Lecture Notes in Computer Science 12363

More information about this subseries at http://www.springer.com/series/7412

Andrea Vedaldi · Horst Bischof ·
Thomas Brox · Jan-Michael Frahm (Eds.)

Computer Vision – ECCV 2020

16th European Conference
Glasgow, UK, August 23–28, 2020
Proceedings, Part XVIII

Springer

Editors
Andrea Vedaldi (iD)
University of Oxford
Oxford, UK

Thomas Brox (iD)
University of Freiburg
Freiburg im Breisgau, Germany

Horst Bischof (iD)
Graz University of Technology
Graz, Austria

Jan-Michael Frahm
University of North Carolina at Chapel Hill
Chapel Hill, NC, USA

ISSN 0302-9743   ISSN 1611-3349 (electronic)
Lecture Notes in Computer Science
ISBN 978-3-030-58522-8   ISBN 978-3-030-58523-5 (eBook)
https://doi.org/10.1007/978-3-030-58523-5

LNCS Sublibrary: SL6 – Image Processing, Computer Vision, Pattern Recognition, and Graphics

This Springer imprint is published by the registered company Springer Nature Switzerland AG
The registered company address is: Gewerbestrasse 11, 6330 Cham, Switzerland

Foreword

Hosting the European Conference on Computer Vision (ECCV 2020) was certainly an exciting journey. From the 2016 plan to hold it at the Edinburgh International Conference Centre (hosting 1,800 delegates) to the 2018 plan to hold it at Glasgow's Scottish Exhibition Centre (up to 6,000 delegates), we finally ended with moving online because of the COVID-19 outbreak. While possibly having fewer delegates than expected because of the online format, ECCV 2020 still had over 3,100 registered participants.

Although online, the conference delivered most of the activities expected at a face-to-face conference: peer-reviewed papers, industrial exhibitors, demonstrations, and messaging between delegates. In addition to the main technical sessions, the conference included a strong program of satellite events with 16 tutorials and 44 workshops.

Furthermore, the online conference format enabled new conference features. Every paper had an associated teaser video and a longer full presentation video. Along with the papers and slides from the videos, all these materials were available the week before the conference. This allowed delegates to become familiar with the paper content and be ready for the live interaction with the authors during the conference week. The live event consisted of brief presentations by the oral and spotlight authors and industrial sponsors. Question and answer sessions for all papers were timed to occur twice so delegates from around the world had convenient access to the authors.

As with ECCV 2018, authors' draft versions of the papers appeared online with open access, now on both the Computer Vision Foundation (CVF) and the European Computer Vision Association (ECVA) websites. An archival publication arrangement was put in place with the cooperation of Springer. SpringerLink hosts the final version of the papers with further improvements, such as activating reference links and supplementary materials. These two approaches benefit all potential readers: a version available freely for all researchers, and an authoritative and citable version with additional benefits for SpringerLink subscribers. We thank Alfred Hofmann and Aliaksandr Birukou from Springer for helping to negotiate this agreement, which we expect will continue for future versions of ECCV.

August 2020

Vittorio Ferrari
Bob Fisher
Cordelia Schmid
Emanuele Trucco

Preface

Welcome to the proceedings of the European Conference on Computer Vision (ECCV 2020). This is a unique edition of ECCV in many ways. Due to the COVID-19 pandemic, this is the first time the conference was held online, in a virtual format. This was also the first time the conference relied exclusively on the Open Review platform to manage the review process. Despite these challenges ECCV is thriving. The conference received 5,150 valid paper submissions, of which 1,360 were accepted for publication (27%) and, of those, 160 were presented as spotlights (3%) and 104 as orals (2%). This amounts to more than twice the number of submissions to ECCV 2018 (2,439). Furthermore, CVPR, the largest conference on computer vision, received 5,850 submissions this year, meaning that ECCV is now 87% the size of CVPR in terms of submissions. By comparison, in 2018 the size of ECCV was only 73% of CVPR.

The review model was similar to previous editions of ECCV; in particular, it was double blind in the sense that the authors did not know the name of the reviewers and vice versa. Furthermore, each conference submission was held confidentially, and was only publicly revealed if and once accepted for publication. Each paper received at least three reviews, totalling more than 15,000 reviews. Handling the review process at this scale was a significant challenge. In order to ensure that each submission received as fair and high-quality reviews as possible, we recruited 2,830 reviewers (a 130% increase with reference to 2018) and 207 area chairs (a 60% increase). The area chairs were selected based on their technical expertise and reputation, largely among people that served as area chair in previous top computer vision and machine learning conferences (ECCV, ICCV, CVPR, NeurIPS, etc.). Reviewers were similarly invited from previous conferences. We also encouraged experienced area chairs to suggest additional chairs and reviewers in the initial phase of recruiting.

Despite doubling the number of submissions, the reviewer load was slightly reduced from 2018, from a maximum of 8 papers down to 7 (with some reviewers offering to handle 6 papers plus an emergency review). The area chair load increased slightly, from 18 papers on average to 22 papers on average.

Conflicts of interest between authors, area chairs, and reviewers were handled largely automatically by the Open Review platform via their curated list of user profiles. Many authors submitting to ECCV already had a profile in Open Review. We set a paper registration deadline one week before the paper submission deadline in order to encourage all missing authors to register and create their Open Review profiles well on time (in practice, we allowed authors to create/change papers arbitrarily until the submission deadline). Except for minor issues with users creating duplicate profiles, this allowed us to easily and quickly identify institutional conflicts, and avoid them, while matching papers to area chairs and reviewers.

Papers were matched to area chairs based on: an affinity score computed by the Open Review platform, which is based on paper titles and abstracts, and an affinity score computed by the Toronto Paper Matching System (TPMS), which is based on the paper's full text, the area chair bids for individual papers, load balancing, and conflict avoidance. Open Review provides the program chairs a convenient web interface to experiment with different configurations of the matching algorithm. The chosen configuration resulted in about 50% of the assigned papers to be highly ranked by the area chair bids, and 50% to be ranked in the middle, with very few low bids assigned.

Assignments to reviewers were similar, with two differences. First, there was a maximum of 7 papers assigned to each reviewer. Second, area chairs recommended up to seven reviewers per paper, providing another highly-weighed term to the affinity scores used for matching.

The assignment of papers to area chairs was smooth. However, it was more difficult to find suitable reviewers for all papers. Having a ratio of 5.6 papers per reviewer with a maximum load of 7 (due to emergency reviewer commitment), which did not allow for much wiggle room in order to also satisfy conflict and expertise constraints. We received some complaints from reviewers who did not feel qualified to review specific papers and we reassigned them wherever possible. However, the large scale of the conference, the many constraints, and the fact that a large fraction of such complaints arrived very late in the review process made this process very difficult and not all complaints could be addressed.

Reviewers had six weeks to complete their assignments. Possibly due to COVID-19 or the fact that the NeurIPS deadline was moved closer to the review deadline, a record 30% of the reviews were still missing after the deadline. By comparison, ECCV 2018 experienced only 10% missing reviews at this stage of the process. In the subsequent week, area chairs chased the missing reviews intensely, found replacement reviewers in their own team, and managed to reach 10% missing reviews. Eventually, we could provide almost all reviews (more than 99.9%) with a delay of only a couple of days on the initial schedule by a significant use of emergency reviews. If this trend is confirmed, it might be a major challenge to run a smooth review process in future editions of ECCV. The community must reconsider prioritization of the time spent on paper writing (the number of submissions increased a lot despite COVID-19) and time spent on paper reviewing (the number of reviews delivered in time decreased a lot presumably due to COVID-19 or NeurIPS deadline). With this imbalance the peer-review system that ensures the quality of our top conferences may break soon.

Reviewers submitted their reviews independently. In the reviews, they had the opportunity to ask questions to the authors to be addressed in the rebuttal. However, reviewers were told not to request any significant new experiment. Using the Open Review interface, authors could provide an answer to each individual review, but were also allowed to cross-reference reviews and responses in their answers. Rather than PDF files, we allowed the use of formatted text for the rebuttal. The rebuttal and initial reviews were then made visible to all reviewers and the primary area chair for a given paper. The area chair encouraged and moderated the reviewer discussion. During the discussions, reviewers were invited to reach a consensus and possibly adjust their ratings as a result of the discussion and of the evidence in the rebuttal.

After the discussion period ended, most reviewers entered a final rating and recommendation, although in many cases this did not differ from their initial recommendation. Based on the updated reviews and discussion, the primary area chair then made a preliminary decision to accept or reject the paper and wrote a justification for it (meta-review). Except for cases where the outcome of this process was absolutely clear (as indicated by the three reviewers and primary area chairs all recommending clear rejection), the decision was then examined and potentially challenged by a secondary area chair. This led to further discussion and overturning a small number of preliminary decisions. Needless to say, there was no in-person area chair meeting, which would have been impossible due to COVID-19.

Area chairs were invited to observe the consensus of the reviewers whenever possible and use extreme caution in overturning a clear consensus to accept or reject a paper. If an area chair still decided to do so, she/he was asked to clearly justify it in the meta-review and to explicitly obtain the agreement of the secondary area chair. In practice, very few papers were rejected after being confidently accepted by the reviewers.

This was the first time Open Review was used as the main platform to run ECCV. In 2018, the program chairs used CMT3 for the user-facing interface and Open Review internally, for matching and conflict resolution. Since it is clearly preferable to only use a single platform, this year we switched to using Open Review in full. The experience was largely positive. The platform is highly-configurable, scalable, and open source. Being written in Python, it is easy to write scripts to extract data programmatically. The paper matching and conflict resolution algorithms and interfaces are top-notch, also due to the excellent author profiles in the platform. Naturally, there were a few kinks along the way due to the fact that the ECCV Open Review configuration was created from scratch for this event and it differs in substantial ways from many other Open Review conferences. However, the Open Review development and support team did a fantastic job in helping us to get the configuration right and to address issues in a timely manner as they unavoidably occurred. We cannot thank them enough for the tremendous effort they put into this project.

Finally, we would like to thank everyone involved in making ECCV 2020 possible in these very strange and difficult times. This starts with our authors, followed by the area chairs and reviewers, who ran the review process at an unprecedented scale. The whole Open Review team (and in particular Melisa Bok, Mohit Unyal, Carlos Mondragon Chapa, and Celeste Martinez Gomez) worked incredibly hard for the entire duration of the process. We would also like to thank René Vidal for contributing to the adoption of Open Review. Our thanks also go to Laurent Charling for TPMS and to the program chairs of ICML, ICLR, and NeurIPS for cross checking double submissions. We thank the website chair, Giovanni Farinella, and the CPI team (in particular Ashley Cook, Miriam Verdon, Nicola McGrane, and Sharon Kerr) for promptly adding material to the website as needed in the various phases of the process. Finally, we thank the publication chairs, Albert Ali Salah, Hamdi Dibeklioglu, Metehan Doyran, Henry Howard-Jenkins, Victor Prisacariu, Siyu Tang, and Gul Varol, who managed to compile these substantial proceedings in an exceedingly compressed schedule. We express our thanks to the ECVA team, in particular Kristina Scherbaum for allowing open access of the proceedings. We thank Alfred Hofmann from Springer who again serve as the publisher. Finally, we thank the other chairs of ECCV 2020, including in particular the general chairs for very useful feedback with the handling of the program.

August 2020   Andrea Vedaldi
                                               Horst Bischof
                                               Thomas Brox
                                               Jan-Michael Frahm

Organization

General Chairs

Vittorio Ferrari	Google Research, Switzerland
Bob Fisher	University of Edinburgh, UK
Cordelia Schmid	Google and Inria, France
Emanuele Trucco	University of Dundee, UK

Program Chairs

Andrea Vedaldi	University of Oxford, UK
Horst Bischof	Graz University of Technology, Austria
Thomas Brox	University of Freiburg, Germany
Jan-Michael Frahm	University of North Carolina, USA

Industrial Liaison Chairs

Jim Ashe	University of Edinburgh, UK
Helmut Grabner	Zurich University of Applied Sciences, Switzerland
Diane Larlus	NAVER LABS Europe, France
Cristian Novotny	University of Edinburgh, UK

Local Arrangement Chairs

Yvan Petillot	Heriot-Watt University, UK
Paul Siebert	University of Glasgow, UK

Academic Demonstration Chair

Thomas Mensink	Google Research and University of Amsterdam, The Netherlands

Poster Chair

Stephen Mckenna	University of Dundee, UK

Technology Chair

Gerardo Aragon Camarasa	University of Glasgow, UK

Tutorial Chairs

Carlo Colombo	University of Florence, Italy
Sotirios Tsaftaris	University of Edinburgh, UK

Publication Chairs

Albert Ali Salah	Utrecht University, The Netherlands
Hamdi Dibeklioglu	Bilkent University, Turkey
Metehan Doyran	Utrecht University, The Netherlands
Henry Howard-Jenkins	University of Oxford, UK
Victor Adrian Prisacariu	University of Oxford, UK
Siyu Tang	ETH Zurich, Switzerland
Gul Varol	University of Oxford, UK

Website Chair

Giovanni Maria Farinella	University of Catania, Italy

Workshops Chairs

Adrien Bartoli	University of Clermont Auvergne, France
Andrea Fusiello	University of Udine, Italy

Area Chairs

Lourdes Agapito	University College London, UK
Zeynep Akata	University of Tübingen, Germany
Karteek Alahari	Inria, France
Antonis Argyros	University of Crete, Greece
Hossein Azizpour	KTH Royal Institute of Technology, Sweden
Joao P. Barreto	Universidade de Coimbra, Portugal
Alexander C. Berg	University of North Carolina at Chapel Hill, USA
Matthew B. Blaschko	KU Leuven, Belgium
Lubomir D. Bourdev	WaveOne, Inc., USA
Edmond Boyer	Inria, France
Yuri Boykov	University of Waterloo, Canada
Gabriel Brostow	University College London, UK
Michael S. Brown	National University of Singapore, Singapore
Jianfei Cai	Monash University, Australia
Barbara Caputo	Politecnico di Torino, Italy
Ayan Chakrabarti	Washington University, St. Louis, USA
Tat-Jen Cham	Nanyang Technological University, Singapore
Manmohan Chandraker	University of California, San Diego, USA
Rama Chellappa	Johns Hopkins University, USA
Liang-Chieh Chen	Google, USA

Yung-Yu Chuang   National Taiwan University, Taiwan
Ondrej Chum   Czech Technical University in Prague, Czech Republic
Brian Clipp   Kitware, USA
John Collomosse   University of Surrey and Adobe Research, UK
Jason J. Corso   University of Michigan, USA
David J. Crandall   Indiana University, USA
Daniel Cremers   University of California, Los Angeles, USA
Fabio Cuzzolin   Oxford Brookes University, UK
Jifeng Dai   SenseTime, SAR China
Kostas Daniilidis   University of Pennsylvania, USA
Andrew Davison   Imperial College London, UK
Alessio Del Bue   Fondazione Istituto Italiano di Tecnologia, Italy
Jia Deng   Princeton University, USA
Alexey Dosovitskiy   Google, Germany
Matthijs Douze   Facebook, France
Enrique Dunn   Stevens Institute of Technology, USA
Irfan Essa   Georgia Institute of Technology and Google, USA
Giovanni Maria Farinella University of Catania, Italy
Ryan Farrell   Brigham Young University, USA
Paolo Favaro   University of Bern, Switzerland
Rogerio Feris   International Business Machines, USA
Cornelia Fermuller   University of Maryland, College Park, USA
David J. Fleet   Vector Institute, Canada
Friedrich Fraundorfer   DLR, Austria
Mario Fritz   CISPA Helmholtz Center for Information Security,
                          Germany
Pascal Fua   EPFL (Swiss Federal Institute of Technology
                          Lausanne), Switzerland
Yasutaka Furukawa   Simon Fraser University, Canada
Li Fuxin   Oregon State University, USA
Efstratios Gavves   University of Amsterdam, The Netherlands
Peter Vincent Gehler   Amazon, USA
Theo Gevers   University of Amsterdam, The Netherlands
Ross Girshick   Facebook AI Research, USA
Boqing Gong   Google, USA
Stephen Gould   Australian National University, Australia
Jinwei Gu   SenseTime Research, USA
Abhinav Gupta   Facebook, USA
Bohyung Han   Seoul National University, South Korea
Bharath Hariharan   Cornell University, USA
Tal Hassner   Facebook AI Research, USA
Xuming He   Australian National University, Australia
Joao F. Henriques   University of Oxford, UK
Adrian Hilton   University of Surrey, UK
Minh Hoai   Stony Brooks, State University of New York, USA
Derek Hoiem   University of Illinois Urbana-Champaign, USA

Timothy Hospedales	University of Edinburgh and Samsung, UK
Gang Hua	Wormpex AI Research, USA
Slobodan Ilic	Siemens AG, Germany
Hiroshi Ishikawa	Waseda University, Japan
Jiaya Jia	The Chinese University of Hong Kong, SAR China
Hailin Jin	Adobe Research, USA
Justin Johnson	University of Michigan, USA
Frederic Jurie	University of Caen Normandie, France
Fredrik Kahl	Chalmers University, Sweden
Sing Bing Kang	Zillow, USA
Gunhee Kim	Seoul National University, South Korea
Junmo Kim	Korea Advanced Institute of Science and Technology, South Korea
Tae-Kyun Kim	Imperial College London, UK
Ron Kimmel	Technion-Israel Institute of Technology, Israel
Alexander Kirillov	Facebook AI Research, USA
Kris Kitani	Carnegie Mellon University, USA
Iasonas Kokkinos	Ariel AI, UK
Vladlen Koltun	Intel Labs, USA
Nikos Komodakis	Ecole des Ponts ParisTech, France
Piotr Koniusz	Australian National University, Australia
M. Pawan Kumar	University of Oxford, UK
Kyros Kutulakos	University of Toronto, Canada
Christoph Lampert	IST Austria, Austria
Ivan Laptev	Inria, France
Diane Larlus	NAVER LABS Europe, France
Laura Leal-Taixe	Technical University Munich, Germany
Honglak Lee	Google and University of Michigan, USA
Joon-Young Lee	Adobe Research, USA
Kyoung Mu Lee	Seoul National University, South Korea
Seungyong Lee	POSTECH, South Korea
Yong Jae Lee	University of California, Davis, USA
Bastian Leibe	RWTH Aachen University, Germany
Victor Lempitsky	Samsung, Russia
Ales Leonardis	University of Birmingham, UK
Marius Leordeanu	Institute of Mathematics of the Romanian Academy, Romania
Vincent Lepetit	ENPC ParisTech, France
Hongdong Li	The Australian National University, Australia
Xi Li	Zhejiang University, China
Yin Li	University of Wisconsin-Madison, USA
Zicheng Liao	Zhejiang University, China
Jongwoo Lim	Hanyang University, South Korea
Stephen Lin	Microsoft Research Asia, China
Yen-Yu Lin	National Chiao Tung University, Taiwan, China
Zhe Lin	Adobe Research, USA

Haibin Ling	Stony Brooks, State University of New York, USA
Jiaying Liu	Peking University, China
Ming-Yu Liu	NVIDIA, USA
Si Liu	Beihang University, China
Xiaoming Liu	Michigan State University, USA
Huchuan Lu	Dalian University of Technology, China
Simon Lucey	Carnegie Mellon University, USA
Jiebo Luo	University of Rochester, USA
Julien Mairal	Inria, France
Michael Maire	University of Chicago, USA
Subhransu Maji	University of Massachusetts, Amherst, USA
Yasushi Makihara	Osaka University, Japan
Jiri Matas	Czech Technical University in Prague, Czech Republic
Yasuyuki Matsushita	Osaka University, Japan
Philippos Mordohai	Stevens Institute of Technology, USA
Vittorio Murino	University of Verona, Italy
Naila Murray	NAVER LABS Europe, France
Hajime Nagahara	Osaka University, Japan
P. J. Narayanan	International Institute of Information Technology (IIIT), Hyderabad, India
Nassir Navab	Technical University of Munich, Germany
Natalia Neverova	Facebook AI Research, France
Matthias Niessner	Technical University of Munich, Germany
Jean-Marc Odobez	Idiap Research Institute and Swiss Federal Institute of Technology Lausanne, Switzerland
Francesca Odone	Università di Genova, Italy
Takeshi Oishi	The University of Tokyo, Tokyo Institute of Technology, Japan
Vicente Ordonez	University of Virginia, USA
Manohar Paluri	Facebook AI Research, USA
Maja Pantic	Imperial College London, UK
In Kyu Park	Inha University, South Korea
Ioannis Patras	Queen Mary University of London, UK
Patrick Perez	Valeo, France
Bryan A. Plummer	Boston University, USA
Thomas Pock	Graz University of Technology, Austria
Marc Pollefeys	ETH Zurich and Microsoft MR & AI Zurich Lab, Switzerland
Jean Ponce	Inria, France
Gerard Pons-Moll	MPII, Saarland Informatics Campus, Germany
Jordi Pont-Tuset	Google, Switzerland
James Matthew Rehg	Georgia Institute of Technology, USA
Ian Reid	University of Adelaide, Australia
Olaf Ronneberger	DeepMind London, UK
Stefan Roth	TU Darmstadt, Germany
Bryan Russell	Adobe Research, USA

Kwang Moo Yi
Zhaozheng Yin
Chang D. Yoo

University of Victoria, Canada
Stony Brook, State University of New York, USA
Korea Advanced Institute of Science and Technology,
    South Korea

Shaodi You
Jingyi Yu
Stella Yu
Stefanos Zafeiriou
Hongbin Zha
Tianzhu Zhang
Liang Zheng
Todd E. Zickler
Andrew Zisserman

University of Amsterdam, The Netherlands
ShanghaiTech University, China
University of California, Berkeley, and ICSI, USA
Imperial College London, UK
Peking University, China
University of Science and Technology of China, China
Australian National University, Australia
Harvard University, USA
University of Oxford, UK

Technical Program Committee

Sathyanarayanan
  N. Aakur
Wael Abd Almgaeed
Abdelrahman
  Abdelhamcd
Abdullah Abuolaim
Supreeth Achar
Hanno Ackermann
Ehsan Adeli
Triantafyllos Afouras
Sameer Agarwal
Aishwarya Agrawal
Harsh Agrawal
Pulkit Agrawal
Antonio Agudo
Eirikur Agustsson
Karim Ahmed
Byeongjoo Ahn
Unaiza Ahsan
Thalaiyasingam Ajanthan
Kenan E. Ak
Emre Akbas
Naveed Akhtar
Derya Akkaynak
Yagiz Aksoy
Ziad Al-Halah
Xavier Alameda-Pineda
Jean-Baptiste Alayrac

Samuel Albanie
Shadi Albarqouni
Cenek Albl
Hassan Abu Alhaija
Daniel Aliaga
Mohammad
  S. Aliakbarian
Rahaf Aljundi
Thiemo Alldieck
Jon Almazan
Jose M. Alvarez
Senjian An
Saket Anand
Codruta Ancuti
Cosmin Ancuti
Peter Anderson
Juan Andrade-Cetto
Alexander Andreopoulos
Misha Andriluka
Dragomir Anguelov
Rushil Anirudh
Michel Antunes
Oisin Mac Aodha
Srikar Appalaraju
Relja Arandjelovic
Nikita Araslanov
Andre Araujo
Helder Araujo

Pablo Arbelaez
Shervin Ardeshir
Sercan O. Arik
Anil Armagan
Anurag Arnab
Chetan Arora
Federica Arrigoni
Mathieu Aubry
Shai Avidan
Angelica I. Aviles-Rivero
Yannis Avrithis
Ismail Ben Ayed
Shekoofeh Azizi
Ioan Andrei Bârsan
Artem Babenko
Deepak Babu Sam
Seung-Hwan Baek
Seungryul Baek
Andrew D. Bagdanov
Shai Bagon
Yuval Bahat
Junjie Bai
Song Bai
Xiang Bai
Yalong Bai
Yancheng Bai
Peter Bajcsy
Slawomir Bak

Mahsa Baktashmotlagh
Kavita Bala
Yogesh Balaji
Guha Balakrishnan
V. N. Balasubramanian
Federico Baldassarre
Vassileios Balntas
Shurjo Banerjee
Aayush Bansal
Ankan Bansal
Jianmin Bao
Linchao Bao
Wenbo Bao
Yingze Bao
Akash Bapat
Md Jawadul Hasan Bappy
Fabien Baradel
Lorenzo Baraldi
Daniel Barath
Adrian Barbu
Kobus Barnard
Nick Barnes
Francisco Barranco
Jonathan T. Barron
Arslan Basharat
Chaim Baskin
Anil S. Baslamisli
Jorge Batista
Kayhan Batmanghelich
Konstantinos Batsos
David Bau
Luis Baumela
Christoph Baur
Eduardo
  Bayro-Corrochano
Paul Beardsley
Jan Bednavr'ik
Oscar Beijbom
Philippe Bekaert
Esube Bekele
Vasileios Belagiannis
Ohad Ben-Shahar
Abhijit Bendale
Róger Bermúdez-Chacón
Maxim Berman
Jesus Bermudez-cameo

Florian Bernard
Stefano Berretti
Marcelo Bertalmio
Gedas Bertasius
Cigdem Beyan
Lucas Beyer
Vijayakumar Bhagavatula
Arjun Nitin Bhagoji
Apratim Bhattacharyya
Binod Bhattarai
Sai Bi
Jia-Wang Bian
Simone Bianco
Adel Bibi
Tolga Birdal
Tom Bishop
Soma Biswas
Mårten Björkman
Volker Blanz
Vishnu Boddeti
Navaneeth Bodla
Simion-Vlad Bogolin
Xavier Boix
Piotr Bojanowski
Timo Bolkart
Guido Borghi
Larbi Boubchir
Guillaume Bourmaud
Adrien Bousseau
Thierry Bouwmans
Richard Bowden
Hakan Boyraz
Mathieu Brédif
Samarth Brahmbhatt
Steve Branson
Nikolas Brasch
Biagio Brattoli
Ernesto Brau
Toby P. Breckon
Francois Bremond
Jesus Briales
Sofia Broomé
Marcus A. Brubaker
Luc Brun
Silvia Bucci
Shyamal Buch

Pradeep Buddharaju
Uta Buechler
Mai Bui
Tu Bui
Adrian Bulat
Giedrius T. Burachas
Elena Burceanu
Xavier P. Burgos-Artizzu
Kaylee Burns
Andrei Bursuc
Benjamin Busam
Wonmin Byeon
Zoya Bylinskii
Sergi Caelles
Jianrui Cai
Minjie Cai
Yujun Cai
Zhaowei Cai
Zhipeng Cai
Juan C. Caicedo
Simone Calderara
Necati Cihan Camgoz
Dylan Campbell
Octavia Camps
Jiale Cao
Kaidi Cao
Liangliang Cao
Xiangyong Cao
Xiaochun Cao
Yang Cao
Yu Cao
Yue Cao
Zhangjie Cao
Luca Carlone
Mathilde Caron
Dan Casas
Thomas J. Cashman
Umberto Castellani
Lluis Castrejon
Jacopo Cavazza
Fabio Cermelli
Hakan Cevikalp
Menglei Chai
Ishani Chakraborty
Rudrasis Chakraborty
Antoni B. Chan

Rozenn Dahyot
Bo Dai
Dengxin Dai
Hang Dai
Longquan Dai
Shuyang Dai
Xiyang Dai
Yuchao Dai
Adrian V. Dalca
Dima Damen
Bharath B. Damodaran
Kristin Dana
Martin Danelljan
Zheng Dang
Zachary Alan Daniels
Donald G. Dansereau
Abhishek Das
Samyak Datta
Achal Dave
Titas De
Rodrigo de Bem
Teo de Campos
Raoul de Charette
Shalini De Mello
Joseph DeGol
Herve Delingette
Haowen Deng
Jiankang Deng
Weijian Deng
Zhiwei Deng
Joachim Denzler
Konstantinos G. Derpanis
Aditya Deshpande
Frederic Devernay
Somdip Dey
Arturo Deza
Abhinav Dhall
Helisa Dhamo
Vikas Dhiman
Fillipe Dias Moreira
  de Souza
Ali Diba
Ferran Diego
Guiguang Ding
Henghui Ding
Jian Ding

Mingyu Ding
Xinghao Ding
Zhengming Ding
Robert DiPietro
Cosimo Distante
Ajay Divakaran
Mandar Dixit
Abdelaziz Djelouah
Thanh-Toan Do
Jose Dolz
Bo Dong
Chao Dong
Jiangxin Dong
Weiming Dong
Weisheng Dong
Xingping Dong
Xuanyi Dong
Yinpeng Dong
Gianfranco Doretto
Hazel Doughty
Hassen Drira
Bertram Drost
Dawei Du
Ye Duan
Yueqi Duan
Abhimanyu Dubey
Anastasia Dubrovina
Stefan Duffner
Chi Nhan Duong
Thibaut Durand
Zoran Duric
Iulia Duta
Debidatta Dwibedi
Benjamin Eckart
Marc Eder
Marzieh Edraki
Alexei A. Efros
Kiana Ehsani
Hazm Kemal Ekenel
James H. Elder
Mohamed Elgharib
Shireen Elhabian
Ehsan Elhamifar
Mohamed Elhoseiny
Ian Endres
N. Benjamin Erichson

Jan Ernst
Sergio Escalera
Francisco Escolano
Victor Escorcia
Carlos Esteves
Francisco J. Estrada
Bin Fan
Chenyou Fan
Deng-Ping Fan
Haoqi Fan
Hehe Fan
Heng Fan
Kai Fan
Lijie Fan
Linxi Fan
Quanfu Fan
Shaojing Fan
Xiaochuan Fan
Xin Fan
Yuchen Fan
Sean Fanello
Hao-Shu Fang
Haoyang Fang
Kuan Fang
Yi Fang
Yuming Fang
Azade Farshad
Alireza Fathi
Raanan Fattal
Joao Fayad
Xiaohan Fei
Christoph Feichtenhofer
Michael Felsberg
Chen Feng
Jiashi Feng
Junyi Feng
Mengyang Feng
Qianli Feng
Zhenhua Feng
Michele Fenzi
Andras Ferencz
Martin Fergie
Basura Fernando
Ethan Fetaya
Michael Firman
John W. Fisher

Matthew Fisher
Boris Flach
Corneliu Florea
Wolfgang Foerstner
David Fofi
Gian Luca Foresti
Per-Erik Forssen
David Fouhey
Katerina Fragkiadaki
Victor Fragoso
Jean-Sébastien Franco
Ohad Fried
Iuri Frosio
Cheng-Yang Fu
Huazhu Fu
Jianlong Fu
Jingjing Fu
Xueyang Fu
Yanwei Fu
Ying Fu
Yun Fu
Olac Fuentes
Kent Fujiwara
Takuya Funatomi
Christopher Funk
Thomas Funkhouser
Antonino Furnari
Ryo Furukawa
Erik Gärtner
Raghudeep Gadde
Matheus Gadelha
Vandit Gajjar
Trevor Gale
Juergen Gall
Mathias Gallardo
Guillermo Gallego
Orazio Gallo
Chuang Gan
Zhe Gan
Madan Ravi Ganesh
Aditya Ganeshan
Siddha Ganju
Bin-Bin Gao
Changxin Gao
Feng Gao
Hongchang Gao

Jin Gao
Jiyang Gao
Junbin Gao
Katelyn Gao
Lin Gao
Mingfei Gao
Ruiqi Gao
Ruohan Gao
Shenghua Gao
Yuan Gao
Yue Gao
Noa Garcia
Alberto Garcia-Garcia
Guillermo
    Garcia-Hernando
Jacob R. Gardner
Animesh Garg
Kshitiz Garg
Rahul Garg
Ravi Garg
Philip N. Garner
Kirill Gavrilyuk
Paul Gay
Shiming Ge
Weifeng Ge
Baris Gecer
Xin Geng
Kyle Genova
Stamatios Georgoulis
Bernard Ghanem
Michael Gharbi
Kamran Ghasedi
Golnaz Ghiasi
Arnab Ghosh
Partha Ghosh
Silvio Giancola
Andrew Gilbert
Rohit Girdhar
Xavier Giro-i-Nieto
Thomas Gittings
Ioannis Gkioulekas
Clement Godard
Vaibhava Goel
Bastian Goldluecke
Lluis Gomez
Nuno Gonçalves

Dong Gong
Ke Gong
Mingming Gong
Abel Gonzalez-Garcia
Ariel Gordon
Daniel Gordon
Paulo Gotardo
Venu Madhav Govindu
Ankit Goyal
Priya Goyal
Raghav Goyal
Benjamin Graham
Douglas Gray
Brent A. Griffin
Etienne Grossmann
David Gu
Jiayuan Gu
Jiuxiang Gu
Lin Gu
Qiao Gu
Shuhang Gu
Jose J. Guerrero
Paul Guerrero
Jie Gui
Jean-Yves Guillemaut
Riza Alp Guler
Erhan Gundogdu
Fatma Guney
Guodong Guo
Kaiwen Guo
Qi Guo
Sheng Guo
Shi Guo
Tiantong Guo
Xiaojie Guo
Yijie Guo
Yiluan Guo
Yuanfang Guo
Yulan Guo
Agrim Gupta
Ankush Gupta
Mohit Gupta
Saurabh Gupta
Tanmay Gupta
Danna Gurari
Abner Guzman-Rivera

JunYoung Gwak
Michael Gygli
Jung-Woo Ha
Simon Hadfield
Isma Hadji
Bjoern Haefner
Taeyoung Hahn
Levente Hajder
Peter Hall
Emanuela Haller
Stefan Haller
Bumsub Ham
Abdullah Hamdi
Dongyoon Han
Hu Han
Jungong Han
Junwei Han
Kai Han
Tian Han
Xiaoguang Han
Xintong Han
Yahong Han
Ankur Handa
Zekun Hao
Albert Haque
Tatsuya Harada
Mehrtash Harandi
Adam W. Harley
Mahmudul Hasan
Atsushi Hashimoto
Ali Hatamizadeh
Munawar Hayat
Dongliang He
Jingrui He
Junfeng He
Kaiming He
Kun He
Lei He
Pan He
Ran He
Shengfeng He
Tong He
Weipeng He
Xuming He
Yang He
Yihui He

Zhihai He
Chinmay Hegde
Janne Heikkila
Mattias P. Heinrich
Stéphane Herbin
Alexander Hermans
Luis Herranz
John R. Hershey
Aaron Hertzmann
Roei Herzig
Anders Heyden
Steven Hickson
Otmar Hilliges
Tomas Hodan
Judy Hoffman
Michael Hofmann
Yannick Hold-Geoffroy
Namdar Homayounfar
Sina Honari
Richang Hong
Seunghoon Hong
Xiaopeng Hong
Yi Hong
Hidekata Hontani
Anthony Hoogs
Yedid Hoshen
Mir Rayat Imtiaz Hossain
Junhui Hou
Le Hou
Lu Hou
Tingbo Hou
Wei-Lin Hsiao
Cheng-Chun Hsu
Gee-Sern Jison Hsu
Kuang-jui Hsu
Changbo Hu
Di Hu
Guosheng Hu
Han Hu
Hao Hu
Hexiang Hu
Hou-Ning Hu
Jie Hu
Junlin Hu
Nan Hu
Ping Hu

Ronghang Hu
Xiaowei Hu
Yinlin Hu
Yuan-Ting Hu
Zhe Hu
Binh-Son Hua
Yang Hua
Bingyao Huang
Di Huang
Dong Huang
Fay Huang
Haibin Huang
Haozhi Huang
Heng Huang
Huaibo Huang
Jia-Bin Huang
Jing Huang
Jingwei Huang
Kaizhu Huang
Lei Huang
Qiangui Huang
Qiaoying Huang
Qingqiu Huang
Qixing Huang
Shaoli Huang
Sheng Huang
Siyuan Huang
Weilin Huang
Wenbing Huang
Xiangru Huang
Xun Huang
Yan Huang
Yifei Huang
Yue Huang
Zhiwu Huang
Zilong Huang
Minyoung Huh
Zhuo Hui
Matthias B. Hullin
Martin Humenberger
Wei-Chih Hung
Zhouyuan Huo
Junhwa Hur
Noureldien Hussein
Jyh-Jing Hwang
Seong Jae Hwang

Sung Ju Hwang
Ichiro Ide
Ivo Ihrke
Daiki Ikami
Satoshi Ikehata
Nazli Ikizler-Cinbis
Sunghoon Im
Yani Ioannou
Radu Tudor Ionescu
Umar Iqbal
Go Irie
Ahmet Iscen
Md Amirul Islam
Vamsi Ithapu
Nathan Jacobs
Arpit Jain
Himalaya Jain
Suyog Jain
Stuart James
Won-Dong Jang
Yunseok Jang
Ronnachai Jaroensri
Dinesh Jayaraman
Sadeep Jayasumana
Suren Jayasuriya
Herve Jegou
Simon Jenni
Hae-Gon Jeon
Yunho Jeon
Koteswar R. Jerripothula
Hueihan Jhuang
I-hong Jhuo
Dinghuang Ji
Hui Ji
Jingwei Ji
Pan Ji
Yanli Ji
Baoxiong Jia
Kui Jia
Xu Jia
Chiyu Max Jiang
Haiyong Jiang
Hao Jiang
Huaizu Jiang
Huajie Jiang
Ke Jiang

Lai Jiang
Li Jiang
Lu Jiang
Ming Jiang
Peng Jiang
Shuqiang Jiang
Wei Jiang
Xudong Jiang
Zhuolin Jiang
Jianbo Jiao
Zequn Jie
Dakai Jin
Kyong Hwan Jin
Lianwen Jin
SouYoung Jin
Xiaojie Jin
Xin Jin
Nebojsa Jojic
Alexis Joly
Michael Jeffrey Jones
Hanbyul Joo
Jungseock Joo
Kyungdon Joo
Ajjen Joshi
Shantanu H. Joshi
Da-Cheng Juan
Marco Körner
Kevin Köser
Asim Kadav
Christine Kaeser-Chen
Kushal Kafle
Dagmar Kainmueller
Ioannis A. Kakadiaris
Zdenek Kalal
Nima Kalantari
Yannis Kalantidis
Mahdi M. Kalayeh
Anmol Kalia
Sinan Kalkan
Vicky Kalogeiton
Ashwin Kalyan
Joni-kristian Kamarainen
Gerda Kamberova
Chandra Kambhamettu
Martin Kampel
Meina Kan

Christopher Kanan
Kenichi Kanatani
Angjoo Kanazawa
Atsushi Kanehira
Takuhiro Kaneko
Asako Kanezaki
Bingyi Kang
Di Kang
Sunghun Kang
Zhao Kang
Vadim Kantorov
Abhishek Kar
Amlan Kar
Theofanis Karaletsos
Leonid Karlinsky
Kevin Karsch
Angelos Katharopoulos
Isinsu Katircioglu
Hiroharu Kato
Zoltan Kato
Dotan Kaufman
Jan Kautz
Rei Kawakami
Qiuhong Ke
Wadim Kehl
Petr Kellnhofer
Aniruddha Kembhavi
Cem Keskin
Margret Keuper
Daniel Keysers
Ashkan Khakzar
Fahad Khan
Naeemullah Khan
Salman Khan
Siddhesh Khandelwal
Rawal Khirodkar
Anna Khoreva
Tejas Khot
Parmeshwar Khurd
Hadi Kiapour
Joe Kileel
Chanho Kim
Dahun Kim
Edward Kim
Eunwoo Kim
Han-ul Kim

Hansung Kim
Heewon Kim
Hyo Jin Kim
Hyunwoo J. Kim
Jinkyu Kim
Jiwon Kim
Jongmin Kim
Junsik Kim
Junyeong Kim
Min H. Kim
Namil Kim
Pyojin Kim
Seon Joo Kim
Seong Tae Kim
Seungryong Kim
Sungwoong Kim
Tae Hyun Kim
Vladimir Kim
Won Hwa Kim
Yonghyun Kim
Benjamin Kimia
Akisato Kimura
Pieter-Jan Kindermans
Zsolt Kira
Itaru Kitahara
Hedvig Kjellstrom
Jan Knopp
Takumi Kobayashi
Erich Kobler
Parker Koch
Reinhard Koch
Elyor Kodirov
Amir Kolaman
Nicholas Kolkin
Dimitrios Kollias
Stefanos Kollias
Soheil Kolouri
Adams Wai-Kin Kong
Naejin Kong
Shu Kong
Tao Kong
Yu Kong
Yoshinori Konishi
Daniil Kononenko
Theodora Kontogianni
Simon Korman

Adam Kortylewski
Jana Kosecka
Jean Kossaifi
Satwik Kottur
Rigas Kouskouridas
Adriana Kovashka
Rama Kovvuri
Adarsh Kowdle
Jedrzej Kozerawski
Mateusz Kozinski
Philipp Kraehenbuehl
Gregory Kramida
Josip Krapac
Dmitry Kravchenko
Ranjay Krishna
Pavel Krsek
Alexander Krull
Jakob Kruse
Hiroyuki Kubo
Hilde Kuehne
Jason Kuen
Andreas Kuhn
Arjan Kuijper
Zuzana Kukelova
Ajay Kumar
Amit Kumar
Avinash Kumar
Suryansh Kumar
Vijay Kumar
Kaustav Kundu
Weicheng Kuo
Nojun Kwak
Suha Kwak
Junseok Kwon
Nikolaos Kyriazis
Zorah Lähner
Ankit Laddha
Florent Lafarge
Jean Lahoud
Kevin Lai
Shang-Hong Lai
Wei-Sheng Lai
Yu-Kun Lai
Iro Laina
Antony Lam
John Wheatley Lambert

Xiangyuan lan
Xu Lan
Charis Lanaras
Georg Langs
Oswald Lanz
Dong Lao
Yizhen Lao
Agata Lapedriza
Gustav Larsson
Viktor Larsson
Katrin Lasinger
Christoph Lassner
Longin Jan Latecki
Stéphane Lathuilière
Rynson Lau
Hei Law
Justin Lazarow
Svetlana Lazebnik
Hieu Le
Huu Le
Ngan Hoang Le
Trung-Nghia Le
Vuong Le
Colin Lea
Erik Learned-Miller
Chen-Yu Lee
Gim Hee Lee
Hsin-Ying Lee
Hyungtae Lee
Jae-Han Lee
Jimmy Addison Lee
Joonseok Lee
Kibok Lee
Kuang-Huei Lee
Kwonjoon Lee
Minsik Lee
Sang-chul Lee
Seungkyu Lee
Soochan Lee
Stefan Lee
Taehee Lee
Andreas Lehrmann
Jie Lei
Peng Lei
Matthew Joseph Leotta
Wee Kheng Leow

Gil Levi
Evgeny Levinkov
Aviad Levis
Jose Lezama
Ang Li
Bin Li
Bing Li
Boyi Li
Changsheng Li
Chao Li
Chen Li
Cheng Li
Chenglong Li
Chi Li
Chun-Guang Li
Chun-Liang Li
Chunyuan Li
Dong Li
Guanbin Li
Hao Li
Haoxiang Li
Hongsheng Li
Hongyang Li
Houqiang Li
Huibin Li
Jia Li
Jianan Li
Jianguo Li
Junnan Li
Junxuan Li
Kai Li
Ke Li
Kejie Li
Kunpeng Li
Lerenhan Li
Li Erran Li
Mengtian Li
Mu Li
Peihua Li
Peiyi Li
Ping Li
Qi Li
Qing Li
Ruiyu Li
Ruoteng Li
Shaozi Li

Sheng Li
Shiwei Li
Shuang Li
Siyang Li
Stan Z. Li
Tianye Li
Wei Li
Weixin Li
Wen Li
Wenbo Li
Xiaomeng Li
Xin Li
Xiu Li
Xuelong Li
Xueting Li
Yan Li
Yandong Li
Yanghao Li
Yehao Li
Yi Li
Yijun Li
Yikang LI
Yining Li
Yongjie Li
Yu Li
Yu-Jhe Li
Yunpeng Li
Yunsheng Li
Yunzhu Li
Zhe Li
Zhen Li
Zhengqi Li
Zhenyang Li
Zhuwen Li
Dongze Lian
Xiaochen Lian
Zhouhui Lian
Chen Liang
Jie Liang
Ming Liang
Paul Pu Liang
Pengpeng Liang
Shu Liang
Wei Liang
Jing Liao
Minghui Liao

Renjie Liao
Shengcai Liao
Shuai Liao
Yiyi Liao
Ser-Nam Lim
Chen-Hsuan Lin
Chung-Ching Lin
Dahua Lin
Ji Lin
Kevin Lin
Tianwei Lin
Tsung-Yi Lin
Tsung-Yu Lin
Wei-An Lin
Weiyao Lin
Yen-Chen Lin
Yuewei Lin
David B. Lindell
Drew Linsley
Krzysztof Lis
Roee Litman
Jim Little
An-An Liu
Bo Liu
Buyu Liu
Chao Liu
Chen Liu
Cheng-lin Liu
Chenxi Liu
Dong Liu
Feng Liu
Guilin Liu
Haomiao Liu
Heshan Liu
Hong Liu
Ji Liu
Jingen Liu
Jun Liu
Lanlan Liu
Li Liu
Liu Liu
Mengyuan Liu
Miaomiao Liu
Nian Liu
Ping Liu
Risheng Liu

Sheng Liu
Shu Liu
Shuaicheng Liu
Sifei Liu
Siqi Liu
Siying Liu
Songtao Liu
Ting Liu
Tongliang Liu
Tyng-Luh Liu
Wanquan Liu
Wei Liu
Weiyang Liu
Weizhe Liu
Wenyu Liu
Wu Liu
Xialei Liu
Xianglong Liu
Xiaodong Liu
Xiaofeng Liu
Xihui Liu
Xingyu Liu
Xinwang Liu
Xuanqing Liu
Xuebo Liu
Yang Liu
Yaojie Liu
Yebin Liu
Yen-Cheng Liu
Yiming Liu
Yu Liu
Yu-Shen Liu
Yufan Liu
Yun Liu
Zheng Liu
Zhijian Liu
Zhuang Liu
Zichuan Liu
Ziwei Liu
Zongyi Liu
Stephan Liwicki
Liliana Lo Presti
Chengjiang Long
Fuchen Long
Mingsheng Long
Xiang Long

Yang Long
Charles T. Loop
Antonio Lopez
Roberto J. Lopez-Sastre
Javier Lorenzo-Navarro
Manolis Lourakis
Boyu Lu
Canyi Lu
Feng Lu
Guoyu Lu
Hongtao Lu
Jiajun Lu
Jiasen Lu
Jiwen Lu
Kaiyue Lu
Le Lu
Shao-Ping Lu
Shijian Lu
Xiankai Lu
Xin Lu
Yao Lu
Yiping Lu
Yongxi Lu
Yongyi Lu
Zhiwu Lu
Fujun Luan
Benjamin E. Lundell
Hao Luo
Jian-Hao Luo
Ruotian Luo
Weixin Luo
Wenhan Luo
Wenjie Luo
Yan Luo
Zelun Luo
Zixin Luo
Khoa Luu
Zhaoyang Lv
Pengyuan Lyu
Thomas Möllenhoff
Matthias Müller
Bingpeng Ma
Chih-Yao Ma
Chongyang Ma
Huimin Ma
Jiayi Ma

K. T. Ma
Ke Ma
Lin Ma
Liqian Ma
Shugao Ma
Wei-Chiu Ma
Xiaojian Ma
Xingjun Ma
Zhanyu Ma
Zheng Ma
Radek Jakob Mackowiak
Ludovic Magerand
Shweta Mahajan
Siddharth Mahendran
Long Mai
Ameesh Makadia
Oscar Mendez Maldonado
Mateusz Malinowski
Yury Malkov
Arun Mallya
Dipu Manandhar
Massimiliano Mancini
Fabian Manhardt
Kevis-kokitsi Maninis
Varun Manjunatha
Junhua Mao
Xudong Mao
Alina Marcu
Edgar Margffoy-Tuay
Dmitrii Marin
Manuel J. Marin-Jimenez
Kenneth Marino
Niki Martinel
Julieta Martinez
Jonathan Masci
Tomohiro Mashita
Iacopo Masi
David Masip
Daniela Massiceti
Stefan Mathe
Yusuke Matsui
Tetsu Matsukawa
Iain A. Matthews
Kevin James Matzen
Bruce Allen Maxwell
Stephen Maybank

Björn Ommer
Mohamed Omran
Elisabeta Oneata
Michael Opitz
Jose Oramas
Tribhuvanesh Orekondy
Shaul Oron
Sergio Orts-Escolano
Ivan Oseledets
Aljosa Osep
Magnus Oskarsson
Anton Osokin
Martin R. Oswald
Wanli Ouyang
Andrew Owens
Mete Ozay
Mustafa Ozuysal
Eduardo Pérez-Pellitero
Gautam Pai
Dipan Kumar Pal
P. H. Pamplona Savarese
Jinshan Pan
Junting Pan
Xingang Pan
Yingwei Pan
Yannis Panagakis
Rameswar Panda
Guan Pang
Jiahao Pang
Jiangmiao Pang
Tianyu Pang
Sharath Pankanti
Nicolas Papadakis
Dim Papadopoulos
George Papandreou
Toufiq Parag
Shaifali Parashar
Sarah Parisot
Eunhyeok Park
Hyun Soo Park
Jaesik Park
Min-Gyu Park
Taesung Park
Alvaro Parra
C. Alejandro Parraga
Despoina Paschalidou

Nikolaos Passalis
Vishal Patel
Viorica Patraucean
Badri Narayana Patro
Danda Pani Paudel
Sujoy Paul
Georgios Pavlakos
Ioannis Pavlidis
Vladimir Pavlovic
Nick Pears
Kim Steenstrup Pedersen
Selen Pehlivan
Shmuel Peleg
Chao Peng
Houwen Peng
Wen-Hsiao Peng
Xi Peng
Xiaojiang Peng
Xingchao Peng
Yuxin Peng
Federico Perazzi
Juan Camilo Perez
Vishwanath Peri
Federico Pernici
Luca Del Pero
Florent Perronnin
Stavros Petridis
Henning Petzka
Patrick Peursum
Michael Pfeiffer
Hanspeter Pfister
Roman Pflugfelder
Minh Tri Pham
Yongri Piao
David Picard
Tomasz Pieciak
A. J. Piergiovanni
Andrea Pilzer
Pedro O. Pinheiro
Silvia Laura Pintea
Lerrel Pinto
Axel Pinz
Robinson Piramuthu
Fiora Pirri
Leonid Pishchulin
Francesco Pittaluga

Daniel Pizarro
Tobias Plötz
Mirco Planamente
Matteo Poggi
Moacir A. Ponti
Parita Pooj
Fatih Porikli
Horst Possegger
Omid Poursaeed
Ameya Prabhu
Viraj Uday Prabhu
Dilip Prasad
Brian L. Price
True Price
Maria Priisalu
Veronique Prinet
Victor Adrian Prisacariu
Jan Prokaj
Sergey Prokudin
Nicolas Pugeault
Xavier Puig
Albert Pumarola
Pulak Purkait
Senthil Purushwalkam
Charles R. Qi
Hang Qi
Haozhi Qi
Lu Qi
Mengshi Qi
Siyuan Qi
Xiaojuan Qi
Yuankai Qi
Shengju Qian
Xuelin Qian
Siyuan Qiao
Yu Qiao
Jie Qin
Qiang Qiu
Weichao Qiu
Zhaofan Qiu
Kha Gia Quach
Yuhui Quan
Yvain Queau
Julian Quiroga
Faisal Qureshi
Mahdi Rad

Filip Radenovic
Petia Radeva
Venkatesh
  B. Radhakrishnan
Ilija Radosavovic
Noha Radwan
Rahul Raguram
Tanzila Rahman
Amit Raj
Ajit Rajwade
Kandan Ramakrishnan
Santhosh
  K. Ramakrishnan
Srikumar Ramalingam
Ravi Ramamoorthi
Vasili Ramanishka
Ramprasaath R. Selvaraju
Francois Rameau
Visvanathan Ramesh
Santu Rana
Rene Ranftl
Anand Rangarajan
Anurag Ranjan
Viresh Ranjan
Yongming Rao
Carolina Raposo
Vivek Rathod
Sathya N. Ravi
Avinash Ravichandran
Tammy Riklin Raviv
Daniel Rebain
Sylvestre-Alvise Rebuffi
N. Dinesh Reddy
Timo Rehfeld
Paolo Remagnino
Konstantinos Rematas
Edoardo Remelli
Dongwei Ren
Haibing Ren
Jian Ren
Jimmy Ren
Mengye Ren
Weihong Ren
Wenqi Ren
Zhile Ren
Zhongzheng Ren

Zhou Ren
Vijay Rengarajan
Md A. Reza
Farzaneh Rezaeianaran
Hamed R. Tavakoli
Nicholas Rhinehart
Helge Rhodin
Elisa Ricci
Alexander Richard
Eitan Richardson
Elad Richardson
Christian Richardt
Stephan Richter
Gernot Riegler
Daniel Ritchie
Tobias Ritschel
Samuel Rivera
Yong Man Ro
Richard Roberts
Joseph Robinson
Ignacio Rocco
Mrigank Rochan
Emanuele Rodolà
Mikel D. Rodriguez
Giorgio Roffo
Grégory Rogez
Gemma Roig
Javier Romero
Xuejian Rong
Yu Rong
Amir Rosenfeld
Bodo Rosenhahn
Guy Rosman
Arun Ross
Paolo Rota
Peter M. Roth
Anastasios Roussos
Anirban Roy
Sebastien Roy
Aruni RoyChowdhury
Artem Rozantsev
Ognjen Rudovic
Daniel Rueckert
Adria Ruiz
Javier Ruiz-del-solar
Christian Rupprecht

Chris Russell
Dan Ruta
Jongbin Ryu
Ömer Sümer
Alexandre Sablayrolles
Faraz Saeedan
Ryusuke Sagawa
Christos Sagonas
Tonmoy Saikia
Hideo Saito
Kuniaki Saito
Shunsuke Saito
Shunta Saito
Ken Sakurada
Joaquin Salas
Fatemeh Sadat Saleh
Mahdi Saleh
Pouya Samangouei
Leo Sampaio
  Ferraz Ribeiro
Artsiom Olegovich
  Sanakoyeu
Enrique Sanchez
Patsorn Sangkloy
Anush Sankaran
Aswin Sankaranarayanan
Swami Sankaranarayanan
Rodrigo Santa Cruz
Amartya Sanyal
Archana Sapkota
Nikolaos Sarafianos
Jun Sato
Shin'ichi Satoh
Hosnieh Sattar
Arman Savran
Manolis Savva
Alexander Sax
Hanno Scharr
Simone Schaub-Meyer
Konrad Schindler
Dmitrij Schlesinger
Uwe Schmidt
Dirk Schnieders
Björn Schuller
Samuel Schulter
Idan Schwartz

William Robson Schwartz
Alex Schwing
Sinisa Segvic
Lorenzo Seidenari
Pradeep Sen
Ozan Sener
Soumyadip Sengupta
Arda Senocak
Mojtaba Seyedhosseini
Shishir Shah
Shital Shah
Sohil Atul Shah
Tamar Rott Shaham
Huasong Shan
Qi Shan
Shiguang Shan
Jing Shao
Roman Shapovalov
Gaurav Sharma
Vivek Sharma
Viktoriia Sharmanska
Dongyu She
Sumit Shekhar
Evan Shelhamer
Chengyao Shen
Chunhua Shen
Falong Shen
Jie Shen
Li Shen
Liyue Shen
Shuhan Shen
Tianwei Shen
Wei Shen
William B. Shen
Yantao Shen
Ying Shen
Yiru Shen
Yujun Shen
Yuming Shen
Zhiqiang Shen
Ziyi Shen
Lu Sheng
Yu Sheng
Rakshith Shetty
Baoguang Shi
Guangming Shi

Hailin Shi
Miaojing Shi
Yemin Shi
Zhenmei Shi
Zhiyuan Shi
Kevin Jonathan Shih
Shiliang Shiliang
Hyunjung Shim
Atsushi Shimada
Nobutaka Shimada
Daeyun Shin
Young Min Shin
Koichi Shinoda
Konstantin Shmelkov
Michael Zheng Shou
Abhinav Shrivastava
Tianmin Shu
Zhixin Shu
Hong-Han Shuai
Pushkar Shukla
Christian Siagian
Mennatullah M. Siam
Kaleem Siddiqi
Karan Sikka
Jae-Young Sim
Christian Simon
Martin Simonovsky
Dheeraj Singaraju
Bharat Singh
Gurkirt Singh
Krishna Kumar Singh
Maneesh Kumar Singh
Richa Singh
Saurabh Singh
Suriya Singh
Vikas Singh
Sudipta N. Sinha
Vincent Sitzmann
Josef Sivic
Gregory Slabaugh
Miroslava Slavcheva
Ron Slossberg
Brandon Smith
Kevin Smith
Vladimir Smutny
Noah Snavely

Roger
    D. Soberanis-Mukul
Kihyuk Sohn
Francesco Solera
Eric Sommerlade
Sanghyun Son
Byung Cheol Song
Chunfeng Song
Dongjin Song
Jiaming Song
Jie Song
Jifei Song
Jingkuan Song
Mingli Song
Shiyu Song
Shuran Song
Xiao Song
Yafei Song
Yale Song
Yang Song
Yi-Zhe Song
Yibing Song
Humberto Sossa
Cesar de Souza
Adrian Spurr
Srinath Sridhar
Suraj Srinivas
Pratul P. Srinivasan
Anuj Srivastava
Tania Stathaki
Christopher Stauffer
Simon Stent
Rainer Stiefelhagen
Pierre Stock
Julian Straub
Jonathan C. Stroud
Joerg Stueckler
Jan Stuehmer
David Stutz
Chi Su
Hang Su
Jong-Chyi Su
Shuochen Su
Yu-Chuan Su
Ramanathan Subramanian
Yusuke Sugano

Masanori Suganuma
Yumin Suh
Mohammed Suhail
Yao Sui
Heung-Il Suk
Josephine Sullivan
Baochen Sun
Chen Sun
Chong Sun
Deqing Sun
Jin Sun
Liang Sun
Lin Sun
Qianru Sun
Shao-Hua Sun
Shuyang Sun
Weiwei Sun
Wenxiu Sun
Xiaoshuai Sun
Xiaoxiao Sun
Xingyuan Sun
Yifan Sun
Zhun Sun
Sabine Susstrunk
David Suter
Supasorn Suwajanakorn
Tomas Svoboda
Eran Swears
Paul Swoboda
Attila Szabo
Richard Szeliski
Duy-Nguyen Ta
Andrea Tagliasacchi
Yuichi Taguchi
Ying Tai
Keita Takahashi
Kouske Takahashi
Jun Takamatsu
Hugues Talbot
Toru Tamaki
Chaowei Tan
Fuwen Tan
Mingkui Tan
Mingxing Tan
Qingyang Tan
Robby T. Tan

Xiaoyang Tan
Kenichiro Tanaka
Masayuki Tanaka
Chang Tang
Chengzhou Tang
Danhang Tang
Ming Tang
Peng Tang
Qingming Tang
Wei Tang
Xu Tang
Yansong Tang
Youbao Tang
Yuxing Tang
Zhiqiang Tang
Tatsunori Taniai
Junli Tao
Xin Tao
Makarand Tapaswi
Jean-Philippe Tarel
Lyne Tchapmi
Zachary Teed
Bugra Tekin
Damien Teney
Ayush Tewari
Christian Theobalt
Christopher Thomas
Diego Thomas
Jim Thomas
Rajat Mani Thomas
Xinmei Tian
Yapeng Tian
Yingli Tian
Yonglong Tian
Zhi Tian
Zhuotao Tian
Kinh Tieu
Joseph Tighe
Massimo Tistarelli
Matthew Toews
Carl Toft
Pavel Tokmakov
Federico Tombari
Chetan Tonde
Yan Tong
Alessio Tonioni

Andrea Torsello
Fabio Tosi
Du Tran
Luan Tran
Ngoc-Trung Tran
Quan Hung Tran
Truyen Tran
Rudolph Triebel
Martin Trimmel
Shashank Tripathi
Subarna Tripathi
Leonardo Trujillo
Eduard Trulls
Tomasz Trzcinski
Sam Tsai
Yi-Hsuan Tsai
Hung-Yu Tseng
Stavros Tsogkas
Aggeliki Tsoli
Devis Tuia
Shubham Tulsiani
Sergey Tulyakov
Frederick Tung
Tony Tung
Daniyar Turmukhambetov
Ambrish Tyagi
Radim Tylecek
Christos Tzelepis
Georgios Tzimiropoulos
Dimitrios Tzionas
Seiichi Uchida
Norimichi Ukita
Dmitry Ulyanov
Martin Urschler
Yoshitaka Ushiku
Ben Usman
Alexander Vakhitov
Julien P. C. Valentin
Jack Valmadre
Ernest Valveny
Joost van de Weijer
Jan van Gemert
Koen Van Leemput
Gul Varol
Sebastiano Vascon
M. Alex O. Vasilescu

Subeesh Vasu
Mayank Vatsa
David Vazquez
Javier Vazquez-Corral
Ashok Veeraraghavan
Erik Velasco-Salido
Raviteja Vemulapalli
Jonathan Ventura
Manisha Verma
Roberto Vezzani
Ruben Villegas
Minh Vo
MinhDuc Vo
Nam Vo
Michele Volpi
Riccardo Volpi
Carl Vondrick
Konstantinos Vougioukas
Tuan-Hung Vu
Sven Wachsmuth
Neal Wadhwa
Catherine Wah
Jacob C. Walker
Thomas S. A. Wallis
Chengde Wan
Jun Wan
Liang Wan
Renjie Wan
Baoyuan Wang
Boyu Wang
Cheng Wang
Chu Wang
Chuan Wang
Chunyu Wang
Dequan Wang
Di Wang
Dilin Wang
Dong Wang
Fang Wang
Guanzhi Wang
Guoyin Wang
Hanzi Wang
Hao Wang
He Wang
Heng Wang
Hongcheng Wang

Hongxing Wang
Hua Wang
Jian Wang
Jingbo Wang
Jinglu Wang
Jingya Wang
Jinjun Wang
Jinqiao Wang
Jue Wang
Ke Wang
Keze Wang
Le Wang
Lei Wang
Lezi Wang
Li Wang
Liang Wang
Lijun Wang
Limin Wang
Linwei Wang
Lizhi Wang
Mengjiao Wang
Mingzhe Wang
Minsi Wang
Naiyan Wang
Nannan Wang
Ning Wang
Oliver Wang
Pei Wang
Peng Wang
Pichao Wang
Qi Wang
Qian Wang
Qiaosong Wang
Qifei Wang
Qilong Wang
Qing Wang
Qingzhong Wang
Quan Wang
Rui Wang
Ruiping Wang
Ruixing Wang
Shangfei Wang
Shenlong Wang
Shiyao Wang
Shuhui Wang
Song Wang

Tao Wang
Tianlu Wang
Tiantian Wang
Ting-chun Wang
Tingwu Wang
Wei Wang
Weiyue Wang
Wenguan Wang
Wenlin Wang
Wenqi Wang
Xiang Wang
Xiaobo Wang
Xiaofang Wang
Xiaoling Wang
Xiaolong Wang
Xiaosong Wang
Xiaoyu Wang
Xin Eric Wang
Xinchao Wang
Xinggang Wang
Xintao Wang
Yali Wang
Yan Wang
Yang Wang
Yangang Wang
Yaxing Wang
Yi Wang
Yida Wang
Yilin Wang
Yiming Wang
Yisen Wang
Yongtao Wang
Yu-Xiong Wang
Yue Wang
Yujiang Wang
Yunbo Wang
Yunhe Wang
Zengmao Wang
Zhangyang Wang
Zhaowen Wang
Zhe Wang
Zhecan Wang
Zheng Wang
Zhixiang Wang
Zilei Wang
Jianqiao Wangni

Anne S. Wannenwetsch
Jan Dirk Wegner
Scott Wehrwein
Donglai Wei
Kaixuan Wei
Longhui Wei
Pengxu Wei
Ping Wei
Qi Wei
Shih-En Wei
Xing Wei
Yunchao Wei
Zijun Wei
Jerod Weinman
Michael Weinmann
Philippe Weinzaepfel
Yair Weiss
Bihan Wen
Longyin Wen
Wei Wen
Junwu Weng
Tsui-Wei Weng
Xinshuo Weng
Eric Wengrowski
Tomas Werner
Gordon Wetzstein
Tobias Weyand
Patrick Wieschollek
Maggie Wigness
Erik Wijmans
Richard Wildes
Olivia Wiles
Chris Williams
Williem Williem
Kyle Wilson
Calden Wloka
Nicolai Wojke
Christian Wolf
Yongkang Wong
Sanghyun Woo
Scott Workman
Baoyuan Wu
Bichen Wu
Chao-Yuan Wu
Huikai Wu
Jiajun Wu

Jialin Wu
Jiaxiang Wu
Jiqing Wu
Jonathan Wu
Lifang Wu
Qi Wu
Qiang Wu
Ruizheng Wu
Shangzhe Wu
Shun-Cheng Wu
Tianfu Wu
Wayne Wu
Wenxuan Wu
Xiao Wu
Xiaohe Wu
Xinxiao Wu
Yang Wu
Yi Wu
Yiming Wu
Ying Nian Wu
Yue Wu
Zheng Wu
Zhenyu Wu
Zhirong Wu
Zuxuan Wu
Stefanie Wuhrer
Jonas Wulff
Changqun Xia
Fangting Xia
Fei Xia
Gui-Song Xia
Lu Xia
Xide Xia
Yin Xia
Yingce Xia
Yongqin Xian
Lei Xiang
Shiming Xiang
Bin Xiao
Fanyi Xiao
Guobao Xiao
Huaxin Xiao
Taihong Xiao
Tete Xiao
Tong Xiao
Wang Xiao

Yang Xiao
Cihang Xie
Guosen Xie
Jianwen Xie
Lingxi Xie
Sirui Xie
Weidi Xie
Wenxuan Xie
Xiaohua Xie
Fuyong Xing
Jun Xing
Junliang Xing
Bo Xiong
Peixi Xiong
Yu Xiong
Yuanjun Xiong
Zhiwei Xiong
Chang Xu
Chenliang Xu
Dan Xu
Danfei Xu
Hang Xu
Hongteng Xu
Huijuan Xu
Jingwei Xu
Jun Xu
Kai Xu
Mengmeng Xu
Mingze Xu
Qianqian Xu
Ran Xu
Weijian Xu
Xiangyu Xu
Xiaogang Xu
Xing Xu
Xun Xu
Yanyu Xu
Yichao Xu
Yong Xu
Yongchao Xu
Yuanlu Xu
Zenglin Xu
Zheng Xu
Chuhui Xue
Jia Xue
Nan Xue

Tianfan Xue
Xiangyang Xue
Abhay Yadav
Yasushi Yagi
I. Zeki Yalniz
Kota Yamaguchi
Toshihiko Yamasaki
Takayoshi Yamashita
Junchi Yan
Ke Yan
Qingan Yan
Sijie Yan
Xinchen Yan
Yan Yan
Yichao Yan
Zhicheng Yan
Keiji Yanai
Bin Yang
Ceyuan Yang
Dawei Yang
Dong Yang
Fan Yang
Guandao Yang
Guorun Yang
Haichuan Yang
Hao Yang
Jianwei Yang
Jiaolong Yang
Jie Yang
Jing Yang
Kaiyu Yang
Linjie Yang
Meng Yang
Michael Ying Yang
Nan Yang
Shuai Yang
Shuo Yang
Tianyu Yang
Tien-Ju Yang
Tsun-Yi Yang
Wei Yang
Wenhan Yang
Xiao Yang
Xiaodong Yang
Xin Yang
Yan Yang

Yanchao Yang
Yee Hong Yang
Yezhou Yang
Zhenheng Yang
Anbang Yao
Angela Yao
Cong Yao
Jian Yao
Li Yao
Ting Yao
Yao Yao
Zhewei Yao
Chengxi Ye
Jianbo Ye
Keren Ye
Linwei Ye
Mang Ye
Mao Ye
Qi Ye
Qixiang Ye
Mei-Chen Yeh
Raymond Yeh
Yu-Ying Yeh
Sai-Kit Yeung
Serena Yeung
Kwang Moo Yi
Li Yi
Renjiao Yi
Alper Yilmaz
Junho Yim
Lijun Yin
Weidong Yin
Xi Yin
Zhichao Yin
Tatsuya Yokota
Ryo Yonetani
Donggeun Yoo
Jae Shin Yoon
Ju Hong Yoon
Sung-eui Yoon
Laurent Younes
Changqian Yu
Fisher Yu
Gang Yu
Jiahui Yu
Kaicheng Yu

Ke Yu
Lequan Yu
Ning Yu
Qian Yu
Ronald Yu
Ruichi Yu
Shoou-I Yu
Tao Yu
Tianshu Yu
Xiang Yu
Xin Yu
Xiyu Yu
Youngjae Yu
Yu Yu
Zhiding Yu
Chunfeng Yuan
Ganzhao Yuan
Jinwei Yuan
Lu Yuan
Quan Yuan
Shanxin Yuan
Tongtong Yuan
Wenjia Yuan
Ye Yuan
Yuan Yuan
Yuhui Yuan
Huanjing Yue
Xiangyu Yue
Ersin Yumer
Sergey Zagoruyko
Egor Zakharov
Amir Zamir
Andrei Zanfir
Mihai Zanfir
Pablo Zegers
Bernhard Zeisl
John S. Zelek
Niclas Zeller
Huayi Zeng
Jiabei Zeng
Wenjun Zeng
Yu Zeng
Xiaohua Zhai
Fangneng Zhan
Huangying Zhan
Kun Zhan

Xiaohang Zhan
Baochang Zhang
Bowen Zhang
Cecilia Zhang
Changqing Zhang
Chao Zhang
Chengquan Zhang
Chi Zhang
Chongyang Zhang
Dingwen Zhang
Dong Zhang
Feihu Zhang
Hang Zhang
Hanwang Zhang
Hao Zhang
He Zhang
Hongguang Zhang
Hua Zhang
Ji Zhang
Jianguo Zhang
Jianming Zhang
Jiawei Zhang
Jie Zhang
Jing Zhang
Juyong Zhang
Kai Zhang
Kaipeng Zhang
Ke Zhang
Le Zhang
Lei Zhang
Li Zhang
Lihe Zhang
Linguang Zhang
Lu Zhang
Mi Zhang
Mingda Zhang
Peng Zhang
Pingping Zhang
Qian Zhang
Qilin Zhang
Quanshi Zhang
Richard Zhang
Rui Zhang
Runze Zhang
Shengping Zhang
Shifeng Zhang

Shuai Zhang
Songyang Zhang
Tao Zhang
Ting Zhang
Tong Zhang
Wayne Zhang
Wei Zhang
Weizhong Zhang
Wenwei Zhang
Xiangyu Zhang
Xiaolin Zhang
Xiaopeng Zhang
Xiaoqin Zhang
Xiuming Zhang
Ya Zhang
Yang Zhang
Yimin Zhang
Yinda Zhang
Ying Zhang
Yongfei Zhang
Yu Zhang
Yulun Zhang
Yunhua Zhang
Yuting Zhang
Zhanpeng Zhang
Zhao Zhang
Zhaoxiang Zhang
Zhen Zhang
Zheng Zhang
Zhifei Zhang
Zhijin Zhang
Zhishuai Zhang
Ziming Zhang
Bo Zhao
Chen Zhao
Fang Zhao
Haiyu Zhao
Han Zhao
Hang Zhao
Hengshuang Zhao
Jian Zhao
Kai Zhao
Liang Zhao
Long Zhao
Qian Zhao
Qibin Zhao

Qijun Zhao
Rui Zhao
Shenglin Zhao
Sicheng Zhao
Tianyi Zhao
Wenda Zhao
Xiangyun Zhao
Xin Zhao
Yang Zhao
Yue Zhao
Zhichen Zhao
Zijing Zhao
Xiantong Zhen
Chuanxia Zheng
Feng Zheng
Haiyong Zheng
Jia Zheng
Kang Zheng
Shuai Kyle Zheng
Wei-Shi Zheng
Yinqiang Zheng
Zerong Zheng
Zhedong Zheng
Zilong Zheng
Bineng Zhong
Fangwei Zhong
Guangyu Zhong
Yiran Zhong
Yujie Zhong
Zhun Zhong
Chunluan Zhou
Huiyu Zhou
Jiahuan Zhou
Jun Zhou
Lei Zhou
Luowei Zhou
Luping Zhou
Mo Zhou
Ning Zhou
Pan Zhou
Peng Zhou
Qianyi Zhou
S. Kevin Zhou
Sanping Zhou
Wengang Zhou
Xingyi Zhou

Yanzhao Zhou
Yi Zhou
Yin Zhou
Yipin Zhou
Yuyin Zhou
Zihan Zhou
Alex Zihao Zhu
Chenchen Zhu
Feng Zhu
Guangming Zhu
Ji Zhu
Jun-Yan Zhu
Lei Zhu
Linchao Zhu
Rui Zhu
Shizhan Zhu
Tyler Lixuan Zhu

Wei Zhu
Xiangyu Zhu
Xinge Zhu
Xizhou Zhu
Yanjun Zhu
Yi Zhu
Yixin Zhu
Yizhe Zhu
Yousong Zhu
Zhe Zhu
Zhen Zhu
Zheng Zhu
Zhenyao Zhu
Zhihui Zhu
Zhuotun Zhu
Bingbing Zhuang
Wei Zhuo

Christian Zimmermann
Karel Zimmermann
Larry Zitnick
Mohammadreza
  Zolfaghari
Maria Zontak
Daniel Zoran
Changqing Zou
Chuhang Zou
Danping Zou
Qi Zou
Yang Zou
Yuliang Zou
Georgios Zoumpourlis
Wangmeng Zuo
Xinxin Zuo

Additional Reviewers

Victoria Fernandez
  Abrevaya
Maya Aghaei
Allam Allam
Christine
  Allen-Blanchette
Nicolas Aziere
Assia Benbihi
Neha Bhargava
Bharat Lal Bhatnagar
Joanna Bitton
Judy Borowski
Amine Bourki
Romain Brégier
Tali Brayer
Sebastian Bujwid
Andrea Burns
Yun-Hao Cao
Yuning Chai
Xiaojun Chang
Bo Chen
Shuo Chen
Zhixiang Chen
Junsuk Choe
Hung-Kuo Chu

Jonathan P. Crall
Kenan Dai
Lucas Deecke
Karan Desai
Prithviraj Dhar
Jing Dong
Wei Dong
Turan Kaan Elgin
Francis Engelmann
Erik Englesson
Fartash Faghri
Zicong Fan
Yang Fu
Risheek Garrepalli
Yifan Ge
Marco Godi
Helmut Grabner
Shuxuan Guo
Jianfeng He
Zhezhi He
Samitha Herath
Chih-Hui Ho
Yicong Hong
Vincent Tao Hu
Julio Hurtado

Jaedong Hwang
Andrey Ignatov
Muhammad
  Abdullah Jamal
Saumya Jetley
Meiguang Jin
Jeff Johnson
Minsoo Kang
Saeed Khorram
Mohammad Rami Koujan
Nilesh Kulkarni
Sudhakar Kumawat
Abdelhak Lemkhenter
Alexander Levine
Jiachen Li
Jing Li
Jun Li
Yi Li
Liang Liao
Ruochen Liao
Tzu-Heng Lin
Phillip Lippe
Bao-di Liu
Bo Liu
Fangchen Liu

Hanxiao Liu
Hongyu Liu
Huidong Liu
Miao Liu
Xinxin Liu
Yongfei Liu
Yu-Lun Liu
Amir Livne
Tiange Luo
Wei Ma
Xiaoxuan Ma
Ioannis Marras
Georg Martius
Effrosyni Mavroudi
Tim Meinhardt
Givi Meishvili
Meng Meng
Zihang Meng
Zhongqi Miao
Gyeongsik Moon
Khoi Nguyen
Yung-Kyun Noh
Antonio Norelli
Jaeyoo Park
Alexander Pashevich
Mandela Patrick
Mary Phuong
Bingqiao Qian
Yu Qiao
Zhen Qiao
Sai Saketh Rambhatla
Aniket Roy
Amelie Royer
Parikshit Vishwas
    Sakurikar
Mark Sandler
Mert Bülent Sarıyıldız
Tanner Schmidt
Anshul B. Shah

Ketul Shah
Rajvi Shah
Hengcan Shi
Xiangxi Shi
Yujiao Shi
William A. P. Smith
Guoxian Song
Robin Strudel
Abby Stylianou
Xinwei Sun
Reuben Tan
Qingyi Tao
Kedar S. Tatwawadi
Anh Tuan Tran
Son Dinh Tran
Eleni Triantafillou
Aristeidis Tsitiridis
Md Zasim Uddin
Andrea Vedaldi
Evangelos Ververas
Vidit Vidit
Paul Voigtlaender
Bo Wan
Huanyu Wang
Huiyu Wang
Junqiu Wang
Pengxiao Wang
Tai Wang
Xinyao Wang
Tomoki Watanabe
Mark Weber
Xi Wei
Botong Wu
James Wu
Jiamin Wu
Rujie Wu
Yu Wu
Rongchang Xie
Wei Xiong

Yunyang Xiong
An Xu
Chi Xu
Yinghao Xu
Fei Xue
Tingyun Yan
Zike Yan
Chao Yang
Heran Yang
Ren Yang
Wenfei Yang
Xu Yang
Rajeev Yasarla
Shaokai Ye
Yufei Ye
Kun Yi
Haichao Yu
Hanchao Yu
Ruixuan Yu
Liangzhe Yuan
Chen-Lin Zhang
Fandong Zhang
Tianyi Zhang
Yang Zhang
Yiyi Zhang
Yongshun Zhang
Yu Zhang
Zhiwei Zhang
Jiaojiao Zhao
Yipu Zhao
Xingjian Zhen
Haizhong Zheng
Tiancheng Zhi
Chengju Zhou
Hao Zhou
Hao Zhu
Alexander Zimin

Contents – Part XVIII

3D Bird Reconstruction: A Dataset, Model, and Shape Recovery from a Single View

Marc Badger(✉)ⓘ, Yufu Wangⓘ, Adarsh Modhⓘ, Ammon Perkesⓘ,
Nikos Kolotourosⓘ, Bernd G. Pfrommerⓘ, Marc F. Schmidtⓘ,
and Kostas Daniilidisⓘ

University of Pennsylvania, Philadelphia, PA 19104, USA
{mbadger,yufu,adarshm,nkolot,pfrommer,kostas}@seas.upenn.edu,
{aperkes,marcschm}@sas.upenn.edu

Abstract. Automated capture of animal pose is transforming how we study neuroscience and social behavior. Movements carry important social cues, but current methods are not able to robustly estimate pose and shape of animals, particularly for social animals such as birds, which are often occluded by each other and objects in the environment. To address this problem, we first introduce a model and multi-view optimization approach, which we use to capture the unique shape and pose space displayed by live birds. We then introduce a pipeline and experiments for keypoint, mask, pose, and shape regression that recovers accurate avian postures from single views. Finally, we provide extensive multi-view keypoint and mask annotations collected from a group of 15 social birds housed together in an outdoor aviary. The project website with videos, results, code, mesh model, and the Penn Aviary Dataset can be found at https://marcbadger.github.io/avian-mesh.

Keywords: Pose estimation · Shape estimation · Birds · Animals · Dataset

1 Introduction

Why Computational Ethology? Accurate measurement of behavior is vital to disciplines ranging from neuroscience and biomechanics to human health and agriculture. Through automated measurement, computational ethology aims to capture complex variation in posture, orientation, and position of multiple individuals over time as they interact with each other and their environment [1]. Pose trajectories contain rich, unbiased, information from which we can extract more abstract features that are relevant to brain function, social interactions, biomechanics, and health. Studying neural functions in the context of natural social

Electronic supplementary material The online version of this chapter (https://doi.org/10.1007/978-3-030-58523-5_1) contains supplementary material, which is available to authorized users.

© Springer Nature Switzerland AG 2020
A. Vedaldi et al. (Eds.): ECCV 2020, LNCS 12363, pp. 1–17, 2020.
https://doi.org/10.1007/978-3-030-58523-5_1

behavior is a critical step toward a deeper understanding how the brain integrates perception, cognition, and learning and memory to produce behavior. Pose trajectories reveal how animals maneuver to negotiate cluttered environments, how animals make decisions while foraging or searching for mates, and how the collective behavior of a group arises from individual decisions. Automated capture of difficult-to-observe behaviors is transforming diverse applications by streamlining the process of extracting quantitative physiological, behavioral, and social data from images and video.

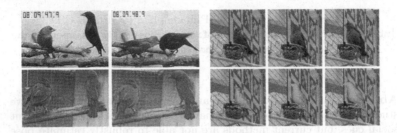

Fig. 1. Visual signals convey important social cues in birds. Motions such as pecking (top left) and wingstrokes (bottom left) drive social behavior in both males and females. Complex wing folding and large changes in body volume when feathers are puffed (upper right) make shape recovery (lower right) a difficult task. Images from [38] and [2].

Why Does Bird Posture Matter? Why Cowbirds? Understanding how the collective behavior of social groups arises from individual interactions is important for studying the evolution of sociality and neural mechanisms behind social behaviors. Although vocalizations are a clear channel for communication in birds, surprisingly, changes in posture, orientation, and position also play an important role in communication. One of the best studied groups from both behavioal and neuroscience perspectives are the brown-headed cowbirds (*Molothrus ater*). **In cowbirds, females influence the behavior of males through a number of visual mechanisms including "wingstrokes", which involve changes in both pose and shape over time** [38] (Fig. 1). Interactions between birds are usually recorded by observing a focal individual's interactions in person in the field. Although insightful, such manual observations contain observer bias, miss interactions between non-focal individuals, and cannot be performed continuously for long periods. Qualitative observations also miss important variation in posture that would be revealed by a quantitative approach. For example, Fig. 1 shows changes in pose and shape that can serve as social cues in cowbirds. The ability to estimate the pose of multiple interacting individuals would transform the study of animal communication [1], as is it beginning to do for humans [13,14,29]. Estimating the pose and shape of birds in a social context, however, presents several challenges.

Fig. 2. Appearance variation across bird identity (top vs bottom) and across viewpoint, time of day, and season (1st column vs. columns 2–4 respectively). The red box within the left image of each panel shows the location of the enlarged crop (right image).

Why is Estimating Bird Pose and Shape Challenging? Recovering shape and pose of birds in the wild is challenging for the following four reasons:

1. Changes in pose and shape are difficult to model in birds.
2. No pose or shape priors are available.
3. Many birds are only visible from a single unoccluded view.
4. Appearance variation in natural settings makes detection difficult.

Shape is particularly difficult to model because birds have highly mobile feathers that allow dramatic changes in both shape (e.g. tail fanning) and perceived body volume (e.g. feather puffing in Fig. 1). Furthermore, when the wings are held next to the body, they are folded in a complex way in which much of the wing surface becomes sandwiched between the top of the wing and the body. These "internal" surfaces cannot be recovered from scans of figurines with folded wings and figurines with wings in intermediate poses are not available. In addition to modeling challenges, cowbirds interact in a complex environment containing extreme variation in illumination and may be heavily occluded either by vegetation by other birds in the group.

Animal posture is often described using joint angles derived from semantic keypoints, joints, or other anatomical locations. This approach is attractive because keypoints are easy to identify and can readily be localized by deep-learning-based software packages such as DeepLabCut [26], DeepPoseKit [10], and LEAP [32]. Under heavy occlusion, however, even multi-view setups frequently do not observe relevant keypoints from more than one view. One solution is to lift the pose from 2D to 3D, but unlike for humans, researchers do not yet have strong species-specific priors for tackling this problem. We overcome the limitations of directly using 2D keypoints and skeletons by fitting a 3D parameterized mesh model with priors learned from a multi-view dataset.

Dataset. With the aim of creating a robust system for estimating the shape and pose of multiple interacting birds over months-long timescales, we recorded the behavior of 15 cowbirds housed together in an outdoor aviary over the course of a single three-month mating season. Our carefully calibrated multi-view dataset contains large variation in (i) bird pose, orientation, and position/depth, (ii) viewpoint across eight cameras, and (iii) appearance across different lighting conditions (time of day and weather) and seasons (Fig. 2). Cowbirds have a nearly textureless appearance and birds move freely and explore all three dimensions of

Fig. 3. The dataset and model. We provide multi-view segmentation masks for over 6300 bird instances, keypoints for 1000 bird instances, the first articulated 3D mesh model of a bird, and a full pipeline for recovering the shape and pose of birds from single views.

their cage, producing a large range of subject depth with respect to the camera. Importantly, both perched and flying birds adopt postures covering a large range of motion in both orientation and pose.

We annotated silhouette and keypoints for 1000 instances and matched these annotations across views. Although 90% of annotated birds were visible from 3 or more cameras, about half of the annotated instances were occluded to some degree. Only 62% of birds had more than one unoccluded view, highlighting the need for a single-view approach.

After collecting keypoint and silhouette ground truth from multiple views, we fit our avian mesh model using a multi-view optimization-based approach to learn a shape space and pose priors. We then use the model and priors to train a neural network to regress pose parameters directly from keypoint and silhouette data. These pose parameters can be used to initialize a single-view optimization procedure to further refine body pose and shape (Fig. 4). We use our dataset for learning instance segmentation and keypoint localization, and for estimating bird pose and shape, but our dataset could also be used in the future for learning Re-ID tasks.

In summary, our contributions are focused around the four challenges mentioned previously:

1. We develop the first parameterized avian mesh model that is capable of capturing the unique pose and shape changes displayed by birds.
2. We fit our mesh model to available multi-view keypoint and silhouette data using an optimization-based approach to obtain an accurate shape space and pose prior.
3. We develop a neural network based pipeline for recovering the shape and pose of birds from a single view.
4. We present a challenging multi-view dataset for studying social behavior in birds. The dataset contains extreme variation in subject appearance and depth and many subjects are fully or partially occluded in all but one view.

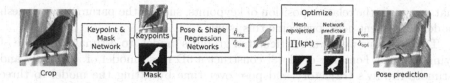

Fig. 4. We estimate the 3D pose and shape of birds from a single view. Given a detection and associated bounding box, we predict body keypoints and a mask. We then predict the parameters of an articulated avian mesh model, which provides a good initial estimate for optional further optimization.

2 Related Work

Human Pose and Shape Estimation. Recent advances in human pose estimation have capitalized on i) powerful 2D joint detectors, ii) 3D pose priors, and iii) low-dimensional articulated 3D shape models of the human body. SMPL [25], the most popular formulation, first deforms a template mesh using shape and pose parameters learned from over 1000 registered body scans of people [4] and then uses linear blend skinning (LBS) to transform mesh vertices given a set of joint angles. In SMPLify, Bogo et al. [3] estimate 3D human pose and shape from single images by fitting SMPL to 2D keypoints. Huang et al. [12] extend SMPLify [3] to the multi-view setting and show a positive effect of silhouette supervision in addition to keypoints. Pavlakos et al. [31] estimate pose and shape directly from predicted keypoints and silhouettes in an end-to-end framework. Recent approaches regress pose and shape directly from images and use adversaries with access to a 3D pose dataset [15], Graph-CNN architectures [21], texture consistency [30], and model-fitting within the training loop [20]. All of the above methods base their approach on parameterized mesh models indicating their critical importance for bridging between observation in 2D and estimation in 3D. In contrast to previous works that rely on 3D scans and SMPL-like models to develop meshes and shape spaces for novel domains such as hands [34], faces [22], and four-legged animals [42], we learn our avian mesh model directly from video data of live birds.

Animal Pose and Shape Estimation. Within biology, most work focuses on isolated animals with no background clutter and few occlusions. Mathis et al. [26] and Pereira et al. [32] recently provided tools for training convolutional neural networks for keypoint localization. Graving et al. [10] localize keypoints on three datasets of fruit flies [32], desert locusts [10], and Grévy's zebras [10]. Günel et al. [11] use a Stacked Hourglass network [27] for 2D keypoint localization in flies and perform pictorial structures and belief propagation message passing [7] to reconstruct 3D pose from 2D detections. Liu and Belhumeur et al. [24] use HOG descriptors and linear SVMs to localize bird parts in the more challenging CUB-200-2011 dataset [37]. All of these works are based on the detection and direct triangulation of 2D keypoints. A fundamental challenge, however, is that any particular keypoint may not be visible from more than one view. Models that constrain the relative position of keypoints, such as the parameterized mesh model we present here, overcome this issue.

Two previous works use articulated graphics models to estimate the pose of flying animals. Fontaine et al. [8] construct a 3D mesh model of a fruit fly and estimate the fly's trajectory and pose over time by fitting the model to three orthogonal views. Breslav [5] create a two-DOF 3D graphics model of a bat and use a Markov Random Field to estimate the 3D pose of bats flying in the wild captured with a multi-view thermal camera setup.

Animal shape estimation is a difficult task. Cashman and Fitzgibbon [6] estimate the shape of dolphins. Ntouskos et al. [28] fit shape primitives to silhouettes of four legged animals. Vincente and Agapito [36] obtain and deform a template mesh using silhouettes from two reference images. Kanazawa et al. [16] learn how animals deform from images by creating an animal-specific model of local stiffness. Kanazawa et al. [17] predict shape, pose, and texture of birds in CUB-200 by deforming a spherical mesh, but do not model pose and thus the locations of wingtips on the mesh are often topologically adjacent to the tail rather than near the shoulders. Zuffi et al. [42] create a realistic, parameterized 3D model (SMAL) from scans of toys by aligning a four-legged template to the scans. They capture shape using PCA coefficients of the aligned meshes and learn a pose prior from a short walking video. Zuffi, Kanazawa, and Black [41] fit the SMAL model to several images of the same animal and then refine the shape to better fit the image data, resulting in capture of both shape and texture (SMALR). Zuffi et al. [40] estimate 3D pose, shape, and texture of zebras in the wild by integrating the SMAL model into an end-to-end network regression pipeline. Their key insight was to first use SMALR to pose an existing horse model and capture a rough texture of the target species. A common feature of these approaches is that they create or leverage a parameterized mesh model. The SMAL model was only trained on four-legged animals so the shape space learned by the model is insufficient for modeling birds, which differ markedly in both limb shape and joint angles. To overcome the lack of a statistical model for birds, we add one additional degree of freedom to each joint and obtain a pose and shape space from multi-view fits to live birds.

Datasets for Animal Pose Estimation. Large-scale object recognition datasets contain many species of animals including dogs, cats, birds, horses, sheep, and more. MS COCO [23] contains 3362 images with bird mask annotations, but no keypoint or pose annotations. The CUB-200 dataset [37] contains 11,788 masks and keypoint instances of birds in the wild. A fruit fly dataset [32] contains 1500 images with centered dorsal views of single flies walking in an arena with a plain white background containing no variation or distractors. The desert locust (800 images) and Grévy's zebras (900 images) include other individuals in the frame, but views are dorsal-only, centered, and narrowly cropped around a focal individual. In contrast our multi-view dataset contains both masks and keypoints of multiple, overlapping subjects and has large variation in relative viewpoint and complex changes in background and lighting.

3   Approach

We adapt a boot-strapped, four-step approach to developing a full pipeline for 3D bird reconstruction from single images (Fig. 5). First we develop a parameterized avian mesh and use a multi-view optimization procedure to fit the model to annotations in our dataset. Because they use information from multiple views, these fits are generally good and do not suffer from ambiguities that can plague pose estimation from single views. It is enticing to deploy this multi-view optimization approach towards our end-goal of estimating the pose and shape of all birds over time, but it is slow (initialization is usually far from the target) and requires multiple views in order to produce realistic poses. Nearly 40% of the birds in our dataset were visible from one or fewer unoccluded views, however, indicating the need for a single-view approach. Second, from the multi-view fits, we extract distributions of shape and pose for birds in the aviary, which we use to create a synthetic dataset on which we train neural networks that regress pose and shape parameters from keypoints and silhouettes in a single view. Third, we train a second network to predict an instance segmentation and keypoints given a detection and corresponding bounding box. Finally, we connect the keypoint and segmentation network to the pose regression network. The full pipeline provides a pose and shape estimate from a single view image, which can be used to initialize further optimization (Fig. 4).

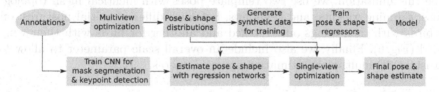

Fig. 5. Overall approach for recovering bird pose and shape from a single view. See Fig. 4 for a detailed view of the final pipeline.

Bird Detection in Full Images. We detect bird instances using a Mask R-CNN pretrained on COCO instance segmentation. We removed weights for non-bird classes (leaving bird and background) and then fine-tuned all layers on our dataset for 15 epochs in PyTorch.

Keypoints and Silhouette Prediction. We train a convolutional neural network to predict keypoints and a silhouette given a detection and corresponding bounding box. We modify the structure of High-Resolution Net (HRNet) [35], which is state-of-the-art for keypoint localization in humans, so that it outputs masks in addition to keypoints. Our modified HRNet achieves 0.46 PCK@05, 0.64 PCK@10, and 0.78 IoU on our dataset.

Skinned Linear Articulated Bird Model. To define an initial mesh, joint locations, and skinning weights, we used an animated 3D mesh of a bird model downloaded from the CGTrader Marketplace website. The model originally contained 18k vertices and 13k faces, but we removed vertices associated with body feathers, eyes, and other fine details to obtain a mesh with 3932 vertices, 5684 faces, and 25 skeletal joints (including a root joint, which is used for camera pose). We use the skinning weights defined in the original file. In addition to skeletal joints, we define 16 mesh keypoints that correspond to the annotated semantic keypoints in our dataset. We obtain keypoint locations by identifying up to four mesh vertices associated with each keypoint and averaging their 3D locations.

To pose the model, we specify a function $M(\alpha, \theta, \gamma, \sigma)$ of bone length parameters $\alpha \in \mathbb{R}^J$ for J joints, pose parameters $\theta \in \mathbb{R}^{3J}$ specifying relative rotation of the joints (and the rotation of the root relative to the global coordinate system) in axis-angle parameterization, global translation inside the aviary γ, and scale σ, that returns a mesh $\mathcal{M} \in \mathbb{R}^{N \times 3}$, with $N = 3932$ vertices. Unlike SMPL [25] and SMAL [42] models, we do not have access to 3D ground truth variation in shape, which prevents the use of shape coefficients drawn from a learned PCA shape space. We mitigate this limitation by including an additional degree of freedom per joint, α_i, that models the distance between parent and child joints, thereby capturing variation in the relative length proportions of the body and limb segments. When birds perch, their wings fold in on themselves and we found that this large deformation is not well modeled by LBS of a single bird mesh model (it is also difficult to capture and register in 3D scans). To overcome this limitation, we use two template poses with identical mesh topology, bones, skinning weights, and keypoints, but with different initial postures: one for birds with their wings outstretched and another for birds with their wings folded (Fig. 6). Finally, we also include an overall scale parameter to allow for consistent 3D multi-view estimation among cameras.

Fig. 6. Our model is capable of capturing both perched and flying poses.

To form the mesh into a given pose, we modify the approach used in SMPL [25] and SMPLify [3] to allow variable bone lengths. Starting with a template mesh $\mathcal{M}^T$ in a canonical pose with joint locations $\mathcal{J} \in \mathbb{R}^{J \times 3}$, we first calculate the position of each joint i relative to its parent as

$$\mathcal{J}_i^o = \mathcal{J}_i - \mathcal{J}_{\text{parent}(i)}. \tag{1}$$

We then multiply this vector by α_i to adjust the distance between the two joints and form a new skeletal shape $\mathcal{J}'$, still in the canonical pose, with joint locations

$$\mathcal{J}'_i = \alpha_i \mathcal{J}^o_i + \sum_{j \in A(i)} \alpha_j \mathcal{J}^o_j, \tag{2}$$

where $A(i)$ is the ordered set of joint ancestors of joint i (i.e. all joints encountered moving along the kinematic tree from joint i to the root). Finally, $\mathcal{J}' = J(\alpha)$ is transformed into the final pose using the global rigid transformation $R_\theta(\cdot)$ defined by pose and root orientation parameters θ, and a LBS function $W(\cdot; \mathcal{M}^T)$ is applied. The final mesh vertices are

$$\mathcal{M} = M(\alpha, \theta, \gamma, \sigma) \overset{\text{def}}{=} \sigma W(R_\theta(J(\alpha)); \mathcal{M}^T) + \gamma. \tag{3}$$

The positions of 3D keypoints are calculated as $P(M(\alpha, \theta, \gamma))$, where $P(\mathcal{M})$: $\mathbb{R}^{N \times 3} \mapsto \mathbb{R}^{K \times 3}$ and K is the number of keypoints. In practice P is simply the average of four selected mesh vertices for each semantic keypoint.

Optimization. To fit our bird model to detected keypoints, we introduce a fitting procedure similar to SMPLify, an optimization-based approach originally described by Bogo et al. [3]. Unlike SMPLify, we capture among individual variation using bone length parameters rather than body shape parameters and we fit to semantic keypoints rather than joint locations. We minimize an objective function with a keypoint reprojection error term and silhouette error term for each camera i, two pose priors, and a prior on the relative 3D distances between joints. Specifically, we minimize:

$$E(\alpha, \theta, \gamma) = \sum_{\text{cam } i} E^{(i)}_{kp}(\cdot; \cdot) + E^{(i)}_{msk}(\cdot; \cdot) + \lambda_\theta E_\theta(\theta) + \lambda_p E_p(\theta) + \lambda_b E_b(\alpha) \tag{4}$$

with

$$E^{(i)}_{kp}(\alpha, \theta, \gamma; K_i, R_i, t_i, \mathcal{P}_i) = \sum_{\text{kpt } k} w_k \rho(\| \Pi_{K_i, R_i, t_i}(P(M(\alpha, \theta, \gamma))_k - \mathcal{P}_{i,k} \|_2)) \tag{5}$$

and

$$E^{(i)}_{msk}(\alpha, \theta, \gamma; K_i, R_i, t_i, \mathcal{S}_i) = \lambda_{msk} \| \mathcal{R}_{K_i, R_i, t_i}(M(\alpha, \theta, \gamma)) - \mathcal{S}_i \|_2. \tag{6}$$

Equation 5 is a weighted reprojection penalty (using the robust Geman-McClure function ρ [9]) between keypoints $\mathcal{P}_i$ and the projected mesh keypoints $\Pi_{K_i, R_i, t_i}(P(M(\alpha, \theta, \gamma)))$ for pinhole projection function $\Pi(x) = K[R|t]x$. The bone lengths, α, are the distances between parent and child joints, θ are the pose parameters, γ is the translation in the global reference frame, K_i, R_i, and t_i are the intrinsics, rotation, and translation, respectively, used in perspective projection for camera i, and $\mathcal{P}_i$ are the detected or annotated 2D keypoint locations in the image. Equation 6 penalizes differences between an annotated mask $\mathcal{S}_i$ and a rendered silhouette $\mathcal{R}_{K_i, R_i, t_i}(M(\alpha, \theta, \gamma))$ obtained using Neural

Mesh Renderer [18]. $E_\theta(\theta) = |\theta - \theta_o|$ is a pose prior that penalizes the L_1 distance from the canonical pose θ_o. $E_p(\theta) = \max(0, \theta - \theta_{\max}) + \max(0, \theta_{\min} - \theta)$ linearly penalizes joint angles outside defined limits $\theta_{\min}$ and $\theta_{\max}$ and $E_b(\alpha) = \max(0, \alpha - \alpha_{\max}) + \max(0, \alpha_{\min} - \alpha)$ penalizes bone lengths outside limits $\alpha_{\min}$ and $\alpha_{\max}$. In the single-view setting, the pose prior (E_θ) and joint angle (E_p) and bone length (E_b) limit losses are disabled and we use the Mahalanobis distance to the distribution of multi-view pose and shape estimates instead. We minimize the objective in 4 using Adam [19] in PyTorch.

Synthetic Data and Pose and Shape Regression. After performing multi-view optimization on 140 3D bird instances in our annotated dataset, we fit a multivariate Gaussian to the estimated pose parameters (pose, viewpoint, and translation). We then sample 100 random points from this distribution for each bird instance, project the corresponding model's visible keypoints onto the camera and render the silhouette, generating 14,000 synthetic instances for training. We keep the bone lengths of the original 140 instances, but add in random noise to the bone lengths for each sample.

We train pose and shape regression networks on the 14,000 synthetic single-view instances supervised by the ground truth pose and shape parameters. For the pose regression network inputs are 2D joint locations and targets are 3D rotations, which are first transformed to the representation proposed by Zhou et al. [39] before computing the L^2 loss. The pose regression network is an MLP with two fully connected layers with the final layer outputting 25∗6+3 translation parameters. The shape regression network takes in a mask and contains one 5×5 convolutional layer followed by four 3×3 convolutional layers and a fully connected layer with 24 outputs, corresponding to the 24 bone lengths. Each convolutional layer is followed by batch normalization and max-pooling layers. Training was performed for 20 epochs using Adam.

4   The Cowbird Dataset

Image Acquisition and Aviary Details. We captured video of 15 individual cowbirds (*Molothrus ater*) in an outdoor aviary from March to June using eight synchronized cameras recording 1920×1200 images at 40 Hz. The aviary is 2.5 m in height and width and is 6 m long. Cameras were positioned in the corners and oriented so that their combined fields view provided maximal coverage of the aviary volume by least four cameras. Intrinsic parameters were estimated for each camera using a standard checkerboard and the camera calibration package in ROS. Extrinsic parameters for camera orientation and translation were estimated online via the TagSLAM package [33] using arrays of fiducial markers permanently attached to the aviary walls.

Dataset Annotation and Statistics. From the above recordings, we exported sets of synchronous frames from 125 "moments" (1000 images) drawn from 10 days uniformly distributed over the recording period (an average of 12.5 uniformly distributed moments each day). On all images, we exhaustively annotated instance segmentation masks for all visible birds, producing over 6355 masks and bounding boxes. On a subset of 18 moments across six of the 10 days we also annotated the locations of 12 semantic keypoints on a total of 1031 masks (Fig. 3). We annotated the bill tip, right and left eyes, neck, nape, right and left wrists, right and left wing tips, right and left feet, and the tail tip. Statistics on the visibility of keypoints (Table S7) and a comparison with other animal datasets (Tables S4, S5) are in the supplementary material.

We manually associated keypoint annotations within each moment across camera views to create 3D instance ID tags. From the 3D instance ID tags, 64%, 26%, and 10% of birds were fully or partially visible from four or more cameras, three cameras, and two or fewer cameras, respectively (Supplementary Table S6). The average width × height of bird masks was 68 × 75 pixels (or ≈5% of image width; the 5th and 95th percentiles of bird max dimensions were 17 × 19 and 239 × 271 pixels, respectively). We provide four types of test/train splits: by moment, by day, by time of day (morning vs. afternoon), and by season (March and April vs May and June). Birds wore colored bands on their legs that, when visible, could provide the true ID of the bird, but we leave the potential application of this dataset to the Re-ID task for future work.

5 Experiments

Detection. We first evaluate the performance of Mask R-CNN on instance segmentation of birds using our dataset. We show excellent generalization (AP = 0.52) when predicting masks on unseen days in the test set (Fig. 7). Further analyses and performance on additional splits of the dataset (e.g. split by time of day or season) are provided in Supplementary Table S1.

Fig. 7. Instance detections made by a fine tuned Mask R-CNN network over a large range of lighting conditions and views. Best viewed in color. (Color figure online)

Multi-view Optimization. We fit our articulated avian mesh model to annotations corresponding to each 3D bird instance in our keypoint dataset. We fit using all keypoint labels from all available views. We present qualitative results in Fig. 8. Our fitting procedure resulted in many plausible results but also in many failure cases, shown in the bottom row of Fig. 8. From the multi-view fits, we obtained a pose and shape space for the mesh model, which we display in the supplementary video. We perform an ablation experiment to investigate the effects of pose priors and joint and bone limits on performance in the single-view setting. For each ablation, we remove the corresponding term from the

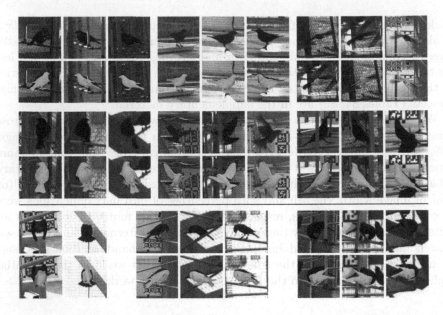

Fig. 8. Multi-view optimization-based fits of the bird mesh to keypoint and mask annotations in our dataset (upper section). Failure cases are shown in the lower section.

objective and report its effect on the accuracy of projected mesh keypoints and silhouettes, which we report in Supplementary Table S3. We measure keypoint accuracy relative to ground truth keypoints using PCK at two thresholds calculated based on the largest dimension of the bounding box and we measure the accuracy of the projected silhouettes using IoU with ground truth masks. We budget 500 iterations for fitting each instance for all settings. The PCK increased as we removed the pose prior and bone limit (but not pose limit) terms from our objective. This increase indicates the model is achieving a better fit to the keypoints, potentially at the cost of producing an unrealistic fit, as might be indicated by the simultaneous decrease in IoU as priors are removed.

Do Silhouettes Improve Multi-view Optimization? We compared fits of the model with and without the silhouette term (Eq. 6) in the objective. The silhouette term improves IoU while only slightly affecting keypoint error (Table 1). More importantly, the silhouette term allows the model to better capture changes in shape produced during feather puffing (Fig. 1).

3D Shape and Pose Recovery from a Single View. Our single-view pipeline produces poses that are consistent across views (Table 2, Supplementary Figure S1). To overcome scale ambiguity, we fix pose and shape and then find the Procrustes transformation (translation, rotation, and scaling) that minimizes keypoint reprojection error in each additional view. We also perform experiments to evaluate the individual components of our full pipeline (Table 3). We first compare pose regression alone (i.e. not optimizing after regression), single-

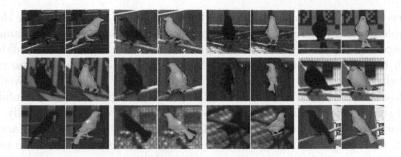

Fig. 9. Regression-based recovery of bird pose and shape from a single view. Each panel shows the input image and refined mesh (see Fig. 4).

Table 1. Ablation study of the silhouette term in the multi-view optimization setting. PCK@05 and PCK@10 denote percent correct keypoints within 5% and 10% of bounding box width, respectively. Silhouettes improve IoU with minimal effect on keypoint error.

	Weight ratio (kpt:mask)	PCK@05	PCK@10	IoU
keypoints only	N/A	0.356	0.631	0.540
keypoints + mask	10:1	0.355	0.637	0.560
keypoints + mask	1:1	0.328	0.618	0.624

view optimization alone (i.e. not initialized by pose regression network), and the full pipeline. Although the regression network alone is less "accurate" than single-view optimization (Table 3), the pose regression network produces good estimates of global pose, which allows optimization to proceed much faster. Additional examples are shown in Fig. 9. Finally, we demonstrate that our model and bone length formulation generalize to similar bird species in the CUB-200 dataset (Supplementary Figure S3).

Table 2. Cross-view PCK and IoU of projected meshes from the single-view pipeline. Values are averaged across all views except the view used to obtain the mesh. Ground truth pipeline input means the keypoint and mask network predictions (Fig. 4) are replaced by ground truth annotations.

Pipeline input	PCK@05	PCK@10	IoU
predictions	0.313	0.632	0.589
ground truth	0.332	0.635	0.586

Table 3. Same-view evaluation of the single-view pipeline and ablations. Regression and optimization are performed using keypoint and mask predictions and evaluated against ground truth. Additional results are presented in Supplementary Table S2.

	PCK@05	PCK@10	IoU
regression	0.104	0.318	0.483
optimization	0.331	0.575	0.641
reg. + opt	0.364	0.619	0.671

Failure Cases. Occasional failures resulted in unnatural poses, which are shown in Supplementary Figure S2. To evaluate the cause of these failures, two annotators inspected the same random sample of 500 crops and rated their confidence in each bird's pose (confident, semi-confident, not-confident). They then rated the predicted keypoints as good or bad for all crops. Finally, they viewed the mesh fits and rated each as a success or failure. We found that 84% of confident, 35% of semi-confident, and 12% of not-confident crops were fit successfully. Bad keypoint detection was responsible for 60% of failures. Even good fits are not perfect, particularly in the tail and feet. Adding more degrees of freedom to the model, such as tail fanning, and annotating additional keypoints on the toes would improve these areas.

6   Conclusions

We present an articulated 3D model that captures changes in pose and shape that have been difficult to model in birds. We provide a novel multi-view dataset with both instance masks and keypoints that contains challenging occlusions and variation in viewpoint and lighting. Our single-view pipeline recovers cross-view-consistent avian pose and shape, and enables robust pose estimation of birds interacting in a social context. We aim to deploy our pipeline in the aviary to better understand how individual interactions drive the formation of avian social networks.

An interesting feature of birds is that variation in a single individual's shape across time can be much larger than overall shape variation among individuals (e.g. due to feather fluffing shown in Fig. 1). In the future, it will be interesting to apply our pipeline to video data and additional species to develop a more nuanced model of how shape varies across time, individuals, and species.

Capturing 3D pose is critical to understanding human and animal health and behavior. Pose data produced by our pipeline will be useful for addressing how flying animals maneuver, negotiate cluttered environments, and make decisions while foraging or searching for mates, and how the collective behavior of a group arises from individual decisions.

Acknowledgements. We thank the diligent annotators in the Schmidt Lab, Kenneth Chaney for compute resources, and Stephen Phillips for helpful discussions. We gratefully acknowledge support through the following grants: NSF-IOS-1557499, NSF-IIS-1703319, NSF MRI 1626008, NSF TRIPODS 1934960.

References

1. Anderson, D.J., Perona, P.: Toward a science of computational ethology. Neuron **84**(1), 18–31 (2014). https://doi.org/10.1016/j.neuron.2014.09.005. http://www.sciencedirect.com/science/article/pii/S0896627314007934
2. Baillie, K.U., Spitzer, S., Crucius, D.: 'Smart aviary' poised to break new ground in behavioral research (2019). https://penntoday.upenn.edu/news/smart-aviary-poised-break-new-ground-behavioral-research

3. Bogo, F., Kanazawa, A., Lassner, C., Gehler, P., Romero, J., Black, M.J.: Keep it SMPL: automatic estimation of 3D human pose and shape from a single image. In: Leibe, B., Matas, J., Sebe, N., Welling, M. (eds.) ECCV 2016. LNCS, vol. 9909, pp. 561–578. Springer, Cham (2016). https://doi.org/10.1007/978-3-319-46454-1_34
4. Bogo, F., Romero, J., Loper, M., Black, M.J.: FAUST: dataset and evaluation for 3D mesh registration. In: Proceedings IEEE Conference on Computer Vision and Pattern Recognition (CVPR), Piscataway, NJ, USA. IEEE, June 2014
5. Breslav, M.: 3D pose estimation of flying animals in multi-view video datasets. Ph.D. thesis, Boston University (2016)
6. Cashman, T., Fitzgibbon, A.: What shape are dolphins? Building 3D morphable models from 2D images. IEEE Trans. Pattern Anal. Mach. Intell. **35**, 232 (2013). https://www.microsoft.com/en-us/research/publication/shape-dolphins-building-3d-morphable-models-2d-images/
7. Felzenszwalb, P.F., Huttenlocher, D.P.: Pictorial structures for object recognition. Int. J. Comput. Vis. **61**(1), 55–79 (2005). https://doi.org/10.1023/B:VISI.0000042934.15159.49
8. Fontaine, E.I., Zabala, F., Dickinson, M.H., Burdick, J.W.: Wing and body motion during flight initiation in drosophila revealed by automated visual tracking. J. Exp. Biol. **212**(9), 1307–1323 (2009). https://doi.org/10.1242/jeb.025379. https://jeb.biologists.org/content/212/9/1307
9. Geman, S., McClure, D.: Statistical methods for tomographic image reconstruction. Bull. Int. Stat. Inst. LI **I**(4), 5–21 (1987)
10. Graving, J.M., et al.: DeepPoseKit, a software toolkit for fast and robust animal pose estimation using deep learning. eLife **8**, e47994 (2019)
11. Günel, S., Rhodin, H., Morales, D., Campagnolo, J., Ramdya, P., Fua, P.: DeepFly3D, a deep learning-based approach for 3D limb and appendage tracking in tethered, adult Drosophila. eLife **8**, e48571 (2019)
12. Huang, Y., et al.: Towards accurate marker-less human shape and pose estimation over time. In: 2017 International Conference on 3D Vision (3DV), pp. 421–430 (2017)
13. Joo, H., et al.: Panoptic studio: a massively multiview system for social motion capture. In: 2015 IEEE International Conference on Computer Vision (ICCV), pp. 3334–3342 (2015)
14. Joo, H., Simon, T., Cikara, M., Sheikh, Y.: Towards social artificial intelligence: nonverbal social signal prediction in a triadic interaction. In: 2019 IEEE/CVF Conference on Computer Vision and Pattern Recognition (CVPR), pp. 10865–10875 (2019)
15. Kanazawa, A., Black, M.J., Jacobs, D.W., Malik, J.: End-to-end recovery of human shape and pose. In: Computer Vision and Pattern Recognition (CVPR) (2018)
16. Kanazawa, A., Kovalsky, S., Basri, R., Jacobs, D.: Learning 3D deformation of animals from 2D images. Comput. Graph. Forum **35**(2), 365–374 (2016). https://doi.org/10.1111/cgf.12838. https://onlinelibrary.wiley.com/doi/abs/10.1111/cgf.12838
17. Kanazawa, A., Tulsiani, S., Efros, A.A., Malik, J.: Learning category-specific mesh reconstruction from image collections. In: Ferrari, V., Hebert, M., Sminchisescu, C., Weiss, Y. (eds.) ECCV 2018. LNCS, vol. 11219, pp. 386–402. Springer, Cham (2018). https://doi.org/10.1007/978-3-030-01267-0_23
18. Kato, H., Ushiku, Y., Harada, T.: Neural 3D mesh renderer. In: 2018 IEEE/CVF Conference on Computer Vision and Pattern Recognition, pp. 3907–3916 (2018)
19. Kingma, D.P., Ba, J.L.: Adam : a method for stochastic optimization (2014)

20. Kolotouros, N., Pavlakos, G., Black, M., Daniilidis, K.: Learning to reconstruct 3D human pose and shape via model-fitting in the loop. In: 2019 IEEE/CVF International Conference on Computer Vision (ICCV), pp. 2252–2261 (2019)
21. Kolotouros, N., Pavlakos, G., Daniilidis, K.: Convolutional mesh regression for single-image human shape reconstruction. In: 2019 IEEE/CVF Conference on Computer Vision and Pattern Recognition (CVPR), pp. 4496–4505 (2019)
22. Li, T., Bolkart, T., Black, M.J., Li, H., Romero, J.: Learning a model of facial shape and expression from 4D scans. ACM Trans. Graph. **36**(6) (2017). https://doi.org/10.1145/3130800.3130813
23. Lin, T.-Y., et al.: Microsoft COCO: common objects in context. In: Fleet, D., Pajdla, T., Schiele, B., Tuytelaars, T. (eds.) ECCV 2014. LNCS, vol. 8693, pp. 740–755. Springer, Cham (2014). https://doi.org/10.1007/978-3-319-10602-1_48
24. Liu, J., Belhumeur, P.N.: Bird part localization using exemplar-based models with enforced pose and subcategory consistency. In: 2013 IEEE International Conference on Computer Vision, pp. 2520–2527 (2013)
25. Loper, M., Mahmood, N., Romero, J., Pons-Moll, G., Black, M.J.: SMPL: a skinned multi-person linear model. ACM Trans. Graph. (Proc. SIGGRAPH Asia) **34**(6), 248:1–248:16 (2015)
26. Mathis, A., et al.: DeepLabCut: markerless pose estimation of user-defined body parts with deep learning. Nat. Neurosci. **21**(9), 1281–1289 (2018)
27. Newell, A., Yang, K., Deng, J.: Stacked hourglass networks for human pose estimation. In: Leibe, B., Matas, J., Sebe, N., Welling, M. (eds.) ECCV 2016. LNCS, vol. 9912, pp. 483–499. Springer, Cham (2016). https://doi.org/10.1007/978-3-319-46484-8_29
28. Ntouskos, V., et al.: Component-wise modeling of articulated objects. In: 2015 IEEE International Conference on Computer Vision (ICCV), pp. 2327–2335 (2015)
29. Pavlakos, G., et al.: Expressive body capture: 3D hands, face, and body from a single image. In: 2019 IEEE/CVF Conference on Computer Vision and Pattern Recognition (CVPR), pp. 10967–10977 (2019)
30. Pavlakos, G., Kolotouros, N., Daniilidis, K.: Texturepose: supervising human mesh estimation with texture consistency. In: 2019 IEEE/CVF International Conference on Computer Vision (ICCV), pp. 803–812 (2019)
31. Pavlakos, G., Zhu, L., Zhou, X., Daniilidis, K.: Learning to estimate 3D human pose and shape from a single color image. In: 2018 IEEE/CVF Conference on Computer Vision and Pattern Recognition, pp. 459–468 (2018)
32. Pereira, T.D., et al.: Fast animal pose estimation using deep neural networks. Nat. Methods **16**, 117–125 (2019)
33. Pfrommer, B., Daniilidis, K.: Tagslam: robust slam with fiducial markers (2019)
34. Romero, J., Tzionas, D., Black, M.J.: Embodied hands: modeling and capturing hands and bodies together. ACM Trans. Graph. **36**(6) (2017). https://doi.org/10.1145/3130800.3130883
35. Sun, K., Xiao, B., Liu, D., Wang, J.: Deep high-resolution representation learning for human pose estimation. In: 2019 IEEE/CVF Conference on Computer Vision and Pattern Recognition (CVPR), pp. 5686–5696 (2019)
36. Vicente, S., Agapito, L.: Balloon shapes: reconstructing and deforming objects with volume from images. In: 2013 International Conference on 3D Vision - 3DV 2013, pp. 223–230 (2013)
37. Wah, C., Branson, S., Welinder, P., Perona, P., Belongie, S.: The Caltech-UCSD birds-200-2011 dataset. Technical report, CNS-TR-2011-001, California Institute of Technology (2011)

38. West, M.J., King, A.P.: Female visual displays affect the development of male song in the cowbird. Nature **334**, 224–246 (1988)
39. Zhou, Y., Barnes, C., Lu, J., Yang, J., Li, H.: On the continuity of rotation representations in neural networks. In: 2019 IEEE/CVF Conference on Computer Vision and Pattern Recognition (CVPR), pp. 5738–5746 (2019)
40. Zuffi, S., Kanazawa, A., Berger-Wolf, T., Black, M.: Three-D safari: learning to estimate zebra pose, shape, and texture from images "in the wild". In: 2019 IEEE/CVF International Conference on Computer Vision (ICCV), pp. 5358–5367 (2019)
41. Zuffi, S., Kanazawa, A., Black, M.J.: Lions and tigers and bears: capturing non-rigid, 3D, articulated shape from images. In: 2018 IEEE/CVF Conference on Computer Vision and Pattern Recognition, pp. 3955–3963 (2018)
42. Zuffi, S., Kanazawa, A., Jacobs, D.W., Black, M.J.: 3D menagerie: modeling the 3D shape and pose of animals. In: 2017 IEEE Conference on Computer Vision and Pattern Recognition (CVPR), pp. 5524–5532 (2017)

We Have So Much in Common: Modeling Semantic Relational Set Abstractions in Videos

Alex Andonian[1]($\boxtimes$), Camilo Fosco[1], Mathew Monfort[1], Allen Lee[1], Rogerio Feris[2], Carl Vondrick[3], and Aude Oliva[1]

[1] Massachusetts Institute of Technology, Cambridge, USA
{andonian,camilolu,mmonfort,allenlee,oliva}@mit.edu
[2] MIT-IBM Watson AI Lab, Cambridge, USA
rsferis@us.ibm.com
[3] Columbia University, New York, USA
vondrick@cs.columbia.edu

Abstract. Identifying common patterns among events is a key capability for human and machine perception, as it underlies intelligent decision making. Here, we propose an approach for learning *semantic relational set abstractions* on videos, inspired by human learning. Our model combines visual features as input with natural language supervision to generate high-level representations of similarities across a set of videos. This allows our model to perform cognitive tasks such as *set abstraction* (which general concept is in common among a set of videos?), *set completion* (which new video goes well with the set?), and *odd one out detection* (which video does not belong to the set?). Experiments on two video benchmarks, Kinetics and Multi-Moments in Time, show that robust and versatile representations emerge when learning to recognize commonalities among sets. We compare our model to several baseline algorithms and show that significant improvements result from explicitly learning relational abstractions with semantic supervision. Code and models are available online (Project website: abstraction.csail.mit.edu).

Keywords: Set abstraction · Video understanding · Relational learning

1 Introduction

Humans are extraordinary at picking out patterns between different events, detecting what they have in common and organize them into abstract categories, a key ability for everyday reasoning. Our goal in this paper is to instantiate this

---

A. Andonian and C. Fosco—Equal contribution.

Electronic supplementary material The online version of this chapter (https://doi.org/10.1007/978-3-030-58523-5_2) contains supplementary material, which is available to authorized users.

A. Vedaldi et al. (Eds.): ECCV 2020, LNCS 12363, pp. 18–34, 2020.
https://doi.org/10.1007/978-3-030-58523-5_2

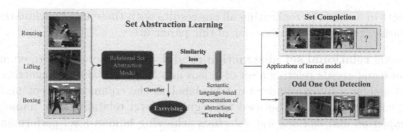

Fig. 1. Semantic Relational Set Abstraction and its Applications: We propose a paradigm to learn the commonalities between events in a set (the *set abstraction*) using a relational video model. Our model is trained to approximate the semantic language-based representation of the abstraction and predict the abstract class shared by the videos. Once trained, the model is able to identify the abstraction that represents a set of videos, select videos that fit this abstraction, and detect when a member of the set does not match the common theme.

ability into a computer vision system. Learning the semantic relational abstraction between a set of events (Fig. 1) allows a model to perform cognitive-level tasks similar to a person abstracting common patterns. If a model has learned that *exercising* can take many forms (*running, weightlifting, boxing*), its feature representation can naturally be used to select which new event is similar to the set, or detect an incompatible exemplar. If the system can enrich the learning of these abstractions with semantic and verbal content, we are one step closer to people's ability to combine visual and contextual relationships to form a deeper understanding of observed events.

In this paper, we propose an approach for learning *semantic relational set abstraction* which recasts a single exemplar recognition task to the task of encoding conceptual relationships shared among a set of videos. We apply our trained model to solve a variety of operations, namely *set abstraction* (what is in common?), *set completion* (which new video goes well with the set?), and *odd one out detection* (which video does not belong to the set?). Additionally, we compare our abstraction model to human performance on a novel *relational event abstraction* task where participants rank a set of query videos according to how closely they align with the abstract semantic relationship found between a set of reference videos. This human baseline provides a strong evaluation metric for determining how well our model can encode semantic relationships and allows us to measure the human ability to abstract common concepts from video sets.

By formulating the new set abstraction task in tandem with a language supervision module, we aim to better approximate human cognitive-level decisions. Importantly, we pioneer a data generation methodology which approximates human behavior in abstraction-related tasks. The semantic relational algorithm allows us to sample training examples that categorize the commonalities between sets of videos in a human understandable way. We root our experiments in event understanding, leveraging the large scale video datasets Kinetics [19] and Multi-

Moments in Time [27], replicating all our results with these two benchmarks. To summarize, the main contributions of this paper are:

1. A novel **relational set abstraction model** which generates representations of *abstract events* relating a set of videos in a language-based geometric space and assigns a human-understandable label to the common concept underlying each set. This can be used for cognitive-level relation tasks, namely *set completion* and *odd one out detection*, achieving human-level performance.
2. A novel paradigm, the **Relational Event Abstraction task**, which measures human performance on event abstraction. Given a set of reference videos representing specific events (e.g. *digging, peeling, unwrapping*), the task involves finding the common abstract concept shared by the videos in the set (e.g. *removing* in this case), and ranking a set of query videos based on how close they align with the abstraction.
3. A **dataset for Relational Event Abstraction** built using a novel *semantic relational algorithmic methodology* rooted in natural language which correlates highly with human results on the *relational event abstraction* task. This allows us to sample a large number of reference and query sets for training and evaluation, avoiding expensive human annotation.

2   Related Work

Concepts Organization. Concepts can be organized in a hierarchical structure (i.e. trees), a chain (i.e. linear organization), or a ring (i.e. perceptual similarities of colors) [34]. Computer vision work on visual classification most often uses trees, with root categories forming the base of taxonomies (i.e. for object [8] and scene classes [39]). Hierarchies can be pre-defined [17,36] or learned [2,10], and can help with transfer learning between categories [23], and class prediction for videos [29,30]. EventNet [40] built an action concept dataset with a hierarchical structure that includes low-level event labels as leaf nodes and increasingly abstract concept labels as parent nodes. They trained a CNN to identify the low-level event labels from the video frames and combine the representation learned from this model with a set of SVMs to predict the higher-level concepts associated with the video. Here, we similarly use a pre-defined relational organization between activities (i.e. *jog, swim* and *weightlift* all share the abstract relation *exercise*) as a tool for learning the abstract semantic relations between sets of videos. While previous works consider a single instance at a time, our goal is to generate representations for *sets* of videos to identify common relationships.

Video Recognition. Two stream convolutional networks (CNNs) [33] combine static images and optical flow. In [11], a recurrent model uses an LSTM to learn temporal relationships between features extracted from each frame. 3D CNNs [35] aim to directly learn motion using 3D convolutional kernels to extract features from a dense sequence of frames. I3D proposes incorporating optical flow with 3D CNNs to form a two stream 3D network [6] "inflated" from 2D filters pre-trained on ImageNet [9]. Temporal Segment [37] and Temporal Relation

Networks [43] model relationships between frames from different time segments while non-local modules [38] capture long-range dependencies. SlowFast Networks [12] combine two streams using dense (fast) and sparse (slow) frame rates to simultaneously learn spatial and temporal information.

Visual Similarity. Prior work has proposed methods for estimating similarity between images and video pairs for retrieval and anomaly detection. Fractal representations have been used to estimate pair-wise image similarity and relationships in order to solve the Odd One Out problem by encoding the spatial transformations needed to convert each image in a set to each other image [24]. Semantic word similarity has been shown to be a good approximation for quantifying visual relationships [41] while Siamese Networks [5] have been used to naturally rank the similarity of sets of images for one-shot image recognition [20]. ViSiL [21] proposes an architecture for learning the similarity between pairs of videos that incorporates both spatial and temporal similarity for video retrieval. Odd-one-out networks [13] learn temporal representations of videos that are used to identify the video in a set that has *out of order* frames. IECO [42] uses ideas from SlowFast Networks [12] to form a two-stream ECO network [44] to learn instance, pose and action representations to estimate video similarity in retrieval.

Learning Semantic Relationships in Visual Data. Different approaches have been proposed for learning semantic relationships between sets of images and videos. Reinforcement learning is used as a method for selecting subsets of data that preserve abstract relationships [28]. A Bayesian model has been used with a conceptual hierarchy formed from ImageNet [9] to identify the concept relation in image sets [18]. Relation Networks have been used to infer object relations [32] and Interaction Networks have helped to identify physical relations in complex systems [3]. We extend the idea of relation learning to videos for learning event relations in sets of varying length rather than object relations between images pairs Laso [1] utilizes similarities, and differences, in object labels of images pairs to improve few shot learning by learning the union, intersection and subtraction between the binary label vectors for each image.

We take a similar approach to learning the intersection operation of the "abstract" labels for sets of videos but we differ in that we operate on sets of varying size and are learning the common semantic abstractions of events shared in the set rather than the common labels found in video pairs. The most similar prior work ranks a set of unseen videos based on their similarity to a provided event description [7] by measuring the semantic correlation between the event and each individual concept in a dictionary of concepts (e.g bike, mountain, etc) with cosine distance on the word embedding generated from a skip-gram model [25] trained on a large text corpus. A set of previously unseen videos is ranked according to the correlation between the video and the provided event using the event-concept correlation and the concept classifier for each video. We utilize word embeddings to capture class relationships as was done in other methods that use similar embeddings to identify semantic visual relations [16, 22, 41]. However, we introduce a relational graph and abstract relational embeddings that are compounded by the embeddings of related classes (see Sect. 3) and apply our approach to the task of recognizing the abstraction between a set of videos.

3 A Dataset for Relational Event Abstraction

Our goal is to categorize the relationships between sets of videos close to human reasoning. We build a dataset for identifying semantic relational event abstractions between sets using word embeddings from natural language [4] and from an abstraction graph where each node represents a semantic relation between its children. Word embeddings, which capture context and word-to-word relationships [25] from a large text corpus, are complementary to our semantic abstraction graph and allow our model to capture relationships not directly encoded into the graph structure. Next, we provide an overview of our approach.

Video Datasets. As an initial step for representing semantic set relationships, we focus on the *relationships between activities in video clips*, using Kinetics [19] and Multi-Moments in Time (M-MiT) datasets [27]. The labels in Kinetics define specific event classes such as *biking through snow* and *cooking on campfire*. The labels in M-MiT are more general such as *bicycling* and *cooking*. The contrast between the class structures allows us to validate our approach to settings with both low-level event categories (Kinetics) and high-level classes (M-MiT).

Semantic Relational Graph. To form our relational graphs, we start with the activity categories provided by the class vocabularies of Kinetics and M-MiT. We assign each category to a synset in the WordNet lexical database [26] that captures the specific meaning of the class label applied to its member videos. Then we extract the hypernym paths of each class in WordNet and add the path, and the members of each path, to our graph. We verify the graph by hand and add any missing relations not captured by the WordNet paths. Building the graph in this way for the full class vocabulary of each dataset allows us to form a trace from each low-level action to high-level categories that capture the abstract relationships between their descendant classes. For example, in the Kinetics graph, the abstraction node for *baking* has two children that share the relation of *baking*, *making a cake* and *baking cookies*, and is a child of the abstraction node *cooking* together with other categories such as *frying* and *cooking chicken* that share the same relation of *cooking*. We do not restrict the nodes in the graph to have a single parent as we are not building a strict hierarchy but rather a directed relational graph where each node represents abstract semantic relations between its descendants. To illustrate, consider the class *sculpting* from M-MiT which is a child of both *carving*, with *peeling* and *shaving*, and *making art* which includes the descendants *drawing* and *painting* (see Fig. 2). Treating the graph as a hierarchy would be incomplete as we would miss out on the full breadth of relations between these different actions.

Category Embeddings. To increase the amount of information given to our model in training and to solidify the relationship between an abstraction node and its children, we generate a semantic embedding vector based on the intuition of distributional semantic word representations in natural language processing [4, 25] for each node. These representations capture contextual word relationships from a large unlabeled corpora. We use the word embeddings generated by the Subword Information Skip Gram (SISG) model [4] which takes into account

Fig. 2. Semantic Relational Graph: Example of a class relation in our semantic graph for M-MiT. A specific video of *a person sculpting* is an exemplar of the category *sculpting*. The set of videos that share the activity *sculpting* is itself a member of multiple sets that capture higher-level abstractions, like *making art* and *carving* which in turn are members of *crafting* and *cutting* respectively.

the morphology of each word. These vectorized relationships are complementary to our semantic abstraction graph and aid in allowing our model to capture additional relationships not directly encoded into the graph structure.

We begin by assigning each leaf node in our graph the average vector of all the words in the class name using SISG. We then consider this the embedding of that node. From here we traverse the graph and assign each node an embedding vector that is the average of the embeddings of all of its direct children. We use this approach for each parent node to ensure that the embeddings of the abstraction nodes are constrained to capture the common relationships among their children while downplaying features that are specific to a single child node. We describe our approach on using these embeddings to train our model in Sect. 4.

Video Embeddings. As a first step in training video models to capture the abstract semantic relationships between different classes described above we assign embedding vectors to each video according to their class associations. For example, a video in the Kinetics dataset with the class *doing aerobics* will be assigned the vector associated with the graph node *doing acrobics*. This ensures that the models described in Sect. 4 learn representations that align with our relational graph. For a video that contains multiple labels in M-MiT we simply take the average vector for all the classes that belong to the video.

Forming a Training Dataset. In order to train models to accurately capture abstract semantic relationships among a set of videos we must first build a dataset consisting of sets of videos that share common abstractions. To do this we iterate through each parent class in our relational graph and sample four videos that belong to any descendant class of the parent. We use four videos to allow a wide class diversity as using more than four videos in a set commonly results in multiple videos belonging to the same class in both M-MiT and Kinetics. This allows us to balance between having a strong training signal (label diversity) while maximizing computational efficiency. We then generate a label set such that we find the lowest common abstraction shared between the members of each subset in the power set of the set of the videos. The abstraction found for a video set of one is simply the label set for that video in the original dataset. In this way we can train a model to find every abstract semantic relationship present in every combination of the videos in the set greatly increasing our training efficiency over training for different set sizes individually. The labels generated for each subset are then paired with their associated embedding vectors. Due to the flexible modularity of our architecture we are able to reduce the subsets to only contain pairs when efficiency is a concern. For this paper we use the full powerset of the input videos to maximize the learning signal for each training step. For Kinetics we generated one million video sets for training and 50k for validation while for M-MiT we generated five million sets for training and 100k for validation. Training and validation videos were all chosen from the associated training and validation sets of each dataset to preserve the original data splits.

3.1 Human Performance on Event Abstraction

We aim to build a model that identifies relationships similar to how humans recognize abstractions between events. First, we collect a human baseline on a video ranking task where the goal is to rank a set of five *query* videos in order of how closely they align with the abstract relationship between a set of *reference* videos. We compare our trained models to human performance in Sect. 5.2.

Collecting a Human Baseline Dataset. Before we collect human baselines we need to build a dataset for ranking videos according to our relational event abstraction paradigm. We begin similar to the approach used for building our training set and iterate through each abstraction node in our relational graph and select a set of N reference videos (where N can be 1, 2, 3 or 4) that share the abstraction. From here we calculate a shared embedding vector that is the average between the vectors for each reference video and the vector of their shared abstraction node. We then sample five query videos from the dataset sorted according to the cosine distance of their embedding vector and this new reference vector. The goal of this approach is to generate a query set that has at least one video closely aligned with the reference set, one that is very different and three videos that have varying levels of similarity to the reference set. This forms a range of videos with a quantifiable metric based on reference set similarity which we can use to evaluate human and model performance.

Collecting Human Performance. To collect human baseline data, we created the Vidrank game (see Supp.) and used Amazon Mechanical Turk to crowdsource annotations. Players were presented with a "Reference" set of 1–4 videos, and an "Unknown" set (Query) of 5 randomly ranked videos. These videos were labeled: "Least Similar", "Less Similar", "Similar", "More Similar", and "Most Similar" based on their position. The task was to drag and rearrange the videos in Unknown to a ranking based on each video's similarity to Reference. To ensure reliable results, we required players to pass "vigilance" rounds, where it was clear that a video should be placed in the "Most Similar" or "Least Similar" position. We collected 40 folds of data for each dataset, 10 questions per fold for a total of 800 human responses[1].

---

[1] Note that these tasks are challenging for humans, who must disregard similarities across scenes, colors, etc. The model can circumvent this problem as it is trained only on event abstractions.

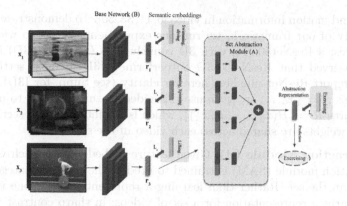

Fig. 3. Set Abstraction Architecture: A set of n videos ($n = 3$ shown) feed into a shared base video model, B. The representations for each video generated by B are combined into each possible subset and fed into the *Set Abstraction Module* (SAM), A, which generates a set-wise representation that is used to identify the common abstractions in each subset. The representations generated by both A and B are contrasted to pretrained semantic embeddings representing the labels for the subset abstractions and the individual videos themselves.

4   Approach

Let x_i be one video, and $\mathcal{X} = \{x_1, \ldots, x_i, \ldots, x_n\}$ be a set of n videos. Given $\mathcal{X}$, our goal is to train a model that correctly predicts the abstract concept that describes all videos in the set and accurately estimates the language model abstraction representation. We will train the model $F(\mathcal{X}; \theta) = (\hat{y}, \hat{e})$ to estimate this category and its semantic word embedding. A naive approach is to first classify each individual video in the set with a traditional video classification network (ignoring learned semantic word embeddings), then use these predictions to look up the lowest common ancestor in the graph. However, this approach is problematic. Firstly, the baseline model will be fragile to errors. If the individual classification model makes a mistake, the set abstraction prediction will be wrong. Secondly, since this does not reason about all elements in the set jointly, this baseline will discard useful information, i.e. the abstract category of a video could change depending on the other videos in the set (which our experiments show). Instead, our approach jointly reasons about all elements in the set.

4.1   Relational Set Abstraction Model

We model $F(\mathcal{X})$ as a deep convolutional network (CNN) that classifies the abstraction of sets of videos. We write F as the composition of two functions A and B: $F(\mathcal{X}) = A(\{B(x) | x \in \mathcal{X}\})$ where A is a set abstraction module and B is a base model network for learning individual features.

Video Feature Network (B). This network estimates visual features for each individual video in the set. Several base networks have been proposed to handle temporal and motion information in video [6,11,33,35]. To demonstrate the wide applicability of our framework, we run our experiments with two widely used architectures: a ResNet50 [14] with 3D convolutions (ResNet50-3D) and I3Ds [6]. We observed that ResNet50-3D outperformed I3D in most settings, and thus only report ResNet numbers here for clarity (see Supp. for I3D). Given a set of videos $\mathcal{X} = \{x_0, \ldots, x_n\}$, we use this video feature network to produce a set of features $\mathcal{R} = \{B(x_0), \ldots, B(x_n)\}$, which is fed into the next section. Note that these weights are shared across each video in the set.

Set Abstraction Module (A). Given feature embeddings of each video, our set abstraction module (SAM) is trained to predict the common category of all the videos in the set. Rather than learning a representation for one video, the network learns a representation for a set of videos, in sharp contrast with previous works. To correctly recognize abstract categories, the model must capture the relationships between elements in the set. However, there can be multiple relationship orders: within a single video, across videos, and across higher-order tuples. We model all these relationships by operating on the power set, which is the set of all subsets. Our approach will learn features for each subset and also combine them to produce the final abstraction. Specifically, let $g_k(r_1, \ldots, r_k)$ be a neural network that accepts k inputs, and produces features to represent those inputs. This g network will be able to capture the kth-order relationship between inputs. Given a set of features $\mathcal{R}$, the model's prediction can be written as,

$$A(\mathcal{R}) = h\Big(\sum_{r_i \in \mathcal{R}} g_1(r_i) + \sum_{r_i, r_j \in \mathcal{R}} g_2(r_i, r_j) + \ldots \Big).$$

Each g computes features to capture the relationships between its inputs. We exert two sources of supervision onto this representation: (1) we train a shared linear classifier to predict the common abstraction between the inputs of g and (2) train a separate linear layer to estimate the word embedding of the abstraction. The representations produced by each g_k are summed together to create an order invariant representation and the abstract category for the entire set is estimated by another network $h(\cdot)$. Figure 3 illustrates this module.

4.2 Learning

The video feature network and the set abstraction module can be trained jointly with stochastic gradient descent. As described in Sect. 5.1, to generate the abstract classification of a set of videos we input the representations computed for each video using the video feature network ($\mathcal{R} = \{B(x_0), \ldots, B(x_n)\}$) into our *set abstraction module* (g) and compute the abstract representations for each set of videos (r_i) in the power set of input videos ($\mathcal{R}^2$), $A(\mathcal{R}) = g(r_i) \ \forall \ r_i \in \mathcal{R}^2$.

We then apply these representations to two linear models (h and e) that capture the abstraction class belonging to the set (h) and the category embedding of the abstraction class (e). The embedding provides additional supervision and ensures that the representations generated by the model adhere to semantically and contextually relevant features.

We train the model by averaging the cross entropy losses, $\mathcal{L}_{ce}$, between the predicted abstraction class $(h(g(r_i)))$ and the ground truth class (a_c) and the mean squared error, $\mathcal{L}_{mse}$, between the generated embedding $(e(g(r_i)))$ and the embedding of the ground truth abstraction class (a_e) for each subset (r_i) in the power set of input videos $(\mathcal{R}^2)$:

$$\mathcal{L}_{total} = \frac{1}{|\mathcal{R}^2|} \sum_{r_i \in \mathcal{R}^2} \mathcal{L}_{ce}\big(h(g(r_i)), a_c\big) + \mathcal{L}_{mse}\big(e(g(r_i)), a_e\big).$$

This loss combines the error from generating both the class and the embedding vector of the abstraction class for each possible set of videos given the input set. By doing this we maximize the supervision signal provided from each video input set without recomputing features for different combinations. In practice we train with 4 videos producing 15 different video sets (we omit the empty set).

5 Experiments

We evaluate our *set abstraction model* on three tasks:

1. **Recognizing set abstractions:** Predict the direct relational abstraction for a set of videos.
2. **Set completion:** Given a set of *reference* videos and a set of *query* videos, select the query that best fits in the reference set.
3. **Finding the odd one out:** Identify the video in a set that does not share the relational abstraction common to the other videos in the set.

Experimental Setup. Our hypothesis is that jointly reasoning about all videos in a set will enable models to more accurately predict set abstractions. We compare our *set abstraction model* (3DResNet50+SAM) against a baseline model that maintains the same base model architecture and is trained for standard classification without a set abstraction module (3DResNet50). As a secondary baseline, we train the same base model for multi-label classification using binary cross entropy loss (3DResNet50+BCE) where the labels consist of all the ancestors of the ground truth class, as well as itself, provided by the dataset. We use the standard training, validation and test splits provided by M-MiT and Kinetics and show evaluation results for each task on the corresponding test set.

Comparison to Previous Work. To the best of our knowledge, there are no existing video-based models that explicitly address the set abstraction task. Thus, we compare our model to the following extensions of similar previous work:

- Relation Networks [31]: we replace SAM with their Relation Module that computes representations over pairs of objects, and use the output of f_ϕ as the abstraction representation. Importantly, this module can only work with pairs of videos, so we cannot compute results on set completion with N=1.

Table 1. Recognizing Set Abstractions: Classification accuracy (percent) of the models evaluated on the set abstraction task. Here, N is the number of elements in the set, and the top$k$ chance level is the sum of the frequency of the top$k$ most frequent abstract nodes presented during evaluation.

Dataset	Model	$N = 2$		$N = 3$		$N = 4$	
		Top1	Top5	Top1	Top5	Top1	Top5
M-MiT	Chance	7.9	16.2	7.9	16.2	7.9	16.2
	3DResNet50	17.1	31.2	22.6	38.8	26.0	42.9
	3DResNet50 (BCE)	3.9	30.0	4.9	34.5	5.1	38.0
	3DResNet50+RN [31]	32.4	65.2	39.2	75.4	44.9	82.1
	3DResNet50+SAM (Ours)	**34.0**	**66.9**	**41.1**	**77.1**	**47.2**	**83.8**
Kinetics	Chance	0.44	2.18	0.44	2.18	0.44	2.18
	3DResNet50	29.9	49.1	22.1	42.8	17.9	40.4
	3DResNet50 (BCE)	2.8	25.0	2.2	22.2	0.5	22.5
	3DResNet50+RN [31]	53.9	83.0	61.6	90.2	66.0	93.8
	3DResNet50+SAM (Ours)	**60.5**	**86.0**	**65.3**	**91.6**	**69.9**	**94.6**

- Odd One Out Networks [13]: although their method works on snippets of a single video and only solves the OOO task to generate a representation, we re-purpose their network ($O3N$) to OOO by extending the number of inputs and training to directly predict the input that does not belong to the set.

Implementation Details. We use PyTorch implementations of 3D ResNet50 [15] as the basis for our video feature networks. Each n-scale relation module in SAM A is a two-layer Multilayer Perceptrons (MLP) with ReLU nonlinearities and 2048 units per layer. All models were optimized using stochastic gradient descent with a momentum term of 0.9 and weight decay of 5e-4. An initial learning rate of 0.001 was decreased by a factor of 10 every 20 epochs of training. Models were trained until convergence ($\sim$50–60 epochs).

5.1   Recognizing Set Abstractions

We first evaluate our model on recognizing the abstract category of a set of N videos. We use our abstraction model to directly predict this category given the set. Our baseline model (3DResNet50) individually predicts the specific class category for each video in the set, then computes the set abstraction class directly from the semantic graph (Sect. 3) based on its predictions. Our multi-label baseline (3DResNet50+BCE) also evaluates each video independently and selects the abstract category with the highest mean probability across videos. Our results suggest that there are significant gains by jointly modeling all elements in the video set. Table 1 quantitatively compares the proposed *set abstraction model* against the baselines and Fig. 4 shows qualitative results. Since the margin of improvement increases with the size of the input set, this suggests that set-based training improves the strength of the learning signal by reducing ambiguity.

Fig. 4. Qualitative Set Abstraction Results. We show results for sets of length 3, and only show the predictions for the individual videos and the entire set to simplify the visualization. Confidence is indicated in parenthesis and ground truth class in brackets.

Table 2. Set Completion: Rank Correlation of our model (3DResNet50+SAM), a baseline (3DResNet50) and human ranking to the ranking achieved using the embedding distance between a video and the abstraction of a *reference* set of size N on the *set completion* task.

Dataset	Model	N = 1	N = 2	N = 3	N = 4	Avg
	Human Baseline	0.547	0.495	0.595	0.541	0.545
M-MiT	3DResNet50	0.388	0.415	0.455	0.463	0.430
	3DResNet50+RN [31]	–	0.483	0.513	0.533	0.489
	3DResNet50+SAM (Ours)	**0.481**	**0.544**	**0.570**	**0.571**	**0.542**
	Human Baseline	0.432	0.653	0.629	0.606	0.58
Kinetics	3DResNet50	0.339	0.421	0.431	0.459	0.413
	3DResNet50+RN [31]	–	0.491	0.487	0.489	0.489
	3DResNet50+SAM (Ours)	**0.523**	**0.627**	**0.659**	**0.606**	**0.604**

5.2 Set Completion

While recognizing the abstract event relationships shared among a set of videos shows that our model has learned to identify common patterns, we aim to solve more cognitive-level tasks. Thus, we apply our model to the complex task of ranking a set of five *query* videos according to how closely they align to the common abstraction found in a set of $N \in [1, 4]$ *reference* videos (see Fig. 5).

First we use the model to generate an abstract representation of the *reference* set of videos using the method from Sect. 4.1. Then we rank each *query* video according to the cosine distance between this abstract representation and their single video feature representations (Sect. 4.1). This distance tells how closely each video is aligned to the *reference* set abstraction found by our model.

To evaluate our results, we correlate our model's ranking order with the ground truth order found by using the natural-language embedding distance (Sect. 3). In this way we compare the performance to both the abstract rela-

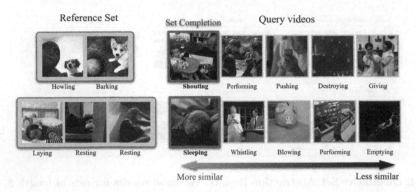

Fig. 5. Set Completion via Learned Abstraction Representations. The videos on the left are the initial reference sets, and the query videos show top ranked retrievals that complete them. The labels of individual videos (not provided to the model) are listed below. The model is able to understand the underlying abstraction regardless of the subject, e.g. choosing the sleeping dog despite all human subjects in the reference.

tionships defined in our *Semantic Relational Graph* and event embeddings associated with the annotated labels of each video. We evaluate the results with, and without, the proposed *abstraction module* (3DResNet50+SAM and 3DResNet50 respectively). For the model without the *abstraction module* we rank the *query* videos using the cosine distance to the average of the feature vectors from each individual *reference* video. Since we are interested in developing models for human-level understanding of abstract relationships, we additionally compare our model results to human performance on the same video ranking task (see Sect. 3). Table 2 summarizes our results. Our abstraction model (3DResNet50+SAM) beats the human baseline on M-MiT when the *reference* set has either 2 or 4 videos and only underperforms the human baseline on Kinetics when the *reference* set has 3 videos. We can see that our model achieves near human performance on M-MIT and surpasses human performance on Kinetics.

5.3 Finding the Odd One Out

Given a set of videos, which video does not belong to the group? With a model trained for set abstraction, we can use its learned internal representation for new tasks, such as identifying the odd one out in a set, without additional training. For example, given a set containing videos of *barking*, *boating* and *flying*, the correct odd one out would be *barking* since *boating* and *flying* are both instances of *traveling*, while *barking* is not.

Given a set of videos, we define a task to identify the odd one out by choosing the video with the largest cosine distance between its video representation and the *abstract* representation of the remaining videos in the set (Sect. 4.1). Intuitively, this method leverages a model's ability to preserve "conceptual" distance in its learned feature space. Table 3 shows that our *set abstraction model*

Fig. 6. Qualitative Results for Odd One Out detection: Given sets of three videos (left) and four videos (right), which one is odd? The odd video detected by our *set abstraction model* is indicated by a red bounding box (probability in parenthesis). Even with a small number of other videos to compare to, the model is able to select the odd video out.

Table 3. Odd One Out detection accuracy: Predict the element that does not belong to the set. The language-enhanced features from the set abstraction network are compared with the features from the corresponding base model.

Dataset	Model	N = 3		N = 4	
		Top-1	Top-2	Top-1	Top-2
	Human Baseline	74.21	–	78.04	–
M-MiT	3DResNet50	49.95	78.02	43.73	68.34
	3DResNet50+RN [31]	36.20	64.30	28.94	50.71
	3DResNet50+O3N [13]	34.10	60.11	35.84	60.71
	3DResNet50+SAM (Ours)	**52.63**	**79.86**	**47.21**	**71.11**
	Human Baseline	87.31	–	85.40	–
Kinetics	3DResNet50	65.15	82.65	69.62	81.48
	3DResNet50+RN [31]	40.11	70.97	30.48	54.41
	3DResNet50+O3N [13]	55.14	80.59	66.00	81.80
	3DResNet50+SAM (Ours)	**85.90**	**92.80**	**83.18**	**91.44**

achieves good performance without additional training by making pairwise distance comparisons on subsets of the input. The abstraction model consistently outperformed the baseline model suggesting that our proposed approach indeed learns stronger set representations than the base model. We show some qualitative examples of our model performance in Fig. 6.

6   Conclusion

A central challenge in computer vision is to learn abstractions of dynamic events. By training models to capture semantic relationships across diverse events and predict common patterns, we show that we can learn rich representations of similarity for a new set of tasks. By rooting our models in language, the model can learn abstractions that are better suited to represent how people form high-level classes. Recognizing abstractions should enable vision systems to summarize high-level patterns for different types of applications. While our focus is on capturing action relationships, future work could take into account abstractions involving scenes, objects and other concepts to provide a larger range of relationships to understand events (e.g. "driving" and "jogging" may both occur on a "road" while "writing" and "drawing" may both use a "pencil").

Acknowledgments. This work was supported by the MIT-IBM Watson AI Lab (to R.F and A.O) and NSF IIS 1850069 (to C.V).

References

1. Alfassy, A., et al.: Laso: label-set operations networks for multi-label few-shot learning. In: The IEEE Conference on Computer Vision and Pattern Recognition (CVPR), June 2019
2. Bannour, H., Hudelot, C.: Hierarchical image annotation using semantic hierarchies. In: Proceedings of the 21st ACM International Conference on Information and Knowledge Management, pp. 2431–2434. ACM (2012)
3. Battaglia, P., Pascanu, R., Lai, M., Jimenez Rezende, D., Kavukcuoglu, K.: Interaction networks for learning about objects, relations and physics. In: Lee, D.D., Sugiyama, M., Luxburg, U.V., Guyon, I., Garnett, R. (eds.) Advances in Neural Information Processing Systems 29, pp. 4502–4510. Curran Associates, Inc. (2016)
4. Bojanowski, P., Grave, E., Joulin, A., Mikolov, T.: Enriching word vectors with subword information. Trans. Assoc. Comput. Linguist. **5**, 135–146 (2017). https://doi.org/10.1162/tacl_a_00051
5. Bromley, J., Guyon, I., LeCun, Y., Säckinger, E., Shah, R.: Signature verification using a "Siamese" time delay neural network. In: Proceedings of the 6th International Conference on Neural Information Processing Systems, NIPS 1993, pp. 737–744. Morgan Kaufmann Publishers Inc., San Francisco (1993)
6. Carreira, J., Zisserman, A.: Quo Vadis, action recognition? A new model and the kinetics dataset. In: Proceedings of the ICCV (2017)
7. Chang, X., Yang, Y., Hauptmann, A.G., Xing, E.P., Yu, Y.L.: Semantic concept discovery for large-scale zero-shot event detection. In: Proceedings of the 24th International Conference on Artificial Intelligence, IJCAI 2015, pp. 2234–2240. AAAI Press (2015)
8. Deng, J., et al.: Large-scale object classification using label relation graphs. In: Fleet, D., Pajdla, T., Schiele, B., Tuytelaars, T. (eds.) ECCV 2014. LNCS, vol. 8689, pp. 48–64. Springer, Cham (2014). https://doi.org/10.1007/978-3-319-10590-1_4
9. Deng, J., Dong, W., Socher, R., Li, L.J., Li, K., Fei-Fei, L.: ImageNet: a large-scale hierarchical image database. In: IEEE Conference on Computer Vision and Pattern Recognition, CVPR 2009, pp. 248–255. IEEE (2009)
10. Deng, J., Satheesh, S., Berg, A.C., Li, F.: Fast and balanced: efficient label tree learning for large scale object recognition. In: Advances in Neural Information Processing Systems, pp. 567–575 (2011)
11. Donahue, J., et al.: Long-term recurrent convolutional networks for visual recognition and description. In: Proceedings of the IEEE Conference on Computer Vision and Pattern Recognition, pp. 2625–2634 (2015)

12. Feichtenhofer, C., Fan, H., Malik, J., He, K.: SlowFast networks for video recognition. In: The IEEE International Conference on Computer Vision (ICCV), October 2019
13. Fernando, B., Bilen, H., Gavves, E., Gould, S.: Self-supervised video representation learning with odd-one-out networks. In: The IEEE Conference on Computer Vision and Pattern Recognition (CVPR), July 2017
14. He, K., Zhang, X., Ren, S., Sun, J.: Deep residual learning for image recognition. CoRR abs/1512.03385 (2015)
15. He, K., Zhang, X., Ren, S., Sun, J.: Deep residual learning for image recognition. In: Proceedings of the IEEE Conference on Computer Vision and Pattern Recognition, pp. 770–778 (2016)
16. Jain, M., van Gemert, J.C., Mensink, T., Snoek, C.G.M.: Objects2action: classifying and localizing actions without any video example. In: The IEEE International Conference on Computer Vision (ICCV), December 2015
17. Jia, Y., Abbott, J.T., Austerweil, J.L., Griffiths, T., Darrell, T.: Visual concept learning: combining machine vision and Bayesian generalization on concept hierarchies. In: Advances in Neural Information Processing Systems, pp. 1842–1850 (2013)
18. Jia, Y., Abbott, J.T., Austerweil, J.L., Griffiths, T., Darrell, T.: Visual concept learning: combining machine vision and Bayesian generalization on concept hierarchies. In: Burges, C.J.C., Bottou, L., Welling, M., Ghahramani, Z., Weinberger, K.Q. (eds.) Advances in Neural Information Processing Systems 26, pp. 1842–1850. Curran Associates, Inc. (2013)
19. Kay, W., et al.: The kinetics human action video dataset. CoRR abs/1705.06950 (2017)
20. Koch, G., Zemel, R., Salakhutdinov, R.: Siamese neural networks for one-shot image recognition. In: ICML Deep Learning Workshop (2015)
21. Kordopatis-Zilos, G., Papadopoulos, S., Patras, I., Kompatsiaris, I.: Visil: fine-grained spatio-temporal video similarity learning. In: The IEEE International Conference on Computer Vision (ICCV), October 2019
22. Lee, H., Seol, J., Lee, S.: Style2vec: representation learning for fashion items from style sets. CoRR abs/1708.04014 (2017)
23. Lim, J.J., Salakhutdinov, R.R., Torralba, A.: Transfer learning by borrowing examples for multiclass object detection. In: Advances in Neural Information Processing Systems, pp. 118–126 (2011)
24. McGreggor, K., Goel, A.: Finding the odd one out: a fractal analogical approach. In: Proceedings of the 8th ACM Conference on Creativity and Cognition, C&C 2011, pp. 289–298. ACM, New York (2011). https://doi.org/10.1145/2069618.2069666
25. Mikolov, T., Sutskever, I., Chen, K., Corrado, G.S., Dean, J.: Distributed representations of words and phrases and their compositionality. In: Burges, C.J.C., Bottou, L., Welling, M., Ghahramani, Z., Weinberger, K.Q. (eds.) Advances in Neural Information Processing Systems 26, pp. 3111–3119. Curran Associates, Inc. (2013)
26. Miller, G.A., Beckwith, R., Fellbaum, C., Gross, D., Miller, K.J.: Introduction to wordnet: an on-line lexical database. Int. J. Lexicography 3(4), 235–244 (1990)
27. Monfort, M., et al.: Multi-moments in time: learning and interpreting models for multi-action video understanding (2019)
28. Muhammad, U.R., Yang, Y., Hospedales, T.M., Xiang, T., Song, Y.Z.: Goal-driven sequential data abstraction. In: The IEEE International Conference on Computer Vision (ICCV), October 2019
29. Nauata, N., Hu, H., Zhou, G.T., Deng, Z., Liao, Z., Mori, G.: Structured label inference for visual understanding. arXiv preprint arXiv:1802.06459 (2018)

30. Nauata, N., Smith, J., Mori, G.: Hierarchical label inference for video classification. arXiv preprint arXiv:1706.05028 (2017)
31. Santoro, A., et al.: A simple neural network module for relational reasoning. In: Advances in Neural Information Processing Systems, pp. 4967–4976 (2017)
32. Santoro, A., et al.: A simple neural network module for relational reasoning. In: Guyon, I., et al. (eds.) Advances in Neural Information Processing Systems 30, pp. 4967–4976. Curran Associates, Inc. (2017)
33. Simonyan, K., Zisserman, A.: Two-stream convolutional networks for action recognition in videos. In: Advances in Neural Information Processing Systems, pp. 568–576 (2014)
34. Tenenbaum, J.B., Kemp, C., Griffiths, T.L., Goodman, N.D.: How to grow a mind: statistics, structure and abstraction. Science **31**, 1279–1285 (2011)
35. Tran, D., Bourdev, L., Fergus, R., Torresani, L., Paluri, M.: Learning spatiotemporal features with 3D convolutional networks. In: Proceedings of the IEEE International Conference on Computer Vision, pp. 4489–4497 (2015)
36. Verma, N., Mahajan, D., Sellamanickam, S., Nair, V.: Learning hierarchical similarity metrics. In: 2012 IEEE Conference on Computer Vision and Pattern Recognition (CVPR), pp. 2280–2287. IEEE (2012)
37. Wang, L., et al.: Temporal segment networks: towards good practices for deep action recognition. In: Leibe, B., Matas, J., Sebe, N., Welling, M. (eds.) ECCV 2016. LNCS, vol. 9912, pp. 20–36. Springer, Cham (2016). https://doi.org/10.1007/978-3-319-46484-8_2
38. Wang, Y., Hoai, M.: Pulling actions out of context: explicit separation for effective combination. In: The IEEE Conference on Computer Vision and Pattern Recognition (CVPR), June 2018
39. Xiao, J., Ehinger, K.A., Hays, J., Torralba, A., Oliva, A.: Sun database: exploring a large collection of scene categories. Int. J. Comput. Vis. **119**(1), 3–22 (2016)
40. Ye, G., Li, Y., Xu, H., Liu, D., Chang, S.F.: EventNet: a large scale structured concept library for complex event detection in video. In: Proceedings of the 23rd ACM International Conference on Multimedia, MM 2015, pp. 471–480. ACM, New York (2015). https://doi.org/10.1145/2733373.2806221
41. Aytar, Y., Shah, M., Luo, J.: Utilizing semantic word similarity measures for video retrieval. In: 2008 IEEE Conference on Computer Vision and Pattern Recognition, pp. 1–8, June 2008. https://doi.org/10.1109/CVPR.2008.4587822
42. Zhao, Z., et al.: Instance-based video search via multi-task retrieval and re-ranking. In: The IEEE International Conference on Computer Vision (ICCV) Workshops, October 2019
43. Zhou, B., Andonian, A., Oliva, A., Torralba, A.: Temporal relational reasoning in videos. In: Ferrari, V., Hebert, M., Sminchisescu, C., Weiss, Y. (eds.) ECCV 2018. LNCS, vol. 11205, pp. 831–846. Springer, Cham (2018). https://doi.org/10.1007/978-3-030-01246-5_49
44. Zolfaghari, M., Singh, K., Brox, T.: ECO: efficient convolutional network for online video understanding. In: Ferrari, V., Hebert, M., Sminchisescu, C., Weiss, Y. (eds.) ECCV 2018. LNCS, vol. 11206, pp. 713–730. Springer, Cham (2018). https://doi.org/10.1007/978-3-030-01216-8_43

Joint Optimization for Multi-person Shape Models from Markerless 3D-Scans

Samuel Zeitvogel$^{(\boxtimes)}$, Johannes Dornheim$^{(\boxtimes)}$, and Astrid Laubenheimer$^{(\boxtimes)}$

Intelligent Systems Research Group (ISRG),
Karlsruhe University of Applied Sciences, Karlsruhe, Germany
{samuel.zeitvogel,johannes.dornheim,astrid.laubenheimer}@hs-karlsruhe.de

Abstract. We propose a markerless end-to-end training framework for parametric 3D human shape models. The training of statistical 3D human shape models with minimal supervision is an important problem in computer vision. Contrary to prior work, the whole training process (i) uses a differentiable shape model surface and (ii) is trained end-to-end by jointly optimizing all parameters of a single, self-contained objective that can be solved with slightly modified off-the-shelf non-linear least squares solvers. The training process only requires a compact model definition and an off-the-shelf 2D RGB pose estimator. No pre-trained shape models are required. For training (iii) a medium-sized dataset of approximately 1000 low-resolution human body scans is sufficient to achieve competitive performance on the challenging FAUST surface correspondence benchmark. The training and evaluation code will be made available for research purposes to facilitate end-to-end shape model training on novel datasets with minimal setup cost.

Keywords: Body shape · Skinning · Subdivision surfaces · Blendshapes

1 Introduction

Statistical human shape models are a prerequisite for a wide variety of tasks such as shape completion, 3D virtual avatar generation, *e.g.* for virtual try-on, gaming, and markerless motion capture.

Conventional approaches [3, 18, 26] for human shape model training employ a two-stage process consisting of a template-to-scan registration step followed by a model parameter estimation step. In the first step, high-quality 3D scans of humans are registered to a common template mesh using additional supervision, *e.g.* hand-picked landmarks [3, 18]. Once the scans are brought into correspondence, the registered meshes (registrations) are manually reviewed for errors.

Electronic supplementary material The online version of this chapter (https://doi.org/10.1007/978-3-030-58523-5_3) contains supplementary material, which is available to authorized users.

© Springer Nature Switzerland AG 2020
A. Vedaldi et al. (Eds.): ECCV 2020, LNCS 12363, pp. 35–51, 2020.
https://doi.org/10.1007/978-3-030-58523-5_3

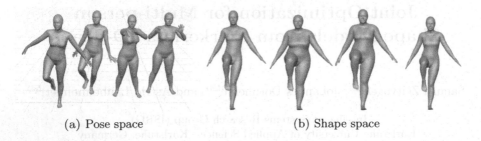

(a) Pose space   (b) Shape space

Fig. 1. Samples from our trained articulated morphable human shape model. A template with $N = 1326$ vertices produces realistic avatars. Shapes are acquired by changing pose-specific parameters (a) and shape-specific parameters (b).

Correct registrations are then used for the training process to produce multi-person articulated human shape models of high quality.

Shape model training is a chicken-and-egg problem. High-quality registrations are best acquired with a good model while the training of a good model requires high-quality registrations. Alternatively, a bootstrapping approach [18] employs a weak model to regularize the registration process and train a better model. This process can be repeated multiple times with alternating optimizers.

In contrast, recent advances in 3D hand model inference on noisy depth images indicate that joint, continuous optimization of data correspondences and model parameters is less likely to converge to bad local minima [33]. Similar work on joint shape model training for hands [21] and humans [37] exists. A differentiable surface model (*e.g.* subdivision surfaces) enables joint, continuous optimization. In this work we follow the best practices for differentiable shape model formulation [21,27,37], objective formulation and joint optimization [11, 12,21,36]. The training and evaluation code[1] will be made available for research purposes to facilitate end-to-end shape model training on novel datasets with minimal setup cost. Our contribution is threefold:

First, we propose a differentiable multi-person articulated human shape model (inspired by [21,26,37]) that can be trained using joint optimization without any 3D supervision. Differentiability is achieved using a Catmull-Clark subdivision surface module [9] with additional benefits: Only a low-poly base mesh is required to generate realistic 3D avatars (see Fig. 1) and the model parameter count is reduced considerably. Similar to the model proposed in [26], the resulting model is compatible with 3D-modelling software and can be computed efficiently.

Secondly, we formulate a single objective for model training that can be minimized with off-the-shelf nonlinear least squares solvers with minor modifications. We employ common best practices to deal with non-euclidean manifolds, robust cost functions, and discrete data-to-model correspondence updates.

---

[1] https://github.com/Intelligent-Systems-Research-Group/JOMS/.

Our third contribution is the application of the aforementioned differentiable multi-person shape model and the proposed optimization procedure to roughly 1000 markerless low-resolution point clouds. The advent of least squares solvers on the GPU [11] enables large scale joint optimization for multi-person shape model training which was previously only considered using alternating optimization methods. We evaluate the reconstruction quality of our approach and benchmark the competitive generalization quality on a challenging shape correspondence benchmark.

2 Related Work

2.1 Human Shape Models

Early work by Blanz and Vetter [4] introduces a morphable shape model for faces using a triangulated mesh structure. Registered meshes in the training set are distilled to a morphable model using principal components analysis (PCA).

Anguelov et al. [3] propose the popular morphable human shape model SCAPE. The model factorizes in a subject-specific shape model and a pose specific shape model. SCAPE does not model the vertex displacement directly but instead relies on triangle transformations. The triangle soup is realigned with an additional least squares estimation step. This approach is also not directly compatible with current software packages, 3D modelling software, and game engines. The required registered meshes are generated in a semi-supervised preprocessing step.

Hirshberg et al. [18] modify SCAPE and incorporate the registration process into the model using alternating optimization. This approach is enhanced by Bogo et al. [5] to incorporate texture information and deal with temporal information that arises by shapes in motion [6]. The Stitched Puppet method [38] transforms SCAPE into a probabilistic graphical model and fits the model to data using a particle-based optimization method.

Loper et al. [26] introduce the SMPL model. This approach produces shape models that are compatible with 3D modelling software. The quality of the model relies heavily on high-quality registrations. Our model formulation extends SMPL and we review a modified version in Sect. 3. SMPL is extended to faces [24], hands [30] and modelling infants [17]. Multiple models can be combined to construct a fully articulated morphable 3D human shape model [20,27].

In contrast to bootstrapping approaches that require registrations, deep learning approaches find shape correspondences without an explicit model [16]. However, many deep learning approaches still rely on templates for training data generation. [10,14,15,23] are trained on large scale synthetic datasets like SURREAL [34]. SURREAL is generated with the help of trained shape models.

2.2 Subdivision Surfaces

Training or fitting mesh models is usually done with variants of nonrigid iterative closest point (NICP) where the data to model correspondence finding and model training is performed in an alternating fashion. To use joint optimization of correspondences and model parameters a differentiable surface representation is required. One way to transform a polygonal mesh into a differentiable surface is the application of subdivision surfaces [9]. Transforming a polygonal mesh into a subdivision surface is easy and can be implemented as a post-processing step.

Cashman and Fitzgibbon [8] train animal shape models on segmented images using a template mesh with subdivision surfaces. Taylor et al. [32] learn a subdivision surface hand model from depth images and cast the whole training process as a single joint optimization problem. They incorporate subdivision surfaces to allow for continuous sliding of the corresponding surface points. Khamis et al. [21] build on this framework to train a morphable articulated hand model from depth images. Taylor et al. [33] use the subdivision surface hand model from [21] to build a hand tracker within a joint continuous optimization framework. Catmull Clark subdivision surfaces [9] have also been considered for human surface models in the context of surface reconstruction [19] and motion capture [37]. The major difference between [37] and our work is that in our work we train a *multi-person* shape model with the addition of shape blend-shapes [26]. The approach outlined in [37] requires existing sparse surface correspondences while our method describes a fully automatic model training pipeline. In contrast to [37], we also provide a quantitative evaluation to show the efficacy of our approach.

2.3   Joint Optimization

In general, alternating optimization is employed due to implementation simplicity and scalability. It is also used in variants of NICP which is in turn required when the surface is not differentiable. Current shape models are trained using forms of bootstrapping with human supervision in the loop.

Taylor et al. [32] compare alternating with joint optimization and report an increased convergence rate and a decreased reconstruction error using joint optimization. For optimization, they linearize the surface at the corresponding point with the tangent plane. The update in the tangent plane is applied to the underlying surface by traversing the mesh and transforming the update direction and magnitude between surface patches accordingly.

Robustified cost functions are prevalent in shape model training to deal with noisy 3D data. Zach [35] proposes a joint optimization scheme for robust bundle adjustment using lifting methods. Lifting methods introduce additional variables and circumvent alternating optimization. Zach and Bourmaud [36] deliver further insight into lifting and gradual refinement for bundle adjustment and recommend using the lifting method when a fast decrease in the objective is preferred and a sensible initial estimation of parameters is known.

Large scale optimization with nonlinear least squares objectives has been made more accessible through open-source optimizers such as Ceres [1] and Optlang [11]. DeVito et al. [11] introduce a GPU solver and show that using generic Gauss-Newton and Levenberg-Marquardt with conjugate gradient as the inner solver is competitive to handcrafted problem-specific solvers for many optimization problems that arise in computer graphics and computer vision.

3   Articulated Morphable Shape Model

The employed statistical shape model is a variation of SMPL [26] that enables joint optimization of all parameters. A differentiable surface model derived from the SMPL pose deformation model is introduced by [37]. They use an articulated person model for joint optimization. The model we propose incorporates the multi-person aspects of [26] and the subdivision surfaces for smooth parametric shape modelling from [21] into the articulated human shape model from [37].

In order to make the following formalism easier to read, the supplementary material contains tables summarizing symbols.

Formalism: The model template consists of a base mesh with $N = 1326$ vertices and $F = 1324$ quadrilateral faces. The underlying skeleton consists of $K = 16$ joints. A root joint is added and the resulting $K + 1$ nodes are connected by K edges (often referred to as bones in the literature). A visualization of the body parts can be found in the supplementary material. Sample meshes can be instantiated from the statistical model using subject-specific parameters $\vec{\beta} \in \mathbb{R}^B$ with $B = 10$ and pose specific parameters $(R, m) \in SO(3)^{K+1} \times \mathbb{R}^3$. $(R, m) = (R_0, R_1, \ldots, R_K, m)$ is separated in global pose parameters R_0 (global rotation), m (global translation) and skeleton pose parameters $R_1, R_2, \cdots, R_K$. The mean shape is denoted by $\bar{T} \in \mathbb{R}^{3N}$ (vectors in $\mathbb{R}^{3N}$ are interpreted as stacked x, y, z coordinates of N vectors in $\mathbb{R}^3$). Shape blend-shapes are denoted by $\mathcal{S} \in \mathbb{R}^{3N \times B}$ and corrective pose blend-shapes by $\mathcal{P} \in \mathbb{R}^{3N \times 9K}$. Each column in $\mathcal{S}$ denotes a shape blend-shape and the columns of $\mathcal{P}$ denote the $9K$ corrective pose blend-shapes. Shape blend-shapes are introduced to model varying shape between different subjects. In contrast to [26], the shape blend-shapes are not enforced to be orthogonal. Corrective pose blend-shape are incorporated to counteract artefacts (*e.g.* surface shrinking near joints during mesh articulation) when applying linear blend skinning Eq. (5). In contrast to [26], our skeleton always binds to the mean shape $\bar{T}$ with the zero pose $(R^*, m^*) = (I, I, \ldots, I, 0) \in SO(3)^{K+1} \times \mathbb{R}^3$.

Blend-Shape Application: Blend-shapes are applied to the mean shape $\bar{T}$ using

$$T = \bar{T} + \mathcal{S}\vec{\beta} + \mathcal{P}\operatorname{vec}(R_1 - R_1^*, R_2 - R_2^*, \ldots, R_K - R_K^*), \tag{1}$$

where $I \in \mathbb{R}^{3 \times 3}$ is the identity matrix and $\operatorname{vec}(\cdot)$ flattens each argument and concatenates the vectors to form a single column vector of the required shape. The resulting vertex positions after the application of all linear blend-shapes are denoted by T. We prescribe a fixed sparsity pattern to the corrective pose blend-shape matrix $\mathcal{P}$ which reduces the parameter count and prevents overfitting. In our implementation, we assign each vertex to two adjacent joints which leads to at most $2 \cdot 9 = 18$ nonzero entries per row in $\mathcal{P}$.

Subject-specific Skeleton: To articulate the shape we use a skeleton that defines a forward kinematic tree. The subject-specific joint locations in the rest pose are given by

$$J = \bar{J} + \mathcal{J}\vec{\beta}, \tag{2}$$

where $\bar{J} \in \mathbb{R}^{3K}$ are the joint locations corresponding to the mean shape $\bar{T}$ and $\mathcal{J} \in \mathbb{R}^{3K \times B}$ are the skeleton basis shapes that are correlated with the respective shape blend-shapes $\mathcal{S}$. We follow [37] for the subject-specific skeleton formalism instead of [26] which instead regresses from $\bar{T} + \mathcal{S}\vec{\beta}$ to J by employing a sparse regression matrix with $3N \cdot 3K$ parameters. Our end-to-end optimization framework prohibits the use of sparsity inducing regularizers (as used by [26]). Additionally, only $(B + 1)3K$ instead of $9NK$ parameters are introduced.

Kinematic Tree: We use the resulting joint locations and a prescribed skeleton topology to construct a transformation for each joint $1 \leq k \leq K$ in the skeleton. This transformation can then be applied to any $x \in \mathbb{R}^3$. The transformation of x w.r.t. the joint indexed by k is denoted by $G'_k : \mathbb{R}^3 \times SO(3)^{K+1} \times \mathbb{R}^{3K} \to \mathbb{R}^3$ with

$$G'_k(x; R, J) = \begin{pmatrix} 1 & 0 & 0 & 0 \\ 0 & 1 & 0 & 0 \\ 0 & 0 & 1 & 0 \end{pmatrix} G_k \begin{pmatrix} x - j_k \\ 1 \end{pmatrix} \text{ where} \tag{3}$$

$$G_0(R, J) = \begin{pmatrix} R_0 & 0 \\ 0^T & 1 \end{pmatrix} \text{ and } G_k(R, J) = \begin{pmatrix} R_k & j_k - j_{A(k)} \\ 0^T & 1 \end{pmatrix} G_{A(k)} \quad \forall_{1 \leq k \leq K}. \tag{4}$$

The position of the joint indexed by k is $j_k \in \mathbb{R}^3$, $A(k)$ denotes the ancestor joint index with respect to joint indexed by k and $j_0 = 0$. Note that the $x - j_k$ transforms x from the joint indexed by k to the root node.

Linear Blend Skinning: The kinematic tree defined above is used to transform the vertex positions T to T' using linear blend skinning [22]:

$$t'_i = m + \sum_{k=1}^{K} w_{k,i} G'_k(t_i; R, J), \quad t_i, t'_i \in \mathbb{R}^3, \forall_{1 \leq i \leq N} \tag{5}$$

where $t_i, t'_i \in \mathbb{R}^3$ denote the i-th vertex of T and T' respectively and $w_{k,i} \in [0, 1]$ is a linear blend skinning weight for vertex t_i transformed by G'_k. All blend skinning weights are denoted by $\mathcal{W} \in [0, 1]^{K \times N}$. The blend skinning weights are constrained by

$$\sum_{k=1}^{K} w_{k,i} = 1 \tag{6}$$

for all vertices indexed by i. After linear blend skinning and global translation m, the resulting vertex positions are denoted by $T' \in \mathbb{R}^{3N}$.

Complete Model Formulation: The whole mesh transformation is denoted by

$$T' = M(R, m, \vec{\beta}; \Theta), \tag{7}$$

where $\Theta = (\bar{T}, \mathcal{S}, \bar{J}, \mathcal{J}, \mathcal{P}, \mathcal{W})$ denote the aggregated model parameters. We convert the quad mesh into a subdivision surface denoted by

$$S(u; T') : \Omega \times \mathbb{R}^{3N} \to \mathbb{R}^3, \tag{8}$$

(a) Hierarchical data setup   (b) OpenPose on synthetic images

Fig. 2. (a) Visualization of the hierarchical input data and latent variables. Top left: Depiction of different persons in the training corpus. Top right: A single person with shape parameters $\vec{\beta}$ shown in different body poses. Bottom left: A fitted model with subject-specific shape coefficients $\vec{\beta}$ and pose-specific parameters (R, m). The 3D point cloud is indicated with filled blue circles. Bottom right: Zoom in on a single point in the point cloud with 3D position p and normal n. The corresponding model surface point is parameterized by $(u, R, m, \vec{\beta})$. Each datapoint induces a residual for the data term E_{Data}. (b) Keypoints: This figure shows an input scan from the Dynamic FAUST dataset [6] with 8 virtual cameras surrounding the scan in the center. Keypoints are extracted and visualized from each synthetic image using OpenPose [7].

where $u = (s, t, f)$ denotes a point on the surface S in local 2D coordinates $(s, t) \in [0, 1]^2$ with patch index $f \in \{1, 2, \cdots, F\}$. More precisely u consists of a 2D bezier patch parameterization [13] and the respective patch index, so $u \in \Omega = [0, 1]^2 \times \{1, 2, \cdots, F\}$. We implement S approximately using the approach outlined in [25] due to its efficiency and simplicity. Implementation details are provided in the supplementary material.

4   Objective and Optimization

To avoid a cluttered multi-index formalism we introduce a number of sets to describe the hierarchical nature of the data and accompanying latent variables (see Fig. 2a).

The training of the human shape requires the estimation $\hat{\Theta}$ of model parameters Θ from given point cloud measurements (scans) of different instances. Here, an instance denotes a 3D-scan of a specific subject in a specific pose. For n scans of Q individual subjects and P measurements per scan, we subsume all latent parameters in the triplet $\Gamma = (\mathcal{B}, \mathcal{R}, \mathcal{U})$, where $\mathcal{B}$ is the set of Q latent shape vectors $\vec{\beta} \in \mathcal{B}$, $\mathcal{R}$ is the set of n latent pose configurations $(R, m) \in \mathcal{R}$ and $\mathcal{U}$ is the set of nP latent surface correspondences $u \in \mathcal{U}$. In total, this leads to an unknown variable count of

$$|\mathcal{B}| + |\mathcal{R}| + |\mathcal{U}| = QB + 3(K + 2)n + 2Pn. \tag{9}$$

The estimation of the model parameters Θ requires the estimation of the latent parameters denoted by $\hat{\Gamma}$. We cast the model learning process as a non-linear least squares problem with unknowns Θ and Γ and propose the following cost function:

Data Term: The main functional is defined as

$$E_{\text{Data}} = \sum_{(p,u,R,m,\vec{\beta}) \in \mathcal{I}_{\text{Data}}} \phi\left(||S(u, M(R, m, \vec{\beta}; \Theta)) - p||^2\right), \qquad (10)$$

where $p \in \mathbb{R}^3$ denotes a point on the 3D scan, $\mathcal{I}_{\text{Data}} \subset \mathbb{R}^3 \times \mathcal{U} \times \mathcal{R} \times \mathcal{B}$ contains all data points with latent variable dependencies and $\phi : \mathbb{R} \to \mathbb{R}$ with $\phi(r) = r^2/(r^2 + \rho^2)$ is the robust Geman-McClure kernel.

Cost Function: Since the data term itself is not sufficient to lead to satisfactory results, additional regularization terms have to be added to constrain the solution space to minimizers which are restricted to non-degenerate shapes. To this end, we introduce additional prior information and regularization terms. Our complete cost function is

$$\begin{aligned}
E = {} & \lambda_{\text{Data}} E_{\text{Data}} && + \lambda_{\text{2D-joint}} E_{\text{2D-joint}} && + \lambda_{\text{2D-surf}} E_{\text{2D-surf}} && + \lambda_{\text{mean}} E_{\text{mean}} \\
& + \lambda_{\text{bshape}} E_{\text{bshape}} && + \lambda_{\text{pshape}} E_{\text{pshape}} && + \lambda_{\text{symm}} E_{\text{symm}} && + \lambda_{\text{symm-skel}} E_{\text{symm-skel}} \\
& + \lambda_{\text{joint}} E_{\text{joint}} && + \lambda_{\text{weights}} E_{\text{weights}} && + \lambda_{\text{convex}} E_{\text{convex}} && + \lambda_{\text{shape}} E_{\text{shape}} \\
& + \lambda_{\text{pose}} E_{\text{pose}} && + \lambda_{\text{ground}} E_{\text{ground}}
\end{aligned}$$

$$(11)$$

and consists of a weighted sum of squared error terms $E_\bullet$ and non-negative scalar hyperparameters $\lambda_\bullet$ that control the impact of each term. The values for the weights $\lambda_\bullet$ are listed in the supplementary material. The remaining section describes the regularization terms in detail.

2D Joint Term: When the model pose and shape is initialized far away from the pose and shape of the scan, the optimization of the data term is likely to end in a local minimum. In order to steer the pose and shape estimation in the right direction, we define a landmark term based on synthetic views of the scans. We apply OpenPose [7] on these scans for 2D keypoint extraction (see Fig. 2b). Those keypoints, which correspond to joints (and have a detection score above 0.5) are used as landmarks. We denote Π as the set of $|\Pi| = 8$ virtual cameras and add a landmark term that penalizes the distance between the model joints projected onto the virtual 2D images and the 2D landmarks:

$$E_{\text{2D-joint}} = \sum_{(Q,\pi,R,m,\vec{\beta}) \in \mathcal{I}_{\text{2D-joint}}} \sum_{k=1}^{K} ||\pi(G'_k(j_k, R, J) + m) - q_k||^2 \qquad (12)$$

where $\pi : \mathbb{R}^3 \to \mathbb{R}^2$ is a camera-specific projection that maps points in world coordinates to points in image coordinates. $\mathcal{I}_{\text{2D-joint}} \subset \mathbb{R}^{2K} \times \Pi \times \mathcal{R} \times \mathcal{B}$ contains the 2D-labels $Q \in \mathbb{R}^{2K}$ with $q_k \in \mathbb{R}^2$ for each joint k in an image with latent dependencies. Even though OpenPose was trained on natural labeled RGB images [7], the predictions work surprisingly well for non-photorealistic synthetic images. The camera setup and the synthetic images with keypoint estimations are depicted in Fig. 2b.

2D Surface Term: OpenPose not only provides keypoints that correspond to joints but also provides keypoints corresponding to landmarks on the surface of the scan (*e.g.* nose and ears). We propose an additional error term for such keypoints. The 2D surface landmark term

$$E_{\text{2D-surf}} = \sum_{(q',u',\pi,R,\boldsymbol{m},\vec{\beta}) \in \mathcal{I}_{\text{2D-surf}}} ||\pi\left(S(u', M(R,\boldsymbol{m},\vec{\beta};\Theta))\right) - q'||^2, \qquad (13)$$

encourages the optimization process to bring the model surface points at $u' \in \Omega$ close to the 2D annotations $q' \in \mathbb{R}^2$ in the image space after the perspective projection π. $\mathcal{I}_{\text{2D-surf}} \subset \mathbb{R}^2 \times \Omega \times \Pi \times \mathcal{R} \times \mathcal{B}$ contains all 2D surface landmarks and links the relevant surface location, camera projection function and latent dependencies.

Smoothing Terms: Self-intersections for mesh-based model-fitting approaches have to be mitigated during optimization. Additionally, we have to deal with missing data (*e.g.* armpits and soles of the feet) and an interpolation scheme has to be adopted in such body regions. To this end, we correlate $\bar{T}$, S and $\mathcal{P}$ so that vertex displacement corresponding to adjacent vertices in the graph of the mesh are similar. When interpreting these variables as vector fields on the mesh we prefer solutions where the vector fields are smooth. One way to encourage this behavior is by exploiting the linear, positive semidefinite Laplace-Beltrami operator from the template shape with vertex positions $\bar{T}^{\text{init}} \in \mathbb{R}^{3N}$. We represent the linear discrete Laplace-Beltrami operator by the matrix $\Delta \in \mathbb{R}^{N \times N}$ and denote $||C||^2_\Delta = C^T \Delta C$ by slight abuse of notation for a provided vector $C \in \mathbb{R}^N$. We estimate Δ with vertex positions $\bar{T}^{\text{init}}$ using the approach from [2]. Our model parameters live mostly in $\mathbb{R}^{3N}$. We make use of Δ by applying it to the N different x, y and z components in the spirit of [31]. To this end, we employ the regularization terms

$$E_{\text{mean}} = ||\bar{T}_x - \bar{T}_x^{\text{init}}||^2_\Delta + ||\bar{T}_y - \bar{T}_y^{\text{init}}||^2_\Delta + ||\bar{T}_z - \bar{T}_z^{\text{init}}||^2_\Delta \qquad (14)$$

$$E_{\text{bshape}} = \sum_{j=1}^{B} \left(||\mathcal{S}_x^{(j)}||^2_\Delta + ||\mathcal{S}_y^{(j)}||^2_\Delta + ||\mathcal{S}_z^{(j)}||^2_\Delta\right) \qquad (15)$$

$$E_{\text{pshape}} = \sum_{j=1}^{9K} \left(||\mathcal{P}_x^{(j)}||^2_\Delta + ||\mathcal{P}_y^{(j)}||^2_\Delta + ||\mathcal{P}_z^{(j)}||^2_\Delta\right) \qquad (16)$$

where $\cdot_x$, $\cdot_y$, $\cdot_z$ refer to the x,y or z component from a vector in $\mathbb{R}^{3N}$ and $\cdot^{(j)}$ denotes the j-th column of the indexed matrix.

Symmetry Term: The body model is split into the left- and right-hand side along the y-z plane. We employ a symmetry term that encourages symmetric shapes with

$$E_{\text{symm}} = ||\bar{T}_x + \bar{T}_x^{\text{mirror}}||^2 + ||\bar{T}_y - \bar{T}_y^{\text{mirror}}||^2 + ||\bar{T}_z - \bar{T}_z^{\text{mirror}}||^2 +$$
$$\sum_{j=1}^{B} \left(||\mathcal{S}_x^{(j)} + \tilde{\mathcal{S}}_x^{(j)}||^2 + ||\mathcal{S}_y^{(j)} - \tilde{\mathcal{S}}_y^{(j)}||^2 + ||\mathcal{S}_z^{(j)} - \tilde{\mathcal{S}}_z^{(j)}||^2\right), \qquad (17)$$

where $\bar{T}^{\mathrm{mirror}}$ and $\tilde{S}$ permute $\bar{T}$ and S, respectively, so that each vertex maps to its mirrored partner. A corresponding term exists for $\bar{J}$ and J with

$$E_{\mathrm{symm\text{-}skel}} = ||\bar{J}_x + \bar{J}_x^{\mathrm{mirror}}||^2 + ||\bar{J}_y - \bar{J}_y^{\mathrm{mirror}}||^2 + ||\bar{J}_z - \bar{J}_z^{\mathrm{mirror}}||^2 +$$
$$\sum_{j=1}^{B} \left(||\mathcal{J}_x^{(j)} + \tilde{\mathcal{J}}_x^{(j)}||^2 + ||\mathcal{J}_y^{(j)} - \tilde{\mathcal{J}}_y^{(j)}||^2 + ||\mathcal{J}_z^{(j)} - \tilde{\mathcal{J}}_z^{(j)}||^2 \right), \tag{18}$$

where $\bar{J}^{\mathrm{mirror}}$ and $\tilde{J}$ are defined accordingly.

Skeleton Consistency Term: Our current formulation does not prevent the optimizer to move the joints to arbitrary locations outside of the surface. To counteract this behavior, we softly constrain the mean joint positions J and skeleton basis shapes $\mathcal{J}$ to deform with the mean shape $\bar{T}$ and blend-shapes S at some specified vertices. We denote the set of manually specified vertex indices for the joint indexed by k with Ring_k (see e.g. [21] or the supplementary material). The joint error term is defined as

$$E_{\mathrm{joint}} = \sum_{k=1}^{K} \left(||j_k - \frac{1}{|\mathrm{Ring}_k|} \sum_{i \in \mathrm{Ring}_k} \bar{t}_i||^2 + \sum_{b=1}^{B} ||\mathcal{J}_k^{(b)} - \frac{1}{|\mathrm{Ring}_k|} \sum_{i \in \mathrm{Ring}_k} S_i^{(b)}||^2 \right), \tag{19}$$

where $S_i^{(b)} \in \mathbb{R}^3$ denotes the i-th vertex position of the b-th blend-shape and $\mathcal{J}_k^{(b)} \in \mathbb{R}^3$ denotes the k-th joint of the b-th skeleton basis shape.

Blend-Skinning Term: We encourage blend-weights W that are close to an initial estimate W^{init} (see the supplementary materials for details) with

$$E_{\mathrm{weights}} = ||W - W^{\mathrm{init}}||_F^2. \tag{20}$$

Convex Combination Term: We introduce another term that softly encourages the convex combination constraint Eq. (6) for the blend-skinning weights W via

$$E_{\mathrm{convex}} = \sum_{i=1}^{N} \left(1 - \sum_{k=1}^{K} w_{k,i} \right)^2. \tag{21}$$

Shape Regularization Term: We encourage the blend-shape coefficients $\vec{\beta}$ to be small using

$$E_{\mathrm{shape}} = \sum_{\vec{\beta} \in \mathcal{B}} ||\vec{\beta}||^2. \tag{22}$$

Pose Regularization Term: We want to discourage unrealistic human postures with the addition of simple pose regularization term

$$E_{\mathrm{pose}} = \sum_{(R,m) \in \mathcal{R}} \sum_{k=1}^{K} ||R_k - \bar{R}_k||_F^2 \tag{23}$$

where $\bar{R}_k \in SO(3)$ denotes the mean rotation of joint k. These are also introduced as variables during optimization and denoted by $\bar{R}$.

Ground Plane Consistency Term: To counteract missing 3D measurements on the soles of the feet we employ a ground plane term to compensate for missing data. We assume a known ground plane perpendicular to the y-axis and offset at height $H \in \mathbb{R}$. We penalize uniformly sampled model surface points $u'' \in \Omega$ that fall below the ground plane via

$$E_{\text{ground}} = \sum_{(u'',R,m,\vec{\beta}) \in \mathcal{I}_{\text{ground}}} \left(S(u'', M(R, m, \vec{\beta}; \Theta))_y - H \right)^2, \qquad (24)$$

where $\mathcal{I}_{\text{ground}} \subset \Omega \times \mathcal{R} \times \mathcal{B}$ contains sampled surface points below the ground plane with latent dependencies for each scan and $\cdot_y$ denotes the y-coordinate of a point in $\mathbb{R}^3$.

Optimization: The objective defined in Eq. (11) is cast as a nonlinear least squares problem and minimized using a truncated variation of Levenberg-Marquardt, where the normal equations are solved approximately using the conjugate gradient method. This leads to the local minimizers

$$\hat{\Theta}, \hat{R} = \arg \min_{\Theta, \bar{R}} \min_{\Gamma} E(\Gamma; \Theta, \bar{R}). \qquad (25)$$

Optimization is performed using the Optlang [11] framework. Several aspects of the optimization procedure require special attention (*e.g.* discrete correspondences updates) and are discussed in the supplementary material. We use a varying hyperparamter schedule for $\lambda_\bullet$, encouraging a rough alignment of body pose before relaxing the constraints on $\mathcal{S}, \mathcal{W}$ and $\mathcal{P}$. At the start of the optimization we mainly rely on detected keypoints. In the later stages, the keypoints are discarded and E_{data} takes precedence.

5   Experimental Evaluation

5.1   Training

We train a female, a male and a unisex model. For training, we test the limits of our approach and use the entire GPU memory (11 GB). We train the models on $n = 911$ scans each where the first 250 scans are acquired from the D-FAUST dataset [6]. D-FAUST contains 10 subjects (5 female and 5 male) and spans a wide variety of body poses. We sample 50 scans per person from this dataset by clustering the pose space with k-means on multi-view OpenPose 3D joint predictions to cover a diverse set of poses. The remaining 661 scans are randomly chosen from the CAESAR dataset [29]. We subsample all 911 scans to 20,000 data points each. This results in a total of $Q = 666$ distinct individuals for each model in the training process. One training process simultaneously optimizes more than $3.6 \cdot 10^7$ latent parameters plus $|\Theta| + 3K$ model parameters (the count of latent parameters Eq. (9) increases with the training set size). This setup requires up to three days of training on an NVIDIA RTX 2080 TI.

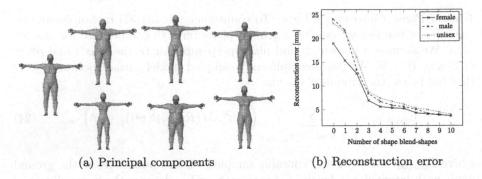

(a) Principal components   (b) Reconstruction error

Fig. 3. (a) Visualization of our trained female shape model, from left to right: Mean shape and the first three PCA components sorted by explained variance in descending order ($\pm 3\sigma$ for the first two principal components and $\pm 5\sigma$ for the third principal component). The model without applied subdivision is overlayed as a wireframe mesh. (b) After training with $B = 10$ shape blend-shapes, we compute the shape and skeleton for each person in the training set and create new blend-shapes using PCA. We vary the number of principal components from 0 to 10 and record the average reconstruction error for each setting.

Fig. 4. Qualitative model fit results on the FAUST dataset: This figure shows 15 different poses in total of all 10 different subjects. The scans are shown in red and the model fits are shown in light blue. (Color figure online)

Model Subspace Evaluation: We analyze the factorization of shape and pose parameters of our models qualitatively. Therefore, we keep the pose parameters (R, m) fixed to a non-canonical pose and vary $\vec{\beta}$ to see if a change in $\vec{\beta}$ leads to obvious changes in posture. Results of this experiment such as shown in Fig. 1b indicate, that shape and pose are factored correctly. In particular changes in $\vec{\beta}$ do not lead to changes in pose.

Principal Components: We orthogonalize the trained shape blend-shapes $\mathcal{S}$ of our female model using PCA and visualize the first three principal components (see Fig. 3a). The first two principal components (PC) correspond to more or less correlated variations in body weight and height. This result is similar to [3] and [28] where the first two components describe variations in gender in addition to weight and height. It differs from the female model in SMPL [26], where the first PC clearly corresponds to body size and the second PC to body weight. We identify two different explanations: (I) Our training set consists of less (approx. one third of [26]) and different (*e.g.* European vs. North American CAESAR dataset) individuals. Therefore, our training set is very likely to represent another statistical distribution. (II) Missing texture information can lead to sliding of correspondences along the surface [5] which has a negative impact on the model training process. We can not easily analyze (I) due to memory constraints on the GPU and the non-availability of the multi-pose dataset. We analyze (II) by benchmarking our approach on the FAUST correspondence challenge in Sect. 5.2.

Model Capacity: We evaluate the reconstruction error by incrementally adding shape blend-shapes to our model (see Fig. 3). After training with $B = 10$ shape blend-shapes, we compute the shape and skeleton for each person in the training set and create new blend-shapes using PCA. We vary the number of principal components from 0 to 10 and record the average reconstruction error for each setting (see Fig. 3b). No retraining is performed. The residuals in E_{Data} are interpreted as the reconstruction error and the euclidean distance $|| \cdot ||_2^2$ is computed for each data point. This error is estimated by sampling $20,000$ points from each scan, computing the distance to the model and averaging the error over all sampled measurements and scans. This error is not sufficient to evaluate the quality of the model but it can give further insight into the training process and model quality. The reconstruction error reduction seems to taper off when using 7 or more principal components. The average reconstruction error for all 10 principal components is 3.7 mm.

5.2 Inference

For inference we keep $\hat{\Theta}$ and $\hat{R}$ fixed and estimate $\hat{\Gamma} = \arg \min_\Gamma E(\Gamma; \hat{\Theta}, \hat{R})$. For evaluation we fit our female, male and unisex models on the FAUST dataset [5] which consists of 10 different persons with 30 scans each.

Qualitative Results: The model-fitting results on 10% of the scans of the FAUST dataset are depicted in Fig. 4. The 2D keypoint terms lead to convincing rough alignments. Coarse body proportions are faithfully reconstructed. Few coarse alignment errors occur due to insufficient pose estimates, mostly stemming from touching body parts or from erroneous head alignment by OpenPose when the scan is not facing at least one of the 8 virtual cameras.

Quantitative Results: We evaluate our approach quantitatively using the FAUST intra-subject challenge and inter-subject challenge. The challenges consist of 100 scan pairs where points on the source scan have to be mapped to corresponding points on the target scan. We evaluate registration results on FAUST using the provided evaluation platform. The average error on the intra-subject challenge and inter-subject challenge for our female model 2.353 cm and 3.234 cm respectively. We compare our results to the current state of the art (see Table 1): In the intra-subject challenge, our approach performs comparably to the listed methods. On the inter-subject challenge our approach is on par with the Stitched Puppet [38] approach which, in contrast to our method, was trained by sampling a pre-trained, strong multi-person shape model. The various variants of the 3D-CODED methods perform worse in their unsupervised settings [15] or require

Table 1. Quantitative results on the FAUST dataset (average error in cm) and prerequisites for model training that were used to achieve the reported performance.

	Intra-subject	Inter-subject	Dependencies
FAUST [5]	0.7	1.1	strong pose prior
Stitched Puppet [38]	1.568	3.126	SCAPE [3]
3D-CODED [15] (unsup)	N/A	4.835	SURR. [34], SMPL [26]
3D-CODED [15] (sup)	1.985	2.878	SURR.[34], SMPL [26]
Deprelle et al. [10]	1.626	2.578	SURR. [34], 3D-CODED [15]
LBS-AE [23]	2.161	4.079	None
Halimi et al. [16]	2.51	N/A	None
Female only (Ours)	2.353	3.234	OpenPose [7]
Male only (Ours)	2.387	3.519	OpenPose [7]
Unisex (Ours)	2.304	3.525	OpenPose [7]
Female and male (Ours)	2.301	3.422	OpenPose [7]

large labeled synthetic training data to outperform model-based approaches [10,15]. Finally, we clearly outperform LBS-AE [23] on the inter-subject challenge. For Halimi et al. [16] no publicly available inter-subject result exists.

6   Discussion and Conclusion

We show that articulated multi-person shape model training can be addressed within a single objective where all parameters are jointly optimized. The proposed method is markerless in the sense that no handcrafted landmarks are required and no pre-existing shape model is required, which might implicitly incorporate expensively generated correspondences. Instead, the landmark term of our objective deals with untextured 3D scans in combination with the output of an off-the-shelf 2D keypoint detector with comparatively low accuracy.

The limiting factor of our approach is the memory of the GPU, which restricts the scalability in terms of the resolution and the amount of training data in comparison to alternating or stochastic optimization methods. On the other hand, the presented results show, that despite the limitations in terms of resolution and variability of the training data, the achieved accuracy in the FAUST correspondence challenge is comparable to strong human shape models that have higher resolution and rely on (semi)-supervised training schemes.

Acknowledgment. We thank A. Bender for the data setup figure. We thank J. Wetzel and N. Link for technical discussion. This work was supported by the German Federal Ministry of Education and Research (BMBF) under Grant 13FH025IX6.

References

1. Agarwal, S., Mierle, K., et al.: Ceres solver. http://ceres-solver.org
2. Alexa, M., Wardetzky, M.: Discrete laplacians on general polygonal meshes. In: ACM Transactions on Graphics (TOG), vol. 30, p. 102. ACM (2011)
3. Anguelov, D., Srinivasan, P., Koller, D., Thrun, S., Rodgers, J., Davis, J.: Scape: shape completion and animation of people. In: ACM Transactions on Graphics (TOG), vol. 24, pp. 408–416. ACM (2005)
4. Blanz, V., Vetter, T.: A morphable model for the synthesis of 3D faces. In: Proceedings of the Conference on Computer Graphics and Interactive Techniques (SIGGRAPH), vol. 99, pp. 187–194 (1999)
5. Bogo, F., Romero, J., Loper, M., Black, M.J.: Faust: dataset and evaluation for 3D mesh registration. In: Proceedings of the IEEE Conference on Computer Vision and Pattern Recognition (CVPR), pp. 3794–3801 (2014)
6. Bogo, F., Romero, J., Pons-Moll, G., Black, M.J.: Dynamic faust: registering human bodies in motion. In: Proceedings of the IEEE Conference on Computer Vision and Pattern Recognition (CVPR), pp. 6233–6242 (2017)
7. Cao, Z., Simon, T., Wei, S.E., Sheikh, Y.: Realtime multi-person 2D pose estimation using part affinity fields. In: Proceedings of the IEEE Conference on Computer Vision and Pattern Recognition (CVPR) (2017)
8. Cashman, T.J., Fitzgibbon, A.W.: What shape are dolphins? Building 3D morphable models from 2D images. Trans. Pattern Anal. Mach. Intell. (PAMI) 35(1), 232–244 (2012)
9. Catmull, E., Clark, J.: Recursively generated b-spline surfaces on arbitrary topological meshes. Comput. Aided Des. 10(6), 350–355 (1978)
10. Deprelle, T., Groueix, T., Fisher, M., Kim, V., Russell, B., Aubry, M.: Learning elementary structures for 3D shape generation and matching. In: Advances in Neural Information Processing Systems (NIPS), pp. 7433–7443 (2019)
11. DeVito, Z., et al.: Opt: a domain specific language for non-linear least squares optimization in graphics and imaging. In: ACM Transactions on Graphics 2017 (TOG) (2017)
12. Engel, J., Koltun, V., Cremers, D.: Direct sparse odometry. Trans. Pattern Anal. Mach. Intell. (PAMI) 40(3), 611–625 (2018)
13. Farin, G.E., Farin, G.: Curves and Surfaces for CAGD: A Practical Guide. Morgan Kaufmann (2002)
14. Genova, K., Cole, F., Maschinot, A., Sarna, A., Vlasic, D., Freeman, W.T.: Unsupervised training for 3D morphable model regression. In: Proceedings of the IEEE Conference on Computer Vision and Pattern Recognition (CVPR), pp. 8377–8386 (2018)
15. Groueix, T., Fisher, M., Kim, V.G., Russell, B.C., Aubry, M.: 3D-coded: 3D correspondences by deep deformation. In: Proceedings of the European Conference on Computer Vision (ECCV), pp. 230–246 (2018)
16. Halimi, O., Litany, O., Rodola, E., Bronstein, A.M., Kimmel, R.: Unsupervised learning of dense shape correspondence. In: Proceedings of the IEEE Conference on Computer Vision and Pattern Recognition (CVPR), pp. 4370–4379 (2019)
17. Hesse, N., et al.: Learning an infant body model from RGB-D data for accurate full body motion analysis. In: Frangi, A.F., Schnabel, J.A., Davatzikos, C., Alberola-López, C., Fichtinger, G. (eds.) MICCAI 2018. LNCS, vol. 11070, pp. 792–800. Springer, Cham (2018). https://doi.org/10.1007/978-3-030-00928-1_89

18. Hirshberg, D.A., Loper, M., Rachlin, E., Black, M.J.: Coregistration: simultaneous alignment and modeling of articulated 3D shape. In: Fitzgibbon, A., Lazebnik, S., Perona, P., Sato, Y., Schmid, C. (eds.) ECCV 2012. LNCS, vol. 7577, pp. 242–255. Springer, Heidelberg (2012). https://doi.org/10.1007/978-3-642-33783-3_18
19. Jaimez, M., Cashman, T.J., Fitzgibbon, A., Gonzalez-Jimenez, J., Cremers, D.: An efficient background term for 3D reconstruction and tracking with smooth surface models. In: Proceedings of the IEEE Conference on Computer Vision and Pattern Recognition (CVPR) (2017)
20. Joo, H., Simon, T., Sheikh, Y.: Total capture: a 3D deformation model for tracking faces, hands, and bodies. In: Proceedings of the IEEE Conference on Computer Vision and Pattern Recognition (CVPR), pp. 8320–8329 (2018)
21. Khamis, S., Taylor, J., Shotton, J., Keskin, C., Izadi, S., Fitzgibbon, A.: Learning an efficient model of hand shape variation from depth images, June 2015
22. Lewis, J.P., Cordner, M., Fong, N.: Pose space deformation: a unified approach to shape interpolation and skeleton-driven deformation. In: Proceedings of the Conference on Computer Graphics and Interactive Techniques, pp. 165–172 (2000)
23. Li, C.L., Simon, T., Saragih, J., Póczos, B., Sheikh, Y.: LBS autoencoder: self-supervised fitting of articulated meshes to point clouds. In: Proceedings of the IEEE Conference on Computer Vision and Pattern Recognition (CVPR), pp. 11967–11976 (2019)
24. Li, T., Bolkart, T., Black, M.J., Li, H., Romero, J.: Learning a model of facial shape and expression from 4D scans. ACM Trans. Graph. (TOG) 36(6), 194 (2017)
25. Loop, C., Schaefer, S.: Approximating Catmull-Clark subdivision surfaces with bicubic patches. ACM Trans. Graph. (TOG) 27(1), 8 (2008)
26. Loper, M., Mahmood, N., Romero, J., Pons-Moll, G., Black, M.J.: SMPL: a skinned multi-person linear model. ACM Trans. Graph. (TOG) 34(6), 248 (2015)
27. Pavlakos, G., et al.: Expressive body capture: 3D hands, face, and body from a single image. In: Proceedings of the IEEE Conference on Computer Vision and Pattern Recognition (CVPR) (2019)
28. Pishchulin, L., Wuhrer, S., Helten, T., Theobalt, C., Schiele, B.: Building statistical shape spaces for 3D human modeling. Pattern Recogn. 67, 276–286 (2017)
29. Robinette, K.M., Daanen, H., Paquet, E.: The CAESAR project: a 3-D surface anthropometry survey. In: International Conference on 3-D Digital Imaging and Modeling, pp. 380–386. IEEE (1999)
30. Romero, J., Tzionas, D., Black, M.J.: Embodied hands: modeling and capturing hands and bodies together. ACM Trans. Graph. (TOG) 36(6), 245 (2017)
31. Sorkine, O., Alexa, M.: As-rigid-as-possible surface modeling. In: Symposium on Geometry Processing, vol. 4 (2007)
32. Taylor, J., et al.: User-specific hand modeling from monocular depth sequences. In: Proceedings of the IEEE Conference on Computer Vision and Pattern Recognition (CVPR), pp. 644–651 (2014)
33. Taylor, J., et al.: Efficient and precise interactive hand tracking through joint, continuous optimization of pose and correspondences. ACM Trans. Graph. (TOG) 35(4), 143 (2016)
34. Varol, G., et al.: Learning from synthetic humans. In: Proceedings of the IEEE Conference on Computer Vision and Pattern Recognition (CVPR) (2017)
35. Zach, C.: Robust bundle adjustment revisited. In: Fleet, D., Pajdla, T., Schiele, B., Tuytelaars, T. (eds.) ECCV 2014. LNCS, vol. 8693, pp. 772–787. Springer, Cham (2014). https://doi.org/10.1007/978-3-319-10602-1_50

36. Zach, C., Bourmaud, G.: Iterated lifting for robust cost optimization. In: Proceedings of the British Machine Vision Conference (BMVC) (2017)
37. Zeitvogel, S., Laubenheimer, A.: Towards end-to-end 3D human avatar shape reconstruction from 4D data. In: International Symposium on Electronics and Telecommunications (ISETC), pp. 1–4. IEEE (2018)
38. Zuffi, S., Black, M.J.: The stitched puppet: a graphical model of 3D human shape and pose. In: Proceedings of the IEEE Conference on Computer Vision and Pattern Recognition (CVPR), pp. 3537–3546 (2015)

Accurate RGB-D Salient Object Detection via Collaborative Learning

Wei Ji[1], Jingjing Li[1], Miao Zhang[1(✉)], Yongri Piao[1], and Huchuan Lu[1,2]

[1] Dalian University of Technology, Dalian, China
weiji.dlut@gmail.com, jingjing.dlut@outlook.com,
{miaozhang,yrpiao,lhchuan}@dlut.edu.cn
[2] Pengcheng Lab, Shenzhen, China
https://github.com/OIPLab-DUT/CoNet

Abstract. Benefiting from the spatial cues embedded in depth images, recent progress on RGB-D saliency detection shows impressive ability on some challenge scenarios. However, there are still two limitations. One hand is that the pooling and upsampling operations in FCNs might cause blur object boundaries. On the other hand, using an additional depth-network to extract depth features might lead to high computation and storage cost. The reliance on depth inputs during testing also limits the practical applications of current RGB-D models. In this paper, we propose a novel collaborative learning framework where edge, depth and saliency are leveraged in a more efficient way, which solves those problems tactfully. The explicitly extracted edge information goes together with saliency to give more emphasis to the salient regions and object boundaries. Depth and saliency learning is innovatively integrated into the high-level feature learning process in a mutual-benefit manner. This strategy enables the network to be free of using extra depth networks and depth inputs to make inference. To this end, it makes our model more lightweight, faster and more versatile. Experiment results on seven benchmark datasets show its superior performance.

1 Introduction

The goal of salient object detection (SOD) is to locate and segment the most attractive and noticeable regions in an image. As a fundamental and pre-processing task, salient object detection plays an important role in various computer vision tasks, e.g., visual tracking [25,52], video SOD [20,58], object detection [14,50], semantic segmentation [37], and human-robot interaction [13].

Recent researches on RGB-D salient object detection have gradually broken the performance bottleneck of traditional methods and RGB-based methods, especially when dealing with complex scenarios like similar foreground and background. However, there are some limitations with the introduction of FCNs [40,51] and depth images. *Firstly*, the emergence of FCNs enables automatic extraction of multi-level and multi-scale features. The high-level features

---

W. Ji and J. Li—Equal contribution.

© Springer Nature Switzerland AG 2020
A. Vedaldi et al. (Eds.): ECCV 2020, LNCS 12363, pp. 52–69, 2020.
https://doi.org/10.1007/978-3-030-58523-5_4

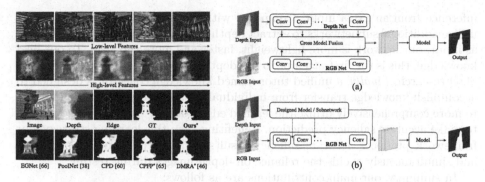

Fig. 1. (Left) First two rows: feature maps in different layers of CNNs. Last two rows: RGB image, depth image, edge map, saliency ground truth (GT) and saliency results of several state-of-the-art methods. * means RGB-D methods. (Right) Two kinds of previous RGB-D SOD network structures. (a) Processing RGB input and depth input separately and then combining the complementary RGB and depth features through cross-modal fusion (e.g. [5–7,23,46]). (b) Using tailor-made depth subnetworks to compensate for RGB representations (e.g. [65,68]).

with rich semantic information can better locate salient objects but the pooling and upsampling operations in FCNs might result in coarse and blur object boundaries (Fig. 1 (left)). The low-level features contain rich local details but suffer from excessive background noises and might cause information chaos. *Secondly*, the spatial layout information from depth images can better express 3D scenes and help locate salient objects. However, previous RGB-D methods either adopted two-stream architectures that process RGB and depth images separately with various cross-modal fusion strategies (Fig. 1a) [5–7,23,46], or utilized subnetworks tailored for depth image to compensate for RGB representations (Fig. 1b) [65,68]. In those methods, the additional depth-networks might lead to high computation and storage cost, and cannot work without depth input, seriously limiting their practical applications.

In this paper, we propose a novel collaborative learning framework (CoNet) to confront the aforementioned limitations. In collaborative learning, multiple group members work together to achieve learning goals through exploratory learning and timely interaction. In our framework, three mutually beneficial collaborators are well-designed from different perspectives of the SOD task, namely edge detection, coarse salient object detection, and depth estimation. *On the one hand*, a edge collaborator is proposed to explicitly extracts edge information from the overabundant low-level features and then goes together with saliency knowledge to jointly assign greater emphasis to salient regions and object boundaries. *On the other hand*, considering the strong consistencies among global semantics and geometrical properties of image regions [54], we innovatively integrate depth and saliency learning into the high-level feature learning process in a mutual-benefit manner. Instead of directly taking depth image as input, this learning strategy enables the network to be free of using an extra depth network to make inference from an extra input. Compared with previous RGB-D models which utilize additional subnetworks to extract depth features and rely on depth images as input, our network is more lightweight, faster and more versatile. To our best knowledge, this is the first attempt to use depth images in such a way in RGB-D SOD research. *Finally*, a unified tutor named knowledge collector is designed to accomplish knowledge transfer from individual collaborators to the group, so as to more comprehensively utilize the learned edge, saliency and depth knowledges to make accurate saliency prediction. Benefiting from this learning strategy, our framework produces accurate saliency results with sharp boundary preserved and simultaneously avoids the reliance on depth images during testing.

In summary, our main contributions are as follows:

- We propose a novel collaborative learning framework (CoNet) where edge, depth, and saliency are leveraged in a different but more efficient way for RGB-D salient object detection. The edge exploitation makes the boundaries of saliency maps more accurate.
- This learning strategy enables our RGB-D network to be free of using an additional depth network and depth input during testing, and thus being more lightweight and versatile.
- Experiment results on seven datasets show the superiority of our method over other state-of-the-art approaches. Moreover, it supports the faster frame rate as it runs at 34 FPS, meeting the needs of real-time prediction (enhances FPS by 55% compared with current best performing method DMRA [46]).

2   Related Work

Early works [10,27,36,45,61] for saliency detection mainly rely on hand-crafted features. [2,3,55] are some comprehensive surveys. Recently, traditional methods have been gradually surpassed by deep learning ones. Among those researches, 2D methods [16,30,32,33,35,41,53,56,57,60,67] based on RGB images have achieved remarkable performance and lone been the mainstream of saliency detection. However, 2D saliency detection appears to make a downgrade when handling complex scenarios due to the lack of spatial information in single RGB image. The introduction of depth images in RGB-D saliency researches [5–7,23,31,46,48,49,65,68] has made great promotions for those complex cases thanks to the embedded rich spatial information of depth images.

The first CNNs-based method [48] for RGB-D SOD uses hand-crafted features extracted from RGB and depth images for training. Then, Chen *et al.* propose to use two-stream models [7,23] to process RGB and depth image separately and then combine cross-modal features to jointly predict saliency. They subsequently design a progressive fusion network [5] to better fuse cross-modal multi-level features and propose a three-stream network [6] which adopts the attention mechanism to adaptively select complement from RGB and depth features. Afterwards, Piao *et al.* [46] utilize residual structure and depth-scale feature fusion module to fuse paired RGB and depth features. The network structures in [5–7,23,46] can be represented as two-stream architectures shown in

Fig. 1a. Another kind of structure is the use of subnetworks tailored for depth images to extract depth features and make compensation for RGB representations [65,68] (Fig. 1b). Zhu *et al.* [68] utilize an auxiliary network to extract depth-induced features and then use them to enhance a pre-trained RGB prior model. In [65], Zhao *et al.* first enhance the depth map by contrast prior and then think of it as an attention map and integrate it with RGB features.

Those methods have some limitations. Using additional depth networks to extract depth features leads to high computation and storage cost. The reliance on depth images as input during testing also severely limits the practical applications of current RGB-D models. Moreover, we found that the boundaries of the produced saliency maps in those methods are a bit coarse and blur. This is mainly because the pooling and upsampling operations in FCNs might lead to the loss of local details and current RGB-D methods have not taken steps to emphasize the boundaries of salient objects.

Some RGB-based SOD methods attempt to enhance the boundary accuracy through adding edge constraints or designing boundary-aware losses. An edge guidance network [66] couples saliency and edge features to better preserve accurate object boundary. Liu *et al.* [38] train their pooling-based network with edge detection task and successfully enhance the details of salient regions. A predict-refine architecture [47] equipped with a hybrid loss segments salient regions and refines the structure with clear boundaries. An attentive feedback module [21] employs a boundary-enhanced loss for learning exquisite boundaries.

In this paper, we propose a novel collaborative learning framework where edge, depth and saliency are leveraged in a different but more efficient way. Different from previous RGB methods using edge supervision [38,66] or boundary-aware losses [21,47], we further combine the learned edge knowledge with saliency knowledge to give extra emphasis to both salient regions and boundaries. For the use of depth, we innovatively integrate it into the high-level feature learning process in a mutual-benefit manner, instead of directly taking depth images as input. Free of using the depth subnetworks and depth input during testing makes our network more lightweight and versatile.

3   Collaborative Learning Framework

3.1   The Overall Architecture

In this paper, we propose a novel CoNet for RGB-D SOD. The overall architecture is shown in Fig. 2. In this framework, three mutually beneficial collaborators, namely edge detection, coarse salient object detection and depth estimation, work together to aid accurate SOD through exploratory learning and timely interaction. From different perspectives of the SOD target, knowledges from edge, depth and saliency are fully exploited in a mutual-benefit manner to enhance the detector's performance. A simplified workflow is given below.

First, a backbone network is used to extract features from original images. Five transition layers and a global guidance module (GGM) are followed to perform feature preprocessing and generate the integrated low-level feature f_l and

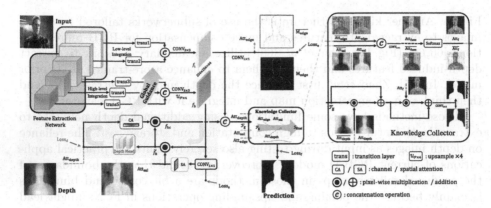

Fig. 2. The overall architecture of our collaborative learning framework. Details of the Global Guidance Module can be found in Fig. 3. Here, $\overline{Att}_* = 1 - Att_*$.

Table 1. Detailed information of the five transition layers in Fig. 2. We show the input size and output size of the feature maps before and after those transition layers, and represent their specific transition operators for better understanding.

Transition	Input size	Transition operators	Output size
trans1	$128 \times 128 \times 64$	$Upsample_{\times 2}$	$256 \times 256 \times 64$
trans2	$64 \times 64 \times 256$	$Upsample_{\times 4}$	$256 \times 256 \times 256$
trans3	$32 \times 32 \times 512$	$Upsample_{\times 2}, Conv_{3\times 3} + BN + PRelu$	$64 \times 64 \times 64$
trans4	$16 \times 16 \times 1024$	$Upsample_{\times 4}, Conv_{3\times 3} + BN + PRelu$	$64 \times 64 \times 64$
trans5	$16 \times 16 \times 2048$	$Upsample_{\times 4}, Conv_{3\times 3} + BN + PRelu$	$64 \times 64 \times 64$

high-level feature f_h (details are shown in Sect. 3.2). Then an edge collaborator is assigned to f_l to extract edge information from the overabundant low-level feature. For the high-level feature f_h, saliency collaborator and depth collaborator work together to jointly enhance the high-level feature learning process of global semantics in a mutual-benefit manner. Finally, all learned knowledges from three collaborators (Att_{edge}, Att_{sal} and Att_{depth}), as well as the integrated low-level and high-level feature (F_g), are uniformly handed to a knowledge collector (KC). Here, acting as a tutor, KC summarizes the learned edge, depth and saliency knowledges and utilizes them to predict accurate saliency results. We elaborate on the three collaborators and the KC in Sect. 3.3.

3.2 Feature Preprocessing

Backbone Network. We use the widely used ResNet [24] suggested by other deep-learning-based methods [15,39,60] as backbone network, where the last fully connected layers are truncated to better fit for the SOD task. As shown in Fig. 2, five side-out features generated from the backbone network are transferred to five transition layers to change their sizes and the number of channels. Detailed parameters are listed in Table 1, and the five output features are defined as $\{f_1, f_2, f_3, f_4, f_5\}$.

Global Guidance Module. In order to obtain richer global semantics and alleviate information dilution in the decoder, a Global Guidance Module (GGM) is applied on high-level features (i.e. f_3, f_4, and f_5)(see Fig. 3). Its key component, global perception module (GPM), takes the progressively integrated feature as input, followed by four parallel dilated convolution operations [62] (kernel size = 3, dilation rates = 1/6/12/18) and one 1×1 traditional convolution operation, to obtain rich global semantics. Benefiting from the dilated convolution [62], the GPM captures affluent multi-scale contextual information without sacrificing image resolution [8,9]. Here, we define the process of GPM as $\widetilde{F} = \Phi(F)$, where F denotes the input feature map and $\widetilde{F}$ means the output feature. In

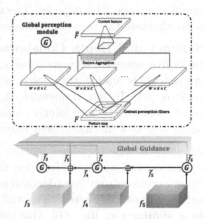

Fig. 3. The architecture of global guidance module (GGM).

GGM, we take the summation of the feature in current layer and the output features of all high-level GPMs as input to alleviate information dilution. Finally, three output features of GPMs are concatenated and an integrated high-level feature f_h is produced, which is computed by:

$$\widetilde{f}_i = \Phi(f_i + \sum_{m=i+1}^{5} \widetilde{f}_m), i = 3, 4, 5, \tag{1}$$

$$f_h = Up(W_h * Concat(\widetilde{f}_3, \widetilde{f}_4, \widetilde{f}_5) + b_h), \tag{2}$$

where $*$ means convolution operation. W_h and b_h are convolution parameters. $Up(\cdot)$ means the upsampling operation.

3.3 Collaborative Learning

Edge Collaborator. Existing 3D methods [5,6,46,48,65] have achieved remarkable performance in locating salient regions, but they still suffer from coarse object boundaries. In our framework, we design an edge collaborator to explicitly extract edge information from the overabundant low-level feature and use this information to give more emphasis to object boundaries.

Specifically, we first formulate this problem by adding edge supervision on the top of integrated low-level feature f_l. The used edge ground truths (GT) (shown in Fig. 1) are derived from saliency GT using canny operator [4]. As shown in Fig. 2, f_l is processed by a 1×1 convolution operation and a softmax function to generate the edge map M_{edge}. Then, binary cross entropy loss (denoted as $loss_e$) is adopted to calculate the difference between M_{edge} and edge GT. As the edge maps M_{edge} in Fig. 2 and Fig. 5 show, edge detection constraint is beneficial for predicting accurate boundaries of salient objects. Additionally, we also transfer the learned edge knowledge before the softmax function (denoted as Att_{edge}) to the knowledge collector (KC), where the edge information is further utilized to emphasize object boundaries. The reason why we use Att_{edge} rather than M_{edge} is to alleviate the negative influence brought by accuracy decrement of M_{edge}.

Saliency and Depth Collaborators. When addressing scene understanding tasks like semantic segmentation and salient object detection, there exist strong consistencies among the global semantics and geometric properties of image regions [54]. In our framework, a saliency collaborator and a depth collaborator work together to jointly enhance the feature learning process of high-level semantics in a mutual-benefit manner.

Stage One: The high-level feature f_h is first processed by a 1×1 convolution operation and a softmax function to predict a coarse saliency map S_{coarse}. Here, binary cross entropy loss (denoted as $loss_s$) is used for training. Then, the learned saliency knowledge acts as a spatial attention map to refine the high-level feature in a similar way like [60]. But different from [60] which considers S_{coarse} as attention map directly, we use the more informative feature map before softmax function (denoted as Att_{sal}) to emphasize or suppress each pixel of f_h. Identify mapping is adopted to alleviate the errors in Att_{sal} to be propagated to depth learning and accelerate network convergence. Formally, this procedure can be defined as:

$$Att_{sal} = W_s * f_h + b_s, \tag{3}$$

$$\widetilde{f_h} = Att_{sal} \odot f_h + f_h, \tag{4}$$

where $\odot$ means element-wise multiplication. $\widetilde{f_h}$ denotes the output saliency-enhanced feature.

Stage Two: As pointed out in previous RGB-D researches [46,65], the spatial information within depth image is helpful for better locating salient objects in a scene. In our network, we innovatively integrate depth learning into the high-level feature learning process, instead of directly taking depth image as input. This learning strategy enables our network to be free of using an extra depth network to make inference from an extra depth input, and thus being more lightweight and versatile. As in Fig. 2, a depth head with three convolution layers (defined as $\Psi(\cdot)$) is first used to make feature $\widetilde{f_h}$ adapt to depth estimation. Then, its output $\Psi(\widetilde{f_h})$ is followed by a 1×1 convolution operation to generate the estimated depth map Att_{depth}. Here, depth images act as GTs for supervision and we use smooth L_1 loss [22] to calculate the difference between Att_{depth} and depth GT, where smooth L_1 loss is a robust L_1 loss proposed in [22] that is less sensitive to outliers than L_2 loss. Formally, the depth loss can be defined as:

$$Loss_d = \frac{1}{W \times H} \sum_{x=1}^{W} \sum_{y=1}^{H} \begin{cases} 0.5 \times |\triangle(x,y)|^2, & \text{if } |\triangle(x,y)| \leq 1, \\ |\triangle(x,y)| - 0.5, & \text{if } \triangle(x,y) < -1 \text{ or } \triangle(x,y) > 1, \end{cases} \tag{5}$$

where W and H denote the width and height of the depth map. $\triangle(x,y)$ means the error between prediction Att_{depth} and the depth GT in each pixel (x,y). Since each channel of a feature map can be considered as a 'feature detector' [59], the depth knowledge Att_{depth} is further employed to learn a channel-wise attention map M_c for choosing useful semantics. Identify mapping operation is also adopted to enhance the fault-tolerant ability. This procedure can be defined as:

$$Att_{depth} = W_d * \Psi(\widetilde{f}_h) + b_d, \tag{6}$$

$$M_c = \sigma(GP(W_c * Att_{depth} + b_c)), \tag{7}$$

$$f_{hc} = M_c \otimes \widetilde{f}_h + \widetilde{f}_h, \tag{8}$$

where w_* and b_* are parameters to be learned. $GP(\cdot)$ means global pooling operation. $\sigma(\cdot)$ is the softmax function. $\otimes$ denotes channel-wise multiplication.

After these two stages, two collaborators can cooperatively generate optimal feature which contains affluent spatial cues and possesses strong ability to distinguish salient and non-salient regions.

Knowledge Collector. In our framework, the KC works as a unified tutor to complete knowledge transfer from individual collaborators to the group.

As illustrated in Fig. 2, all knowledges learned from three collaborators (i.e. Att_{edge}, Att_{sal}, and Att_{depth}) and the concatenated multi-level feature $F_g = Concat(f_l, f_{hc})$ are uniformly transferred to the KC. Those information are comprehensively processed in a triple-attention manner to give more emphasis to salient regions and object boundaries. In Fig. 2, we show a detailed diagram with visualized attention maps for better understanding. To be specific, Att_{edge} and Att_{sal} are first concatenated together to jointly learn a fused attention map Att_f, where the locations and boundaries of the salient objects are considered uniformly. Then, F_g is in turn multiplied with the depth attention map Att_{depth} and the fused attention map Att_f, which significantly enhances the contrast between salient and non-salient areas. Ablation analysis shows the ability of the KC to enhance the performance significantly.

There is a vital problem worth thinking about. The quality of Att_{depth} and Att_f might lead to irrecoverable inhibition of salient areas. Therefore, we add several residual connection operations [24] to the KC to retain the original features. Formally, this process can be defined as:

$$Att_f = \sigma(W_f * Concat(Att_{sal}, Att_{edge}) + b_f), \tag{9}$$

$$\widetilde{F}_g = Att_{depth} \odot F_g + F_g, \tag{10}$$

$$F = Att_f \odot \widetilde{F}_g + \widetilde{F}_g. \tag{11}$$

In the end, F is followed by a 1×1 convolution operation and an upsampling operation to generate the final saliency map S_{final}. Here, binary cross entropy loss (denoted as $loss_f$) is used to calculate the difference between S_{final} and saliency GT. Thus, the total loss L can be represented as:

$$L = \lambda_e Loss_e + \lambda_s Loss_s + \lambda_d Loss_d + \lambda_f Loss_f, \tag{12}$$

where $Loss_e$, $Loss_s$, and $Loss_f$ are cross entropy loss and $Loss_d$ is a smooth L_1 loss. In this paper, we set $\lambda_e = \lambda_s = \lambda_f = 1$ and $\lambda_d = 3$.

4   Experiments

4.1   Dataset

To evaluate the performance of our network, we conduct experiments on seven widely used benchmark datasets.

DUT-D [46]: contains 1200 images with 800 indoor and 400 outdoor scenes paired with corresponding depth images. This dataset contains many complex scenarios. **NJUD** [28]: contains 1985 stereo images (the latest version). They are gathered from the Internet, 3D movies and photographs taken by a Fuji W3 stereo camera. **NLPR** [44]: includes 1000 images captured by Kinect under different illumination conditions. **SIP** [19]: contains 929 salient person samples with different poses and illumination conditions. **LFSD** [34]: is a relatively small dataset with 100 images captured by Lytro camera. **STEREO** [43]: contains 797 stereoscopic images downloaded from the Internet. **RGBD135** [11]: consists of seven indoor scenes and contains 135 images captured by Kinect.

For training, we split 800 samples from DUT-D, 1485 samples from NJUD, and 700 samples from NLPR as in [5,6,46]. The remaining images and other public datasets are all for testing to comprehensively evaluate the generation abilities of models. To reduce overfitting, we augment the training set by randomly flipping, cropping and rotating those images.

4.2   Experimental Setup

Evaluation Metrics. We adopt 6 widely used evaluation metrics to verify the performance of various models, including precision-recall (PR) curve, mean F-measure (F_β) [1], mean absolute error (MAE) [3], weighted F-measure (F_β^w) [42] and recently proposed S-measure (S) [17] and E-measure (E) [18]. Saliency maps are binarized using a series of thresholds and then pairs of precision and recall are computed to plot the PR curve. The F-measure is a harmonic mean of average precision and average recall. Here, we calculate the mean F-measure which uses adaptive threshold to generate binary saliency map. The MAE represents the average absolute difference between the saliency map and ground truth. Weighted F-measure intuitively generalizes F-measure by alternating the way to calculate the Precision and Recall. S-measure contains two terms: object-aware and region-aware structural similarities. E-measure jointly captures image level statistics and local pixel matching information. Details of those evaluation metrics can refer to [55]. For MAE, lower value is better. For others, higher is better.

Implementation Details. We implement our proposed framework using the Pytorch toolbox and train it with a GTX 1080 Ti GPU. All training and test images are uniformly resized to 256×256. Our network is trained in an end-to-end manner using the standard SGD optimizer, and it converges after 50 epochs with batch size of 2. The momentum, weight decay and learning rate are set as 0.9, 0.0005 and 1e−10, respectively. Any post-processing procedure (e.g., CRF [29]) is not applied in this work. The model size of our network has only 167.6M and the inference speed for a 256×256 image only takes 0.0290 s (34FPS).

4.3 Ablation Analysis

Overview of Performance. We show the quantitative and qualitative results of different modules of our proposed network in Table 2 and Fig. 4. The backbone network (denoted as B) is constructed by directly concatenating low-level feature f_l and high-level feature f_h without using GGM for prediction. Comparison of the results (a) and (b) shows that adding our GGM can more effectively extract rich semantic features and prevent information dilution in the decoding stage.

After introducing edge supervision (E), the boundaries of the saliency maps are sharper (b vs c in Fig. 4). The edge maps (M_{edge}) in Fig. 2 and Fig. 5 also show the ability of our network in explicitly extracting object boundaries. By adding additional saliency supervision on f_h (denoted as S), the performance can be further improved. However, by comparing (d) and (e), we can see that our mutual-benefit learning style between saliency collaborator and depth collaborator $(S_{SA}$ and $D_{CA})$ can further improve the detector's ability to locate salient objects. This also verifies the strong correlation between saliency and depth. Finally, by using our proposed KC, all learned edge, depth and saliency knowledges from three collaborators can be effectively summarized and utilized to give more emphasis to salient regions and object boundaries, improving the average MAE performance on two datasets by nearly 9.6% points. By comparing (e) and (f) in Fig. 4, we can also see that salient regions in (f) are more consistent with

Table 2. Quantitative results of the ablation analysis on two benchmark datasets. B means the backbone network. E and S represent edge supervision and saliency supervision respectively. $S_{SA} + D_{CA}$ means our mutual-benefit learning strategy between depth and saliency. +KC means adding our knowledge collector on (e).

Indexes	Modules	NJUD		NLPR	
		$F_\beta\uparrow$	$MAE\downarrow$	$F_\beta\uparrow$	$MAE\downarrow$
(a)	B	0.831	0.065	0.797	0.050
(b)	B+GGM	0.839	0.060	0.813	0.044
(c)	(b)+E	0.851	0.056	0.825	0.041
(d)	(b)+E+S	0.857	0.054	0.833	0.038
(e)	(b)+E+S_{SA}+D_{CA}	0.864	0.051	0.841	0.035
(f)	(e)+KC	**0.872**	**0.047**	**0.848**	**0.031**

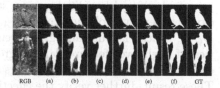

RGB   (a)   (b)   (c)   (d)   (e)   (f)   GT

Fig. 4. Visual saliency maps of ablation analysis. The meaning of the indexes (a)–(f) can refer to Table 2.

the saliency GT and the object boundaries are explicitly highlighted benefiting from the comprehensive knowledge utilization. Those advances demonstrate that using our collaborative learning strategy is beneficial for accurate saliency prediction. We list some numerical results here for better understanding. The Root Mean Squared Error (RMSE) of depth prediction on NJUD and NLPR datasets are 0.3684 and 0.4696, respectively. The MAE scores of edge prediction are 0.053 and 0.044, respectively.

The Interactions Between Collaborators

Saliency and Edge. To explore the correlation between saliency and edge, we gradually add edge supervision (E) and saliency supervision (S_l) on the low-level feature f_l. From the quantitative results in Table 3, we can see that adding edge supervision can explicitly extract clear boundary information and significantly enhance the detection performance, especially for the F-measure scores. However, when adding S_l on f_l, the performances on both datasets decrease.

This is partly because the low-level features contain too much information and are relatively too coarse to predict saliency, and partly because the two tasks are to some extent incompatible, in which one is for highlighting the boundaries and another.is for highlighting the whole salient objects. Hence, it is optimal to only add edge supervision on the low-level feature.

Table 3. Ablation analysis of the interactions between three collaborators. The meaning of indexes (b)–(f) can refer to Table 2. $+S_l$ means adding saliency supervision on low-level feature. D means depth supervision.

Modules	NJUD		NLPR	
	$F_\beta \uparrow$	$MAE \downarrow$	$F_\beta \uparrow$	$MAE \downarrow$
Saliency & Edge				
(b)	0.839	0.060	0.813	0.044
(b)+E (c)	0.851	0.056	0.825	0.041
(b)+E+S_l	0.835	0.062	0.807	0.044
Saliency & Depth				
(c)	0.851	0.056	0.825	0.041
(c)+S (d)	0.857	0.054	0.833	0.038
(c)+S+D	0.859	0.054	0.835	0.037
(c)+S+D_{CA}	0.861	0.053	0.837	0.036
(c)+S_{SA}+D_{CA} (e)	0.864	0.051	0.841	0.035
Saliency & Edge & Depth				
(e)	0.864	0.051	0.841	0.035
(e)+Att_{edge}	0.868	0.049	0.846	0.032
(e)+Att_{sal}	0.866	0.049	0.844	0.033
(e)+Att_{edge}+Att_{sal}	0.869	0.048	0.846	0.031
(e)+Att_{edge}+Att_{sal}+Att_{depth} (f)	0.872	0.047	0.848	0.031

Saliency and Depth. In order to verify the effectiveness of the proposed mutual-benefit learning strategy on high-level feature f_h, we gradually add two collaborators and their mutual-benefit operations to the baseline model (c). As shown in Table 3, adding saliency supervision (S) and adding depth supervision (D) are all beneficial for extracting more representative high-level semantic features. In addition, by gradually introducing our proposed mutual-benefit learning strategy between two collaborators (S_{SA} and D_{CA}), spatial layouts and global semantics of high-level feature can be greatly enhanced, which consequently brings addition These results further verify the effectiveness o work.

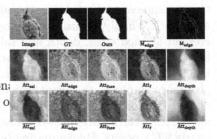

Fig. 5. Internal results in the knowledge collector. The results of another sample can be seen in Fig. 2. Here, $\overline{F} = 1 - F$.

Saliency, Edge and Depth. In our knowledge collector, all knowledge learned from three collaborators are summarized and utilized in a triple-attention manner. As the visualized attention maps in Fig. 2 and Fig. 5 show, the edge knowledge (Att_{edge}) can help highlight object boundaries, and the depth and saliency knowledge (Att_{depth} and Att_{sal}) can also be used to emphasize salient regions and suppress non-salient regions. We can see from Table 3 that both Att_{edge} and Att_{sal} are beneficial for enhancing the feature representation and improving the F-measure and MAE performance. In our framework, we adopt a better strategy that Att_{edge} and Att_{sal} are concatenated together to jointly emphasize salient objects and their boundaries. Finally, by comparing the results in the last two lines of Table 3, we can see that by further utilizing the learned depth knowledge, the detector's performance can be

Table 4. Quantitative comparisons on seven benchmark datasets. The best three results are shown in **blue**, red, and green fonts respectively.

Dataset	Metric	DES [11]	LHM [44]	DCMC [12]	MB [69]	CDCP [70]	DF [48]	CTMF [23]	PDNet [68]	MPCI [7]	TANet [6]	PCA [5]	CPFP [65]	DMRA [46]	Ours
DUT-D [46]	$E\uparrow$	0.733	0.767	0.712	0.691	0.794	0.842	0.884	0.861	0.855	0.866	0.858	0.854	0.927	**0.941**
	$S\uparrow$	0.659	0.568	0.499	0.607	0.687	0.730	0.834	0.799	0.791	0.808	0.801	0.749	0.888	**0.918**
	$F_\beta^w\uparrow$	0.386	0.350	0.290	0.464	0.530	0.542	0.690	0.650	0.636	0.712	0.696	0.644	0.858	**0.896**
	$F_\beta\uparrow$	0.668	0.659	0.406	0.577	0.633	0.748	0.792	0.757	0.753	0.779	0.760	0.736	0.883	**0.908**
	$MAE\downarrow$	0.280	0.174	0.243	0.156	0.159	0.145	0.097	0.112	0.113	0.093	0.100	0.099	0.048	**0.034**
NJUD [28]	$E\uparrow$	0.421	0.722	0.796	0.643	0.751	0.818	0.864	0.890	0.878	0.893	0.896	0.894	0.908	**0.912**
	$S\uparrow$	0.413	0.530	0.703	0.534	0.673	0.735	0.849	0.883	0.859	0.878	0.877	0.878	0.886	**0.894**
	$F_\beta^w\uparrow$	0.241	0.311	0.506	0.369	0.522	0.552	0.732	0.798	0.749	0.812	0.811	0.837	0.853	**0.856**
	$F_\beta\uparrow$	0.165	0.625	0.715	0.492	0.618	0.744	0.788	0.832	0.813	0.844	0.844	0.850	**0.872**	**0.872**
	$MAE\downarrow$	0.448	0.201	0.167	0.202	0.181	0.151	0.085	0.062	0.079	0.061	0.059	0.053	0.051	**0.047**
NLPR [44]	$E\uparrow$	0.735	0.772	0.684	0.814	0.785	0.838	0.869	0.876	0.871	0.916	0.916	0.924	**0.942**	0.936
	$S\uparrow$	0.582	0.591	0.550	0.714	0.724	0.769	0.860	0.835	0.855	0.886	0.873	0.888	0.899	**0.907**
	$F_\beta^w\uparrow$	0.259	0.320	0.265	0.574	0.524	0.524	0.691	0.659	0.688	0.789	0.772	0.820	0.845	**0.850**
	$F_\beta\uparrow$	0.583	0.520	0.328	0.637	0.591	0.682	0.723	0.740	0.729	0.795	0.794	0.822	**0.855**	0.848
	$MAE\downarrow$	0.301	0.119	0.196	0.089	0.114	0.099	0.056	0.064	0.059	0.041	0.044	0.036	**0.031**	**0.031**
STEREO [43]	$E\uparrow$	0.451	0.781	0.838	0.693	0.801	0.844	0.870	0.903	0.890	0.911	0.905	0.897	0.920	**0.923**
	$S\uparrow$	0.473	0.567	0.745	0.579	0.727	0.763	0.853	0.874	0.856	0.877	0.880	0.871	0.886	**0.908**
	$F_\beta^w\uparrow$	0.277	0.369	0.551	0.445	0.595	0.576	0.727	0.799	0.747	0.811	0.810	0.818	0.850	**0.871**
	$F_\beta\uparrow$	0.223	0.716	0.761	0.572	0.680	0.761	0.786	0.833	0.812	0.849	0.845	0.827	0.868	**0.885**
	$MAE\downarrow$	0.417	0.179	0.150	0.178	0.149	0.142	0.087	0.064	0.080	0.060	0.061	0.054	0.047	**0.041**
SIP [19]	$E\uparrow$	0.742	0.722	0.787	0.715	0.721	0.794	0.824	0.802	0.886	0.893	0.898	0.899	0.863	**0.909**
	$S\uparrow$	0.616	0.523	0.684	0.624	0.597	0.651	0.716	0.691	0.833	0.835	0.844	0.850	0.806	**0.858**
	$F_\beta^w\uparrow$	0.352	0.286	0.426	0.474	0.411	0.411	0.551	0.503	0.726	0.762	0.777	0.798	0.750	**0.814**
	$F_\beta\uparrow$	0.646	0.593	0.646	0.573	0.494	0.672	0.684	0.620	0.795	0.809	0.824	0.818	0.819	**0.842**
	$MAE\downarrow$	0.300	0.182	0.186	0.163	0.224	0.186	0.139	0.166	0.086	0.075	0.071	0.064	0.085	**0.063**
LFSD [34]	$E\uparrow$	0.475	0.742	0.842	0.631	0.737	0.841	0.851	0.872	0.840	0.845	0.846	0.867	**0.899**	0.897
	$S\uparrow$	0.440	0.558	0.754	0.538	0.658	0.796	0.796	0.845	0.787	0.801	0.800	0.828	0.847	**0.862**
	$F_\beta^w\uparrow$	0.278	0.379	0.605	0.401	0.524	0.645	0.700	0.738	0.668	0.723	0.720	0.779	0.814	**0.819**
	$F_\beta\uparrow$	0.228	0.708	0.815	0.543	0.634	0.810	0.781	0.824	0.779	0.794	0.794	0.813	**0.849**	0.848
	$MAE\downarrow$	0.415	0.211	0.155	0.218	0.199	0.142	0.120	0.109	0.132	0.111	0.112	0.088	0.075	**0.071**
RGBD135 [11]	$E\uparrow$	0.786	0.850	0.674	0.798	0.806	0.801	0.907	0.915	0.899	0.916	0.909	0.927	**0.945**	**0.945**
	$S\uparrow$	0.627	0.577	0.470	0.661	0.706	0.685	0.863	0.868	0.847	0.858	0.845	0.874	0.901	**0.910**
	$F_\beta^w\uparrow$	0.301	0.372	0.173	0.516	0.484	0.397	0.694	0.731	0.656	0.745	0.718	0.794	0.849	**0.856**
	$F_\beta\uparrow$	0.689	0.857	0.228	0.588	0.583	0.566	0.765	0.800	0.750	0.782	0.763	0.819	0.857	**0.861**
	$MAE\downarrow$	0.289	0.097	0.194	0.102	0.119	0.130	0.055	0.050	0.064	0.045	0.049	0.037	0.029	**0.027**

further improved. We visualize all internal results of the KC in Fig. 5 for better understanding.

4.4 Comparison with State-of-the-Arts

We compare results from our method with various state-of-the-art approaches on seven public datasets. For fair comparisons, the results from competing methods are generated by authorized codes or directly provided by authors.

Quantitative Evaluation. Table 4 shows the quantitative results of our method over other 13 RGB-D ones on seven benchmark datasets. We can see that our proposed collaborative learning framework achieves superior performance. Noted that our method avoids the reliance on depth images and only takes RGB image as input in the testing stage. To comprehensively verify the effectiveness of our model, we additionally conduct comparisons with 9 state-of-the-art RGB methods on three public datasets. Results in Table 5 consistently show that our

Table 5. Quantitative comparisons with state-of-the-art 2D methods.

Dataset	Metric	DSS [26]	Amulet [63]	R³Net [15]	PiCANet [39]	PAGRN [64]	EGNet [66]	PoolNet [38]	BASNet [47]	CPD [60]	Ours
NJUD [28]	$S\uparrow$	0.807	0.843	0.837	0.847	0.829	0.871	0.872	0.872	0.876	**0.894**
	$F_\beta^w\uparrow$	0.678	0.758	0.736	0.768	0.746	0.812	0.816	0.839	0.834	**0.856**
	$MAE\downarrow$	0.108	0.085	0.092	0.071	0.081	0.057	0.057	0.055	0.054	**0.047**
NLPR [44]	$S\uparrow$	0.816	0.848	0.798	0.834	0.844	0.861	0.867	0.890	0.887	**0.907**
	$F_\beta^w\uparrow$	0.614	0.716	0.611	0.707	0.707	0.760	0.771	0.834	0.820	**0.850**
	$MAE\downarrow$	0.076	0.062	0.101	0.053	0.051	0.046	0.046	0.036	0.036	**0.031**
STEREO [43]	$S\uparrow$	0.841	0.881	0.855	0.868	0.851	0.897	0.898	0.896	0.899	**0.908**
	$F_\beta^w\uparrow$	0.718	0.811	0.752	0.774	0.792	0.847	0.849	**0.873**	0.865	0.871
	$MAE\downarrow$	0.087	0.062	0.084	0.062	0.067	0.045	0.045	0.042	0.042	**0.041**

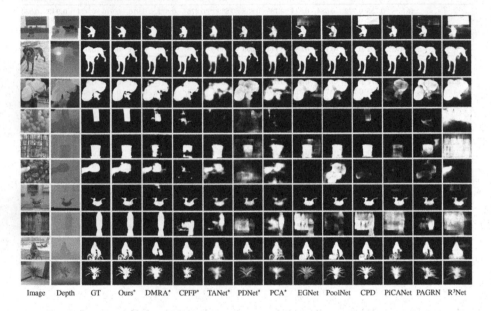

Image Depth GT Ours* DMRA* CPFP* TANet* PDNet* PCA* EGNet PoolNet CPD PiCANet PAGRN R³Net

Fig. 6. Visual comparisons of our method with other state-of-the-art CNNs-based methods in some representative scenes. * means RGB-D methods.

method also achieves comparable results compared to 2D methods. The PR curves in Fig. 7 also verify the superiority of our method.

Qualitative Evaluation. Figure 6 shows some representative samples of results comparing our method with some top-ranking CNNs-based RGB and RGB-D approaches. For the complex scenes with lower-contrast (the 4th and 5th rows) or multiple objects (the 8th row), our method can better locate the salient objects thanks to the useful spatial information in depth image and sufficient extraction and utilization of edge information. Thus, our method can produce accurate saliency results with sharp boundaries preserved.

Complexity Evaluation. We also compare the model size and run time (Frame Per Second, FPS) of our method with 11 representative models in Table 6.

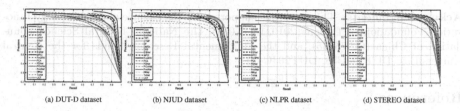

(a) DUT-D dataset	(b) NJUD dataset	(c) NLPR dataset	(d) STEREO dataset

Fig. 7. The PR curves of our method compared to other state-of-the-art approaches on four datasets.

Table 6. Complexity comparisons of various methods. The best three results are shown in **blue**, red, and green fonts respectively. FPS means frame per second.

Types	Methods	Years	Size	FPS	NJUD [28] $F_\beta^w \uparrow$	NJUD [28] $MAE \downarrow$	NLPR [44] $F_\beta^w \uparrow$	NLPR [44] $MAE \downarrow$
	DSS	2017'CVPR	447.3MB	22	0.678	0.108	0.614	0.076
	Amulet	2017'ICCV	**132.6 MB**	16	0.758	0.085	0.716	0.062
2D	PiCANet	2018'CVPR	197.2 MB	7	0.768	0.071	0.707	0.053
	PoolNet	2019'CVPR	278.5 MB	32	0.816	0.057	0.771	0.046
	CPD	2019'CVPR	183 MB	**62**	0.834	0.054	0.820	0.036
	PCA	2018'CVPR	533.6 MB	15	0.811	0.059	0.772	0.044
	TANet	2019'TIP	951.9 MB	14	0.812	0.061	0.789	0.041
3D	MPCI	2019'PR	929.7 MB	19	0.749	0.079	0.688	0.059
	PDNet	2019'ICME	192 MB	19	0.798	0.062	0.659	0.064
	CPFP	2019'CVPR	278 MB	6	0.837	0.053	0.820	0.036
	DMRA	2019'ICCV	238.8 MB	22	0.853	0.051	0.845	**0.031**
*	Ours		167.6 MB	34	**0.856**	**0.047**	**0.850**	0.031

Thanks to the well-designed depth learning strategy, our network is free of using extra depth networks and depth inputs to make inference. It can also be seen that our method achieves outstanding scores with a smaller model size and higher FPS (enhances FPS by 55% compared to current best performing RGB-D model DMRA). Those results confirm that our model is suitable for the pre-processing task in terms of model size and running speed.

5   Conclusion

In this work, we propose a novel collaborative learning framework for accurate RGB-D salient object detection. In our framework, three mutually beneficial collaborators, i.e., edge detection, coarse salient object detection and depth estimation, jointly accomplish the SOD task from different perspectives. Benefiting from the well-designed mutual-benefit learning strategy between three collaborators, our method can produce accurate saliency results with sharp boundaries preserved. Free of using extra depth subnetworks and depth inputs during testing also makes our network more lightweight and versatile. Experiment results on seven benchmark datasets show that our method achieves superior performance over 22 state-of-the-art RGB and RGB-D methods.

Acknowledgements. This work was supported by the Science and Technology Innovation Foundation of Dalian (2019J12GX034), the National Natural Science Foundation of China (61976035), and the Fundamental Research Funds for the Central Universities (DUT19JC58, DUT20JC42).

References

1. Achanta, R., Hemami, S.S., Estrada, F.J., Süsstrunk, S.: Frequency-tuned salient region detection. In: CVPR, pp. 1597–1604 (2009)
2. Borji, A., Cheng, M.M., Jiang, H., Li, J.: Salient object detection: a benchmark. TIP **24**(12), 5706–5722 (2015)
3. Borji, A., Sihite, D.N., Itti, L.: Salient object detection: a benchmark. In: Fitzgibbon, A., Lazebnik, S., Perona, P., Sato, Y., Schmid, C. (eds.) ECCV 2012. LNCS, vol. 7573, pp. 414–429. Springer, Heidelberg (2012). https://doi.org/10.1007/978-3-642-33709-3_30
4. Canny, J.: A computational approach to edge detection. TPAMI **8**(6), 679–698 (1986)
5. Chen, H., Li, Y.: Progressively complementarity-aware fusion network for RGB-D salient object detection. In: CVPR, pp. 3051–3060 (2018)
6. Chen, H., Li, Y.: Three-stream attention-aware network for RGB-D salient object detection. TIP **28**(6), 2825–2835 (2019)
7. Chen, H., Li, Y., Su, D.: Multi-modal fusion network with multi-scale multi-path and cross-modal interactions for RGB-D salient object detection. PR **86**, 376–385 (2019)
8. Chen, L.C., Papandreou, G., Kokkinos, I., Murphy, K., Yuille, A.L.: Deeplab: semantic image segmentation with deep convolutional nets, atrous convolution, and fully connected CRFs. TPAMI **40**(4), 834–848 (2018)
9. Chen, L.-C., Zhu, Y., Papandreou, G., Schroff, F., Adam, H.: Encoder-decoder with atrous separable convolution for semantic image segmentation. In: Ferrari, V., Hebert, M., Sminchisescu, C., Weiss, Y. (eds.) ECCV 2018. LNCS, vol. 11211, pp. 833–851. Springer, Cham (2018). https://doi.org/10.1007/978-3-030-01234-2_49
10. Cheng, M.M., Zhang, G.X., Mitra, N.J., Huang, X., Hu, S.M.: Global contrast based salient region detection. TPAMI **37**(3), 409–416 (2011)
11. Cheng, Y., Fu, H., Wei, X., Xiao, J., Cao, X.: Depth enhanced saliency detection method. In: ICIMCS, pp. 23–27 (2014)
12. Cong, R., Lei, J., Zhang, C., Huang, Q., Cao, X., Hou, C.: Saliency detection for stereoscopic images based on depth confidence analysis and multiple cues fusion. SPL **23**(6), 819–823 (2016)
13. Craye, C., Filliat, D., Goudou, J.F.: Environment exploration for object-based visual saliency learning. In: ICRA, pp. 2303–2309 (2016)
14. Dai, J., Li, Y., He, K., Sun, J.: R-FCN: object detection via region-based fully convolutional networks. In: NIPS, pp. 379–387 (2016)
15. Deng, Z., et al.: R^3net: recurrent residual refinement network for saliency detection. In: IJCAI, pp. 684–690 (2018)
16. Fan, D.-P., Cheng, M.-M., Liu, J.-J., Gao, S.-H., Hou, Q., Borji, A.: Salient objects in clutter: bringing salient object detection to the foreground. In: Ferrari, V., Hebert, M., Sminchisescu, C., Weiss, Y. (eds.) ECCV 2018. LNCS, vol. 11219, pp. 196–212. Springer, Cham (2018). https://doi.org/10.1007/978-3-030-01267-0_12
17. Fan, D.P., Cheng, M.M., Liu, Y., Li, T., Borji, A.: Structure-measure: a new way to evaluate foreground maps. In: ICCV, pp. 4558–4567 (2017)

18. Fan, D.P., Gong, C., Cao, Y., Ren, B., Cheng, M.M., Borji, A.: Enhanced-alignment measure for binary foreground map evaluation. In: IJCAI, pp. 698–704 (2018)
19. Fan, D.P., et al.: Rethinking RGB-D salient object detection: models, datasets, and large-scale benchmarks. arXiv preprint arXiv:1907.06781 (2019)
20. Fan, D.P., Wang, W., Cheng, M.M., Shen, J.: Shifting more attention to video salient object detection. In: CVPR, pp. 8554–8564 (2019)
21. Feng, M., Lu, H., Ding, E.: Attentive feedback network for boundary-aware salient object detection. In: CVPR, pp. 1623–1632 (2019)
22. Girshick, R.: Fast R-CNN. In: ICCV, pp. 1440–1448 (2015)
23. Han, J., Chen, H., Liu, N., Yan, C., Li, X.: CNNs-based RGB-D saliency detection via cross-view transfer and multiview fusion. IEEE Trans. Syst. Man Cybern. **48**(11), 3171–3183 (2018)
24. He, K., Zhang, X., Ren, S., Sun, J.: Deep residual learning for image recognition. In: CVPR, pp. 770–778 (2016)
25. Hong, S., You, T., Kwak, S., Han, B.: Online tracking by learning discriminative saliency map with convolutional neural network. In: ICML, pp. 597–606 (2015)
26. Hou, Q., Cheng, M.M., Hu, X., Borji, A., Tu, Z., Torr, P.H.S.: Deeply supervised salient object detection with short connections. In: CVPR, pp. 815–828 (2017)
27. Itti, L., Koch, C., Niebur, E.: A model of saliency-based visual attention for rapid scene analysis. TPAMI **20**(11), 1254–1259 (1998)
28. Ju, R., Ge, L., Geng, W., Ren, T., Wu, G.: Depth saliency based on anisotropic center-surround difference. In: ICIP, pp. 1115–1119 (2014)
29. Krähenbühl, P., Koltun, V.: Efficient inference in fully connected CRFs with gaussian edge potentials. In: NIPS, pp. 109–117 (2011)
30. Lee, G., Tai, Y.W., Kim, J.: Deep saliency with encoded low level distance map and high level features. In: CVPR, pp. 660–668 (2016)
31. Li, G., Zhu, C.: A three-pathway psychobiological framework of salient object detection using stereoscopic technology. In: ICCVW, pp. 3008–3014 (2017)
32. Li, G., Yu, Y.: Visual saliency based on multiscale deep features. In: CVPR, pp. 5455–5463 (2015)
33. Li, G., Yu, Y.: Visual saliency detection based on multiscale deep CNN features. TIP **25**(11), 5012–5024 (2016)
34. Li, N., Ye, J., Ji, Y., Ling, H., Yu, J.: Saliency detection on light field. TPAMI **39**(8), 1605–1616 (2017)
35. Li, X., et al.: Deepsaliency: multi-task deep neural network model for salient object detection. TIP **25**(8), 3919–3930 (2016)
36. Li, Y., Hou, X., Koch, C., Rehg, J.M., Yuille, A.L.: The secrets of salient object segmentation. In: CVPR, pp. 280–287 (2014)
37. Lin, G., Milan, A., Shen, C., Reid, I.D.: RefineNet: multi-path refinement networks for high-resolution semantic segmentation. In: CVPR, pp. 5168–5177 (2017)
38. Liu, J.J., Hou, Q., Cheng, M.M., Feng, J., Jiang, J.: A simple pooling-based design for real-time salient object detection. In: CVPR, pp. 3917–3926 (2019)
39. Liu, N., Han, J., Yang, M.H.: PicaNet: learning pixel-wise contextual attention for saliency detection. In: CVPR, pp. 3089–3098 (2018)
40. Long, J., Shelhamer, E., Darrell, T.: Fully convolutional networks for semantic segmentation. In: CVPR, pp. 3431–3440 (2015)
41. Luo, Z., Mishra, A.K., Achkar, A., Eichel, J.A., Li, S., Jodoin, P.M.: Non-local deep features for salient object detection. In: CVPR, pp. 6593–6601 (2017)
42. Margolin, R., Zelnik-Manor, L., Tal, A.: How to evaluate foreground maps. In: CVPR, pp. 248–255 (2014)

43. Niu, Y., Geng, Y., Li, X., Liu, F.: Leveraging stereopsis for saliency analysis. In: CVPR, pp. 454–461 (2012)
44. Peng, H., Li, B., Xiong, W., Hu, W., Ji, R.: RGBD salient object detection: a benchmark and algorithms. In: Fleet, D., Pajdla, T., Schiele, B., Tuytelaars, T. (eds.) ECCV 2014. LNCS, vol. 8691, pp. 92–109. Springer, Cham (2014). https://doi.org/10.1007/978-3-319-10578-9_7
45. Perazzi, F., Krähenbühl, P., Pritch, Y., Hornung, A.: Saliency filters: contrast based filtering for salient region detection. In: CVPR, pp. 733–740 (2012)
46. Piao, Y., Ji, W., Li, J., Zhang, M., Lu, H.: Depth-induced multi-scale recurrent attention network for saliency detection. In: ICCV (2019)
47. Qin, X., Zhang, Z., Huang, C., Gao, C., Dehghan, M., Jagersand, M.: BasNet: boundary-aware salient object detection. In: CVPR, pp. 7479–7489 (2019)
48. Qu, L., He, S., Zhang, J., Tian, J., Tang, Y., Yang, Q.: RGBD salient object detection via deep fusion. TIP **26**(5), 2274–2285 (2017)
49. Ren, J., Gong, X., Yu, L., Zhou, W., Yang, M.Y.: Exploiting global priors for RGB-D saliency detection. In: CVPRW, pp. 25–32 (2015)
50. Ren, S., He, K., Girshick, R.B., Sun, J.: Faster R-CNN: towards real-time object detection with region proposal networks. In: NIPS 2015, pp. 91–99 (2015)
51. Simonyan, K., Zisserman, A.: Very deep convolutional networks for large-scale image recognition. In: ICLR (2015)
52. Smeulders, A.W.M., Chu, D.M., Cucchiara, R., Calderara, S., Dehghan, A., Shah, M.: Visual tracking: an experimental survey. TPAMI **36**(7), 1442–1468 (2014)
53. Wang, L., Lu, H., Ruan, X., Yang, M.H.: Deep networks for saliency detection via local estimation and global search. In: CVPR, pp. 3183–3192 (2015)
54. Wang, P., Shen, X., Lin, Z., Cohen, S., Price, B., Yuille, A.: Towards unified depth and semantic prediction from a single image. In: CVPR, pp. 2800–2809 (2015)
55. Wang, W., Lai, Q., Fu, H., Shen, J., Ling, H.: Salient object detection in the deep learning era: an in-depth survey. arXiv preprint arXiv:1904.09146 (2019)
56. Wang, W., Shen, J.: Deep visual attention prediction. TIP **27**(5), 2368–2378 (2018)
57. Wang, W., Shen, J., Dong, X., Borji, A.: Salient object detection driven by fixation prediction. In: CVPR, pp. 1711–1720 (2018)
58. Wang, W., Shen, J., Xie, J., Cheng, M.M., Ling, H., Borji, A.: Revisiting video saliency prediction in the deep learning era. TPAMI 1 (2019)
59. Woo, S., Park, J., Lee, J.-Y., Kweon, I.S.: CBAM: convolutional block attention module. In: Ferrari, V., Hebert, M., Sminchisescu, C., Weiss, Y. (eds.) ECCV 2018. LNCS, vol. 11211, pp. 3–19. Springer, Cham (2018). https://doi.org/10.1007/978-3-030-01234-2_1
60. Wu, Z., Su, L., Huang, Q.: Cascaded partial decoder for fast and accurate salient object detection. In: CVPR, pp. 3907–3916 (2019)
61. Yan, Q., Xu, L., Shi, J., Jia, J.: Hierarchical saliency detection. In: CVPR, pp. 1155–1162 (2013)
62. Yu, F., Koltun, V.: Multi-scale context aggregation by dilated convolutions. In: ICLR (2016)
63. Zhang, P., Wang, D., Lu, H., Wang, H., Ruan, X.: Amulet: aggregating multi-level convolutional features for salient object detection. In: ICCV, pp. 202–211 (2017)
64. Zhang, X., Wang, T., Qi, J., Lu, H., Wang, G.: Progressive attention guided recurrent network for salient object detection. In: CVPR, pp. 714–722 (2018)
65. Zhao, J., Cao, Y., Fan, D., Cheng, M., LI, X., Zhang, L.: Contrast prior and fluid pyramid integration for RGBD salient object detection. In: CVPR (2019)
66. Zhao, J., Liu, J., Fan, D.P., Cao, Y., Yang, J., Cheng, M.M.: EGNet: edge guidance network for salient object detection. In: ICCV (2019)

67. Zhao, R., Ouyang, W., Li, H., Wang, X.: Saliency detection by multi-context deep learning. In: CVPR, pp. 1265–1274 (2015)
68. Zhu, C., Cai, X., Huang, K., Li, T.H., Li, G.: PDNet: prior-model guided depth-enhanced network for salient object detection. In: ICME, pp. 199–204 (2019)
69. Zhu, C., Li, G., Guo, X., Wang, W., Wang, R.: A multilayer backpropagation saliency detection algorithm based on depth mining. In: Felsberg, M., Heyden, A., Krüger, N. (eds.) CAIP 2017. LNCS, vol. 10425, pp. 14–23. Springer, Cham (2017). https://doi.org/10.1007/978-3-319-64698-5_2
70. Zhu, C., Li, G., Wang, W., Wang, R.: An innovative salient object detection using center-dark channel prior. In: ICCVW, pp. 1509–1515 (2017)

Finding Your (3D) Center: 3D Object Detection Using a Learned Loss

David Griffiths(✉)📧, Jan Boehm📧, and Tobias Ritschel

University College London, London, UK
{david.griffiths.16,j.boehm,t.ritschel}@ucl.ac.uk

Abstract. Massive semantically labeled datasets are readily available for 2D images, however, are much harder to achieve for 3D scenes. Objects in 3D repositories like ShapeNet are labeled, but regrettably only in isolation, so without context. 3D scenes can be acquired by range scanners on city-level scale, but much fewer with semantic labels. Addressing this disparity, we introduce a new optimization procedure, which allows training for 3D detection with raw 3D scans while using as little as 5% of the object labels and still achieve comparable performance. Our optimization uses two networks. A *scene network* maps an entire 3D scene to a set of 3D object centers. As we assume the scene not to be labeled by centers, no classic loss, such as Chamfer can be used to train it. Instead, we use another network to emulate the loss. This *loss network* is trained on a small labeled subset and maps a non-centered 3D object in the presence of distractions to its own center. This function is very similar – and hence can be used instead of – the gradient the supervised loss would provide. Our evaluation documents competitive fidelity at a much lower level of supervision, respectively higher quality at comparable supervision. Supplementary material can be found at: dgriffiths3.github.io.

Keywords: 3D learning · 3D point clouds · 3D object detection · Unsupervised

1 Introduction

We can reason about one 3D chair as we do about a 2D chair image, however, we cannot yet machine-understand a point cloud of a 3D room as we would do for a 2D room image. For 2D images, massive amounts of manual human labeling have enabled amazing state-of-the-art object detectors [10,19,26,34]. We also have massive repositories of clean 3D objects [3] which we can classify thanks to deep 3D point processing [25]. But we do not have, despite commendable efforts [6,28], and probably might never have, 3D scene labeling at the extent of 2D images. We hence argue that progress in 3D understanding even more critically depends on reducing the amount of supervision required.

Electronic supplementary material The online version of this chapter (https://doi.org/10.1007/978-3-030-58523-5_5) contains supplementary material, which is available to authorized users.

© Springer Nature Switzerland AG 2020
A. Vedaldi et al. (Eds.): ECCV 2020, LNCS 12363, pp. 70–85, 2020.
https://doi.org/10.1007/978-3-030-58523-5_5

While general unsupervised detection is an elusive goal, we suggest taking a shortcut: while we do not have labeled 3D scenes, we do have labeled 3D objects. The key idea in this work is to first teach a *loss network* everything that can be learned from seeing snippets of labeled objects. Next, we use this network to learn a *scene network* that explores the relation of objects within scenes, but without any scene labels, i.e., on raw scans.

After reviewing previous work, we will show how this cascade of networks is possible when choosing a slightly more primitive loss than the popular Chamfer loss and we propose two network architectures to implement it. Results show how a state-of-the-art, simple, fast and feed-forward 3D detection network can achieve similar Chamfer distance and mAP@.25 scores to a supervised approach, but with only 5% of the labels.

2 Previous Work

2D object detection has been addressed by deep-learning based approaches like Fast R-CNN [10], YOLO [26], SSD [19] or the stacked hourglass architecture [21] with great success.

In early work Song and Xiao [29] have extended sliding window-detection to a 3D representation using voxels with templates of Hough features and SVM classifiers. This approach was later extended to deep templates [30]. Both approaches use fully-supervised training on object locations given by bounding boxes. We compare to such a sliding window approach using a point-based deep template. Hou et al. [14] complement a voxel-based approach with color 2D image information which more easily represents finer details.

Karpathy et al. [15] detect objects by over-segmenting the scene and classifying segments as objects based on geometric properties such as compactness, smoothness, etc. Similarly, Chen et al. [4] minimize other features to 3D-detect objects in street scans.

While modern software libraries make voxels simple to work with, they are limited in the spatial extent of scenes they can process, and the detail of the scene they can represent. Qi et al. [24] were first to suggest an approach to work on raw point clouds. 3D object detection in point clouds is investigated by Qi et al. [23] and Engelcke et al. [7] map the scene to votes, then those votes are clustered and each cluster becomes a proposal. The vectors pointing from a seed to a vote are similar to the loss network gradients proposed in our method, but for VoteNet, this is part of the architecture during training and testing while for us these vectors are only part of the training. Finally, VoteNet is trained fully supervised with object positions. The idea of merging 2D images and 3D processing is applicable to point clouds as well, as shown by Ku et al. [16] and Qi et al. [22].

Zhou and Tuzel [35] question the usefulness of points for 3D detection and have suggested to re-sample point clouds to voxels again. Also Chen et al. [5] show how combining point inputs, volume convolutions and point proposals can lead to good results. For a survey on 3D detection, also discussing trade-offs of points and voxels, see the survey by Griffiths and Boehm [11].

Our architecture is inspired by Fast R-CNN [10,27], which regresses object proposals in one branch, warps them into a canonical frame and classifies them in the other branch. Recently, Yang et al. [33] have shown how direct mapping of a point cloud to bounding boxes is feasible. Feng et al. [8] train a network with supervision that makes multiple proposals individually and later reasons about their relation. Also, Zhou et al. [34] first work on center points for object representation alone and later regress the 2D bounding box and all other object features from image content around the points. All these works tend to first extract proposals in a learned fashion and then reason about their properties or relations in a second, learned step. We follow this design for the scene network, but drive its learning in an entirely different, unsupervised, way. Finally, all of these works require only one feed-forward point cloud network pass, a strategy we will follow as well.

Unrelated to 2D or 3D detection, Adler and Öktem [1] have proposed to replace the gradient computation in an optimization problem by a neural network. In computer vision, this idea has been used to drive light field [9] or appearance synthesis [20]. We take this a step further and use a network to emulate the gradients in a very particular optimization: the training of another network.

3   OurApproach

We learn two networks: a *scene network* and a *loss network* (Fig. 1). The first (Fig. 1, bottom) is deployed, while the second (Fig. 1, top) is only used in training.

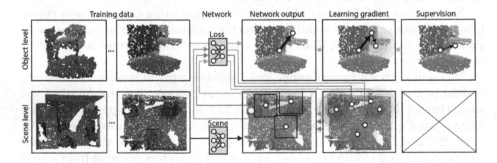

Fig. 1. Our approach proceeds in two steps of training **(row)** with different training data (column one and two), networks (column three), outputs (column four), gradients (column five) and supervision (column six). Object level training **(first row)** data comprises of 3D scene patches with known objects that are not centered. The loss network maps off-center scenes to their center (big black arrow). Its learning follows the gradient of a quadratic potential (orange field) that has the minimum at the offset that would center the object. This offset is the object-level supervision, as seen in the last column. The scene network **(second row)** is trained to map a scene to all object centers, here for three chairs. The gradient to train the scene network is computed by running the loss network from the previous step once for each object (here three times: blue, pink, green). Note, that there is no scene-level supervision (cross). (Color figure online)

The *scene network* maps 3D scenes to sets of 3D object centers. The input data is a 3D point cloud. The output is a fixed sized list of 3D object centers. We assume a feed-forward approach, that does not consider any proposals [14,21, 29,30] or voting [22,23], but directly regresses centers from input data [33,34].

The *loss network* emulates the loss used to train the scene network. The input data is again a 3D point cloud, but this time of a single object, displaced by a random amount and subject to some other distortions. Output is not the scalar loss, but the gradient of a Mean Squared Error loss function.

In the following, we will first describe the training (Sect. 3.1) before looking into the details of both the scene and loss network implementation (Sect. 3.2).

3.1 Training

The key contribution of our approach is a new way of training. We will first look into a classic baseline with scene-level supervision, then introduce a hypothetical oracle that solves almost the same problem and finally show how this problem can be solved without scene-level supervision by our approach.

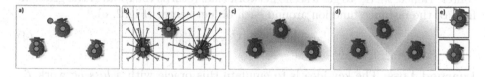

Fig. 2. a) A 2D scene with three chair objects, supervised by centers (orange) and their predictions (blue). b) The same scene, with the vector field of the oracle ∇ shown as arrows. c) A 2D Slice through a 6D cost function. d) A 2D Slice through an alternative cost function, truncated at the Voronoi cell edges. The oracle is the gradient of this. e) The simple task of the loss network: given a chair not in the center (top), regress an offset such that it becomes centered. (Color figure online)

Supervised. Consider learning the parameters θ of a scene network $\mathcal{S}_\theta$ which regresses object centers $\mathcal{S}_\theta(\mathbf{x}_i) = \hat{\mathbf{c}}$ from a scene $\mathbf{x}_i$. The scene is labeled by a set of 3D object centers $\mathbf{c}_i$ (Fig. 2, a). This is achieved by minimizing the expectation

$$\arg \min_{\theta} \mathbb{E}_i[H(\mathcal{S}_\theta(\mathbf{x}_i) - \mathbf{c}_i)], \tag{1}$$

using a two-sided Chamfer loss between the label point set $\mathbf{c}_i$ and a prediction $\hat{\mathbf{c}}_i$

$$H(\hat{\mathbf{c}}, \mathbf{c}) = \mathbb{E}_i[\min_j ||\hat{\mathbf{c}}_i - \mathbf{c}_j||_2^2] + \mathbb{E}_i[\min_j ||\mathbf{c}_i - \hat{\mathbf{c}}_j||_2^2]. \tag{2}$$

Under H, the network is free to report centers in any order, and ensures all network predictions are close to a supervised center (precision) and all supervised centers are close to at least one network prediction (recall) (Fig. 3).

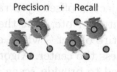

Precision + Recall

Fig. 3. Chamfer loss.

In this work, we assume the box center supervision $\mathbf{c}_i$ to not be accessible. Tackling this, we will first introduce an oracle solving a similar problem.

Oracle. Consider, instead of supervision, an *oracle* function $\nabla(\mathbf{x})$ which returns for a 3D scene $\mathbf{p}$ *the smallest offset by which we need to move the scene so that the world center falls onto an object center* (Fig. 2, b). Then, learning means to

$$\arg\min_\theta \mathbb{E}_{i,j}[||\nabla(\underbrace{\mathbf{x}_i \ominus \mathcal{S}_\theta(\mathbf{x}_i)_j}_{\mathbf{y}_{\theta,i,j}})||_2^2], \tag{3}$$

where $\mathbf{x} \ominus \mathbf{d}$ denotes shifting a scene $\mathbf{x}$ by an offset $\mathbf{d}$. The relation between Eq. 1 and Eq. 3 is intuitive: knowing the centers is very similar to pointing to the nearest center from every location. It is, however, not quite the same. It assures every network prediction would map to a center, but does not assure, that there is a prediction for every center. We will need to deal with this concern later, by assuring space is well covered, so that there are enough predictions such that at least one maps to every center. We will denote a scene i shifted to be centered around object j by a *scene network* with parameters θ as $\mathbf{y}_{\theta,i,j}$.

Every location that maps to itself, i.e., a *fixed point* [31] of ∇, is an object center. Hence, we try to get a scene network that returns the roots of the gradient field of the distance function around each object center (Fig. 2, c):

$$\arg\min_\theta \mathbb{E}_{i,j}[||\nabla(\mathbf{y}_{\theta,i,j})||_2^2]. \tag{4}$$

Learned Loss. The key idea is to emulate this oracle with a *loss* network $\mathcal{L}_\phi$ having parameters ϕ as in

$$\arg\min_\theta \mathbb{E}_{i,j}[||\mathcal{L}_\phi(\mathbf{y}_{\theta,i,j})||_2^2]. \tag{5}$$

The loss network does not need to understand any global scene structure, it only locally needs to *center* the scene around the nearest object (Fig. 2, d). This task can be learned by working on local 3D object *patches*, without scene-level supervision. So we can train the loss network on any set of objects $\mathbf{o}_k$, translated by a known offset $\mathbf{d}_k$ using

$$\arg\min_\phi \mathbb{E}_k[||\mathbf{d}_k - \mathcal{L}_\phi(\mathbf{o}_k \ominus \mathbf{d}_k)||_2]. \tag{6}$$

As the loss network is local, it is also only ever trained on 3D patches. These can be produced in several different ways: sampling of CAD models, CAD models with simulated noise, by pasting simulated results on random scene pieces, etc. In our experiments, we use a small labeled scene subset to extract objects as follows: we pick a random object center and a 3D box 1 m size such that at least point representing an object surface is present in the box. Hence the center of the object is offset by a random, but known $\mathbf{d}_k$ we regress and subject to natural clutter. Note, that the box does not, and does not have to, strictly cover the entire object – which are of different sizes – but has to be just large enough to guess the center. Algorithm 1 demonstrates how the *loss network* output can be used to provide *scene network* supervision.

Algorithm 1: $\mathcal{L}$: loss network, $\mathcal{S}$: scene network, k: proposal count, n 3D patch point count, m scene point count.

$\mathcal{L}_\phi : \mathbb{R}^{n \times 3} \to \mathbb{R}^3$;
$\mathcal{S}_\theta : \mathbb{R}^{m \times 3} \to \mathbb{R}^{k \times 3}$;
crop : $\mathbb{R}^{m \times 3} \to \mathbb{R}^{n \times 3}$;
while *loss training* **do**
    $x = \texttt{sampleScene}()$;
    $o = \texttt{randObjectCenter}()$;
    $d = \texttt{randOffset}()$;
    $p = \texttt{crop}(x \ominus (o + d))$;
    $\nabla = \frac{\partial}{\partial \phi} \|\mathcal{L}_\phi(p) - d\|_2^2$;
    $\phi = \texttt{optimizer}(\phi, \nabla)$;
end
while *scene training* **do**
    $x = \texttt{sampleScene}()$;
    $c = \mathcal{S}_\theta(x)$;
    for $i = 1 \ldots k$ **do**
        $p = \texttt{crop}(x \ominus c_i)$;
        $\nabla_i = \mathcal{L}_\phi(p)$;
    end
    $\theta = \texttt{optimizer}(\theta, \nabla)$;
end

Varying Object Count. The above was assuming the number of objects n_c to be known. It did so when assuming a vector of a known dimension as supervision in Eq. 1 and did so, when assuming the oracle Eq. 3 and its derivations were returning gradient vectors of a fixed size. In our setting this number is unknown. We address this by bounding the number of objects and handling occupancy i.e., a weight indicting if an object is present or not, at two levels.

First, we train an occupancy branch $\mathcal{O}_\phi$ of the loss network that classifies occupancy of a single patch, much like the loss network regresses the center. We define space to be *occupied*, if the 3D patch contains any points belonging to the given objects surface. This branch is trained on the same patches as the loss network plus an equal number of additional 3D patches that do not contain any objects i.e. occupancy is zero.

Second, the occupancy branch is used to support the training of the scene network which has to deal with the fact that the number of actual centers is lower than the maximal number of centers. This is achieved by ignoring the gradients to the scene networks parameters θ if the occupancy network reports the 3D patch about a center to not contain an object of interest. So instead of Eq. 5, we learn

$$\arg \min_\theta \mathbb{E}_{i,j}[\mathcal{O}_\phi(\mathbf{y}_{\theta,i,j}) \mathcal{L}_\phi(\mathbf{y}_{\theta,i,j})]. \tag{7}$$

The product in the sum is zero for centers of 3D patches that the loss network thinks, are not occupied and hence should not affect the learning.

Overlap. When neither object centers nor their count are known, there is nothing to prevent two network outputs to map to the same center. While such duplicates can to some level be addressed by non-maximum suppression as a (non-differentiable) post-process to testing, we have found it essential to already prevent them (differentiable) from occurring when training the scene network. Without doing so, our training degenerates to a single proposal.

To this end, we avoid *overlap*. Let $v(q_1, q_2)$ be a function that is zero if the bounding boxes of the object in the scene centers do not overlap, one if they are identical and otherwise be the ratio of intersection. We then optimize

$$\arg\min_{\theta} c_1(\theta) = \mathbb{E}_{i,j,k}\left[\mathcal{O}_\phi(\mathbf{y}_{\theta,i,j})\mathcal{L}_\phi(\mathbf{y}_{\theta,i,j}) + v(\mathbf{y}_{\theta,i,j},\mathbf{y}_{\theta,i,k})\right]. \tag{8}$$

We found that in case of a collision instead of mutually repelling all colliding objects, it can be more effective if out of multiple colliding objects, the collision acts on all but one winner object (winner-takes-all). To decide the winner, we again use the gradient magnitude: if multiple objects collide, the one that is already closest to the target i.e., the one with the smallest gradient, remains unaffected ($v = 0$) and takes possession of the target, while all others adapt.

Additional Features. For other object properties such as size, orientation, class of object, etc. we can proceed in two similar steps. First, we know the object-level property vector $\mathbf{q}$, so we can train a *property branch* denoted $\mathcal{P}_\theta$ that shares parameters θ with the loss network to regresses the property vector from the same displaced 3D patches as in Eq. 6

$$\arg\min_{\phi} \mathbb{E}_k[\|\mathbf{q}_k - \mathcal{P}_\phi(\mathbf{o}_k \ominus \mathbf{d}_k)\|_1]. \tag{9}$$

For scene-level learning we extend the scene network by a branch $\mathcal{T}_\theta$ to emulate what the property network had said about the 3D patch at each center, but now with global context and on a scene-level

$$\arg\min_{\theta} c_1(\theta) + \alpha \cdot \mathbb{E}_{i,j}[|\mathcal{T}_\theta(\mathbf{y}_{\theta,i,j}) - \mathcal{P}_\phi(\mathbf{y}_{\theta,i,j})|_1]. \tag{10}$$

For simplicity, we will denote occupancy just as any other object property and assume it to be produced by $\mathcal{T}$ just, that it has a special meaning in training as defined in Eq. 7. We will next detail the architecture of all networks.

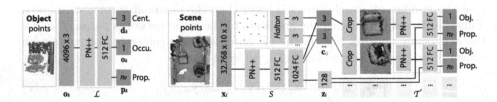

Fig. 4. The object (**left**) and scene (**right**) network. Input denoted orange, output blue, trainable parts yellow, hard-coded parts in italics. Please see Sect. 3.2 for a details. (Color figure online)

3.2 Network

Both networks are implemented using PointNet++ [25] optimized using ADAM. We choose particularly simple designs and rather focus on the analysis of changes from the levels of supervision we enable.

Loss and Occupancy Network. The loss network branches $\mathcal{L}$ and $\mathcal{O}$ share parameters ϕ and both map 4,096 3D points to a 3D displacement vector, occupancy and other scalar features (left in Fig. 4).

Scene Network. The scene network branches $\mathcal{S}$ and $\mathcal{T}$ jointly map a point cloud to a vector of 3D object centers and property vectors (including occupancy), sharing parameters θ. The box branch $\mathcal{S}$ first generates positions, next the scene is cropped around these positions and each 3D patch respectively fed into a small PointNet++ encoder $\mathcal{M}$ to produce crop specific local feature encodings. Finally, we concatenate the global scene latent code $\mathcal{S}_z$ with the respective local latent code $\mathcal{M}_z$ and pass it through the scene property branch $\mathcal{T}$ MLP.

The scene property branch is trained sharing all weights across all instances for all objects. This is intuitive, as deciding that e.g., a chair's orientation is the same for all chairs (the back rest is facing backwards), can at the same time be related to global scene properties (alignment towards a table).

Instead of learning the centers, we learn the residual relative to a uniform coverage of 3D space such that no object is missed during training. The Hammersley pattern [32] assures that, no part of 3D space is left uncovered.

We assume a fixed number of 32,768 input points for one scene. Note, that we do not use color as input, a trivial extension. Each MLP sub-layer is an MLP consisting of 3 fully-connected layers where layer 1 has 512 hidden states and the final layer contains the branch specific output nodes.

Post-process. Our scene network returns a set of oriented bounding boxes with occupancy. To reduce this soft answer to a set of detected objects, e.g., to compute mAP metrics, we remove all bounding boxes with occupancy below a threshold τ_o, which we set to 0.9 in all our results.

In the evaluation, the same will be done for our ablations SLIDING and SUPERVISED, just that these also require additional non-maximum suppression (NMS) as they frequently propose boxes that overlap. To construct a final list of detections, we pick the proposal with maximal occupancy and remove any overlapping proposal with IoU $>.25$ and repeat until no proposals remain.

4 Evaluation

We compare to different variants of our approach under different metrics and with different forms of supervision as well as to other methods.

4.1 Network

Data Sets

We consider two large-scale sets of 3D scanned scenes: Stanford 2D-3D-S dataset (S3D) [2] and ScanNet [6]. From both we extract, for each scene, the list of object centers and object features for all objects of one class.

We split the dataset in three parts (Fig. 5): First, the test dataset is the official test dataset (pink in Fig. 5). The remaining training data is split into two parts: a labeled, and and unlabeled part. The labeled part (orange in Fig. 5) has all 3D scenes with complete annotations on them. The unlabeled part (blue in Fig. 5) contains only raw 3D point cloud without annotation. Note, that the labeled data is a subset of the unlabeled data, not a different set.

We call the ratio of labeled over unlabeled data the *label ratio*. To more strictly evaluate transfer across data sets, we consider ScanNet completely unlabeled. All single-class results are reported for the class `chair`.

Metrics. Effectiveness is measured using the Chamfer distance (less is better) also used as a loss in Eq. 1 and the established mean Average Precision mAP@.25, (more is better) of a x% bounding box overlap test. X is chosen at 25%.

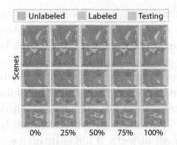

Fig. 5. Label ratio. (Color figure online)

Methods. We consider the following *three* methods: SUPERVISED is the supervised approach define by Eq. 1. This method can be trained only on the labeled part of the training set. SLIDING window is an approach that applies our loss network, trained on the labeled data, to a dense regular grid of 3D location in every 3D scene to produce a heat map from which final results are generated by NMS. OURS is our method. The loss network is trained on the labeled data (orange in Fig. 5). The scene network is trained on the unlabeled data (blue in Fig. 5), which includes the labeled data (but without accessing the labels) as a subset.

4.2 Results

Effect of Supervision. The main effect to study is the change of 3D detection quality in respect to the level of supervision. In Table 1, different rows show different label ratios. The columns show Chamfer error and mAP@.25 for the class `chair` trained and tested on S3D.

Table 1. Chamfer error (less is better) and mAP@.25 (more is better) (**columns**), as a function of supervision (**rows**) in units of label ratio on the S3D class `chair`. Right, the supervision-quality-relation plotted as a graph for every method (color).

Ratio	Chamfer error			mAP			
	SUP	SLI	OUR	SUP	SLI	OUR	
1%	1.265	.850	**.554**	.159	**.473**	.366	
5%	.789	.577	**.346**	.352	.562	**.642**	
25%	.772	.579	**.274**	.568	.573	**.735**	
50%	.644	.538	**.232**	.577	.589	**.773**	
75%	.616	.437	**.203**	.656	.592	**.785**	
100%	.557	.434	**.178**	.756	.598	**.803**	

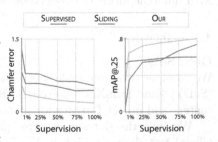

We notice that across all levels of supervision, OUR approach performs better in Chamfer error and mAP than SLIDING window using the same object training or SUPERVISED training of the same network. It can further be seen, how all methods improve with more labels. Looking at a condition with only 5% supervision, OUR method can perform similar to a SUPERVISED method that had 20× the labeling effort invested. At this condition, our detection is an acceptable .642, which SUPERVISED will only beat when at least 75% of the dataset is labeled. It could be conjectured, that the scene network does no more than emulating to slide a neural-network object detector across the scene. If this was true, SLIDING would be expected to perform similar or better than OURS, which is not the case. This indicates, that the scene network has indeed learned something not known at object level, something about the relation of the global scene without ever having labels on this level.

Figure 6 plots the rank distribution (horizontal axis) of Chamfer distances (vertical axis) for different methods (colors) at different levels of supervision (lightness). We see that OUR method performs well across the board, SUPERVISED has a more steep distribution compared to SLIDING, indicating it produces good as well as bad results, while the former is more uniform. In terms of supervision scalability, additional labeling invested into our method (brighter shades of yellow) result in more improvements to the right side of the curve, indicating, additional supervision is reducing high error-responses while already-good answers remain.

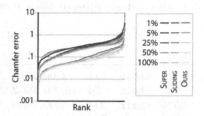

Fig. 6. Error distribution. (Color figure online)

Transfer Across Data Sets. So far, we have only considered training and testing on S3D. In Table 2, we look into how much supervision scaling would transfer to another data set, ScanNet. Remember, that we treat ScanNet as unlabeled, and hence, the loss network will be strictly only trained on objects from S3D. The three first rows in Table 2 define the conditions compared here: a loss network always trained on S3D, a scene network trained on either S3D or ScanNet and testing all combinations on both data sets.

Table 2. Transfer across data sets: Different rows show different levels of supervision, different columns indicate different methods and metrics. The plot on the right visualizes all methods in all conditions quantified by two metrics. Training either on S3D or on ScanNet. The metrics again are Chamfer error (also the loss) and mAP@.25. Colors in the plot correspond to different training, dotted/solid to different test data.

Loss:	S3D		S3D				
Scene:	S3D		SCANNET				
Test:	S3D		ScanNet		S3D		ScanNet
Ratio	Err.	mAP	Err.	mAP	Err.	mAP	Err.	mAP
1%	0.554	.366	1.753	.112	0.579	.296	0.337	.548
5%	0.346	.642	0.727	.138	0.466	.463	0.703	.599
50%	0.232	.773	0.588	.380	0.447	.497	0.258	.645
100%	0.178	.803	0.789	.384	0.336	.555	0.356	.661

Column two and three in Table 2 and the dotted violet line in the plot, iterate the scaling of available label data we already see in Table 1 when training and testing on S3D. Columns four and five, show a method trained on S3D but tested on ScanNet. We find performance to be reduced, probably, as the domain of ScanNet is different from the one of S3D. If we include the unlabeled scenes of ScanNet in the training, as seen in columns six to nine, the quality increases again, to competitive levels, using only S3D labels and 0% of the labels available for ScanNet.

Table 3. Performance of the loss network for different label ratio (**rows**) on different test data and according to different metrics (**columns**). [1] *Class not present in ScanNet.*

Ratio	Class	#Sce	#Obj	S3D Err.	S3D Acc.	SCANNET Err.	SCANNET Acc.
1%	chair	11	2400	.0085	.853	.0099	.843
5%	chair	54	16,000	.0052	.936	.0075	.899
25%	chair	271	47,200	.0049	.949	.0071	.907
50%	chair	542	121,191	.0046	.953	.0069	.902
75%	chair	813	162,000	.0045	.955	.0065	.920
100%	chair	1084	223,980	.0043	.960	.0068	.911
5%	table	54	5060	.0078	.921	—[1]	—[1]
5%	bcase	54	4780	.0093	.819	—[1]	—[1]
5%	column	54	2780	.0100	.855	—[1]	—[1]

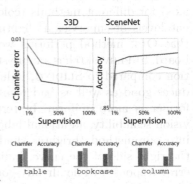

Table 3 further illustrate the loss network: how good are we at finding vectors that point to an object center? We see that the gradient error and the confidence error, both go down moderately with more labels when training and testing on S3D (violet). The fact that not much is decreasing in the loss network, while the scene network keeps improving, indicates the object task can be learned from little data, and less object-level supervision is required than what can be learned on a scene level, still. We further see, that the loss network generalizes between data sets from the fact that it is trained on S3D (violet curve) but when tested on ScanNet (green curve) goes down, too.

Besides seeing how quality scales with the amount of labeled supervision for training the loss network, it is also relevant to ask what happens when the amount of unlabeled training data for the scene network is increased while holding the labeled data fixed. This is analyzed in Table 4. Here we took our loss network and trained it at 5% label ratio on S3D and tested on ScanNet. Next, the scene network was trained, but on various number of scenes from

Table 4. Chamfer error and mAP@.25 reported for varying the number of scenes.

#Sce	OUR Err.	OUR mAP
66	.643	.079
330	.509	.242
1648	.506	.360
3295	.457	.412
4943	.435	.479
6590	.407	.599

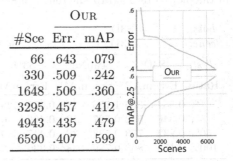

ScanNet, which, as we said, is considered unlabeled. The number of scenes changes over columns, resp. along the horizontal axis in the plot. We see that without investing any additional labeling effort, the scene network keeps increasing substantially, indicting what was learned on a few labeled S3D objects can enable understanding the structure of ScanNet.

Different Classes. Table 1 was analyzing the main axis of contribution: different levels of supervision but for a single class. This has shown that at around a label ratio of 5% OUR method performs similar to a SUPERVISED one. Holding the label ration of 5% fixed and repeating the experiment for other classes, is summarized in Table 5. We see, that the relation between SUPERVISED, SLIDING and OURS is retained across classes.

Table 5. Chamfer error (less is better) and mAP@.25 precision (more is better) **(columns)**, per class **(rows)** at a supervision of 5% labeling ratio.

	Chamfer error			mAP					
Class	SUP	SLI	OUR	SUP	SLI	OUR			
chair	0.789	0.577	.346	.352	.562	.642			
table	1.144	1.304	.740	.282	.528	.615			
bookcase	1.121	1.427	.979	.370	.298	.640			
column	0.900	2.640	.838	.490	.353	.654			

Comparison to Other Work. In Table 6 we compare our approach to other methods. Here, we use 20% of ScanNet V2 for testing and the rest for training. Out of the training data, we train our approach once with 100% labeled and once with only 5% labeled. Other methods were trained at 100% label ratio.

Table 6. Performance (mAP(%) with IoU threshold .25) of different methods **(rows)** on all classes **(columns)** of ScanNet V2. [1] 5 images. [2] Only xyz. [3] Their ablation; similar to our backbone.

Method		cabinet	bed	chair	sofa	table	door	window	bookshelf	picture	counter	desk	curtain	fridge	curtain	toilet	sink	bathtub	other	mAP
3DSIS[1]	[14]	19.8	69.7	66.2	71.8	36.1	30.6	10.9	27.3	0.0	10.0	46.9	14.1	53.8	36.0	87.6	43.0	84.3	16.2	40.2
3DSIS[2]	[14]	12.8	63.1	66.0	46.3	26.9	8.0	2.8	2.3	0.0	6.9	33.3	2.5	10.4	12.2	74.5	22.9	58.7	7.1	25.4
MTML	[17]	32.7	80.7	64.7	68.8	57.1	41.8	39.6	58.8	18.2	0.4	18.0	81.5	44.5	100.0	100.0	44.2	100.0	36.4	54.9
VoteNet	[23]	36.3	87.9	88.7	89.6	58.8	47.3	38.1	44.6	7.8	56.1	71.7	47.2	45.4	57.1	94.9	54.7	92.1	37.2	58.7
BoxNet[3]	[17]						No per-class information available													45.4
3D-BoNet	[33]	58.7	88.7	64.3	80.7	66.1	52.2	61.2	83.6	24.3	55.0	72.4	62.0	51.2	100.0	90.9	75.1	100.0	50.1	68.7
Ours 100%		43.0	70.8	58.3	16.0	44.6	28.0	13.4	58.2	4.9	69.9	74.0	75.0	36.0	58.9	79.0	47.0	77.9	48.2	50.2
Ours 5%		38.1	68.9	58.9	88.8	42.5	21.1	9.0	53.2	6.8	53.9	68.0	62.3	26.5	45.6	69.9	40.4	66.9	48.0	48.3

We see that our approach provides competitive performance, both at 100% of the labels, as well as there is only a small drop when reducing supervision by factor 20×. Our mAP at 100% of the labels is better than both variants (with and without color) of 3DSIS [14] from 2018 and similar to MTML [17] from 2019. VoteNet [23] and 3D-BoNet [33] are highly specialized architectures from 2019 that have a higher mAP. We have included BoxNet from Qi et al. [23], an ablation they include as a vanilla 3D detection approach that is similar to what we work with. We achieve similar even slightly better performance, yet at 5% of the supervision. In some categories, our approach wins over all approaches. We conclude that a simple backbone architecture we use is no contribution and cannot win over specialized ones, but that it also is competitive to the state-of-the-art. We should note here, as we do not carry out Semantic instance segmentation in our network, we did not test on the official test ScanNet benchmark test set. Instead, we reserve 20% of the labeled training scenes for testing.

Qualitative Results. Figure 7 shows qualitative example results of our approach.

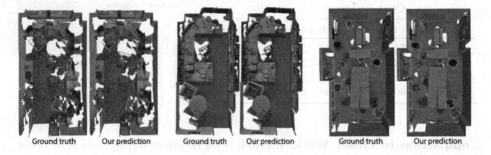

Fig. 7. Qualitative results of our approach and the ground truth for `chair` on S3D.

Computational Efficiency. Despite the additional complexity in training, at deployment, out network is a direct and fast forward architecture, mapping a point cloud to bounding boxes. Finding 20 proposals in 32,768 points takes 189 ms, while the supervised takes the same amount of time, with the small overhead of a NMS (190 ms) on a Nvidia RTX 2080Ti. Our CPU implementation of sliding window requires 4.7 s for the same task on a i7-6850K CPU @ 3.60 GHz. All results are computed with those settings.

5   Discussion

How can OURS *be better than the* SUPERVISED*?* It is not obvious why at 100% label ratio in Table 1, the SUPERVISED architecture performs at an mAP of .756 while OURS remains slightly higher at an mAP of .803. This is not just variance of the mAP estimation (computed across many identical objects and scenes).

A possible explanation for this difference is, that our training is no drop-in replacement for supervised training. Instead, it optimizes a different loss (truncation to the nearest object and collision avoidance) that might turn out to be better suited for 3D detection than what it was emulating in the beginning. We, for example, do not require NMS. As our training does not converge without those changes to the architecture, some of the effects observed might be due to differences in architecture and not due to the training. We conjecture future work might consider exploring different losses, involving truncation and collision, even when labels are present.

Why Hammersley? Other work has reasoned about what intermediate points to use when processing point clouds. When voting [23], the argument is, that the centers of bounding boxes are not part of the point set, and hence using a point set that is any subset of the input is not a good solution. While we do not vote, we also have chosen not to use points of the scene as the initial points. We also refrain from using any improved sampling of the surface, such as Poisson disk [13] sampling as we do not seek to cover any particular instance but space in general, covered by scenes uniformly.

How can the scene network be "better" than the loss network? As the loss network is only an approximation to the true loss, one might ask, how a scene network, trained with this loss network, can perform better than the loss network alone, e.g., how can it, consistently (Tables 1, 2, 4 and 5), outperform SLIDINGWINDOW?

Let us assume, that a scene network trained by a clean supervision signal can use global scene structure to solve the task. If now the supervision signal would start to be corrupted by noise, recent work has shown for images [18] or point clouds [12], that a neural network trained under noise will converge to a result that is very similar to the clean result: under $\mathcal{L}_2$ it will converge to the mean of the noise, under $\mathcal{L}_1$ to its median, etc. The amount of variance of that noise does not influence the result, what matters is that the noise is unbiased. In our case, this means if we were to have supervision by noisy bounding boxes, that would not change anything, except that the scene network training would converge slower but still to the mean or median of that noise distribution, which is, the correct result. So what was done in our training, by using a network to approximate the loss, means to just introduce another form of noise into the training.

6 Conclusion

We have suggested a novel training procedure to reduce the 3D labeling effort required to solve a 3D detection task. The key is to first learn a loss function on a small labeled local view of the data (objects), which is then used to drive a second learning procedure to capture global relations (scenes). The way to enlightenment here is to "find your center": the simple task of taking any piece of 3D scene and shifting it so it becomes centered around the closest object. Our analysis indicates that the scene network actually understands global scene structure not accessible to a sliding window. Our network achieves state of the art results, executes in a fraction of a second on large point clouds with typically only 5% of the labeling effort. We have deduced what it means exactly to learn the loss function, the new challenges associated with this problem and proposed several solutions to overcome these challenges.

In future work, other tasks might benefit from similar decoupling of supervision labels and a learned loss, probably across other domains or modalities.

References

1. Adler, J., Öktem, O.: Solving ill-posed inverse problems using iterativedeep neural networks. Inverse Prob. **33**(12) (2017)
2. Armeni, I., Sax, S., Zamir, A.R., Savarese, S.: Joint 2D–3D-semantic data for indoor scene understanding. arXiv:1702.01105 (2017)
3. Chang, A.X., et al.: ShapeNet: an information-rich 3D model repository. arXiv:1512.03012 (2015)
4. Chen, X., et al.: 3D object proposals for accurate object class detection. In: NIPS (2015)
5. Chen, Y., Liu, S., Shen, X., Jia, J.: Fast point R-CNN. In: ICCV (2019)
6. Dai, A., Chang, A.X., Savva, M., Halber, M., Funkhouser, T., Nießner, M.: Scan-Net: richly-annotated 3D reconstructions of indoor scenes. In: CVPR (2017)
7. Engelcke, M., Rao, D., Wang, D.Z., Tong, C.H., Posner, I.: Vote3deep: fast object detection in 3D point clouds using efficient convolutional neural networks. In: ICRA (2017)
8. Feng, M., Gilani, S.Z., Wang, Y., Zhang, L., Mian, A.: Relation graph network for 3D object detection in point clouds. arXiv:1912.00202 (2019)
9. Flynn, J., et al.: Deepview: view synthesis with learned gradient descent. In: CVPR (2019)
10. Girshick, R.: Fast R-CNN. In: ICCV (2015)
11. Griffiths, D., Boehm, J.: A review on deep learning techniques for 3D senseddata classification. Remote Sens. **11**(12) (2019)
12. Hermosilla, P., Ritschel, T., Ropinski, T.: Total denoising: unsupervised learning of 3D point cloud cleaning. In: Proceedings of the IEEE International Conference on Computer Vision, pp. 52–60 (2019)
13. Hermosilla, P., Ritschel, T., Vázquez, P.P., Vinacua, À., Ropinski, T.: Monte Carlo convolution for learning on non-uniformly sampled point clouds. ACM Trans. Graph. (Proc. SIGGRAPH Asia) **37**(6), 1–12 (2018)
14. Hou, J., Dai, A., Nießner, M.: 3D-SIS: 3D semantic instance segmentation of RGB-D scans. In: CVPR (2019)
15. Karpathy, A., Miller, S., Fei-Fei, L.: Object discovery in 3D scenes via shape analysis. In: ICRA (2013)
16. Ku, J., Mozifian, M., Lee, J., Harakeh, A., Waslander, S.L.: Joint 3D proposal generation and object detection from view aggregation. In: IROS (2018)
17. Lahoud, J., Ghanem, B., Pollefeys, M., Oswald, M.R.: 3D instance segmentation via multi-task metric learning. In: ICCV (2019)
18. Lehtinen, J., et al.: Noise2noise: learning image restoration without clean data. arXiv:1803.04189 (2018)
19. Liu, W., et al.: SSD: single shot multibox detector. In: Leibe, B., Matas, J., Sebe, N., Welling, M. (eds.) ECCV 2016. LNCS, vol. 9905, pp. 21–37. Springer, Cham (2016). https://doi.org/10.1007/978-3-319-46448-0_2

20. Maximov, M., Leal-Taixe, L., Fritz, M., Ritschel, T.: Deep appearance maps. In: ICCV (2019)
21. Newell, A., Yang, K., Deng, J.: Stacked hourglass networks for human pose estimation. In: Leibe, B., Matas, J., Sebe, N., Welling, M. (eds.) ECCV 2016. LNCS, vol. 9912, pp. 483–499. Springer, Cham (2016). https://doi.org/10.1007/978-3-319-46484-8_29
22. Qi, C.R., Chen, X., Litany, O., Guibas, L.J.: ImVoteNet: boosting 3D object detection in point clouds with image votes. In: arXiv preprint arXiv:2001.10692 (2020)
23. Qi, C.R., Litany, O., He, K., Guibas, L.J.: Deep hough voting for 3D object detection in point clouds. In: ICCV (2019)
24. Qi, C.R., Su, H., Mo, K., Guibas, L.J.: PointNet: deep learning on point sets for 3D classification and segmentation. In: Proceedings of the IEEE Conference on Computer Vision and Pattern Recognition, pp. 652–660 (2017)
25. Qi, C.R., Yi, L., Su, H., Guibas, L.J.: PointNet++: deep hierarchical feature learning on point sets in a metric space. arXiv:1706.02413 (2017)
26. Redmon, J., Divvala, S., Girshick, R., Farhadi, A.: You only look once: unified, real-time object detection. In: CVPR (2016)
27. Shi, S., Wang, X., Li, H.: PointRCNN: 3D object proposal generation and detection from point cloud. In: CVPR (2019)
28. Song, S., Lichtenberg, S.P., Xiao, J.: Sun RGB-D: a RGB-D scene understanding benchmark suite. In: CVPR (2015)
29. Song, S., Xiao, J.: Sliding shapes for 3D object detection in depth images. In: Fleet, D., Pajdla, T., Schiele, B., Tuytelaars, T. (eds.) ECCV 2014. LNCS, vol. 8694, pp. 634–651. Springer, Cham (2014). https://doi.org/10.1007/978-3-319-10599-4_41
30. Song, S., Xiao, J.: Deep sliding shapes for amodal 3D object detection in RGB-D images. In: CVPR (2016)
31. Weisstein, E.: Fixed point (2020). http://mathworld.wolfram.com/FixedPoint.html
32. Weisstein, E.: Hammersley point set (2020). http://mathworld.wolfram.com/HammersleyPointSet.html
33. Yang, B., et al.: Learning object bounding boxes for 3D instance segmentation on point clouds. arXiv:1906.01140 (2019)
34. Zhou, X., Wang, D., Krähenbühl, P.: Objects as points. arXiv:1904.07850 (2019)
35. Zhou, Y., Tuzel, O.: VoxelNet: end-to-end learning for point cloud based 3D object detection. In: CVPR (2018)

Collaborative Training Between Region Proposal Localization and Classification for Domain Adaptive Object Detection

Ganlong Zhao, Guanbin Li[✉], Ruijia Xu, and Liang Lin

School of Data and Computer Science, Sun Yat-sen University, Guangzhou, China
{zhaoglong,xurj3}@mail2.sysu.edu.cn, liguanbin@mail.sysu.edu.cn,
linliang@ieee.org

Abstract. Object detectors are usually trained with large amount of labeled data, which is expensive and labor-intensive. Pre-trained detectors applied to unlabeled dataset always suffer from the difference of dataset distribution, also called domain shift. Domain adaptation for object detection tries to adapt the detector from labeled datasets to unlabeled ones for better performance. In this paper, we are the first to reveal that the region proposal network (RPN) and region proposal classifier (RPC) in the endemic two-stage detectors (e.g., Faster RCNN) demonstrate significantly different transferability when facing large domain gap. The region classifier shows preferable performance but is limited without RPN's high-quality proposals while simple alignment in the backbone network is not effective enough for RPN adaptation. We delve into the consistency and the difference of RPN and RPC, treat them individually and leverage high-confidence output of one as mutual guidance to train the other. Moreover, the samples with low-confidence are used for discrepancy calculation between RPN and RPC and minimax optimization. Extensive experimental results on various scenarios have demonstrated the effectiveness of our proposed method in both domain-adaptive region proposal generation and object detection. Code is available at https://github.com/GanlongZhao/CST_DA_detection.

Keywords: Domain adaptation · Object detection · Transfer learning

1 Introduction

Benefiting from massively well-labeled data, deep convolutional neural networks have recently shown unparalleled advantages in various visual tasks, e.g., image recognition and object detection. Unfortunately, such data is usually prohibitive in many real-world scenarios. The problem becomes extremely serious for object detection, since it requires more precise object-level annotations. A common solution for this problem is to transfer the pretrained model from label-rich domain(i.e. source domain) to the other(i.e. target domain), but this often suffers from performance degradation due to domain gap.

© Springer Nature Switzerland AG 2020
A. Vedaldi et al. (Eds.): ECCV 2020, LNCS 12363, pp. 86–102, 2020.
https://doi.org/10.1007/978-3-030-58523-5_6

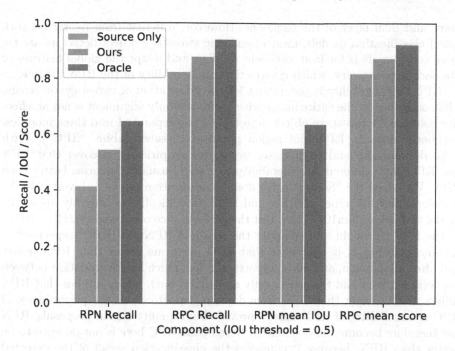

Fig. 1. Comparison of domain adaptation performance between "Source Only" model, "Ours" and the "Oracle" model w.r.t "Recall", "Average IOU" and "Mean Score". Threshold of IOU is set to 0.5 for "Recall" calculation. Experiment is conducted on adapting models trained on Sim10k [14] to Cityscapes [4].

Various kinds of methods have been proposed to overcome the domain gap. Most of them are based on adversarial training, and can be separated into two categories: feature-level and pixel-level domain adaptation. Feature-level domain adaptation tries to align the feature distributions from the two domains by adversarial training, while pixel-level domain adaptation uses GANs [10] to generate target-like images from source domain with labels unchanged.

There have been some research works focusing on domain adaptive object detection with feature-level or pixel-level adaptation techniques. Chen *et al.* [3] attempt to reduce domain discrepancy in both image level and instance level. Saito *et al.* [25] leverage weak global feature alignment and strong local alignment to mitigate the performance degradation caused by distinct scene layouts and different combinations of objects. Zhu *et al.* [34] propose to bridge the domain gap through effective region mining and region-based domain alignment. Noted that most of the previous research works on domain adaptive object detection focus on bridging the whole-image representations and thus perform alignment in the backbone branch of a detector. Though Zhu *et al.* [34] propose a selective adaptation framework based on region patches generated by the RPN branch, the loss gradient of the region classifier is just propagated to the backbone without adapting the RPN itself. Saito *et al.* [25] conduct feature alignment only on the lower and final layer of the backbone. However, different from neural network based classification models, most of endemic two-stage object detectors are far more complex. It is far from sufficient to align and adapt the global features of the backbone network, which ignores the transferability of the RPN module.

RPN transferability is paramount for the adaptation of two-stage detectors, while adapting on the entire image with backbone-only alignment is not an effective solution. A two-stage object detector can be separated into three modules: backbone network, RPN and region proposal classifier (abbr. "RPC"). With large domain gap and complicated scene, we empirically discover that RPN and RPC show different transferability, i.e., RPC usually performs better than RPN. We adopt the "Source Only" domain adaptation model to investigate the transferability difference of RPN and RPC. Specifically, we directly apply the model trained on Sim10k [14] to test the performance on Cityscapes [4], take 0.5 as the IOU threshold and compute the recall of RPN and RPC respectively[1]. As shown in Fig. 1, it is obvious that RPC performs better than RPN before and after adaptation, and more importantly, has much less degradation between oracle(green bar) and the source-only model(red bar), which implies that RPN suffers much severer than RPC from domain gap. However, the performance of RPC is also limited if RPN fails to provide high-quality region proposals. RPN has therefore become the bottleneck. Noted that RPC here is not doomed to be better than RPN, because it considers the classification recall of the detected region proposals (even if the RPN detection is accurate, the accuracy of the proposal classification is still uncertain).

On the other hand, it is noteworthy that, in some kinds of two-stage object detectors (e.g., Faster RCNN), there is no gradient flow between RPN and RPC. A natural idea is to take them as two individual branches from backbone. If we consider RPN as a kind of foreground/background classifier, it can be regarded as a coarse and fast RPC classifying each anchor across the feature map. Similarly, if we sum up the output of RPC stream to background and foreground scores, it performs just like a fine-grained and selective RPN. Based on the above discussion, we propose a novel domain adaptation method on Faster RCNN using collaborative training between RPN and RPC. It can also be easily generalized to other two-stage detectors. Specifically, we first apply collaborative self-training between RPN and RPC, which leverages the high-confident output of one to train the other. Besides, we introduce focal loss [17] in our method to impose more weight on ROIs of high-confidence and improve stability by removing the threshold selection. Second, ROIs of low-confidence that are ignored in the first part are used to calculate the foreground/background discrepancy between RPN and RPC. To improve the detector's transferability, the backbone network is trained to minimize the discrepancy while RPN and RPC try to maximize it. We verify its effectiveness under different adaptation scenarios.

To sum up, this work has the following contributions: (1) We are the first to reveal the significance of exploring the transferability of RPN module for domain-

---

[1] Noted that the recall computation of RPC here refers to the proportion of detected ROIs (not GT objects) having correct label prediction.

adaptive object detection. Simple alignment in backbone can not guarantee that the RPC receives high quality proposals. (2) From the perspective of treating RPN and RPC independently, we derive a collaborative self-training method that can propagate the loss gradient through the whole network and mutually enhance each other. (3) To the best of our knowledge, we are the first to adapt Maximum Classifier Discrepancy, MCD [26] to two-stage object detection framework for domain adaptation by focusing on ambiguous ROIs and show its effectiveness.

2   Related Works

Object Detection. Object detection has been around for a long time and is now an important research topic in computer vision. The development of convolutional neural networks has greatly advanced the performance of object detection. CNN-based detectors can be mainly divided into two categories: one-stage detectors and two-stage detectors. Although one-stage detectors such as YOLO [22] and SSD [18] have notably higher efficiency and have become popular paradigms, two-stage detectors like Fast RCNN [9], Faster RCNN [23] and Mask RCNN [11] are still widely adopted for their much higher performance. Faster RCNN [23] is a classical two-stage object detector and is commonly used as a baseline for domain adaptation. However, object detectors suffer from domain gap when being applied to an unseen domain. Generally the backbone network pre-trained with ImageNet [5] is fine-tuned on large amount of object-level labeled data for detection together with RPN and RPC. Unfortunately, such annotated data is usually prohibitive in target domains.

Domain Adaptation. Domain adaptation aims to utilize the labeled source domain data and unlabeled target domain data to boost performance on the latter. Domain adaptation on classification has been widely studied with technical paradigms like subspace alignment [6], asymmetric metric learning [16] and covariance matrix alignment [29]. A typical approach for domain adaptation is to reduce domain gap by making features or images from the two domains indistinguishable. Some methods try to minimize the distance between features of the two domains by resorting to MMD [1] or $\mathcal{H}\Delta\mathcal{H}$ [26], while some of the other works employ adversarial training with gradient reverse layer [7] or use generative adversarial networks [2,33]. Besides, entropy minimization has also been applied to domain adaptation for classification [19] and segmentation [31]. Domain-adaptive object detection has a completely different framework from image classification and semantic segmentation. It includes both object proposal detection and region-level classification. Therefore, when designing a domain-adaptive detection algorithm, it is far from sufficient to simply consider the alignment of the backbone network features, and it is necessary to consider the transferability of the algorithm in both RPN and RPC.

Domain Adaptation for Object Detection. In the past few years there has been some research in domain adaptation for object detection [12,15,24, 30,32]. Raj *et al.* [21] first proposed a domain adaptation method on RCNN with subspace alignment. Chen *et al.* [3] used a global-level and an instance-level alignment method respectively for the global and regional features. Inoue *et al.* [13] proposed a weakly-supervised framework with pseudo label. Saito *et al.* [25] pointed out that global alignment on complicated scenes and layouts might lead to negative transfer, and they proposed to employ a weak global alignment in the final layer of the backbone network, which puts more emphasis on images that are globally similar, and a strong local alignment in lower layer of the backbone. Zhu *et al.* [34] utilized RPN proposals to mine the discriminative regions of fixed size, which are pertinent to object detection, and focused on the alignment of those regions. These methods mainly focus on the alignment of the backbone stream regardless of the RPN transferability.

3 Method

3.1 Framework Overview

Given one labeled dataset from the source domain and an unlabeled one from the target domain, our task is to train a detector to obtain the best performance on the target dataset. For simplicity, both datasets share the same label space. Traditionally, feature-level domain adaptation methods try to extract the domain-invariant feature from both datasets, neglecting the adaptation of the main modules (i.e., RPN and RPC) besides the backbone stream.

The architecture of our model is illustrated in Fig. 3. The blue region on the top includes the modules in Faster RCNN, in which RPN generates and sends ROIs to the head of RPC for ROI-pooling. The yellow part is a domain discriminator that tries to determine which domain the input features originate from. It takes the backbone feature as input and outputs a domain prediction map of the same size. Besides, RPN prediction is used to highlight the foreground anchors in discriminator loss calculation. The red part consists of the proposed collaborative self-training scheme and the discrepancy maximization/minimization between RPN and RPC. ROIs with high-confidence of RPN (RPC) are used to train RPC (RPN), while ambiguous ROIs are used for discrepancy optimization. We illustrate the mutual complementary relationship of the proposed collaborative training and MCD optimization in Fig. 2. Among them, for collaborative training, the higher the confidence level of the ROI, the greater the weight will be given when calculating the loss function, while the opposite is true for MCD optimization. The lower the confidence level of ROI, the larger the sample weight will be. The two curves are implemented by tailor-designed polynomial functions in our experiment.

3.2 Collaborative Self-training

Generally a prevailing two-stage detector can be separated into three parts: the backbone F, RPN and RPC. Backbone F plays the role of feature representation and extracts the feature of the whole image. Then RPN takes the feature as

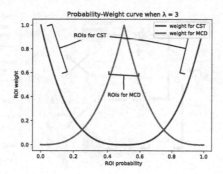

Fig. 2. The weight of ROI w.r.t its probability in loss calculation. Blue curve is used for collaborative self-training, and the red curve is for MCD. (Color figure online)

input, and predicts the foreground/background score of each anchor across the feature map. ROI pooling is applied to the anchors of high foreground probability for feature extraction and further sent to the RPC. Finally, RPC performs category prediction and regression of size and position of bounding boxes.

Although RPC performs regional classification based on the resulted proposals of the RPN module, it does not back propagate gradient to RPN during training. RPN filters out anchors with low foreground probability before feeding the rest to the RPC module. If the proposal filtering operation is removed and the RPN module performs ROI pooling at each anchor, the RPC module can be considered equivalent to the RPN. Ideally, the outputs of RPN and RPC should also be consistent. Those anchors with high background score in RPC should have low RPN foreground probability. Similarly, anchors having high score with non-background classes should also have high RPN foreground probability.

The core motivation of this paper is to improve the performance of object detection in the target domain by fully exploiting the domain-adaptive capability of the RPN module. Now we have accessed to the labeled image x_s and its annotation y_s from an annotated source dataset $\{X_s, Y_s\}$, as well as the unlabeled image x_t drawn from an unlabeled target dataset $\{X_t\}$. For labeled source image x_s, we introduce supervised training loss of Faster RCNN, which is calculated as follows [3]:

$$L_{det} = L_{rpn}(x_s, y_s) + L_{cls}(x_s, y_s). \tag{1}$$

As we do not have accessed to the annotations $\{y_t\}$, we mutually train the two modules of RPN and RPC by leveraging the output of one module to train the other. Given a target image x_t, feature f_t is first extracted by the feature extractor F. Based on f_t, RPN predicts the score (i.e., foreground and background probability) s_{rpn} for each ROI r_t while RPC outputs its class probability distribution s_{cls}, including the score of the background category and several foreground ones. For those ROIs with high-confident s_{cls}, we reuse them to update RPN, the loss of which is calculated as:

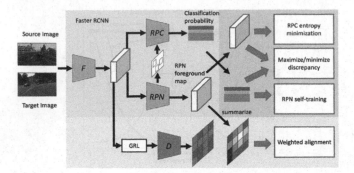

Fig. 3. Network architecture of our method. GRL stands for gradient reverse layer. From top to bottom: entropy minimization on RPC probability weighted by corresponding foreground score in RPN foreground map; discrepancy calculation on ambiguous ROIs shared in both RPN and RPC; RPN self-training using high-confident RPC detection results; domain discriminator loss weighted by summarized foreground probability in the same position.

$$L_{rpn_t} = f_w(s_{cls})L_{rpn}(x_t, \hat{y}_t), \tag{2}$$

where f_w can be defined as any function that decreases when s_{cls} is uncertain on the foreground/background classification. For simplicity, we define it as:

$$f_w(s_{cls}) = (|1 - 2s_{cls}^{bg}|)^{\lambda}, \tag{3}$$

where s_{cls}^{bg} is the background score in s_{cls}. λ controls the weight on samples of low-confidence. Besides, $\hat{y}_t$ in Eq. 2 refers to the pseudo label which contains ROIs with high-confident s_{cls}, including both foreground and background region proposals. It does not need to contain every object in the original image nor the specific class. In Faster RCNN, RPN is trained in a selective way. Assuming that the feature extractor forwards a feature map of spatial size $H \times W$, there will generally be $H \times W \times 9$ anchors and a prediction map of equal size. Only a small portion of anchors are referenced in L_{rpn} calculation and missing labels do not hurt the performance of RPN without extra processing.

On the other side, we can perform similar operations in the RPC module. Since RPN focuses on foreground/background classification and can not provide anchors with category-level pseudo labels which is necessary for RPC training, we adopt entropy minimization [19] for RPC, and adaptively assign higher weight to the samples of high-confidence in the calculation of the loss function. Based on entropy minimization, RPC is trained to output high-confident s_{cls} for ROIs with high-confident s_{rpn}. Similar to the RPN module, we define:

$$L_{cls_t} = f_w(s_{rpn})E(s_{cls}), \tag{4}$$

$$E(s_{cls}) = -\sum_{c \in C} s_{cls}^c log(s_{cls}^c), \tag{5}$$

$$f_w(s_{rpn}) = |1 - 2s_{rpn}^{fg}|^\lambda, \tag{6}$$

where C includes the background and all foreground classes. s_{cls}^c denotes the predicted probability of class c in s_{cls} while s_{rpn}^{fg} refers to the output foreground probability of the RPN module.

3.3 Maximize Discrepancy Classifier on Detectors

As described above, the loss term calculation of each ROI is multiplied by an adaptive weight during the collaborative self-training process. As shown in Fig. 2, the weight value of RPN update depends on the relevant output score of the RPC branch, and vice versa. We design the weight function and guide the training process to focus more on ROIs with high-confidence. In this section, we creatively introduce a customized maximizing discrepancy classifier, i.e., MCD [26], and point out that those ROIs with low-confidence can also be effectively leveraged to improve the model adaptation. MCD is a method originally proposed for domain adaptive image classification, which utilizes task-specific classifiers to align the distributions of source and target. It works by first separating the network into the feature extractor and classifier, and duplicating the latter. During training, the two classifiers learn to maximize the prediction discrepancy between themselves while the feature extractor tries to minimize it. MCD theoretically pointed out that by minimizing and maximizing discrepancy, the transferability of the model can be effectively improved. We borrow it here and formulate the two-stage classification process in detection to satisfy its setting, which further complements the collaborative self-training. Specifically, we regard RPN and RPC as two foreground/background classifiers without duplication but weight ROIs in an opposite way. As the red curve shown in Fig. 2, we assign higher weight to ROIs with low-confidence when performing MCD loss calculation. The discrepancy between RPN and RPC is defined as:

$$s_{cls}^{fg} = \sum_{c \in C} s_{cls}^c, \tag{7}$$

$$L_{discrepancy}(s_{cls}, s_{rpn}) = |s_{cls}^{fg} - s_{rpn}^{fg}|, \tag{8}$$

where C is the set of foreground categories. $L_{discrepancy}$ measures the foreground/background discrepancy between RPN and RPC based on their predictions. In addition, we define the weight function as:

$$f_w(s_{cls}, s_{rpn}) = (2min(|s_{cls}^{fg}|, |1 - s_{cls}^{fg}|, |s_{rpn}^{fg}|, |1 - s_{rpn}^{fg}|))^\lambda. \tag{9}$$

$f_w(s_{cls}, s_{rpn})$ is set to obtain a larger value when both s_{cls}^{fg} and s_{rpn}^{fg} are around 0.5 (i.e., of low-confident prediction), and a smaller value when either of them approaches 0 or 1. It is introduced here to mitigate the negative impact of noisy RPN prediction in the calculation of MCD loss, which is defined as,

$$L_{MCD} = f_w(s_{cls}, s_{rpn})L_{discrepancy}(s_{cls}, s_{rpn}). \tag{10}$$

The feature extractor F is trained to minimize the L_{MCD} while RPN and RPC try to maximize it alternately. As shown in Fig. 4, each curve represents a decision boundary of a specific classifier. In this case, they are replaced by RPN and RPC. Samples between two decision boundaries are more likely to be wrongly classified while those far from decision boundaries are more similar to the source domain and thus the output of which can be regarded as reliable pseudo labels.

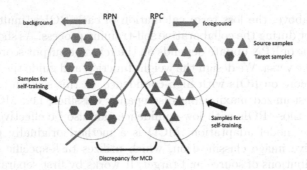

Fig. 4. Illustration of MCD principle. Samples far from the two decision boundaries of RPN and RPC tend to be reliable while others between the two boundaries are utilized for discrepancy computation.

3.4   RPN Weighted Alignment

Feature alignment between domains is a basic strategy often used by domain adaptive algorithms, and its effectiveness is widely recognized. As shown in Fig. 3, we further introduce a domain discriminator to achieve cross-domain feature alignment, and verify that it stabilize and is complementary to the previously introduced collaborative self-training paradigm. However, due to the diversity of object category combinations and the scene complexity in object detection, simply aligning the whole image might lead to failure. Some works have attempted to solve this problem, for example Saito *et al.* [25] use a weak global alignment and Zhu *et al.* [34] try to align the ROIs after region mining. We instead design a fine-grained and more flexible alignment with RPN score, i.e., the discriminator focusing on the regions of higher foreground probability.

Specifically, we design RPN weighted alignment as a local domain discriminator [25]. Discriminator D takes the feature of size $H \times W \times C$ from the backbone as input, and outputs a $H \times W$ probability map, the value of which denotes the domain probability of the specific position. We scale the RPN foreground map f to the same spatial size ($H \times W \times 9$) and weight the loss as follows:

$$L_{adv_s} = \frac{1}{HW} \sum_{w=1}^{W} \sum_{h=1}^{H} (D(F(x_s))_{wh}^2 \sum_{i=1}^{9} (f_i)_{wh}), \tag{11}$$

$$L_{adv_t} = \frac{1}{HW} \sum_{w=1}^{W} \sum_{h=1}^{H} ((1 - D(F(x_t))_{wh})^2 \sum_{i=1}^{9} (f_i)_{wh}. \tag{12}$$

f_i is the i-th channel of the RPN foreground map f with size $H \times W$, which represents the probability that the i-th anchor box defined at each position belongs to the foreground. $(f_i)_{wh}$ and $D(F(x))_{wh}$ are the element of f_i and $D(F(x))$ at position (w, h). The foreground map f might be rough at the beginning, but it will continue to be optimized as the collaborative self-training iterates and become an effective complement to the collaborative self-training at the backbone.

3.5 Overall Objective

The detection loss of original Faster RCNN consists of localization loss L_{rpn} calculated on RPN and classification loss L_{cls} on RPC. For source image, loss is defined as follows:

$$L_s = L_{rpn} + L_{cls} + L_{adv_s}. \tag{13}$$

For the target domain, it is slightly different due to L_{MCD}. we define the backbone loss and the loss of RPN and RPC as:

$$L_{t_{backbone}} = L_{adv_t} + \alpha L_{rpn_t} + \beta L_{cls_t} + \gamma L_{MCD}, \tag{14}$$

$$L_{t_{RPN,RPC}} = \alpha L_{rpn_t} + \beta L_{cls_t} - \gamma L_{MCD}, \tag{15}$$

and the loss for the discriminator is:

$$L_D = L_{adv_s} + L_{adv_t}. \tag{16}$$

α, β, γ control the trade-off between the detection loss of the source image and other losses. As [7], we adopt GRL (gradient reverse layer), which flips the sign of gradients in back-propagation, to implement the adversarial loss.

4 Experiment

We adopt unsupervised domain adaptation protocol and use four datasets in our experiments for evaluation, including Cityscapes [4], FoggyCityscapes [27], Sim10k [14] and KITTI [8]. Both images and annotations are provided for the source domain, while only images are available for target domain at training. As with [3] and [34], we evaluate our method on three kinds of domain shifts.

4.1 Implement Details

Following [3,25,34], we adopt Faster RCNN [23] with VGG16 [28] backbone and ROI-alignment [11] in all our experiments. We resize the input images so that the length of the shorter size is 600 and keep the aspect ratio following [25]. Our model is trained with three steps using SGD with 0.9 momentum and 0.001 learning rate. We first pre-train the model using the source dataset, followed by 10,000 iterations with L_{det} calculated on the source domain and L_{adv} on both domains, and finally train the network with all loss terms in Sect. 3.5 for 6,000 iterations. Without specific notation, we set α as 0.1, β as 0.05, γ as 0.1. For simplicity, pseudo boxes with confidence under 0.9 are discarded in RPN self-training and $f_w(s_{cls})$ is set to 1. λ is set to 2 in L_{MCD} and 5 in other loss terms. We implement all methods with Pytorch [20]. The architecture of domain discriminator follows [25].

We compare our method with four baselines: Faster RCNN [23], domain adaptive Faster RCNN (DA-Faster) [3], Strong Weak alignment (SWDA) [25], and selective cross-domain alignment (SCDA) [34]. The Faster RCNN model is only trained with the source images and labels without referring to the target data, which is also referred to as the source only model.

4.2 Domain Adaptation for Detection

Normal to Foggy. Cityscapes [4] is used as the source domain while Fog-gyCityscapes [27] as the target. In both domains, we use the training set for pre-training and adaptation without augmentation, and evaluate our model on the validation set for 8 classes. The results are reported in Table 1. As shown in the table, our method outperforms the existing state-of-the-art by 1.6% w.r.t mAP. Besides, our method outperforms existing methods in class *car*, which is the most common object in target domain.

Synthetic to Real. Sim10k [14] is used as the source domain and Cityscapes as the target domain. Similar to [3,25,34], we evaluate the detection performance on *car* in Cityscapes validation set. The results of our method is reported in Table 2. Our method outperforms the existing state-of-the-art method by 1.5% w.r.t mAP.

Cross Camera Adaptation. We used KITTI [8] as source domain and Cityscapes as the target domain for evaluation. The results under this scenario is reported in Table 3. Our method improves the existing best method by 1.1% w.r.t mAP.

Table 1. AP(%) from Cityscapes to FoggyCityscapes

Method	Person	Rider	Car	Truck	Bus	Train	Motobike	Bicycle	mAP
Faster RCNN [23]	29.7	32.2	44.6	16.2	27.0	9.1	20.7	29.7	26.2
DA-Faster [3]	25.0	31.0	40.5	22.1	35.3	20.2	20.0	27.1	27.6
SWDA [25]	29.9	42.3	43.5	24.5	36.2	**32.6**	30.0	35.3	34.3
SCDA [34]	**33.5**	38	48.5	**26.5**	39	23.3	28	33.6	33.8
Proposed	32.7	**44.4**	**50.1**	21.7	**45.6**	25.4	**30.1**	**36.8**	**35.9**

Table 2. AP on "Car"(%) from Sim10k to Cityscapes.

Method	AP on "Car"
Faster RCNN [23]	34.57
DA-Faster [3]	38.97
SWDA [25]	40.10
SCDA [34]	43.05
Proposed	44.51

Table 3. AP on "Car"(%) from KITTI to Cityscapes.

Method	AP on "Car"
Faster RCNN [23]	34.9
DA-Faster [3]	38.5
SWDA [25]	–
SCDA [34]	42.5
Proposed	43.6

5 Ablation Study

Effectiveness of RPN Adaptation. As one of our core motivations is to explore the significance of the RPN module for domain adaptation. We verify the superiority of collaborative self-training by analyzing the quality of region proposals generated by different adaptation models. We first define a metric called *proposal coverage*. Given a ground truth bounding box, we define the largest IOU with all detected proposals as its proposal coverage. For each ground truth in the target domain, we calculate the proposal coverage and count the distribution for each detection model. We conduct experiments on the adaptation from Sim10k to the Cityscapes dataset for comparison. Firstly, in order to verify that the naive local alignment in the backbone branch is not sufficient for domain adaptive object detection, we adopt a non-weighted local alignment method as naive alignment for comparison, in which a domain discriminator is applied to every position of the feature map. It can be implemented by removing the f_i weighting strategy in Eq. 11 and Eq. 12. We compare the proposal coverage distribution of four different detection models, including the source-only model, naive alignment model, our RPN adaptation with collaborative self-training and the Oracle model (i.e., the performance upper bound which refers to the annotations of the target domain). As shown in Fig. 5, our method greatly improves the quality of the generated object proposals in the target domain and its proposal coverage distribution is much more closer to the Oracle model, when compared with the naive alignment. Specifically, our method greatly reduces the boxes with proposal coverage = 0 compared with both the

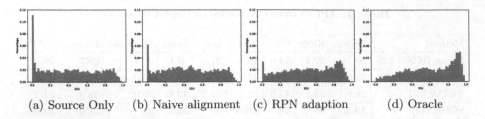

(a) Source Only   (b) Naive alignment   (c) RPN adaption   (d) Oracle

Fig. 5. Proposal coverage distribution of different proposal generation models. "Naive alignment" stands for non-weighted local alignment and "Oracle" refers to the model trained with labeled target dataset. RPN adaption is trained using our proposed collaborative self-training paradigm.

source only and naive alignment models. It also obviously improves the quality of proposals with IOU ≥ 0.5 while the naive alignment mainly changes the distribution of IOU < 0.5 w.r.t the source-only baseline. This shows that our method can effectively improve the accuracy of the generated proposals, and therefore bring about significant numerical performance improvements (Table 4). Noted that the performance benefit may become more apparent as the IOU threshold setting to a higher value.

Effectiveness of Different Components. We evaluate the contribution of different components by designing several variants of our model. The results are reported in Table 4. All experiments are conducted on the adaptation from Sim10k to Cityscapes dataset. Hyper-parameter setting of all model variants remain the same as described in Sect. 4.1. As shown in the table, incorporating the core module, i.e., collaborative self-training (CST), to the baseline source only model can bring significant performance gain of 7.76% w.r.t AP, increasing AP from 34.57% to 42.33%. This reveals that the domain adaptation capability can indeed be improved by effectively mining the complementary advantages of RPN and RPC. On the other hand, applying a naive local alignment only results in 2.46% performance improvement, which proves that simple alignment on the backbone branch is far from sufficient for domain adaptive object detection. Nevertheless, the proposed weighted local alignment still outperforms the naive alignment by 1.28% w.r.t AP even without RPN self-training. It is worth noting that using MCD alone does not significantly improve the baseline (36.42% VS 34.57%) because most of the uncertain ROIs are filtered out by RPN. Last but not least, as can be seen from the last two rows of the table, CST is complementary to the RPN weighted alignment and MCD, our entire model gains an additional 2.18% AP when compared to the CST-only version. In general, all three proposed components make their own contributions compared with the source-only model, which overall improve the baseline by 9.94% w.r.t AP.

Visualization of Detection Results. The detection results after adaptation are illustrated in Fig. 6. Our model can accurately localize and classify objects under different kinds of domain shifts. We also visualize the weight used in the local alignment calculation. It is obvious that RPN weighted method can

Fig. 6. Detection result on target domain. From left to right: Sim10k to Cityscapes; Kitti to Cityscapes; Cityscapes to FoggyCityscapes. The second row shows the weight inferred from our RPN weighted local domain discriminator. Brighter colors indicate higher attention. Apparently, regions with considered objects (e.g. car) are of higher weight in loss calculation. (Color figure online)

Table 4. AP on "Car" from Sim10k to Cityscapes of different method variants to demonstrate the effectiveness of the proposed algorithm. "Local" represents naive local alignment and "CST" refers to the collaborative self-training.

Local	Weight local	CST	MCD	AP on Car
				34.57
✓				37.03
	✓			38.31
		✓		42.33
			✓	36.42
	✓	✓		43.08
	✓	✓	✓	44.51

effectively suppress non-critical parts of the image. As shown in the Figure, although the sky and roads occupy most of the area of the image, the inferred weight map shows that these areas have little effect on distinguishing objects of different domains, which is consistent with our optimization goal.

6   Conclusion

In this paper, we are the first to empirically reveal that the RPN and RPC module in the endemic two-stage detectors (e.g., Faster RCNN) demonstrate significantly different transferability when facing large domain gap. Base on this observation, we design a collaborative self-training method for RPN and RPC to train each other with ROIs of high-confidence. Moreover, a customized maximizing discrepancy classifier is introduced to effectively leverage ROIs with low-confidence to further increase the accuracy and generalization of the detection model. Experimental results demonstrated that our method significantly improves the transferability and outperforms existing methods in various domain adaptation scenarios.

Acknowledgements. This work was supported in part by the Guangdong Basic and Applied Basic Research Foundation (2020B1515020048), in part by the National Natural Science Foundation of China (61976250, 61702565, U1811463), in part by the National High Level Talents Special Support Plan (Ten Thousand Talents Program), in part by the Fundamental Research Funds for the Central Universities (18lgpy63). This work was also sponsored by Meituan-Dianping Group.

References

1. Borgwardt, K.M., Gretton, A., Rasch, M.J., Kriegel, H.P., Schölkopf, B., Smola, A.J.: Integrating structured biological data by kernel maximum mean discrepancy. Bioinformatics **22**(14), e49–e57 (2006)
2. Bousmalis, K., Silberman, N., Dohan, D., Erhan, D., Krishnan, D.: Unsupervised pixel-level domain adaptation with generative adversarial networks. In: Proceedings of the IEEE Conference on Computer Vision and Pattern Recognition, pp. 3722–3731 (2017)
3. Chen, Y., Li, W., Sakaridis, C., Dai, D., Van Gool, L.: Domain adaptive faster R-CNN for object detection in the wild. In: Proceedings of the IEEE Conference on Computer Vision and Pattern Recognition, pp. 3339–3348 (2018)
4. Cordts, M., et al.: The cityscapes dataset for semantic urban scene understanding. In: Proceedings of the IEEE Conference on Computer Vision and Pattern Recognition, pp. 3213–3223 (2016)
5. Deng, J., Dong, W., Socher, R., Li, L.J., Li, K., Fei-Fei, L.: ImageNet: a large-scale hierarchical image database. In: 2009 IEEE Conference on Computer Vision and Pattern Recognition, pp. 248–255. IEEE (2009)
6. Fernando, B., Habrard, A., Sebban, M., Tuytelaars, T.: Unsupervised visual domain adaptation using subspace alignment. In: Proceedings of the IEEE International Conference on Computer Vision, pp. 2960–2967 (2013)
7. Ganin, Y., et al.: Domain-adversarial training of neural networks. J. Mach. Learn. Res. **17**(1), 2096–2130 (2016)
8. Geiger, A., Lenz, P., Urtasun, R.: Are we ready for autonomous driving? The KITTI vision benchmark suite. In: 2012 IEEE Conference on Computer Vision and Pattern Recognition, pp. 3354–3361. IEEE (2012)
9. Girshick, R.: Fast R-CNN. In: Proceedings of the IEEE International Conference on Computer Vision, pp. 1440–1448 (2015)
10. Goodfellow, I., et al.: Generative adversarial nets. In: Advances in Neural Information Processing Systems, pp. 2672–2680 (2014)
11. He, K., Gkioxari, G., Dollár, P., Girshick, R.: Mask R-CNN. In: Proceedings of the IEEE International Conference on Computer Vision, pp. 2961–2969 (2017)
12. Hoffman, J., et al.: CyCADA: cycle-consistent adversarial domain adaptation. arXiv preprint arXiv:1711.03213 (2017)
13. Inoue, N., Furuta, R., Yamasaki, T., Aizawa, K.: Cross-domain weakly-supervised object detection through progressive domain adaptation. In: Proceedings of the IEEE Conference on Computer Vision and Pattern Recognition, pp. 5001–5009 (2018)

14. Johnson-Roberson, M., Barto, C., Mehta, R., Sridhar, S.N., Rosaen, K., Vasudevan, R.: Driving in the matrix: can virtual worlds replace human-generated annotations for real world tasks? arXiv preprint arXiv:1610.01983 (2016)
15. Kim, T., Jeong, M., Kim, S., Choi, S., Kim, C.: Diversify and match: a domain adaptive representation learning paradigm for object detection. In: Proceedings of the IEEE Conference on Computer Vision and Pattern Recognition, pp. 12456–12465 (2019)
16. Kulis, B., Saenko, K., Darrell, T.: What you saw is not what you get: domain adaptation using asymmetric kernel transforms. In: CVPR 2011, pp. 1785–1792. IEEE (2011)
17. Lin, T.Y., Goyal, P., Girshick, R., He, K., Dollár, P.: Focal loss for dense object detection. In: Proceedings of the IEEE International Conference on Computer Vision, pp. 2980–2988 (2017)
18. Liu, W., et al.: SSD: single shot multibox detector. In: Leibe, B., Matas, J., Sebe, N., Welling, M. (eds.) ECCV 2016. LNCS, vol. 9905, pp. 21–37. Springer, Cham (2016). https://doi.org/10.1007/978-3-319-46448-0_2
19. Long, M., Zhu, H., Wang, J., Jordan, M.I.: Unsupervised domain adaptation with residual transfer networks. In: Advances in Neural Information Processing Systems, pp. 136–144 (2016)
20. Paszke, A., et al.: Automatic differentiation in PyTorch (2017)
21. Raj, A., Namboodiri, V.P., Tuytelaars, T.: Subspace alignment based domain adaptation for RCNN detector. arXiv preprint arXiv:1507.05578 (2015)
22. Redmon, J., Divvala, S., Girshick, R., Farhadi, A.: You only look once: unified, real-time object detection. In: Proceedings of the IEEE Conference on Computer Vision and Pattern Recognition, pp. 779–788 (2016)
23. Ren, S., He, K., Girshick, R., Sun, J.: Faster R-CNN: towards real-time object detection with region proposal networks. In: Advances in Neural Information Processing Systems, pp. 91–99 (2015)
24. RoyChowdhury, A., et al.: Automatic adaptation of object detectors to new domains using self-training. In: Proceedings of the IEEE Conference on Computer Vision and Pattern Recognition, pp. 780–790 (2019)
25. Saito, K., Ushiku, Y., Harada, T., Saenko, K.: Strong-weak distribution alignment for adaptive object detection. arXiv (2018)
26. Saito, K., Watanabe, K., Ushiku, Y., Harada, T.: Maximum classifier discrepancy for unsupervised domain adaptation. In: Proceedings of the IEEE Conference on Computer Vision and Pattern Recognition, pp. 3723–3732 (2018)
27. Sakaridis, C., Dai, D., Van Gool, L.: Semantic foggy scene understanding with synthetic data. Int. J. Comput. Vis. **126**(9), 973–992 (2018)
28. Simonyan, K., Zisserman, A.: Very deep convolutional networks for large-scale image recognition. arXiv preprint arXiv:1409.1556 (2014)
29. Sun, B., Feng, J., Saenko, K.: Return of frustratingly easy domain adaptation. In: Thirtieth AAAI Conference on Artificial Intelligence (2016)
30. Tzeng, E., Burns, K., Saenko, K., Darrell, T.: SPLAT: semantic pixel-level adaptation transforms for detection. arXiv preprint arXiv:1812.00929 (2018)
31. Vu, T.H., Jain, H., Bucher, M., Cord, M., Pérez, P.: ADVENT: adversarial entropy minimization for domain adaptation in semantic segmentation. In: Proceedings of the IEEE Conference on Computer Vision and Pattern Recognition, pp. 2517–2526 (2019)

32. Wang, T., Zhang, X., Yuan, L., Feng, J.: Few-shot adaptive faster R-CNN. In: Proceedings of the IEEE Conference on Computer Vision and Pattern Recognition, pp. 7173–7182 (2019)

33. Yoo, D., Kim, N., Park, S., Paek, A.S., Kweon, I.S.: Pixel-level domain transfer. In: Leibe, B., Matas, J., Sebe, N., Welling, M. (eds.) ECCV 2016. LNCS, vol. 9912, pp. 517–532. Springer, Cham (2016). https://doi.org/10.1007/978-3-319-46484-8_31

34. Zhu, X., Pang, J., Yang, C., Shi, J., Lin, D.: Adapting object detectors via selective cross-domain alignment. In: Proceedings of the IEEE Conference on Computer Vision and Pattern Recognition, pp. 687–696 (2019)

Two Stream Active Query Suggestion
for Active Learning in Connectomics

Zudi Lin[1]([⊠]), Donglai Wei[1], Won-Dong Jang[1], Siyan Zhou[1], Xupeng Chen[2],
Xueying Wang[1], Richard Schalek[1], Daniel Berger[1], Brian Matejek[1],
Lee Kamentsky[3], Adi Peleg[4], Daniel Haehn[5], Thouis Jones[6], Toufiq Parag[7],
Jeff Lichtman[1], and Hanspeter Pfister[1]

[1] Harvard University, Cambridge, USA
linzudi@g.harvard.edu
[2] New York University, New York, USA
[3] MIT, Cambridge, USA
[4] Google, Mountain View, USA
[5] University of Massachusetts Boston, Boston, USA
[6] Broad Institute, Cambridge, USA
[7] Comcast Research, Cambridge, USA

Abstract. For large-scale vision tasks in biomedical images, the labeled
data is often limited to train effective deep models. Active learning is a
common solution, where a query suggestion method selects representative unlabeled samples for annotation, and the new labels are used to
improve the base model. However, most query suggestion models optimize their learnable parameters only on the limited labeled data and
consequently become less effective for the more challenging unlabeled
data. To tackle this, we propose a *two-stream active* query suggestion
approach. In addition to the supervised feature extractor, we introduce
an unsupervised one optimized on all raw images to capture diverse image
features, which can later be improved by fine-tuning on new labels. As
a use case, we build an end-to-end active learning framework with our
query suggestion method for 3D synapse detection and mitochondria
segmentation in connectomics. With the framework, we curate, to our
best knowledge, the largest connectomics dataset with dense synapses
and mitochondria annotation. On this new dataset, our method outperforms previous state-of-the-art methods by 3.1% for synapse and 3.8%
for mitochondria in terms of region-of-interest proposal accuracy. We also
apply our method to image classification, where it outperforms previous
approaches on CIFAR-10 under the same limited annotation budget. The
project page is https://zudi-lin.github.io/projects/#two_stream_active.

X. Chen, L. Kamentsky, A. Peleg, D. Haehn, T. Jones and T. Parag—Works were done
at Harvard University.

Electronic supplementary material The online version of this chapter (https://
doi.org/10.1007/978-3-030-58523-5_7) contains supplementary material, which is available to authorized users.

Keywords: Active learning · Connectomics · Object detection · Semantic segmentation · Image classification

1 Introduction

Deep convolutional neural networks (CNNs) have advanced many areas in computer vision. Despite their success, CNNs need a large amount of labeled data to learn their parameters. However, for detection and segmentation tasks, dense annotations can be costly. Further, in the biomedical image domain, annotations need to be conducted by domain experts after years of training. Thus, under the limited annotation budget, it is critical to effectively select a subset of unlabeled data for annotation to train deep learning models.

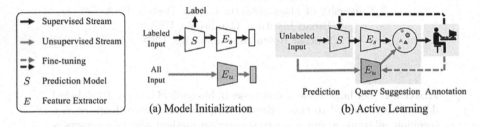

(a) Model Initialization   (b) Active Learning

Fig. 1. Two-stream active query suggestion. Active learning methods transform unlabeled data into a feature space to suggest informative queries and improve the base model S. Previous methods optimize their feature extractor (E_s) only on the labeled data. We propose a second one (E_u) trained *unsupervisedly* on all data to capture diverse image features, which can later be updated by fine-tuning with new annotations.

Active learning is a common solution that iteratively improves the prediction model by suggesting informative queries for human annotation to increase labels. There are three main categories of query suggestion approaches that have been explored for CNNs: uncertainty-based [42,44,52], expected model change-based [53], and clustering-based methods [43]. However, all these methods use features extracted from CNNs that are trained on the labeled set (Fig. 1a, →). For example, core-set [45] uses the last feature space before the classification layer to find representative queries, and learning-loss [53] takes multiple features maps to estimate the loss of the model prediction. Therefore, these methods can be biased towards the feature distribution of the small labeled set. Notably, in many biomedical image applications, the labeled dataset is far from representative of the whole dataset due to its vast quantity and great diversity.

To address this challenge, we propose a *two-stream active clustering* method to improve query suggestion by introducing an additional *unsupervised* feature extractor to capture the image statistics of the whole dataset (Fig. 1a,→). During active learning, we combine features extracted by both the supervised and unsupervised streams from the unlabeled data (Fig. 1b). The unsupervised stream can better select representative samples based on image features even when the supervised model makes wrong predictions. Given new annotations, we can further finetune the unsupervised feature extractor to make the embedding space more discriminative. For the clustering module, we show that combining the features from both streams in a *hierarchical* manner achieves significantly better query suggestion performance than directly concatenating the feature vectors.

We test our method in the field of *connectomics*, where the goal is to reconstruct the wiring diagram of neurons to enable new insights into the workings of the brain [26,31]. Recent advances in electron microscopy (EM) allow researchers to collect brain images at nanometer resolution and petabyte scale [21,56]. One crucial task is to detect and segment biological structures like synapses and mitochondria for a deeper understanding of neural anatomy and activation patterns [2] (Fig. 2a-b). However, most labeled connectomics datasets [11,30] are only a few gigavoxels in size, hundreds of times smaller than the unlabeled volume needed for down-stream biological analysis (Fig. 2c).

With our *two-stream active clustering* as the key component, we build an end-to-end framework with a base model and an annotation workflow. Before active learning, our base model achieves state-of-the-art results on public benchmarks. Besides, our annotation workflow reduces interactive error correction time by 26%, as shown by a controlled user study. With this framework, we finished the dense annotation of synapse objects and mitochondria semantic mask for a $(50 \ \mu m)^3$ EM image volume (300 gigavoxels) in the rat visual cortex, called EM-R50, which is over 100× larger than existing datasets. For the evaluation of active learning approaches on this connectomics dataset, our method improves the performance of previous state-of-the-art methods by 3.1% for synapses and 3.8% for mitochondria, respectively, in terms of the accuracy of the region-of-interest (ROI) proposals. We further perform ablation studies to examine the importance of different framework components and hyper-parameters. To demonstrate its broader impact, we also benchmark our method on natural image classification (CIFAR-10), which outperforms previous state-of-the-art methods by over 2% under a limited annotation budget ≈5% of the total training images.

Contributions. First, we introduce a novel active learning method that incorporates information from an unsupervised model to improve the effectiveness of query suggestions. Second, our method achieves state-of-the-art results for detection and segmentation tasks on connectomics datasets and image classification on CIFAR-10. Third, we release the code and a densely annotated connectomics dataset (100× bigger than current datasets) to facilitate future researches.

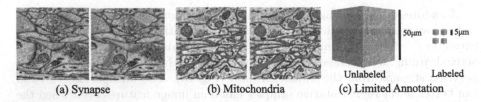

(a) Synapse　　　　　　(b) Mitochondria　　　　(c) Limited Annotation

Fig. 2. Two essential vision tasks in connectomics: (a) object detection of synapses to quantify the neuronal connectivity strength and (b) semantic segmentation of mitochondria to estimate the neuronal activity level. (c) However, the terabyte-level test data can be 100× larger than the training data, making active learning necessary.

2　Related Work

Synapse Detection and Mitochondria Segmentation. Synapse detection and mitochondria segmentation are two popular tasks in connectomics. Due to the complex shapes, bounding box-based detection [38] and segmentation [12] methods can have poor performance. Thus, most previous works for biomedical vision directly predict the semantic segmentation of the object and generate bounding-box proposals for the detection task via post-processing.

For synapse detection, previous approaches focus on segmenting the synaptic cleft region using hand-crafted image features [2,17,19,24,25,37] or learned features [39]. To further predict synapse polarity, the direction of information transmission among neurons, recent works apply random forest classifiers [23,49], neural networks [5,14,18,34], and combinations [7]. For mitochondria segmentation, earlier works leverage various image processing techniques and manually-designed image features [27,30,32,36,46,50]. Recent methods employ 2D or 3D fully convolutional network architectures [6,33] to regress the semantic mask.

In this paper, we adopt the 3D U-Net [40] model for both synapse detection and mitochondria semantic segmentation. Incorporating recent deep learning techniques including residual blocks [13] and squeeze-and-excitation blocks [16], our model achieves top performance on public connectomics benchmarks.

Active Learning. Active learning methods iteratively query human annotators to obtain new informative samples to label and then improve the base model. Transductive active learning [54] aims to improve the later step by training the base model on the additional unlabeled data with pseudo labels. The similarity graph among samples [3,35,57] is often used to generate pseudo labels from manual annotations. We focus on the former step to suggest better queries [45], where traditional methods use uncertainty-based sampling [42,52], and diversity-based optimization [9,10]. Tailored for neural networks, recent works explore ideas of maximizing feature coverage [43], margin-based sampling [55], expected error-based selection [53] and adversarial learning [8].

Besides the image classification task, active learning has been applied to object detection for different image domains [1,4,48]. Roy *et al.* [41] formulates the detection task as a structured prediction with novel margin sampling techniques and Vijayanarasimhan *et al.* [51] scales up the labeling process with crowdsourcing. Kao *et al.* [20] proposes location-aware measures for query suggestion. Instead of solely using the feature extractor optimized on the labeled set, our key insights are to improve query suggestions with unsupervised image information and fine-tune the learned feature extractor to distinguish ambiguous samples.

3   Active Learning Framework Overview

Our active learning framework for large-scale vision tasks in connectomics has three components: base model, query suggestion, and annotation (Fig. 3). We here describe our base model and annotation workflow, leaving the query suggestion method for Sect. 4. Further details are in the supplementary document.

Overview. During active learning on unlabeled images, the base model first predicts dense probability map and generates regions of interest (ROIs). Then the proposed query suggestion method extracts features for all ROIs and suggests queries through the two-stream clustering for annotation. With the new annotation, in addition to fine-tuning the base model, we further fine-tune the proposed query suggestion model to make it more discriminative in query suggestion.

Model Prediction. The base model handles two tasks: synapse detection and mitochondria segmentation. The irregular shapes make it hard to directly predict 3D bounding boxes for synapses, while the vast volume quantity makes it infeasible to conduct pixel-wise annotation for mitochondria. Therefore, following common practice for biomedical images, we first predict a dense semantic probability map and apply connected component labeling with post-processing to generate ROI proposals. We thus unify two different tasks as judging the correctness of ROIs in active learning. Since finding false positives from proposals is more efficient than locating false negatives in the vast volume, we re-balance the weights between foreground and background pixels to ensure a high recall.

In Fig. 3a, we show an example for synapse detection. Each synapse instance has a pre-synaptic (purple) and a post-synaptic (cyan) segment, and we predict a three-channel probability map representing pre- and post-synaptic regions and their union. We align extracted ROIs to a reference orientation to normalize its rotation variation. To this end, we select the 2D slice with the biggest area from the 3D instance, apply the principal component analysis (PCA) of the mask, and rotate the instance to align its first principal component to the vertical direction. For synapse, we further make sure the pre-synaptic segment (gray) is on the left.

Annotation. We focus on the correctness of ROIs instead of the pixel-level correctness of the mask. During annotation, an annotator judges the ROI to be

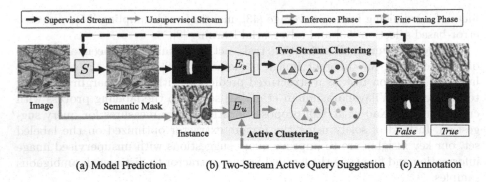

Fig. 3. Overview of our active learning framework. (a) The base model S predicts semantic masks, which are post-processed to generate ROIs. We align them to the same orientation for better clustering. (b) Our method adds an additional stream of *unsupervised* feature extracted by E_u. We apply hierarchical clustering to partition the unlabeled data and suggest cluster centers as queries for annotation. (c) Annotators provide *True or False* annotations for query samples that are used to fine-tune both the based mode S (black dashed line) and the proposed E_u (red dashed line). (Color figure online)

correct if the mask within covers more than half of the ground truth mask in this ROI. In practice, thanks to the performance of the base model, annotators find most predicted instances are unambiguously right or wrong. We built a browser-based labeling interface, where annotators can click on each suggested query to change its label, *e.g.*, from *True* to *False* (Fig. 3c). For better judgment, we display both image patches and predicted instance masks to annotators. For annotation efficiency, we display the selected query samples in a grouped manner using our clustering results instead of a random order (Sect. 6.3).

4   Two-Stream Active Query Suggestion

Compared to previous methods, our *two-stream active query suggestion* introduces an additional unsupervised feature extractor that is trained on all images to capture dataset statistics. We then cluster unlabeled data with the two-steam features and use cluster centers as query samples for annotation (Sect. 4.1). With new annotations, we further fine-tune the image feature extractor to adjust the feature space for better query suggestions in the next round (Sect. 4.2).

4.1   Two-Stream Clustering

For the task of deciding the correctness of ROI proposals, we use the predicted mask from the base model for the supervised stream and its corresponding raw image for the unsupervised stream. We first apply the feature extractor to reduce the feature dimension for each stream. Then, we fuse the two-stream clustering results to partition the unlabeled data into smaller subsets where the samples share similar features to make the cluster centers more representative.

Feature Extraction Network. We train the feature extraction model through self-supervision. Specifically, we train a variational auto-encoder (VAE) [22] to regress the input through a bottleneck network structure (Fig. 4a) and use the embedded features as the low-dimensional representations. The VAE model consists of an encoder network E and a decoder network D. We use several convolutional layers for the encoder, followed by a fully connected layer to predict both the mean and standard deviation of the low-dimensional embedding vector. For the decoder, we use deconvolution (or transposed convolution) layers to learn to reconstruct the input image from the embedding vector.

The loss function for the VAE is the sum of the ℓ_1 reconstruction loss and the KL-divergence between the distribution of the predicted embedding feature and the standard normal distribution. In practice, we use samples with a fixed patch size where the mask or image is rotated and aligned when training the VAE. We then only use the VAE mean vector as the extracted feature.

Feature Fusion. Given the extracted image and mask features, we propose two designs of clustering architectures to fuse such two-stream information: late-fusion clustering and hierarchical clustering (Fig. 4b). Inspired by the two-stream architecture designs for video action recognition [47], we design a *late-fusion* strategy to directly concatenate image and mask features and feed into a clustering module C. We expect that with the combined features from two streams, the clustering method can better distinguish ambiguous samples.

Another strategy is the *hierarchical clustering* that clusters the mask features and the image features *sequentially*. The intuition behind the design is that since the embedding spaces for both extractors can be very different (*e.g.*, dimension, and distance scale), the hierarchical approach can alleviate the needs for rebalancing. In the hierarchical clustering, the members of each of the N mask clusters separated in the first round are further divided into M sub-clusters by applying the k-means algorithm on the unsupervised image VAE embedding space, which yields MN clusters in total. We show in Sect. 6.2 that conditioning the image clustering on the mask features can prevent the image features, which are of high dimension than mask features, from dominating the results. Therefore hierarchical clustering can better suggest queries compare to late-fusion.

Query Suggestion. Given the clustering result from either late-fusion or hierarchical clustering, we run an additional round of clustering (e.g., k-means) with Q clusters and use the samples with minimum distances to each cluster center as queries presented to the annotators. Thus the annotator needs to annotate in total MNQ samples. In the ablation study (Sect. 6.2), we will examine the query suggestion effectiveness and efficiency with different hyper-parameter choices.

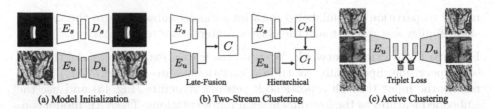

(a) Model Initialization   (b) Two-Stream Clustering   (c) Active Clustering

Fig. 4. Architectures for the two-stream active query suggestion model. (a) For model initialization, we train the supervised (E_s) and unsupervised (E_u) feature extractors using VAEs. (b) For two-stream clustering, we compare two design choices to combine E_u and E_s features in an either parallel (late-fusion) or hierarchical manner. The block C_i denotes the clustering algorithm. (c) For active clustering, we fine-tune E_u with triplet loss to encourage the learning of discriminative features.

4.2   Active Clustering

Since the encoders are learned in an unsupervised manner, we expect that with new annotations, we can improve the encoder to encourage the learning of more discriminative features. Therefore we adaptively adjust the embedding space of the encoder with new labels to make the clustering module *active*.

Triplet Loss. We employ the triplet loss [15] to incorporate new label information into the encoder. Suppose that we have a set of labeled positive and negative instances. After randomly select one positive sample $\mathbf{x}_P$ and one negative sample $\mathbf{x}_N$ as anchors, we hope that the third sample $\mathbf{x}$ becomes close to $\mathbf{x}_P$ and distant from $\mathbf{x}_N$ if it is positive, and vice versa. This can encourage the encoders to learn more discriminative features and facilitate query suggestions since samples share the same label are closer while different classes are projected further apart. Following Hoffer *et al.* [15], we calculate the distances as

$$d_P = ||\phi(\mathbf{x}) - \phi(\mathbf{x}_P)||_2, \qquad d_N = ||\phi(\mathbf{x}) - \phi(\mathbf{x}_N)||_2, \qquad (1)$$

where $\phi(\mathbf{x})$ indicates features extracted from the encoder. We then define the loss function for adjusting the feature extractor as

$$L_{\text{Triplet}}(\mathbf{x}, \mathbf{x}_P, \mathbf{x}_N) = \left\| \frac{e^{d_P}}{e^{d_P} + e^{d_N}}, \frac{e^{d_N}}{e^{d_P} + e^{d_N}} - 1 \right\|_2^2 \qquad (2)$$

to minimize d_P and maximize d_N. Incorporating the triplet loss enables the active adjusting of the feature space to be more discriminative, which can further improve the effectiveness of query suggestion.

4.3   Learning Strategy

Inference Phase. (Figure 3, solid line) For both synapses and mitochondria, the base model was initially trained on a small manually labeled volume of size

$(5\ \mu m)^3$, comparable to public benchmark datasets. We conduct sliding-window prediction for the large test volume of size $(50\ \mu m)^3$ and use the connected component algorithm to generate ROI candidates for active learning. The VAE of the mask encoder E_s model is trained on the aligned patches with predicted object masks, while the image encoder E_u is trained with image patches uniformly sampled from the whole volume to capture diverse texture information.

Fine-Tuning Phase. (Figure 3, dashed line) For active clustering, the E_u is initialized by fine-tuning it with labeled patches from the small labeled volume. Then the queries are generated with two-stream hierarchical clustering by successively using the latent spaces of both E_s and E_u. After query annotation, we fine-tune the image encoder with new ground truth labels and apply it for future iterations of query suggestion. For non-active clustering, we conduct the same hierarchical clustering but use the original E_u trained under a totally unsupervised setting. In both cases, the new query samples are used to fine-tune and improve the base model as a standard active learning practice.

5   EM-R50 Connectomics Dataset

With our two-stream active query suggestion method, we annotated, to the best of our knowledge, the largest 3D EM connectomics image volume datasets with dense synapses object and mitochondria mask annotation. Specifically, we imaged a tissue block from Layer II/III in the primary visual cortex of an adult rat at a resolution of $8 \times 8 \times 30\ nm^3$ using a multi-beam scanning electron microscope. After stitching and aligning the images on multi-CPU clusters, we obtained a final volume of 50 μm cube. We also apply deflickering, frame interpolation, and image de-striping techniques to improve image quality.

Annotation Quantity. All ROIs are annotated by three neuroscience experts, and we take the majority decision mapped back to the volume as the final label. In total, we obtain around 104K synapses, and mitochondria mask that occupies around 7.6% of the voxels. Compared to benchmarks like CREMI [11] and Lucchi [28] in connectomics, EM-R50 dataset is over 150× larger in image volume size, and over 100× larger in terms of the number of synapse instance.

Instance Diversity. To exhibit instance diversity, we show 3D meshes of all synapses and mitochondria within a subvolume (Fig. 5b). We use representative 2D slices for each 3D ROIs during annotation, and we show the variation of instance shape and orientation (Fig. 5c).

6   Experiments on Connectomics Datasets

We first benchmark our query suggestion method against others on the EM-R50 dataset for the ROI-level accuracy for synapse and mitochondria. Then we examine the design choices of the proposed method and the whole active learning pipeline through ablation studies, public benchmarks, and user studies.

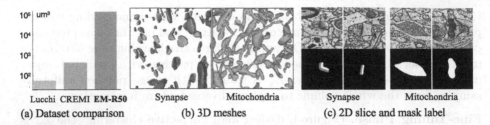

Fig. 5. EM-R50 connectomics dataset with dense synapse and mitochondria annotation. (a) We compare the size of the densely annotated image volume with other connectomics datasets (log-scale). To visualize the diversity of instance shape and orientation, we show (b) 3D meshes of synapses and mitochondria within a sub-volume, and (c) sample 2D image patches with corresponding mask annotations.

6.1   Comparing with State-of-the-Art Methods

Dataset and Metric. We randomly sample a subset of ROIs for synapses and mitochondria from EM-R50 for the benchmark experiments. The number of samples in the training-test split is 28.7K-10K and 20K-5K for synapse and mitochondria, respectively. We use the ROI proposal accuracy of the base model after fine-tuning as the evaluation metric for active learning methods.

Methods in Comparison. We compare our method with random uniform sampling, core-set [43], and learning-loss [53] approaches under a two-iteration scenario. For all the methods, after generating queries with a fixed annotation budget (1,280 instances, ≈5% of the training set), we use new labels to fine-tune the base network and evaluate the updated network prediction.

Results on Synapse. The initial accuracy of the network on the active learning test split is 0.811. We fine-tune the network using the instances suggested by different methods from the training set, where it has an initial accuracy of 0.762. To prevent overfitting, we construct mini-batches from the initial labeled volume with a ratio of 0.25. After first-round fine-tuning, the accuracy of the synapse proposals is increased to 0.892 for the test set, which outperforms random uniform sampling, core-set, and learning-loss (Table 1, Round 1). For our method, the new annotations are used to fine-tune the image encoder E_u. After another round of active clustering and fine-tuning the base model, the test accuracy is increased to 0.926, which shows that our method outperforms previous approaches significantly on synapse detection (Table 1, Round 2).

Results on Mitochondria. Since other structures like lysosome and artifacts caused by tissue staining and imaging look similar to mitochondria, the initial accuracy on the unlabeled ROIs is only 0.57. After applying our approach, the accuracy is improved to 0.809 on the test set, which outperforms core-set [43] method by 4.2% and the learning-loss [53] by 3.8% (Table 1). We observe that the accuracy improvement is relatively small for mitochondria at the second active

Table 1. Active learning performance comparison on our EM-R50 connectomics benchmark. Our two-stream query suggestion approach significantly out-perform previous methods in terms of the ROI proposal accuracy (higher is better).

Method	Synapse		Mitochondria	
	Round 1	Round 2	Round 1	Round 2
Random	0.824	0.871	0.704	0.749
Core-set [43]	0.847	0.895	0.726	0.767
Learning-loss[a] [53]	0.832	0.889	0.724	0.771
Two-stream (ours)	**0.892**	**0.926**	**0.802**	**0.809**

[a] Please check Sect. S-1 in the supplementary document for model details.

learning round. This may result from the intrinsic ambiguity of the mitochondria feature, where even human experts are not confident about the label.

Discussion. Despite the state-of-the-art performance of core-set [43] and learning-loss [53] on natural image benchmarks, those methods are less effective in handling the connectomics tasks due to two reasons. First, both methods use features from the supervised model, which can hardly capture the images features in the large unlabeled volume. Second, for the learning-loss approach, estimating the prediction loss with the global-average-pooling (GAP) module can ignore the useful structure information of objects. Nevertheless, we also compare the CIFAR-10 image classification benchmark (Sect. 7), where the methods are optimally tuned by the authors, for an even fairer comparison.

6.2 Ablation Analysis of Two-Stream Active Query Suggestion

In this part, we validate our design choices of the proposed *two-stream active clustering* module through ablation studies. Since the goal of query suggestion in active learning is to find the most "representative" samples for annotation, we perform the experiments to evaluate how different hyper-parameter and design choices influence the accuracy of annotation under a limited label budget.

Table 2. Comparison of design choices for two-stream clustering. We compute the object detection accuracy by assigning the labels of the cluster centers to other cluster members. The number of candidates per cluster, Q, is fixed to 5.

Description	Random		One-stream		Two-stream			
			Mask-only	Image-only	Late-fusion	Hierarchical		
E_s clusters (N)	–	–	128	256	1	1	–	64	128	64	32
E_u clusters (M)	–	–	1	1	128	256	–	2	2	4	8
Total num. (MN)	–	–	128	256	128	256	256	128	256	256	256
Annotation ratio (%)	2.23	4.46	2.23	4.46	2.23	4.46	4.46	2.23	4.46	4.46	4.46
Accuracy	0.767	0.772	0.805	0.819	0.420	0.578	0.738	0.821	0.826	**0.846**	0.814

Table 3. Comparison of design choices for active clustering. We show the accuracy w/ or w/o fine-tuning feature extractors. Fine-tuning only E_u shows the best performance while fine-tuning E_s can confuse the encoder, which leads to worse performance.

Active encoder	None	E_s	E_u	E_u and E_s
Accuracy	0.846	0.830	**0.880**	0.871

Dataset and Metric. We use the synapse benchmark dataset above to perform the ablation study. Suppose that after sliding window inference of the detection model, we have N proposed instances with an accuracy of p. Here p is the number of correct predictions over the total number of ROIs. By fixing the annotation budget s, the baseline accuracy is defined by the expectation of the accuracy that can be achieved by random one-by-one annotation, which is $p(1 - \frac{s}{N}) + \frac{s}{N}$. For example, with an initial accuracy of 0.7, randomly annotating 10% of the instances can improve the overall accuracy by 3%, since 70% of the queries are positive, and no errors can be corrected by annotating them. Then for evaluating the proposed methods, after annotating the cluster representatives in the clustering module, we assign the major representative labels to all samples in that cluster and calculate the label accuracy of the ROIs.

Effect of Two-Stream Clustering. We examine the active learning method accuracy with respect to the number of clusters and clustering architectures. Note that we fix the number of representatives $Q = 5$. Initially, for the instance proposals generated by the detection model, we assign 'correct' labels to all instances, and the accuracy is 0.762. As shown in Table 2, both with manual annotation of 4.46% of the data, random annotation can increase the accuracy by $4.46\% \times (1 - 0.762) \approx 0.01$, while our clustering module can increase the label accuracy by around 0.08 in absolute value. Besides, combining two-stream information with late fusion performs worse than the 'mask only' design. This is because the dimension of image embedding space is 1,000 to achieve reasonable reconstruction performance, which is much larger than the mask embedding space (20). Image embedding tends to dominate the result with direct concatenation and clustering using the same distance metric.

Effect of Active Clustering. We examine the effect of active clustering for the feature extractors E_u and E_s. There are three choices of the architectures, fine-tuning E_s only, fine-tuning E_u only, as well as fine-tuning both E_s and E_u. As indicated in Table 3, fine-tuning only E_s decreases the accuracy, because add supervision can distort the shape priors learned by the mask VAE; fine-tuning only E_u have a significant improvement over the static hierarchical baseline; fine-tuning both E_s and E_u decreases the E_u only performance, which further indicate that the shape information learned in E_u by self-supervision already contains distinguishable information that can be extracted from object masks. Therefore, we only fine-tuning E_u.

Table 4. Pixel-level evaluation on public connectomics datasets. For synapse, ours ranks 1st among results in publications on the CREMI dataset (left). For mitochondria, ours is on-par with state-of-the-art methods on the Lucchi dataset (right).

Synapse	CREMI ↓	ADGT ↓	ADF ↓		Mitochondria	VOC ↑
DTU1 [14]	72.21	106.31	38.11		Cheng [6]	0.942
DTU2 [14]	67.56	109.67	**25.46**		Lucchi [29]	**0.948**
Base model (**Ours**)	**63.92**	**97.64**	30.19		Base model (**Ours**)	0.937

6.3   Ablation Analysis of Active Learning Pipeline

Besides the evaluation of the proposed query suggestion method above, we examine the performance of the other two modules in the whole pipeline.

Model Prediction: Pixel-Level Evaluation. We provide pixel-level evaluations of the base model[1] to show its effectiveness on small benchmark datasets and indicate the necessity of active learning on large datasets. For synaptic cleft, we evaluate on the CREMI Challenge dataset [11], which contains 3 training and 3 test volumes of the size $1250 \times 1250 \times 125$ voxels. The results are evaluated by two scores: the average distance of any predicted cleft voxel to its closest ground-truth cleft voxel (ADGT) for penalizing false positives and the average distance of any ground-truth cleft voxel to its closest predicted cleft voxel (ADF) for penalizing false negatives. The final ranking criterion (CREMI score) is the mean of ADGT and ADF over the three test volumes. For mitochondria, we evaluate the model on the Lucchi dataset [30], which contains 1 training and 1 test volumes of size $1024 \times 768 \times 165$ voxels. We use the standard VOC score, which is the average of the Jaccard index of the foreground and background pixels.

For synaptic cleft, our proposed model outperforms previous leading methods by 5% and ranks 1st among published results on the public leaderboard (Table 4, left). For mitochondria, our model achieves comparable results to the previous state-of-the-art methods [6, 29], with ∼1% difference (Table 4, right). The results suggest that our base model is strong enough to enable a fair comparison of the following active learning methods on the large-scale benchmark dataset.

Model Prediction: Recall. As objects are sparse in the images, correcting false positive is much easier than finding false negatives. Thus we rebalance the loss and reject batches without foreground pixels with a 95% probability to heavily penalize false negatives as in Heinrich et al. [14]. In return, the instance-level recall for synapses on a fully labeled validation volume is 0.94 (IoU threshold is 0.5), which is adequate for the ROI-based active learning experiments.

Annotation: Query Display Order. To speed up the annotation, we sort the suggested query samples by their cluster indices and distance from their cluster centers. Such cluster-based query display order potentially allows participants to scan and identify false predictions faster, as patches with similar features are

---

[1] Architecture details are shown in Fig. S-1 in the supplementary document.

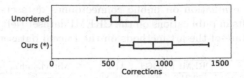

Fig. 6. User study on annotation throughput. The box plots show the median and interquartile range of the number of annotated instances in a fixed time frame of 30 min.

grouped closer than those in a random ordering. For evaluation, we performed a user study with novices as a single factor between-subjects experiment.

From the EM-R50 dataset, we randomly select 2.1K synapses, with 211 are false predictions. We recruited 20 novice participants and asked them to annotate as many synapses as possible within the 30-min time frame after a 10-min proper instruction on the task. Each participant was randomly assigned to either our clustering method or random ordering of the synapses.

Our clustering method allows study participants to annotate synapse with higher throughput, 930 ± 237 synapses, compared to the random order, 670 ± 224 (Fig. 6). Besides the efficiency improvement, the cluster-based query display order leads to a slight average accuracy improvement: for users with clustering 0.728 ± 0.087 compared to the random ordering with 0.713 ± 0.114.

7   Application to Natural Image Classification

The proposed two-stream active query suggestion can be applied to image classification in the active learning setting. Instead of the predicted mask encoded by a VAE model, we use the class label prediction as the supervised stream feature.

Dataset and Metric. CIFAR-10 has 60K images of size 32×32 pixels, with 50K for training and 10K for testing. Each image has a label from one of the ten classes. For evaluation, we use the top-1 classification accuracy on the test split.

Methods in Comparison. We use the same training protocol as Yoo et al. [53] for a fair comparison. For query suggestion methods, we compare with random uniform sampling, core-set [43], and learning-loss [53] approaches. For the active learning pipeline, we run a five-round comparison. We first uniformly sample 1K samples from the training set as the initial pool. After training the classification model, we apply different query suggestion approaches and label additional 1K samples from the unlabeled pool. Then we train the model from scratch again and conduct another round of query suggestion and labeling. We iterate the process until the total number of labeled samples reaches 5K.

Implementation Details. For classification, we use the same ResNet-18 model as Yoo et al. [53]. During training, we apply data augmentation, including random crop and horizontal flip, and image normalization. During active learning,

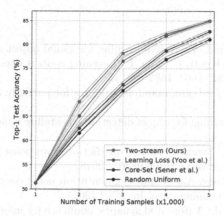

Fig. 7. Active learning results on the CIFAR-10 dataset. The accuracy improvement of our approach over previous state-of-the-art methods is most significant when training with a limited number of samples (2k and 3k out of total 50k images), similar to the annotation budget for EM-R50 ($\approx$5%). Mean and standard deviation are estimated from 5 runs. We also show that the accuracy saturates after ten iterations of query suggestion (Fig. S-4 in the supplementary material).

the number of training epochs is 200, and the mini-batch size is 64. The learning rate of the SGD optimizer is initially 0.1 and decreased to 0.01 after 160 epochs. For indicating the effectiveness of the unsupervised stream, we only use the two-stream clustering module of our query suggestion method. We pre-train the unsupervised stream feature with a VAE on all the training images with a latent dimension of 32. At the clustering phase, we fix the number of clusters at the output space and VAE latent space to be 50 and 20, respectively.

Results. Our proposed method outperforms the random uniform sampling and core-set methods, and is higher or comparable to the recent learning-loss approach (Fig. 7). Empirically, when the number of training samples is around 5% of the whole dataset (*i.e.*, 2K, and 3K out of 50K training images), our method achieves 2–3% improvement upon the learning-loss approach.

8   Conclusion

In this paper, we demonstrate the effectiveness of our proposed two-stream active query suggestion method for large-scale vision tasks in connectomics under the active learning setting. Besides the state-of-the-art results on the connectomics data, we show its applicability to a natural image classification benchmark. We evaluate each module of our active learning pipeline through public benchmarks, ablation studies, and user studies. As a use case, we build a connectomics dataset from a $(50 \, \mu m)^3$ cubic tissue with dense annotation of synapse and mitochondria.

Acknowledgment. This work has been partially supported by NSF award IIS-1835231 and NIH award 5U54CA225088-03.

References

1. Abramson, Y., Freund, Y.: Active learning for visual object detection (2006)
2. Becker, C., Ali, K., Knott, G., Fua, P.: Learning context cues for synapse segmentation. IEEE TMI **32**, 1864–1877 (2013)
3. Belkin, M., Niyogi, P.: Using manifold stucture for partially labeled classification. In: NIPS (2003)
4. Bietti, A.: Active learning for object detection on satellite images. Technical report, Caltech (2012)
5. Buhmann, J., et al.: Synaptic partner prediction from point annotations in insect brains. arXiv preprint arXiv:1806.08205 (2018)
6. Cheng, H.C., Varshney, A.: Volume segmentation using convolutional neural networks with limited training data. In: ICIP (2017)
7. Dorkenwald, S., et al.: Automated synaptic connectivity inference for volume electron microscopy. Nat. Methods **14**(4), 435 (2017)
8. Ducoffe, M., Precioso, F.: Adversarial active learning for deep networks: a margin based approach. In: ICML (2018)
9. Dutt Jain, S., Grauman, K.: Active image segmentation propagation. In: CVPR (2016)
10. Freytag, A., Rodner, E., Denzler, J.: Selecting influential examples: active learning with expected model output changes. In: Fleet, D., Pajdla, T., Schiele, B., Tuytelaars, T. (eds.) ECCV 2014. LNCS, vol. 8692, pp. 562–577. Springer, Cham (2014). https://doi.org/10.1007/978-3-319-10593-2_37
11. Funke, J., Saalfeld, S., Bock, D., Turaga, S., Perlman, E.: Circuit reconstruction from electron microscopy images (2016). https://cremi.org
12. He, K., Gkioxari, G., Dollár, P., Girshick, R.: Mask R-CNN. In: Proceedings of the IEEE International Conference on Computer Vision, pp. 2961–2969 (2017)
13. He, K., Zhang, X., Ren, S., Sun, J.: Deep residual learning for image recognition. In: CVPR (2016)
14. Heinrich, L., Funke, J., Pape, C., Nunez-Iglesias, J., Saalfeld, S.: Synaptic cleft segmentation in non-isotropic volume electron microscopy of the complete *Drosophila* brain. In: Frangi, A.F., Schnabel, J.A., Davatzikos, C., Alberola-López, C., Fichtinger, G. (eds.) MICCAI 2018. LNCS, vol. 11071, pp. 317–325. Springer, Cham (2018). https://doi.org/10.1007/978-3-030-00934-2_36
15. Hoffer, E., Ailon, N.: Deep metric learning using triplet network. In: International Workshop on Similarity-Based Pattern Recognition (2015)
16. Hu, J., Shen, L., Sun, G.: Squeeze-and-excitation networks. In: CVPR (2018)
17. Huang, G.B., Plaza, S.: Identifying synapses using deep and wide multiscale recursive networks. arXiv preprint arXiv:1409.1789 (2014)
18. Huang, G.B., Scheffer, L.K., Plaza, S.M.: Fully-automatic synapse prediction and validation on a large data set. arXiv preprint arXiv:1604.03075 (2016)
19. Jagadeesh, V., Anderson, J., Jones, B., Marc, R., Fisher, S., Manjunath, B.: Synapse classification and localization in electron micrographs. Pattern Recogn. Lett. **43**, 17–24 (2014)
20. Kao, C.-C., Lee, T.-Y., Sen, P., Liu, M.-Y.: Localization-aware active learning for object detection. In: Jawahar, C.V., Li, H., Mori, G., Schindler, K. (eds.) ACCV 2018. LNCS, vol. 11366, pp. 506–522. Springer, Cham (2019). https://doi.org/10.1007/978-3-030-20876-9_32
21. Kasthuri, N., et al.: Saturated reconstruction of a volume of neocortex. Cell **162**(3), 648–661 (2015)

22. Kingma, D.P., Welling, M.: Auto-encoding variational bayes. ICLR (2013)
23. Kreshuk, A., Funke, J., Cardona, A., Hamprecht, F.A.: Who is talking to whom: synaptic partner detection in anisotropic volumes of insect brain. In: Navab, N., Hornegger, J., Wells, W.M., Frangi, A.F. (eds.) MICCAI 2015. LNCS, vol. 9349, pp. 661–668. Springer, Cham (2015). https://doi.org/10.1007/978-3-319-24553-9_81
24. Kreshuk, A., Koethe, U., Pax, E., Bock, D.D., Hamprecht, F.A.: Automated detection of synapses in serial section transmission electron microscopy image stacks. PLoS One **9**(2), e87351 (2014)
25. Kreshuk, A., et al.: Automated detection and segmentation of synaptic contacts in nearly isotropic serial electron microscopy images. PLoS One **6**(10), e24899 (2011)
26. Lichtman, J.W., Sanes, J.R.: Ome sweet ome: what can the genome tell us about the connectome? Curr. Opin. Neurobiol. **18**(3), 346–353 (2008)
27. Lucchi, A., Li, Y., Fua, P.: Learning for structured prediction using approximate subgradient descent with working sets. In: CVPR (2013)
28. Lucchi, A., Li, Y., Smith, K., Fua, P.: Structured image segmentation using kernelized features. In: Fitzgibbon, A., Lazebnik, S., Perona, P., Sato, Y., Schmid, C. (eds.) ECCV 2012. LNCS, vol. 7573, pp. 400–413. Springer, Heidelberg (2012). https://doi.org/10.1007/978-3-642-33709-3_29
29. Lucchi, A., et al.: Learning structured models for segmentation of 2D and 3D imagery. IEEE TMI **34**, 1096–1110 (2015)
30. Lucchi, A., Smith, K., Achanta, R., Knott, G., Fua, P.: Supervoxel-based segmentation of mitochondria in EM image stacks with learned shape features. IEEE TMI **31**, 474–486 (2012)
31. Morgan, J.L., Lichtman, J.W.: Why not connectomics? Nat. Methods **10**(6), 494 (2013)
32. Narasimha, R., Ouyang, H., Gray, A., McLaughlin, S.W., Subramaniam, S.: Automatic joint classification and segmentation of whole cell 3D images. Pattern Recogn. **42**, 1067–1079 (2009)
33. Oztel, I., Yolcu, G., Ersoy, I., White, T., Bunyak, F.: Mitochondria segmentation in electron microscopy volumes using deep convolutional neural network. In: Bioinformatics and Biomedicine (2017)
34. Parag, T., et al.: Detecting synapse location and connectivity by signed proximity estimation and pruning with deep nets. arXiv preprint arXiv:1807.02739 (2018)
35. Parag, T., Ciresan, D.C., Giusti, A.: Efficient classifier training to minimize false merges in electron microscopy segmentation. In: ICCV (2015)
36. Perez, A.J., et al.: A workflow for the automatic segmentation of organelles in electron microscopy image stacks. Front. Neuroanat. **8**, 126 (2014)
37. Plaza, S.M., Parag, T., Huang, G.B., Olbris, D.J., Saunders, M.A., Rivlin, P.K.: Annotating synapses in large EM datasets. arXiv preprint arXiv:1409.1801 (2014)
38. Ren, S., He, K., Girshick, R., Sun, J.: Faster R-CNN: towards real-time object detection with region proposal networks. In: Advances in Neural Information Processing Systems, pp. 91–99 (2015)
39. Roncal, W.G., et al.: VESICLE: volumetric evaluation of synaptic interfaces using computer vision at large scale. arXiv preprint arXiv:1403.3724 (2014)
40. Ronneberger, O., Fischer, P., Brox, T.: U-net: convolutional networks for biomedical image segmentation. In: Navab, N., Hornegger, J., Wells, W.M., Frangi, A.F. (eds.) MICCAI 2015. LNCS, vol. 9351, pp. 234–241. Springer, Cham (2015). https://doi.org/10.1007/978-3-319-24574-4_28
41. Roy, S., Namboodiri, V.P., Biswas, A.: Active learning with version spaces for object detection. arXiv preprint arXiv:1611.07285 (2016)

42. Scheffer, T., Decomain, C., Wrobel, S.: Active hidden Markov models for information extraction. In: Hoffmann, F., Hand, D.J., Adams, N., Fisher, D., Guimaraes, G. (eds.) IDA 2001. LNCS, vol. 2189, pp. 309–318. Springer, Heidelberg (2001). https://doi.org/10.1007/3-540-44816-0_31
43. Sener, O., Savarese, S.: Active learning for convolutional neural networks: a core-set approach. In: ICLR (2018)
44. Settles, B.: Active learning literature survey. Technical report, University of Wisconsin-Madison Department of Computer Sciences (2009)
45. Settles, B.: Active learning literature survey. 2010. Computer Sciences Technical Report (2014)
46. Seyedhosseini, M., Ellisman, M.H., Tasdizen, T.: Segmentation of mitochondria in electron microscopy images using algebraic curves. In: ISBI, pp. 860–863. IEEE (2013)
47. Simonyan, K., Zisserman, A.: Two-stream convolutional networks for action recognition in videos. In: NIPS (2014)
48. Sivaraman, S., Trivedi, M.M.: Active learning for on-road vehicle detection: a comparative study. Mach. Vis. Appl. **25**, 599–611 (2014)
49. Staffler, B., Berning, M., Boergens, K.M., Gour, A., van der Smagt, P., Helmstaedter, M.: SynEM, automated synapse detection for connectomics. Elife (2017)
50. Vazquez-Reina, A., Gelbart, M., Huang, D., Lichtman, J., Miller, E., Pfister, H.: Segmentation fusion for connectomics. In: ICCV (2011)
51. Vijayanarasimhan, S., Grauman, K.: Large-scale live active learning: training object detectors with crawled data and crowds. IJCV **108**, 97–114 (2014)
52. Wang, K., Zhang, D., Li, Y., Zhang, R., Lin, L.: Cost-effective active learning for deep image classification. IEEE TCSVT **27**, 2591–2600 (2017)
53. Yoo, D., Kweon, I.S.: Learning loss for active learning. In: Proceedings of the IEEE Conference on Computer Vision and Pattern Recognition, pp. 93–102 (2019)
54. Yu, K., Bi, J., Tresp, V.: Active learning via transductive experimental design. In: Proceedings of the 23rd International Conference on Machine Learning, pp. 1081–1088 (2006)
55. Zhang, Y., Lease, M., Wallace, B.C.: Active discriminative text representation learning. In: AAAI (2017)
56. Zheng, Z., et al.: A complete electron microscopy volume of the brain of adult drosophila melanogaster. Cell **174**, 730–743 (2018)
57. Zhu, X., Ghahramani, Z., Lafferty, J.D.: Semi-supervised learning using Gaussian fields and harmonic functions. In: ICML (2003)

Pix2Surf: Learning Parametric 3D Surface Models of Objects from Images

Jiahui Lei[1]([✉]), Srinath Sridhar[2], Paul Guerrero[3], Minhyuk Sung[3], Niloy Mitra[3,4], and Leonidas J. Guibas[2]

[1] Zhejiang University, Hangzhou, China
lei_jiahui@zju.edu.cn
[2] Stanford University, Stanford, USA
[3] Adobe Research, London, UK
[4] University College London, London, UK

Abstract. We investigate the problem of learning to generate 3D parametric surface representations for novel object instances, as seen from one or more views. Previous work on learning shape reconstruction from multiple views uses discrete representations such as point clouds or voxels, while continuous surface generation approaches lack multi-view consistency. We address these issues by designing neural networks capable of generating *high-quality parametric 3D surfaces* which are also *consistent* between views. Furthermore, the generated 3D surfaces preserve accurate image pixel to 3D surface point *correspondences*, allowing us to lift texture information to reconstruct shapes with rich geometry *and* appearance. Our method is supervised and trained on a public dataset of shapes from common object categories. Quantitative results indicate that our method significantly outperforms previous work, while qualitative results demonstrate the high quality of our reconstructions.

Keywords: 3D reconstruction · Multi-view · Single-view · Parametrization

1 Introduction

Reconstructing the 3D shape of an object from one or more views is an important problem with applications in 3D scene understanding, robotic navigation or manipulation, and content creation. Even with multi-view images, the problem can be challenging when camera baselines are large, or when lighting and occlusions are inconsistent across the views. Recent developments in supervised deep learning have demonstrated the potential to overcome these challenges.

https://geometry.stanford.edu/projects/pix2surf.

Electronic supplementary material The online version of this chapter (https://doi.org/10.1007/978-3-030-58523-5_8) contains supplementary material, which is available to authorized users.

Fig. 1. Pix2Surf learns to generate a continuous parametric 3D surface of an object seen in one or more views. Given a single image, we can reconstruct a continuous partial 3D shape (top row). When multiple views are available, we can aggregate the views to form a set of consistent 3D surfaces (bottom row). Our reconstructions preserve 2D pixel to 3D shape correspondence that allows the transport of textures, even from real images (last column).

Ideally, a multi-view surface reconstruction algorithm should have the following desirable **3C** properties: surface **continuity, multi-view consistency** and 2D-3D **correspondence**. First, it should be able to reconstruct high-quality shapes that can be readily used in downstream applications. While much progress has been made in learning shape representations such as point clouds [9,15,23,37], volumetric grids [7,39,40], and meshes [42,45], their geometric quality is limited by the discrete nature of the underlying representation. Therefore, representations such as implicit functions [6,32,35], and *UV* surface parametrizations [8,13] are preferable, since they can represent a **continuous surface** at arbitrary resolution. Second, the algorithm should be able to reconstruct objects from a sparse set of views while ensuring that the combined shape is **consistent** across the views. Recent approaches exploit geometric constraints to solve this problem but require additional supervision through knowledge of the exact camera geometry [5]. Finally, the algorithm should provide accurate **correspondences** between 2D pixels and points on the 3D shape, so as to accurately transport object properties (e.g., texture) directly from 2D and support aggregation across views. While some extant methods satisfy a subset of these properties, we currently lack any method that has all of them.

In this paper, we present **Pix2Surf**, a method that learns to reconstruct *continuous* and *consistent* 3D surface from single or multiple views of novel object instances, while preserving accurate 2D–3D *correspondences*. We build upon recent work on category-specific shape reconstruction using *Normalized Object Coordinate Space (NOCS)* [37,41], which reconstructs the 3D point cloud as a *NOCS map* – an object-centered depth map – in a canonical space that is in accurate correspondence with image pixels. Importantly, NOCS maps do not require knowledge of camera geometry. However, these maps do not directly encode the underlying surface of the object. In this paper, we present a method that incorporates a representation of the underlying surface by predicting a continuous parametrization that maps a learned *UV parameter space* to 3D NOCS coordinates, similar in spirit to AtlasNet [13]. Unlike AtlasNet, however, our approach also provides accurate 2D–3D correspondences and an emergent learned chart that can be used to texture the object directly from the input image.

When multiple views of an object are available, we also present a version of Pix2Surf that is capable of reconstructing an object by predicting an atlas, i.e., view-specific charts assembled to form the final shape. While in the NOCS approach [37] individual views can also be directly aggregated since they live in the same canonical space, this naïve approach can lead to discontinuities at view boundaries. Instead, for view-consistent reconstruction, we aggregate multiple views at the feature level and explicitly enforce consistency during training.

Extensive experiments and comparisons with previous work show that Pix2Surf is capable of reconstructing high-quality shapes that are consistent within and across views. In terms of reconstruction error, we outperform state-of-the-art methods while maintaining the 3C properties. Furthermore, accurate 2D–3D correspondences allow us to texture the reconstructed shape with rich color information as shown in Fig. 1. In summary, the primary contributions of our work are:

- a method to generate a set of **continuous** parametric 3D surfaces representing the shape of a novel object observed from single or multiple views;
- the unsupervised extraction of a learned UV parametrization that retains accurate 2D to 3D surface point **correspondences**, allowing lifting of texture information from the input image; and
- a method to **consistently** aggregate such parametrizations across different views, using multiple charts.

Emergent Properties: A notable emergent property of our network is that the learned UV parametrization domains are consistent across different views of the same object (i.e., corresponding pixels in different views have similar UV coordinates) – and even across views of related objects in the same class. This is despite the UV domain maps only being indirectly supervised for consistency, through 3D reconstruction.

Scope: In this work, our focus is on *continuity, consistency, and 2D image–3D surface correspondences*. We focus on the case when the multi-view images have little overlap, a setting where traditional stereo matching techniques fail. Our method only requires supervision for the input views and their corresponding NOCS maps but does not require camera poses or ground truth UV parametrization. We note that the generated surfaces need not be watertight, and continuity at the seams between views is not guaranteed.

2   Related Work

There is a large body of work on object reconstruction which we categorize broadly based on the underlying shape representation.

Voxels: The earliest deep-learning-based methods predict a voxel representation of an object's shape. Many of these methods are trained as generative models for 3D shapes, with a separate image encoder to obtain the latent code for a given image [12]. Later methods use more efficient data structures, such as octrees [34,38,43] to alleviate the space requirements of explicit voxels. Multiple views can also be aggregated into a voxel grid using a recurrent network [7]. Several methods use supervision in the form of 2D images from different viewpoints, rather than a 3D shape, to perform both single-view and multi-view voxel reconstruction [14,17,40,47]. These methods usually use some form of a differentiable voxel renderer to obtain a 2D image that can be compared to the ground truth image. The quality gap of these methods to their counterparts that use 3D supervision is still quite large. Voxels only allow for a relatively coarse representation of a shape, even with the more efficient data representations. Additionally, voxels do not explicitly represent an object's surface prompting the study of alternative representations.

Point Clouds: To recover the point cloud of an object instance from a single view, methods with 3D supervision [9,23] and without 3D supervision [15] have been proposed. These methods encode the input image into a latent code thus losing correspondences between the image pixels and the output points. Some methods establish a coarse correspondence implicitly by estimating the camera parameters, but this is typically inaccurate. A recent method reconstructs a point cloud of a shape from multiple views [5], but requires ground truth camera parameters. A large body of monocular or stereo depth estimation methods obtain a point cloud for the visible parts of the scene in an image, but do not attempt to recover the geometry of individual object instances in their local coordinate frames [3]. NOCS [37,41] obtains exact correspondences between 2D pixels and 3D points by predicting the 3D coordinates of each pixel in a canonical coordinate frame. NOCS can even be extended to reconstruct unseen parts of an object [37] (X-NOCS). All these approaches that output point clouds do not describe the connectivity of a surface, which has to be extracted separately – a classical and difficult geometry problem. We extend NOCS to directly recover continuous surfaces and consistently handle multiple views.

Implicit Functions: Poisson Surface Reconstruction [20,21] has long been the gold standard for recovering an implicit surface from a point cloud. More recently, data-driven methods have been proposed that model the implicit function with a small MLP [6,29,32], with the implicit function representing the occupancy probability or the distance to the surface. These methods can reconstruct an implicit function directly from a single image, but do not handle multiple views and do not establish a correspondence between pixels and the 3D space. PiFU [35] and DISN [46] are more recent methods that establish a correspondence between pixels and 3D space and use per-pixel features to parameterize an implicit function. Both single and multiple views can be handled, but the methods either require ground truth camera poses as input [35], or use a network to get a coarse approximation of the camera poses, giving only approximate correspondences [46]. Some recent works integrate the neural rendering with deep implicit functions [27,30], but they depend on the known camera information. Furthermore, to obtain an explicit surface from an implicit function, an expensive post-processing step is needed, such as Marching Cubes [28] or ray tracing.

Parametric Surfaces or Templates: Several methods attempt to directly reconstruct a parametric representation of a shape's surface. These parametric representations range from class-specific templates [16, 22], general structured templates [11], or more generic surface representations, such as meshes or continuous functions. Pixel2Mesh and its sequel [42, 45] deform a genus-zero mesh based on local image features at each vertex, obtained by projecting the vertices to the image plane(s). Camera parameters are assumed to be known for this projection. 3DN [44] deforms a given source mesh to approximate a single target image, using global features for both the source and the target, without establishing correspondences to the target pixels. Several methods use 2D images instead of 3D meshes as supervisory signal [18,19,26,33] using differentiable mesh renderers. This makes it easier to collect training data, but the accuracy of these methods still lags behind methods with 3D supervision. Atlas-Net [13] represents shapes with continuous 2D patches that can be inferred from a single input image, or from a video clip [24]. Mesh DeformNet [31] introduces topology modification to AtlasNet. Similar to AtlasNet, we use a 2D patch as a UV parametrization, but we handle multiple non-adjacent views and establish correspondences between 2D pixels and 3D surface points.

3   Preliminaries

We build our approach upon two previous ideas that we describe below.

(X-)NOCS: Normalized object coordinate space (NOCS) is a canonicalized unit container space used for category-level reasoning of object pose, size, and shape [37,41]. Instances from a given object category are *normalized* for their position, orientation, and size, thus disentangling intra-category shape variation from the exact pose and size of instances. NOCS maps (see Fig. 2) are perspective projections of the 3D NOCS shape onto a specific camera and can be interpreted as **object-centered depth maps** that simultaneously encode mask and partial shape of the object. When used to predict 3D point cloud from images, NOCS maps retain correspondences from 2D pixels to 3D points, and can be used to transport image texture directly to 3D. X-NOCS is an extension of NOCS maps to also encode the occluded parts of a shape [37]. However, using NOCS maps for reconstruction results in a discontinuous point cloud.

Surface Parametrization: A two-manifold surface in 3D can be mapped to a 2D plane (*chart*) parametrized by two coordinates (u, v). This UV parametrization of a 3D surface is widely used in computer graphics and, more recently, in 3D shape reconstruction [13,22]. The parameterization can be limited in expressing complex shapes, depending on the functional formulation used.

For example, in typical CAD settings, low-degree polynomial or rational functions are used to represent the mappings. In our case, instead, we use a fully connected network to overcome the limitation of expressibility. A single map, however, still lacks the ability to describe complicated shapes with high-genus topology. Thus, multiple charts are often used, where multiple 2D planar patches are mapped by separate maps to a 3D surface – effectively partitioning the surfaces into parts, each of which is the image of a different map in the chart. We show how a single chart can be used

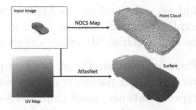

Fig. 2. Given a single image, X-NOCS [37] reconstructs a point cloud preserving pixel–3D correspondece. AtlasNet [13] learns shape as a continuous surface.

for 3D shape reconstruction while multiple charts allow consistent reconstruction over multiple views while still preserving pixel to 3D correspondence.

4   Pix2Surf: Pixels to Surface

Our goal is to predict a continuous and consistent parametric 3D surface for a novel object instance observed from one or more views. The word "continuous" parametric surfaces in our method refers to parametric C^0 continuity (similar to AtlasNet [13]), i.e., any continuous trajectory over the chart space maps to a continuous curve in 3D space. Additionally, we would like to preserve correspondences between 2D pixels and 3D surface points. We first describe our approach for reconstructing a 3D surface from a single image using a single chart, and then generalize it to multiple views using an atlas.

4.1   Single-View Single-Chart Pix2Surf

At inference time, the single-view version of Pix2Surf takes an RGB image of an object observed from an arbitrary camera as input. We use a CNN to extract image features that compactly encode object shape. The features are then processed by two branches: (1) the **NOCS-UV branch** is a CNN that estimates a mask, a learned UV map, and a NOCS map and (2) the **Surface Parametrization (SP) branch** is an MLP that generates a continuous 3D surface. This single-view, single-chart architecture is shown in Fig. 3.

(1) NOCS-UV Branch: Similar to X-NOCS [37], we predict the NOCS map and mask that encode the partial shape of the object observed in the image. We use an encoder-decoder architecture building on top of SegNet [1] and VGG [36]. Our network uses skip connections and shares pool indices between the encoder and the decoder. The predicted NOCS maps and masks are the same size as the input image. During training, the object mask is supervised with a binary

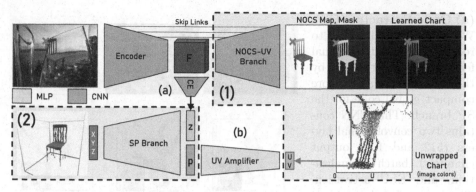

Fig. 3. Single-View Single-chart Pix2Surf network architecture. The input image is processed using an encoder-decoder architecture to predict a NOCS map, object mask, and a learned chart (top right two channels, color coded). The Surface Parameterization branch takes sampled and amplified chart coordinates **p** and a latent image code **z** to predict the final continuous surface. Unwrapped chart (bottom right) refers to a visualization of foreground colors using the predicted two-channel learned chart (top right) as coordinate. The colors of the input image can be transported to all intermediate steps ($\times$ and arrows). (Color figure online)

cross entropy loss and the NOCS map is supervised with an L^2 loss. Note that the NOCS map here is *not* our final 3D output, but acts as an intermediate supervision signal for the network.

Emergence of a Chart: Different from previous work, we predict a 2-channel output in additional to the NOCS map and mask. These 2 channels are not explicitly supervised during training, so the network can predict any value between 0 and 1. However, when jointly trained with the other branches, we observe the **emergence** of a *learned chart* in these 2 channels (see Fig. 4). The network discovers how to unwrap an object shape onto a flat surface. **Remarkably, this learned chart is (almost) consistent across multiple views and even across instances.** During reconstruction, each image pixel's learned chart coordinates are passed on to the SP branch. We show that using the learned chart coordinates is superior to using arbitrary UV coordinates like AtlasNet [13], or alternatively using the original image coordinates (Image2Surf, Sect. 5.1). Additionally, we preserve exact correspondences between input image pixels and the learned chart.

(a) Code Extractor (CE): We use a small CNN to reduce the high dimensional feature map extracted by the encoder to make a more compact global code for the SP branch. This CNN contains two convolutional layers (512 and 1024 output channels), batch normalization, and ELU activation. The output is a latent code z of size 1024 and is passed to the SP branch.

(b) UV Amplifier: Before we use the learned chart coordinates as an input to the SP branch, we process each UV coordinate with a *UV amplifier* MLP. The motivation for this comes from the information imbalance the two inputs to the

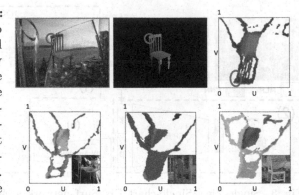

Fig. 4. Given an object image (row 1, col 1), our network predicts a 2-channel image without explicit supervision (row 1, col 2, color coded). Remarkably, the output of these two channels visualized in a *UV* space (row 1, col 3) show that the network has learned to unwrap the 3D shape onto a plane (corresponding patches shown in red). This unwrapping is consistent over multiple views, and even across multiple object instances (last row). For more unwrapped charts of cars and airplanes please see supplementary Fig. S.3. (Color figure online)

SP branch – one input is the global latent code z which has 1024 dimensions, while the UV coordinates would have only 2 dimensions. To overcome this, we *amplify* the UV coordinates to p (256 dimensions) using a 3-layer MLP that progressively amplifies the 2 coordinates (2, 64, 128, 256). This allows the SP branch to make use of the image and UV information in a more balanced manner.

(2) SP Branch: Similar to AtlasNet [13], our surface parametrization (SP) branch takes the global latent code z from the code extractor (CE) and the amplified coordinates p as input and produces a continuous 3D position as the output. Note that the learned chart coordinates can be continuously sampled at inference time. The continuity of the output 3D surface emerges from our use of a continuous MLP mapping function between the uv coordinates and the output 3D positions [13]. Our SP branch is a MLP with 9 layers and skip connection every 2 layers (input: 1024+256, intermediate: 512, last: 3). Since we train on canonically oriented ShapeNet models, the predicted 3D surface also lies within the canonical NOCS container [41].

Our approach has three key differences to AtlasNet. First, we use a UV amplifier to transform the 2D *UV* coordinates to higher dimensions allowing better information balancing. Second, the learned chart is in direct correspondence with the pixels of the input image (see Fig. 4). This allows us to transport appearance information directly from the image to the 3D surface. Third, our sampling of the *UV* chart is learned by a network (NOCS-UV branch) instead of uniform sampling, which enables us to reconstruct complex topologies. Our inference processing allows us to sample any continuous point in the learned chart space within the predicted object mask allowing the generation of **continuous textured 3D surface**.

Training: The encoder and decoder CNNs are first initialized by training them on the NOCS map and mask prediction tasks using only the L^2 and BCE losses. Subsequently, we jointly train the NOCS-UV and SP branches, code extractor, and UV amplifier end-to-end. The joint loss is given as,

$$\mathcal{L}_I = w_1 \left(w_n \, \mathcal{L}_n + w_m \, \mathcal{L}_m \right) + w_2 \, \mathcal{L}_s, \tag{1}$$

where $\mathcal{L}_n$ and $\mathcal{L}_m$ are the L^2 NOCS map and BCE losses respectively, w_n, w_m are the weights for the NOCS map and mask prediction respectively, and w_1, w_2 are the weights for the NOCS-UV and SP branches respectively. For the SP branch we supervise on K points sampled randomly from within the foreground mask. For each sampled point, a corresponding amplified chart coordinate **p** is predicted without any supervision. This is concatenated with the global latent code **z** to predict the final 3D surface position. Empirically, we found the best hyperparameters to be: $K = 4096, w_1 = 0.1, w_2 = 0.9, w_n = 0.7, w_m = 0.3$. The loss for the SP branch is given as, $L_s = \frac{1}{K} \sum_{i=1}^{K} \|\mathbf{x}_i - \hat{\mathbf{x}}_i\|_2$, where $\mathbf{x}$ and $\hat{\mathbf{x}}$ are the ground truth and predicted 3D surface position obtained from the 3D ShapeNet models (same as ground truth NOCS map values). During inference, we can predict a continuous 3D surface for any given image and its learned chart coordinate. Please see the supplementary document for more details on inference and final 3D model generation.

4.2 Multi-view Atlas Pix2Surf

The method described above is suitable when we have a single view of the object. When multiple views are available, we could naively extend the single view network and combine the generated surfaces using a union operation. However, this leads to sharp discontinuities (Fig. 5). To overcome this issue, we propose a generalization of our single-view single-chart method to consistently aggregate 2D surface information from multiple views, using an *atlas* i.e., a separate learned chart (UV map) for each view. Figure 6 shows an overview of our multi-view network. This design shares similarities with the single view network but has additional multi-view consistency which is enforced both at the feature level

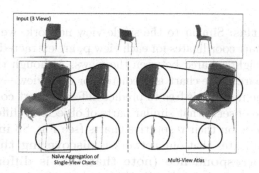

Fig. 5. Given 3 views, naïve aggregation of individual charts leads to discontinuities or double surfaces (left). Our multi-view atlas method produces more consistent surfaces (right), for instance, at the legs and backrest.

through a feature pooling step, and using a consistency loss for better 3D surface generation.

Multi-view Feature Pooling: The goal of this step is to promote multi-view information sharing at the *feature level* (see Fig. 6). Different from the single view network, the latent codes z_i extracted for each view i (using a shared encoder and code extractor) are maxpooled into a common shared multi-view latent code z_m. Intuitively, this shared latent code captures the most salient information from each view.

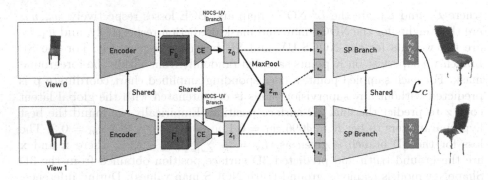

Fig. 6. Multi-view atlas network architecture. The multi-view network allows multiple charts to be consistently aggregated. This network has two main features: (1) the MaxPool operation to pool features between views, and (2) a multi-view consistency loss $\mathcal{L}_C$ that ensures corresponding points produce 3D surface points that are nearby. Only two views are shown in this figure, but we use multiple views during training and inference. The encoder, NOCS-UV branch, CE branch, and SP branches share weights.

Atlas: Similar to the single view network, we learn a chart for each view. The chart coordinates for each view p_i are extracted using the NOCS-UV branch with weights shared between the views. Although the NOCS-UV branch weights are shared, one chart is predicted for each view – thus, we have an atlas. Note that the network is free to predict different chart coordinates for each view. However, we observe that similar parts of objects in different images map to similar locations on their respective charts (see Fig. S3 in supplementary document). This indicates that our network is **discovering the notion of image cross-view correspondence (note that this is different from 2D-3D correspondence).** As in the single-view network, chart coordinates are passed through a shared UV amplifier.

We concatenate the shared latent code z_m to each of the per-view latent codes z_i. This concatenated multi-view code and the learned per-view chart coordinates p_i are passed to the SP branch. The UV amplifier, code extractor and structure of the learned UV map are similar to the single view network.

Multi-view Loss: In addition to the L^2 loss function on the 3D surface generated by the SP branch, we also have a multi-view consistency loss. This loss enforces corresponding points on multiple views to predict similar 3D surface positions. To obtain correspondence information at training time, we sample a random set of foreground points within the mask and find the exact match of the ground truth NOCS values of that pixel in the other input views. Note that this correspondence information is *not* provided as additional supervision – the ground truth NOCS maps already contain this information since corresponding points multiple views have the same NOCS position. Given these correspondences, the multi-view consistency loss for a pair of views is given as, $\mathcal{L}_C = \frac{1}{|\mathcal{P}|} \sum_{(i,j) \in \mathcal{P}} \|\mathbf{x}_i - \mathbf{x}_j\|_2$, where $\mathbf{x}_{i,j}$ are the paired predicted xyz from two different views and the set $\mathcal{P}$ contains all matched correspondence pair from these two views. During training, within each mini-batch, we sample multiple views per object and compute the loss for all possible pairs.

Training: The multi-view network is trained similar to the single view model. The NOCS-UV branch is first trained and subsequently the whole network is trained end-to-end. The loss function we use is $\mathcal{L}_M = \mathcal{L}_I + \frac{w_3}{a} \sum_{j=0}^{a} \mathcal{L}_C$, where a denotes the number of pairs of views within that batch, and w_3 is the correspondence loss weight empirically set to 0.9. We set w_n, w_m to 0.1 inside $\mathcal{L}_I$. Please see the supplementary document for more details on inference and final 3D model generation.

5 Experiments

We present extensive experimental comparison of Pix2Surf with several recent single- and multi-view reconstruction methods, and validate our design choices. We do so by focusing on the 3C properties (consistency, correspondence and continuity) for visible surface reconstruction (Sect. 5.1). Since we learn a strong prior over shapes, we can also estimate surfaces that are hidden in the input image (Sect. 5.2). For the details of training, inference, and evaluation metrics, and also ablations, more comparisons, and results with real images, refer to the supplementary document.

Dataset: For quantitative comparisons, we use ShapeNetPlain [37] dataset which consists of 5 random views for each shape in ShapeNet [4] with a white background. For additional robustness to the background found in real-world images, we train Pix2Surf on ShapeNetCOCO [37] which consists of 20 random views of each ShapeNet object with a random background from MS COCO [25]. We use this dataset for all qualitative results and for real-world results. Each shape category is trained separately in all experiments.

Experimental Setting: We follow the experimental setup of X-NOCS [37]. Our ground truth for each input image is the point cloud represented by the NOCS map (or X-NOCS map for hidden surface) provided in the dataset [37]. The outputs of all methods are converted to a NOCS map (using the ground truth camera pose) allowing us to compute metrics even for partial shapes. Multi-view experiments use all 5 views in the dataset to reconstruct a surface, using the same dataset as the single-view experiments. All metrics are computed per-view

Table 1. Visible surface reconstruction. We compare our method to a baseline and three state-of-the-art methods evaluating reconstruction accuracy and the 3C properties. The top half of the table shows single-view reconstruction, the bottom half is multi-view reconstruction. Note how Pix2Surf is close to the top performance in each of the metrics, while all other methods have significant shortcomings. The **Recons. Error** and **Correspond. Error, Consistency Error** are all multiplied by 10^3.

	Recons. Error ↓				Correspond. Error ↓				Consistency Error ↓				Disconti. Score ↑			
	car	chair	plane	avg.	car	chair	plane	avg.	car	chair	plane	avg.	car	chair	plane	avg.
Im.2Surf	2.23	3.81	2.66	2.90	**8.49**	9.54	8.76	8.93	13.08	12.55	10.75	12.13	0.46	0.39	0.35	0.40
X-NOCS	2.25	2.95	2.08	2.43	12.82	8.63	8.93	10.13	18.93	12.00	10.59	13.84	0.59	**0.47**	0.59	0.55
AtlasNet	**1.54**	3.36	3.15	2.68	–	–	–	–	–	–	–	–	0.68	0.39	0.64	0.57
Pix2Surf	1.67	**1.91**	**1.61**	**1.73**	9.52	**5.79**	7.19	**7.50**	**12.72**	**7.75**	**8.48**	**9.65**	**0.69**	0.43	**0.65**	**0.59**
X-NOCS	2.89	2.80	2.19	2.63	14.30	9.48	8.95	10.91	22.18	14.26	11.65	16.03	**0.67**	**0.48**	0.54	0.56
P2M++	2.88	5.59	3.24	3.90	–	–	–	–	–	–	–	–	0.67	0.36	0.63	0.55
Pix2Surf	**1.41**	**1.78**	**1.38**	**1.52**	**8.49**	**5.84**	**7.06**	**7.13**	**10.98**	**6.65**	**7.50**	**8.38**	0.66	0.43	**0.66**	**0.58**

and then averaged up, making the single- and multi-view values comparable in our quantitative experiments.

Metrics: We quantify the quality of reconstructed surfaces with several metrics. The reconstruction error of predictions is computed as the Chamfer distance [2,10] between the estimated surface and the ground truth NOCS map (interpreted as a point cloud). To obtain points on a reconstructed surface, we convert it into a NOCS map using the ground truth camera pose.

In addition to the accuracy of reconstructed surfaces, we quantify the 3C properties of a surface with the following metrics. The 2D–3D correspondence error measures the accuracy of the estimated correspondence between input pixels and 3D points on the reconstructed surface. The error for each foreground pixel is the distance between the estimated 3D location of the pixel and the ground truth location. Unlike the Chamfer distance, this uses the 2D–3D correspondence to compare points. We average over all foreground pixels to obtain the correspondence error of a surface. The **multi-view consistency error** was defined in Sect. 4.2 as the 3D distance between corresponding points in different views. We average the distance for a given point over all pairs of views that contain the point. Corresponding points are found based on the ground truth NOCS map. The continuity is measured based on the surface connectivity. While the continuity of Pix2Surf is induced by our use of a continous MLP as mapping from uv space to 3D space [13], the mapping from the *input image* to the 3D space should *not* be C^0-continuous everywhere, due to self occlusions and boundaries of the 3D shape. The reconstructed surface should have the same C^0 discontinuities as the ground truth surface. We define a C^0 discontinuity as large difference in the 3D locations of the neighboring pixels in a NOCS map (above a threshold of 0.05). We take a statistical approach to measure the surface connectivity, by computing a histogram over the 3D distances between neighboring pixels that are discontinuous. The discontinuity score is the correlation of this histogram to a histogram of the ground truth surface. A higher score indicates a distribution of discontinuities that is more similar to the ground truth surface. Note that continuity is a property induced from method design itself, and the score can penalize the over-smooth case from methods that produces continuous prediction.

5.1   Visible Surface Reconstruction

We compare the quality of single- and multi-view reconstructions to one baseline [Image2Surf (single-view)], and three state-of-the-art methods [AtlasNet [13] (single-view), X-NOCS [37] (single- and multi-view), Pixel2Mesh++ [45] (multiview)]. Note that Pix2Surf deals with a more challenging problem compared to AtlasNet and Pixel2Mesh++: (1) we predict 2D–3D correspondences (AtlasNet does not), and (2) we do not require camera geometry information as input (Pixel2Mesh++ does). In this section, we only focus on reconstructing *visible* surfaces, but we also report hidden surface generation in the next section.

The single-view performance of each method in all of our metrics is shown in the first four rows of Table 1, and the multi-view performance in the last three rows. Metrics are comparable across single- and multi-view methods. For each of the four metrics, we show the performance on each dataset category, and an average over all categories.

Image2Surf: This baseline is similar to Pix2Surf, but takes image UV coordinates (normalized by predicted mask) as input to the UV amplifier instead of the learned UV chart, i.e., the input image is the chart. We observe that it is hard for the network to learn depth discontinuities, resulting in over-smoothed occlusion boundaries (see supplementary document). The over-smoothing is reflected in a high reconstruction error, and particularly low discontinuity correlation score. This comparison justifies our design to include a learned UV chart.

X-NOCS: This is a state-of-the-art reconstruction method that predicts a 3D point cloud, i.e., a 3D point for each foreground pixel. Since X-NOCS has no notion of surface connectivity, there is no coordination between neighboring points, resulting in poor reconstruction accuracy and noisy output point clouds (see Fig. 7).

Note that the output point cloud from X-NOCS can capture the right discontinuity. However, it can only produce discrete noisy point cloud instead of continuous surfaces.

AtlasNet: This method also uses an explicit surface parametrization, giving it a low reconstruction error on the Car category. However, since the

Fig. 7. Our results (left) compared with surface-agnostic X-NOCS (right), visualized with image connectivity. Pix2Surf produces significantly smoother results.

parametrization is not learned and has a fixed layout and connectivity, the reconstruction error increases significantly for categories with more complex shapes and topologies, such as Chair and Airplane. Correspondence and multi-view consistency are not evaluated, since AtlasNet lacks pixel-to-point correspondences and works only for a single view.

Pixel2Mesh++: This method deforms a given starting mesh in a coarse-to-fine approach to approximate an object shown from multiple views. In each refinement step, a mesh vertex is deformed based on a small image neighborhood around the projection of the vertex in each view. Unlike in our method, ground truth camera positions need to be known for this projection. The fixed connectivity and topology of the starting mesh results in a higher reconstruction error. Since correspondence and multi-view consistency are trivial given a ground truth camera model, we do not evaluate these properties.

Unlike the previous methods, **Pix2Surf** learns a continuous parametrization of the surface that does not have a fixed topology or connectivity. This gives us more flexibility to approximate complex surfaces, for instance, to correctly place holes that can model C^0 discontinuities. This explains our high discontinuity correlation scores which also benefits the accuracy of reconstruction and 2D-3D correspondence. In the multi-view setting, Pix2Surf shares information across the views, improving the overall reconstruction accuracy. For example, surfaces that are only visible at a very oblique angle in one view can benefit from additional views. Our use of a consistency loss additionally ensures an improvement of the multi-view consistency over the baselines, and a lower consistency error compared to single view Pix2Surf (Fig. 5). We observe that Pix2Surf is the only method that has top performance on all quality metrics (reconstruction and 3 C properties), all other methods reconstruct surfaces that fall short in at least some of the metrics.

5.2   Hidden Surface Generation

Since Pix2Surf learns a strong prior of the shapes it was trained on, we can generate plausible estimates for surfaces in parts of the object that are not directly visible in the image (see Fig. 8). Similar to X-NOCS, we represent a 3D object with two layers: a visible layer that we reconstruct in the experiments described previously, and a hidden layer denoting the last intersection of camera rays [37]. Pix2Surf can be easily extended to reconstruct hidden surface farthest from the camera by adding additional output channels to the NOCS-UV branch. The rest of the architecture remains the same with the learned UV parametrization additionally also learning about the hidden surface. In Table 2, we show our performance when jointly reconstructing the visible and hidden surfaces from an image. We compare to both the single- and multi-view version of X-NOCS on all categories. The improvement in accuracy for our method shows that hidden surfaces benefits from our learned parametrization as well. Comparing the performance of the visible surface reconstruction to Table 1, we see that the joint reconstruction of visible and hidden surfaces does not significantly decrease the reconstruction accuracy of the visible surfaces.

Table 2. We compare the reconstruction error of visible and hidden surfaces (trained jointly) for Pix2Surf and X-NOCS [single view (sv.) and multi-view (mv.)]. The learned parametrization of Pix2Surf also benefits from hidden surface generation, and the additional reconstruction of the hidden surface does not adversely affect the accuracy of the visible surfaces.

Fig. 8. Pix2Surf can reconstruct both visible (textured) and hidden parts (color coded). (Color figure online)

	Visible Error ↓				Hidden Error ↓			
	Car	Chair	Plane	**Avg.**	Car	Chair	Plane	**Avg.**
X-NOCS (sv.)	2.25	2.95	2.08	2.43	1.86	3.34	2.25	2.48
X-NOCS (mv.)	2.89	2.80	2.19	2.63	3.11	3.32	2.03	2.82
Pix2Surf	**1.66**	**2.01**	**1.66**	**1.78**	**1.52**	**2.47**	**1.77**	**1.92**

6   Conclusion

We have presented Pix2Surf, a method for predicting 3D surface from a single- or multi-view images. Compared with the previous work, Pix2Surf simultaneously achieves three properties in the prediction: **continuity** of the surface, **consistency** across views, and pixel-level **correspondences** from the images to the 3D shape. By attaining these properties, our method enables the generation of high-quality parametric surfaces, readily integrating the output surfaces from multi-views, and lifting texture information from images to the 3D shape. In future work, we will explore ways of guaranteeing continuity even across different views and improving the quality of mapped textures. Another interesting direction is to exploit the intermediate learned chart as a container for material properties. A longer-term goal would be to investigate how the network can generalize across multiple categories.

Acknowledgement. We thank the anonymous reviewers for their comments and suggestions. This work was supported by a Vannevar Bush Faculty Fellowship, NSF grant IIS-1763268, grants from the Stanford GRO Program, the SAIL-Toyota Center for AI Research, AWS Machine Learning Awards Program, UCL AI Center, and a gift from the Adobe.

References

1. Badrinarayanan, V., Kendall, A., Cipolla, R.: SegNet: a deep convolutional encoder-decoder architecture for image segmentation. IEEE Trans. Pattern Anal. Mach. Intell. **39**, 2481–2495 (2017)
2. Barrow, H.G., Tenenbaum, J.M., Bolles, R.C., Wolf, H.C.: Parametric correspondence and chamfer matching: two new techniques for image matching. In: Proceedings of the 5th International Joint Conference on Artificial Intelligence, IJCAI 1977, vol. 2, pp. 659–663. Morgan Kaufmann Publishers Inc., San Francisco (1977). http://dl.acm.org/citation.cfm?id=1622943.1622971
3. Bhoi, A.: Monocular depth estimation: a survey. arXiv preprint arXiv:1901.09402 (2019)

4. Chang, A.X., et al.: ShapeNet: an information-rich 3D model repository. arXiv preprint arXiv:1512.03012 (2015)
5. Chen, R., Han, S., Xu, J., Su, H.: Point-based multi-view stereo network. In: Proceedings of ICCV (2019)
6. Chen, Z., Zhang, H.: Learning implicit fields for generative shape modeling. In: Proceedings of CVPR (2019)
7. Choy, C.B., Xu, D., Gwak, J.Y., Chen, K., Savarese, S.: 3D-R2N2: a unified approach for single and multi-view 3D object reconstruction. In: Leibe, B., Matas, J., Sebe, N., Welling, M. (eds.) ECCV 2016. LNCS, vol. 9912, pp. 628–644. Springer, Cham (2016). https://doi.org/10.1007/978-3-319-46484-8_38
8. Deprelle, T., Groueix, T., Fisher, M., Kim, V.G., Russell, B.C., Aubry, M.: Learning elementary structures for 3D shape generation and matching. In: Proceedings of NeurIPS (2019)
9. Fan, H., Su, H., Guibas, L.: A point set generation network for 3D object reconstruction from a single image. In: Proceedings of CVPR (2017)
10. Fan, H., Su, H., Guibas, L.J.: A point set generation network for 3D object reconstruction from a single image. In: Proceedings of the IEEE Conference on Computer Vision and Pattern Recognition, pp. 605–613 (2017)
11. Genova, K., Cole, F., Vlasic, D., Sarna, A., Freeman, W.T., Funkhouser, T.: Learning shape templates with structured implicit functions. In: Proceedings of ICCV (2019)
12. Girdhar, R., Fouhey, D.F., Rodriguez, M., Gupta, A.: Learning a predictable and generative vector representation for objects. In: Leibe, B., Matas, J., Sebe, N., Welling, M. (eds.) ECCV 2016. LNCS, vol. 9910, pp. 484–499. Springer, Cham (2016). https://doi.org/10.1007/978-3-319-46466-4_29
13. Groueix, T., Fisher, M., Kim, V.G., Russell, B.C., Aubry, M.: AtlasNet: a papier-mâché approach to learning 3D surface generation. In: Proceedings of CVPR (2018)
14. Henzler, P., Mitra, N., Ritschel, T.: Escaping plato's cave using adversarial training: 3D shape from unstructured 2D image collections. In: Proceedings of ICCV (2019)
15. Insafutdinov, E., Dosovitskiy, A.: Unsupervised learning of shape and pose with differentiable point clouds. In: Proceedings of NeurIPS (2018)
16. Kanazawa, A., Tulsiani, S., Efros, A.A., Malik, J.: Learning category-specific mesh reconstruction from image collections. In: Proceedings of ECCV (2018)
17. Kar, A., Häne, C., Malik, J.: Learning a multi-view stereo machine. In: Proceedings of NeurIPS (2017)
18. Kato, H., Harada, T.: Learning view priors for single-view 3D reconstruction. In: Proceedings of CVPR (2019)
19. Kato, H., Ushiku, Y., Harada, T.: Neural 3D mesh renderer. In: Proceedings of CVPR (2018)
20. Kazhdan, M., Bolitho, M., Hoppe, H.: Poisson surface reconstruction. In: Proceedings of the Eurographics Symposium on Geometry Processing (2006)
21. Kazhdan, M., Hoppe, H.: Screened poisson surface reconstruction. ACM Trans. Graph. (ToG) 32, 1–13 (2013)
22. Kulkarni, N., Gupta, A., Tulsiani, S.: Canonical surface mapping via geometric cycle consistency. In: Proceedings of ICCV (2019)
23. Lin, C.H., Kong, C., Lucey, S.: Learning efficient point cloud generation for dense 3D object reconstruction. In: Proceedings of AAAI (2018)
24. Lin, C.H., et al.: Photometric mesh optimization for video-aligned 3D object reconstruction. In: Proceedings of CVPR (2019)

25. Lin, T.-Y., et al.: Microsoft COCO: common objects in context. In: Fleet, D., Pajdla, T., Schiele, B., Tuytelaars, T. (eds.) ECCV 2014. LNCS, vol. 8693, pp. 740–755. Springer, Cham (2014). https://doi.org/10.1007/978-3-319-10602-1_48

26. Liu, S., Li, T., Chen, W., Li, H.: Soft rasterizer: a differentiable renderer for image-based 3D reasoning (2019)

27. Liu, S., Saito, S., Chen, W., Li, H.: Learning to infer implicit surfaces without 3D supervision. In: Advances in Neural Information Processing Systems, pp. 8295–8306 (2019)

28. Lorensen, W.E., Cline, H.E.: Marching cubes: a high resolution 3D surface construction algorithm. In: SIGGRAPH (1987)

29. Mescheder, L., Oechsle, M., Niemeyer, M., Nowozin, S., Geiger, A.: Occupancy networks: learning 3D reconstruction in function space. In: Proceedings of CVPR (2019)

30. Niemeyer, M., Mescheder, L., Oechsle, M., Geiger, A.: Differentiable volumetric rendering: learning implicit 3D representations without 3D supervision. In: Proceedings of the IEEE/CVF Conference on Computer Vision and Pattern Recognition, pp. 3504–3515 (2020)

31. Pan, J., Han, X., Chen, W., Tang, J., Jia, K.: Deep mesh reconstruction from single RGB images via topology modification networks. In: Proceedings of the IEEE International Conference on Computer Vision, pp. 9964–9973 (2019)

32. Park, J.J., Florence, P., Straub, J., Newcombe, R., Lovegrove, S.: DeepSDF: learning continuous signed distance functions for shape representation. In: Proceedings of CVPR (2019)

33. Petersen, F., Bermano, A.H., Deussen, O., Cohen-Or, D.: Pix2vex: image-to-geometry reconstruction using a smooth differentiable renderer. arXiv preprint arXiv:1903.11149 (2019)

34. Richter, S.R., Roth, S.: Matryoshka networks: predicting 3D geometry via nested shape layers. In: Proceedings of CVPR (2018)

35. Saito, S., Huang, Z., Natsume, R., Morishima, S., Kanazawa, A., Li, H.: PIFu: pixel-aligned implicit function for high-resolution clothed human digitization. In: Proceedings of ICCV (2019)

36. Simonyan, K., Zisserman, A.: Very deep convolutional networks for large-scale image recognition. arXiv preprint arXiv:1409.1556 (2014)

37. Sridhar, S., Rempe, D., Valentin, J., Bouaziz, S., Guibas, L.J.: Multiview aggregation for learning category-specific shape reconstruction. In: Proceedings of NeurIPS (2019)

38. Tatarchenko, M., Dosovitskiy, A., Brox, T.: Octree generating networks: efficient convolutional architectures for high-resolution 3D outputs. In: Proceedings of ICCV (2017)

39. Tatarchenko, M., Richter, S.R., Ranftl, R., Li, Z., Koltun, V., Brox, T.: What do single-view 3D reconstruction networks learn? In: Proceedings of CVPR (2019)

40. Tulsiani, S., Zhou, T., Efros, A.A., Malik, J.: Multi-view supervision for single-view reconstruction via differentiable ray consistency. In: Proceedings of CVPR (2017)

41. Wang, H., Sridhar, S., Huang, J., Valentin, J., Song, S., Guibas, L.J.: Normalized object coordinate space for category-level 6D object pose and size estimation. In: Proceedings of CVPR (2019)

42. Wang, N., Zhang, Y., Li, Z., Fu, Y., Liu, W., Jiang, Y.G.: Pixel2Mesh: generating 3D mesh models from single RGB images. In: Proceedings of ECCV (2018)

43. Wang, P.S., Liu, Y., Guo, Y.X., Sun, C.Y., Tong, X.: Adaptive O-CNN: a Patch-based Deep Representation of 3D Shapes. In: SIGGRAPH Asia (2018)

44. Wang, W., Ceylan, D., Mech, R., Neumann, U.: 3DN: 3D deformation network. In: Proceedings of CVPR (2019)
45. Wen, C., Zhang, Y., Li, Z., Fu, Y.: Pixel2Mesh++: multi-view 3D mesh generation via deformation. In: Proceedings of ICCV (2019)
46. Xu, Q., Wang, W., Ceylan, D., Mech, R., Neumann, U.: DISN: deep implicit surface network for high-quality single-view 3D reconstruction. In: Proceedings of NeurIPS (2019)
47. Yan, X., Yang, J., Yumer, E., Guo, Y., Lee, H.: Perspective transformer nets: learning single-view 3D object reconstruction without 3D supervision. In: Proceedings of NeurIPS (2016)

6D Camera Relocalization in Ambiguous Scenes via Continuous Multimodal Inference

Mai Bui[1]([⊠]), Tolga Birdal[2], Haowen Deng[1,3], Shadi Albarqouni[1,4],
Leonidas Guibas[2], Slobodan Ilic[1,3], and Nassir Navab[1,5]

[1] Technical University of Munich, Munich, Germany
mai.bui@tum.de
[2] Stanford University, Stanford, USA
[3] Siemens AG, Munich, Germany
[4] ETH Zurich, Zurich, Switzerland
[5] Johns Hopkins University, Baltimore, USA

Abstract. We present a multimodal camera relocalization framework
that captures ambiguities and uncertainties with continuous mixture
models defined on the manifold of camera poses. In highly ambiguous
environments, which can easily arise due to symmetries and repetitive
structures in the scene, computing one plausible solution (what most
state-of-the-art methods currently regress) may not be sufficient. Instead
we predict multiple camera pose hypotheses as well as the respective
uncertainty for each prediction. Towards this aim, we use Bingham distributions, to model the orientation of the camera pose, and a multivariate
Gaussian to model the position, with an end-to-end deep neural network.
By incorporating a Winner-Takes-All training scheme, we finally obtain a
mixture model that is well suited for explaining ambiguities in the scene,
yet does not suffer from mode collapse, a common problem with mixture density networks. We introduce a new dataset specifically designed
to foster camera localization research in ambiguous environments and
exhaustively evaluate our method on synthetic as well as real data on
both ambiguous scenes and on non-ambiguous benchmark datasets. We
plan to release our code and dataset under multimodal3dvision.github.io.

1 Introduction

Camera relocalization is the term for determining the 6-DoF rotation and translation parameters of a camera with respect to a known 3D world. It is now
a key technology in enabling a multitude of applications such as augmented
reality, autonomous driving, human computer interaction and robot guidance,
thanks to its extensive integration in simultaneous localization and mapping

Electronic supplementary material The online version of this chapter (https://
doi.org/10.1007/978-3-030-58523-5_9) contains supplementary material, which is available to authorized users.

© Springer Nature Switzerland AG 2020
A. Vedaldi et al. (Eds.): ECCV 2020, LNCS 12363, pp. 139–157, 2020.
https://doi.org/10.1007/978-3-030-58523-5_9

Fig. 1. In a highly ambiguous environment, similar looking views can easily confuse current camera pose regression models and lead to incorrect localization results. Instead, given a query RGB image, our aim is to predict the possible modes as well as the associated uncertainties, which we model by the parameters of Bingham and Gaussian mixture models.

(SLAM) [17,22,64], structure from motion (SfM) [68,72], metrology [5] and visual localization [57,69]. For decades, vision scholars have worked on finding the unique solution of this problem [38,39,39,60,66,75,76]. However, this trend is now witnessing a fundamental challenge. A recent school of thought has begun to point out that for highly complex and ambiguous real environments, obtaining a single solution for the location and orientation of a camera is simply not sufficient. This has led to a paradigm shift towards estimating a range of solutions, in the form of full probability distributions [1,6,7] or at least solutions that estimate the uncertainty in orientation estimates [40,50]. Thanks to advances in state-of-the-art machine learning, this important problem can now be tackled via data driven algorithms that are able to discover multi-modal and complex distributions, targeting the task at hand.

In this paper, we devise a multi-hypotheses method, depicted in Fig. 1, for learning continuous mixture models on manifold valued rotations (parameterized by quaternions) and Euclidean translations that can explain uncertainty and ambiguity in 6DoF camera relocalization, while avoiding mode collapse [65]. In particular, we propose a direct regression framework utilizing a combination of antipodally symmetric Bingham [3] and Gaussian probability distributions in order to deal with rotational and translational uncertainties respectively. Together, they are well suited to the geometric nature of $SE(3)$ pose representations. Using said distributions, we aim to build novel variational models that enable estimation of full covariances on the discrete modes to be predicted. For better exploration of the variational posterior, we extend the established particle based training approaches [49,49,63] to mixtures of Gaussians and Binghams. While these techniques only optimize the location of individual hypothesis to cover diverse modes, we additionally learn to predict associated variances on the manifold. We can then approximate the manifold valued posterior of our problem in a continuous fashion. Note that, to the best of our knowledge, such continuous distributions for multi-modal modeling of the network posteriors formed by the 6D pose parameters have not been explored previously. Our synthetic and real experiments demonstrate promising performance both under ambiguities and non-ambiguous scenes. Our method is also flexible in the sense that it can be used with a wide variety of backbone architectures. In a nutshell, our contributions are:

1. We provide a general framework for continuously modelling conditional density functions on quaternions using Bingham distributions, while explaining the translational uncertainty with multi-modal Gaussians.
2. We propose a new training scheme for deep neural networks that enables the inference of a diverse set of modes and related concentration parameters as well as the prior weights for the mixture components.
3. We exhaustively evaluate our method on existing datasets, demonstrating the validity of our approach. Additionally, we create a new highly ambiguous camera relocalization dataset, which we use to showcase the quality of results attained by our algorithm and provide a thorough study on camera localization in ambiguous scenes.

2   Prior Art

6D camera relocalization is a very well studied topic with a vast literature [10,12, 13,15,24,43,52,69]. Our work considers the uncertainty aspect and this is what we focus on here: We first review the uncertainty estimation in deep networks, and subsequently move to uncertainty in 6D pose and relocalization.

Characterizing the Posterior in Deep Networks. Typical CNNs [36,70] are over-confident in their predictions [33,77]. Moreover, these networks tend to approximate the conditional averages of the target data [8]. These undesired properties render the immediate outputs of those networks unsuitable for the quantification of calibrated uncertainty. This has fostered numerous works as we will summarize in the following. *Mixture Density Networks (MDN)* [8] is the pioneer to model the conditional distribution by predicting the parameters of a mixture model. Yet, it is repeatedly reported that optimizing for general mixture models suffers from mode collapse and numerical instabilities [20,49]. These issues can to a certain extent be alleviated by using Dropout [27] as a Bayesian approximation, but even for moderate dimensions these methods still face difficulties in capturing multiple modes. Instead, the more tailored *Winner Takes All* (WTA) [25,34] as well as *Relaxed-WTA* (RWTA) [63] try to capture the multimodal posterior in the K-best hypotheses predictions of the network. *Evolving-WTA* (EWTA) [49] further avoids the inconsistencies related to the WTA losses. Though, a majority of these works consider only low dimensional posterior with the assumption of a Euclidean space, whereas we consider a 7D non-Euclidean highly non-convex posterior.

Uncertainty in 6D. Initial attempts to capture the uncertainty of camera relocalization involved Random Forests (RF) [14]. Valentin *et al.* [73] stored GMM components at the leaves of a scene coordinate regression forest [69]. The modes are obtained via a mean shift procedure, and the covariance is explained by a 3D Gaussian. A similar approach later considered the uncertainty in object coordinate labels [11]. It is a shortcoming of RF that both of these approaches require hand crafted depth features. Moreover, their uncertainty is on the correspondences and not on the final camera pose. Thus a costly RANSAC [26] is required to propagate the uncertainty in the leaves to the camera pose.

Only recently, Manhardt et al. [50] localized a camera against a known object under rotational ambiguities arising due to symmetries or self-occlusions. They extended the RTWA [63] to deal with the 3D rotations using quaternions. This method can only yield discrete hypotheses not continuous density estimates. Similarly, the pose estimation network of Pitteri et al. [58] explicitly considered axis-symmetric objects whose pose cannot be uniquely determined. Likewise, Corona et al. [19] addressed general rotational symmetries. All of these works require extensive knowledge about the object and cannot be extended to the scenario of localizing against a scene without having a 3D model. Note that the latter two works cannot handle the case of self-symmetry and [19] additionally requires a dataset of symmetry-labeled objects, an assumption unlikely to be fulfilled in real applications.

Bayesian PoseNet [40] was one of the first works to model uncertainty for the 6D relocalization problem. It leveraged Dropout [27] to sample the posterior as a way to enable approximate probabilistic pose inference. Although in theory this method can generate discrete samples from the multi-modal distribution, in practice, as we will demonstrate, the Monte Carlo scheme tends to draw samples around a single mode. This method also suffers from the large errors associated to PoseNet [43] itself. The successive VidLoc [18] adapted MDNs [8] to model and predict uncertainty for the 6D relocalization problem. Besides the reported issues of MDNs, VidLoc incorrectly modeled the rotation parameters using Gaussians and lacked the demonstrations of uncertainty on rotations. Contrarily, in this work we devise a principled method using Bingham distributions [3] that are well suited to the double covering nature of unit quaternions. HydraNet [56] provided calibrated uncertainties on the $SO(3)$ group, but assumed a unimodal posterior that is centered on the naive $\mathbb{R}^4$-mean of predicted quaternions.

Our work is inspired by [59], where a variational auto-encoder [44] is learnt to approximate the posterior of $SO(2)$ modeled by von Mises mixtures [51]. Though, it is not trivial to tweak and generalize [59] to the continuous, highly multi-modal and multi-dimensional setting of 6D camera relocaliztion. This is what we precisely contribute in this work. Note that we are particularly interested in the *aleatoric* uncertainty (noise in the observations) and leave the *epistemic* (noise in the model) part as a future work [42].

3   The Bingham Distribution

Derived from a zero-mean Gaussian, the Bingham distribution [3] (BD) is an antipodally symmetric probability distribution conditioned to lie on $\mathbb{S}^{d-1}$ with probability density function (PDF) $\mathcal{B} : \mathbb{S}^{d-1} \to \mathbb{R}$:

$$\mathcal{B}(\mathbf{x}; \boldsymbol{\Lambda}, \mathbf{V}) = (1/F) \exp(\mathbf{x}^T \mathbf{V} \boldsymbol{\Lambda} \mathbf{V}^T \mathbf{x}) = (1/F) \exp\left(\sum_{i=1}^{d} \lambda_i (\mathbf{v}_i^T \mathbf{x})^2\right) \quad (1)$$

where $\mathbf{V} \in \mathbb{R}^{d \times d}$ is an orthogonal matrix ($\mathbf{V}\mathbf{V}^T = \mathbf{V}^T\mathbf{V} = \mathbf{I}_{d \times d}$) describing the orientation, $\mathbf{\Lambda} \in \mathbb{R}^{d \times d}$ is called the *concentration matrix* with $0 \geq \lambda_1 \geq \cdots \geq \lambda_{d-1}$: $\mathbf{\Lambda} = \mathrm{diag}([0\ \lambda_1\ \lambda_2\ \ldots\ \lambda_{d-1}])$.

It is easy to show that adding a multiple of the identity matrix $\mathbf{I}_{d \times d}$ to $\mathbf{V}$ does not change the distribution [3]. Thus, we conveniently force the first entry of $\mathbf{\Lambda}$ to be zero. Moreover, since it is possible to swap columns of $\mathbf{\Lambda}$, we can build $\mathbf{V}$ in a sorted fashion. This allows us to obtain *the mode* very easily by taking the first column of $\mathbf{V}$. Due to its antipodally symmetric nature, the mean of the distribution is always zero. F in Eq (1) denotes the *the normalization constant* dependent only on $\mathbf{\Lambda}$ and is of the form:

$$F \triangleq |S_{d-1}| \cdot {}_1F_1\left(1/2, d/2, \mathbf{\Lambda}\right), \tag{2}$$

where $|S_{d-1}|$ is the surface area of the d-sphere and ${}_1F_1$ is a confluent hypergeometric function of matrix argument [37,46]. The computation of F is not trivial. In practice, following Glover [31], this quantity as well as its gradients are approximated by tri-linear interpolation using a pre-computed look-up table over a predefined set of possible values in $\mathbf{\Lambda}$, lending itself to differentiation [45,47].

Relationship to Quaternions. The antipodal symmetry of the PDF makes it amenable to explain the topology of quaternions, i. e., $\mathcal{B}(\mathbf{x}; \cdot) = \mathcal{B}(-\mathbf{x}; \cdot)$ holds for all $\mathbf{x} \in \mathbb{S}^{d-1}$. In 4D when $\lambda_1 = \lambda_2 = \lambda_3$, one can write $\mathbf{\Lambda} = \mathrm{diag}([1, 0, 0, 0])$. In this case, Bingham density relates to the dot product of two quaternions $\mathbf{q}_1 \in \mathbb{H}_1 \triangleq \mathbf{x}$ and the mode of the distribution, say $\bar{\mathbf{q}}_2 \in \mathbb{H}_1$. This induces a metric of the form: $d_{\mathrm{bingham}} = d(\mathbf{q}_1, \bar{\mathbf{q}}_2) = (\mathbf{q}_1 \cdot \bar{\mathbf{q}}_2)^2 = \cos(\theta/2)^2$.

Bingham distributions have been extensively used to represent distributions on unit quaternions ($\mathbb{H}_1$) [4,7,29,30,46]; however, to the best of our knowledge, never for the problem we consider here.

Constructing a Bingham Distribution on a Given Mode. Creating a Bingham distribution on any given mode $\mathbf{q} \in \mathbb{H}_1$ requires finding a set of vectors orthonormal to $\mathbf{q}$. This is a frame bundle $\mathbb{H}_1 \to \mathcal{F}\mathbb{H}_1$ composed of four unit vectors: the mode and its orthonormals. We follow Birdal *et al.* [7] and use the *parallelizability* of unit quaternions to define the orthonormal basis $\mathbf{V} : \mathbb{H}_1 \mapsto \mathbb{R}^{4 \times 4}$:

$$\mathbf{V}(\mathbf{q}) \triangleq \begin{bmatrix} q_1 & -q_2 & -q_3 & q_4 \\ q_2 & q_1 & q_4 & q_3 \\ q_3 & -q_4 & q_1 & -q_2 \\ q_4 & q_3 & -q_2 & -q_1 \end{bmatrix}. \tag{3}$$

It is easy to verify that the matrix valued function $\mathbf{V}(\mathbf{q})$ is orthonormal for every $\mathbf{q} \in \mathbb{H}_1$. $\mathbf{V}(\mathbf{q})$ further gives a convenient way to represent quaternions as matrices paving the way to linear operations, such as quaternion multiplication or orthonormalization without the Gram-Schmidt.

Relationship to Other Representations. Note that geometric [2] or measure theoretic [23], there are multitudes of ways of defining probability distributions

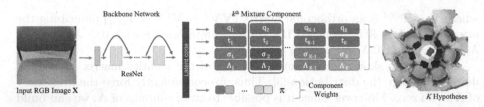

Fig. 2. Forward pass of our network. For an input RGB image we predict K camera pose hypotheses as well as Bingham concentration parameters, Gaussian variances and component weights to obtain a mixture model. (Color figure online)

on the Lie group of 6D rigid transformations [35]. A choice would be to define Gaussian distribution on the Rodrigues vector (or exponential coordinates) [54] where the geodesics are straight lines [53] or the use of Concentrated Gaussian distributions [9] on matrices of SE(3). However, as our purpose is not tracking but direct regression, in this work we favor quaternions as continuous and minimally redundant parameterizations without singularities [16, 32] and use the Bingham distribution that is well suited to their topology. We handle the redundancy $\mathbf{q} \equiv -\mathbf{q}$ by mapping all the rotations to the northern hemisphere.

4 Proposed Model

We now describe our model for uncertainty prediction following [59]. We consider the situation where we observe an input image $\mathbf{X} \in \mathbb{R}^{W \times H \times 3}$ and assume the availability of a predictor function $\mu_{\mathbf{\Gamma}}(\mathbf{X}) : \mathbb{R}^{W \times H \times 3} \mapsto \mathbb{H}_1$ parameterized by $\mathbf{\Gamma} = \{\mathbf{\Gamma}_i\}$. Note that predicting entities that are non-Euclidean easily generalizes to prediction of Euclidean quantities such as translations e.g. $\mathbf{t} \in \mathbb{R}^3$. For the sake of conciseness and clarity, we will omit the translations and concentrate on the rotations. Translations modeled via Gaussians will be precised later on.

The Unimodal Case. We momentarily assume that $\mu_{\mathbf{\Gamma}}(\cdot)$, or short $\mu(\cdot)$, can yield the correct values of the absolute camera rotations $\mathbf{q}_i \in \mathbb{H}_1$ with respect to a common origin, admitting a non-ambiguous prediction, hence a posterior of single mode. We use the predicted rotation to set the most likely value (mode) of a BD:

$$p_{\mathbf{\Gamma}}(\mathbf{q} \mid \mathbf{X}; \mathbf{\Lambda}) = (1/F) \exp\left(\mathbf{q}^{\top} \mathbf{V}_{\mu} \mathbf{\Lambda} \mathbf{V}_{\mu}^{\top} \mathbf{q}\right), \tag{4}$$

and let $\mathbf{q}_i$ differ from this value up to the extent determined by $\mathbf{\Lambda} = \{\lambda_i\}$. For the sake of brevity we use $\mathbf{V}_{\mu} \equiv \mathbf{V}(\mu(\mathbf{X}))$, the orthonormal basis aligned with the predicted quaternion $\mu(\mathbf{X})$ and as defined in Eq (3).

While for certain applications, fixing $\mathbf{\Lambda}$ can work, in order to capture the variation in the input, it is recommended to adapt $\mathbf{\Lambda}$ [59]. Thus, we introduce it among the unknowns. To this end we define the function $\mathbf{\Lambda}_{\mathbf{\Gamma}}(\mathbf{X})$ or in short $\mathbf{\Lambda}_{\mathbf{\Gamma}}$ for computing the concentration values depending on the current image and the parameters $\mathbf{\Gamma}$. Our final model for the unimodal case reads:

$$p_{\mathbf{\Gamma}}(\mathbf{q}\,|\,\mathbf{X}) = \frac{\exp\left(\mathbf{q}^{\top}\mathbf{V}(\mu(\mathbf{X}))\mathbf{\Lambda_{\Gamma}}(\mathbf{X})\mathbf{V}(\mu(\mathbf{X}))^{\top}\mathbf{q}\right)}{F(\mathbf{\Lambda_{\Gamma}}(\mathbf{X}))} = \frac{\exp\left(\mathbf{q}^{\top}\mathbf{V}_{\mu}\mathbf{\Lambda_{\Gamma}}\mathbf{V}_{\mu}^{\top}\mathbf{q}\right)}{F(\mathbf{\Lambda_{\Gamma}})} \quad (5)$$

The latter follows from the short-hand notations and is included for clarity. Given a collection of observations i.e., images $\mathcal{X} = \{\mathbf{X}_i\}$ and associated rotations $\mathbf{Q} = \{\mathbf{q}_i\}$, where $i = 1, ..., N$, the parameters of $\mu_{\mathbf{\Gamma}}(\mathbf{X})$ and $\mathbf{\Lambda_{\Gamma}}(\mathbf{X})$ can be obtained simply by maximizing the log-likelihood:

$$\mathbf{\Gamma}^{\star} = \arg\max_{\mathbf{\Gamma}} \log \mathcal{L}_u(\mathbf{\Gamma}|\mathcal{X},\mathbf{Q}) \quad (6)$$

$$\log \mathcal{L}_u(\mathbf{\Gamma}|\mathcal{X},\mathbf{Q}) = \sum_{i=1}^{N} \mathbf{q}_i^{\top}\mathbf{V}_{\mu}\mathbf{\Lambda_{\Gamma}}\mathbf{V}_{\mu}^{\top}\mathbf{q}_i - \sum_{i=1}^{N} \log F(\mathbf{\Lambda_{\Gamma}}). \quad (7)$$

Note once again that $\mathbf{\Lambda_{\Gamma}} \equiv \mathbf{\Lambda_{\Gamma}}(\mathbf{X}_i)$ and $\mathbf{V}_{\mu} \equiv \mathbf{V}(\mu(\mathbf{X}_i))$. If $\mathbf{\Lambda_{\Gamma}}$ were to be fixed as in [59], the term on the right would have no effect and minimizing that loss would correspond to optimizing the Bingham log-likelihood. To ensure $0 \geq \lambda_1 \geq \cdots \geq \lambda_{d-1}$, we parameterize $\boldsymbol{\lambda}$ by λ_1 and the positive offsets $e_2, ..., e_{d-1}$ such that $\lambda_k = \lambda_{k-1} - e_k$ where $k = 2, ..., d-1$. This allows us to make an ordered prediction from the network.

Extension to Finite Bingham Mixture Models (BMM). Ambiguities present in the data requires us to take into account the multimodal nature of the posterior. To achieve this, we now extend the aforementioned model to Bingham Mixture Models [61]. For the finite case, we use K different components associated with K mixture weights $\pi_j(\mathbf{X}, \mathbf{\Gamma})$ for $j = 1, ..., K$. With each component being a Bingham distribution, we can describe the density function as

$$P_{\mathbf{\Gamma}}(\mathbf{q}\,|\,\mathbf{X}) = \sum_{j=1}^{K} \pi_j(\mathbf{X}, \mathbf{\Gamma})p_{\mathbf{\Gamma}j}(\mathbf{q}\,|\,\mathbf{X}), \quad (8)$$

where $p_{\mathbf{\Gamma}j}(\mathbf{q}\,|\,\mathbf{X})$ are the K component distributions and $\pi_j(\mathbf{X}, \mathbf{\Gamma})$ the mixture weights s.t. $\sum_j \pi_j(\mathbf{X}, \mathbf{\Gamma}) = 1$. The model can again be trained by maximizing the log-likelihood, but this time of the mixture model [71,74]

$$\mathbf{\Gamma}^{\star} = \arg\max_{\mathbf{\Gamma}} \log \mathcal{L}_m(\mathbf{\Gamma}|\mathcal{X},\mathbf{Q}) \quad (9)$$

$$\log \mathcal{L}_m(\mathbf{\Gamma}|\mathcal{X},\mathbf{Q}) = \sum_{i=1}^{N} \log \sum_{j=1}^{K} \pi_j(\mathbf{X}_i, \mathbf{\Gamma})p_{\mathbf{\Gamma}j}(\mathbf{q}_i\,|\,\mathbf{X}_i). \quad (10)$$

5 Deeply Modeling $\mu(\cdot)$ and $\Lambda(\cdot)$

Following up on the recent advances, we jointly model $\mu(\cdot)$ and $\Lambda(\cdot)$ by a deep residual network [36]. $\mathbf{\Gamma}$ denotes the entirety of the trainable parameters. On the output we have **fourteen** quantities per density: four for the mode quaternion, three for translation, three for $\mathbf{\Lambda}$ the Bingham concentration, three for variances of the multivariate Gaussian and one for the weight $\pi_j(\cdot)$. In total our K mixture components result in $K \times 14$ output entities. Our architecture is shown in Fig. 2 and we provide further details in the suppl. document. While a typical way to train our network is through simultaneously regressing the output variables, this is known to severely harm the accuracy [63]. Instead we exploit modern approaches to training in presence of ambiguities as we detail in what follows.

MHP Training Scheme. Due to the increased dimensionality, in practice training our variational network in an unconstrained manner is likely to suffer from mode collapse, where all the heads concentrate around the same prediction. To avoid this and obtain a diverse set of modes, instead of training all branches equally by maximizing the log-likelihood of the mixture model, we follow the multi-hypotheses schemes of [49,63] and train our model using a WTA loss function, for each branch maximizing the log-likelihood of a unimodal distribution,

$$\mathbf{\Gamma}^\star = \arg\max_{\mathbf{\Gamma}} \sum_{i=1}^{N} \sum_{j=1}^{K} w_{ij} \log \mathcal{L}_u(\mathbf{\Gamma}|\mathbf{X}_i, \mathbf{q}_i)), \tag{11}$$

according to the associated weights w_{ij} for each of the k hypotheses. In this work, we compute the weights w_{ij} during training following RWTA [63] as

$$w_{ij} = \begin{cases} 1 - \epsilon, & \text{if } j = \arg\min_k |\mathbf{q}_i - \hat{\mathbf{q}}_{ik}| \\ \frac{\epsilon}{K-1}, & \text{otherwise} \end{cases}, \tag{12}$$

where $\hat{\mathbf{q}}_{ik}$ is the predicted mode of a single Bingham distribution. Note that WTA [34] would amount to updating only the branch of the best hypothesis and EWTA [49] the top k branches closest to the ground truth. However, for our problem, we found RWTA to be a more reliable machinery. Finally, to obtain the desired continuous distribution, we train the weights of our Bingham mixture model using the following loss function:

$$\mathcal{L}_\pi(\mathbf{\Gamma}|\mathcal{X}, \mathbf{Q}) = \sum_{i=1}^{N} \sum_{j=1}^{K} \sigma(\hat{\pi}_j(\mathbf{X}_i, \mathbf{\Gamma}), y_{ij}), \tag{13}$$

where σ is the cross-entropy, $\hat{\pi}(\mathbf{X}, \mathbf{\Gamma})$ the predicted weight of the neural network and y_{ij} the associated label of the mixture model component given as

$$y_{ij} = \begin{cases} 1, & \text{if } j = \arg\min_k |\mathbf{q}_i - \hat{\mathbf{q}}_{ik}| \\ 0, & \text{otherwise} \end{cases}. \tag{14}$$

Our final loss, therefore, consists of the weighted likelihood for a unimodal distribution of each branch and the loss of our mixture weights, $\mathcal{L}_\pi(\mathbf{\Gamma}|\mathcal{X}, \mathbf{Q})$.

Incorporating Translations. We model translations $\{\mathbf{t}_i \in \mathbb{R}^3\}_i$ by the standard Gaussian distributions with covariances $\{\mathbf{\Sigma}_i \in \mathbb{R}^{3\times3} \succeq 0\}_i$. Hence, we use

Table 1. Evaluation in non-ambiguous scenes, displayed is the median rotation and translation error. (Numbers for MapNet on the Cambridge Landmarks dataset are taken from [67]). BPN depicts Bayesian-PoseNet [13]. *Uni* and *BMDN* refer to our unimodal version and Bingham-MDN respectively.

Dataset	7-Scenes							Cambridge Landmarks				
[° / m]	Chess	Fire	Heads	Office	Pumpkin	Kitchen	Stairs	Kings	Hospital	Shop	St. Marys	Street
PoseNet	4.48/0.13	**11.3**/0.27	13.0/0.17	5.55/0.19	4.75/0.26	5.35/0.23	12.4/0.35	**1.04**/0.88	**3.29**/3.2	**3.78**/0.88	**3.32**/1.57	25.5/20.3
MapNet	**3.25/0.08**	11.69/0.27	13.2/0.18	**5.15/0.17**	**4.02/0.22**	4.93/0.23	12.08/0.3	1.89/1.07	3.91/1.94	4.22/1.49	4.53/2.0	-
BPN	7.24/0.37	13.7/0.43	**12.0**/0.31	8.04/0.48	7.08/0.61	7.54/0.58	13.1/ 0.48	4.06/1.74	5.12/2.57	7.54/1.25	8.38/2.11	-
VidLoc	-/0.18	-/**0.26**	-/0.14	-/0.26	-/0.36	-/0.31	-/**0.26**	-	-	-	-	-
Uni	4.97/0.1	12.87/0.27	14.05/**0.12**	7.52/0.2	7.11/0.23	8.25/**0.19**	13.1/0.28	1.77/0.88	3.71/**1.93**	4.74/**0.8**	6.19/1.84	**24.1**/16.8
BMDN	4.35/0.1	11.86/0.28	12.76/**0.12**	6.55/0.19	6.9/**0.22**	8.08/0.21	**9.98**/0.31	2.08/**0.83**	3.64/2.16	4.93/0.92	6.03/**1.37**	36.9/**9.7**

the ordinary MDNs [8] to handle them. Yet, once again, during training we apply the MHP scheme explained above to avoid mode collapse and diversify the predictions. In practice, we first train the network to predict the translation and its variance. Then, intuitively, recovering the associated rotation should be an easier task, after which we fine-tune the network on all components of the distribution. Such split has already been shown to be prosperous in prior work [21].

Inference. Rather than reporting the conditional average which can result in label blur, we propose to obtain a single best estimate according to the weighted mode, where we choose the best mixture component according to its mixture weight and pick the mode as a final prediction.

We finally measure the uncertainty of the prediction according to the entropy of the resulting Bingham and Gaussian distributions, given as

$$H_B = \log F - \mathbf{\Lambda}\frac{\nabla F(\mathbf{\Lambda})}{F}, \quad \text{and} \quad H_G = \frac{c}{2} + \frac{c}{2}\log(2\pi) + \frac{1}{2}\log(|\mathbf{\Sigma}|), \quad (15)$$

respectively, where $c = 3$ the dimension of the mean vector of the Gaussian. For a given image we first normalize the entropy values over all pose hypotheses, and finally obtain a measure of (un)certainty as the sum of both rotational (H_B) and translational (H_G) normalized entropy.

Implementation Details. We implement our method in Python using PyTorch library [55]. Following the current state-of-the-art direct camera pose regression methods, we use a *ResNet-34* [36] as our backbone network architecture, followed by fully-connected layers for rotation and translation, respectively. The predicted quaternions are normalized during training. We provide further details of training in the supplementary material.

6 Experimental Evaluation

When evaluating our method we consider two cases: (1) camera relocalization in non-ambiguous scenes, where our aim is to not only predict the camera pose, but the posterior of both rotation and translation that can be used to associate each pose with a measure of uncertainty; (2) we create a highly ambiguous environment, where similar looking images are captured from very different viewpoints.

We show the problems current regression methods suffer from in handling such scenarios and in contrast show the merit of our proposed method.

Error Metrics. Note that, under ambiguities a best mode is unlikely to exist. In those cases, as long as we can generate a hypothesis that is close to the Ground Truth (GT), our network is considered successful. For this reason, in addition to the standard metrics and the weighted mode, we will also speak of the so called *Oracle* error, assuming an oracle that is able to choose the best of all predictions: the one closest to the GT. In addition, we report the *Self-EMD* (SEMD) [49], the earth movers distance [62] of turning a multi-modal distribution into a unimodal one. With this measure we can evaluate the diversity of predictions, where the unimodal distribution is chosen as the predicted mode of the corresponding method. Note that this measure by itself does not give any indication about the accuracy of the prediction.

Datasets. In addition to the standard datasets of 7-Scenes [69] and Cambridge Landmarks [43], we created synthetic as well as real datasets, that are specifically designed to contain repetitive structures and allow us to assess the real benefits of our approach. For synthetic data we render table models from 3DWarehouse[1] and create camera trajectories, e.g. a circular movement around the object, such that ambiguous views are ensured to be included in our dataset. Specifically we use a *dining table* and a *round table* model with discrete modes of ambiguities. In addition, we create highly ambiguous real scenes using Google Tango and the graph-based SLAM approach RTAB-Map [48]. We acquire RGB and depth images as well as distinct ground truth camera trajectories for training and testing. We also reconstruct those scenes. However, note that only the RGB images and corresponding camera poses are required to train our model and the reconstructions are used for visualization only. In particular, our training and test sets consist of 2414 and 1326 frames, respectively. Note that our network sees a single pose label per image. We provide further details, visualizations and evaluations in our supplementary material.

Baselines and SoTA. We compare our approach to current state-of-the-art direct camera pose regression methods, PoseNet [41] and MapNet [13], that output a single pose prediction. More importantly, we assess our performance against two state-of-the-art approaches, namely BayesianPoseNet [40] and Vid-Loc [18], that are most related to our work and predict a distribution over the pose space by using dropout and mixture density networks, respectively. We further include the *unimodal* predictions as well as BMMs trained using mixture density networks [8,28] as baselines. We coin the latter Bingham-MDN or in short *BMDN*.

6.1 Evaluation in Non-ambiguous Scenes

We first evaluate our method on the publicly available 7-Scenes [69] and Cambridge Landmarks [43] datasets. As most of the scenes contained in these datasets

---

[1] https://3dwarehouse.sketchup.com/.

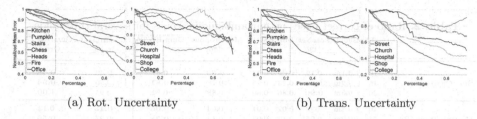

(a) Rot. Uncertainty   (b) Trans. Uncertainty

Fig. 3. Uncertainty evaluation on the 7-Scenes and Cambridge Landmarks datasets, showing the correlation between predicted uncertainty and pose error. Based on the entropy of our predicted distribution uncertain samples are gradually removed. We observe that as we remove the uncertain samples the overall error drops indicating a strong correlation between our predictions and the actual erroneous estimations.

do not show highly ambiguous environments, we consider them to be non-ambiguous. Though, we can not guarantee that some ambiguous views might arise in these datasets as well, such as in the *Stairs* scene of the 7-Scenes dataset. Both datasets have extensively been used to evaluate camera pose estimation methods. Following the state-of-the-art, we report the median rotation and translation errors, the results of which can be found in Table 1. In comparison to methods that output a single pose prediction (*e.g.* PoseNet [41] and MapNet [13]), our methods achieves similar results. Yet, our network provides an additional piece of information that is the uncertainty. On the other hand, especially in translation our method outperforms uncertainty methods, namely BayesianPoseNet [40] and VidLoc [18], on most scenes.

Uncertainty Evaluation. One benefit of our method is that we can use the resulting variance of the predicted distribution as a measure of uncertainty in our predictions. The resulting correlation between pose error and uncertainty can be seen in Fig. 3, where we gradually remove the most uncertain predictions and plot the mean error for the remaining samples. The inverse correlation between the actual errors vs our confidence shows that whenever our algorithm labels a prediction as uncertain it is also likely to be a bad estimate.

It has been shown that current direct camera pose regression methods still have difficulties in generalizing to views that differ significantly from the camera trajectories seen during training [67]. However, we chose to focus on another problem these methods have to face and analyze the performance of direct regression methods in a highly ambiguous environment. In this scenario even similar trajectories can confuse the network and easily lead to wrong predictions, for which our method proposes a solution.

6.2 Evaluation in Ambiguous Scenes

We start with quantitative evaluations on our synthetic as well as real scenes before showing qualitative results. We compare our method to PoseNet and BayesianPoseNet, which we refer to as MC-Dropout. In comparison, we replace

Table 2. Ratio of correct poses on our ambiguous scenes for several thresholds.

	Threshold	PoseNet [43]	Uni.	BMDN	MC-Dropout [40]	Ours-RWTA	MC-Dropout Oracle	Ours-RWTA Oracle
Blue Chairs (A)	10° / 0.1m	0.19	0.29	0.24	**0.39**	0.35	0.40	**0.58**
	15° / 0.2m	0.69	0.73	0.75	0.78	**0.81**	0.90	**0.94**
	20° / 0.3m	**0.90**	0.86	0.80	0.88	0.82	0.95	**1.00**
Meeting Table (B)	10° / 0.1m	0.0	0.02	0.01	0.04	**0.05**	**0.13**	0.12
	15° / 0.2m	0.05	0.12	0.07	0.13	**0.28**	0.27	**0.56**
	20° / 0.3m	0.10	0.19	0.10	0.22	**0.39**	0.32	**0.78**
Staircase (C)	10° / 0.1m	0.14	0.11	0.04	0.13	**0.18**	**0.27**	0.19
	15° / 0.2m	0.45	0.48	0.15	0.32	**0.50**	**0.54**	0.53
	20° / 0.3m	0.60	0.62	0.25	0.49	**0.68**	0.70	**0.74**
Staircase Extended (D)	10° / 0.1m	0.07	0.06	0.06	0.02	**0.09**	**0.16**	0.09
	15° / 0.2m	0.31	0.26	0.21	0.14	**0.39**	**0.45**	0.40
	20° / 0.3m	0.49	0.41	0.32	0.31	**0.58**	0.64	0.64
Seminar Room (E)	10° / 0.1m	**0.37**	0.11	0.06	0.18	0.35	**0.46**	0.36
	15° / 0.2m	0.81	0.36	0.23	0.57	**0.83**	**0.85**	0.83
	20° / 0.3m	0.90	0.57	0.40	0.78	**0.95**	0.90	**0.95**
Average	10° / 0.1m	0.15	0.12	0.08	0.15	**0.20**	**0.28**	0.27
	15° / 0.2m	0.46	0.39	0.28	0.39	**0.56**	0.60	**0.65**
	20° / 0.3m	0.60	0.53	0.37	0.54	**0.68**	0.70	**0.82**

the original network architecture by a ResNet, that has been shown to improve the performance of direct camera pose regression methods [13].

Quantitative Evaluations. Due to the design of our synthetic table scenes, we know that there are two and four possible modes for each image in *dining* and *round* table scenes respectively. Hence, we analyze the predictions of our model by computing the accuracy of correctly detected modes of the true posterior. A mode is considered as found if there exists one pose hypothesis that falls into a certain rotational (5°) and translational (10% of the diameter of GT camera trajectory) threshold of it. In the dining-table, MC-Dropout obtains an accuracy of 50%, finding one mode for each image, whereas the accuracy of Ours-RWTA on average achieves 96%. On round-table, our model shows an average detection rate of 99.1%, in comparison to 24.8% of MC-Dropout.

On our real scenes, we report the recall, where a pose is considered to be correct if both the rotation and translation errors are below a pre-defined threshold. Table 2 shows the accuracy of our baseline methods in comparison to ours for various thresholds. Especially on our *Meeting Table* scene, it can be seen that the performance of direct camera pose regression methods that suffer from mode collapse drops significantly due to the presence of ambiguities in the scene. Thanks to the diverse mode predictions of Ours-RWTA, which is indicated by the high Oracle accuracy as well as the high SEMD shown in Table 3, we are able to improve upon our baseline predictions. Further, by a semi-automatic labeling procedure detailed in our suppl. material, we are able to extract GT modes for the *Blue Chairs* and *Meeting Table* scenes. This way, we can evaluate the entire set of predictions against the GT. Table 4 shows the percentage of correctly detected modes for our method in comparison to MC-Dropout when evaluating

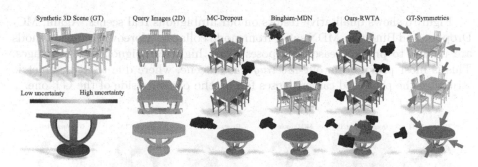

Low uncertainty   High uncertainty

Fig. 4. Qualitative results on our synthetic *dining* and *round table* datasets. Camera poses are colored by uncertainty. Viewpoints are adjusted for best perception.

Table 3. SEMD of our method and MC-Dropout indicating highly diverse predictions by our method in comparison to the baseline.

(a)   (b)   (c)

Fig. 5. Bingham distributions plotted on the unit sphere: **(a)** low uncertainty, **(b)** higher uncertainty and **(c)** the mixtures of Ours-RWTA.

Method/Scene	A	B	C	D	E
MC-Dropout	0.06	0.11	0.13	0.26	0.10
Ours-RWTA	1.19	2.13	2.04	3.81	1.70

with these GT modes. The results support our qualitative observations, that MC-Dropout suffers from mode collapse such that even with increasing threshold, the number of detected modes does not increase significantly.

Qualitative Evaluations. Qualitative results of our proposed model on our synthetic table datasets are shown in Fig. 4. MC-Dropout as well as our finite mixture model, *Bingham-MDN*, suffer from mode collapse. In comparison, the proposed MHP model is able to capture plausible, but diverse modes as well as associated uncertainties.

In contrast to other methods that obtain an uncertainty value for one prediction, we obtain uncertainty values for each hypothesis. This way, we could easily remove non-meaningful predictions, that for example can arise in the WTA and RWTA training schemes. Resulting predicted Bingham distributions are visualized in Fig. 5, by marginalizing over the angle component.

Table 4. Ratio of correctly detected modes for various translational thresholds (in meters). A and B denote *Blue Chairs* and *Meeting Table* scenes.

Scene	Method	0.1	0.2	0.3	0.4
A	MC-Dropout	0.11	0.15	0.16	0.16
	Ours-RWTA	0.36	0.79	0.80	0.80
B	MC-Dropout	0.04	0.07	0.09	0.11
	Ours-RWTA	0.10	0.43	0.63	0.73

Figure 6 shows qualitative results on our ambiguous real scenes. Again, MC-Dropout and Bingham-MDN suffer from mode collapse. Moreover, these methods are unable to predict reasonable poses given highly ambiguous query images. This is most profound in our *Meeting Table* scene, where due to its symmetric structure the predicted camera poses fall on the opposite side of the GT one.

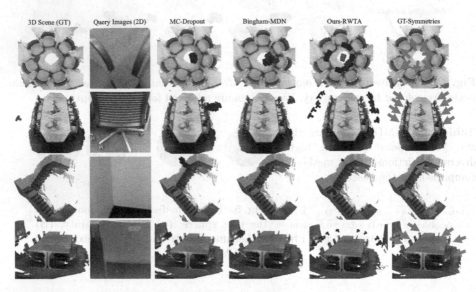

Fig. 6. Qualitative results in our ambiguous dataset. For better visualization, camera poses have been pruned by their uncertainty values.

7 Conclusion

We have presented a novel method dealing with problems of direct camera pose regression in highly ambiguous environments where a unique solution to the 6DoF localization might be nonexistent. Instead, we predict camera pose hypotheses as well as associated uncertainties that finally produce a mixture model. We use the Bingham distribution to model rotations and multivariate Gaussian distribution to obtain the position of a camera. In contrast to other methods like MC-Dropout [40] or mixture density networks our training scheme is able to avoid mode collapse. Thus, we can obtain better mode predictions and improve upon the performance of camera pose regression methods in ambiguous environments while retaining the performance in non-ambiguous ones.

Acknowledgements. This project is supported by Bavaria California Technology Center (BaCaTeC), Stanford-Ford Alliance, NSF grant IIS-1763268, Vannevar Bush Faculty Fellowship, Samsung GRO program, the Stanford SAIL Toyota Research, and the PRIME programme of the German Academic Exchange Service (DAAD) with funds from the German Federal Ministry of Education and Research (BMBF).

References

1. Arun Srivatsan, R., Xu, M., Zevallos, N., Choset, H.: Probabilistic pose estimation using a Bingham distribution-based linear filter. Int. J. Robot. Res. **37**(13–14), 1610–1631 (2018)
2. Barfoot, T.D., Furgale, P.T.: Associating uncertainty with three-dimensional poses for use in estimation problems. IEEE Trans. Robot. **30**(3), 679–693 (2014)
3. Bingham, C.: An antipodally symmetric distribution on the sphere. Ann. Stat. 1201–1225 (1974)
4. Birdal, T., Arbel, M., Şimşekli, U., Guibas, L.: Synchronizing probability measures on rotations via optimal transport. In: Proceedings of the IEEE Conference on Computer Vision and Pattern Recognition (2020)
5. Birdal, T., Bala, E., Eren, T., Ilic, S.: Online inspection of 3D parts via a locally overlapping camera network. In: 2016 IEEE Winter Conference on Applications of Computer Vision (WACV), pp. 1–10. IEEE (2016)
6. Birdal, T., Simsekli, U.: Probabilistic permutation synchronization using the Riemannian structure of the Birkhoff polytope. In: Proceedings of the IEEE Conference on Computer Vision and Pattern Recognition, pp. 11105–11116 (2019)
7. Birdal, T., Simsekli, U., Eken, M.O., Ilic, S.: Bayesian pose graph optimization via Bingham distributions and tempered geodesic MCMC. In: Advances in Neural Information Processing Systems, pp. 308–319 (2018)
8. Bishop, C.M.: Mixture density networks (1994)
9. Bourmaud, G., Mégret, R., Arnaudon, M., Giremus, A.: Continuous-discrete extended Kalman filter on matrix lie groups using concentrated Gaussian distributions. Jo. Math. Imaging Vis. **51**(1), 209–228 (2015)
10. Brachmann, E., et al.: DSAC-differentiable RANSAC for camera localization. In: Proceedings of the IEEE Conference on Computer Vision and Pattern Recognition (2017)
11. Brachmann, E., Michel, F., Krull, A., Ying Yang, M., Gumhold, S., et al.: Uncertainty-driven 6D pose estimation of objects and scenes from a single RGB image. In: Proceedings of the IEEE Conference on Computer Vision and Pattern Recognition, pp. 3364–3372 (2016)
12. Brachmann, E., Rother, C.: Learning less is more-6D camera localization via 3D surface regression. In: Proceedings of the IEEE Conference on Computer Vision and Pattern Recognition, pp. 4654–4662 (2018)
13. Brahmbhatt, S., Gu, J., Kim, K., Hays, J., Kautz, J.: Geometry-aware learning of maps for camera localization. In: Proceedings of the IEEE Conference on Computer Vision and Pattern Recognition, pp. 2616–2625 (2018)
14. Breiman, L.: Random forests. Mach. Learn. **45**(1), 5–32 (2001)
15. Bui, M., Albarqouni, S., Ilic, S., Navab, N.: Scene coordinate and correspondence learning for image-based localization. In: British Machine Vision Conference (BMVC) (2018)
16. Busam, B., Birdal, T., Navab, N.: Camera pose filtering with local regression geodesics on the Riemannian manifold of dual quaternions. In: Proceedings of the IEEE International Conference on Computer Vision Workshops, pp. 2436–2445 (2017)
17. Cadena, C., et al.: Past, present, and future of simultaneous localization and mapping: toward the robust-perception age. IEEE Trans. Robot. **32**(6), 1309–1332 (2016)

18. Clark, R., Wang, S., Markham, A., Trigoni, N., Wen, H.: VidLoc: a deep spatio-temporal model for 6-DoF video-clip relocalization. In: Proceedings of the IEEE Conference on Computer Vision and Pattern Recognition (2017)
19. Corona, E., Kundu, K., Fidler, S.: Pose estimation for objects with rotational symmetry. In: 2018 IEEE/RSJ International Conference on Intelligent Robots and Systems (IROS), pp. 7215–7222. IEEE (2018)
20. Cui, H., et al.: Multimodal trajectory predictions for autonomous driving using deep convolutional networks. In: 2019 International Conference on Robotics and Automation (ICRA), pp. 2090–2096. IEEE (2019)
21. Deng, H., Birdal, T., Ilic, S.: 3D local features for direct pairwise registration. In: The IEEE Conference on Computer Vision and Pattern Recognition (CVPR), June 2019
22. Durrant-Whyte, H., Bailey, T.: Simultaneous localization and mapping: part I. IEEE Robot. Autom. Mag. **13**(2), 99–110 (2006)
23. Falorsi, L., de Haan, P., Davidson, T.R., Forré, P.: Reparameterizing distributions on lie groups. arXiv preprint arXiv:1903.02958 (2019)
24. Feng, W., Tian, F.P., Zhang, Q., Sun, J.: 6D dynamic camera relocalization from single reference image. In: Proceedings of the IEEE Conference on Computer Vision and Pattern Recognition, pp. 4049–4057 (2016)
25. Firman, M., Campbell, N.D., Agapito, L., Brostow, G.J.: DiverseNet: when one right answer is not enough. In: Proceedings of the IEEE Conference on Computer Vision and Pattern Recognition, pp. 5598–5607 (2018)
26. Fischler, M.A., Bolles, R.C.: Random sample consensus: a paradigm for model fitting with applications to image analysis and automated cartography. Commun. ACM **24**(6), 381–395 (1981)
27. Gal, Y., Ghahramani, Z.: Dropout as a Bayesian approximation: representing model uncertainty in deep learning. In: International Conference on Machine Learning, pp. 1050–1059 (2016)
28. Gilitschenski, I., Sahoo, R., Schwarting, W., Amini, A., Karaman, S., Rus, D.: Deep orientation uncertainty learning based on a Bingham loss. In: International Conference on Learning Representations (2020)
29. Glover, J., Kaelbling, L.P.: Tracking the spin on a ping pong ball with the quaternion Bingham filter. In: 2014 IEEE International Conference on Robotics and Automation (ICRA), pp. 4133–4140, May 2014
30. Glover, J., Bradski, G., Rusu, R.B.: Monte Carlo pose estimation with quaternion kernels and the Bingham distribution. In: Robotics Science System (2012)
31. Glover, J.M.: The quaternion Bingham distribution, 3D object detection, and dynamic manipulation. Ph.D. thesis, Massachusetts Institute of Technology (2014)
32. Grassia, F.S.: Practical parameterization·of rotations using the exponential map. J. Graph. Tools **3**(3), 29–48 (1998)
33. Guo, C., Pleiss, G., Sun, Y., Weinberger, K.Q.: On calibration of modern neural networks. In: Proceedings of the 34th International Conference on Machine Learning, vol. 70, pp. 1321–1330. JMLR. org (2017)
34. Guzman-Rivera, A., Batra, D., Kohli, P.: Multiple choice learning: learning to produce multiple structured outputs. In: Advances in Neural Information Processing Systems, pp. 1799–1807 (2012)
35. Haarbach, A., Birdal, T., Ilic, S.: Survey of higher order rigid body motion interpolation methods for keyframe animation and continuous-time trajectory estimation. In: 2018 Sixth International Conference on 3D Vision (3DV), pp. 381–389. IEEE (2018). https://doi.org/10.1109/3DV.2018.00051

36. He, K., Zhang, X., Ren, S., Sun, J.: Deep residual learning for image recognition. In: Proceedings of the IEEE Conference on Computer Vision and Pattern Recognition, pp. 770–778 (2016)
37. Herz, C.S.: Bessel functions of matrix argument. Ann. Math. **61**(3), 474–523 (1955). http://www.jstor.org/stable/1969810
38. Hinterstoisser, S., et al.: Model based training, detection and pose estimation of texture-less 3D objects in heavily cluttered scenes. In: Lee, K.M., Matsushita, Y., Rehg, J.M., Hu, Z. (eds.) ACCV 2012. LNCS, vol. 7724, pp. 548–562. Springer, Heidelberg (2013). https://doi.org/10.1007/978-3-642-37331-2_42
39. Horaud, R., Conio, B., Leboulleux, O., Lacolle, B.: An analytic solution for the perspective 4-point problem. In: Proceedings CVPR 1989: IEEE Computer Society Conference on Computer Vision and Pattern Recognition. IEEE (1989)
40. Kendall, A., Cipolla, R.: Modelling uncertainty in deep learning for camera relocalization. In: 2016 IEEE International Conference on Robotics and Automation (ICRA), pp. 4762–4769. IEEE (2016)
41. Kendall, A., Cipolla, R., et al.: Geometric loss functions for camera pose regression with deep learning. In: Proceedings of CVPR, vol. 3, p. 8 (2017)
42. Kendall, A., Gal, Y.: What uncertainties do we need in Bayesian deep learning for computer vision? In: Advances in Neural Information Processing Systems (2017)
43. Kendall, A., Grimes, M., Cipolla, R.: PoseNet: a convolutional network for real-time 6-DoF camera relocalization. In: Proceedings of the IEEE International Conference on Computer Vision, pp. 2938–2946 (2015)
44. Kingma, D.P., Welling, M.: Auto-encoding variational bayes. arXiv preprint arXiv:1312.6114 (2013)
45. Kume, A., Wood, A.T.: Saddlepoint approximations for the bingham and fisher-bingham normalising constants. Biometrika **92**(2), 465–476 (2005)
46. Kurz, G., Gilitschenski, I., Julier, S., Hanebeck, U.D.: Recursive estimation of orientation based on the Bingham distribution. In: 2013 16th International Conference on Information Fusion (FUSION), pp. 1487–1494. IEEE (2013)
47. Kurz, G., et al.: Directional statistics and filtering using libdirectional. arXiv preprint arXiv:1712.09718 (2017)
48. Labbé, M., Michaud, F.: Rtab-map as an open-source lidar and visual simultaneous localization and mapping library for large-scale and long-term online operation. J. Field Robot. **36**(2), 416–446 (2019)
49. Makansi, O., Ilg, E., Cicek, O., Brox, T.: Overcoming limitations of mixture density networks: a sampling and fitting framework for multimodal future prediction. In: Proceedings of the IEEE Conference on Computer Vision and Pattern Recognition, pp. 7144–7153 (2019)
50. Manhardt, F., et al.: Explaining the ambiguity of object detection and 6D pose from visual data. In: International Conference of Computer Vision. IEEE/CVF (2019)
51. Mardia, K.V., Jupp, P.E.: Directional Statistics. Wiley, Hoboken (2009)
52. Massiceti, D., Krull, A., Brachmann, E., Rother, C., Torr, P.H.: Random forests versus neural networks–what's best for camera localization? In: 2017 IEEE International Conference on Robotics and Automation (ICRA). IEEE (2017)
53. Morawiec, A., Field, D.: Rodrigues parameterization for orientation and misorientation distributions. Philos. Mag. A **73**(4), 1113–1130 (1996)
54. Murray, R.M.: A Mathematical Introduction to Robotic Manipulation. CRC Press, Boca Raton (1994)
55. Paszke, A., et al.: Automatic differentiation in PyTorch. In: NIPS Autodiff Workshop (2017)

56. Peretroukhin, V., Wagstaff, B., Giamou, M., Kelly, J.: Probabilistic regression of rotations using quaternion averaging and a deep multi-headed network. arXiv preprint arXiv:1904.03182 (2019)
57. Piasco, N., Sidibé, D., Demonceaux, C., Gouet-Brunet, V.: A survey on visual-based localization: on the benefit of heterogeneous data. Pattern Recogn. **74**, 90–109 (2018)
58. Pitteri, G., Ramamonjisoa, M., Ilic, S., Lepetit, V.: On object symmetries and 6D pose estimation from images. In: 3D Vision (3DV). IEEE (2019)
59. Prokudin, S., Gehler, P., Nowozin, S.: Deep directional statistics: pose estimation with uncertainty quantification. In: Proceedings of the European Conference on Computer Vision (ECCV), pp. 534–551 (2018)
60. Qi, C.R., Litany, O., He, K., Guibas, L.J.: Deep hough voting for 3D object detection in point clouds. In: The IEEE International Conference on Computer Vision (ICCV), October 2019
61. Riedel, S., Marton, Z.C., Kriegel, S.: Multi-view orientation estimation using Bingham mixture models. In: 2016 IEEE International Conference on Automation, Quality and Testing, Robotics (AQTR), pp. 1–6. IEEE (2016)
62. Rubner, Y., Tomasi, C., Guibas, L.J.: The earth mover's distance as a metric for image retrieval. Int. J. Comput. Vis. **40**(2), 99–121 (2000)
63. Rupprecht, C., et al.: Learning in an uncertain world: representing ambiguity through multiple hypotheses. In: Proceedings of the IEEE International Conference on Computer Vision, pp. 3591–3600 (2017)
64. Salas-Moreno, R.F., Newcombe, R.A., Strasdat, H., Kelly, P.H., Davison, A.J.: SLAM++: simultaneous localisation and mapping at the level of objects. In: Proceedings of the IEEE Conference on Computer Vision and Pattern Recognition, pp. 1352–1359 (2013)
65. Salimans, T., Goodfellow, I., Zaremba, W., Cheung, V., Radford, A., Chen, X.: Improved techniques for training GANs. In: Advances in Neural Information Processing Systems, pp. 2234–2242 (2016)
66. Sattler, T., Havlena, M., Radenovic, F., Schindler, K., Pollefeys, M.: Hyperpoints and fine vocabularies for large-scale location recognition. In: Proceedings of the IEEE International Conference on Computer Vision, pp. 2102–2110 (2015)
67. Sattler, T., Zhou, Q., Pollefeys, M., Leal-Taixe, L.: Understanding the limitations of CNN-based absolute camera pose regression. In: Proceedings of the IEEE Conference on Computer Vision and Pattern Recognition, pp. 3302–3312 (2019)
68. Schonberger, J.L., Frahm, J.M.: Structure-from-motion revisited. In: Proceedings of the IEEE Conference on Computer Vision and Pattern Recognition (2016)
69. Shotton, J., Glocker, B., Zach, C., Izadi, S., Criminisi, A., Fitzgibbon, A.: Scene coordinate regression forests for camera relocalization in RGB-D images. In: Proceedings of the IEEE Conference on Computer Vision and Pattern Recognition, pp. 2930–2937 (2013)
70. Simonyan, K., Zisserman, A.: Very deep convolutional networks for large-scale image recognition. CoRR abs/1409.1556 (2014)
71. Suvrit, S., Ley, C., Verdebout, T.: Directional statistics in machine learning: a brief review. In: Applied Directional Statistics. Chapman and Hall/CRC (2018)
72. Ullman, S.: The interpretation of structure from motion. Proc. Roy. Soc. London. Ser. B. Biol. Sci. **203**(1153), 405–426 (1979)
73. Valentin, J., Nießner, M., Shotton, J., Fitzgibbon, A., Izadi, S., Torr, P.H.: Exploiting uncertainty in regression forests for accurate camera relocalization. In: Proceedings of the IEEE Conference on Computer Vision and Pattern Recognition, pp. 4400–4408 (2015)

74. Yamaji, A.: Genetic algorithm for fitting a mixed bingham distribution to 3D orientations: a tool for the statistical and paleostress analyses of fracture orientations. Island Arc **25**(1), 72–83 (2016)
75. Zakharov, S., Shugurov, I., Ilic, S.: DPOD: 6D pose object detector and refiner. In: The IEEE International Conference on Computer Vision (ICCV) (2019)
76. Zeisl, B., Sattler, T., Pollefeys, M.: Camera pose voting for large-scale image-based localization. In: Proceedings of the IEEE International Conference on Computer Vision, pp. 2704–2712 (2015)
77. Zolfaghari, M., Çiçek, Ö., Ali, S.M., Mahdisoltani, F., Zhang, C., Brox, T.: Learning representations for predicting future activities. arXiv:1905.03578 (2019)

Modeling Artistic Workflows for Image Generation and Editing

Hung-Yu Tseng[1(✉)], Matthew Fisher[2], Jingwan Lu[2], Yijun Li[2],
Vladimir Kim[2], and Ming-Hsuan Yang[1]

[1] University of California, Oakland, USA
htseng6@ucmerced.edu
[2] Adobe Research, San Jose, USA

Abstract. People often create art by following an artistic workflow involving multiple stages that inform the overall design. If an artist wishes to modify an earlier decision, significant work may be required to propagate this new decision forward to the final artwork. Motivated by the above observations, we propose a generative model that follows a given artistic workflow, enabling both multi-stage image generation as well as multi-stage image editing of an existing piece of art. Furthermore, for the editing scenario, we introduce an optimization process along with learning-based regularization to ensure the edited image produced by the model closely aligns with the originally provided image. Qualitative and quantitative results on three different artistic datasets demonstrate the effectiveness of the proposed framework on both image generation and editing tasks.

1 Introduction

Creating artwork from scratch is a herculean task for people without years of artistic experience. For novices to the world of art, it would be more feasible to accomplish this task if there are clear creation steps to follow. Take a watercolor painting for example. One may be guided to first sketch the outline with pencils, then fill out areas with large brushes, and finalize details such as the color gradient and shadow with small brushes. At each stage, some aspects (i.e., variations) of the overall design are determined to carry forward to the final piece of art.

Inspired by these observations, we aim to model workflows for creating art, targeting two relevant artistic applications: multi-stage artwork creation and multi-stage artwork editing. As shown in Fig. 1, multi-stage artwork generation guides the user through the creation process by starting from the first stage then

---

H.-Y. Tseng—Work done during HY's internship at Adobe Research.

Electronic supplementary material The online version of this chapter (https://doi.org/10.1007/978-3-030-58523-5_10) contains supplementary material, which is available to authorized users.

A. Vedaldi et al. (Eds.): ECCV 2020, LNCS 12363, pp. 158–174, 2020.
https://doi.org/10.1007/978-3-030-58523-5_10

selecting the variation at each subsequent creation stage. In the multi-stage art-work editing, we are given a final piece of artwork and infer all the intermediate creation stages, enabling the user to perform different types of editing on various stages and propagate them forward to modify the final artwork.

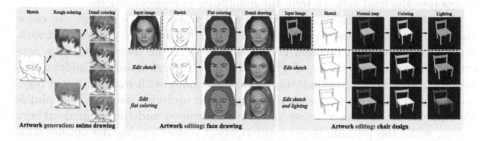

Fig. 1. We model the sequential creation stages for a given artistic workflow by learning from examples. At test time, our framework can guide the user to create new artwork by sampling different variations at each stage (left), and infer the creation stages of existing artwork to enable the user to perform natural edits by exploring variations at different stages (middle and right).

Existing artwork creation approaches use conditional generative adversarial networks (conditional GANs) [20,28,49] to produce the artwork according to user-provided input signals. These methods can take user inputs such as a sketch image [7] or segmentation mask [34,43] and perform a single-step generation to synthesize the final artwork. To make the creation process more tractable, recent frameworks adopt a multi-step generation strategy to accomplish the generation tasks such as fashion simulation [38] and sketch-to-image [12]. However, these approaches typically do not support editing existing artwork. To manipulate an existing artwork image without degrading the quality, numerous editing schemes [4,31,36,46,47] have been proposed in the past decade. Nevertheless, these methods either are designed for specific applications [31,36,46] or lack flexible controls over the editing procedure because of the single-stage generation strategy [4,47].

In this paper, we develop a conditional GAN-based framework that 1) synthesizes novel artwork via multiple creation stages, and 2) edits existing artwork at various creation stages. Our approach consists of an artwork generation module and a workflow inference module. The artwork generation module learns to emulate each artistic stage by a series of multi-modal (i.e., one-to-many) conditional GAN [49] networks. Each network in the artwork generation module uses a stage-specific latent representation to encode the variation presented at the corresponding creation stage. At test time, the user can determine the latent representation at each stage sequentially for the artwork generation module to synthesize the desired artwork image.

To enable editing existing artwork, we also design an inference module that learns to sequentially infer the corresponding images at all intermediate stages.

We assume a one-to-one mapping from the final to intermediate stages, and use a series of uni-modal conditional GANs [20] to perform this inference. At test time, we predict the stage-specific latent representations from the inferred images at all intermediate stages. Depending on the desired type of edit, the user can edit any stage to manipulate the stage-specific image or latent representation and regenerate the final artwork from the manipulated representations.

We observe that directly applying our workflow inference module can cause the reconstructed image to differ slightly from the initially provided artwork. Such a reconstruction problem is undesirable since the user expects the generated image to be unchanged when no edits are performed. To address this problem, we design an optimization procedure along with learning-based regularization to refine the reconstructed image. This optimization aims to minimize the appearance difference between the reconstructed and the original artwork image, while the learning-based regularization seeks to guide the optimization process and alleviate overfitting.

We collect three datasets with different creation stages to demonstrate the use cases of our approach: face drawing, anime drawing, and chair design. We demonstrate the creation process guided by the proposed framework and present editing results made by artists. For quantitative evaluations, we measure the reconstruction error and Fréchet inception distance (FID) [14] to validate the effectiveness of the proposed optimization and learning-based regularization scheme. We make the code and datasets public available to stimulate the future research.[1]

In this work, we make the following three contributions:

- We propose an image generation and editing framework which models the creation workflow for a particular type of artwork.
- We design an optimization process and a learning-based regularization function for the reconstruction problem in the editing scenario.
- We collect three different datasets containing various design stages and use them to evaluate the proposed approach.

2 Related Work

Generative Adversarial Networks (GANs). GANs [2,5,13,22,23] model the real image distribution via adversarial learning schemes. Typically, these methods encode the distribution of real images into a latent space by learning the mapping from latent representations to generated images. To make the latent representation more interpretable, the InfoGAN [8] approach learns to disentangle the latent representations by maximizing the mutual information. Similar to the FineGAN [37] and VON [50] methods, our approach learns to synthesize an image via multiple stages of generation, and encode different types of variation into separate latent spaces at various stages. Our framework extends these approaches to also enables image editing of different types of artwork.

---

[1] https://github.com/hytseng0509/ArtEditing.

Conditional GANs. Conditional GANs learn to synthesize the output image by referencing the input context such as text descriptions [44], scene graphs [42], segmentation masks [15,34], and images [20]. According to the type of mapping from the input context to the output image, conditional GANs can be categorized as uni-modal (one-to-one) [20,48] or multi-modal (one-to-many) [17,29,32,49]. Since we assume there are many possible variations involved for the generation at each stage of the artwork creation workflow, we use the multi-modal conditional GANs to synthesize the next-stage image, and utilize the uni-modal conditional GANs to inference the prior-stage image.

Image Editing. Image editing frameworks enable user-guided manipulation without degrading the realism of the edited images. Recently, deep-learning-based approaches have made significant progress on various image editing tasks such as colorization [19,26,45,46], image stylization [16,31], image blending [18], image inpainting [33,35], layout editing [30], and face editing [6,9,36]. Unlike these task-specific methods, the task-agnostic iGAN [47] and GANPaint [4] models map the variation in the training data onto a low-dimensional latent space using GAN models. Editing can be conducted by manipulating the representation in the learned latent space. Different from iGAN and GANPaint, we develop a multi-stage generation method to model different types of variation at various stages.

Optimization for Reconstruction. In order to embed an existing image to the latent space learned by a GAN model, numerous approaches [10,25,49] propose to train an encoder to learn the mapping from images to latent representations. However, the generator sometimes fails to reconstruct the original image from the embedded representations. To address this problem, optimization-based methods are proposed in recent studies. Abdal et al. [1] and Bau et al. [4] adopt the gradient descent scheme to optimize the latent representations and modulations for the feature activations, respectively. The goal is to minimize the appearance distance between the generated and original images. We also utilize the optimization strategy to reconstruct existing artwork images. In addition, we introduce a learning-based regularization function to guide the optimization process.

Regularizations for Deep Learning. These approaches [11,24,27,39–41] aim to prevent the learning function from overfitting to a specific solution. Particularly, the weight decay scheme [24] regularizes by constraining the magnitude of learning parameters during the training phase. Nevertheless, regularization methods typically involve hyper-parameters that require meticulous hand-tuning to ensure the effectiveness. The MetaReg [3] method designs a learning-to-learn algorithm to automatically find the hyper-parameters of the weight decay regularization to address the domain generalization problem. Our proposed learning-based regularization is trained with a similar strategy but different objectives to alleviate the overfitting problem described in Sect. 3.2.

3 Method

Our approach is motivated by the sequential creation stages of artistic work-flows. We build a model that enables a user to 1) follow the creation stages to generate novel artwork and 2) conduct edits at different stages. Our framework is composed of an artwork generation and a workflow inference module. As shown in Fig. 2(a), the artwork generation module learns to model the creations stages of the artist workflow. To enable editing an existing piece of art, the workflow inference module is trained to sequentially infer the corresponding images at all creation stages. When editing existing artwork, it is important that the artwork remains as close as possible to the original artwork, and only desired design deci-sions are altered. To enable this, we design an optimization process together with a learning-based regularization that allows faithful reconstruction of the input image. We provide the implementation and training details for each component in the proposed framework as supplemental material.

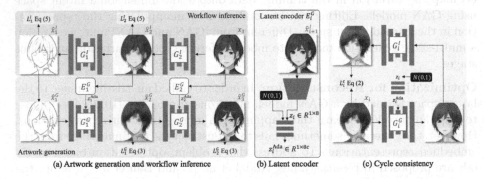

(a) Artwork generation and workflow inference   (b) Latent encoder   (c) Cycle consistency

Fig. 2. Overview of the proposed framework. (a) Given N creation stages ($N = 3$ in this example), our approach consists of $N-1$ workflow inference networks and $N-1$ artwork generation networks. The workflow inference module produces the intermedi-ate results of the input artwork at all creation stages. The artwork generation module computes the latent representation z and transformation parameter z^{Ada} for each stage, then reconstructs the input artwork images from these transformation parameters. (b) The latent encoder E_i^G extracts the stage-specific latent representation z from the example, and computes the transformation parameters z^{Ada} for the AdaIN normal-ization layers (c channels). (c) We introduce a cycle consistency loss for each stage to prevent the artwork generation model (which accounts for detail coloring in this example) from memorizing the variation determined at the previous stages (sketching and flat coloring).

3.1 Artwork Generation and Workflow Inference

Preliminaries. The proposed approach is driven by the number of stages in the training dataset and operates in a supervised setting with aligned training data.

Denoting N as the number of stages, the training dataset is comprised of a set of image groups $\{(x_1, x_2, \cdots, x_N)\}$, where x_N denotes the artwork image at the final stage. We construct the proposed framework with $N - 1$ workflow inference models $\{G_i^I\}_{i=1}^N$ as well as $N - 1$ artwork generation models $\{(E_i^G, G_i^G)\}_{i=1}^N$. We show an example of 3 stages in Fig. 2(a). Since the proposed method is based on the observation that artists sequentially determine a design factor (i.e., variation) at each stage, we assume that the generation from the image in the prior stage to the later one is multi-modal (i.e., one-to-many mapping), while the inference from the final to the previous stages is uni-modal (i.e., one-to-one mapping).

Artwork Generation. The artwork generation module aims to mimic the sequential creation stages of the artistic workflow. Since we assume the generation from the prior stages to the following ones is multi-modal, we construct a series of artwork generation networks by adopting the multi-modal conditional GAN approach in BicycleGAN [49] and the network architecture of MUNIT [17]. As shown in Fig. 2(a) and (b), each artwork generation model contains two components: latent encoder E_i^G and generator G_i^G. The latent encoder E_i^G encodes the variation presented at the i-th stage in a stage-specific latent space. Given an input image x_i and the corresponding next-stage image x_{i+1}, the latent encoder E_i^G extracts the stage-specific latent representation z_i from the image x_{i+1}, and computes the transformation parameter z_i^{Ada}. The generator G_i^G then takes the current-stage image x_i as input and modulates the activations through the AdaIN normalization layers [49] with the transformation parameter z_i^{Ada} to synthesize the next-stage image $\hat{x}_{i+1}^G$, namely

$$\hat{x}_{i+1}^G = G_i^G(x_i, E_i^G(x_{i+1})) \quad i \in \{1, 2, \cdots, N - 1\}. \tag{1}$$

We utilize the objective introduced in the BicycleGAN [49], denoted as L_i^{bicycle}, for training the generation model. The objective L_i^{bicycle} is detailed in the supplementary material.

Ideally, the artwork generation networks corresponding to a given stage would encode only new information (i.e., incremental variation), preserving prior design decisions from earlier stages. To encourage this property, we impose a cycle consistency loss to enforce the generation network to encode the variation presented at the current stage only, as shown in Fig. 2(c). Specifically, we use the inference model G_i^I to map the generated next-stage image back to the current stage. The mapped image should be identical to the original image x_i at the current stage, namely

$$L_i^c = \|G_i^I(G_i^G(x_i, E_i^G(z_i))) - x_i\|_1 \quad z_i \sim N(0, 1). \tag{2}$$

Therefore, the overall training objective for the artwork generation model at the i-th stage is

$$L_i^G = L_i^{\text{bicycle}} + \lambda^c L_i^c, \tag{3}$$

where λ^c controls the importance of the cycle consistency.

Workflow Inference. To enable the user to edit the input artwork x_N at different creation stages, our inference module aims to hallucinate the corresponding images at all previous stages. For the i-th stage, we use a unimodal conditional GAN network [20] to generate the image at i-th stage from the image at $(i+1)$-th stage, namely

$$\hat{x}_i^I = G_i^I(x_{i+1}) \quad i \in \{1, 2, \cdots, N-1\}. \tag{4}$$

During the training phase, we apply the hinge version of GAN loss [5] to ensure the realism of the generated image $\hat{x}_i^I$. We also impose an ℓ_1 loss between the synthesized image $\hat{x}_i^I$ and the ground-truth image x_i to stabilize and accelerate the training. Hence the training objective for the inference network at the i-th stage is

$$L_i^I = L_i^{\text{GAN}}(\hat{x}_i^I) + \lambda^1 \|\hat{x}_i^I - x_i\|_1, \tag{5}$$

where λ^1 controls the importance of the ℓ_1 loss.

Test-Time Inference. As shown in Fig. 2(a), given an input artwork image x_N, we sequentially obtain the images at all previous stages $\{\hat{x}_i^I\}_{i=1}^N$ using the workflow inference module (blue block). We then use the artwork generation module (green block) to extract the latent representations $\{z_i\}_{i=1}^{N-1}$ from the inferred images $\{\hat{x}_i^I\}_{i=1}^N$, and compute the transformation parameters $\{z_i^{\text{Ada}}\}_{i=1}^{N-1}$. Combining the first-stage image $x_1^G = x_1^I$ and the transformation parameters $\{z_i^{\text{Ada}}\}_{i=1}^{N-1}$, the generation module consecutively generates the images $\{\hat{x}_i^G\}_{i=2}^N$ at the following stages. The user can choose the stage to manipulate based on the type of edit desired. Edits at the i-th stage can be performed by either manipulating the latent representation z_i or directly modifying the image x_i^G. For example, in Fig. 2(a), the user can choose to augment the representation z_1 to adjust the flat coloring. After editing, the generation module generates the new artwork image at the final stage.

Fig. 3. Motivation of the AdaIN optimization and learning-based regularization. The proposed AdaIN optimization and the learning-based regularization are motivated by the observations that 1) using the computed transformation parameters z^{Ada} in Fig. 2 cannot well reconstruct the original input image (red outline in 1-st row)), and 2) the AdaIN optimization may degrade the quality of the editing results (yellow outline in 2-nd row). (Color figure online)

3.2 Optimization for Reconstruction

As illustrated in Sect. 3.1, the artwork generation module would ideally reconstruct the input artwork image (i.e., $\hat{x}_N^G = x_N$) from the transformation parameters $\{z_i^{\text{Ada}}\}_{i=1}^{N-1}$ before the user performs an edit. However, the reconstructed image $\hat{x}_N^G$ may be slightly different from the input image x_N, as shown in the first row of Fig. 3. Therefore, we adopt an AdaIN optimization algorithm to optimize the transformation parameters $\{z_i^{\text{Ada}}\}_{i=1}^{N}$ of the AdaIN normalization layers in the artwork generation models. The goal of the AdaIN optimization is to minimize the appearance distance between the reconstructed and input image.

While this does improve the reconstruction of the input image, we observe that the optimization procedure causes the generation module to memorize input image details, which degrades the quality of some edited results, as shown in the second row of Fig. 3. To mitigate this memorization, we propose a learning-based regularization to improve the AdaIN optimization.

AdaIN Optimization. The AdaIN optimization approach aims to minimize the appearance distance between the reconstructed image $\hat{x}_N^G$ and the input artwork image x_N. There are many choices for what to optimize to improve reconstruction: we could optimize the parameters in the generation models or the extracted representations $\{z_i\}_{i=1}^{N}$. Optimizing model parameters is inefficient because of the large number of parameters to be updated. On the other hand, we find that optimizing the extracted representation is ineffective, as validated in Sect. 4.3. As a result, we choose to optimize the transformation parameters $\{z_i^{\text{Ada}}\}_{i=1}^{N}$ of the AdaIN normalization layers in the generation models, namely the AdaIN optimization. Note that a recent study [1] also adopts a similar strategy.

We conduct the AdaIN optimization for each stage sequentially. The transformation parameter at the early stage is optimized and then fixed for the optimization at the later stages. Except for the last stage (i.e., $i = N - 1$) that uses the input artwork image x_N, the inferred image x_{i+1}^I by the inference model serves as the reference image x^{ref} for the optimization. For each stage, we first use the latent encoder E_i^G to compute the transformation parameter z_i^{Ada} from the reference image for generating the image. Since there are four AdaIN normalization layers with c channels in each artwork generation model, the dimension of the transformation parameter is $1 \times 8c$ (a scale and a bias term for each channel). Then we follow the standard gradient descent procedure to optimize the transformation parameters with the goal of minimizing the loss function L^{Ada} which measures the appearance distance between the synthesized image $\hat{x}_i^G$ by the generator G_i^G and the reference image x^{ref}. The loss function L^{Ada} is a combination of the pixel-wise ℓ_1 loss and VGG-16 perceptual loss [21], namely

$$L^{\text{Ada}}(\hat{x}_i^G, x^{\text{ref}}) = \|\hat{x}_i^G - x^{\text{ref}}\|_1 + \lambda^p L^p(\hat{x}_i^G, x^{\text{ref}}), \tag{6}$$

where λ_p is the importance term. We summarize the AdaIN optimization in Algorithm 1. Note that in practice, we optimize the incremental term δ_i^{Ada} for the transformation parameter z_i^{Ada}, instead of updating the parameter itself.

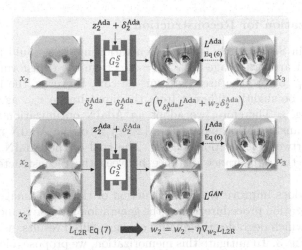

Fig. 4. Training process for learning-based regularization. For the i-th stage ($i = 2$ in this example), we optimize the hyper-parameter w_i for the weight decay regularization (orange text) by involving the AdaIN optimization in the training process: after the incremental term δ_i^{Ada} is updated via one step of AdaIN optimization and the weight decay regularization (blue arrow), the generation model should achieve improved reconstruction as well as maintain the quality of the editing result (green block). Therefore, we use the losses $L^{\text{Ada}}, L^{\text{GAN}}$ computed from the updated parameter $\tilde{\delta}_i^{\text{Ada}}$ to optimize the hyper-parameter w_i (red arrow). (Color figure online)

Learning-Based Regularization. Although the AdaIN optimization scheme addresses the reconstruction problem, it often degrades the quality of editing operations, as shown in the second row of Fig. 3. This is because the AdaIN optimization causes overfitting (memorization of the reference image x^{ref}). The incremental term δ_i^{Ada} for the transformation parameter z_i^{Ada} is updated to extreme values to achieve better reconstruction, so the generator becomes sensitive to the change (i.e., editing) on the input image and produces unrealistic results.

To address the overfitting problem, we use weight decay regularization [24] to constrain the magnitude of the incremental term δ_i^{Ada}, as shown in Line 6 in Algorithm 1. However, it is difficult to find a general hyper-parameter setting $w_i \in R^{1 \times 8c}$ for different generation stages of various artistic workflows. Therefore, we propose a learning algorithm to optimize the hyper-parameter w_i. The core idea is that updating the incremental term δ_i^{Ada} with the regularization $w_i \delta_i^{\text{Ada}}$ should 1) improve the reconstruction and 2) maintain the realism of edits on an input image. We illustrate the proposed algorithm in Fig. 4. In each iteration of training at the i-th stage, we sample an image pair (x_i, x_{i+1}) and an additional input image x_i' from the training dataset. The image x_i' serves as the edited image of x_i. We first use the latent encoder E_i^G to extract the transformation parameter z_i^{Ada} from the next-stage image x_{i+1}. As shown in the grey block of Fig. 4, we then update the incremental term from δ_i^{Ada} to $\tilde{\delta}_i^{\text{Ada}}$ via one step of the AdaIN optimization and the weight decay regularization. With the

---

Algorithm 1: AdaIN optimization at i-th stage

---

1 **Require:** reference image $x^{\text{ref}} = x_N$ or $x^{\text{ref}} = \hat{x}^I_{i+1}$, input image $\hat{x}^G_i$, learning
   rate α, iterations T, regularization parameter w_i
2 $z^{\text{Ada}}_i = E^G_i(x^{\text{ref}})$, $\delta^{\text{Ada}}_i = \mathbf{0} \in R^{1 \times 8c}$
3 **while** $t = \{1, \ldots, T\}$ **do**
4   $\quad \hat{x}^G_{i+1} = G^G_i(\hat{x}^G_i, z^{\text{Ada}}_i + \delta^{\text{Ada}}_i)$
5   $\quad L^{\text{Ada}} = \|\hat{x}^G_{i+1} - x^{\text{ref}}\|_1 + \lambda^p L^p(\hat{x}^G_{i+1}, x^{\text{ref}})$
6   $\quad \delta^{\text{Ada}}_i = \delta^{\text{Ada}}_i - \alpha \left(\nabla_{\delta^{\text{Ada}}_i} L_{\text{Ada}} + w_i \delta^{\text{Ada}}_i \right)$
7 **end**
8 **Return:** $z^{\text{Ada}}_i + \delta^{\text{Ada}}_i$

---

updated incremental term $\tilde{\delta}^{\text{Ada}}_i$, we use the loss function L^{Ada} to measure the reconstruction quality, and use the GAN loss to evaluate the realism of editing results, namely

$$L^{\text{L2R}} = L^{\text{Ada}}(G^G_i(x_i, z^{\text{Ada}}_i + \tilde{\delta}^{\text{Ada}}_i), x_{i+1}) \\ + \lambda^{\text{GAN}} L^{\text{GAN}}(G^G_i(x'_i, z^{\text{Ada}}_i + \tilde{\delta}^{\text{Ada}}_i)). \tag{7}$$

Finally, since the loss L^{L2R} indicates the efficacy of the weight decay regularization, we optimize the hyper-parameter w_i by

$$w_i = w_i - \eta \nabla_{w_i} L^{\text{L2R}}, \tag{8}$$

where η is the learning rate of the training algorithm for the proposed learning-based regularization.

4 Experimental Results

4.1 Datasets

To evaluate our framework, we manually process face drawing, anime drawing, and chair design datasets. We describe details in the supplementary material.

4.2 Qualitative Evaluation

Generation. We present the generation results at all stages in Fig. 5. In this experiment, we use the testing images at the first stage as inputs, and randomly sample various latent representation $z \in \{z_i\}^{N-1}_{i=1}$ at each stage of the proposed artwork generation module. The generation module sequentially synthesizes the final result via multiple stages. It successfully generates variations by sampling different random latent codes at different stages. For example, when generating anime drawings, manipulating the latent code at the final stage produces detailed color variations, such as modifying the saturation or adding the highlights to the hair regions.

Editing. Figure 6 shows the results of editing the artwork images at different stages. Specifically, after the AdaIN optimization reconstructs the testing image at the final stage (first row), we re-sample the representations $z \in \{z_i\}_{i=1}^{N-1}$ at various stages. Our framework is capable of synthesizing the final artwork such that its appearance only changes with respect to the stage with re-sampled latent code. For example, for editing face drawings, re-sampling representations at the flat coloring stage only affects hair color, while maintaining the haircut style and details.

To evaluate the interactivity of our system, we also asked professional artists to edit some example sketches (Fig. 7). First, we use the proposed framework to infer the initial sketch from the input artwork image. Given the artwork image and the corresponding sketch, we asked an artist to modify the sketch manually. For the edited sketch (second row), we highlight the edits with the red outlines. This experiment confirms that the proposed framework enables the artists to adjust only some stages of the workflow, controlling only desired aspects of the final synthesized image. Additional artistic edits are shown in Fig. 1.

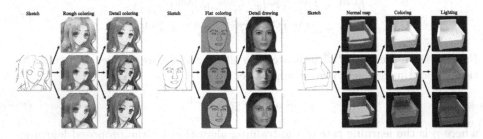

Fig. 5. Results of image generation from the first stage. We use the first-stage testing images as input and randomly sample the latent representations to generate the image at the final stage.

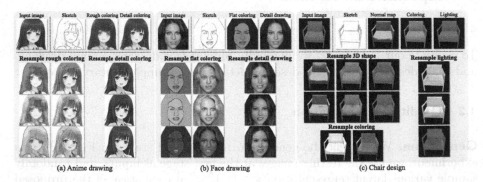

Fig. 6. Re-sampling latent representation at each stage. After we use the AdaIN optimization process to reconstruct the input image (1st row), we edit the reconstructed image by re-sampling the latent representations at various stages.

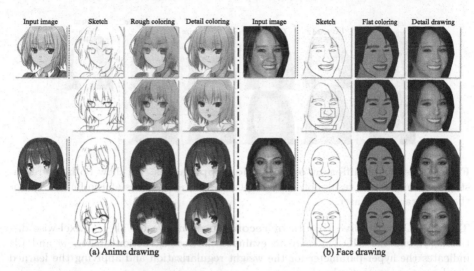

| Input image | Sketch | Rough coloring | Detail coloring | Input image | Sketch | Flat coloring | Detail drawing |

(a) Anime drawing   (b) Face drawing

Fig. 7. Results of artistic editing. Given an input artwork image, we ask the artist to edit the inferred sketch image. The synthesis model then produces the corresponding edited artwork. The first row shows the input artwork and inferred images, and the red outlines indicate the edited regions. (Color figure online)

AdaIN Optimization and Learning-Based Regularization. Figure 8 presents the results of the AdaIN optimization and the proposed learning-based regularization. As shown in the first row, optimizing representations z fails to refine the reconstructed images due to the limited capacity of the low-dimensional latent representation. In contrast, the AdaIN optimization scheme minimizes the perceptual difference between the input and reconstructed images. We also demonstrate how the optimization process influences the editing results in the second row. Although the AdaIN optimization resolves the reconstruction problem, it leads to overfitting and results in unrealistic editing results synthesized by the generation model. By utilizing the proposed learning-based regularization, we address the overfitting problem and improve the quality of the edited images.

4.3   Quantitative Evaluation

Evaluation Metrics. We use the following evaluation metrics:

- Reconstruction error: Given the input artwork x_N and the reconstructed image $\hat{x}_N^G$, we use the ℓ_1 distance $\|\hat{x}_N^G - x_N\|$ to evaluate the reconstruction quality.
- FID: We use the FID [14] score to measure the realism of generated images $\hat{x}_N^G$. A smaller FID score indicates better visual quality.

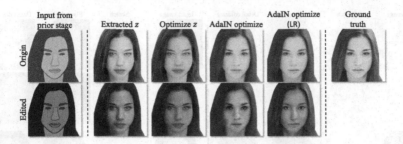

Fig. 8. Results of different optimization approaches. We show both the reconstruction and editing results of various optimization approaches at the final stage for the face drawing dataset.

Table 1. Quantitative results of reconstruction. We use the $\ell 1$ pixel-wise distance ($\downarrow$) and the FID ($\downarrow$) score to evaluate the reconstruction ability. w and LR indicates the hyper-parameter for the weight regularization and applying the learned regularization, respectively.

Optimization	w	Face		Anime		Chair	
		$\ell 1$	FID	$\ell 1$	FID	$\ell 1$	FID
None	–	0.094	39.78	0.127	36.73	0.074	129.2
z	0	0.104	40.70	0.126	45.66	0.068	107.0
AdaIN	0	**0.040**	**34.61**	**0.042**	**26.56**	**0.009**	**46.48**
AdaIN	10^{-3}	0.043	35.78	0.056	29.14	0.019	53.08
AdaIN	10^{-2}	0.053	39.19	0.097	46.31	0.049	83.58
AdaIN	LR	0.045	33.28	0.070	34.16	0.018	49.44

Reconstruction. As shown in Sect. 3.2, we conduct the AdaIN optimization for each stage sequentially to reconstruct the testing image at the final stage. We use both the reconstruction error and FID score to evaluate several baseline methods and the AdaIN optimization, and show the results in Table 1. Results on the 2-nd and 3-rd rows demonstrate that the AdaIN optimization is more effective than optimizing the latent representations $\{z_i\}_{i=1}^{N-1}$. On the other hand, applying stronger weight decay regularization (i.e., $w_i = 10^{-2}$) diminishes the reconstruction ability of the AdaIN optimization. By applying the weight decay regularization with learned hyper-parameter w (i.e., LR), we achieve comparable reconstruction performance in comparison to the optimization without regularization.

Editing. In this experiment, we investigate how various optimization methods influence the quality of edited images. For each testing final-stage image, we first use different optimization approaches to refine the reconstructed images. We then conduct the editing by re-sampling the latent representation z_i at a randomly chosen stage. We adopt the FID score to measure the quality of the edited images and show the results in Table 2. As described in Sect. 3.2, applying the AdaIN optimization causes overfitting that degrades the quality of the edited images. For instance, applying the AdaIN optimization increases the FID score from 38.68 to 44.28 on the face drawing dataset. One straightforward solution to alleviate this issue is to apply strong weight decay regularizations (i.e., $w = 10^{-2}$). However, according to the results in 5-th row of Table 1, such strong regularizations reduce the reconstruction effectiveness of the AdaIN optimization. Combining the results in Table 1 and Table 2, we conclude that applying the regularization with the learned hyper-parameter w not only mitigates overfitting but also maintains the efficacy of the AdaIN optimization. We conduct more analysis of the proposed learning-based regularization in the supplementary materials.

4.4 Limitations

The proposed framework has several limitations (see supplemental material for visual examples). First, since the model learns the multi-stage generation from a training dataset, it fails to produce appealing results if the style of the input image is significantly different from images in the training set. Second, the uni-modal inference assumption may not be correct. In practice, the mapping from later stages to previous ones can also be multi-modal. For instance, the style of the pencil sketches by various artists may be different. Finally, artists may not follow a well-staged workflow to create artwork in practice. However, our main goal is to provide an example workflow to make the artwork creation and editing more feasible, especially for the users who may not be experts in that type of artwork.

Table 2. Quantitative results of editing. We use the FID ($\downarrow$) score to evaluate the quality of the edited images $\hat{x}_N^G$ synthesized by the proposed framework. w and LR indicates the hyper-parameter for the weight regularization and applying the learned regularization, respectively.

Optimization	w	Face	Anime	Chair
None	–	38.68 ± 0.44	$\mathbf{35.59 \pm 0.12}$	128.4 ± 1.50
AdaIN	0	44.28 ± 0.45	37.40 ± 0.36	97.90 ± 1.20
AdaIN	10^{-3}	41.75 ± 0.49	38.95 ± 0.59	$\mathbf{91.68 \pm 4.23}$
AdaIN	10^{-2}	$\mathbf{38.57 \pm 0.94}$	38.07 ± 0.54	99.36 ± 7.23
AdaIN	LR	39.40 ± 0.21	35.73 ± 0.26	95.25 ± 0.73

5   Conclusions

In this work, we introduce an image generation and editing framework that models the creation stages of an artistic workflow. We also propose a learning-based regularization for the AdaIN optimization to address the reconstruction problem for enabling non-destructive artwork editing. Qualitative results on three different datasets show that the proposed framework 1) generates appealing artwork images via multiple creation stages and 2) synthesizes the editing results made by the artists. Furthermore, the quantitative results validate the effectiveness of the AdaIN optimization and the learning-based regularization.

We believe there are many exciting areas for future research in this direction that could make creating high-quality artwork both more accessible and faster. We would like to study video sequences of artists as they create artwork to automatically learn meaningful workflow stages that better align with the artistic process. This could further enable the design of editing tools that more closely align with the operations artists currently perform to iterate on their designs.

Acknowledgements. This work is supported in part by the NSF CAREER Grant #1149783.

References

1. Abdal, R., Qin, Y., Wonka, P.: Image2StyleGAN: how to embed images into the styleGAN latent space? In: ICCV (2019)
2. Arjovsky, M., Chintala, S., Bottou, L.: Wasserstein GAN. In: ICML (2017)
3. Balaji, Y., Sankaranarayanan, S., Chellappa, R.: MetaReg: towards domain generalization using meta-regularization. In: NIPS (2018)
4. Bau, D., et al.: Semantic photo manipulation with a generative image prior. ACM TOG (Proc. SIGGRAPH) **38**(4), 59 (2019)
5. Brock, A., Donahue, J., Simonyan, K.: Large scale GAN training for high fidelity natural image synthesis. In: ICLR (2019)
6. Chang, H., Lu, J., Yu, F., Finkelstein, A.: PairedCycleGAN: asymmetric style transfer for applying and removing makeup. In: CVPR (2018)
7. Chen, W., Hays, J.: SketchyGAN: towards diverse and realistic sketch to image synthesis. In: CVPR (2018)
8. Chen, X., Duan, Y., Houthooft, R., Schulman, J., Sutskever, I., Abbeel, P.: InfoGAN: interpretable representation learning by information maximizing generative adversarial nets. In: NIPS (2016)
9. Cheng, Y.C., Lee, H.Y., Sun, M., Yang, M.H.: Controllable image synthesis via SegVAE. In: ECCV (2020)
10. Donahue, J., Simonyan, K.: Large scale adversarial representation learning. In: NIPS (2019)
11. Ghiasi, G., Lin, T.Y., Le, Q.V.: DropBlock: a regularization method for convolutional networks. In: NIPS (2018)
12. Ghosh, A., et al.: Interactive sketch & fill: multiclass sketch-to-image translation. In: CVPR (2019)
13. Goodfellow, I., et al.: Generative adversarial nets. In: NIPS (2014)
14. Heusel, M., Ramsauer, H., Unterthiner, T., Nessler, B., Hochreiter, S.: GANs trained by a two time-scale update rule converge to a local nash equilibrium. In: NIPS (2017)
15. Huang, H.-P., Tseng, H.-Y., Lee, H.-Y., Huang, J.-B.: Semantic view synthesis. In: Vedaldi, A., Bischof, H., Brox, T., Frahm, J.-M. (eds.) ECCV 2020. LNCS, vol. 12357, pp. 592–608. Springer, Cham (2020). https://doi.org/10.1007/978-3-030-58610-2_35

16. Huang, X., Belongie, S.: Arbitrary style transfer in real-time with adaptive instance normalization. In: ICCV (2017)
17. Huang, X., Liu, M.Y., Belongie, S., Kautz, J.: Multimodal unsupervised image-to-image translation. In: ECCV (2018)
18. Hung, W.C., Zhang, J., Shen, X., Lin, Z., Lee, J.Y., Yang, M.H.: Learning to blend photos. In: ECCV (2018)
19. Iizuka, S., Simo-Serra, E., Ishikawa, H.: Let there be color!: joint end-to-end learning of global and local image priors for automatic image colorization with simultaneous classification. ACM TOG (Proc. SIGGRAPH) **35**(4), 110 (2016)
20. Isola, P., Zhu, J.Y., Zhou, T., Efros, A.A.: Image-to-image translation with conditional adversarial networks. In: CVPR (2017)
21. Johnson, J., Alahi, A., Fei-Fei, L.: Perceptual losses for real-time style transfer and super-resolution. In: Leibe, B., Matas, J., Sebe, N., Welling, M. (eds.) ECCV 2016. LNCS, vol. 9906, pp. 694–711. Springer, Cham (2016). https://doi.org/10.1007/978-3-319-46475-6_43
22. Karras, T., Aila, T., Laine, S., Lehtinen, J.: Progressive growing of GANs for improved quality, stability, and variation. In: ICLR (2018)
23. Karras, T., Laine, S., Aila, T.: A style-based generator architecture for generative adversarial networks. In: CVPR (2019)
24. Krogh, A., Hertz, J.A.: A simple weight decay can improve generalization. In: NIPS (1992)
25. Larsen, A.B.L., Sønderby, S.K., Larochelle, H., Winther, O.: Autoencoding beyond pixels using a learned similarity metric. arXiv preprint arXiv:1512.09300 (2015)
26. Larsson, G., Maire, M., Shakhnarovich, G.: Learning representations for automatic colorization. In: Leibe, B., Matas, J., Sebe, N., Welling, M. (eds.) ECCV 2016. LNCS, vol. 9908, pp. 577–593. Springer, Cham (2016). https://doi.org/10.1007/978-3-319-46493-0_35
27. Larsson, G., Maire, M., Shakhnarovich, G.: FractalNet: ultra-deep neural networks without residuals. In: ICML (2017)
28. Lee, H.Y., Tseng, H.Y., Huang, J.B., Singh, M.K., Yang, M.H.: Diverse image-to-image translation via disentangled representations. In: ECCV (2018)
29. Lee, H.Y., et al.: DRIT++: diverse image-to-image translation via disentangled representations. IJCV 1–16 (2020)
30. Lee, H.Y., et al.: Neural design network: graphic layout generation with constraints. In: ECCV (2020)
31. Li, Y., Liu, M.Y., Li, X., Yang, M.H., Kautz, J.: A closed-form solution to photo-realistic image stylization. In: ECCV (2018)
32. Mao, Q., Lee, H.Y., Tseng, H.Y., Ma, S., Yang, M.H.: Mode seeking generative adversarial networks for diverse image synthesis. In: CVPR (2019)
33. Nazeri, K., Ng, E., Joseph, T., Qureshi, F., Ebrahimi, M.: EdgeConnect: generative image inpainting with adversarial edge learning. arXiv preprint arXiv:1901.00212 (2019)
34. Park, T., Liu, M.Y., Wang, T.C., Zhu, J.Y.: Semantic image synthesis with spatially-adaptive normalization. In: CVPR (2019)
35. Pathak, D., Krahenbuhl, P., Donahue, J., Darrell, T., Efros, A.A.: Context encoders: feature learning by inpainting. In: CVPR (2016)
36. Portenier, T., Hu, Q., Szabo, A., Bigdeli, S.A., Favaro, P., Zwicker, M.: FaceShop: deep sketch-based face image editing. ACM TOG (Proc. SIGGRAPH) **37**(4), 99 (2018)
37. Singh, K.K., Ojha, U., Lee, Y.J.: FineGAN: unsupervised hierarchical disentanglement for fine-grained object generation and discovery. In: CVPR (2019)

38. Song, S., Zhang, W., Liu, J., Mei, T.: Unsupervised person image generation with semantic parsing transformation. In: CVPR (2019)
39. Srivastava, N., Hinton, G., Krizhevsky, A., Sutskever, I., Salakhutdinov, R.: Dropout: a simple way to prevent neural networks from overfitting. JMLR **15**(1), 1929–1958 (2014)
40. Tseng, H.Y., Chen, Y.W., Tsai, Y.H., Liu, S., Lin, Y.Y., Yang, M.H.: Regularizing meta-learning via gradient dropout. arXiv preprint arXiv:2004.05859 (2020)
41. Tseng, H.Y., Lee, H.Y., Huang, J.B., Yang, M.H.: Cross-domain few-shot classification via learned feature-wise transformation. In: ICLR (2020)
42. Tseng, H.Y., Lee, H.Y., Jiang, L., Yang, W., Yang, M.H.: RetrieveGAN: image synthesis via differentiable patch retrieval. In: ECCV (2020)
43. Wang, T.C., Liu, M.Y., Zhu, J.Y., Tao, A., Kautz, J., Catanzaro, B.: High-resolution image synthesis and semantic manipulation with conditional GANs. In: CVPR (2018)
44. Zhang, H., et al.: StackGAN++: realistic image synthesis with stacked generative adversarial networks. TPAMI **41**(8), 1947–1962 (2018)
45. Zhang, R., Isola, P., Efros, A.A.: Colorful image colorization. In: Leibe, B., Matas, J., Sebe, N., Welling, M. (eds.) ECCV 2016. LNCS, vol. 9907, pp. 649–666. Springer, Cham (2016). https://doi.org/10.1007/978-3-319-46487-9_40
46. Zhang, R., et al.: Real-time user-guided image colorization with learned deep priors. ACM TOG (Proc. SIGGRAPH) **9**(4) (2017)
47. Zhu, J.-Y., Krähenbühl, P., Shechtman, E., Efros, A.A.: Generative visual manipulation on the natural image manifold. In: Leibe, B., Matas, J., Sebe, N., Welling, M. (eds.) ECCV 2016. LNCS, vol. 9909, pp. 597–613. Springer, Cham (2016). https://doi.org/10.1007/978-3-319-46454-1_36
48. Zhu, J.Y., Park, T., Isola, P., Efros, A.A.: Unpaired image-to-image translation using cycle-consistent adversarial networks. In: ICCV (2017)
49. Zhu, J.Y., et al.: Toward multimodal image-to-image translation. In: NIPS (2017)
50. Zhu, J.Y., et al.: Visual object networks: image generation with disentangled 3D representations. In: NIPS (2018)

A Large-Scale Annotated Mechanical Components Benchmark for Classification and Retrieval Tasks with Deep Neural Networks

Sangpil Kim[1]([envelope]), Hyung-gun Chi[1], Xiao Hu[1], Qixing Huang[2], and Karthik Ramani[1]

[1] Purdue University, West Lafayette, IN 47907, USA
{kim2030,chi45,hu440,ramani}@purdue.edu
[2] The University of Texas at Austin, Austin, TX 78712, USA
huangqx@cs.utexas.edu

Abstract. We introduce a large-scale annotated mechanical components benchmark for classification and retrieval tasks named Mechanical Components Benchmark (MCB): a large-scale dataset of 3D objects of mechanical components. The dataset enables data-driven feature learning for mechanical components. Exploring the shape descriptor for mechanical components is essential to computer vision and manufacturing applications. However, not much attention has been given on creating annotated mechanical components datasets on a large scale. This is because acquiring 3D models is challenging and annotating mechanical components requires engineering knowledge. Our main contributions are the creation of a large-scale annotated mechanical component benchmark, defining hierarchy taxonomy of mechanical components, and benchmarking the effectiveness of deep learning shape classifiers on the mechanical components. We created an annotated dataset and benchmarked seven state-of-the-art deep learning classification methods in three categories, namely: (1) point clouds, (2) volumetric representation in voxel grids, and (3) view-based representation.

Keywords: Deep learning · Mechanical components · Benchmark · 3D objects · Classification · Retrieval

1 Introduction

The application of machine learning is highlighted recently due to the improved effectiveness of the deep neural networks [12,16,21,32,36,37]. Along with deep

S. Kim and H. Chi—These authors made an equal contribution.

Electronic supplementary material The online version of this chapter (https://doi.org/10.1007/978-3-030-58523-5_11) contains supplementary material, which is available to authorized users.

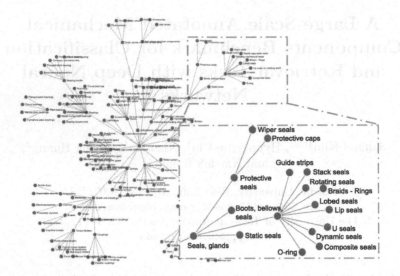

Fig. 1. The hierarchy taxonomy of mechanical components based on the International Classification for Standards.

neural networks, data driven algorithms and the creation of a large-scale datasets [4,8,26,47] have led to a series of breakthroughs in computer vision [6,14,45] and graphics [9,23,33]. The development of ImageNet [8], which used the class hierarchical structure from WordNet [31] to maximize the dataset coverage, showed that a well-structured and annotated dataset is crucial for developing geometric feature descriptors. The pre-trained deep neural network descriptors using ImageNet have been widely used to extract low-dimensional representations that are used in tasks of object detection [25,28,37], semantic segmentation [5,15,34,39], image caption generator [19,46], and image retrieval [20,43].

The creation of a large-scale mechanical components dataset with well-organized hierarchical classes and annotations is needed for developing and benchmarking geometric feature descriptors in the manufacturing industry [10, 18]. Geometric features extracted from the descriptors are fundamental cues to retrieve objects given the query object and classifying objects given image, volumetric representation, or point clouds.

However, in the manufacturing, design, and supply chain areas, the classification of mechanical components with deep neural networks has not been addressed due to the lack of large-scale annotated datasets. Without a standardized dataset, it is difficult to develop and compare learning algorithms on mechanical components [42].

Creating a large-scale mechanical component dataset is challenging due to the significant difficulty of collecting 3D CAD models of mechanical components. Different from common-object datasets [4,32,42,47], the accessibility of most mechanical components is limited because of proprietary and ownership issues with specially designed models. Products and manufacturing models are held by

Fig. 2. Randomly sampled mechanical components from the MCB.

companies for commercial usages, resulting in a deficiency of open-source components datasets. The inconsistency and incompatibility of mechanical components from available sources require massive effort on filtering and annotating the data. Also, annotating mechanical components is harder than common objects since it demands more knowledge and expertise from annotators to properly annotate engineering components.

To resolve this difficulty, we established a hierarchical semantic taxonomy as a guideline based on the International Classification for Standards (ICS) published by the International Organization for Standardization (ISO). The tree structure of our proposed hierarchical taxonomy is depicted in Fig. 1. Details are provided in the supplementary document. To collect annotations, we developed a web application which reduce the difficulty of filtering and annotating (Fig. 4). This application supports controllable viewing, displaying meta-information, annotation, and filtering parts by viewing multiple parts as a tabular to visually see the consistent shape features within the same class rather than viewing each individual part. These functionalities make the benchmark creation faster and more accurate for fine-grained categories.

Furthermore, we benchmark seven state-of-the-art shape descriptors to analyze the properties of mechanical components. Seven methods are carefully selected from three different 3D object representations: (1) point clouds, (2) voxel girds, and (3) view-based. From the benchmark results, the input representation is not the core factor that determines the performance. DLAN [11], which uses voxel grids, and PointCNN [24], which uses a point cloud input representation that focuses on local shape features, perform relatively well on both retrieval and classification tasks. The view-based methods are not robust on unseen orientation in shape retrieval tasks, which is also observed in common object retrieval tasks [41]. However, the descriptors [11,36,44] show significantly different results from common object classification tasks, which indirectly indicates that topological and geometrical characteristics of mechanical components are different from common objects. We report micro- and macro-precision, recall, F-score, mAP, and NDCG to evaluate retrieval tasks. For classification tasks, we report accuracy per class, accuracy per instance, F1 score, and average precision.

Our main contributions are as follows:

1. We create a hierarchical taxonomy based on the International Classification for Standards.
2. We annotate and collect a large-scale mechanical components benchmark. The mechanical components are annotated by mechanical engineering experts.
3. We benchmark seven deep learning-based classifiers and analyze their performances with the mechanical components dataset.
4. We develop a web-based framework which has a viewing, annotating, and filtering feature to guide annotators.

Each contribution above provides a significant benefit to the computer vision community and opportunities for researchers to develop new algorithms for the mechanical components.

2 Related Works

We will explain existing common-object and mechanical components datasets for 3D geometric data-driven learning methods. We summarized the overview of these datasets in Table 1. The reviews of shape classification and retrieval are detailed in Sect. 5.

Table 1. Comparison table of the MCB dataset with other datasets. CO and MC stands for common objects and mechanical components, respectively. ShapeNetCore, ShapeNetSem, and PartNet use models from the ShapeNet.

Dataset	# Class	# Models	Type
ModelNet [47]	40	12,311	CO
ShapeNet [4]	3,135	+3,000,000	CO
ShapeNetCore	55	51,300	CO
ShapeNetSem	270	12,000	CO
PrincetonSB [42]	92	6,670	CO
PartNet [32]	24	26,671	CO
ABC [11]	N/A	+1,000,000	MC
AAD [2]	9	180	MC
ESB [18]	45	867	MC
MCB (Ours)	68	58,696	MC

Large-Scale 3D Object Datasets. The Princeton Shape Benchmark (PSB) [42] is an early work that collected and annotated 3D objects for shape matching and classification benchmarking. It collected 3D polygonal models from the World Wide Web and classified them based on the method of construction, such as man-made and natural objects. ShapeNet [4] is a large-scale dataset of high-quality 3D models of objects, which are widely used in various tasks such as instance segmentation [32], shape retrieval [41], and shape reconstruction [7]. ModelNet [47] consists of two datasets (a 10-class dataset and a 40-class dataset) and demonstrates a comprehensive clean collection of 3D CAD models of objects. PartNet [32] is a fine-grained, instance-level, hierarchical parts dataset. It used 3D objects from ShapeNet and was annotated by 66 annotators.

Engineering Shape Datasets. Engineering shape datasets has been developed to improve the shape-based retrieval of 3D data [17]. The Engineering Shape Benchmark (ESB) [18] is an annotated engineering shape dataset. It proposed an approach that defines the class by mechanical part's name—not by functionality—and benchmarked analytical shape descriptor. However, the number of models in the ESB dataset is not sufficient for training a robust feature extractor, and classes are only classified by their shape, which limits the usage of the dataset. The Actual Artifacts Dataset (AAD) [2] consists of four datasets with a total around 700 models and provides several classifications for engineering artifacts selected from the National Design Repository (NDR) [38]. Recently, A Big Cad (ABC) Model Dataset [22] proposed one million Computer-Aided Design (CAD) models dataset without annotations.

3 Properties of Mechanical Components

Mechanical components, shown in Fig. 2, have sharp edges, well-defined surfaces, and high genus, which distinguishes them from common objects. Since the shape of mechanical parts represents their physical functions, the functionality of machine elements is sensitive to small details, resulting in the difficulty in annotation. Therefore, mechanical components are often categorized by their detailed shape, whereas common objects are mainly identified by their general shape. The shape and location of detail features often determine the function of engineering parts.

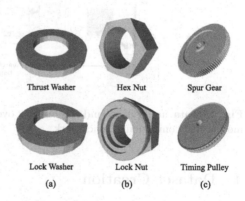

Thrust Washer   Hex Nut   Spur Gear

Lock Washer   Lock Nut   Timing Pulley

(a)   (b)   (c)

Fig. 3. Examples of detail features making categorical changes.

The shape of the detail features and the function of engineering parts are usually interdependent. For example, the only difference in shape between a thrust washer and a lock washer is the split detail feature, as seen in Fig. 3 (a), but they possess distinct functionality. A thrust washer spreads fastener loads, while a split lock washer uses the split feature to lock a nut and bolt in place. In another case, a hex nut and a lock nut share a hexagonal shape. However, the lock nut has an additional circular feature that houses a nylon insert, as seen in

Fig. 3 (b). A hex nut mates with a bolt to fasten materials together, and while the lock nut performs a similar function, the nylon insert keeps the nut from coming loose from the bolt. In another application, a spur gear transfers power to other toothed surfaces, while a timing pulley transfers power to a timing belt. Both parts' shapes and functions are similar, but small details in tooth shape and design differentiate the two parts, as seen in Fig. 3 (c). In contrast, changing the shape of common objects, like using longer legs on chairs, may not change the function of the object. Because these characteristics do not appear in common object datasets [4,47], the existing shape descriptors [24,36,41] need to be benchmarked on MCB to explore the shape descriptors on mechanical components. This is because recently-proposed deep neural networks descriptors are developed to capture the features from the common objects but not validated on the mechanical components. In this sense, Koch *et al.* [22] created a large CAD model dataset and benchmarked surface normal estimation, but they could not benchmark object classification or shape retrieval because they are not labeled.

An annotated benchmark dataset such as MCB can link the shape to the particular representation inside product data management systems of CAD kernels. Our work opens up ways for implementation of fine-grained searches with features of mechanical components, semantic text, and mechanical meta-data.

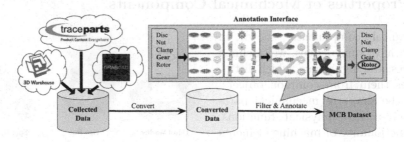

Fig. 4. Data acquisition and annotation overview for the creation of a large-scale mechanical components benchmark.

4   Dataset Creation

For the dataset creation, we first elaborate on the acquisition of mechanical components and explain how we annotated them. We acquire models from online 3D CAD repositories. To effectively annotate, CAD models are filtered and annotated using web-based tools. We define classes by following the field "Mechanical Systems and Components" of the International Classification Standard (ICS).

4.1   Data Acquisition

We collect mechanical components from online large 3D CAD repositories: TraceParts[1], 3D Warehouse[2], and GrabCAD[3]. 3D Warehouse and GrabCAD are large online open repositories for professional designers, engineers, manufacturers, and students to share CAD models. They provide numerous CAD models with various classes, including mechanical components. The models from TraceParts are industry standard components and shape variation within class is small. By merging the models from different sources, we obtained 163K mechanical components before annotation and purification as shown in Table 2.

4.2   Acquired Dataset Purification

We developed a web-based platform to manage large-scale dataset functioning, collecting, viewing, filtering, and annotating data. The overview of the platform is available in Fig. 4. Web-based applications have the advantage that users are free of installation and can easily access to the platform from any computer with internet connection. This accessibility accelerated the annotation process. We utilized the tool to trigger the scrapper collecting CAD models, also filtering and annotating the data with intuitive user interfaces, which is available in Fig. 4. A dataset managing platform visualizes multi-view images of each engineering part, which gives users a more comprehensive understanding of the mechanical part during filtering and annotating. The data creation pipeline consists of consecutive three steps.

Step 1: *Conversion/Multi-view image generation*. The file format conversion process is necessary to create a unified file format dataset, since collected CAD models consist of various formats such as STL, STEP, and OFF. The converter module in the platform converts file format into OBJ format and captures projected images from multiple viewpoints. For 3D data conversion, we used Open Asset Import Library (Assimp) [1] and Gmsh [13]. We used projected images for annotating engineering parts.

Step 2: *Filtering*. We filter scrapped CAD models by deleting broken and duplicated models and capturing wrongly categorized models for re-annotation. Meta-information (i.e. file size, file name, search keyword, and data source) tagged in the process of scrapping step helps users to filter the data. Eight ME experts manually filtered out the duplicates with our annotation tool. We group objects with similar meta-information and these experts manually removed duplicates. An annotation interface presents the models in a table format rather than one model at a time. Several models can be

Table 2. The number of data before and after filtering.

Data source	#Data	
	Before	After
GrabCAD	22,703	5,301
3D Warehouse	20,478	12,737
TraceParts	120,665	40,658
Total	163,216	58,696

---

[1] https://www.traceparts.com/.
[2] https://3dwarehouse.sketchup.com/.
[3] https://grabcad.com/.

viewed at one time, which increases the speed of filtering and makes identifying duplicate models easier. A quantitative comparison of the dataset between before and after filtering is shown in Table 2.

Step 3: *Annotation*. After filtering, we re-annotate the missed categorized models to the correct category based on the tagged information and multi-view image. We use a 3D viewer in the annotation interface to present a close-up look when the multi-view image does not provide enough information for annotation. Some of the models do not belong to any of our mechanical components categories but are still relevant to engineering parts. We defined these models as miscellaneous and labeled them into new part categories as needed.

5   Experiments

To analyze the behavior of learning algorithms developed for common objects works on mechanical components, we benchmarked classification and retrieval tasks with three different representations: point could, projected views, and voxel grids. We use two NVIDIA GeForce RTX 2080Ti GPUs, i9-9900k CPU, and 64GB RAM for the experiments. We carefully choose seven state-of-the-art shape classification algorithms from three different 3D shape representations: point cloud, multi-view, and voxel grids as the benchmark methods. In point cloud method, we use PointCNN [24], PointNet++ [36], and SpiderCNN [48]. For the multi-view based, we use MVCNN [44] and RotationNet [20]. DLAN [11] and VRN [3] are used to evaluate voxel grids representation. For training each method, we use the code and the hyper-parameters from seven deep-learning algorithm papers. We use 2,048 points density for point cloud, $32 \times 32 \times 32$ grid for voxel grids, and $3 \times 224 \times 224$ resolution for image-based representations. We follow the original papers for the input data processing and training procedures. For all the benchmark datasets, we randomly split the datasets into train and test set as 80% and 20%, respectively. Training is conducted for each method to prevent initialization variation and report the best results.

Point Clouds. A point cloud is a collection of points in Euclidean space. PointCNN [24] relaxes irregularity of point clouds by approximating the transformation matrix with multi-layer perception, which simultaneously weights and permutes the input features for point cloud data feature learning. PointNet [35] learns a set of optimization functions for selecting feature points that contain meaningful content, which canonicalizes the input point clouds and aggregates all feature points to capture global point cloud features. PointNet++ [35] is an advanced version of PointNet. This work focused on recognizing fine-grained patterns with a hierarchical neural network which iteratively applied on a nested partitioned point set. SpiderCNN [48] proposes a convolutional layer, which is a product of a step function that captures local geodesic information and a Taylor polynomial to convolve in point cloud.

Projected Views. View-based methods [20,43,44] extract features of 3D shape representations by observing multi-view images of an object and jointly estimating their poses. Their method successfully works for object classification and shape retrieval tasks, but performed poorly on unknown orientation models. Su *et al.* [43] uses a collection of multiple views of 3D objects, which is effective for learning their representations. MVCNN [44] further improves Su *et al.* with cross-modal distillation and adversarial inputs with a differentiable renderer.

Voxel Grids. Three-dimensional objects can be discredited and represented in voxel grids, and voxel-based classifiers use voxel grids as their inputs. DLAN [11] proposes Rotation Normalized Grids (RNGs), which are samples of oriented point sets rotated by PCA for shape retrieval. Multiple blocks of RNGs are converted into local features with 3D convolution, and these features are aggregated with average pooling as object representations. VSL [27] learns the probabilistic manifold of the underlying structure of voxelized 3D shapes with an auto-encoder in an unsupervised manner. VoxNet [30] converts point clouds into voxels in voxel grids and extracts features with a 3D convolution layer for the classification tasks. VRN [3] uses a series of 3D convolutional layers to extract features for classifying objects compactly.

Table 3. The statistics of Mechanical Components Benchmark dataset.

Class	#Models	Class	#Models	Class	#Models
Articulations eyelets& joints	1,632	Impeller	145	Socket	858
Bearing accessories	107	Keys and keyways splines	4,936	Spacers	113
Bushes	764	Knob	644	Split pins	472
Cap nuts	225	Lever	1,032	Spring washers	55
Castle nuts	226	Locating pins	55	Springs	328
Castor	99	Locknuts	254	Square	72
Chain drives	100	Lockwashers	434	Square nuts	53
Clamps	155	Nozzle	154	Standard fitting	764
Collars	52	Plain guidings	49	Studs	4,089
Conventional rivets	3,806	Plates circulate plates	365	Switch	173
Convex washer	91	Plugs	169	T-nut	101
Cylindrical pins	1,895	Pulleys	121	T-shape fitting	338
Elbow fitting	383	Radial contact ball bearings	1,199	Taper pins	1,795
Eye screws	1,131	Right angular gearings	60	Tapping screws	2,182
Fan	213	Right spur gears	430	Threaded rods	1,022
Flange nut	53	Rivet nut	51	Thrust washers	2,333
Flanged block bearing	404	Roll pins	1,597	Toothed	47
Flanged plain bearings	110	Screws& bolts \ w countersunk head	2,452	Turbine	85
Grooved pins	2,245	Screws& bolts \ w cylindrical head	3,656	Valve	94
Helical geared motors	732	Screws& bolts \ w hexagonal head	7,058	Washer bolt	912
Hexagonal nuts	1,039	Setscrew	1,334	Wheel	243
Hinge	54	Slotted nuts	78	Wingnuts	50
Hook	119	Snap rings	609	Total	58,696

5.1   Statistics of the Dataset

MCB has a total number of 58,696 mechanical components with 68 classes. The exact name of the types and amount of data in each category are shown in Table 3. Objects from TraceParts are aligned, but the objects from the other two sources (30 % of the objects) are not consistently oriented. We did not perform additional alignments as many object classes do not possess consistent orientations due to a variety of continuous/discrete symmetries. On the other hand, having unaligned models in shape classification and retrieval tasks helps to evaluate the generalization of the shape descriptors [40]. Unlike 3D Warehouse and GrabCAD that provide data from general usages, TraceParts stores data from the manufacturing companies. The CAD models from manufacturing companies show a tiny variation because they follow the parameterized catalogs for standardization. Therefore, to see the effect of data that has dense distribution and orientation invariance, we built two datasets for the experiment:

- Dataset A (MCB): Aggregated data from TraceParts(see footnote 1), 3D Warehouse(see footnote 2), and GrabCAD(see footnote 3)
- Dataset B: Aggregated data from 3D Warehouse and GrabCAD.

The dataset A has the same statistics with the original MCB dataset, and the dataset B has 18,038 data with 25 classes. The detailed statistics of the dataset B is explained in supplementary material.

Table 4. Summary table of evaluation metrics of shape retrieval benchmark for seven deep learning methods. They are grouped by their representation types. Each *, †, and ⊠ symbol indicates the method point cloud, volumetric, and image, respectively.

Dataset	Method	micro					macro				
		P@N	R@N	F1@N	mAP	NDCG@N	P@N	R@N	F1@N	mAP	DCG@N
A	PointCNN* [24]	**0.892**	**0.892**	**0.690**	**0.889**	**0.898**	0.869	**0.797**	**0.833**	0.886	0.854
	PointNet++* [36]	0.778	0.778	0.613	0.794	0.754	0.772	0.678	0.712	0.803	0.746
	SpiderCNN* [48]	0.839	0.839	0.669	0.867	0.793	0.844	0.741	0.776	0.877	0.812
	MVCNN⊠ [44]	0.579	0.579	0.488	0.657	0.487	0.667	0.552	0.585	0.735	0.641
	RotationNet⊠ [20]	0.688	0.699	0.508	0.805	0.683	0.784	0.652	0.683	0.815	0.735
	DLAN† [11]	0.840	0.840	0.568	0.879	0.828	**0.878**	0.786	0.820	0.880	0.845
	VRN† [3]	0.537	0537	0.402	0.653	0.519	0.646	0.480	0.507	0.664	0.576
B	PointCNN*	0.905	0.905	**0.676**	**0.913**	0.899	0.895	0.829	0.853	**0.909**	**0.871**
	PointNet++*	0.847	0.847	0.657	0.892	0.798	0.873	0.799	0.823	0.903	0.846
	SpiderCNN*	0.779	0.779	0.609	0.829	0.728	0.782	0.698	0.719	0.841	0.757
	MVCNN⊠	0.786	0.786	0.609	0.831	0.742	0793	0.719	0.741	0.852	0.776
	RotationNet⊠	0.529	0.529	0.434	0.607	0.454	0.560	0.466	0.483	0.647	0.540
	DLAN†	**0.912**	**0.912**	0.674	0.908	**0.925**	**0.903**	**0.830**	**0.854**	0.902	0.870
	VRN†	0.607	0.607	0.460	0.628	0.613	0.565	0.468	0.484	0.619	0.534

5.2   Retrieval Benchmark

At each entry, we calculate scores of the precision-recall curve in the retrieval results: precision, recall, F1-score, mAP, and Normalized Discounted Cumulative

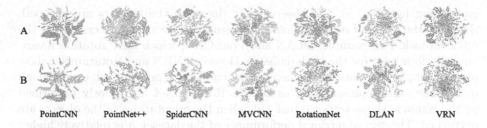

Fig. 5. t-SNE[29] plots of seven different deep neural networks trained with the dataset A and B. We set perplexity as 40 and iterate 300 times.

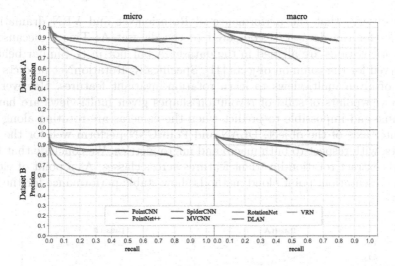

Fig. 6. Precision-recall curve plots for retrieval with seven different methods. The PointCNN shows best retrieval results for the dataset A, and DLAN shows best retrieval results for the dataset B.

Gain (NDCG). In shape retrieval, NDCG has a heavier tail at high ranks, which means that it does not discount lower ranks as much as mAP does [41]. Therefore, NDCG has a better ability to show the ration between the real performance and ideal performance to evaluate the metrics.

Since each object has a different number of positive retrievals, the score table metrics are referred to as P@N, R@N, F1@N, and NDCG@N, where the N refers to the total retrieval list length of each object, which varies across queries. The macro-averaged version presents the performance of the dataset combining the result of each category. The micro-averaged version treats each query, and the retrieval result equally treats cross groups. Therefore, it eventually has the same P@N and R@N.

The summary results of all tested methods are given in Table 4. Corresponding precision-recall curves are given in Fig. 6. To see the similarity of the geometric features from the descriptors, we perform t-distributed stochastic neighbor embedding (see Fig. 5). We observe that the more the clusters are grouped, the more the retrieval results enhanced. Orientation invariance is crucial for the retrieval task. For example, DLAN and PointCNN, which have rotation invariance, perform best for the both datasets. However, VRN and RotationNet show poor results for the dataset B where the orientation is not aligned, even though it uses the same representation as DLAN. RotationNet also poorly performed on the common shape retrieval task [40] when the orientations of the objects are perturbed. The overall retrieval performance of the dataset A is relatively higher than the dataset B. Micro has slightly better results on P@N, while much better results on R@N show that the metrics have better performance in cross-category testing.

We observe that the performance of RotationNet and VRN dramatically decreases for the dataset B compared to the dataset A. This is because the object orientations are aligned in the dataset A but not in B. Similar behavior is observed for the common objects [41]. Specifically, RotationNet predicts view orders of given multi-views to learn rotation-invariant features. However, the camera viewpoints of solid of revolution shapes given multi-views are hard to determine and impossible to predict when the cameras are rotating along with the center axis of the object. DLAN and PointCNN perform well for the both datasets, with respect to both macro and micro metrics. We conclude that these methods extract rotation-invariant features across classes. As a point of view in data representation, point cloud methods show stable performance for the both datasets.

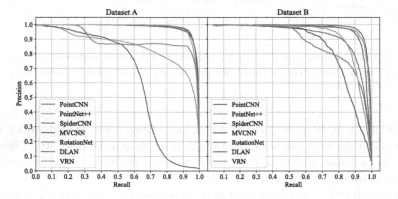

Fig. 7. Precision and Recall curve plots of classification task. Left plot shows the PR curve of the dataset A and right plot shows the PR curve the dataset B. The RotationNet shows the best performance in terms of accuracy.

5.3   Classification Benchmark

For the classification task, we measure four metrics, mean accuracy over objects, average accuracy per class, F1-score and average precision (AP) and plotted precision-recall curves. We use the macro method for F1 and AP calculation.

Table 5. Benchmark results of the seven classification models which were trained and evaluated on our mechanical engineering part benchmark. We trained five times per model and reported the highest result. Each *, †, and ⊠ symbol indicates the method: point cloud, volumetric, and image representation, respectively.

Method	Acc. over object (%)		Acc. over class (%)		F1-score		Average Precision	
	A	B	A	B	A	B	A	B
PointCNN* [24]	93.89	93.67	81.85	86.80	83.86	88.63	**90.13**	**93.86**
PointNet++* [36]	87.45	93.91	73.68	87.97	74.59	88.32	73.45	91.33
SpiderCNN* [48]	93.59	89.31	79.70	79.29	81.30	80.72	86.64	82.47
MVCNN⊠ [44]	64.67	79.17	80.47	84.09	69.69	77.69	79.82	86.66
RotationNet⊠ [20]	**97.35**	**94.73**	**90.79**	**89.70**	**92.29**	**91.05**	87.58	84.87
DLAN† [11]	93.53	91.38	82.97	84.21	83.81	83.88	89.80	90.14
VRN† [3]	93.17	85.44	80.34	70.15	81.48	73.01	85.72	77.36

AP metrics are used to compare the network performance across the dataset A and B. F1-score is the harmonic mean of the precision and recall. The benchmark results for the datasets A and B are available in Table 5 and Fig. 7. Additionally, to compare the performance between common objects and mechanical objects, we provide classification performance on MondelNet40 in Table 6.

Unlike the retrieval task, RotationNet outperforms the other methods for the both datasets (see Table 5 and Table 6). The performance of MVCNN drops significantly on the mechanical components compared to the common objects which is ModelNet40. On the other hand, the accuracy of RotationNet drops slightly. The major differences between MVCNN and RotationNet are estimating correspondence between each image and view order during

Table 6. Classification accuracy on ModelNet40. Each *, †, and ⊠ symbol indicates the method is based on point cloud representation, volumetric representation, and image representation, respectively.

Method	Acc. over object (%)
PointCNN* [24]	92.2
PointNet++* [36]	91.9
SpiderCNN* [48]	92.4
MVCNN⊠ [44]	95.0
RotationNet⊠ [20]	**97.37**
DLAN† [11]	84.0
VRN† [3]	95.5

training. This correspondence estimation relaxes rotation variant property by implicitly learning mapping function between each view of the object and camera view point. In point cloud methods, PointCNN shows the best performance on both datasets, and SpiderCNN perform better for the dataset A than B. PointCNN performs best for the AP (see Table 5). This is because mechanical components are sensitive to the local changes and PointCNN leverages spatially-local correlation. In the same sense, DLAN performs better on mechanical components due to oriented point sets. However, VRN performance drops on the mechanical components benchmark since voxel grids are orientation variant.

From our benchmark result, capturing local features and having orientation invariance are crucial for developing mechanical components classifier. Although

RotationNet shows 97% accuracy over an object, the accuracy over the class, which is 90.79%, is not good enough to utilize in the industry. For the deep learning application for mechanical components in the industry, a deeper understanding of mechanical parts is required. The classification result of the 'Flanged plain bearings' class shows a low accuracy, which is under 87% for every network. This value is relatively lower than the accuracy of other classes (see appendix). This result shows the limitation of existing 3D object classification algorithms in terms of extracting local features. The general shape of the bearing is almost similar to thick washers or rings. Therefore, if the network cannot capture the local difference of ring-shaped object, it is hard to distinguish these objects.

We experiment how point cloud density affects the results in point cloud base algorithms. We perform five different densities: 128, 256, 512, 1,024, and 2,048 points on three point cloud classification methods [24,36,48] for the classification task for the dataset B. From our experiment results, the performance increases as the density of the point cloud increases, as shown in line plots in Fig. 8. However, the enhancement of results saturates as the point cloud density grows and the performance of SpiderCNN [48] decreases even the density increases

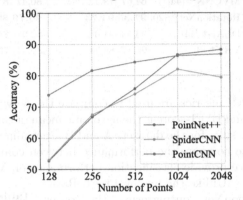

Fig. 8. Classification results of five different point cloud densities.

from 1,025 to 2,048. PointNet++ [36] is the most sensitive in the density of the point cloud, and PointCNN [24] is the least vulnerable in the variation of the point cloud density.

6   Conclusion

We propose a large-scale mechanical component benchmark with annotations. For the creation of the dataset, we develop an annotation framework that enhances the efficacy of the annotation and filtering processes. We perform shape classification and retrieval experiments with seven deep-learning shape classification methods which are designed to classify common objects. We find that view-based and voxel grid presentation-based methods perform poorly on random orientation of mechanical components. However, DLAN, a voxel-based method, performs well on random orientation since it has orientation invariance. The creation of MCB and experimental results can be used for the development of data-driven algorithms of mechanical components. For the future work, the development on dedicated data-driven feature descriptor for mechanical components will be conducted.

Acknowledgment. We wish to give a special thanks to the reviewers for their invaluable feedback. Additionally, we thank TraceParts for providing CAD models of mechanical components. This work is partially supported by NSF under the grants FW-HTF 1839971, OIA 1937036, and CRI 1729486. We also acknowledge the Feddersen Chair Funds. Any opinions, findings, and conclusions or recommendations expressed in this material are those of the authors and do not necessarily reflect the views of the funding agency.

References

1. Open asset import library. http://www.assimp.org/. Accessed 10 Nov 2019
2. Bespalov, D., Ip, C.Y., Regli, W.C., Shaffer, J.: Benchmarking CAD search techniques. In: Proceedings of the 2005 ACM Symposium on Solid and Physical Modeling, pp. 275–286. ACM (2005)
3. Brock, A., Lim, T., Ritchie, J.M., Weston, N.: Generative and discriminative voxel modeling with convolutional neural networks. arXiv preprint arXiv:1608.04236 (2016)
4. Chang, A.X., et al.: ShapeNet: an information-rich 3D model repository. Technical report arXiv:1512.03012 [cs.GR], Stanford University – Princeton University – Toyota Technological Institute at Chicago (2015)
5. Chen, L.-C., Zhu, Y., Papandreou, G., Schroff, F., Adam, H.: Encoder-decoder with atrous separable convolution for semantic image segmentation. In: Ferrari, V., Hebert, M., Sminchisescu, C., Weiss, Y. (eds.) ECCV 2018. LNCS, vol. 11211, pp. 833–851. Springer, Cham (2018). https://doi.org/10.1007/978-3-030-01234-2_49
6. Choi, C., Kim, S., Ramani, K.: Learning hand articulations by hallucinating heat distribution. In: Proceedings of the IEEE International Conference on Computer Vision, pp. 3104–3113 (2017)
7. Choy, C.B., Xu, D., Gwak, J.Y., Chen, K., Savarese, S.: 3D-R2N2: a unified approach for single and multi-view 3D object reconstruction. In: Leibe, B., Matas, J., Sebe, N., Welling, M. (eds.) ECCV 2016. LNCS, vol. 9912, pp. 628–644. Springer, Cham (2016). https://doi.org/10.1007/978-3-319-46484-8_38
8. Deng, J., Dong, W., Socher, R., Li, L.J., Li, K., Fei-Fei, L.: Imagenet: a large-scale hierarchical image database. In: 2009 IEEE Conference on Computer Vision and Pattern Recognition, pp. 248–255. IEEE (2009)
9. Ellis, K., Ritchie, D., Solar-Lezama, A., Tenenbaum, J.: Learning to infer graphics programs from hand-drawn images. In: Advances in Neural Information Processing Systems, pp. 6059–6068 (2018)
10. Farag, A., Ali, A., Graham, J., Farag, A., Elshazly, S., Falk, R.: Evaluation of geometric feature descriptors for detection and classification of lung nodules in low dose CT scans of the chest. In: 2011 IEEE International Symposium on Biomedical Imaging: from Nano to Macro, pp. 169–172. IEEE (2011)
11. Furuya, T., Ohbuchi, R.: Diffusion-on-manifold aggregation of local features for shape-based 3D model retrieval. In: Proceedings of the 5th ACM on International Conference on Multimedia Retrieval, pp. 171–178. ACM (2015)
12. Furuya, T., Ohbuchi, R.: Deep aggregation of local 3D geometric features for 3D model retrieval. In: BMVC, pp. 121–1 (2016)
13. Geuzaine, C., Remacle, J.F.: Gmsh: a 3-D finite element mesh generator with built-in pre-and post-processing facilities. Int. J. Numer. Methods Eng. **79**(11), 1309–1331 (2009)

14. Goodfellow, I., et al.: Generative adversarial nets. In: Advances in Neural Information Processing Systems, pp. 2672–2680 (2014)
15. He, K., Gkioxari, G., Dollár, P., Girshick, R.: Mask R-CNN. In: Proceedings of the IEEE International Conference on Computer Vision, pp. 2961–2969 (2017)
16. Huang, J., Kwok, T.H., Zhou, C.: Parametric design for human body modeling by wireframe-assisted deep learning. Comput.-Aided Des. **108**, 19–29 (2019)
17. Iyer, N., Jayanti, S., Lou, K., Kalyanaraman, Y., Ramani, K.: Three-dimensional shape searching: state-of-the-art review and future trends. Comput.-Aided Des. **37**(5), 509–530 (2005)
18. Jayanti, S., Kalyanaraman, Y., Iyer, N., Ramani, K.: Developing an engineering shape benchmark for CAD models. Comput.-Aided Des. **38**(9), 939–953 (2006)
19. Jia, X., Gavves, E., Fernando, B., Tuytelaars, T.: Guiding the long-short term memory model for image caption generation. In: The IEEE International Conference on Computer Vision (ICCV), December 2015
20. Kanezaki, A., Matsushita, Y., Nishida, Y.: Rotationnet: joint object categorization and pose estimation using multiviews from unsupervised viewpoints. In: Proceedings of the IEEE Conference on Computer Vision and Pattern Recognition, pp. 5010–5019 (2018)
21. Kim, S., Winovich, N., Chi, H.G., Lin, G., Ramani, K.: Latent transformations neural network for object view synthesis. Vis. Comput. 1–15 (2019)
22. Koch, S., et al.: ABC: a big CAD model dataset for geometric deep learning. In: Proceedings of the IEEE Conference on Computer Vision and Pattern Recognition, pp. 9601–9611 (2019)
23. Kulkarni, T.D., Whitney, W.F., Kohli, P., Tenenbaum, J.: Deep convolutional inverse graphics network. In: Advances in Neural Information Processing Systems, pp. 2539–2547 (2015)
24. Li, Y., Bu, R., Sun, M., Wu, W., Di, X., Chen, B.: PointCNN: Convolution on X-transformed points. In: Advances in Neural Information Processing Systems, pp. 820–830 (2018)
25. Lin, T.Y., Dollár, P., Girshick, R., He, K., Hariharan, B., Belongie, S.: Feature pyramid networks for object detection. In: Proceedings of the IEEE Conference on Computer Vision and Pattern Recognition, pp. 2117–2125 (2017)
26. Lin, T.Y., et al.: Microsoft COCO: common objects in context. In: Fleet, D., Pajdla, T., Schiele, B., Tuytelaars, T. (eds.) ECCV 2014. LNCS, vol. 8693, pp. 740–755. Springer, Cham (2014). https://doi.org/10.1007/978-3-319-10602-1_48
27. Liu, S., Giles, L., Ororbia, A.: Learning a hierarchical latent-variable model of 3D shapes. In: 2018 International Conference on 3D Vision (3DV), pp. 542–551. IEEE (2018)
28. Liu, W., et al.: SSD: single shot multibox detector. In: Leibe, B., Matas, J., Sebe, N., Welling, M. (eds.) ECCV 2016. LNCS, vol. 9905, pp. 21–37. Springer, Cham (2016). https://doi.org/10.1007/978-3-319-46448-0_2
29. van den Maaten, L., Hinton, G.: Visualizing data using t-SNE. J. Mach. Learn. Res. **9**(Nov), 2579–2605 (2008)
30. Maturana, D., Scherer, S.: Voxnet: a 3D convolutional neural network for real-time object recognition. In: 2015 IEEE/RSJ International Conference on Intelligent Robots and Systems (IROS), pp. 922–928. IEEE (2015)
31. Miller, G.A.: WordNet: An Electronic Lexical Database. MIT Press, Cambridge (1998)
32. Mo, K., et al.: Partnet: a large-scale benchmark for fine-grained and hierarchical part-level 3D object understanding. In: Proceedings of the IEEE Conference on Computer Vision and Pattern Recognition, pp. 909–918 (2019)

33. Monti, F., Boscaini, D., Masci, J., Rodola, E., Svoboda, J., Bronstein, M.M.: Geometric deep learning on graphs and manifolds using mixture model CNNs. In: The IEEE Conference on Computer Vision and Pattern Recognition (CVPR), July 2017
34. Paszke, A., Chaurasia, A., Kim, S., Culurciello, E.: ENET: A deep neural network architecture for real-time semantic segmentation. arXiv preprint arXiv:1606.02147 (2016)
35. Qi, C.R., Su, H., Mo, K., Guibas, L.J.: Pointnet: deep learning on point sets for 3D classification and segmentation. In: Proceedings of the IEEE Conference on Computer Vision and Pattern Recognition, pp. 652–660 (2017)
36. Qi, C.R., Yi, L., Su, H., Guibas, L.J.: Pointnet++: deep hierarchical feature learning on point sets in a metric space. In: Advances in Neural Information Processing Systems, pp. 5099–5108 (2017)
37. Redmon, J., Divvala, S., Girshick, R., Farhadi, A.: You only look once: unified, real-time object detection. In: Proceedings of the IEEE Conference on Computer Vision and Pattern Recognition, pp. 779–788 (2016)
38. Regli, W.C., et al.: National design repository project: a status report. In: International Joint Conferences on Artificial Intelligence (IJCAI), Seattle, WA, pp. 4–10 (2001)
39. Ronneberger, O., Fischer, P., Brox, T.: U-Net: convolutional networks for biomedical image segmentation. In: Navab, N., Hornegger, J., Wells, W.M., Frangi, A.F. (eds.) MICCAI 2015. LNCS, vol. 9351, pp. 234–241. Springer, Cham (2015). https://doi.org/10.1007/978-3-319-24574-4_28
40. Savva, M., et al.: Shrec16 track: largescale 3D shape retrieval from shapenet core55. In: Proceedings of the Eurographics Workshop on 3D Object Retrieval, pp. 89–98 (2016)
41. Savva, M., et al.: Large-scale 3D shape retrieval from shapenet core55: Shrec'17 track. In: Proceedings of the Workshop on 3D Object Retrieval, pp. 39–50. Eurographics Association (2017)
42. Shilane, P., Min, P., Kazhdan, M., Funkhouser, T.: The Princeton shape benchmark. In: Proceedings Shape Modeling Applications, pp. 167–178. IEEE (2004)
43. Su, H., Maji, S., Kalogerakis, E., Learned-Miller, E.: Multi-view convolutional neural networks for 3D shape recognition. In: Proceedings of the IEEE International Conference on Computer Vision, pp. 945–953 (2015)
44. Su, J.-C., Gadelha, M., Wang, R., Maji, S.: A deeper look at 3D shape classifiers. In: Leal-Taixé, L., Roth, S. (eds.) ECCV 2018. LNCS, vol. 11131, pp. 645–661. Springer, Cham (2019). https://doi.org/10.1007/978-3-030-11015-4_49
45. Szegedy, C., Ioffe, S., Vanhoucke, V., Alemi, A.A.: Inception-v4, inception-resnet and the impact of residual connections on learning. In: Thirty-First AAAI Conference on Artificial Intelligence (2017)
46. Vinyals, O., Toshev, A., Bengio, S., Erhan, D.: Show and tell: a neural image caption generator. In: Proceedings of the IEEE Conference on Computer Vision and Pattern Recognition, pp. 3156–3164 (2015)
47. Wu, Z., et al.: 3D shapenets: a deep representation for volumetric shapes. In: Proceedings of the IEEE Conference on Computer Vision and Pattern Recognition, pp. 1912–1920 (2015)
48. Xu, Y., Fan, T., Xu, M., Zeng, L., Qiao, Y.: SpiderCNN: deep learning on point sets with parameterized convolutional filters. In: Ferrari, V., Hebert, M., Sminchisescu, C., Weiss, Y. (eds.) ECCV 2018. LNCS, vol. 11212, pp. 90–105. Springer, Cham (2018). https://doi.org/10.1007/978-3-030-01237-3_6

Hidden Footprints: Learning Contextual Walkability from 3D Human Trails

Jin Sun[✉], Hadar Averbuch-Elor, Qianqian Wang, and Noah Snavely

Cornell Tech, New York, NY 10044, USA
{jinsun,hadarelor,qw246,snavely}@cornell.edu

Abstract. Predicting where people can walk in a scene is important for many tasks, including autonomous driving systems and human behavior analysis. Yet learning a computational model for this purpose is challenging due to semantic ambiguity and a lack of labeled data: current datasets only tell you where people *are*, not where they *could be*. We tackle this problem by leveraging information from existing datasets, without additional labeling. We first augment the set of valid, labeled walkable regions by propagating person observations between images, utilizing 3D information to create what we call *hidden footprints*. However, this augmented data is still sparse. We devise a training strategy designed for such sparse labels, combining a class-balanced classification loss with a contextual adversarial loss. Using this strategy, we demonstrate a model that learns to predict a walkability map from a single image. We evaluate our model on the Waymo and Cityscapes datasets, demonstrating superior performance compared to baselines and state-of-the-art models.

Keywords: Scene understanding · Context · Human analysis

1 Introduction

Walking, one of our most common daily activities, requires complex scene understanding and planning. Consider the image in Fig. 1(a). We have no trouble imagining where a person might walk in this scene—on sidewalks, in crosswalks, or into a subway entrance. Given an image, predicting where people might walk is critical for engineering robust autonomous systems like self-driving cars that must anticipate human behaviors. Such a capability could also be used in urban planning and public policy making to assess walkability in a neighborhood [11].

At first glance, one might think that predicting where people might appear in an image is equivalent to reasoning about semantics, since pedestrians are associated with sidewalks but not streets, etc. Perhaps we can simply convert a semantic segmentation into a prediction of walkable regions. However, we find

Electronic supplementary material The online version of this chapter (https://doi.org/10.1007/978-3-030-58523-5_12) contains supplementary material, which is available to authorized users.

© Springer Nature Switzerland AG 2020
A. Vedaldi et al. (Eds.): ECCV 2020, LNCS 12363, pp. 192–207, 2020.
https://doi.org/10.1007/978-3-030-58523-5_12

(a)   (b)   (c)

Fig. 1. (a) Imagine an urban scene like this one. Where could people walk in such a scene? The answer is not simply 'crosswalk' or 'sidewalk'—scene context plays a significant role. For example, people can be standing on sidewalks and not on crosswalks while waiting for a green light (b, top). A road can be open to walk, or only partially walkable (e.g., when occupied by cows, as in (b, bottom)). (c) We propose to predict where people could walk (highlighted in green) by learning from observing human behavior in weakly annotated data. (Color figure online)

empirically that semantic categories are a surprisingly poor predictor of where humans might walk in a given scene. Common semantic categories like road and sidewalk alone do not fully explain affordances that are indicated by other visual cues, such as crosswalks, curb ramps, and red lights. And some cases require even more complex reasoning: 1) one may walk in front of a parked car, but not a moving one; 2) a sidewalk can become unwalkable due to temporary construction work; 3) during special events, a street that is usually unsafe for walking might be closed to traffic and hence walkable. A single semantic label is insufficient for reasoning about these cases, and labeling scenes with such fine-grained semantic granularity as 'street during a special event' would be prohibitively expensive.

In this work, we propose to predict *contextual walkability*—a map of where human *could* be walking on under the current scene's conditions—to distinguish from measuring a generic static 'walkable' map defined by semantics or geometry. We seek to design a computational model to predict such a map given a street image. Rather than learning from a pre-defined set of semantic categories and detecting where objects *are*, we address this novel task of predicting where they *could be* by taking a data-driven approach where we learn from direct observation of human behavior. While this approach allows for a flexible modeling of context, this task is challenging because the visual cues for predicting where people *could* be are significantly more ambiguous. Furthermore, no such labeled dataset exists to the best of our knowledge.

Fortunately, we are witnessing a surge of street scene datasets collected for autonomous driving, such as the Waymo [31], nuScenes [2], and Argoverse [3] datasets. In such datasets, a car drives around urban areas and continuously captures sensory data, including images. These datasets also provide 2D and 3D people labels, along with the 3D poses of the vehicle and its cameras. We

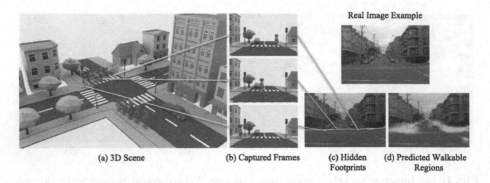

(a) 3D Scene   (b) Captured Frames   (c) Hidden   (d) Predicted Walkable
                                                        Footprints   Regions

Fig. 2. When a person is observed walking through a scene (a), a sequence of images (b) of that person defines a 3D trail, where each position indicates a valid location for people to walk in the scene. We call such virtual trails *hidden footprints* (c). While such trails provide partial annotation of walkable regions, our trained model can generalize to other plausible locations (d).

leverage such data for learning contextual walkability models *without additional labels*. Our approach, illustrated in Fig. 2, relies on two key insights:

- We can significantly augment walkable region labels in a scene by re-projecting all 3D people annotations into a given reference frame using the provided camera poses (Sect. 3). Such "hidden footprints", which capture true walkable regions, not only expand the labeling of walkable regions, but also facilitate the learning of purely contextual cues because no real person is present at those projected locations.
- However, the number of people we observe in a scene is still limited: many walkable regions that share similar contextual cues may be missing from the ground-truth. We thus train a dense prediction neural network using a novel training strategy to recover a walkability distribution from under-sampled ground-truth labels (Sect. 4). Our training objective guides a model to not only be faithful to the original ground-truth labels, but also to be able to expand beyond to contextually similar regions in the same scene.

Combining these two insights yields a model that can predict a broad distribution with good coverage of where people could walk, trained solely from an initially sparse and incomplete set of true labels.

We evaluate our approach on popular datasets that feature pedestrians (Sect. 5). Our model significantly outperforms alternative approaches, in terms of covering true walkable locations and classification performance on a manually labeled test set. In addition, our model generalizes surprisingly well to the Cityscapes dataset, even with no fine-tuning. Finally, we show that our hidden footprints framework is flexible and can be adapted to predict additional maps, such as expected pedestrian walking directions.

2   Related Work

Our work addresses contextually-correct placement of objects within a target image. Sun and Jacobs [26] learn a representation of context for locating missing objects in an image, such as curb ramps on sidewalks that ought to be present but are not. Like us, Chien et al. [5] detect non-existent pedestrians. However, unlike our 3D-aware data collection approach, they use images with labeled pedestrians and remove a subset of them using image inpainting. Lee et al. [18] use semantic maps to jointly predict the locations and shapes of potential people. Our work doesn't require semantic inputs.

Several prior works focus on synthesizing pedestrians in images. Pedestrian-Synthesis-GAN [25] trains a GAN on pedestrian bounding boxes filled with random noise. The network learns to synthesize the missing pedestrian based on nearby regions. Lee et al. [19] address the related problem of cutting an object from one video, e.g., a walking pedestrian, and pasting it into a user-specified region in another video. Hong et al. [14] use semantic maps to infer a person's shape and appearance inside user-provided bounding boxes.

Inserting an object into a photo is useful for data augmentation, and determining where to insert an object is often an essential step. ST-GAN [21] uses a GAN framework equipped with a spatial transformer network to output a natural composite restricted to the space of target geometric manipulations. Dwibedi et al. [9] propose a simple approach to synthesizing training images for instance detection tasks. Huang et al. [15] focus on detecting pedestrians in unexpected scenarios and use GANs to select a realistic subset of synthetically generated images. Interactive techniques for object insertion are also common. Lalonde et al. [17] built an interactive system that, given a target image and a desired 3D location in the scene, retrieves object instances for insertion. Our model can aid such methods by providing good candidate object locations.

Our work is also related to human-centric scene analysis. Kitani et al. [16] combine semantic scene understanding and optimal control theory to forecast the activities of people in an image. Fouhey et al. [10] use human actions as a cue for 3D scene understanding. A key subproblem of human analysis is to infer affordances, i.e., potential interactions of people with environments. Gupta et al. [13] manually associate human actions with exemplar poses and use 3D correlation between poses and scenes to obtain valid locations for these actions in the scene. Wang et al. [30] use human poses extracted from TV sitcoms to learn affordances. Li et al. [20] further encourage physically feasible solutions by leveraging geometric knowledge from voxel representations. Chuang et al. [6] train a model with semantic maps to reason about affordances that respect both physical rules and social norms. During inference, our model takes a single image as input and requires no additional information such as 3D or semantics.

3   Propagating Hidden Footprints

In this section, we describe how to obtain an augmented set of valid walkable locations in a scene, exploiting the nature of data collection in recent datasets.

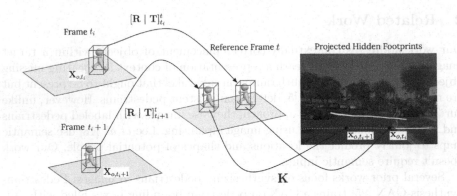

Fig. 3. *Propagating 3D annotations to new frames.* Each labeled person o is represented by the 3D centroid, $\mathbf{X}_{o,t_i} \in \mathbb{R}^3$, of the bottom face of its 3D cuboid (representing the averaged feet location) in the coordinate frame of frame t_i. This point is transformed to the reference frame t's coordinate systems by the 3D rotation and translation $[\mathbf{R} \mid \mathbf{T}]_{t_i}^t$ relating these two frames. The transformed centroid is then projected to the image plane via the camera intrinsics matrix $\mathbf{K}$. The process continues for $i + 1, i + 2, \ldots, T$ until the end of the sequence. The final reference image plane contains the accumulation of these virtual 2D footprints from the whole sequence.

We focus on the Waymo Open Dataset [31], but any dataset that provides 3D labels and camera parameters, such as the nuScenes [2], could be used as well.

In autonomous driving datasets, street scenes are often captured in units of driving sequences, that is, sequences of frames each with a known 3D pose in a world coordinate system. The 3D location of each visible object (e.g., cars, persons, signs, and cyclists) is annotated. Using camera poses (and associated intrinsics), any 3D location can be projected to any frame in the same sequence.

We observe that when a person is walking during the capture, the "footprints" they leave are not only valid locations for a person to walk on for the *current* frame, but also a valid location for *any other frame* in the same scene.

Therefore, we project all person locations, across the entire sequence, from 3D coordinates to 2D coordinates in all frames. This results in a massive expansion of walkable location labels, and, most importantly, these locations are all real, not hallucinated. Note that we do not project labels across sequences, as walkability might change over longer periods of time (e.g., from day to night). We further assume that walkability conditions remain the same during the duration of the relative short sequences (around 20 s in Waymo).

Figure 3 illustrates our process for propagating these "hidden footprints". In particular, we generate a footprint map L for each frame t from all time steps and all objects in the same driving sequence:

$$L_t(\mathbf{x}) = \sum_{i=1}^{T} \sum_{o=1}^{O} g(\mathbf{x}_{o,t_i} - \mathbf{x}; \sigma), \quad \text{where} \quad \mathbf{x}_{o,t_i} = \mathbf{\Pi}_{t_i}^t(\mathbf{X}_{o,t_i}), \quad (1)$$

T is the total number of frames in the sequence, O is the number of objects (e.g., people) present in the scene, $\mathbf{x}$ is a 2D location in L_t in the reference frame t, $g(\cdot; \sigma)$ is a Gaussian kernel function with standard deviation σ, and $\Pi_{t_i}^t$ is a transformation from frame t_i to image coordinates in frame t, consisting of a camera rotation $\mathbf{R}$ and translation $\mathbf{T}$, followed by an application of the camera intrinsics matrix $\mathbf{K}$ such that $\Pi_{t_i}^t = \mathbf{K}[\mathbf{R} \mid \mathbf{T}]_{t_i}^t$. In other words, $\Pi_{t_i}^t$ maps a 3D point $\mathbf{X}_{o,t_i}$ to a 2D point $\mathbf{x}_{o,t_i}$. (For simplicity, we do not distinguish between Cartesian and homogeneous coordinates.) The accumulated map shows the observed contextual walkability in the reference scene. The benefits of our hidden footprints are two-fold:

Vastly More Labels. The propagated locations represent a significant expansion of the labels from existing pedestrians. All these locations are real and valid, not artificially augmented. Such abundant labeled data greatly improve a learned model's generalization power.

Easier to Learn Human Context. Since the propagated locations usually do not contain an existing person in the reference frame, no special treatment is necessary in order to learn a context-only model for human walkability, and hence a network trained on these augmented labels will not simply learn to detect people. In prior work, an object of interest has to be either inpainted with background [5] or blocked [26] to learn a context-only model.

By propagating hidden footprints, we obtain a large expansion in the available training labels for learning contextual walkable locations. However, it still does not cover *all* possible walkable locations. For example, there might be people walking on a sidewalk on one side of a street in a given sequence, while the other side is entirely empty. Both are likely to be equally valid walkable locations given their contextual similarity. A good model should predict both sides of the sidewalk as contextually walkable. In the next section we describe a training strategy to obtain such a model.

4 Contextual Walkability from Partial Observations

Propagating hidden footprints greatly increases the number of observations, alleviating, but not completely solving, our prediction task. Our goal is to learn the latent underlying distribution of where a person can walk in a scene from partial observations. We propose a training strategy with two losses that each focuses on different regions where ground-truth labels and predictions disagree. Our approach is illustrated in Fig. 4.

In particular, we treat positive labels and negative labels differently. For positive ground-truth locations (Sect. 4.1), we discourage the model from predicting false negatives by using a class-balanced classification loss. For negative ground-truth locations (Sect. 4.2), we are more forgiving of discrepancies due to the fact that ground-truth is severely under-sampled (and hence many negatives in the ground-truth could in reality be walkable). We use deep visual features extracted at those ambiguous locations as a proxy to measure whether they are from the

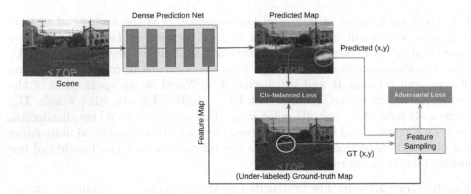

Fig. 4. We model the underlying ground-truth distribution by training a dense prediction network with two losses. The first is a *class-balanced classification loss* that encourages the network to predict positives (blue circle) for ground-truth positive regions (white circle). The second is an *adversarial feature loss* used for determining hard false positive regions (orange circle). These regions will not be penalized if they share similar visual context with the true positive regions. (Color figure online)

same distribution as ground-truth locations—i.e., contextually similar. Together, these terms encourage the model to make faithful predictions for ground-truth locations and educated guesses in uncertain regions.

4.1   Class-Balanced Classification Loss for a Global Prediction

We treat the prediction of contextual walkability of an image as a binary classification problem: given a location in an image, is it possible for a person to be walking there? This setting is used in previous context modeling work [5,26].

Directly using a standard classification loss such as the cross-entropy loss is problematic, as it will be dominated by the overwhelmingly large and uncertain negative labels; Waymo images have many times more background pixels than human ones. We reduce such bias by adding class weights to the ground-truth labels. In particular, for each image, we re-balance the positive and negative samples by the reciprocal of their label frequencies:

$$CBL(p, y) = -c_{pos} \cdot y \cdot \log(p) - c_{neg} \cdot (1 - y) \cdot \log(1 - p), \qquad (2)$$

where p is the model's prediction, y is a binary 1 or 0 label, and c_{pos} and c_{neg} are class weights for positive and negative labels, respectively. We convert the footprint map L from Eq. 1 into a binary map by assigning non-zero values to 1. By adding a higher weight to positive label locations, we encourage the model to make correct predictions on them. Note that our overall training strategy is not limited to the specific choice of the class-rebalanced loss. We tried alternatives such as the Focal Loss [22], but found that they gave similar performance.

4.2   Adversarial Feature Loss for Hard False Positives

For this loss, we only investigate possible false positives and use an adversarial approach [12] to handle regions predicted as walkable by our model but where the ground-truth says otherwise. We use features extracted from the same prediction model as a proxy to measure contextual similarity and encourage the model to produce positives at locations that are visually similar to known positives.

More concretely, after the model outputs a prediction map, we sample locations where the predicted score is high but the ground-truth label is negative. Ground-truth positives are sampled as well. We then extract features at these locations, and feed them into a discriminator network. While the classification loss penalizes false positives equally, the adversarial loss is lower for false positives that share similar features with ground-truth locations. By training the discriminator to distinguish whether features are from the same distribution, gradients flow back to the walkability prediction network to encourage it to predict high scores more confidently on plausible locations.

To sample features, we explored a few distribution sampling methods, including max sampling and rejection sampling. To balance efficiency and effectiveness, we adopt a sampling strategy that selects the top $K\%$ scored regions and uniformly samples N locations among them. Though the sampling operation is not differentiable, gradients from the discriminator flow back to the model through the sampled features, and so the whole pipeline can be trained end-to-end.

The class-balanced classification loss and the adversarial feature loss work together to encourage the network to faithfully predict ground-truth positive labels, while also expanding to locations that share similar visual context.

Discussion. Compared to our proposed training strategy, there are alternatives for this task. Mean Squared Error (MSE) is a popular loss for modeling unimodal probability maps (e.g., [24]). We find that models trained with MSE produce visually noisier output. This may be due to the multi-modal nature of walkability maps. The training can also be done by an adversarial-only loss [5], without any additional direct regression. We find that training is unstable without regression. Our training can also be seen as a soft handling of unlabeled samples, compared to pseudo-labeling used in semi-supervised settings [4]. We empirically compare the proposed training strategy with the alternatives in the next section.

5   Evaluation

In this section, we evaluate how well our proposed approach works as a contextual walkability model. Just as our problem lacks full ground-truth for training, a proper evaluation is also non-trivial. To help provide a comprehensive view of performance, our evaluation considers several aspects: 1) quality of predictions over the whole test set measured by the region *expansion multiplier* over the labeled data; 2) a standard classification metric on a subset with manual ground-truth labels; 3) quality of predictions compared to semantics based reasoning and alternatives, and 4) generalization power to another dataset. The results

Scene   Hidden Footprints   Predicted Regions

Fig. 5. *Example results on Waymo.* For each scene (left), we illustrate the projected footprints (center) and our walkability predictions (right). Red boxes are existing pedestrians. Predictions are overlaid on top of the input images for visualization purposes. Greener regions correspond to a higher likelihood for walkability. While projected footprints greatly expand the sparse existing human positions in a scene, our model predicts even more regions, yielding a nice coverage of locations such as sidewalks and crosswalks. Note that our model is not biased towards any specific configuration of predicted map, and is robust to day/night changes. (Color figure online)

suggest that our method learns a good universal walkability model. It greatly expands positive predictions from limited labels, makes reasonable additional predictions based on contextual similarity, and is able to generalize to unseen street images from another dataset. Many qualitative examples are provided in the supplemental material.

5.1   Implementation Details

Network. All models are trained using a generic U-Net–style dense prediction structure with an ImageNet pretrained ResNet-50 network as the backbone. Note that our method is not limited to any particular architecture. The network takes a 640 × 480 pixels image and produces a 160 × 120 single-channel map, where at each position a score represents the likelihood of a person being able to walk there. Features sampled from the layer before the last single-channel output layer are used in the adversarial feature loss. We pick the top 1% regions and sample 10 features per image. A three-layer MLP network is used as the discriminator,

Table 1. Quantitative comparisons between our proposed training (hidden footprints + class-balanced loss + adversarial feature loss) and alternatives. Best results are in bold and second best are underlined. Except Raw GT, all others are trained with propagated footprints. The numbers in the entire test set experiment are ratios, normalized by ground-truth number of locations.

	Raw GT	MSE	Adversarial	BCE	RBCE	Ours
Entire Test Set						
Pred Total (TP+FP) ↑	0.80	2.12	**22.15**	0.24	3.13	<u>8.18</u>
Pred Valid (TP) ↑	0.06	0.37	<u>0.60</u>	0.13	0.53	**0.69**
Missing (FN) ↓	0.94	0.63	<u>0.39</u>	0.87	0.47	**0.31**
Expansion	0.73	1.75	21.55	0.11	2.60	7.48
50-Scenes Manually Labeled Set						
Mean Average Precision ↑	0.111	0.145	0.173	0.119	<u>0.190</u>	**0.269**

which takes a 256-D feature vector and predicts a single score. During inference, we generate three fixed crops of an image and merge predicted maps.

Training. We train networks using Adam with betas set to 0.5 and 0.999 and with a batch size of 10. Images are resized to 480px in height and then randomly cropped along the width dimension to 640×480. The initial learning rate is 10^{-4} and is reduced by 10% when the validation loss reaches a plateau. Loss weights are 1.0 for both the class-balanced classification loss and the adversarial feature loss.

Baselines. We compare with models trained on hidden footprints with different losses: MSE, standard cross-entropy (BCE), Reweighted BCE, and adversarial loss-only. In addition, we compare with a model trained without our footprint propagation (Raw GT). The baselines also provide insights on ablations of our design choices: our method combines the hidden footprints + class-balanced classification loss + adversarial feature loss. All adversarial training is done with a WGAN loss with gradient penalty. More details are in the supplementary material.

5.2 Data

The Waymo Open Dataset [31] contains over 200k images with 3D labels of four object categories, including pedestrians. The data is divided into multiple capture sequences where each segment is a 20 s recording of a vehicle driving in diverse conditions. We adopt the official train/val splits and use front camera scenes with at least one pedestrian. Our final training and test sets contain about 80k and 2k images, respectively. For each frame, we store 3D labels, camera intrinsics and poses. Footprints are propagated using the procedure described in Sect. 3.

Table 2. Quantitative comparison with semantic based approaches on the 50-scenes manually labeled set. Best result is in bold.

	Raw GT	RBCE	Ours	S-Semantic	M-Semantic	Semantic-Net
mAP ↑	0.111	0.190	**0.269**	0.087	0.105	0.201

5.3 Results

Effect of Propagating Hidden Footprints. In the raw Waymo dataset, there are on average 3.8 pedestrians per scene. With our proposed footprints propagation, we obtain on average 362 valid locations per image, resulting in a 116× increase. We show qualitative examples of projected labels in Fig. 5. While the visible pedestrians are scarce in the images, the propagated footprints have a much larger, yet still valid, coverage.

Next we show that out proposed training strategy encourages a model to make predictions that have even greater coverage over plausible regions, all from limited supervision that is severely under-sampled in the original data.

Quality of Walkability Predictions: Region Expansion Ratio. Knowing that the ground-truth walkable regions are not exhaustive, a good prediction model should fully cover groundtruth positive locations, and also expand to additional locations that are also likely to be positive. We measure the quality of such an expanded prediction by computing the following metrics on the test set: true positives (TP), false negatives (FN), and expansion ratio compared to ground-truth. The expansion ratio is the quotient of the number of predicted locations not covered by ground-truth and ground-truth locations.

Table 1 top part shows that our method achieves a balance in expanding predicted regions while reducing missing regions, compared to alternative approaches. The model trained with adversarial loss–only (Adversarial) has the largest expansion ratio as there is no other loss restricting the number of positive predicted labels. But the expansion ratio only tells one side of the story: how good are those expanded regions?

Quality of Predictions: Classification Metric. A model that predicts all regions to be positive would have the best expansion score, but that is clearly undesirable. In order to quantify whether the expanded regions are indeed good, we report standard metrics on a subset of the test data containing manually labelled positive regions. Specifically, we labeled 50 randomly selected scenes. For each image, we draw a walkability map of where a pedestrian might walk under normal conditions. For instance, we would not label the whole road (even though people might occasionally jaywalk). Example labeled scenes are provided in the supplemental material. With these fully labeled images, we can report standard classification metrics to further evaluate the quality of our predictions.

Table 1 bottom part shows the mean Average Precision (mAP) of all methods on the manually labeled set. By looking at the table as a whole, it is now clear that the adversarial loss–only model produces largely false positives. Fur-

Table 3. Quantitative comparison with alternative approaches on the 100-scenes manually labeled set. Best result is in bold.

	Where & Who [28]	D-NP [5]	Inpainting [33]	Ours
mAP ↑	0.149	0.142	0.107	**0.246**

thermore, the benefit of the propagated footprints is clearly demonstrated by comparing MSE (0.145 mAP) to Raw GT (0.111 mAP). Our full model significantly outperforms others (including ablated alternatives) on the manually labeled set. It shows a good balance between expansion and precision.

Comparison with Semantic Segmentation. In this experiment, we test the claim that walkability predictions cannot be solved by semantic segmentation alone. We consider the following alternatives to our method, ordered by the richness of semantic information introduced into reasoning. 1) *S-Semantic*: estimate the conditional probability of a location being walkable if it is in a single specific semantic class—we choose 'sidewalk' as it is empirically the most predictive class. 2) *M-Semantic*: estimate the conditional probability of a location being walkable given its predicted semantic class (from multiple classes). 3) *Semantic-Net*: given a semantic segmentation as input, train a neural network to predict walkability.

Table 2 shows mAP scores on the manually labeled set. We use a state-of-the-art semantic segmentation network model, HRNet [27,29], that is pre-trained on street scenes. The generated semantic maps are reasonable upon visual inspection. All semantics-based approaches are trained with our propagated footprints. The clear performance advantage of our method suggests that predicting walkability is not trivially a downstream application of semantic segmentation. An interesting future direction is to explore combining our approach with semantics.

Comparison with Alternative Walkability Predictions. Here, we compare to three alternative walkability prediction methods:

Where & Who [28]: This method generates person compositions given an RGB image and its semantic layout. We use the authors' provided model trained on the COCO dataset [23]. The semantic layouts are obtained from Detectron2 [32].

D-NP [5]: This method predicts human poses from empty scenes. We adapt it to predict a heatmap of people's feet. We use the same U-Net structure as ours for fair comparison. D-NP has an additional discriminator to classify (scene, map) pairs, adopted from DCGAN as in [5]. We train the whole pipeline on the Waymo dataset with inpainted scenes in which people are removed.

Inpainting [33]: For a stronger baseline, we use a state-of-the-art image inpainting algorithm pre-trained on the Paris StreetView data [8] to remove existing people. The original locations of people are then used as the ground-truth walkable regions. With the empty scenes and walkable regions, we train a regression network with the same U-Net structure as ours and with a 2D binary cross-entropy loss.

Input image   Prediction of Lee *et al.* [18]   Our Prediction

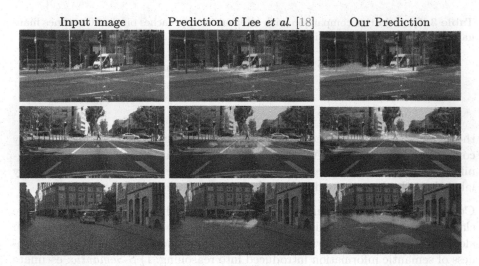

Fig. 6. *Generalization to Cityscapes.* This figure compares predicted walkability maps produced by a semantics-based pedestrian insertion method [18] and by our Waymo-trained model on the Cityscapes dataset. Locations predicted by our method are diverse and extensive.

Table 3 shows results on our Waymo manually labeled set, for which we include an additional set of scenes such that there are in total 100 scenes with walkability maps. Our full model outperforms all others by a significant margin.

Generalization to Cityscapes. We also evaluate our Waymo-trained model on the Cityscapes datasets [7] *without finetuning*, demonstrating its generalization ability. Cityscapes contains 5,000 images and we evaluate on the official validation set. In addition to RGB images, the dataset also provides a fine-grained semantic label map of common classes (e.g., building, bicycle, and ground).

We use a semantics-based metric to measure the similarity between our predicted walkable locations and the ground-truth person locations. To this end, we sample pedestrian feet locations both from our predicted map, as well as from the ground-truth. At each location, we accumulate the local surrounding areas' ground-truth semantic labels and store the most frequent label as where the person is standing on. A histogram of those labels is collected by sampling locations over all images in the validation set. This shows distribution of walkable regions according to semantic categories. For comparison, we also sample predicted person locations predicted from Lee *et al.* [18], which was trained on the Cityspaces dataset, and collect a histogram from those locations. Note that unlike in [18], our method does not require semantic maps as inputs.

To measure the discrepancy of the distributions between the predicted walking locations and ground-truth, we compute the Kullback–Leibler (KL) divergence. The KL-divergence between our model (trained on Waymo, and not fine-tuned on Cityscapes) and the ground-truth is 0.42. This distance is significantly smaller than the distance of 0.78 achieved by the Cityscapes-trained

Fig. 7. Predicted walking directions in Waymo scenes on valid walkable regions. Directions are colored following a common optical flow visualization scheme [1].

model of [18]. Figure 6 shows qualitative comparisons. Taking a closer look at a few selected semantic categories, 50% of the predictions from the model proposed in [18] are in the road, whereas our method places 18% of the people in streets, compared to 21% of the ground-truth people. In contrast, our model predicts 26% of the pedestrians on regions already covered by cars, although in ground-truth 11% of people has feet locations on car segments (and 8% in [18]). We attribute this behaviour to the fact that people can be walking behind cars in our training data, thus our model may predict occluded regions. In the supplemental material we show a full distribution of the semantic categories associated with our walkability maps vs. those of ground-truth labeled people.

5.4   Predicting Walking Directions

So far we have focused on modeling the *locations* where people walk in a scene. Here, we show that our hidden footprint propagation setting is adaptable to a richer family of predictions. In particular, we show a preliminary study of predicting a person's walking *directions*. For each person in the dataset, we track their trajectory, compute 3D walking directions at each time step, and then project this direction to all frames using the procedure described in Sect. 3. We train a neural network to predict a pixel-wise map where at each location it outputs a unit-length directional vector. Figure 7 shows qualitative results of our predicted walking directions.

6   Conclusion

We proposed a new method for predicting where people can potentially walk in a scene. To circumvent the challenges of insufficient labels, we first project existing human labels from each frame to each other frame by utilizing 3D information. We then devise a training strategy to encourage the network to learn the true underlying distribution from under-sampled ground-truth labels. Our experiments show that our model can expand ground-truth human walkable locations in the Waymo Open Dataset, and can also outperform state-of-the-art human context models, including on unobserved datasets such as Cityscapes.

Acknowledgements. This research was supported in part by the generosity of Eric and Wendy Schmidt by recommendation of the Schmidt Futures program.

References

1. Baker, S., Scharstein, D., Lewis, J., Roth, S., Black, M.J., Szeliski, R.: A database and evaluation methodology for optical flow. Int. J. Comput. Vis. **92**(1), 1–31 (2011)
2. Caesar, H., et al.: nuscenes: A multimodal dataset for autonomous driving. arXiv preprint arXiv:1903.11027 (2019)
3. Chang, M.F., et al.: Argoverse: 3D tracking and forecasting with rich maps. In: The IEEE Conference on Computer Vision and Pattern Recognition (CVPR), June 2019
4. Chapelle, O., Scholkopf, B., Zien, A.: Semi-supervised learning (chapelle, o. et al., eds.; 2006)[book reviews]. IEEE Trans. Neural Netw. **20**(3), 542 (2009)
5. Chien, J.T., Chou, C.J., Chen, D.J., Chen, H.T.: Detecting nonexistent pedestrians. In: Proceedings of International Conference on Computer Vision Workshops, pp. 182–189 (2017)
6. Chuang, C.Y., Li, J., Torralba, A., Fidler, S.: Learning to act properly: predicting and explaining affordances from images. In: Proceedings of the IEEE Conference on Computer Vision and Pattern Recognition, pp. 975–983 (2018)
7. Cordts, M., et al.: The cityscapes dataset for semantic urban scene understanding. In: Proceedings of the IEEE Conference on Computer Vision and Pattern Recognition, pp. 3213–3223 (2016)
8. Doersch, C., Singh, S., Gupta, A., Sivic, J., Efros, A.A.: What makes Paris look like Paris? ACM Trans. Graph. (SIGGRAPH) **31**(4), 101:1–101:9 (2012)
9. Dwibedi, D., Misra, I., Hebert, M.: Cut, paste and learn: surprisingly easy synthesis for instance detection. In: Proceedings of the IEEE International Conference on Computer Vision, pp. 1301–1310 (2017)
10. Fouhey, D.F., Delaitre, V., Gupta, A., Efros, A.A., Laptev, I., Sivic, J.: People watching: human actions as a cue for single view geometry. In: Fitzgibbon, A., Lazebnik, S., Perona, P., Sato, Y., Schmid, C. (eds.) ECCV 2012. LNCS, vol. 7576, pp. 732–745. Springer, Heidelberg (2012). https://doi.org/10.1007/978-3-642-33715-4_53
11. Frank, L.D., et al.: The development of a walkability index: application to the neighborhood quality of life study. Br. J. Sports Med. **44**(13), 924–933 (2010)
12. Goodfellow, I., et al.: Generative adversarial nets. In: Advances in Neural Information Processing Systems, pp. 2672–2680 (2014)
13. Gupta, A., Satkin, S., Efros, A.A., Hebert, M.: From 3D scene geometry to human workspace. In: CVPR 2011, pp. 1961–1968. IEEE (2011)
14. Hong, S., Yan, X., Huang, T.S., Lee, H.: Learning hierarchical semantic image manipulation through structured representations. In: Advances in Neural Information Processing Systems, pp. 2708–2718 (2018)
15. Huang, S., Ramanan, D.: Expecting the unexpected: training detectors for unusual pedestrians with adversarial imposters. In: Proceedings of the IEEE Conference on Computer Vision and Pattern Recognition, pp. 2243–2252 (2017)
16. Kitani, K.M., Ziebart, B.D., Bagnell, J.A., Hebert, M.: Activity forecasting. In: Fitzgibbon, A., Lazebnik, S., Perona, P., Sato, Y., Schmid, C. (eds.) ECCV 2012. LNCS, vol. 7575, pp. 201–214. Springer, Heidelberg (2012). https://doi.org/10.1007/978-3-642-33765-9_15

17. Lalonde, J.F., Hoiem, D., Efros, A.A., Rother, C., Winn, J., Criminisi, A.: Photo clip art. ACM Trans. Graph. (TOG) **26**, 3 (2007)
18. Lee, D., Liu, S., Gu, J., Liu, M.Y., Yang, M.H., Kautz, J.: Context-aware synthesis and placement of object instances. In: Advances in Neural Information Processing Systems, pp. 10393–10403 (2018)
19. Lee, D., Pfister, T., Yang, M.H.: Inserting videos into videos. In: Proceedings of the IEEE Conference on Computer Vision and Pattern Recognition, pp. 10061–10070 (2019)
20. Li, X., Liu, S., Kim, K., Wang, X., Yang, M.H., Kautz, J.: Putting humans in a scene: learning affordance in 3D indoor environments. In: Proceedings of the IEEE Conference on Computer Vision and Pattern Recognition, pp. 12368–12376 (2019)
21. Lin, C.H., Yumer, E., Wang, O., Shechtman, E., Lucey, S.: ST-GAN: spatial transformer generative adversarial networks for image compositing. In: Proceedings of the IEEE Conference on Computer Vision and Pattern Recognition, pp. 9455–9464 (2018)
22. Lin, T.Y., Goyal, P., Girshick, R., He, K., Dollár, P.: Focal loss for dense object detection. In: Proceedings of the IEEE International Conference on Computer Vision, pp. 2980–2988 (2017)
23. Lin, T.-Y., et al.: Microsoft COCO: common objects in context. In: Fleet, D., Pajdla, T., Schiele, B., Tuytelaars, T. (eds.) ECCV 2014. LNCS, vol. 8693, pp. 740–755. Springer, Cham (2014). https://doi.org/10.1007/978-3-319-10602-1_48
24. Newell, A., Yang, K., Deng, J.: Stacked hourglass networks for human pose estimation. In: Leibe, B., Matas, J., Sebe, N., Welling, M. (eds.) ECCV 2016. LNCS, vol. 9912, pp. 483–499. Springer, Cham (2016). https://doi.org/10.1007/978-3-319-46484-8_29
25. Ouyang, X., Cheng, Y., Jiang, Y., Li, C.L., Zhou, P.: Pedestrian-synthesis-GAN: Generating pedestrian data in real scene and beyond. arXiv preprint arXiv:1804.02047 (2018)
26. Sun, J., Jacobs, D.W.: Seeing what is not there: learning context to determine where objects are missing. In: Proceedings of the IEEE Conference on Computer Vision and Pattern Recognition, pp. 5716–5724 (2017)
27. Sun, K., Xiao, B., Liu, D., Wang, J.: Deep high-resolution representation learning for human pose estimation. In: CVPR (2019)
28. Tan, F., Bernier, C., Cohen, B., Ordonez, V., Barnes, C.: Where and who? Automatic semantic-aware person composition. In: 2018 IEEE Winter Conference on Applications of Computer Vision (WACV), pp. 1519–1528. IEEE (2018)
29. Wang, J., et al.: Deep high-resolution representation learning for visual recognition. CoRR p. abs/1908.07919 (2019)
30. Wang, X., Girdhar, R., Gupta, A.: Binge watching: scaling affordance learning from sitcoms. In: Proceedings of the IEEE Conference on Computer Vision and Pattern Recognition, pp. 2596–2605 (2017)
31. Waymo: Waymo Open Dataset: An autonomous driving dataset (2019)
32. Wu, Y., Kirillov, A., Massa, F., Lo, W.Y., Girshick, R.: Detectron2 (2019). https://github.com/facebookresearch/detectron2
33. Xie, C., et al.: Image inpainting with learnable bidirectional attention maps. In: Proceedings of the IEEE International Conference on Computer Vision, pp. 8858–8867 (2019)

Self-supervised Learning of Audio-Visual Objects from Video

Triantafyllos Afouras[1]([⊠]), Andrew Owens[2], Joon Son Chung[1,3], and Andrew Zisserman[1]

[1] University of Oxford, Oxford, UK
afourast@robots.ox.ac.uk
[2] University of Michigan, Ann Arbor, USA
[3] Naver Corporation, Seongnam-si, South Korea

Abstract. Our objective is to transform a video into a set of discrete audio-visual objects using self-supervised learning. To this end, we introduce a model that uses attention to localize and group sound sources, and optical flow to aggregate information over time. We demonstrate the effectiveness of the audio-visual object embeddings that our model learns by using them for four downstream speech-oriented tasks: (a) multi-speaker sound source separation, (b) localizing and tracking speakers, (c) correcting misaligned audio-visual data, and (d) active speaker detection. Using our representation, these tasks can be solved entirely by training on unlabeled video, without the aid of object detectors. We also demonstrate the generality of our method by applying it to non-human speakers, including cartoons and puppets. Our model significantly outperforms other self-supervised approaches, and obtains performance competitive with methods that use supervised face detection.

1 Introduction

When humans organize the visual world into objects, hearing provides cues that affect the perceptual grouping process. We group different image regions together not only because they look alike, or move together, but also because grouping them together helps us explain the *causes* of co-occurring audio signals.

In this paper, our objective is to replicate this organizational capability, by designing a model that can ingest raw video and transform it into a set of *discrete audio-visual objects*. The network is trained using only self-supervised learning from audio-visual cues. We demonstrate this capability on videos containing talking heads.

This organizational task must overcome a number of challenges if it is to be applicable to raw videos in the wild: (i) there are potentially many visually

Electronic supplementary material The online version of this chapter (https://doi.org/10.1007/978-3-030-58523-5_13) contains supplementary material, which is available to authorized users.

Fig. 1. We learn through self-supervision to represent a video as a set of discrete *audio-visual objects*. Our model groups a scene into object instances and represents each one with a feature embedding. We use these embeddings for speech-oriented tasks that typically require object detectors: (a) multi-speaker source separation, (b) speaker localization, (c) synchronizing misaligned audio and video, and (d) active speaker detection. Using our representation, these tasks can be solved without any labeled data, and on domains where off-the-shelf detectors are not available, such as cartoons and puppets. Please see our webpage for videos: http://www.robots.ox.ac.uk/~vgg/research/avobjects.

similar sound generating objects in the scene (multiple heads in our case), and the model must correctly attribute the sound to the actual sound source; (ii) these objects may move over time; and (iii) there can be multiple other objects in the scene (clutter) as well.

To address these challenges, we build upon recent works on self-supervised audio-visual localization. These include video methods that find motions temporally synchronized with audio onsets [13,40,44], and single-frame methods [6,31,46,52] that find regions that are likely to co-occur with the audio. However, their output is a typically a "heat map" that indicates whether a given pixel is likely (or unlikely) to be attributed to the audio; they do not group a scene into *discrete objects*; and, if only using semantic correspondence, then they cannot distinguish which, of several, object instances is making a sound.

Our first contribution is to propose a network that addresses all three of these challenges; it is able to use synchronization cues to detect sound sources, group them into distinct instances, and track them over time as they move. Our second contribution is to demonstrate that object embeddings obtained from this network facilitate a number of audio-visual downstream tasks that have previously required hand-engineered supervised pipelines.

As illustrated in Fig. 1, we demonstrate that the embeddings enable: (a) multi-speaker sound source separation [2,20]; (b) detecting and tracking talking heads; (c) aligning misaligned recordings [12,15]; and (d) detecting active speakers, i.e. identifying which speaker is talking [13,50]. In each case, we significantly outperform other self-supervised localization methods, and obtain comparable (and in some cases better) performance to prior methods that are trained using stronger supervision, despite the fact that we learn to perform them entirely from a raw audio-visual signal.

The trained model, which we call the Look Who's Talking Network (LWTNet), is essentially "plug and play" in that, once trained on unlabeled data (without preprocessing), it can be applied directly to other video material. It can easily be fine-tuned for other audio-visual domains: we demonstrate this functionality on active speaker detection for non-human speakers, such as animated characters in *The Simpsons* and puppets in *Sesame Street*. This demonstrates the generality of the model and learning framework, since this is a domain where off-the-shelf supervised methods, such as methods that use face detectors, cannot transfer without additional labeling.

2   Related Work

Sound Source Localization. Our task is closely related to the *sound source localization* problem, i.e. finding the location in a video that is the source of a sound. Early work performed localization [7,22,34,39] and segmentation [37] by doing inference on simple probabilistic models, such as methods based on canonical correlation analysis.

Recent efforts learn audio and video representations using self-supervised learning [13,40,44] with *synchronization* as the proxy task: the network has to predict whether video and audio are temporally aligned (or synthetically shifted). Owens and Efros [44] show via heat-map visualizations that their network often attends to sound sources, but do not quantitatively evaluate their model. Recent work [38] added an attention mechanism to this model. Other work has detected sound-making objects using *correspondence* cues [6,31,35,36,46,48,52,54], e.g. by training a model to predict whether audio and a single video frame come from the same (or different) videos. Since these models do not use motion and are trained only to find the correspondence between object appearance and sound, they would not be able to identify which of several objects of the same category is the actual source of a sound. In contrast, our goal is to obtain discrete audio-visual objects from a scene, even when they bellong to the same category (e.g. multiple talking heads). In a related line of work, [25] distill visual object detectors into an audio model using stereo sound, while [27] use spatial information in a scene to convert mono sound to stereo.

Active Speaker Detection (ASD). Early work on active speaker detection trained simple classifiers on hand-crafted feature sets [16]. Later, Chung and Zisserman [13] used synchronization cues to solve the active speaker detection problem. They used a hand-engineered face detection and tracking pipeline to select candidate speakers, and ran their model only on cropped faces. In contrast, our model learns to do ASD entirely from unlabeled data. Chung *et al.*[11] extended the pipeline by enrolling speaker models from visible speaking segments. Recently, Roth *et al.* [50] proposed an active speaker detection dataset and evaluated a variety of supervised methods for it.

Source Separation. In recent years, researchers have proposed a variety of methods for separating the voices of multiple speakers in a scene [2,20,23,44]. These methods either only handle a single on-screen speaker [44] or use hand-engineered, supervised face detection pipelines. Afouras *et al.* [2] and Ephrat

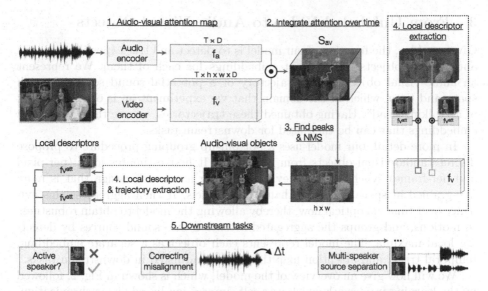

Fig. 2. The Look Who's Talking Network (LWTNet): (1) Computes an audio-visual attention map S_{av} by solving a synchronization task, (2) accumulates attention over time, (3) selects *audio-visual objects* by computing the N highest peaks with non-maximum suppression (NMS) from the accumulated attention map, each corresponding to a trajectory of the pixel over time; (4) for every audio-visual object, it extracts embedding vectors from a spatial window ρ, using the local attention map S_{av} to select visual features, and (5) provides the audio-visual objects as inputs to downstream tasks.

et al. [20], for example, detect and track faces and extract visual representations using off-the-shelf packages. In contrast, we use our model to separate multiple speakers entirely via self-supervision.

Other recent work has explored separating the sounds of musical instruments and other sound-making objects. Gao *et al.* [26,28] use semantic object detectors trained on instrument categories, while [51,58] do not explicitly group a scene into objects and instead either pool the visual features or produce a per-pixel map that associates each pixel with a separated audio source. Recently, [57] added motion information from optical flow. We, too, use flow in our model, but instead of using it as a *cue* for motion, we use it to integrate information from moving objects over time [24,47] in order to track them. In concurrent work [36] propose a model that groups and separates sound sources.

Representation Learning. In recent years, researchers have proposed a variety of self-supervised learning methods for learning representations from images [10,18,32,33,41,43,55,56], videos [29,30] and multimodal data [5,40,42, 45,46]. Often the representation learned by these methods is a feature set (e.g., CNN weights) that can be adapted to downstream tasks by fine-tuning. By contrast, we learn an additional *attention mechanism* that can be used to group discrete objects of interest for downstream speech tasks.

3 From Unlabeled Video to Audio-Visual Objects

Given a video, the function of our model is to detect and track (possibly several) audio-visual objects, and extract embeddings for each of them. We represent an audio-visual object as the trajectory of a potential sound source through space and time, which in the domain that we experiment on is often the track of a "talking head". Having obtained these trajectories, we use them to extract embeddings that can be then used for downstream tasks.

In more detail, our model uses a bottom-up grouping procedure to propose discrete audio-visual objects from raw video. It first estimates local (per-pixel and per-frame) synchronization evidence, using a network design that is more fine-grained in space and time than prior models. It then aggregates this evidence over time via optical flow, thereby allowing the model to obtain robustness to motions, and groups the aggregated attention into sound sources by detecting local maxima. The model represents each object as a separate embedding, temporal track, and attention map that can be adjusted in downstream tasks.

We will now give an overview of the model, which is shown in Fig. 2, followed by the learning framework which uses self-supervision based on synchronization. For architecture details, please refer to the the arXiv version.

3.1 Estimating Audio-Visual Attention

Before we group a scene into sound sources, we estimate a per-pixel attention map that picks out the regions of a video whose motions have a high degree of synchronization with the audio. We propose an attention mechanism that provides highly localized spatio-temporal attention, and which is sensitive to speaker motion. As in [6,31], we estimate audio-visual attention via a multimodal embedding (Fig. 2, step 1). We learn vector embeddings for each audio clip and embedding vectors for each pixel, such that if a pixel's vector has a high dot product with that of the audio, then it is likely to belong to that sound source. For this, we use a two-stream architecture similar to those in other sound-source localization work [6,31,52], with a network backbone similar to [11].

Video Encoder. Our video feature encoder is a spatio-temporal VGG-M [9] with a 3D convolutional layer first, followed by a stack of 2D convolutions. Given a $T \times H \times W \times 3$ input RGB video, it extracts a video embedding map $f_v(x, y, t)$ with dimensions $T \times h \times w \times D$.

Audio Encoder. The audio encoder is a VGG-M network operating on log-mel spectrograms, treated as single-channel images. Given an audio segment, it extracts a D-dimensional embedding $f_a(t)$ for every corresponding video frame t.

Computing Fine-Grained Attention Maps. For each space-time pixel, we ask: how correlated is it with the events in the audio? To estimate this, we measure the similarity between the audio and visual features at every spatial location. For every space-time feature vector $f_v(x, y, t)$, we compute the cosine similarity with the audio feature vector $f_a(t)$:

$$S_{av}(x, y, t) = f_v(x, y, t) \cdot f_a(t), \tag{1}$$

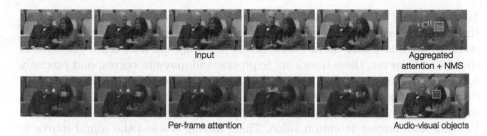

Fig. 3. **Intermediate representations from our model.** We show the per-frame attention maps $S_{av}(t)$, the aggregated attention map S_{av}^{tr} and the two highest scoring extracted audio-visual objects. We show the audio-visual objects for a single frame, with a square of constant width.

where we first l_2 normalize both features. We refer to the result, $S_{av}(x, y, t)$, as the *audio-visual attention map*.

3.2 Extracting Audio-Visual Objects

Given the audio-visual evidence, we parse a video into object representations.

Integrating Evidence Over Time. Audio-visual objects may only intermittently make sounds. Therefore, we need to integrate sparse attention evidence over time. We also need to group and track sound sources *between* frames, while accounting for camera and object motion. To make our model more robust to these motions, we aggregate information over time using optical flow (Fig. 2, step 2). We extract dense optical flow for every frame, chain the flow values together to obtain long-range tracks, and average the attention scores over these tracks. Specifically, if $\mathcal{T}(x, y, t)$ is the tracked location of pixel (x, y) from frame 1 to the later frame t, we compute the score:

$$S_{av}^{tr}(x, y) = \frac{1}{T} \sum_{t=1}^{T} S_{av}(\mathcal{T}(x, y, t), t), \tag{2}$$

where we perform the sampling using bilinear interpolation. The result is a 2D map containing a score for the future trajectory of every pixel of the initial frame through time. Note that any tracking method can be used in place of optical flow (e.g. with explicit occlusion handling); we use optical flow for simplicity.

Grouping a Scene into Instances. To obtain discrete audio-visual objects, we detect spatial local maxima (peaks) on the temporally aggregated synchronization maps, and apply non-maximum suppression (NMS). More specifically, we find peaks in the time-averaged synchronization map, $S_{av}^{tr}(x, y)$, and sort them in decreasing order; we then choose the peaks greedily, each time suppressing the ones that are within a $\rho \times \rho$ box. The selected peaks can be now viewed as distinct audio-visual objects. Examples of the intermediate representations extracted at the steps described so far are shown in Fig. 3.

Extracting Object Embeddings. Now that the sound sources have been grouped into distinct audio-visual objects, we can extract feature embeddings for each one of them that we can use in downstream tasks. Before extracting these features, we locate the position of the sound source in each frame. A simple strategy for this would be to follow the object's optical flow track throughout the video. However, these tracks are imprecise and may not correspond precisely to the location of the sound source. Therefore, we "snap" to the track location to the nearest peak in the attention map. More specifically, in frame t, we search in an area of $\rho \times \rho$ centered on the tracked location $\mathcal{T}(x, y, t)$, and select the pixel location with largest attention value. Then, having tracked the sound source in each frame, we select the corresponding spatial feature vector from the visual feature map f_v (Fig. 2, step 4). These per-frame embedding features, $f_v^{att}(t)$, can then be used to solve downstream tasks (Sect. 4). One can equivalently view this procedure as an audio-visual attention mechanism that operates on f_v.

3.3   Learning the Attention Map

Training our model amounts to learning the attention map S_{av} on which the audio-visual objects are subsequently extracted. We obtain this map by solving a self-supervised audio-visual synchronization task [13,40,44]: we encourage the embedding at each pixel to be correlated with the true audio and uncorrelated with shifted versions of it. We estimate the synchronization evidence for each frame by aggregating the per-pixel synchronization scores. Following common practice in multiple instance learning [6], we measure the per-frame evidence by the maximum spatial response:

$$S_{av}^{att}(t) = \max_{x,y} S_{av}(x, y, t). \tag{3}$$

We maximize the similarity between a video frame's true audio track while minimizing that of N shifted (i.e. misaligned) versions of the audio. Given visual features f_v and true audio a_i, we sample N other audio segments from the same video clip: $a_1, a_2, ..., a_N$, and minimize the contrastive loss [15,43]:

$$\mathcal{L} = -\log \frac{\exp(S_{av}^{att}(v, a_i))}{\exp(S_{av}^{att}(v, a_i)) + \sum_{j=1}^{N} \exp(S_{av}^{att}(v, a_j))}. \tag{4}$$

For the negative examples, we select all audio features (except for the true example) in a temporal window centered on the video frame.

In addition to the synchronization task, we also consider the *correspondence* task of Arandjelović and Zisserman [6], which chooses negatives audio samples from random video clips. Since this problem can be solved with even a single frame, it results in a model that is less sensitive to motion.

4   Applications of Audio-Visual Object Embeddings

We use our learned audio-visual objects for a variety of applications.

4.1   Audio-Visual Object Detection and Tracking

We can use our model for spatially localizing speakers. To do this, we use the tracked location of an audio-visual object in each frame.

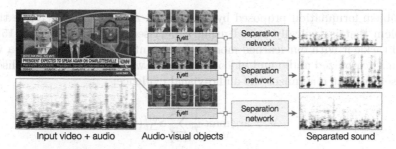

Input video + audio   Audio-visual objects   Separated sound

Fig. 4. Multi-speaker separation. We isolate the sound of each speaker's voice by combining our audio-visual objects with a network similar to [2]. Given a spectrogram of a noisy sound mixture, the network isolates the voice of each speaker, using the visual features provided by their audio-visual object.

4.2 Active Speaker Detection

For every frame in our video, our model can locate potential speakers and decide whether or not they are speaking. In our setting, this can be viewed as deciding whether an audio-visual object has strong evidence of synchronization in a given frame. For every tracked audio-visual object, we extract the visual features $f_v^{att}(t)$ (Sect. 3.2) for each frame t. We then obtain a score that indicates how strong the audio-visual correlation for frame t is, by computing the dot product: $f_v^{att}(t) \cdot f_a(t)$. Following previous work [13], we threshold the result to make a binary decision (active speaker or not).

4.3 Multi-speaker Source Separation

Our audio-visual objects can also be used for separating the voices of speakers in a video. We consider the *multi-speaker* separation problem [2,20]: given a video with multiple people speaking on-screen (*e.g.*, a television debate show), we isolate the sound of each speaker's voice from the audio stream. We note that this problem is distinct from on/off-screen audio separation [44], which requires only a single speaker to be on-screen.

We train an additional network that, given a waveform containing an audio mixture and an audio-visual object, isolates the speaker's voice (Fig. 4, full details in the the arXiv version of the paper). We use an architecture that is similar to [2], but conditions on our self-supervised representations instead of detections from a face detector. More specifically, the method of [2] runs a face detection and tracking system on a video, computes CNN features on each crop, and then feeds those to a source separation network. We, instead, simply provide the same separation network with the embedding features $f_v^{att}(t)$.

4.4 Correcting Audio-Visual Misalignment

We can also use our model to correct misaligned audio-visual data—a problem that often occurs in the recording and television broadcast process. We follow the problem formulation proposed by Chung and Zisserman [13]. While this is a problem that is typically solved using supervised face detection [13,15], we instead tackle it with our learned model. During inference, we are given a video with unsynchronized audio and video tracks, and we shift the audio to discover the offset $\hat{\Delta}t$ that maximizes the audio-visual evidence:

$$\hat{\Delta}t = \arg\max_{\Delta t} \frac{1}{T} \sum_{t=1}^{T} S_{\Delta t}^{att}(t), \tag{5}$$

where $S_{\Delta t}^{att}(t)$ is the synchronization score of frame t after shifting the audio by Δt. This can be estimated efficiently by recomputing the dot products in Eq. 1.

In addition to treating this alignment procedure as a stand-alone application, we also use it as a preprocessing step for our other applications (a common practice in other speech analysis work [2]). When given a test video, we first compute the optimal offset $\hat{\Delta}t$, and use it to shift the audio accordingly. We then recompute $S_{av}(t)$ from the synchronized embeddings.

5   Experiments

5.1   Datasets

Human Speech. We evaluate our model on the Lip Reading Sentences (LRS2 and LRS3) datasets and the Columbia active speaker dataset. LRS2 [1] and LRS3 [3] are audio-visual speech datasets containing 224 and 475 h of videos respectively, along with ground truth face tracks of the speakers. The Columbia dataset [8] contains footage from an 86-minute panel discussion, where multiple individuals take turns in speaking, and contains approximate bounding boxes and active speaker labels, *i.e.* whether a visible face is speaking at a given point in time. All datasets provide (pseudo-)ground truth bounding boxes obtained via face detection, which we use for evaluation. We resample all videos to a resolution of $H \times W = 270 \times 480$ pixels before feeding them to our model, which outputs $h \times w = 18 \times 31$ attention maps. We train all models on LRS2, and use LRS3 and Columbia only for evaluation.

Non-human Speakers. To evaluate our method on non-human speakers, we collected television footage from *The Simpsons* and *Sesame Street* shows (Table 3a). For testing, we obtained ASD and speaker localization labels, using the VIA tool [19]: we asked human annotators to label frames that they believed to contain an active speaker and to localize them. For every dataset, we create a *single-head* and a *multi-head* set, where clips are constrained to contain a single active speaker or multiple heads (talking or not) respectively. We provide dataset statistics in Table 3a and more details in the the arXiv version of the paper.

5.2   Training Details

Audio-Visual Object Detection Training. To make training easier, we follow [40] and use a simple learning curriculum. At the beginning of training, we

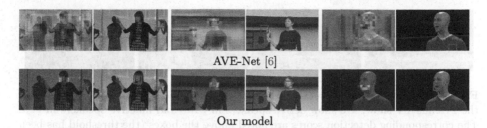

AVE-Net [6]

Our model

Fig. 5. Talking head detection and tracking on LRS3 datasets. For each of the 4 examples, we show the audio-visual attention score on every spatial location for the depicted frame, and a bounding box centered on the largest value, indicating the speaker location. Please see our webpage for video results.

Fig. 6. Handling motion: Talking head detection and tracking on continuous scenes from the validation set of LRS2. Despite the significant movement of the speakers and the camera, our method accurately tracks them.

sample negatives from random video clips, then switch to shifted audio tracks later in training. To speed up training, we also begin by taking the mean dot product (Eq. 3), and then switch to the maximum. We set ρ to 100 pixels.

Source Separation Training. Training takes place in two steps: we first train our model to produce audio-visual objects by solving a synchronization problem. Then, we train the multi-speaker separation network on top of these learned representations. We follow previous work [2,20] and use a mix-and-separate learning procedure. We create synthetic videos containing multiple talking speakers by 1) selecting two or three videos at random from the training set, depending on the experiment, 2) summing their waveforms together, and 3) vertically concatenating the video frames together. The model is then tasked with extracting a number of talking heads equal to the number of mixed videos and predicting an original corresponding waveform for each.

Non-human Model Training. We fine-tune the best model from LRS2 separately on each of the two datasets with non-human speakers. The lip motion for non-human speakers, such as the motion of a puppet's mouth, is only loosely correlated with speech, suggesting that there is less of an advantage to obtaining our negative examples from temporally shifted audio. We therefore sample our negative audio examples from other video clips rather than from misaligned audio (Sect. 3.3) when computing attention maps (Fig. 7).

Fig. 7. Active speaker detection on the Columbia dataset, and an example from the *Friends* TV show. We show active speakers in blue and inactive speakers in red. The corresponding detection scores are noted above the boxes (the threshold has been subtracted so that positive scores indicate active speakers). (Color figure online)

Fig. 8. Active speaker detection for non-human speakers. We show the top 2 highest-scoring audio-visual objects in each scene, along with the aggregated attention map. Please see our webpage for video results.

5.3   Results

1. Talking Head Detection and Tracking. We evaluate how well our model is able to localize speakers, i.e. talking heads (Table 1a). First, we evaluate two simple baselines: the *random* one, which selects a random pixel in each frame and the *center* one, which always selects the center pixel. Next, we compared with two recent sound source localization methods: Owens and Efros [44] and AVE-Net [6]. Since these methods require input videos that are longer than most of the videos in the test set of LRS2, we only evaluate them on LRS3. We also perform several ablations of our model: To evaluate the benefit of integrating the audio-visual evidence over flow trajectories, we create a variation of our model called *No flow* that, instead, computes the attention S_{av}^{tr} by globally pooling over time throughout the video. Finally, we also consider a variation of this model that uses a larger NMS window ($\rho = 150$).

We found that our method obtains very high accuracy, and that it significantly outperforms all other methods. AVE-Net solves a correspondence task that doesn't require motion information, and uses a single video frame as input. Consequently, it does not take advantage of informative motion, such as moving

Table 1. (a): **Talking head detection and tracking accuracy**. A detection is considered correct if it lies within the true bounding box. (b): **Active speaker detection accuracy** on the Columbia dataset [8]. F1 Scores (%) for each speaker, and the overall average.

Method	LRS2	LRS3
Random	2.8%	2.9%
Center	23.9%	25.9%
Owens & Efros [44]	-	24.8%
AVE-Net [6]	-	58.1%
No flow	98.4%	94.2%
No flow + large NMS	98.8%	97.2%
Full model	**99.6%**	**99.7%**

Method	Speaker					
	Bell	Boll	Lieb	Long	Sick	Avg.
Chakravarty [8]	82.9	65.8	73.6	86.9	81.8	80.2
Shahid [53]	87.3	96.4	92.2	83.0	87.2	89.2
SyncNet [13]	93.7	83.4	86.8	97.7	86.1	89.5
Ours	92.6	82.4	88.7	94.4	95.9	**90.8**

lips. As can be seen in Fig. 5, the localization maps produced by AVE-Net [6] are less precise, as it only loosely associates appearance of a person to speech, and won't consistently focus on the same region. Owens and Efros [44], by contrast, has a large temporal receptive field, which results in temporally imprecise predictions, causing very large errors when the subjects are moving. The *No flow* baseline fails to track the talking head well outside the NMS area, and its accuracy is consequently lower on LRS3. Enlarging the NMS window partially alleviates this issue, but the accuracy is still lower than that of our model. We note that the LRS2 test set contains very short clips (usually 1–2 seconds long) with predominantly static speakers, which explains why using flow does not provide an advantage. We show some challenging examples with significant speaker and camera motion in Fig. 6. Please refer to the the arXiv version of the paper for further analysis of camera and speaker motion.

2. Active Speaker Detection. Next, we ask how well our model can determine *which* speaker is talking. Following previous work that uses supervised face detection [14,53], we evaluate our method on the Columbia dataset [8]. For each video clip, we extract 5 audio-visual objects (an upper bound on the number of speakers), each of which has an ASD score indicating the likelihood that it is a sound source (Sect. 4.2). We then associate each ground truth bounding box with the audio-visual object whose trajectory follows it the closest. For comparison with existing work, we report the F1 measure (the standard for this dataset) per individual speaker as well as averaged over all speakers. For calculating the F1 we set the ASD threshold to the one that yields the Equal Error Rate (EER) for the pretext task on the LRS2 validation set. As shown in Table 1b, our model outperforms all previously reported results on this dataset, even though (unlike other methods) it does not use labeled face bounding boxes for training.

3. Multi-speaker Source Separation. To evaluate our model on speaker separation, we follow the protocol of [2]. We create synthetic examples from the test set of LRS2, using only videos that are between 2 − 5 seconds long, and evaluate performance using Signal-to-Distortion-Ratio (SDR) [21] and Perceptual Evaluation of Speech Quality (PESQ, varies between 0 and 4.5) [49] (higher is better

Table 2. (a): **Source separation** on LRS2. #Spk indicates the number of speakers. The WER on the ground truth signal is 20.0%. (b): **Audio-visual synchronization** accuracy (%) evaluation for a given number of input frames.

Method \ # Spk.	SDR 2	SDR 3	PESQ 2	PESQ 3	WER % 2	WER % 3
Mixed input	-0.3	-3.4	1.7	1.5	91.0	97.2
Conv.-Sync [2]	11.3	7.5	3.0	2.5	30.3	43.5
Ours Frozen	10.7	7.0	3.0	2.5	30.7	44.2
Oracle-BB	10.8	7.1	2.9	2.5	30.9	44.9
Small-NMS	10.6	6.8	3.0	2.5	31.2	44.7
Full	10.8	7.2	3.0	2.6	30.4	42.0

Method	Input frames 5	7	9	11	13	15
SyncNet [13]	75.8	82.3	87.6	91.8	94.5	96.1
PM [15]	88.1	93.8	96.4	97.9	98.7	99.1
Ours	78.8	87.1	92.1	94.8	96.3	97.3

for both). We also assess the intelligibility of the output by computing the Word Error Rate (WER, lower is better) between the transcriptions obtained with the Google Cloud speech recognition system. Following [3], we train and evaluate separate models for 2 and 3 speakers, though we note that if the number of speakers were unknown, it could be estimated using active speaker detection.

For comparison, we implement the model of Afouras *et al.* [2], and train it on the same data. For extracting visual features to serve as its input, we use a state-of-the-art audio-visual synchronization model [15], rather than the lip-reading features from Afouras *et al.* [4]. We refer to this model as *Conversation-Sync*. This model uses bounding boxes from a well-engineered face detection system, and thus represents an approximate upper limit on the performance of our self-supervised model. Our main model for this experiment is trained end-to-end and uses $\rho = 150$. We also performed a number of ablations: a model that freezes the pretrained audio-visual features and a model with a smaller $\rho = 100$.

We observed (Table 2a) that our self-supervised model obtains results close to those of [2], which is based on supervised face detection. We also asked how much error is introduced by lack of face detection. In this direction we extract the local visual descriptors using tracks obtained with face detectors instead of our audio-visual object tracks. This model, *Oracle-BB*, obtains results similar to ours, suggesting that the quality of our face localization is high.

4. Correcting Misaligned Visual and Audio Data. We use the same metric as [15] to evaluate on LRS2. The task is to determine the correct audio-to-visual offset within a ±15 frame window. An offset is considered correct if it is within 1 video frame from the ground truth. The distances are averaged over 5 to 15 frames. We compare our method to two state-of-the-art synchronization methods: SyncNet [13] and the state-of-the-art Perfect Match [15]. We note that [15] represents an approximate upper limit to what we would expect our method to achieve, since we are using a similar network and training objective; the major difference is that we use our audio-visual objects instead of image crops from a face detector. The results (Table 2b) show that our self-supervised model obtains comparable accuracy to these supervised methods.

5. Generalization to Non-human Speakers. We evaluate the LWTNet model's generalization to non-human speakers using the *Simpsons* and *Sesame*

Table 3. (a): **Label statistics** for non-human test sets. S is *single head* and M *multi-head*. (b): **Non-human speaker evaluation** *for ASD and localization tasks on* Simpsons *and* Sesame Street. *MN: MobileNet; RN: ResNet50.*

Source	Type	Clips	Frames
The Simpsons	S	41	87
The Simpsons	M	582	251
Sesame Street	S	57	120
Sesame Street	M	143	424

	Loc. Acc		ASD AP			
	Single-head		Single-head		Multi-head	
Method	Simp.	Ses.	Simp.	Ses.	Simp.	Ses.
Random	8.7	16.0	-	-	-	-
Center	62.0	80.1	-	-	-	-
RetinaFace RN	47.7	61.2	40.0	46.8	-	-
RetinaFace MN	72.1	70.2	60.4	52.4	-	-
Ours	**98.8**	**81.0**	**98.7**	**72.2**	**85.5**	**55.6**

Street datasets described in Sect. 5.1. The results of our evaluation are summarized in Table 3b. Since supervised speech analysis methods are often based on face detection systems, we compare our method's performance to off-the-shelf face detectors, using the *single-head* subset. As a face detector baseline, we use the state-of-the-art RetinaFace [17] detector, with both the MobileNet and ResNet-50 backbones. We report localization accuracy (as in Table 1a) and Average Precision (AP). It is clear that our model outperforms the face detectors in both localization and retrieval performance for both datasets.

The second evaluation setting is detecting active speakers in videos from the *multi-head* test set. As expected, our model's performance decreases in this more challenging scenario; however, the AP for both datasets indicates that our method can be useful for retrieving the speaker in this entirely new domain. We show qualitative examples of ASD on the *multi-head* test sets in Fig. 8.

6   Conclusion

In this paper, we have proposed a unified model that learns from raw video to detect and track speakers. The embeddings learned by the model are effective for many downstream speech analysis tasks, such as source separation and active speaker detection, that in previous work required supervised face detection.

Acknowledgements. We thank V. Kalogeiton for generous help with the annotations and the *Friends* videos, A. A. Efros for helpful discussions, L. Momeni, T. Han and Q. Pleple for proofreading, A. Dutta for help with VIA, and A. Thandavan for infrastructure support. This work is funded by the UK EPSRC CDT in AIMS, DARPA Medifor, and a Google-DeepMind Graduate Scholarship.

References

1. Afouras, T., Chung, J.S., Senior, A., Vinyals, O., Zisserman, A.: Deep audio-visual speech recognition. IEEE PAMI (2019)
2. Afouras, T., Chung, J.S., Zisserman, A.: The conversation: deep audio-visual speech enhancement. In: INTERSPEECH (2018)

3. Afouras, T., Chung, J.S., Zisserman, A.: LRS3-TED: a large-scale dataset for visual speech recognition. In: arXiv preprint arXiv:1809.00496 (2018)
4. Afouras, T., Chung, J.S., Zisserman, A.: My lips are concealed: audio-visual speech enhancement through obstructions. In: INTERSPEECH (2019)
5. Arandjelović, R., Zisserman, A.: Look, listen and learn. In: Proceedings of ICCV (2017)
6. Arandjelović, R., Zisserman, A.: Objects that sound. In: Ferrari, V., Hebert, M., Sminchisescu, C., Weiss, Y. (eds.) ECCV 2018. LNCS, vol. 11205, pp. 451–466. Springer, Cham (2018). https://doi.org/10.1007/978-3-030-01246-5_27
7. Barzelay, Z., Schechner, Y.Y.: Harmony in motion. In: 2007 IEEE Conference on Computer Vision and Pattern Recognition (2007)
8. Chakravarty, P., Tuytelaars, T.: Cross-modal supervision for learning active speaker detection in video. In: Leibe, B., Matas, J., Sebe, N., Welling, M. (eds.) ECCV 2016. LNCS, vol. 9909, pp. 285–301. Springer, Cham (2016). https://doi.org/10.1007/978-3-319-46454-1_18
9. Chatfield, K., Simonyan, K., Vedaldi, A., Zisserman, A.: Return of the devil in the details: Delving deep into convolutional nets. arXiv preprint arXiv:1405.3531 (2014)
10. Chen, T., Kornblith, S., Norouzi, M., Hinton, G.: A simple framework for contrastive learning of visual representations. ICML (2020)
11. Chung, J.S., Lee, B.J., Han, I.: Who said that?: Audio-visual speaker diarisation of real-world meetings. In: Interspeech (2019)
12. Chung, J.S., Nagrani, A., Zisserman, A.: VoxCeleb2: deep speaker recognition. In: INTERSPEECH (2018)
13. Chung, J.S., Zisserman, A.: Out of time: automated lip sync in the wild. In: Chen, C.-S., Lu, J., Ma, K.-K. (eds.) ACCV 2016. LNCS, vol. 10117, pp. 251–263. Springer, Cham (2017). https://doi.org/10.1007/978-3-319-54427-4_19
14. Chung, J.S., Zisserman, A.: Signs in time: encoding human motion as a temporal image. In: Workshop on Brave New Ideas for Motion Representations, ECCV (2016)
15. Chung, S.W., Chung, J.S., Kang, H.G.: Perfect match: improved cross-modal embeddings for audio-visual synchronisation. In: Proceedings of ICASSP, pp. 3965–3969. IEEE (2019)
16. Cutler, R., Davis, L.: Look who's talking: speaker detection using video and audio correlation. In: 2000 IEEE International Conference on Multimedia and Expo. ICME 2000. Proceedings. Latest Advances in the Fast Changing World of Multimedia (Cat. No. 00TH8532), vol. 3, pp. 1589–1592. IEEE (2000)
17. Deng, J., Guo, J., Yuxiang, Z., Yu, J., Kotsia, I., Zafeiriou, S.: Retinaface: Single-stage dense face localisation in the wild. In: arxiv (2019)
18. Doersch, C., Gupta, A., Efros, A.A.: Unsupervised visual representation learning by context prediction. In: Proceedings of ICCV, pp. 1422–1430 (2015)
19. Dutta, A., Zisserman, A.: The VIA annotation software for images, audio and video. In: Proceedings of the 27th ACM International Conference on Multimedia. MM 2019. ACM, New York (2019)
20. Ephrat, A., et al.: Looking to listen at the cocktail party: a speaker-independent audio-visual model for speech separation. ACM Trans. Graph. (TOG) 37(4), 112 (2018)
21. Févotte, C., Gribonval, R., Vincent, E.: BSS EVAL toolbox user guide. IRISA Technical Report 1706 (2005). http://www.irisa.fr/metiss/bsseval/
22. Fisher III, J.W., Darrell, T., Freeman, W.T., Viola, P.A.: Learning joint statistical models for audio-visual fusion and segregation. In: NeurIPS (2000)

23. Gabbay, A., Ephrat, A., Halperin, T., Peleg, S.: Seeing through noise: visually driven speaker separation and enhancement. In: Proceedings of ICASSP, pp. 3051–3055. IEEE (2018)
24. Gadde, R., Jampani, V., Gehler, P.V.: Semantic video CNNs through representation warping. In: Proceedings of ICCV, pp. 4463–4472 (2017)
25. Gan, C., Zhao, H., Chen, P., Cox, D., Torralba, A.: Self-supervised moving vehicle tracking with stereo sound. In: Proceedings of the IEEE International Conference on Computer Vision, pp. 7053–7062 (2019)
26. Gao, R., Feris, R., Grauman, K.: Learning to separate object sounds by watching unlabeled video. In: Ferrari, V., Hebert, M., Sminchisescu, C., Weiss, Y. (eds.) ECCV 2018. LNCS, vol. 11207, pp. 36–54. Springer, Cham (2018). https://doi.org/10.1007/978-3-030-01219-9_3
27. Gao, R., Grauman, K.: 2.5D visual sound. In: CVPR (2019)
28. Gao, R., Grauman, K.: Co-separating sounds of visual objects. arXiv preprint arXiv:1904.07750 (2019)
29. Han, T., Xie, W., Zisserman, A.: Video representation learning by dense predictive coding. In: Workshop on Large Scale Holistic Video Understanding, ICCV (2019)
30. Han, T., Xie, W., Zisserman, A.: Memory-augmented dense predictive coding for video representation learning. In: ECCV (2020)
31. Harwath, D., Recasens, A., Surís, D., Chuang, G., Torralba, A., Glass, J.: Jointly discovering visual objects and spoken words from raw sensory input. In: Ferrari, V., Hebert, M., Sminchisescu, C., Weiss, Y. (eds.) ECCV 2018. LNCS, vol. 11210, pp. 659–677. Springer, Cham (2018). https://doi.org/10.1007/978-3-030-01231-1_40
32. He, K., Fan, H., Wu, Y., Xie, S., Girshick, R.: Momentum contrast for unsupervised visual representation learning. In: CVPR (2020)
33. Hénaff, O.J., et al.: Data-efficient image recognition with contrastive predictive coding. In: ICML (2020)
34. Hershey, J., Movellan, J.: Audio-vision: locating sounds via audio-visual synchrony. In: NeurIPS, vol. 12 (1999)
35. Hu, D., Nie, F., Li, X.: Deep multimodal clustering for unsupervised audiovisual learning. In: Proceedings of the IEEE/CVF Conference on Computer Vision and Pattern Recognition (CVPR), June 2019
36. Hu, D., Wang, Z., Xiong, H., Wang, D., Nie, F., Dou, D.: Curriculum audiovisual learning. arXiv preprint arXiv:2001.09414 (2020)
37. Izadinia, H., Saleemi, I., Shah, M.: Multimodal analysis for identification and segmentation of moving-sounding objects. IEEE Trans. Multimedia 15(2), 378–390 (2012)
38. Khosravan, N., Ardeshir, S., Puri, R.: On attention modules for audio-visual synchronization. arXiv preprint arXiv:1812.06071 (2018)
39. Kidron, E., Schechner, Y.Y., Elad, M.: Pixels that sound. In: Proceedings of CVPR (2005)
40. Korbar, B., Tran, D., Torresani, L.: Co-training of audio and video representations from self-supervised temporal synchronization. CoRR (2018)
41. Misra, I., van der Maaten, L.: Self-supervised learning of pretext-invariant representations. In: CVPR (2020)
42. Nagrani, A., Chung, J.S., Albanie, S., Zisserman, A.: Disentangled speech embeddings using cross-modal self-supervision. In: Proceedings of ICASSP, pp. 6829–6833. IEEE (2020)
43. Oord, A.v.d., Li, Y., Vinyals, O.: Representation learning with contrastive predictive coding. arXiv preprint arXiv:1807.03748 (2018)

224   T. Afouras et al.

44. Owens, A., Efros, A.A.: Audio-visual scene analysis with self-supervised multisensory features. In: Ferrari, V., Hebert, M., Sminchisescu, C., Weiss, Y. (eds.) ECCV 2018. LNCS, vol. 11210, pp. 639–658. Springer, Cham (2018). https://doi.org/10.1007/978-3-030-01231-1_39

45. Owens, A., Isola, P., McDermott, J., Torralba, A., Adelson, E.H., Freeman, W.T.: Visually indicated sounds. In: Computer Vision and Pattern Recognition (CVPR) (2016)

46. Owens, A., Wu, J., McDermott, J.H., Freeman, W.T., Torralba, A.: Learning sight from sound: ambient sound provides supervision for visual learning. Int. J. Comput. Vis. (2018)

47. Pfister, T., Charles, J., Zisserman, A.: Flowing convnets for human pose estimation in videos. In: Proceedings of ICCV (2015)

48. Ramaswamy, J., Das, S.: See the sound, hear the pixels. In: Proceedings of the IEEE/CVF Winter Conference on Applications of Computer Vision (WACV), March 2020

49. Rix, A.W., Beerends, J.G., Hollier, M.P., Hekstra, A.P.: Perceptual evaluation of speech quality (PESQ)-a new method for speech quality assessment of telephone networks and codecs. In: Proceedings of ICASSP, vol. 2, pp. 749–752. IEEE (2001)

50. Roth, J., et al.: AVA-ActiveSpeaker: An audio-visual dataset for active speaker detection. arXiv preprint arXiv:1901.01342 (2019)

51. Rouditchenko, A., Zhao, H., Gan, C., McDermott, J., Torralba, A.: Self-supervised audio-visual co-segmentation. In: Proceedings of ICASSP, pp. 2357–2361. IEEE (2019)

52. Senocak, A., Oh, T.H., Kim, J., Yang, M.H., Kweon, I.S.: Learning to localize sound source in visual scenes. In: Proceedings of CVPR (2018)

53. Shahid, M., Beyan, C., Murino, V.: Voice activity detection by upper body motion analysis and unsupervised domain adaptation. In: The IEEE International Conference on Computer Vision (ICCV) Workshops, October 2019

54. Tian, Y., Shi, J., Li, B., Duan, Z., Xu, C.: Audio-visual event localization in unconstrained videos. In: Ferrari, V., Hebert, M., Sminchisescu, C., Weiss, Y. (eds.) ECCV 2018. LNCS, vol. 11206, pp. 252–268. Springer, Cham (2018). https://doi.org/10.1007/978-3-030-01216-8_16

55. Tian, Y., Krishnan, D., Isola, P.: Contrastive multiview coding. arXiv preprint arXiv:1906.05849 (2019)

56. Wang, X., Gupta, A.: Unsupervised learning of visual representations using videos. In: Proceedings of ICCV, pp. 2794–2802 (2015)

57. Zhao, H., Gan, C., Ma, W.C., Torralba, A.: The sound of motions. In: Proceedings of ICCV (2019)

58. Zhao, H., Gan, C., Rouditchenko, A., Vondrick, C., McDermott, J., Torralba, A.: The sound of pixels. In: Ferrari, V., Hebert, M., Sminchisescu, C., Weiss, Y. (eds.) ECCV 2018. LNCS, vol. 11205, pp. 587–604. Springer, Cham (2018). https://doi.org/10.1007/978-3-030-01246-5_35

GAN-Based Garment Generation Using Sewing Pattern Images

Yu Shen$^{(\boxtimes)}$, Junbang Liang, and Ming C. Lin

University of Maryland, College Park, USA
yushen@umd.edu
https://gamma.umd.edu/researchdirections/virtualtryon/garmentgeneration/

Abstract. The generation of realistic apparel model has become increasingly popular as a result of the rapid pace of change in fashion trends and the growing need for garment models in various applications such as virtual try-on. For such application requirements, it is important to have a general cloth model that can represent a diverse set of garments. Previous studies often make certain assumptions about the garment, such as the topology or *suited* body shape. We propose a unified method using the generative network. Our model is applicable to different garment topologies with different sewing patterns and fabric materials. We also develop a novel image representation of garment models, and a reliable mapping algorithm between the general garment model and the image representation that can regularize the data representation of the cloth. Using this special intermediate image representation, the generated garment model can be easily retargeted to another body, enabling garment customization. In addition, a large garment appearance dataset is provided for use in garment reconstruction, garment capturing, and other applications. We demonstrate that our generative model has high reconstruction accuracy and can provide rich variations of virtual garments.

1 Introduction

The generation of realistic garment is one of the most important steps during the garment design and manufacturing process. Usually, a garment model needs to be manually designed by an experienced designer—this step can be time-consuming and labor-intensive. The efficiency can be dramatically improved if a garment model can be generated automatically. The generation of garment model can also benefit certain virtual-reality applications such as the virtual try-on system. As e-commerce becomes more prevalent in the apparel industry, a rich and realistic virtual try-on system can considerably improve the user experience during online shopping, where garment model generation plays a central role.

Electronic supplementary material The online version of this chapter (https://doi.org/10.1007/978-3-030-58523-5_14) contains supplementary material, which is available to authorized users.

© Springer Nature Switzerland AG 2020
A. Vedaldi et al. (Eds.): ECCV 2020, LNCS 12363, pp. 225–247, 2020.
https://doi.org/10.1007/978-3-030-58523-5_14

However, there are many challenges in automatically generating garment models. First, garments usually have different topologies, especially for fashion apparel, that make it difficult to design a universal generation pipeline. Moreover, it is often not straightforward for the general garments design to be retargeted onto another body shape, making it difficult for customization. Some previous work started to address this problem using either user-assisted input [15] or cloths with fixed topology such as a T-shirt or a skirt [31].

We propose a learning-based parametric generative model to overcome the above difficulties. Given garment sewing patterns and human body shapes as inputs, we compute the *displacement image* on the UV space of the human body as a unified representation of the garment mesh. Different sizes and topologies of garments are represented by different values and geometric adjacencies in the image. The 2D displacement image, as the representation of the 3D garment mesh data, is given as input into a conditional Generative Adversarial Network (GAN) for latent space learning. Using this 2D representation for the garment mesh, on one hand, we can transform the irregular 3D mesh data to regular image data where a traditional CNN can easily learn; on the other hand, we can extract the relative geometric information with respect to the human body, enabling straightforward garment retargeting onto a different human body.

Our network can generate a series of garment models that meet the constraints of inputs with various appearances. The generated garments can be easily retargeted to another body shape using our 2D representation, while other generative methods [15,31] need to rerun the generative network and cannot ensure the same appearance as the original.

To train such a generative model, a sufficient amount of garment data is needed. However, there is no publicly available garment dataset which provides the appearances of the garments undergoing different motions and on varying human body shapes. Therefore, we generate a large dataset with different garment geometries for this specific task. We employ physically-based simulation of different garment meshes and fabric materials. Together with different human body motions, we can obtain a large variety of garment appearances on the body.

Overall, our contributions include:

- The first image-based garment generative model (Sect. 5) which supports most garment topologies and patterns (Sect. 6.3), human body shapes and sizes (Sect. 6.5), and garment materials (Appendix 2).
- A novel image representation for garments (Sect. 4) that can transfer to/from general 3D garment models with little information loss (Sect. 6.2), enabling garment retargeting (Sect. 6.5).
- A large garment appearance dataset for training (Appendix 2).

2   Related Work

In this section, we survey related works in garment modeling, garment retargeting, and generative networks.

2.1 Garment Modeling

Garment model generation has attracted attention these days due to its importance in both real-world and virtual garment design application. Although professional tools, such as Marvelous Designer [2018], can help design high-quality garment models, it may take an excessive amount of time to use it. Several studies have addressed this issue by introducing an automatic generation pipeline to improve the efficiency. Assuming different priors, most previous studies lie in three categories: sketch-based, image-based, and depth-based.

Sketch-Based Methods. Generating garment models with sketches is one of the most popular ways. Turquin *et al.* [28] and Decaudin *et al.* [9] developed some of the early work in this area. They used grid and geometric methods to generate garment models with sketches. Later, Robson *et al.* [26] proposed a context-aware method to make the generated garment model more realistic based on a set of observations on key factors which could affect the shapes of garments. Jung *et al.* [17] proposed a method to model 3D developable surfaces with a multi-view sketch input. Recently, Huang *et al.* [15] proposed a realistic 3D garment generation algorithm based on front and back image sketches. Wang *et al.* [31] proposed an algorithm that can achieve retargeting easily.

In addition, a common limitation of these methods is the domain knowledge requirement on garment sketching, while our method does not require any domain knowledge.

Image-Based or Depth-Based Methods. Other information such as images can also be used to generate a garment model. Bradley *et al.* [5] and Zhou *et al.* [34] researched early on garment modeling using multi-view images and single-view image, respectively. Jeong *et al.* [16] created the garment model with a single photograph by detecting the landmark points of the garment. Yang [32] made a full use of garment and human body databases to generate the garment models from image. Daněřek *et al.*'s [8] method can estimate the 3D garment shape from a single image using deep neural networks. Recently, Tex2Shape [1], PIFu [27], DeepHuman [33], Gabeur et al. [11] proposed models for detailed clothed full-body geometry reconstruction. MGN [4] predicts body shape and clothing, layered on top of the SMPL [21] model from a few (typically 1–8) frames of a video. Depth information can also be useful. Chen *et al.* [7] proposed a method to generate garment models given an RGBD sequence of a worn garment.

However, these methods require photos or depth images from a real garment, which means they cannot generate a garment model from size parameters only. In contrast, our model is able to generate 3D garment meshes directly from sewing patterns and sizing parameters by using the generative network.

2.2 Garment Retargeting

Retargeting the garment model from one body to another is often needed due to different body shapes. Retargeting can save computational costs if it can be done efficiently. Brouet *et al.* [6] introduced a fully automatic method for

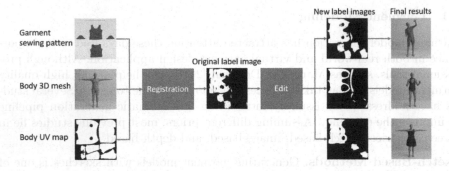

Fig. 1. Label image generation process. We first generate the label image with the pattern configuration registered on the body mesh and mapped to the body UV map. Then we can edit the original label image to new, different label images, which will lead to different garment topologies in the final results.

design-preserving transfer of garments among characters with different body shapes. In contrast, Guan *et al.* [13] used a learning-based architecture to replace the expensive simulation process and present retargeting examples. GarNet [14] presented a two-stream architecture to fit a 3D garment template to a 3D body. TailorNet [24] predicts clothing deformation given the pose and shape of human and garment model.

In our method, by making use of the image representation of the garment, we can easily retarget one generated garment model from one body shape to another, without additional computations.

2.3   Generative Network

Generative networks have been becoming more popular due to their impressive performance. There are several well-known generative networks, such as Generative Adversarial Network (GAN) [12] and Variational Auto-Encoder (VAE) [10]. With the development of the neural network research, new variants of generative networks have been proposed, such as Pix2PixHD [29] based on GAN or VQ-VAE [25] based on VAE. In our algorithm, we design the network architecture based on the Pix2PixHD network architecture due its high accuracy and efficiency.

3   Method Overview

Our objective for this work is to develop a GAN-based generator that creates different types of garment meshes, given the garment design (or sewing) patterns. The overall pipeline is shown in Fig. 1.

First, we unify the common garment pattern configurations to a body mask that shows the region of garment coverage. To do this, we mark the sizes of each pattern pieces from the 2D sewing pattern and register each piece to its

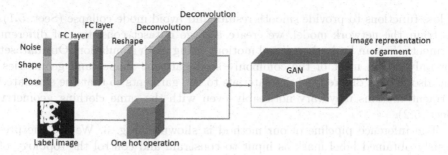

Fig. 2. Our network architecture. We first encode one dimensional input to match the sizes to the label image (upper branch). It is then concatenated with one hot labelled image (bottom branch) and fed into the GAN network. Finally, the network outputs the image representation of the garment (right).

corresponding body part. We can then obtain the label map by coloring the covered body part according to the registration. As an auxiliary step, we may edit the label image to vary the sizes and the connectivity of different parts, leading to different garment styles and topologies in the final results.

We model the garment mesh using a 2D image representation in the UV space of the corresponding human body (Fig. 11), which shares the same space as the label map that we obtained from the pattern input. This step regularizes the input mesh onto a CNN-friendly format that is independent to the original mesh resolution. We compute the correspondence between 3D points of the mesh and the 2D pixels of the image using non-rigid ICP and a Voronoi diagram, as later discussed in Sect. 4.

We then train a deep GAN to learn the distribution of the representative images. We use a state-of-the-art conditional GAN to learn a mapping between a topology label mask and the final image representation, conditioned on the human pose, shape and a random noise, as shown in Fig. 2. We define a set

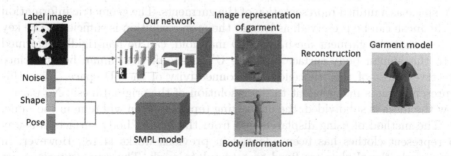

Fig. 3. Our inference pipeline. The upper branch generates the image representation of the garment, while the bottom branch generates the body mesh. Finally, we recover the garment mesh by decoding the image representation of the garment given the body mesh.

of loss functions to provide smooth results and avoid mode collapse (Sect. 5.1). To train the network model, we create a large dataset consisting of different garments, human body shapes and motions using cloth simulation. Our dataset not only covers most of the commonly seen garment shapes and geometries, but also assigns different fabric materials to the garments so that the simulated garment motions may vary noticeably even with the same clothing geometry (Sect. 5.2).

The inference pipeline of our method is shown in Fig. 3. We use the previously obtained label mask as input to constrain and control the topology of the output mesh. Given the label mask, we can generate a set of different image representations of the garment by varying the human pose and shape parameters, as well as the noise vector. As the last step, we recover the 3D garment mesh using its image representation and the corresponding human body. The final garment mesh can naturally fit onto the given human body shape due to the nature of our representation model (See Sect. 4), and can provide realistic details depending on the body pose and shape.

4 Garment Representation in UV Space

As stated before, there are several challenges involved in modeling garments. First, garment meshes are graph data with nonuniform structures. Different meshes usually have different numbers of vertices and connections. It is difficult to set up a uniform and vectorized graph representation for all garments. Also, in contrast to other graph data, subdivision does not change the geometric information of the mesh. Graph representation cannot easily account for this ambiguity or redundancy of the mesh. Next, there are many kinds of garments that have different topologies. Shoulder style can provide a large variety of garment looks, not to mention the difference between skirts and pants. This makes high-level parameterization (e.g., sleeve length) impossible without predefined classification.

To overcome these difficulties, we employ displacement maps on human body UV space as a unified representation of the garments. The geometric information of the mesh can be preserved, as long as the map resolution is sufficient. The key idea is that the garment mesh, as a 2D manifold, can be non-rigidly deformed onto the human body surface, and the UV space of the human body surface preserves most of the adjacency and connectivity of the 3D space. Also, this representation is independent to the resolution of the original mesh. No matter how the mesh is subdivided, the underlying representation will remain the same.

The method of using displacements from the human body surface as a way to represent clothes has been adopted in previous works [4,18]. However, in their work, the clothes are fixed to a template mesh. The representations are thus forced to be separated into a set of different clothes, since they have different templates. In contrast, we do not rely on specific clothing templates. Our model not only unifies different cloth types, but also generates clothes with new topologies.

4.1 Encoding Process

To create a displacement map of a certain garment, we first use non-rigid ICP [2] to register the cloth surface to the body surface, which makes the cloth tight-fit to the body. We then subdivide the cloth surface according to the Voronoi regions of body vertices to assign garment surface to body vertices. Finally, for each point on the body UV map, we compute the corresponding 3D position on the body surface, match it to the point on the cloth with the interpolated normal vector (of the garment surface that is assigned to the region), and fill in pixel value of the map using the displacement.

Specifically, we first register the cloth surface $\mathcal{G} = (\mathcal{V}_G, \mathcal{E}_G)$ to the body surface $\mathcal{B} = (\mathcal{V}_B, \mathcal{E}_B)$ by optimization:

$$\mathbf{X} = \arg\min_{\mathbf{X}} E(\mathbf{X}) = \sum_{v_i \in \mathcal{V}_G} d^2(\mathcal{B}, \mathbf{X}_i v_i) + \alpha \sum_{(\mathbf{v}_i, \mathbf{v}_j) \in \mathcal{E}_G} \|(\mathbf{X}_i - \mathbf{X}_j)\mathbf{D}\|_F^2 \quad (1)$$

where $\mathbf{X}$ is the set of affine matrices for all garment vertices, α and $\mathbf{D} = diag(1, 1, 1, \gamma)$ are importance weights, and $d()$ is the distance between a point to a mesh. We set α and γ to a small value (typically 0.1) to encourage non-rigidity so that the cloth is mapped onto the body surface without large global rigid transformation. Note that after the non-rigid ICP, there may still be some vertices that are far from the body surface because of the topology constraint (e.g. dresses). We then design an algorithm to create a correspondence mapping of the surfaces between the cloth and the body.

As a preliminary step, we create the correspondence between each face of the cloth mesh and the vertices of the body mesh according to the Euclidean distance. First, we subdivide the registered cloth mesh using the 3D Voronoi diagram of the body surface. Voronoi regions [20] of the body vertices cuts the garment surface into convex polygons, which can be easily triangulated. Given that computing the analytical intersection to Voronoi regions is challenging, we achieve the subdivision by repeatedly check if an edge of the cloth mesh belongs to multiple Voronoi regions:

$$\exists e = (\mathbf{v}_0, \mathbf{v}_1) \in \mathcal{E}_G : V_{min}(\mathcal{V}_B, \mathbf{v}_0) \bigcap V_{min}(\mathcal{V}_B, \mathbf{v}_1) = \emptyset \quad (2)$$

where $V_{min}(\mathbf{V}, \mathbf{u})$ computes the subset of $\mathbf{V}$ that has the closest distance to $\mathbf{u}$:

$$V_{min}(\mathbf{V}, \mathbf{u}) = \{\mathbf{v} \in \mathbf{V} : \forall \mathbf{v}' \in \mathbf{V}, d(\mathbf{v}', \mathbf{u}) \geq d(\mathbf{v}, \mathbf{u})\} \quad (3)$$

If so, we subdivide the edge using the perpendicular bisector plane of the two vertices selected from $V_{min}(\mathcal{V}_B, \mathbf{v}_0)$ and $V_{min}(\mathcal{V}_B, \mathbf{v}_1)$, and its adjacent faces accordingly. Finally, we ensure that each face of the cloth mesh belongs to only one Voronoi region.

The next step is to match each subdivided face of the cloth mesh to the UV space of the corresponding Voronoi region. The intersection of the Voronoi region of a vertex and the body surface is bounded by the perpendicular bisector planes of each of its adjacent edges. We refer it as the 'Voronoi surface' of a vertex. Instead of further subdividing the cloth face into smaller faces and mapping them to different UV regions of the Voronoi surface, we iterate each pixel of the UV regions and shoot a ray out of the surface. To ensure an even sampling, the direction of each ray is computed by interpolating between the normal direction of face, edge and vertex (See description with figure in Appendix 5). An intersection of the ray and the cloth face creates a match between a pixel of the UV space and a point on the cloth surface. We enforce that the pixels on the edge of the pattern cuts are positioned on the body edge in 3D space. This ensures that their ray directions are the same, resulting in that the adjacent pairs of faces that are separated in the UV map have their common edge mapped onto the same garment edge in the 3D space, thus preserving connectivity. This property is used to reconstruct the 3D cloth mesh from the representation, as discussed in Sect. 4.2.

The quality of our mapping algorithm depends heavily on the load balance of the Voronoi regions. This is why we perform non-rigid ICP as pre-processing: it prevents loss of reconstruction details when the garment pieces are far from the body surface. Nonetheless, the non-rigid ICP may still not be able to handle extreme cases such as complex stacked garment layers. When multiple faces overlap on the same region, we choose the garment vertices that are the farthest from the body surface. This will result in smoother and simpler reconstructed garments in these challenging cases.

4.2   Decoding Process

Decoding the image representation back to the 3D cloth mesh is straightforward. Since adjacent pixels of the UV space correspond to adjacent points in the 3D space, we can simply connect adjacent pixels together to form the mesh. The only problem is that the connectivity will be lost when the cloth is cut into different UV regions. We solve this problem by ensuring that the two edges at different sides of the cut boundary are mapped to the same garment edge, as discussed in the encoding process. After fusing the duplicated 3D edge, the surface will be faithfully reconstructed.

5   Latent Space Learning

We apply a GAN-based model to learn the latent space of the representation image. Our network structure is shown in Fig. 2.

Since the pixel values in the representation image are related to the human body pose and shape, we add them as the conditional input in the network. Additionally, we provide a label map that indicates the overall topology of the garment to further constrain the generated image. The noise vector here serves mostly as the encoded detailed appearance, such as wrinkles and tightness of the cloth. We re-format the label image to one-hot version, and concatenate it with the encoded features of the other 1D input. Currently we only have binary information for the garment label map, but we can also support labels of different garment parts, as long as the corresponding data is provided. We use

Pix2PixHD [30] as our backbone network, but other state-of-the-art methods can also work in practice.

5.1   Loss Functions and Training Process

Because we cannot simply enumerate every possible garment and simulate them on every possible human pose, the trained model can easily have mode collapse problems, which is not ideal. To deal with this problem, we use a two-phase learning process. First, we train the model with the usual GAN loss and the feature loss:

$$\mathcal{L} = \mathcal{L}_{GAN} + \lambda_0 \mathcal{L}_{feat} \tag{4}$$

$$\mathcal{L}_{GAN} = \|D(I_{real}) - 1)\|_1 + \|D(I_{fake})\|_1 + \|D(G(I_{fake})) - 1\|_1 \tag{5}$$

$$\mathcal{L}_{feat} = \|D^*(I_{real}) - D^*(I_{fake})\|_1 + \|VGG^*(I_{real}) - VGG^*(I_{fake})\|_1 \tag{6}$$

In the above equations, $\mathcal{L}$ is the total loss, $\mathcal{L}_{GAN}$ is the GAN loss, and $\mathcal{L}_{feat}$ is the feature loss. $D()$ is the discriminator, $G()$ is the generator, and $VGG()$ is the pretrained VGG network. I_{real} and I_{fake} are the real and the fake images. D^* and VGG^* means the concatenation of the activations in all layers. After the first phase, the network can learn a conditional mapping between the input label and the output image, but it lacks variation from the noise vector.

Next, we fine-tune the model using the GAN loss and the new smoothness loss only:

$$\mathcal{L} = \mathcal{L}_{GAN} + \lambda_1 \mathcal{L}_{smooth} \tag{7}$$

$$\mathcal{L}_{smooth} = \|\frac{\partial I_{fake}}{\partial x}\|_1 + \|\frac{\partial I_{fake}}{\partial y}\|_1 \tag{8}$$

where $\mathcal{L}_{smooth}$ is introduced to the GAN model to enforce the smoothness of the representation image. Since the paired supervision from the feature loss is removed, the model will gradually become diverse to include more plausible but unseen results. We show later in our experiments that our learned model can generate clothing styles that are not from the training dataset.

5.2   Data Preparation

To learn the network model with high accuracy and variety, a large dataset depicting the joint distribution between the garment geometry and the human body is required. Previous datasets such as Bhatnagar et al. [4] or Liang et al. [19] have limited garment styles and body motions and are thus not suitable for our needs. Therefore, we propose a physics-based simulated dataset to represent most common garment types, human motions, and cloth materials. We sample different human motions and body shapes using the Moshed CMU

MoCap dataset [22]. Our garments are obtained from various online sources, which we will make public with the dataset. We initialize the human to a T-pose and dress the body with each of the garments. Then we use the cloth simulator [23] to generate the cloth motion along with the body motion. We notice that the cloth material of the garment can significantly alter the appearance, so we also vary the material parameters during data generation. For quantitative details, please refer to Sect. 6.3. We show examples of different garment data in Appendix 2.

6   Experimental Analysis

In this section, we will first introduce the implementation details of our method. Next, we show the effectiveness and performance of the key parts of our method by various experiments, including garment reconstruction, clothing style generation, and garment retargeting.

6.1   Implementation Details

We collected 104 types of garment models, each with 10 materials, and chose one random body motion sequence out of the 10 most commonly seen sequences. Then we dressed the garment on the body and simulated it using a cloth simulator [23] to generate a series of garment meshes with different poses, thereby generating 104 * 10 * 250 = 260,000 garment (split into 80%/20% for training/test). After that, we applied the representation transfer process on those garment instances and generated the image representation as well as the label mask. Next, we fed the images together with body shapes, poses, and the label images to the network for training. In practice, we randomly chose 2 materials in each epoch, to reduce the training time while making full use of the whole dataset.

We set λ_1 to 500, and the learning rate to 0.0002. We trained the model on an Nvidia GTX 1080 GPU. It took around 4 hours to train for each epoch, and we trained our model for 20 epochs in total.

6.2   Garment Reconstruction

Image representation of garments is one of the key contributions for the entire pipeline. We show the accuracy of the representation transfer process on our training data both qualitatively and quantitatively.

By transferring the 3D mesh of the garment to its 2D image representation and transferring back to a 3D mesh, we were able to recover the original 3D garment mesh. We randomly chose 5 different types of garments from the entire training dataset, chose 1 instance in each type, and generated the 3D mesh pair. The first row of Fig. 4 shows the original garments, while the second row shows the results of the recovered garments. As shown in the figure, our method is able to retain most of the original information when transferring between the

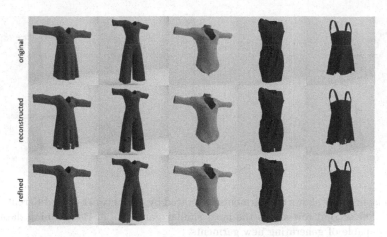

Fig. 4. Comparison between original mesh (first row), reconstructed mesh (second row), and refined mesh (third row). Our method is able to retain most of the original information, independent of the topology or the geometry of the garment mesh. The refined meshes indicate that the post-process is able to fix the small holes and gaps on the reconstructed meshes.

3D mesh and 2D image representation, under different types and topologies of garments. There might be small gaps or holes on the reconstructed meshes because of the resolution differences between two representations. We performed post-processing on the reconstructed meshes to resolve these small gaps/holes, as shown in the third row of Fig. 4. The post-processing method that we used is Ball Pivoting [3] on incomplete regions.

Since the regenerated vertices and edges of garments are aligned with those of the body mesh, it is inadequate to only compare the Euclidean distance of vertices of the original and reconstructed garment meshes. Assume we have mesh $M_1 = \{V_1, E_1, F_1\}$ and mesh $M_2 = \{V_2, E_2, F_2\}$, we define a mesh-based reconstruction error as the average distance from each point in V_1 to M_2, and each point in V_2 to M_1, shown as follows:

$$d_m = \frac{\sum_{p_1 \in V_1} dist(p_1, M_2) + \sum_{p_2 \in V_2} dist(p_2, M_1)}{\|V_1\| + \|V_2\|}$$

where $dist(p, M)$ is the smallest distance from point p to the surface of mesh M. We randomly sampled 6,000 garment instances from all the 260,000 garment instances in our training dataset, calculated the reconstruction error for each sample and computed the error distribution. The average percentage error is less than 1%, with the largest being less than 1.4%. The error distribution is shown in Appendix 6. Our method is robust to all garment topologies, materials, body poses, and shapes.

Fig. 5. The first row shows the garments generated by our network with different design patterns. The second row shows the most similar garments in the training data. Our model is capable of generating new garments.

6.3 Clothing Style Generation

In this section, we demonstrate the generalization ability of our method. Specifically, we did the experiments in the following steps. First, we fed new label images and body information not included in the training data to the network, and obtained the image representation result. The output further went through the reconstruction algorithm and the post-processing and was finally transformed to the refined 3D garment. The generated image representation was further searched for its nearest neighbor in the training data using L1 distance. We retrieved the original mesh of the nearest neighbor for comparison.

In Fig. 5, we show the generated garments in the first row with different topologies or patterns. There are cases including a single-shoulder dress (the first column) and a backless dress (the last column), showing that our model is able to generate garments of varying topology. The second row shows the nearest neighbors in the training dataset. The geometric differences between the generated meshes and the nearest neighbors are significant, which means that our network can generalize to unseen topologies.

6.4 Interpolation Results

We did the interpolation experiment to show the effectiveness of our method. In the experiment, we chose two garments, generated the intermediate label images and fed them into our method. We show the interpolation results between two specific cases in Fig. 6. As shown in the figure, the garment changes smoothly from the leftmost style to the rightmost style, showing that our learned latent space is smooth and compact.

Fig. 6. Interpolation results between two specific cases. As shown in the figure, the garment changes smoothly from the leftmost style to the rightmost style, showing that our learned latent space is smooth and compact.

6.5   Garment Retargeting

Ease of retargeting is an important property in garment generation. In this experiment, we first generated a garment model with a specific body shape, then retargeted the generated garment to different body shapes. We show some of the retargeting results in Fig. 7, which are qualitatively as good as the results of Wang *et al.* [31]. We found that both algorithms can retain the appearance of the original garment retargeted onto bodies of different shapes and sizes. However, in their method, an additional Siamese network needs to be trained to achieve the retargeting goal [31], while our method can retarget the garment directly from the generated image representation and the new body shape – requiring less computation and demonstrating greater ease. Our method can also naturally ensure the consistency of the garment style by the definition of our image presentation. Other works, such as Brouet *et al.* [6], which is based on an optimization framework, or Guan *et al.* [13], which uses a learning-based architecture, are more computationally expensive than our method in the retargeting process.

6.6   Garment Generation Methods Comparison

There are methods that can generate garments through sketches, *e.g.*, Huang *et al.* [15] and Wang *et al.* [31]. Thanks to the information contained in the sketches, Huang *et al.*'s method can generate textures of garments, and Wang *et al.*'s method can generate garments with realistic wrinkles. However, our method only needs label images instead of full sketches. Also, our method can generate garments with different topologies given our image representation of garments, while these methods can only support at most three types of topologies.

In addition, a recent work Tex2Shape [1] can generate the combined body and garment mesh from photographs. However, it can only reconstruct the entire body mesh with garments as a whole and is not able to separate the garment apart, while our method generates a stand-alone garment mesh. Moreover, Tex2Shape reconstructs the result with the same topology as the body mesh, so it can only handle body-like garments. In contrast, our method uses an extra label image to provide sewing information to the network, and reconstructs the garment mesh by training the network to assemble and stitch different pieces together, thereby applicable to generate garments of varying topologies.

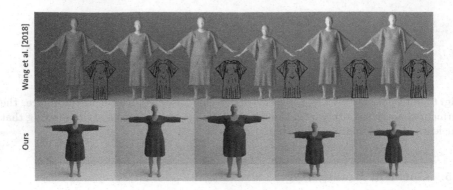

Fig. 7. Retargeting results for different body shapes and sizes, compared with Wang *et al.* [31]. The retargeting qualities are nearly the same qualitatively, *i.e.* both algorithms can retain the appearance of the original garment retargeted onto bodies of different shapes and sizes. However, an additional Siamese network is needed in their retargeting process, while our method retargets the cloth directly from the image representation, thereby requiring less computation than [31].

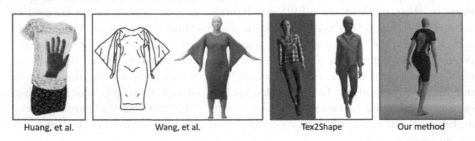

Fig. 8. Output comparison. Huang *et al.* [15] generate garment model with texture. Wang *et al.* [31] generate garments with realistic wrinkles as the sketch. Tex2Shape [1] generates combined body and garment models. Our method generates garments with various topologies.

We show the outputs of the three methods mentioned above and our method in Fig. 8. Huang *et al.* [15] generate garment model with texture. Wang *et al.* [31] generate garments with realistic wrinkles as the sketch. Tex2Shape [1] generates combined body and garment model. Our method can generate garments with various topologies. Also, we show the different characteristics of different methods in Table 1. Because different methods have different characteristics and focus on different aspects, specific inputs would require different methods.

6.7 Performance

Our network inference (Sect. 5) takes about 369 msec on average, which is around 16.4% of the entire process. Garment reconstruction (Sect. 4.2) takes about 1,303 msec on average, around 57.9%. Post-processing refinement takes the last 25.7%, nearly 576 msec on average. Overall, our method takes 2,248 msec on average.

Table 1. Characteristic comparisons of different methods

Characteristics	Huang et al. [15]	Wang et al. [31]	Tex2Shape [1]	ours
Input sketch	YES	YES	NO	NO
Input photograph	NO	NO	YES	NO
Input body pose or shape	NO	NO	NO	YES
Input garment sewing pattern	NO	NO	NO	YES
Use geometric representation	YES	NO	NO	NO
Use GAN	NO	YES	YES	YES
Use body UV map	NO	NO	YES	YES
Infer body pose or shape	NO	YES	YES	NO
Generate texture	YES	NO	NO	NO
Generate wrinkles	NO	YES	NO	NO
Generate body model	NO	NO	YES	NO
Topology supported	Limited	Limited	Limited	Various

Since the image resolution in our method is fixed to 512 * 512, the variation in image processing time is insignificant. It is possible to further accelerate the performance of our algorithm. Please refer to Appendix 8.

7   Conclusion

We presented a learning-based parametric generative model, which is the first garment generative model that can support any type of garment material, body shape, and most garment topologies. To offer this capability, we propose a special image representation of the garment model. Our method also makes garment retargeting much easier. In addition, a large garment dataset will be made available for further research in this area.

Limitation and Future Work: Currently our method does not automatically generate fabric textures. In addition, due to the intermediate image representation of the garment, our method cannot generate multi-layer garment models, e.g., multi-layer lace skirts. This problem offers new research challenges. Our network can be further used as an extension of existing garment datasets because of its applicability and generalizability to unseen topologies. The generated 3D garments can also be used in user-driven fashion design and apparel prototyping.

Acknowledgment. This work is supported in part by Elizabeth Stevinson Iribe Professorship and National Science Foundation.

1   Samples of Garment Sewing Pattern

We show garment sewing pattern samples in Fig. 9, including a dress, pants, a shirt, and a skirt. Since sewing patterns offer common information about the garments, they are generally available.

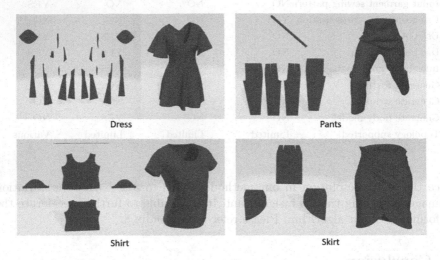

Fig. 9. Garment sewing pattern samples. We show 4 cases here, including a dress, pants, a shirt, and a skirt. Since sewing patterns offer common information about the garments, they are generally available.

2   Example Meshes from Our Garment Dataset

In Fig. 10, we show that our garment dataset consists of clothing on different human body poses/sizes/shapes and of varying garment topologies, patterns, and materials. We sample ten human motions from the CMU MoCap dataset, including motions of walking, running, climbing and dancing. As stated in the main text, we have over 100 different garment types in the dataset, including dresses, t-shirts, pants, skirts and swimsuits. We use different material parameters and material space scales to control the sizes of the garments. Given this large and diverse dataset, our network can successfully disentangle different parts of the body label to generate garments with topologies totally different with those in the training dataset, while keeping a visually plausible result.

Fig. 10. Examples meshes from our garment dataset. The dataset includes several common garment topologies and materials, as well as various human poses. The last two columns show the same garment pattern with different materials. The wrinkle appearances of the two sequences are different.

3   One-Hot Version of the Label Image

In Fig. 2, we first transfer the label image to one-hot version by the one-hot operation. One-hot version of the label image is an array of binary images, where each pixel on the i-th binary image represents whether the label ID of that pixel on the original image equals to i. We use one-hot to support different garment components (e.g., shirts+jacket+pants) in future work. One-hot version of label image can decouple different class IDs and will be easier for the network to learn. Currently, we use it to differentiate between garment and other pixels. Overall, it is an extensible data format.

4   Data Format Transfer Process

Figure 11 shows the data format transfer process. The garment model and the image representation of the garment can transfer to each other using the body mesh and UV map, as discussed in Sect. 4.

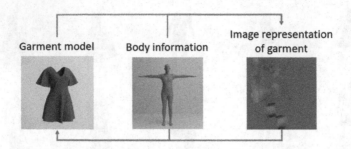

Fig. 11. Data format transfer process. The garment model and the image representation of the garment can transfer to each other using the body mesh.

5   Point Matching Between Body and Garment

We show the mapping process from body surface pixels to the garment surface in Fig. 12. Within the Voronoi region of a body vertex, the ray direction of a pixel (brown) is interpolated between the vertex normal (black) and the face normal (gray), according to the barycentric coordinates.

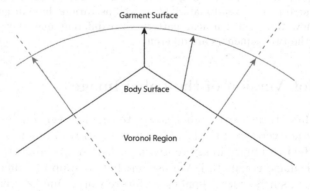

Fig. 12. Mapping from body surface pixels to the garment surface. Within the Voronoi region of a body vertex, the ray direction of a pixel (brown) is interpolated between the vertex normal (black) and the face normal (gray), according to the barycentric coordinates. (Color figure online)

6   Reconstruction Results Under Different Conditions

We show in Fig. 14 that our algorithm introduced in Sec. 4 is robust to any kind of garment input. We tested our algorithm with garments of different topologies on different human bodies. Different cloth materials are expressed with different sizes and detailed wrinkles based on the geometry. Our method can also retain the material information faithfully.

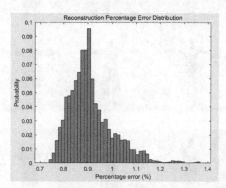

Fig. 13. Distribution of reconstruction error d_m (in percentage w.r.t. the garment height) over 6,000 randomly selected garment instances. The error is relatively small across all types of garments, with the largest being less than 1.4% and most within 1%.

Figure 13 shows the distribution of d_m in our training dataset. The error is relatively small across all types of garments, with the largest being less than 1.4% and most of them within 1%.

7   Garment Retargeting

In Sect. 6.5, we show the garment retargeting results using only a T-pose. In Fig. 15, we show more cases with different poses. As shown in the figure, our method can retarget garments with different topologies, patterns, and materials to bodies with different shapes, sizes, and poses.

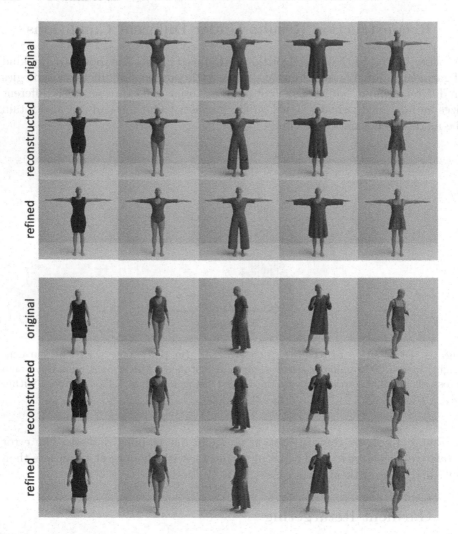

Fig. 14. Reconstructed mesh results under different human poses and shapes, garment topologies, sizes, and materials. Our data transfer method is able to map any 3D mesh to its 2D image representation with little information loss.

8   Performance

Our method takes about 2,248 msec on average. For garment generation, 2 s would be quite acceptable if the quality is good enough, while the garments designed manually usually take much longer. Also, there is room for performance improvement and parallelization of the post-processing after the network inference. More importantly, we use a resolution of 512 * 512 for the displacement map, so there are up to 512*512 vertices and 511 * 511 * 3 edges in our reconstructed mesh, with resolution much higher than other works, thus tak-

Fig. 15. Garment retargeting results. Our method can retarget garments with different topologies, patterns, and materials to bodies with different shapes, sizes, and poses.

ing slightly longer. As needed, the implementation of our method can be much improved by reducing the resolution of the displacement map.

References

1. Alldieck, T., Pons-Moll, G., Theobalt, C., Magnor, M.: Tex2shape: detailed full human body geometry from a single image, pp. 2293–2303 (2019)
2. Amberg, B., Romdhani, S., Vetter, T.: Optimal step nonrigid ICP algorithms for surface registration. In: 2007 IEEE Conference on Computer Vision and Pattern Recognition, pp. 1–8. IEEE (2007)
3. Bernardini, F., Mittleman, J., Rushmeier, H., Silva, C., Taubin, G.: The ball-pivoting algorithm for surface reconstruction. IEEE Trans. Vis. Comput. Graph. **5**(4), 349–359 (1999)
4. Bhatnagar, B.L., Tiwari, G., Theobalt, C., Pons-Moll, G.: Multi-garment net: learning to dress 3D people from images. In: IEEE International Conference on Computer Vision (ICCV). IEEE, October 2019
5. Bradley, D., Popa, T., Sheffer, A., Heidrich, W., Boubekeur, T.: Markerless garment capture. ACM Trans. Graph. **27**(3), 99 (2008). https://doi.org/10.1145/1360612. 1360698

6. Brouet, R., Sheffer, A., Boissieux, L., Cani, M.: Design preserving garment transfer. ACM Trans. Graph. **31**(4), 36:1–36:11 (2012). https://doi.org/10.1145/2185520.2185532
7. Chen, X., Zhou, B., Lu, F., Wang, L., Bi, L., Tan, P.: Garment modeling with a depth camera. ACM Trans. Graph. **34**(6), 203:1–203:12 (2015). https://doi.org/10.1145/2816795.2818059
8. Danerek, R., Dibra, E., Öztireli, A.C., Ziegler, R., Gross, M.H.: Deepgarment: 3D garment shape estimation from a single image. Comput. Graph. Forum **36**(2), 269–280 (2017). https://doi.org/10.1111/cgf.13125
9. Decaudin, P., Julius, D., Wither, J., Boissieux, L., Sheffer, A., Cani, M.: Virtual garments: a fully geometric approach for clothing design. Comput. Graph. Forum **25**(3), 625–634 (2006). https://doi.org/10.1111/j.1467-8659.2006.00982.x
10. Doersch, C.: Tutorial on variational autoencoders. CoRR abs/1606.05908 (2016). http://arxiv.org/abs/1606.05908
11. Gabeur, V., Franco, J.S., Martin, X., Schmid, C., Rogez, G.: Moulding humans: non-parametric 3D human shape estimation from single images. In: Proceedings of the IEEE International Conference on Computer Vision, pp. 2232–2241 (2019)
12. Goodfellow, I.J., et al.: Generative adversarial nets. In: Advances in Neural Information Processing Systems 27: Annual Conference on Neural Information Processing Systems 2014, December 8–13 2014, Montreal, Quebec, Canada, pp. 2672–2680 (2014). http://papers.nips.cc/paper/5423-generative-adversarial-nets
13. Guan, P., Reiss, L., Hirshberg, D.A., Weiss, A., Black, M.J.: DRAPE: dressing any person. ACM Trans. Graph. **31**(4), 35:1–35:10 (2012). https://doi.org/10.1145/2185520.2185531
14. Gundogdu, E., Constantin, V., Seifoddini, A., Dang, M., Salzmann, M., Fua, P.: Garnet: a two-stream network for fast and accurate 3d cloth draping. In: Proceedings of the IEEE International Conference on Computer Vision, pp. 8739–8748 (2019)
15. Huang, P., Yao, J., Zhao, H.: Automatic realistic 3D garment generation based on two images. In: 2016 International Conference on Virtual Reality and Visualization (ICVRV) (2016)
16. Jeong, M., Han, D., Ko, H.: Garment capture from a photograph. J. Vis. Comput. Animation **26**(3–4), 291–300 (2015). https://doi.org/10.1002/cav.1653
17. Jung, A., Hahmann, S., Rohmer, D., Bégault, A., Boissieux, L., Cani, M.: Sketching folds: developable surfaces from non-planar silhouettes. ACM Trans. Graph. **34**(5), 155:1–155:12 (2015). https://doi.org/10.1145/2749458
18. Lähner, Z., Cremers, D., Tung, T.: DeepWrinkles: accurate and realistic clothing modeling. In: Ferrari, V., Hebert, M., Sminchisescu, C., Weiss, Y. (eds.) ECCV 2018. LNCS, vol. 11208, pp. 698–715. Springer, Cham (2018). https://doi.org/10.1007/978-3-030-01225-0_41
19. Liang, J., Lin, M.C.: Shape-aware human pose and shape reconstruction using multi-view images. In: Proceedings of the IEEE International Conference on Computer Vision, pp. 4352–4362 (2019)
20. Lin, M.C.: Efficient collision detection for animation and robotics. Ph.D. thesis, Department of Electrical Engineering and Computer Science (1993)
21. Loper, M., Mahmood, N., Romero, J., Pons-Moll, G., Black, M.J.: SMPL: a skinned multi-person linear model. ACM Trans. Graph. **34**(6), 248:1–248:16 (2015). https://doi.org/10.1145/2816795.2818013
22. Loper, M.M., Mahmood, N., Black, M.J.: MoSh: motion and shape capture from sparse markers. ACM Trans. Graph. (Proc. SIGGRAPH Asia) **33**(6), 220:1–220:13 (2014). https://doi.org/10.1145/2661229.2661273

23. Narain, R., Samii, A., O'Brien, J.F.: Adaptive anisotropic remeshing for cloth simulation. ACM Trans. Graph. **31**(6), 152:1–152:10 (2012). https://doi.org/10.1145/2366145.2366171

24. Patel, C., Liao, Z., Pons-Moll, G.: Tailornet: predicting clothing in 3D as a function of human pose, shape and garment style. In: Proceedings of the IEEE/CVF Conference on Computer Vision and Pattern Recognition, pp. 7365–7375 (2020)

25. Razavi, A., van den Oord, A., Vinyals, O.: Generating diverse high-fidelity images with VQ-VAE-2. In: Advances in Neural Information Processing Systems, pp. 14866–14876 (2019)

26. Robson, C., Maharik, R., Sheffer, A., Carr, N.: Context-aware garment modeling from sketches. Comput. Graph. **35**(3), 604–613 (2011). https://doi.org/10.1016/j.cag.2011.03.002

27. Saito, S., Huang, Z., Natsume, R., Morishima, S., Kanazawa, A., Li, H.: Pifu: pixel-aligned implicit function for high-resolution clothed human digitization. In: Proceedings of the IEEE International Conference on Computer Vision, pp. 2304–2314 (2019)

28. Turquin, E., Cani, M., Hughes, J.F.: Sketching garments for virtual characters. In: 34. International Conference on Computer Graphics and Interactive Techniques, SIGGRAPH 2007, San Diego, California, USA, August 5–9, 2007, Courses, p. 28 (2007). https://doi.org/10.1145/1281500.1281539

29. Wang, T., Liu, M., Zhu, J., Tao, A., Kautz, J., Catanzaro, B.: High-resolution image synthesis and semantic manipulation with conditional GANs. In: 2018 IEEE Conference on Computer Vision and Pattern Recognition, CVPR 2018, Salt Lake City, UT, USA, June 18–22, 2018, pp. 8798–8807 (2018). http://openaccess.thecvf.com/content_cvpr_2018/html/Wang_High-Resolution_Image_Synthesis_CVPR_2018_paper.html

30. Wang, T.C., Liu, M.Y., Zhu, J.Y., Tao, A., Kautz, J., Catanzaro, B.: High-resolution image synthesis and semantic manipulation with conditional GANs. In: Proceedings of the IEEE Conference on Computer Vision and Pattern Recognition (2018)

31. Wang, T.Y., Ceylan, D., Popovic, J., Mitra, N.J.: Learning a shared shape space for multimodal garment design. CoRR abs/1806.11335 (2018). http://arxiv.org/abs/1806.11335

32. Yang, S., et al.: Physics-inspired garment recovery from a single-view image. ACM Trans. Graph. (TOG) **37**(5), 170 (2018)

33. Zheng, Z., Yu, T., Wei, Y., Dai, Q., Liu, Y.: Deephuman: 3D human reconstruction from a single image. In: Proceedings of the IEEE International Conference on Computer Vision, pp. 7739–7749 (2019)

34. Zhou, B., Chen, X., Fu, Q., Guo, K., Tan, P.: Garment modeling from a single image. Comput. Graph. Forum **32**(7), 85–91 (2013). https://doi.org/10.1111/cgf.12215

Style Transfer for Co-speech Gesture Animation: A Multi-speaker Conditional-Mixture Approach

Chaitanya Ahuja[1]([⊠]) ⓘ, Dong Won Lee[1], Yukiko I. Nakano[2],
and Louis-Philippe Morency[1] ⓘ

[1] Carnegie Mellon University, Pittsburgh, PA, USA
{cahuja,dongwonl,morency}@cs.cmu.edu
[2] Seikei University, Musashino, Tokyo, Japan
y.nakano@st.seikei.ac.jp

Abstract. *How can we teach robots or virtual assistants to gesture naturally? Can we go further and adapt the gesturing style to follow a specific speaker?* Gestures that are naturally timed with corresponding speech during human communication are called co-speech gestures. A key challenge, called gesture style transfer, is to learn a model that generates these gestures for a speaking agent 'A' in the gesturing style of a target speaker 'B'. A secondary goal is to simultaneously learn to generate co-speech gestures for multiple speakers while remembering what is unique about each speaker. We call this challenge style preservation. In this paper, we propose a new model, named Mix-StAGE, which trains a single model for multiple speakers while learning unique style embeddings for each speaker's gestures in an end-to-end manner. A novelty of Mix-StAGE is to learn a mixture of generative models which allows for conditioning on the unique gesture style of each speaker. As Mix-StAGE disentangles style and content of gestures, gesturing styles for the same input speech can be altered by simply switching the style embeddings. Mix-StAGE also allows for style preservation when learning simultaneously from multiple speakers. We also introduce a new dataset, Pose-Audio-Transcript-Style (PATS), designed to study gesture generation and style transfer. Our proposed Mix-StAGE model significantly outperforms the previous state-of-the-art approach for gesture generation and provides a path towards performing gesture style transfer across multiple speakers. Link to code, data and videos: http://chahuja.com/mix-stage.

Keywords: Gesture animation · Style transfer · Co-speech gestures

1 Introduction

Nonverbal behaviours such as body posture, hand gestures and head nods play a crucial role in human communication [41,55]. Pointing at different objects, mov-

Electronic supplementary material The online version of this chapter (https://doi.org/10.1007/978-3-030-58523-5_15) contains supplementary material, which is available to authorized users.

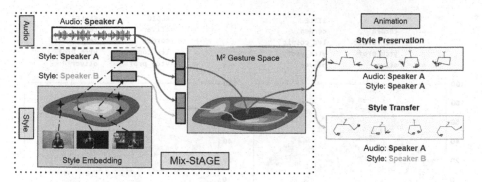

Fig. 1. Overview of co-speech gesture generation and gesture style transfer/preservation task. The models learns a style embedding for each speaker, which can be mapped to a gesture space with either the same speaker's audio to generate style preserved gestures or a different speaker's audio to generate style transferred gestures.

ing hands up-down in emphasis, and describing the outline of a shape are some of the many gestures that co-occur with the verbal and vocal content of communication. These are known as co-speech gestures [27,38]. When creating new robots or embodied virtual assistants designed to communicate with humans, it is important to generate *naturalistic* looking gestures that are meaningful with the speech [6]. Some recent works have proposed speaker-specific gesture generation models [11,13,16,18] that are both trained and tested on the same speaker. The intuition behind this prior work is that co-speech gestures are idiosyncratic [38,57]. There is an unmet need to learn generative models that are able to learn to generate gestures simultaneously from multiple speakers (→ in Fig. 1) while at the same time remembering what is unique for each speaker's gesture style. These models should not simply remember the "average" speaker. A bigger technical challenge is to be able to transfer gesturing style of speaker 'B' to speaker 'A' (→ in Fig. 1).

The gesturing style can defined along two dimensions which is a result of (a) the speaker's idiosyncrasy (or speaker-level style), and (b) due to some more general attributes such as standing versus sitting, or the body orientation such as left versus right (or attribute-level style). For both gesture style types, the generation model needs to be able to learn the diversity and expressivity [8,42] present in the gesture space, within and amongst speakers. The gesture distribution is likely to have multiple modes, some of them shared among speakers and some distinct to a speaker's prototypical gestures.

In this paper, we introduce the Mixture-Model guided Style and Audio for Gesture Generation (or Mix-StAGE) approach which trains a single model for multiple speakers while learning unique style embeddings for each speaker's gestures in an end-to-end manner (see Fig. 1). We use this model to perform two tasks for gesture generation conditioned on the input audio signal, (1) **style preservation** which ensures that while learning from multiple speakers we are still able to preserve unique gesturing styles of each speaker, and (2) **style**

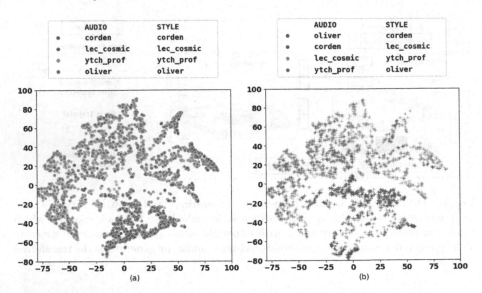

Fig. 2. t-SNE [37] representation of the Multi-mode Multimodal Gesture Space (Sect. 4.1). Each color represents a style, which is fixed for both plots. The plot on the left visualizes the gesture space generated from the audio content and style of the same speaker. The plot on the right shows the generated gesture space where the audio content and style are not from the same speaker. It can be observed that a similar gesture space is occupied by each speaker's style even when the audio content is not of their own.

transfer where generated gestures are from a new style that was not the same as the source of the speech. A novelty of Mix-StAGE is to learn a mixture of generative models which allows for conditioning on the unique gesture style of each speaker. Our experiments study the impact of multiple speakers on both style transfer and preservation. Our study focuses on the non-verbal components of speech asking the research question if we can predict gestures without explicitly modeling verbal signals. We also introduce a new dataset, Pose-Audio-Transcript-Style (PATS), designed to study gesture generation and style transfer.

2   Related Work

Speech Driven Gesture Generation: For prosody-driven head motion generation [49] and body motion generation [31,32], Hidden Markov Models were used to predict a sequence of frames. Chiu & Marsella [12] proposed a two-step process: predicting gesture labels from speech signal using conditional random fields (CRFs) and converting the label to gesture motion using Gaussian process latent variable models (GPLVMs). More recently, an LSTM network was applied to MFCC features extracted from speech to predict a sequence of frames for gestures [22] and body motions [1,50]. Generative adversarial networks (GAN) were used to generate head motions [47] and body motions [16]. Gestures driven by an audio signal [18] is the closest approach to our task of style preservation but it uses models trained on single speakers unlike our multi-speaker models.

Disentanglement and Transfer of Style : Style extraction and transfer have been studied in context of image artistic style [17,26], factorizing foreground and background in videos [15,54], disentanglement in speech [9,20,56]. These approaches were extended to translation between properties of style such as map edges and real photos using paired samples [25]. Paired data limits the variety of attributes of source and target, which encouraged unsupervised domain translation for images [58,59] and videos[7]. Style was disentangled from content using a shared latent space [33], a cycle consistency loss [58] and contrastive learning [39]. Cycle consistency losses were shown to limit diversity in the generated outputs as opposed to a weak consistency loss [24] and shared content space [29]. Cycle consistency in cross-domain translation assumes reversibility (i.e. domain A can be translated to domain B and vice-versa). These assumptions are violated in cross-modal translation [36] and style control [56] tasks where information in modality B (e.g. pose) is a subset of that in modality B (e.g. audio). Style transfer for pose has been studied in context of generating dance moves based on the content of the audio [30] or walking styles [52]. Generated dance moves are conditioned on both the style and content of the audio (i.e. kind of music like ballet or hip-hop), unlike co-speech gesture generation which requires only the content and not the style of the audio (i.e. speaker specific style like identity or fundamental frequency). Co-speech gesture styles have been studied in context of speaker personalities [40], but requires a long annotation process to create a profile for each speaker. To our knowledge, this is the first fully data-driven approach that learns gesture style transfer for multiple speakers in a co-speech gesture generation setting.

3 Stylized Co-speech Gesture Animation

We define the problem of stylized co-speech gesture animation with two main goals, (1) generation of an animation which represents the gestures that would co-occur with the spoken segment and (2) modification of the style of these gestures. Figure 1 shows the first goal ($\rightarrow$) exemplified with the style preservation scenario, while the second goal ($\rightarrow$) exemplifies with the style transfer scenario.

Formally, given a sequence of T audio frames $\mathbf{X}_a \sim F_a$ and i^{th} speaker's style $S(i)$, the goal is to predict a sequence of T frames of 2-D poses $\mathbf{Y}_p \sim F_p$. Here F_a and F_p are the marginal distributions of the content of input audio and style of output pose sequences. To control pose generation by both style and audio, we learn a joint distribution over pose, audio and style $F_{p,a,s}$ which can be broken down into 3 parts

$$F_{p,a,s} = F_{p|\Phi} F_{\Phi|a,s} \cdot F_s \cdot F_a \quad (1)$$

where $F_{\Phi|a,s}$ is the distribution of the gesture space Φ conditioned on the audio and style of pose (Fig. 1). We discuss the modelling of $F_{p|\Phi} F_{\Phi|a,s}$, F_a, and F_s in Sect. 4.1, 4.2 and 4.3 respectively.

4   Mix-StAGE: Mixture-Model guided Style and Audio for Gesture Generation

Figure 3 shows an overview of our Mix-StAGE model, including the training inference pathways. A first component of our Mix-StAGE model is the audio encoder E_a^c, which takes as input the spoken audio X_a. During training, we also have the pose sequence of the speaker Y_p. This pose sequence is decomposed into content and style, with two specialized encoders E_p^c and E_p^s. During training, the style for the pose sequence can either be concatenated with the audio or the pose content.

The pose sequences for multiple speakers are represented as a distribution with multiple modes [21]. To decode from this multi-mode multimodal gesture space, we use a common generator G with multiple sub-generators (or CMix-GAN) conditioned on input audio and style to decode both these embeddings to output pose $\mathbf{Y}_p$.

Our loss function comprises of a mode-regularization loss (Sect. 4.1) to ensure that audio and style embedding can sample from the appropriate marginal distribution of poses, a joint loss (Sect. 4.2) to ensure latent distribution matching for content in a cross-modal translation task, a style consistency loss (Sect. 4.3) to ensure that the correct style is being generated and an adversarial loss (Sect. 4.4) that matches the generated pose distribution to the target pose distribution.

4.1   M²GS: Multi-mode Multimodal Gesture Space

Humans perform different styles of gestures, where each style consists of different kinds of gestures (i.e beat, metaphorical, emblematic, iconic and so on) [38]. Learning pose generators for multiple speakers, each with their own style of gestures, presents a distribution with multiple modes. These gestures have a tendency of switching from one mode to the other over time, which depends on style embeddings and content of the audio.

To prevent mode collapse [4] we propose the use of mixture-model guided sub-generators [5,21,23], each learning a different mode of M^2 gesture space $F_{\Phi|a,s}$.

$$\hat{\mathbf{Y}}_p = \sum_{m=1}^{M} \phi_m G_m(\mathbf{Z}) = G(\mathbf{Z}) \qquad (2)$$

where $\mathbf{Z} \in \{\mathbf{Z}_{a \to p}, \mathbf{Z}_{p \to p}\}$ are cross-modal and self-modal latent spaces respectively. They are defined as $\mathbf{Z}_{a \to p} = \left[E_a^c(\mathbf{X}_a), E_p^s(\mathbf{Y}_p) \otimes \mathbf{S} \right]$ and $\mathbf{Z}_{p \to p} = \left[E_p^c(\mathbf{Y}_p), E_p^s(\mathbf{Y}_p) \otimes \mathbf{S} \right]$ where $\mathbf{S}$ is the style embedding matrix (See Sect. 4.3) and $\otimes$ is argmax for style ID followed by matrix multiplication. Pose sequence $\hat{Y}_p \sim F_{p|\Phi} F_{\Phi|a,s}$ represents the pose probability distribution conditioned on audio and style. $G_m \sim F_{p|a,s}^m \ \forall m \in [1, 2, \ldots M]$ are sub-generator functions with corresponding mixture-model priors $\Phi = \{\phi_1, \phi_2, \ldots \phi_M\}$. These mixture model priors represent the M^2 gesture space and are estimated at inference time conditioned on the input audio and style.

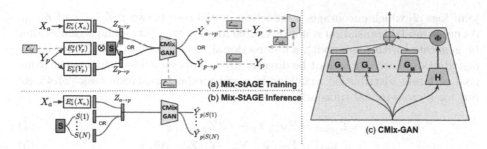

Fig. 3. (a) Overview of the proposed model Mix-StAGE in training mode, where audio X_a and pose Y_p are fed as inputs to learn a style embedding and concurrently generate a gesture animation. **S** represents the style matrix, which is multiplied with a separately encoded pose. $\otimes$ represents argmax for style ID followed by matrix multiplication. Discriminator D is used for adversarial training. All the loss functions are represented with dashed lines. (b) Mix-StAGE in inference mode, where any speaker's style embedding can be used on an input audio X_a to generate gesture style-transferred or style-preserved animations (c) CMix-GAN generator: a visual representation of the conditional Mix-GAN model, where the $\oplus$ represents a weighted sum of the model priors Φ with the generated outputs by the sub-generators.

Estimating Mixture Model Priors: During training, we partition poses Y_p into M clusters using an unsupervised approach, Lloyd's algorithm [35]. While other unsupervised clustering methods [43] can also be used at this stage, we choose Lloyd's algorithm for its simplicity and speed. Each of these clusters represent samples from probability distributions $\{F_{p|a,s}^1, F_{p|a,s}^2, \ldots F_{p|a,s}^M\}$. If a sample belongs to the mth cluster, $\phi_m = 1$, otherwise $\phi_m = 0$, making Φ a sequence of one-hot vectors. While training the generator G with loss function $\mathcal{L}_{rec}$, if a sample belongs to the distribution $F_{p|a,s}^m$, only parameters of sub-generator G_m are updated. Hence, each sub-generator learns different components of the true distribution, which are combined using Eq. 2 to give the generated pose.

At inference time, we do not have the true values of mixture-model priors Φ. As mixture model priors modulate based on the style of the speaker and audio content at any given moment, we jointly learn a classification network $H \sim F_{\Phi|a,s}$ to estimate values of Φ in form of a mode regularization loss function

$$\mathcal{L}_{mix} = \mathbb{E}_{\Phi,\mathbf{Z}} \text{CCE}(\Phi, H(\mathbf{Z})) \tag{3}$$

where CCE is categorical cross-entropy.

4.2 Joint Space of Style, Audio and Pose

A set of marginal distributions F_a and F_s are learnt by our content encoders E_a^c and E_p^c, which together define the joint distribution of the generated poses: $F_{p,a,s}$. Since both cross-modal $\mathbf{Z}_{a \to p}$ and self-modal $\mathbf{Z}_{p \to p}$ latent spaces are designed to represent the same underlying content distribution, they should be consistent with each other. Using the same generator G for decoding both of these embeddings [34] yields content invariant generator. We enforce a reconstruction and joint loss [2] which encourages a reduction in distance between $\mathbf{Z}_{a \to p}$ and $\mathbf{Z}_{p \to p}$. As cross-modal translation is not reversible for this task (i.e. audio signal cannot be generated with pose input), a bi-directional reconstruction loss [29] for latent distribution matching cannot be directly used. This joint loss achieves the same goal of latent distribution matching in a uni-modal translation task [24,45,46] but for a cross-modal translation task.

$$\mathcal{L}_{joint} = \mathbb{E}_{\mathbf{Y}_p} \|\mathbf{Y}_p - G(\mathbf{Z}_{p \to p})\|_1 \tag{4}$$

$$\mathcal{L}_{rec} = \mathbb{E}_{\mathbf{Y}_p, \mathbf{X}_a} \|\mathbf{Y}_p - G(\mathbf{Z}_{a \to p})\|_1 \tag{5}$$

4.3 Style Embedding

We represent style as a collection of embeddings $S(i) \in \mathbf{S} \sim F_s$, where $S(i)$ is the style of the ith speaker in the style matrix $\mathbf{S}$. Style space and embeddings are conceptually similar to the GST (Global Style Token) layer [56] which decomposes the audio embedding space into a set of basis vectors or style tokens, but only one out of the two modalities in the stylized audio generation task [36,56] have both style and content. In our case, both audio and pose have style and content. To ensure that generator G is attending only to style of pose while ignoring style of the audio, a style consistency loss is enforced on input $\mathbf{Y}_p$ and generated $\hat{\mathbf{Y}}_p$.

$$\mathcal{L}_{id} = \mathbb{E}_{Y \in \{\mathbf{Y}_p, \hat{\mathbf{Y}}_p\}} \mathrm{CCE}\left(\mathrm{Softmax}\left(E_p^s(Y)\right), \mathbf{ID}\right) \tag{6}$$

where $\mathbf{ID}$ is a one-hot vector denoting the speaker level style.

4.4 Total Loss with Adversarial Training

To alleviate the challenge of overly smooth generation caused by L1 reconstruction and joint losses in Eq. 4,5, we use the generated pose sequence $\hat{\mathbf{Y}}^p$ as a signal for the adversarial discriminator D [18]. The discriminator tries to classify the true pose $\mathbf{Y}^p$ from the generated pose $\hat{\mathbf{Y}}^p$, while the generator learns to fool the discriminator by generating realistic poses. This adversarial loss [19] is written as:

$$\mathcal{L}_{adv} = \mathbb{E}_{\mathbf{Y}_p} \log D\left(\mathbf{Y}_p\right) + \mathbb{E}_{\mathbf{X}_a, \mathbf{Y}_p} \log\left(1 - D(G\left(\left[E_a^c(\mathbf{X}_a), E_p^s(\mathbf{X}_p)\right]\right)))\right) \tag{7}$$

The model is jointly trained to optimize the overall loss function:

$$\max_{D} \min_{E_a^c, E_p^c, E_p^s, G} \mathcal{L}_{mix} + \mathcal{L}_{joint} + \mathcal{L}_{rec} + \lambda_{id}\mathcal{L}_{id} + \mathcal{L}_{adv} \tag{8}$$

where λ_{id} controls the weight of the style consistency loss term.

4.5 Network Architectures

Our proposed approach can work with any temporal network, giving it the flexibility of incorporating domain dependent or pre-trained temporal models.

Table 1. Style Preservation: Objective metrics for pose generation of single-speaker and multi-speaker models as indicated in the columns. Each row refers to the number of speakers the model was trained, with the average performance indicated at the top. The scores for common individual speakers are also indicated below alongside. For detailed results on other speakers please refer to the supplementary. Bold numbers indicate $p < 0.1$ in a bootstrapped two sided t-test.

No. of Speakers	Speaker	Single-Speaker Models				Multi-Speaker Models					
		S2G[18]		CMix-GAN		MUNIT[24]		StAGE		Mix-StAGE	
		PCK	F1	PCK	F1	PCK	F1	PCK	F1	PCK	F1
	Mean	0.25	0.08	0.26	**0.27**	0.24	0.06	**0.36**	0.21	0.34	0.22
2	Corden	0.30	0.05	0.32	0.21	0.25	0.06	0.36	0.21	0.34	0.24
	lec_cosmic	0.19	0.12	0.19	0.33	0.15	0.19	0.20	0.48	0.24	0.49
	Mean	0.37	0.18	0.37	0.27	0.22	0.05	0.38	**0.34**	0.39	**0.35**
4	Corden	0.30	0.05	0.32	0.21	0.24	0.07	0.35	0.27	0.35	0.30
	lec_cosmic	0.19	0.12	0.19	0.33	0.19	0.16	0.18	0.23	0.20	0.19
	Mean	0.36	0.14	0.37	0.26	0.31	0.21	0.38	**0.32**	0.40	**0.33**
8	Corden	0.30	0.05	0.32	0.21	0.23	0.03	0.32	0.28	0.36	0.27
	lec_cosmic	0.19	0.12	0.19	0.33	0.13	0.09	0.23	0.34	0.24	0.32

In our experiments we use a Temporal Convolution Network (TCN) module for both content and style encoders. The style space is a matrix $\mathbf{S} \in \mathbb{R}^{N \times D}$ where N is the number of speakers and D is the length of the style embeddings. The generator $G(.)$ consists of a 1D version of U-Net [18,44] followed by M TCNs as sub-generator functions. The discriminator is also a TCN module with lower capacity than the generators. A more detailed architecture can be found in the supplementary.

5 Experiments

Our experiments are divided into 2 sections, (1) **Style Preservation:** Generating co-speech gestures for multiple speakers with their own individualistic style, (2) **Style Transfer:** Generating co-speech gestures with content (or audio) of a speaker and gesture style of another speaker. Additionally, style transfer can be speaker-level as well as attribute-level. We choose visually distinguishable attribute-level styles: (1) body orientation, (2) gesture frequency, (3) primary arm function and (4) sitting/standing posture. We start by describing the baseline models followed by the evaluation metrics, which we will use to compare our model. We end this section with the description of our proposed dataset.

5.1 Baseline Models

Single-Speaker Models: These models are not designed to perform style transfer and hence are not included for those experiments.

- **Speech2Gesture** [18]: The closest work to co-speech gesture generation is one that only generates individualistic styles. We use the pre-trained models available from their code-base to render the videos. For rest of the speakers in PATS, we replicate their model, hyper-parameters and train speaker specific models.
- **CMix-GAN** (variant of our model): As an ablation, we remove the style embedding module and style consistency losses from our model Mix-StAGE . Hence, a separate model is required to be trained for each speaker for style preservation experiments.

Multi-speaker Models

- **MUNIT** [24]: The closest work to our style-transfer task is MUNIT which takes multiple domains of images (i.e. uni-modal). We modify the encoders and decoders to domain specific architectures (i.e. 1D convolutions for audio instead of 2D convolutions for images) while retaining the loss functions.
- **StAGE** (variant of our model): As an ablation, we fix the number of sub-generators in our model Mix-StAGE to one. This is equivalent to setting $M = 1$ in Eq. 2.

5.2   Evaluation Metrics

Human Perceptual Study: We conduct a human perceptual study on Amazon Mechanical Turk (AMT) for co-speech gesture generation (or style preservation) and style transfer (speaker-level and attribute-level) and measure preferences in two aspects of the generated animations, (1) **naturalness**, and (2) **style transfer correctness** for animation generation with content (i.e. audio) of speaker A and style of speaker B. We show a pair of videos with skeletal animations to the annotators. One of the animations is from the ground-truth set, while the other is generated using our proposed model. The generated animation could either have the same style or a different style as the original speaker. With unlimited time, the annotator has to answer two questions, (1) Which of the videos has more natural gestures? and (2) Do these videos have the same attribute-level style (or speaker-level style)? The first question is a real vs. fake perceptual study against the ground truth, while the second question measures how often the algorithm is able to visually preserve or transfer style (attribute or individual level). We run this study for randomly selected 100 pairs of videos from the held-out set.

Probability of Correct Keypoints (PCK): To measure the accuracy of the gesture generation, PCK [3,51] is used to evaluate all models. PCK values are averaged over $\alpha = 0.1, 0.2$ as suggested in [18].

Mode Classification F1: Correctness of shape of a gesture can be quantified by measuring the number of times the model has sampled from the correct mode

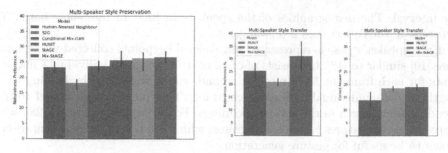

(a) Style Preservation Naturalness   (b) Style Transfer Natu- (c) Style Transfer Cor-
                                        ralness   rectness

Fig. 4. Perceptual Study for speaker-level style preservation in (a) and speaker level style transfer in (b), (c). We have naturalness preference for both style transfer and preservation, and style transfer correctness scores for style transfer. Higher is better. Error bars calculated for $p < 0.1$ using a bootstrapped two sided t-test.

of the pose distribution. Formally, we use the true $(\mathbf{Y}_p)$ and generated $(\hat{\mathbf{Y}}^p)$ pose to find the closest cluster $\hat{m}$ and m respectively. If $m = \hat{m}$, the generated pose was sampled from the correct mode. F1 score of this M-class classification problem is defined as Mode Classification F1, or simply F1.

Inception Score (IS): Generated pose sequences with the audio of speaker A and style of speaker B does not have a ground truth reference. To quantitatively measure the correctness and diversity of generated pose sequence we use the inception score [48]. For generative tasks such as image generation, this metric has been used with a pre-trained classification network such as Inception Model [53]. In our case, the generated samples are not images, but a set of 2D keypoints. Hence, we train a network which classifies a sequence of poses to its corresponding speaker which estimates the conditional likelihood to calculate IS scores.

5.3   Pose-Audio-Transcript-Style (PATS) Dataset

Gesture styles, which may be defined by attributes such as type, frequency, orientation of the body, is representative of the idiosyncrasies of the speaker [14]. We create a new dataset, Pose-Audio-Transcript-Style (PATS), to study various styles of gestures for a large number of speakers in diverse settings.

PATS contains pose sequences aligned with corresponding audio signals and transcripts[1] for 25 speakers (including 10 speakers from [18]) to offer a total of 251 h of data, with a mean of 10.7 s and a standard deviation of 13.5 s

---

[1] While transcripts are a part of this dataset, they are ignored for the purposed of this work.

per interval. The demographics of the speakers include 15 talk show hosts, 5 lecturers, 3 YouTubers, and 2 televangelists.

Each speaker's pose is represented via skeletal keypoints collected via Open-Pose [10] similar to [18]. It consists of of 52 coordinates of an individual's major joints for each frame at 15 frames per second, which we rescale by holding the length of each individual's shoulder constant. This prevents the model from encoding limb length in the style embeddings. Following prior work [18,28], we represent audio features as mel spetrograms, which is a rich input representation shown to be useful for gesture generation.

6   Results and Discussion

We group our results and discussions in (1) a first set of experiments studying style preservation (when output gesture styles are the same the original speaker) and (2) a second set of experiments studying transfer of gesture styles.

6.1   Gesture Animation and Style Preservation

To understand the impact of adding more speakers, we select a random sample of 8 speakers for the largest 8-speaker multi-speaker model, and train smaller 4-speaker and 2-speaker models where the speakers trained are always a subset of the speakers that were trained in a larger model. This allows to compare the performance on the same two initial speakers which are '*Corden*' and '*lec_cosmic*' in our case[2]. We also compare with single-speaker models trained and tested on one speaker at a time.

Impact of Training with Multiple Speakers. Results from Table 1 show that multi-speaker models outperform single-speaker models especially for pose accuracy (i.e. PCK), shape and timing (i.e. F1). We find that increasing the number of speakers could sometimes reduce the performance of individual speakers but the overall performance generally shows improvement.

Comparison with Previous Baselines. To compare with prior baselines, we focus first on the subjective evaluation shown in Fig. 4a, since it is arguably the most important metric. The results show consistent improvements on the naturalness rating for our proposed model Mix-StAGE and also our single-speaker variant CMix-GAN over the previous state of the art approach S2G [18]. We also observe that multi-speaker models perform better than single speaker-models. In Table 1, we show similar quantitative improvements of Mix-StAGE and CMix-GAN over S2G for both PCK and F1 scores.

---

[2] The complete set of speakers used in our experiments are listed in the supplementary.

Table 2. Style Transfer: Inception scores for style transfer on multi-speaker models (indicated in each row). Columns on the left refer to the speaker-level style transfer task while those on the right refer to the specific attribute-level style task. Bold numbers indicate $p < 0.1$ in a bootstrapped two sided t-test.

Model	Number of Speakers			Attributes			
	2 Speakers	4 Speakers	8 Speakers	Sitting vs Standing	Gesture Frequency	Body Orientation	Primary Arm Func.
MUNIT [24]	1.11	1.90	2.06	1.10	2.49	1.05	3.32
StAGE	2.17	**2.85**	3.89	1.68	4.38	**6.81**	3.14
Mix-StAGE	**2.61**	**2.85**	**4.48**	**3.08**	**4.50**	6.69	**3.32**

Impact of Multiple Generators for Decoding. Mix-StAGE's gesture space models multiple modes, as seen in Fig. 2. Its importance is shown in Table 1 where models with single generators as the decoder (i.e. S2G, MUNIT and StAGE) showed lower F1 scores, most likely due to mode collapse while training. Multiple generators in CMix-GAN and Mix-StAGE boost F1 scores as compared to other models in the single-speaker and multi-speaker regimes respectively. A similar trend was observed in the perceptual study in Fig. 4.

We also study the impact of the number of generators (hyperparameter M) in our Mix-StAGE model. While for small number of speakers (i.e. 2 speakers) a single generator is good enough, the positive effect of multiple generators can be observed as the number of speakers increase (see Table 1). We also vary $M \in \{1, 2, 4, 8, 12\}$ and observe that improvements seem to plateau at $M = 8$ with only marginal improvements for larger number of sub-generators. For the ablation study we refer the readers to the supplementary.

Attribute-Level Style Preservation in Multi-speaker Models. We also study style preservation for attributes in Sect. 5 as a perceptual study in Fig. 6. We observe that humans deem animations generated by Mix-StAGE significantly more natural in most cases. High scores ranging 60–90% for style preservation correctness, with Mix-StAGE outperforming others, are observed for pairs of speakers in Fig. 6b. This indicates that style preservation may be a relatively easy task as compared to style transfer for multi-speaker models. With this, we now shift our focus to style transfer.

6.2 Style Transfer

Speaker-Level Style Transfer. To study our capability to transfer style of a specific speaker to a new speaker, we will compare the gesture spaces between the original speakers and the transferred speakers. Figure 2a shows that each original speaker occupies different regions in the M^2 gesture space. Using our Mix-StAGE model to transfer style, we can see the new gesture space in Fig. 2b. For the transferred speakers the 2 spaces look quite similar. For instance,

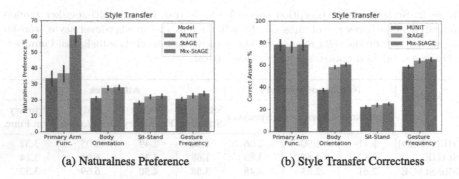

(a) Naturalness Preference   (b) Style Transfer Correctness

Fig. 5. A visualization of the perceptual human study for attribute-level style transfer with (a) naturalness preference, and (b) style transfer correctness scores for the generated animations for a different style than the speaker. Higher is better. Error bars calculated for $p < 0.1$ using a bootstrapped two sided t-test.

'*Corden*' style (a speaker in our dataset) is represented by the color blue in Fig. 2a and occupies the lower region of the gesture space. When Mix-StAGE generates co-speech gestures using audio of '*Oliver*' and the style of '*Corden*', it occupies a subset of '*Corden's*' region in the gesture space, also represented by blue in Fig. 2b. We see a similar trend for styles of '*Oliver*' and '*ytch_prof*'. This is an indication of a successful style transfer across different speakers. We note the lack of clean separation in the gesture space among different styles as there could common gestures across multiple speakers.

For the perceptual study, we want to know if humans can distinguish the generated speaker styles. For this, we show human annotators two videos: a ground truth video in a specific style, and a generated video which is either from the style of the same speaker or a different speaker. Annotators have to decide if this is the same style or not. We use the 4-speaker model for this experiment. Figure 4b shows naturalness preference and 4c shows percentage of the time style was transferred correctly. Our model Mix-StAGE performs best in both cases. This trend is corroborated with higher inception scores in Table 2.

Impact of Number of Speakers for Style Transfer. In Table 2, we observe that increasing the number of speakers used for training also increases the average inception score for the stylized gesture generations. This is a welcome effect as it indicates increases in the diversity and the accuracy of the generations.

Attribute-Level Style Transfer in Multi-speaker Models. We study four common attributes of gesture style which are also visually distinguishable by humans: (1) sitting vs. standing, (2) high vs low gesture frequency, (3) left vs right body orientation and (4) left vs right primary arm. Speakers were selected carefully to represent each extremes of these four attributes. We run a perceptual study similar to the one for speaker-level styles. However, we ask the annotators

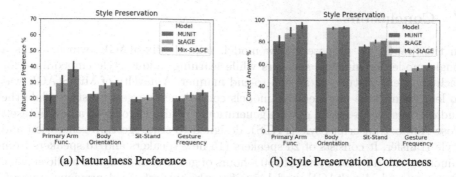

(a) Naturalness Preference   (b) Style Preservation Correctness

Fig. 6. A visualization of the perceptual human study for attribute-level style preservation with (a) naturalness preference, and (b) style preservation correctness scores for the generated animations for the same style as the speaker. Higher is better. Error bars calculated for $p < 0.1$ using a bootstrapped two sided t-test.

to judge if the attribute is the same in both of the videos (e.g. are both the people gesturing with the same arm?). Results from Fig. 5 show that Mix-StAGE generates more (or similar) number of natural gestures with the correct attribute-level style compared to the other baselines. We also observe that it is harder for humans to determine if a person is standing or sitting, which we suspect is due to the missing waistline in the animation.

For a visual understanding of the generated gestures and stylized gestures, we plot a style-content heatmap in Fig. 7, where columns represent generations for a specific style, while rows represent different speaker's audio as input. These heatmaps show that gestures are consistent across audio inputs but different between styles. Accuracy and diversity of style transfer is corroborated by inception scores in Table 2.

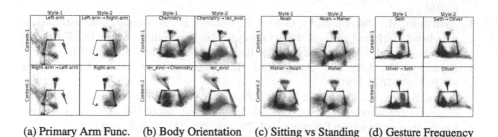

(a) Primary Arm Func.   (b) Body Orientation   (c) Sitting vs Standing   (d) Gesture Frequency

Fig. 7. Style-Content Heatmaps for attribute-level style transfer. Each column represents the same style, while rows have input audio from different speakers. These heatmaps show that gestures are consistent across audio inputs but different between styles. Red regions correspond to the motion of the right arm, while blue corresponds to the left. (Color figure online)

7   Conclusions

In this paper, we propose a new model, named Mix-StAGE, which learns a single model for multiple speakers while learning unique style embeddings for each speaker's gestures in an end-to-end manner. A novelty of Mix-StAGE was to learn a mixture of generative models conditioned on gesture style while the audio drives the co-speech gesture generation. We also introduced a new dataset, Pose-Audio-Transcript-Style (PATS), designed to study gesture generation and style transfer. It consists of 25 speakers (15 new speakers and 10 speakers from Ginosar et al. [18]) for a total of 250+ hours of gestures and aligned audio signals. Our proposed Mix-StAGE model significantly outperformed previous state-of-the-art approach for gesture generation and provided a path towards performing gesture style transfer across multiple speakers. We also demonstrated, through human perceptual studies, that the generated animations by our model are more natural whilst being able to retain or transfer style.

Acknowledgements. This material is based upon work partially supported by the National Science Foundation (Awards #1750439 #1722822), National Institutes of Health and the InMind project. Any opinions, findings, and conclusions or recommendations expressed in this material are those of the author(s) and do not necessarily reflect the views of National Science Foundation or National Institutes of Health, and no official endorsement should be inferred.

References

1. Ahuja, C., Ma, S., Morency, L.P., Sheikh, Y.: To react or not to react: end-to-end visual pose forecasting for personalized avatar during dyadic conversations. In: 2019 International Conference on Multimodal Interaction, pp. 74–84. ACM (2019)
2. Ahuja, C., Morency, L.P.: Language2pose: natural language grounded pose forecasting. In: 2019 International Conference on 3D Vision (3DV), pp. 719–728. IEEE (2019)
3. Andriluka, M., Pishchulin, L., Gehler, P., Schiele, B.: 2D human pose estimation: new benchmark and state of the art analysis. In: Proceedings of the IEEE Conference on computer Vision and Pattern Recognition, pp. 3686–3693 (2014)
4. Arjovsky, M., Chintala, S., Bottou, L.: Wasserstein GAN. arXiv preprint arXiv:1701.07875 (2017)
5. Arora, S., Ge, R., Liang, Y., Ma, T., Zhang, Y.: Generalization and equilibrium in generative adversarial nets (GANs). In: Proceedings of the 34th International Conference on Machine Learning, vol. 70, pp. 224–232. JMLR. org (2017)
6. Bailenson, J.N., Yee, N., Merget, D., Schroeder, R.: The effect of behavioral realism and form realism of real-time avatar faces on verbal disclosure, nonverbal disclosure, emotion recognition, and copresence in dyadic interaction. Presence: Teleoperators Virtual Environ. **15**(4), 359–372 (2006)
7. Bansal, A., Ma, S., Ramanan, D., Sheikh, Y.: Recycle-GAN: unsupervised video retargeting. In: Ferrari, V., Hebert, M., Sminchisescu, C., Weiss, Y. (eds.) ECCV 2018. LNCS, vol. 11209, pp. 122–138. Springer, Cham (2018). https://doi.org/10.1007/978-3-030-01228-1_8

8. Bergmann, K., Kopp, S.: Increasing the expressiveness of virtual agents: autonomous generation of speech and gesture for spatial description tasks. In: Proceedings of The 8th International Conference on Autonomous Agents and Multiagent Systems-Volume 1, pp. 361–368 (2009)
9. Bian, Y., Chen, C., Kang, Y., Pan, Z.: Multi-reference tacotron by intercross training for style disentangling, transfer and control in speech synthesis. arXiv preprint arXiv:1904.02373 (2019)
10. Cao, Z., Hidalgo, G., Simon, T., Wei, S.E., Sheikh, Y.: Openpose: realtime multi-person 2D pose estimation using part affinity fields. arXiv preprint arXiv:1812.08008 (2018)
11. Cassell, J., Vilhjálmsson, H.H., Bickmore, T.: Beat: the behavior expression animation toolkit. In: Prendinger, H., Ishizuka, M. (eds.) Life-Like Characters, pp. 163–185. Springer, Heidelberg (2004)
12. Chiu, C.C., Marsella, S.: Gesture generation with low-dimensional embeddings. In: Proceedings of the 2014 International Conference on Autonomous Agents and Multi-agent Systems, pp. 781–788 (2014)
13. Chiu, C.-C., Morency, L.-P., Marsella, S.: Predicting co-verbal gestures: a deep and temporal modeling approach. In: Brinkman, W.-P., Broekens, J., Heylen, D. (eds.) IVA 2015. LNCS (LNAI), vol. 9238, pp. 152–166. Springer, Cham (2015). https://doi.org/10.1007/978-3-319-21996-7_17
14. Davis, R.O., Vincent, J.: Sometimes more is better: agent gestures, procedural knowledge and the foreign language learner. Br. J. Educ. Technol. **50**(6), 3252–3263 (2019)
15. Denton, E.L., et al.: Unsupervised learning of disentangled representations from video. In: Advances in Neural Information Processing Systems, pp. 4414–4423 (2017)
16. Ferstl, Y., Neff, M., McDonnell, R.: Multi-objective adversarial gesture generation. In: Motion, Interaction and Games, p. 3. ACM (2019)
17. Gatys, L.A., Ecker, A.S., Bethge, M.: A neural algorithm of artistic style. arXiv preprint arXiv:1508.06576 (2015)
18. Ginosar, S., Bar, A., Kohavi, G., Chan, C., Owens, A., Malik, J.: Learning individual styles of conversational gesture. In: Proceedings of the IEEE Conference on Computer Vision and Pattern Recognition, pp. 3497–3506 (2019)
19. Goodfellow, I., et al.: Generative adversarial nets. In: Advances in Neural Information Processing Systems, pp. 2672–2680 (2014)
20. Gurunath, N., Rallabandi, S.K., Black, A.: Disentangling speech and non-speech components for building robust acoustic models from found data. arXiv preprint arXiv:1909.11727 (2019)
21. Hao, G.Y., Yu, H.X., Zheng, W.S.: Mixgan: learning concepts from different domains for mixture generation. arXiv preprint arXiv:1807.01659 (2018)
22. Hasegawa, D., Kaneko, N., Shirakawa, S., Sakuta, H., Sumi, K.: Evaluation of speech-to-gesture generation using bi-directional LSTM network. In: Proceedings of the 18th International Conference on Intelligent Virtual Agents (IVA18), pp. 79–86 (2018)
23. Hoang, Q., Nguyen, T.D., Le, T., Phung, D.: MGAN: training generative adversarial nets with multiple generators (2018)
24. Huang, X., Liu, M.-Y., Belongie, S., Kautz, J.: Multimodal unsupervised image-to-image translation. In: Ferrari, V., Hebert, M., Sminchisescu, C., Weiss, Y. (eds.) ECCV 2018. LNCS, vol. 11207, pp. 179–196. Springer, Cham (2018). https://doi.org/10.1007/978-3-030-01219-9_11

25. Isola, P., Zhu, J.Y., Zhou, T., Efros, A.A.: Image-to-image translation with conditional adversarial networks. In: Proceedings of the IEEE Conference on Computer Vision and Pattern Recognition, pp. 1125–1134 (2017)
26. Johnson, J., Alahi, A., Fei-Fei, L.: Perceptual losses for real-time style transfer and super-resolution. In: Leibe, B., Matas, J., Sebe, N., Welling, M. (eds.) ECCV 2016. LNCS, vol. 9906, pp. 694–711. Springer, Cham (2016). https://doi.org/10.1007/978-3-319-46475-6_43
27. Kendon, A.: Gesture and speech: two aspects of the process of utterance. In: Key, M.R. (ed.) Nonverbal Communication and Language, pp. 207–227 (1980)
28. Kucherenko, T., Hasegawa, D., Henter, G.E., Kaneko, N., Kjellström, H.: Analyzing input and output representations for speech-driven gesture generation. arXiv preprint arXiv:1903.03369 (2019)
29. Lee, H.Y., et al.: Drit++: Diverse image-to-image translation via disentangled representations. arXiv preprint arXiv:1905.01270 (2019)
30. Lee, H.Y., et al.: Dancing to music. In: Advances in Neural Information Processing Systems, pp. 3581–3591 (2019)
31. Levine, S., Krähenbühl, P., Thrun, S., Koltun, V.: Gesture controllers. ACM Trans. Graph. **29**(4), 124:1–124:11 (2010)
32. Levine, S., Theobalt, C., Koltun, V.: Real-time prosody-driven synthesis of body language. ACM Trans. Graph. **28**(5), 172:1–172:10 (2009)
33. Liu, M.Y., Breuel, T., Kautz, J.: Unsupervised image-to-image translation networks. In: Advances in Neural Information Processing Systems, pp. 700–708 (2017)
34. Liu, M.Y., Tuzel, O.: Coupled generative adversarial networks. In: Advances in Neural Information Processing Systems, pp. 469–477 (2016)
35. Lloyd, S.: Least squares quantization in PCM. IEEE Trans. Inf. Theory **28**(2), 129–137 (1982)
36. Ma, S., Mcduff, D., Song, Y.: Neural TTS stylization with adversarial and collaborative games (2018)
37. van den Maaten, L., Hinton, G.: Visualizing data using t-SNE. J. Mach. Learn. Res. **9**(Nov), 2579–2605 (2008)
38. McNeill, D.: Hand and mind: What gestures reveal about thought. University of Chicago Press (1992)
39. Nagrani, A., Chung, J.S., Albanie, S., Zisserman, A.: Disentangled speech embeddings using cross-modal self-supervision. In: ICASSP 2020–2020 IEEE International Conference on Acoustics, Speech and Signal Processing (ICASSP), pp. 6829–6833. IEEE (2020)
40. Neff, M., Kipp, M., Albrecht, I., Seidel, H.P.: Gesture modeling and animation based on a probabilistic re-creation of speaker style. ACM Trans. Graph. (TOG) **27**(1), 1–24 (2008)
41. Obermeier, C., Kelly, S.D., Gunter, T.C.: A speaker's gesture style can affect language comprehension: ERP evidence from gesture-speech integration. Soc. Cogn. Affective Neurosci. **10**(9), 1236–1243 (2015)
42. Pelachaud, C.: Studies on gesture expressivity for a virtual agent. Speech Commun. **51**(7), 630–639 (2009)
43. Reynolds, D.A.: Gaussian mixture models. Encyclopedia Biometrics **741** (2009)
44. Ronneberger, O., Fischer, P., Brox, T.: U-Net: convolutional networks for biomedical image segmentation. In: Navab, N., Hornegger, J., Wells, W.M., Frangi, A.F. (eds.) MICCAI 2015. LNCS, vol. 9351, pp. 234–241. Springer, Cham (2015). https://doi.org/10.1007/978-3-319-24574-4_28

45. Rosca, M., Lakshminarayanan, B., Warde-Farley, D., Mohamed, S.: Variational approaches for auto-encoding generative adversarial networks. arXiv preprint arXiv:1706.04987 (2017)
46. Royer, A., et al.: XGAN: unsupervised image-to-image translation for many-to-many mappings. In: Singh, R., Vatsa, M., Patel, V.M., Ratha, N. (eds.) Domain Adaptation for Visual Understanding, pp. 33–49. Springer, Cham (2020). https://doi.org/10.1007/978-3-030-30671-7_3
47. Sadoughi, N., Busso, C.: Novel realizations of speech-driven head movements with generative adversarial networks, pp. 6169–6173 (2018). https://doi.org/10.1109/ICASSP.2018.8461967
48. Salimans, T., Goodfellow, I., Zaremba, W., Cheung, V., Radford, A., Chen, X.: Improved techniques for training GANs. In: Advances in Neural Information Processing Systems, pp. 2234–2242 (2016)
49. Sargin, M.E., Yemez, Y., Erzin, E., Tekalp, A.M.: Analysis of head gesture and prosody patterns for prosody-driven head-gesture animation. IEEE Trans. Pattern Anal. Mach. Intell. **30**, 1330–1345 (2008). https://doi.org/10.1109/TPAMI.2007.70797
50. Shlizerman, E., Dery, L., Schoen, H., Kemelmacher, I.: Audio to body dynamics. Proceedings/CVPR, IEEE Computer Society Conference on Computer Vision and Pattern Recognition. IEEE Computer Society Conference on Computer Vision and Pattern Recognition (2018)
51. Simon, T., Joo, H., Matthews, I., Sheikh, Y.: Hand keypoint detection in single images using multiview bootstrapping. In: Proceedings of the IEEE Conference on Computer Vision and Pattern Recognition, pp. 1145–1153 (2017)
52. Smith, H.J., Cao, C., Neff, M., Wang, Y.: Efficient neural networks for real-time motion style transfer. Proc. ACM Comput. Graph. Interactive Tech. **2**(2), 1–17 (2019)
53. Szegedy, C., Vanhoucke, V., Ioffe, S., Shlens, J., Wojna, Z.: Rethinking the inception architecture for computer vision. In: Proceedings of the IEEE Conference on Computer Vision and Pattern Recognition, pp. 2818–2826 (2016)
54. Villegas, R., Yang, J., Hong, S., Lin, X., Lee, H.: Decomposing motion and content for natural video sequence prediction. arXiv preprint arXiv:1706.08033 (2017)
55. Wagner, P., Malisz, Z., Kopp, S.: Gesture and speech in interaction: an overview (2014)
56. Wang, Y., et al.: Style tokens: Unsupervised style modeling, control and transfer in end-to-end speech synthesis. arXiv preprint arXiv:1803.09017 (2018)
57. Xu, J., Gannon, P.J., Emmorey, K., Smith, J.F., Braun, A.R.: Symbolic gestures and spoken language are processed by a common neural system. Proc. Natl. Acad. Sci. **106**(49), 20664–20669 (2009)
58. Zhu, J.Y., Park, T., Isola, P., Efros, A.A.: Unpaired image-to-image translation using cycle-consistent adversarial networks. In: Proceedings of the IEEE International Conference on Computer Vision, pp. 2223–2232 (2017)
59. Zhu, J.Y., et al.: Toward multimodal image-to-image translation. In: Advances in Neural Information Processing Systems, pp. 465–476 (2017)

An LSTM Approach to Temporal 3D Object Detection in LiDAR Point Clouds

Rui Huang$^{(\boxtimes)}$, Wanyue Zhang, Abhijit Kundu, Caroline Pantofaru,
David A. Ross, Thomas Funkhouser, and Alireza Fathi

Google Research, Mountain View, US
huangrui@google.com

Abstract. Detecting objects in 3D LiDAR data is a core technology for
autonomous driving and other robotics applications. Although LiDAR
data is acquired over time, most of the 3D object detection algorithms
propose object bounding boxes independently for each frame and neglect
the useful information available in the temporal domain. To address this
problem, in this paper we propose a sparse LSTM-based multi-frame
3d object detection algorithm. We use a U-Net style 3D sparse convolution network to extract features for each frame's LiDAR point-cloud.
These features are fed to the LSTM module together with the hidden
and memory features from last frame to predict the 3d objects in the
current frame as well as hidden and memory features that are passed
to the next frame. Experiments on the Waymo Open Dataset show that
our algorithm outperforms the traditional frame by frame approach by
7.5% mAP@0.7 and other multi-frame approaches by 1.2% while using
less memory and computation per frame. To the best of our knowledge,
this is the first work to use an LSTM for 3D object detection in sparse
point clouds.

Keywords: 3D object detection · LSTM · Point cloud

1 Introduction

3D object detection is one of the fundamental tasks in computer vision. Given
observations of a scene with a 3D sensor (e.g., LiDAR), the goal is to output semantically labeled 3D oriented bounding boxes for all objects in every
observation. This task is critical for autonomous driving, object manipulation,
augmented reality, and many other robot applications.

Although almost all robot sensors capture data continuously (LiDAR, RGB-
D video, RGB video, etc.), most 3D object detection algorithms consider only
one "frame" of input sensor data when making bounding box predictions. Histor-
ically, multi-frame data has not been widely available (e.g. the Kitti 3D Object
Detection Challenge [15] provides only one LiDAR sweep for each scene). However, after datasets with multi-frame sequences of LiDAR were released [3,5,42],

© Springer Nature Switzerland AG 2020
A. Vedaldi et al. (Eds.): ECCV 2020, LNCS 12363, pp. 266–282, 2020.
https://doi.org/10.1007/978-3-030-58523-5_16

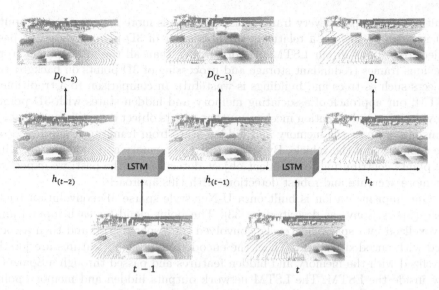

Fig. 1. Our method consumes a sequence of point clouds as input. At each time step, the proposed LSTM module combines the point cloud features from the current frame with the hidden and memory features from the previous frame to predict the 3d objects in the current frame together with the hidden and memory features that are passed to the next frame. For memory efficiency we pass only the hidden feature points that have high score to the next frame (pink in the images). D_t and h_t represent the 3d object detections and hidden features in frame t respectively. (Color figure online)

most 3D object detection algorithms still work frame by frame. Among the algorithms with reported results on the nuScenes and Waymo object detection tasks, we find that only Ngiam et al. [34] and Hu et al. [21] consider multiple frames as input, and they both use simple methods based on reusing seed points or concatenating of input data for multiple frames.

In this paper, we investigate a new method as depicted in Fig. 1 that utilizes the temporal sequence of LiDAR data acquired by autonomous vehicle for 3D object detection. Our approach is to use the memory of an LSTM to encode information about objects detected in previous frames in a way that can assist object detection in the current frame. Specifically, we represent the memory and hidden state of the LSTM as 64-dimensional features associated with 3D points observed in previous frames. At each frame, we use an LSTM architecture to combine features from these memory and hidden state 3D point clouds with features extracted from the latest observed 3D point cloud to produce bounding box predictions for the current frame and update the memory and hidden state for the next frame.

The rationale for this approach is that the LSTM memory can represent everything known about object detections in the past in a concise set of features associated with a sparse set of 3D positions (ideally near past object detections). In comparison to previous methods that concatenate input point clouds from multiple timesteps at every frame, this approach is more memory and compute efficient, as we include a relatively small number of 3D points related to past object detections in our LSTM memory rather than all the input points from previous frames (redundant storage and processing of 3D points on background objects such as trees and buildings is wasteful). In comparison to a traditional LSTM, our approach of associating memory and hidden states with 3D points provides a spatial attention mechanism that assists object detection and enables transformation of the memory and hidden state from frame to frame based on the egomotion of the vehicle. By associating the memory and hidden state with 3D points contributing to confident object detections in the past, we expect to get more accurate and robust detections with this approach.

Our implementation is built on a U-Net style sparse 3D convolution backbone (SparseConv) as described in [33]. The point cloud for each input frame is voxelized into sparse 3d voxels, convolved on a sparse grid, and then associated with encoded features. Then, the encoded point cloud features are jointly voxelized with the memory and hidden features and passed through a SparseConv inside the LSTM. The LSTM network outputs hidden and memory point cloud features that will be passed to the next frame. Furthermore, the predicted hidden features are fed into the 3d detection head to generate per point object bounding box proposals (center, size, rotation, and confidence), which are further processed with a graph convolution to smooth per point predictions in local neighborhoods and non maximum suppression (NMS) to select a highly confident and non-overlapping set of proposed bounding boxes. The hidden and memory state (point features) only keep track of features in locations that have high objectness score. This enables us to be more memory efficient and to be able to aggregate and reason about information in a long sequence.

Experiments show that this method outperforms frame by frame detection models by 7.5% mAP@0.7 and beats a strong multi-frame concatenation baseline model by 1.2%. Our model achieves 6.8% better results than a baseline that refines predicted bounding boxes using the classical combination of frame by frame detection, Hungarian assignment, and Kalman filtering [44].

Our key contributions are summarized below:

- We propose the first LSTM-based sequential point cloud processing framework for 3D object detection. It provides a significant performance boost over a single frame state-of-the-art 3D SparseConv model. Furthermore, our model outperforms a strong baseline based on concatenating multi-frame data.
- We propose a 3D Sparse Conv LSTM where a small 3d sparse U-Net replaces the fully connected layer in vanilla LSTM. Our model has explicit memory to facilitate reasoning across long sequence of point clouds. Compared to point-based methods, our voxel-based module is effective and efficient in fusing accumulated memory and input data in multiple scales, while maintaining a constant memory footprint regardless of the sequence length in inference time.

2   Related Work

3D Object Detection. A common approach to 3D object detection is to utilize ideas that have been successful for 2D object detection [6,26,27,32,40,41,47]. For instance, Frustum-PointNet [36] uses 2D detectors on RGB images and point clouds from the depth sensor. However, the search space for potential objects is limited in the 3D viewing frustum extended from 2D regions. MV3D [6] deploys a multi-view fusion network for features extracted from the bird-eye view, Lidar range view and RGB images. Building on soft voxelization, Zhou et al. [52] fuse features based on Cartesian coordinate, perspective coordinate and the output of a shared fully connected layer from LiDAR points.

Another class of methods [19,35,37,39,48,50,51] propose networks that directly consume the 3d point cloud as input. Shi et al. [39] propose a bottom-up approach to directly generate 3D bounding box proposals from the point cloud, followed by a sub-network for refinement. VoteNet [35] uses PointNet++ [37] backbone to vote for object centers. The votes are then clustered to produce the final bounding box proposals.

There is an increasing trend to convert point clouds to regular grids where 3d convolution can be conveniently constructed. VoxelNet [53] partitions point clouds into voxels but this method is computationally expensive. Some of the previous works attempt to solve this issue by making use of the sparsity pattern of 3D points [12,16,17,33,38]. SparseConv [18,33] is exceptionally efficient as convolutions are restricted to active sites and sparsity pattern is preserved even after layers of convolutions. Our work uses a sparse voxel U-Net as the backbone as described in [33].

Spatio-Temporal Methods. Various ways to make use of the temporal information are experimented for different vision tasks such as prediction in video data and modeling the human motion dynamics [7,22,23,45,46]. In addition, there are various LSTM based methods for object detection in video [7,14,24,45]. Among those, Xiao et al. [45] introduce a spatio-temporal memory module (STMM) to model temporal appearance and motion changes of objects. Teng et al. [43] explore detecting objects in streaming video using weak supervision by tracking and optical flow.

For LiDAR point clouds, [11,31,49] use ConvGRU or ConvLSTM to process the bird-eye view projection. Luo et al. [30] explore the synergy of 4 tasks for autonomous driving: detection, tracking, motion forecasting and motion planning. By concatenating multiple frames of input, 2d convolution is performed on voxels to forecast the next n frames. Inspired by Luo et al. [30], Casas et al. [4] jointly tackle detection and motion estimation by adding the rasterized map to provide the environment context for more accurate forecasting. MeteorNet [29] processes point clouds directly and proposes direct grouping and chained grouping to find nearest spatio-temporal neighbourhours. This method is applied to semantic segmentation, classification and flow estimation. Choy et al. [9] augment 3D data with the time axis and build sparse 4D convolutions using non-conventional kernel shapes. Our approach is distinct from the above as we propose a 3d sparse LSTM model that consumes sparse 3d data and performs 3d sparse operations and merging to perform 3d object detection.

The closest related work to ours is PointRNN [13], which adapts RNNs for predicting scene flow on multi-frame point clouds. It proposes a point-rnn function to aggregate the past state and the current input based on the point coordinates. Our approach is different as we conduct 3D sparse convolution on adjacent voxels which avoids the expensive step of finding nearest neighbors for each point in the point cloud. Besides, no permutation invariant aggregation is needed by our method. Furthermore, we focus on 3d object detection in a sequence while PointRNN [13] focuses on scene flow.

3D Scene Flow and Object Tracking. Some researchers have focused on the related problem of predicting scene flow (3D motion vector per point) from pairs of input point clouds in adjacent frames. Since the magnitude and direction of movement of points provides a cue for object detection, these two tasks could provide mutual context for each other. For instance, Behl et al. [1] extract xyz object coordinates from 4 RGB images and incorporate detection and instance segmentation cues from 2D to improve scene flow. PointFlowNet [2] gets rid of the reliance on 2D images by using an encoder-decoder model to tackle flow, object location and motion in conjunction. FlowNet3D [10] consumes point clouds directly by using a Set Conv Layer to down-sample points, a flow embedding layer to aggregate features from two point clouds and a Set UpConv layer to get a per-point estimation of the translation vectors. While these methods are loosely related to ours, they are aimed at predicting flow for the entire scene and do not aim at improving 3d object detection.

Some of the previous works [5,8,44] have focused on 3D tracking of objects. However, these algorithms mainly focus on generating multi-frame tracks and do not necessarily result in a more accurate per frame 3d object detection. Wang et al. [44] detect objects in every frame and then use the Hungarian algorithm to associate objects and Kalman filter to aggregate predictions across frames. We use this method as a baseline and compare our 3d object detection accuracy with theirs in the results section.

3   Method

The architecture of our method is shown in Fig. 2. In each frame, we extract point features from the input point cloud by feeding it into a 3d sparse voxel conv U-Net as described in [33]. We feed the extracted features together with the memory and hidden features from previous frame to our proposed 3d sparse conv LSTM which processes them and outputs the hidden and memory features that will be consumed by the next frame. In the mean time, an object detection head is applied to the hidden features to produce 3d object proposals in each frame. The proposals are then passed through a graph convolution stage and then non-maximum suppression to output the detected 3d objects for each frame.

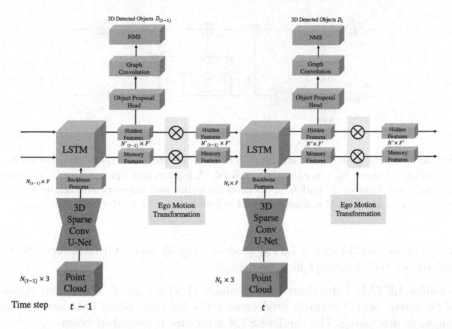

Fig. 2. Overview of the temporal detection framework: A sequence of point clouds are processed by a Sparse Conv U-Net backbone in each frame. The 3d sparse LSTM fuses the backbone feature at the current time step t with the hidden and memory feature at the previous time step $t - 1$ to produce hidden and memory feature at time step t. Object proposals are generated from the hidden feature and refined using a graph convolution network. Farthest point sampling and non-maximum suppression are applied to the proposed 3d objects to produce the final detected 3d objects.

3.1 3D Sparse Conv U-Net

To extract point features from the input point cloud, we use a U-Net shaped backbone as described in [33]. The input to our feature extractor is a point cloud as a $N \times 3$ tensor (points with their xyz position). The network first voxelizes the point cloud into sparse 3d voxels. If multiple points fall in the same voxel, the voxel feature would be the average of the xyz location of those points. The netowrk encoder consists of several blocks of 3d sparse convolution layers where each block is followed by a 3d max pooling. The U-Net decoder upsamples the spatial resolution gradually with several blocks of sparse convolution layers and upsampling with skip connections from the encoder layers. The extracted voxel features are de-voxelized back to the points to output a $N \times F$ tensor where F is the extracted feature dimension.

3.2 3D Sparse Conv LSTM

We use an LSTM based on 3d sparse conv network to leverage the temporal information in the sequence of LiDAR frames. Here we first review the basic

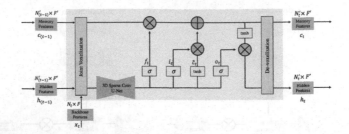

Fig. 3. 3D sparse conv LSTM structure. The backbone feature x_t, memory feature c_{t-1} and hidden feature h_{t-1} are jointly voxelized. A lightweight SparseConv U-Net takes the concatenation of x_t and h_{t-1} to produce gates and memory candidate. Output features from the LSTM are de-voxelized before being sent to the next time step.

notation for our 3d sparse LSTM module (Fig. 3) and then introduce the key differences and challenges in more details.

Vanilla LSTM: Long short term memory (LSTM) [20] is a common variant of recurrent neural network used extensively for time series data and natural language processing. The vanilla LSTM structure is described below:

$$f_t = \sigma(W_f \cdot [h_{t-1}, x_t] + b_f) \tag{1}$$

$$i_t = \sigma(W_i \cdot [h_{t-1}, x_t] + b_i) \tag{2}$$

$$\tilde{c}_t = tanh(W_c \cdot [h_{t-1}, x_t] + b_c) \tag{3}$$

$$c_t = f_t \times c_{t-1} + i_t \times \tilde{c}_t \tag{4}$$

$$o_t = \sigma(W_o[h_{t-1}, x_t] + b_o) \tag{5}$$

$$h_t = o_t \times tanh(c_t) \tag{6}$$

The input feature at current time step x_t and the hidden feature at the previous time step h_{t-1} are concatenated before being transformed by a fully connected layer with weight matrix W and bias b. The transformed feature is activated by either sigmoid (σ) or tanh function to produce input gate (i_t), forget gate (f_t), output gate (o_t) and cell memory candidate ($\tilde{c}_t$) for the current time step. The cell memory c_t is updated from $\tilde{c}_t$ and the cell memory at previous time step c_{t-1}, where $\times$ denotes element-wise multiplication.

LSTM on Sparse Point Clouds: In our context, x_t of size $N_t \times F$ is the point cloud features that is extracted using our 3d sparse backbone, h_{t-1} and c_{t-1} of size $N'_{t-1} \times F'$ are hidden and memory point features respectively. We subsample the hidden and memory point features and only keep a subset (N'_{t-1}) of the points that have high semantic scores (obtained from the pre-trained single frame detection model).

In order for the LSTM to be able to fuse multiple 3d sparse tensor features, we replace the fully connected layer in vanilla LSTM with a lightweight 3d sparse conv U-Net structure to produce gates and cell memory candidate. This approach ensures that the LSTM has enough capacity to conduct sequential reasoning in the 3d sparse space and avoid the expensive nearest neighbor search in point-based methods [13,28,29].

Joint Voxelization: Due to the object motion in the scene (even though that we compensate for the egomotion), x_t and h_{t-1} (or c_{t-1}) will not align in the 3D space. Our solution to this problem is to jointly voxelize the three point clouds, namely x_t, h_{t-1} and c_{t-1}. The resulting three voxel grids are then concatenated in feature dimension. If one of the sparse point features has no point in a voxel but the other ones do, we will pad the voxel features of the one with the missing point with zeros.

In other words, since the point clouds are non-overlapping in some region, after joint voxelization there will be empty voxels inserted into each voxel grids in non-overlapping region. This means that the joint sparse voxel representation covers the union of the spatial extent of all participating point clouds which is still extremely sparse.

3.3 Object Detection

The proposal head takes the voxelized hidden features h_t of size $N \times F'$ from the LSTM at each time step to independently generate per voxel bounding box proposals (center, rotation, height, length and width). The predictions are then de-voxelized to produce per point bounding box predictions, taking into account each point's position offset within the voxel. During de-voxelization we transfer the prediction associated with each voxel to all the points that fall inside it. The head is implemented with 3 layers of sparse convolutions for each attribute.

As described in [33], we construct a graph on top of the per point predictions, where each point (node) is connected to its K nearest neighbors with similar object center predictions. The predicted object attributes are propagated based on a predicted weight per point. The weight determines the significance of each point in comparison with its neighbors. The bounding box prediction loss is applied both before and after the propagation.

During the inference time, we sample a subset of high score and farthest predictions and then apply non-maximum suppression to output the final 3d object detection results.

3.4 Training

We first train a single frame backbone which we use to extract the encoded point features for each frame. The proposed LSTM module processes the encoded point features x_t together with hidden and memory point features from previous frame (h_{t-1} and c_{t-1}) and outputs hidden and memory features h_t and c_t for the current frame. The object detection head takes h_t as input and outputs the 3d detected objects in frame t. The 3d box regression and classification losses are applied to the outputs in every frame. Our algorithm operates in the local coordinate frame which means the features from previous frames are transformed to the current frame to compensate for egomotion.

As described in [33], we adopt a hybrid of regression and classification losses for bounding box prediction. Each bounding box is represented by height, length, width, center location and a 3×3 rotation matrix. Instead of computing a separate loss for each of the parameters, we use an integrated box corner loss which can be back propagated and get all the individual attributes updated at once. We first calculate the 8 box corners in a differentiable way, then apply Huber loss on the distance between ground-truth and predicted boxes. The benefit of doing this is the ease of training as we do not have to tune multiple individual losses.

We use a dynamic classification loss as described in [33]. At each step, we classify the predictions that have more than 70% IOU with their corresponding ground-truth object as positive and the rest of the predictions as classified as negative. As the model gets better in box prediction, there will be more positive predicted boxes over time.

4   Experimental Results

We perform a series of experiments to evaluate the performance of our LSTM network in comparison to the alternative approaches. Furthermore, we study the effects of our design decisions through ablation studies.

Fig. 4. Example sequences in Waymo Open Dataset. Each frame is colored differently. A few fast moving cars and two walking pedestrians are shown over three frames. (Color figure online)

Dataset: We use the recently released Waymo Open Dataset [42] for our experiments. It contains 1000 sequences (798 training and 202 validation) captured in major US cities under diverse weather conditions and times of the day. Each sequence (Fig. 4) has approximately 200 frames at a frame rate of 100 ms. Each frame has multiple LiDARs and cameras with annotated 3D and 2D bounding box labels for vehicles, pedestrians, cyclists, signs. In our experiments, only LiDAR point clouds and 3D bounding box labels for vehicles with 5+ points are used for training and evaluation. The Waymo Open Dataset is a larger scale dataset in comparison to the previous self-driving car datasets such as Kitti dataset [15]. The 20 s sequence for each scene enables the training and evaluation of the temporal 3D object detection task on point clouds in challenging and realistic autonomous driving scenarios.

Table 1. 3D object detection results on Waymo Open dataset *validation* set. Unless noted otherwise, the models are using single frame. †:re-implemented by [34].

Model	mAP@0.7IoU
StarNet [34]	53.7
PointPillars† [25]	57.2
MVF [52]	62.9
U-Net	56.1
U-Net + Kalman Filter [44]	56.8
Concatenation (4 frames)	62.4
Ours (4 frames)	**63.6**

Experiment Details: We follow the metric used in almost all self-driving car datasets which is the mean average precision (mAP) metric for the 7 degree-of-freedom 3D boxes at intersection over union (IoU) threshold of 0.7 for vehicles.

For the object detection backbone, the encoder contains 6 blocks each with two 3D SparseConv layers, with output feature dimensions of 64, 96, 128, 160, 192, 224, 256. The decoder has the same structure in reversed order with skip connections from encoder layers.

We use a lightweight 3D sparse U-Net for LSTM that has one encoder block (of 128 dimensions), max pooling, one bottleneck block (of 128 dimensions), unpooling, and one decoder block (of 256 dimensions). Models are trained on 20 synced GPUs with a batch size of 2 (which means effectively a batch size of 40). We train the model with 0.1 initial learning rate. After 25k steps, we decay the learning rate every 7k steps by the factors of [0.3, 0.1, 0.01, 0.001, 0.0001]. We use a voxel size of [0.2 m, 0.2 m, 0.2 m]. The LSTM module uses the sub-sampled point cloud features that are computed by the backbone as well as hidden and memory features that it receives from the previous frame. Hidden feature points in previous steps are accumulated during the sequence. In practice, this only slightly increase the number of non-empty voxels at each step.

4.1 Object Detection Results

We show our results on Waymo Open Dataset in Table 1. Our first baseline (U-Net) is a single frame model built on our sparse 3D convolution U-Net backbone (without the LSTM) which achieves 56.1% mAP at IoU 0.7. Our second baseline combines the single frame detector with AB3DMOT [44], which deploys a

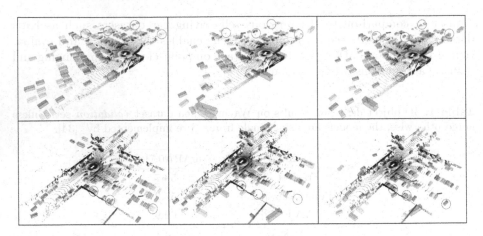

Fig. 5. We compare our sparse LSTM 3d object detection results with the one frame 3d detection baseline. Left: ground truth labels; Middle: single frame predictions; Right: LSTM predictions. Misaligned (arrows) and missing (circles) vehicles are highlighted.

combination of 3D Kalman Filter and Hungarian algorithm. Kalman filter is a classical way for tracking objects which we use to update measurements based on prior from previous frames[1]. We build this baseline by applying the Kalman filter on top of the single frame detector. Based on our experiments, this method achieves 0.7% gain in comparison to the single frame baseline.

Our third baseline feeds the concatenation of 4 frames into our U-Net backbone (same as the first baseline, but with 4 frames of input). We concatenate the points in the feature dimension after applying ego-motion transformation. Since points in different frames do not align, we use zero padding to offset the features. This is more flexible than Luo et al. [30]'s early fusion with 1D convolution, and more memory and compute efficient than their late fusion since the backbone runs only once. In comparison to the U-Net baseline, this gives rise to a 6.3% increase of mAP to 62.4%. Finally, in the last row we show our proposed LSTM model (4-frames) results with the best performance of 63.6% mAP@0.7.

We report the results of other single-frame detectors for comparison. StarNet [34] is a point-based detector based on sampling instead of learned proposals. It achieves 53.7% on the validation dataset. PointPillars [25] organizes point clouds into regular vertical columns and detects objects using 2D CNN. It achieves 57.2% mAP (re-implemented by [34]). MVF [52] has the state-of-the-art single frame results on the Waymo Open Dataset. However, their method is not directly comparable to ours since they perform significant data augmentation. Regardless, our focus is on how an LSTM can be used to improve the results of any method, which is largely orthogonal to any particular choice of single-frame baseline. The results demonstrate its effectiveness (7.5% improvement in mAP@0.7 over the single frame model with the same U-Net backbone).

---

[1] Using code released by [44].

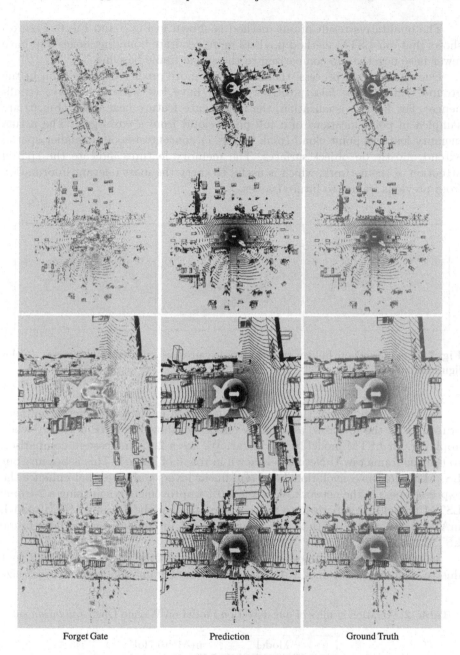

| Forget Gate | Prediction | Ground Truth |

Fig. 6. Visualization of the forget gate of our proposed LSTM module, prediction of object detection and the ground truth label. The gate is visualized as a heatmap, where a high value (red) intuitively means *not* forgetting the hidden features at the location. We included ground truth boxes in this figure for clarity. The points within vehicles are mostly in high value, while the buildings and other objects are mostly in blue. The color in the prediction represents semantic classes of vehicles (red), background (black). (Color figure online)

The qualitative result of our method is shown in Fig. 5 and Fig. 6. Figure 5 shows that our LSTM method predicts more accurate bounding boxes and have fewer false negatives in comparison to the single frame baseline.

Due to the NMS process, there are often more predicted boxes than in the ground truth. These false positives usually have low semantics scores (confidence). For better visualization, the forget gate feature heat maps (Fig. 6) are sampled in point locations of a full point cloud from a voxel grid. The actual memory features point cloud (pink in Fig. 7) concentrates on a smaller spatial extend, mostly on object surfaces. The memory features indicate the spatial attention of the network, which is useful to carry the most relevant information from previous frames to future frames.

Fig. 7. Locations where hidden and memory features are selected (pink points). (Color figure online)

In Table 2, we present the results of our LSTM model with different number of frames. In the first row, as a sanity check, we show the mAP accuracy when applying our LSTM model to one frame. We see a 2.6% increase in comparison to the one frame raw U-Net model shown in row 3 of Table 1. This is because our LSTM model has convolutional and non-linear layers and gates that enhance the expressiveness of the network. We see a 1.0% improvement when using a 2-frame LSTM model in comparison to the 1-frame one. For the 4-frame LSTM model, mAP reaches 63.6%, with a 4.9% improvement in comparison to the 1-frame LSTM model.

In order to take 7 frames as input, we decrease the batch size from 2 to 1 due to memory constraints. Compared with the 4 frame model with batch size

Table 2. Ablation studies of our detection model on Waymo Open *validation* set

Model	mAP@0.7IoU
1 frame	58.7
2 frames	59.7
4 frames	**63.6**
4 frames (batch size 1)	62.3
7 frames (batch size 1)	63.3

1, the performance increases by 1.0%. Overall, the ablation study shows that LSTM with hidden and memory features over longer sequences results in higher 3d object detection accuracy.

5 Discussion

We have proposed an LSTM approach for detecting 3D objects in a sequence of LiDAR point cloud observations. Our method leverages memory and hidden state features associated with 3D points from previous object detections, which are transformed according to vehicle egomotion at each timestep. The backbone for our LSTM is a sparse 3D convolution network that co-voxelizes the input point cloud, memory, and hidden state at each frame. Experiments on Waymo Open Dataset demonstrate that our algorithm achieves the state-of-the-art results and outperforms a single frame baseline by 7.5%, a multi-frame object detection baseline by 1.2%, and a multi-frame object tracking baseline by 6.8%. In the future, we would like to also predict scene flow and use it to better transform the memory and hidden states in our LSTM, and we would like to study how our LSTM can be used to improve a variety of other single-frame object detectors.

Memory Efficiency: Our proposed model is more memory efficient in comparison to previous temporal models that concatenate the point clouds from multiple frames [25,30]. A method that concatenates M frames needs to apply the 3d network at each frame to M times more number of points, while our 3d network is only applied to the points in the current frame plus a small set of features coming from last frame. Please note that our sub-sampled hidden and memory feature points (we sample 30k points in each frame out of 180k LiDAR points) are lightweight in comparison to passing full size point features from previous frames.

Computation Efficiency: In comparison to the single frame model, the LSTM module adds a very small computational overhead. LSTM runs in a stream and is able to reuse the intermediate tensors that are computed in the previous time steps. The only additional overhead to per-frame computation is 3 sparse conv blocks which is a small fraction (10% of the parameters) of the single frame network that uses 15 sparse conv blocks. Note that our single frame feature extractor runs in 19 ms on a Titan V GPU. Given that the lidar input arrives at 10 hz, our network is still able to run in real-time within its 100 ms computation budget. Therefore, it adds only a small overhead while it gains 7.5% in comparison to our single frame method. In comparison to the 4-frame concatenation baseline, the LSTM approach is more efficient. Concatenation reduces the sparsity and results in feeding a denser set of voxels to the network. We show that not only LSTM method is more efficient but also it achieves 1.2% better result.

References

1. Behl, A., Hosseini Jafari, O., Karthik Mustikovela, S., Abu Alhaija, H., Rother, C., Geiger, A.: Bounding boxes, segmentations and object coordinates: how important

is recognition for 3D scene flow estimation in autonomous driving scenarios? In: Proceedings of the IEEE International Conference on Computer Vision, pp. 2574–2583 (2017)

2. Behl, A., Paschalidou, D., Donné, S., Geiger, A.: PointFlowNet: learning representations for rigid motion estimation from point clouds. In: Proceedings of the IEEE Conference on Computer Vision and Pattern Recognition, pp. 7962–7971 (2019)

3. Caesar, H., et al.: nuScenes: a multimodal dataset for autonomous driving. arXiv preprint arXiv:1903.11027 (2019)

4. Casas, S., Luo, W., Urtasun, R.: IntentNet: learning to predict intention from raw sensor data. In: Conference on Robot Learning, pp. 947–956 (2018)

5. Chang, M.F., et al.: Argoverse: 3D tracking and forecasting with rich maps. In: Proceedings of the IEEE Conference on Computer Vision and Pattern Recognition, pp. 8748–8757 (2019)

6. Chen, X., Ma, H., Wan, J., Li, B., Xia, T.: Multi-view 3D object detection network for autonomous driving. In: Proceedings of the IEEE Conference on Computer Vision and Pattern Recognition, pp. 1907–1915 (2017)

7. Chen, X., Yu, J., Wu, Z.: Temporally identity-aware SSD with attentional LSTM. IEEE Trans. Cybern. 50(6), 2674–2686 (2019)

8. Chiu, H.k., Prioletti, A., Li, J., Bohg, J.: Probabilistic 3D multi-object tracking for autonomous driving. arXiv preprint arXiv:2001.05673 (2020)

9. Choy, C., Gwak, J., Savarese, S.: 4D spatio-temporal convnets: Minkowski convolutional neural networks. In: Proceedings of the IEEE Conference on Computer Vision and Pattern Recognition, pp. 3075–3084 (2019)

10. Dosovitskiy, A., et al.: FlowNet: learning optical flow with convolutional networks. In: Proceedings of the IEEE International Conference on Computer Vision, pp. 2758–2766 (2015)

11. El Sallab, A., Sobh, I., Zidan, M., Zahran, M., Abdelkarim, S.: YOLO4D: a spatio-temporal approach for real-time multi-object detection and classification from LiDAR point clouds. In: NIPS 2018 Workshop MLITS (2018)

12. Engelcke, M., Rao, D., Wang, D.Z., Tong, C.H., Posner, I.: Vote3Deep: fast object detection in 3D point clouds using efficient convolutional neural networks. In: 2017 IEEE International Conference on Robotics and Automation (ICRA), pp. 1355–1361. IEEE (2017)

13. Fan, H., Yang, Y.: PointRNN: point recurrent neural network for moving point cloud processing. arXiv preprint arXiv:1910.08287 (2019)

14. Feng, Y., Ma, L., Liu, W., Luo, J.: Spatio-temporal video re-localization by warp LSTM. In: Proceedings of the IEEE Conference on Computer Vision and Pattern Recognition, pp. 1288–1297 (2019)

15. Geiger, A., Lenz, P., Urtasun, R.: Are we ready for autonomous driving? the KITTI vision benchmark suite. In: 2012 IEEE Conference on Computer Vision and Pattern Recognition, pp. 3354–3361. IEEE (2012)

16. Graham, B.: Sparse 3D convolutional neural networks. In: Xie, X., Jones, M.W.G.K.L.T. (eds.) Proceedings of the British Machine Vision Conference (BMVC), pp. 150.1–150.9. BMVA Press, September 2015. https://doi.org/10.5244/C.29.150

17. Graham, B., Engelcke, M., van der Maaten, L.: 3D semantic segmentation with submanifold sparse convolutional networks. In: Proceedings of the IEEE Conference on Computer Vision and Pattern Recognition, pp. 9224–9232 (2018)

18. Graham, B., van der Maaten, L.: Submanifold sparse convolutional networks. arXiv preprint arXiv:1706.01307 (2017)

19. Groueix, T., Fisher, M., Kim, V.G., Russell, B.C., Aubry, M.: AtlasNet: a papier-mâché approach to learning 3D surface generation. arXiv preprint arXiv:1802.05384 (2018)
20. Hochreiter, S., Schmidhuber, J.: Long short-term memory. Neural Comput. 9(8), 1735–1780 (1997)
21. Hu, P., Ziglar, J., Held, D., Ramanan, D.: What you see is what you get: exploiting visibility for 3D object detection. In: Proceedings of the IEEE/CVF Conference on Computer Vision and Pattern Recognition, pp. 11001–11009 (2020)
22. Huang, L., Yan, P., Li, G., Wang, Q., Lin, L.: Attention embedded spatio-temporal network for video salient object detection. IEEE Access 7, 166203–166213 (2019)
23. Kanazawa, A., Zhang, J.Y., Felsen, P., Malik, J.: Learning 3D human dynamics from video. In: Computer Vision and Pattern Regognition (CVPR) (2019)
24. Kang, K., et al.: Object detection in videos with tubelet proposal networks. In: Proceedings of the IEEE Conference on Computer Vision and Pattern Recognition, pp. 727–735 (2017)
25. Lang, A.H., Vora, S., Caesar, H., Zhou, L., Yang, J., Beijbom, O.: PointPillars: fast encoders for object detection from point clouds. In: Proceedings of the IEEE Conference on Computer Vision and Pattern Recognition, pp. 12697–12705 (2019)
26. Li, B., Zhang, T., Xia, T.: Vehicle detection from 3d lidar using fully convolutional network. arXiv preprint arXiv:1608.07916 (2016)
27. Liang, M., Yang, B., Wang, S., Urtasun, R.: Deep continuous fusion for multi-sensor 3D object detection. In: Proceedings of the European Conference on Computer Vision (ECCV), pp. 641–656 (2018)
28. Liu, X., Qi, C.R., Guibas, L.J.: FlowNet3D: learning scene flow in 3D point clouds. In: Proceedings of the IEEE Conference on Computer Vision and Pattern Recognition, pp. 529–537 (2019)
29. Liu, X., Yan, M., Bohg, J.: MeteorNet: deep learning on dynamic 3D point cloud sequences. In: Proceedings of the IEEE International Conference on Computer Vision, pp. 9246–9255 (2019)
30. Luo, W., Yang, B., Urtasun, R.: Fast and furious: real time end-to-end 3D detection, tracking and motion forecasting with a single convolutional net. In: Proceedings of the IEEE conference on Computer Vision and Pattern Recognition, pp. 3569–3577 (2018)
31. McCrae, S., Zakhor, A.: 3D object detection using temporal lidar data. In: IEEE International Conference on Image Processing (ICIP) (2020)
32. Meyer, G.P., Laddha, A., Kee, E., Vallespi-Gonzalez, C., Wellington, C.K.: LaserNet: an efficient probabilistic 3D object detector for autonomous driving. In: Proceedings of the IEEE Conference on Computer Vision and Pattern Recognition, pp. 12677–12686 (2019)
33. Najibi, M., et al.: DOPS: learning to detect 3D objects and predict their 3D shapes. In: Proceedings of the IEEE Conference on Computer Vision and Pattern Recognition (2020)
34. Ngiam, J., et al.: StarNet: targeted computation for object detection in point clouds. arXiv preprint arXiv:1908.11069 (2019)
35. Qi, C.R., Litany, O., He, K., Guibas, L.J.: Deep Hough voting for 3D object detection in point clouds. In: Proceedings of the IEEE International Conference on Computer Vision, pp. 9277–9286 (2019)
36. Qi, C.R., Liu, W., Wu, C., Su, H., Guibas, L.J.: Frustum pointnets for 3D object detection from RGB-D data. In: Proceedings of the IEEE Conference on Computer Vision and Pattern Recognition, pp. 918–927 (2018)

37. Qi, C.R., Yi, L., Su, H., Guibas, L.J.: PointNet++: deep hierarchical feature learning on point sets in a metric space. In: Advances in Neural Information Processing Systems, pp. 5099–5108 (2017)
38. Riegler, G., Osman Ulusoy, A., Geiger, A.: OctNet: Learning deep 3D representations at high resolutions. In: Proceedings of the IEEE Conference on Computer Vision and Pattern Recognition, pp. 3577–3586 (2017)
39. Shi, S., Wang, X., Li, H.: PointRCNN: 3D object proposal generation and detection from point cloud. In: Proceedings of the IEEE Conference on Computer Vision and Pattern Recognition, pp. 770–779 (2019)
40. Simon, M., et al.: Complexer-YOLO: real-time 3D object detection and tracking on semantic point clouds. In: Proceedings of the IEEE Conference on Computer Vision and Pattern Recognition Workshops (2019)
41. Simon, M., Milz, S., Amende, K., Gross, H.-M.: Complex-YOLO: an Euler-region-proposal for real-time 3D object detection on point clouds. In: Leal-Taixé, L., Roth, S. (eds.) ECCV 2018. LNCS, vol. 11129, pp. 197–209. Springer, Cham (2019). https://doi.org/10.1007/978-3-030-11009-3_11
42. Sun, P., et al.: Scalability in perception for autonomous driving: Waymo open dataset. arXiv pp. arXiv-1912 (2019)
43. Teng, E., Falcão, J.D., Huang, R., Iannucci, B.: ClickBAIT: click-based accelerated incremental training of convolutional neural networks. In: 2018 IEEE Applied Imagery Pattern Recognition Workshop (AIPR), pp. 1–12. IEEE (2018)
44. Weng, X., Kitani, K.: A Baseline for 3D Multi-Object Tracking. arXiv:1907.03961 (2019)
45. Xiao, F., Jae Lee, Y.: Video object detection with an aligned spatial-temporal memory. In: Proceedings of the European Conference on Computer Vision (ECCV), pp. 485–501 (2018)
46. Xu, Z., et al.: Unsupervised discovery of parts, structure, and dynamics. arXiv preprint arXiv:1903.05136 (2019)
47. Yang, B., Luo, W., Urtasun, R.: PIXOR: real-time 3D object detection from point clouds. In: Proceedings of the IEEE Conference on Computer Vision and Pattern Recognition, pp. 7652–7660 (2018)
48. Yang, Y., Feng, C., Shen, Y., Tian, D.: FoldingNet: point cloud auto-encoder via deep grid deformation. In: Proceedings of the IEEE Conference on Computer Vision and Pattern Recognition, pp. 206–215 (2018)
49. Yin, J., Shen, J., Guan, C., Zhou, D., Yang, R.: LiDAR-based online 3D video object detection with graph-based message passing and spatiotemporal transformer attention. In: Proceedings of the IEEE/CVF Conference on Computer Vision and Pattern Recognition (CVPR), June 2020
50. Zhao, H., Jiang, L., Fu, C.W., Jia, J.: PointWeb: enhancing local neighborhood features for point cloud processing. In: Proceedings of the IEEE Conference on Computer Vision and Pattern Recognition, pp. 5565–5573 (2019)
51. Zhao, Y., Birdal, T., Deng, H., Tombari, F.: 3D point capsule networks. In: Proceedings of the IEEE Conference on Computer Vision and Pattern Recognition, pp. 1009–1018 (2019)
52. Zhou, Y., et al.: End-to-end multi-view fusion for 3D object detection in LiDAR point clouds. arXiv preprint arXiv:1910.06528 (2019)
53. Zhou, Y., Tuzel, O.: VoxelNet: end-to-end learning for point cloud based 3D object detection. In: Proceedings of the IEEE Conference on Computer Vision and Pattern Recognition, pp. 4490–4499 (2018)

Monotonicity Prior for Cloud Tomography

Tamar Loeub[1][(✉)], Aviad Levis[2][(✉)], Vadim Holodovsky[1][(✉)],
and Yoav Y. Schechner[1][(✉)]

[1] Viterbi Faculty of Electrical Engineering, Technion - Israel Institute of Technology,
Haifa, Israel
tamarloeub@gmail.com, vholod@ef.technion.ac.il, yoav@ee.technion.ac.il
[2] Computing and Mathematical Sciences Department,
California Institute of Technology, Pasadena, CA, USA
aviad.levis@gmail.com

Abstract. We introduce a differentiable monotonicity prior, useful to
express signals of monotonic tendency. An important natural signal
of this tendency is the optical extinction coefficient, as a function of
altitude in a cloud. Cloud droplets become larger as vapor condenses
on them in an updraft. Reconstruction of the volumetric structure of
clouds is important for climate research. Data for such reconstruction is
multi-view images of each cloud taken simultaneously. This acquisition
mode is expected by upcoming future spaceborne imagers. We achieve
three-dimensional volumetric reconstruction through stochastic scattering tomography, which is based on optimization of a cost function. Part of
the cost is the monotonicity prior, which helps to improve the reconstruction quality. The stochastic tomography is based on Monte-Carlo (MC)
radiative transfer. It is formulated and implemented in a coarse-to-fine
form, making it scalable to large fields.

Keywords: Scattering · Regularization · Physics-based vision

1 Introduction

Clouds have a key role of the climate system. They are part of the water cycle
(the source of freshwater supply), and account for $\approx 2/3$ of the Earth's albedo.
Inaccurate models of cloud properties yield the largest errors in climate predictions. To put matters in perspective, an error of $\approx 1\%$ in cloud properties yields
errors in current climate predictions, that are comparable to climatic effects of
man-made greenhouse gases. To reduce climate prediction uncertainties, it is
required to model clouds far better than the state-of-the-art. Models, naturally,
need observations to set parameters and statistics. Observations require both relevant raw data and proper ways to extract information from this data, regarding
cloud structure.

Electronic supplementary material The online version of this chapter (https://
doi.org/10.1007/978-3-030-58523-5_17) contains supplementary material, which is
available to authorized users.

© Springer Nature Switzerland AG 2020
A. Vedaldi et al. (Eds.): ECCV 2020, LNCS 12363, pp. 283–299, 2020.
https://doi.org/10.1007/978-3-030-58523-5_17

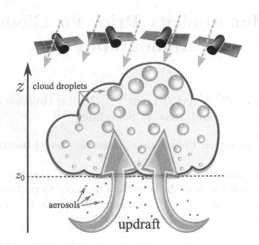

Fig. 1. As air containing water vapor ascends, humidity starts condensing at the condensation level z_0, forming a cloud base. Condensation is initiated by aerosols (condensation nuclei), which are much smaller than water droplets. In the cloud, condensation increases the size (radius) of the droplets, with altitude. The clouds are irradiated by the sun and observed by multiple cameras. From the images, 3D volumetric information is sought.

However, current observations cannot meet this need. Cloud observations on a planetary scale rely on 2D satellite images to provide raw data. Then, analysis of the data assumes the atmosphere is layered. In other words, the prevailing assumption underlying analysis and modeling is that clouds are homogeneous horizontally for infinite extents, having only 1D vertical variations. This is termed the *plane parallel* model. Analysis is based on 1D radiative transfer (RT) across layers. However, clouds are naturally 3D volumetric objects having spatial heterogeneity, and light propagates in them according to 3D-RT. This discrepancy yields significant errors and uncertainties the smaller the clouds are. While such clouds are small, they are numerous.

For these reasons, it is important to derive the volumetric structure of small clouds. This requires devising 3D computed tomography (CT) based on multiple scattering [1, 8, 15, 16, 19, 25, 31–33] and passive imaging. Scattering-based CT stands contrary to traditional CT, which relies on linear image formation models [3, 17, 18, 40, 57, 59] and treats scattering as a disturbance. It has recently been shown [14] that scattering-based bio-medical CT can have advantages relative to linear X-ray CT, in terms of radiation dose, chemical analysis and elimination of moving parts.

This paper advances scattering-based tomography of small clouds. To enable imaging over wide scales, the assumed imaging setup is a formation of orbiting satellites, designed to image cloud fields simultaneously from multiple directions (Fig. 1). This is in accordance to a space mission planned to provide raw image data dedicated for tomographic analysis [54]. CT analysis relies on optimized fitting of an image formation model to data, while complying with priors. In this paper, we introduce and use a differentiable monotonicity prior. It is helpful for expressing the nature of convective clouds. Consequently, the prior helps in lowering the recovery error and reaching faster convergence.

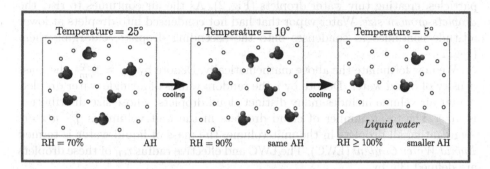

Fig. 2. An illustration of an isolated air parcel, demonstrating cloud-droplet condensation, as described in Sect. 2. The absolute humidity is AH, and the relative humidity is RH. The right-most stage is at saturation or super-saturation

Moreover, our algorithm is scalable to large fields. It is a stochastic approach, based on MC RT, where accuracy is a matter of time to simulate transport of photons. We develop the recovery in a coarse-to-fine (hierarchical) approach, to speed convergence and handling of large fields. This is contrary to prior atmospheric scattering CT, which is based on discrete ordinates [31,32]. The latter uses high quality RT which is memory-limited: the domain size is severely limited by memory resources, and not designed for grid coarsening.

Proper real raw spaceborne data is expected to become available in a few years [54]. Therefore, we test the methods using simulations. The simulations emulate realistic photon (Poisson) noise, by using realistic photon counts per pixel in the raw images.

2   Droplets in Convective Clouds

To form priors on an object, it is helpful to know its nature. We put to use a cloud trend. First, we note three terms: saturation, absolute humidity and relative humidity. *Absolute humidity* quantifies the density of water vapor in air ($[gr/m^3]$). *Saturation* is a state at which vapor is at equilibrium with liquid water. *Relative humidity* is the ratio of absolute humidity of an air parcel, to the parcel's absolute humidity at saturation. By definition at saturation, the relative humidity is 100%. For an unsaturated *isolated air parcel*, the absolute humidity is fixed, but relative humidity increases as temperature decreases.

Convective clouds require three ingredients: humidity (water vapor), ascending air current (updraft) and condensation nuclei (aerosols). The ascending air lifts water vapor and aerosols. Ascent leads to cooling of air with altitude z. A lower temperature increases the relative humidity. At some altitude, cooling yields a critical temperature at which relative humidity reaches (or surpasses) 100%. This altitude is termed *condensation level*, or *cloud-base*, z_0. Above this level, the vapor reaches over-saturation and thus starts to condense on the aerosol particles, creating tiny water droplets (Fig. 2). As the air continues to rise, the *droplets grow in size*. Water vapor that had not condensed into droplets at lower altitude continues to condenses onto the ascending droplets, enlarging them (Fig. 1).

We now formulate the above mathematically. Denote by $\rho_w \approx 1 \frac{gr}{cm^3}$ the mass density of liquid water. Denote by r the random radius of a single water droplet. A unit air volume includes many distinct cloud droplets, whose size distribution is $n(r)$. The total number of cloud droplets in the unit volume is $\int_0^\infty n(r)dr$. Aggregating all droplets in the unit volume, the mass of liquid water is termed *Liquid Water Content* (LWC). The LWC and effective radius r_{eff} of these droplets are defined [35] by

$$\text{LWC} = \frac{4}{3}\rho_w \int_0^\infty \pi r^3 n(r)dr, \quad r_{eff} = \frac{\int_0^\infty \pi r^3 n(r)dr}{\int_0^\infty \pi r^2 n(r)dr} . \tag{1}$$

Intuitively, r_{eff} is a weighted sum of the droplets radiuses, according to the geometric cross section of each droplet. In nature, both r_{eff} and LWC tend to *increases with altitude* inside a cloud [34,39,60]. A model for this at $z > z_0$ is

$$r_{eff} \propto (z - z_0)^{\frac{1}{3}}, \qquad \text{LWC} \propto (z - z_0). \tag{2}$$

The relations in (2), however, are approximate. They assume, as mentioned in the beginning of this section, that air parcels are *isolated*, exchanging neither material nor heat with surrounding parcels. This assumption is common in atmospheric models, and is termed the *adiabatic* approximation. Real clouds experience mixing with surrounding air parcels, evaporation and in some cases, precipitation. Therefore, we do not impose Eq. (2) as a constraint, but it leads us to regularization that tolerates deviation (Sect. 4).

3   Radiative Transfer and Memory Loads

Many methods in remote sensing of the atmosphere assume a *plane parallel medium* [23,36,41], in which the atmosphere is assumed to be layered. There, radiance varies essentially only vertically [10,27], i.e. the model is 1D-RT. However, we seek recovery of clouds as they are: 3D volumetric and heterogeneous. This is consistent with recent work in computer vision [2,19,31]. In a 3D heterogeneous medium, the forward model is 3D-RT. Attempting 3D-RT is computationally complex. Compounding complexity is pursuit of a 3D *inverse problem*: recovering an unknown 3D scattering medium. Solving such an inverse problem requires repeated, iterative calculations: 3D-RT calculations are to be repeated over a hypothetical medium, which varies as iterations progress.

One approach for 3D-RT forward model calculations is deterministic, epitomized by the Spherical Harmonics Discrete Ordinate Method (SHDOM) [11]. As in analysis of reflection in computer vision [4,50], spherical harmonics help in efficient analysis of angular scattering. SHDOM is fast and accurate for small domain sizes, when many radiometric quantities (i.e. pixels for image synthesis) are sought. However, its memory use is very expensive, as we explain. SHDOM uses a discrete representation of angles. Let N_{voxels} be the number of voxels and N_{angles} be the number of angles. Typically, $N_{\text{angles}} \approx 500$. To reach accuracy, SHDOM [11] relies on an adaptive subdivision of voxels and a finite number of spherical harmonics. This leads memory requirements to inflate by a factor C. In a typical cloudy scene, $C \approx 75$. Hence, memory load is $\approx C N_{\text{angles}} N_{\text{voxels}} \approx 4 \cdot 10^4 \cdot N_{\text{voxels}}$. This expense inhibits scalability to large domains. Moreover, subdivision is enforced, because SHDOM is based on assumptions of cloud voxels having high internal uniformity and small optical depth. Consequently, SHDOM is not amenable to hierarchical coarse-to-fine analysis.

Contrary to SHDOM, stochastic forward model RT done by MC [38,58] is not prone to these problems. In MC, simulated photons independently roam the domain and interact randomly (Sect. 5). Angles are not discretized in memory structure per voxel. Hence, memory consumption is $\approx N_{\text{voxels}}$. MC is highly *parallelizable* due to photon independence [48]. Accuracy is controlled (as is runtime) by the number of photons simulated. The variance of the RT result decreases linearly with the number of photons sampled, and does not require adaptive spatial subdivision. It is, therefore, scalable to large domains. Using MC as the forward model along with our hierarchical coarse-to-fine approach, enables solving large scale inverse scattering problems.

4 Inverse Problem

We seek to recover the volumetric optical parameters of a scattering medium, based on two dimensional images, i.e. perform scatter-based tomography. A scattering medium domain is $\mathcal{M} \subset \mathbb{R}^3$. Consider an infinitesimal volume element, voxel v, around location $\boldsymbol{x} \in \mathcal{M}$. The extinction coefficient of the medium at a point $\boldsymbol{x}$ is denoted by $\beta(\boldsymbol{x})$. When multiple particle types exist in a medium, each has its own extinction coefficient. For example, if a volume has air molecules and cloud droplets, then their respective extinction coefficients are $\beta^{\text{a}}(\boldsymbol{x})$ and $\beta^{\text{c}}(\boldsymbol{x})$. The total extinction coefficient at $\boldsymbol{x}$ is

$$\beta(\boldsymbol{x}) = \beta^{\text{a}}(\boldsymbol{x}) + \beta^{\text{c}}(\boldsymbol{x}). \tag{3}$$

The field $\beta^{\text{a}}(\boldsymbol{x})$ is known from auxiliary measurements of the atmosphere, e.g. through radiosondes (weather balloons). Therefore, the computer vision task is to recover the unknown cloud extinction β_v^{c} for each voxel $v \in [1, \dots, N_{\text{voxels}}]$. The vector of unknowns is $\boldsymbol{\beta} = \left[\beta_1^{\text{c}}, \beta_2^{\text{c}}, \dots, \beta_{N_{\text{voxels}}}^{\text{c}} \right]$.

Images are obtained from several viewpoints, in different locations and angles. The radiance measurements are indexed by $d \in [1, \dots, N_{\text{detectors}}]$. They are concatenated into a data vector i^{measured}. We fit to i^{measured} a corresponding model image set, concatenated to a vector, $i(\beta)$. Data fitting is quantified by the difference between measurements i^{measured} and the corresponding model image set $i(\beta)$, using

$$F(\beta) = \frac{1}{2}\|i(\beta) - i^{\text{measured}}\|_2^2 . \tag{4}$$

The field β is estimated by optimization

$$\hat{\beta} = \arg\min_{\beta \geq 0} [F(\beta) + R(\beta)] . \tag{5}$$

Here $R(\beta)$ is a regularization term, which expresses priors on the medium. The optimization problem in Eq. (5) is solved using stochastic gradient descent (SGD). Sect. 5 describes the image formation model i^{measured}, its implications to $F(\beta)$ and the optimization problem in Eq. (5). We now discuss the regularization term.

Regularization

Priors are useful to better constrain results to more physically plausible solutions and aid convergence and quality performance. Recall Sect. 2. For a given droplet size distribution $n(r)$, the extinction coefficient [35] by cloud droplets is

$$\beta^c = \int_0^\infty Q_{\text{eff}}(r)\pi r^2 n(r)dr. \tag{6}$$

Here $Q_{\text{eff}}(r)$ is an extinction efficiency factor. For cloud droplets[1], $Q_{\text{eff}}(r) \approx 2$ thus insensitive to r. Combining Eqs. (1, 6),

$$\beta^c(x) = \frac{3}{2\rho_w} \frac{\text{LWC}(x)}{r_{\text{eff}}(x)}. \tag{7}$$

Then, considering Eqs. (2, 7), for $z > z_0$,

$$\beta^c(x) \propto (z - z_0)^{\frac{2}{3}}. \tag{8}$$

As written in Sect. 2, Eq. (2) is an approximation, thus we use Eq. (8) to form a loose prior: $\beta^c(x)$ tends to *monotonically increase* with z, within a cloud. The trend of vertical increase in $\beta^c(x)$ only *applies* to voxels in which a cloud resides: outside cloud voxels, $\beta^c(x) \equiv 0$. Hence the monotonicity preference is nulled outside a cloud.

For a possible expression for such a prior, define $\mathbb{1}\{\text{statement}\}$ to be a binary column vector. Each element corresponds to a voxel v and satisfies

$$\mathbb{1}\{\text{statement}\}_v = \begin{cases} 1 & \text{statement is true in } v \\ 0 & \text{otherwise} \end{cases}. \tag{9}$$

---

[1] The factor $Q_{\text{eff}}(r)$ depends on the wavelength λ. For clouds, typically $r \gg \lambda$. Then, it is reasonable to neglect the dependency of Q_{eff} on r and λ [35].

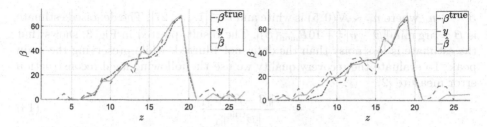

Fig. 3. A 1D example illustrating the denoising capability of the monotonicity prior R_{mon}, when applied to a non-negative signal whose origin tends to be monotonous or null. [Left] The input y noise is $\sim \mathcal{N}(0,5)$. Using Eq. (13), the error in $\hat{\beta}$ is $\epsilon = 15\%$. [Right] The input y noise is $\sim \mathcal{N}(0,10)$. The error in $\hat{\beta}$ is $\epsilon = 20\%$.

Further, let $\mathbf{D}_z$ be a matrix expressing discrete vertical differentiation, in lieu of continuous differentiation $\partial \beta^c(\boldsymbol{x})/\partial z$. Then, consider this term:

$$R_{\text{mon}} = -\left[\mathbb{1}\left\{\mathbf{D}_z\boldsymbol{\beta}^c > 0\right\}\right]^{\mathsf{T}} \mathbb{1}\left\{\boldsymbol{\beta}^c > 0\right\}, \tag{10}$$

where T denotes transposition.

Consider voxels where $\boldsymbol{\beta}^c$ increases vertically. These voxels decrease R_{mon}. This lowers the optimization cost (5), if R_{mon} is used for regularization. Voxels which are not in a cloud or have vertical decrease of the extinction coefficient do not contribute to decreasing of R_{mon}, meaning essentially a higher cost.

Equation (10) is not differentiable with respect to $\boldsymbol{\beta}^c$. To enable practical optimization based on SGD, we use a softer version. For an arbitrary vector $\mathbf{q}$, let $\tanh(\mathbf{q})$ be a diagonal matrix, where the diagonal is defined by the hyperbolic-tangent of each element of $\mathbf{q}$. Let $\mathbf{1}$ be a column vector, of all whose elements are ones. We use

$$R_{\text{mon}}(\boldsymbol{\beta}^c) = -\mathbf{1}^{\mathsf{T}} \tanh\left(c_1\mathbf{D}_z\boldsymbol{\beta}^c\right)\tanh\left(c_2\boldsymbol{\beta}^c\right)\mathbb{1}\left\{\boldsymbol{\beta}^c > 0\right\}. \tag{11}$$

The constants $c_1, c_2 \geq 0$ control the slope of the hyperbolic-tangent functions.

In addition, we let the field $\boldsymbol{\beta}^c$ be spatially smooth. This is expressed by penalizing the energy of the Laplacian-filtered field $\boldsymbol{\beta}^c$. The 3D Laplacian operator is represented by matrix $\mathbf{A}$. Overall, the regularization term is

$$R(\boldsymbol{\beta}^c) = \frac{1}{2}\mu_1\|\mathbf{A}\boldsymbol{\beta}^c\|_2^2 + \mu_2 R_{\text{mon}}(\boldsymbol{\beta}^c), \tag{12}$$

where $\mu_1, \mu_2 \geq 0$ control the weight of the regularization components. For the gradient of the regularization term, see the supplementary material.

Example of Simple Use of the Monotonicity prior. We illustrate here the effect of the prior on a 1D denoising problem, unrelated to scattering tomography. The original signal $\beta^{\text{true}}(z) \in [0, 70]$, is a representative vertical profile of a cloud, containing 27 samples. The signal $\beta^{\text{true}}(z)$ monotonically increases within part of the domain. The monotonic increase ends with an abrupt fall. The signal has null value in other parts of the domain, as seen in Fig. 3. The noisy input is $y_v =$

$\beta_v^{\text{true}} + n_v$ where $n_v \sim \mathcal{N}(0,5)$ is white and $v \in [1, \ldots, 27]$. The denoised estimate is $\hat{\beta} = \arg\min[\|\boldsymbol{\beta} - \boldsymbol{y}\|_2^2 + 10R_{\text{mon}}(\boldsymbol{\beta})]$. The result, plotted in Fig. 3, shows that the estimate is less noisy than the data, not blurred, while preserving the main peak. To evaluate the recovery quality we use the following local reconstruction error measure [2]

$$\epsilon = \frac{\|\hat{\boldsymbol{\beta}} - \boldsymbol{\beta}^{\text{true}}\|_1}{\|\boldsymbol{\beta}^{\text{true}}\|_1}. \tag{13}$$

The mean error over 1,000 realizations of $\boldsymbol{y}$ is $\epsilon = 16\%$, down from 18% without the prior. For a higher noise level, $n_v \sim \mathcal{N}(0,10)$, $\epsilon = 32\%$ down from 37%.

5   The Forward Model

The image formation model $\boldsymbol{i}(\boldsymbol{\beta})$ is known in the literature [9,13–15,19,38,48,49, 52]. We provide it to make the paper self-contained and provide the background for deriving the data-fitting gradient $\partial F(\boldsymbol{\beta})/\partial\boldsymbol{\beta}$. Terms relating to volumetric scattering are often used in dehazing or defogging computer vision literature [5, 22,28,42–46,53,55,56].

In an infinitesimal volume, the interaction can be scattering or absorption. Relative to the incoming radiance, the energy scattered in an infinitesimal volume is set by the coefficient $\varpi\beta(\boldsymbol{x})$. Here $\varpi \in [0,1]$, the *single-scattering albedo*, is the ratio of the scattering coefficient to the total extinction coefficient $\beta(\boldsymbol{x})$. The single-scattering albedo of air molecules and cloud droplets are denote by ϖ^{a}, and ϖ^{c} respectively.

Denote direction by $\boldsymbol{\omega}, \boldsymbol{\omega}' \in \mathbb{S}^2$, where $\mathbb{S}^2$ is the unit sphere. Scattered light is distributed angularly relative to the radiation incident on point $\boldsymbol{x}$, from direction $\boldsymbol{\omega}$. When scattering particles are oriented randomly and uniformly in a voxel, the probability density for scattering to direction $\boldsymbol{\omega}$ is expressed by a *phase function* $f(\theta)$, where $\theta = \arccos(\boldsymbol{\omega} \cdot \boldsymbol{\omega}')$ is the scattering angle. The phase functions of air molecules and cloud droplets are denoted by $f^{\text{a}}(\theta)$, and $f^{\text{c}}(\theta)$ respectively.

To describe the forward model by MC RT, we draw some definitions and notations from [14,19]. MC RT aggregates random photon paths. Denote by L_d the set of all possible photon paths via $\mathcal{M}$, connecting a source to a detector d. The source in our case-study is the sun, located at a distant location $\boldsymbol{x}_\odot$. A specific photon path, denoted by $\mathcal{L} = (\boldsymbol{x}_0, \boldsymbol{x}_1, \ldots, \boldsymbol{x}_B)$ is a sequence of B interaction points. One way for photons to contribute a signal to a detector d is for a photon to perform a random walk (via $\mathcal{L}$) from the source to a terminal location $\boldsymbol{x}_B$ at a camera aperture. There, the photon deposits a terminal intensity $I_B(\mathcal{L})$ at a pixel d. A clear signal at detector d is the expectation of a photon to reach d, multiplied by the number of photons (N_{phot}) generated at the source,

$$i_d(\boldsymbol{\beta}) = N_{\text{phot}} \int_{L_d} P(\mathcal{L}) I_B(\mathcal{L}) d\mathcal{L}. \tag{14}$$

Here $P(\mathcal{L})$ is the probability density of a general path $\mathcal{L}$. However, a random walk from the source to a camera pixel is a relatively rare event. For efficient rendering, we sample light paths in the medium using Backward MC.

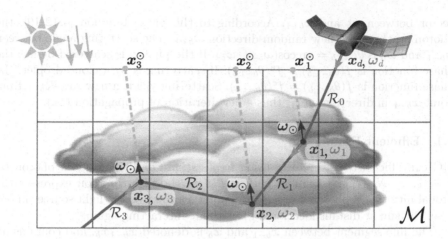

Fig. 4. Backward MC simulated imaging. This illustration demonstrates a photon path from a detector through the medium. The contribution to i_d is the sum of signals along the yellow lines towards the sun (local estimation).

The Backward MC process, illustrated in Fig. 4, back-propagates mathematically photons from a detector to the source. Denote the direction to the source (sun) by $\boldsymbol{\omega}_\odot$. From each camera pixel d, photons are back-projected in a corresponding direction $\boldsymbol{\omega}_d$ and then traced through the medium, as follows:
(i) Initially, a photon has associated intensity I_0. The photon is back projected along a ray denoted $\mathcal{R}_0$, which has a direction $\boldsymbol{\omega}_d$ and starts at detector (pixel) d at location $\boldsymbol{x}_d$, meaning, $\boldsymbol{x}_0 = \boldsymbol{x}_d$.
Per iteration b:
(ii) On ray $\mathcal{R}_b$, a random location $\boldsymbol{x}_{b+1}$ is sampled. The photon propagates to $\boldsymbol{x}_{b+1}$. This random distance, along $\mathcal{R}_b$, is sampled as described in [19]. It is a function of the spatially varying extinction coefficient $\beta(\boldsymbol{x})$.
(iii) If $\boldsymbol{x}_b \notin \mathcal{M}$, the photon is terminated. If $\boldsymbol{x}_b \notin \mathcal{M}$ while $\mathcal{R}_b \| \boldsymbol{\omega}_\odot$, the photon contributes intensity to pixel d.
(iv) If $\boldsymbol{x}_b \in \mathcal{M}$, then the photon interacts there with a particle. The type of particle (air molecule or cloud droplet) is sampled randomly based on the ratio of extinction coefficients $\beta^{\mathrm{a}}(\boldsymbol{x}_b)$, $\beta^{\mathrm{c}}(\boldsymbol{x}_b)$. If the scattering event at $\boldsymbol{x}_b$ is due to air, the single scattering albedo is $\varpi_b = \varpi^{\mathrm{a}}$. If the scattering event at $\boldsymbol{x}_b$ is due to a cloud droplet, the single scattering albedo is $\varpi_b = \varpi^{\mathrm{c}}$. The intensity carried by the photon is attenuated to $I_b = \varpi_b I_{b-1}$, hence

$$I_b = I_0 \prod_{b'=1}^{b} \varpi_{b'}. \tag{15}$$

If I_b is lower than a threshold, the photon can be stochastically terminated [20].
(v) The extinction coefficient at $\boldsymbol{x}_b$ is determined by the type of particle that caused the scattering. Meaning, $\beta(\boldsymbol{x}_b) = \beta^{\mathrm{a}}(\boldsymbol{x}_b)$ if the particle is an air molecule, and $\beta(\boldsymbol{x}_b) = \beta^{\mathrm{c}}(\boldsymbol{x}_b)$ if the particle is a cloud droplet. Denote $\boldsymbol{\omega}_b$ as the unit

vector between $\boldsymbol{x}_b$ and $\boldsymbol{x}_{b+1}$. According to the phase function [6,12,19], the photon scatters to a new random direction $\boldsymbol{\omega}_{b+1}$. The scattering angle between $\boldsymbol{\omega}_{b-1}$ and $\boldsymbol{\omega}_b$ is $\theta_{b-1,b} = \arccos(\boldsymbol{\omega}_{b-1} \cdot \boldsymbol{\omega}_b)$. If the photon is scattered by air, the phase function is $f(\theta_{b-1,b}) = f^{\mathrm{a}}(\theta_{b-1,b})$. If scattering is by a cloud droplet, the phase function is $f(\theta_{b-1,b}) = f^{\mathrm{c}}(\theta_{b-1,b})$. Scattering yields a new ray, $\mathcal{R}_{b+1}$ from point $\boldsymbol{x}_{b+1}$ in direction $\boldsymbol{\omega}_{b+1}$, thus a new iteration of propagation (ii).

5.1   Efficient Rendering

MC can efficiently estimate i_d using *local estimation* [35]. Index a photon by $p \in [1, \ldots, N_{\mathrm{phot}}]$. Its random path is $\mathcal{L}(d, p)$. Local estimation expresses the probability that a photon scatters from point $\boldsymbol{x}_b$ towards the light source (in our case, the sun at distant location $\boldsymbol{x}_{\odot}$), without interacting again.

The line segment between $\boldsymbol{x}_{b-1}$ and $\boldsymbol{x}_b$ is denoted $\overline{\boldsymbol{x}_{b-1}\boldsymbol{x}_b}$, and contains all points satisfying $\{\boldsymbol{x} = \tilde{\alpha}\boldsymbol{x}_{b-1} + (1 - \tilde{\alpha})\boldsymbol{x}_b, \forall \tilde{\alpha} \in [0, 1]\}$. The transmittance of the medium on this line segment is

$$a\left(\overline{\boldsymbol{x}_{b-1}\boldsymbol{x}_b}\right) = \exp\left\{-\int_{\boldsymbol{x}_{b-1}}^{\boldsymbol{x}_b} [\beta^{\mathrm{a}}(\boldsymbol{x}) + \beta^{\mathrm{c}}(\boldsymbol{x})]\, d\boldsymbol{x}\right\}. \qquad (16)$$

As illustrated in Fig. 4, the line between $\boldsymbol{x}_b$ and $\boldsymbol{x}_{\odot}$ intersects with the top of the atmosphere (TOA). This intersection is denoted by $\boldsymbol{x}_b^{\odot}$. Analogously, define the line segment $\overline{\boldsymbol{x}_b\boldsymbol{x}_b^{\odot}}$. Along this segment, the transmittance is $a\left(\overline{\boldsymbol{x}_b\boldsymbol{x}_b^{\odot}}\right)$. The scattering angle to the sun is denoted by $\theta_{b,\odot}$. Using Eq. (15, 16), the local estimation contribution of photon p at $\boldsymbol{x}_b$ is

$$I_b^{\mathrm{le}}[\mathcal{L}(d, p)] = I_b f(\theta_{b,\odot})\, a\left(\overline{\boldsymbol{x}_b\boldsymbol{x}_b^{\odot}}\right). \qquad (17)$$

The MC estimate of $i_d(\boldsymbol{\beta})$ is then

$$i_d(\boldsymbol{\beta}) \approx \sum_{p=1}^{N_{\mathrm{phot}}} \sum_{b=1}^{B} I_b^{\mathrm{le}}\left[\mathcal{L}(d, p)\right]. \qquad (18)$$

5.2   Gradient of the Model-Fit

The problem (4, 5) is solved iteratively using SGD. This requires, specifically, estimation of the gradient of $F(\boldsymbol{\beta})$. A component of the gradient in voxel v is given by

$$\frac{\partial F(\boldsymbol{\beta})}{\partial \beta_v^{\mathrm{c}}} = \sum_{d=1}^{N_{\mathrm{detectors}}} \left[i_d(\boldsymbol{\beta}) - i_d^{\mathrm{measured}}\right] \frac{\partial i_d(\boldsymbol{\beta})}{\partial \beta_v^{\mathrm{c}}}. \qquad (19)$$

We now assess the Jacobian $\partial i_d(\boldsymbol{\beta})/\partial \beta_v^{\mathrm{c}}$, i.e., how image pixels change in response to an infinitesimal perturbation of the medium $\boldsymbol{\beta}$. We describe here the essence of the Jacobian. For more details and mathematical derivations, see the supplementary material.

The domain of voxel v is denoted by $\mathcal{V}_v$. The line segment $\overline{\boldsymbol{x}_{b-1}\boldsymbol{x}_b}$ traverses several voxels. Denote by $l_{b-1,v}$ the length of the intersection of $\mathcal{V}_v$ with the line segment. Similarly, denote by $l_v^\odot$ the length of the intersection of $\mathcal{V}_v$ with the line segment and $\overline{\boldsymbol{x}_b\boldsymbol{x}_b^\odot}$. Then,

$$\frac{\partial i_d(\boldsymbol{x})}{\partial \beta_v^c} \approx - \sum_{p=1}^{N_{\text{phot}}} \sum_{b=1}^{B} I_b^{\text{le}} \left[\mathcal{L}(d,p)\right] \cdot \left(l_v^\odot + l_{b-1,v} - \begin{cases} \frac{1}{\beta_v^c} & \boldsymbol{x}_b \in \mathcal{V}_v, \ \&\ \text{scatter} \\ & \text{is by a cloud droplet} \\ 0 & \text{else} \end{cases} \right).$$

(20)

Based on Eqs. (19, 20) we solve Eq. (5) using SGD.

6 Coarse to Fine

We take several measures to enhance the estimation speed. We use a coarse-to-fine approach, spatially and in RT quality. RT quality is coarsened by using a small number of MC photons when estimating the gradient. The estimate is then noisy (coarse), per iteration. However, this way, SGD iterations are fast, and significantly advance the minimization (5) in its initial stages [51]. For more details and figures regarding specific simulations and performances, see the supplementary material. After a while, however, SGD in the coarse quality yields diminishing returns. Therefore, we use the result based on the coarse quality as initialization for a later stage, where the number of photons increases, leading to finer quality, better accuracy yet slower iterations.

Spatial coarse-to-fine is similar to common analysis in image analysis. Here, $\mathcal{M}$ is first represented in a coarse voxel grid. Thus the vector of unknown extinctions coefficients has a small dimension. Furthermore, the modeled and measured images are represented in a coarse resolution. The use of a small number of pixels and voxels enables significant shortening of runtime while significantly advancing the minimization (5) in its initial stages. We use the result obtained in the coarse voxels as initialization for the next stage: there, both $\mathcal{M}$ and the images are represented by smaller elements, leading to finer spatial resolution, yet slower iterations. This process is repeated in additional stages, where a result obtained in a coarser resolution initializes iterations on a finer-resolution grid. The number of voxels and pixels used in each stage of our examples are specified in the supplementary material.

We used additional speedup measures. Using the data images, we perform space-carving [29] of $\mathcal{M}$. This creates a photo-hull within which clouds potentially reside, pruning many voxels from the optimization. To correspond to the voxel resolution at the intermediate resolution steps described above, the photo-hull map is coarsened by dilation, before subsampling.

Furthermore, Eq. (19) requires two MC simulations: one is for estimating $i_d(\boldsymbol{\beta})$ (forward model); the other is for estimating its Jacobian $\partial i_d(\boldsymbol{\beta})/\partial\boldsymbol{\beta}$. Per iteration we use the same paths to sample both fields, saving about half the runtime. We found that possible bias of the gradient is not a major concern to the inverse problem, as each iteration samples a new random set.

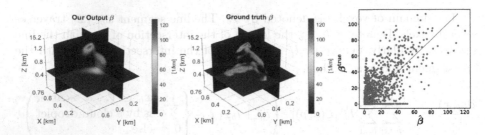

Fig. 5. 3D slices of the medium from scene C. [Left] Our recovered medium. [Middle] The true medium. [Right] scatter plot of 25% of the data points, randomly selected. The red line represents ideal reconstruction, where $\hat{\beta} = \beta^{\text{true}}$. Some areas are not well estimated. This is because they are occluded from the viewpoints of the cameras, by higher and denser parts of the cloud. (Color figure online)

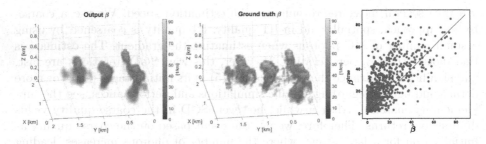

Fig. 6. Scene F. [Left] Our recovered medium. [Middle] The true medium. [Right] Scatter plot of 10% of the data points, randomly selected. The red line represents ideal reconstruction, where $\hat{\beta} = \beta^{\text{true}}$. (Color figure online)

7   Simulations

Currently, there is no adequate gear for simultaneous multi-view imaging of small clouds from space. However, proper orbiting cameras and data are upcoming in a few years [54]. This paper develops methods in anticipation for the upcoming spaceborne technological advancement. At such a stage, testing is by simulations, as in several recent computer vision papers [15,19,31,32] about scattering-CT. As in [31,32] a realistically complex scene is created by use a *Large Eddy Simulation (LES)* [7,37]. This is a tool used by atmospheric scientists as a standard for computationally generating cloud fields from first principles. We deal with two public domain scenes [37], with results shown in Figs. 5, 6.

C has a single cloud in a 0.76 km × 0.76 km × 1.52 km domain having 38 × 38 × 38 voxels (54, 872 unknowns). Cameras have 76 × 76 pixels, each having 30 m ground-level footprint (nadir view). In rendering of ground-truth data, $N_{\text{phot}} = 8192$ photons. Cameras are at $z = 1.9$ km. One camera is 1.8° off the zenith. The other eight reside on a ring ≈ 29° around the zenith.

F has several clouds in a 1.72 km × 1.72 km × 1.08 km domain having 86 × 86 × 27 voxels (199, 692 unknowns). Cameras have 86 × 86 pixels, each having 20 m ground-level footprint, $N_{\text{phot}} = 4096$ photons. One camera is $0.4°$ off the zenith at $z = 2.16\,\text{km}$. The other eight are at $z = 1.86\,\text{km}$ on a ring $\approx 33°$ around the zenith.

In both C and F, the sun is at the zenith, voxel are $20 \times 20 \times 40\,\text{m}^3$ large, and the background surface is Lambertian having albedo 0.05, emulating a dark ocean. Each domain has background $\beta = 0.01\,\text{km}^{-1}$ wherever $\beta^c = 0$. The domain is observed by nine perspective cameras. To reduce the effect of air molecules (which are not our object of interest), sensing focused on a red-colored wavelength 632 nm.

The values of N_{phot} are comparable to the full-well of common machine vision cameras. Due to the randomness of MC rendering, this photon count yields typical Poissonian noise resembling photon noise in cameras. The MC kernel we used is by Mitsuba [21].

After estimation of the photo-hull (Sect. 6), tomographic estimation was initialized by $\beta(\boldsymbol{x}) = 2\,\text{km}^{-1}$. We use the adaptive moment estimation method (Adam) [26], with the following parameters: first moment decay $\eta_1 = 0.9$, second moment decay $\eta_2 = 0.999$, $\varepsilon = 10^{-8}$, and step size is $\alpha = 0.3$ for all scenes.

Reconstruction quality is assessed by two measures [2] which compare the estimated extinction field to the ground-truth. One of them is ϵ from Eq. (13). The other is

$$\delta = \frac{\|\hat{\beta}\|_1 - \|\beta^{\text{true}}\|_1}{\|\beta^{\text{true}}\|_1}, \tag{21}$$

which is a relative bias of the overall extinction.

Results

In the small scene C, the results are displayed in Fig. 5. Without using the monotonicity prior, meaning only using a smoothness prior, the results were $\delta = -11.5\%$ and $\epsilon = 87\%$ for a runtime of about 6.5 h to convergence. Our approach, with the monotonicity prior, yielded $\delta - 5.5\%$ and $\epsilon = 84.5\%$ for the same runtime. Results of the cloud field (scene F) are displayed in Fig. 6. Without a monotonicity prior, $\delta = -34.2\%$ and $\epsilon = 54.8\%$, after 127 h. Afterwards, the error measures increased. The monotonicity prior improved results to $\delta = -28.2\%$ and $\epsilon = 44.7\%$. Results reached convergence after 134 h.

Comparing our results to previous work, running a code based on [15] for scene C, after a fixed 6.5 h runtime, yielded $\epsilon = 196\%$ and $\delta = 14\%$, and code based on [19] yielded $\epsilon = 174\%$ and $\delta = -9.6\%$, after the same runtime. To compare the SHDOM-based tomography approach [31] (and [32], which uses the same algorithm yet solves for micro-physics), we used an improved code [30] and solved for scene C. The deterministic approach resulted in more accurate output, having $\delta = -3.6\%$ and $\epsilon = 72\%$. Moreover, the SHDOM-based tomography code [30] ran for about an hour while our method ran for about 6.5 h. Recalling Sect. 3, this outcome is not surprising, as the deterministic SHDOM is inherently faster and more accurate than random walks of MC. Yet, this approach has a fundamental challenge to scale to large fields, contrary to MC. In accordance, for larger scenes, containing a 100s of thousands of voxels, like scene **F**, [31] reported $\epsilon = 70\%$ and $\delta = 30\%$, compared to our results in scene **F**.

Overall, the most significant cost reduction is achieved within minutes, using the coarse scales. The finest scales are much slower, yet enable a significant improvement of the results. See the supplementary for more details.

8   Conclusions

Spaceborne data of simultaneous imaging of cloud fields, in resolution that is required for scattering-CT is not yet accessible by an experiment. However, a future spaceborne mission [54] is planned to provide such data. Hence, priors and algorithms developed for wide field scattering-CT should become useful in several years, enabling scientists to draw new information and understanding of nature. Moreover, advancement of scattering CT should be useful also to some modalities of bio-medical imaging and material engineering.

To enhance speed, good initialization can be obtained in the future by a neural-network. Current differential neural renderers (DNRs) are set for opaque, reflective objects [24,47]. A DNR based on 3D, volumetric RT in large heterogeneous scattering media requires large training sets that are currently unavailable. Once suitable scattering training data and DNRs become available, it may efficiently initialize our process, which is proper physics-based MC RT analysis. The latter is needed as the results should be trusted by physicists studying nature.

We believe that the corase-to-fine stochastic approach taken in this paper has significant potential. MC can excellently use technologies of parallel computations. The calculations per photon are mutually independent of other photons. Since most of the calculation in MC rendering are similar, it is worthwhile to adapt such a method to GPU, to accelerate the runtime. Encouragingly, the most recent version of Mitsuba [48], states an implementation on GPU. For these reasons, the developments of scattering tomography are timely.

Acknowledgments. We thank Ilan Koren, Eshkol Eytan Liebeskind, and Tom Dror-Schwartz for useful discussions. We thank Johanan Erez, Ina Talmon and Daniel Yagodin for technical support. Yoav Schechner is the Mark and Diane Seiden Chair in Science at the Technion. He is a Landau Fellow - supported by the Taub Foundation. His work was conducted in the Ollendorff Minerva Center. Minvera is funded through the BMBF. This research is funded by the European Research Council (ERC) under the European Unions Horizon 2020 research and innovation program (grant agreement No 810370: CloudCT). Aviad Levis is a Zuckerman Postdoctoral Fellow.

References

1. Aides, A., Levis, A., Holodovsky, V., Schechner, Y.Y., Althausen, D., Vainiger, A.: Distributed sky imaging radiometry and tomography. In: ICCP, pp. 1–12. IEEE (2020)
2. Aides, A., Schechner, Y.Y., Holodovsky, V., Garay, M.J., Davis, A.B.: Multi sky-view 3D aerosol distribution recovery. Opt. Express **21**(22), 25820–25833 (2013)

3. Alterman, M., Schechner, Y.Y., Vo, M., Narasimhan, S.G.: Passive tomography of turbulence strength. In: Fleet, D., Pajdla, T., Schiele, B., Tuytelaars, T. (eds.) ECCV 2014. LNCS, vol. 8692, pp. 47–60. Springer, Cham (2014). https://doi.org/10.1007/978-3-319-10593-2_4

4. Basri, R., Jacobs, D.W.: Lambertian reflectance and linear subspaces. IEEE Trans. Pattern Anal. Mach. Intell. **2**, 218–233 (2003)

5. Berman, D., Treibitz, T., Avidan, S.: Air-light estimation using haze-lines. In: ICCP, pp. 1–9. IEEE (2017)

6. Binzoni, T., Leung, T.S., Gandjbakhche, A.H., Ruefenacht, D., Delpy, D.: The use of the Henyey-Greenstein phase function in Monte Carlo simulations in biomedical optics. Phys. Med. Biol. **51**(17), N313 (2006)

7. Chung, D., Matheou, G.: Large-eddy simulation of stratified turbulence. Part i: a vortex-based subgrid-scale model. J. Atmos. Sci. **71**(5), 1863–1879 (2014)

8. Davis, C., Emde, C., Harwood, R.: A 3-D polarized reversed Monte Carlo radiative transfer model for Millimeter and submillimeter passive remote sensing in cloudy atmospheres. IEEE Trans. Geosci. Remote Sens. **43**(5), 1096–1101 (2005)

9. Deng, X., Jiao, S., Bitterli, B., Jarosz, W.: Photon surfaces for robust, unbiased volumetric density estimation. ACM Trans. Graph. **38**(4), 46 (2019)

10. Diner, D.J., Martonchik, J.V.: Atmospheric transmittance from spacecraft using multiple view angle imagery. Appl. Opt. **24**(21), 3503–3511 (1985)

11. Evans, K.F.: The spherical harmonics discrete ordinate method for three-dimensional atmospheric radiative transfer. J. Atmos. Sci. **55**(3), 429–446 (1998)

12. Frisvad, J.R.: Importance sampling the Rayleigh phase function. JOSA **28**(12), 2436–2441 (2011)

13. Georgiev, I., Misso, Z., Hachisuka, T., Nowrouzezahrai, D., Křivánek, J., Jarosz, W.: Integral formulations of volumetric transmittance. ACM Trans. Graph. **38**(6), 1–17 (2019)

14. Geva, A., Schechner, Y.Y., Chernyak, Y., Gupta, R.: X-ray computed tomography through scatter. In: Ferrari, V., Hebert, M., Sminchisescu, C., Weiss, Y. (eds.) Computer Vision – ECCV 2018. LNCS, vol. 11218, pp. 37–54. Springer, Cham (2018). https://doi.org/10.1007/978-3-030-01264-9_3

15. Gkioulekas, I., Levin, A., Zickler, T.: An evaluation of computational imaging techniques for heterogeneous inverse scattering. In: Leibe, B., Matas, J., Sebe, N., Welling, M. (eds.) ECCV 2016. LNCS, vol. 9907, pp. 685–701. Springer, Cham (2016). https://doi.org/10.1007/978-3-319-46487-9_42

16. Gkioulekas, I., Zhao, S., Bala, K., Zickler, T., Levin, A.: Inverse volume rendering with material dictionaries. ACM Trans. Graph. **32**(6), 162 (2013)

17. Gorbunov, M.E., Sokolovky, S., Bengtsson, L.: Space refractive tomography of the atmosphere: modeling of direct and inverse problems (1996)

18. Gregson, J., Krimerman, M., Hullin, M.B., Heidrich, W.: Stochastic tomography and its applications in 3D imaging of mixing fluids. ACM Trans. Graph. **31**(4), 1–52 (2012)

19. Holodovsky, V., Schechner, Y.Y., Levin, A., Levis, A., Aides, A.: In-situ multi-view multi-scattering stochastic tomography. In: ICCP, pp. 1–12. IEEE (2016)

20. Iwabuchi, H.: Efficient Monte Carlo methods for radiative transfer modeling. J. Atmos. Sci. **63**(9), 2324–2339 (2006)

21. Jakob, W.: Mitsuba renderer (2010). http://www.mitsuba-renderer.org

22. Kaftory, R., Schechner, Y.Y., Zeevi, Y.Y.: Variational distance-dependent image restoration. In: CVPR, pp. 1–8. IEEE (2007)

23. Kalashnikova, O.V., Garay, M.J., Davis, A.B., Diner, D.J., Martonchik, J.V.: Sensitivity of multi-angle photo-polarimetry to vertical layering and mixing of absorbing aerosols: quantifying measurement uncertainties. J. Quant. Spectrosc. Radiat. Transf. **112**(13), 2149–2163 (2011)
24. Kato, H., Ushiku, Y., Harada, T.: Neural 3D mesh renderer. In: CVPR, pp. 3907–3916. IEEE (2018)
25. Khungurn, P., Schroeder, D., Zhao, S., Bala, K., Marschner, S.: Matching real fabrics with micro-appearance models. ACM Trans. Graph. **35**(1), 1–1 (2015)
26. Kingma, D.P., Ba, J.: Adam: a method for stochastic optimization. arXiv preprint arXiv:1412.6980 (2014)
27. Kokhanovsky, A.A.: Light Scattering Media Optics. Springer, Heidelberg (2004)
28. Kratz, L., Nishino, K.: Factorizing scene albedo and depth from a single foggy image. In: ICCV, pp. 1701–1708. IEEE (2009)
29. Kutulakos, K.N., Seitz, S.M.: A theory of shape by space carving. Int. J. Comput. Vis. **38**(3), 199–218 (2000)
30. Levis, A., Aides, A.: pyshdom (2019). https://github.com/aviadlevis/pyshdom
31. Levis, A., Schechner, Y.Y., Aides, A., Davis, A.B.: Airborne three-dimensional cloud tomography. In: ICCV, pp. 3379–3387. IEEE (2015)
32. Levis, A., Schechner, Y.Y., Davis, A.B.: Multiple-scattering microphysics tomography. In: CVPR, pp. 6740–6749. IEEE (2017)
33. Levis, A., Schechner, Y.Y., Davis, A.B., Loveridge, J.: Multi-view polarimetric scattering cloud tomography and retrieval of droplet size. arXiv preprint arXiv:2005.11423 (2020)
34. Lu, M.-L., et al.: Aerosol-cloud relationships in continental shallow cumulus. J. Geophys. Res. **113**, D15201 (2008). https://doi.org/10.1029/2007JD009354
35. Marshak, A., Davis, A.: 3D Radiative Transfer in Cloudy Atmospheres. Springer, Heidelberg (2005)
36. Martonchik, J.V., et al.: Techniques for the retrieval of aerosol properties over land and ocean using multiangle imaging. IEEE Trans. Geosci. Remote Sens. **36**(4), 1212–1227 (1998)
37. Matheou, G., Chung, D.: Large-eddy simulation of stratified turbulence. Part ii: application of the stretched-vortex model to the atmospheric boundary layer. J. Atmos. Sci. **71**(12), 4439–4460 (2014)
38. Mayer, B.: Radiative transfer in the cloudy atmosphere. In: EPJ Web of Conferences, vol. 1, pp. 75–99. EDP Sciences (2009)
39. McFarlane, S.A., Grabowski, W.W.: Optical properties of shallow tropical cumuli derived from ARM ground-based remote sensing. Geophys. Res. Lett. **34**, L06808 (2007). https://doi.org/10.1029/2006GL028767
40. Messer, H., Zinevich, A., Alpert, P.: Environmental sensor networks using existing wireless communication systems for rainfall and wind velocity measurements. Instrum. Meas. Mag. **15**(2), 32–38 (2012)
41. Mobley, C.D.: Light and Water: Radiative Transfer in Natural Waters. Academic Press, Cambridge (1994)
42. Narasimhan, S.G., Ramamoorthi, R., Nayar, S.K.: Analytic rendering of multiple scattering in participating media. Technical report, Columbia University (2004)
43. Narasimhan, S.G., Gupta, M., Donner, C., Ramamoorthi, R., Nayar, S.K., Wann-Jensen, H.: Acquiring scattering properties of participating media by dilution. ACM Trans. Graph. **25**(3), 1003–1012 (2006)
44. Narasimhan, S.G., Nayar, S.K.: Vision and the atmosphere. Int. J. Comput. Vis. **48**(3), 233–254 (2002)

45. Narasimhan, S.G., Nayar, S.K.: Interactive (de) weathering of an image using physical models. In: IEEE Workshop on Color and Photometric Methods in Computer Vision, vol. 6, p. 1. France (2003)
46. Narasimhan, S.G., Nayar, S.K., Sun, B., Koppal, S.J.: Structured light in scattering media. In: ICCV, vol. 1, pp. 420–427. IEEE (2005)
47. Nguyen-Phuoc, T.H., Li, C., Balaban, S., Yang, Y.: RenderNet: a deep convolutional network for differentiable rendering from 3D shapes. In: NeurIPS, pp. 7891–7901 (2018)
48. Nimier-David, M., Vicini, D., Zeltner, T., Jakob, W.: Mitsuba 2: a retargetable forward and inverse renderer. ACM Trans. Graph. **38**(6) (2019). https://doi.org/10.1145/3355089.3356498
49. Novák, J., Georgiev, I., Hanika, J., Jarosz, W.: Monte Carlo methods for volumetric light transport simulation. In: Computer Graphics Forum, vol. 37, pp. 551–576. Wiley Online Library (2018)
50. Okabe, T., Sato, I., Sato, Y.: Spherical harmonics vs. haar wavelets: basis for recovering illumination from cast shadows. In: CVPR, vol. 1, pp. 50–57. IEEE (2004)
51. Pfeiffer, G.T., Sato, Y.: On stochastic optimization methods for Monte Carlo least-squares problems. arXiv preprint arXiv:1804.10079 (2018)
52. Pharr, M., Jakob, W., Humphreys, G.: Physically Based Rendering: from Theory to Implementation. Morgan Kaufmann, Burlington (2016)
53. Schaul, L., Fredembach, C., Süsstrunk, S.: Color image dehazing using the near-infrared. In: ICIP, pp. 1629–1632. IEEE (2009)
54. Schilling, K., Schechner, Y.Y., Koren, I.: CloudCT - computed tomography of clouds by a small satellite formation. In: IAA symposium on Small Satellites for Earth Observation (2019)
55. Sheinin, M., Schechner, Y.Y.: The next best underwater view. In: CVPR, pp. 3764–3773. IEEE (2016)
56. Spier, O., Treibitz, T., Gilboa, G.: In situ target-less calibration of turbid media. In: ICCP, pp. 1–9. IEEE (2017)
57. Vainiger, A., Schechner, Y.Y., Treibitz, T., Avni, A., Timor, D.S.: Optical wide-field tomography of sediment resuspension. Opt. Express **27**(12), A766–A778 (2019)
58. Villefranque, N., et al.: A path-tracing Monte Carlo library for 3-D radiative transfer in highly resolved cloudy atmospheres. J. Adv. Model. Earth Syst. **11**(8), 2449–2473 (2019)
59. Wright, T.E., Burton, M., Pyle, D.M., Caltabiano, T.: Scanning tomography of SO_2 distribution in a volcanic gas plume. Geophys. Res. Lett. **35**, L17811 (2008). https://doi.org/10.1029/2008GL034640
60. Zhang, S., Xue, H., Feingold, G.: Vertical profiles of droplet effective radius in shallow convective clouds. Atmos. Chem. Phys. **11**(10), 4633–4644 (2011)

Learning Trailer Moments in Full-Length Movies with Co-Contrastive Attention

Lezi Wang[1]([✉]), Dong Liu[2], Rohit Puri[3], and Dimitris N. Metaxas[1]

[1] Rutgers University, New Brunswick, USA
{lw462,dnm}@cs.rutgers.edu
[2] Netflix, Los Gatos, USA
dongl@netflix.com
[3] Twitch, San Francisco, USA
rohipur@twitch.tv

Abstract. A movie's key moments stand out of the screenplay to grab an audience's attention and make movie browsing efficient. But a lack of annotations makes the existing approaches not applicable to movie key moment detection. To get rid of human annotations, we leverage the officially-released trailers as the weak supervision to learn a model that can detect the key moments from full-length movies. We introduce a novel ranking network that utilizes the Co-Attention between movies and trailers as guidance to generate the training pairs, where the moments highly corrected with trailers are expected to be scored higher than the uncorrelated moments. Additionally, we propose a Contrastive Attention module to enhance the feature representations such that the comparative contrast between features of the key and non-key moments are maximized. We construct the first movie-trailer dataset, and the proposed Co-Attention assisted ranking network shows superior performance even over the supervised(The term "supervised" refers to the approach with access to the manual ground-truth annotations for training.) approach. The effectiveness of our Contrastive Attention module is also demonstrated by the performance improvement over the state-of-the-art on the public benchmarks.

Keywords: Trailer moment detection · Video highlight detection · Co-contrastive attention · Weak supervision · Video feature augmentation

1 Introduction

"Just give me five great moments and I can sell that movie." – Irving Thalberg (Hollywood's first great movie producer).

L. Wang—This work was done when Lezi Wang worked as an intern at Netflix.
R. Puri—This work was done when Rohit Puri was with Netflix.

Electronic supplementary material The online version of this chapter (https://doi.org/10.1007/978-3-030-58523-5_18) contains supplementary material, which is available to authorized users.

A. Vedaldi et al. (Eds.): ECCV 2020, LNCS 12363, pp. 300–316, 2020.
https://doi.org/10.1007/978-3-030-58523-5_18

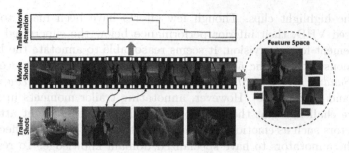

Fig. 1. We leverage the trailer shots to estimate the attention scores of individual shots in the full-length movie, which indicate the "trailerness" of the shots and can be used as weak supervision to model the contrastive relation between the key and non-key moments in the feature space.

Movie is made of moments [34], while not all of the moments are equally important. In the spirit of the quote above, some key moments are known as coming attraction or preview, which can not only grab an audience's attention but also convey the movie's theme.

The importance of detecting the key moments is two-fold. First, key moments migrate the content overwhelming. There are millions of movies produced in human history [11]. A full-length movie typically lasts two or three hours, making it incredibly time-consuming for consumers to go through many of them. The key moments in the form of short video clips can make the movie browsing efficient, where audiences can quickly get the theme by previewing those short clips with story highlightings. Second, for the purpose of movie promotion, the well-selected moments can attract audience to the movie, where the key moments are usually drawn from the most exciting, funny, or otherwise noteworthy parts of the film but in abbreviated form and usually without spoilers[1].

A popular form of key moments in the movie industry is the trailer, which is a short preview of the full-length movie and contains the significant shots selected by professionals in the field of cinematography. In this paper, we focus on moments in the movie trailer and try to answer an important question regarding *Movie Trailer Moment Detection* (MTMD) – can we learn a vision model to detect trailer moments in full-length movies automatically?

The MTMD problem is related to the existing line of research on *Video Highlight Detection* (VHD), a task of extracting highlight clips from videos. Recently, deep learning has become a dominant approach to this task, which formulates it as a problem of learning a ranking model to score the human-labeled highlight clips higher than the non-highlight. Given video clips, the deep spatial-temporal features are extracted as the input to train the ranking model [3, 13, 24, 36, 38]. However, the existing VHD approaches cannot be directly applied to MTMD due to the following reasons.

First, there is no labeled data available for MTMD. To train a robust VHD model, it requires extensive supervision where the annotators must manually

---

[1] https://en.wikipedia.org/wiki/Trailer_(promotion).

identify the highlight clips. Though few efforts have been made to conduct unsupervised VHD, their inferior performance below the supervised indicates the requirement for supervision. It seems reasonable to annotate the highlights which demonstrate specific actions (e.g., *"Skiing"*, *"Skating"*) or events (e.g., *"Making Sandwich"*, *"Dog Show"*) as in the VHD datasets like *Youtube Highlight* [31] and TVSum [30]. However, annotating trailer moments in movies is much more challenging as the selection of trailer moments might attribute to various factors such as emotion, environment, story-line, or visual effects, which requires the annotators to have specialized domain knowledge. To resolve this issue, we create the supervision signal by matching moments between the trailers and the corresponding movies, as shown in Fig. 1. Specifically, we propose a *Co-Attention* module to measure the coherence between the shots from trailers and movies, through which a set of the best and worst matched shots from the movies are discovered as weakly labeled positive and negative samples.

Second, the existing VHD approaches treat the individual short clips in the long videos separately without exploring their relations. In fact, the trailer moments follow certain common patterns and should be distinguishable from the non-trailer moments. Taking action movies as an example, although different movies tell different stories, their trailer moments always contain shots with intensive motion activities. To incorporate such prior into MTMD, we propose a *Contrastive Attention module* to enforce the feature representations of the trailer moments to be highly correlated while at the same time encourage the high contrast between the trailer and non-trailer moments. In this way, the features of trailer moments can form a compact clique in the feature space and stand out from the features of the non-trailer moments.

We integrate the two modules, i.e., *Co-Attention* and *Contrastive Attention*, into the state-of-the-art 3D CNN architecture that can be employed as a feature encoder with a scoring function to produce the ranking score for each shot in the movie. We dub the integrated network *CCANet: Co-Contrastive Attention Network*. To support this study and facilitate researches in this direction, we construct TMDD, a Trailer Moment Detection Dataset, which contains 150 movies and their official trailers. The total length of these videos is over 300 h. We conduct experiments on TMDD, and our CCANet shows promising results, even outperforming the supervised approaches. We also demonstrate that our proposed Contrastive Attention module significantly achieves marginal performance-boosting over the state-of-the-art on the public VHD benchmarks, including Youtube Highlight [31] and TVSum [30].

In summary, we make the following contributions:

- We propose CCANet that can automatically detect trailer moments from full-length movies without the need of human annotation.
- We propose a Contrastive Attention to constrain the feature representations such that the contrastive relation can be well exploited in the feature space.
- To our best knowledge, we are the first to collect a trailer moment detection dataset to facilitate this research direction.
- Our approach shows the superior performance over the state-of-the-art on the public benchmarks, outperforming the existing best approach by 13%.

2 Related Works

Studies on movie and trailer have been on the increase interests in computer vision research because of their rich content [11]. Several efforts have been made to analyze movies or trailers from different angles. A growing line of research is trying to understand the semantics in movies via audio-visual information together with the plot, subtitles, sentiment, and scripts [1,4,12,19,22,29,33,37]. The works [1,22,37] focus on understanding the relationships of movie characters. Zhu et al. [44] proposed an approach to match movie shots and scripts so as to understand high-level storylines. Tapaswi et al. [33] developed a movie Q&A benchmark, proposing a way to understand movies via visual question answering. Chu et al. [4] use machine learning approaches to construct emotional arcs of the visual or audio signals, cluster the type of arcs, and predict audience engagement. Besides the studies on movies, there are also efforts trying to understand the trailers. The works in [12,19] attempt to generate trailers for user-uploaded videos by learning from structures of movies. Smith et al. [29] present a heuristic system to fuse the multi-modality to select the candidate shots for trailer creation and the analysis is preformed on horror movies. In [27,43], the genre classification problem is investigated by using the trailers to represent the movie content. For this purpose, datasets with several thousand trailers have been constructed. These works are all based on the movie or trailers separately without considering their correspondence. As a pioneering work, Huang et al. [11] propose an approach to bridge trailers and movies, allowing the knowledge learned from trailers to be transferred to full-length movie analysis. However, a dataset consisting of full-length movies and the key moment annotations is still unavailable, which motivates us to collect TMDD to facilitate this research direction.

Video highlight detection has been studied a lot for sports videos [25,32,40]. Recently, *supervised video highlight detection* has been applied to Internet videos [31] and first-person videos [42]. The Video2GIF approach [8] learns to construct a GIF for a video from the user-created GIF-Video pairs. The supervised highlight detection requires human-labeled training pairs, which are expensive to obtain. Recently, several efforts have been made for *unsupervised video highlight detection*, which does not require manual annotations. These approaches can be further divided into *domain-agnostic* or *domains-specific* approaches. The domain-agnostic approaches operate uniformly on any video containing different semantic concepts. The approach in [18] is based on motion intensity. Works [20,23] are to train a set of video category classifiers and then detect highlights based on the classifier scores or spatial-temporal gradients. In contrast, the domain-specific approaches train highlight detectors on a collection of videos containing the same concept. In [41], Yang et al.propose a category-aware reconstruction loss for unsupervised domain-specific highlight detection.

A very recent work [39] is proposed to get rid of human annotations by leveraging the video duration as the supervision to train highlight detectors. The key insight is that the clips from the shorter user-generated videos are more likely to be the highlights than those from longer videos since users tend to be more focused on the content when capturing shorter videos [39]. While the insight does not apply to movie domain. As shown in Fig. 2, the duration of the trailer and non-trailer shots is similar statistically, which severely mutes the duration signal.

Inspired by the fact that movies come with trailers, we tackle the annotation problem by leveraging the trailer moments to generate the super-vision. A Co-Attention module is proposed to measure the coherence between the shots from trailers and movies. Different from the existing Pseudo-Label approach, which offline predicts the labels [15, 26], our Co-Attention module is updated in the learning process, where training is in an end-to-end fashion.

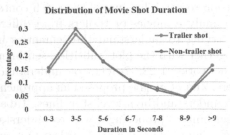

Fig. 2. Duration distribution for the trailer and non-trailer shots indicates that the duration of the two kinds of shots is similar.

3 Approach

We develop CCANet with two goals: 1) with the weak-supervision from the publicly available trailers, the network is trained without human labeling; 2) we incorporate the "contrastive" relation into the learning process so that the trailer moment can be distinguishable from the non-trailer. We first describe how we construct the Trailer Moment Detection Dataset (TMDD) in Sect. 3.1. Then we present the CCANet in Sect. 3.2, consisting of the *Co-Attention* for learning the trailer moments and the *Contrastive Attention* for feature augmentation.

3.1 Trailer Moment Detection Dataset

We aim to detect the key moments in movies using the publicly available trailers as supervision. However, the existing movie or trailer related benchmarks [27, 43] are not appropriate for this task. They collect the trailers or the movie posters for genre classification without full movies provided. Recently, Huang *et al.*[11] learn the vision models from both movies and trailers by proposing a Large-Scale Movie and Trailer Dataset (LSMTD). However, LSMTD is not publicly available. Moreover, due to the different purposes of learning a semantic model for movie understanding, LSMTD has no ground-truth for MTMD evaluation. To this end, we construct a new dataset, named *Trailer Moment Detection Dataset* (TMDD).

TMDD contains 150 movies in full length paired with their official trail-ers. The movies are split into three domains according to the genre, including *"Action"*, *"Drama"*, and *"Sci-Fi"*. Each domain has 50 movie-trailer pairs. We train an MTMD model for each domain, which accounts for the intuition that the key moments are highly domain-dependent, e.g., a fighting moment might be crucial in *"Action"* movie but not in romantic *"Drama"*.

We define a movie moment as a shot that consists of consecutive frames in one camera recording time [28]. We apply the shot boundary detection [28] to

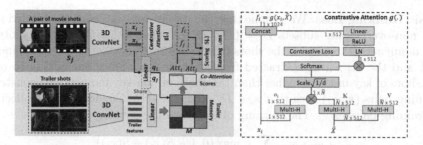

Fig. 3. Left: overview of the proposed CCANet. We use the Co-Attention between the trailer and movie as the weak supervision and propose Contrastive Attention to augment the feature representations such that the trailer shots can stand out from the non-trailer shots in the feature space. **Right:** the details of Contrastive Attention module. $\otimes$ indicates matrix multiplication and "Concat" stands for vector concatenation.

segment movies and trailers into multiple shots. Overall, the TMDD contains 263,837 movie shots and 15,790 trailer shots. Hence, MTMD on this dataset is a quite challenging task as the *true positives* only take ~6% if we regard all trailer shots as the key moments. To our best knowledge, this is the first and largest dataset that has ever been built for MTMD.

To build the ground-truth without the requirement of experts annotating the key moments, we conduct visual similarity matching between trailers and movies at the shot-level and then manually verify the correctness of the matches. The shots occurring both in trailers and full-length movies are regarded as the ground-truth key moments in the movie. Notably, the annotations obtained in this way are only for performance evaluation but not for training the model. In the next section, we present our approach of leveraging the trailers to learn the movie key moments without human annotations needed.

3.2   CCANet for Trailer Moment Detection

We integrate the Co-Attention and Contrastive Attention modules into a unified CCANet, as shown in Fig. 3(Left). Our goal is to learn a scoring function $S(\cdot)$ that predicts the "trailerness" score of a movie shot given its feature as input, where the feature is extracted from the individual shot by a 3D ConvNet [9]. At test time, movie shots can be ranked based on the predicted scores, and the top-ranked shots are deemed as the key moments that can be applied to create trailers. Specifically, instead of relying on human annotations to create the pairwise shots for learning the $S(\cdot)$, we create shot pairs based on the Co-Attention scores *Att* between trailers and movies. Additionally, the Contrastive Attention module is proposed to augment the 3D features so as to explore the relations between the trailer and non-trailer shots. The details are descried below.

Learning Trailer Moments via Co-Attention . We leverage the Co-Attention between movies and trailers to modify the basic ranking loss for MTMD.

Basic Ranking Loss. We assume that the movie dataset D can be divided into two non-overlapping subsets $D = \{D^+, D^-\}$, where D^+ contains the shots of key moments, D^- contains the shots of non-key moment. Let s_i refer to a movie shot and the 3D feature extracted from shot s_i is x_i. Since our goal is to rank the shots of key moment higher than the shots of non-key moment, we construct training pairs (s_i, s_j) such that $s_i \in D^+$ and $s_j \in D^-$. We denote the collection of training pairs as $\mathcal{P}$. The learning objective is the ranking loss:

$$\mathcal{L}_{Rank} = \sum_{(s_i, s_j) \in \mathcal{P}} \max\big(0, 1 - S(x_i) + S(x_j)\big). \tag{1}$$

Co-Attention Between Trailer and Movie. Let T refers to a set of N_t shots in a trailer. We encode each $t_i \in T$ into a 3D feature. As shown in Fig. 3(Left), a linear layer is applied to map the shot features into a memory $M \in \mathbb{R}^{N_t \times d}$, where d is the dimension of the memory vector $m_\tau \in M$. Given the feature x_i of shot s_i from a full movie, we generate the query q_i by applying the linear layer to x_i. The Co-Attention can be calculated as the maximal convolution activation between the query q_i and the vectors in M:

$$Att_i = \max_{\tau \in N_t}(q_i \circledast m_\tau). \tag{2}$$

The Co-Attention score Att_i measures the coherence of shot s_i in the movie to all shots in the trailer T. A large Att_i value indicates that the shot s_i is highly correlated to the trailer and therefore is a potential key moment in the movie.

Ranking Loss with Co-Attention. The ranking loss in Eq. (1) assumes that we have annotations for constructing the training set D^+ and D^-. However, it requires extensive human efforts and domain knowledge to annotate them. To achieve the learning goal without access to human annotations, we leverage the trailer to predict the attention score Att_i and use it as a "soft label" to measure the importance of shot s_i in the full movie. Additionally, as shown in Fig. 3(Left), we introduce a *Contrastive Attention* module $g(\cdot)$ (described in the next section and illustrated by Fig. 3(Right)) to augment the feature x_i of shot s_i into f_i. With the soft labels and augmented features, we can rewrite the learning objective as follows:

$$\mathcal{L}_{Rank} = \sum_{(s_i, s_j) \in \mathcal{P}} w_{ij} \max\big\{0, 1 - \sigma[S(f_i) - S(f_j)]\big\}$$
$$\text{where} \quad w_{ij} = \lambda(\exp(|Att_i - Att_j|) - 1), \tag{3}$$
$$\text{and} \quad \sigma = \text{sgn}(Att_i - Att_j),$$

where λ is a scaling factor and w_{ij} is introduced as a variable to identify the validness of a pair $(s_i, s_j) \in \mathcal{P}$ to the loss. The underlying intuition is that we assign a large weight to the contrastive pair where the difference between Att_i and Att_j is significant and therefore, should be treated as a confident training sample. The variable σ is used to determine the order of the predicted scores based on their Co-Attention values.

It is worth noting that our approach module is different from the existing app-roach of *learning with Pseudo-Label* (PL). In PL, labels are collected offline from the highly confident predictions made by the model. While our Co-Attention module updates the label predictions in the end-to-end training process.

Augmenting Features via Contrastive Attention. As shown in Fig. 3 (Right), we draw inspiration from the attention mechanism [35] to exploit the contrastive relation among shots. Given a target shot s_i and an auxiliary shot set $\tilde{S}$ with $\tilde{N}$ shots, we extract a 3D visual feature $x_i \in \mathbb{R}^d$ and a feature set $\tilde{X} \in \mathbb{R}^{\tilde{N} \times d}$, respectively. We apply $\tilde{X}$ as the supportive set to augment x_i to be $f_i = g(x_i, \tilde{X}) \in \mathbb{R}^{2d}$. We aim to make the attention contrastive such that the features of key moments can form a compact clique in the feature space and stand out from the features of the non-key moments. Specifically, the attention $A \in \mathbb{R}^{1 \times \tilde{N}}$ between x_i and each $x_j \in \tilde{X}$ is computed as:

$$A(x_i, \tilde{X}) = \text{softmax}(\frac{o_i^T K}{\sqrt{d}}), \tag{4}$$

where we use linear layers to map x_i and $\tilde{X}$ to a query vector o_i and key matrix K respectively, and d is the output channel number of the linear layers. The attention score is used to weight the contribution of shots in $\tilde{S}$ to augmenting s_i. We apply another linear layer to map $\tilde{X}$ to a value matrix V. Then the Contrastive Attention for augmenting x_i to be f_i is formulated as:

$$f_i = \text{concat}\big[x_i, \text{Linear}(\text{ReLu}(A(x_i, \tilde{X}) \cdot V))\big]. \tag{5}$$

Now we describe how to construct the auxiliary shot set $\tilde{S}$ for a specific s_i and how to regularize the feature augmentation discussed above. Inspired by our intuition that the cross-video key moments share common patterns and the key and non-key moments in the same video are supposed to be contrastive, we choose both common key moments and non-key moments to construct $\tilde{S}$. In particular, given a shot s_i in a mini-batch during training, we collect all the key moment shot *across videos* as well as the non-key moment shots surrounding s_i in the *same video* into $\tilde{S}$ (More details can be found in the supplementary material). The key and non-key moment shots in the supportive set $\tilde{S}$ are denoted by $\tilde{S}^+$ and $\tilde{S}^-$ respectively, and we propose the following loss as a regularizer to explicitly impose the contrastive relation between the key and non-key moments:

$$\mathcal{L}_C = -\sum_i \theta_i \log \frac{\sum_{j \in \tilde{S}+} \theta_j \exp(o_i^T k_j)}{\sum_{j \in \tilde{S}+} \theta_j \exp(o_i^T k_j) + \sum_{j \in \tilde{S}-} (1 - \theta_j) \exp(o_i^T k_j)} \tag{6}$$

where k_j is the j-th vector in the embedding key matrix K as in Eq. (4), and θ_i is a confidence weight indicating the reliability of the soft label for the shot, defined as a function of the Co-Attention score Att_i:

$$\theta_i = \frac{1}{1 + \exp(-\gamma(Att_i - \epsilon))} \tag{7}$$

where we empirically choose values of γ and ϵ to be $0.65 \times \max(Att_i)$ and 100 respectively. Equation (7) approximately maps the Co-Attention score to values

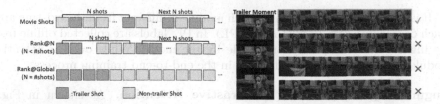

Fig. 4. Left: Rank@N. We calculate AP within every N consecutive shots in a full-length movie and average them as the overall performance metric, offering a local-view on the ranking performance. The top row lists trailer (blue) and non-trailer (grey) shots in a movie along the timeline before ranking. The middle and bottom illustrate the ideal Rank@N results. **Right:** the "hard" annotation brings about ambiguity in the labels. A trailer shot and its four visually similar movie shots are shown. The movie shot marked by the green border is labeled as positive and the rest shots are negative. (Color figure online)

of 0 or 1, which is a differentiable function and can be incorporated into the back-propagation of the learning process.

Finally, we combine the Co-Attention ranking loss Eq. (3) and the contrastive loss Eq. (6) as the training objective of CCANet:

$$\mathcal{L} = \mathcal{L}_{Rank} + \mathcal{L}_C. \tag{8}$$

4   Experiment Results

4.1   Movie Key Moment Detection Results

Dataset and Evaluation Metric. We evaluate our CCANet on the constructed dataset TMDD. Under a specific movie genre containing 50 movies, we randomly split the movies into the training and test set, containing 45 and 5 movies respectively. In the experiment, we repeat the split three times and report the average across three runs as the final result. During test, the movie shots are ranked based on the predicted score and then compared with the human-verified "key moment" ground-truth obtained by matching shots between trailers and movies as described in Sect. 3.1.

For the evaluation metric, we calculate Average Precision (AP) on each test video to measure the shot ranking performance. In order to get a fine-grain local view on the ranking performance on each video, we adapt AP to a Rank@N metric which can be illustrated in Fig. 4(Left). As seen, we examine the ranking AP within every N consecutive shots in the movie and average them across the entire movie as the performance metric. Rank@Global is equivalent to AP where N equals to the number of shots in the movie. We calculate the results on each movie and average them across all test movies as the overall performance.

Feature Extraction. The 3D CNN [9] (S3D) with a ResNet-34 [10] backbone pre-trained on Kinetics-400 dataset [3] are used to compute the input features.

Table 1. The trailer moment detection results on TMDD. "Sup", "PL" and "CoA" denote the different approaches, including fully-supervised, Pseudo-Label and our Co-Attention, with the basic 3D features [9]. The "Sup+CA", "PL+CA" and "CCANet" denote that the shot features in 'Sup', "PL" and "CoA" are augmented with our proposed Contrastive Attention module. The terms "Act", "Dra" and "ScF" refer to the movie categories, i.e., *Action, Drama* and *Sci-Fi*, and "Avg" indicates the "Average" result across categories. The subscripts "10", "20" and "GL" indicate different evaluation metrics of Rank@10,Rank@20 and Rank@Global.

	Act_{10}	Dra_{10}	ScF_{10}	Avg_{10}	Act_{20}	Dra_{20}	ScF_{20}	Avg_{20}	Act_{GL}	Dra_{GL}	ScF_{GL}	Avg_{GL}
Sup	0.691	0.603	0.562	0.619	0.558	0.507	0.363	0.476	0.153	0.158	0.141	0.151
Sup+CA	**0.725**	0.641	0.589	0.652	0.583	0.524	0.382	0.496	0.171	0.163	0.153	0.162
PL	0.681	0.591	0.515	0.596	0.542	0.506	0.361	0.460	0.218	0.191	0.169	0.193
PL+CA	0.714	0.625	0.548	0.629	0.577	0.539	0.372	0.486	0.269	0.215	**0.212**	0.232
CoA	0.695	0.667	0.556	0.639	0.574	0.540	0.397	0.504	0.228	0.221	0.176	0.208
CCANet	0.723	**0.692**	**0.591**	**0.669**	**0.612**	**0.562**	**0.428**	**0.534**	**0.271**	**0.246**	0.210	**0.242**

We use the output after the global pooling of the final convolution layer and a shot is represented by a feature of 512 dimensions, same as the work [39]. Specifically, a feature vector is extracted from a snippet covering 16 consecutive frames. The snippet features are averaged to represent the shot, where a snippet belongs to specific shot if >70% frames of the snippet are covered by the shot.

Implementation Details. We implement our model with PyTorch[2]., and optimize the loss with Adam optimizer [14] for 50 epochs. We use a batch size of 2048 and set the base learning rate to 0.001. With a single NVIDIA K80 gpu, the total feature extraction time for a 4-second shot is 0.05s. After extracting features, the time to train a ranking model for *Drama* movies is one hour, which contains $480K$ snippets in a total duration of ~100 h. At test time, it takes 0.04 s to score a batch of snippets after feature extraction.

Comparison Baselines. We compare our CCANet to two baselines, where the training and inference settings such as learning rate, batch size and so on, follow the same practice as CCANet.

- **Fully Supervised MTMD.** We assume the annotated trailer shots are accessible. Then we can perform supervised training as the VHD approaches described in Sect. 1. The movie shots annotated as trailer moment are the positive samples. For each positive sample, we sample 20 negative (non-trailer) shots, forming a set of pairs to train the ranking model as in Eq. (1).
- **Weakly Supervised MTMD with Pseudo Label.** We also compare CCANet to a weakly supervised approach using the Pseudo Label, which does not require access to manual annotations. We offline calculate the visual similarity between trailer and movie shots. The movie shots having the high similarity to the trailer are regarded as the positive samples, and those with low similarity as the negatives.

---

[2] https://pytorch.org/.

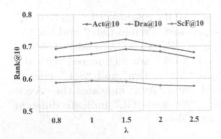

Fig. 5. Performance variance with respect to λ. We change the λ value and report the performance of the proposed CCANet. The evaluation metric is Rank@10.

Table 2. *Rank@Global* of the proposed CoA approach with different feature encoding strategies.

	Act_{GL}	Dra_{GL}	ScF_{GL}
CoA	0.228	0.221	0.208
+*FeaAug*	0.255	0.230	0.236
+*FeaAug*+$\mathcal{L}_C$	**0.271**	**0.246**	**0.242**

Results. Table 1 presents the trailer moment detection results of different approaches. As seen, by using our Co-Attention (CoA) module alone, our approach substantially outperforms the two baselines. Notably, CoA achieves ∼6% Rank@Global margin over the supervised approach. The trend is that the Rank@N drops as N increases, and Rank@Global is the lowest compared to $N=10, 20$. The performance drop is attributed to the fact that increasing N involves more negative samples for ranking. Especially, the fully-supervised approach drops the most at the global ranking metric. An explanation is that it suffers from the "hard" annotations provided by annotators. The "hard" means that a movie shot is considered as a positive sample only when it is an exact trailer moment. As shown in Fig. 4 (Right), only the shot at the top is annotated as the trailer shot (positive sample) as it is an exact match to the trailer while the other three are regarded as negative samples. Forcing those movie shots to be separable largely in the feature space brings the ambiguity to train the model. Our CoA module tackles this problem by assigning the soft labels to the data and a training pair with the closer attention scores contributes less to the loss calculation.

We also apply the proposed Contrastive Attention module to augmenting the features in all comparison approaches. In Table 1, the models with augmented features show superior performance over their origins with the 3D features only [9]. The results validate that exploring the relations among different shots can enhance the feature representation and boost the performance.

Impact of Parameter λ. In Eq. (3), we introduce a heuristic parameter λ to weight the validness of a training pair. The impact of λ to CCANet's performance is shown in Fig. 5, where we report the results measured by Rank@10 and choose the value leading to the best performance. As can be seen, the performance is not sensitive to the value variation of λ and we set the value of $\lambda = 1.5$ as default.

Ablation Study. In Table 2, we perform ablation study to examine our key contribution of Contrastive Attention by evaluating the CoA approach with three variants of shot feature encoding: 1) CoA uses the 3D feature only [9];

2) CoA+*FeaAug* augments features as Eq. (5) without contrastive loss $\mathcal{L}_C$; 3) CoA+*FeaAug*+$\mathcal{L}_C$ is our CCANet. The ~2% performance gain from +*FeaAug* over CoA shows the importance of exploring the relations among clips for feature encoding. Further, our CCANet consistently improves CoA+*FeaAug*. Our interpretation is that the loss $\mathcal{L}_C$ is introduced to guide the attention to be contrastive, encouraging the features of trailer shots to form a compact clique in the feature space and more distinguishable from the features of the non-trailer shots. As a result, it relieves the difficulty of learning the rank model and make CCANet achieve the best performance.

4.2   Video Highlight Detection Results

We also evaluate the proposed Contrastive Attention[3] on the VHD benchmarks, demonstrating its effectiveness. VHD has a similar goal to MTMD, aiming to detect the highlight moments in video which are supposed to be noticeable among the non-highlight moments, which naturally manifest the contrastive relations. We follow the work [39] to choose two challenging public video highlight detection datasets including YouTube Highlights [31] and TVSum [30]. The trained highlight detectors are domain-specific [39].

Datasets. YouTube Highlights dataset [31] contains six domain-specific categories: *surfing, skating, skiing, gymnastics, parkour, and dog.* Each domain consists of ~100 videos and the total duration is ~1430 min. Each video is divided into multiple clips and humans annotate whether a clip contains a specific category. TVSum [30] is collected from YouTube using 10 queries and consists of 50 videos in total from domains such as *changing vehicle tire, grooming an animal, making sandwiches, parade,* etc. (see Table 4). We follow the works [20,39] to average the frame-level scores to obtain the shot-level scores, and then select the top 50% shots from each video to build the ground-truth. Finally, the highlights selected by our approach are compared with the ground-truth.

Evaluation Metric and Baselines. We follow the works in [31,39], using the mean Average Precision (mAP) and mAP at top-5 to evaluate the highlight detection results on Youtube Highlights [31] and TVSum [30], respectively. We compare with eleven state-of-the-art approaches, which are categorized into unsupervised and supervised approaches. Those previous works' results are reported by the original papers. Specifically, We compare with the unsupervised approaches of RRAE [41], MBF [5], CVS [21], SG [16], DeSum-Net(DSN) [20], VESD [2] and LM [39]. In particular, the latest approach LM [39] uses the duration signal as the supervision to train a ranking model and training data contains around 10M Instagram videos.

We also include the supervised approaches, e.g. KVS [23], seqDPP [6], Sub-Mod [7], CLA, GIFs [8] and LSVM [31]. The latent SVM (LSVM) [31] has the same supervised ranking loss as ours, but LSVM uses the classic visual features while our features are augmented by the Contrastive Attention module.

---

[3] Our Co-Attention module is not applicable for the VHD task since there are no video pairs in VHD as the trailer-movie pairs in MTMD.

Table 3. The highlight detection mAP on YouTube Highlight dataset. Avg. is the average mAP over all the domains. Our approach outperforms all the baselines.

	RRAE [41]	GIFs [8]	LSVM [31]	CLA [39]	LM [39]	Ours
dog	0.49	0.308	0.6	0.502	0.579	**0.633**
gymnastic	0.35	0.335	0.41	0.217	0.417	**0.825**
parkour	0.5	0.54	0.61	0.309	**0.67**	0.623
skating	0.25	0.554	**0.62**	0.505	0.578	0.529
skiing	0.22	0.328	0.36	0.379	0.486	**0.745**
surfing	0.49	0.541	0.61	0.584	0.651	**0.793**
Avg.	0.383	0.464	0.536	0.416	0.564	**0.691**

Table 4. The highlight detection top-5 mAP score on TVSum [30]. The '-' means that mAP value is not provided in the original paper.

	MBF [5]	KVS [23]	CVS [21]	SG [16]	DSN [20]	VESD [2]	seqDPP [6]	SubMod [7]	CLA [39]	LM [39]	Ours
Vehicle tire	0.295	0.353	0.328	0.423	-	-	-	-	0.294	0.559	**0.613**
Vehicle unstuck	0.357	0.441	0.413	0.472	-	-	-	-	0.246	0.429	**0.546**
Grooming animal	0.325	0.402	0.379	0.475	-	-	-	-	0.590	0.612	**0.657**
Making sandwich	0.412	0.417	0.398	0.489	-	-	-	-	0.433	0.540	**0.608**
Parkour	0.318	0.382	0.354	0.456	-	-	-	-	0.505	**0.604**	0.591
Parade	0.334	0.403	0.381	0.473	-	-	-	-	0.491	0.475	**0.701**
Flash mob	0.365	0.397	0.365	0.464	-	-	-	-	0.430	0.432	**0.582**
Beekeeping	0.313	0.342	0.326	0.417	-	-	-	-	0.517	**0.663**	0.647
Bike tricks	0.365	0.419	0.402	0.483	-	-	-	-	0.578	0.691	**0.656**
Dog show	0.357	0.394	0.378	0.466	-	-	-	-	0.382	0.626	**0.681**
Average	0.345	0.398	0.372	0.462	0.424	0.423	0.447	0.461	0.447	0.563	**0.628**

Results on Youtube Highlights. Table 3 shows the results on YouTube Highlights dataset [31]. All the baseline results are quoted from the original papers. Our approach achieves the best performance and substantially improves those baselines with a large margin. Notably, our approach outperforms the following best performing LM [39] by 12.7% and CLA by 27.5%, with relative gains of 23% and 66%, where both LM and CLA models are trained on the additional 10M Instagram videos. We also achieve 15.5% performance gain over the LSVM. The LSVM [31] trains a ranking model with domain-specific manually annotated data, but its basic visual feature is limited to capture the feature distribution. Our proposed Contrastive Attention module explicitly models the relations between highlights and non-highlights so that highlight feature can form a compact clique in the feature space and stand out from the features of the non-highlights, leading to a more robust ranking model.

Results on TVSum. Table 4 shows the results on TVSum dataset [30]. Our approach outperforms all the baselines by a noticeable margin. In particular, our results achieve 6.5% mAP higher than the following best performing approach

LM [39]. Regarding the supervised approaches, we also outperform SubMod [7] by 16.7%, where the SubMod [7] proposes an adapted submodular function with structured learning for the highlight detection.

4.3 Understanding the Co-Contrastive Attention

Co-Attention Between Trailer and Movie Shots. We examine the Co-Attention scores between the trailer and movie shots. In Fig. 6, the score achieves the highest when the trailer moments exactly comes from the movie shots. Our model assigns reasonable high scores to the shots which are visually similar to the trailer moment.

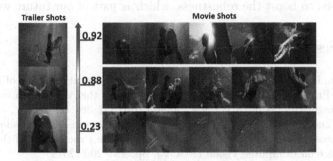

Fig. 6. The normalized Co-Attention scores between Trailer and Movie shots.

Feature Augmented by Contrastive Attention. In Fig. 7, we plot the UMAP embedding [17] of the basic 3D features and the augmented features with the Contrastive Attention on domains of *"Surfing"* and *"Gymnastic"* from Youtube Highlights [31] dataset. As can be seen, the augmented highlight and non-highlight features are more separable in the feature space, which eases the difficulty of learning a robust model for highlight detection, resulting in the performance improvement in both domains.

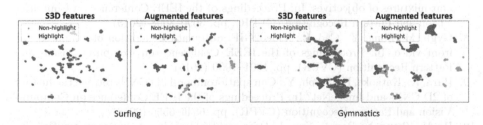

Fig. 7. The UMAP visualization of features over domains *"Surfing"* and *"Gymnastics"* (best view in color, zoom in).

5   Conclusion

In this work, we propose the CCANet to address the problem of learning the trailer moments from movies. Our approach utilizes Co-Attention scores as supervision, which does not require expensive human-annotations. Additionally, we introduce the Contrastive Attention module to augment the video features, equipping the model with the capacity of capturing the contrastive relation between the trailer and non-trailer moments. To evaluate our approach, we are the first to collect the dataset, TMDD. The effectiveness of our approach is demonstrated by the performance gain not only on our collected data but also on the public benchmarks. The results on TMDD also demonstrate there is a large room for improvements in trailer moment detection, e.g. multi-modality might be used to boost the robustness, which is part of our future work.

References

1. Bamman, D., O'Connor, B., Smith, N.A.: Learning latent personas of film characters. In: Proceedings of the 51st Annual Meeting of the Association for Computational Linguistics (Volume 1: Long Papers), pp. 352–361 (2013)
2. Cai, S., Zuo, W., Davis, L.S., Zhang, L.: Weakly-supervised video summarization using variational encoder-decoder and web prior. In: Proceedings of the European Conference on Computer Vision (ECCV), pp. 184–200 (2018)
3. Carreira, J., Zisserman, A.: Quo vadis, action recognition? a new model and the kinetics dataset. In: Proceedings of the IEEE Conference on Computer Vision and Pattern Recognition (CVPR), pp. 6299–6308 (2017)
4. Chu, E., Roy, D.: Audio-visual sentiment analysis for learning emotional arcs in movies. In: 2017 IEEE International Conference on Data Mining (ICDM), pp. 829–834. IEEE (2017)
5. Chu, W.S., Song, Y., Jaimes, A.: Video co-summarization: video summarization by visual co-occurrence. In: Proceedings of the IEEE Conference on Computer Vision and Pattern Recognition (CVPR), pp. 3584–3592 (2015)
6. Gong, B., Chao, W.L., Grauman, K., Sha, F.: Diverse sequential subset selection for supervised video summarization. In: Advances in Neural Information Processing Systems (NeurIPs), pp. 2069–2077 (2014)
7. Gygli, M., Grabner, H., Van Gool, L.: Video summarization by learning submodular mixtures of objectives. In: Proceedings of the IEEE Conference on Computer Vision and Pattern Recognition (CVPR), pp. 3090–3098 (2015)
8. Gygli, M., Song, Y., Cao, L.: Video2GIF: automatic generation of animated gifs from video. In: Proceedings of the IEEE Conference on Computer Vision and Pattern Recognition (CVPR), pp. 1001–1009 (2016)
9. Hara, K., Kataoka, H., Satoh, Y.: Can spatiotemporal 3D CNNs retrace the history of 2D CNNs and ImageNet? In: Proceedings of the IEEE Conference on Computer Vision and Pattern Recognition (CVPR), pp. 6546–6555 (2018)
10. He, K., Zhang, X., Ren, S., Sun, J.: Deep residual learning for image recognition. In: Proceedings of the IEEE Conference on Computer Vision and Pattern Recognition (CVPR), pp. 770–778 (2016)
11. Huang, Q., Xiong, Y., Xiong, Y., Zhang, Y., Lin, D.: From trailers to storylines: an efficient way to learn from movies. arXiv preprint arXiv:1806.05341 (2018)

12. Kang, H.W., Matsushita, Y., Tang, X., Chen, X.Q.: Space-time video montage. In: 2006 IEEE Computer Society Conference on Computer Vision and Pattern Recognition (CVPR), vol. 2, pp. 1331–1338. IEEE (2006)
13. Karpathy, A., Toderici, G., Shetty, S., Leung, T., Sukthankar, R., Fei-Fei, L.: Large-scale video classification with convolutional neural networks. In: Proceedings of the IEEE Conference on Computer Vision and Pattern Recognition (CVPR), pp. 1725–1732 (2014)
14. Kingma, D.P., Ba, J.: Adam: a method for stochastic optimization. arXiv preprint arXiv:1412.6980 (2014)
15. Lee, D.H.: Pseudo-label: the simple and efficient semi-supervised learning method for deep neural networks. In: Workshop on Challenges in Representation Learning, ICML, vol. 3, p. 2 (2013)
16. Mahasseni, B., Lam, M., Todorovic, S.: Unsupervised video summarization with adversarial LSTM networks. In: Proceedings of the IEEE Conference on Computer Vision and Pattern Recognition (CVPR), pp. 202–211 (2017)
17. McInnes, L., Healy, J., Melville, J.: UMAP: uniform manifold approximation and projection for dimension reduction. arXiv preprint arXiv:1802.03426 (2018)
18. Mendi, E., Clemente, H.B., Bayrak, C.: Sports video summarization based on motion analysis. Comput. Electr. Eng. **39**(3), 790–796 (2013)
19. Oosterhuis, H., Ravi, S., Bendersky, M.: Semantic video trailers. arXiv preprint arXiv:1609.01819 (2016)
20. Panda, R., Das, A., Wu, Z., Ernst, J., Roy-Chowdhury, A.K.: Weakly supervised summarization of web videos. In: Proceedings of the IEEE International Conference on Computer Vision (ICCV), pp. 3657–3666 (2017)
21. Panda, R., Roy-Chowdhury, A.K.: Collaborative summarization of topic-related videos. In: Proceedings of the IEEE Conference on Computer Vision and Pattern Recognition (CVPR), pp. 7083–7092 (2017)
22. Park, S.B., Kim, Y.W., Uddin, M.N., Jo, G.S.: Character-net: character network analysis from video. In: Proceedings of the 2009 IEEE/WIC/ACM International Joint Conference on Web Intelligence and Intelligent Agent Technology, vol. 01. pp. 305–308. IEEE Computer Society (2009)
23. Potapov, D., Douze, M., Harchaoui, Z., Schmid, C.: Category-specific video summarization. In: Fleet, D., Pajdla, T., Schiele, B., Tuytelaars, T. (eds.) ECCV 2014. LNCS, vol. 8694, pp. 540–555. Springer, Cham (2014). https://doi.org/10.1007/978-3-319-10599-4_35
24. Qiu, Z., Yao, T., Mei, T.: Learning spatio-temporal representation with pseudo-3D residual networks. In: proceedings of the IEEE International Conference on Computer Vision (ICCV), pp. 5533–5541 (2017)
25. Rui, Y., Gupta, A., Acero, A.: Automatically extracting highlights for TV baseball programs. In: Proceedings of the Eighth ACM International Conference on Multimedia, pp. 105–115. ACM (2000)
26. Shi, W., Gong, Y., Ding, C., MaXiaoyu Tao, Z., Zheng, N.: Transductive semi-supervised deep learning using min-max features. In: Proceedings of the European Conference on Computer Vision (ECCV), pp. 299–315 (2018)
27. Simões, G.S., Wehrmann, J., Barros, R.C., Ruiz, D.D.: Movie genre classification with convolutional neural networks. In: 2016 International Joint Conference on Neural Networks (IJCNN), pp. 259–266. IEEE (2016)
28. Smeaton, A.F., Over, P., Doherty, A.R.: Video shot boundary detection: seven years of TRECVid activity. Comput. Vis. Image Underst. (CVIU) **114**(4), 411–418 (2010)

29. Smith, J.R., Joshi, D., Huet, B., Hsu, W., Cota, J.: Harnessing AI for augmenting creativity: application to movie trailer creation. In: Proceedings of the 25th ACM International Conference on Multimedia, pp. 1799–1808 (2017)
30. Song, Y., Vallmitjana, J., Stent, A., Jaimes, A.: TVSum: summarizing web videos using titles. In: Proceedings of the IEEE Conference on Computer Vision and Pattern Recognition (ICCV), pp. 5179–5187 (2015)
31. Sun, M., Farhadi, A., Seitz, S.: Ranking domain-specific highlights by analyzing edited videos. In: Fleet, D., Pajdla, T., Schiele, B., Tuytelaars, T. (eds.) ECCV 2014. LNCS, vol. 8689, pp. 787–802. Springer, Cham (2014). https://doi.org/10.1007/978-3-319-10590-1_51
32. Tang, H., Kwatra, V., Sargin, M.E., Gargi, U.: Detecting highlights in sports videos: cricket as a test case. In: 2011 IEEE International Conference on Multimedia and Expo (ICME), pp. 1–6. IEEE (2011)
33. Tapaswi, M., Zhu, Y., Stiefelhagen, R., Torralba, A., Urtasun, R., Fidler, S.: MovieQA: understanding stories in movies through question-answering. In: Proceedings of the IEEE Conference on Computer Vision and Pattern Recognition (ICCV), pp. 4631–4640 (2016)
34. Thomson, D.: Moments That Made the Movies. Thames & Hudson (2014). https://books.google.com/books?id=_vNFngEACAAJ
35. Vaswani, A., et al.: Attention is all you need. In: Guyon, I., et al. (eds.) Advances in Neural Information Processing Systems (NeurIPS), vol. 30, pp. 5998–6008. Curran Associates Inc. (2017)
36. Wang, X., Girshick, R., Gupta, A., He, K.: Non-local neural networks. In: Proceedings of the IEEE Conference on Computer Vision and Pattern Recognition (CVPR), pp. 7794–7803 (2018)
37. Weng, C.Y., Chu, W.T., Wu, J.L.: RoleNet: movie analysis from the perspective of social networks. IEEE Trans. Multimedia 11(2), 256–271 (2009)
38. Xie, S., Sun, C., Huang, J., Tu, Z., Murphy, K.: Rethinking spatiotemporal feature learning for video understanding. In: Proceedings of the European Conference on Computer Vision ECCV (2018)
39. Xiong, B., Kalantidis, Y., Ghadiyaram, D., Grauman, K.: Less is more: learning highlight detection from video duration. In: Proceedings of the IEEE Conference on Computer Vision and Pattern Recognition (CVPR), pp. 1258–1267 (2019)
40. Xiong, Z., Radhakrishnan, R., Divakaran, A., Huang, T.S.: Highlights extraction from sports video based on an audio-visual marker detection framework. In: 2005 IEEE International Conference on Multimedia and Expo (ICME), p. 4. IEEE (2005)
41. Yang, H., Wang, B., Lin, S., Wipf, D., Guo, M., Guo, B.: Unsupervised extraction of video highlights via robust recurrent auto-encoders. In: Proceedings of the IEEE International Conference on Computer Vision (ICCV), pp. 4633–4641 (2015)
42. Yao, T., Mei, T., Rui, Y.: Highlight detection with pairwise deep ranking for first-person video summarization. In: Proceedings of the IEEE Conference on Computer Vision and Pattern Recognition (CVPR), pp. 982–990 (2016)
43. Zhou, H., Hermans, T., Karandikar, A.V., Rehg, J.M.: Movie genre classification via scene categorization. In: Proceedings of the 18th ACM International Conference on Multimedia, pp. 747–750. ACM (2010)
44. Zhu, Y., et al.: Aligning books and movies: towards story-like visual explanations by watching movies and reading books. In: Proceedings of the IEEE International Conference on Computer Vision (ICCV), pp. 19–27 (2015)

Preserving Semantic Neighborhoods
for Robust Cross-Modal Retrieval

Christopher Thomas$^{(\boxtimes)}$ and Adriana Kovashka

University of Pittsburgh, Pittsburgh, PA 15260, USA
{chris,kovashka}@cs.pitt.edu

Abstract. The abundance of multimodal data (e.g. social media posts) has inspired interest in cross-modal retrieval methods. Popular approaches rely on a variety of metric learning losses, which prescribe what the proximity of image and text should be, in the learned space. However, most prior methods have focused on the case where image and text convey redundant information; in contrast, real-world image-text pairs convey complementary information with little overlap. Further, images in news articles and media portray topics in a visually diverse fashion; thus, we need to take special care to ensure a meaningful image representation. We propose novel within-modality losses which encourage semantic coherency in both the text and image subspaces, which does not necessarily align with visual coherency. Our method ensures that not only are paired images and texts close, but the expected image-image and text-text relationships are also observed. Our approach improves the results of cross-modal retrieval on four datasets compared to five baselines.

1 Introduction

Vision-language tasks such as image captioning [2,27,58] and cross-modal generation and retrieval [40,60,63] have seen increased interest in recent years. At the core of methods in this space are techniques to bring together images and their corresponding pieces of text. However, most existing cross-modal retrieval methods only work on data where the two modalities (images and text) are well aligned, and provide fairly redundant information. As shown in Fig. 1, captioning datasets such as COCO contain samples where the overlap between images and text is significant (both image and text mention or show the same objects). In this setting, cross-modal retrieval means finding the manifestation of a single concept in two modalities (e.g. learning embeddings such that the word "banana" and the pixels for "banana" project close by in a learned space).

In contrast, real-world news articles contain image and text pairs that cover the same topic, but show complementary information (protest signs vs information about the specific event; guns vs discussion of rights; rainbow flag vs LGBT rights). While a human viewer can still guess which images go with which text, the alignment between image and text is abstract and symbolic. Further, images in news articles are ambiguous *in isolation*. We show in Fig. 2 that an image

© Springer Nature Switzerland AG 2020
A. Vedaldi et al. (Eds.): ECCV 2020, LNCS 12363, pp. 317–335, 2020.
https://doi.org/10.1007/978-3-030-58523-5_19

Fig. 1. Image-text pairs from COCO [25] and Politics [49]. Traditional image captions (top) are descriptive of the image, while we focus on the more challenging problem of aligning images and text with a non-literal complementary relationship (bottom).

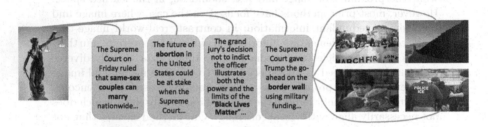

Fig. 2. The image on the left symbolizes justice and may be paired with text about a variety of subjects (e.g. abortion, same sex marriage). Similarly, text regarding immigration (right) may be paired with visually dissimilar images. Our approach enforces that *semantically* similar content (images on the right) is close in the learned space. To discover such content, we use semantic neighbors of the text and their paired images. (Color figure online)

might illustrate multiple related texts (shown in green), and each text in turn could be illustrated with multiple visually distant images (e.g. the four images on the right-hand side could appear with the border wall text). Thus, we must first resolve any ambiguities in the image, and figure out "what it means".

We propose a metric learning approach where we use the semantic relationships between text segments, to guide the embedding learned for corresponding images. In other words, to understand what an image "means", we look at what articles it appeared with. Unlike prior approaches, we capture this information not only across modalities, but within the image modality itself. If texts y_i and y_j are semantically similar, we learn an embedding where we explicitly encourage their paired images x_i and x_j to be similar, using a new unimodal loss. Note that in general x_i and x_j need not be similar in the original visual space (Fig. 2). In addition, we encourage texts y_i and y_j, who were close in the unimodal space, to remain close.

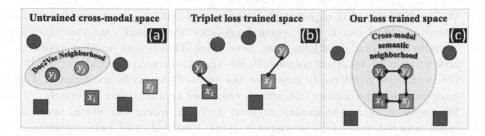

Fig. 3. We show how our method enforces cross-modal semantic coherence. Circles represent text and squares images. In (a), we show the untrained cross-modal space. Note y_i and y_j are neighbors in Doc2Vec space and thus semantically similar. (b) shows the space after triplet loss training. y_i and x_i, and y_j and x_j, are now close as desired, but y_i and y_j have moved apart, and x_i and x_j remain distant. (c) shows our loss's effect. Now, all semantic neighbors (both images and text) are pulled closer.

Our novel loss formulation explicitly encourages *within-modality semantic coherence.* Figure 3 shows the effect. On the left, we show the proximity of samples before cross-modal learning; specifically, while two texts are close in the document space, their paired articles may be far from the texts. In the middle, we show the effect of using a standard triplet loss, which pulls image-text pairs close, but does not necessarily preserve the similarity of related articles; they are now further than they used to be in the original space. In contrast, on the right, we show how our method brings paired images and text closer, while also preserving a semantically coherent region, i.e. the texts remained close.

In our approach, we use neighborhoods in the original text document space, to compute semantic proximity. We also experiment with an alternative approach where we compute neighborhoods using the visual space, then guide the corresponding texts to be close. This approach is a variant of ours, and is novel in the sense that it uses proximity in one unimodal space, to guide the other space/modality. While unimodal losses based on visual similarity are helpful over a standard cross-modal loss (e.g. triplet loss), our main approach is superior.

Next, we compare to a method [52] which utilizes the *set* of text annotations available for an image in COCO, to guide the structure of the learned space. We show that when these ground-truth annotations are available, using them to compute neighborhoods in the textual space is the most reliable. However, on many datasets, such sets of annotations (more than one for the same image) are not available. We show that our approach offers a comparable alternative.

Finally, we test the contribution of our additional losses using PVSE [48], a state-of-the-art visual semantic embedding model, as a backbone. We show that our proposed loss further improves the performance of this model.

To summarize, our contributions are as follows.

- We preserve relationships in the original *semantic* space. Because images do not clearly capture semantics, we use the semantic space (from text) to guide the image representation, through a unimodal (within-modality) loss.

- We perform detailed experimental analysis of our proposed loss function, including ablations, on four recent large-scale image-text datasets. One [3] contains multimodal articles from New York Times, and another contains articles from far-left/right media [49]. We also conduct experiments on [25,43]. Our approach significantly improves the state-of-the-art in most cases. The more abstract the dataset/alignment, the more beneficial our approach.
- We tackle a new cross-modal retrieval problem where the visual space is much less concrete. This scenario is quite practical, and has applications ranging from automatic caption generation for news images, to detection of fake multimodal articles (i.e. detecting whether an image supports the text).

2   Related Work

Cross-Modal Learning. A fundamental problem in cross-modal inference is the creation of a shared semantic manifold on which multiple modalities may be represented. The goal is to learn a space where content about related semantics (e.g. images of "border wall" and text about "border wall") projects close by, regardless of which modality it comes from. Many image-text embedding methods rely on a two-stream architecture, with one stream handling visual content (e.g. captured by a CNN) and the other stream handling textual content (e.g. through an RNN). Both streams are trained with paired data, e.g. an image and its captions, and a variety of loss functions are used to encourage both streams to produce similar embeddings for paired data. Recently, purely attention-based approaches have been proposed [6,26]. One common loss used to train retrieval models is triplet loss, which originates in the (single-modality) metric learning literature, e.g. for learning face representations [42]. In cross-modal retrieval, the triplet loss has been used broadly [9,32,34,38,57,66]. Alternative choices include angular loss [51], N-pairs loss [47], hierarchical loss [11], and clustering loss [36].

While single-modality losses like triplet, angular and N-pairs have been used across and within modalities, they are not sufficient for cross-modal retrieval. These losses do not ensure that the general semantics of the text are preserved in the new cross-modal space; thus, the cross-modal matching task might distort them too much. This phenomenon resembles forgetting [14,24] but in the cross-modal domain. Our method preserves within-modal structure, and a similar effect can be achieved by leveraging category labels as in [5,31,50,64]; however, such labels are not available in the datasets we consider, nor is it clear how to define them, since matches lie beyond the presence of objects. Importantly, classic retrieval losses do not tackle the complementary relationship between images and text, which makes the space of topically related images more visually diffuse. In other words, two images might depict substantially *different visual* content but nonetheless be *semantically related.*

Note that we do not propose a new *model* for image-text alignment, but instead propose cross-modal embedding *constraints* which can be used to train any such model. For example, we compare to Song et al. [48]'s recent polysemous visual semantic embedding (PVSE) model, which uses global and local features to compute self-attention residuals. Our loss improves [48]'s performance.

Our work is also related to cross-modal distillation [10, 12, 15, 46], which transfers supervision across modalities, but none of these approaches exploit the semantic signal that text neighborhoods carry to constrain the visual representations. Finally, [1, 22, 61] detect different types of image-text relationships (e.g. parallel, complementary) but do not retrieve across modalities.

Metric Learning approaches learn distance metrics which meaningfully measure the similarity of objects. These can be broadly categorized into: 1) sampling-based methods [17, 18, 28, 29, 36, 44, 45, 53, 54, 56, 59], which intelligently choose easy/hard samples or weight samples; or 2) loss functions [8, 11, 16, 42, 47, 51, 55] which impose intuitions regarding neighborhood structure, data separation, etc. Our method relates to the second category. Triplet loss [20, 42] takes into account the *relative* similarity of positives and negatives, such that positive pairs are closer to each other than positives are to negatives. [62] generalize triplet loss by fusing it with classification loss. [37] integrate all positive and negative pairs within a minibatch, such that all pair combinations are updated jointly. Similarly, [47]'s N-pair loss pushes multiple negatives away in each triplet. [52] propose a structural loss, which pulls multiple text paired with the same image together, but requires more than one ground truth caption per image (which most datasets lack). In contrast, our approach pulls semantically similar images *and* text together and only requires a single caption per image. More recently, [51] propose an angular loss which leverages the triangle inequality to constrain the angle between points within triplets. We show how cross-modal complementary information (semantics paired with diverse visuals) can be leveraged to improve the learned embedding space, regardless of the specific loss used.

3   Method

Consider two image-text pairs, $\{x_i, y_i\}$ and $\{x_j, y_j\}$. To ground the "meaning" of the images, we use proximity in a generic, pre-trained textual space between the texts y_i and y_j. If y_i and y_j are semantically close, we expect that they will also be relatively close in the learned space, and further, that x_i and x_j will be close also. We observed that, while intuitive, this expectation does not actually hold in the learned cross-modal space. The problem becomes more severe when image and paired text do not exhibit literal alignment, as shown in Fig. 1, because images paired via text neighbors could be visually different.

We describe how several common existing loss functions tackle cross-modal retrieval, and discuss their limitations. We then propose two constraints which pull within-modality semantic neighbors close to each other. Fig. 4 illustrates how our approach differs from standard metric learning losses.

3.1   Problem Formulation and Existing Approaches

We assume a dataset $\mathcal{D} = \{\mathbf{I}, \mathbf{T}\}$ of n image-text pairs, where $\mathbf{I} = \{x_1, x_2, \ldots, x_n\}$ and $\mathbf{T} = \{y_1, y_2, \ldots, y_n\}$ denote the set of paired images and

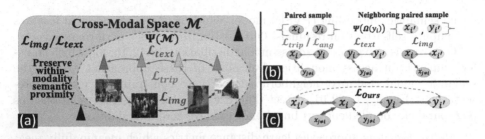

Fig. 4. (a): $\mathcal{L}_{text}$ and $\mathcal{L}_{img}$ pull semantic neighbors of the same modality closer. The images are visually distinct, but semantically similar. (b): Pull connections are shown in green, and push in red. $\mathcal{L}_{trip}$ and $\mathcal{L}_{ang}$ operate cross-modally, but impose no within-modality constraints. (c): $\mathcal{L}_{ours}$ (which combines all three losses above) exploits the paired nature of the data to enforce the expected inter/intra-modal relationships. Solid lines indicate connections that our loss enforces but triplet/angular do not. (Color figure online)

text, respectively. By pairs, we mean y_i is text related to or co-occurring with image x_i. Let f_I denote a convolutional neural network which projects images into the joint space and f_T a recurrent network which projects text. We use the notational shorthand $f_T(y) = y$ and $f_I(x) = x$. The goal of training f_I and f_T is to learn a cross-modal manifold $\mathcal{M}$ where semantically similar samples are close. At inference time, we wish to retrieve a ground-truth paired text given an input image, or vice versa. One common technique is triplet loss [42] which posits that paired samples should be closer to one another than they are to non-paired samples. Let $\mathcal{T} = (x_i^a, y_i^p, y_j^n)$ denote a triplet of samples consisting of an anchor (a), positive or paired sample (p), and negative or non-paired sample (n) chosen randomly such that $i \neq j$. Let m denote a margin. The triplet loss $\mathcal{L}_{trip}$ is then:

$$\mathcal{L}_{trip}(\mathcal{T}) = \left[\|x_i^a - y_i^p\|_2^2 - \|x_i^a - y_j^n\|_2^2 + m\right]_+ \tag{1}$$

This loss is perhaps the most common one used in cross-modal retrieval tasks, but it has some deficiencies. For example, the gradient of the triplet loss wrt. each point only considers two points, but ignores their relationship with the third one; for example, $\frac{\partial \mathcal{L}_{trip}}{\partial x_i^a} = 2\left(y_j^n - y_i^p\right)$. This allows for degenerate cases, so angular loss $\mathcal{L}_{ang}$ [51] accounts for the angular relationship of all three points:

$$\mathcal{L}_{ang}(\mathcal{T}) = \left[\|x_i^a - y_i^p\|_2^2 - 4\tan^2\alpha\|y_j^n - \mathcal{C}_i\|_2^2\right]_+ \tag{2}$$

where $\mathcal{C}_i = (x_i^a + y_i^p)/2$ is the center of a circle through anchor and positive.

One challenging aspect of these losses is choosing a good negative term in the triplet. If the negative is too far from the anchor, the loss becomes 0 and no learning occurs. In contrast, if negatives are chosen too close, the model may have difficulty converging to a reasonable solution as it continuously tries to move samples to avoid overlap with the negatives. How to best sample triplets to avoid these issues is an active area of research [8]. One recent technique, the

N-pairs loss [47], proposes that instead of a single negative sample being used, all negatives within the minibatch should be used. The N-pairs loss $\mathcal{L}_{ang}^{NP}$ pushes the anchor and positive embedding away from *multiple* negatives simultaneously:

$$\mathcal{L}_{ang}^{NP}(\mathcal{T}) = \sum_{y_j \in \text{minibatch}, \; j \neq i} \mathcal{L}_{ang}\left(x_i^a, y_i^p, y_j^n\right) \tag{3}$$

The symmetric constraint [65] can also be added to explicitly account for bidirectional retrieval, i.e. text-to-image, by swapping the role of images and text to form symmetric triplets $\mathcal{T}_{sym} = (y_i^a, x_i^p, x_i^n)$:

$$\mathcal{L}_{ang}^{NP+SYM}(\mathcal{T}, \mathcal{T}_{sym}) = \mathcal{L}_{ang}^{NP}(\mathcal{T}) + \mathcal{L}_{ang}^{NP}(\mathcal{T}_{sym}) \tag{4}$$

Limitations. While these loss functions have been used for cross-modal retrieval, they do not take advantage of several unique aspects of the multi-modal setting. Only the dashed pull/push connections in Fig. 4 (c) are part of triplet/angular loss. The solid connections are intuitive, but only enforced in our novel formulation. We argue the lack of explicit *within-modality* constraints allows discontinuities within the space for semantically related content from the same modality.

3.2   Our Proposed Loss

The text domain provides a semantic fingerprint for the image-text pair, since vastly dissimilar visual content may still be semantically related (e.g. image of White house, image of protest), while similar visual content (e.g. crowd in church, crowd at mall) could be semantically unrelated. We thus use the text domain to constrain within-modality semantic locality for both images and text.

To measure ground-truth semantic similarity, we pretrain a Doc2Vec [23] model Ω on the train set of text. Specifically, let d be the document embedding of article y_i, T denote the number of words in y_i, w_t represent the embedding learned for word t, $p(\cdot)$ be the probability of the given word, and k denote the look-around window. Ω learns word embeddings and document embeddings which maximize the average log probability: $\frac{1}{T}\sum_{t=1}^{T} \log p\left(w_t | d, w_{t-k}, \ldots, w_{t+k}\right)$. After training Ω, we use iterative backpropagation to compute the document embedding which maximizes the log probability for every article in the dataset: $\Omega(\mathbf{T}) = \{\Omega(y_1), \ldots, \Omega(y_n)\}$.

Because Doc2Vec has been shown to capture latent topics within text documents well [35], we seek to enforce that locality originally captured in $\Omega(\mathbf{T})$'s space also be preserved in the cross-modal space $\mathcal{M}$. Let

$$\Psi\left(\Omega(y_i)\right) = \langle x_{i'}, y_{i'} \rangle \tag{5}$$

denote a nearest neighbor function over $\Omega(\mathbf{T})$, where $\langle \cdot, \cdot \rangle$ is an image-text pair in the train set randomly sampled from the $k = 200$ nearest neighbors to y_i, and $i \neq i'$. $\Psi\left(\Omega(y_i)\right)$ thus returns an image-text pair semantically related to y_i.

We formulate two loss functions to enforce within-modality semantic locality in $\mathcal{M}$. The first, $\mathcal{L}_{text}$, enforces locality of the text's projections:

$$T'_{text} = \left(y_i^a, y_{i'}^p, y_j^n\right)$$

$$\mathcal{L}_{text}\left(T'_{text}\right) = \mathcal{L}_{ang}\left(T'_{text}\right) \tag{6}$$

$$\mathcal{L}_{ang}\left(T'_{text}\right) = \left[\|y_i^a - y_{i'}^p\|_2^2 - 4\tan^2\alpha\|y_j^n - \mathcal{C}_i\|_2^2\right]_+$$

where y_j^n is the negative sample chosen randomly such that $i \neq j$ and $\mathcal{C}_i = \left(y_i^a + y_i^p\right)/2$. $\mathcal{L}_{text}$ is the most straightforward transfer of semantics from $\Omega(\mathbf{T})$'s space to the joint space: nearest neighbors in Ω should remain close in $\mathcal{M}$.

As Fig. 4 (c) shows, $\mathcal{L}_{text}$ also indirectly causes semantically related images to move closer in $\mathcal{M}$: there is now a weak connection between x_i and $x_{i'}$ through the now-connected y_i and $y_{i'}$. To directly ensure smoothness and semantic coherence between x_i and $x_{i'}$, we propose a second constraint, $\mathcal{L}_{img}$:

$$T'_{img} = \left(x_i^a, x_{i'}^p, x_j^n\right)$$

$$\mathcal{L}_{img}\left(T'_{img}\right) = \mathcal{L}_{ang}\left(T'_{img}\right) \tag{7}$$

$$\mathcal{L}_{ang}\left(T'_{img}\right) = \left[\|x_i^a - x_{i'}^p\|_2^2 - 4\tan^2\alpha\|x_j^n - \mathcal{C}_i\|_2^2\right]_+$$

where x_j^n is the randomly chosen negative sample such that $i \neq j$ and $\mathcal{C}_i = \left(x_i^a + x_i^p\right)/2$. Note that x_i and $x_{i'}$ are often not going to be neighbors in the original visual space. We use N-pairs over all terms to maximize discriminativity, and symmetric loss to ensure robust bidirectional retrieval:

$$\mathcal{L}_{ang}^{OURS}\left(T, T_{sym}, T'_{text}, T'_{img}\right) =$$

$$\mathcal{L}_{ang}^{NP+SYM}\left(T, T_{sym}\right) + \alpha\mathcal{L}_{text}^{NP}\left(T'_{text}\right) + \beta\mathcal{L}_{img}^{NP}\left(T'_{img}\right) \tag{8}$$

where α, β are hyperparameters controlling the importance of each constraint.

Second Variant. We also experiment with a variant of our method where the nearest neighbor function in Eq. 5 (computed in Doc2Vec space) is replaced with one that computes nearest neighbors in the space of visual (e.g. ResNet) features. Now $x_i, x_{i'}$ are neighbors in the original visual space before cross-modal training, and $y_i, y_{i'}$ are their paired articles (which may not be neighbors in the original Doc2Vec space). We denote this method as OURS (Img NNs) in Table 1, and show that while it helps over a simple triplet- or angular-based baseline, it is inferior to our main method variant described above.

Discussion. At a low level, our method combines three angular losses. However, note that our losses in Eq. 6 and Eq. 7 do not exist in prior literature. While [52] leverages ground-truth neighbors (sets of neighbors provided together for the same image sample in a dataset), we are not aware of prior work that estimates neighbors. Importantly, we are not aware of prior work that uses the text space to construct a loss over the image space, as Eq. 7 does. We show that the choice of space in which semantic coherency is computed is important; doing this in the original textual space is superior than using the original image space. We show the contribution of both of these losses in our experiments.

3.3 Implementation Details

All methods use a two-stream architecture, with the image stream using a ResNet-50 [19] architecture initialized with ImageNet features, and the text stream using Gated Recurrent Units [7] with hidden state size 512. We use image size 224×224 and random horizontal flipping, and initialize all non-pretrained learnable weights via Xavier init. [13]. Text models are initialized with word embeddings of size 200 learned on the target dataset. We apply a linear transformation to each model's output features ($\mathbb{R}^{2048 \times 256}$ for image, $\mathbb{R}^{512 \times 256}$ for text) to get the final embedding, and perform L_2 normalization. We use Adam [21] with minibatch size 64, learning rate $1.0e-4$, and weight decay $1e-5$. We decay the learning rate by a factor of 0.1 after every 5 epochs of no decrease in val. loss. We use a train-val-test split of 80-10-10. For Doc2Vec, we use [41] with $d \in \mathbb{R}^{200}$ and train using distributed memory [23] for 20 epochs with window $k = 20$, ignoring words that appear less than 20 times. We use hierarchical softmax [33] to compute $p(\cdot)$. To efficiently compute approximate nearest neighbors for Ψ, we use [30]; our method adds negligible computational overhead as neighbors are computed prior to training. We choose $\alpha = 0.3, \beta = 0.1$ for $\mathcal{L}_{trip}^{OURS}$, and $\alpha = 0.2, \beta = 0.3$ for $\mathcal{L}_{ang}^{OURS}$, on a held-out val. set.

4 Experiments

We compare our method to five baselines on four recent large-scale datasets. Our results consistently demonstrate the superiority of our approach at bidirectional retrieval. We also show our method better preserves within-modality semantic locality by keeping neighboring images and text closer in the joint space.

4.1 Datasets

Two datasets feature challenging indirect relations between image and text, compared to standard captioning data. These also exhibit longer text paired with images: 59 and 18 words on average, compared to 11 in COCO.

Politics. [49] consists of images paired with news articles. In some cases, multiple images were paired with boilerplate text (website headliner, privacy policy) due to failed data scraping. We removed duplicates using MinHash [4]. We were left with 246,131 unique image-text pairs. Because the articles are lengthy, we only use the first two sentences of each. [49] do not perform retrieval.

GoodNews. [3] consists of ~466k images paired with their captions. All data was harvested from the New York Times. Captions often feature abstract or indirect text in order to relate the image to the article it appeared with. The method in [3] takes image and text as input, hence cannot serve as a baseline.

We also test on two large-scale standard image captioning datasets, where the relationship between image and text is typically more direct:

Table 1. We show retrieval results for image to text (**I→T**) and text to image (**T→I**) on all datasets. The best method per group is shown in bold.

| | Img-Text Non-Literal | | | | Img-Text Literal | | | |
| | Politics [49] | | GoodNews [3] | | ConcCap [43] | | COCO [25] | |
Method	I→T	T→I	I→T	T→I	I→T	T→I	I→T	T→I
Ang+NP+Sym	0.6270	0.6216	0.8704	0.8728	0.7687	0.7695	**0.6976**	**0.6964**
Ours (Img NNs)	0.6370	0.6378	0.8840	0.8852	0.7636	0.7666	0.6819	0.6876
Ours	**0.6467**	**0.6492**	**0.8849**	**0.8865**	**0.7760**	**0.7835**	0.6900	0.6885
PVSE	0.6246	0.6199	0.8724	0.8709	0.7746	0.7809	0.6878	0.6892
PVSE+Ours	**0.6264**	**0.6314**	**0.8867**	**0.8864**	**0.7865**	**0.7924**	**0.6932**	**0.6925**
Trip+NP+Sym	0.4742	0.4801	0.7203	0.7216	**0.5413**	0.5332	**0.4957**	**0.4746**
Ours (Trip)	**0.4940**	**0.4877**	**0.7390**	**0.7378**	0.5386	**0.5394**	0.4790	0.4611

COCO. [25] is a large dataset containing numerous annotations, such as objects, segmentations, and captions. The dataset contains ∼120k images with captions. Unlike our other datasets, COCO contains more than one caption per image, with each image paired with four to seven captions.

Conceptual Captions. [43] is composed of ∼3.3M image-text pairs. The text comes from automatically cleaned alt-text descriptions paired with images harvested from the internet and has been found to represent a much wider variety of style and content compared to COCO.

4.2   Baselines

We compare to N-Pairs Symmetric Angular Loss (Ang+NP+Sym, a combination of [47,51,65], trained with $\mathcal{L}_{ang}^{NP+SYM}$). For a subset of results, we also replace the angular loss with the weaker but more common triplet loss (Trip+NP+Sym). We show the result of choosing to enforce coherency within the image and text modalities by using images rather than text; this is the second variant of our method, denoted Ours (Img NNs).

We also compare our approach against the deep structure preserving loss [52] (Struc), which enforces that captions paired with the same image are closer to each other than to non-paired captions.

Finally, we show how our approach can improve the performance of a state-of-the-art cross-modal retrieval model. PVSE [48] uses both images and text to compute a self-attention residual before producing embeddings.

4.3   Quantitative Results

We formulate a cross-modal retrieval task such that given a query image or text, the embedding of the paired text/image must be closer to the query embedding than non-paired samples also of the target modality. We sample random (non-paired) samples from the test set, along with the ground-truth paired sample. We then compute Recall@1 within each task: that is, whether the ground truth paired sample is closer to its cross-modal embedding than the non-paired embeddings. For our most challenging datasets (GoodNews and Politics), we use a 5-way task. For COCO and Conceptual Captions, we found this task to be too simple and that all methods easily achieved very high performance due to the literal image-text relationship. Because we wish to distinguish meaningful performance differences between methods, we used a 20-way task for Conceptual Captions and a 100-way task for COCO. Task complexities were chosen based on the baseline's performance, before our method's results were computed.

We report the results in Table 1. The first and second group of results all use angular loss, while the third set use triplet loss. We observe that our method significantly outperforms all baselines tested for both directions of cross-modal retrieval for three of the four datasets. Our method achieves a 2% relative boost in accuracy (on average across both retrieval tasks) vs. the strongest baseline on GoodNews, and a 4% boost on Politics. We also observe recall is much worse for all tasks on the Politics dataset compared to GoodNews, likely because the images and article text are much less well-aligned. The performance gap seems small but note that given the figurative use of images in these datasets, often there may not be a clear ground-truth answer. In Fig. 2, Themis may be constrained to be close to protestors or border wall. At test time, the ground-truth text paired with Themis may be about the Supreme Court, but one of the "incorrect" answers could be about immigration or freedom, which still make sense. Our method keeps more neighbors closer to the query point as shown next, thus may retrieve plausible, but technically "incorrect" neighbors for a query.

Importantly, we see that while the variant of our method using neighborhoods computed in image space (OURS Img NNs) does outperform ANG+NP+SYM, it is weaker than our main method variant (OURS). We also observe that when adding our loss on top of the PVSE model [48], accuracy of retrieval improves. In other words, our loss is complementary to advancements accomplished by network model-based techniques such as attention.

Our method outperforms the baselines on ConcCap also, but not on COCO, since COCO is the easiest, least abstract of all datasets, with the most literal image-text alignment. Our approach constrains neighboring texts and their images to be close, and for datasets where matching is on a more abstract, challenging level, the benefit of neighbor information outweighs the disadvantage of this inexact similarity. However, for more straightforward tasks (e.g. in COCO), it may introduce noise. For example, for caption "a man on a bicycle with a banana", the model may pull that image and text closer to images with a banana in a bowl of fruit. Overall, our approach of enforcing within-modality semantic neighborhoods substantially improves cross-view retrieval, particularly when the relationship between image and text is complementary, rather than redundant.

To better ground our method's performance in datasets typically used for retrieval, we also conducted an experimented on Flickr30K [39]. Since that

Table 2. We show retrieval results for image to text $(I{\to}T)$ and text to image $(T{\to}I)$ on COCO using [52]'s loss vs. ours. GT requires multiple Ground Truth captions per image, while NN uses Nearest Neighbors. The best method per row is shown in bold, while the best method which does not require a set of neighboring text is underlined.

		Ours (Trip)	Struc (GT, Text)	Struc (NN$_\Omega$, Text)	Struc (NN$_\Omega$, Img)
COCO	I→T	<u>0.4790</u>	**0.4817**	0.4635	0.4752
	T→I	<u>0.4611</u>	**0.4867**	0.4594	0.4604

Table 3. We test how well each method preserves the semantic neighborhood (see text) of Ω in $\mathcal{M}$. Higher values are better. Best method is shown in bold.

Method	GoodNews [3]		Politics [49]	
	I	T	I	T
Trip+NP+Sym	0.1183	0.1294	0.1135	0.1311
Ours (Trip)	**0.1327**	**0.1426**	**0.1319**	**0.1483**
Ang+NP+Sym	0.1032	0.1131	0.1199	0.1544
Ours (Ang)	**0.1270**	**0.1376**	**0.1386**	**0.1703**

dataset does not exhibit image-text complementarity, we do not expect our method to improve performance, but it should not significantly reduce it. We compared the original PVSE against PVSE with our novel loss. We observed that our method slightly outperformed the original PVSE, on both text-to-image and image-to-text retrieval (0.5419 and 0.5559 for ours, vs 0.5405 and 0.5539 for PVSE).

In Table 2, we show a result comparing our method to Deep Structure Preserving Loss [52]. Since this method requires a *set* of annotations (captions) for an image, i.e. it requires *ground-truth neighbor relations* for texts, we can only apply it on COCO. In the first column, we show our method. In the second, we show [52] using ground-truth neighbors. Next, we show using [52] with *estimated neighbors*, as in our method. We see that as expected, using estimated rather than ground-truth text neighbors reduces performance (third vs. second columns). When estimated neighbors are used in [52]'s structural constraint, our method performs better (third vs. first columns). Interestingly, we observe that defining [52]'s structural constraint in image rather than text space is better (fourth vs. third columns). In both cases, neighborhoods are computed in *text* space (Eq. 5). This may be because the structural constraint, which requires the group of neighbors to be closer together than to others, is too strict for estimated text neighbors. That is, the constraint may require the text embeddings to lose useful discriminativity to be closer to neighboring text. Neighboring images are likely to be much more visually similar in COCO than in GoodNews or Politics as they will contain the same objects.

We next test how well each method preserves the *semantic neighborhood* given by Ω, i.e. Doc2Vec space. We begin by computing the embeddings in

Table 4. We show an ablation of our method where we remove either component of our loss. The best method is shown in **bold** and the best ablation is underlined.

Method	GoodNews [3]		Politics [49]	
	I→T	T→I	I→T	T→I
OURS (ANG)	**0.8849**	**0.8865**	**0.6467**	**0.6492**
OURS (ANG)-$\mathcal{L}_{text}$	<u>0.8786</u>	0.8813	0.6387	<u>0.6467</u>
OURS (ANG)-$\mathcal{L}_{img}$	0.8782	<u>0.8817</u>	<u>0.6390</u>	0.6413

Fig. 5. Uncurated results showing image/text samples that our method keeps closest in $\mathcal{M}$ compared to the baseline, i.e. pairs where $\frac{d_{ours}(s_1,s_2)}{d_{baseline}(s_1,s_2)}$ is smallest. Our method keeps semantically related images and text closer in the space, relative to the baseline. While the images are not visually similar, they are semantically similar (EU and Merkel; judge's gavel and Supreme Court).

$\mathcal{M}$ (cross-modal space) for all test samples. For each such sample s_i (either image or text), we compute $\Psi_{\mathcal{M}}(s_i)$, that is, we retrieve the neighbors (of the same modality as s_i) in $\mathcal{M}$. We next retrieve the neighbors of s_i in Ω, $\Psi_{\Omega}(s_i)$, described in Sect. 3.2. For each sample, we compute $|\Psi_{\mathcal{M}}(s_i) \cap \Psi_{\Omega}(s_i)| / |\Psi_{\Omega}(s_i)|$, i.e. the percentage of the nearest neighbors of the sample in Ω which are also its neighbors in $\mathcal{M}$. That is, we measure how well each method preserves within-modality semantic locality through the number of neighbors in Doc2Vec space which remain neighbors in the learned space. We consider the 200 nearest neighbors. We report the result for competitive baselines in Table 3. We find that our constraints are, indeed, preserving within-modality semantic locality, as sample proximity in Ω is more preserved in $\mathcal{M}$ with our approach than without it, i.e. we better reconstruct the semantic neighborhood of Ω in $\mathcal{M}$. We believe this allows our model to ultimately perform better at cross-modal retrieval.

We finally test the contribution of each component of our proposed loss. We test two variants of our method, where we remove either $\mathcal{L}_{text}$ or $\mathcal{L}_{img}$. We present our results in Table 4. In every case, *combining* our losses for our full method performs the best, suggesting that each loss plays a complementary role in enforcing semantic locality for its target modality.

4.4 Qualitative Results

In this section, we present qualitative results illustrating how our constraints both improve semantic proximity and demonstrate superior retrieval results.

Fig. 6. We show cross-modal retrieval results on Politics [49] using our method and the strongest baseline. We bold text aligning with the image. For text retrieval, ours returns more relevant (and semantically consistent) results. For image retrieval, our method exhibits more consistency (e.g. drug images are marijuana, immigration images show arrests), while the baseline returns more inconsistent and irrelevant images.

Semantic Proximity: In Fig. 5, we perform an experiment to discover what samples our constraints affect the most. We randomly sampled 10k image-image and text-text neighbor pairs (in Ω) and computed their distance in $\mathcal{M}$ using features from our method vs. the baseline ANG+NP+SYM. Small ratios indicate the samples were closer in $\mathcal{M}$ using our method, relative to the baseline, while larger indicate the opposite. We show the samples with the *two smallest* ratios for images and text. We observe that visually dissimilar, but semantically similar images have the smallest ratio (e.g. E.U. flag and Merkel, Judge's gavel and Supreme Court), which suggests our $\mathcal{L}_{img}$ constraint has moved the samples closer than the baseline places them. For text, we observe articles about the same issue are brought closer even though specifics differ.

Cross-Modal Retrieval Results: In Fig. 6 we show the top-3 results for a set of queries, retrieved by our method vs. ANG+NP+SYM. We observe increased semantic homogeneity in the returned samples compared with the baseline. For example, images retrieved for "drugs" using our method consistently feature marijuana, while the baseline returns images of pills, smoke, and incorrect retrievals; "wall" results in consistent images of the border wall; "immigration" features arrests. For text retrieval, we find that our method consistently performs better at recognizing public figures and returning related articles.

5   Conclusions

We proposed a novel loss function which improves semantic coherence for cross-modal retrieval. Our approach leverages a latent space learned on text alone, in order to enforce proximity within samples of the same modality, in the learned cross-modal space. We constrain text and image embeddings to be close in joint space if they or their partners were close in the unimodal text space. We experimentally demonstrate that our approach significantly improves upon several state-of-the-art loss functions on multiple challenging datasets. We presented qualitative results demonstrating increased semantic homogeneity of retrieval results. Applications of our method include improving retrieval of abstract, non-literal text, visual question answering over news and multimodal media, news curation, and learning general-purpose robust visual-semantic embeddings.

Acknowledgements. This material is based upon work supported by the National Science Foundation under Grant No. 1718262. It was also supported by Adobe and Amazon gifts, and an NVIDIA hardware grant. We thank the reviewers and AC for their valuable suggestions.

References

1. Alikhani, M., Sharma, P., Li, S., Soricut, R., Stone, M.: Clue: cross-modal coherence modeling for caption generation. In: Proceedings of the Annual Meeting of the Association for Computational Linguistics (ACL) (2020)
2. Anderson, P., et al.: Bottom-up and top-down attention for image captioning and visual question answering. In: Proceedings of the IEEE Conference on Computer Vision and Pattern Recognition (CVPR) (2018)
3. Biten, A.F., Gomez, L., Rusinol, M., Karatzas, D.: Good news, everyone! context driven entity-aware captioning for news images. In: Proceedings of the IEEE Conference on Computer Vision and Pattern Recognition (CVPR) (2019)
4. Broder, A.: On the resemblance and containment of documents. In: Proceedings of the Compression and Complexity of Sequences (1997)
5. Carvalho, M., Cadène, R., Picard, D., Soulier, L., Thome, N., Cord, M.: Cross-modal retrieval in the cooking context: learning semantic text-image embeddings. In: ACM SIGIR Conference on Research and Development in Information Retrieval (2018)
6. Chen, Y.C., et al.: UNITER: learning universal image-text representations. In: Proceedings of the European Conference on Computer Vision (ECCV) (2020)
7. Cho, K., van Merrienboer, B., Bahdanau, D., Bengio, Y.: On the properties of neural machine translation: encoder-decoder approaches. In: Eighth Workshop on Syntax, Semantics and Structure in Statistical Translation (SSST-8) (2014)
8. Duan, Y., Zheng, W., Lin, X., Lu, J., Zhou, J.: Deep adversarial metric learning. In: Proceedings of the IEEE Conference on Computer Vision and Pattern Recognition (CVPR) (2018)
9. Faghri, F., Fleet, D.J., Kiros, J.R., Fidler, S.: VSE++: improved visual-semantic embeddings. In: British Machine Vision Conference (BMVC) (2018)
10. Frome, A., et al.: DeVISE: a deep visual-semantic embedding model. In: Advances in Neural Information Processing Systems (NIPS) (2013)

11. Ge, W.: Deep metric learning with hierarchical triplet loss. In: Proceedings of the European Conference on Computer Vision (ECCV) (2018)
12. Girdhar, R., Tran, D., Torresani, L., Ramanan, D.: Distinit: learning video representations without a single labeled video. In: Proceedings of the IEEE International Conference on Computer Vision (ICCV) (2019)
13. Glorot, X., Bengio, Y.: Understanding the difficulty of training deep feedforward neural networks. In: Proceedings of the International Conference on Artificial Intelligence and Statistics (AISTATS) (2010)
14. Goodfellow, I.J., Mirza, M., Da Xiao, A.C., Bengio, Y.: An empirical investigation of catastrophic forgeting in gradient-based neural networks. In: Proceedings of International Conference on Learning Representations (ICLR) (2014)
15. Gupta, S., Hoffman, J., Malik, J.: Cross modal distillation for supervision transfer. In: Proceedings of the IEEE Conference on Computer Vision and Pattern Recognition (CVPR) (2016)
16. Hadsell, R., Chopra, S., LeCun, Y.: Dimensionality reduction by learning an invariant mapping. In: Proceedings of the IEEE Conference on Computer Vision and Pattern Recognition (CVPR) (2006)
17. Han, X., Leung, T., Jia, Y., Sukthankar, R., Berg, A.C.: MatchNet: unifying feature and metric learning for patch-based matching. In: Proceedings of the IEEE Conference on Computer Vision and Pattern Recognition (CVPR) (2015)
18. Harwood, B., Kumar, B., Carneiro, G., Reid, I., Drummond, T., et al.: Smart mining for deep metric learning. In: Proceedings of the IEEE International Conference on Computer Vision (ICCV) (2017)
19. He, K., Zhang, X., Ren, S., Sun, J.: Deep residual learning for image recognition. In: Proceedings of the IEEE Conference on Computer Vision and Pattern Recognition (CVPR) (2016)
20. Hoffer, E., Ailon, N.: Deep metric learning using triplet network. In: Feragen, A., Pelillo, M., Loog, M. (eds.) SIMBAD 2015. LNCS, vol. 9370, pp. 84–92. Springer, Cham (2015). https://doi.org/10.1007/978-3-319-24261-3_7
21. Kingma, D.P., Ba, J.: Adam: a method for stochastic optimization (2015)
22. Kruk, J., Lubin, J., Sikka, K., Lin, X., Jurafsky, D., Divakaran, A.: Integrating text and image: determining multimodal document intent in Instagram posts. In: Empirical Methods in Natural Language Processing (EMNLP) (2019)
23. Le, Q., Mikolov, T.: Distributed representations of sentences and documents. In: International Conference on Machine Learning (ICML) (2014)
24. Li, Z., Hoiem, D.: Learning without forgetting. IEEE Trans. Pattern Anal. Mach. Intell. 40(12), 2935–2947 (2017)
25. Lin, T.-Y., et al.: Microsoft COCO: common objects in context. In: Fleet, D., Pajdla, T., Schiele, B., Tuytelaars, T. (eds.) ECCV 2014. LNCS, vol. 8693, pp. 740–755. Springer, Cham (2014). https://doi.org/10.1007/978-3-319-10602-1_48
26. Lu, J., Batra, D., Parikh, D., Lee, S.: ViLBERT: pretraining task-agnostic visiolinguistic representations for vision-and-language tasks. In: Advances in Neural Information Processing Systems (NeurIPS) (2019)
27. Lu, J., Yang, J., Batra, D., Parikh, D.: Neural baby talk. In: Proceedings of the IEEE Conference on Computer Vision and Pattern Recognition (CVPR) (2018)
28. Lu, J., Xu, C., Zhang, W., Duan, L.Y., Mei, T.: Sampling wisely: deep image embedding by top-k precision optimization. In: Proceedings of the IEEE International Conference on Computer Vision (ICCV) (2019)
29. Lu, J., Hu, J., Tan, Y.P.: Discriminative deep metric learning for face and kinship verification. IEEE Trans. Image Process. 26(9), 4269–4282 (2017)

30. Malkov, Y.A., Yashunin, D.A.: Efficient and robust approximate nearest neighbor search using hierarchical navigable small world graphs. IEEE Trans. Pattern Anal. Mach. Intell. (2016)
31. Marin, J., et al.: Recipe1M+: a dataset for learning cross-modal embeddings for cooking recipes and food images. IEEE Trans. Pattern Anal. Mach. Intell. (2019)
32. Mithun, N.C., Paul, S., Roy-Chowdhury, A.K.: Weakly supervised video moment retrieval from text queries. In: The IEEE Conference on Computer Vision and Pattern Recognition (CVPR) (2019)
33. Morin, F., Bengio, Y.: Hierarchical probabilistic neural network language model. In: International Workshop on Artificial Intelligence and Statistics (AISTATS) (2005)
34. Murrugarra-Llerena, N., Kovashka, A.: Cross-modality personalization for retrieval. In: Proceedings of the IEEE Conference on Computer Vision and Pattern Recognition (CVPR) (2019)
35. Niu, L., Dai, X., Zhang, J., Chen, J.: Topic2Vec: learning distributed representations of topics. In: International Conference on Asian Language Processing (IALP) (2015)
36. Oh Song, H., Jegelka, S., Rathod, V., Murphy, K.: Deep metric learning via facility location. In: Proceedings of the IEEE Conference on Computer Vision and Pattern Recognition (CVPR) (2017)
37. Oh Song, H., Xiang, Y., Jegelka, S., Savarese, S.: Deep metric learning via lifted structured feature embedding. In: Proceedings of the IEEE Conference on Computer Vision and Pattern Recognition (CVPR) (2016)
38. Pang, K., et al.: Generalising fine-grained sketch-based image retrieval. In: The IEEE Conference on Computer Vision and Pattern Recognition (CVPR) (2019)
39. Plummer, B.A., Wang, L., Cervantes, C.M., Caicedo, J.C., Hockenmaier, J., Lazebnik, S.: Flickr30k entities: collecting region-to-phrase correspondences for richer image-to-sentence models. In: Proceedings of the IEEE International Conference on Computer Vision (ICCV) (2015)
40. Reed, S., Akata, Z., Yan, X., Logeswaran, L., Schiele, B., Lee, H.: Generative adversarial text to image synthesis. In: International Conference on Machine Learning (ICML) (2016)
41. Řehůřek, R., Sojka, P.: Software framework for topic modelling with large corpora. In: Proceedings of the LREC 2010 Workshop on New Challenges for NLP Frameworks (2010)
42. Schroff, F., Kalenichenko, D., Philbin, J.: FaceNet: a unified embedding for face recognition and clustering. In: Proceedings of the IEEE Conference on Computer Vision and Pattern Recognition (CVPR) (2015)
43. Sharma, P., Ding, N., Goodman, S., Soricut, R.: Conceptual captions: a cleaned, hypernymed, image alt-text dataset for automatic image captioning. In: Proceedings of the Annual Meeting of the Association for Computational Linguistics (ACL) (2018)
44. Shrivastava, A., Gupta, A., Girshick, R.: Training region-based object detectors with online hard example mining. In: Proceedings of the IEEE Conference on Computer Vision and Pattern Recognition (CVPR) (2016)
45. Simo-Serra, E., Trulls, E., Ferraz, L., Kokkinos, I., Fua, P., Moreno-Noguer, F.: Discriminative learning of deep convolutional feature point descriptors. In: Proceedings of the IEEE International Conference on Computer Vision (ICCV) (2015)
46. Socher, R., Ganjoo, M., Manning, C.D., Ng, A.: Zero-shot learning through cross-modal transfer. In: Advances in Neural Information Processing Systems (NIPS) (2013)

47. Sohn, K.: Improved deep metric learning with multi-class N-pair loss objective. In: Advances in Neural Information Processing Systems (NIPS) (2016)
48. Song, Y., Soleymani, M.: Polysemous visual-semantic embedding for cross-modal retrieval. In: Proceedings of the IEEE Conference on Computer Vision and Pattern Recognition (CVPR) (2019)
49. Thomas, C., Kovashka, A.: Predicting the politics of an image using Webly supervised data. In: Advances in Neural Information Processing Systems (NeurIPS) (2019)
50. Wang, B., Yang, Y., Xu, X., Hanjalic, A., Shen, H.T.: Adversarial cross-modal retrieval. In: ACM International Conference on Multimedia (2017)
51. Wang, J., Zhou, F., Wen, S., Liu, X., Lin, Y.: Deep metric learning with angular loss. In: Proceedings of the IEEE International Conference on Computer Vision (ICCV) (2017)
52. Wang, L., Li, Y., Lazebnik, S.: Learning deep structure-preserving image-text embeddings. In: Proceedings of the IEEE Conference on Computer Vision and Pattern Recognition (CVPR) (2016)
53. Wang, X., Gupta, A.: Unsupervised learning of visual representations using videos. In: Proceedings of the IEEE International Conference on Computer Vision (ICCV) (2015)
54. Wang, X., Han, X., Huang, W., Dong, D., Scott, M.R.: Multi-similarity loss with general pair weighting for deep metric learning. In: Proceedings of the IEEE Conference on Computer Vision and Pattern Recognition (CVPR) (2019)
55. Weinberger, K.Q., Saul, L.K.: Distance metric learning for large margin nearest neighbor classification. J. Mach. Learn. Res. 10(Feb), 207–244 (2009)
56. Wu, C.Y., Manmatha, R., Smola, A.J., Krahenbuhl, P.: Sampling matters in deep embedding learning. In: Proceedings of the IEEE International Conference on Computer Vision (ICCV) (2017)
57. Ye, K., Kovashka, A.: Advise: symbolism and external knowledge for decoding advertisements. In: Proceedings of the European Conference on Computer Vision (ECCV) (2018)
58. You, Q., Jin, H., Wang, Z., Fang, C., Luo, J.: Image captioning with semantic attention. In: Proceedings of the IEEE Conference on Computer Vision and Pattern Recognition (CVPR) (2016)
59. Yuan, Y., Yang, K., Zhang, C.: Hard-aware deeply cascaded embedding. In: Proceedings of the IEEE International Conference on Computer Vision (ICCV) (2017)
60. Zhang, H., et al.: StackGAN: text to photo-realistic image synthesis with stacked generative adversarial networks. In: Proceedings of the IEEE International Conference on Computer Vision (ICCV) (2017)
61. Zhang, M., Hwa, R., Kovashka, A.: Equal but not the same: understanding the implicit relationship between persuasive images and text. In: British Machine Vision Conference (BMVC) (2018)
62. Zhang, X., Zhou, F., Lin, Y., Zhang, S.: Embedding label structures for fine-grained feature representation. In: Proceedings of the IEEE Conference on Computer Vision and Pattern Recognition (CVPR) (2016)
63. Zhang, Z., Xie, Y., Yang, L.: Photographic text-to-image synthesis with a hierarchically-nested adversarial network. In: Proceedings of the IEEE Conference on Computer Vision and Pattern Recognition (CVPR) (2018)
64. Zhen, L., Hu, P., Wang, X., Peng, D.: Deep supervised cross-modal retrieval. In: Proceedings of the IEEE/CVF Conference on Computer Vision and Pattern Recognition (CVPR) (2019)

65. Zhou, S., Wang, J., Wang, J., Gong, Y., Zheng, N.: Point to set similarity based deep feature learning for person re-identification. In: Proceedings of the IEEE Conference on Computer Vision and Pattern Recognition (CVPR) (2017)
66. Zhu, B., Ngo, C.W., Chen, J., Hao, Y.: R2GAN: cross-modal recipe retrieval with generative adversarial network. In: Proceedings of the IEEE Conference on Computer Vision and Pattern Recognition (CVPR) (2019)

Large-Scale Pretraining for Visual Dialog:
A Simple State-of-the-Art Baseline

Vishvak Murahari[1(✉)], Dhruv Batra[1,2], Devi Parikh[1,2], and Abhishek Das[1,2]

[1] Georgia Institute of Technology, Atlanta, USA
{vishvak.murahari,dbatra,parikh,abhshkdz}@gatech.edu
[2] Facebook AI Research, Menlo Park, USA

Abstract. Prior work in visual dialog has focused on training deep neural models on VisDial in isolation. Instead, we present an approach to leverage pretraining on related vision-language datasets before transferring to visual dialog. We adapt the recently proposed ViLBERT model for multi-turn visually-grounded conversations. Our model is pretrained on the Conceptual Captions and Visual Question Answering datasets, and finetuned on VisDial. Our best single model outperforms prior published work by >1% absolute on NDCG and MRR.

Next, we find that additional finetuning using "dense" annotations in VisDial leads to even higher NDCG – more than 10% over our base model – but hurts MRR – more than 17% below our base model! This highlights a trade-off between the two primary metrics – NDCG and MRR – which we find is due to dense annotations not correlating well with the original ground-truth answers to questions.

Keywords: Vision & Language · Visual dialog

1 Introduction

Recent years have seen incredible progress in Visual Dialog [1–22], spurred in part by the initial efforts of Das *et al.* [2] in developing a concrete task definition – given an image, dialog history consisting of a sequence of question-answer pairs, and a follow-up question about the image, to predict a free-form natural language answer to the question – along with a large-scale dataset and evaluation metrics. The state-of-the-art on the task has improved by more than 20% absolute ($\sim$54% $\rightarrow$ $\sim$74% NDCG) and the original task has since been extended to challenging domains, *e.g.* video understanding [23], navigation assistants [24–26].

While this is promising, much of this progress has happened in isolation, wherein sophisticated neural architectures are trained and benchmarked solely on the VisDial dataset. This is limiting – since there is a significant amount of shared abstraction and visual grounding in related tasks in vision and language

Electronic supplementary material The online version of this chapter (https://doi.org/10.1007/978-3-030-58523-5_20) contains supplementary material, which is available to authorized users.

A. Vedaldi et al. (Eds.): ECCV 2020, LNCS 12363, pp. 336–352, 2020.
https://doi.org/10.1007/978-3-030-58523-5_20

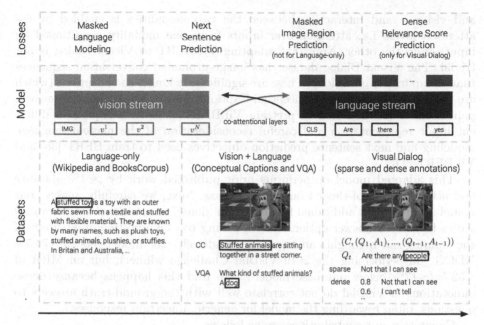

Fig. 1. First, the language stream of our model is pretrained on English Wikipedia and the BooksCorpus [27] datasets with the masked language modeling (MLM) and next sentence prediction (NSP) losses. Next, the entire model is trained on the Conceptual Captions [28] and VQA [29] datasets with the masked image region (MIR), MLM and NSP losses. Finally, we finetune the model on sparse annotations from VisDial [2] with the MIR, MLM and NSP losses, and optionally finetune on dense annotations.

(*e.g.* captioning, visual question answering) that can benefit Visual Dialog – and wasteful – since it is expensive and dissatisfying to have to collect a large-scale dataset for every new task. In this work, we explore an approach to pretrain our model on other related vision and language datasets and then transfer to Visual Dialog (Fig. 1).

Our work is inspired by prior work in transfer learning in computer vision and natural language understanding where large models [30–40] are pretrained on large datasets [27,41,42] with simple yet powerful self-supervised objectives to learn powerful representations that are then transferred to downstream tasks, leading to state-of-the-art results on a variety of benchmarks [41,43]. Recent work has extended this to vision and language tasks [44–50], leading to compelling results in Visual Question Answering [29], Commonsense Reasoning [51], Natural Language Visual Reasoning [52], Entailment [53], Image-Text Retrieval [54,55], Referring Expressions [56], and Vision-Language Navigation [57].

In this work, we adapt ViLBERT [44] to Visual Dialog. ViLBERT uses two Transformer-based[34] encoders, one for each of the two modalities – language and vision – and interaction between the two modalities is enabled by co-attention layers *i.e.* attention over inputs from one modality conditioned on inputs from the other. Note that adapting ViLBERT to Visual Dialog is not trivial. The Visual Dialog dataset has image-grounded conversation sequences that are up to 10 rounds long. These are significantly longer than captions (which are ≤2 sentences) from the Conceptual Captions dataset [28] or question-answer pairs from VQA [29] used to pretrain ViLBERT, and thus requires a different input representation and careful reconsideration of the masked language modeling and next sentence prediction objectives used to train BERT [35] and ViLBERT [44].

This adapted model outperforms prior published work by > 1% absolute and achieves state-of-the-art on Visual Dialog. Next, we carefully analyse our model and find that additional finetuning on 'dense' annotations[1] *i.e.* relevance scores for all 100 answer options corresponding to each question on a subset of the training set, highlights an interesting trade-off – the model gets to ∼74.5% NDCG (outperforming the 2019 VisDial Challenge winner), but an MRR of ∼52% (∼17% below our base model!). We find this happens because dense annotations in VisDial do not correlate well with the ground-truth answers to questions, often rewarding the model for generic, uncertain responses.

Concretely, our contributions are as follows:

- We introduce an adaptation of the ViLBERT [44] model for Visual Dialog, thus making use of the large-scale Conceptual Captions [28] and Visual Question Answering (VQA) [29] datasets for pretraining and learning powerful visually-grounded representations before finetuning on VisDial [2]. Since captioning and VQA differ significantly from Visual Dialog in input size (≤2 sentence descriptions *vs.* ≤10 question-answer rounds), this requires rethinking the input representation to learn additional segment embeddings representing questions-answer pairs. Our adapted model improves over prior published work by >1% and sets a new state-of-the-art.
- We next finetune our model on dense annotations *i.e.* relevance scores for all 100 answer options corresponding to each question on a subset of the training set, leading to even higher NDCG – more than 10% over our base model – but hurting MRR – more than 17% below our base model! This highlights a stark trade-off between the two primary metrics for this task – NDCG and MRR. Through qualitative and quantitative results, we show that this happens because dense annotations do not correlate well with the original ground-truth answers, often rewarding the model for generic, uncertain responses.
- Our PyTorch [58] code is publicly available[2] to encourage further work in large-scale transfer learning for VisDial.

---

[1] Publicly available on visualdialog.org/data.
[2] github.com/vmurahari3/visdial-bert/.

2 Related Work

Our work is related to prior work in visual dialog [1–22], and self-supervised pretraining and transfer learning in computer vision and language [30–40].

Visual Dialog. Das *et al.* [2] and de Vries *et al.* [1] introduced the task of Visual Dialog – given an image, dialog history consisting of a sequence of question-answer pairs, and a follow-up question, predict a free-form natural language answer to the question – along with a dataset, evaluation metrics, and baseline models. Follow-up works on visual dialog have explored the use of deep reinforcement learning [3,4,17], knowledge transfer from discriminative to generative decoders [5], conditional variational autoencoders [6], generative adversarial networks [7], attention mechanisms for visual coreference resolution [9,11], and modeling the questioner's theory of mind [10]. Crucially, all of these works train and evaluate on the VisDial dataset *in isolation*, without leveraging related visual grounding signals from other large-scale datasets in vision and language. We devise a unified model that can be pretrained on the Conceptual Captions [28] and VQA [29] datasets, and then transferred and finetuned on VisDial.

Self-supervised Learning in Vision and Language. Building on the success of transfer learning in natural language understanding [33–40] leading to state-of-the-art results on a broad set of benchmarks [41,43], recent work has extended this to vision and language tasks [44–50]. These works pretrain single [45,48,49] or two [44,46]-stream Transformer [34]-based models with self-supervised objectives, such as next-sentence prediction and masked language/image modeling, on large-scale image-text datasets and have led to compelling results in Visual Question Answering [29], Commonsense Reasoning [51], Natural Language Visual Reasoning [52], Entailment [53], Image-Text Retrieval [54,55], and Referring Expressions [56], and Vision-Language Navigation [57].

3 Adapting ViLBERT [44] for Visual Dialog

Lu *et al.* [44] introduced ViLBERT[3], which extended BERT [35] to a two-stream multi-modal architecture for jointly modeling visual and linguistic inputs. Interaction between the two modalities was enabled through co-attention layers, *i.e.* attending to one modality conditioned on the other – attention over language conditioned on visual input, and attention over image regions conditioned on linguistic input. This was operationalized as swapping the key and value matrices between the visual and linguistic Transformer [34] blocks. We next discuss our changes to adapt it for Visual Dialog followed by our training pipeline.

Input Representation. Recall that the model gets image I, dialog history (including image caption C) $H = (C, (Q_1, A_1), ..., (Q_{t-1}, A_{t-1}))$, question Q_t, and a list of 100 answer options $A_t = \{A_t^{(1)}, A_t^{(2)}, ..., A_t^{(100)}\}$ as input, and is asked to return a sorting of A_t. We concatenate the t rounds of dialog history

---

[3] Along with code released at github.com/jiasenlu/ViLBERT_beta.

and follow-up question Q_t, with each question and answer separated by a <SEP> token. The overall input to the language stream is represented as:

$$\text{<CLS>} \; C \; \text{<SEP>} \; Q_1 \; \text{<SEP>} \; A_1 \; \text{<SEP>}, ..., \text{<SEP>} \; Q_t \; \text{<SEP>} \; A_t \; \text{<SEP>} \qquad (1)$$

Similar to Wolf et al. [59], we use different segment embeddings for questions and answers to help the model distinguish between the two and understand question and answer boundaries in the input. Captions and answers share the same segment embeddings. To represent the image, we follow [44,60] and extract object bounding boxes and their visual features for top-36 detected objects in the image from a Faster R-CNN [61] (with a ResNet-101 [30] backbone) object detection network pretrained on the Visual Genome dataset [42]. The feature vector for each detected object is computed as mean-pooled convolutional features from the regions of that object. A 5-d feature vector, consisting of normalized top-left and bottom-right object coordinates, and the fraction of image area covered, is projected to the same dimensions as the feature vector for the detected object, and added to it, giving us the final visual features $\{v_1, ..., v_{36}\}$. The beginning of this image region sequence (consisting of object detection features) is demarcated by an IMG token with mean-pooled features from the entire image. The overall input to ViLBERT can be written as the following sequence:

$$\text{<IMG>} \; v_1, ..., v_{36} \; \text{<CLS>} \; C \; \text{<SEP>} \; Q_1 \; \text{<SEP>} \; A_1 \; \text{<SEP>}, ..., \text{<SEP>} \; Q_t \; \text{<SEP>} \; A_t \; \text{<SEP>} \quad (2)$$

3.1 Pretraining on Conceptual Captions [28]

To pretrain the model, we follow [44] and train on the Conceptual Captions (CC) dataset, which is a large corpus (with ∼3M samples) of aligned image-caption pairs. During pretraining, the sum of the *masked language modeling* (MLM) loss [35] and the *masked image region* (MIR) loss is optimized. To compute the MLM loss, a set of tokens in the input sequence are masked and the model is trained to predict these tokens given context. We mask around 15% of the tokens in the input sequence. For the MIR loss, similar to the MLM loss, we zero out 15% of the image features and the model learns to predict the semantic category of the masked out object (out of 1601 classes from Visual Genome [42,60]).

3.2 Pretraining on VQA [29]

The VQA dataset is quite related to Visual Dialog in that it can be interpreted as independent visually-grounded question-answer pairs with no dialog history, and thus is a natural choice for further pretraining prior to finetuning on VisDial. Similar to Lu et al. [44], we pretrain on VQA by learning a small decoder – a two-layer MLP – on top of the element-wise product between the image and text representations to predict a distribution over 3129 answers.

3.3 Finetuning on Visual Dialog [2]

To finetune on Visual Dialog, we use the MLM loss along with the next sentence prediction (NSP) and MIR losses. For MLM, we mask 10% of the tokens in the dialog sequence. For MIR, similar to pretraining, we mask 15% of the image features. Note that the discriminative task in visual dialog is to identify the ground-truth answer from a list of 100 answer options consisting of popular, nearest neighbors, and random answers from the dataset. We achieve this through the NSP loss. The NSP head is trained to predict 1 when the ground-truth answer is appended to the input sequence, and 0 when a negative answer sampled from the remaining answer options is appended to it. Each image in VisDial has 10 rounds of dialog, leading to 10 sets of positive and negative samples for the NSP loss per mini-batch. Since these are fairly correlated samples, we randomly sub-sample 2 out of these 20 during training. At test time, we use log-probabilities from the NSP head to rank the 100 answer options per round.

3.4 Finetuning with Dense Annotations

The authors of [2] recently released dense annotations[4] *i.e.* relevance scores for all 100 answer options from A_t corresponding to the question on a subset of the training set. These relevance scores range from 0 to 1 and are calculated as the ratio of number of human annotators who marked a particular answer option as correct to the total number of human annotators ($= 4$). So 1 means that the answer option was considered correct by 4 human annotators. In our final stage of training, we utilize these dense annotations to finetune our model. Concretely, we use the NSP head to predict likelihood scores $\hat{\ell}_t^{(i)}$ for each answer option $A_t^{(i)}$ at round t, normalize these to form a probability distribution over the 100 answers $\hat{y}_t = [\hat{y}_t^{(1)}, ..., \hat{y}_t^{(100)}]$, and then compute a cross-entropy (CE) loss against the normalized ground-truth relevance scores y_t, given by $-\sum_i y_t^{(i)} \log \hat{y}_t^{(i)}$.

4 Experiments

To compare to previous research, we conduct experiments on VisDial v1.0 [2]. The dataset contains human-human dialogs on $\sim$130k COCO [62]-like images. We follow the original splits and use $\sim$120k for training, $\sim$2k for validation, and $\sim$8k for testing. We next describe the various settings we experiment with.

Evaluation Metrics. We use metrics introduced in [2]. Specifically, given the predicted ranking of 100 answer options from a model at each round, we compute retrieval metrics – mean rank (MR) of the ground-truth answer, mean reciprocal rank (MRR), and recall@k ($k = \{1, 5, 10\}$). Additionally, along with the release of dense annotations, *i.e.* relevance scores $\in [0, 1]$ for all 100 answer options, a new metric – NDCG – was introduced. NDCG accounts for multiple correct answers in the option set and penalizes low-ranked but correct answer options.

---

[4] Publicly available on visualdialog.org/data.

4.1   Language-Only

We begin with a 'blind' setting, where given the dialog history and follow-up question, and without access to the image, the model is tasked with predicting the answer. We do not use the ViLBERT formulation for these experiments, and finetune the BERT model released in [35] and pretrained on BooksCorpus [27] and English Wikipedia. For the MLM loss, we mask 15% of tokens and sub-sample 8 out of 20 sequences per mini-batch during training. We experiment with two variants – training only with NSP, and training with both NSP and MLM. See Table 3 for language-only results (marked 'L-only'). This setting helps us benchmark gains coming from switching to Transformer [34]-based architectures before the added complexity of incorporating visual input.

Varying Number of Dialog Rounds. We train ablations of our language-only model (with NSP and MLM losses) where we vary the number of rounds in dialog history, starting from 0, where the input sequence only contains the follow-up question and answer, to 2, 4, and 6 and 10 rounds of dialog history (Table 1).

Zero-Shot and 'Cheap' Finetuning. We report performance for ablations of our NSP+MLM model with no/minimal training in Table 2. First, we do a zero-shot test where we initialize BERT with weights from Wikipedia and BooksCorpus pretraining and simply run inference on VisDial. Second, with the same initialization, we freeze all layers and finetune only the MLM and NSP loss heads.

4.2   Finetuning on VisDial

We finetune ViLBERT on VisDial with four different weight initializations – 1) with randomly initialized weights, 2) from the best language-only weights (from Sect. 4.1) for the language stream (visual stream and co-attention layers initialized randomly), 3) from a model pretrained on CC [28] (as described in Sect. 3.1) and 4) from a model pretrained on CC [28] +VQA [29] (as described in Sect. 3.2). 1) helps us benchmark improvements due to pretraining, 2) helps us benchmark performance if the model learns visual grounding solely from VisDial, 3) quantifies effects of learning visual grounding additionally from CC, and 4) helps us quantify improvements with additional exposure to visually-grounded question-answering data. See Table 3 for results.

4.3   Finetuning with Dense Annotations

Finally, we finetune our best model from Sect. 4.2 – marked 'w/ CC+VQA' in Table 3 – on dense annotations, as described in Sect. 3.4. Note that computing the CE loss requires a separate forward pass for each of the 100 answer options, since dialog history, question, answer are all concatenated together before passing as input. This is memory-expensive, and so in practice, we sub-sample and only use 80 options, and use gradient accumulation to (artificially) construct a larger mini-batch. Finetuning with the CE loss only leads to significant improvements on NDCG but hurts other metrics (see Table 3). We discuss and analyse this in more detail later. But to control for this 'metric-overfitting', we also train a variant with both the CE and NSP losses.

5   Results

We list findings from all experiments described in Sect. 4 below.

Table 1. Performance of the NSP + MLM language-only model on VisDial v1.0 val as the number of dialog history rounds is varied

# history rounds	NDCG ↑	MRR ↑	R@1 ↑	R@5 ↑	R@10 ↑	MR ↓
0	50.54	54.29	38.88	72.67	83.09	5.90
2	53.69	61.31	46.83	78.96	88.15	4.51
4	55.10	62.83	48.36	80.61	89.57	4.19
6	55.69	63.73	49.31	**81.13**	**90.06**	**4.04**
10	**57.22**	**64.10**	**50.05**	81.09	90.00	4.16

Table 2. Performance of the NSP + MLM language-only model on VisDial v1.0 val with no/minimal training (described in Sect. 4.1)

Model	NDCG ↑	MRR ↑	R@1 ↑	R@5 ↑	R@10 ↑	MR ↓
No training	11.63	6.88	2.63	7.17	11.30	46.90
Loss heads only	19.69	9.81	3.42	10.44	18.85	31.38

- **Language-only performs well.** The language-only model gets to 57.22 on NDCG and 64.10 on MRR (Table 3), which is already competitive with several prior published works (Table 4). These trends are consistent with high human performance on VisDial [2] with just language (question and dialog history) – 48.5 on MRR – which further improves to 63.5 on MRR with image.
- **Increasing dialog history rounds helps.** We report performance of the language-only model as a function of dialog history rounds in Table 1 and Fig. 2a. Note that the change in performance from including 0 to 4 rounds of dialog history (+4.56 on NDCG, +8.54 on MRR) is much more than from 4 to 10 dialog history rounds (+2.12 on NDCG, +1.27 on MRR). Thus, performance continues to go up with increasing dialog history rounds but starts to plateau with ≥4 history rounds. We believe these improvements are largely indicative of the Transformer's ability to model long-term dependencies.
- **Zero-shot model performs poorly.** Running inference with the language-only model pretrained on BooksCorpus [27] and Wikipedia without any fine-tuning on VisDial only gets to 11.63 on NDCG and 6.88 on MRR (Table 2).

Finetuning the loss heads with all other layers frozen leads to an improvement of ~8 NDCG points over this. This low performance can be attributed to significantly longer sequences in VisDial than the model was pretrained with.

- **VQA initialization helps more than random or CC initialization.** Finetuning ViLBERT on VisDial with weights initialized from VQA pretraining gets to 64.82 on NDCG and 68.97 on MRR, ~3 points better than random initialization on NDCG and ~2 points better than CC pretraining (Table 3). We believe poorer transfer from CC is because both VQA and VisDial have images from COCO and are more closely related tasks than captioning on CC.
- **Dense annotations boost NDCG, hurt MRR.** Finetuning with the CE loss leads to 74.47 on NDCG – a ~10% improvement over the 'w/ CC + VQA' base model – but 50.74 on MRR, a ~17% decline below the base model (Table 4). This is a surprising finding! We carefully analyze this behavior in Sect. 6.
- **Ensembling does not improve performance.** We trained 3 models initialized with different random seeds for each of the 3 variants ('w/ CC + VQA', 'CE' and 'CE + NSP') and aggregated results by averaging the normalized scores from the 3 models. We did not observe any significant improvement.

Table 3. Results on VisDial v1.0 val (with 95% CI). ↑ indicates higher is better.

	Model	NDCG ↑	MRR ↑	R@1 ↑	R@5 ↑	R@10 ↑	MR ↓
L-only	NSP	55.80 ±0.9	63.37 ±0.5	49.28 ±0.7	80.51 ±0.5	89.22 ±0.4	4.32 ±0.1
	NSP + MLM	57.22 ±0.9	64.10 ±0.5	50.05 ±0.7	81.09 ±0.5	90.00 ±0.4	4.16 ±0.1
+vision	Random init	61.88 ±0.9	67.04 ±0.5	53.51 ±0.7	83.94 ±0.5	92.27 ±0.4	3.55 ±0.1
	w/ L-only	62.08 ±0.9	67.73 ±0.5	54.67 ±0.7	84.02 ±0.5	92.07 ±0.4	3.58 ±0.1
	w/ CC [28]	62.99 ±0.9	68.64 ±0.5	55.55 ±0.7	85.04 ±0.5	92.98 ±0.4	3.36 ±0.1
	w/ CC [28]+VQA [29]	**64.82** ±0.9	**68.97** ±0.5	**55.78** ±0.7	**85.34** ±0.5	**93.11** ±0.4	**3.35** ±0.1
+dense	CE	**75.10** ±1.1	52.12 ±0.6	39.84 ±0.7	64.93 ±0.7	80.47 ±0.5	6.26 ±0.1
	CE + NSP	69.11 ±1.0	65.76 ±0.5	53.30 ±0.7	80.77 ±0.5	90.00 ±0.4	4.33 ±0.1

We report results from the Visual Dialog evaluation server[5] for our best models – 'w/ CC + VQA', 'CE' and 'CE + NSP' – on the unseen test-std split in Table 4. We compare against prior published results and top entries from the leaderboard. Our models outperform prior results and set a new state-of-the-art – ViLBERT with CC + VQA pretraining on MRR, R@k, MR metrics, and further finetuning with a CE loss on dense annotations on NDCG. Finally, adding NSP loss along with CE (as in Sect. 4.3) offers a balance between optimizing metrics that reward both sparse (original ground-truth answers) and dense annotations.

---

[5] evalai.cloudcv.org/web/challenges/challenge-page/161/leaderboard/483.

Table 4. Results on VisDial v1.0 test-std. ↑ indicates higher is better. ↓ indicates lower is better. † denotes ensembles. Best single-model results are **bolded** and best ensemble results are underlined. ⋆ denotes the winning team of the 2019 Visual Dialog Challenge.

	Model	NDCG ↑	MRR ↑	R@1 ↑	R@5 ↑	R@10 ↑	MR ↓
Published Results	GNN [12]	52.82	61.37	47.33	77.98	87.83	4.57
	CorefNMN [9]	54.70	61.50	47.55	78.10	88.80	4.40
	RvA [11]	55.59	63.03	49.03	80.40	89.83	4.18
	HACAN [19]	57.17	64.22	50.88	80.63	89.45	4.20
	NMN [9]	58.10	58.80	44.15	76.88	86.88	4.81
	DAN [14]	57.59	63.20	49.63	79.75	89.35	4.30
	DAN† [14]	59.36	64.92	51.28	81.60	90.88	3.92
	ReDAN [15]	61.86	53.13	41.38	66.07	74.50	8.91
	ReDAN+† [15]	64.47	53.74	42.45	64.68	75.68	6.64
	DualVD [22]	56.32	63.23	49.25	80.23	89.70	4.11
	FGA [13]	56.93	66.22	52.75	82.92	91.08	3.81
	FGA† [13]	57.20	69.30	55.65	86.73	94.05	3.14
	DL-61 [20]	57.32	62.20	47.90	80.43	89.95	4.17
	DL-61† [20]	57.88	63.42	49.30	80.77	90.68	3.97
	MReal - BDAI⋆ [21]	74.02	52.62	40.03	68.85	79.15	6.76
Leaderboard Entries	LF	45.31	55.42	40.95	72.45	82.83	5.95
	HRE	45.46	54.16	39.93	70.45	81.50	6.41
	MN	47.50	55.49	40.98	72.30	83.30	5.92
	MN-Att	49.58	56.90	42.43	74.00	84.35	5.59
	LF-Att	51.63	60.41	46.18	77.80	87.30	4.75
	MS ConvAI	55.35	63.27	49.53	80.40	89.60	4.15
	USTC-YTH	56.47	61.44	47.65	78.13	87.88	4.65
	UET-VNU	57.40	59.50	45.50	76.33	85.82	5.34
	square	60.16	61.26	47.15	78.73	88.48	4.46
	MS D365 AI	64.47	53.73	42.45	64.68	75.68	6.63
Ours	Random init	60.40	65.53	51.03	83.45	91.83	3.60
	w/ CC [28]+VQA [29]	63.87	**67.50**	**53.85**	**84.68**	**93.25**	**3.32**
	CE	**74.47**	50.74	37.95	64.13	80.00	6.28
	CE + NSP	68.08	63.92	50.78	79.53	89.60	4.28

6 Analysis

As described in Sect. 5, finetuning on dense annotations leads to a significant increase in NDCG, but hurts the other 5 metrics – MRR, R@1, R@5, R@10 and MR – which depend on the original sparse annotations in VisDial *i.e.* follow-up answers provided in human-human dialog.

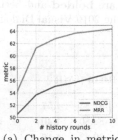

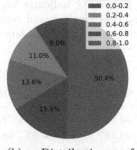

Relevance Score	w/ CC + VQA	CE	CE + NSP
0.0 − 0.2		6.47	14.88	10.79
0.2 − 0.4		4.77	11.11	6.62
0.4 − 0.6		4.02	8.49	4.86
0.6 − 0.8		3.12	6.63	3.77
0.8 − 1.0		1.95	3.26	2.21

(a) Change in metrics for varying number of rounds in dialog history for the language-only model trained with NSP + LM

(b) Distribution of dense annotation relevance scores for ground-truth answers in VisDial v1.0 val. ∼50% ground-truth answers have relevance scores < 0.8, and ∼10% have scores < 0.2

(c) Mean rank (lower is better) of the GT answers on VisDial v1.0 val split across model variants and ranges of relevance scores

Fig. 2. Analysis plots of dense annotations in Visdial v1.0 val split and dialog history ablations.

We begin by visualizing the distribution of dense relevance scores for these sparse ground-truth (GT) answers in Fig. 2b and observe that ∼50% GT answers have relevance ≤0.8, and ∼30% have relevance ≤0.6. Thus, there is some degree of misalignment between dense and sparse annotations – answers originally provided during human-human dialog in VisDial were not always judged to be relevant by all humans during the post-hoc dense annotation phase.

Why are GT and Dense Annotations Misaligned? We notice that many questions with discrepancy between GT and dense annotations are somewhat subjective. For *e.g.*, in row 1, round 7 (Fig. 5), Q: 'what color is the chair?', the GT answer is 'black' but the chair is in shadow and it is difficult to accurately identify its color. And thus, we expect to see variance when multiple humans are polled for the answer. Instead, the GT answer is just one sample from the human answer distribution, not necessarily from its peak. In general, the dense annotations seem less wrong than GT (as they are sourced by consensus) since they are safer – often resolving to answers like 'I cannot tell' when there is uncertainty/subjectivity – but also uninformative – not conveying additional information *e.g.* 'I think 3 but they are occluded so it is hard to tell' – since such nuanced answers are not part of the list of answer options in VisDial [2].

Model Performance on GT *vs*. Dense annotations. Table 2c shows mean ranks of these GT answers as predicted by three model variants – ViLBERT w/ CC + VQA, CE, and CE + NSP – grouped by dense relevance scores. The 'CE' model gets worse mean ranks than 'w/ CC + VQA' for all GT answers, since it is no longer trained with these GT answers during dense annotation finetuning. The CE model assigns low mean ranks to GT answers with higher relevance

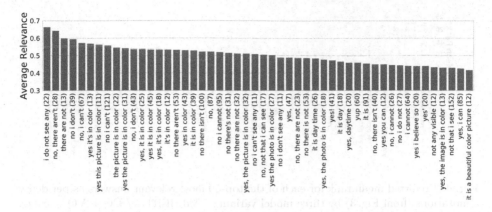

Fig. 3. Mean relevance scores and counts for top-50 most-relevant answers from VisDial v1.0 val dense annotations. These contain several sets of paraphrases – {"yes it's in color", "yes this picture is in color", "the picture is in color", "yes the picture is in color", "yes, it is in color", "yes it is in color", "yes, it's in color", "yes in color"}, *etc.* and have a bias towards binary answers (Color figure online)

scores (≥ 0.8), which translates to a high NDCG score (Table 3). But it assigns poor mean ranks to GT answers with relatively lower relevance scores (≤ 0.8), and since ~50% GT answers have relevance scores ≤ 0.8, this hurts MRR, R@k, MR for the CE model (Table 3).

Next, we consider the top-50 most-relevant answer options (occurring ≥ 10 times) as per dense annotations in VisDial v1.0 val (not restricting ourselves to only GT answers). Figure 3 shows the mean relevance scores for this set, and Fig. 4 shows the mean ranks assigned to these answers by our models. The CE model gets better mean ranks in this set compared to Base, leading to high NDCG.

Qualitative Examples. Finally, we present uniformly sampled example answer predictions on VisDial v1.0 val from our models along with the ground-truth dialog sequences in Fig. 5 and present additional samples in the appendix. In these examples, consistent with the Visual Dialog task definition [2], at every round of dialog, the model gets the image, ground-truth human dialog history (including caption), and follow-up question as input, and predicts the answer. Specifically, the model ranks 100 answer options. Here we show the top-1 prediction.

We make a few observations. 1) The Base model is surprisingly accurate, *e.g.* in row 2, round 1 (Fig. 5), Q: 'can you see any people?', predicted answer: 'part of a person', in row 2, round 10, Q: 'anything else interesting about the photo?', predicted answer: 'the dog is looking up at the person with his tongue out'. 2) The CE model often answers with generic responses (such as 'I cannot tell'), especially for questions involving some amount of subjectivity/uncertainty, *e.g.* in row 1, round 7, Q: 'what color is the chair?', predicted answer: 'I cannot tell' (the chair seems to be in shadow in the image), in row 2, round 7, Q: 'does the dog look happy?', predicted answer: 'I can't tell' (subjective question). 3) This

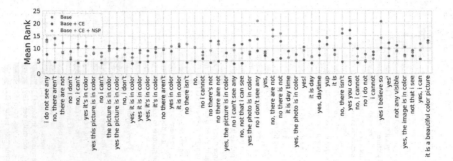

Fig. 4. Predicted mean rank for each of the top-50 most relevant answers as per dense annotations (from Fig. 3) by three model variants – ViLBERT w/ CC + VQA (called 'Base'), CE, and CE + NSP. The CE model gets lower mean ranks for most answers in this set compared to Base. This leads to significantly higher NDCG, as reported in Table 3 and Table 4, but low MRR, since these relevant answers as per dense annotations do not correlate well with the set of original ground-truth answers, as shown in Fig. 2b

Image + Caption	GT Dialog	Base Model	Base Model + CE	Base Model + CE + NSP
a chair sitting on the beach with boats in view offshore	Q: are there any people? A: yes Q: is the picture in color? A: yes Q: how many people are there? A: i can't tell Q: what are they doing? A: sitting on the boats Q: is this near a beach? A: yes Q: are they big boats? A: 1 is Q: what color is the chair? A: white Q: what kind of chair is it? A: beach Q: what is it made out of? A: wood Q: is it a nice day? A: overcast	Q: are there any people? A: no Q: is the picture in color? A: yes Q: how many people are there? A: 1 Q: what are they doing? A: sitting on the boats Q: is this near a beach? A: yes Q: are they big boats? A: no Q: what color is the chair? A: black Q: what kind of chair is it? A: beach Q: what is it made out of? A: metal Q: is it a nice day? A: yes	Q: are there any people? A: no Q: is the picture in color? A: yes, it is Q: how many people are there? A: 0 Q: what are they doing? A: i cannot tell Q: is this near a beach? A: yes Q: are they big boats? A: nope Q: what color is the chair? A: i cannot tell Q: what kind of chair is it? A: i can't tell Q: what is it made out of? A: i can't tell Q: is it a nice day? A: yes	Q: are there any people? A: no Q: is the picture in color? A: yes Q: how many people are there? A: 1 Q: what are they doing? A: sitting on the boats Q: is this near a beach? A: yes Q: are they big boats? A: no Q: what color is the chair? A: black Q: what kind of chair is it? A: beach Q: what is it made out of? A: i can't tell Q: is it a nice day? A: yes
a brown and white dog riding a skateboard	Q: can you see any people? A: part of a person Q: is the photo in color? A: yes Q: man or woman? A: i can't tell Q: how many dogs? A: just 1 Q: is the dog big or small? A: medium Q: inside or outside? A: hard to tell but i think outside Q: does the dog look happy? A: kind of Q: can you see the ground? A: yes Q: what color is the skateboard? A: black Q: anything else interesting about the photo? A: the dog is looking up at the person with his tongue out	Q: can you see any people? A: part of a person Q: is the photo in color? A: yes Q: man or woman? A: can't see his face Q: how many dogs? A: 1 Q: is the dog big or small? A: big dog Q: inside or outside? A: outside Q: does the dog look happy? A: yes Q: can you see the ground? A: yes Q: what color is the skateboard? A: black Q: anything else interesting about the photo? A: the dog is looking up at the person with his tongue out	Q: can you see any people? A: yes Q: is the photo in color? A: yes Q: man or woman? A: can't tell Q: how many dogs? A: 1 Q: is the dog big or small? A: big dog Q: inside or outside? A: outside Q: does the dog look happy? A: i can't tell Q: can you see the ground? A: yes Q: what color is the skateboard? A: black Q: anything else interesting about the photo? A: no	Q: can you see any people? A: yes Q: is the photo in color? A: yes Q: man or woman? A: can't tell Q: how many dogs? A: 1 Q: is the dog big or small? A: it looks like a large dog Q: inside or outside? A: outside Q: does the dog look happy? A: yes Q: can you see the ground? A: yes Q: what color is the skateboard? A: black Q: anything else interesting about the photo? A: the dog is looking up at the person with his tongue out
a group of giraffes is standing in a savannah	Q: how many giraffes are there? A: 3 Q: are they all adults? A: yes Q: do you see grass? A: yes Q: are there any trees? A: yes Q: is it sunny? A: yes Q: do the trees have any leaves on them? A: yes Q: are there any other animals visible? A: no Q: are the giraffes heads up? A: yes Q: is the sky clear? A: yes Q: does it look dry? A: yes	Q: how many giraffes are there? A: there are 4 giraffes Q: are they all adults? A: yes Q: do you see grass? A: yes Q: are there any trees? A: yes Q: is it sunny? A: yes Q: do the trees have any leaves on them? A: yes Q: are there any other animals visible? A: no Q: are the giraffes heads up? A: yes Q: is the sky clear? A: yes Q: does it look dry? A: yes	Q: how many giraffes are there? A: 3 Q: are they all adults? A: yes Q: do you see grass? A: yes Q: are there any trees? A: yes Q: is it sunny? A: yes Q: do the trees have any leaves on them? A: yes Q: are there any other animals visible? A: no Q: are the giraffes heads up? A: yes they are Q: is the sky clear? A: yes Q: does it look dry? A: yes	Q: how many giraffes are there? A: 3 Q: are they all adults? A: yes Q: do you see grass? A: yes Q: are there any trees? A: yes Q: is it sunny? A: yes Q: do the trees have any leaves on them? A: yes Q: are there any other animals visible? A: no Q: are the giraffes heads up? A: yes Q: is the sky clear? A: yes Q: does it look dry? A: yes

Fig. 5. Qualitative samples for three model variants – ViLBERT w/ CC + VQA (called 'Base'), Base + CE, and Base + CE + NSP

also highlights a consequence of misalignment between ground-truth and dense annotations. While the ground-truth answer provides *one* reasonable response for the question asked, it is answerer-specific to quite an extent and there may be other correct answers (annotated in the dense annotations). A negative effect of this misalignment is that when finetuned on dense annotations (CE), the model gets rewarded for generic answers (*e.g.* 'cannot tell'). While being able to capture and reason about uncertainty is a desirable property models should have, it would be more helpful if these agents can convey more information with appropriate qualifiers (*e.g.* 'I think 3 but they are occluded so it is hard to tell') than a blanket 'I cannot tell'. We aim to study this in future work.

7 Implementation

We use the BERT$_{BASE}$ model [35] for the linguistic stream. We use 6 layers of Transformer blocks (with 8 attention heads and a hidden state size of 1024) for the visual stream. The co-attention layers connect the 6 Transformer layers in the visual stream to the last 6 Transformer layers in the linguistic stream. We train on dialog sequences with atmost 256 tokens as most sequences had atmost 256 tokens. During inference, we truncate longer sequences by removing rounds starting from round 1 (we keep the caption). We set all loss coefficients to 1. We use a batch size of 128 for language-only experiments and 80 for other experiments. We use Adam [63] and linearly increase learning rate from 0 to $2e^{-5}$ over $10k$ iterations and decay to $1e^{-5}$ over $200k$ iterations. Our code is available at github.com/vmurahari3/visdial-bert/.

8 Conclusion

We introduce a model for Visual Dialog that enables pretraining on large-scale image-text datasets before transferring and finetuning on VisDial. Our model is an adaptation of ViLBERT [44], and our best single model is pretrained on BooksCorpus [27], English Wikipedia (at the BERT stage), and on Conceptual Captions [28], VQA [29] (at the ViLBERT stage), before finetuning on Vis-Dial, optionally with dense annotations. Our model outperforms prior published results by > 1% absolute on NDCG and MRR, achieving state-of-the-art results, and providing a simple baseline for future 'pretrain-then-transfer' approaches.

Through careful analysis of our results, we find that the recently released dense annotations for the task do not correlate well with the original ground-truth dialog answers, leading to a trade-off when models optimize for metrics that take into account these dense annotations (NDCG) *vs.* the original sparse annotations (MRR). This opens up avenues for future research into better evaluation metrics.

Finally, note that our model is discriminative – it can pick a good answer from a list of answer options – but cannot generate an answer. In the future, we aim to develop robust decoding techniques, based on decoding strategies for transformer-based models introduced in [33,64], for a strong generative model.

Acknowledgments. The Georgia Tech effort was supported in part by NSF, AFRL, DARPA, ONR YIPs, ARO PECASE, Amazon. AD was supported by fellowships from Facebook, Adobe, Snap Inc. Views and conclusions contained herein are those of the authors and should not be interpreted as necessarily representing the official policies or endorsements, either expressed or implied, of the U.S. Government, or any sponsor.

References

1. de Vries, H., Strub, F., Chandar, S., Pietquin, O., Larochelle, H., Courville, A.: GuessWhat?! visual object discovery through multi-modal dialogue. In: CVPR (2017)
2. Das, A., et al.: Visual dialog. In: CVPR (2017)
3. Strub, F., De Vries, H., Mary, J., Piot, B., Courville, A., Pietquin, O.: End-to-end optimization of goal-driven and visually grounded dialogue systems, arXiv preprint arXiv:1703.05423 (2017)
4. Das, A., Kottur, S., Moura, J.M., Lee, S., Batra, D.: Learning cooperative visual dialog agents with deep reinforcement learning. In: ICCV (2017)
5. Lu, J., Kannan, A., Yang, J., Parikh, D., Batra, D.: Best of both worlds: transferring knowledge from discriminative learning to a generative visual dialog model. In: NIPS (2017)
6. Massiceti, D., Siddharth, N., Dokania, P.K., Torr, P.H.: FLIPDIAL: a generative model for two-way visual dialogue. In: CVPR (2018)
7. Wu, Q., Wang, P., Shen, C., Reid, I., van den Hengel, A.: Are you talking to me? reasoned visual dialog generation through adversarial learning. In: CVPR (2018)
8. Jain, U., Lazebnik, S., Schwing, A.G.: Two can play this game: visual dialog with discriminative question generation and answering. In: CVPR (2018)
9. Kottur, S., Moura, J.M., Parikh, D., Batra, D., Rohrbach, M.: Visual coreference resolution in visual dialog using neural module networks. In: ECCV (2018)
10. Lee, S.-W., Gao, T., Yang, S., Yoo, J., Ha, J.-W.: Large-scale answerer in questioner's mind for visual dialog question generation. In: ICLR (2019)
11. Niu, Y., Zhang, H., Zhang, M., Zhang, J., Lu, Z., Wen, J.-R.: Recursive visual attention in visual dialog. In: CVPR (2019)
12. Zheng, Z., Wang, W., Qi, S., Zhu, S.-C.: Reasoning visual dialogs with structural and partial observations. In: CVPR (2019)
13. Schwartz, I., Yu, S., Hazan, T., Schwing, A.G.: Factor graph attention. In: CVPR (2019)
14. Kang, G.-C., Lim, J., Zhang, B.-T.: Dual attention networks for visual reference resolution in visual dialog. In: EMNLP (2019)
15. Gan, Z., Cheng, Y., Kholy, A.E., Li, L., Liu, J., Gao, J.: Multi-step reasoning via recurrent dual attention for visual dialog. In: ACL (2019)
16. Kottur, S., Moura, J.M., Parikh, D., Batra, D., Rohrbach, M.: CLEVR-dialog: a diagnostic dataset for multi-round reasoning in visual dialog. In: NAACL (2019)
17. Murahari, V., Chattopadhyay, P., Batra, D., Parikh, D., Das, A.: Improving generative visual dialog by answering diverse questions. In: EMNLP (2019)
18. Shekhar, R., et al.: Beyond task success: a closer look at jointly learning to see, ask, and guesswhat. In: NAACL (2019)
19. Yang, T., Zha, Z.-J., Zhang, H.: Making history matter: gold-critic sequence training for visual dialog. arXiv preprint arXiv:1902.09326 (2019)
20. Guo, D., Xu, C., Tao, D.: Image-question-answer synergistic network for visual dialog. In: CVPR (2019)

21. Qi, J., Niu, Y., Huang, J., Zhang, H.: Two causal principles for improving visual dialog, arXiv preprint arXiv:1911.10496 (2019)
22. Jiang, X., Yu, J., et al.: DualVD: an adaptive dual encoding model for deep visual understanding in visual dialogue. In: AAAI (2020)
23. Alamri, H., et al.: Audio visual scene-aware dialog. In: CVPR (2019)
24. de Vries, H., Shuster, K., Batra, D., Parikh, D., Weston, J., Kiela, D.: Talk the walk: navigating new york city through grounded dialogue, arXiv preprint arXiv:1807.03367 (2018)
25. Nguyen, K., Daumé III, H.: Help, anna! visual navigation with natural multimodal assistance via retrospective curiosity-encouraging imitation learning. In: EMNLP (2019)
26. Thomason, J., Murray, M., Cakmak, M., Zettlemoyer, L.: Vision-and-dialog navigation (2019)
27. Zhu, Y., Kiros, R., et al.: Aligning books and movies: towards story-like visual explanations by watching movies and reading books. In: ICCV (2015)
28. Sharma, P., Ding, N., Goodman, S., Soricut, R.: Conceptual captions: a cleaned, hypernymed, image alt-text dataset for automatic image captioning. In: ACL (2018)
29. Antol, S., et al.: VQA: visual question answering. In: ICCV (2015)
30. He, K., Zhang, X., Ren, S., Sun, J.: Deep residual learning for image recognition. In: CVPR (2016)
31. Simonyan, K., Zisserman, A.: Very deep convolutional networks for large-scale image recognition. In: ICLR (2015)
32. Krizhevsky, A., Sutskever, I., Hinton, G.: ImageNet classification with deep convolutional neural networks. In: NIPS (2012)
33. Radford, A., Narasimhan, K., Salimans, T., Sutskever, I.: Improving language understanding with unsupervised learning (2018)
34. Vaswani, A., et al.: Attention is all you need. In: NIPS (2017)
35. Devlin, J., Chang, M.-W., Lee, K., Toutanova, K.: BERT: pre-training of deep bidirectional transformers for language understanding. In: NAACL (2019)
36. Liu, Y., et al.: RoBERTa: a robustly optimized BERT pretraining approach, arXiv preprint arXiv:1907.11692 (2019)
37. Lan, Z., Chen, M., Goodman, S., Gimpel, K., Sharma, P., Soricut, R.: ALBERT: a lite BERT for self-supervised learning of language representations, arXiv preprint arXiv:1909.11942 (2019)
38. Yang, Z., Dai, Z., Yang, Y., Carbonell, J., Salakhutdinov, R., Le, Q.L.: XLNET: generalized autoregressive pretraining for language understanding, arXiv preprint arXiv:1906.08237 (2019)
39. Raffel, C., et al.: Exploring the limits of transfer learning with a unified text-to-text transformer, arXiv preprint arXiv:1910.10683 (2019)
40. Zhang, Y., et al.: DialoGPT: large-scale generative pre-training for conversational response generation, arXiv preprint arXiv:1911.00536 (2019)
41. Russakovsky, O., et al.: ImageNet large scale visual recognition challenge. IJCV (2015)
42. Krishna, R., et al.: Visual genome: connecting language and vision using crowd-sourced dense image annotations. IJCV (2017)
43. Wang, A., Singh, A., Michael, J., Hill, F., Levy, O., Bowman, S.R.: GLUE: a multi-task benchmark and analysis platform for natural language understanding, arXiv preprint arXiv:1804.07461 (2018)
44. Lu, J., Batra, D., Parikh, D., Lee, S.: ViLBERT: pretraining task-agnostic visiolinguistic representations for vision-and-language tasks. In: NeurIPS (2019)

45. Li, L.H., Yatskar, M., Yin, D., Hsieh, C.-J., Chang, K.-W.: VisualBERT: a simple and performant baseline for vision and language, arXiv preprint arXiv:1908.03557 (2019)
46. Tan, H., Bansal, M.: LXMERT: learning cross-modality encoder representations from transformers, arXiv preprint arXiv:1908.07490 (2019)
47. Chen, Y.-C., et al.: UNITER: Learning UNiversal Image-TExt Representations, arXiv preprint arXiv:1909.11740 (2019)
48. Li, G., Duan, N., Fang, Y., Jiang, D., Zhou, M.: Unicoder-VL: a universal encoder for vision and language by cross-modal pre-training, arXiv preprint arXiv:1908.06066 (2019)
49. Su, W., et al.: VL-BERT: pre-training of generic visual-linguistic representations, arXiv preprint arXiv:1908.08530 (2019)
50. Sun, C., Myers, A., Vondrick, C., Murphy, K., Schmid, C.: VideoBERT: a joint model for video and language representation learning, arXiv preprint arXiv:1904.01766 (2019)
51. Zellers, R., Bisk, Y., Farhadi, A., Choi, Y.: From recognition to cognition: visual commonsense reasoning. In: CVPR (2019)
52. Suhr, A., Zhou, S., Zhang, A., Zhang, I., Bai, H., Artzi, Y.: A corpus for reasoning about natural language grounded in photographs. In: ACL (2019)
53. Xie, N., Lai, F., Doran, D., Kadav, A.: Visual entailment: a novel task for fine-grained image understanding, arXiv preprint arXiv:1901.06706 (2019)
54. Young, P., Lai, A., Hodosh, M., Hockenmaier, J.: From image descriptions to visual denotations: new similarity metrics for semantic inference over event descriptions. In: TACL (2014)
55. Lee, K.-H., Chen, X., Hua, G., Hu, H., He, X.: Stacked cross attention for image-text matching. In: ECCV (2018)
56. Kazemzadeh, S., Ordonez, V., Matten, M., Berg, T.L.: ReferItGame: referring to objects in photographs of natural scenes. In: EMNLP (2014)
57. Hao, W., Li, C., Li, X., Carin, L., Gao, J.: Towards learning a generic agent for vision-and-language navigation via pre-training. In: CVPR (2020)
58. Paszke, A., et al.: Pytorch: an imperative style, high-performance deep learning library. In: Wallach, H., Larochelle, H., Beygelzimer, A., d'Alché-Buc, F., Fox, E., Garnett, R. (eds.) Advances in Neural Information Processing Systems, vol. 32, pp. 8024–8035, Curran Associates Inc. (2019)
59. Wolf, T., Sanh, V., Chaumond, J., Delangue, C.: TransferTransfo: a transfer learning approach for neural network based conversational agents, arXiv preprint arXiv:1901.08149 (2019)
60. Anderson, P., et al.: Bottom-up and top-down attention for image captioning and visual question answering, arXiv preprint arXiv:1707.07998 (2017)
61. Ren, S., He, K., Girshick, R., Sun, J.: Faster R-CNN: towards real-time object detection with region proposal networks. In: NIPS (2015)
62. Lin, T.-Y., et al.: Microsoft COCO: common objects in context. In: ECCV (2014)
63. Kingma, D., Ba, J.: Adam: a method for stochastic optimization. In: ICLR (2015)
64. Lewis, M., et al.: Bart: denoising sequence-to-sequence pre-training for natural language generation, translation, and comprehension, arXiv preprint arXiv:1910.13461 (2019)

Learning to Generate Grounded Visual Captions Without Localization Supervision

Chih-Yao Ma[1,3]([✉]), Yannis Kalantidis[2], Ghassan AlRegib[1], Peter Vajda[3], Marcus Rohrbach[3], and Zsolt Kira[1]

[1] Georgia Tech, Atlanta, Georgia
{cyma,alregib,zkira}@gatech.edu
[2] NAVER LABS Europe, Grenoble, France
yannis.kalantidis@naverlabs.com
[3] Facebook, California, USA
{cyma,vajdap,mrf}@fb.com

Abstract. When automatically generating a sentence description for an image or video, it often remains unclear how well the generated caption is grounded, that is whether the model uses the correct image regions to output particular words, or if the model is hallucinating based on priors in the dataset and/or the language model. The most common way of relating image regions with words in caption models is through an attention mechanism over the regions that are used as input to predict the next word. The model must therefore learn to predict the attentional weights without knowing the word it should localize. This is difficult to train without grounding supervision since recurrent models can propagate past information and there is no explicit signal to force the captioning model to properly ground the individual decoded words. In this work, we help the model to achieve this via a novel cyclical training regimen that forces the model to localize each word in the image *after* the sentence decoder generates it, and then reconstruct the sentence from the localized image region(s) to match the ground-truth. Our proposed framework only requires learning one extra fully-connected layer (the localizer), a layer that can be removed at test time. We show that our model significantly improves grounding accuracy without relying on grounding supervision or introducing extra computation during inference, for both image and video captioning tasks. Code is available at https://github.com/chihyaoma/cyclical-visual-captioning.

Keywords: Image captioning · Video captioning · Self-supervised learning · Visual grounding

Y. Kalantidis—Work done while at Facebook.

Electronic supplementary material The online version of this chapter (https://doi.org/10.1007/978-3-030-58523-5_21) contains supplementary material, which is available to authorized users.

1   Introduction

Vision and language tasks such as visual captioning or question answering, combine linguistic descriptions with data from real-world scenes. Deep learning models have achieved great success for such tasks, driven in part by the development of attention mechanisms that focus on various objects in the scene while generating captions. The resulting models, however, are known to have poor grounding performance [19], leading to undesirable behaviors (such as object hallucinations [26]), despite having high captioning accuracy. That is, they often do not correctly associate generated words with the appropriate image regions (*e.g.,* objects) in the scene, resulting in models that lack interpretability.

Several existing approaches have tried to improve the grounding of captioning models. One class of methods generate sentence *templates* with slot locations explicitly tied to specific image regions. These slots are then filled in by visual concepts identified by off-the-shelf object detectors [20]. Other methods have developed specific grounding or attention modules that aim to *attend* to the correct region(s) for generating visually groundable word. Such methods, however, rely on explicit supervision for optimizing the grounding or attention modules [19,46] and require bounding box annotations for each visually groundable word.

In this work, we propose a novel cyclical training regimen that is able to significantly improve grounding performance without any grounding annotations (Fig. 1). An important insight of our work is that current models use attention mechanisms conditioned on the hidden features of recurrent modules such as LSTMs, which leads to effective models with high accuracy but entangle grounding and decoding. Since LSTMs are effective at propagating information across the decoding process, the network does not necessarily need to associate particular decoded words with their corresponding image region(s). However, for a captioning model to be visually grounded, the model has to predict attentional weights without knowing the word to localize.

Based on this insight, we develop a cyclical training regimen to force the network to ground individual decoded words: *decoding → localization → reconstruction*. Specifically, the model of the decoding stage can be any state-of-the-art captioning model; in this work, we follow GVD [46] to extend the widely used Up-Down model [2]. At the localization stage, each word generated by the first decoding stage is localized through a *localizer*, and the resulting grounded image region(s) are then used to reconstruct the ground-truth caption in the final stage. Both decoding and reconstruction stages are trained using a standard cross-entropy loss. Important to our method, both stages share the same decoder, thereby causing the localization stage to guide the decoder to improve its attention mechanism. Our method is simple and only adds a fully-connected layer to perform localization. During inference, we only use the (shared) decoder, thus we do not add any computational cost.

To compare with the state-of-the-art [46], we evaluate our proposed method on the challenging Flickr30k Entities image captioning dataset [24] and the ActivityNet-Entities video captioning dataset [46] on both captioning and

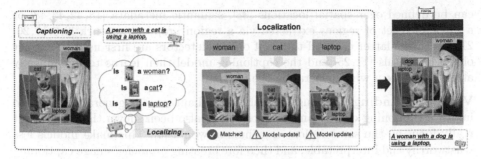

Fig. 1. Visual captioning models are often not visually-grounded. As humans, after generating a caption, we perform localization to check whether the generated caption is visually-grounded. If the localized image region is incorrect, we will correct it. Same goes for training a model. We would like to update the model accordingly. However, without the ground-truth grounding annotation, how does the model know the localized region is incorrect? How can the model then be updated? To overcome this issue, we propose to perform *localization* and *reconstruction* to regularize the captioning model to be visually-grounded without relying on the grounding annotations.

grounding. In addition to the existing grounding metrics that calculate the grounding accuracy for each object class [46], we further include a grounding metric that compute grounding accuracy for each generated sentence. We achieve around 18% relative improvements averaged over grounding metrics in terms of bridging the gap between the unsupervised baseline and supervised methods on Flickr30k Entites and around 34% on ActivityNet-Entities. We further find that our method can even outperform the supervised method on infrequent words, owing to its self-supervised nature. In addition, we also conduct human evaluation on visual grounding to further verify the improvement of the proposed method.

Contributions Summary. We propose object re-localization as a form of self-supervision for grounded visual captioning and present a cyclical training regimen that re-generates sentences after re-localizing the objects conditioned on each word, implicitly imposing grounding consistency. We evaluate our proposed approach on both image and video captioning tasks. We show that the proposed training regime can boost grounding accuracy over a state-of-the-art baseline, enabling grounded models to be trained without bounding box annotations, while retaining high captioning quality across two datasets and various experimental settings.

2   Related Work

Visual Captioning. Neural models for visual captioning have received significant attention recently [2,8,20,21,23,28,29,32,36,37]. Most current state-of-the-art models contain attention mechanisms, allowing the process to focus on subsets of the image when generating the next word. These attention mechanisms can be defined over spatial locations [38], semantic metadata [18,42,43,48] or a predefined set of regions extracted via a region proposal network [2,6,17,20, 21,44]. In the latter case, off-the-shelf object detectors are first used to extract object proposals [11,25] and the captioning model then learns to dynamically attend over them when generating the caption.

Visual Grounding. Although attention mechanisms are generally shown to improve captioning quality and metrics, it has also been shown that they don't really focus on the same regions as a human would [5]. This make models less trustworthy and interpretable, and therefore creating *grounded* image captioning models, *i.e.,* models that accurately link generated words or phrases to specific regions of the image, has recently been an active research area. A number of approaches have been proposed, *e.g.,* for grounding phrases or objects from image descriptions [7,14,27,41,45,46], grounding visual explanations [12], visual co-reference resolution for actors in video [28], or improving grounding via human supervision [30]. Recently, Zhou et al. [46] presented a model with self-attention based context encoding and direct grounding supervision that achieves state-of-the-art results in both the image and video tasks. They exploit ground-truth bounding box annotations to significantly improve the visual grounding accuracy. In contrast, we focus on reinforcing the visual grounding capability of the existing captioning model via a cyclical training regimen without using bounding box annotations and present a method that can increase grounding accuracy while maintaining comparable captioning performance with state-of-the-arts. Another closely related work is [27] where the authors focus on grounding or localizing a given textual phrase in an image. This creates a critical difference, as the ground-truth caption is provided during both training and test time and renders the training regimen proposed in [27] not applicable to visual captioning tasks.

Cyclical Training. Cycle consistency [4,10,40,49] has been used recently in a wide range of domains, including machine translation [10], unpaired image-to-image translation [49], visual question answering [31], question answering [33], image captioning [4], video captioning [9,39], captioning and drawing [15] as well as domain adaptation [13]. While the cyclical training regime has been explored vastly in both vision and language domains, it has not yet been used for enforcing the *visual grounding* capability of a captioning model.

3   Method

Notation. For the visual captioning task we have pairs of target sentences and images (or videos). Each image (or video) is represented by spatial feature map(s) extracted by a ResNet-101 model and a bag of regions obtained from Faster-RCNN [25] as $R = [r_1, r_2, ..., r_N] \in \mathbb{R}^{d \times N}$. The target sentence is represented as a sequence of one-hot vectors $y_t^* \in \mathbb{R}^s$, where T is the sentence length, $t \in 1, 2, ..., T$, and s is the dictionary size.

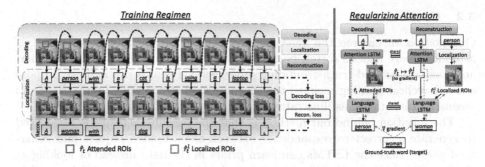

Fig. 2. (Left) Proposed cyclical training regimen: *decoding → localization → reconstruction*. The decoder attends to the image regions and sequentially generates each of the output words. The localizer then uses the generated words as input to locate the image regions. Finally, the shared decoder during the reconstruction stage uses the localized image regions to regenerate a sentence that matches with the ground-truth sentence. (Right) Because of the shared Attention LSTM, Language LSTM, and equal word inputs, this training regimen regularizes the attention mechanism inside the Attention LSTM so that the attended ROIs get closer to the less biased and better localized ROIs $\hat{r}_t \mapsto \hat{r}_t^l$.

3.1 Baseline

We reimplemented the model used in GVD [46] without self-attention for region feature encoding [21,34] as our baseline[1]. It is an extension of the state-of-the-art Up-Down [2] model with the *grounding-aware region encoding* (see Appendix).

Specifically, our baseline model uses two LSTM modules: Attention LSTM and Language LSTM. The Attention LSTM identifies which visual representation in the image is needed for the Language LSTM to generate the next word. It encodes the global image feature v_g, previous hidden state output of the Language LSTM h_{t-1}^L, and the previous word embedding e_{t-1} into the hidden state h_t^A.

$$h_t^A = LSTM_{Attn}([v_g; h_{t-1}^L; e_{t-1}]), \quad e_{t-1} = W_e y_{t-1}, \tag{1}$$

where $[;]$ denotes concatenation, and W_e are learned parameters. We omit the Attention LSTM input hidden and cell states to avoid notational clutter in the exposition.

The Language LSTM uses the hidden state h_t^A from the Attention LSTM to dynamically attend on the bag of regions R for obtaining visual representations of the image $\hat{r}_t$ to generate a word y_t.

$$z_{t,n} = W_{aa}tanh(W_a h_t^A + r_n), \quad \alpha_t = \text{softmax}(z_t), \quad \hat{r}_t = R\alpha, \tag{2}$$

where W_{aa} and W_a are learned parameters. The conditional probability distribution over possible output words y_t is computed as:

$$h_t^L = LSTM_{Lang}([\hat{r}_t, h_t^A]), \quad p(y_t|y_{1:t-1}) = \text{softmax}(W_o h_t^L), \tag{3}$$

where $y_{1:t-1}$ is a sequence of outputs $(y_1, ..., y_{t-1})$. We refer the Language LSTM and the output logit layer as the complete language decoder.

---

[1] We removed self-attention because we found that removing it slightly improved both captioning and grounding accuracy in our implementation.

3.2   Overview

Our goal is to enforce the generated caption to be visually grounded, *i.e.*, attended image regions correspond specifically to individual words being generated, *without* ground-truth grounding supervision. Towards this end, we propose a novel cyclical training regimen that is comprised of *decoding, localization,* and *reconstruction* stages, as illustrated in Fig. 2.

The intuition behind our method is that the baseline network is not forced to generate a correct correspondence between the attended objects and generated words, since the LSTMs can learn priors in the data instead of looking at the image or propagate information forward which can subsequently be used to generate corresponding words in future time steps. The proposed cyclical training regimen, in contrast, aims at enforcing visual grounding to the model by requiring the language decoder (Eq. 3) to rely on the localized image regions $\hat{r}_t^l$ to reconstruct the ground-truth sentence, where the localization is conditioned *only* on the generated word from the decoding stage. Our cyclical method can therefore be done without using any annotations of the grounding itself.

Specifically, let $\boldsymbol{y}_t^d = \mathcal{D}^d(\hat{\boldsymbol{r}}_t; \theta_d)$ be the initial language decoder with parameters θ_d (Eq. 3), trained to sequentially generate words $\boldsymbol{y}_t^d$. Let $\hat{\boldsymbol{r}}_t^l = \mathcal{G}(\boldsymbol{y}_t^d, \boldsymbol{R}; \theta_g)$ define a *localizer* unit with parameters θ_g, that learns to map (ground) each generated word $\boldsymbol{y}_t^d$ to region(s) in the image $\boldsymbol{R}$. Finally, let $\boldsymbol{y}_t^l = \mathcal{D}^l(\hat{\boldsymbol{r}}_t^l; \theta_l)$ be a second decoder, that is required to reconstruct the ground-truth caption using the localized region(s), instead of the attention computed by the decoder itself. We define the cycle:

$$\boldsymbol{y}_t^l = \mathcal{D}^l(\mathcal{G}(\mathcal{D}^d(\hat{\boldsymbol{r}}_t; \theta_d), \boldsymbol{R}; \theta_g); \theta_l), \quad \theta_d = \theta_l, \tag{4}$$

where $\mathcal{D}^d$ and $\mathcal{D}^l$ share parameters. Although parameters are shared, the inputs for the two language decoders differ, leading to unique LSTM hidden state values during a run. Note that the Attention LSTMs and logit layers in the two stages also share parameters, though they are omitted for clarity.

Through cyclical joint training, both $\mathcal{D}^d$ and $\mathcal{D}^l$ are required to generate the same ground-truth sentence. They are both optimized to maximize the likelihood of the correct caption:

$$\theta^* = \arg\max_{\theta_d} \sum \log p(\boldsymbol{y}_t^d; \theta_d) + \arg\max_{\theta_l} \sum \log p(\boldsymbol{y}_t^l; \theta_l), \tag{5}$$

During training, the localizer regularizes the region attention of the reconstructor and the effect is further propagated to the baseline network in the decoding stage, since the parameters of Attention LSTM and Language LSTM are shared for both decoding and reconstruction stages. Note that the gradient from reconstruction loss will not backprop to the decoder $\mathcal{D}^d$ in the first decoding stage since the generated words used as input to the localizer are leafs in the computational graph. The network is implicitly regularized to update its attention mechanism to match with the localized image regions $\hat{r}_t \mapsto \hat{r}_t^l$. In Sect. 4.4, we demonstrate that the localized image regions $\hat{r}_t^l$ indeed have higher attention accuracy than $\hat{r}_t$ when using ground-truth words as inputs for the localizer, which drives the attention mechanism and helps the attended region $\hat{r}_t$ to be more visually grounded.

3.3 Cyclical Training

We now describe each stage of our cyclical model, also illustrated in Fig. 3.

Decoding. We first use the baseline model presented in Sect. 3.1 to generate a sequence of words $\boldsymbol{y} = [\boldsymbol{y}_1^d, \boldsymbol{y}_2^d, ..., \boldsymbol{y}_T^d]$, where T is the ground-truth sentence length.

Localization. Following the decoding process, a localizer $\mathcal{G}$ is then learned to localize the image regions from each generated word y_t.

$$e_t = \boldsymbol{W}_e \boldsymbol{y}_t^d, \quad z_{t,n}^l = (\boldsymbol{W}_l e_t)^\top \boldsymbol{r}_n \quad \text{and} \quad \beta_t = \text{softmax}(\boldsymbol{z}_t^l), \qquad (6)$$

where e_t is the embedding for the word generated during decoding stage at step t, $\boldsymbol{r}_n$ is the image representation of a region proposal, and $\boldsymbol{W}_e$ and $\boldsymbol{W}_l$ are the learned parameters. Based on the localized weights β_t, the localized region representation can be obtained by $\hat{r}_t^l = \boldsymbol{R}\beta$. Our localizer essentially is a linear layer, and we have experimented with non-linear layers but found it performed worse (see Appendix for comparison).

Reconstruction. Finally, the shared language decoder $\mathcal{D}^l$ relies on the localized region representation $\hat{r}_t^l$ to generate the next word. The probability over possible output words is:

$$h_t^L = LSTM_{Lang}([\hat{r}_t^l; h_t^A]), \quad p(\boldsymbol{y}_t^l|\boldsymbol{y}_{1:t-1}^l) = \text{softmax}(\boldsymbol{W}_o h_t^L), \qquad (7)$$

Given the target ground truth caption $\boldsymbol{y}_{1:T}^*$ and our proposed captioning model parameterized with θ, we minimize the following cross-entropy losses:

$$\mathcal{L}_{CE}(\theta) = -\lambda_1 \underbrace{\sum_{t=1}^{T} log(p_\theta(\boldsymbol{y}_t^*|\boldsymbol{y}_{1:t-1}^*))\mathbb{1}_{(\boldsymbol{y}_t^*=\boldsymbol{y}_t^d)}}_{\text{decoding loss}} - \lambda_2 \underbrace{\sum_{t=1}^{T} log(p_\theta(\boldsymbol{y}_t^*|\boldsymbol{y}_{1:t-1}^*))\mathbb{1}_{(\boldsymbol{y}_t^*=\boldsymbol{y}_t^l)}}_{\text{reconstruction loss}}$$

$$(8)$$

where λ_1 and λ_2 are weighting coefficient selected on the validation split.

Note that before entering the proposed cyclical training regimen, the decoder was first pre-trained until convergence to make sure the generated caption, which used as inputs to the localizer, are reasonable. The cyclical training regimen mainly serves as a regularization method.

4   Experiments

Datasets. We use the Flickr30k Entities image dataset [24] and the ActivityNet-Entities video dataset [46] to provide a comparison with the state-of-the-art [46].

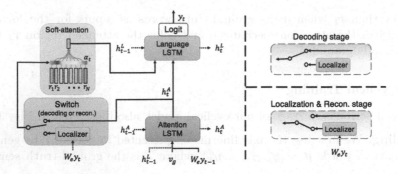

Fig. 3. Proposed model architecture (left) and how the model operates during decoding, localization, and reconstruction stages (right). During the decoding stage, the soft-attention module uses the hidden state of the Attention LSTM to compute attention weights on image regions. During the localization and reconstruction stage, the soft-attention module instead uses the generated word from decoding stage to compute attention weights on image regions.

Flickr30k Entities contains 275k annotated bounding boxes from 31k images associated with natural language phrases. Each image is annotated with 5 crowd-sourced captions. ActivityNet-Entities contains 15k videos with 158k spatially annotated bounding boxes from 52k video segments.

4.1   Evaluation Metrics

Captioning Evaluation Metrics. We measure captioning performance using common language metrics, *i.e.,* BLEU [22], METEOR [3], CIDEr [35], and SPICE [1].

Grounding Evaluation Metrics. Following the grounding evaluation from GVD [46], we measure the attention accuracy on generated sentences, denoted by $F1_{all}$ and $F1_{loc}$. In $F1_{all}$, a region prediction is considered correct if the object word[2] is correctly predicted and also correctly localized. We also compute $F1_{loc}$, which only considers correctly-predicted object words. Please see illustration of the grounding metrics in Appendix.

In the original formulation, the precision and recall for the two F1 metrics are computed **for each object class**, and it is set to zero if an object class has never been predicted. The scores are computed for each object class and averaged over the total number of classes. Such metrics are extremely stringent as captioning models are generally biased toward certain words in the vocabulary, given the long-tailed distribution of words. In fact, both the baseline and proposed method only generate about 45% of the annotated object words within the val set in Flickr30k Entities. The grounding accuracy of the other 55% of the classes are therefore zero, making the averaged grounding accuracy seemingly low.

---

[2] The object words are words in the sentences that are annotated with corresponding image regions.

Table 1. Performance comparison on the Flickr30k Entities **test set**. *: our results are averaged **across five runs**. Only numbers reported by multiple runs are considered to be bolded (Note that since supervised methods are used as upper bound, their numbers are not bolded.).

Method	Grounding supervision	Captioning Evaluation					Grounding Evaluation			
		B@1	B@4	M	C	S	F1$_{all}$	F1$_{loc}$	F1$_{all\_per\_sent}$	F1$_{loc\_per\_sent}$
ATT-FCN [43]		64.7	19.9	18.5	-	-	-	-	-	-
NBT [20]		69.0	27.1	21.7	57.5	15.6	-	-	-	-
Up-Down [2]		69.4	27.3	21.7	56.6	16.0	4.14	12.3	-	-
GVD (w/o SelfAttn) [46]		69.2	26.9	22.1	60.1	16.1	3.97	11.6	-	-
GVD [46]	✓	69.9	27.3	22.5	62.3	16.5	7.77	22.2	-	-
Baseline*	✓	69.0	26.8	22.4	61.1	16.8	8.44 (+100%)	22.78 (+100%)	27.37 (+100%)	63.19 (+100%)
Baseline*		69.1	26.0	22.1	59.6	16.3	4.08 (+0%)	11.83 (+0%)	13.20 (+0%)	31.83 (+0%)
Cyclical*		**69.9**	**27.4**	**22.3**	**61.4**	**16.6**	**4.98** (+21%)	**13.53** (+16%)	**15.03** (+13%)	**35.54** (+12%)

Measuring Grounding per Generated Sentence. Instead of evaluating grounding on each object class (which might be less intuitive), we include a new grounding evaluation metric *per sentence* to directly reflect the grounding measurement of each generated sentence. The metrics are computed against a pool of object words and their ground-truth bounding boxes (GT bbox) collected across five GT captions on Flickr30k Entities (and one GT caption on ActivityNet-Entities). We use the same Prec$_{all}$, Rec$_{all}$, Prec$_{loc}$, and Rec$_{loc}$ as defined previously, but their scores are averaged on each of the generated sentence. As a result, the F1$_{loc\_per\_sent}$ measures the F1 score only on the generated words. The model will not be punished if some object words are not generated, but it also needs to maintain diversity to achieve high captioning performance.

4.2   Implementation and Training Details

Region Proposal and Spatial Features. Following GVD [46], we extracted 100 region proposals from each image (video frame) and encode them via the *grounding-aware region encoding*.

Training. We train the model with ADAM optimizer [16]. The initial learning rate is set to $1e-4$. Learning rates automatically drop by 10x when the CIDEr score is saturated. The batch size is 32 for Flickr30k Entities and 96 for ActivityNet-Entities. We learn the word embedding layer from scratch for fair comparisons with existing work [46]. Please see the Appendix for additional training and implementation details.

4.3   Captioning and Grounding Performance Comparison

Flickr30k Entities. We first compare the proposed method with our baseline with or without grounding supervision on the Flickr30k Entities test set (see Table 1). To train the supervised baseline, we train the attention mechanism as well as add the region classification task using the ground-truth grounding annotation, similar to GVD [46]. We train the proposed baselines and our method

Table 2. Performance comparison on the ActivityNet-Entities **val set**. *: our results are averaged **across five runs**. Only numbers reported by multiple runs are considered to be bolded.

Method	Grounding supervision	Captioning Evaluation						Grounding Evaluation			
		B@1	B@4	M	C	S	$F1_{all}$	$F1_{loc}$	$F1_{all\_per\_sent}$	$F1_{loc\_per\_sent}$
GVD [46]		23.0	2.27	10.7	44.6	13.8	0.28	1.13	-	-
GVD (w/o SelfAttn) [46]		23.2	2.28	10.9	45.6	15.0	3.70	12.7	-	-
GVD [46]	✓	23.9	2.59	11.2	47.5	15.1	7.11	24.1	-	-
Baseline*	✓	23.1	2.13	10.7	45.0	14.6	7.30 (+100%)	25.02 (+100%)	17.88 (+100%)	60.23 (+100%)
Baseline*		23.2	2.22	10.8	45.9	**15.1**	3.75 (+0%)	12.00 (+0%)	9.41 (+0%)	31.68 (+0%)
Cyclical*		**23.7**	**2.45**	**11.1**	**46.4**	14.8	**4.71** (+26%)	**15.84** (+29%)	**11.73** (+38%)	**41.56** (+43%)

on the training set and choose the best performing checkpoints based on their CIDEr score on the val set. Unlike previous work, our experimental results are reported by averaging across five runs on the test set. We report only the mean of the five runs to keep the table uncluttered.

When compared to the existing state of the arts, our proposed baselines achieve comparable captioning evaluation performances and better grounding accuracy. Using the resulting supervised baseline as the upper bound, our proposed method with cyclical training statistically achieves around 15 to 20% relative grounding accuracy improvements for both $F1_{all}$ and $F1_{loc}$ and 10 to 15% for $F1_{all}\_per\_sent$ and $F1_{loc}\_per\_sent$ without utilizing any grounding annotations or additional computation during inference.

ActivityNet-Entities. We adapt our proposed baselines and method to the ActivityNet-Entities video dataset (see Table 2 for results on the validation set and Appendix for results on the test set). We can see that our baseline again achieved comparable performance to the state of the arts. The proposed method then significantly improved the grounding accuracy around 25% to 30% relative grounding accuracy improvements for both $F1_{all}$ and $F1_{loc}$ and around 40% for $F1_{all}\_per\_sent$ and $F1_{loc}\_per\_sent$. Interestingly, we observed that baseline with grounding supervision does not improve the captioning accuracy, different from the observation in previous work [46].

4.4 Quantitative Analysis

Are Localized Image Regions Better than Attended Image Regions *During Training*? Given our intuition described in Sect. 3, we expect the decoder to be regularized to update its attention mechanism to match with the localized image regions $\hat{r}_t \mapsto \hat{r}_t^l$ during training. This indicates that the localized image regions should be more accurate than the attended image regions by the decoder and drives the update on attention mechanism. To verify this, we compute the attention accuracy for both decoder and localizer over ground-truth sentences following [27,47]. The attention accuracy for localizer is 20.4% and is higher than the 19.3% from the decoder at the end of training, which confirms our hypothesis on how cyclical training helps the captioning model to be more grounded.

#	Grounding supervision	Captioning eval.				Grounding eval.	
		M	C	S	F1$_{all}$	F1$_{loc}$	F1$_{loc\_per\_sent}$
Unrealistically perfect object detector							
Baseline	✓	25.3	76.5	22.3	23.19	52.83	90.76
Baseline		25.2	76.3	22.0	20.82	48.74	77.81
Cyclical		**25.8**	**80.2**	**22.7**	**25.27**	**54.54**	81.56
Grounding-biased object detector							
Baseline	✓	21.3	53.3	15.5	8.23	23.95	66.96
Baseline		**21.2**	**52.4**	**15.4**	5.95	17.51	42.84
Cyclical		**21.2**	52.0	**15.4**	**6.87**	**19.65**	**50.25**

Table 3. Comparison when using *better* object detector on Flickr30k Entities **test** set (see Appendix for complete version).

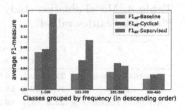

Fig. 4. Average F1$_{all}$-score per class as a function of class frequency.

Grounding Performance When Using a *Better* Object Detector. In Table 1 and 2 we showed that our proposed method significantly improved the grounding accuracy for both image and video captioning. These experimental settings follow the widely used procedure for visual captioning systems: extract regional proposal features and generate visual captions by attending to those extracted visual features. One might ask, what if we have a better object detector that can extract robust visual representation that are better aligned with the word embeddings? Will visual grounding still an issue for captioning?

To answer this, we ran two sets of experiments (Table 3): **(1) *Perfect* object detector**: we replace the ROIs by ground-truth bbox and represent the new ROIs by learning embedding features directly from ground-truth object words associated with each ground-truth bbox. This experiment gives an estimate of the captioning and grounding performance if we have (almost) perfect ROI representations (though unrealistic). We can see that the fully-supervised method achieves an F1$_{all}$ of only 23%, which further confirms the difficulty of the metric and the necessity of our grounding metric on a per sentence level (note that F1$_{loc\_per\_sent}$ shows 90%). We can also see that baseline (unsup.) still leaves room for improvement on grounding performance. Surprisingly, our method improved both captioning and grounding accuracy and surpasses the fully-supervised baseline except on the F1$_{loc\_per\_sent}$. We find that it is because the baseline (sup.) overfits to the training set, while ours is regularized from the cyclical training. Also, our generated object words are more diverse, which is important for F1$_{all}$ and F1$_{loc}$. **(2) *Grounding-biased* object detector**: we extract ROI features from an object detector pre-trained on Flickr30k. Thus, the ROI features and their associated object predictions are biased toward the annotated object words but do not generalize to predict diverse captions compared to the original object detector trained from Visual Genome, resulting in lower captioning performance. We can see that our proposed method still successfully improves grounding and maintains captioning performance in this experiment setting as well.

How Does the Number of Annotations Affect Grounding Performance? In Fig. 4, we present the average F1-score on the Flickr30k Entities

Table 4. Model ablation study on the Flickr30k Entities val set.

#	Captioning eval.			Grounding eval.	
	M	C	S	F1$_{all}$	F1$_{loc}$
Baseline (Unsup.)	**22.3**	62.1	16.0	4.18	11.9
Cyclical	22.2	**62.2**	**16.2**	**5.63**	**14.6**
- Attention consistency	**22.3**	61.8	**16.2**	4.19	11.3
- Localizer using h^A	22.2	61.8	16.1	4.58	11.3

Table 5. Human evaluation on grounding on the Flickr30k Entities val set.

Method	Human grounding eval.
	%
About equal	47.1
Cyclical is better	28.1
Baseline is better	24.8

val set when grouping classes according to their frequency of appearance in the training set[4]. We see that, unsurprisingly, the largest difference in grounding accuracy between the supervised and our proposed cyclical training is for the 50 most frequently appearing object classes, where enough training data exists. As the number of annotated boxes decreases, however, the difference in performance diminishes, and cyclical training appears to be more robust. Overall, we see that the supervised method is biased towards frequently appearing objects, while grounding performance for the proposed approach is more balanced among classes.

Should We Explicitly Make Attended Image Regions to be Similar to Localized Image Regions? One possible way to regularize the attention mechanism of the decoder is to explicitly optimize $\hat{r}_t \mapsto \hat{r}_t^l$ via KL divergence over two soft-attention weights α_t and β_t. The experimental results are shown in Table 4 (*Attention consistency*). We use a single run unsupervised baseline with a fix random seed as baseline model for ablation study. We can see that when explicitly forcing the attended regions to be similar to the localized regions, both the captioning performance and the grounding accuracy remain similar to the baseline (unsup.). We conjecture that this is due to the noisy localized regions at the initial training stage. When forcing the attended regions to be similar to noisy localized regions, the Language LSTM will eventually learn to not rely on the attended region at each step for generating sequence of words. To verify, we increase the weight for attention consistency loss and observed that it has lower grounding accuracy (F1$_{all}$ = 3.2), but the captioning will reach similar performance while taking 1.5× longer to reach convergence.

Is Using Only the Generated Word for Localization Necessary? Our proposed localizer (Eq. 6 and Fig. 3) relies on purely the word embedding representation to locate the image regions. This forces the localizer to rely only on the word embedding without biasing it with the memorized information from the Attention LSTM. As shown in the Table 4 (localizer using h^A), although this achieves comparable captioning performance, it has lower grounding accuracy improvement compared to our proposed method.

---

[4] We group the 460 object classes in 10 groups, sorted by the number of annotated bounding boxes.

Fig. 5. Generated captions and corresponding visual grounding regions with comparison between baseline (left) and proposed approach (right). Our proposed method is able to generate more descriptive sentences while selecting the correct regions for generating the corresponding words.

4.5   Human Evaluation on Grounding

We conduct a human evaluation on the perceptual quality of the grounding. We asked 10 human subjects (not familiar with the proposed method) to pick the best among two grounded regions (by baseline and Cyclical) for each word. The subjects have three options to choose from: 1) grounded region A is better, 2) grounded region B is better, and 3) they are about the same (see Appendix for illustration).

Each of the human subjects were given 25 images, each with a varying number of groundable words. Each image was presented to two different human subjects in order to be able to measure inter-rater agreement. For this study, we define a word to be groundable if it is either a noun or verb. The image regions are selected based on the region with the maximum attention weight in α_t for each word. The order of approaches was randomized for each sentence.

Our experiment on the Flickr30k Entities val set is shown in Table 5: 28.1% of words are more grounded by Cyclical, 24.8% of words are more grounded by baseline, and 47.1% of words are similarly grounded. We also measured inter-rater agreement between each pair of human subjects: 72.7% of ratings are the same, 4.9% of ratings are the opposite, and 22.4% of ratings could be ambiguous (*e.g.,* one chose A is better, the other chose they are about the same).

We would also like to note that the grounded words judged to be similar largely consisted of very easy or impossible cases. For example, words like *mountain, water, street, etc,* are typically rated to be "about the same" since they usually have many possible boxes and is very easy for both models to ground the words correctly. On the other hand, for visually ungroundable cases, *e.g.,* *stand* appears a lot and the subject would choose *about the same* since the image does not cover the fact that the person's feet are on the ground. We see that

Fig. 6. Examples with correct grounding (top) as well as failure cases (bottom).

the human study results follow the grounding results presented in the paper and show an improvement in grounding accuracy for the proposed method over a strong baseline. The improvement is achieved without grounding annotations or extra computation at test time.

4.6   Qualitative Analysis

We additionally present some qualitative results comparing the baseline (Unsup.) and the proposed method in Fig. 5. To avoid clutter, only the visually-relevant object words in the generated sentences are highlighted with colors. Each high-lighted word has a corresponding image region annotated on the original image. The image regions are selected based on the region with the maximum attention weight in α_t. We can see that our proposed method significantly outperforms the baseline (Unsup.) in terms of both the quality of the generated sentence and grounding accuracy. In Fig. 6, we show a number of correct and incorrect examples of our proposed method. We observe that while the model is able to generate grounded captions for the images, it may sometimes overlook the semantic meaning of the generated sentences (*e.g.*, "*A young girl [...] walking down a brick wall*"), or the spatial relationship between the objects ("*A man [...] is holding up a flag*").

5   Conclusion

Working from the intuition that typical attentional mechanisms in the visual captioning task are not forced to ground generated words since recurrent models can propagate past information, we devise a novel cyclical training regime to explicitly force the model to ground each word without grounding annotations. Our method only adds a fully-connected layer during training, which can be removed during inference, and we show thorough quantitative and qualitative results demonstrating around 20% or 30% relative improvements in visual grounding accuracy over existing methods for image and video captioning tasks.

Acknowledgments. Chih-Yao Ma and Zsolt Kira were partly supported by DARPA's Lifelong Learning Machines (L2M) program, under Cooperative Agreement HR0011-18-2-0019, as part of their affiliation with Georgia Tech. We thank Chia-Jung Hsu for her valuable and artistic help on the figures.

References

1. Anderson, P., Fernando, B., Johnson, M., Gould, S.: SPICE: semantic propositional image caption evaluation. In: Leibe, B., Matas, J., Sebe, N., Welling, M. (eds.) ECCV 2016. LNCS, vol. 9909, pp. 382–398. Springer, Cham (2016). https://doi.org/10.1007/978-3-319-46454-1_24
2. Anderson, P., et al.: Bottom-up and top-down attention for image captioning and visual question answering. In: Proceedings of the IEEE Conference on Computer Vision and Pattern Recognition (CVPR), p. 6 (2018)
3. Banerjee, S., Lavie, A.: Meteor: An automatic metric for MT evaluation with improved correlation with human judgments. In: Proceedings of the ACL Workshop on Intrinsic and Extrinsic Evaluation Measures for Machine Translation and/or Summarization, vol. 29, pp. 65–72 (2005)
4. Chen, X., Lawrence Zitnick, C.: Mind's eye: a recurrent visual representation for image caption generation. In: Proceedings of the IEEE Conference on Computer Vision and Pattern Recognition (CVPR), pp. 2422–2431 (2015)
5. Das, A., Agrawal, H., Zitnick, L., Parikh, D., Batra, D.: Human attention in visual question answering: do humans and deep networks look at the same regions? Comput. Vis. Image Underst. **163**, 90–100 (2017)
6. Das, P., Xu, C., Doell, R.F., Corso, J.J.: A thousand frames in just a few words: lingual description of videos through latent topics and sparse object stitching. In: Proceedings of the IEEE Conference on Computer Vision and Pattern Recognition (CVPR), pp. 2634–2641 (2013)
7. Deng, C., Wu, Q., Wu, Q., Hu, F., Lyu, F., Tan, M.: Visual grounding via accumulated attention. In: Proceedings of the IEEE Conference on Computer Vision and Pattern Recognition (CVPR), pp. 7746–7755 (2018)
8. Donahue, J., et al.: Long-term recurrent convolutional networks for visual recognition and description. In: Proceedings of the IEEE Conference on Computer Vision and Pattern Recognition (CVPR), pp. 2625–2634 (2015)
9. Duan, X., Huang, W., Gan, C., Wang, J., Zhu, W., Huang, J.: Weakly supervised dense event captioning in videos. In: Advances in Neural Information Processing Systems (NeurIPS), pp. 3063–3073 (2018)
10. He, D., et al.: Dual learning for machine translation. In: Advances in Neural Information Processing Systems (NeurIPS), pp. 820–828 (2016)
11. He, K., Gkioxari, G., Dollár, P., Girshick, R.: Mask R-CNN. In: 2017 IEEE International Conference on Computer Vision (ICCV), pp. 2980–2988. IEEE (2017)

12. Hendricks, L.A., Hu, R., Darrell, T., Akata, Z.: Grounding visual explanations. In: Ferrari, V., Hebert, M., Sminchisescu, C., Weiss, Y. (eds.) ECCV 2018. LNCS, vol. 11206, pp. 269–286. Springer, Cham (2018). https://doi.org/10.1007/978-3-030-01216-8_17

13. Hosseini-Asl, E., Zhou, Y., Xiong, C., Socher, R.: Augmented cyclic adversarial learning for low resource domain adaptation. In: Proceedings of the International Conference on Learning Representations (ICLR) (2019)

14. Hu, R., Xu, H., Rohrbach, M., Feng, J., Saenko, K., Darrell, T.: Natural language object retrieval. In: Proceedings of the IEEE Conference on Computer Vision and Pattern Recognition (CVPR), pp. 4555–4564 (2016)

15. Huang, Q., Zhang, P., Wu, D., Zhang, L.: Turbo learning for captionbot and drawingbot. In: Advances in Neural Information Processing Systems (NeurIPS), pp. 6456–6466 (2018)

16. Kingma, D.P., Ba, J.: Adam: a method for stochastic optimization. In: Proceedings of the International Conference on Learning Representations (ICLR) (2015)

17. Kulkarni, G., et al.: Babytalk: understanding and generating simple image descriptions. IEEE Trans. Pattern Anal. Mach. Intell. **35**(12), 2891–2903 (2013)

18. Li, Y., Yao, T., Pan, Y., Chao, H., Mei, T.: Jointly localizing and describing events for dense video captioning. In: Proceedings of the IEEE Conference on Computer Vision and Pattern Recognition (CVPR), pp. 7492–7500 (2018)

19. Liu, C., Mao, J., Sha, F., Yuille, A.: Attention correctness in neural image captioning. In: Thirty-First AAAI Conference on Artificial Intelligence (2017)

20. Lu, J., Yang, J., Batra, D., Parikh, D.: Neural baby talk. In: Proceedings of the IEEE Conference on Computer Vision and Pattern Recognition (CVPR), pp. 7219–7228 (2018)

21. Ma, C.Y., Kadav, A., Melvin, I., Kira, Z., AlRegib, G., Graf, H.P.: Attend and interact: higher-order object interactions for video understanding. In: Proceedings of the IEEE Conference on Computer Vision and Pattern Recognition (CVPR) (2018)

22. Papineni, K., Roukos, S., Ward, T., Zhu, W.J.: Bleu: a method for automatic evaluation of machine translation. In: Proceedings of the 40th Annual Meeting on Association for Computational Linguistics, pp. 311–318. Association for Computational Linguistics (2002)

23. Park, J.S., Rohrbach, M., Darrell, T., Rohrbach, A.: Adversarial inference for multi-sentence video description. In: Proceedings of the IEEE Conference on Computer Vision and Pattern Recognition (CVPR) (2019)

24. Plummer, B.A., Wang, L., Cervantes, C.M., Caicedo, J.C., Hockenmaier, J., Lazebnik, S.: Flickr30k entities: collecting region-to-phrase correspondences for richer image-to-sentence models. In: Proceedings of the IEEE International Conference on Computer Vision (ICCV), pp. 2641–2649 (2015)

25. Ren, S., He, K., Girshick, R., Sun, J.: Faster R-CNN: towards real-time object detection with region proposal networks. In: Advances in Neural Information Processing Systems (NeurIPS), pp. 91–99 (2015)

26. Rohrbach, A., Hendricks, L.A., Burns, K., Darrell, T., Saenko, K.: Object hallucination in image captioning. In: Proceedings of the 2018 Conference on Empirical Methods in Natural Language Processing (EMNLP), pp. 4035–4045 (2018)

27. Rohrbach, A., Rohrbach, M., Hu, R., Darrell, T., Schiele, B.: Grounding of textual phrases in images by reconstruction. In: Leibe, B., Matas, J., Sebe, N., Welling, M. (eds.) ECCV 2016. LNCS, vol. 9905, pp. 817–834. Springer, Cham (2016). https://doi.org/10.1007/978-3-319-46448-0_49

28. Rohrbach, A., Rohrbach, M., Tang, S., Joon Oh, S., Schiele, B.: Generating descriptions with grounded and co-referenced people. In: Proceedings of the IEEE Conference on Computer Vision and Pattern Recognition (CVPR), pp. 4979–4989 (2017)
29. Rohrbach, A., et al.: Movie description. Int. J. Comput. Vis. **123**(1), 94–120 (2017)
30. Selvaraju, R.R., Lee, S., Shen, Y., Jin, H., Batra, D., Parikh, D.: Taking a hint: Leveraging explanations to make vision and language models more grounded. In: Proceedings of the IEEE International Conference on Computer Vision (ICCV) (2019)
31. Shah, M., Chen, X., Rohrbach, M., Parikh, D.: Cycle-consistency for robust visual question answering. In: Proceedings of the IEEE Conference on Computer Vision and Pattern Recognition (CVPR) (2019)
32. Shetty, R., Rohrbach, M., Anne Hendricks, L., Fritz, M., Schiele, B.: Speaking the same language: Matching machine to human captions by adversarial training. In: Proceedings of the IEEE International Conference on Computer Vision (ICCV), pp. 4135–4144 (2017)
33. Tang, D., et al.: Learning to collaborate for question answering and asking. In: Proceedings of the 2018 Conference of the North American Chapter of the Association for Computational Linguistics: Human Language Technologies, Volume 1 (Long Papers), vol. 1, pp. 1564–1574 (2018)
34. Vaswani, A., et al.: Attention is all you need. In: Advances in Neural Information Processing Systems (NeurIPS), pp. 5998–6008 (2017)
35. Vedantam, R., Lawrence Zitnick, C., Parikh, D.: Cider: consensus-based image description evaluation. In: Proceedings of the IEEE Conference on Computer Vision and Pattern Recognition (CVPR), pp. 4566–4575 (2015)
36. Venugopalan, S., Anne Hendricks, L., Rohrbach, M., Mooney, R., Darrell, T., Saenko, K.: Captioning images with diverse objects. In: Proceedings of the IEEE Conference on Computer Vision and Pattern Recognition (CVPR), pp. 5753–5761 (2017)
37. Venugopalan, S., Rohrbach, M., Donahue, J., Mooney, R., Darrell, T., Saenko, K.: Sequence to sequence-video to text. In: Proceedings of the IEEE International Conference on Computer Vision (ICCV), pp. 4534–4542 (2015)
38. Vinyals, O., Toshev, A., Bengio, S., Erhan, D.: Show and tell: a neural image caption generator. In: Proceedings of the IEEE Conference on Computer Vision and Pattern Recognition (CVPR), pp. 3156–3164 (2015)
39. Wang, B., Ma, L., Zhang, W., Liu, W.: Reconstruction network for video captioning. In: Proceedings of the IEEE Conference on Computer Vision and Pattern Recognition (CVPR), pp. 7622–7631 (2018)
40. Wang, F., Huang, Q., Guibas, L.J.: Image co-segmentation via consistent functional maps. In: Proceedings of the IEEE International Conference on Computer Vision (ICCV), December 2013
41. Xiao, F., Sigal, L., Jae Lee, Y.: Weakly-supervised visual grounding of phrases with linguistic structures. In: Proceedings of the IEEE Conference on Computer Vision and Pattern Recognition (CVPR), pp. 5945–5954 (2017)
42. Yao, T., Pan, Y., Li, Y., Qiu, Z., Mei, T.: Boosting image captioning with attributes. In: Proceedings of the IEEE International Conference on Computer Vision (ICCV), pp. 22–29 (2017)
43. You, Q., Jin, H., Wang, Z., Fang, C., Luo, J.: Image captioning with semantic attention. In: Proceedings of the IEEE Conference on Computer Vision and Pattern Recognition (CVPR), pp. 4651–4659 (2016)

44. Zanfir, M., Marinoiu, E., Sminchisescu, C.: Spatio-temporal attention models for grounded video captioning. In: Asian Conference on Computer Vision, pp. 104–119 (2016)
45. Zhang, Y., Niebles, J.C., Soto, A.: Interpretable visual question answering by visual grounding from attention supervision mining. In: 2019 IEEE Winter Conference on Applications of Computer Vision (WACV), pp. 349–357. IEEE (2019)
46. Zhou, L., Kalantidis, Y., Chen, X., Corso, J.J., Rohrbach, M.: Grounded video description. In: Proceedings of the IEEE Conference on Computer Vision and Pattern Recognition (CVPR) (2019)
47. Zhou, L., Louis, N., Corso, J.J.: Weakly-supervised video object grounding from text by loss weighting and object interaction. In: British Machine Vision Conference (BMVC) (2018)
48. Zhou, L., Xu, C., Koch, P., Corso, J.J.: Watch what you just said: image captioning with text-conditional attention. In: Proceedings of the on Thematic Workshops of ACM Multimedia 2017, pp. 305–313. ACM (2017)
49. Zhu, J.Y., Park, T., Isola, P., Efros, A.A.: Unpaired image-to-image translation using cycle-consistent adversarial networks. In: Proceedings of the IEEE International Conference on Computer Vision (ICCV), pp. 2223–2232 (2017)

Neural Hair Rendering

Menglei Chai(✉), Jian Ren, and Sergey Tulyakov

Snap Inc., Santa Monica, USA
cmlatsim@gmail.com

Abstract. In this paper, we propose a generic neural-based hair rendering pipeline that can synthesize photo-realistic images from virtual 3D hair models. Unlike existing supervised translation methods that require model-level similarity to preserve consistent structure representation for both real images and fake renderings, our method adopts an unsupervised solution to work on arbitrary hair models. The key component of our method is a shared latent space to encode appearance-invariant structure information of both domains, which generates realistic renderings conditioned by extra appearance inputs. This is achieved by domain-specific pre-disentangled structure representation, partially shared domain encoder layers and a structure discriminator. We also propose a simple yet effective temporal conditioning method to enforce consistency for video sequence generation. We demonstrate the superiority of our method by testing it on a large number of portraits and comparing it with alternative baselines and state-of-the-art unsupervised image translation methods.

Keywords: Neural rendering · Unsupervised image translation

1 Introduction

Hair is a critical component of human subjects. Rendering virtual 3D hair models into realistic images has been long studied in computer graphics, due to the extremely complicated geometry and material of human hair. Traditional graphical rendering pipelines try to simulate every aspect of natural hair appearance, including surface shading, light scattering, semi-transparent occlusions, and soft shadowing. This is usually achieved by leveraging physics-based shading models of hair fibers, global illumination rendering algorithms, and artistically designed material parameters. Given the extreme complexity of the geometry and associated lighting effects, such a direct approximation of physical hair appearance requires a highly detailed 3D model, carefully tuned material parameters, and a huge amount of rendering computation. However, for interactive application scenarios that require efficient feedback and user-friendly interactions, such as games and photo editing softwares, it is often too expensive and unaffordable.

Electronic supplementary material The online version of this chapter (https://doi.org/10.1007/978-3-030-58523-5_22) contains supplementary material, which is available to authorized users.

© Springer Nature Switzerland AG 2020
A. Vedaldi et al. (Eds.): ECCV 2020, LNCS 12363, pp. 371–388, 2020.
https://doi.org/10.1007/978-3-030-58523-5_22

With the recent advances in generative adversarial networks, it becomes natural to formulate hair rendering as a special case of the conditional image generation problem, with the hair structure controlled by the 3D model, while realistic appearance synthesized by neural networks. In the context of image-to-image translation, one of the major challenges is how to bridge both the source and target domains for proper translation. Most existing hair generation methods fall into the supervised category, which demands enough training image pairs to provide direct supervision. For example, sketch-based hair generation methods [28, 34, 49] construct training pairs by synthesizing user sketches from real images. While several such methods are introduced, rendering 3D hair models with the help of neural networks do not receive similar treatment. The existing work on this topic [66] requires real and fake domains considerably overlap, such that the common structure is present in both domains. This is achieved at the cost of a complicated strand-level high-quality model, which allows for extracting edge and orientation maps that serve as the common representations of hair structures between real photos and fake models. However, preparing such a high-quality hair model is itself expensive and non-trivial even for a professional artist, which significantly restricts the application scope of this method.

In this paper, we propose a generic neural-network-based hair rendering pipeline that provides efficient and realistic rendering of a generic low-quality 3D hair model borrowing the material features extracted from an arbitrary reference hair image. Instead of using a complicated strand-level model to match real-world hairs like [66], we allow users to use any type of hair model requiring only the isotropic structure of hair strands be properly represented. Particularly, we adopt sparse polygon strip meshes which are much more widely used in interactive applications [65]. Given the dramatic difference between such a coarse geometry and real hair, we are not able to design common structure representations at the model level. Therefore, supervised image translation methods will be infeasible due to the lack of paired data.

To bridge the domains of real hair images and low-quality virtual hair models in an unsupervised manner, we propose to construct a shared latent space between both real and fake domains, which encodes a common structural representation from distinct inputs of both domains and renders the realistic hair image from this latent space with the appearance conditioned by an extra reference. This is achieved by 1) different domain structure encodings used as the network inputs, to pre-disentangle geometric structure and chromatic appearance for both real hair images and 3D models; 2) a UNIT [39]-like architecture adopted to enable common latent space by partially sharing encoder weights between two auto-encoder branches that are trained with in-domain supervision; 3) a structure discriminator introduced to further match the distribution of the encoded structure features; 4) supervised reconstruction enforced on both branches to guarantee all necessary structure information is kept in the shared feature space. In addition, to enable temporally-smooth animation rendering, we introduce a simple yet effective temporal condition method with single image training data only, utilizing the exact hair model motion fields. We demonstrate the effectiveness of the pipeline and each key component by extensively testing on a large amount of diverse human portraits and various hair models. We also compare our method with general unsupervised image translation methods, and show that due to the limited sampling ability on the synthetic hair domain, all existing methods fail to produce convincing results.

2   Related Work

Image-to-image translation aims at converting images from one domain to another while keeping the structure of the source image unchanged. The literature contains various methods performing this task in different settings. Paired image-to-image translation methods [27,64] operate when pairs of images in both domains are available. For example, semantic labels to scenes [8,48,64], edges to objects [54], and image super-resolution [29,33]. However, paired data are not always available in many tasks. Unsupervised image-to-image translation tackles a setting in which paired data is not available, while sampling from two domains is possible [12,26,39,40,55,58,73]. Clearly, unpaired image-to-image translation is an ill-posed problem for there are numerous ways an image can be transformed to a different domain. Hence, recently proposed methods introduce constraints to limit the number of possible transformations. Some studies enforce certain domain properties [1,55], while other concurrent works apply cycle-consistency to transform images between different domains [31,69,73]. Our work differs from existing studies that we focus on a specific challenging problem, which is the realistic hair generation, where we want to translate manually designed hair models from the domain of rendered images to the domain of real hair. For the purpose of controllable hair generation, we leverage rendered hair structure and arbitrary hair appearance to synthesize diverse realistic hairstyles. The further difference in our work compared to the image-to-image translation papers is unbalanced data. The domain of images containing real hair is far more diverse than that of rendered hair, making it even more challenging for classical image-to-image translation works to address the problem.

Neural style transfer is related to image-to-image translation in a way that image style is changed while content is maintained [9,16,20,25,36–38,63]. Style in this case is represented by unique style of an artist [16,63] or is copied from an example image provided by the user. Our work follows the research idea from example-guide style transfer that hairstyle is obtained from reference real image. However, instead of changing the style of a whole image, our aim is to keep the appearance of the human face and background unchanged, while having full control over the hair region. Therefore, instead of following exiting works that inject style features into image generation networks directly [25,48], we propose a new architecture that combines only hair appearance features and latent features that encodes image content and adapted hair structure for image generation. This way we can achieve the goal that only the style of the hair region is manipulated according to the provided exemplar image.

Domain Adaptation addresses the domain-shift problem that widely exists between the source and target domains [53]. Various feature-based methods have been proposed to tackle the problem [13,17,18,32,62]. Recent works on adversarial learning for the embedded feature alignment between source and target domains achieve better results than previous studies [14,15,22,41,60,61]. Efforts using domain adaptation for both classification and pixel-level prediction tasks have gained significantly progress [1,10,60]. In this work, we follow the challenging setting of unsupervised domain adaptation that there is no corresponding annotation between source and target domains. We aim at learning an embedding space that only contains hair structure information for both rendered and real domains. Considering the domain gap, instead of using original images as input, we use rendered and real structure map as inputs to the encoders, which contain both domain-specific layers and shared layers, to obtain latent features. The adaptation is achieved by adversarial training and image reconstruction.

Hair Modeling, Rendering, and Generation share a similar goal with our paper, which is synthesizing photo-realistic hair images. With 3D hair models manually created [65,70], captured [19,23,42,47,71], or reconstructed from images [3–6,24,72], traditional graphical hair rendering methods focus on improving rendering quality and performance by either more accurately modeling the special hair material and lighting behaviours [11,43,44,68], or approximating certain aspects of rendering pipeline to reduce the computation complexity [45,50,52,67,74]. However, the extremely huge computation cost for realistic hair rendering usually prohibits them to be directly applied in real-time applications. Utilizing the latest advances in GANs, recent works [28,34,46,49,59] achieved impressive progress on conditioned hair image generation as supervised image-to-image translation. A GAN-based hair rendering method [66] proposes to perform conditioned 3D hair rendering by starting with a common structure representation and progressively enforce various conditions. However, it requires the hair model to be able to generate consistent representation (strand orientation map) with real images, which is challenging for low-quality mesh-based models, and cannot achieve temporally smooth results.

3   Approach

Let h be the target 3D hair model, with camera parameters c and hair material parameters m, we formulate the *traditional graphic rendering pipeline* as $R_t(h, m, c)$. Likewise, our *neural network-based rendering pipeline* is defined as $R_n(h, r, c)$, with a low-quality hair model h and material features extracted from an arbitrary reference hair image r.

3.1   Overview of Network Architecture

The overall system pipeline is shown in Fig. 1, which consists of two parallel branches for both domains of real photo (i.e., *real*) and synthetic renderings (i.e., *fake*), respectively.

On the encoding side, the *structure adaptation subnetwork*, which includes a real encoder E_r and a fake encoder E_f, achieves cross-domain structure embedding e. Similar to UNIT [39], we share the weights of the last few ResNet layers

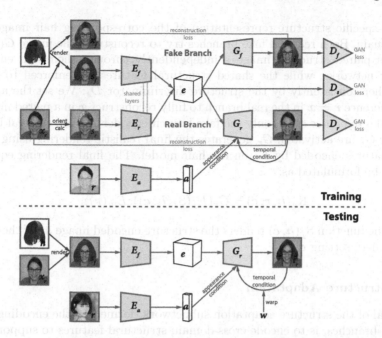

Fig. 1. The overall pipeline of our neural hair rendering framework. We use two branches to encode hair structure features, one for the real domain and the other for the fake domain. A domain discriminator is applied to the outputs from both encoders, to achieve domain invariant features. We also use two decoders to reconstruct images for two domains. The decoder in the real domain is different from the one in the fake domain, for it is conditioned on a reference image. Additionally, to generate consistent videos, we apply a temporal condition on the real branch. During inference, we use the encoder in the fake branch to get hair structure features from a 3D hair model and use the generator in the real branch to synthesized an appearance conditioned image.

in E_r and E_f to extract consistent structural representation from two domains. In addition, a structure discriminator D_s is introduced to match the high-level feature distributions between two domains to enforce the shared latent space further to be domain invariant.

On the decoding side, the *appearance rendering subnetwork*, consisting of G_r and G_f for the real and fake domain respectively, is attached after the shared latent space e to reconstruct the images in the corresponding domain. Each decoder owns its exclusive domain discriminator D_r and D_f to ensure the reconstruction matches the domain distribution, besides the reconstruction losses. The hair appearance is conditioned in an asymmetric way that G_r accepts the extra condition of material features extracted from a reference image r by using material encoder E_m, while the unconditional decoder G_f is asked to memorize the appearance, which is made on purpose for training data generation (Sect. 4.1).

At the *training stage*, all these networks are jointly trained using two sets of image pairs (s, x) for both real and fake domains, where s represents a domain-specific structure representation of the corresponding hair image x in this domain. Both real and fake branches try to reconstruct the image $G(E(x))$ from its paired structure image s independently through their own encoder-decoder networks, while the shared structural features are enforced to match each other consistently by the structure discriminator D_s. We set the appearance reference $r = x$ in the real branch to fully reconstruct x in a paired manner.

At the *inference stage*, only the fake branch encoder E_f and the real branch decoder G_r are activated. G_r generates the final realistic rendering using structural features encoded by E_f on the hair model. The final rendering equation R_n can be formulated as:

$$R_n(h, r, c) = G_r(E_f(S_f(h, c)), E_m(r)), \tag{1}$$

where the function $S_f(h, c)$ renders the structure encoded image s_f of the model h in camera setting c.

3.2   Structure Adaptation

The goal of the structure adaptation subnetwork, formed by the encoding parts of both branches, is to encode cross-domain structural features to support final rendering. Since the inputs to both encoders are manually disentangled structure representation (Sect. 4.1), the encoded features $E(s)$ only contain structural information of the target hair. Moreover, as the appearance information is either conditioned by extra decoder input in a way that non-spatial-varying structural information is leaked (the real branch) or simple enough to be memorized by the decoder (the fake branch) (Sect. 3.3), the encoded features should also include all the structural information necessary to reconstruct x.

E_r and E_f share a similar network structure: five downsampling convolution layers followed by six ResBlks. The last two ResBlks are weight-sharing to enforce the shared latent space. D_s follows PatchGAN [27] to distinguish between the latent feature maps from both domains:

$$\mathcal{L}_{D_s} = \mathbb{E}_{s_r}[\log(D_s(E_r(s_r)))] + \mathbb{E}_{s_f}[\log(1 - D_s(E_f(s_f)))]. \tag{2}$$

3.3   Appearance Rendering

The hair appearance rendering subnetwork decodes the shared cross-domain hair features into the real domain images. The decoders G_r and G_f have different network structures and do not share weights since the neural hair rendering is a unidirectional translation that aims to map the rendered 3D model in the fake domain to real images in the real domain. Therefore, G_f is required to make sure the latent features e encode all necessary information from the input 3D model, instead of learning to render various appearance. On the other hand, G_r is designed in a way to accept arbitrary inputs for realistic image generation.

Specifically, the unconditional decoder G_f starts with two ResBlks, and then five consecutive upsampling transposed convolutional layers followed by one final convolutional layer. G_r adopts a similar structure as G_f, with each transposed convolutional layer replaced with a SPADE [48] ResBlk to use appearance feature maps a_{r,s_r} at different scales to condition the generation. Assuming the binary hair mask of the reference and the target images are m_r and m_s, the appearance encoder E_m extracts the appearance feature vector on $r \times m_r$ with five downsampling convolutional layers and an average pooling. This feature vector $E_m(r)$ is then used to construct the feature map a_{r,s_r} by duplicating it spatially in the target hair mask m_s as follows:

$$a_{r,s_r}(p) = \begin{cases} E_m(r), & \text{if } m_{s_r}(p) = 1, \\ 0, & \text{if } m_{s_r}(p) = 0. \end{cases} \tag{3}$$

To make sure the reconstructed real image $G_r(E_r(s_r), a_{r,s_r})$ and the reconstructed fake image $G_f(E_f(s_f))$ belong to their respective distributions, we apply two domain specific discriminator D_r and D_f for the real and fake domain respectively. The adversarial losses write as:

$$\mathcal{L}_{D_r} = \mathbb{E}_{x_r}[\log(D_r(x_r))] + \mathbb{E}_{s_r, r}[\log(1 - D_r(G_r(E_r(s_r), a_{r,s_r})))], \tag{4}$$

$$\mathcal{L}_{D_f} = \mathbb{E}_{x_f}[\log(D_f(x_f))] + \mathbb{E}_{s_f}[\log(1 - D_f(G_f(E_f(s_f))))]. \tag{5}$$

We also adopt perceptual losses to measure high-level feature distance utilizing the paired data:

$$\mathcal{L}_p = \sum_{l=0}^{L} \|\Psi_l(G_r(E_r(s_r), a_{r,s_r})) - \Psi_l(x_r)\|_1 \\ + \|\Psi_l(G_f(E_f(s_f))) - \Psi_l(x_f)\|_1, \tag{6}$$

where $\Psi_l(i)$ computes the activation feature map of input image i at the lth selected layer of VGG-19 [56] pre-trained on ImageNet [51].

Finally, we have the overall training objective as:

$$\min_{E,G} \max_{D} (\lambda_s \mathcal{L}_{D_s} + \lambda_g(\mathcal{L}_{D_r} + \mathcal{L}_{D_f}) + \lambda_p \mathcal{L}_p). \tag{7}$$

3.4 Temporal Conditioning

The aforementioned rendering network is able to generate plausible single-frame results. However, despite the hair structure is controlled by smoothly-varying inputs of s_f with the appearance conditioned by a fixed feature map a_{r,s_r}, the spatially-varying appearance details are still generated in a somewhat arbitrary manner which tends to flicker in time (Fig. 5). Fortunately, with the availability of the 3D model, we can calculate the exact hair motion flow w^t for each pair of frames $t-1$ and t, which can be used to warp image i from $t-1$ to t as $\mathrm{W}(i, w^t)$. We utilize this dense correspondences to enforce temporal smoothness.

Let $I = \{i^0, i^1, \ldots, i^T\}$ be the generated result sequence, we achieve this temporal conditioning by simply using the warped result of the previous frame

$W(i^{t-1}, w^t)$ as an additional condition, stacked with the appearance feature map a_{r,s_r}, to the real branch decoder G_r when generating the current frame i^t.

We achieve temporally consistent by changing the real branch decoder only with temporally finetuning. During temporal training, we fix all other networks and use the same objective as Eq. 7, but randomly (50% of chance) concatenate x_r into the condition inputs to the SPADE ResBlks of G_r^t. The generation pipeline of the real branch now becomes $G_r^t(E_r(s_r), a_{r,s_r}, x_r)$, so that the network learns to preserve the consistency if the previous frame is inputted as the temporal condition, or generate randomly from scratch if the condition is zero.

Finally, we have the rendering equation for sequential generation:

$$i^t = R_n(h, r, c^t) = \begin{cases} G_r(E_f(s_f^t), a_{r,s_f^t}), & \text{if } t = 0, \\ G_r^t(E_f(s_f^t), a_{r,s_f^t}, W(i^{t-1}, w^t)). & \text{if } t > 0, \end{cases} \quad (8)$$
$$s_f^t = S_f(h, c^t).$$

4   Experiments

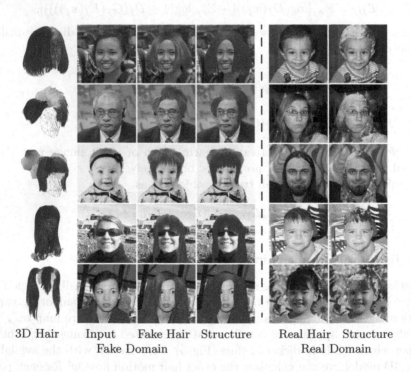

3D Hair   Input   Fake Hair   Structure   Real Hair   Structure
            Fake Domain   Real Domain

Fig. 2. Training data preparation. For the fake domain (left), we use hair model and input image to generate fake rendering and model structure map. For the real domain (b), we generate image structure map for each image.

4.1 Data Preparation

To train the proposed framework, we generate a dataset that includes image pairs (s, x) for both real and fake domains. In each domain, $s \rightarrow x$ indicates the mapping from structure to image, where s encodes only the structure information, and x is the corresponded image that conforms to the structure condition.

Real Domain. We adopt the widely used FFHQ [30] portrait dataset to generate the training pairs for the real branch, given it contains diverse hairstyles on shapes and appearances. To prepare real data pairs, we use original portrait photos from FFHQ as x_r, and generate s_r to encode only structure information from hair. However, obtaining s_r is a non-trivial process since hair image also contains material information, besides structural knowledge. To fully disentangle structure and material, and construct a universal structural representation s of all real hair, we apply a dense pixel-level orientation map in the hair region, which is formulated as $s_r = S_r(x_r)$, calculated with oriented filter kernels [47]. Thus, we can obtain s_r that only consists of local hair strand flow structures. Example generated pairs are presented in Fig. 2b.

For the purpose of training and validation, we randomly select 65, 000 images from FFHQ as training, and use the remaining 5, 000 images for testing. For each image x_r, we perform hair segmentation using off-the-shelf model [4], and calculate s_r for the hair region.

Fake Domain. There are multiple ways to model and render virtual hair models. From coarse to fine, typical virtual hair models range from a single rigid shape, coarse polygon strips representing detached hair wisps, to a large number of thin hair fibers that mimic real-world hair behaviors. Due to various granularity of the geometry, the structural representation is hardly shared with each other or real hair images. In our experiments, all the hair models we used are polygon strips based considering this type of hair model is widely adopted in real-time scenarios for it is efficient to render and flexible to be animated. To generate s_f for a given hair model h and specified camera parameters c, we use smoothly varying color gradient as texture to render h into a color image that embeds the structure information of the hair geometry, such that $s_f = S_f(h, c)$. As for x_f, we use traditional graphic rendering pipeline to render h with a uniform appearance color and simple diffuse shading, so that the final synthetic renderings have a consistent appearance that can be easily disentangled without any extra condition, and keep all necessary structural information to verify the effectiveness of the encoding step. Example pairs are shown in Fig. 2a.

For the 3D hair used for fake data pairs, we create five models (leftmost column in Fig. 2). The first four models are used for training, and the last one is used to evaluate the generalization capability of the network, for the network has never seen it. All these models consist of 10 to 50 polygon strips, which is sparse enough for real-time applications. We use the same training set from the real domain to form training pairs. Each image is overlaid by one of the four 3D hair models according to the head position and pose. Then the image with the fake hair model is used to generate x_f through rendering the hair model with

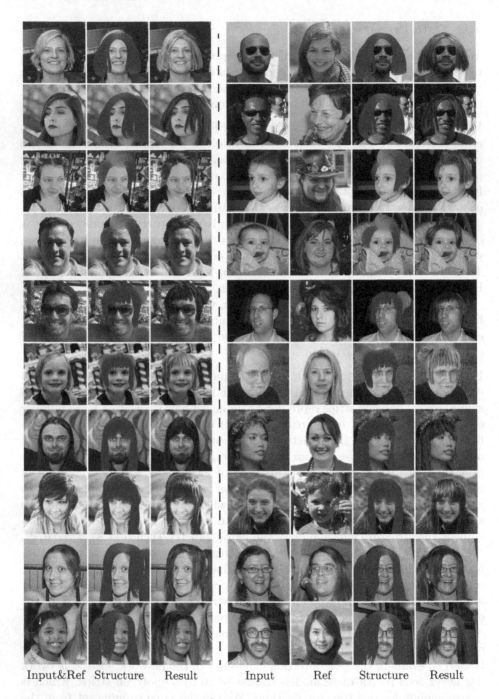

| Input&Ref | Structure | Result | Input | Ref | Structure | Result |

Fig. 3. Results for the hair models used in this study (2 rows per model). We visualize examples where the input and the reference image are the same (left), and the input and the reference are different images (right). In the former case the method copies appearance from another image.

simple diffuse shading, and s_f by exporting color textures that encode surface tangent of the mesh. We strictly use the same shading parameters, including lighting and color, to enforce a uniform appearance of hair that can be easily disentangled by the networks.

4.2   Implementation Details

We apply a two-stage learning strategy. During the first stage, all networks are trained jointly following Eq. 7 for the single-image renderer R_n. After that, we temporally fine-tune the decoder G_r of the real branch, to achieve temporally-smooth renderer R_n^t, by introducing the additional temporal condition as detailed in Sect. 3.4. To make the networks of both stages consistent, we keep the same condition input dimensions, including appearance and temporal, but set the temporal condition to zero during the first stage. During the second stage, we set it to zero with 50% of chance. The network architecture discussed in Sect. 3 is implemented using PyTorch. We adopt Adam solver with a learning rate set to 0.0001 for the first stage, and 0.00001 for the fine-tuning stage. The training resolution of all images is 512×512, with the mini-batch size set to 4. For the loss functions, weights λ_p, λ_s, and λ_g are set to 10, 1, and 1, respectively. All experiments are conducted on a workstation with 4 Nvidia Tesla P100 GPUs. During test time, rendering a single frame takes less than 1 second, with structure encoding less than 200 ms and final generation less than 400 ms.

4.3   Qualitative Results

We present visual hair rendering results from two settings in Fig. 3. The left three columns in Fig. 3 show that the reference image r is the same as x_r. By applying a hair model, we can modify human hair shape but keep the original hair appearance and orientation. The right four columns show that the reference image is different from x_r, therefore, both structure and appearance of hair from x_r can be changed at the same time to render the hair with a new style. We also demonstrate our video results in Fig. 5 (please click the image to watch video results online), where we adopt 3D face tracking [2] to guide the rigid position of the hair model, and physics-based hair simulation method [7] to generate secondary hair motion. These flexible applications demonstrate that our method can be easily applied to modify hair and generate novel high-quality hair images.

4.4   Comparison Results

To the best of our knowledge, there is no previous work that tackles the problem of neural hair rendering; thus, a direct comparison is not feasible. However, in light of our methods aim to bridge two different domains without ground-truth image pairs, which is related to unsupervised image translation, we compare our network with state-of-the-art unpaired image translation studies. It is important

Table 1. Quantitative comparison results. We compare our method against commonly adopted image-to-image translation frameworks, reporting Fréchet Inception Distance (FID, lower the better), Intersection over Union (IoU, higher the better) and pixel accuracy (Accuracy, higher the better). Additionally we report ablation studies by first removing the structural discriminator (w/o SD) followed by removing both the structural discriminator and the shared latent space (w/o SL and SD).

Method	FID ↓	IoU(%) ↑	Accuracy(%) ↑
Graphic Renderer	98.62	55.77	86.17
CycleGAN [73]	107.11	46.46	84.06
UNIT [39]	116.79	30.89	84.27
DRIT [35]	174.39	30.69	65.80
w/o SL and SD	94.25	80.10	93.89
w/o SD	77.09	86.60	96.35
Ours	**57.08**	**86.74**	**96.45**

to stress that although our hair rendering translation falls into the range of image translation problems, there exist fundamental differences compared to the generic unpaired image translation formulations for the following two reasons.

First and foremost, compared with translation between two domains, such as painting styles, or seasons/times of the day, which have roughly the same amount of images for two domains and enough representative training images can be sampled to provide nearly-uniform domain coverage, our real/fake domains have dramatically different sizes–it is easy to collect a huge amount of real human portrait photos with diverse hairstyles to form the real domain. Unfortunately, for the fake domain, it is impossible to reach the same variety since it would require manually designing *every* possible hair shape and appearance to describe the distribution of the whole domain of rendered fake hair. Therefore, we focus on a realistic assumption that only a limited set of such models are available for training and testing, such that we use four 3D models for training and one for testing, which is far from being able to produce variety in the fake domain.

Second, as a deterministic process, hair rendering should be conditioned strictly on both geometric shape and chromatic appearance, which can be hardly achieved with unconditioned image translation frameworks.

With those differences bearing in mind, we show the comparison between our method and three unpaired image translation studies, including CycleGAN [73], DRIT [35], and UNIT [39]. For the training of these methods, we use the same sets of images, x_r and x_f, for both real and fake domains, and the default hyperparameters reported by the original papers. Additionally, we compare with the images generated by the traditional graphic rendering pipeline. We denote the method as **Graphic Renderer**. Finally, we report two ablation studies to evaluate the soundness of the network and the importance of each step: 1) we first remove the structural discriminator (termed as **w/o SD**); 2) we then additionally remove the shared latent space (termed as **w/o SL and SD**).

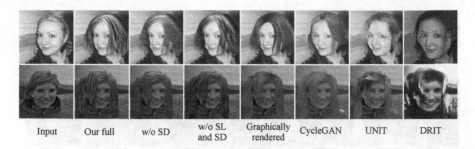

| Input | Our full | w/o SD | w/o SL and SD | Graphically rendered | CycleGAN | UNIT | DRIT |

Fig. 4. Visual comparisons. We show selected visual comparisons against commonly adopted image-to-image translation methods as well as visualize ablation results. Our method synthesizes more realistic hair images compared to other approaches.

Quantitative Comparison. For quantitative evaluation, we adopt FID (Fréchet Inception Distance) [21] to measure the distribution distance between two domains. Moreover, inspired by the evaluation protocol from existing work [8,64], we apply a pre-trained hair segmentation model [57] on the generated images to get the hair mask, and compare it with the ground truth. Intuitively, the segmentation model should predict the hair mask that similar to the ground-truth for the realistic synthesized images. To measure the segmentation accuracy, we use both Intersection-over-Union (IoU) and pixel accuracy (Accuracy).

The quantitative results are reported in Table 1. Our method significantly outperforms the state-of-the-art unpaired image translation works and graphic rendering approach by a large margin for all three evaluation metrics. The low FID score proves our method can generate high-fidelity hair images that contain similar hair appearance distribution as images from the real domain. The high IoU and Accuracy demonstrate the ability of the network to minimize the structure gap between real and fake domains so that the synthesized images can follow the manually designed structure. Furthermore, the ablation analysis in Table 1 shows both shared encoder layers and the structural discriminator are essential parts of the network, for the shared encoder layers help the network to find a common latent space that embeds hair structural knowledge, while the structural discriminator forces the hair structure features to be domain invariant.

Qualitative Comparison. The qualitative comparison of different methods is shown in Fig. 4. It can be easily seen that our generated images have much higher quality than the synthesized images created by other state-of-the-art unpaired image translation methods, for they have clearer hair mask, follow hair appearance from reference images, maintain the structure from hair models, and look like natural hair. Compared with the ablation methods (Fig. 4c and d), our full method (Fig. 4b) can follow the appearance from reference images (Fig. 4a) by generating hair with similar orientation.

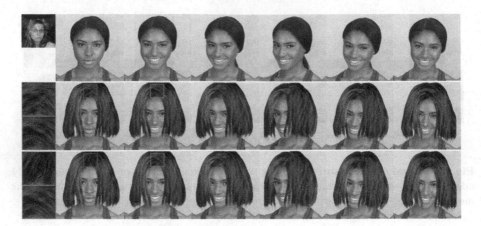

Fig. 5. Video results and comparisons. Top row: the first image is the appearance reference image and others are continuous input frames; middle row: generated hair images with temporal conditioning; bottom row: generated hair images without temporal conditioning. We show two zoom-in hair regions for each result. By applying temporal conditioning, our model synthesizes hair images with consistent appearance, while not using temporal conditioning leads to hair appearance flickering as indicated by blue and green boxes. *Click the image to play the video results and comparisons.* (Color figure online)

We also show the importance of temporal conditioning (Sect. 3.4) in Fig. 5. The temporal conditioning helps us generate consistent and smooth video results, for hair appearance and orientation are similar between continuous frames. Without temporal conditioning, the hair texture could be different between frames, as indicated by blue and green boxes, which may result in flickering for the synthesized video. Please refer to the supplementary video for more examples.

5 Conclusions

We propose a neural-based rendering pipeline for general virtual 3D hair models. The key idea of our method is that instead of enforcing model-level representation consistency to enable supervised paired training, we relax the strict requirements on the model and adopt an unsupervised image translation framework. To bridge the gap between real and fake domains, we construct a shared latent space to encode a common structure feature space for both domains, even if their inputs are dramatically different. In this way, we can encode a virtual hair model into such a structure feature, and switch it into the real generator to produce realistic rendering. The conditional real generator not only allows flexible appearance conditioning but can also be used to introduce temporal conditioning to generate smooth sequential results.

Our method has several limitations. First, the current method does not change the input. A smaller fake hair won't be able to fully occlude the original one in the input image. It is possible to do face inpainting to remove the excessive hair regions to fix this issue. Second, when the lighting/material of the appearance reference is dramatically different from the input, the result may look unnatural. Better reference selection would help to make the results better. Third, the current method simply blends the generated hair onto the input, which causes blending artifacts in some results especially when the background is complicated. A simple solution is to train a supervised boundary refinement network to achieve better blending quality.

References

1. Bousmalis, K., Silberman, N., Dohan, D., Erhan, D., Krishnan, D.: Unsupervised pixel-level domain adaptation with generative adversarial networks. In: CVPR, pp. 95–104 (2017)
2. Cao, C., Chai, M., Woodford, O.J., Luo, L.: Stabilized real-time face tracking via a learned dynamic rigidity prior. ACM Trans. Graph. **37**(6), 233:1–233:11 (2018)
3. Chai, M., Luo, L., Sunkavalli, K., Carr, N., Hadap, S., Zhou, K.: High-quality hair modeling from a single portrait photo. ACM Trans. Graph. **34**(6), 204:1–204:10 (2015)
4. Chai, M., Shao, T., Wu, H., Weng, Y., Zhou, K.: AutoHair: fully automatic hair modeling from a single image. ACM Trans. Graph. **35**(4), 116:1–116:12 (2016)
5. Chai, M., Wang, L., Weng, Y., Jin, X., Zhou, K.: Dynamic hair manipulation in images and videos. ACM Trans. Graph. **32**(4), 75:1–75:8 (2013)
6. Chai, M., Wang, L., Weng, Y., Yu, Y., Guo, B., Zhou, K.: Single-view hair modeling for portrait manipulation. ACM Trans. Graph. **31**(4), 116:1–116:8 (2012)
7. Chai, M., Zheng, C., Zhou, K.: A reduced model for interactive hairs. ACM Trans. Graph. **33**(4), 124:1–124:11 (2014)
8. Chen, Q., Koltun, V.: Photographic image synthesis with cascaded refinement networks. In: ICCV, pp. 1520–1529 (2017)
9. Chen, T.Q., Schmidt, M.: Fast patch-based style transfer of arbitrary style. CoRR abs/1612.04337 (2016)
10. Chen, Y., Chen, W., Chen, Y., Tsai, B., Wang, Y.F., Sun, M.: No more discrimination: cross city adaptation of road scene segmenters. In: ICCV, pp. 2011–2020 (2017)
11. d'Eon, E., François, G., Hill, M., Letteri, J., Aubry, J.: An energy-conserving hair reflectance model. Comput. Graph. Forum **30**(4), 1181–1187 (2011)
12. Dundar, A., Liu, M., Wang, T., Zedlewski, J., Kautz, J.: Domain stylization: a strong, simple baseline for synthetic to real image domain adaptation. CoRR abs/1807.09384 (2018)
13. Fernando, B., Habrard, A., Sebban, M., Tuytelaars, T.: Unsupervised visual domain adaptation using subspace alignment. In: ICCV, pp. 2960–2967 (2013)
14. Ganin, Y., Lempitsky, V.S.: Unsupervised domain adaptation by backpropagation. In: ICML, vol. 37, pp. 1180–1189 (2015)
15. Ganin, Y., et al.: Domain-adversarial training of neural networks. J. Mach. Learn. Res. **17**, 59:1–59:35 (2016)
16. Gatys, L.A., Ecker, A.S., Bethge, M.: Image style transfer using convolutional neural networks. In: CVPR, pp. 2414–2423 (2016)
17. Gong, B., Shi, Y., Sha, F., Grauman, K.: Geodesic flow kernel for unsupervised domain adaptation. In: CVPR, pp. 2066–2073 (2012)

18. Gopalan, R., Li, R., Chellappa, R.: Domain adaptation for object recognition: an unsupervised approach. In: ICCV, pp. 999–1006 (2011)
19. Herrera, T.L., Zinke, A., Weber, A.: Lighting hair from the inside: a thermal approach to hair reconstruction. ACM Trans. Graph. **31**(6), 146:1–146:9 (2012)
20. Hertzmann, A., Jacobs, C.E., Oliver, N., Curless, B., Salesin, D.: Image analogies. In: SIGGRAPH, pp. 327–340 (2001)
21. Heusel, M., Ramsauer, H., Unterthiner, T., Nessler, B., Hochreiter, S.: Gans trained by a two time-scale update rule converge to a local nash equilibrium. In: NIPS, pp. 6626–6637 (2017)
22. Hoffman, J., et al.: CyCADA: cycle-consistent adversarial domain adaptation. ICML, vol. 80, pp. 1994–2003 (2018)
23. Hu, L., Ma, C., Luo, L., Li, H.: Robust hair capture using simulated examples. ACM Trans. Graph. **33**(4), 126:1–126:10 (2014)
24. Hu, L., Ma, C., Luo, L., Li, H.: Single-view hair modeling using a hairstyle database. ACM Trans. Graph. **34**(4), 125:1–125:9 (2015)
25. Huang, X., Belongie, S.J.: Arbitrary style transfer in real-time with adaptive instance normalization. In: ICCV, pp. 1510–1519 (2017)
26. Huang, X., Liu, M., Belongie, S.J., Kautz, J.: Multimodal unsupervised image-to-image translation. In: ECCV, vol. 11207, pp. 179–196 (2018)
27. Isola, P., Zhu, J., Zhou, T., Efros, A.A.: Image-to-image translation with conditional adversarial networks. In: CVPR, pp. 5967–5976 (2017)
28. Jo, Y., Park, J.: SC-FEGAN: face editing generative adversarial network with user's sketch and color. In: ICCV, pp. 1745–1753 (2019)
29. Johnson, J., Alahi, A., Fei-Fei, L.: Perceptual losses for real-time style transfer and super-resolution. In: Leibe, B., Matas, J., Sebe, N., Welling, M. (eds.) ECCV 2016. LNCS, vol. 9906, pp. 694–711. Springer, Cham (2016). https://doi.org/10.1007/978-3-319-46475-6_43
30. Karras, T., Laine, S., Aila, T.: A style-based generator architecture for generative adversarial networks. In: CVPR, pp. 4401–4410 (2019)
31. Kim, T., Cha, M., Kim, H., Lee, J.K., Kim, J.: Learning to discover cross-domain relations with generative adversarial networks. In: ICML, vol. 70, pp. 1857–1865 (2017)
32. Kulis, B., Saenko, K., Darrell, T.: What you saw is not what you get: domain adaptation using asymmetric kernel transforms. In: CVPR, pp. 1785–1792 (2011)
33. Ledig, C., et al.: Photo-realistic single image super-resolution using a generative adversarial network. In: CVPR, pp. 105–114 (2017)
34. Lee, C., Liu, Z., Wu, L., Luo, P.: MaskGAN: towards diverse and interactive facial image manipulation. CoRR abs/1907.11922 (2019)
35. Lee, H., Tseng, H., Huang, J., Singh, M., Yang, M.: Diverse image-to-image translation via disentangled representations. ECCV, vol. 11205, pp. 36–52 (2018)
36. Li, C., Wand, M.: Precomputed real-time texture synthesis with markovian generative adversarial networks. In: Leibe, B., Matas, J., Sebe, N., Welling, M. (eds.) ECCV 2016. LNCS, vol. 9907, pp. 702–716. Springer, Cham (2016). https://doi.org/10.1007/978-3-319-46487-9_43
37. Li, Y., Wang, N., Liu, J., Hou, X.: Demystifying neural style transfer. In: IJCAI, pp. 2230–2236 (2017)
38. Li, Y., Fang, C., Yang, J., Wang, Z., Lu, X., Yang, M.: Diversified texture synthesis with feed-forward networks. In: CVPR, pp. 266–274 (2017)
39. Liu, M., Breuel, T., Kautz, J.: Unsupervised image-to-image translation networks. In: NeurIPS, pp. 700–708 (2017)

40. Liu, M., et al.: Few-shot unsupervised image-to-image translation. In: ICCV, pp. 10550–10559 (2019)
41. Liu, M., Tuzel, O.: Coupled generative adversarial networks. In: NIPS, pp. 469–477 (2016)
42. Luo, L., Li, H., Rusinkiewicz, S.: Structure-aware hair capture. ACM Trans. Graph. **32**(4), 76:1–76:12 (2013)
43. Marschner, S.R., Jensen, H.W., Cammarano, M., Worley, S., Hanrahan, P.: Light scattering from human hair fibers. ACM Trans. Graph. **22**(3), 780–791 (2003)
44. Moon, J.T., Marschner, S.R.: Simulating multiple scattering in hair using a photon mapping approach. ACM Trans. Graph. **25**(3), 1067–1074 (2006)
45. Moon, J.T., Walter, B., Marschner, S.: Efficient multiple scattering in hair using spherical harmonics. ACM Trans. Graph. **27**(3), 31 (2008)
46. Olszewski, K., et al.: Intuitive, interactive beard and hair synthesis with generative models. In: CVPR, pp. 7446–7456 (2020)
47. Paris, S., et al.: Hair photobooth: geometric and photometric acquisition of real hairstyles. ACM Trans. Graph. **27**(3), 30 (2008)
48. Park, T., Liu, M., Wang, T., Zhu, J.: Semantic image synthesis with spatially-adaptive normalization. In: CVPR, pp. 2337–2346 (2019)
49. Qiu, H., Wang, C., Zhu, H., Zhu, X., Gu, J., Han, X.: Two-phase hair image synthesis by self-enhancing generative model. Comput. Graph. Forum **38**(7), 403–412 (2019)
50. Ren, Z., Zhou, K., Li, T., Hua, W., Guo, B.: Interactive hair rendering under environment lighting. ACM Trans. Graph. **29**(4), 55:1–55:8 (2010)
51. Russakovsky, O., et al.: Imagenet large scale visual recognition challenge. IJCV **115**(3), 211–252 (2015)
52. Sadeghi, I., Pritchett, H., Jensen, H.W., Tamstorf, R.: An artist friendly hair shading system. ACM Trans. Graph. **29**(4), 56:1–56:10 (2010)
53. Saenko, K., Kulis, B., Fritz, M., Darrell, T.: Adapting visual category models to new domains. In: Daniilidis, K., Maragos, P., Paragios, N. (eds.) ECCV 2010. LNCS, vol. 6314, pp. 213–226. Springer, Heidelberg (2010). https://doi.org/10.1007/978-3-642-15561-1_16
54. Sangkloy, P., Lu, J., Fang, C., Yu, F., Hays, J.: Scribbler: controlling deep image synthesis with sketch and color. In: CVPR, pp. 6836–6845 (2017)
55. Shrivastava, A., Pfister, T., Tuzel, O., Susskind, J., Wang, W., Webb, R.: Learning from simulated and unsupervised images through adversarial training. In: CVPR, pp. 2242–2251 (2017)
56. Simonyan, K., Zisserman, A.: Very deep convolutional networks for large-scale image recognition. In: ICLR (2015)
57. Svanera, M., Muhammad, U.R., Leonardi, R., Benini, S.: Figaro, hair detection and segmentation in the wild. In: ICIP, pp. 933–937 (2016)
58. Taigman, Y., Polyak, A., Wolf, L.: Unsupervised cross-domain image generation. In: ICLR (2017)
59. Tan, Z., et al.: MichiGAN: multi-input-conditioned hair image generation for portrait editing. ACM Trans. Graph. **39**(4), 95:1–95:13 (2020)
60. Tsai, Y., Hung, W., Schulter, S., Sohn, K., Yang, M., Chandraker, M.: Learning to adapt structured output space for semantic segmentation. In: CVPR, pp. 7472–7481 (2018)
61. Tzeng, E., Hoffman, J., Saenko, K., Darrell, T.: Adversarial discriminative domain adaptation. In: CVPR, pp. 2962–2971 (2017)
62. Tzeng, E., Hoffman, J., Zhang, N., Saenko, K., Darrell, T.: Deep domain confusion: Maximizing for domain invariance. CoRR abs/1412.3474 (2014)

63. Ulyanov, D., Lebedev, V., Vedaldi, A., Lempitsky, V.S.: Texture networks: feed-forward synthesis of textures and stylized images. In: ICML, vol. 48, pp. 1349–1357 (2016)
64. Wang, T., Liu, M., Zhu, J., Tao, A., Kautz, J., Catanzaro, B.: High-resolution image synthesis and semantic manipulation with conditional gans. In: CVPR, pp. 8798–8807 (2018)
65. Ward, K., Bertails, F., Kim, T., Marschner, S.R., Cani, M., Lin, M.C.: A survey on hair modeling: styling, simulation, and rendering. IEEE Trans. Vis. Comput. Graph. **13**(2), 213–234 (2007)
66. Wei, L., Hu, L., Kim, V.G., Yumer, E., Li, H.: Real-time hair rendering using sequential adversarial networks. In: ECCV, vol. 11208, pp. 105–122 (2018)
67. Xu, K., Ma, L., Ren, B., Wang, R., Hu, S.: Interactive hair rendering and appearance editing under environment lighting. ACM Trans. Graph. **30**(6), 173 (2011)
68. Yan, L., Tseng, C., Jensen, H.W., Ramamoorthi, R.: Physically-accurate fur reflectance: modeling, measurement and rendering. ACM Trans. Graph. **34**(6), 185:1–185:13 (2015)
69. Yi, Z., Zhang, H.R., Tan, P., Gong, M.: DualGAN: unsupervised dual learning for image-to-image translation. In: ICCV, pp. 2868–2876 (2017)
70. Yuksel, C., Schaefer, S., Keyser, J.: Hair meshes. ACM Trans. Graph. **28**(5), 166 (2009)
71. Zhang, M., Chai, M., Wu, H., Yang, H., Zhou, K.: A data-driven approach to four-view image-based hair modeling. ACM Trans. Graph. **36**(4), 156:1–156:11 (2017)
72. Zhou, Y., et al.: HairNet: single-view hair reconstruction using convolutional neural networks. In: ECCV, vol. 11215, pp. 249–265 (2018)
73. Zhu, J., Park, T., Isola, P., Efros, A.A.: Unpaired image-to-image translation using cycle-consistent adversarial networks. In: ICCV, pp. 2242–2251 (2017)
74. Zinke, A., Yuksel, C., Weber, A., Keyser, J.: Dual scattering approximation for fast multiple scattering in hair. ACM Trans. Graph. **27**(3), 32 (2008)

JNR: Joint-Based Neural Rig Representation for Compact 3D Face Modeling

Noranart Vesdapunt[(✉)], Mitch Rundle, HsiangTao Wu, and Baoyuan Wang

Microsoft Cloud and AI, Pittsburgh, USA
{noves,mitchr,musclewu,baoyuanw}@microsoft.com

Abstract. In this paper, we introduce a novel approach to learn a 3D face model using a joint-based face rig and a neural skinning network. Thanks to the joint-based representation, our model enjoys some significant advantages over prior blendshape-based models. First, it is very compact such that we are orders of magnitude smaller while still keeping strong modeling capacity. Second, because each joint has its semantic meaning, interactive facial geometry editing is made easier and more intuitive. Third, through skinning, our model supports adding mouth interior and eyes, as well as accessories (hair, eye glasses, etc.) in a simpler, more accurate and principled way. We argue that because the human face is highly structured and topologically consistent, it does not need to be learned entirely from data. Instead we can leverage prior knowledge in the form of a human-designed 3D face rig to reduce the data dependency, and learn a compact yet strong face model from only a small dataset (less than one hundred 3D scans). To further improve the modeling capacity, we train a skinning weight generator through adversarial learning. Experiments on fitting high-quality 3D scans (both neutral and expressive), noisy depth images, and RGB images demonstrate that its modeling capacity is on-par with state-of-the-art face models, such as FLAME and Facewarehouse, even though the model is 10 to 20 times smaller. This suggests broad value in both graphics and vision applications on mobile and edge devices.

Keywords: Face modeling · 3D face reconstruction · GANs

1 Introduction

Parametric face models are popular representations for 3D face reconstruction and face tracking, where the face geometry is represented in a lower dimensional space than raw vertices. 3D Morphable Face Model (3DMM) proposed in 1999 [10] is still the dominating representation, although there are a few more

Electronic supplementary material The online version of this chapter (https://doi.org/10.1007/978-3-030-58523-5_23) contains supplementary material, which is available to authorized users.

© Springer Nature Switzerland AG 2020
A. Vedaldi et al. (Eds.): ECCV 2020, LNCS 12363, pp. 389–405, 2020.
https://doi.org/10.1007/978-3-030-58523-5_23

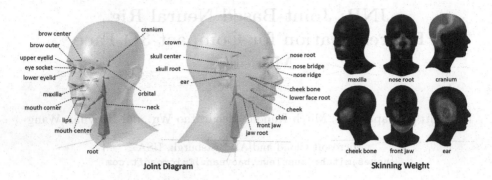

Fig. 1. An illustration of 52 joints in our joint-based model with hand-painted skinning weight. Our joints are defined hierarchically and semantically by following human anatomy without any training data. Our model is very compact, i.e., only a single mesh (5k vertices) with 9k floating points for skinning weight.

recent models such as BFM [4] and FLAME [31] which showed stronger modeling capacity with more training data. They all advocate learning the face model from a large number of high quality 3D scans. Collecting a large scan dataset, however, comes with a high cost along with the need for complicated and labor-intensive post processing such as cleanup, retopologizing and labeling to ensure accurate point to point correspondence before learning the face model. The previous state-of-the-art FLAME [31] model was learned from 33k registered scans which is obviously not easy to scale.

Model size is another important aspect to democratizing 3D face applications, which unfortunately has not received much attention. A cheap, efficient solution is required to support low memory 3D face reconstruction and tracking, especially on edge devices. Even on the cloud, efficiency is compelling as a means of saving cost. The highest capacity model from FLAME was shown with 300 identity basis which is over 4 million floating points, and while BFM [4] has fewer identity basis, the larger number of vertices increases the size to 32 million floating points. The challenging issues of collecting data and reducing model size lead us to explore a strong prior that can reduce the need for both of them.

A few recent works [40,43] propose to learn a face model from massive wild images or videos. Although impressive results were shown, those models were primarily designed for RGB input. Moreover, the face model is implicitly embedded within the trained network, so it is unclear how it can be extracted to serve other purposes, such as fitting to depth data and interactive editing (such as ZEPETO [3] App). With the ubiquity of depth cameras (e.g., Microsoft Azure Kinect), fitting 3D face models to depth or point cloud data may be an increasingly important problem for both face reconstruction and tracking [44].

In this paper, we propose to use a new face representation that leverages a human designed, joint-based, parametric 3D face rig with learned skinning weights to model the shape variations. At the heart of our model, we bring human prior knowledge into the modeling which significantly reduces the data

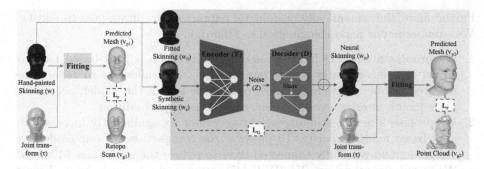

Fig. 2. Our pipeline has 3 stages. (1) Optimize joint transformation (τ) and hand-painted skinning weight on retopologized scans to generate learned linear skinning weight, and synthetic skinning weight training data. (2) Train an autoencoder to output neural skinning weight then finetune the decoder to cover entire Z distribution. (3) Fix the decoder then optimize τ and Z on point cloud with the same loss as (1), but correspondences are built by ICP. Note that only (3) is needed at test time.

that would otherwise be needed to learn everything from scratch. Although joint-based representations receive little attention in literature, they are standard in the gaming and movie industries. To build our model, we used a standard low-poly base mesh whose topology supports anatomically plausible deformations with minimal artifacts. We created a deformation skeleton and skinned the mesh to it using hand-painted skinning weights (see Fig. 1). Our model can be replicated easily using the details provided in supplementary materials, and it needs to be created only once. It can then be used in applications just like any other face models (FLAME [31], BFM [4], Facewarehouse [13], etc). Our technique is also compatible with joint-based face models publicly available on the internet. The model capacity is defined by the number of joints, their bind poses and the skinning weights, which are typically fixed. The skinning weights, however, do not need to be fixed and are a key to further increasing the representation capacity.

To learn skinning weights, we collected 94 high quality scans and retopologized them to our face topology. We took 85 of our scans as training data, then fit one fixed global skinning weight matrix across all the scans through an iterative optimizer. This shows noticeable improvement over the hand-painted skinning weights, and with this model we achieve lower error than FLAME 49 while being 30 times smaller. We then further increased our model capacity by training a neural network to encode the skinning weight space into 50 parameters. We test the effectiveness of the neural skinning weights on BU-3DFE neutral and we can approximate FLAME 300 scan-to-mesh error, while staying 20 times more compact. Our model also handles expressions out of the box, but we added a subset of expression blendshapes from [14] to further increase capacity. Our final model is on-par with FLAME 300 while remaining 10 times more compact (evaluate on expressive scans from BU-3DFE). Our fitted face mesh enjoys the benefit of joint model as it facilitates accurate placement of mouth interior, eyes and accessories.

Furthermore, the semantically meaningful parameters facilitate post-fit editing. We summarize our main contribution as following:

1. We propose a new 3D face representation based on a joint-based face rig model that incorporates strong human prior. This allow us to reduce the required number of training scans, reduce the size of the model, accurately place mouth interior, eyes and hair, as well as support editing.
2. We propose a neural skinning weight generator to significantly increase our joint-based model capacity. To preserve the compactness of our model, we design a group-wise fully connected layer to reduce the model size 51 times.
3. We tested our model on BU-3DFE [47], retopologized scans, Azure Kinect depth dataset, 2D images with 2D landmarks, and retargeting. Results show that our model can achieve similar error with FLAME 300 [31] with 20 times smaller for neutral scans, and 10 times smaller for expressive scans.

2 Related Work

Linear 3D Face Modeling. In 1999, Blanz and Vetter [10] are the first to introduce morphable model by learning PCA from 200 scans. Ten years later, Basel Face Model (BFM) [4] was built on another 200 scans using a nonrigid extension of iterative closest point algorithm (NICP) [6]. Recently, Large Scale Facial Model (LSFM) [11] built a large scale version of BFM by learning from 10k scans. Wu et al. [45] added a joint for jaw rotation to significantly reduce the number of required blendshapes. Inspire by [45], FLAME [31], added joints to allow eye, jaw and head rotation. FLAME was trained with 33k scans, and is considered to be the current state-of-the-art. While the trend is towards larger amounts of training data for building face models, we believe that it is hard to democratize such methods due to the high cost of data collection as well as privacy concerns. The authors of [41] propose to learn a geometry and texture correction in a PCA space based of 3DMM. JALI [20] proposes a face model specifically for speaking movement. Combined Face & Head Model (CFHM) [34] designed a combination of Liverpool-York Head Model (LYHM) [18] and LSFM [11] to achieve a face model that can perform well on both head and face region. York Ear Model [17] proposes an ear modeling. Muscle-based model [9] seeks to explain human movement anatomically similar to us, but express in blendshapes and jaw joint. We define our model anatomically purely in joints and introduce neural skinning weight to increase model capacity.

Nonlinear 3D Face Modeling. Fernandez et al. [5] designs a convolutional encoder to decouple latent representation and apply a multilinear model as a decoder to output depth images. Non-linear 3DMM [43] proposes to learn face model by convolutional neural network on 300W-LP [39] with RGB input and the outputs are position map and texture in UV space. Face Model Learning from Videos (FML) [40] learns face model on fully-connected layers to output shape and texture in the vertex space instead of UV space, and is designed to tackle the lack of training data through unsupervised learning on a large-scale video dataset (VoxCeleb2 [16]). FML is the closest work to ours in term of addressing limited 3D scan datasets, but it was not designed for point cloud fitting.

While autoencoder family [19,28,30,37] achieves impressive error on retopologized mesh, which we call known-correspondence, it is unclear how to accommodate other tasks. We consider FLAME as a more complete face model as it can use NICP to fit unknown correspondence, and the authors demonstrate 2D fitting and retargeting in their paper. To make a fair comparison, we retrained COMA on our 85 training scans, but COMA [37] does not generalize to BU-3DFE at all. COMA [37] and Li et al. [30] require 20k scans to achieve their results. The model size is also larger, e.g., Dai et al. [19]'s decoder alone is 17.8M Floats and Jiang et al. [28]'s encoder alone is 14.0M Floats, while our entire model is only 0.2M Floats.

Personalized 3D Face Modeling. High quality 3D face modeling is already possible if the model is for a specific person. Past works [27,46] capture images from multiple views and apply multi-view stereo reconstruction to obtain 3D face mesh. Using videos instead of multiview [22,23] is also popular for reducing the cost of camera rig. Using only internet photos is also possible in [38] where the authors personalize normal maps to obtain user's mesh. Our work indirectly creates personalized 3D face models using the neural skinning weights.

Skinning Weight Learning. Bailey et al. [8] and Neuroskinning [32] propose to learn skinning weight to replace manual effort in painting the skinning weight for character animation. Although the idea of using neural network in Neuroskinning is similar to ours, the goal is different as we wish to increase model capacity for face modeling. Moreover, Neuroskinning takes a 3D model as an input, while we train our network as a GAN. Perhaps a direct comparison would be inputting both 3D scan and joint transformation into our network, but the network will not generalize well due to the small amount of scans in training set. The network size is also directly counted towards face model size, so Neuroskinning's network design would defeat the purpose of compactness of our joint model.

3   Methodology

We first define the plain joint-based model with linear blend skinning where the weights are fixed, then introduce an advanced version where we use a network to generate skinning weights adaptively though optimizing the latent code in Sect. 3.2. Figure 2 illustrates the overview of our system pipeline.

3.1   Joint-Based Rigging Model

Joint-Based Face Representation. We use vertex based LBS similar to FLAME [31] with $N = 5236$ vertices, but $K = 52$ joints. This can be viewed as replacing the blendshapes in FLAME with 47 more joints which are defined as:

$$M_k = B^{-1} * \tau * M_p \tag{1}$$

where B is a fixed 4×4 bind pose matrix and the fitting variables are rigid transformation τ, which includes Euler rotation (R), translation (T) and scaling (S). Our joint are defined hierarchically, where the global joint transformation M_k is defined by recursive multiplications from the parent joint M_p, to the root joint M_{p0} (an identity matrix). We first define the root joint to allow global pose, and design all the other joints hierarchically such that each level covers a smaller area of the mesh and therefore finer detail. We then evaluate and visualize the fitting error on high quality scans and adjust the current joints and/or add smaller joints, and repeat the process until we reach desired level of morphing detail. As we only allow specific transformation on each joint, and we enforce bilateral symmetry in most cases, we end up with only 105 variables as opposed to 468 (9 DOF $\times$ 52 joints). The full details can be found in supplementary.

Linear Blend Skinning. We apply M_k and LBS on each vertex.

$$v' = \sum_{k=1}^{K} (w_k * v * M_k) \tag{2}$$

Skinning weights (w) are defined in a 2-dimensional $N * K$ matrix, and v is the vertex position. We first initialize w with skinning weights created manually using the weight painting tool in Blender. Even with some expertise in the area, it is difficult to produce good results in regions influenced by many joints, and quite intractable to find weights that work well across a wide range of identities. Therefore, we took a data-driven approach and regressed w over a set of ground-truth meshes, rather than tweaking them manually to fit all kinds of scans. As we design our joint to mainly model the local geometry variations, more than half of the joints only affect small regions of the face. While Dictionary 3DMM [21] uses sparse PCA to enforce local sparsity, our skinning weights w are naturally very sparse. Similar to symmetry constraint in 3DMM [33], our joints are also bilaterally symmetrical, so too are the skinning weights. Sparsity and symmetry reduce the number of variables from $N * K = 272,272$ to 8,990 floating points (30.3$\times$). Figure 1 shows a sample of our skinning weights. As we only need 9k floats to represent skinning weights, we optimize them as unknown parameters and fit them across all the scans in the training set. As a result, the learned skinning weights do not produce sharp artifacts that would be difficult to fix manually.

Losses. In order to eliminate the correspondences error, we use the commercial tool Wrap3 [2] to manually register all the scans into our topology. Once we establish the dense point-point correspondences across all the scans and our model, the fitting is conducted by minimizing RMSE:

$$L_v = \frac{1}{N} \sqrt{\sum_{n=1}^{N} (v_{p,n} - v_{g,n})^2} \tag{3}$$

Where v_p is the predicted vertex and v_g is the ground-truth vertex. When fitting to depth test data, we use NICP to build the correspondences between predicted mesh and point cloud. Similar to mean face regularization, we initialize local transforms to identity transformation (I) and add a diminishing regularization term to limit joint transformations while NICP converges.

$$L_m = \frac{1}{K} \sum_{k=1}^{K} (|R_k| + |T_k| + |S_k - 1|) \tag{4}$$

We manually limit the transformations to keep deformations within anatomically plausible bounds. This is similar to PCA coefficient limit, as FLAME also generates artifacts if the coefficients exceed $[-2, 2]$.

$$L_x = \frac{1}{K} \sum_{k=1}^{K} \begin{cases} |x_k - x_{max}|, & \text{if } x_k > x_{max} \\ |x_k - x_{min}|, & \text{if } x_k < x_{min} \\ 0, & \text{if } x_{min} \leq x_k \leq x_{max} \end{cases} \tag{5}$$

Where x is a substituted annotation for R, S, T, and we have L_x on all of them. We also follow the standard regularization in $[13, 22, 38, 42, 46]$.

$$L_p = \frac{1}{N} \sqrt{\sum_{n=1}^{N} (\nabla^2(v_{p,n}) - \nabla^2(v_{g,n}))^2} \tag{6}$$

Where ∇^2 is a Laplacian operator between v_p and v_g to ensures the overall topology does not deviate too much. Our final loss is the summation of all losses.

$$L_r = L_v + \lambda_m L_m + \lambda_x (L_R + L_T + L_S) + \lambda_p L_p \tag{7}$$

We empirically set the weight for each loss with $\lambda_m = 0.03$, $\lambda_x = \lambda_p = 0.3$.

3.2 Neural Skinning Weight

Model Capacity. As we seek to further increase the capacity of our joint-based model, we could consider applying free form deformation or corrective blendshapes on top of the fitting result, but that would nullify the ability afforded by the joint-based rig to accessorize the mesh. Although a local region expression blendshapes is tolerable, identity blendshapes that effect the mesh globally will also break the accessorizing. We could also add more joints, either manually or through learning, but this begins to dilute the semantic meaning of the joints, and furthermore can lead to deformation artifacts caused by too many joints in close proximity. Instead, we chose to fit the skinning weights to each test subject thereby creating person-specific skinning weights. While we limit our skinning weight to be sparse and symmetrical, the 9k skinning weight parameters still have too much freedom, which without sufficient regularization can result in artifacts. This motivates us to use a lower dimensional latent space to model the skinning weights generation. This approach is similar in spirit to GANFit [24] who optimizes the latent code of a GAN [25] network.

Skinning Weight Synthesis. To train the skinning weight generator, we needed a set of ground-truth skinning weights. As our joint-based model generates meshes by joint transformation and skinning weight, if we perturb joint transformations, there should exist another set of skinning weights that result in about the same mesh. We generated 32k joint transformations by iteratively fitting joint transformations and skinning weights on scans. We saved these intermediate joint transformations and corresponding meshes during the fitting. We then froze the joint transformations and optimized skinning weights until they converged on each mesh. We end generating 32k sets of skinning weights (w_s).

Network Design. After generating the training data, we first follow COMA [37] by building an autoencoder to encode a latent vector. We remove the encoder at test time and freeze the decoder, then fit the latent vector to generate skinning weights for each test scan. We found that the fitted latent vector during testing could be very different from training because the skinning weight training data are quite closed (because they are generated from only a small number of scans), thus the latent vector does not need to use all the space. We then remove the encoder and fine-tune the decoder with random noise (Z), similar to Generative Adversarial Network (GAN) to increase the coverage of Z space. We found that pre-training the autoencoder is important as training only the decoder in GAN style is not stable. We choose fully connected (fc) layers with swish activation [35] as our network block. While graph convolution [29] is a compelling alternative, there is no obvious design for connecting skinning weights across different joints, so we let the network learn these connections.

The encoder E reduces the dimension from 8,990 to 50 (Fig. 2), and the discriminator C maps the dimension from 50 to 1. While the encoder and discriminator do not have to be compacted because we do not use them at test time, our decoder (D) maps Z from 50 to 8,990, and is suffering the large number of parameters, especially on the late fc layers. Reducing the feature dimension of these fc layers causes the network to struggle to learn, so we keep the dimension large, but reuse the fc layer weight on different portions of the feature. To be specific, let X denote the input dimension into fc layer, and Y is the output dimension. Instead of learning $X * Y$ weights, we split X into n groups of $\frac{X}{n}$ inputs and only learn a single weight $\frac{X}{n} * \frac{Y}{n}$ to reuse on all the n groups. This reduces the number of parameters by a factor of n^2. We call this: group-wise fully connected layer inspired by group-wise convolution from MobileNet [26]. This can also be viewed as reshaping the 1D input X into 2D shape of $n * \frac{X}{n}$, and apply 2D convolution with $\frac{Y}{n}$ filters of kernel size $1 * \frac{X}{n}$.

Losses. We follow WGAN [7] where the discriminator loss on batch size (m) is:

$$L_C = \frac{1}{m} \sum_{i=1}^{m} (C(J_i) - C(D(Z_i))) \tag{8}$$

We also add sparsity loss to the generator to prevent our neural skinning weights (w_s) from deviating too far from the learned linear skinning weights from Sect. 3.1. We use the sparsity loss weight: $\lambda_0 = 0.05$ and adversarial loss weight: $\lambda_1 = 0.1$ throughout the experiments. Our generator loss is:

$$L_G = \sum_{i=1}^{m}(|J_i - D(Z_i)| + \lambda_0|D(Z_i)| - \lambda_1 C(D(Z_i))) \qquad (9)$$

The generator and discriminator loss remain the same on both autoencoder and generator. Only the input changes from skinning weight (w_s) to noise (Z).

3.3 Expression Modeling

Because our joints have semantic meaning, we can use 12 of them to model a subset of facial expressions directly. However, to make sure our model has enough capacity to model all other expressions (such as cheek puff), we simply add 24 of the 47 blendshapes from [14] as each of them represents one specific expression. The full list of blendshapes can be found in the supplemental material. To apply expression blendshapes on top of the neural skinning weight model, we fit joint transformation and neural skinning weight iteratively, then fit expression coefficients, and repeat the whole process for a few cycles. We use the loss function from Eq. 7 through out every stage.

4 Experimental Setup

4.1 Datasets

Retopologized Scans: We collected 94 scans, then manually retopologized them into our template. We then split 85 for training, 2 for validation and 7 for testing. These 85 scans were used to learn a global skinning weight in Sect. 3.1. For neural skinning weights, we fit joint transformation with hand-painted skinning weights for 3k iterations, and saved these joint transformations on intermediate iterations. We did not use every iteration because if the iterations were too closed, the joint transformation had similar values. We then fit the global skinning weights across all the scans and refit joint transformations, then repeated the process for 5 cycles, resulting in 8k joint transformations. We sparsely perturbed them by 5% for 3 more times to collect 32k joint transformations in total. We fit skinning weights on these joint transformations until converge on each of the corresponding retopologized scan. We split these 32k skinning weights into 29k training set, 680 validation set, and 2.4k test set.

Separate Testing Scans: For external comparison, we evaluate the identity part of our model on BU-3DFE [47] 100 neutral scans, and evaluate the expressiveness of our model on 480 BU-3DFE scans (10 male and 10 female with all the scans except neutral). We setup a small rig with 1 Microsoft Azure Kinect (for depth) and 3 point grey cameras (for high-quality texture) and used it to collect, fit and visualize 70 scans to test the effectiveness of our model.

4.2   Evaluation Metric

We use point-to-point error on the retopologized test set as we already have the correspondences between predicted mesh and ground-truth scan. For BU-3DFE, we follow scan-to-mesh distance from FLAME. Unlike FLAME, our joint-based model automatically aligns with the root as part of fitting, so we do not need an extra alignment process.

4.3   Training Configuration

Our joint-based model was created by adding joints and painting skinning weights to a generic base mesh in Blender. We exported the model to .npz file and fit the parameters on Tensorflow. We use AdamOptimizer throughout the training. We fit joint transformations using a learning rate of 10^{-3} for 3k iterations. We fit the global skinning weights across the training set using a learning rate of 10^{-4} and batch size of 85 for 3k iterations. We fit joint transformations and skinning weights iteratively for 5 cycles. On a single Nvidia V100, fitting joint transformations took 1 min per scan ($\sim$1.5 h on training set), and fitting the skinning weights for 5 cycles took $\sim$20 h. Neural skinning weights were trained using a learning rate of 10^{-3} during the first phase of training the autoencoder, and learning rate of 10^{-4} during the second phase of fine-tuning the decoder with a batch size of 85 for 30k iterations. It took $\sim$40 h to train neural skinning weights on a V100 GPU. At test time, we built the correspondence on point clouds using NICP. We fit identity by fitting joint transformation for 500 iterations and Z for 500 iterations for 2 cycles, then fit expressions for 500 iterations, then repeated the iterative fitting between identity and expressions for 2 cycles. The fitting took $\sim$2 min with a Nvidia 1080Ti.

4.4   Neural Skinning Weight Architecture

Our neural skinning weights consists of encoder, decoder and discriminator (Fig. 2). The basic building block is fc layer and swish activation. We use both fc layer weight and bias on the encoder and discriminator, but only use fc layer weight on the decoder. We remove fc layer bias from the decoder because if we initialize the input noise Z as zero, our block will always output zero. We then designed the final skinning weights to be a summation between learned linear skinning weights from Sect. 3.1, and the residual from our neural network. At test time, we first set Z to zero and fit joint transformations to convergence, then fit Z to generate a new skinning weight and repeat the cycle. Another reason for removing fc layer bias is to decrease the network size by half.

As we do not use encoder and discriminator during the testing, they can be large. Each fc layer in our encoder reduces the dimensions by half from 8990 to 50. Our discriminator dimension goes from 50 to 256 in 4 layers, and another 4 layers to decrease the dimension to 1. The decoder contains 6 layers where the first 3 fc layers increase the dimension from 50 to 250, and the last 3 group-wise fc layers dimensions are 500 $(n = 2)$, 1100 $(n = 5)$ and 8990 $(n = 10)$, where n is the number of groups.

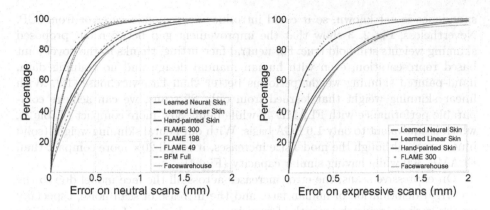

Fig. 3. External comparison on BU-3DFE. Our model is much more compact than the counterpart model with about the same scan-to-mesh error. The improvement of each skinning weight learning step also holds true.

5   Results

5.1   Ablation Study

Known Correspondence. We conduct our experiments in a control environment on high quality scans with known correspondence to eliminate external factors. Table 1 shows that, in this control environment, hand-painted skinning weight already achieves a reasonable error (i.e. 0.41 mm), while learning the linear skinning weight across training scans generalizes to the test set and further reduces the error. However, with our learned neural skinning, the error is reduced drastically from 0.41 to 0.11, which indicates that our neural skinning indeed increases joint-based model capacity very significantly. This is because each of the scan can have their personalized skinning weight, instead of just a single skinning weight that is shared across all the scans. This is a clear conclusion on high quality, known correspondence dataset.

Table 1. Point-to-point error on known correspondence test set. Our learned skinning weights demonstrate improvement of model capacity on every step.

Model	Error (mm)
(1) Hand-painted skinning	0.41 ± 0.02
(2) Learned Linear skinning	0.34 ± 0.02
(3) Learned Neural skinning	0.11 ± 0.01

Unknown Correspondence. To compare to external methods, we study our model on BU-3DFE scans which are noisier than our training set. The correspondence is not known, so it could introduce correspondence error from ICP. Nevertheless, Fig. 2, 3 show that the improvement gap between our proposed skinning weights still hold true. For neutral face fitting, thanks to the novel joint based representation, even with human manual design and no training data, hand-painted skinning weight performs better than Facewarehouse [12]. With linear skinning weight that learned from only 85 scans, we can achieve comparable performance with FLAME 49 while being 30× more compact (Table 2), which is equivalent to only 1.6 PCA basis. With our neural skinning weight (solid blue curve), although the model size increases, it is still 20× more compact than FLAME 300 while having similar capacity (Fig. 4).

On expressive scans, the error increases across all the face model due to the non-rigid deformation of human face, and the increase of scan noise, especially on the surface inside the mouth of surprise scans. In spite of error increase, we observe similar improvement on our propose skinning weight (Table 3 and Fig. 3). Our neural skinning model supports 6 expression in BU-3DFE out of the box and can achieve 0.239 mm error. Adding expression blendshapes improves the error to be comparable with FLAME 300 while remaining 10× more compact. We also outperform Nonlinear 3DMM [43] on BU-3DFE with Chamfer distance of 0.00083, 0.00075, 0.00077 for Nonlinear 3DMM, FLAME and ours respectively.

Table 2. Model size and RMS error on BU-3DFE neutral scans. Our neural skinning model is comparable to FLAME 300 with 20× smaller model size. We only count the identity part of the model in #Float.

Model	#Float	vs (3)	vs (1)	Error (mm)
(1) FLAME 300 [31]	4.52M	0.16×	1×	0.158 ± 0.172
(2) FLAME 49 [31]	738K	1×	6.1×	–
(3) Facewarehouse [13]	1.73M	0.43×	2.61	0.437 ± 0.474
(4) BFM full [4]	31.8M	0.02×	0.14×	–
(5) Hand-painted skinning	24.7K	29.9×	183×	0.375 ± 0.387
(6) Learned linear skinning	24.7K	29.9×	183×	0.338 ± 0.346
(7) Learned neural skinning	225.4K	3.3×	20.1×	0.153 ± 0.206

5.2 Applications

As we proof the effectiveness and compactness of our model in the Sect. 5.1, in this section, we show our results in real world applications.

Noisy Depth Fitting. We demonstrate our result on consumer rig. We use depth from a single Microsoft Azure Kinect [1] which suffers from multipass issue (cause the nose to shrink), missing parts due to IR absorbant material, and the depth itself is noisy. As our model can recover the full head from single depth map (the missing area will be automatically imply from the joint-based

Table 3. Model size and RMS error on BU-3DFE expressive scans. Our neural skinning model is comparable to FLAME 300 with 10× smaller model size.

Model	#Float	vs (1)	Error (mm)
(1) FLAME 300 [31]	6.03M	1×	0.211 ± 0.261
(2) Facewarehouse [13]	79.42M	0.08×	0.558 ± 0.670
(3) Hand-painted skinning	401.7K	15×	0.432 ± 0.518
(4) Learned linear skinning	401.7K	15×	0.405 ± 0.494
(5) Learned neural skinning	602.4K	10×	0.218 ± 0.289

model prior), we use 3 point-grey cameras only for extracting full frontal texture. Figure 6 illustrates the effectiveness of our fitting result. Capturing the subject takes a split second and fitting takes 2 min on our setting.

RGB Image Fitting. Like any previous face models, our model also supports 2D image based 3D face reconstruction. We conducted such an experiment on RGB images from Facewarehouse [13]. We use 2D landmarks as the only loss to fit the geometry. Figure 7 shows a few fitting results. As we can see, our model can fit reasonably well for even large facial expressions (Fig. 5).

5.3 Model Benefit

Editing and Accessorizing. As our model is designed with comprehensible list of joints, artist can tweak the joints to edit facial mesh. Adding accessory (e.g., hat, glass) is also possible, as most of the industrial computer graphic software (e.g., Blender, Maya) supports skinning weight transfer in just a few click. The artist can bind the new accessory to one of our joint and transfer skinning weight from the closest vertex. Changing hairstyle, attaching teeth, beard, whisker and mustache are also doable as well. The pipeline figure for adding accessories, and more demo video can be found in supplemental material.

Retargeting. As our joints and expression blendshapes have semantic meaning, we can transfer the expression and pose into puppet (Fig. 7) or other face scans.

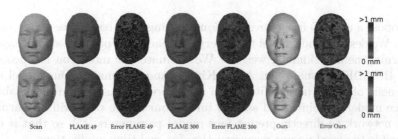

Fig. 4. Visual comparison between FLAME and neural skinning model on BU-3DFE.

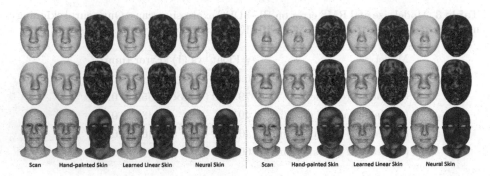

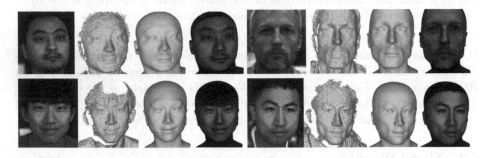

Fig. 5. Visualization of each step of improvement. Top 2 rows are BU-3DFE (top row scans are female), last row is our retopologized scan (right is female).

Fig. 6. Top: result on expressive scan from BU-3DFE (scan, result, error map). Bottom: result on Azure Kinect with point-grey cameras. Images are RGB (only for texture), depth (from Kinect), our fitted geometry, raycasted texture on our fitted geometry.

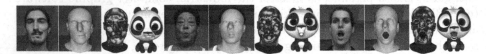

Fig. 7. 3D face fitting from single 2D image by 2D landmark loss. Note that scan is only used for evaluation. The order is input, result, error map, and retargeting.

6   Conclusion

We propose a new 3D face representation by using joint-based face rig as the face model. We designed our model to be very compact, yet, preserve strong capacity by learning neural skinning weights. We evaluate our model on retopologized scans, BU-3DFE, visualize on Azure Kinect, and 2D images. Our model enjoy the benefit of facial mesh editing and accessorizing. As we can reasonably fit our face model to point cloud and 2D image, our future work will be learning a neural network to directly predict face model parameters [15], so that it could be possible to speed up our reconstruction for real-time applications.

References

1. Microsoft azure kinect. https://azure.microsoft.com/en-us/services/kinect-dk/
2. R3ds wrap 3. https://www.russian3dscanner.com/
3. Zepeto. https://zepeto.me/
4. Paysan, P., Knothe, R., Amberg, B., Romdhani, S., Vetter, T.: A 3D Face Model for Pose and Illumination Invariant Face Recognition (2009)
5. Abrevaya, V.F., Wuhrer, S., Boyer, E.: Multilinear autoencoder for 3D face model learning. In: 2018 IEEE Winter Conference on Applications of Computer Vision (WACV), pp. 1–9, March 2018. https://doi.org/10.1109/WACV.2018.00007
6. Amberg, B., Romdhani, S., Vetter, T.: Optimal step nonrigid ICP algorithms for surface registration. In: CVPR. IEEE Computer Society (2007). http://dblp.uni-trier.de/db/conf/cvpr/cvpr2007.html#AmbergRV07
7. Arjovsky, M., Chintala, S., Bottou, L.: Wasserstein generative adversarial networks. In: Precup, D., Teh, Y.W. (eds.) Proceedings of the 34th International Conference on Machine Learning. Proceedings of Machine Learning Research, vol. 70, pp. 214–223. PMLR, International Convention Centre, Sydney, Australia, 06–11 August 2017. http://proceedings.mlr.press/v70/arjovsky17a.html
8. Bailey, S.W., Otte, D., Dilorenzo, P., O'Brien, J.F.: Fast and deep deformation approximations. ACM Trans. Graph. (TOG) **37**(4), 1–12 (2018)
9. Bao, M., Cong, M., Grabli, S., Fedkiw, R.: High-quality face capture using anatomical muscles. In: The IEEE Conference on Computer Vision and Pattern Recognition (CVPR), June 2019
10. Blanz, V., Vetter, T.: A morphable model for the synthesis of 3D faces. In: Proceedings SIGGRAPH, pp. 187–194 (1999)
11. Booth, J., Roussos, A., Ponniah, A., Dunaway, D., Zafeiriou, S.: Large scale 3D morphable models. Int. J. Comput. Vis. **126**(2–4), 233–254 (2018). https://doi.org/10.1007/s11263-017-1009-7
12. Cao, C., Weng, Y., Lin, S., Zhou, K.: 3D shape regression for real-time facial animation. ACM Trans. Graph. **32**(4), 1–10 (2013)
13. Cao, C., Weng, Y., Zhou, S., Tong, Y., Zhou, K.: Facewarehouse: a 3D facial expression database for visual computing. IEEE Trans. Visual. Comput. Graph. **20**(3), 413 425 (2014)
14. Chaudhuri, B., Vesdapunt, N., Wang, B.: Joint face detection and facial motion retargeting for multiple faces. In: IEEE Conference on Computer Vision and Pattern Recognition (CVPR) (2019)
15. Chaudhuri, B., Vesdapunt, N., Wang, B.: Joint face detection and facial motion retargeting for multiple faces. In: The IEEE Conference on Computer Vision and Pattern Recognition (CVPR), June 2019
16. Chung, J.S., Nagrani, A., Zisserman, A.: Voxceleb2: deep speaker recognition. In: INTERSPEECH (2018)
17. Dai, H., Pears, N., Smith, W.: A data-augmented 3D morphable model of the ear. In: 2018 13th IEEE International Conference on Automatic Face Gesture Recognition (FG 2018), pp. 404–408, May 2018. https://doi.org/10.1109/FG.2018.00065
18. Dai, H., Pears, N., Smith, W.A.P., Duncan, C.: A 3D morphable model of craniofacial shape and texture variation. In: The IEEE International Conference on Computer Vision (ICCV), October 2017
19. Dai, H., Shao, L.: Pointae: point auto-encoder for 3D statistical shape and texture modelling. In: Proceedings of the IEEE/CVF International Conference on Computer Vision (ICCV), October 2019

20. Edwards, P., Landreth, C., Fiume, E., Singh, K.: Jali: an animator-centric viseme model for expressive lip synchronization. ACM Trans. Graph. **35**(4), 127:1–127:11 (2016). https://doi.org/10.1145/2897824.2925984

21. Ferrari, C., Lisanti, G., Berretti, S., Bimbo, A.D.: Dictionary learning based 3D morphable model construction for face recognition with varying expression and pose. In: 2015 International Conference on 3D Vision. IEEE, October 2015. https://doi.org/10.1109/3dv.2015.63

22. Garrido, P., Valgaerts, L., Wu, C., Theobalt, C.: Reconstructing detailed dynamic face geometry from monocular video. ACM Trans. Graph. **32**, 158:1–158:10 (2013). (Proceedings of SIGGRAPH Asia 2013). https://doi.org/10.1145/2508363.2508380

23. Garrido, P., et al.: Reconstruction of personalized 3D face rigs from monocular video. ACM Trans. Graph. **35**(3), 28:1–28:15 (2016). (Presented at SIGGRAPH 2016)

24. Gecer, B., Ploumpis, S., Kotsia, I., Zafeiriou, S.: GANFIT: generative adversarial network fitting for high fidelity 3D face reconstruction. In: CVPR (2019). http://arxiv.org/abs/1902.05978

25. Goodfellow, I., et al.: Generative adversarial nets. In: Ghahramani, Z., Welling, M., Cortes, C., Lawrence, N.D., Weinberger, K.Q. (eds.) Advances in Neural Information Processing Systems, vol. 27, pp. 2672–2680. Curran Associates, Inc. (2014). http://papers.nips.cc/paper/5423-generative-adversarial-nets.pdf

26. Howard, A.G., et al.: Mobilenets: efficient convolutional neural networks for mobile vision applications (2017). http://arxiv.org/abs/1704.04861, cite arxiv:1704.04861

27. Huynh, L., et al.: Mesoscopic facial geometry inference using deep neural networks. In: Proceedings of the IEEE Conference on Computer Vision and Pattern Recognition (CVPR), Salt Lake City, UT (2018)

28. Jiang, Z.H., Wu, Q., Chen, K., Zhang, J.: Disentangled representation learning for 3D face shape. In: Proceedings of the IEEE/CVF Conference on Computer Vision and Pattern Recognition (CVPR), June 2019

29. Kipf, T.N., Welling, M.: Semi-supervised classification with graph convolutional networks. In: Proceedings of the 5th International Conference on Learning Representations. ICLR 2017 (2017). https://openreview.net/forum?id=SJU4ayYgl

30. Li, K., Liu, J., Lai, Y.K., Yang, J.: Generating 3D faces using multi-column graph convolutional networks. Comput. Graph. Forum **38**(7), 215–224 (2019). https://doi.org/10.1111/cgf.13830

31. Li, T., Bolkart, T., Black, M.J., Li, H., Romero, J.: Learning a model of facial shape and expression from 4D scans. ACM Trans. Graph. **36**(6), 194:1–194:17 (2017)

32. Liu, L., Zheng, Y., Tang, D., Yuan, Y., Fan, C., Zhou, K.: Neuroskinning: automatic skin binding for production characters with deep graph networks. ACM Trans. Graph. (TOG) **38**(4), 1–12 (2019)

33. Patel, A., Smith, W.: Simplification of 3D morphable models. In: Proceedings of the International Conference on Computer Vision, pp. 271–278 (2011). https://doi.org/10.1109/ICCV.2011.6126252, International Conference on Computer Vision; Conference date: 06-11-2011 Through 13-11-2011

34. Ploumpis, S., Wang, H., Pears, N., Smith, W.A.P., Zafeiriou, S.: Combining 3D morphable models: a large scale face-and-head model. In: The IEEE Conference on Computer Vision and Pattern Recognition (CVPR), June 2019

35. Ramachandran, P., Zoph, B., Le, Q.V.: Searching for activation functions. ArXiv abs/1710.05941 (2017)

36. Ramachandran, P., Zoph, B., Le, Q.V.: Searching for activation functions (2017)

37. Ranjan, A., Bolkart, T., Sanyal, S., Black, M.J.: Generating 3D faces using convolutional mesh autoencoders. In: Ferrari, V., Hebert, M., Sminchisescu, C., Weiss, Y. (eds.) ECCV 2018. LNCS, vol. 11207, pp. 725–741. Springer, Cham (2018). https://doi.org/10.1007/978-3-030-01219-9_43
38. Roth, J., Tong, Y., Liu, X.: Adaptive 3D face reconstruction from unconstrained photo collections. In: Proceedings of the IEEE Conference on Computer Vision and Pattern Recognition (CVPR) (2016)
39. Sagonas, C., Tzimiropoulos, G., Zafeiriou, S., Pantic, M.: 300 faces in-the-wild challenge: the first facial landmark localization challenge. In: IEEE International Conference on Computer Vision Workshops (ICCVW) (2013)
40. Tewari, A., et al.: Fml: face model learning from videos. In: The IEEE Conference on Computer Vision and Pattern Recognition (CVPR), June 2019
41. Tewari, A., et al.: Self-supervised multi-level face model learning for monocular reconstruction at over 250 Hz. In: Proceedings of the IEEE Conference on Computer Vision and Pattern Recognition (CVPR) (2018)
42. Tewari, A., et al.: Self-supervised multi-level face model learning for monocular reconstruction at over 250 Hz. In: 2018 IEEE Conference on Computer Vision and Pattern Recognition, CVPR 2018, Salt Lake City, UT, USA, 18–22 June 2018, pp. 2549–2559 (2018). https://doi.org/10.1109/CVPR.2018.00270, http://openaccess.thecvf.com/content_cvpr_2018/html/Tewari_Self-Supervised_Multi-Level_Face_CVPR_2018_paper.html
43. Tran, L., Liu, X.: Nonlinear 3D face morphable model. In: Proceedings of the IEEE Conference on Computer Vision and Pattern Recognition (CVPR) (2018)
44. Weise, T., Bouaziz, S., Li, H., Pauly, M.: Realtime performance-based facial animation. In: ACM SIGGRAPH 2011 Papers, pp. 77:1–77:10. SIGGRAPH 2011 (2011)
45. Wu, C., Bradley, D., Gross, M., Beeler, T.: An anatomically-constrained local deformation model for monocular face capture. ACM Trans. Graph. 35(4), 115:1–115:12 (2016). https://doi.org/10.1145/2897824.2925882
46. Wu, C., Shiratori, T., Sheikh, Y.: Deep incremental learning for efficient high-fidelity face tracking. ACM Trans. Graph. 37(6), 234:1–234:12 (2018). https://doi.org/10.1145/3272127.3275101
47. Zhang, Z., et al.: Multimodal spontaneous emotion corpus for human behavior analysis. In: The IEEE Conference on Computer Vision and Pattern Recognition (CVPR), June 2016

On Disentangling Spoof Trace for Generic Face Anti-spoofing

Yaojie Liu$^{(\boxtimes)}$, Joel Stehouwer$^{(\boxtimes)}$, and Xiaoming Liu$^{(\boxtimes)}$

Michigan State University, East Lansing, MI 48823, USA
{liuyaoj1,stehouw7,liuxm}@msu.edu

Abstract. Prior studies show that the key to face anti-spoofing lies in the subtle image pattern, termed "spoof trace", *e.g.*, color distortion, 3D mask edge, Moiré pattern, and many others. Designing a generic anti-spoofing model to estimate those spoof traces can improve both generalization and interpretability. Yet, this is a challenging task due to the diversity of spoof types and the lack of ground truth. This work designs a novel adversarial learning framework to disentangle the spoof traces from input faces as a hierarchical combination of patterns. With the disentangled spoof traces, we unveil the live counterpart from spoof face, and synthesize realistic new spoof faces after a proper geometric correction. Our method demonstrates superior spoof detection performance on both seen and unseen spoof scenarios while providing visually-convincing estimation of spoof traces. Code is available at https://github.com/yaojieliu/ECCV20-STDN.

1 Introduction

In recent years, the vulnerability of face biometric systems has been widely recognized and brought increasing attention to the vision community due to various physical and digital attacks. There are various physical and digital attacks, such as face morphing [13,52,55], face adversarial attacks [14,20,44], face manipulation attacks (*e.g.*, deepfake, face swap) [9,45], and face spoofing (*i.e.*, presentation attacks) [5,19,40], that can be used to attack the biometric systems. Among all these attacks, face spoofing is the only physical attack to deceive the systems, where attackers present faces from spoof mediums, such as photograph, screen, mask and makeup, instead of a live human. These spoof mediums can be easily manufactured by ordinary people, therefore posing a huge threat to applications such as mobile face unlock, building access control, and transportation security. Therefore, face biometric systems need to be reinforced with face anti-spoofing techniques before performing face recognition tasks.

Electronic supplementary material The online version of this chapter (https://doi.org/10.1007/978-3-030-58523-5_24) contains supplementary material, which is available to authorized users.

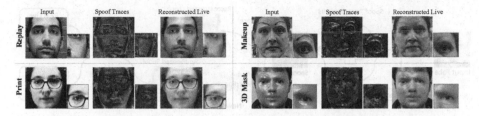

Fig. 1. Our approach can detect spoof faces, disentangle the spoof traces, and reconstruct the live counterparts. It can recognize diverse traces from various spoof types (*e.g.*, Moiré pattern in replay, artificial eyebrow and wax in makeup, color distortion in print, and specular highlights in 3D mask). Zoom in for details. (Color figure online)

Face anti-spoofing[1] has been studied for over a decade, and one of the most common approaches is based on texture analysis [6,7,37]. Researchers notice that presenting faces from spoof mediums introduces special texture differences, such as color distortions, unnatural specular highlights, and Moiré patterns. Those texture differences are inherent within spoof mediums and thus hard to remove or camouflage. Early works build feature extractor with classifier, such as LBP+SVM and HOG+SVM [17,26]. Recent works leverage deep learning techniques and show great progress [4,29,31,41,51].

However, there are two limitations in deep learning approaches. First, most works concern either print/replay or 3D mask alone, while a real-world scenario may encounter more spoof types at the same time. Second, most approaches formulate face anti-spoofing simply as a classification/regression problem. Although a few methods [24,29,51] attempt to offer insights via fixation, saliency, or noise analysis, there is little understanding on what the exact differences are between live and spoof, and what patterns the classifier's decision is based upon. We regard the face spoof detection for *all* spoof types as **generic face anti-spoofing**, and term the patterns differentiating spoof with live as **spoof trace**. Shown in Fig. 1, this work aims to equip generic face anti-spoofing models with the ability to explicitly extract the spoof traces from the input faces. We term this process as **spoof trace disentanglement**. This is a challenging objective due to spoof traces diversity and the lack of ground truth. However, we believe that solving this problem can bring several benefits:

1. Binary classification for face anti-spoofing would harvest any cue that helps classification, which might include spoof-irrelevant cues such as lighting, and thus hinder generalization. Spoof trace disentanglement explicitly tackles the most fundamental cue in spoofing, upon which the classification can be grounded and witnesses better generalization.
2. With the trend of pursuing explainable AI [1,3], it is desirable to generate the patterns that support its decision. Spoof trace serves as a good visual explana-

---
[1] As most face recognition systems are based on a monocular camera, this work only concerns monocular face anti-spoofing methods, and terms as face anti-spoofing hereafter for simplicity.

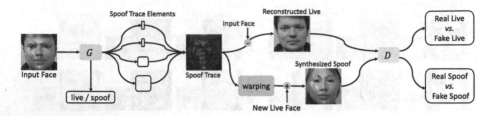

Fig. 2. Overview of the proposed Spoof Trace Disentanglement Network (STDN).

tion of the model's decision. Certain properties (*e.g.*, severity, methodology) of spoof attacks could potentially be revealed based on the traces.
3. Spoof traces are good sources for synthesizing realistic spoof samples. High-quality synthesis can address the issue of limited training data for the minority spoof types, such as special 3D masks and makeup.

Shown in Fig. 2, we propose a Spoof Trace Disentanglement Network (STDN) to tackle this problem. Given only the binary labels of live *vs.* spoof, STDN adopts an overall GAN training strategy. The generator takes input faces, detect the spoof faces, and disentangles the spoof traces as a combination of multiple elements. With the spoof traces, we can reconstruct the live counterpart from the spoof and synthesize new spoof from the live. To correct possible geometric discrepancy during spoof synthesis, we propose a novel 3D warping layer to deform spoof traces toward the target face. We deploy multiscale discriminators to improve the fidelity of both the reconstructed live and synthesized spoof. Moreover, the synthesized spoof are further utilized to train the generator in a supervised fashion, thanks to disentangled spoof traces as ground truth for the synthesized sample.

In summary, the main contributions are as follows:

- We for the first time study spoof trace for generic face anti-spoofing;
- We propose a novel model to disentangle spoof traces hierarchically;
- We utilize the spoof traces to synthesize new data and enhance the training;
- We achieve SOTA performance and provide convincing visualization.

2   Related Work

Face Anti-spoofing has been studied for more than a decade and its development can be roughly divided into three stages. In early years, researchers leverage the spontaneous motion, such as eye blinking, to detect simple print photograph or static replay attacks [25,35]. However, simple counter attacks would fail those methods, such as print with eye holes, and video replaying. Later, researchers focus on texture differences between live and spoof. Researchers extract handcrafted features from faces, *e.g.*, LBP [6,17,18,33], HoG [26,50], SIFT [37] and SURF [7], and train a classifier such as SVM and LDA. Recently, deep learning demonstrates significant improvements over the conventional methods. [16,27,36,49] train a deep neural network to do binary classification between live and spoof. [4,29,31,41,51] propose to learn additional information, such as face depth map and rPPG signal. With the latest approaches become saturating on several benchmarks, researchers start to explore more challenging cases, such as few-shot/zero-shot face anti-spoofing [31,38,54], domain adaptation in face anti-spoofing [41,42], *etc.*

This work aims to disentangling the spoof traces from the input faces. [12,24, 43] are the first few to estimate the different traces. However, they formulate the traces as low-intensity noises, which is limited to print/replay attacks with no convincing visual results. In contrast, we explore and visualize spoof traces for a wide range of spoof attacks, and evaluate the proposed method on challenging cases (*e.g.*, zero-shot face anti-spoofing).

Disentanglement Learning is often adopted to better represent complex data representation. DR-GAN [46,47] disentangles face into identity and pose vectors for pose-invariant face recognition and synthesis. [53] disentangles the representations of appearance, canonical, and pose features for gait recognition. [28] disentangles the representation of a 3D face into identity, expressions, poses, albedo, and illuminations. To solve the problem of image synthesis, [15] disentangles an image into appearance and shape with U-Net and Variational Auto Encoder (VAE). In this work, we intend to disentangle features with different geometry and scales. We leverage outputs from different layers to represent features at different scales, and propose a novel warping layer to align the geometry.

3 Spoof Trace Disentanglement Network

3.1 Problem Formulation

Let the domain of live faces be denoted as $\mathcal{L} \subset \mathbb{R}^{N \times N \times 3}$ and spoof faces as $\mathcal{S} \subset \mathbb{R}^{N \times N \times 3}$, where N is the image size. We intend to obtain not only the correct prediction (live *vs.* spoof) of the input face, but also a convincing estimation of the spoof traces. Without the guidance of ground truth spoof traces, our key idea is to find a minimum change that transfers an input face to the live domain:

$$\arg\min_{\hat{\mathbf{I}}} \|\mathbf{I} - \hat{\mathbf{I}}\|_F \ s.t. \ \mathbf{I} \in (\mathcal{S} \cup \mathcal{L}) \text{ and } \hat{\mathbf{I}} \in \mathcal{L}, \tag{1}$$

where $\mathbf{I}$ is the input face from either domain, $\hat{\mathbf{I}}$ is the target face in the live domain, and $\mathbf{I} - \hat{\mathbf{I}}$ is defined as the spoof trace. For an input live face $\mathbf{I}_{\text{live}}$, the spoof traces should be 0 as it's already in $\mathcal{L}$. For an input spoof face $\mathbf{I}_{\text{spoof}}$, this L-2 regularization on spoof traces is also preferred, as there is no paired solution for the domain transfer and we hope the spoof traces to be bounded. Based on [24, 37], spoof traces can be partitioned into multiple elements based on scales: global traces, low-level traces, and high-level traces. Global traces, such as color balance bias and range bias, can be efficiently modeled by a single value. The color biases here only refer to those created by the interaction between spoof mediums and

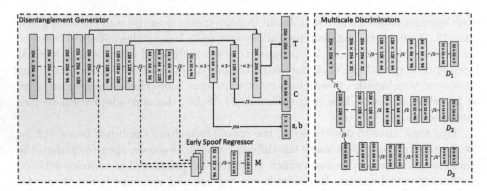

Fig. 3. The proposed STDN architecture. Except the last layer, each conv and transpose conv is concatenated with a Leaky ReLU layer and a batch normalization layer. /2 denotes a downsampling by 2, and ×2 denotes an upsampling by 2.

the capturing camera, and the model is expected to ignore those spoof-irrelevant color variations. Low-level traces consist of smooth content patterns, such as makeup strokes, and specular highlights. High-level traces include sharp patterns and high-frequency texture, such as mask edges and Moiré pattern. Denoted as $G(\cdot)$, the spoof trace disentanglement is formulated as a coarse-to-fine spoof effect build-up:

$$
\begin{aligned}
G(\mathbf{I}) &= \mathbf{I} - \hat{\mathbf{I}} \\
&= \mathbf{I} - ((1-\mathbf{s})\mathbf{I} - \mathbf{b} - \lfloor \mathbf{C} \rfloor_N - \mathbf{T}) \\
&= \mathbf{s}\mathbf{I} + \mathbf{b} + \lfloor \mathbf{C} \rfloor_N + \mathbf{T},
\end{aligned}
\tag{2}
$$

where $\mathbf{s}, \mathbf{b} \in \mathbb{R}^{1 \times 1 \times 3}$ represent color range bias and balance bias, $\mathbf{C} \in \mathbb{R}^{L \times L \times 3}$ denotes the smooth content patterns ($L < N$ to enforce the smoothness), $\lfloor \cdot \rfloor$ is the resizing operation, and $\mathbf{T} \in \mathbb{R}^{N \times N \times 3}$ is the high-level texture patterns. Compared to the single layer representation [24], this disentangled representation $\{\mathbf{s}, \mathbf{b}, \mathbf{C}, \mathbf{T}\}$ can largely improve disentanglement quality and suppress unwanted artifacts due to its coarse-to-fine process.

As shown in Fig. 3, Spoof Trace Disentanglement Network (STDN) consists of a generator and multiscale discriminators. They are jointly optimized to disentangle the spoof trace elements $\{\mathbf{s}, \mathbf{b}, \mathbf{C}, \mathbf{T}\}$ from the input faces. In the rest of this section, we discuss the details of the generator, face reconstruction and synthesis, the discriminators, and the training steps and losses used in STDN.

3.2   Disentanglement Generator

Spoof trace disentanglement is implemented via the generator. The disentanglement generator adopts an encoder-decoder as the backbone network. The encoder progressively downsamples the input face $\mathbf{I} \in \mathbb{R}^{256 \times 256 \times 3}$ to a latent feature tensor $\mathbf{F} \in \mathbb{R}^{32 \times 32 \times 96}$ via conv layers. The decoder upsamples the feature tensor $\mathbf{F}$ with transpose conv layers back to the input face size. To properly disentangle each spoof trace element, we leverage the natural upscaling property of the decoder structure: $\mathbf{s}, \mathbf{b}$ have the lowest spatial resolution and thus are disentangled in the very beginning of the decoder; $\mathbf{C}$ is extracted in the middle of the decoder with the size of 64; $\mathbf{T}$ is accordingly estimated in the last layer of the decoder. Similar to U-Net [39], we apply the short-cut connection between encoder and decoder to leak the high-frequency details for a high-quality estimation.

Unlike typical GAN scenarios where the generator only takes data from the source domain, our generator takes data from both source (spoof) and target (live) domains, and requires high accuracy in distinguishing two domains. Although the spoof traces should be significantly different between the two domains, they solely are not perfect hint for classification as the intensity of spoof traces varies from type to type. For this objective, we additionally introduce an Early Spoof Regressor (ESR) to enhance discriminativeness of the generator. ESR takes the bottleneck features $\mathbf{F}$ and outputs a $\mathbf{0/1}$ map $\mathbf{M} \in \mathbb{R}^{16 \times 16}$, where $\mathbf{0}$ means live and $\mathbf{1}$ means spoof. Moreover, we purposely make the encoder much heavier than the decoder, $i.e.$, more channels and deeper layers. This benefits the classification since ESR can better leverage the features learnt for spoof trace disentanglement.

In the testing phase, we use the average of the output from ESR and the intensity of spoof traces for classification:

$$\text{score} = \frac{1}{2K^2}\|\mathbf{M}\|_1 + \frac{\alpha_0}{2N^2}\|G(\mathbf{I})\|_1, \tag{3}$$

where α_0 is the weight for the spoof trace, $K=16$ is the size of $\mathbf{M}$, and $N=256$ is the image size.

3.3 Reconstruction and Synthesis

There are two ways to use the spoof traces:

- **Reconstruction:** obtaining the live face counterpart from the input as $\hat{\mathbf{I}} = \mathbf{I} - G(\mathbf{I})$;
- **Synthesis:** obtaining a new spoof face by applying the spoof traces $G(\mathbf{I}_i)$ disentangled from face image $\mathbf{I}_i$ to a live face $\mathbf{I}_j$.

To note that, spoof traces contain shape-dependent content associated with the original spoof face. Directly combining them with a new face with different shape or pose may result in poor alignment and strong visual implausibility. Hence, we propose an online 3D warping layer to correct the shape discrepancy. With ground truth traces, the synthesized spoof enable *supervised* training for the generator.

Online 3D Warping Layer. The spoof traces for face i can be expressed as:

$$G_i = G(\mathbf{I}_i)[\mathbf{p}_0], \tag{4}$$

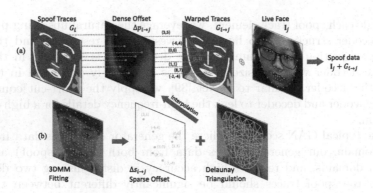

Fig. 4. 3D warping pipeline. (a) Given the corresponding dense offset, we warp the spoof trace and add them to the target live face to create a new spoof. E.g. pixel (x, y) with offset $(3, 5)$ is warped to pixel $(x + 3, y + 5)$ in the new image. (b) To obtain a dense offsets from the spare offsets of the selected face shape vertices, Delaunay triangulation interpolation is adopted.

where $\mathbf{p}_0 = \{(0,0), (0,1), ..., (255, 255)\} \in \mathbb{R}^{256 \times 256 \times 2}$ enumerates pixel locations in $\mathbf{I}_i$. To warp the spoof trace, a dense offset $\Delta\mathbf{p}_{i \to j} \in \mathbb{R}^{256 \times 256 \times 2}$ is required to indicate the offset value from face i to face j. The warped traces can be denoted as:

$$G_{i \to j} = G(\mathbf{I}_i)[\mathbf{p}_0 + \Delta\mathbf{p}_{i \to j}], \tag{5}$$

$\Delta\mathbf{p}_{i \to j}$ can be fractional numbers, and the sampling of fractional pixel locations is implemented via bilinear interpolation. During data preparation, we use [30] to fit 3DMM model and extract the 2D locations of Q selected vertices for each face as:

$$\mathbf{s} = \{(x_0, y_0), (x_1, y_1), ..., (x_N, y_N)\} \in \mathbb{R}^{Q \times 2}, \tag{6}$$

A sparse offset on the corresponding vertices can then be computed between face i and j as $\Delta\mathbf{s}_{i \to j} = \mathbf{s}_j - \mathbf{s}_i$. We select $Q = 140$ vertices to cover the face region so that they can represent non-rigid deformation, due to pose and expression. To convert the sparse offset $\Delta\mathbf{s}_{i \to j}$ to the dense offset $\Delta\mathbf{p}_{i \to j}$, we apply a triangulation interpolation:

$$\Delta\mathbf{p}_{i \to j} = \text{Tri}(\mathbf{p}_0, \mathbf{s}_i, \Delta\mathbf{s}_{i \to j}), \tag{7}$$

where $\text{Tri}(\cdot)$ is the bilinear interpolation operation based on Delaunay triangulation. Since the pixel values in the warped face are a linear combination of pixel values of the triangulation vertices, this whole process is differentiable. This process is illustrated in Fig. 4. Compared to previous methods [11, 29] that use offline face swapping or pre-computed dense offset, our warping layer only requires a sparse set of vertex locations, which is differentiable and computationally efficient.

Creating "harder" Samples. We can manipulate the spoof traces via tuning $\{s, b, C, T\}$, such as diminishing or amplifying certain element. Diminishing one or a few elements in $\{s, b, C, T\}$ would make the faces "less spoofed" as spoof traces are weakened. Those "less spoofed" data can be regarded as *harder* samples and may benefit the generalization. *E.g.*, removing the color distortion s may force the generator to explore high-level texture patterns. In this work, we randomly set one element from $\{s, b, C, T\}$ to be zero when synthesizing a new spoof face. Compared with other methods, such as brightness and contrast change [32], reflection and blurriness effect [51], or 3D distortion [21], our approach can introduce more realistic and effective data samples, as shown in Sect. 4.

3.4   Multi-scale Discriminators

Motivated by [48], we adopt 3 discriminators D_1, D_2, and D_3 at different resolutions (*i.e.*, 256, 128, and 64). The faces are resized to corresponding resolutions and sent to discriminators. D_1, working at the highest scale, processes the fine texture. D_2, working at the middle scale, focuses on the content pattern mostly in C. D_3, working at the lowest scale, focuses on global elements as the higher-frequency details are erased. We adopt the structure of fully convolutional network, similar to PatchGAN [23], and each discriminator consists of 10 conv with 3 downsampling. It outputs a 2-channel map, where the first channel compares the reconstructed live with the real live, and the second channel compares the synthesized spoof with real spoof.

3.5   Training Steps and Loss Functions

We utilize multiple loss functions in three training steps.

ESR Loss: M for live should be zero, and for spoof should be one:

$$L_{ESR} = \frac{1}{K^2}(\mathbb{E}_{i\sim\mathcal{L}}[\|\mathbf{M}_i\|_1] + \mathbb{E}_{i\sim\mathcal{S}\cup\hat{\mathcal{S}}}[\|\mathbf{M}_i - 1\|_1]), \tag{8}$$

where $\hat{\mathcal{S}}$ denotes synthesized spoof faces and $K = 16$ is the size of $\mathbf{M}$.

Adversarial Loss: We employ the LSGANs [34] on reconstructed live and synthesized spoof. For G:

$$L_G = \sum_{n=1,2,3} \{\mathbb{E}_{i\sim\mathcal{S}}[(D_n^1(\mathbf{I}_i - G_i) - 1)^2] + \mathbb{E}_{i\sim\mathcal{L}, j\sim\mathcal{S}}[(D_n^2(\mathbf{I}_i + G_{j\to i}) - 1)^2]\}, \tag{9}$$

and for D:

$$L_D = \sum_{n=1,2,3} \{\mathbb{E}_{i\sim\mathcal{L}}[(D_n^1(\mathbf{I}_i) - 1)^2] + \mathbb{E}_{i\sim\mathcal{S}}[(D_n^2(\mathbf{I}_i) - 1)^2]$$
$$+ \mathbb{E}_{i\sim\mathcal{S}}[(D_n^1(\mathbf{I}_i - G_i(x)))^2] + \mathbb{E}_{i\sim\mathcal{L}, j\sim\mathcal{S}}[D_n^2(\mathbf{I}_i + G_{j\to i})^2]\}. \tag{10}$$

where D_n^1 and D_n^2 denote the first and second channel of discriminator D_n.

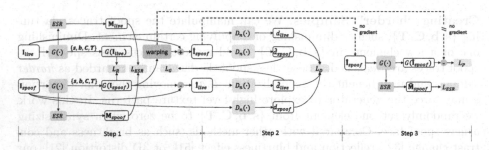

Fig. 5. The three training steps of STDN. Each mini-batch includes the same number of live and spoof samples.

Regularizer Loss: In Eq. 1, the task regularizes the intensity of spoof traces while satisfying certain domain conditions. This regularizer loss is denoted as:

$$L_R = \beta\, \mathbb{E}_{\mathbf{x}\sim\mathcal{L}}[\|G(\mathbf{I}_i)\|_2^2] + \mathbb{E}_{\mathbf{i}\sim\mathcal{S}}[\|G(\mathbf{I}_i)\|_2^2], \tag{11}$$

where $\beta > 1$ is a weight to further compress the traces of live faces to be zero.

Pixel Loss: Synthesized spoof data come with ground truth spoof traces. Therefore we can enable a supervised pixel loss for the generator to disentangle the exact spoof traces that were added to the live faces:

$$L_P = \mathbb{E}_{\mathbf{i}\sim\mathcal{L},\mathbf{j}\sim\mathcal{S}}[\|G(\lceil\mathbf{I}_i + G_{j\to i}\rceil) - \lceil G_{j\to i}\rceil\|_1], \tag{12}$$

where $\lceil\cdot\rceil$ is the `stop_gradient` operation. In this loss, we regard the traces $G_{j\to i}$ as ground truth, and the `stop_gradient` operation can prevent changing $G_{j\to i}$ to minimize the loss.

Training Steps and Total Loss: Figure 5 shows the 3 training steps: generator step, discriminator step, and extra supervision step. In the generator step, live faces $\mathbf{I}_{live}$ and spoof faces $\mathbf{I}_{spoof}$ are fed to the generator to disentangle the spoof traces. The spoof traces are used to reconstruct the live counterpart $\hat{\mathbf{I}}_{live}$ and synthesize new spoof $\hat{\mathbf{I}}_{spoof}$. The generator is updated with respect to adversarial loss L_G, ESR loss L_{ESR}, and regularizer loss L_R:

$$L = \alpha_1 L_G + \alpha_2 L_{ESR} + \alpha_3 L_R. \tag{13}$$

For the discriminator step, $\mathbf{I}_{live}$, $\mathbf{I}_{spoof}$, $\hat{\mathbf{I}}_{live}$, and $\hat{\mathbf{I}}_{spoof}$ are fed into the discriminators $D_n(\cdot), n = \{1,2,3\}$. The discriminators are supervised with adversarial loss L_D to compete with the generator. For the extra supervision step, $\mathbf{I}_{live}$ and $\hat{\mathbf{I}}_{spoof}$ are fed into the generator with ground truth label and trace to enable pixel loss L_P and ESR loss L_{ESR}:

$$L = \alpha_4 L_{ESR} + \alpha_5 L_P, \tag{14}$$

α_1-α_5 are the weights to balance the multitask training. We execute 3 steps in every iteration, and reduce the learning rate for discriminator step by half.

Table 1. Known spoof detection on: (a) OULU-NPU (b) SiW (c) SiW-M Protocol I.

(a)

Protocol	Method	APCER (%)	BPCER (%)	ACER (%)
1	STASN[51]	1.2	2.5	1.9
	Auxiliary [29]	1.6	1.6	1.6
	DeSpoof [24]	1.2	1.7	1.5
	Ours	**0.8**	1.3	**1.1**
2	Auxiliary [29]	2.7	2.7	2.7
	GRADIANT [8]	3.1	1.9	2.5
	STASN[51]	4.2	0.3	2.2
	Ours	**2.3**	1.6	**1.9**
3	DeSpoof [24]	4.0 ± 1.8	3.8 ± 1.2	3.6 ± 1.6
	Auxiliary [29]	2.7 ± 1.3	3.1 ± 1.7	2.9 ± 1.5
	STASN[51]	4.7 ± 3.9	0.9 ± 1.2	2.8 ± 1.6
	Ours	**1.6 ± 1.6**	4.0 ± 5.4	**2.8 ± 3.3**
4	Auxiliary [29]	9.3 ± 5.6	10.4 ± 6.0	9.5 ± 6.0
	STASN[51]	6.7 ± 10.6	8.3 ± 8.4	7.5 ± 4.7
	DeSpoof [24]	5.1 ± 6.3	6.1 ± 5.1	5.6 ± 5.7
	Ours	**2.3 ± 3.6**	5.2 ± 5.4	**3.8 ± 4.2**

(b)

Protocol	Method	APCER (%)	BPCER (%)	ACER (%)
1	Auxiliary[29]	3.6	3.6	3.6
	STASN[51]	–	–	1.0
	Meta-FAS-DR[54]	0.5	0.5	0.5
	Ours	**0.0**	**0.0**	**0.0**
2	Auxiliary[29]	0.6 ± 0.7	0.6 ± 0.7	0.6 ± 0.7
	Meta-FAS-DR[54]	0.3 ± 0.3	0.3 ± 0.3	0.3 ± 0.3
	STASN[51]			0.3 ± 0.1
	Ours	**0.0 ± 0.0**	**0.0 ± 0.0**	**0.0 ± 0.0**
3	STASN[51]	–	–	12.1 ± 1.5
	Auxiliary[29]	8.3 ± 3.8	8.3 ± 3.8	8.3 ± 3.8
	Meta-FAS-DR[54]	**8.0 ± 5.0**	**7.4 ± 5.7**	**7.7 ± 5.3**
	Ours	8.3 ± 3.3	7.5 ± 3.3	7.9 ± 3.3

(c)

Metrics(%)	Replay	Print	3D Mask Half	Silic.	Trans.	Paper	Manne.	Makeup Ob.	Im.	Cos.	Partial Attacks Funny.	Papergls.	Paper	Overall
ACER(%)														
Auxiliary[29]	5.1	5.0	5.0	10.2	5.0	9.8	6.3	19.6	5.0	26.5	5.5	5.2	5.0	6.3
Ours	**3.2**	**3.1**	**3.0**	**9.0**	**3.0**	**3.4**	**4.7**	**3.0**	**3.0**	**24.5**	**4.1**	**3.7**	**3.0**	**4.1**
EER(%)														
Auxiliary[29]	4.7	0.0	1.6	10.5	4.6	10.0	6.4	12.7	0.0	19.6	7.2	7.5	0.0	6.6
Ours	**2.1**	**2.2**	**0.0**	**7.2**	**0.1**	**3.9**	**4.8**	**0.0**	**0.0**	**19.6**	**5.3**	**5.4**	**0.0**	**4.8**
TDR@FDR=0.5(%)														
Ours	90.1	76.1	80.7	71.5	62.3	74.4	85.0	100.0	100.0	33.8	49.6	30.6	97.7	70.4

4 Experiments

We first introduce the setup, and present the experiment results. Next, we quantitatively evaluate the spoof traces by performing a spoof medium classification, and conduct an ablation study on each design. Finally, we provide visualization on the spoof trace disentanglement and new spoof synthesis.

4.1 Experimental Setup

Databases. We conduct experiments on three major databases: Oulu-NPU [8], SiW [29], and SiW-M [31]. Oulu-NPU and SiW include print/replay attacks, while SiW-M includes 13 spoof types. We follow all the testing protocols and compare with SOTA methods. Similar to most prior works, we only use the face region for training and testing.

Evaluation Metrics. Two standard metrics are used in this work for comparison: EER and APCER/BPCER/ACER [22]. We also report True Detection Rate (TDR) at a given False Detection Rate (FDR). This metric describes the spoof detection rate at a strict tolerance to live errors, which is widely used to evaluate systems in real-world applications [2]. In this work, we report TDR at FDR = 0.5%.

Parameter Setting. STDN is implemented in Tensorflow with an initial learning rate of $1e$-4. We train in total $150,000$ iterations with a batch size of 8, and decrease the learning rate by a ratio of 10 every $45,000$ iterations. We initialize

Table 2. The evaluation on SiW-M Protocol II: unknown spoof detection. **Bold** indicates the best score in each protocol. Red indicates protocols that our method improves over 50% than SOTA.

Methods	Replay	Print	3D Mask					Makeup			Partial Attacks			Average
			Half	Silic.	Trans.	Paper	Manne.	Ob.	Im.	Cos.	Fun.	Papergls.	Paper	
						APCER(%)								
LBP+SVM [8]	19.1	15.4	40.8	20.3	70.3	0.0	4.6	96.9	35.3	11.3	53.3	58.5	0.6	32.8 ± 29.8
Auxiliary[29]	23.7	7.3	27.7	18.2	97.8	8.3	16.2	100.0	18.0	16.3	91.8	72.2	0.4	38.3 ± 37.4
DTL [31]	**1.0**	**0.0**	0.7	24.5	58.6	0.5	3.8	**73.2**	13.2	12.4	17.0	17.0	0.2	17.1 ± 23.3
Ours	1.6	**0.0**	**0.5**	**7.2**	**9.7**	0.5	**0.0**	96.1	**0.0**	21.8	**14.4**	**6.5**	**0.0**	**12.2 ± 26.1**
						BPCER(%)								
LBP+SVM [8]	22.1	21.5	21.9	21.4	20.7	23.1	22.9	21.7	12.5	22.2	18.4	20.0	22.9	21.0 ± 2.9
Auxiliary[29]	10.1	6.5	10.9	11.6	6.2	7.8	9.3	11.6	9.3	7.1	6.2	8.8	10.3	**8.9 ± 2.0**
DTL [31]	18.6	11.9	29.3	12.8	13.4	8.5	23.0	11.5	9.6	16.0	21.5	22.6	16.8	16.6 ± 6.2
Ours	14.0	14.6	13.6	18.6	18.1	8.1	13.4	**10.3**	**9.2**	17.2	27.0	35.5	11.2	16.2 ± 7.6
						ACER(%)								
LBP+SVM [8]	20.6	18.4	31.3	21.4	45.5	11.6	13.8	59.3	23.9	16.7	35.9	39.2	11.7	26.9 ± 14.5
Auxiliary[29]	16.8	6.9	19.3	14.9	52.1	8.0	12.8	55.8	13.7	**11.7**	49.0	40.5	**5.3**	23.6 ± 18.5
DTL [31]	9.8	6.0	15.0	18.7	36.0	4.5	13.4	48.1	11.4	14.2	19.3	19.8	8.5	16.8 ± 11.1
Ours	**7.8**	7.3	**7.1**	**12.9**	**13.9**	**4.3**	**6.7**	53.2	**4.6**	19.5	20.7	21.0	5.6	**14.2 ± 13.2**
						EER(%)								
LBP+SVM [8]	20.8	18.6	36.3	21.4	37.2	7.5	14.1	51.2	19.8	16.1	34.4	33.0	7.9	24.5 ± 12.9
Auxiliary[29]	14.0	4.3	11.6	**12.4**	24.6	7.8	10.0	72.3	10.1	9.4	21.4	**18.6**	4.0	17.0 ± 17.7
DTL [31]	10.0	**2.1**	14.4	18.6	26.5	5.7	9.6	50.2	10.1	13.2	**19.8**	20.5	8.8	16.1 ± 12.2
Ours	**7.6**	3.8	**8.4**	13.8	14.5	**5.3**	**4.4**	**35.4**	**0.0**	19.3	21.0	20.8	1.6	**12.0 ± 10.0**
						TDR@FDR=0.5(%)								
Ours	45.0	40.5	45.7	36.7	11.7	40.9	74.0	0.0	67.5	16.0	13.4	9.4	62.8	35.7 ± 23.9

the weights with $[0, 0.02]$ normal distribution. $\{\alpha_1, \alpha_2, \alpha_3, \alpha_4, \alpha_5, \beta\}$ are set to be $\{1, 100, 1e - 3, 50, 1, 1e - 4\}$. α_0 is empirically determined from the training or validation set. We use open source face alignment [10] and 3DMM fitting [30] to crop the face and provide 140 landmarks.

4.2 Anti-spoofing for Known Spoof Types

Oulu-NPU [8] is a common benchmark due to its high quality and challenging testing. Shown in Table 1(a), our approach achieves the best performance in all four protocols. Specifically, we demonstrate significant improvement in protocol 1 and 4, reducing the ACER by 30% and 32% relative to prior works. We notice, in protocol 3 and 4, the performances of camera 6 are much lower than those of cameras 1-5: the ACER for camera 6 are 9.5% and 8.6%, while the average ACER for the other cameras are 1.7% and 3.1%. Compared to other cameras, camera 6 has stronger sensor noises and STDN recognizes them as unknown spoof traces, which leads to an increasing BPCER. Separating sensor noises from spoof traces can be an important future research.

SiW [29] Compared to Oulu, SiW includes fewer cameras but more spoof mediums and environment variations, such as pose, illumination, and expression. The comparisons are shown in Table 1(b). We outperform the previous works on the first two protocols and have a competitive performance on protocol 3. Protocol 3 requires the model to be trained on one spoof attack (print or replay) and tested on the other. Shown in Fig. 8, the traces of print and replay are significantly different, which may prevent the model from generalizing well.

SiW-M [31] contains a large amount of spoof types, including print, replay, 3D mask, makeup, and partial attacks. To use SiW-M, we randomly split the

Table 3. Confusion matrices of spoof mediums classification based on spoof traces. The left table is 3-class classification, and the right is 5-class classification. The results are compared with the previous method [24]. Green represents improvement over [24]. Red represents performance drop.

Label \ Predict	Live	Print	Replay
Live	60(+1)	0(−1)	0
Print	3(+3)	108(+20)	9(−23)
Replay	1(−12)	11(+3)	108(+9)

Label \ Predict	Live	Print1	Print2	Replay1	Replay2
Live	56(−4)	1(+1)	1(+1)	1(+1)	1(+1)
Print1	0	43(+2)	11(+9)	3(−8)	3(−3)
Print2	0	9(−25)	48(+37)	1(−8)	2(−4)
Replay1	1(−9)	2(−1)	3(+3)	51(+38)	3(−28)
Replay2	1(−7)	2(−5)	2(+2)	3(−3)	52(+13)

data into train/test set with a ratio of 60% and 40%, and the results are shown in Table 1(c). Compared to one of the best anti-spoofing models [29], our method outperforms on all spoof types as well as the overall performance, which demonstrates the superiority of our anti-spoofing on known spoof attacks.

4.3   Anti-spoofing for Unknown Spoof Types

Another important aspect of anti-spoofing model is to generalize to the unknown/unseen. SiW-M comes with the testing protocol to evaluate the performance of unknown attack detection. Shown in Table 2, STDN achieves significant improvement over the previous best model by relatively 24.8% on the overall EER and 15.5% on the overall ACER. This is especially noteworthy because DTL was specifically designed for detecting unknown spoof types, while our proposed approach shines in *both known and unknown spoof detection*. Specifically, we reduce the EERs of transparent mask, mannequin head, impersonation makeup and partial paper attack relatively by 45.3%, 54.2%, 100.0%, 81.8%, respectively. Among all, obfuscation makeup is the most challenging one, where we predict almost all the spoof samples as live. This is due to the fact that such makeup looks very similar to the live faces, while being dissimilar to any other spoof types. Once we obtain a few samples, our model can quickly recognize the spoof traces on the eyebrow and cheek, and successfully detect the attack (0% in Table 1(c)). However, with the TDR = 35.7% at FDR = 0.5%, the proposed method is still far from applicable in practices when dealing with unknown spoof types, which warrant future research.

4.4   Spoof Traces Classification

To quantitatively evaluate the spoof trace, we perform a spoof medium classification on the disentangled spoof traces. After convergence, we fix STDN and apply a simple CNN to classify the spoof mediums given the estimated spoof traces. We follow the same setting in [24] on Oulu-NPU Protocol 1. Shown in Table 3, our 3-class model and 5-class model can achieve classification accuracy of 92.0% and 83.3% respectively. Compared to [24], we improve 10% on the 3-class model and 29% on the 5-class model. In addition, we train the same CNN on the original

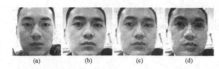

| | | | |
| (a) | (b) | (c) | (d) |

Fig. 6. Live reconstruction comparison: (a) live, (b) spoof, (c) ESR+D-GAN, (d) ESR+GAN.

Table 4. Quantitative ablation study of components in our approach.

Method	APCER (%)	BPCER (%)	ACER (%)
ESR	0.8	4.3	2.6
ESR+GAN	1.5	2.7	2.1
ESR+D-GAN	0.8	2.4	1.6
ESR+GAN+L_P	0.8	8.2	4.5
ESR+D-GAN+L_P	0.8	1.3	1.1

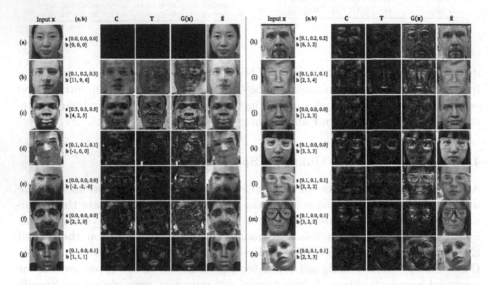

Fig. 7. Examples of spoof trace disentanglement on SiW-M. The (a)-(n) items are live, print, replay, half mask, silicone mask, paper mask, transparent mask, obfuscation makeup, impersonation makeup, cosmetic makeup, paper glasses, partial paper, funny eye glasses, and mannequin head. The first column is the input face, the 2nd-4th columns are the spoof trace elements $\{\mathbf{s}, \mathbf{b}, \mathbf{C}, \mathbf{T}\}$, the 5th column is the overall spoof traces, and the last column is the reconstructed live. (Color figure online)

images for spoof medium classification, and the classification accuracy is 86.3% (3-class) and 80.6% (5-class). This demonstrates that STDN distills significant information to distinguish different spoof mediums.

4.5   Ablation Study

In this section, we show the importance of each design on Oulu Protocol 1. Our baseline is the encoder with ESR (denoted as ESR), which is a conventional regression model. To validate the effectiveness of GAN, we report the results of ESR+GAN. In this case, the generator outputs a single-layer spoof trace with the input size, instead of the proposed four elements. To demonstrate the effectiveness of disentangled 4-element spoof trace, we change the single layer to the proposed $\{\mathbf{s}, \mathbf{b}, \mathbf{C}, \mathbf{T}\}$, denoted as ESR+D-GAN. In addition, we evaluate the effect

Fig. 8. Examples of the spoof data synthesis. (a) The source spoof samples $\mathbf{I}_i$. (b) The disentangled spoof traces $G(\mathbf{I}_i)$. (c) The target live faces $\mathbf{I}_j$. (d) The synthesized spoof $\mathbf{I}_j + G_{i \to j}$.

of synthesized data via enabling training step 3, denoted as ESR+GAN+L_P and ESR+D-GAN+L_P (*i.e.*, our final approach).

Shown in Table 4, the baseline achieves a decent performance of ACER 2.6%. Using a generative model can improve the ACER from 2.6% to 2.1%, while a proper disentanglement can improve to 1.6%. Shown in Fig. 6, ESR+D-GAN produces higher-quality spoof traces than ESR+GAN. If feeding bad-quality spoof samples in the training step 3, it would increase the error rate from 2.1% to 4.5%. But if feeding the good-quality spoof samples, it can achieve a significant improvement from 1.6% to 1.1%.

4.6 Visualization

As shown in Fig. 7, we successfully disentangle various spoof traces. *E.g.*, strong color distortion shows up in print/replay attacks (Fig. 7b–c). Moiré patterns in the replay attack are well detected (Fig. 7c). For makeup attacks (Fig. 7h–j), the fake eyebrows, lipstick, artificial wax, and cheek shade are clearly detected. The folds and edges in paper-crafted mask (Fig. 7f) are well detected. Although our method cannot provide a convincing estimation for a few spoof types (*e.g.*, funny eye glasses in Fig. 7m), the model effectively focuses on the correct region and disentangles parts of the traces.

Additionally, we show some examples of spoof synthesis using the disentangled spoof traces in Fig. 8. The spoof traces can be precisely transferred to a new face without changing the identity of the target face. Thanks to the proposed 3D warping layer, the geometric discrepancy between the source spoof trace and the target face is corrected during the synthesis. These two figures demonstrate that our approach disentangles visually convincing spoof traces that help face anti-spoofing.

5   Conclusions

This work proposes a network (STDN) to tackle a challenging problem of disentangling spoof traces from faces. With the spoof traces, we reconstruct the live faces as well as synthesize new spoofs. To correct the geometric discrepancy in synthesis, we propose a 3D warping layer to deform the traces. The disentanglement not only improves the SOTA of both known and unknown anti-spoofing, but also provides visual evidence to support the model's decision.

Acknowledgment. This research is based upon work supported by the Office of the Director of National Intelligence (ODNI), Intelligence Advanced Research Projects Activity (IARPA), via IARPA R&D Contract No. 2017-17020200004. The views and conclusions contained herein are those of the authors and should not be interpreted as necessarily representing the official policies or endorsements, either expressed or implied, of the ODNI, IARPA, or the U.S. Government. The U.S. Government is authorized to reproduce and distribute reprints for Governmental purposes notwithstanding any copyright annotation thereon.

References

1. Explainable Artificial Intelligence (XAI). https://www.darpa.mil/program/explainable-artificial-intelligence
2. IARPA research program Odin). https://www.iarpa.gov/index.php/research-programs/odin
3. Arrieta, A.B., et al.: Explainable artificial intelligence (XAI): concepts, taxonomies, opportunities and challenges toward responsible AI. Inf. Fusion **58**, 82–115 (2020)
4. Atoum, Y., Liu, Y., Jourabloo, A., Liu, X.: Face anti-spoofing using patch and depth-based CNNs. In: IJCB. IEEE (2017)
5. Bigun, J., Fronthaler, H., Kollreider, K.: Assuring liveness in biometric identity authentication by real-time face tracking. In: International Conference on Computational Intelligence for Homeland Security and Personal Safety (CIHSPS). IEEE (2004)
6. Boulkenafet, Z., Komulainen, J., Hadid, A.: Face anti-spoofing based on color texture analysis. In: ICIP. IEEE (2015)
7. Boulkenafet, Z., Komulainen, J., Hadid, A.: Face antispoofing using speeded-up robust features and fisher vector encoding. Signal Process. Lett. **24**(2), 141–145 (2016)
8. Boulkenafet, Z., Komulainen, J., Li, L., Feng, X., Hadid, A.: OULU-NPU: a mobile face presentation attack database with real-world variations. In: FG. IEEE (2017)
9. Boylan, J.F.: Will deep-fake technology destroy democracy? In: The New York Times (2018)
10. Bulat, A., Tzimiropoulos, G.: How far are we from solving the 2D & 3D face alignment problem? (and a dataset of 230,000 3D facial landmarks). In: ICCV. IEEE (2017)
11. Chang, H., Lu, J., Yu, F., Finkelstein, A.: Paired cycleGAN: asymmetric style transfer for applying and removing makeup. In: CVPR, IEEE (2018)
12. Chen, C., Xiong, Z., Liu, X., Wu, F.: Camera trace erasing. In: CVPR (2020)
13. Dale, K., Sunkavalli, K., Johnson, M.K., Vlasic, D., Matusik, W., Pfister, H.: Video face replacement. In: TOG. ACM (2011)

14. Deb, D., Zhang, J., Jain, A.K.: Advfaces: adversarial face synthesis. arXiv preprint arXiv:1908.05008 (2019)
15. Esser, P., Sutter, E., Ommer, B.: A variational U-Net for conditional appearance and shape generation. In: CVPR. IEEE (2018)
16. Feng, L., et al.: Integration of image quality and motion cues for face anti-spoofing: a neural network approach. J. Visual Commun. Image Represent. **38**(2016), 451–460 (2016)
17. de Freitas Pereira, T., Anjos, A., De Martino, José M., Marcel, S.: LBP- TOP based countermeasure against face spoofing attacks. In: Park, J.-I. Kim, J. (eds.) ACCV 2012. LNCS, vol. 7728, pp. 121-132. Springer, Heidelberg (2013). https://doi.org/10.1007/978-3-642-37410-4_11
18. de Freitas Pereira, T., Anjos, A., De Martino, J.M., Marcel, S.: Can face anti-spoofing countermeasures work in a real world scenario? In: ICB. IEEE (2013)
19. Frischholz, R.W., Werner, A.: Avoiding replay-attacks in a face recognition system using head-pose estimation. In: International SOI Conference. IEEE (2003)
20. Goodfellow, I.J., Shlens, J., Szegedy, C.: Explaining and harnessing adversarial examples. arXiv preprint arXiv:1412.6572 (2014)
21. Guo, J., Zhu, X., Xiao, J., Lei, Z., Wan, G., Li, S.Z.: Improving face anti-spoofing by 3D virtual synthesis. arXiv preprint arXiv:1901.00488 (2019)
22. ISO/IEC JTC 1/SC 37 Biometrics: Information technology biometric presentation attack detection part 1: Framework. International organization for standardization. https://www.iso.org/obp/ui/iso (2016)
23. Isola, P., Zhu, J.Y., Zhou, T., Efros, A.A.: Image-to-image translation with conditional adversarial networks. In: CVPR. IEEE (2017)
24. Jourabloo, A., Liu, Y., Liu, X.: Face de-spoofing: anti-spoofing via noise modeling. In: Ferrari, V., Hebert, M., Sminchisescu, C., Weiss, Y. (eds.) ECCV 2018. LNCS, vol. 11217, pp. 297–315. Springer, Cham (2018). https://doi.org/10.1007/978-3-030-01261-8_18
25. Kollreider, K., Fronthaler, H., Faraj, M.I., Bigun, J.: Real-time face detection and motion analysis with application in "liveness" assessment. TIFS **2**(3), 548–558 (2007)
26. Komulainen, J., Hadid, A., Pietikäinen, M.: Context based face anti-spoofing. In: BTAS. IEEE (2013)
27. Li, L., Feng, X., Boulkenafet, Z., Xia, Z., Li, M., Hadid, A.: An original face anti-spoofing approach using partial convolutional neural network. In: Sixth International Conference on Image Processing Theory, Tools and Applications (IPTA). IEEE (2016)
28. Liu, F., Zeng, D., Zhao, Q., Liu, X.: Disentangling features in 3D face shapes for joint face reconstruction and recognition. In: CVPR. IEEE (2018)
29. Liu, Y., Jourabloo, A., Liu, X.: Learning deep models for face anti-spoofing: binary or auxiliary supervision. In: CVPR. IEEE (2018)
30. Liu, Y., Jourabloo, A., Ren, W., Liu, X.: Dense face alignment. In: ICCV Workshops. IEEE (2017)
31. Liu, Y., Stehouwer, J., Jourabloo, A., Liu, X.: Deep tree learning for zero-shot face anti-spoofing. In: CVPR. IEEE (2019)
32. Liu, Y., Stehouwer, J., Jourabloo, A., Atoum, Y., Liu, X.: Presentation attack detection for face in mobile phones. In: Rattani, A., Derakhshani, R., Ross, A. (eds.) Selfie Biometrics. ACVPR, pp. 171–196. Springer, Cham (2019). https://doi.org/10.1007/978-3-030-26972-2_8
33. Määttä, J., Hadid, A., Pietikäinen, M.: Face spoofing detection from single images using micro-texture analysis. In: IJCB. IEEE (2011)

422   Y. Liu et al.

34. Mao, X., Li, Q., Xie, H., Lau, R.Y., Wang, Z., Paul Smolley, S.: Least squares generative adversarial networks. In: ICCV. IEEE (2017)
35. Pan, G., Sun, L., Wu, Z., Lao, S.: Eyeblink-based anti-spoofing in face recognition from a generic webcamera. In: ICCV. IEEE (2007)
36. Patel, K., Han, H., Jain, A.K.: Cross-database face antispoofing with robust feature representation. In: You, Z., et al. (eds.) CCBR 2016. LNCS, vol. 9967, pp. 611–619. Springer, Cham (2016). https://doi.org/10.1007/978-3-319-46654-5_67
37. Patel, K., Han, H., Jain, A.K.: Secure face unlock: spoof detection on smartphones. TIFS **11**(10), 2268–2283 (2016)
38. Qin, Y., et al.: Learning meta model for zero-and few-shot face anti-spoofing. arXiv preprint arXiv:1904.12490 (2019)
39. Ronneberger, O., Fischer, P., Brox, T.: U-net: convolutional networks for biomedical image segmentation. In: International Conference on Medical image computing and computer-assisted intervention. Springer (2015)
40. Schuckers, S.A.: Spoofing and anti-spoofing measures. Information Security technical report (2002)
41. Shao, R., Lan, X., Li, J., Yuen, P.C.: Multi-adversarial discriminative deep domain generalization for face presentation attack detection. In: CVPR. IEEE (2019)
42. Shao, R., Lan, X., Yuen, P.C.: Regularized fine-grained meta face anti-spoofing. arXiv preprint arXiv:1911.10771 (2019)
43. Stehouwer, J., Jourabloo, A., Liu, Y., Liu, X.: Noise modeling, synthesis and classification for generic object anti-spoofing. In: CVPR. IEEE (2020)
44. Szegedy, C., et al.: Intriguing properties of neural networks. arXiv preprint arXiv:1312.6199 (2013)
45. Thies, J., Zollhofer, M., Stamminger, M., Theobalt, C., Nießner, M.: Face2face: Real-time face capture and reenactment of RGB videos. In: CVPR. IEEE (2016)
46. Tran, L., Yin, X., Liu, X.: Disentangled representation learning GAN for pose-invariant face recognition. In: CVPR. IEEE (2017)
47. Tran, L., Yin, X., Liu, X.: Representation learning by rotating your faces. IEEE Trans. Pattern Anal. Mach. Intell. **41**(12), 3007–3021 (2019)
48. Wang, T.C., Liu, M.Y., Zhu, J.Y., Tao, A., Kautz, J., Catanzaro, B.: High-resolution image synthesis and semantic manipulation with conditional GANs. In: CVPR. IEEE (2018)
49. Yang, J., Lei, Z., Li, S.Z.: Learn convolutional neural network for face anti-spoofing. arXiv preprint arXiv:1408.5601 (2014)
50. Yang, J., Lei, Z., Liao, S., Li, S.Z.: Face liveness detection with component dependent descriptor. In: ICB. IEEE (2013)
51. Yang, X., et al.: Face anti-spoofing: Model matters, so does data. In: CVPR. IEEE (2019)
52. Zakharov, E., Shysheya, A., Burkov, E., Lempitsky, V.: Few-shot adversarial learning of realistic neural talking head models. arXiv preprint arXiv:1905.08233 (2019)
53. Zhang, Z., et al.: Gait recognition via disentangled representation learning. In: CVPR. IEEE (2019)
54. Zhao, C., Qin, Y., Wang, Z., Fu, T., Shi, H.: Meta anti-spoofing: Learning to learn in face anti-spoofing. arXiv preprint arXiv:1904.12490 (2019)
55. Zollhöfer, M., et al.: State of the art on monocular 3D face reconstruction, tracking, and applications. Comput. Graph. Forum **37**(2), 523–550 (2018)

Streaming Object Detection
for 3-D Point Clouds

Wei Han[1(✉)], Zhengdong Zhang[1], Benjamin Caine[1], Brandon Yang[1],
Christoph Sprunk[2], Ouais Alsharif[2], Jiquan Ngiam[1], Vijay Vasudevan[1],
Jonathon Shlens[1], and Zhifeng Chen[1]

[1] Google Brain, Mountain View, USA
{weihan,zhangzd}@google.com
[2] Waymo, LLC, Mountain View, USA

Abstract. Autonomous vehicles operate in a dynamic environment,
where the speed with which a vehicle can perceive and react impacts the
safety and efficacy of the system. LiDAR provides a prominent sensory
modality that informs many existing perceptual systems including object
detection, segmentation, motion estimation, and action recognition. The
latency for perceptual systems based on point cloud data can be dominated by the amount of time for a complete rotational scan (e.g. 100 ms).
This built-in data capture latency is artificial, and based on treating the
point cloud as a camera image in order to leverage camera-inspired architectures. However, unlike camera sensors, most LiDAR point cloud data
is natively a *streaming* data source in which laser reflections are sequentially recorded based on the precession of the laser beam. In this work,
we explore how to build an object detector that removes this artificial
latency constraint, and instead operates on native streaming data in
order to significantly reduce latency. This approach has the added benefit of reducing the peak computational burden on inference hardware
by spreading the computation over the acquisition time for a scan. We
demonstrate a family of streaming detection systems based on sequential
modeling through a series of modifications to the traditional detection
meta-architecture. We highlight how this model may achieve competitive
if not superior predictive performance with state-of-the-art, traditional
non-streaming detection systems while achieving significant latency gains
(e.g. $1/15^{th}$–$1/3^{rd}$ of peak latency). Our results show that operating on
LiDAR data in its native streaming formulation offers several advantages
for self driving object detection – advantages that we hope will be useful
for any LiDAR perception system where minimizing latency is critical
for safe and efficient operation.

1   Introduction

Self-driving cars are typically equipped with an array of sensors to robustly identify objects across highly variable environmental conditions [3,7,13,54]. In turn,

Electronic supplementary material The online version of this chapter (https://
doi.org/10.1007/978-3-030-58523-5_25) contains supplementary material, which is
available to authorized users.

driving in the real world requires responding to this large array of data with minimal latency to maximize the opportunity for safe and effective navigation [28].

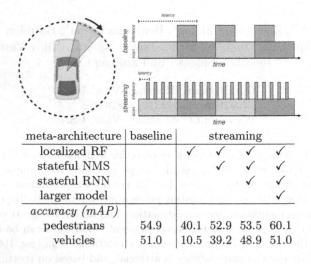

meta-architecture	baseline	streaming			
localized RF		✓	✓	✓	✓
stateful NMS			✓	✓	✓
stateful RNN				✓	✓
larger model					✓
accuracy (mAP)					
pedestrians	54.9	40.1	52.9	53.5	60.1
vehicles	51.0	10.5	39.2	48.9	51.0

Fig. 1. Streaming object detection pipelines computation to minimize latency without sacrificing accuracy. LiDAR accrues a point cloud incrementally based on a rotation around the z axis. Instead of artificially waiting for a complete point cloud scene based on a 360° rotation (*baseline*), we perform inference on subsets of the rotation to pipeline computation (*streaming*). Gray boxes indicate the duration for a complete rotation of a LiDAR (e.g. 100 ms [13,53]). Green boxes denote inference time. The expected latency for detection – defined as the time between a measurement and a detection decreases substantially in a streaming architecture (dashed line). At 100 ms scan time, the expected latency reduces from ∼120 ms (baseline) versus ∼30 ms (streaming), i.e. >3× (see text for details). The table compares the detection accuracy (mAP) for the baseline on pedestrians and vehicles to several streaming variants [32]. (Color figure online)

LiDAR represents one of the most prominent sensory modalities in SDC systems [7,54] informing object detection [58–60,64], region segmentation [33,36] and motion estimation [27,62]. Existing approaches to LiDAR-based perception derive from a family of camera-based approaches [5,9,19,25,39,49], requiring a complete 360° scan of the environment. This artificial requirement to have the complete scan limits the minimum latency a perception system can achieve, and effectively inserts the LiDAR scan period into the latency[1]. Unlike CCD cameras, many LiDAR systems are *streaming* data sources, where data arrives sequentially as the laser rotates around the z axis [2,21].

---

[1] LiDAR typically operates with a 5–20 Hz scan rate. We focus on 10 Hz (i.e. 100 ms period), because several prominent academic datasets employ this scan rate [13,53], however our results may be equally applied across the range of scan rates available.

Object detection in LiDAR-based perception systems [58–60,64] presents a unique and important opportunity for re-imagining LiDAR-based meta architectures in order to significantly minimize latency. In particular, streaming LiDAR data permits the design of meta-architectures which operate on the data as it arrives, in order to pipeline the sensory readout with the inference computation to significantly reduce latency (Fig. 1).

In this work, we propose a series of modifications to standard meta architectures that may generically adapt an object detection system to operate in a streaming manner. This approach combines traditional elements of single-stage detection systems [39,48] as well as design elements of sequence-based learning systems [6,18,26]. The goal of this work is to show how we can modify an existing object detection system – with a minimum set of changes and the addition of new operations – to efficiently and accurately emit detections as data arrives (Fig. 1). We find that this approach matches or exceeds the performance of several baseline systems [32,43], while substantially reducing latency. For instance, a family of streaming models on pedestrian detection achieves up to 60.1 mAP compared to 54.9 mAP for a baseline non-streaming model, while reducing the expected latency >3×. In addition, the resulting model better utilizes computational resources by pipelining the computation throughout the duration of a LiDAR scan. We demonstrate through this work that designing architectures to leverage the native format of LiDAR data achieves substantial latency gains for perception systems generically that may improve the safety and efficacy of SDC systems.

2 Related Work

2.1 Object Detection in Camera Images

Object detection has a long history [8,10,12,35,55]. The re-emergence of convolutional neural networks (CNN) for computer vision [30,31] inspired the field to harness both the rich image features and final training objective of a CNN model for object detection [50]. In particular, the features of a CNN trained on an image classification task proved sufficient for providing reasonable candidate locations for objects [15]. Subsequent work demonstrated that a single CNN may be trained in an end-to-end fashion to sub-serve for both stages of an object detection system [14,49]. The resulting two-stage systems, however, suffered from relatively poor computational performance [24]. This inspired researchers to pursue one stage object detection systems [39,48] at the cost of lower predictive performance [39,48].

2.2 Object Detection in Videos

Strategies to tackle object detection in videos include breaking up the problem into a computationally-heavy detection phase and a computationally-light tracking phase [37], and building blended CNN recurrent architectures for providing memory between time steps of each frame [38,41]. Recent methods have also explored the potential to persist and update a memory of the scene [4,22]. This problem reflects possibly the closest set of methods relevant to our proposed work. predictions.

In our work, we examine the possibility of dividing a *single frame* into slices that can be processed in a streaming fashion. A time step in our setup correspond to a slice of one frame. An object may appear across multiple slices, but generally, each slice contains distinct objects. This may require similar architectures to video object detection (e.g., convolutional LSTMs) in order to provide a memory and state of earlier slices for refining or merging detections [44,56,57].

2.3 Object Detection in Point Clouds

The prominence of LiDAR systems in self-driving cars necessitated the application of object detection to point cloud data [3,13]. Much work has employed object detection systems originally designed for camera images to point cloud data by projecting such data from a Bird's Eye View (BEV) [32,40,59,60] (but see [42]) or a 3-D voxel grid [58,64]. Alternatively, some methods have re-purposed two stage object detector design with a region-proposal stage but replacing the feature extraction operations [45,52,61].

In parallel, others have pursued replacing a discretization operation with a featurization based on native point-cloud data [46,47]. Such methods have led to methods for building detection systems that blend aspects of point cloud featurization and traditional object detectors on cameras [32,43,63] to achieve favorable performance for a given computational budget.

3   Methods

3.1   Streaming LiDAR Inputs

A LiDAR system for an SDC measures the distance to objects by shining multiple lasers at fixed inclinations and measuring the reflectance in a sensor [7,54]. The lasers precess around the z axis, and make a complete rotation at a 5–20 Hz scan rate. Typically, SDC perception systems artificially wait for a complete 360° rotation before processing the data.

This work simulates a streaming system with the Waymo Open Dataset [53] by artificially manipulating the point cloud data[2]. The native format of point cloud data are *range images* whose resolution in height and width correspond to the number of lasers and the rotation speed and laser pulse rate [42]. In this work, we artificially slice the input range image into n vertical strips along the image width to experiment with streaming detection models.

---

[2] KITTI [13] is the most popular LiDAR detection dataset, however this dataset provides annotations within a 90° frustum. The Waymo Open Dataset provides a completely annotated 360° point cloud which is necessary to demonstrate the efficacy of the streaming architecture across all angular orientations.

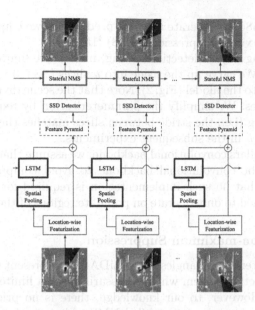

Fig. 2. Diagram of streaming detection architecture. A streaming object detection system processes a spatially restricted slice of the scene. We introduce two stateful components: Stateful NMS (red) and a LSTM (blue) between input slices. Detections produced by the model are denoted in green boxes. Dashed line denotes feature pyramid uniquely employed in [32]. (Color figure online)

3.2   Streaming Object Detection

This work introduces a meta-architecture for adapting object detection systems for point clouds to operate in a streaming fashion. We employ two models as a baseline to demonstrate how the proposed changes to the meta-architecture are generic, and may be employed in notably different detection systems.

We first investigate PointPillars [32] as a baseline model because it provides competitive performance in terms of predictive accuracy and computational budget. The model divides the x-y space into a top-down 2D grid, where each grid cell is referred to as a *pillar*. The points within each non-zero pillar are featurized using a variant of a multi-layer perceptron architecture designed for point cloud data [46,47]. The resulting d-dimensional point cloud features are scattered to the grid and a standard multi-scale, convolutional feature pyramid [34,64] is computed on the spatially-arranged point cloud features to result in a global activation map. The second model investigated is StarNet [43]. StarNet is an entirely point-based detection system which uses sampling instead of a learned region proposal to operate on targeted regions of a point cloud. StarNet avoids usage of global information, but instead targets the computational demand to regions of interest, resulting in a locally targeted activation map. See the Appendix for architecture details for both models. For both PointPillars and StarNet, the resulting activation map is regressed on to a 7-dimensional target parameterizing the 3D bounding box as well as a classification logit [58]. Ground truth labels are assigned to individual anchors based on intersection-over-union

(IoU) overlap [32,58]. To generate the final predictions, we employ oriented, 3-D multi-class non-maximal suppression (NMS) [15].

In the streaming object detection setting, models are limited to a restricted view of the scene. We carve up the scene into n slices (Sect. 3.1) and only supply an individual slice to the model (Fig. 2). Note that the scene division is performed in polar coordinates. We simplify the parameterization by requiring that slices are non-overlapping (i.e., the stride between slices matches the slice width). We explore a range of n in the subsequent experiments.

For the PointPillars convolutional backbone, we assume that sparse operators are employed in the convolutional backbone to avoid computation on empty pillars [17]. Note that no such implementation is required for StarNet because the model is designed to only operate on populate regions of the point cloud [43].

3.3 Stateful Non-maximum Suppression

Objects which subtend large angles of the LiDAR scan present unique challenges to a steaming detection system which necessarily have a limited range of sensor input (Table 1). However, to our knowledge, there is no prior work on NMS with state. Hence, we explore a modified NMS technique that maintains *states for the streaming setting*. Generically, NMS with state may take into account detections in the previous k slices to determine if a new detection is indeed unique. Therefore, detections from the current slice can be suppressed by those in previous slices. Stateful NMS does not require a complete LiDAR rotation and may likewise operate in a streaming fashion. Furthermore. In our experiments, for a broad range of n, we found that $k = 1$ achieves as good of performance as $k = n - 1$ which would correspond to a global NMS available to a non-streaming system. We explore the selection of k in Sect. 4.2.

3.4 Adding State with Recurrent Architectures

Given a restricted view of the point cloud scene, streaming models can be limited by the context available to make predictions. To increase the amount of context that the model has access to, we consider augmenting the baseline model to maintain a recurrent memory across consecutive slices. This memory may be placed in the intermediate representations of the network.

We select a standard single layer LSTM as our recurrent architecture, although any other RNN architecture may suffice [23]. The LSTM is inserted after the final global representation from either baseline model before regressing on to the detection targets. For instance, in the PointPillars baseline model, this corresponds to inserting the LSTM after the convolutional representation. The memory input is then the spatial average pooling for all activations of the final convolutional layer before the feature pyramid[3]. Note that the LSTM memory does not hard code the spatial location of the slice it is processing.

---

[3] The final convolutional layer corresponds to the third convolutional layer in Point-Pillars [32] after three strided convolutions. When computing the feature pyramid, note that the output of the LSTM is only added the corresponding outputs of the third convolutional layer.

Based on earlier preliminary experiments, we employed LSTM with 128 hidden dimensions and 256 output dimensions. The output of the LSTM is summed with the final activation map by broadcasting across all spatial dimensions in the hidden representation. More sophisticated elaborations are possible, but are not explored in this work [37,38,41,44,56,57].

4   Results

We present all results on the Waymo Open Dataset [53]. All models are trained with Adam [29] using the Lingvo machine learning framework [51] built on top of TensorFlow [1][4]. No data augmentation is used. We perform hyper-parameter tuning through cross-validated studies and final evaluations on the corresponding test datasets. In our experiments, we explore how the proposed meta-architecture for streaming 3-D object detection compares to a standard object detection system, i.e. PointPillars [32] and StarNet [43]. All experiments use the first return from the medium range LiDAR (labeled TOP) in the Waymo Open Dataset, ignoring the four short range LiDARs for simplicity. This results in slightly lower baseline and final accuracy as compared to previous results [43,53,63].

We begin with a simple, generic modification to the baseline architecture by limiting its spatial view to a single slice. This modification permits the model to operate in a streaming fashion, but suffers from a severe performance degradation. We address these issues through stateful variants of non-maximum suppression (NMS) and recurrent architectures (RNN). We demonstrate that the resulting streaming meta-architecture restores competitive performance with favorable computation and latency benefits. Importantly, such a meta-architecture may be applied generically to most other point cloud detection models [40,42,43,58–60,64].

4.1   Spatially Localized Detection Degrades Baseline Models

To build a baseline model for streaming, we modify existing models by restricting the operation to a slice of point cloud (Fig. 2). The model may in principle operate with a reduced latency since inference can proceed before a complete LiDAR rotation (Fig. 1).

Two free parameters that govern a localized receptive field are the angle of restriction and the stride between subsequent inference operations. To simplify this analysis, we parameterize the angle based on the number of slices n where the angular width is $360°/n$. We specify the stride to match the angle such that each inference performs inference on non-overlapping slices; overlapping slices is an option, but requires recomputing features on the same points. We compare the performance of the streaming models against the baselines in Fig. 3.

As the number of slices n grows, the streaming model receives a decreasing fraction of the scene and the predictive performance monotonically decreases

---

[4] Code available at http://github.com/tensorflow/lingvo.

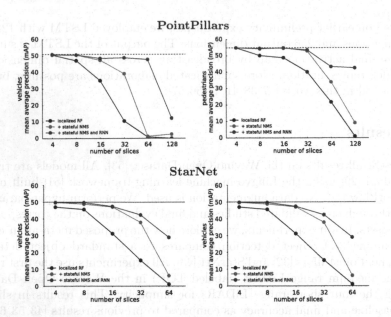

Fig. 3. Streaming object detection may achieve comparable performance to a non-streaming baseline. Mean average precision (mAP) versus the number of slices n within a single rotation for (left) vehicles and (right) pedestrians for (top) PointPillars [32] and (bottom) StarNet [43]. The solid black line (localized rf) corresponds to the modified baseline that operates on a spatially restricted region. Dashed lines corresponds to baseline models that processes the entire laser spin (vehicle = 51.0%; pedestrian = 54.9%). Each curve represents a streaming architecture (see text). (Color figure online)

(black solid line). This reduction in accuracy may be expected because less sensory data is available for each inference operation. All objects sizes appear to be severely degraded by the localized receptive field, although there is a slight negative trend for increasing sizes (Table 1). The negative trend is consistent with the observation that vehicle mAP drops off faster than pedestrians, as vehicles are larger and will more frequently cross slices boundaries. For instance, at $n = 16$, the mean average precision is 35.0% and 48.8% for vehicles and pedestrians, respectively. A large number of slices $n = 128$ severely degrades model performance for both types. Beyond only observing fewer points with smaller slices, we observe that even at a modest 8 slices the mAP drops compared to the baseline for both pedestrians and vehicles. Our hypothesis is that when NMS only operates per slice, there are false positives or false negatives created on any object that crosses a border of two consecutive slices. As a result, we next turn to investigating ways in which we can improve NMS in the intra-frame scenario to restore performance compared to the baseline while retaining the streaming model's latency benefits.

Table 1. Localized receptive field leads to duplicate detections. Vehicle detection performance (mAP) across subtended angle of ground truth objects for localized receptive field and stateful NMS ($n = 32$ slices) for [32]. Red indicates the percent drop from baseline. **Localized receptive field leads to duplicate detections.** Vehicle detection performance (mAP) across subtended angle of ground truth objects for localized receptive field and stateful NMS ($n = 32$ slices) for [32]. Red indicates the percent drop from baseline.

	0–5°	5–15°	15–25°	25–35°	>35°
Baseline	12.6	30.9	47.2	69.1	83.4
Localized RF	2.6	6.6	9.2	13.1	14.0
	−79%	−79%	−80%	−81%	−83%
+Stateful NMS	9.0	24.1	36.0	54.9	68.0
	−28%	−22%	−23%	−20%	−18%
+Stateful NMS	10.8	29.9	46.0	65.3	82.1
and RNN	−14%	−3%	−3%	−6%	−2%

Table 2. Stateful NMS achieves comparable performance gains as global NMS. Table entries report the mAP for detection on vehicles (car) and pedestrians (ped) [32]. Localized indicates the mAP for a spatially restricted receptive field. Global NMS boosts performance significantly but is a non-streaming heuristic. Stateful NMS achieves comparable results but is amenable to a streaming architecture.

Slices		Localized	Global	Stateful
16	Car	35.0	47.5	47.4
	Ped	48.8	54.8	54.9
32	Car	10.5	39.2	39.0
	Ped	40.1	53.1	52.9

4.2 Adding State to Non-maximum Suppression Boosts Performance

NMS provides a standard heuristic method to remove highly overlapping predictions [12, 15]. With a spatially localized receptive field, NMS has no ability to suppress overlapping detections arising from distinct slices of the LiDAR spin.

We can verify this failure mode by modifying NMS to operate over the concatenation of all detections across *all* n slices in a complete rotation of the LiDAR. We term this operation *global* NMS. (Note that global NMS does not enable a streaming detection system because a complete LiDAR rotation is required to finish inference.) We compute the detection accuracy for global NMS as a point of reference to measure how many of the failures of a spatially localized receptive field can be rescued. Indeed, Table 2 (localized vs. global) indicates that applying global NMS improves predictive performance significantly.

We wish to develop a new form of NMS that achieves the same performance as global NMS but may operate in a streaming fashion. We construct a simple form of *stateful* NMS that stores detections from the previous slice of the LiDAR scene and use the previous slice's detections to rule out overlapping detections (Sect. 3.3). Stateful NMS does not require a complete LiDAR rotation and may operate in a streaming fashion. Table 2 (global vs. stateful) indicates that stateful NMS provides predictive performance that is comparable to a global NMS operation within ± 0.1 mAP. Indeed, stateful NMS boosts performance of the spatially restricted receptive field across all ranges of slices for both baseline models (Fig. 3, red curve). This observation is also consistent with the fact that a large fraction of the failures across object size are largely systematically recovered (Table 1), suggesting that indeed duplicate detections across boundaries hamper model performance. These results suggest that this generic change to introduce state into the NMS heuristic may recover much of the performance drop due to a spatially localized receptive field.

4.3 Adding a Learned Recurrent Model Further Boosts Performance

Adding stateful NMS substantially improves streaming detection performance, however, the network backbone and learned components of the system only operate on the current slice of the scene and the outputs of the previous slice.

This section examines how a recurrent model may improve predictive performance by providing access to (1) more than just the previous LiDAR slice, and (2) the lower-level features of the model, instead of just the outputs. In particular, we provide an alteration to the meta-architecture by adding a recurrent layer to the global representation of the network featurization (Fig. 2). We investigated variations of recurrent architectures and hyper-parameters, and settled on a simple single layer LSTM, but most RNN's produces similar results.

The blue curve of Fig. 3 shows the results of adding a recurrent bottleneck on top of a spatially localized receptive field and a stateful NMS for both baseline architectures. Generically, we observe that predictive performance increases across the entire range of n. The performance gains are more notable across increasing number of slices n, as well as all vehicle object sizes (Table 1). These results are expected given that we observed systematic failure introduced for large objects that subtend multiple slices (Fig. 3). For instance, with PointPillars at $n = 32$ slices, the mAP for vehicles is boosted from 39.2% to 48.9% through the addition of a learned, recurrent architecture (Fig. 3, blue curves). These results also demonstrate how objects close to the scanner may be divided across slices, thus leading to higher error, since the object size can be approximately measured by the angle subtended in the θ direction. Specifically, our streaming model only introduces a 1.3 mAP drop for close-by objects ($>35°$), and 1.8 drop in mAP for faraway objects ($<5°$). Overall, this streaming model is only marginally lower than the mAP of the original model that has access to the complete LiDAR scan.

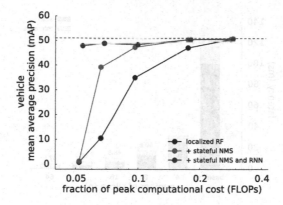

Fig. 4. Fraction of peak computational demand (FLOPs) versus detection accuracy (vehicle mAP) across varying number of slices n for [32]. Note the logarithmic scale on the x-axis. Each curve represents streaming architecture (see text). Dashed line indicates non-streaming baseline. (Color figure online)

Although a streaming model sacrifices slightly in terms of predictive performance, we expect that the resulting streaming model would realize substantial gains in terms of computational demand and latency. In the following section we explore this question in detail.

4.4   Streaming Detection Reduces Latency and Peak Computation

We define the **end-to-end** latency as the duration of time between the earliest observation of an object (i.e. the earliest time at which reflecting LiDAR points are received) and the identification of a localized, labeled object. For a non-streaming model the latency corresponds to the summation of the time for a complete (worst-case) 360° rotation and the subsequent inference operation on the complete LiDAR scene. The latency in a streaming detection system should be improved for two reasons: (1) the computational demand for processing a fraction of the LiDAR spin should be roughly $\frac{1}{n}$ of the complete scene and (2) the inference operation does not require artificially waiting for the full LiDAR spin and may be pipelined. In this section, we focus our analysis on [32] because of previously reported latencies.

To test the first point, we compute the theoretical peak computational demand (in FLOPS) for running single frame inference on the baseline model as well as streaming models. Figure 4 compares the peak FLOPS versus the detection accuracy across varying slices n for each of the 3 streaming architectures. We display the compute of each approach in terms of the *fraction* of peak FLOPS required for the non-streaming baseline model (Note the log x-axis). We observe that a model with a localized receptive field (black curve) reveals a trade-off in accuracy versus the amount of computational demand. However, subsequent models with stateful NMS (red curve) and stateful NMS and RNN (blue curve) require much fewer peak FLOPS to achieve most of the detection

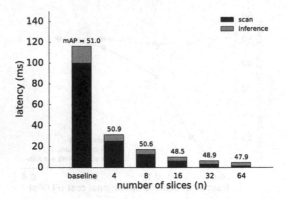

Fig. 5. Worst-case latency from the initial measurement of the vehicle to the detection for non-streaming (baseline) and streaming detection model (stateful NMS and RNN), broken down by phase. Scan phase latency is based on 10 Hz LiDAR period [53]. Inference phase latency estimated from baseline GPU implementation [32]. Numbers on top of bar are vehicle mAP.

performance of the baseline. Furthermore, the stateful NMS and RNN achieves nearly baseline predictive performance across a wide range of slices n with a computational cost of roughly $\frac{1}{n}$. Thus, the streaming model requires less peak computational demand with minimal degradation in predictive performance.

Note that latency is very heavily determined by the hardware, as well as the rotational period of the LiDAR system. To estimate reasonable speed up gains, we test these ideas on a previously reported implementation speed for the PointPillars model [32] (Sect. 6), and employ the rotational period of the Waymo Open Dataset (i.e. 100 ms for 10 Hz) [53]. Figure 5 plots the latency versus the detection accuracy across the streaming model variants; we assume that the scan time is equivalent to the worst case delay between the first measurement of the object and the triggering of inference (the end of the slice): e.g., with 4 slices and 10 Hz period, each slice triggers inference every 25 ms. We estimate inference latency by scaling the floating point computation time [32] by the fraction of point cloud scene $\frac{1}{n}$ observed within an angular wedge necessary for inference.

The streaming models reduce end-to-end latency significantly in comparison to a non-streaming baseline model. In fact, we observe that for $n = 8$, a streaming model with a stateful NMS and RNN achieves competitive accuracy with the non-streaming model, but with $\frac{1}{15}$ of the latency (17 ms vs 116 ms). We take these results as a conceptual proof that a streaming detection system may significantly reduce end-to-end latency without significantly sacrificing predictive accuracy.

4.5   Increasing Model Size Exceeds Baseline with Fewer FLOPs

This section explores whether the computational resources that are freed up by switching to streaming settings may instead be harnessed (with a still reduced

PointPillars

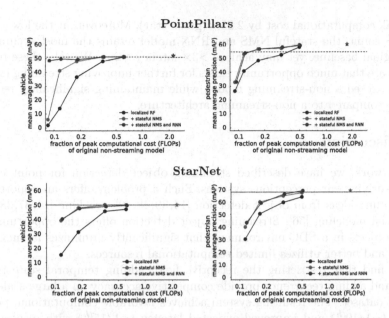

StarNet

Fig. 6. A larger streaming model may exceed the baseline performance, but with a fraction of the peak computational budget. Results presented for vehicle (left) and pedestrian (right) detection on the larger (top) PointPillars [32] and (bottom) StarNet [43] relative to the original non-streaming baseline. Green star indicates the relative peak FLOPS and accuracy of the larger non-streaming model. Computational cost varies inversely with the number of slices(n) from 4 to 64. (Color figure online)

latency) to boost performance above and beyond the baseline model. Broadly speaking, we attempt to double the computational cost of each baseline network in order to attempt to boost the overall performance. For instance, for Point-Pillars we increase the size of the feature pyramid by systematically increasing the spatial resolution of the activation map. Specifically, we increase the top-down view grid size from 384 to 512 for pedestrian models, and from 512 to 784 for vehicle models, resulting in models with 1.75× and 2.23× relative expense, respectively. For StarNet, we increase all hidden units by 1.44×.

Figure 6 shows the predictive accuracy of the resulting both streaming baseline models on vehicles (left) and pedestrians (right) across slices n in comparison to the non-streaming baseline for both architectures. As a point of reference, we show the relative peak computational cost of these larger models when run in a non-streaming node (green star). The x-axis measures logarithmically the peak computational demand of the resulting model expressed as a fraction of the non-streaming baseline model. Importantly, we observe that even though the peak computational demand is a small fraction of the non-streaming baseline model, the predictive performance matches or exceeds the non-streaming baseline model (e.g. 60.1 mAP versus 54.9 mAP for pedestrians with PointPillars). Note that in order to achieve these gains in the non-streaming model requires increasing the peak computational cost by 2.25× (green star). Moreover, at the lower peak FLOPS count, the stateful NMS and RNN model retains the most accuracy of the original baseline, yet may achieve >3× latency gains. We take these results to indicate that much opportunity exists for further improving streaming models to well exceed a non-streaming model, while maintaining significantly reduced latency compared to a non-streaming architecture.

5   Discussion

In this work, we have described streaming object detection for point clouds for self-driving car perceptions systems. Such a problem offers an opportunity for blending ideas from object detection [14,39,48,49], tracking [11,37,38] and sequential modeling [56]. Streaming object detection offers the opportunity to detect objects in a SDC environment that significantly minimizes latency (e.g. 3-15×) and better utilizes limited computational resources.

We find that restricting the receptive field, adding temporal state to the NMS, and adding recurrence provide competitive accuracy on a large-scale self-driving dataset. The resulting system achieves favorable computational performance ($\sim 1/10^{th}$) and improved expected latency ($\sim 1/15^{th}$) with respect to a baseline non-streaming system. Such gains provide headroom to scale up the system to surpass baseline performance (60.1 vs 54.9 mAP) while maintaining a peak computational budget far below a non-streaming model.

This work offers opportunity for further improving this methodology, or application to new streaming sensors (e.g. high-resolution cameras that emit rasterized data in slices). While this work focuses on streaming models for a single frame, it is possible to also extend the models to incorporate data across multiple frames. We note that a streaming model may be amendable towards tracking problems since it already incorporates state. Finally, we have explored meta-architecture changes with respect to two competitive object detection baselines [32,43]. We hope our work will encourage further research on other point cloud based perception systems to test their efficacy in a streaming setting [32,40,42,43,58–60,64].

Acknowledgements. We thank the larger teams at Google Brain and Waymo for their help and support. We also thank Chen Wu, Pieter-jan Kindermans, Matthieu Devin and Junhua Mao for detailed comments on the project and manuscript.

Appendix

A   Architecture and Training Details

A.1   Training data

There are actually two different training settings in the paper: (1) for the non-streaming models and (2) for the streaming models with stateful NMS. In the first settings, there is no RNN so we use the original full scene data. In the second setting, we only use sliced data to train the streaming models with LSTM (Tables 3 and 4).

Table 3. PointPillars detection baseline [32].

Operation	Stride	# In	# Out	Activation	Other
Streaming detector					
Featurizer MLP		12	64		With max pooling
Convolution block	1	64	64	ReLU	Layers=4
Convolution block	2	64	128	ReLU	Layers=6
Convolution block	2	128	256	ReLU	Layers=6
LSTM		256	256		Hidden=128
Deconvolution 1	1	64	128	ReLU	
Deconvolution 2	2	128	128	ReLU	
Deconvolution 3	4	256	128	ReLU	
Convolution/Detector	1	384	16		Kernel=3×3
Convolution block					
(S, C_{in}, C_{out}, L)					
Convolution 1	S	C_{in}	C_{out}	ReLU	Kernel=3×3
Convolution $2, ..., L-1$	1	C_{out}	C_{out}	ReLU	Kernel=3×3

Normalization Batch normalization before ReLU for every
convolution and deconvolution layer
Optimizer Adam [29] ($\alpha = 0.001$, $\beta_1 = 0.9$, $\beta_2 = 0.999$)
Parameter updates 40,000 - 80,000
Batch size 64
Weight initialization Xavier-Glorot[16]

Table 4. StarNet detection baseline [43].

Operation	# In	# Out	Activation	Other
Streaming detector				
Linear	4	64	ReLU	
StarNet block $\times$ 5	64	64	ReLU	Final feature is the concat of all layers
LSTM	384	384		Hidden=128
Detector	384	16		
StarNet block				
Max-Concat	64	128		
Linear	128	256	ReLU	
Linear	256	64	ReLU	

Normalization Batch normalization before ReLU
Optimizer Adam [29] ($\alpha = 0.001$, $\beta_1 = 0.9$, $\beta_2 = 0.999$)
Parameter updates 100,000
Batch size 64
Weight initialization Kaiming Uniform [20]

References

1. Abadi, M., et al.: Tensorflow: a system for large-scale machine learning. In: 12th {USENIX} Symposium on Operating Systems Design and Implementation ({OSDI} 2016), pp. 265–283 (2016)
2. Ackerman, E.: Lidar that will make self-driving cars affordable [news]. IEEE Spectr. **53**(10), 14–14 (2016)
3. Caesar, H., et al.: nuscenes: a multimodal dataset for autonomous driving. arXiv preprint arXiv:1903.11027 (2019)
4. Chai, Y.: Patchwork: A patch-wise attention network for efficient object detection and segmentation in video streams. In: IEEE Conference on Computer Vision and Pattern Recognition (2019)
5. Chen, L.C., Papandreou, G., Kokkinos, I., Murphy, K., Yuille, A.L.: Deeplab: semantic image segmentation with deep convolutional nets, atrous convolution, and fully connected CRFs. IEEE Trans. Pattern Anal. Mach. Intell. **40**(4), 834–848 (2017)
6. Chiu, C.C., et al.: State-of-the-art speech recognition with sequence-to-sequence models. In: 2018 IEEE International Conference on Acoustics, Speech and Signal Processing (ICASSP), pp. 4774–4778. IEEE (2018)
7. Cho, H., Seo, Y.W., Kumar, B.V., Rajkumar, R.R.: A multi-sensor fusion system for moving object detection and tracking in urban driving environments. In: 2014 IEEE International Conference on Robotics and Automation (ICRA), pp. 1836–1843. IEEE (2014)
8. Dean, T., Ruzon, M.A., Segal, M., Shlens, J., Vijayanarasimhan, S., Yagnik, J.: Fast, accurate detection of 100,000 object classes on a single machine. In: Proceedings of the IEEE Conference on Computer Vision and Pattern Recognition, pp. 1814–1821 (2013)
9. Dosovitskiy, A., et al.: Flownet: learning optical flow with convolutional networks. In: Proceedings of the IEEE International Conference on Computer Vision,. pp. 2758–2766 (2015)
10. Everingham, M., Van Gool, L., Williams, C.K., Winn, J., Zisserman, A.: The pascal visual object classes (voc) challenge. Int. J. Comput. Vis. **88**(2), 303–338 (2010)
11. Feichtenhofer, C., Pinz, A., Zisserman, A.: Detect to track and track to detect. In: Proceedings of the IEEE International Conference on Computer Vision, pp. 3038–3046 (2017)
12. Felzenszwalb, P.F., Girshick, R.B., McAllester, D., Ramanan, D.: Object detection with discriminatively trained part-based models. IEEE Trans. Pattern Anal. Mach. Intell. **32**(9), 1627–1645 (2010)
13. Geiger, A., Lenz, P., Stiller, C., Urtasun, R.: Vision meets robotics: the kitti dataset. Int. J. Robot. Res. **32**(11), 1231–1237 (2013)
14. Girshick, R.: Fast R-CNN. In: Proceedings of the IEEE International Conference on Computer Vision, pp. 1440–1448 (2015)
15. Girshick, R., Donahue, J., Darrell, T., Malik, J.: Rich feature hierarchies for accurate object detection and semantic segmentation. In: Proceedings of the IEEE Conference on Computer Vision and Pattern Recognition, pp. 580–587 (2014)
16. Glorot, X., Bengio, Y.: Understanding the difficulty of training deep feedforward neural networks. In: Proceedings of the Thirteenth International Conference on Artificial Intelligence and Statistics, pp. 249–256 (2010)
17. Graham, B., van der Maaten, L.: Submanifold sparse convolutional networks. CoRR abs/1706.01307 (2017). http://arxiv.org/abs/1706.01307

18. Graves, A.: Sequence transduction with recurrent neural networks. arXiv preprint arXiv:1211.3711 (2012)
19. He, K., Gkioxari, G., Dollár, P., Girshick, R.: Mask R-CNN. In: Proceedings of the IEEE International Conference on Computer Vision, pp. 2961–2969 (2017)
20. He, K., Zhang, X., Ren, S., Sun, J.: Delving deep into rectifiers: Surpassing human-level performance on imagenet classification. In: Proceedings of the IEEE International Conference on Computer Vision, pp. 1026–1034 (2015)
21. Hecht, J.: Lidar for self-driving cars. Opt. Photonics News **29**(1), 26–33 (2018)
22. Henriques, J.F., Vedaldi, A.: Mapnet: an allocentric spatial memory for mapping environments. In: IEEE Conference on Computer Vision and Pattern Recognition (2018)
23. Hochreiter, S., Schmidhuber, J.: Long short-term memory. Neural Comput. **9**(8), 1735–1780 (1997)
24. Huang, J., et al.: Speed/accuracy trade-offs for modern convolutional object detectors. In: Proceedings of the IEEE Conference on Computer Vision and Pattern Recognition, pp. 7310–7311 (2017)
25. Ilg, E., Mayer, N., Saikia, T., Keuper, M., Dosovitskiy, A., Brox, T.: Flownet 2.0: evolution of optical flow estimation with deep networks. In: Proceedings of the IEEE Conference on Computer Vision and Pattern Recognition, pp. 2462–2470 (2017)
26. Jaitly, N., Sussillo, D., Le, Q.V., Vinyals, O., Sutskever, I., Bengio, S.: A neural transducer. arXiv preprint arXiv:1511.04868 (2015)
27. Jeon, H.H., Ko, Y.H.: Lidar data interpolation algorithm for visual odometry based on 3D–2D motion estimation. In: 2018 International Conference on Electronics, Information, and Communication (ICEIC), pp. 1–2. IEEE (2018)
28. Kim, J., Kim, H., Lakshmanan, K., Rajkumar, R.R.: Parallel scheduling for cyber-physical systems: analysis and case study on a self-driving car. In: Proceedings of the ACM/IEEE 4th International Conference on Cyber-physical Systems, pp. 31–40. ACM (2013)
29. Kingma, D.P., Ba, J.: Adam: a method for stochastic optimization. arXiv preprint arXiv:1412.6980 (2014)
30. Krizhevsky, A., Hinton, G.: Learning multiple layers of features from tiny images. University of Toronto, Technical report (2009)
31. Krizhevsky, A., Sutskever, I., Hinton, G.E.: Imagenet classification with deep convolutional neural networks. In: Advances in Neural Information Processing Systems (2012)
32. Lang, A.H., Vora, S., Caesar, H., Zhou, L., Yang, J., Beijbom, O.: Pointpillars: fast encoders for object detection from point clouds. arXiv preprint arXiv:1812.05784 (2018)
33. Lim, K.L., Drage, T., Bräunl, T.: Implementation of semantic segmentation for road and lane detection on an autonomous ground vehicle with lidar. In: 2017 IEEE International Conference on Multisensor Fusion and Integration for Intelligent Systems (MFI), pp. 429–434. IEEE (2017)
34. Lin, T.Y., Dollár, P., Girshick, R., He, K., Hariharan, B., Belongie, S.: Feature pyramid networks for object detection. In: Proceedings of the IEEE Conference on Computer Vision and Pattern Recognition (2017)
35. Lin, T.-Y., et al.: Microsoft COCO: common objects in context. In: Fleet, D., Pajdla, T., Schiele, B., Tuytelaars, T. (eds.) ECCV 2014. LNCS, vol. 8693, pp. 740–755. Springer, Cham (2014). https://doi.org/10.1007/978-3-319-10602-1_48

36. Lindner, P., Richter, E., Wanielik, G., Takagi, K., Isogai, A.: Multi-channel lidar processing for lane detection and estimation. In: 2009 12th International IEEE Conference on Intelligent Transportation Systems, pp. 1–6. IEEE (2009)
37. Liu, M., Zhu, M.: Mobile video object detection with temporally-aware feature maps. In: Proceedings of the IEEE Conference on Computer Vision and Pattern Recognition, pp. 5686–5695 (2018)
38. Liu, M., Zhu, M., White, M., Li, Y., Kalenichenko, D.: Looking fast and slow: Memory-guided mobile video object detection. arXiv preprint arXiv:1903.10172 (2019)
39. Liu, W., et al.: SSD: single shot multibox detector. In: Leibe, B., Matas, J., Sebe, N., Welling, M. (eds.) ECCV 2016. LNCS, vol. 9905, pp. 21–37. Springer, Cham (2016). https://doi.org/10.1007/978-3-319-46448-0_2
40. Luo, W., Yang, B., Urtasun, R.: Fast and furious: real time end-to-end 3D detection, tracking and motion forecasting with a single convolutional net. In: Proceedings of the IEEE Conference on Computer Vision and Pattern Recognition, pp. 3569–3577 (2018)
41. McIntosh, L., Maheswaranathan, N., Sussillo, D., Shlens, J.: Recurrent segmentation for variable computational budgets. In: Proceedings of the IEEE Conference on Computer Vision and Pattern Recognition Workshops, pp. 1648–1657 (2018)
42. Meyer, G.P., Laddha, A., Kee, E., Vallespi-Gonzalez, C., Wellington, C.K.: Lasernet: an efficient probabilistic 3D object detector for autonomous driving. arXiv preprint arXiv:1903.08701 (2019)
43. Ngiam, J., et al.: Starnet: targeted computation for object detection in point clouds. arXiv preprint arXiv:1908.11069 (2019)
44. Pinheiro, P., Collobert, R.: Recurrent convolutional neural networks for scene labeling. In: Xing, E.P., Jebara, T. (eds.) Proceedings of the 31st International Conference on Machine Learning. Proceedings of Machine Learning Research, vol. 32, pp. 82–90. PMLR, Bejing, China, 22–24 June 2014
45. Qi, C.R., Liu, W., Wu, C., Su, H., Guibas, L.J.: Frustum pointnets for 3D object detection from RGB-D data. In: Proceedings of the IEEE Conference on Computer Vision and Pattern Recognition, pp. 918–927 (2018)
46. Qi, C.R., Su, H., Mo, K., Guibas, L.J.: Pointnet: deep learning on point sets for 3D classification and segmentation. In: Proceedings of the IEEE Conference on Computer Vision and Pattern Recognition, pp. 652–660 (2017)
47. Qi, C.R., Yi, L., Su, H., Guibas, L.J.: Pointnet++: deep hierarchical feature learning on point sets in a metric space. In: Advances in Neural Information Processing Systems, pp. 5099–5108 (2017)
48. Redmon, J., Divvala, S., Girshick, R., Farhadi, A.: You only look once: unified, real-time object detection. In: Proceedings of the IEEE Conference on Computer Vision and Pattern Recognition, pp. 779–788 (2016)
49. Ren, S., He, K., Girshick, R., Sun, J.: Faster R-CNN: towards real-time object detection with region proposal networks. In: Advances in Neural Information Processing Systems, pp. 91–99 (2015)
50. Sermanet, P., Eigen, D., Zhang, X., Mathieu, M., Fergus, R., LeCun, Y.: Overfeat: integrated recognition, localization and detection using convolutional networks. arXiv preprint arXiv:1312.6229 (2013)
51. Shen, J., et al.: Lingvo: a modular and scalable framework for sequence-to-sequence modeling. arXiv preprint arXiv:1902.08295 (2019)
52. Shi, S., Wang, X., Li, H.: PointRCNN: 3D object proposal generation and detection from point cloud. In: Proceedings of the IEEE Conference on Computer Vision and Pattern Recognition, pp. 770–779 (2019)

53. Sun, P., et al.: Scalability in perception for autonomous driving: Waymo open dataset. In: Proceedings of IEEE Conference on Computer Vision and Pattern Recognition (CVPR) (2020)
54. Thrun, S., et al.: Stanley: the robot that won the darpa grand challenge. J. Field Robot. **23**(9), 661–692 (2006)
55. Uijlings, J.R., Van De Sande, K.E., Gevers, T., Smeulders, A.W.: Selective search for object recognition. Int. J. Comput. Vis. **104**(2), 154–171 (2013)
56. Wu, Y., et al.: Google's neural machine translation system: bridging the gap between human and machine translation. arXiv preprint arXiv:1609.08144 (2016)
57. Xingjian, S., Chen, Z., Wang, H., Yeung, D.Y., Wong, W.K., Woo, W.C.: Convolutional LSTM network: a machine learning approach for precipitation nowcasting. In: Advances in Neural Information Processing Systems, pp. 802–810 (2015)
58. Yan, Y., Mao, Y., Li, B.: Second: sparsely embedded convolutional detection. Sensors **18**(10), 3337 (2018)
59. Yang, B., Liang, M., Urtasun, R.: Hdnet: exploiting HD maps for 3D object detection. In: Conference on Robot Learning, pp. 146–155 (2018)
60. Yang, B., Luo, W., Urtasun, R.: Pixor: real-time 3D object detection from point clouds. In: Proceedings of the IEEE Conference on Computer Vision and Pattern Recognition, pp. 7652–7660 (2018)
61. Yang, Z., Sun, Y., Liu, S., Shen, X., Jia, J.: Ipod: intensive point-based object detector for point cloud. arXiv preprint arXiv:1812.05276 (2018)
62. Zhang, J., Singh, S.: Visual-lidar odometry and mapping: low-drift, robust, and fast. In: 2015 IEEE International Conference on Robotics and Automation (ICRA), pp. 2174–2181. IEEE (2015)
63. Zhou, Y., et al.: End-to-end multi-view fusion for 3D object detection in lidar point clouds. In: Conference on Robot Learning (CoRL) (2019)
64. Zhou, Y., Tuzel, O.: Voxelnet: end-to-end learning for point cloud based 3D object detection. In: Proceedings of the IEEE Conference on Computer Vision and Pattern Recognition, pp. 4490–4499 (2018)

NAS-DIP: Learning Deep Image Prior with Neural Architecture Search

Yun-Chun Chen[✉], Chen Gao, Esther Robb, and Jia-Bin Huang

Virginia Tech, Blacksburg, USA
ycchen918@gmail.com

Abstract. Recent work has shown that the structure of deep convolutional neural networks can be used as a structured image prior for solving various inverse image restoration tasks. Instead of using hand-designed architectures, we propose to search for neural architectures that capture stronger image priors. Building upon a generic U-Net architecture, our core contribution lies in designing new search spaces for (1) an upsampling cell and (2) a pattern of cross-scale residual connections. We search for an improved network by leveraging an existing neural architecture search algorithm (using reinforcement learning with a recurrent neural network controller). We validate the effectiveness of our method via a wide variety of applications, including image restoration, dehazing, image-to-image translation, and matrix factorization. Extensive experimental results show that our algorithm performs favorably against state-of-the-art learning-free approaches and reaches competitive performance with existing learning-based methods in some cases.

1 Introduction

Convolutional neural networks (CNNs) have been successfully applied to various computer vision tasks. Apart from visual recognition tasks, CNNs have also demonstrated strong performance in restoration and synthesis problems. The reason behind these successful stories is often attributed to the ability of CNNs to *learn priors* from large-scale datasets (i.e., the priors are embedded in the *parameters/weights* of the trained network). In contrast to existing supervised learning paradigms that require learning the network parameters from labeled datasets, recent studies have discovered that the *structure* of the network by itself is sufficient to capture rich low-level image statistics [69,78]. Such structured image priors encoded in the network architecture are critical for image restoration (e.g., single image super-resolution [44–46], image denoising [12,49,82], joint filtering [50]), and image synthesis (e.g., image-to-image translation [37,38,47,48,91]) tasks.

Y.-C. Chen and C. Gao—Equal Contribution.

Electronic supplementary material The online version of this chapter (https://doi.org/10.1007/978-3-030-58523-5_26) contains supplementary material, which is available to authorized users.

© Springer Nature Switzerland AG 2020
A. Vedaldi et al. (Eds.): ECCV 2020, LNCS 12363, pp. 442–459, 2020.
https://doi.org/10.1007/978-3-030-58523-5_26

Super-Res   Denoising   Inpainting   Dehazing   Translation

Fig. 1. Applications. We propose to *learn* deep image prior using a neural architecture search. The resulting network can be applied to solve various inverse image problems *without* pre-training the model with a large-scale dataset with ground truth. Through extensive experimental evaluations, we show that our model compares favorably against existing hand-crafted CNN models for learning-free image restoration tasks. In some cases, our model even reaches competitive performance when compared with recent learning-based models.

While learning-free methods [78] have demonstrated competitive performance on image restoration tasks when compared with learning-based approaches [44, 46], only conventional network architectures such as ResNet [31] or U-Net [68] have been evaluated. There are two important aspects of network designs for these image restoration problems. First, while the design of an encoder has been extensively studied [31, 35, 43, 71], the design of a decoder [80, 81] (the upsampling cell in particular) receives considerably less attention. Second, as the spatial resolution of the features is progressively reduced along the path of the feature encoder, it is crucial for the network to recover feature maps with higher spatial resolution. U-Net [68] is one popular design to address this issue by concatenating the encoded features at the corresponding encoder layer with the features in the decoder when performing a sequence of up-convolutions. Such skip connection patterns, however, are manually designed and fixed for each task at hand.

Our Work. In this paper, we propose to *search* for both (1) the upsampling cells in the decoder and (2) the skip connection patterns between the encoder and the decoder (i.e., cross-level feature connections). To achieve this, we develop new search spaces for the two components. First, to search for the upsampling cell, we decompose a typical upsampling operation into two steps: i) ways of changing the spatial resolution (e.g., bilinear or bicubic [20]) and ii) ways of feature transformation (e.g., 2D convolution or 2D transposed convolution [86]). Second, to search for the cross-level connection patterns, we propose to search connection patterns shared across different feature levels in an encoder-decoder network.

Motivated by the Neural Architecture Search (NAS) algorithm [24, 26, 92] which has been shown effective in discovering networks with top performance in a large search space, we leverage reinforcement learning (RL) with a recurrent neural network (RNN) controller [24, 26, 92] and use the PSNR as the reward to guide the architecture search. By simultaneously searching in the two developed search spaces, our method is capable of discovering a CNN architecture that captures stronger structured image priors for the task of interest. We show the applicability of our method through four *learning-free* image restoration tasks, including single image super-resolution, image denoising, image inpainting, and image dehazing, and a *learning-based* unpaired image-to-image translation problem (see Fig. 1). Our experimental results demonstrate that searching for both the upsampling cell and the cross-level feature connections results in performance improvement over conventional neural architectures.

Our Contributions. First, we present a decomposition based on several commonly used upsampling operators that allows us to search for a novel upsampling cell for each task. Second, we develop a search space that consists of patterns of cross-level feature connections in an encoder-decoder architecture. Third, extensive evaluations on a variety of image restoration and synthesis tasks demonstrate that our proposed algorithm compares favorably against existing learning-based methods in some cases and achieves state-of-the-art performance when compared with existing learning-free approaches.

2   Related Work

Upsampling Cell. The design of the upsampling cell can be categorized into two groups: 1) non-learnable parameter-based methods and 2) learnable parameter-based approaches. *Non-learnable* parameter-based methods use interpolation to resize the feature maps from lower spatial resolutions to higher spatial resolutions. Example operators include bilinear/bicubic interpolation [20,62], nearest neighbor upsampling [9,39,62], and depth-to-space upsampling [46]. *Learnable* parameter-based approaches *learn* the mappings between feature maps of lower spatial resolutions and higher ones. Among the design choices, 2D transposed convolution is one of the popular choices for various dense prediction tasks, including semantic segmentation [56], optical flow estimation [21], depth prediction [15], image restoration [49,75], and synthesis [62] problems. Recent advances include bilinear additive upsampling [80,81] and CARAFE [79].

In contrast to these methods that manually design the upsampling operations, we develop a search space for the upsampling cell by decoupling several existing upsampling operations into methods of changing the spatial resolution and methods of feature transformation. We then adopt a Neural Architecture Search algorithm [24,26,92] (i.e., reinforcement learning with an RNN controller) to *automatically* discover the optimal upsampling cell for each individual task. We further demonstrate that the discovered upsampling cells can be transferred across tasks with favorable performance compared with the base network architectures.

NAS Applications and Search Space. NAS algorithms have been successfully applied to various tasks, including image classification [65,66,92,93], semantic segmentation [14,52,61], object detection [24], image restoration [17,74], and image generation [26]. In the context of object detection, NAS-FPN [24] develops a search space that allows the model to learn pyramidal representations by merging cross-scale features for improved detection of multiple objects with different scales and locations. In semantic segmentation, several methods focus on searching for the encoder architecture [52], the Atrous Spatial Pyramid Pooling module [14], or the decoder cell for a compact architecture [61]. In image restoration tasks, existing algorithms aim at discovering a better encoder structure for single image super-resolution [17] or asymmetric encoder-decoder architecture for image inpainting and image denoising [74]. In image generation problems, Auto-GAN [26] adopts a progressive scheme to search for a better generative adversarial network (GAN) architecture [27]. When searching for the decoder architecture, other methods focus on searching for more compact architectures [61] or optimizing cell structures with hand-designed upsampling cells and feature connections [26, 74].

Our focus differs from these methods in two aspects. First, we develop a search space for the *upsampling cell*, allowing us to discover an optimal upsampling choice for each task at hand. Second, we search for a pattern of *cross-level feature connections* and share it across different feature levels in an encoder-decoder architecture. Our cross-level feature connections are different from those in NAS-FPN [24] in that we aim at recovering feature maps with higher spatial resolution in the decoder, whereas NAS-FPN [24] aims at learning pyramidal feature representations for object detection. By simultaneously searching for both the upsampling cell and the cross-level feature connections, our searched architecture achieves the state-of-the-art performance when compared with existing learning-free approaches on a variety of image restoration tasks, and also reaches competitive results when compared with existing learning-based algorithms.

NAS Algorithms. NAS methods can be grouped into several categories depending on the search algorithm. The primary methods are evolution [5,53, 59,65,66,72,73,83], reinforcement learning [7,13,77,90,92,93], and differentiable search [2,54]. Evolutionary search leverages evolutionary algorithms to discover network structures by randomly mutating high-performing candidates from a population of architectures. RL-based approaches adopt either Q-learning [7,90] or policy gradient [13,77,92,93] strategies to train a recurrent network which outputs a sequence of symbols describing a network architecture [92] or a repeatable cell structure [93]. Differentiable search methods develop a continuous search space and optimize the network architecture using gradient descent, providing time efficiency at the expense of being memory intensive.

In this work, we follow the RL-based approaches and adopt an RL-based search algorithm with an RNN controller [26,92] to search for the upsampling cell and the cross-level connection patterns. We note that the search algorithm is *not* limited to RL-based approaches. Other alternatives, such as evolutionary-based methods or differentiable architecture search algorithms, can also be applied. The focus of our paper lies in the design of the two search spaces. We leave the development of the search algorithm as future work.

Test-Time Training. Training CNNs on *testing data* has been shown as an effective approach for improving performance [55,57,76]. Our work also trains a

CNN at the test time. Our method differs in that we do not rely on any *external* data at the training stage.

3 Proposed Method

In this section, we first provide an overview of the proposed method. We then describe the two developed search spaces for the upsampling cell and the cross-scale residual connections, respectively.

3.1 Method Overview

In contrast to existing learning-based methods that learn directly from large-scale datasets (Fig. 2a), recent studies [78] have shown that by randomly mapping noise to a degenerated (e.g., noisy, low-resolution, or occluded) image, the *untrained* CNN can solve the image restoration problems with competitive performance (Fig. 2b). To discover network structures that capture stronger image priors, we consider the task of searching for an upsampling cell and a pattern of cross-scale residual connections *without* learning from paired data. To achieve this, we present an algorithm consisting of two steps. As shown in Fig. 2c, we first apply reinforcement learning with an RNN controller [92] (using the PSNR as the reward) to search for the best-performing network structure f_θ^* on a held-out training set (blue block). After the network architecture search step, for each image in the test set, we randomly reinitialize the weights of the best-performing network structure f_θ^* and optimize the mapping from the random noise to a degenerated image (green block).

Searching for the Best-Performing Network Structure. Given an image $x \in \mathbb{R}^{H \times W \times 3}$ in the training set, we first generate a *degenerated* version x_0 by adding noise, downsampling, or dropping certain pixels from x depending on the task of interest. That is, for image denoising, $x_0 \in \mathbb{R}^{H \times W \times 3}$ denotes the noisy version of x, for single image super-resolution, $x_0 \in \mathbb{R}^{\frac{H}{r} \times \frac{W}{r} \times 3}$ denotes the low-resolution version of x where r represents the downsampling ratio, and for image inpainting, $x_0 \in \mathbb{R}^{H \times W \times 3}$ denotes the occluded version of x. We then sample a noise image $z \in \mathbb{R}^{H \times W \times C}$ and enforce the searched network f_θ to map the noise image z to the denoised, high-resolution, or inpainted version of x_0, i.e., map the noise image z to x.

To achieve this, we follow DIP [78] and optimize different objectives for different tasks. Please refer to the supplementary material for details.

As the ground-truth images in the training set are available, we can rank each of the searched network structure by computing the Peak Signal to Noise Ratio (PSNR) between the ground-truth image and the network's output (i.e., $f_\theta(z)$) and determine the best-performing network structure f_θ^* for the training set.

Determining the Optimal Stopping Point t^*. We note that the optimal stopping point t^* (i.e., the number of iterations required) depends on the network

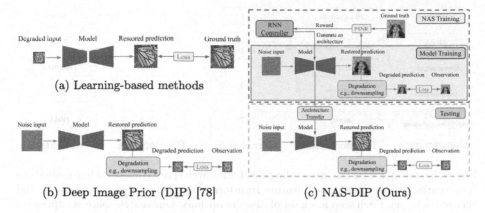

(a) Learning-based methods

(b) Deep Image Prior (DIP) [78]   (c) NAS-DIP (Ours)

Fig. 2. Overview of the main workflow and comparison. (a) Learning-based methods, e.g., super-resolution models such as LapSRN [44] or VDSR [41]. Given a dataset with labeled input/output pairs (e.g., low-resolution observation and the corresponding high-resolution image), this class of methods trains a deep CNN model to learn the mapping between the degraded input and its corresponding ground truth for the task of interest (e.g., super-resolution, denoising, etc.). (b) Learning-free methods. Given a noise input and an input observation, the DIP method [78] optimizes the model to produce a restored image that matches the given input observation after the specified image degradation model (e.g., downsampling). Here, the weights of the CNN model are *randomly initialized* (i.e., no need to train the model on a labeled dataset). (c) NAS-DIP (Ours). As DIP leverages CNN architectures as structured image priors, we explore ways to *learn* such priors. Specifically, we develop two search spaces (one for the upsampling cell and the other for the cross-level feature connections) and leverage existing neural architecture search techniques (an RNN-based controller trained with reinforcement learning) with PSNR as our reward to search for an improved network structure on a held-out training set (blue block). After the network architecture search, we then transfer the best-performing architecture and optimize the model the same way as DIP (green block). (Color figure online)

structure. Since we have a held-out training set, we are able to estimate the best stopping point for each randomly generated network structure. We then rank all the sampled network structures by measuring the differences (i.e., computing the PSNR) between the recovered image and its corresponding ground truth (i.e., the original image from the training set). After that, we apply the best-performing network structure to the test set and report the results recorded at the optimal stopping point t^*.

Testing with the Searched Network Structure f_θ^*. After searching on the training set, we apply the best-performing network structure f_θ^* with *random initialization* on each image in the test set at a time for optimization with t^* iterations using the same objective as DIP [78] for different tasks.

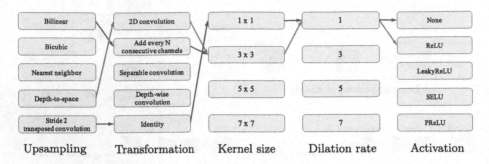

Fig. 3. Search space for the upsampling cell. Our search space consists of five main steps (i.e., spatial feature upsampling, feature transformation, kernel size, dilation rate, and activation layer). Each step has a set of discrete options. Our search space is expressive and covers many existing designs of upsampling operations. For example, the blue path indicates the bilinear additive upsampling operation [80,81]. The red path corresponds to the stride 2 transposed convolution used in [56,86]. The green path represents the sub-pixel convolution [46,70]. Searching in this space allows us to discover new and effective upsampling layers for capturing stronger image priors. (Color figure online)

3.2 Search Space for the Upsampling Layer

We develop a search space for the upsampling layer by decomposing existing upsampling operations based on two steps: 1) methods of changing the spatial resolution of the feature map and 2) methods of feature transformation. Our search space of operations for changing the spatial resolution includes: bilinear upsampling, bicubic upsampling [20], nearest-neighbor interpolation, depth-to-space [46,70], and stride 2 transposed convolution [21,56,86]. Our search space of operations for feature transformation includes: 2D convolution, add every N consecutive channels [80,81], separable convolution [28,80,81], depth-wise convolution [23,28,29], and identity. To relax the degree of freedom during the network architecture search, our search space allows the operations that contain learnable parameters to search for the kernel size, the dilation rate, and whether to include an activation at the end. Our search space of operations for the activation function includes none, ReLU [60], LeakyReLU [84], SELU [42], and PReLU [30]. Figure 3 presents our developed search space for the upsampling cell. By decomposing several commonly used upsampling operators, our search space is more flexible and allows us to discover a novel upsampling cell for each task. Newly developed spatial upsampling operators (e.g., CARAFE [79]) can also be incorporated in our search space easily in the future.

3.3 Cross-Scale Residual Connections

For the cross-scale residual connections, we develop a search space that covers cross-level feature connections. In contrast to U-Net [68] which *concatenates feature maps* at the same feature level, we adopt *residual connections* [31]. This allows us to merge feature maps extracted from different feature levels without

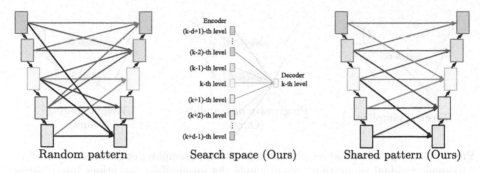

Fig. 4. Illustration of the cross-level feature connections. (*Left*) U-Net architecture with a random pattern of cross-level feature connections. Without any constraint, the search space is large. (*Middle*) Our proposed search space for the cross-level feature connections. To constrain the search space, we enforce the connection pattern to depend only on the level difference and share the pattern across different feature levels, e.g., each feature level in the decoder receives feature maps from two levels lower, the same level, and one level higher. With this constraint, the size of the search space is significantly reduced. (*Right*) U-Net architecture with the pattern of cross-level feature connections shown in the middle example shared across different feature levels.

the need to pre-define the number of input channels for each layer in the decoder because we design the number of input channels to always be the same.

Sharing Cross-Level Patterns. The search space for the cross-scale residual connections can be extremely large. Assuming that the network depth is d (i.e., there are d feature levels in the encoder and d feature levels in the decoder), the number of possible connection patterns is 2^{d^2}. To constrain the search space, we enforce the connection patterns to depend only on *the difference of feature levels* (e.g., connecting all the feature maps to the feature maps one level higher). For the k-th feature level in the decoder, it can receive feature maps from the $(k - d + 1)$-th feature level up to the $(k + d - 1)$-th feature level (i.e., there are in total 2^{2d-1} possible connection patterns). With this constraint, we then search for a pattern of feature connections and *share this pattern across different feature levels* in the decoder. Figure 4 illustrates the main idea of the proposed cross-level residual connections. The size of the search space can be significantly reduced from 2^{d^2} (without any constraints on the connection patterns) to 2^{2d-1}.

Progressive Upsampling. For the cross-scale upsampling operation, as illustrated in Fig. 5, we propose to *decouple* the 4× upsampling operation in the left into two consecutive 2× upsampling operations in the middle, and share the weights with the 2× upsampling operations in the right at the same feature level. This allows us to define 2× upsampling operations between each consecutive feature level only, and all other possible upsampling scales can be achieved by decoupling them into a series of 2× upsampling operations. The cross-scale downsampling connections can be achieved similarly.

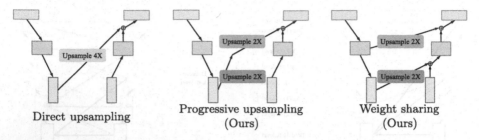

Fig. 5. Decomposition and weight sharing of the upsampling operations. To achieve cross-scale residual connections, we decouple the upsampling operations into a series of 2× upsampling operations, e.g., a 4× upsampling operation (*left*) can be realized by two consecutive 2× upsampling operations (*middle*). We adopt *weight sharing* for the upsampling operation at the same feature level, i.e., the weights of the 2× upsampling operations in the middle example are shared with those in the right example. After the cross-level feature upsampling, we add all the input feature maps at the same feature level. The resulting feature map then becomes the input of the decoder at the next layer. We note that the same level feature connections always exist in our model. We do not visualize them in this figure due to presentation clarity.

4   Experimental Results

In this section, we first describe the implementation details. We then present the quantitative and visual comparisons to existing methods as well as the ablation study. More visual results, implementation details, and the searched architectures are provided in the supplementary material.

4.1   Implementation Details

Here, we provide the implementation details regarding the neural architecture search, model training, and testing.

Neural Architecture Search. We implement our model using PyTorch. The network architecture of our RNN controller is the same as [26]. To create the training data for searching for the network architecture, we randomly sample 100 images from the DIV2K [1] training set. To search for the optimal network structure for the task of interest, the neural architecture search process is composed of two alternating phases. For the first phase, the RNN controller first samples a candidate network architecture with random initialization. We then optimize the sampled candidate model on the held-out training set (i.e., model training in our NAS-DIP framework). For the second phase, we first compute the PSNR between the restored prediction and the corresponding ground truth as the reward and use reinforcement learning to update the RNN controller (i.e., NAS training in our NAS-DIP framework). The training time required for each task varies. Specifically, finding the optimal network structure for the super-resolution task takes about 3 days, denoising about 3 days, and inpainting about 5 days (using an NVIDIA V100 GPU with 12 GB memory). We refer the reader

Table 1. Comparison with existing methods on image restoration tasks. For (a), (b), and (c), we report the average PSNR results. For (d), we follow the evaluation protocol in [6] and report the mean squared error (MSE). (a) Results of single image super-resolution on the Set5 [10] and Set14 [87] datasets with 2×, 4×, and 8× scaling factors. (b) Results of image inpainting on the dataset provided by [33] (*left*) and image denoising on the BM3D dataset [18] (*right*). (c) Comparison with Deep Decoder [32] on the dataset provided by [32] on single image super-resolution and image inpainting tasks. (d) Comparison with Latent Convolutional Models [6] on the Bedrooms dataset of LSUN [85]. For super-resolution in (c) and (d), we report the results with 4× scaling factor. Marker * indicates that the method uses the ground truth to get the best PSNR results. The **bold** and underlined numbers indicate the top two results, respectively.

(a)

Type	Method	Set5 [10]			Set14 [87]		
		2×	4×	8×	2×	4×	8×
Learning based	LapSRN [44]	37.52	31.54	26.14	33.08	28.19	24.44
	VDSR [41]	37.53	31.35	25.93	33.05	28.02	24.26
	EDSR [51]	38.11	32.46	26.96	33.92	28.80	24.91
	RDN [89]	38.24	32.47	-	34.01	28.81	-
	RCAN [88]	38.27	32.63	<u>27.31</u>	<u>34.12</u>	28.87	<u>25.23</u>
	SAN [19]	<u>38.31</u>	<u>32.64</u>	27.22	34.07	<u>28.92</u>	25.14
	EBRN [64]	**38.35**	**32.79**	**27.45**	**34.24**	**29.01**	**25.44**
Learning free	Bicubic [20]	33.66	28.44	24.37	30.24	26.05	23.09
	Glasner et al. [25]	-	28.84	-	-	26.46	-
	TV prior [8]	-	28.85	24.87	-	26.42	23.48
	RED [67]	-	30.23	25.56	-	27.36	23.89
	DeepRED [58]	-	30.72	26.04	-	27.63	24.28
	SelfExSR [36]	**36.60**	30.34	25.49	**32.24**	27.41	23.92
	DIP* [78]	33.19	29.89	25.88	29.80	27.00	24.15
	Ours	35.32	<u>30.81</u>	<u>26.41</u>	31.58	<u>27.84</u>	<u>24.59</u>
	Ours*	<u>35.90</u>	**31.09**	**27.03**	<u>31.89</u>	**28.37**	**25.17**

(b)

Method	Inpainting	Denoising
Papyan et al. [63]	31.19	-
DIP [78]	33.48	30.43
SGLD [16]	<u>34.51</u>	<u>30.81</u>
Ours	**34.72**	**31.42**

(c)

Method	SR 4×	Inpainting
Deep Decoder [32]	25.8	31.9
DIP [78]	<u>26.9</u>	<u>32.3</u>
Ours	**27.4**	**33.1**

(d)

Method	SR 4× ↓	Inpainting ↓
PGAN [40]	0.0183	0.0097
GLO [11]	0.0069	0.0085
LCM [6]	0.0071	0.0065
DIP [78]	<u>0.0057</u>	<u>0.0063</u>
Ours	**0.0054**	**0.0060**

to the supplementary material for the details of the neural architecture search process.

Testing Details. When applying the searched model for testing, there is a hyperparameter (number of iterations) that one can set to obtain the final prediction results. An early method [78] uses the ground truth of the test image and the PSNR to select the number of iterations with the best performance. However, this scheme may not be practical as the ground-truth image is not often available. To address this issue, we use the same training set for NAS training to find an optimal number of iterations that allow the model to reach the best performance. Specifically, we select model prediction at 4,500 iterations for super-resolution, 3,500 iterations for denoising, and 9,000 iterations for inpainting.

4.2 Quantitative Comparison

We validate the effectiveness of our searched model architecture by evaluating its performance when used as a deep image prior to solve various inverse problems

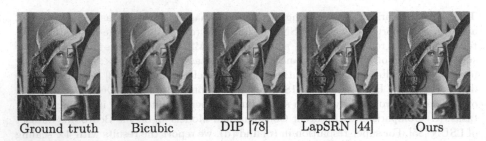

Ground truth   Bicubic   DIP [78]   LapSRN [44]   Ours

Fig. 6. Qualitative results of single image super-resolution. We present visual comparisons with learning-free methods (i.e., bicubic and DIP [78]) and a learning-based approach (i.e., LapSRN [44]) with 8× scaling factor.

in image restoration. In each task, we compare with state-of-the-art learning-free methods on benchmark datasets.

Single Image Super-Resolution. Following DIP [78], we adopt two standard benchmarks: the Set5 [10] and Set14 [87] datasets. We compare our approach with existing learning-free methods [8, 20, 25, 36, 58, 67, 78] and learning-based approaches [19, 41, 44, 51, 64, 88, 89] on three different upsampling scales (i.e., 2×, 4×, and 8× upsampling). We use the evaluation code provided by DIP [78]. Table 1a presents the experimental results.

Results on all three upsampling scales show that our algorithm performs favorably against state-of-the-art *learning-free* methods. Our results show that for a larger upsampling scale (e.g., 8× upsampling), our method achieves competitive or even better performance when compared with existing learning-based methods [41, 44] on both datasets. This is interesting because our model has *never seen any paired low-/high-resolution image pair*. The results also highlight the importance of our searched architecture, resulting in significant performance boost over existing methods that use hand-crafted priors (e.g., DIP [78]).

Image Denoising and Inpainting. We adopt the BM3D dataset [18] to evaluate the image denoising task. For a fair comparison, we follow DIP [78] and average the output of our network using an exponential sliding window (moving average) to obtain the final result. For evaluating the performance of the inpainting task, we follow DIP [78] and use the dataset provided by Heide et al. [33]. Here, we compare our results with DIP [78], Papyan et al. [63], and SGLD [16] using the 50% missing pixels setting. Table 1b reports the experimental results. Similarly, our method compares favorably against all competing approaches on both tasks.

Comparisons to Recent CNN Designs. There have been several recent methods that design the CNN architecture for improved performance on image restoration tasks. We first follow the same experimental setting as DeepDecoder [32] and evaluate our method on the 4× super-resolution and inpainting tasks. Table 1c reports the experimental results. Next, we follow the evaluation

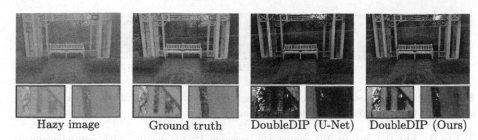

| Hazy image | Ground truth | DoubleDIP (U-Net) | DoubleDIP (Ours) |

Fig. 7. Qualitative results of image dehazing. We present the visual comparisons with DoubleDIP [22] on the O-HAZE dataset [4].

protocol in Latent Convolutional Models [6] and report our results on the 4× super-resolution and inpainting tasks in Table 1d.

From the extensive quantitative evaluations, we show that our model with both the searched upsampling cells and the cross-scale residual connections can serve as a stronger structured image prior to existing manual CNN architecture designs.

4.3 Visual Comparison

Here, we show sample qualitative results of several image restoration tasks and compare them with the state-of-the-art approaches. We refer the reader to review the full resolution results to better perceive visual quality improvement.

Figure 6 presents the visual results of single image super-resolution. Generally, using our model as an image prior results in clearly visible improvement in terms of the visual quality. This improvement highlights the strength of *learning* a stronger structured image prior through neural architecture search.

In addition to standard image restoration tasks, we also experiment with *model transferability* to two different tasks. We use the dehazing application in DoubleDIP [22] and the matrix factorization task in CompMirror [3] for demonstration.

For dehazing, we follow the official implementation by DoubleDIP [22] for generating the dehazed results. To generate our results, we swap the standard U-Net model in DoubleDIP [22] with our searched model searched in the *denoising* task. Figure 7 shows an example of visual comparison with DoubleDIP [22] on the O-HAZE dataset [4].[1] Our results show that using our model produces dehazed images with better visual quality.

For matrix factorization, we use the official implementation by CompMirror [3] to generate the factorized results. To generate our results, we replace the upsampling layer in the CompMirror [3] model with our searched upsampling

---

[1] We originally plan to conduct a quantitative evaluation using the O-HAZE dataset [4]. Unfortunately, using the provided source code and email correspondences with the authors, we were still unable to reproduce the results of DoubleDIP on this dataset. We thus did not report quantitative results on dehazing in this work.

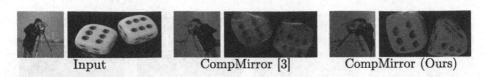

| Input | CompMirror [3] | CompMirror (Ours) |

Fig. 8. Qualitative results of the matrix factorization. We present the visual comparisons with CompMirror [3] on the matrix factorization task.

Table 2. Ablation study. We report the average PSNR results with comparisons to our variant methods. For single image super-resolution, we report the results on the Set 14 dataset [87] with 4× and 8× scaling factors. For image inpainting, we report the results using the dataset provided by [33]. For image denoising, we use the BM3D dataset [18] for evaluation. We denote "S-U" for searching upsampling and "S-C" for searching connection. The numbers in the parenthesis denote the performance gain over DIP [78]. The **bold** and underlined numbers indicate the top two results, respectively.

Method	Search upsampling	Search connection	SR 4×	SR 8×	Inpainting	Denoising
DIP [78]	-	-	27.00	24.15	33.48	30.43
Ours w/o S-C	✓	-	27.54 (+0.54)	24.44 (+0.29)	34.01 (+0.53)	31.08 (+0.65)
Ours w/o S-U	-	✓	27.32 (+0.32)	24.29 (+0.14)	34.59 (+1.11)	31.16 (+0.73)
Ours	✓	✓	**27.84** (+0.84)	**24.59** (+0.44)	**34.72** (+1.24)	**31.42** (+0.99)

layer searched in the *denoising* task. Figure 8 presents an example of visual comparison with CompMirror [3]. Our results show that using our model produces smoother factorized images with fewer visual artifacts.

4.4 Ablation Study

We conduct an ablation study to isolate the contributions from individual components. Specifically, we aim to understand how much performance improvement can be attributed to each of our two technical contributions. As our method builds upon the U-Net architecture of DIP [78], we use their results as the baseline.

We report the results of our variant methods in Table 2. Our results demonstrate that searching for an upsampling cell and a pattern of cross-level residual connections consistently helps improve the performance over DIP [78] across multiple tasks. We also observe that the upsampling cell is particularly important for the single image super-resolution task. On the other hand, introducing the searched cross-scale residual connections offers a larger performance boost over upsampling cell for both inpainting and denoising tasks. The model with both components shows the best performance, highlighting the complementary nature of these two components.

4.5 Image-to-image Translation

We also explore transferring the searched model (from the denoising task) to a different problem. Specifically, we aim to test if our searched model generalizes

Fig. 9. Qualitative results of the unpaired image-to-image translation task. We present the visual comparisons with CycleGAN [91] on the Winter → Summer (*left*) and the Summer → Winter (*right*) translation tasks.

Table 3. Quantitative results of unpaired image-to-image translation on the Summer ↔ Winter dataset. (*Left*) The FID scores. (*Right*) The user study results.

Method	FID ↓		Method	User study	
	Summer → Winter	Winter → Summer		Summer → Winter	Winter → Summer
CycleGAN (U-Net of [91])	78.62	73.91	CycleGAN (U-Net of [91])	20.7%	27.82%
CycleGAN (U-Net of [78])	79.74	74.83	CycleGAN (Ours)	79.3%	72.18%
CycleGAN (Ours)	76.22	71.98			

well to image-to-image translation tasks. We take the official PyTorch implementation of CycleGAN [91] and train the Summer ↔ Winter translation. We compare our results with the standard U-Net based model of CycleGAN [91]. Figure 9 shows one sample result for each of the translation directions. To quantify the performance, we also compute the FID score [34] and perform a user study. We report the results in Table 3. Both objective (FID score) and subjective (user study) results indicate that our searched model improves the performance over the base CycleGAN [91] model.

5   Conclusions

In this paper, we propose to use neural architecture search techniques to discover stronger structured image priors captured by the CNN architecture. The core technical contributions of our work lie in (1) the search space design for the upsampling layer commonly used in a decoder and (2) the cross-level feature connections between the encoder and the decoder. We build our network design upon the standard U-Net architecture and search for an optimal upsampling cell and a pattern of cross-level feature connections for each task of interest. We validate the effectiveness of our model on four image restoration tasks, one matrix factorization application, and an unpaired image-to-image translation problem. Through extensive experimental evaluations, our results show consistent performance improvement over conventional network designs.

References

1. Agustsson, E., Timofte, R.: Ntire 2017 challenge on single image super-resolution: dataset and study. In: CVPRW (2017)

2. Ahmed, K., Torresani, L.: MaskConnect: connectivity learning by gradient descent. In: Ferrari, V., Hebert, M., Sminchisescu, C., Weiss, Y. (eds.) ECCV 2018. LNCS, vol. 11209, pp. 362–378. Springer, Cham (2018). https://doi.org/10.1007/978-3-030-01228-1_22

3. Aittala, M., et al.: Computational mirrors: blind inverse light transport by deep matrix factorization. In: NeurIPS (2019)

4. Ancuti, C.O., Ancuti, C., Timofte, R., De Vleeschouwer, C.: O-HAZE: a dehazing benchmark with real hazy and haze-free outdoor images. In: CVPRW (2018)

5. Angeline, P.J., Saunders, G.M., Pollack, J.B.: An evolutionary algorithm that constructs recurrent neural networks. Trans. Neural Netw. 5, 54–65 (1994)

6. Athar, S., Burnaev, E., Lempitsky, V.: Latent convolutional models. In: ICLR (2019)

7. Baker, B., Gupta, O., Naik, N., Raskar, R.: Designing neural network architectures using reinforcement learning. In: ICLR (2017)

8. Beck, A., Teboulle, M.: Fast gradient-based algorithms for constrained total variation image denoising and deblurring problems. TIP 18, 2419–2434 (2009)

9. Berthelot, D., Schumm, T., Metz, L.: Began: boundary equilibrium generative adversarial networks. arXiv (2017)

10. Bevilacqua, M., Roumy, A., Guillemot, C., Alberi-Morel, M.L.: Low-complexity single-image super-resolution based on nonnegative neighbor embedding. In: BMVC (2012)

11. Bojanowski, P., Joulin, A., Lopez-Paz, D., Szlam, A.: Optimizing the latent space of generative networks. In: ICML (2018)

12. Burger, H.C., Schuler, C.J., Harmeling, S.: Image denoising: can plain neural networks compete with BM3D? In: CVPR (2012)

13. Cai, H., Chen, T., Zhang, W., Yu, Y., Wang, J.: Efficient architecture search by network transformation. In: AAAI (2018)

14. Chen, L.C., et al.: Searching for efficient multi-scale architectures for dense image prediction. In: NeurIPS (2018)

15. Chen, Y.C., Lin, Y.Y., Yang, M.H., Huang, J.B.: CrDoCo: pixel-level domain transfer with cross-domain consistency. In: CVPR (2019)

16. Cheng, Z., Gadelha, M., Maji, S., Sheldon, D.: A Bayesian perspective on the deep image prior. In: CVPR (2019)

17. Chu, X., Zhang, B., Ma, H., Xu, R., Li, J., Li, Q.: Fast, accurate and lightweight super-resolution with neural architecture search. arXiv (2019)

18. Dabov, K., Foi, A., Egiazarian, K.: Video denoising by sparse 3D transform-domain collaborative filtering. In: European Signal Processing Conference (2007)

19. Dai, T., Cai, J., Zhang, Y., Xia, S.T., Zhang, L.: Second-order attention network for single image super-resolution. In: CVPR (2019)

20. Dong, C., Loy, C.C., He, K., Tang, X.: Image super-resolution using deep convolutional networks. TPAMI 38, 295–307 (2015)

21. Dosovitskiy, A., et al.: FlowNet: learning optical flow with convolutional networks. In: ICCV (2015)

22. Gandelsman, Y., Shocher, A., Irani, M.: "Double-DIP": unsupervised image decomposition via coupled deep-image-priors. In: CVPR (2019)

23. Gao, H., Wang, Z., Ji, S.: ChannelNets: compact and efficient convolutional neural networks via channel-wise convolutions. In: NeurIPS (2018)

24. Ghiasi, G., Lin, T.Y., Le, Q.V.: NAS-FPN: learning scalable feature pyramid architecture for object detection. In: CVPR (2019)

25. Glasner, D., Bagon, S., Irani, M.: Super-resolution from a single image. In: ICCV (2009)

26. Gong, X., Chang, S., Jiang, Y., Wang, Z.: AutoGAN: neural architecture search for generative adversarial networks. In: ICCV (2019)
27. Goodfellow, I., et al.: Generative adversarial nets. In: NeurIPS (2014)
28. Guo, J., Li, Y., Lin, W., Chen, Y., Li, J.: Network decoupling: from regular to depthwise separable convolutions. In: BMVC (2018)
29. Guo, Y., Li, Y., Wang, L., Rosing, T.: Depthwise convolution is all you need for learning multiple visual domains. In: AAAI (2019)
30. He, K., Zhang, X., Ren, S., Sun, J.: Delving deep into rectifiers: surpassing human-level performance on ImageNet classification. In: ICCV (2015)
31. He, K., Zhang, X., Ren, S., Sun, J.: Deep residual learning for image recognition. In: CVPR (2016)
32. Heckel, R., Hand, P.: Deep decoder: Concise image representations from untrained non-convolutional networks. In: ICLR (2019)
33. Heide, F., Heidrich, W., Wetzstein, G.: Fast and flexible convolutional sparse coding. In: CVPR (2015)
34. Heusel, M., Ramsauer, H., Unterthiner, T., Nessler, B., Hochreiter, S.: GANs trained by a two time-scale update rule converge to a local Nash equilibrium. In: NeurIPS (2017)
35. Huang, G., Liu, Z., Van Der Maaten, L., Weinberger, K.Q.: Densely connected convolutional networks. In: CVPR (2017)
36. Huang, J.B., Singh, A., Ahuja, N.: Single image super-resolution from transformed self-exemplars. In: CVPR (2015)
37. Huang, X., Liu, M.-Y., Belongie, S., Kautz, J.: Multimodal unsupervised image-to-image translation. In: Ferrari, V., Hebert, M., Sminchisescu, C., Weiss, Y. (eds.) ECCV 2018. LNCS, vol. 11207, pp. 179–196. Springer, Cham (2018). https://doi.org/10.1007/978-3-030-01219-9_11
38. Isola, P., Zhu, J.Y., Zhou, T., Efros, A.A.: Image-to-image translation with conditional adversarial networks. In: CVPR (2017)
39. Jia, X., Chang, H., Tuytelaars, T.: Super-resolution with deep adaptive image resampling. arXiv (2017)
40. Karras, T., Aila, T., Laine, S., Lehtinen, J.: Progressive growing of gans for improved quality, stability, and variation. In: ICLR (2018)
41. Kim, J., Kwon Lee, J., Mu Lee, K.: Accurate image super-resolution using very deep convolutional networks. In: CVPR (2016)
42. Klambauer, G., Unterthiner, T., Mayr, A., Hochreiter, S.: Self-normalizing neural networks. In: NeurIPS (2017)
43. Krizhevsky, A., Sutskever, I., Hinton, G.E.: ImageNet classification with deep convolutional neural networks. In: NeurIPS (2012)
44. Lai, W.S., Huang, J.B., Ahuja, N., Yang, M.H.: Deep Laplacian pyramid networks for fast and accurate super-resolution. In: CVPR (2017)
45. Lai, W.S., Huang, J.B., Ahuja, N., Yang, M.H.: Fast and accurate image super-resolution with deep Laplacian pyramid networks. TPAMI 41, 2599–2613 (2018)
46. Ledig, C., et al.: Photo-realistic single image super-resolution using a generative adversarial network. In: CVPR (2017)
47. Lee, H.-Y., Tseng, H.-Y., Huang, J.-B., Singh, M., Yang, M.-H.: Diverse image-to-image translation via disentangled representations. In: Ferrari, V., Hebert, M., Sminchisescu, C., Weiss, Y. (eds.) ECCV 2018. LNCS, vol. 11205, pp. 36–52. Springer, Cham (2018). https://doi.org/10.1007/978-3-030-01246-5_3
48. Lee, H.-Y., et al.: DRIT++: diverse image-to-image translation via disentangled representations. Int. J. Comput. Vis. 128(10), 2402–2417 (2020). https://doi.org/10.1007/s11263-019-01284-z

49. Lefkimmiatis, S.: Non-local color image denoising with convolutional neural networks. In: CVPR (2017)
50. Li, Y., Huang, J.-B., Ahuja, N., Yang, M.-H.: Deep joint image filtering. In: Leibe, B., Matas, J., Sebe, N., Welling, M. (eds.) ECCV 2016. LNCS, vol. 9908, pp. 154–169. Springer, Cham (2016). https://doi.org/10.1007/978-3-319-46493-0_10
51. Lim, B., Son, S., Kim, H., Nah, S., Mu Lee, K.: Enhanced deep residual networks for single image super-resolution. In: CVPR (2017)
52. Liu, C., et al.: Auto-DeepLab: hierarchical neural architecture search for semantic image segmentation. In: CVPR (2019)
53. Liu, H., Simonyan, K., Vinyals, O., Fernando, C., Kavukcuoglu, K.: Hierarchial representations for efficient architecture search. In: ICLR (2018)
54. Liu, H., Simonyan, K., Yang, Y.: Darts: differentiable architecture search. In: ICLR (2019)
55. Liu, Y.L., Lai, W.S., Yang, M.H., Chuang, Y.Y., Huang, J.B.: Learning to see through obstructions. In: CVPR (2020)
56. Long, J., Shelhamer, E., Darrell, T.: Fully convolutional networks for semantic segmentation. In: CVPR (2015)
57. Luo, X., Huang, J., Szeliski, R., Matzen, K., Kopf, J.: Consistent video depth estimation. ACM Trans. Graph. (Proc. ACM SIGGRAPH) (2020)
58. Mataev, G., Milanfar, P., Elad, M.: DeepRED: deep image prior powered by red. In: ICCVW (2019)
59. Miikkulainen, R., et al.: Evolving deep neural networks. In: Artificial Intelligence in the Age of Neural Networks and Brain Computing (2019)
60. Nair, V., Hinton, G.E.: Rectified linear units improve restricted Boltzmann machines. In: ICML (2010)
61. Nekrasov, V., Chen, H., Shen, C., Reid, I.: Fast neural architecture search of compact semantic segmentation models via auxiliary cells. In: CVPR (2019)
62. Odena, A., Dumoulin, V., Olah, C.: Deconvolution and checkerboard artifacts. Distill 1, e3 (2016)
63. Papyan, V., Romano, Y., Sulam, J., Elad, M.: Convolutional dictionary learning via local processing. In: ICCV (2017)
64. Qiu, Y., Wang, R., Tao, D., Cheng, J.: Embedded block residual network: a recursive restoration model for single-image super-resolution. In: ICCV (2019)
65. Real, E., Aggarwal, A., Huang, Y., Le, Q.V.: Regularized evolution for image classifier architecture search. In: ICML (2018)
66. Real, E., et al.: Large-scale evolution of image classifiers. In: ICML (2017)
67. Romano, Y., Elad, M., Milanfar, P.: The little engine that could: regularization by denoising (RED). SIAM J. Imaging Sci. 10, 1804–1844 (2017)
68. Ronneberger, O., Fischer, P., Brox, T.: U-Net: convolutional networks for biomedical image segmentation. In: Navab, N., Hornegger, J., Wells, W.M., Frangi, A.F. (eds.) MICCAI 2015. LNCS, vol. 9351, pp. 234–241. Springer, Cham (2015). https://doi.org/10.1007/978-3-319-24574-4_28
69. Saxe, A.M., Koh, P.W., Chen, Z., Bhand, M., Suresh, B., Ng, A.Y.: On random weights and unsupervised feature learning. In: ICML (2011)
70. Shi, W., et al.: Real-time single image and video super-resolution using an efficient sub-pixel convolutional neural network. In: CVPR (2016)
71. Simonyan, K., Zisserman, A.: Very deep convolutional networks for large-scale image recognition. In: ICLR (2015)
72. Stanley, K.O., D'Ambrosio, D.B., Gauci, J.: A hypercube-based encoding for evolving large-scale neural networks. Artif. Life 15, 185–212 (2009)

73. Stanley, K.O., Miikkulainen, R.: Evolving neural networks through augmenting topologies. Evol. Comput. **10**, 99–127 (2002)
74. Suganuma, M., Ozay, M., Okatani, T.: Exploiting the potential of standard convolutional autoencoders for image restoration by evolutionary search. In: ICML (2018)
75. Sun, Q., Ma, L., Joon Oh, S., Van Gool, L., Schiele, B., Fritz, M.: Natural and effective obfuscation by head inpainting. In: CVPR (2018)
76. Sun, Y., Wang, X., Liu, Z., Miller, J., Efros, A.A., Hardt, M.: Test-time training for out-of-distribution generalization. In: ICML (2020)
77. Tan, M., Chen, B., Pang, R., Vasudevan, V., Le, Q.V.: MnasNet: platform-aware neural architecture search for mobile. In: CVPR (2019)
78. Ulyanov, D., Vedaldi, A., Lempitsky, V.: Deep image prior. In: CVPR (2018)
79. Wang, J., Chen, K., Xu, R., Liu, Z., Loy, C.C., Lin, D.: CARAFE: content-aware reassembly of features. In: ICCV (2019)
80. Wojna, Z., et al.: The devil is in the decoder: classification, regression and GANs. Int. J. Comput. Vis. **127**(11), 1694–1706 (2019). https://doi.org/10.1007/s11263-019-01170-8
81. Wojna, Z., et al.: The devil is in the decoder. In: BMVC (2017)
82. Xie, J., Xu, L., Chen, E.: Image denoising and inpainting with deep neural networks. In: NeurIPS (2012)
83. Xie, L., Yuille, A.: Genetic CNN. In: ICCV (2017)
84. Xu, B., Wang, N., Chen, T., Li, M.: Empirical evaluation of rectified activations in convolutional network. In: ICMLW (2015)
85. Yu, F., Seff, A., Zhang, Y., Song, S., Funkhouser, T., Xiao, J.: LSUN: construction of a large-scale image dataset using deep learning with humans in the loop. arXiv (2015)
86. Zeiler, M.D., Taylor, G.W., Fergus, R.: Adaptive deconvolutional networks for mid and high level feature learning. In: ICCV (2011)
87. Zeyde, R., Elad, M., Protter, M.: On single image scale-up using sparse-representations. In: Boissonnat, J.-D., et al. (eds.) Curves and Surfaces 2010. LNCS, vol. 6920, pp. 711–730. Springer, Heidelberg (2012). https://doi.org/10.1007/978-3-642-27413-8_47
88. Zhang, Y., Li, K., Li, K., Wang, L., Zhong, B., Fu, Y.: Image super-resolution using very deep residual channel attention networks. In: Ferrari, V., Hebert, M., Sminchisescu, C., Weiss, Y. (eds.) ECCV 2018. LNCS, vol. 11211, pp. 294–310. Springer, Cham (2018). https://doi.org/10.1007/978-3-030-01234-2_18
89. Zhang, Y., Tian, Y., Kong, Y., Zhong, B., Fu, Y.: Residual dense network for image super-resolution. In: CVPR (2018)
90. Zhong, Z., Yan, J., Wu, W., Shao, J., Liu, C.L.: Practical block-wise neural network architecture generation. In: CVPR (2018)
91. Zhu, J.Y., Park, T., Isola, P., Efros, A.A.: Unpaired image-to-image translation using cycle-consistent adversarial networks. In: ICCV (2017)
92. Zoph, B., Le, Q.V.: Neural architecture search with reinforcement learning. In: ICLR (2017)
93. Zoph, B., Vasudevan, V., Shlens, J., Le, Q.V.: Learning transferable architectures for scalable image recognition. In: CVPR (2018)

Learning to Learn in a Semi-supervised Fashion

Yun-Chun Chen[1]($\boxtimes$), Chao-Te Chou[1], and Yu-Chiang Frank Wang[1,2]

[1] Graduate Institute of Communication Engineering, National Taiwan University,
Taipei, Taiwan
{b03901148,b03901096,ycwang}@ntu.edu.tw
[2] ASUS Intelligent Cloud Services, Taipei, Taiwan

Abstract. To address semi-supervised learning from both labeled and unlabeled data, we present a novel meta-learning scheme. We particularly consider that labeled and unlabeled data share disjoint ground truth label sets, which can be seen tasks like in person re-identification or image retrieval. Our learning scheme exploits the idea of leveraging information from labeled to unlabeled data. Instead of fitting the associated class-wise similarity scores as most meta-learning algorithms do, we propose to derive semantics-oriented similarity representations from labeled data, and transfer such representation to unlabeled ones. Thus, our strategy can be viewed as a self-supervised learning scheme, which can be applied to fully supervised learning tasks for improved performance. Our experiments on various tasks and settings confirm the effectiveness of our proposed approach and its superiority over the state-of-the-art methods.

1 Introduction

Recent advances of deep learning models like convolutional neural networks (CNNs) have shown encouraging performance in various computer vision applications, including image retrieval [70,75,89] and person re-identification (re-ID) [9,21,41,42,96]. Different from recognizing the input as a particular category, the above tasks aim at learning feature embeddings, making instances of the same type (e.g., object category) close to each other while separating those of distinct classes away. Similar tasks such as image-based item verification [46], face verification [72], face recognition [17,62,68], and vehicle re-ID [16,73,83] can all be viewed as the tasks of this category.

Existing methods for image matching generally require the collection of a large number of labeled data, and tailor algorithms to address the associated tasks (e.g., image retrieval [70,75] and person re-ID [21,41,96,97]). However, the assumption of having a sufficient amount of labeled data available during training may not be practical. To relax the dependency of manual supervision, several

Electronic supplementary material The online version of this chapter (https://doi.org/10.1007/978-3-030-58523-5_27) contains supplementary material, which is available to authorized users.

© Springer Nature Switzerland AG 2020
A. Vedaldi et al. (Eds.): ECCV 2020, LNCS 12363, pp. 460–478, 2020.
https://doi.org/10.1007/978-3-030-58523-5_27

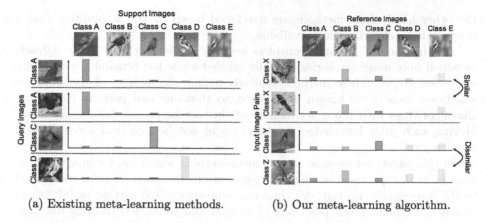

(a) Existing meta-learning methods.   (b) Our meta-learning algorithm.

Fig. 1. Illustration of learning class-wise similarity. (a) Standard meta-learning methods for visual classification compute class-wise similarity *scores* between the query image and those in the support set, where the two sets share *overlapping* ground truth labels. (b) To deal with training data with non-overlapping labels, our meta-learning scheme derives semantics-oriented similarity representations in a learning-to-learn fashion, allowing the determination of pairwise relationship between images with unseen labels.

semi-supervised methods for image retrieval [78,82,92] and person re-ID [39,85] are proposed. These methods focus on learning models from datasets where each category is partially labeled (i.e., some data in *each* category are labeled, while the rest in that category remain unlabeled). Thus, they choose to use the models learned from the labeled data to assign pseudo labels to the unlabeled ones [82, 85,92], or adopt ensemble learning techniques to enforce the predictions of the unlabeled data to be consistent across multiple networks [78]. Despite significant progress having been reported, these methods *cannot* be directly applied to scenarios where novel objects or persons are present.

To deal with instances of unseen categories for image matching purposes, one can approach such problem in two different ways. The majority of existing methods focuses on the cross-dataset (domain adaptation [10,31]) setting, where one dataset is fully labeled (i.e., source domain dataset) while the other one remains unlabeled. (i.e., target domain dataset) [18,81,91]. Existing methods for this category typically assume that there is a domain gap between the two datasets. These methods either leverage adversarial learning strategies to align feature distributions between the two datasets [18,81], or aim at assigning pseudo labels for each unlabeled image in the target dataset through predicting class-wise similarity scores from models trained on the source (labeled) dataset [91]. By carefully selecting hyperparameters such as the prediction score threshold, one can determine whether or not a given image pair from the target dataset is of the same category. However, the class-wise similarity scores are computed based on a network trained on the source dataset, which might not generalize well to the target dataset, especially when their labels are non-overlapping. On the other hand, these methods are developed based on the assumption that a large-scale labeled dataset is available.

Another line of research considers learning models from a *single* dataset, in which only some categories are fully labeled while the remaining classes are unlabeled [87,88]. These methods typically require the number of classes of the unlabeled data to be known in advance, so that one can perform clustering-like algorithms with the exact number of clusters for pseudo label assignment. Having such prior knowledge, however, might not be practical for real-world applications.

In this paper, we propose a novel meta-learning algorithm for image matching in a semi-supervised setting, with applications to image retrieval and person re-ID. Specifically, we consider the same semi-supervised setting as [87,88], in which the ground truth label sets of labeled and unlabeled training data are *disjoint*. Our meta-learning strategy aims at exploiting and leveraging class-wise similarity representation across labeled and unlabeled training data, while such similarity representation is derived by a learning-to-learn fashion. The resulting representations allow our model to relate images with pseudo labels in the unlabeled set (e.g., Fig. 1b). This is very different from existing meta-learning for visual classification methods (like few-shot learning), which typically assume that the support and query sets share the same label set and focus on fitting the associated class-wise similarity scores (e.g., Fig. 1a). Our learning scheme is realized by learning to match randomly selected labeled data pairs, and such concepts can be applied to observe both labeled and unlabeled data for completing the semi-supervised learning process.

The contributions of this paper are summarized as follows:

- We propose a meta-learning algorithm for image matching in semi-supervised settings, where labeled and unlabeled data share non-overlapping categories.
- Our learning scheme aims at deriving semantics-oriented similarity representation across labeled and unlabeled sets. Since pseudo labels can be automatically assigned to the unlabeled training data, our approach can be viewed as a self-supervised learning strategy.
- With the derivation of semantics-oriented similarity representations, our learning scheme can be applied to fully supervised settings and further improves the performance.
- Evaluations on four datasets in different settings confirm that our method performs favorably against existing image retrieval and person re-ID approaches.

2   Related Work

Semi-supervised Learning. Semi-supervised learning for visual analysis has been extensively studied in the literature. Most of the existing methods focus on image classification and can be categorized into two groups depending on the learning strategy: 1) labeling-based methods and 2) consistency-based approaches. Labeling-based methods focus on assigning labels to the unlabeled images through pseudo labeling [38], label propagation [49], or leveraging regularization techniques for performing the above label assignment [25]. Consistency-based approaches, on the other hand, exploit the idea of cycle consistency [8,11,101] and adopt ensemble learning algorithms to enforce the predictions of the unlabeled samples to be consistent across multiple models [3,5,37,51,56,59,74]. In addition to image classification, another line of research focuses on utilizing annotation-free images to improve the performance of semantic segmentation [33,67]. These methods adopt generative adversarial networks (GANs) [24] to generate images conditioned on the class labels to enhance the learning of feature representations for the unlabeled images [67], or develop a fully convolutional discriminator to generate dense probability maps that indicate the confidence of correct segmentation for each pixel in the unlabeled images [33].

To match images of the same category, a number of methods for image retrieval [78,82,92] and person re-ID [19,22,32,39,44,45,85] also consider learning models in semi-supervised settings. Methods for semi-supervised image retrieval can be grouped into two categories depending on the adopted descriptors: 1) hand-crafted descriptor based methods and 2) approaches based on trainable descriptors. The former typically focuses on optimizing the errors on the labeled set and leverages a regularizer to maximize the information entropy between labeled and unlabeled sets [82]. Trainable descriptor based approaches either utilize a graph to model the relationship between labeled and unlabeled sets [92] or leverage a GAN [24] to learn triplet-wise information from both labeled and unlabeled data [78]. Similarly, methods for semi-supervised person re-ID also aim at relating labeled and unlabeled images through dictionary learning [44], multi-feature learning [22], pseudo labeling with regularizers [32], or considering complex relationships between labeled and unlabeled images [19]. While promising performance has been shown, these methods cannot be directly applied to scenarios where datasets contain labeled and unlabeled images with non-overlapping category labels, which are practical in many real-world applications.

To tackle this issue, two recent methods for semi-supervised person re-ID are proposed [87,88]. These methods either combine K-means clustering and multi-view clustering [87], or develop a self-paced multi-view clustering algorithm [88] to assign pseudo labels to images in the unlabeled set. However, these methods require the number of identities of the unlabeled set to be known in advance. Our work does not need such prior knowledge. As noted above, we approach such problems and assign pseudo labels to the unlabeled data by learning their semantics-oriented similarity representations, which are realized in a unique learning-to-learn fashion.

Meta-learning. The primary objective of meta-learning is to enable a base learning algorithm which observes data with particular properties to adapt to similar tasks with new concepts of interest. Few-shot learning [65,69] and neural architecture search [7] are among the popular applications of meta-learning.

Existing meta-learning algorithms can be grouped into three categories based on the learning task: 1) initialization-based methods, 2) memory-based approaches, and 3) metric-based algorithms. Initialization-based methods focus on learning an optimizer [2,12,30,57] or learning to initialize the network parameters so that the models can rapidly adapt to novel classes or new tasks [23]. Memory-based approaches leverage memory-augmented models (e.g., the hidden activations in a recurrent network or external memory) to retain the learned knowledge [35,53,60], and associate the learned knowledge with the newly encountered tasks for rapid generalization. Metric-based algorithms aim at learning a feature embedding with proper distance metrics for few-shot [65,69] or one-shot [76] classification.

Similar to metric-based meta-learning algorithms, our method also aims at learning a feature embedding. Our method differs from existing meta-learning for visual classification approaches in that we learn a feature embedding from both labeled and unlabeled data with disjoint label sets. Moreover, both the support and query sets share the same label set in most other meta-learning approaches, while the label sets of our meta-training and meta-validation sets are disjoint.

3   Proposed Method

3.1   Algorithmic Overview

We first describe the setting of our semi-supervised learning task, and define the notations. When matching image pairs in the tasks of image retrieval and person re-ID, we assume that our training set contains a set of N_L labeled images $X_L = \{x_i^L\}_{i=1}^{N_L}$ with the corresponding labels $Y_L = \{y_i^L\}_{i=1}^{N_L}$, and a set of N_U unlabeled images $X_U = \{x_j^U\}_{j=1}^{N_U}$. For the labeled data, each $x_i^L \in \mathbb{R}^{H \times W \times 3}$ and $y_i^L \in \mathbb{R}$ denote the ith image and the associated label, respectively. As for $x_j^U \in \mathbb{R}^{H \times W \times 3}$, it is the jth unlabeled image in X_U. Note that the class number of the labeled set is denoted as C_L, while that of the unlabeled set is *unknown*. We assume the label sets of the labeled and unlabeled sets are *disjoint*.

The goal of this work is to learn a feature embedding model by jointly observing the above labeled and unlabeled sets, with the learned features can be applied for matching images for tasks of retrieval and re-ID. As shown in Fig. 2, our proposed algorithm comprises two learning phases: 1) meta-learning with labeled training data and 2) meta semi-supervised learning on both labeled and unlabeled training sets. For the first phase (i.e., Fig. 2a), we first partition the labeled set X_L into a meta-training set $M_T = \{x_k^{M_T}\}_{k=1}^{N_{M_T}}$ and a meta-validation set $M_V = \{x_l^{M_V}\}_{l=1}^{N_{M_V}}$, with disjoint labels for M_T and M_V (from Y_L). The numbers of images for M_T and M_V are denoted as N_{M_T} and N_{M_V}, and the numbers of classes for M_T and M_V are denoted as C_{M_T} and C_{M_V}, respectively, summing up as C_L (i.e., $C_L = C_{M_T} + C_{M_V}$). Our model F takes images x from M_T and M_V as inputs, and learns feature representations $f = F(x) \in \mathbb{R}^d$ (d is the dimension of f) for input images. Our model then derives semantics-oriented similarity

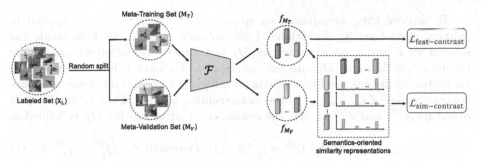

(a) Meta-learning with the labeled training set X_L.

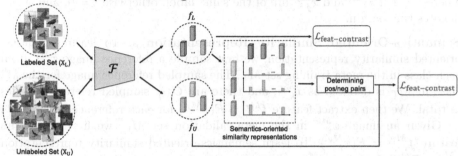

(b) Meta semi-supervised learning on both labeled set X_L and unlabeled set X_U.

Fig. 2. Overview of the proposed meta-learning framework. (a) Our model first takes labeled data and learns semantics-oriented similarity representation in a learning-to-learn fashion (i.e., joint learning of $\mathcal{L}_{\text{feat}-\text{contrast}}$ and $\mathcal{L}_{\text{sim}-\text{contrast}}$). (b) The learned concept of matching semantics information allows us to learn from both labeled and unlabeled data, determining positive/negative pairs for computing $\mathcal{L}_{\text{feat}-\text{contrast}}$ for all the training data pairs. Note that the meta-training and meta-validation sets in (a) do *not* share the same labels, *neither* do the labeled and unlabeled training sets in (b).

representation $s \in \mathbb{R}^{C_{M_T}}$ (C_{M_T} denotes the dimension of s) for each image in M_V. In the second meta-learning stage (i.e., Fig. 2b), we utilize the learned concept of semantics-oriented similarity representation to guide the learning of the unlabeled set. The details of each learning phase are elaborated in the following subsections.

As for testing, our model takes a query image as input and extracts its feature $f \in \mathbb{R}^d$, which is applied to match gallery images via nearest neighbor search.

3.2 Meta-learning on X_L

Motivated by [91], we exploit the idea of leveraging information from class-wise similarity to guide the learning of the unlabeled data. In our work, we choose to implicitly learn semantics-oriented similarity representation instead of explicit label-specific representation in a learning-to-learn fashion, so that the learned representation can be applied for describing the unlabeled images.

To achieve this, we advance an episodic learning paradigm as applied in existing meta-learning algorithms [4,23]. In each episode, we first divide the labeled set L into a meta-training set M_T and a meta-validation set M_V, where the labels of M_T and M_V are *not* overlapped. To learn a feature embedding for matching images, we follow existing methods [14,26] and introduce a feature contrastive loss $\mathcal{L}_{\text{feat-contrast}}$ in the meta-training set M_T. That is, given a pair of images $x_i^{M_T}$ and $x_j^{M_T}$ in M_T, the feature contrastive loss for M_T is defined as

$$\mathcal{L}_{\text{feat-contrast}}(M_T; F) = t \cdot \| f_i^{M_T} - f_j^{M_T} \| + (1-t) \cdot \max(0, \phi - \| f_i^{M_T} - f_j^{M_T} \|), \quad (1)$$

where $t = 1$ if $x_i^{M_T}$ and $x_j^{M_T}$ are of the same label, otherwise $t = 0$, and $\phi > 0$ denotes the margin.

Semantics-Oriented Similarity Representation s. To learn semantics-oriented similarity representation, we first sample a reference image $\hat{x}^{M_T}$ from each class in the meta-training set M_T. The sampled reference image for the kth class in M_T is denoted as $\hat{x}_k^{M_T}$, and there are C_{M_T} sampled reference images in total. We then extract feature $\hat{f}^{M_T} = F(\hat{x}^{M_T})$ for each reference image $\hat{x}^{M_T}$.

Given an image $x_i^{M_V}$ in the meta-validation set M_V, we first extract its feature $f_i^{M_V} = F(x_i^{M_V})$. To learn semantics-oriented similarity representation $s_i^{M_V} \in \mathbb{R}^{C_{M_T}}$ for image $x_i^{M_V}$, we compute the class-wise similarity scores between $f_i^{M_V}$ and all reference features $\hat{f}^{M_T}$ sampled from the meta-training set. The kth entry of the semantics-oriented similarity representation $s_i^{M_V}$ is defined as

$$s_i^{M_V}(k) = \text{sim}(f_i^{M_V}, \hat{f}_k^{M_T}), \quad (2)$$

where $\text{sim}(f_i^{M_V}, \hat{f}_k^{M_T})$ denotes the similarity between feature $f_i^{M_V}$ and the sampled reference feature $\hat{f}_k^{M_T}$. We note that we do not limit the similarity measurement in the above equation. For example, we compute the cosine similarity for image retrieval and calculate the ℓ_2 distance for person re-ID.

To achieve the learning of semantics-oriented similarity representation, we utilize the ground truth label information from the meta-validation set, and develop a similarity contrastive loss $\mathcal{L}_{\text{sim-contrast}}$. Specifically, given an image pair $x_i^{M_V}$ and $x_j^{M_V}$ in M_V, the associated similarity contrastive loss is defined as

$$\mathcal{L}_{\text{sim-contrast}}(M_V; F) = t \cdot \| s_i^{M_V} - s_j^{M_V} \| + (1-t) \cdot \max(0, \phi - \| s_i^{M_V} - s_j^{M_V} \|), \quad (3)$$

where $t = 1$ if $x_i^{M_V}$ and $x_j^{M_V}$ are of the same category, otherwise $t = 0$. $\phi > 0$ denotes the margin.

By repeating the above procedure across multiple episodes until the convergence of the meta-validation loss (i.e., the similarity contrastive loss), our model carries out the learning of semantics-oriented similarity representation in a learning-to-learn fashion, without fitting particular class label information. Utilizing such representation allows our model to realize joint learning of labeled and unlabeled data, as discussed next.

3.3 Meta Semi-supervised Learning on X_L and X_U

In the semi-supervised setting, learning from labeled data X_L simply follows the standard feature contrastive loss $\mathcal{L}_{\text{feat}-\text{contrast}}(X_L; F)$. To jointly exploit labeled and unlabeled data, we advance the aforementioned meta-training strategy and start from randomly sampling C_{M_T} categories from the labeled set X_L. For each sampled class, we then randomly sample one reference image $\hat{x}_k^L$ and extract its feature $\hat{f}_k^L = F(\hat{x}_k^L)$, where $\hat{x}_k^L$ denotes the sampled reference image of the kth sampled class. Namely, there are C_{M_T} sampled reference images in total.

Next, given an image x_i^U in the unlabeled set U, we extract its feature $f_i^U = F(x_i^U)$, followed by computing the semantics-oriented similarity representation $s_i^U \in \mathbb{R}^{C_{M_T}}$ between f_i^U and all features $\hat{f}^L$ of the above sampled reference images from the labeled set. It is worth repeating that our learning scheme is very different from existing methods [91], which focus on fitting class-wise similarity scores on the entire labeled set. Instead, we only compute the similarity scores between features of sampled classes. This is the reason why we view our representation to be semantics-oriented instead of class-specific (as [91] does).

Now, we are able to measure the similarity between semantics-oriented similarity representations s_i^U and s_j^U, with a threshold ψ to determine whether the corresponding input images x_i^U and x_j^U are of the same class ($t = 1$) or not ($t = 0$). Namely,

$$\begin{cases} t = 1, & \text{if } \|s_i^U - s_j^U\| < \psi, \\ t = 0, & \text{otherwise.} \end{cases} \tag{4}$$

The above process can be viewed as assigning pseudo positive/negative labels for the unlabeled data X_U, allowing us to compute the feature contrastive loss $\mathcal{L}_{\text{feat}-\text{contrast}}(X_U; F)$ on any image pair from the unlabeled set.

4 Experiments

We present quantitative and qualitative results in this section. In all of our experiments, we implement our model using PyTorch and train our model on a single NVIDIA TITAN RTX GPU with 24 GB memory. The performance of our method can be possibly further improved by applying pre/post-processing methods, attention mechanisms, or re-ranking techniques. However, such techniques are not used in all of our experiments.

4.1 Datasets and Evaluation Metrics

We conduct experiments on four public benchmarks, including the CUB-200 [77], Car196 [36], Market-1501 [96], and DukeMTMC-reID [58] datasets.

Datasets. For image retrieval, we adopt the CUB-200 [77] and Car196 [36] datasets. The CUB-200 dataset [77] is a fine-grained bird dataset containing $11,788$ images of 200 bird species. Following existing methods [27,54,55], we use the first 100 categories with $5,864$ images for training and the remaining

100 categories with 5,924 images for testing. The Car196 [36] dataset is a fine-grained car dataset consisting of 16,189 images with 196 car categories. Following [27,54,55], we use the first 98 categories with 8,054 images for training while the remaining 98 categories with 8,131 images are used for testing.

As for person re-ID, we consider the Market-1501 [96] dataset, which contains 32,668 labeled images of 1,501 identities captured by 6 camera views. This dataset is partitioned into a training set of 12,936 images from 751 identities, and a test set of 19,732 images from the other 750 identities. We also have the DukeMTMC-reID [58] dataset which is composed of 36,411 labeled images of 1,404 identities collected from 8 camera views. We utilize the benchmarking training/test split, where the training set consists of 16,522 images of 702 identities, and the test set contains 19,889 images of the other 702 identities.

Evaluation Metrics. Following recent image retrieval methods [27,55], we use the Recall@K (R@K) metric and the normalized mutual information (NMI) [15] with cosine similarity for evaluating image retrieval performance. For person re-ID, we adopt the standard single-shot person re-ID setting [43] and use the average cumulative match characteristic (CMC) and the mean Average Precision (mAP) with Euclidean distance as similarity measurements.

4.2 Evaluation of Semi-Supervised Learning Tasks

4.2.1 Image Retrieval

Implementation Details and Settings. Following [52,55], we adopt an ImageNet-pretrained Inception-v1 [71] to serve as the backbone of our model F. A fully connected layer with ℓ_2 normalization is added after the pool5 layer to serve as the feature embedding layer. All images are resized to $256 \times 256 \times 3$ in advance. During the first stage of meta-learning, we set the batch sizes of the meta-training and meta-validation sets to 32 and 64, respectively. We use the Adam optimizer to train our model for 600 epochs. The initial learning rate is set to 2×10^{-5} and the momentum is set to 0.9. The learning rate is decreased by a factor of 10 for every 150 epochs. The margin ϕ is set to 0.3. As for the meta semi-supervised learning stage, we set the batch size of the labeled set to 32, the batch size of the unlabeled set to 64, and the initial learning rate to 1×10^{-5}. Similarly, the learning rate is decayed by a factor of 10 for every 150 epochs. We train our model for another 600 epochs. The similarity threshold ψ is set to 0.01. We evaluate our method with three different label ratios, i.e., 25%, 50%, and 75% of the categories are labeled, while the remaining categories are unlabeled.

Results. We compare our method with existing fully supervised and unsupervised methods. Table 1 reports the results recorded at Recall@1, 2, 4, and 8, and NMI on the CUB-200 [77] and Car196 [36] datasets. We note that while the results of our method (semi-supervised setting) are not directly comparable to those of fully supervised and unsupervised approaches, their results can be viewed as upper (for fully supervised methods) and lower (for unsupervised approaches) bounds of our results.

Table 1. Results of semi-supervised image retrieval. The bold and underlined numbers indicate top two results, respectively.

Method	Supervision	CUB-200 [77]					Car196 [36]				
		R@1	R@2	R@4	R@8	NMI	R@1	R@2	R@4	R@8	NMI
Triplet [84]	Supervised	35.9	47.7	59.1	70.0	49.8	45.1	57.4	69.7	79.2	52.9
Lifted [55]		46.9	59.8	71.2	81.5	56.4	59.9	70.4	79.6	87.0	_57.8_
Clustering [54]		48.2	61.4	71.8	81.9	59.2	58.1	70.6	80.3	87.8	–
Smart+ [27]		_49.8_	_62.3_	_74.1_	_83.3_	_59.9_	_64.7_	_76.2_	_84.2_	_90.2_	–
Angular [80]		**53.6**	**65.0**	**75.3**	**83.7**	**61.0**	**71.3**	**80.7**	**87.0**	**91.8**	**62.4**
Exemplar [20]	Unsupervised	38.2	50.3	62.8	75.0	45.0	36.5	48.1	59.2	71.0	35.4
NCE [86]		39.2	51.4	63.7	75.8	45.1	_37.5_	_48.7_	59.8	71.5	35.6
DeepCluster [6]		42.9	54.1	65.6	76.2	53.0	32.6	43.8	57.0	69.5	_38.5_
MOM [34]		_45.3_	_57.8_	_68.6_	_78.4_	_55.0_	35.5	48.2	_60.6_	_72.4_	**38.6**
Instance [90]		**46.2**	**59.0**	**70.1**	**80.2**	**55.4**	**41.3**	**52.3**	**63.6**	**74.9**	35.8
Ours	Semi-supervised (25%)	48.4	60.3	71.7	81.0	55.9	54.5	66.8	77.2	85.1	48.6
	Semi-supervised (50%)	_50.5_	_61.1_	_72.3_	_82.9_	_57.6_	_62.2_	_73.8_	_83.0_	_89.4_	_55.0_
	Semi-supervised (75%)	**51.0**	**62.3**	**73.4**	**83.0**	**59.3**	**65.9**	**76.6**	**84.4**	**90.1**	**57.7**

The results on both datasets show that our method performs favorably against all competing unsupervised approaches and achieves competitive or even better performance when compared with fully supervised methods.

4.2.2 Person re-ID

Implementation Details and Settings. Following [87], our model F employs an ImageNet-pretrained ResNet-50 [28]. All images are resized to $256 \times 128 \times 3$ in advance. During the first stage of meta-learning, we set the batch sizes of the meta-training and meta-validation sets to 32 and 128, respectively. We use the Adam optimizer to train our model for 600 epochs. The initial learning rate is set to 2×10^{-3}, and is decayed by a factor of 10 for every 150 epochs. The momentum and the margin ϕ are set to 0.9 and 0.3, respectively. As for the meta semi-supervised learning stage, we set the batch size of the labeled set to 32, the batch size of the unlabeled set to 128, and the initial learning rate to 2×10^{-5}. Similarly, the learning rate is decreased by a factor of 10 for every 150 epochs. We train our model for another 600 epochs. The similarity threshold ψ is set to 0.5. Also following [87], we evaluate our method with three different label ratios, i.e., $\frac{1}{3}$, $\frac{1}{6}$, and $\frac{1}{12}$ of the person IDs are fully labeled, while the remaining person IDs are unlabeled.

Results. We compare our method with unsupervised approaches [43,96], semi-supervised methods [87,88], a fully supervised approach [93], and a cross-dataset person re-ID method [91]. Similarly, the results of fully supervised/unsupervised methods can be regarded as the upper/lower bounds of our results. For the

Table 2. Results of semi-supervised person re-ID. The bold numbers indicate the best results.

Method	Supervision	Backbone	Market-1501 [96]		DukeMTMC-reID [58]	
			Rank 1	mAP	Rank 1	mAP
LOMO [43]	Unsupervised	–	27.2	8.0	12.3	4.8
BOW [96]		–	35.8	14.8	17.1	8.3
AlignedReID [93]	Supervised	ResNet-50	89.2	72.8	79.3	65.6
MVC [87]	Semi-supervised ($\frac{1}{12}$)	ResNet-50	46.6	–	34.8	–
Ours		ResNet-50	**56.7**	**32.4**	**44.9**	**24.4**
MVC [87]	Semi-supervised ($\frac{1}{6}$)	ResNet-50	60.0	–	43.8	–
Ours		ResNet-50	**70.8**	**46.4**	**56.6**	**33.6**
MVC [87]	Semi-supervised ($\frac{1}{3}$)	ResNet-50	72.2	48.7	52.9	33.6
SPMVC [88]		ResNet-50	71.5	53.2	58.5	37.4
MAR [91]		ResNet-50	69.9	46.4	–	–
Ours		ResNet-50	**80.3**	**58.7**	**67.5**	**46.3**

cross-dataset person re-ID method [91], we use their official implementation[1] with their default hyperparameter settings, and set the labeled set as their source domain and the unlabeled set as their target domain. Table 2 compares the rank 1 and mAP scores on the Market-1501 [96] and DukeMTMC-reID [58] datasets.

From this table, when comparing to semi-supervised learning methods, i.e., MVC [87] and SPMVC [88], our method consistently outperforms their results by large margins on all three evaluated label ratios of both datasets. When comparing to a fully-supervised method, e.g., AlignedReID [93], our method achieves 90% and 85% of their results recorded at rank 1 on the Market-1501 [96] and DukeMTMC-reID [58] datasets, respectively, using relatively fewer labeled information, i.e., only $\frac{1}{3}$ of the person IDs are labeled. From these results, we show that under the same experimental setting, our method achieves the state-of-the-art performance, while resulting in comparable results compared to fully supervised approaches.

4.3 Evaluation of Supervised Learning Tasks

4.3.1 Evaluation of Limited Labeled Data

In addition to evaluating the performance of our semi-supervised learning, we now apply our meta-learning strategy to the labeled training set only and see whether our learning-to-learn strategy would benefit such scenario.

Implementation Details. All images are resized to $256 \times 128 \times 3$ in advance. We set the batch sizes of the meta-training and meta-validation sets to 32 and 128, respectively. We use the Adam optimizer to train our model for 600 epochs. The initial learning rate is set to 2×10^{-3}, and is decayed by a factor of 10 for every 150 epochs. The momentum and margin ϕ are set to 0.9 and 0.3, respectively.

---

[1] https://github.com/KovenYu/MAR.

Table 3. Results of fully-supervised person re-ID with limited training data. The bold and underlined numbers indicate the top two results, respectively.

Method	Market-1501 [96]											
	$\frac{1}{3}$ of the IDs are available				$\frac{1}{6}$ of the IDs are available				$\frac{1}{12}$ of the IDs are available			
	Rank 1	Rank 5	Rank 10	mAP	Rank 1	Rank 5	Rank 10	mAP	Rank 1	Rank 5	Rank 10	mAP
DDML [47]	72.1	86.9	91.4	45.5	62.3	81.1	86.5	35.4	49.6	70.1	78.0	24.8
Triplet hard [62]	72.6	87.5	92.0	49.9	61.4	81.2	87.0	_38.2_	47.5	69.2	77.7	25.6
Triplet+HDML [98]	73.2	88.6	92.3	48.0	62.0	80.7	86.9	35.3	48.0	70.0	78.9	25.2
AlignedReID [93]	73.3	88.1	92.1	47.7	62.2	81.3	87.0	36.1	49.7	71.4	78.8	25.8
MGN [79]	74.1	88.2	92.1	50.8	62.3	81.4	86.7	38.1	_50.6_	71.6	79.8	27.4
BoT [48]	74.8	_89.7_	_93.5_	51.8	60.6	80.3	86.5	36.7	47.1	69.0	77.8	24.2
PyrNet [50]	_74.9_	_89.7_	92.7	_52.1_	_63.2_	_81.8_	_87.2_	37.8	50.2	_71.7_	_79.9_	_28.4_
Ours	**77.0**	**90.8**	**93.9**	**54.0**	**66.0**	**85.2**	**90.3**	**41.2**	**53.4**	**74.9**	**82.2**	**29.2**
	DukeMTMC-reID [58]											
PyrNet [50]	59.7	75.8	80.5	40.6	51.6	68.0	73.5	31.9	39.8	56.6	63.5	_21.2_
Triplet hard [62]	60.6	76.5	82.3	40.1	51.8	_69.1_	75.5	30.1	40.0	_58.2_	_65.4_	20.3
DDML [47]	60.6	75.0	79.1	36.7	51.5	67.3	73.2	29.6	40.1	57.4	64.3	20.1
MGN [79]	60.6	75.1	80.3	_41.2_	51.4	66.3	72.7	_31.9_	39.7	56.3	63.4	20.8
BoT [48]	61.9	_78.5_	_83.8_	40.5	_52.4_	68.9	_75.9_	31.8	_40.7_	57.6	64.3	21.1
AlignedReID [93]	_62.5_	77.1	82.2	40.3	51.3	68.2	75.6	29.6	40.5	58.1	_65.4_	20.5
Ours	**64.9**	**80.9**	**85.6**	**44.8**	**54.0**	**70.9**	**76.7**	**32.1**	**42.6**	**60.5**	**67.3**	**22.2**

Table 4. Results of fully-supervised person re-ID. The bold and underlined numbers indicate the top two results, respectively.

Method	Market-1501 [96]			DukeMTMC-reID [58]		
	Rank 1	Rank 5	mAP	Rank 1	Rank 5	mAP
Part-Aligned [95]	81.0	92.0	63.4	–	–	–
PAN [99]	82.8	93.5	63.4	7 1.6	83.9	51.5
MGCAM [66]	83.8	–	74.3	–	–	–
TriNet [29]	84.9	94.2	69.1	–	–	–
JLML [40]	85.1	–	65.5	–	–	–
PoseTransfer [100]	87.7	–	68.9	78.5	–	56.9
PSE [61]	87.7	94.5	69.0	79.8	89.7	62.0
CamStyle [100]	88.1	–	68.7	75.3	–	53.5
DPFL [13]	88.9	92.3	73.1	79.2	–	60.6
AlignedReID [93]	89.2	95.9	72.8	79.3	89.7	65.6
DML [94]	89.3	–	70.5	–	–	–
DKP [63]	90.1	96.7	75.3	80.3	89.5	63.2
DuATM [64]	91.4	97.1	76.6	81.8	90.2	68.6
RDR [1]	92.2	97.9	81.2	85.2	**93.9**	72.8
SPReID [66]	93.7	97.6	83.4	85.9	92.9	73.3
BoT [48]	_94.5_	_98.2_	_85.9_	_86.4_	_93.6_	_76.4_
AlignedReID [93] + Ours	91.1	96.3	78.1	81.7	91.0	67.7
BoT [48] + Ours	**94.8**	**98.3**	**86.1**	**86.6**	**93.9**	**76.8**

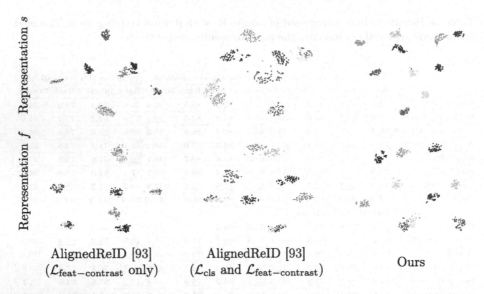

Representation s

Representation f

AlignedReID [93]
($\mathcal{L}_{\text{feat}-\text{contrast}}$ only)

AlignedReID [93]
($\mathcal{L}_{\text{cls}}$ and $\mathcal{L}_{\text{feat}-\text{contrast}}$)

Ours

Fig. 3. Visual comparisons of the learned representations on Market-1501. (*Top row*) visualizes the semantics-oriented similarity representation. (*Bottom row*) visualizes the feature representation f. Note that selected samples of 20 identities are illustrated, each in a specific color. Comparing to AlignedReID, our model only learns semantics similarity and achieves comparable/improved performances. (Color figure online)

Results. We adopt the Market-1501 [96] and DukeMTMC-reID [58] datasets for performance evaluations and compare our method with a number of supervised approaches [47,48,50,62,79,93]. Table 3 presents the experimental results. The results show that our method consistently performs favorably against all competing approaches, demonstrating sufficient re-ID ability can be exhibited by our proposed method even when only limited labeled data are observed.

Visualization of the Learned Representations. To demonstrate that our model benefits from learning semantics-oriented similarity representation s, we select 20 person IDs and visualize both semantics-oriented similarity representation s and the learned feature representation f on the Market-1501 [96] test set via t-SNE in Fig. 3, in which we compare our approach with AlignedReID [93] and its variant method.

We observe that without learning the semantics-oriented similarity representation, AlignedReID [93] and its variant method cannot separate the representation s well. Our method, on the other hand, learns semantics-oriented similarity representations from the labeled set in a learning-to-learn fashion. The learned similarity representation s allows our model to guide the learning of the unlabeled set, resulting in a well-separated space for the feature representation f.

4.3.2   Extension to Fully-Supervised Learning Tasks

Finally, to show that our formulation is not limited to semi-supervised learning settings, we apply our learning algorithm to fully-supervised setting on the Market-1501 [96] and DukeMTMC-reID [58] datasets.

Results. We initialize our model from AlignedReID [93] and BoT [48], respectively, and apply our meta-learning strategy on the entire training set, i.e., there are two variant methods: (1) AlignedReID [93] + Ours and (2) BoT [48] + Ours. As shown in Table 4, our method further improves the performance of AlignedReID [93] and BoT [48] on both datasets, respectively, comparing favorably against existing fully-supervised learning methods.

4.4   Limitations and Potential Issues

We observe that our method is memory intensive as learning from the unlabeled set requires larger batch size to increase the likelihood of selecting positive image pairs (sampling a negative pair is easier than sampling a positive pair). On the other hand, our learning algorithm is suitable for solving tasks where the categories are visually similar.

5   Conclusions

We presented a meta-learning algorithm for semi-supervised learning with applications to image retrieval and person re-ID. We consider the training schemes in which labeled and unlabeled data share non-overlapping categories. Our core technical novelty lies in learning semantics-oriented similarity representation from the labeled set in a learning-to-learn fashion, which can be applied to semi-supervised settings without knowing the number of classes of the unlabeled data in advance. Our experiments confirmed that our method performs favorably against state-of-the-art image retrieval and person re-ID approaches in semi-supervised settings. We also verified that our algorithm can be applied to supervised settings for improved performance, which further exhibits the effectiveness and applicability of our learning algorithm.

Acknowledgments. This paper is supported in part by the Ministry of Science and Technology (MOST) of Taiwan under grant MOST 109-2634-F-002-037.

References

1. Almazan, J., Gajic, B., Murray, N., Larlus, D.: Re-id done right: towards good practices for person re-identification. arXiv (2018)
2. Andrychowicz, M., et al.: Learning to learn by gradient descent by gradient descent. In: NeurIPS (2016)
3. Athiwaratkun, B., Finzi, M., Izmailov, P., Wilson, A.G.: There are many consistent explanations of unlabeled data: why you should average. In: ICLR (2019)

4. Balaji, Y., Sankaranarayanan, S., Chellappa, R.: MetaReg: towards domain generalization using meta-regularization. In: NeurIPS (2018)
5. Berthelot, D., Carlini, N., Goodfellow, I., Papernot, N., Oliver, A., Raffel, C.: Mixmatch: a holistic approach to semi-supervised learning. In: NeurIPS (2019)
6. Caron, M., Bojanowski, P., Joulin, A., Douze, M.: Deep clustering for unsupervised learning of visual features. In: Ferrari, V., Hebert, M., Sminchisescu, C., Weiss, Y. (eds.) Computer Vision – ECCV 2018. LNCS, vol. 11218, pp. 139–156. Springer, Cham (2018). https://doi.org/10.1007/978-3-030-01264-9_9
7. Chen, Y.C., Gao, C., Robb, E., Huang, J.B.: NAS-DIP: learning deep image prior with neural architecture search. In: Vedaldi, A., et al. (eds.) ECCV 2020. LNCS, vol. 12363, pp. 459–476. Springer, Heidelberg (2020)
8. Chen, Y.-C., Huang, P.-H., Yu, L.-Y., Huang, J.-B., Yang, M.-H., Lin, Y.-Y.: Deep semantic matching with foreground detection and cycle-consistency. In: Jawahar, C.V., Li, H., Mori, G., Schindler, K. (eds.) ACCV 2018. LNCS, vol. 11363, pp. 347–362. Springer, Cham (2019). https://doi.org/10.1007/978-3-030-20893-6_22
9. Chen, Y.C., Li, Y.J., Du, X., Wang, Y.C.F.: Learning resolution-invariant deep representations for person re-identification. In: AAAI (2019)
10. Chen, Y.C., Lin, Y.Y., Yang, M.H., Huang, J.B.: CrDoCo: pixel-level domain transfer with cross-domain consistency. In: CVPR (2019)
11. Chen, Y.C., Lin, Y.Y., Yang, M.H., Huang, J.B.: Show, match and segment: joint weakly supervised learning of semantic matching and object co-segmentation. TPAMI (2020)
12. Chen, Y., et al.: Learning to learn without gradient descent by gradient descent. In: ICML (2017)
13. Cheng, D., Gong, Y., Zhou, S., Wang, J., Zheng, N.: Person re-identification by multi-channel parts-based CNN with improved triplet loss function. In: CVPR (2016)
14. Chopra, S., Hadsell, R., LeCun, Y.: Learning a similarity metric discriminatively, with application to face verification. In: CVPR (2005)
15. Christopher, D.M., Prabhakar, R., Hinrich, S.: Introduction to Information Retrieval. Cambridge University Press, Cambridge (2008)
16. Chu, R., Sun, Y., Li, Y., Liu, Z., Zhang, C., Wei, Y.: Vehicle re-identification with viewpoint-aware metric learning. In: ICCV (2019)
17. Deng, J., Guo, J., Xue, N., Zafeiriou, S.: ArcFace: additive angular margin loss for deep face recognition. In: CVPR (2019)
18. Deng, W., Zheng, L., Ye, Q., Kang, G., Yang, Y., Jiao, J.: Image-image domain adaptation with preserved self-similarity and domain-dissimilarity for person re-identification. In: CVPR (2018)
19. Ding, G., Zhang, S., Khan, S., Tang, Z., Zhang, J., Porikli, F.: Feature affinity based pseudo labeling for semi-supervised person re-identification. TMM **21**, 2891–2902 (2019)
20. Dosovitskiy, A., Fischer, P., Springenberg, J.T., Riedmiller, M., Brox, T.: Discriminative unsupervised feature learning with exemplar convolutional neural networks. TPAMI **38**, 1734–1747 (2015)
21. Farenzena, M., Bazzani, L., Perina, A., Murino, V., Cristani, M.: Person re-identification by symmetry-driven accumulation of local features. In: CVPR (2010)
22. Figueira, D., Bazzani, L., Minh, H.Q., Cristani, M., Bernardino, A., Murino, V.: Semi-supervised multi-feature learning for person re-identification. In: AVSS (2013)

23. Finn, C., Abbeel, P., Levine, S.: Model-agnostic meta-learning for fast adaptation of deep networks. In: ICML (2017)
24. Goodfellow, I., et al.: Generative adversarial nets. In: NeurIPS (2014)
25. Grandvalet, Y., Bengio, Y.: Semi-supervised learning by entropy minimization. In: NeurIPS (2005)
26. Hadsell, R., Chopra, S., LeCun, Y.: Dimensionality reduction by learning an invariant mapping. In: CVPR (2006)
27. Harwood, B., Kumar, B., Carneiro, G., Reid, I., Drummond, T., et al.: Smart mining for deep metric learning. In: ICCV (2017)
28. He, K., Zhang, X., Ren, S., Sun, J.: Deep residual learning for image recognition. In: CVPR (2016)
29. Hermans, A., Beyer, L., Leibe, B.: In defense of the triplet loss for person re-identification. arXiv (2017)
30. Hochreiter, S., Younger, A.S., Conwell, P.R.: Learning to learn using gradient descent. In: Dorffner, G., Bischof, H., Hornik, K. (eds.) ICANN 2001. LNCS, vol. 2130, pp. 87–94. Springer, Heidelberg (2001). https://doi.org/10.1007/3-540-44668-0_13
31. Hoffman, J., et al.: CyCADA : cycle-consistent adversarial domain adaptation. In: ICML (2018)
32. Huang, Y., Xu, J., Wu, Q., Zheng, Z., Zhang, Z., Zhang, J.: Multi-pseudo regularized label for generated data in person re-identification. TIP **28**, 1391–1403 (2018)
33. Hung, W.C., Tsai, Y.H., Liou, Y.T., Lin, Y.Y., Yang, M.H.: Adversarial learning for semi-supervised semantic segmentation. In: BMVC (2018)
34. Iscen, A., Tolias, G., Avrithis, Y., Chum, O.: Mining on manifolds: metric learning without labels. In: CVPR (2018)
35. Kaiser, Ł., Nachum, O., Roy, A., Bengio, S.: Learning to remember rare events. ICLR (2018)
36. Krause, J., Stark, M., Deng, J., Fei-Fei, L.: 3D object representations for fine-grained categorization. In: ICCVW (2013)
37. Laine, S., Aila, T.: Temporal ensembling for semi-supervised learning. In: ICLR (2017)
38. Lee, D.H.: Pseudo-label: the simple and efficient semi-supervised learning method for deep neural networks. In: ICMLW (2013)
39. Li, J., Ma, A.J., Yuen, P.C.: Semi-supervised region metric learning for person re-identification. Int. J. Comput. Vis. **126**(8), 855–874 (2018). https://doi.org/10.1007/s11263-018-1075-5
40. Li, W., Zhu, X., Gong, S.: Person re-identification by deep joint learning of multi-loss classification. In: IJCAI (2017)
41. Li, Y.J., Chen, Y.C., Lin, Y.Y., Du, X., Wang, Y.C.F.: Recover and identify: a generative dual model for cross-resolution person re-identification. In: ICCV (2019)
42. Li, Y.J., Chen, Y.C., Lin, Y.Y., Wang, Y.C.F.: Cross-resolution adversarial dual network for person re-identification and beyond. arXiv (2020)
43. Liao, S., Hu, Y., Zhu, X., Li, S.Z.: Person re-identification by local maximal occurrence representation and metric learning. In: CVPR (2015)
44. Liu, X., Song, M., Tao, D., Zhou, X., Chen, C., Bu, J.: Semi-supervised coupled dictionary learning for person re-identification. In: CVPR (2014)
45. Liu, Yu., Song, G., Shao, J., Jin, X., Wang, X.: Transductive centroid projection for semi-supervised large-scale recognition. In: Ferrari, V., Hebert, M., Sminchisescu, C., Weiss, Y. (eds.) ECCV 2018. LNCS, vol. 11209, pp. 72–89. Springer, Cham (2018). https://doi.org/10.1007/978-3-030-01228-1_5

46. Liu, Z., Luo, P., Qiu, S., Wang, X., Tang, X.: DeepFashion: powering robust clothes recognition and retrieval with rich annotations. In: CVPR (2016)
47. Lu, J., Hu, J., Tan, Y.P.: Discriminative deep metric learning for face and kinship verification. TIP **26**, 4269–4282 (2017)
48. Luo, H., Gu, Y., Liao, X., Lai, S., Jiang, W.: Bag of tricks and a strong baseline for deep person re-identification. In: CVPRW (2019)
49. Luo, Y., Zhu, J., Li, M., Ren, Y., Zhang, B.: Smooth neighbors on teacher graphs for semi-supervised learning. In: CVPR (2018)
50. Martinel, N., Luca Foresti, G., Micheloni, C.: Aggregating deep pyramidal representations for person re-identification. In: CVPRW (2019)
51. Miyato, T., Maeda, S., Koyama, M., Ishii, S.: Virtual adversarial training: a regularization method for supervised and semi-supervised learning. TPAMI **41**, 1979–1993 (2018)
52. Movshovitz-Attias, Y., Toshev, A., Leung, T.K., Ioffe, S., Singh, S.: No fuss distance metric learning using proxies. In: ICCV (2017)
53. Munkhdalai, T., Yu, H.: Meta networks. In: ICML (2017)
54. Oh Song, H., Jegelka, S., Rathod, V., Murphy, K.: Deep metric learning via facility location. In: CVPR (2017)
55. Oh Song, H., Xiang, Y., Jegelka, S., Savarese, S.: Deep metric learning via lifted structured feature embedding. In: CVPR (2016)
56. Qiao, S., Shen, W., Zhang, Z., Wang, B., Yuille, A.: Deep co-training for semi-supervised image recognition. In: Ferrari, V., Hebert, M., Sminchisescu, C., Weiss, Y. (eds.) ECCV 2018. LNCS, vol. 11219, pp. 142–159. Springer, Cham (2018). https://doi.org/10.1007/978-3-030-01267-0_9
57. Ravi, S., Larochelle, H.: Optimization as a model for few-shot learning. In: ICLR (2017)
58. Ristani, E., Solera, F., Zou, R., Cucchiara, R., Tomasi, C.: Performance measures and a data set for multi-target, multi-camera tracking. In: Hua, G., Jégou, H. (eds.) ECCV 2016. LNCS, vol. 9914, pp. 17–35. Springer, Cham (2016). https://doi.org/10.1007/978-3-319-48881-3_2
59. Sajjadi, M., Javanmardi, M., Tasdizen, T.: Regularization with stochastic transformations and perturbations for deep semi-supervised learning. In: NeurIPS (2016)
60. Santoro, A., Bartunov, S., Botvinick, M., Wierstra, D., Lillicrap, T.: Meta-learning with memory-augmented neural networks. In: ICML (2016)
61. Saquib Sarfraz, M., Schumann, A., Eberle, A., Stiefelhagen, R.: A pose-sensitive embedding for person re-identification with expanded cross neighborhood re-ranking. In: CVPR (2018)
62. Schroff, F., Kalenichenko, D., Philbin, J.: FaceNet: a unified embedding for face recognition and clustering. In: CVPR (2015)
63. Shen, Y., Xiao, T., Li, H., Yi, S., Wang, X.: End-to-end deep kronecker-product matching for person re-identification. In: CVPR (2018)
64. Si, J., et al.: Dual attention matching network for context-aware feature sequence based person re-identification. In: CVPR (2018)
65. Snell, J., Swersky, K., Zemel, R.: Prototypical networks for few-shot learning. In: NeurIPS (2017)
66. Song, C., Huang, Y., Ouyang, W., Wang, L.: Mask-guided contrastive attention model for person re-identification. In: CVPR (2018)
67. Souly, N., Spampinato, C., Shah, M.: Semi supervised semantic segmentation using generative adversarial network. In: ICCV (2017)

68. Sun, Y., Chen, Y., Wang, X., Tang, X.: Deep learning face representation by joint identification-verification. In: NeurIPS (2014)
69. Sung, F., Yang, Y., Zhang, L., Xiang, T., Torr, P.H., Hospedales, T.M.: Learning to compare: Relation network for few-shot learning. In: CVPR (2018)
70. Swets, D.L., Weng, J.J.: Using discriminant eigenfeatures for image retrieval. TPAMI **18**, 831–836 (1996)
71. Szegedy, C., et al.: Going deeper with convolutions. In: CVPR (2015)
72. Taigman, Y., Yang, M., Ranzato, M., Wolf, L.: DeepFace: closing the gap to human-level performance in face verification. In: CVPR (2014)
73. Tang, Z., et al.: PAMTRI: pose-aware multi-task learning for vehicle re-identification using highly randomized synthetic data. In: ICCV (2019)
74. Tarvainen, A., Valpola, H.: Mean teachers are better role models: weight-averaged consistency targets improve semi-supervised deep learning results. In: NeurIPS (2017)
75. Tieu, K., Viola, P.: Boosting image retrieval. IJCV **56**, 17–36 (2004). https://doi.org/10.1023/B:VISI.0000004830.93820.78
76. Vinyals, O., Blundell, C., Lillicrap, T., Wierstra, D., et al.: Matching networks for one shot learning. In: NeurIPS (2016)
77. Wah, C., Branson, S., Welinder, P., Perona, P., Belongie, S.: The Caltech-UCSD Birds-200-2011 dataset (2011)
78. Wang, G., Hu, Q., Cheng, J., Hou, Z.: Semi-supervised generative adversarial hashing for image retrieval. In: Ferrari, V., Hebert, M., Sminchisescu, C., Weiss, Y. (eds.) ECCV 2018. LNCS, vol. 11219, pp. 491–507. Springer, Cham (2018). https://doi.org/10.1007/978-3-030-01267-0_29
79. Wang, G., Yuan, Y., Chen, X., Li, J., Zhou, X.: Learning discriminative features with multiple granularities for person re-identification. In: ACM MM (2018)
80. Wang, J., Zhou, F., Wen, S., Liu, X., Lin, Y.: Deep metric learning with angular loss. In: ICCV (2017)
81. Wang, J., Zhu, X., Gong, S., Li, W.: Transferable joint attribute-identity deep learning for unsupervised person re-identification. In: CVPR (2018)
82. Wang, J., Kumar, S., Chang, S.F.: Semi-supervised hashing for large-scale search. TPAMI **34**, 2393–2406 (2012)
83. Wang, P., et al.: Vehicle re-identification in aerial imagery: dataset and approach. In: ICCV (2019)
84. Weinberger, K.Q., Saul, L.K.: Distance metric learning for large margin nearest neighbor classification. JMLR (2009)
85. Wu, Y., Lin, Y., Dong, X., Yan, Y., Bian, W., Yang, Y.: Progressive learning for person re-identification with one example. TIP **28**, 2872–2881 (2019)
86. Wu, Z., Xiong, Y., Yu, S.X., Lin, D.: Unsupervised feature learning via non-parametric instance discrimination. In: CVPR (2018)
87. Xin, X., Wang, J., Xie, R., Zhou, S., Huang, W., Zheng, N.: Semi-supervised person re-identification using multi-view clustering. Pattern Recognit. **88**, 285–297 (2019)
88. Xin, X., Wu, X., Wang, Y., Wang, J.: Deep self-paced learning for semi-supervised person re-identification using multi-view self-paced clustering. In: ICIP (2019)
89. Xu, J., Shi, C., Qi, C., Wang, C., Xiao, B.: Unsupervised part-based weighting aggregation of deep convolutional features for image retrieval. In: AAAI (2018)
90. Ye, M., Zhang, X., Yuen, P.C., Chang, S.F.: Unsupervised embedding learning via invariant and spreading instance feature. In: CVPR (2019)
91. Yu, H.X., Zheng, W.S., Wu, A., Guo, X., Gong, S., Lai, J.H.: Unsupervised person re-identification by soft multilabel learning. In: CVPR (2019)

92. Zhang, J., Peng, Y.: SSDH: semi-supervised deep hashing for large scale image retrieval. TCSVT **29**, 212–225 (2017)
93. Zhang, X., et al.: AlignedReID: surpassing human-level performance in person re-identification. arXiv (2017)
94. Zhang, Y., Xiang, T., Hospedales, T.M., Lu, H.: Deep mutual learning. In: CVPR (2018)
95. Zhao, L., Li, X., Zhuang, Y., Wang, J.: Deeply-learned part-aligned representations for person re-identification. In: ICCV (2017)
96. Zheng, L., Shen, L., Tian, L., Wang, S., Wang, J., Tian, Q.: Scalable person re-identification: a benchmark. In: ICCV (2015)
97. Zheng, L., Yang, Y., Hauptmann, A.G.: Person re-identification: past, present and future. arXiv (2016)
98. Zheng, W., Chen, Z., Lu, J., Zhou, J.: Hardness-aware deep metric learning. In: CVPR (2019)
99. Zheng, Z., Zheng, L., Yang, Y.: Pedestrian alignment network for large-scale person re-identification. TCSVT **29**, 3037–3045 (2018)
100. Zhong, Z., Zheng, L., Zheng, Z., Li, S., Yang, Y.: Camera style adaptation for person re-identification. In: CVPR (2018)
101. Zhu, J.Y., Park, T., Isola, P., Efros, A.A.: Unpaired image-to-image translation using cycle-consistent adversarial networks. In: ICCV (2017)

FeatMatch: Feature-Based Augmentation for Semi-supervised Learning

Chia-Wen Kuo[1], Chih-Yao Ma[1], Jia-Bin Huang[2], and Zsolt Kira[1(✉)]

[1] Georgia Tech, Atlanta, USA
{albert.cwkuo,cyma,zkira}@gatech.edu
[2] Virginia Tech, Blacksburg, USA
jbhuang@vt.edu

Abstract. Recent state-of-the-art semi-supervised learning (SSL) methods use a combination of image-based transformations and consistency regularization as core components. Such methods, however, are limited to simple transformations such as traditional data augmentation or convex combinations of two images. In this paper, we propose a novel learned feature-based refinement and augmentation method that produces a varied set of complex transformations. Importantly, these transformations also use information from both within-class and across-class prototypical representations that we extract through clustering. We use features already computed across iterations by storing them in a memory bank, obviating the need for significant extra computation. These transformations, combined with traditional image-based augmentation, are then used as part of the consistency-based regularization loss. We demonstrate that our method is comparable to current state of art for smaller datasets (CIFAR-10 and SVHN) while being able to scale up to larger datasets such as CIFAR-100 and mini-Imagenet where we achieve significant gains over the state of art (*e.g.*, absolute 17.44% gain on mini-ImageNet). We further test our method on DomainNet, demonstrating better robustness to out-of-domain unlabeled data, and perform rigorous ablations and analysis to validate the method. Code is available here: https://sites.google.com/view/chiawen-kuo/home/featmatch.

Keywords: Semi-supervised learning · Feature-based augmentation · Consistency regularization

1 Introduction

Driven by large-scale datasets such as ImageNet as well as computing resources, deep neural networks have achieved strong performance on a wide variety of tasks. Training these deep neural networks, however, requires millions of labeled

Electronic supplementary material The online version of this chapter (https://doi.org/10.1007/978-3-030-58523-5_28) contains supplementary material, which is available to authorized users.

© Springer Nature Switzerland AG 2020
A. Vedaldi et al. (Eds.): ECCV 2020, LNCS 12363, pp. 479–495, 2020.
https://doi.org/10.1007/978-3-030-58523-5_28

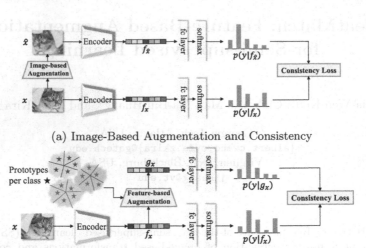

(a) Image-Based Augmentation and Consistency

(b) Feature-Based Augmentation and Consistency

Fig. 1. Consistency regularization methods are the most successful methods for semi-supervised learning. The main idea of these methods is to enforce consistency between the predictions of different *transformations* of an input image. (*a*) Image-based augmentation method generate different views of an input image via data augmentation, which are limited to operations in the image space as well as operations within a single instance or simple convex combination of two instances. (*b*) We propose an additional learned feature-based augmentation that operates in the abstract feature space. The learned feature refinement and augmentation module is capable of leveraging information from other instances, within or outside of the same class.

examples that are expensive to acquire and annotate. Consequently, numerous methods have been developed for semi-supervised learning (SSL), where a large number of unlabeled examples are available alongside a smaller set of labeled data. One branch of the most successful SSL methods [3,4,15,18,21,24,25] uses image-based augmentation [6,8,12,33] to generate different *transformations* of an input image, and consistency regularization to enforce invariant representations across these transformations. While these methods have achieved great success, the data augmentation methods for generating different transformations are limited to transformations in the image space and fail to leverage the knowledge of other instances in the dataset for diverse transformations.

In this paper, we propose novel feature-based refinement and augmentation that addresses the limitations of conventional image-based augmentation described above. Specifically, we propose a module that learns to refine and augment input image features via soft-attention toward a small set of representative prototypes extracted from the image features of other images in the dataset. The comparison between image-based augmentation and our proposed feature-based refinement and augmentation is shown in Fig. 1. Since the proposed module is learned and carried out in the feature space, diverse and abstract transformations of input images can be applied, which we validate in Sect. 4.4. Our approach only requires minimum computation via maintaining a memory bank and using k-means clustering to extract prototypes.

We demonstrate that adding our proposed feature-based augmentation along with conventional image-based augmentations, when used for consistency regularization, achieves significant gains. We test our method on standard SSL datasets such as SVHN and CIFAR-10, and show that our method, despite its simplicity, compares favorably against state-of-art methods in all cases. Further, through testing on CIFAR-100 and mini-ImageNet, we show that our method is scalable to larger datasets and outperformed the current best methods by significant margins. For example, we outperformed the closest state of the art by an absolute **17%** on mini-ImageNet with 4k labels. We also propose another realistic setting on DomainNet [20] to test the robustness of our proposed method under the case where the unlabeled samples are partially coming from shifted domains, in which we improved **23%** over supervised baseline and **12%** over semi-supervised baseline when 50% unlabeled samples are all coming from shifted domains. Finally, we conduct thorough ablations and thorough analysis to highlight that the method does, in fact, perform varied complex transformations in feature space (as evidenced by t-SNE and nearest neighbor image samples). To summarize, our key contributions include:

- We develop a learned feature-based refinement and augmentation module to transform input image features in the abstract feature space by leveraging a small set of representative prototypes of all classes in the dataset.
- We propose a memory bank mechanism to efficiently extract prototypes from images of the entire dataset with minimal extra computations.
- We demonstrate thorough results across four standard SSL datasets and also propose a realistic setting where the unlabeled data partially come from domains shifted from the target labeled set.
- We perform in-depth analysis of the prototype representations extracted and used for each instance, as well as what transformations the proposed feature-based refinement and augmentation module learns.

2   Related Works

Consistency Regularization Methods. Current state-of-the-art SSL methods mostly fall into this category. The key insight of this branch of methods is that the prediction of a deep model should be consistent across different *semantic-preserving transformations* of the same data. Consistency regularization methods regularize the model to be invariant to textural or geometric changes of an image. Specifically, given an input image x and a network composed of a feature encoder $f_x = Enc(x)$ and a classifier $p_x = Clf(f_x)$, we can generate the pseudo-label of the input image by $p_x = Clf(Enc(x))$. Furthermore, given a data augmentation module $AugD(\cdot)$, we can generate an augmented copy of x by $\hat{x} = AugD(x)$. A consistency loss $\mathcal{H}$, typically KL-Divergence loss,

Table 1. Comparison to other SSL methods with consistency regularization.

	ReMixMatch [3]	MixMatch [4]	Mean Teacher [25]	ICT [29]	PLCB [1]	FeatMatch (Ours)
Feature-Based Augmentation	–	–	–	–	–	✓
Image-Based Augmentation	✓	✓	✓	✓	✓	✓
Temporal Ensembling	✓	✓	✓			–
Self-Supervised Loss	✓	–	–	–	–	–
Alignment of Class Distribution	✓	–	–	–	✓	–

is then applied on the model predictions of $\hat{x}$ to enforce consistent prediction: $\mathcal{L}_{con} = \mathcal{H}(p, Clf(Enc(\hat{x})))$.

Image-Based Augmentation. The core to consistency-based methods is how to generate diverse but reasonable transformations of the same data. A straightforward answer is to incorporate data augmentation, which has been widely used in the training of a deep model to increase data diversity and prevent overfitting. For example, [4,15,24,25] use traditional data augmentation to generate different transformations of semantically identical images. Data augmentation method randomly perturbs an image in terms of its texture, eg. brightness, hue, sharpness, or its geometry, eg. rotation, translation, or affine transform. In addition to data augmentation, Miyato et al. [18] and Yu et al. [31] perturbed images along the adversarial direction, and Qiao et al. [21] use multiple networks to generate different views (predictions) of the same data. Recently, several works propose data augmentation modules for supervised learning or semi-supervised learning, where the augmentation parameters can either be easily tuned [8], found by RL-training [7], or decided by the confidence of network prediction [3].

Mixup [12,32,32,33], similar to data augmentation, is another effective way of increasing data diversity. It generates new training samples by a convex combination of two images and their corresponding labels. It has been shown that models trained with Mixup is robust toward out-of-distribution data [9] and is beneficial for the uncertainty calibration of a network [26]. Given two images x_1 and x_2 and their labels (or pseudo labels) y_1 and y_2, they are mixed by a randomly sampled ratio r by $\hat{x} = r \cdot x_1 + (1 - r) \cdot x_2$ and $\hat{y} = r \cdot y_1 + (1 - r) \cdot y_2$. This has been done in feature space as well [28]. A standard classification loss $\mathcal{H}(\cdot)$ is then applied on the prediction of the mixed sample $\hat{x}$ and the mixed label $\hat{y}$ by $\mathcal{L}_{mix} = \mathcal{H}(\hat{y}, Clf(Enc((\hat{x})))$. Originally, Mixup methods were developed for supervised learning. ICT [29] and MixMatch [4] introduce Mixup into semi-supervised learning by using the pseudo-label of the unlabeled data. Furthermore, by controlling the mixing ratio r to be greater than 0.5 as proposed by [4], we can make sure that the mixed sample is closer to x_1. Therefore, we can separate the mixed data into labeled mixed batch $\hat{\mathcal{X}}$ if x_1 is labeled, and unlabeled mixed batch $\hat{\mathcal{U}}$ if x_1 is unlabeled. Different loss weights can then be applied to modulate the strength of regularization from the unlabeled data.

3   Feature-Based Augmentation and Consistency

Image-based augmentation has been shown to be an effective approach to generate different views of an image for consistency-based SSL methods. However, conventional image-based augmentation has the following two limitations: (1) Operate in image space, which limits the possible transformations to textural or geometric within images, and (2) Operate within a single instance, which fails to transform data with the knowledge of other instances, either within or outside of the same class. Some recent works that utilize Mixup only partially address the second limitation of conventional data augmentation since mixup operates only between two instances. On the other hand, Manifold Mixup [28] approaches the first limitation by performing Mixup in the feature space but is limited to a simple convex combination of two samples.

We instead propose to address these two limitations simultaneously. We proposed a novel method that refines and augments image features in the abstract feature space rather than image space. To efficiently leverage the knowledge of other classes, we condense the information of each class into a small set of prototypes by performing clustering in the feature space. The image features are then refined and augmented through information propagated from prototypes of all classes. We hypothesize that this feature refinement/augmentation can further improve the feature representations, and these refined features can produce better pseudo-labels than features without the refinement (See Sect. 4.4 for our analysis on this hypothesis). The feature refinement and augmentation are learned via a lightweight attention network for the representative prototypes and optimized end-to-end with other objectives such as classification loss. A consistency loss can naturally be applied between the prediction from the original features and the refined features to regularize the network as shown in Fig. 1b.

The final model seamlessly combines our novel feature-based augmentation with conventional image-based augmentation for consistency regularization, which is applied to data augmented from both sources. Despite the simplicity of the method, we find this achieves significant performance improvement. In summary, we compare our method with other highly relevant SSL works in Table 1.

3.1   Prototype Selection

In order to efficiently leverage the knowledge of other classes for feature refinement and augmentation, we propose to compactly represent the information of each class by clustering in the feature space. To select representative prototypes from the dataset, we propose to use K-Means clustering in the feature space to extract p_k cluster means as prototypes for *each class*. However, there are two technical challenges: (1) in an SSL setting, most images are unlabeled; (2) even if all the labels are available, it is still computationally expensive to extract features of all the images from the entire dataset before running K-Means.

To tackle these issues, as shown in Fig. 2, we collect features f_{xi} and pseudo-labels $\hat{y}_i$ already generated by the network at every iteration of the training loop, *i.e.*, no extra computation needed. In the recording loop, the pairs of pseudo

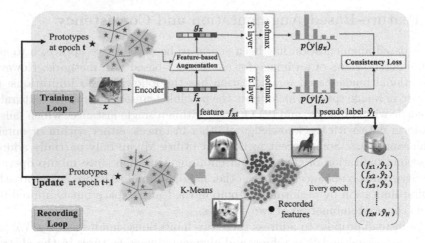

Fig. 2. A prototype recording loop that runs alongside the model training loop. The image features f_{xi} as well as their pseudo labels $\hat{y}_i$ already generated at each iteration of the training loop are collected and recorded in a memory bank as $(f_{xi}, \hat{y}_i)$ pairs. Once the training loop goes over the whole dataset, the recording loop will run K-Means to extract prototypes for each class, update the prototypes for feature-based augmentation, and clear the memory bank.

label and features are detached from the computation graph and pushed into a memory bank for later usage. The prototypes are extracted by K-Means at every epoch when we go over the whole dataset. Finally, the feature refinement and augmentation module updates the prototypes with the newly extracted ones in the training loop. Even though the prototypes are extracted from the feature computed from the model a few iterations ago, as training progresses and the model gradually converges, the extracted prototypes fall on the correct cluster and are diverse enough to compactly represent the feature distribution per class. More analyses can be found in Sect. 4.4. Similar idea is concurrently explored in self-supervised learning by He et al. [10,30].

3.2   Learned Feature Augmentation

With a set of prototypes selected by the process described above, we propose a learned feature refinement and augmentation module via soft-attention [27] toward the set of selected prototypes. The proposed module refines and augments input image features in the feature space by leveraging the knowledge of prototypes, either within or outside of the same class, as shown in Fig. 3. The lightweight feature refinement and augmentation module composed of three fully connected layers is jointly optimized with other objectives and hence learns a reasonable feature-based augmentation to aid classification. We provide further analysis and discussion in Sect. 4.4.

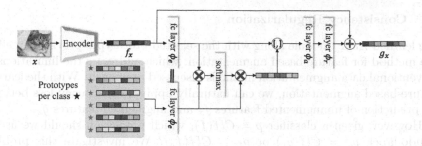

Fig. 3. Feature-Based Augmentation: The input image features are augmented by attention using extracted prototype features (Eq. 1), where the colors of $\star$ represent the classes of prototypes. The prototype features are calculated via a weighted sum using the attention weights, concatenated with the image features, and then undergo a fc layer ϕ_a (Eq. 2) to produce attention features f_a. Finally, we use the attention features to refine and augment the input image features with a residual connection (Eq. 3).

Inspired by the attention mechanism [27], each input image feature *attends* to prototype features via attention weights computed by dot product similarity. The prototype features are then weighted summed by the attention weights and then fed back to the input image feature via residual connect for feature augmentation and refinement. Specifically, for an input image with extracted features f_x and the i-th prototype features $f_{p,i}$, we first project them into an embedding space by a learned function ϕ_e as $e_x = \phi_e(f_x)$ and $e_{p,i} = \phi_e(f_{p,i})$ respectively. We compute an attention weight w_i between e_x and $e_{p,i}$ as:

$$w_i = \mathrm{softmax}(e_x^T e_{p,i}), \tag{1}$$

where $\mathrm{softmax}(\cdot)$ normalizes the dot product similarity scores across all prototypes. The information aggregated from the prototypes and passed to the image features for feature refinement and augmentation can then be expressed as a sum of prototype features weighted by the attention weights:

$$f_a = \mathrm{relu}(\phi_a([e_x, \sum_i w_i e_{p,i}])), \tag{2}$$

where ϕ_a is a learnable function, and $[\cdot, \cdot]$ is a concatenation operation along the feature dimension. Finally, the input image features f_x is refined via a residual connection as:

$$g_x = \mathrm{relu}(f_x + \phi_r(f_a)), \tag{3}$$

where g_x are the refined features of f_x, and ϕ_r is a learnable function.

The attention mechanism described above can be trivially generalized to multi-head attention as in [27]. In practice, we use multi-head attention, instead of single head for slightly better results. For simplicity, we define the feature refinement and augmentation process $AugF(\cdot)$ described above as $g_x = AugF(f_x)$.

3.3 Consistency Regularization

The learned $AugF$ module along with the selected prototypes provides an effective method for feature-based augmentation, which addresses the limitations of conventional data augmentation methods discussed previously. With the learned feature-based augmentation, we can naturally apply a consistency loss between the prediction of unaugmented features f_x and augmented features g_x.

However, given a classifier $p = Clf(f)$, which prediction should we use as pseudo-label, $p_g = Clf(g_x)$ or $p_f = Clf(f_x)$? We investigate this problem in Sect. 4.4 and find that $AugF$ is able to *refine* the input features for better representation, thus generating better pseudo-labels. Therefore, we compute pseudo-label p_g on the refined feature g_x by $p_g = Clf(g_x)$. The feature-based consistency loss can be computed as: $\mathcal{L}_{con} = \mathcal{H}(p_g, Clf(f_x))$. We can easily extend $\mathcal{L}_{con}$ to work seamlessly with traditional augmentation methods, *i.e.*, traditional data augmentation and Mixup. For simplicity, we will illustrate with only data augmentation, but Mixup can be easily adapted. Inspired by Berthelot et al. [3], we generate a weakly augmented image x and its strongly augmented copy $\hat{x}$. The pseudo-label is computed on the weakly augmented image x that undergoes feature-based augmentation and refinement for better pseudo-labels as $p_g = Clf(AugF(Enc(x)))$. We can then compute two consistency losses on the strongly augmented data $\hat{x}$, one with $AugF$ applied and the other without:

$$\mathcal{L}_{con\text{-}g} = \mathcal{H}(p_g, Clf(AugF(Enc(\hat{x})))) \tag{4}$$

$$\mathcal{L}_{con\text{-}f} = \mathcal{H}(p_g, Clf(Enc(\hat{x}))) \tag{5}$$

Since the pseudo-label p_g is computed on the image undergoing weak data augmentation and feature-based augmentation, the regularization signal of $\mathcal{L}_{con\text{-}g}$ and $\mathcal{L}_{con\text{-}f}$ comes from both image-based and feature-based augmentation.

3.4 Total Loss

Consistency regularization losses $\mathcal{L}_{con\text{-}g}$ and $\mathcal{L}_{con\text{-}f}$ in Eq. 4 and 5 are applied on unlabeled data. For labeled image x with label y, a regular classification loss can be applied:

$$\mathcal{L}_{clf} = \mathcal{H}(y, Clf(AugF(Enc(x)))) \tag{6}$$

Therefore, the total loss can be written as: $\mathcal{L}_{clf} + \lambda_g \mathcal{L}_{con\text{-}g} + \lambda_f \mathcal{L}_{con\text{-}f}$. Where λ_g and λ_f are weights for $\mathcal{L}_{con\text{-}g}$ and $\mathcal{L}_{con\text{-}f}$ losses respectively.

4 Experiments

4.1 Datasets

Standard SSL Datasets. We conduct experiments on commonly used SSL datasets: SVHN [19], CIFAR-10 [14], CIFAR-100 [14], and mini-ImageNet [22]. Following the standard approach in SSL, we randomly choose a certain number

Table 2. Comparison on CIFAR-100 and mini-imageNet. Numbers represent error rate in three runs. For fair comparison, we use the same model as other methods: CNN-13 for CIFAR-100 and ResNet-18 for mini-ImageNet.

Method	CIFAR-100		mini-ImageNet	
	# Labeled samples		# Labeled samples	
	4,000	10,000	4,000	10,000
Π-model [24]	–	39.19 ± 0.36	–	–
SNTG [17]	–	37.97 ± 0.29	–	–
SSL with Memory [5]	–	34.51 ± 0.61	–	–
Deep Co-Training [21]	–	34.63 ± 0.14	–	–
Weight Averaging [2]	–	33.62 ± 0.54	–	–
Mean Teacher [25]	45.36 ± 0.49	36.08 ± 0.51	72.51 ± 0.22	57.55 ± 1.11
Label Propagation [13]	43.73 ± 0.20	35.92 ± 0.47	70.29 ± 0.81	57.58 ± 1.47
PLCB [1]	37.55 ± 1.09	32.15 ± 0.50	56.49 ± 0.51	46.08 ± 0.11
FeatMatch (Ours)	**31.06 ± 0.41**	**26.83 ± 0.04**	**39.05 ± 0.06**	**34.79 ± 0.22**

of labeled samples as a small labeled set and discard the labels for the remaining data to form a large unlabeled set. Our proposed method is tested under various amounts of labeled samples. SVHN is a dataset of 10 digits, which has about 70k training samples. CIFAR-10 and CIFAR-100 are natural image datasets with 10 and 100 classes respectively. Both dataset contains 50k training samples. For mini-ImageNet, we follow [1,13] to construct the mini-ImageNet training set. Specifically, given a predefined list of 100 classes [22] from ILSVRC [23], 500 samples are selected randomly for each class, thus forming a training set of 50k samples. The samples are center-cropped and resized to 84 × 84 resolution. We then follow the same standard procedure and construct a small labeled set and a large unlabeled set from the 50k training samples.

SSL Under Domain Shift. In another realistic setting, we argue that the unlabeled data may come from a domain different from that of the target labeled data. For instance, given a small set of labeled natural images of animals, the large unlabeled set may also contain paintings of animals. To investigate the effect of domain shift in the unlabeled set, we proposed a new SSL task based on the DomainNet dataset [20], which contains 345 classes of images coming from six domains: Clipart, Infograph, Painting, Quickdraw, Real, and Sketch.

We use the *Real* domain as our target. Five percent of the data from the Real domain are kept as the target labeled set, and the rest are the target unlabeled set. We select *Clipart, Painting, Sketch,* and *Quickdraw* as shifted domains. To modulate the level of domain shift in the unlabeled data, we propose a parameter r_u that controls the ratio of unlabeled data coming from the target Real domain or the shifted domains. Specifically, r_u percent of target Real unlabeled set is replaced with data uniformly drawn from the shifted domains. By formulating the problem this way, the amount of unlabeled data remains constant. The only

Table 3. Comparison between the image-based baseline with our proposed feature-based augmentation method on DomainNet with 1) unlabeled data coming from the same domain as the labeled target ($r_u = 0\%$), and 2) half of unlabeled data coming from the same domain as the labeled target and the other half from shifted domains ($r_u = 50\%$). Numbers are error rates across 3 runs.

Method (5% labeled samples)	$r_u = 0\%$	$r_u = 50\%$
(Semi-supervised) Baseline	56.63 ± 0.17	65.82 ± 0.07
FeatMatch (Ours)	40.66 ± 0.60	54.01 ± 0.66
Supervised baseline (5% labeled samples, lower bound)	77.25 ± 0.52	
Supervised baseline (100% labeled samples, upper bound)	31.91 ± 0.15	

factor that affects the performance of the proposed method is the ratio between in-domain data and shifted domain data in the unlabeled set.

We randomly reserve 1% of data from the Real domain as the validation set. The final result is reported on the test set of the Real domain, with the model selected on the reserved validation set. The images are center-cropped and resized to 128×128 resolution, and the model we use is the standard ResNet-18 [11]. There are around 120k training samples, which is more than twice larger than the standard SSL datasets such as CIFAR-10 and CIFAR-100. For a fair comparison, we fix *all* hyper-parameters across experiments of different r_u to truly assess the robustness of proposed methods toward domain shift in the unlabeled data.

Hyper-parameters. We tune the hyper-parameters on CIFAR-10 with 250 labels with a validation set held-out from the training set. Our method is not sensitive to the hyper-parameters, which are kept fixed across *all* the datasets and settings. Please see the supplementary for more implementation details and the values of hyper-parameters.

4.2   Results

We first show our results on CIFAR-100 and mini-ImageNet with 4k and 10k labels in Table 2. Our method consistently improves over state of the arts by large margins, with about absolute 5% on CIFAR-100 with 4k labels and 17% on mini-ImageNet with 4k labels.

In Table 3, we show our results on the larger dataset of DomainNet setting, which contains unlabeled data coming from other shifted domains. It can be clearly seen that in the setting of $r_u = 50\%$, where 50% of the unlabeled data are coming from other shifted domains, the performance drops by a large margin compared with the setting of $r_u = 0\%$, where all the unlabeled data are coming from the same domain as the target labeled set. Nevertheless, our proposed feature-based augmentation method improves over supervised baseline by absolute 36% error rate when $r_u = 0\%$ and 23% when $r_u = 50\%$. When compared to the conventional image-based augmentation baseline, we improves by 12% when $r_u = 50\%$ and 16% when $r_u = 0\%$.

Table 4. Comparison on CIFAR-10 and SVHN. Numbers represent error rate across three runs. The results reported in the first block with CNN-13 model [15,18] are from the original paper. The results reported in the second block with wide ResNet (WRN) are reproduced by [3,4].

Method	Model (param.)	CIFAR-10			SVHN		
		# Labeled samples			# Labeled samples		
		250	1,000	4,000	250	1,000	4,000
SSL with Memory [5]	CNN-13 (3M)	–	–	11.91± 0.22	8.83	4.21	–
Deep Co-Training [21]		–	–	8.35 ± 0.06	–	3.29 ± 0.03	–
Weight Averaging [2]		–	15.58 ± 0.12	9.05 ± 0.21	–	–	–
ICT [29]		–	15.48 ± 0.78	7.29 ± 0.02	4.78 ± 0.68	3.89 ± 0.04	–
Label Propagation [13]		–	16.93 ± 0.70	10.61 ± 0.28	–	–	–
SNTG [17]		–	18.41 ± 0.52	9.89 ± 0.34	4.29 ± 0.23	3.86 ± 0.27	–
PLCB [1]		–	6.85 ± 0.15	5.97 ± 0.15	–	–	–
Π-model [24]	WRN (1.5M)	53.02 ± 2.05	31.53 ± 0.98	17.41 ± 0.37	17.65 ± 0.27	8.60 ± 0.18	5.57 ± 0.14
PseudoLabel [16]		49.98 ± 1.17	30.91 ± 1.73	16.21 ± 0.11	21.16 ± 0.88	10.19 ± 0.41	5.71 ± 0.07
Mixup [33]		47.43 ± 0.92	25.72 ± 0.66	13.15 ± 0.20	39.97 ± 1.89	16.79 ± 0.63	7.96 ± 0.14
VAT [18]		36.03 ± 2.82	18.68 ± 0.40	11.05 ± 0.31	8.41 ± 1.01	5.98 ± 0.21	4.20 ± 0.15
Mean Teacher [25]		47.32 ± 4.71	17.32 ± 4.00	10.36 ± 0.25	6.45 ± 2.43	3.75 ± 0.10	3.39 ± 0.11
MixMatch [4]		11.08 ± 0.87	7.75 ± 0.32	6.24 ± 0.06	3.78 ± 0.26	3.27 ± 0.31	2.89 ± 0.06
ReMixMatch [3]		**6.27 ± 0.34**	**5.73 ± 0.16**	5.14 ± 0.04	**3.10 ± 0.50**	**2.83 ± 0.30**	**2.42 ± 0.09**
FeatMatch (Ours)		7.50 ± 0.64	**5.76 ± 0.07**	**4.91 ± 0.18**	**3.34 ± 0.19**	**3.10 ± 0.06**	2.62 ± 0.08

Table 5. Ablation study on CIFAR-10 with various amount of labeled samples.

Experiment	Image-Based Augmentation	Feature-Based Augmentation	$\mathcal{L}_{con-f}$	$\mathcal{L}_{con-g}$	$\mathcal{L}_{con}$	#Labeled samples		
						250	1,000	4,000
Baseline	✔	–	–	–	✔	19.55 ± 1.58	9.04 ± 1.00	6.08 ± 0.16
w/o $\mathcal{L}_{con-f}$	✔	✔	–	✔	–	18.57 ± 3.19	8.38 ± 0.35	6.09 ± 0.16
w/o $\mathcal{L}_{con-g}$	✔	✔	✔	–	–	8.19 ± 1.74	6.07 ± 0.46	5.16 ± 0.30
FeatMatch (Ours)	✔	✔	✔	✔	–	7.90 ± 0.49	5.94 ± 0.16	5.00 ± 0.21

In Table 4, we show the comparison of our method with other SSL methods on standard CIFAR-10 and SVHN datasets. Our method achieves comparable results with the current state of the art, ReMixMatch, even though 1) we start from a lower baseline and 2) our method is much simpler (*e.g.*, no class distribution alignment and no self-supervised loss), as compared in Table 1. Our proposed feature-based augmentation method is complementary to image-based methods and can be easily integrated to further improve the performance.

4.3 Ablation Study

In the ablation study, we are interested in answering the following questions: 1) what is the effectiveness of the two proposed consistency losses – $\mathcal{L}_{con-f}$ (Eq. 5)

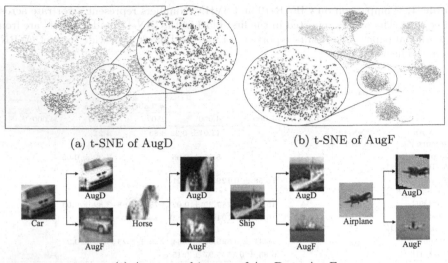

(a) t-SNE of AugD   (b) t-SNE of AugF

(c) Augmented images of AugD vs. AugF

Fig. 4. (*a*) We jointly compute and plot t-SNE of input unaugmented image features (dimmer color) and image-based augmented features (brighter color). (*b*) We also jointly compute and plot t-SNE of input unaugmented image features (dimmer color) and feature-based augmented features (brighter color) with the exact same t-SNE parameters with (a). (*c*) To concretely visualize the augmented feature, we find their nearest image neighbor in the feature space and compare against the image-based augmentation method side by side. (Color figure online)

and $\mathcal{L}_{con\text{-}g}$ (Eq. 4). 2) how much of the improvement is from the proposed feature-based augmentation method over the image-based augmentation baseline? For the image-based augmentation baseline, the $AugF$ module is completely removed and thus the consistency regularization comes only from image-based augmentation. This is also the same image-based augmentation baseline that our final model with feature-based augmentation builds upon. The ablation study is conducted on CIFAR-10 with various amount of labeled samples (Table 5).

We can see from Table 5 that our image-based augmentation baseline achieves good results but only on cases where there are more labeled samples. We conjecture this is because the aggressive data augmentation applied to training images makes the training unstable. Nevertheless, our baseline performance is still competitive with respect to other image-based augmentation methods in Table 4 (though slightly worse than MixMatch). By adding our proposed $AugF$ module($\mathcal{L}_{con\text{-}f}$ and $\mathcal{L}_{con\text{-}g}$) for feature refinement and augmentation on top of the image-based augmentation baseline, the performance improves over baseline consistently, especially for 250 labels.

We can also see that $\mathcal{L}_{con\text{-}f}$ plays a more important role than $\mathcal{L}_{con\text{-}g}$, though our final model with both loss terms achieves the best result. In both $\mathcal{L}_{con\text{-}f}$ and $\mathcal{L}_{con\text{-}g}$, the pseudo-labels are computed from the features undergone feature-

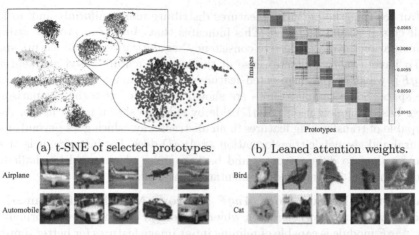

(a) t-SNE of selected prototypes.   (b) Leaned attention weights.

(c) Nearest image neighbors of prototypes

Fig. 5. (*a*) In the t-SNE plot, the extracted prototypes (◇) fall on the correct clusters and are able to compactly represent the cluster. (*b*) We visualize the learned attention weights from a batch of images toward prototypes. The images and prototypes are sorted by their classes for ease of illustration. As can be seen, images have higher attention weights to the prototypes with the same class. (*c*) We find the prototypes' nearest image neighbors in the feature space. The prototypes compactly represent a diverse sets of images in each class.

based augmentation. The only difference is which prediction we're driving to match the pseudo-label: 1) the prediction from the feature undergone both $AugD$ and $AugF$ (by $\mathcal{L}_{con\text{-}g}$ loss), or 2) the prediction from the feature undergone only $AugD$ (by $\mathcal{L}_{con\text{-}f}$ loss)? As claimed in Sect. 3.3 and analyzed in Sect. 4.4, $AugF$ is able to refine input image features for better representation and pseudo-labels of higher quality. Therefore, matching the slightly worse prediction from the feature undergone only $AugD$ (by $\mathcal{L}_{con\text{-}f}$ loss) induces a stronger consistency regularization. This explains why $\mathcal{L}_{con\text{-}f}$ improves performance more crucially.

4.4 Analysis

What augmentation does $AugF$ Learn? We compare the feature distribution via t-SNE 1) between input unaugmented image features and image-based augmented features in Fig. 4a, and 2) between input unaugmented image features and feature-based augmented features in Fig. 4b. In Fig. 4a, some local small clusters are captured by t-SNE and can be found in the zoomed subfigure. This indicates that $AugD$ can only perturb data locally, and fail to produce stronger augmentation for more effective consistency regularization in the feature space. In Fig. 4b, we can see $AugF$ indeed learns to augment and refine features. Furthermore, the learned augmentation preserves semantic meaning as the augmented features still fall in the correct cluster. In the zoomed figure, we can see that the perturbed features distribute more uniformly and no local small clusters could be found. This indicates that $AugF$ can produce stronger augmentation for more effective consistency regularization in the feature space.

To have a more concrete sense of the learned feature-based augmentation ($AugF$), we show the augmented feature's nearest image neighbor in the feature space. Some sample results are shown in Fig. 4c, with the comparison to image-based augmentation ($AugD$) side by side. As shown in the figure, $AugF$ is capable of transforming features in an abstract way, which goes beyond simple textural and geometric transformation as $AugD$ does. For instance, it is able to augment data to different poses and backgrounds, which could be challenging for conventional image-based augmentation methods.

What Other Reason Does $AugF$ Improve Model Performance? We hypothesize that one other reason why our method can improve performance is that $AugF$ module is capable of refining input image features for better representation by the extracted prototypes, and thus provides better pseudo-labels. The consistency regularization losses then drive the network's prediction to match the target pseudo-labels of higher quality, leading to overall improvement. With this hypothesis, we expect classification accuracy to be higher for features after feature refinement. To verify, we remove $\mathcal{L}_{con\text{-}f}$ loss and retrain. The accuracy of pseudo-labeling from the features refined by $AugF$ is on average $0.5 - 1.0\%$ higher. This confirms our hypothesis that $\mathcal{L}_{con\text{-}f}$ drives the feature encoder to learn a better feature representation refined by $AugF$.

The reader may wonder: why doesn't $AugF$ learn a shortcut solution of identity mapping to minimize $\mathcal{L}_{con\text{-}f}$ and $\mathcal{L}_{con\text{-}g}$? As can be seen from Fig. 4, $AugF$ does *not* learn an identity mapping. Although learning an identity mapping may be a shortcut solution for minimizing $\mathcal{L}_{con\text{-}f}$ and $\mathcal{L}_{con\text{-}g}$, it is not the case for the classification loss $\mathcal{L}_{clf}$ (Eq. 6). This finding implicitly confirms our hypothesis that there is extra information from the prototypes that $AugF$ can leverage to refine the feature representation for higher (pseudo-label) classification accuracy.

What does Aug do Internally? In Fig. 5a and c, we can see that even though our proposed prototype extraction method only uses simple K-Means to extract prototypes of each class based on potentially noisy pseudo-labels, and features recorded several iterations ago, our prototype selection method can still successfully extract a diverse set of prototypes per class. Moreover, in Fig. 5b, the attention mechanism inside $AugF$ learns to attend to prototypes that belong to the same class with the input image feature. Note that there is no loss term specific for $AugF$, as it is simply jointly optimized with the standard classification and consistency regularization loss from semi-supervised learning.

5   Conclusion

We introduce a method to jointly learn a classifier and feature-based refinement and augmentations which can be used within existing consistency-based SSL methods. Unlike traditional image-based transformations, our method can learn complex, feature-based transformations as well as incorporate information from class-specific prototypical representations extracted in an efficient manner (specifically using a memory bank). Using this method, we show comparable results as the current state of the art for smaller datasets such as CIFAR-10 and SVHN, and significant improvements on datasets with a large number of categories (*e.g.*, 17.44% absolute improvement on mini-ImageNet). We also demonstrate increased robustness to out-of-domain unlabeled data, which is an important real-world problem, and perform ablations and analysis to demonstrate the learned feature transformation and extracted prototypical representations.

Acknowledgement. This work was funded by DARPA's Learning with Less Labels (LwLL) program under agreement HR0011-18-S-0044 and DARPA's Lifelong Learning Machines (L2M) program under Cooperative Agreement HR0011-18-2-0019.

References

1. Arazo, E., Ortego, D., Albert, P., O'Connor, N.E., McGuinness, K.: Pseudo-labeling and confirmation bias in deep semi-supervised learning. arXiv preprint arXiv:1908.02983 (2019)
2. Athiwaratkun, B., Finzi, M., Izmailov, P., Wilson, A.G.: Improving consistency-based semi-supervised learning with weight averaging. arXiv preprint arXiv:1806.05594, **2** (2018)
3. Berthelot, D., et al.: Remixmatch: semi-supervised learning with distribution alignment and augmentation anchoring. In: Proceedings of International Conference on Learning Representations (ICLR) (2020)
4. Berthelot, D., Carlini, N., Goodfellow, I., Papernot, N., Oliver, A., Raffel, C.A.: Mixmatch: a holistic approach to semi-supervised learning. In: Advances in Neural Information Processing Systems, pp. 5050–5060 (2019)
5. Chen, Y., Zhu, X., Gong, S.: Semi-supervised deep learning with memory. In: Ferrari, V., Hebert, M., Sminchisescu, C., Weiss, Y. (eds.) ECCV 2018. LNCS, vol. 11205, pp. 275–291. Springer, Cham (2018). https://doi.org/10.1007/978-3-030-01246-5_17
6. Cubuk, E.D., Zoph, B., Mane, D., Vasudevan, V., Le, Q.V.: Autoaugment: learning augmentation policies from data. arXiv preprint arXiv:1805.09501 (2018)
7. Cubuk, E.D., Zoph, B., Mane, D., Vasudevan, V., Le, Q.V.: Autoaugment: learning augmentation strategies from data. In: Proceedings of the IEEE Conference on Computer Vision and Pattern Recognition, pp. 113–123 (2019)
8. Cubuk, E.D., Zoph, B., Shlens, J., Le, Q.V.: Randaugment: Practical automated data augmentation with a reduced search space. arXiv preprint arXiv:1909.13719 (2019)
9. Guo, H., Mao, Y., Zhang, R.: Mixup as locally linear out-of-manifold regularization. In: Proceedings of the AAAI Conference on Artificial Intelligence, vol. 33, pp. 3714–3722 (2019)
10. He, K., Fan, H., Wu, Y., Xie, S., Girshick, R.: Momentum contrast for unsupervised visual representation learning. arXiv preprint arXiv:1911.05722 (2019)
11. He, K., Zhang, X., Ren, S., Sun, J.: Deep residual learning for image recognition. In: Proceedings of the IEEE Conference on Computer Vision and Pattern Recognition, pp. 770–778 (2016)

12. Hendrycks, D., Mu, N., Cubuk, E.D., Zoph, B., Gilmer, J., Lakshminarayanan, B.:AugMix: a simple data processing method to improve robustness and uncertainty. In: Proceedings of the International Conference on Learning Representations (ICLR) (2020)
13. Iscen, A., Tolias, G., Avrithis, Y., Chum, O.: Label propagation for deep semi-supervised learning. In: Proceedings of the IEEE Conference on Computer Vision and Pattern Recognition, pp. 5070–5079 (2019)
14. Krizhevsky, A., Hinton, G., et al.: Learning multiple layers of features from tiny images. Technical report. Citeseer (2009)
15. Laine, S., Aila, T.: Temporal ensembling for semi-supervised learning. In: Proceedings of International Conference on Learning Representations (ICLR) (2017)
16. Lee, D.H.: Pseudo-label: the simple and efficient semi-supervised learning method for deep neural networks. In: Workshop on Challenges in Representation Learning, ICML, vol. 3, p. 2 (2013)
17. Luo, Y., Zhu, J., Li, M., Ren, Y., Zhang, B.: Smooth neighbors on teacher graphs for semi-supervised learning. In: Proceedings of the IEEE Conference on Computer Vision and Pattern Recognition, pp. 8896–8905 (2018)
18. Miyato, T., Maeda, S., Koyama, M., Ishii, S.: Virtual adversarial training: a regularization method for supervised and semi-supervised learning. IEEE Trans. Pattern Anal. Mach. Intell. **41**(8), 1979–1993 (2018)
19. Netzer, Y., Wang, T., Coates, A., Bissacco, A., Wu, B., Ng, A.Y.: Reading digits in natural images with unsupervised feature learning (2011)
20. Peng, X., Bai, Q., Xia, X., Huang, Z., Saenko, K., Wang, B.: Moment matching for multi-source domain adaptation. In: Proceedings of the IEEE International Conference on Computer Vision, pp. 1406–1415 (2019)
21. Qiao, S., Shen, W., Zhang, Z., Wang, B., Yuille, A.: Deep co-training for semi-supervised image recognition. In: Ferrari, V., Hebert, M., Sminchisescu, C., Weiss, Y. (eds.) ECCV 2018. LNCS, vol. 11219, pp. 142–159. Springer, Cham (2018). https://doi.org/10.1007/978-3-030-01267-0_9
22. Ravi, S., Larochelle, H.: Optimization as a model for few-shot learning. In: International Conference on Learning Representations (ICLR) (2017)
23. Russakovsky, O., et al.: ImageNet large scale visual recognition challenge. Int. J. Comput. Vis. **115**(3), 211–252 (2015)
24. Sajjadi, M., Javanmardi, M., Tasdizen, T.: Regularization with stochastic transformations and perturbations for deep semi-supervised learning. In: Advances in Neural Information Processing Systems, pp. 1163–1171 (2016)
25. Tarvainen, A., Valpola, H.: Mean teachers are better role models: weight-averaged consistency targets improve semi-supervised deep learning results. In: Advances in Neural Information Processing Systems, pp. 1195–1204 (2017)
26. Thulasidasan, S., Chennupati, G., Bilmes, J.A., Bhattacharya, T., Michalak, S.: On mixup training: improved calibration and predictive uncertainty for deep neural networks. In: Advances in Neural Information Processing Systems, pp. 13888–13899 (2019)
27. Vaswani, A., et al.: Attention is all you need. In: Advances in Neural Information Processing Systems, pp. 5998–6008 (2017)
28. Verma, V., Lamb, A., Beckham, C., Courville, A., Mitliagkis, I., Bengio, Y.: Manifold mixup: encouraging meaningful on-manifold interpolation as a regularizer. Stat, **1050**, 13 (2018)
29. Verma, V., Lamb, A., Kannala, J., Bengio, Y., Lopez-Paz, D.: Interpolation consistency training for semi-supervised learning. arXiv preprint arXiv:1903.03825 (2019)

30. Wu, Z., Xiong, Y., Yu, S.X., Lin, D.: Unsupervised feature learning via non-parametric instance discrimination. In: Proceedings of the IEEE Conference on Computer Vision and Pattern Recognition, pp. 3733–3742 (2018)
31. Yu, B., Wu, J., Ma, J., Zhu, Z.: Tangent-normal adversarial regularization for semi-supervised learning. In: Proceedings of the IEEE Conference on Computer Vision and Pattern Recognition, pp. 10676–10684 (2019)
32. Yun, S., Han, D., Oh, S.J., Chun, S., Choe, J., Yoo, Y.: Cutmix: regularization strategy to train strong classifiers with localizable features. In: Proceedings of the IEEE International Conference on Computer Vision, pp. 6023–6032 (2019)
33. Zhang, H., Cisse, M., Dauphin, Y.N., Lopez-Paz, D.: mixup: beyond empirical risk minimization. In: Proceedings of International Conference on Learning Representations (ICLR) (2018)

RadarNet: Exploiting Radar for Robust Perception of Dynamic Objects

Bin Yang[1,2]([✉]), Runsheng Guo[3], Ming Liang[1], Sergio Casas[1,2],
and Raquel Urtasun[1,2]

[1] Uber Advanced Technologies Group, Pittsburgh, USA
{byang10,ming.liang,sergio.casas,urtasun}@uber.com
[2] Univeristy of Toronto, Toronto, Canada
[3] University of Waterloo, Waterloo, Canada
r9guo@edu.uwaterloo.ca

Abstract. We tackle the problem of exploiting Radar for perception in
the context of self-driving as Radar provides complementary information to other sensors such as LiDAR or cameras in the form of Doppler
velocity. The main challenges of using Radar are the noise and measurement ambiguities which have been a struggle for existing simple input or
output fusion methods. To better address this, we propose a new solution that exploits both LiDAR and Radar sensors for perception. Our
approach, dubbed RadarNet, features a voxel-based early fusion and an
attention-based late fusion, which learn from data to exploit both geometric and dynamic information of Radar data. RadarNet achieves state-
of-the-art results on two large-scale real-world datasets in the tasks of
object detection and velocity estimation. We further show that exploiting
Radar improves the perception capabilities of detecting faraway objects
and understanding the motion of dynamic objects.

Keywords: Radar · Autonomous driving · Object detection

1 Introduction

Self-driving vehicles (SDVs) have to perceive the world around them in order
to interact with the environment in a safe manner. Perception systems typically
detect the objects of interest and track them over time in order to estimate their
motion. Despite many decades of research, perception systems have not achieved
the level of reliability required to deploy self-driving vehicles at scale without
safety drivers.

B. Yang and R. Guo—Equal contribution. Work done during RG's internship at Uber
ATG.

Electronic supplementary material The online version of this chapter (https://
doi.org/10.1007/978-3-030-58523-5_29) contains supplementary material, which is
available to authorized users.

Recent 3D perception systems typically exploit cameras [5,36,42], LiDAR [17,34,45], or their combination [6,20,32] to achieve high-quality 3D object detection. While cameras capture rich appearance features, LiDAR provides direct and accurate 3D measurements. The sparsity of LiDAR measurements (e.g., at long range) and the sensor's sensitivity to weather (e.g., fog, rain and snow) remain open challenges. In addition to detecting and recognizing objects, estimating their velocities is also of vital importance. In some safety critical situations, for example a child running out of occlusion in front of the SDV, the SDV needs to estimate velocities from a single measurement cycle in order to avoid collision. This estimation is often inaccurate (or even impossible) when using LiDAR or cameras alone as they provide static information only. While for pedestrians we may infer the motion from its pose with large uncertainty, for rigid objects like vehicles we can not make reasonable predictions from their appearance alone.

An appealing solution is to use sensors that are robust to various weather conditions and can provide velocity estimations from a single measurement. This is the case of Radar, which uses the Doppler effect to compute the radial velocities of objects relative to the SDV. Radar brings its own challenges, as the data is very sparse (typically much more so than LiDAR), the measurements are ambiguous in terms of position and velocity, the readings lack tangential information and often contain false positives. As a result, previous methods either focus on the ADAS by fusing Radar with cameras [4,8,29,30], where the performance requirements are relatively low; or fuse Radar data at the perception output level (e.g., tracks) [7,9,10], thus failing to fully exploit the complementary information of the sensors.

In this paper, we take a step forward in this direction and design a novel neural network architecture, dubbed *RadarNet*, which can exploit both LiDAR and Radar to provide accurate detections and velocity estimates for the actors in the scene. Towards this goal, we propose a multi-level fusion scheme that can fully exploit both geometric and dynamic information of Radar data. In particular, we first fuse Radar data with LiDAR point clouds via a novel *voxel-based early fusion* approach to leverage the Radar's long sensing range. Furthermore, after we get object detections, we fuse Radar data again via an *attention-based late fusion* approach to leverage the Radar's velocity readings. The proposed attention module captures the uncertainties in both detections and Radar measurements and plays an important role in transforming the 1D radial velocities from Radar to accurate 2D object velocity estimates.

We demonstrate the effectiveness of RadarNet on two large-scale driving datasets, where it surpasses the previous state-of-the-art in both 3D object detection and velocity estimation. We further show that exploiting Radar brings significant improvements in perceiving dynamic objects, improving both motion estimation and long range detection.

2   Related Work

Exploiting LiDAR for Perception: As a high-quality 3D sensor, LiDAR has been widely used for 3D object detection in self-driving. Previous methods mainly differ in two aspects: the detection architecture and the input representation. While single-stage detectors [17,45,47] have the advantages of simplicity and fast inference, two-stage methods [6,15,34] are often superior in producing precisely localized bounding boxes. Different representations of LiDAR point clouds have been proposed: 3D voxel grids [18], range view (RV) projections [6,19,27], bird's eye view (BEV) projections [44,45,47], and point sets [32,34,35] are amongst the most popular. While 3D voxel grids are slow and wasteful to process due to the size of the volume which is mainly sparse, range view projections are dense representations by nature. However, RV images suffer from the large variance in object size and shape due to the projection. BEV projections achieve a better trade-off between accuracy and speed. Voxel features represented with either simple statistics [6,45] or learned representations [17] have been proposed. In this paper, we use a single-stage detector with BEV representation for its simplicity, effectiveness and efficiency.

Fig. 1. LiDAR and Radar sensor data: We show LiDAR data in white, dynamic Radar returns (with radial velocity) in red, static Radar returns in yellow, and object labels in blue. (Color figure online)

Exploiting Radar for Perception: Radar has long been used in ADAS for adaptive cruise control and collision avoidance due to its cost and robustness to severe weather conditions. Recently, Radar has been exploited in many other applications, spanning across free space estimation [25,38], object detection [4,8,29], object classification [13,31,43] and segmentation [25,33,38]. However, most of these methods treat Radar as another 3D sensor, ignoring its high-fidelity velocity information. In contrast, we exploit both Radar's geometric and dynamic information thanks to a novel specialized fusion mechanism for each type of information.

Sensor Fusion with Radar: In many self-driving perception systems, Radar data has been fused at the perception output level in the form of object tracks [7,9,10]. Kalman Filter [39] or IMM [2] trackers are popular approaches to digest Radar data, and the resulting tracks are then fused with object tracks from other

Table 1. Hardware comparison between LiDAR and Radar sensors.

Sensor modality	Detection range	Range accuracy	Azimuth resolution	Velocity accuracy
LiDAR	100 m	2 cm	0.1°–0.4°	–
Radar	250 m	10 cm near range	3.2°–12.3° near range	0.1 km/h
		40 cm far range	1.6° far range	

sensors. However, sensor fusion is not exploited during the process of generating those object tracks. Recent works also look at fusion between Radar and cameras within the perception system. Different Radar representations are proposed to facilitate fusion: spectrogram images [22], sparse locations in image space [29], pseudo-image by projecting to image space [4,30], BEV representation [28] and object detections [16]. However, these methods do not have high accuracy in terms of 3D perception. Instead, here we choose to fuse Radar with LiDAR and design a multi-level fusion mechanism that outperforms the state-of-the-art in self-driving.

3 Review of LiDAR and Radar Sensors

We first provide a review of LiDAR and Radar sensors and introduce our notation. We hope this short review can help readers better understand the intuitions behind our model designs, which will be described in the next section.

LiDAR (light detection and ranging) sensors can be divided into three main types: spinning LiDAR, solid state LiDAR, and flash LiDAR. In this paper we focus on the most common type: spinning LiDAR. This type of LiDAR emits and receives laser light pulses in 360° and exploits the time of flight (ToF) to calculate the distance to the obstacles. As a result, LiDAR data is generated as a continuous stream of point clouds. We denote each LiDAR point as a vector $P = (x, y, z, t)$, encoding the 3D position and the capture timestamp. In practice we often divide the LiDAR data into consecutive 360° sweeps for frame-wise point cloud processing. LiDAR is the preferred sensor for most self-driving vehicles due to its accurate 3D measurements. The main drawbacks are its sensitivity to dirt (which leads to poor performance in fog, rain and snow), cold (that causes exhaust plumes) as well as the lack of reflectivity of certain materials (such as windows and certain paints). Furthermore, its density decreases with range, making long range detection challenging.

Radar (radio detection and ranging) sensors work similarly as LiDAR, but transmit electromagnetic waves to sense the environment. The Radar outputs can be organized in three different levels: raw data in the form of time-frequency spectrograms, clusters from applying DBSCAN [12] or CFAR [37] on raw data, and tracks from performing object tracking on the clusters. From one representation to the next, the data sparsity and abstraction increases, while the noise in the data decreases. In this paper we focus on the mid-level data form, Radar

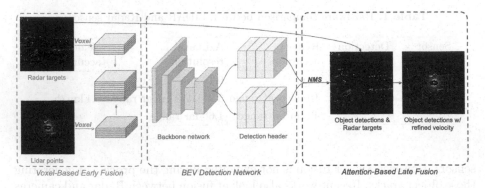

Fig. 2. RadarNet: Multi-level LiDAR and Radar fusion is performed for accurate 3D object detection and velocity estimation.

clusters, for its good balance between information richness and noise. In the following we refer to these clusters as Radar targets. We denote each Radar target as a vector $Q = (\mathbf{q}, v_{\parallel}, m, t)$, where $\mathbf{q} = (x, y)$ is the 2D position in BEV, $v_{\parallel}$ is a scalar value representing the radial velocity, m is a binary value indicating whether the target is moving or not, and t is the capture timestamp. The main advantages of Radar are that it provides instantaneous velocity measurements and is robust to various weather conditions. However, its drawbacks are also significant. It has a low resolution and thus it is difficult to detect small objects. There are ambiguities (a modulo function) in range and velocity due to Radar aliasing, as well as false positive detections from clutter and multi-path returns. It is also worth noting that the objects' real-world velocities (2D vectors in BEV) are ambiguous given only the radial velocity. Therefore we need to additionally estimate the tangential velocity or the 2D velocity direction in order to properly utilize the radial velocity.

We compare LiDAR and Radar data both quantitatively and qualitatively. We visualize both sensors' data from the nuScenes dataset [3] in Fig. 1, and we compare their technical specifications in Table 1. Note that LiDAR outperforms Radar in both accuracy and resolution by over an order of magnitude. The accurate 3D surface measurements makes LiDAR the first choice for high-precision 3D object detection. Radar can provide complementary information in two aspects: more observations at long range and instantaneous velocity evidence from the Doppler effect. We thus argue that since these sensors are very complementary, their combination provides a superior solution for self-driving.

4   Exploiting LiDAR and Radar for Robust Perception

In this section we present our novel approach to 3D perception, involving 3D object detection and velocity estimation. We refer the reader to Fig. 2 for an illustration of the overall architecture of our approach. To fully exploit the complementary information of the two sensor modalities and thereby benefit both

object detection and velocity estimation, we propose two sensor fusion mechanisms, namely *early fusion* and *late fusion*, that operate at different granularities.
More specifically, while early fusion learns joint representations from both sensor
observations, late fusion refines object velocities via an attention-based association and aggregation mechanism between object detections and Radar targets.

4.1 Exploiting Geometric Information via Early Fusion

LiDAR Voxel Representation: We take multiple sweeps of LiDAR point
clouds (those within the past 0.5 seconds) as input so that the model has enough
information to infer the objects' motion while still being able to run in real-time.
All point cloud sweeps are transformed to the ego-vehicle's centric coordinates at
the current frame. Note that this is easy to do as sensors are calibrated and the
vehicle pose is estimated by the localization system. Following FAF [26], we adopt
a bird's eye view (BEV) representation and concatenate multiple height slices
and sweeps together along the channel dimension. We use a weighted occupancy
value as each voxel's feature representation. Specifically, for each voxel, if no
point falls in it, the voxel's value is 0. If one or more points $\{(x_i, y_i, z_i), i =
1 \dots N\}$ fall into it, the voxel's value is defined as $\sum_i (1 - \frac{|x_i - a|}{dx/2})(1 - \frac{|y_i - b|}{dy/2})(1 -
\frac{|z_i - c|}{dz/2})$, where (a, b, c) is the voxel's center and (dx, dy, dz) is the voxel's size.

Radar Voxel Representation: Similar to how we accumulate multiple sweeps
of LiDAR data, we also take multiple cycles of Radar data as input, in the
same coordinate system as LiDAR. We keep only the (x, y) position of Radar
targets and ignore the height position as it is often inaccurate (if it ever exists).
As a result, each cycle of Radar data can be voxelized as one BEV image. We
concatenate multiple cycles along the channel dimension and use a motion-aware
occupancy value as the feature for each voxel. Specifically, for each BEV voxel, if
no Radar target falls into it, the voxel's value is 0. If at least one moving Radar
target (i.e., $m = 1$) falls into it, the voxel's value is 1. If all Radar targets falling
into it are static, the voxel's value is -1.

Early Fusion: We use the same BEV voxel size for LiDAR and Radar data.
Thus their voxel representations have the same size in BEV space. We perform
early fusion by concatenating them together along the channel dimension.

4.2 Detection Network

We adopt a single-stage anchor-free BEV object detector with additional velocity
estimation in the detection header.

Backbone Network: We adopt the same backbone network architecture as
PnPNet [21]. The backbone network is composed of three initial convolution
layers, three consecutive multi-scale inception blocks [40], and a feature pyramid

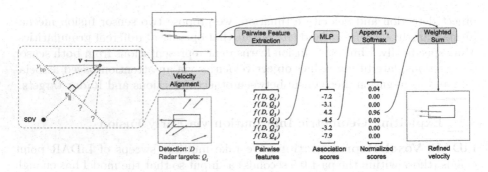

Fig. 3. Attention-based late fusion of object detection and Radar targets: In the figure we show an example of fusing Radar with one detection, while in practice this is applied to all detections in parallel. We first align the radial velocities of Radar targets with the detection's motion direction, then predict pairwise association scores for all detection-Radar pairs. The refined velocity is computed as a weighted sum of all Radar evidences as well as the original velocity estimate. (Color figure online)

network [23]. The three initial convolution layers down-sample the voxel input by 4 and output 64-D feature maps. The inception block consists of three branches, each with a down-sampling ratio of 1×, 2× and 4× implemented by stride of the first convolution. The number of convolution layers in each branch is 2, 4 and 6, and the number of feature channels in each branch is 32, 64 and 96. The feature pyramid network merges multi-scale feature maps from the inception block into one, with 256 channels for each layer. The final output of the backbone network is a 256-D feature map with a 4× down-sampling ratio compared to the voxel input.

Detection Header: We apply a fully-convolutional detection header [24] for anchor-free dense detection, which consists of a classification branch and a regression branch, each with 4 convolution layers and 128 channels. The detection is parameterized as $D = (c, x, y, w, l, \theta, \mathbf{v})$, which represents the confidence score, the object's center position in BEV, its width, length and orientation, and its 2-D velocity $\mathbf{v} = (v_x, v_y)$ in BEV. The classification branch predicts the confidence score c, while the regression branch predicts all the other terms $(x - p_x, y - p_y, w, l, \cos(\theta), \sin(\theta), m, v_x, v_y)$, where p_x and p_y are the 2D coordinates of every voxel center and m is an additional term that indicates the probability of moving. During inference, we set the 2-D velocity to $(0, 0)$ if the predicted probability of moving is smaller than 50%.

4.3   Exploiting Dynamic Information via Late Fusion

While early fusion exploits the position and density information of Radar targets, late fusion is designed to explicitly exploit the Radar's radial velocity evidence. Due to the lack of tangential information, the actual object velocity vector is ambiguous given the radial velocity alone. To address this issue, we propose to use the velocity estimation in object detections to align the radial velocity, which is simply back-projecting the radial velocity to the motion direction of the detection. We refer the reader to Fig. 3 for an illustration. It is thus apparent that the radial velocity is more confident when the angle between the radial direction and motion direction is small, as when it is close to 90°, a very small variance in radial velocity will be exaggerated by back-projection.

Given a set of object detections and Radar targets, the key of fully exploiting Radar data lies in solving the following two tasks: (1) *association* of each Radar target with the correct object detection for velocity alignment; (2) *aggregation* to combine the velocity estimates from detection and associated Radar targets robustly. Both tasks are non-trivial to solve. The association is not a one-to-one mapping as there are many objects without any associated Radar targets, and there are also objects with multiple Radar targets. False positives and noisy positions of Radar targets also make association difficult. For the aggregation problem, it is hard to estimate the uncertainty of the Radar velocity as it also depends on the associated detection.

In this paper, we propose an attention-based mechanism that learns from data to both associate and aggregate. This is illustrated in Fig. 3. Specifically, given pairwise features defined between each object detection and Radar target, we first compute pairwise association scores via a learnable matching function. We then aggregate each detection with all Radar targets according to the normalized association scores to get the refined velocity estimate. Note that late fusion is performed on *dynamic* Radar targets only.

Pairwise Detection-Radar Association: Given an object detection denoted as $D = (c, x, y, w, l, \theta, \mathbf{v})$ and a Radar target denoted as $Q = (\mathbf{q}, v_{\parallel}, m, t)$, we first define their pairwise feature as follows:

$$f(D, Q) = (f^{\text{det}}(D), f^{\text{det-radar}}(D, Q)) \tag{1}$$

$$f^{\text{det}}(D) = (w, l, \|\mathbf{v}\|, \frac{v_x}{\|\mathbf{v}\|}, \frac{v_y}{\|\mathbf{v}\|}, \cos(\gamma)) \tag{2}$$

$$f^{\text{det-radar}}(D, Q) = (\mathrm{d}x, \mathrm{d}y, \mathrm{d}t, v^{\text{bp}}) \tag{3}$$

$$v^{\text{bp}} = \min(50, \frac{v_{\parallel}}{\cos(\phi)}) \tag{4}$$

where $(\cdot, \cdot)$ indicates the concatenation operator, γ is the angle between D's motion direction and D's radial direction, ϕ is the angle between D's motion direction and Q's radial direction, v^{bp} is the back-projected radial velocity (capped $50\,\text{m/s}$ to avoid very large values), and $(\mathrm{d}x, \mathrm{d}y, \mathrm{d}t)$ are the offsets in BEV positions and timestamps of D and Q.

We then compute the pairwise association score by feeding the above feature to a learnable matching function:

$$s_{i,j} = \text{MLP}_{\text{match}}(f(D_i, Q_j)) \tag{5}$$

In our case the matching function is parameterized as a Multi-Layer Perceptron (MLP) with five layers with 32, 64, 64, 64 and 1 channels respectively.

Velocity Aggregation: We compute the association scores for all detections and Radar target pairs and refine the velocity estimate of each detection D_i by aggregating information from all Radar targets. Towards this goal, we first normalize the association scores of all Radar targets to sum to 1. We append an additional score of 1 before normalization to handle cases with no association.

$$\mathbf{s}_i^{\text{norm}} = \texttt{softmax}((1, s_{i,:})) \tag{6}$$

We then refine the velocity magnitude by summing all the candidates (the detection itself as well as all Radar targets) weighted by their normalized scores:

$$v_i' = \mathbf{s}_i^{\text{norm}} \cdot (\|\mathbf{v}_i\|, v_{i,:}^{\text{bp}})^\top \tag{7}$$

The 2D velocity estimate is then computed as the refined velocity magnitude:

$$\mathbf{v}' = v' \cdot \left(\frac{v_x}{\|\mathbf{v}\|}, \frac{v_y}{\|\mathbf{v}\|}\right) \tag{8}$$

where the detection index i is omitted for brevity.

4.4   Learning and Inference

We trained the proposed LiDAR and Radar fusion model with a multi-task loss defined as a weighted sum of the detection loss, velocity loss on the detection output, as well as the velocity loss on the late fusion output:

$$\mathcal{L} = (\mathcal{L}_{\text{cls}}^{\text{det}} + \alpha \cdot \mathcal{L}_{\text{reg}}^{\text{det}}) + \beta \cdot (\mathcal{L}_{\text{cls}}^{\text{velo}} + \mathcal{L}_{\text{reg}}^{\text{velo}}) + \delta \cdot \mathcal{L}_{\text{reg}}^{\text{velo\_attn}} \tag{9}$$

where $\mathcal{L}_{\text{cls}}^{\text{det}}$ is the cross-entropy loss on classification score c, $\mathcal{L}_{\text{reg}}^{\text{det}}$ is the smooth ℓ_1 loss summed over the position, size and orientation terms, $\mathcal{L}_{\text{cls}}^{\text{velo}}$ is the cross-entropy loss on moving probability m, $\mathcal{L}_{\text{reg}}^{\text{velo}}$ is the smooth ℓ_1 loss on $\mathbf{v}$, and $\mathcal{L}_{\text{reg}}^{\text{velo\_attn}}$ is the smooth ℓ_1 loss on $\mathbf{v}'$. α, β and δ are scalars that balance different tasks. Note that we do not require explicit supervision to learn object and Radar association, which is an advantage of the attention-based late fusion module where the association is implicitly learned.

We use the Adam optimizer [14] with batch normalization [11] after every convolution layer and layer normalization [1] after every fully-connected layer (except for the final output layer). For detection we use hard negative mining. $\mathcal{L}_{\text{reg}}^{\text{det}}$, $\mathcal{L}_{\text{cls}}^{\text{velo}}$ and $\mathcal{L}_{\text{reg}}^{\text{velo}}$ are computed on positive samples only, and $\mathcal{L}_{\text{reg}}^{\text{velo\_attn}}$ is computed on true positive detections only. We apply the same post-processing to generate final detections during training and testing phases, where the top 200 detections per class are kept and NMS is applied thereafter.

5   Experimental Evaluation

5.1   Datasets and Evaluation Metrics

nuScenes: We validate the proposed method on the nuScenes dataset [3]. This dataset contains sensor data from 1 LiDAR and 5 Radars, with object labels at 2 Hz. Velocity labels are computed as finite difference between consecutive frames. Since we focus on dynamic objects, we evaluate on two challenging object classes: cars and motorcycles, as their velocities have high variance. We follow the official training/validation split with 700/150 logs each. We report the model performance on object detection and velocity estimation. Average Precision (AP) is used as the detection metric, which is defined on center distance in BEV between the detection and the label. The final AP is averaged over four different distance thresholds (0.5 m, 1 m, 2 m and 4 m). Average Velocity Error (AVE) is used as the velocity metric, which is computed as the ℓ_2 velocity error averaged over all true positive detections (2 m threshold). Cars are evaluated 50 m range, while motorcycles are evaluated 40 m range. Labels with 0 LiDAR and Radar points are ignored.

DenseRadar: One advantage of Radar over LiDAR is its longer sensing range. To showcase this, we further evaluate our model on a self-collected dataset, called *DenseRadar*, with vehicle labels 100 m range for 5002 snippets. Velocity labels are estimated by fitting a kinematic bicycle model to the trajectory, which produces smoother velocities compared with the finite difference procedure employed in nuScenes. We use similar metrics as nuScenes. For detection we compute AP at 0.7 IoU in BEV. For velocity we report Average Dynamic Velocity Error (ADVE) on *dynamic* objects only. We make a training/validation split with 4666/336 logs each.

5.2   Implementation Details

We train a two-class model on nuScenes with a shared backbone network and class-specific detection headers. Global data augmentation is used during training, with random translations from $[-1, 1]$m in the X and Y axes and $[-0.2, 0.2]$m in the Z axis, random scaling from $[0.95, 1.05]$, random rotation from $[-45°, 45°]$ along the Z axis, and random left-right and front-back flipping. We do not apply augmentation at test time. To alleviate the class imbalance, we duplicate training frames that contain motorcycles by 5 times. The model is trained for 25 epochs with a batch size of 32 frames on 8 GPUs. We use an input voxel size 0.125 m in the X and Y axes, 0.2 m in the Z axis. We use $\alpha = 1$ and $\beta = \delta = 0.1$. Hyper-parameter tuning is conducted on the train-detect/train-track split.

We train a single-class model on DenseRadar. Since the dataset is much larger, we do not apply data augmentation. We use an input voxel resolution 0.2 m in all three axes due to the extra computation due to the longer detection range. We use $\alpha = 1$ and $\beta = \delta = 0.5$. The model is trained for 1.5 epochs.

Table 2. Comparison with the state-of-the-art on nuScenes validation set.

Method	Input	Cars			Motorcycles		
		AP↑	AP@2m↑	AVE↓	AP↑	AP@2m↑	AVE↓
MonoDIS [36]	I	47.8	64.9	–	28.1	37.7	–
PointPillar [17]	L	70.5	76.1	0.269	20.0	22.8	0.603
PointPillar+ [41]	L	76.7	80.5	0.209	35.0	38.6	0.371
PointPainting [41]	L+I	78.8	82.9	0.206	44.4	48.1	0.351
3DSSD [46]	L	81.2	85.8	0.188	36.0	39.9	0.356
CBGS [48]	L	82.3	85.9	0.230	50.6	52.4	0.339
RadarNet (Ours)	L+R	**84.5**	**87.9**	**0.175**	**52.9**	**55.6**	**0.269**

5.3   Comparison with the State-of-the-Art

We compare our LiDAR and Radar fusion model with other state-of-the-art perception models on nuScenes and show the evaluation results in Table 2. Specifically, we compare with the camera-based method MonoDIS [36], the LiDAR-based methods PointPillar [17], PointPillar+ [41], 3DSSD [46], CBGS [48], and the LiDAR and camera fusion method PointPainting [41]. RadarNet outperforms all methods significantly in both detection AP and velocity error. Compared with the second best on cars/motorcycles, our model shows an absolute gain of 2.2%/2.3% in detection AP and a relative reduction of 7%/21% in velocity error.

5.4   Ablation Study

We conduct an ablation study on the nuScenes and DenseRadar datasets to validate the effectiveness of our two-level fusion scheme. To better verify the advantage of the proposed attention-based late fusion, we build a strong baseline with carefully designed heuristics. Recall that our attention-based late fusion consists of two steps: association and aggregation. As a counterpart, we build the baseline fusion method by replacing each step with heuristics. In particular, for each detection candidate, we first use a set of rules to determine the Radar targets associated with it. Given a set of associated Radar targets (if any), we then take the median of their aligned velocities (by back-projecting to the motion direction of the detection) as the estimate from Radar and average it with the initial velocity estimate of the detection. If there are no associated Radar targets, we keep the original detection velocity.

Below we define the set of rules we designed for determining the associated Radar targets. Given the features in Eq. 2 and Eq. 3, a Radar target is considered as associated if it meets all of the following conditions:

$$\sqrt{(dx)^2 + (dy)^2} < 3 \text{ m} \tag{10}$$

$$\gamma < 40° \tag{11}$$

$$\|\mathbf{v}\| > 1 \text{ m/s} \tag{12}$$

$$v^{bp} < 30 \text{ m/s} \tag{13}$$

Table 3. Ablation study on nuScenes validation set.

Model	LiDAR	Radar		Cars		Motorcycles	
		Early	Late	AP@2m↑	AVE↓	AP@2m↑	AVE↓
LiDAR	✓	–	–	87.6	0.203	53.7	0.316
Early	✓	✓	–	+0.3	−2%	+1.9	−0%
Heuristic	✓	✓	Heuristic	+0.3	−9%	+1.9	−4%
RadarNet	✓	✓	Attention	**+0.3**	**−14%**	**+1.9**	**−15%**

Table 4. Ablation study on DenseRadar validation set.

Model	LiDAR	Radar		Vehicles AP ↑			ADVE ↓
		Early	Late	0–40	40–70	70–100	
LiDAR	✓	–	–	95.4	88.0	77.5	0.285
Early	✓	✓	–	+0.3	+0.5	+0.8	−3%
Heuristic	✓	✓	Heuristic	+0.3	+0.5	+0.8	−6%
RadarNet	✓	✓	Attention	**+0.3**	**+0.5**	**+0.8**	**−19%**

We define these rules to filter out unreliable Radar targets, and the thresholds are chosen via cross-validation.

Evaluation on nuScenes: We show ablation results on nuScenes in Table 3. Note that our LiDAR only model already achieves state-of-the-art performance. Adding early fusion improves detection of motorcycles by 1.9% absolute AP, as the LiDAR observations are sparse and therefore Radar data serves as additional evidence. Early fusion does not affect the velocity performance much as only density information is exploited at present. When it comes to late fusion, our approach achieves over 14% velocity error reduction, significantly outperforming the heuristic baseline especially in motorcycles, where we typically have few Radar targets and therefore more noise.

Evaluation on DenseRadar: Ablation results on DenseRadar are depicted in Table 4. We show detection APs in near range (0–40), mid range (40–70) and long range (70–100) respectively. Early fusion helps long-distance object detection, bringing 0.8% absolute gain in the 70–100 range detection AP. When late fusion is added, larger improvements are achieved than on nuScenes (from 14% to 19%). Two reasons may account for this: (1) DenseRadar uses higher-end Radar sensors that produce denser returns; (2) we evaluate in longer range (100 m vs. 50 m), which is more challenging and therefore there is more room for improvement. However, the heuristic baseline still gets lower than 10% gain, showing the advantage of the proposed attention-based mechanism which can learn from noisy data.

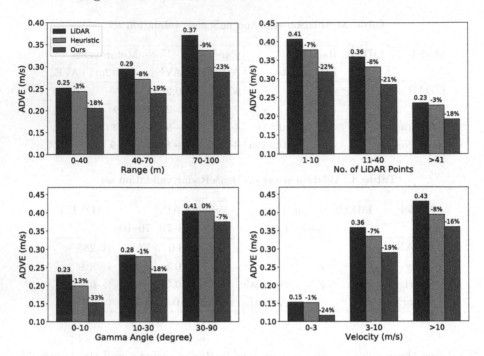

Fig. 4. Fine-grained evaluation of velocity estimation on DenseRadar validation set.

5.5   Fine-Grained Analysis

To better understand in which aspects the velocity estimation performance is improved by exploiting Radar we conduct fine-grained evaluation on the larger-scale DenseRadar dataset with respect to different subsets of object labels. In particular, we create different subsets of labels by varying the object distance to the ego vehicle, number of observed LiDAR points, angle γ between motion direction and radial direction, and the velocity magnitude.

We compare three model variants: LiDAR only, our model with heuristic late fusion and our model in Fig. 4. From the results we see that the heuristic model brings negligible gains when $\gamma > 10°$ or $\|\mathbf{v}\| < 3$ m/s. This justifies the $40°$ and 1 m/s thresholds in our heuristics as these are cases where Radar data contain large uncertainty. In contrast, our attention-based model consistently and significantly outperforms the heuristic model under all conditions, showing its effectiveness in capturing sensor uncertainties and exploiting both sensors.

5.6   Qualitative Results

In Fig. 5 we show the learned detection and Radar associations. Results are shown in sequence for each object to illustrate the temporal change in the association. From the results we observe that: (1) the association is sparse in that only relevant Radar targets are associated; (2) the association is quite robust to

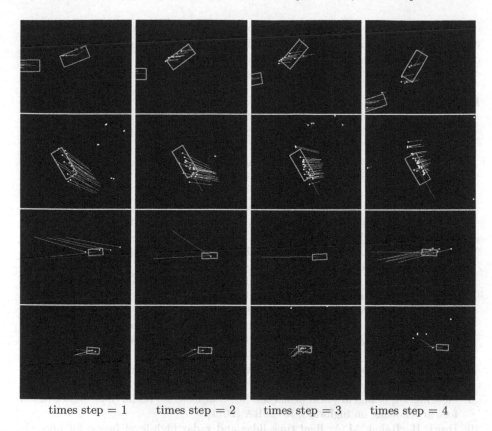

times step = 1   times step = 2   times step = 3   times step = 4

Fig. 5. Qualitative Results: Visualization of learned detections and Radar associations for cars (row 1 & 2) and motorcycles (row 3 & 4) on nuScenes validation set. Each row corresponds to the same object across time. We draw object detections in cyan, Radar targets within past 0.5 s in white, and associated Radar targets with > 0.1 normalized score in yellow. (Color figure online)

noisy locations of the Radar targets; (3) the model captures the uncertainty of Radar targets very well. For example, when the radial direction is near tangential to the object's motion direction, the model tends to not associate any Radar targets as in such cases the Radar evidence is often very unreliable.

6   Conclusion

We have proposed a new method to exploit Radar in combination with LiDAR for robust perception of dynamic objects in self-driving. To exploit geometric information from Radar, we use a voxel-based early fusion approach, which is shown to improve long-distance object detection due to Radar's longer sensing range. To exploit dynamic information, we propose an attention-based late fusion approach, which addresses the critical problem of associating Radar targets and objects without ground-truth association labels. By learning to associate and aggregate information, a significant performance boost in velocity estimation is observed under various conditions.

References

1. Ba, J.L., Kiros, J.R., Hinton, G.E.: Layer normalization. arXiv preprint arXiv:1607.06450 (2016)
2. Blom, H.A., Bar-Shalom, Y.: The interacting multiple model algorithm for systems with Markovian switching coefficients. IEEE Trans. Autom. Control **33**, 780–783 (1988)
3. Caesar, H., et al.: nuScenes: ldataset for autonomous driving. In: CVPR (2020)
4. Chadwick, S., Maddetn, W., Newman, P.: Distant vehicle detection using radar and vision. In: ICRA (2019)
5. Chen, X., Kundu, K., Zhang, Z., Ma, H., Fidler, S., Urtasun, R.: Monocular 3D object detection for autonomous driving. In: CVPR (2016)
6. Chen, X., Ma, H., Wan, J., Li, B., Xia, T.: Multi-view 3D object detection network for autonomous driving. In: CVPR (2017)
7. Cho, H., Seo, Y.W., Kumar, B.V., Rajkumar, R.R.: A multi-sensor fusion system for moving object detection and tracking in urban driving environments. In: 2014 IEEE International Conference on Robotics and Automation (ICRA), pp. 1836–1843. IEEE (2014)
8. Danzer, A., Griebel, T., Bach, M., Dietmayer, K.: 2D car detection in radar data with pointNets. In: ITSC (2019)
9. Göhring, D., Wang, M., Schnürmacher, M., Ganjineh, T.: Radar/lidar sensor fusion for car-following on highways. In: ICRA (2011)
10. Hajri, H., Rahal, M.C.: Real time lidar and radar high-level fusion for obstacle detection and tracking with evaluation on a ground truth. Int. J. Mech. Mechatron. Eng. (2018)
11. Ioffe, S., Szegedy, C.: Batch normalization: accelerating deep network training by reducing internal covariate shift. In: ICML (2015)
12. Kellner, D., Klappstein, J., Dietmayer, K.: Grid-based DBSCAN for clustering extended objects in radar data. In: IEEE Intelligent Vehicles Symposium (2012)
13. Kim, S., Lee, S., Doo, S., Shim, B.: Moving target classification in automotive radar systems using convolutional recurrent neural networks. In: 26th European Signal Processing Conference (EUSIPCO) (2018)
14. Kingma, D., Ba, J.: Adam: a method for stochastic optimization. In: ICLR (2015)
15. Ku, J., Mozifian, M., Lee, J., Harakeh, A., Waslander, S.: Joint 3D proposal generation and object detection from view aggregation. In: IROS (2018)
16. Kuang, H., Liu, X., Zhang, J., Fang, Z.: Multi-modality cascaded fusion technology for autonomous driving. In: 4th International Conference on Robotics and Automation Sciences (ICRAS) (2020)
17. Lang, A.H., Vora, S., Caesar, H., Zhou, L., Yang, J., Beijbom, O.: PointPillars: fast encoders for object detection from point clouds. In: CVPR (2019)
18. Li, B.: 3D fully convolutional network for vehicle detection in point cloud. In: IROS (2017)
19. Li, B., Zhang, T., Xia, T.: Vehicle detection from 3D lidar using fully convolutional network. In: RSS (2016)

20. Liang, M., Yang, B., Wang, S., Urtasun, R.: Deep continuous fusion for multi-sensor 3D object detection. In: Ferrari, V., Hebert, M., Sminchisescu, C., Weiss, Y. (eds.) ECCV 2018. LNCS, vol. 11220, pp. 663–678. Springer, Cham (2018). https://doi.org/10.1007/978-3-030-01270-0_39
21. Liang, M., et al.: Object trajectory evolution for end-to-end perception and prediction. In: CVPR (2020)
22. Lim, T.Y., et al.: Radar and camera early fusion for vehicle detection in advanced driver assistance systems. In: Machine Learning for Autonomous Driving Workshop at the 33rd Conference on Neural Information Processing Systems (2019)
23. Lin, T.Y., Dollár, P., Girshick, R., He, K., Hariharan, B., Belongie, S.: Feature pyramid networks for object detection. In: CVPR (2017)
24. Lin, T.Y., Goyal, P., Girshick, R., He, K., Dollár, P.: Focal loss for dense object detection. In: ICCV (2017)
25. Lombacher, J., Laudt, K., Hahn, M., Dickmann, J., Wöhler, C.: Semantic radar grids. In: IEEE Intelligent Vehicles Symposium (IV) (2017)
26. Luo, W., Yang, B., Urtasun, R.: Fast and furious: real time end-to-end 3D detection, tracking and motion forecasting with a single convolutional net. In: CVPR (2018)
27. Meyer, G.P., Laddha, A., Kee, E., Vallespi-Gonzalez, C., Wellington, C.K.: LaserNet: an efficient probabilistic 3D object detector for autonomous driving. In: CVPR (2019)
28. Meyer, M., Kuschk, G.: Deep learning based 3D object detection for automotive radar and camera. In: 16th European Radar Conference (EuRAD) (2019)
29. Nabati, R., Qi, H.: RRPN: radar region proposal network for object detection in autonomous vehicles. In: ICIP (2019)
30. Nobis, F., Geisslinger, M., Weber, M., Betz, J., Lienkamp, M.: A deep learning-based radar and camera sensor fusion architecture for object detection. In: Sensor Data Fusion: Trends, Solutions, Applications (SDF) (2019)
31. Patel, K., Rambach, K., Visentin, T., Rusev, D., Pfeiffer, M., Yang, B.: Deep learning-based object classification on automotive radar spectra. In: RadarConf (2019)
32. Qi, C.R., Liu, W., Wu, C., Su, H., Guibas, L.J.: Frustum PointNets for 3D object detection from RGB-D data. In: CVPR (2018)
33. Schumann, O., Hahn, M., Dickmann, J., Wöhler, C.: Semantic segmentation on radar point clouds. In: FUSION (2018)
34. Shi, S., Wang, X., Li, H.: PointRCNN: 3D object proposal generation and detection from point cloud. In: CVPR (2019)
35. Shi, W., Rajkumar, R.: Point-GNN: graph neural network for 3D object detection in a point cloud. In: CVPR (2020)
36. Simonelli, A., Bulo, S.R., Porzi, L., López-Antequera, M., Kontschieder, P.: Disentangling monocular 3D object detection. In: ICCV (2019)
37. Skolnik, M.I.: Radar Handbook, 2nd edn. McGrawHill, LOndon (1990)
38. Sless, L., El Shlomo, B., Cohen, G., Oron, S.: Road scene understanding by occupancy grid learning from sparse radar clusters using semantic segmentation. In: Proceedings of the IEEE International Conference on Computer Vision Workshops (2019)
39. Sun, S.L., Deng, Z.L.: Multi-sensor optimal information fusion Kalman filter. Automatica 40(6), 1017–1023 (2004)
40. Szegedy, C., et al.: Going deeper with convolutions. In: CVPR (2015)
41. Vora, S., Lang, A.H., Helou, B., Beijbom, O.: PointPainting: sequential fusion for 3D object detection. In: CVPR (2020)

42. Weng, X., Kitani, K.: Monocular 3D object detection with pseudo-lidar point cloud. In: ICCVW (2019)
43. Wöhler, C., Schumann, O., Hahn, M., Dickmann, J.: Comparison of random forest and long short-term memory network performances in classification tasks using radar. In: Sensor Data Fusion: Trends, Solutions, Applications (SDF) (2017)
44. Yan, Y., Mao, Y., Li, B.: Second: sparsely embedded convolutional detection. Sensors 18, 3337 (2018)
45. Yang, B., Luo, W., Urtasun, R.: PIXOR: real-time 3D object detection from point clouds. In: CVPR (2018)
46. Yang, Z., Sun, Y., Liu, S., Jia, J.: 3DSSD: point-based 3D single stage object detector. In: CVPR (2020)
47. Zhou, Y., Tuzel, O.: VoxelNet: end-to-end learning for point cloud based 3D object detection. In: CVPR (2018)
48. Zhu, B., Jiang, Z., Zhou, X., Li, Z., Yu, G.: Class-balanced grouping and sampling for point cloud 3D object detection. arXiv preprint arXiv:1908.09492 (2019)

Seeing the Un-Scene: Learning Amodal Semantic Maps for Room Navigation

Medhini Narasimhan[1,2]([⊠]), Erik Wijmans[1,3], Xinlei Chen[1], Trevor Darrell[2], Dhruv Batra[1,3], Devi Parikh[1,3], and Amanpreet Singh[1]

[1] Facebook AI Research, Menlo Park, USA
[2] University of California, Berkeley, Berkeley, USA
medhini@berkeley.edu
[3] Georgia Institute of Technology, Atlanta, USA

Abstract. We introduce a learning-based approach for room navigation using semantic maps. Our proposed architecture learns to predict top-down belief maps of regions that lie beyond the agent's field of view while modeling architectural and stylistic regularities in houses. First, we train a model to generate amodal semantic top-down maps indicating beliefs of location, size, and shape of rooms by learning the underlying architectural patterns in houses. Next, we use these maps to predict a point that lies in the target room and train a policy to navigate to the point. We empirically demonstrate that by predicting semantic maps, the model learns common correlations found in houses and generalizes to novel environments. We also demonstrate that reducing the task of room navigation to point navigation improves the performance further.

Keywords: Embodied AI · Room navigation

1 Introduction

Humans have an uncanny ability to seamlessly navigate in unseen environments by quickly understanding their surroundings. Consider the example in Fig. 1. You're visiting a friend's home for the first time and you want to go to the kitchen. You're in the entryway and you look around to observe a bedroom in one direction and the dining room in the other direction. A possible, but tedious solution is to head in a random direction and exhaustively search the space until you end up in the kitchen. Another option, and most probably the one you'd pick, is to walk towards the dining room as you are more likely to find the kitchen near the dining room rather than the bedroom. We believe that there are underlying architectural principles which govern the design of houses

M. Narasimhan—Work done while an intern at Facebook AI Research.

Electronic supplementary material The online version of this chapter (https://doi.org/10.1007/978-3-030-58523-5_30) contains supplementary material, which is available to authorized users.

© Springer Nature Switzerland AG 2020
A. Vedaldi et al. (Eds.): ECCV 2020, LNCS 12363, pp. 513–529, 2020.
https://doi.org/10.1007/978-3-030-58523-5_30

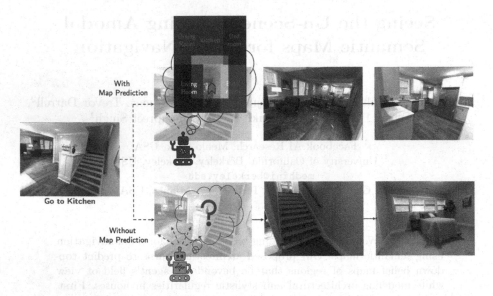

Fig. 1. We train an agent that can navigate to rooms in unseen environments by predicting amodal semantic maps. Here, the agent is placed in the entryway and is asked to navigate to the kitchen. It recognizes a dining room and predicts that the kitchen must be near, even though it has not seen it yet, and heads in that direction. It updates its semantic belief maps as it sees more of the house. Our agent learns to model architectural and stylistic regularities in houses, which helps it find its way in a novel house.

and our prior knowledge of house layouts helps us to navigate effectively in new spaces. We also improve our predictions of how the house is laid out as we walk around and gather more information. The goal of this work is to elicit a similar behaviour in embodied agents by enabling them to predict regions which lie beyond their field of view. The agent models correlations between the appearance and architectural layout of houses to efficiently navigate in unseen scenes.

Currently, there exist two paradigms for the navigation problem (1) Classical path planning based methods: SLAM-based approaches which first build a geometric map and then use path planning with localization for navigation [18,25,36]. (2) Learning-based methods: A policy is learned for a specific task in an environment [17,46,47]. In this work, we introduce a learning-based method that unlike previous approaches, predicts an intermediate representation that captures the model's current belief of the *semantic layout of the house*, beyond the agent's field of view. [43,46] represent priors in the form of knowledge graphs or probabilistic relation graphs which capture room location priors. However, we also want to learn other correlations such as estimating the shape of a room by observing parts of it. We choose to model correlations as semantic maps indicating the location, shape, and size of rooms which lie beyond the agent's field of view. This offers a more flexible representation.

In this work, we develop a novel technique to dynamically generate amodal semantic top-down maps of rooms by learning architectural regularities in houses and to use these predicted maps for room navigation. We define the task of room navigation as navigating to the nearest room of the specified type. For e.g., navigating to the bedroom closest to the starting point. We train an agent using supervision to predict regions on a map that lie beyond its field of view which forces it to develop beliefs about where a room might be present before navigating to it. The agent constantly updates these beliefs as it steps around the environment. The learned beliefs help the agent navigate in novel environments.

Contributions. (1) We introduce a novel learning-based approach for room navigation via amodal prediction of semantic maps. The agent learns architectural and stylistic regularities in houses to predict regions beyond its field of view. (2) Through carefully designed ablations, we show that our model trained to predict semantic maps as intermediate representations achieves better performance on unseen environments compared to a baseline which doesn't explicitly generate semantic top-down maps. (3) To evaluate our approach, we introduce the room navigation task and dataset in the Habitat platform [32].

2   Related Work

Navigation in Mobile Robotics. Conventional solutions to the navigation problem in robotics are comprised of two main steps: (1) mapping the environment and simultaneously localizing the agent in the generated map (2) path planning towards the target using the generated map. Geometric solutions to the mapping problem include (i) structure from motion and (ii) simultaneous localization and mapping (SLAM) [4,7,13,16,18,36]. Various SLAM algorithms have been developed for different sensory inputs available to the agent. Using the generated map, a path can be computed to the target location via several path planning algorithms [25]. These approaches fall under the passive SLAM category where the maps are generated ahead of time. On the other hand, active SLAM research focuses on dynamically controlling the camera for building spatial representations of the environment. Some works formulate active SLAM as Partially Observable Markov Decision Process and use either Bayesian Optimization [27] or Reinforcement Learning [23] to plan trajectories that lead to accurate maps. [8] and [35] use Rao-Blackwellized Particle Filters to choose the set of actions that maximize the information gain and minimize the uncertainty of the predicted maps.

A less studied yet actively growing area of SLAM research is incorporating semantics into SLAM [6,29,37,40]. [40] and [6] use semantic features for improved localization performance and for performing SLAM on dynamic scenes. [3] leverages conceptual spatial knowledge in the form of object co-occurrences for active object search. On the other hand, our work learns semantic maps of room locations for the task of Room Navigation. [11] summarizes the various ways of representing semantic information and using it for indoor navigation.

However, all of these techniques rely on sensor data which is highly susceptible to noise [10] and have no mechanism to learn or update beliefs across environments. The limitations and open challenges in SLAM have been outlined in [7]. This motivates learning based methods for navigation, which we describe next.

Learning Based Methods for Navigation. With the motivation of generalizing to novel environments and learning semantic cues, a number of end-to-end learning based approaches have been developed in the recent past [14,17,28,31, 47]. [31] use a topological graph for navigation and [14] propose a memory based policy that uses attention to exploit spatio-temporal dependencies. [28] jointly learn the goal-driven reinforcement learning problem with auxiliary depth prediction tasks. [42] introduce the RoomNav task in the House3D simulation platform and train a policy using deep deterministic policy gradient [26] to solve the same. [10] focus on building a task-agnostic exploration policy and demonstrate its usefulness for downstream navigation tasks. Most relevant to our work is Cognitive Mapping and Planning (CMP) [17]. It uses a differentiable mapper to learn an egocentric map of the environment and a differentiable planner that uses these maps alongside the goal to output navigational actions. These maps only indicate free space and contain no semantic information. On the other hand, we predict semantic top-down maps indicating the location, shape, and size of rooms in the house. Furthermore, unlike the maps in CMP which indicate regions within the agent's field of view, our maps model architectural regularities in houses and contain unseen regions which lie beyond the agent's field of view. [46] uses spatial and visual relationships between objects represented as a knowledge graph [24] for semantic navigation. The agent doesn't learn new priors or update existing beliefs during training. [43] estimates priors as probabilistic relationship graphs over semantic entities and uses these graphs for planning. However, their graphs don't capture information regarding size and shape of rooms and patterns in houses. Contrary to previous approaches, our work does not rely on pre-constructed maps or knowledge graphs representing priors. We dynamically learn amodal semantic belief maps which model architectural and stylistic regularities in houses. Further, the agent updates its beliefs as it moves around in an environment.

Vision-Language Navigation (VLN). A different but related task, language guided visual navigation was introduced by [2]. In VLN, an agent follows language instructions to reach a goal in a home. For *e.g.*, "Walk up the rest of the stairs, stop at the top of the stairs near the potted plant". There are multiple works which attempt to solve this problem [15,38,39]. The room navigation task introduced here is different in that the agent doesn't receive any language based instructions, just the final goal in the form of a room type. In [2], the agent "teleports" from one location to another on sparse pre-computed navigation graph and can never collide with anything. Our Room Navigation task starts from the significantly more realistic setting as in Point Navigation [32] where the agent takes the low-level actions such as move-forward (0.25m), turn-left/right (10°) and learns to avoid collisions. Compared to VLN, the navigation in our task is significantly more challenging. We'd like to highlight that the methods devel-

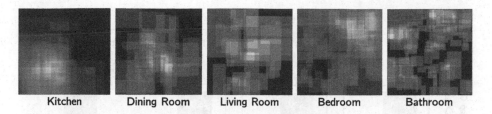

Fig. 2. Architectural patterns in Matterport 3D. All homes are scaled and aligned such that the kitchen is in the bottom left corner. The maps confirm the existence of structure in house layouts which can be leveraged by an agent trying to navigate to a room in a new environment.

oped for VLN aren't directly applicable to room navigation as they all rely on intermediate goals in the form of language based instructions.

3   Room Navigation Task

The agent is spawned at a random starting location and orientation in a *novel* environment and is tasked with navigating to a given target room – *e.g.*, "Kitchen." If there exist multiple rooms of same type, the agent needs to navigate to the room closest to its starting location. We ensure that the agent never starts in a room of the target room type *i.e.* if the agent is in a bedroom the target room cannot be a bedroom. Similar to [32], with each step the agent receives an RGB image from a single color vision sensor, depth information from the depth sensor and GPS+Compass that provides the current position and orientation *relative* to the start position. When there's no GPS information available we only need egomotion, which most robotics platforms provide via IMU sensors or odometry. As in [32, 41], the agent does not have access to any ground truth floor plan map and must navigate using only its sensors. Unlike point navigation [32], room navigation is a semantic task, so GPS+Compass is insufficient to solve the task and only helps in preventing the agent from going around in circles.

4   Room Navigation Using Amodal Semantic Maps

We develop a room navigation framework with an explicit mapping strategy to predict amodal semantic maps by learning underlying correlations in houses, and navigating using these maps. Our approach assumes there are architectural and stylistic regularities in houses which allow us to predict regions of a house which lie beyond our field of view. For instance, if we are in the kitchen we can guess that the dining room is adjacent to it and we would be right in most cases.[1] We first verify the existence of such correlations by scaling and aligning all the homes in the Matterport 3D[9] dataset such that the kitchen is in the bottom-left corner. As shown in Fig. 2, we can observe that the concentration of dinning

---

[1] We acknowledge that these regularities likely vary across geographies and cultures.

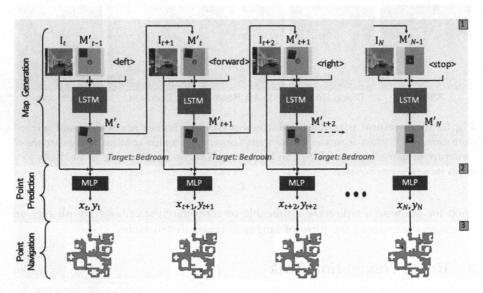

Fig. 3. Room Navigation using Amodal Semantic Maps: Our room navigation framework consists of 3 components (1) **Map Generation:** We predict egocentric crops of top down maps indicating rooms in a house. We use a sequence-to-sequence network which takes as input the current image I_t, the previous semantic map M'_{t-1}, and previous a_{t-1} and predicts the current semantic map M'_t. (2) **Point Prediction:** We feed the predicted maps (M'_t) along with I_t and target room ID tr to a network to predict a target point $P_t^{\text{pred}} = (x_t, y_t)$ that lies in the target room. The predicted and ground truth points are shown in green and red respectively on the top down map. (3) **Point Navigation:** We use a pre-trained point navigation policy to navigate to P_t^{pred}. The process is repeated for N steps until the agent calls stop or $N = 500$, whichever happens first. (Color figure online)

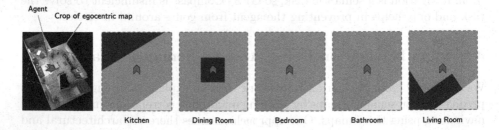

Fig. 4. Egocentric crops of Semantic Top Down Maps. We show the regions in a $26\,\text{m} \times 26\,\text{m}$ crop of the ground truth semantic maps w.r.t. agent's location and orientation. Agents are trained to predict these semantic maps, including regions they have not seen yet, based on regions of the house they have seen so far. Regions colored ■ lie inside the specified room. ■ denotes regions inside the house and not in the room and ■ denotes regions outside the house. (Color figure online)

rooms is close to the kitchens and the bedrooms are in the opposite corner, away from the kitchens. We believe there exist similar and more subtle correlations (e.g.size of the kitchen could be indicative of the number of bedrooms in a house) which our agent can automatically learn while predicting amodal semantic top down maps of regions. We now provide an overview of our approach followed by a detailed explanation of each of its sub-components.

Overview. Our room navigation framework is outlined in Fig. 3. The agent is spawned in a random location and is asked to navigate to a specified target room, $tr \in \mathcal{R}$, where $\mathcal{R}$ is the set of all possible target rooms. With each step t, the agent receives RGB image I_t, Depth D_t and GPS+Compass information. The agent predicts egocentric crops of top down semantic maps indicating the rooms which lie in and beyond its field of view. An example of these maps is shown in Fig. 4. To generate the maps, the agent uses a sequence to sequence (Seq2Seq) network which takes as input the RGB Image from the current time-step I_t, the predicted semantic maps from the previous time step $M_{t-1,r}^{\mathrm{pred}} \forall r \in \mathcal{R}$, and the previous action a_{t-1} and predicts semantic maps $M_{t,r}^{\mathrm{pred}}$ for the current time step. The predicted semantic maps $M_{t,r}^{\mathrm{pred}}$ are fed to a point prediction network which predicts a target point P_t^{pred} lying inside the target room. The agent navigates to the predicted point using a point navigation policy $\pi_{nav}(P_t^{\mathrm{pred}}, D_t)$. The agent updates its beliefs of the predicted semantic maps and the predicted point as it steps around the environment. The episode is deemed successful if the agent stops inside the target room before 500 steps.

Next, we describe the three main components of our model architecture: Map Generation, Point Prediction and Point Navigation, as shown in Fig. 3.

4.1   Map Generation

We model the correlations in houses by learning to predict amodal semantic top-down maps. The agent uses the information it has gathered so far to determine where the different rooms in the house are present even before visiting these regions. The maps are crops of top-down maps indicating the type, location, size and shape of rooms and are egocentric to the agent. Figure 4 shows an example of ground truth maps. The maps have 3 classes: (1) Regions which lie outside the house, (2) Regions which lie in the house but outside the target room, (3) Regions which lie in the target room. We hypothesize that the agent will learn architectural regularities in houses and correlations between the RGB images and layouts in order to predict these maps. We train a map generation network f_{map}, to predict egocentric crops of top down semantic maps for each room $r \in \mathcal{R}$. f_{map} consists of a sequence to sequence network f_{seq} and a decoder network f_{dec}. At each time step t, f_{seq} takes in as input a concatenation of learned representations of the current RGB frame $f_i(I_t)$, the previous action $f_{act}(a_{t-1})$, and the semantic map of the previous time step $f_m(M_{t-1,r})$. f_{seq} outputs a latent representation $h_{t,r}$. $h_{t,r}$ is passed through a parameterized decoder network f_{dec} that resembles the decoder in [30]. f_{dec} upsamples the latent representation using multiple transpose convolutions layers to produce output $M_{t,r}^{\text{pred}} \forall r \in \mathcal{R}$. Following [5], during training, we uniformly choose to set $M_{t,r}^{\text{input}}$ to be $M_{t-1,r}^{\text{pred}}$ (predicted map from previous time step) 50% of the time and $M_{t-1,r}^{\text{GT}}$ (ground truth map from the previous time step) the rest of the time. At each step, we use the ground truth semantic map $M_{t,r}^{\text{GT}}$ to train f_{seq} and f_{dec} with cross entropy loss, $\mathcal{L}_{map}$. Eq. 1–5 describe the exact working of f_{map}. The agent continuously generates maps at each step until it calls stop or reaches end of the episode.

$$\mu \sim \text{Uniform}(0,1) \tag{1}$$

$$M_{t,r}^{\text{input}} = \begin{cases} M_{t-1,r}^{\text{GT}}, \text{if } \mu > 0.5 \text{ or } t = 0, \\ M_{t-1,r}^{\text{pred}}, \text{otherwise} \end{cases} \tag{2}$$

$$h_{t,r} = f_{seq}(f_m(M_t^{\text{input}}), f_{act}(a_{t-1}), f_i(I_t)) \tag{3}$$

$$M_{t,r}^{\text{pred}} = f_{dec}(h_{t,r}) \tag{4}$$

$$\mathcal{L}_{map} = \sum_{r \in \mathcal{R}} \text{CrossEntropy}(M_{t,r}^{\text{pred}}, M_{t,r}^{\text{GT}}) \tag{5}$$

During inference, we feed a random image as the input semantic map for the first time step and use $M_{t,r}^{\text{pred}}$ as $M_{t,r}^{\text{input}}$ for all consecutive steps.

4.2   Point Prediction

The maps predicted by f_{map} are amodal – the agent predicts regions that it has not seen yet. They are however *crops* – the agent does not predict the layout of the entire house. These crops are egocentric to the agent and the target room may not always appear in these maps. For *e.g.*, consider Fig. 4, there exists a bathroom in the house but this region does not appear inside the crop as it does not fall inside the crop w.r.t. the agent's current location. Inspired by the recent progress in point navigation [32,41], we reduce the room navigation problem to point navigation. We train a network f_{point} to predict a target point $P_t^{\text{pred}} = (x'_t, y'_t)$ that lies in the target room tr, at each step t of the agent. Similar to Sect. 4.1, we learn representations $f_i(I_t)$ of the RGB image I_t, $f_m(M_{t,r}^{\text{pred}}) \forall r \in \mathcal{R}$ of the predicted semantic maps $M_{t,r}^{\text{pred}}$, and $f_{emb}(tr)$ which is a one-hot embedding of the target room ID tr. The predicted semantic map representations for the different rooms are combined using Eq. 6,

$$g_M = f_m(M_{t,r}^{\text{pred}}) \odot f_{emb}(tr) \tag{6}$$

where $\odot$ represents element-wise multiplication. These are then concatenated with the target room ID tr and fed to a multilayer perceptron (MLP) f_{point} which outputs P_t^{pred} as described in Eq. 7. f_{point} is trained using mean square loss w.r.t. a ground truth target point in the target room, P^{GT}, as shown in Eq. 8. Sect. 5 describes how this point is chosen.

$$P_t^{\text{pred}} = (x'_t, y'_t) = f_{point}(g_M, f_i(I_t), f_{emb}(tr)) \tag{7}$$

$$\mathcal{L}_{point} = \text{MSELoss}(P_t^{\text{pred}}, P^{\text{GT}}) \tag{8}$$

'During inference, the agent predicts a point every $k = 6$ steps. Once the agent completes $K = 60$ steps, the target point is simply fixed and no longer updated. The episode terminates if the agent calls stop or reaches $N = 500$ steps.

4.3   Point Navigation

At this stage, we have reduced room navigation to point navigation where the agent needs to navigate to the predicted target point P_t^{pred}. Following the approach in [41], we train a point navigation policy using Proximal Policy Optimization (PPO) [34] on the dataset of point navigation episodes in [32]. The policy, described in Eq. 9, is parameterized by a 2-layer LSTM with a 512-dimensional hidden state. It takes three inputs: the previous action a_{t-1}, the predicted target point P_t^{pred}, and an encoding of the Depth input $f_d(D_t)$. We only feed Depth input to the point navigation policy as this was found to work best [32]. The LSTM's output is used to produce a softmax distribution over the action space and an estimate of the value function.

$$a_t = \pi_{nav}(P_t, f_d(D_t), a_{t-1}) \tag{9}$$

The agent receives terminal reward $r_T = 2.5$, and shaped reward $r_t(a_t) = -\Delta_{\text{geo\_dist}} - 0.01$, where $\Delta_{\text{geo\_dist}} = d_t - d_{t-1}$ is the change in geodesic distance to the goal by performing action a_t. We then use this pre-trained policy, π_{nav}, to navigate to the predicted point, P_t^{pred}. We also fine-tune π_{nav} on the points predicted by our model and this improves the performance.

To recap, f_{map} generates the semantic map of the space, f_{point} acts as a high-level policy and predicts a point, and the low level point navigation controller π_{nav} predicts actions to navigate to this point.

4.4   Implementation Details

Map Generation. The image and map representations, $f_i(I_t) \in \mathbb{R}^{256}$ and $f_m(M_{t,r}^{\text{input}}) \in \mathbb{R}^{256}$, are obtained by first embedding the input RGB image I_t and semantic map M_t^{input} using a ResNet50 [19] pre-trained on ImageNet [12] followed by a fully connected layer. The action representation $f_{act}(a_t) \in \{0,1\}^{32}$ is a one-hot embedding of the action a_t.

All three are concatenated to form a 544 dimensional vector which is reduced to a vector of size 512 using a linear layer. This is fed through f_{seq}, a 2 layer LSTM [20] which outputs a hidden state $h_{t,r} \in \mathbb{R}^{512}$. This is passed through f_{dec} which consists of 5 transpose convolutions interleaved with BatchNorm [21] and followed by ReLU. The map generation network is trained using data collected from shortest path trajectories between source and target points in the house. We experimented with multiple semantic map crop sizes ranging 20 m 40 m and 26 m to work the best.

Point Prediction. The input RGB image I_t and semantic maps $M_{t,r}^{input}$ are embedded using a pre-trained ResNet50 [19], same as before, but $f_m(M_{t,r}^{input})$ is now a 32 dimensional vector using a fully-connected layer. The target room embedding $f_{emb}(tr) \in \mathbb{R}^{32}$ is a 32 dimensional one-hot encoding of the target room tr. The generated maps for each room type $f_m(M_{t,r}^{input})$ are multiplied with target embedding $f_{emb}(tr)$ and concatenated to form $g_M \in \mathbb{R}^{160}$. This is concatenated with the RGB image to form a 416 dimensional vector which is passed through f_{point} to obtain P_t^{pred}. We compute a goal vector relative to the agent's current location which is used by the point navigation module. Similar to map generation, the point prediction network is also trained using data collected from a shortest path trajectories.

Point Navigation Policy. The Depth encoding $f_d(D_t)$ is based on ResNeXt [45] with the number of output channels at every layer reduced by half. As in [41], we replace every BatchNorm layer [21] with GroupNorm [44] to account for highly correlated inputs seen in on-policy RL. As in [41], we use PPO with Generalized Advantage Estimation (GAE) [33] to train the policy network. We set the discount factor γ to 0.99 and the GAE parameter τ to 0.95. Each worker collects (up to) 128 frames of experience from 4 agents running in parallel (all in different environments) and then performs 2 epochs of PPO with 2 mini-batches per epoch. We use Adam [22] with a learning rate of 2.5×10^{-4}. We use DD-PPO [41] to train 64 workers on 64 GPUs.

5 Room Navigation Dataset

Simulator and Datasets. We conduct our experiments in Habitat [32], a 3D simulation platform for embodied AI research. We introduce the room navigation task in the Habitat API and create a dataset of room navigation episodes using scenes from Matterport 3D [9]. We use Matterport 3D as it is equipped with room category and boundary annotations and hence is best suited for the task of Room Navigation. It consists of 61 scenes for training, 11 for validation, and 18 for testing. We only use the subset of 90 buildings which are houses and exclude others such as spas as those locations do not have common room categories with the majority of the dataset. We extract a subset of these scenes which contain at least one of the following room types: Bathroom, Bedroom, Dining Room, Kitchen, Living Room on the first floor. We only use the first floor of the house because, (1) In Matterport3d, the bounding boxes of rooms on different floors overlap at times, e.g. the box for a room on the first floor often overlaps with the room right above it on the second floor, making it hard to sample points which lie on the same floor, (2) The floors are uneven, making it difficult to distinguish between the different levels of the house.

Our dataset is comprised of 2.6 million episodes in 32 train houses, 200 episodes in 4 validation houses, and 500 episodes in 10 test houses.

Episode Specification. An episode starts with the agent being initialized at a starting position and orientation that are sampled at random from all navigable positions of the environment [32]. The target room is chosen from R if it is present in the house and is navigable from the starting position. We ensure the start position is not in the target room and has a geodesic distance of at 4 m and at 45 m from the target point in the room. During the episode, the agent is allowed to take up to 500 actions. After each action, the agent receives a set of observations from the active sensors. Statistics of the room navigation episodes can be found in the supplementary.

Evaluation Metric. Similar to [1], we design two evaluation metrics for room navigation - RoomNav Success weighted by (normalized inverse) Path Length (*RoomNav SPL*) and *Success*. An episode is considered a success if the agent 0.2 m inside the bounds of the specified target room. We 0.2 m as the room boundaries in Matterport 3D sometimes lie outside the room and this factor ensures the agent has indeed stepped inside the room. As shown in Fig. 5, we compute the geodesic distance from the source point to all the navigable points that

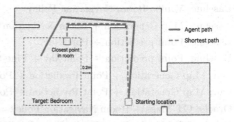

Fig. 5. SPL for Room Navigation. We compute the geodesic distance from source point to every point inside the room, 0.2 m of the bounds of the room. The point with the smallest geodesic distance is chosen as the *ground truth target point* and the SPL is computed w.r.t. this point.

0.2 m within the bounds of the room and choose a ground-truth target point P^{GT} that is closest to the agent's start position, *i.e.* has the shortest geodesic distance. RoomNav SPL is similar to the SPL defined in [1]. Let S indicate 'success', l be the length of the shortest geodesic distance between start point and P^{GT} defined above, and p be the length of the agent's path, then RoomNav SPL $= S \frac{l}{\max(l,p)}$. To achieve an SPL of 1, the agent must enter the nearest target room, and call stop when it has 0.2 m into the room.

Agent. As in [32], the agent is modeled as a cylinder with 0.2 m and height 1.5 m. The actions and the sensors are the same as in [32].

6   Results

We design baselines to evaluate the effectiveness of each component of our proposed room navigation framework and to validate our approach. We also report oracle results using ground-truth annotations to establish an upper-bound on the scores that can be achieved by our model.

Table 1 shows the RoomNav SPL and Success scores on the room navigation validation and test sets (for selected baselines). Our room navigation framework described in Sect. 4 achieves an SPL of 0.31 on validation and 0.29 on the test set. Fine-tuning the point navigation policy on points predicted by the point prediction network improves the SPL to 0.35 on validation and 0.33 on test, making this our best performing model.

Table 1. RoomNav-SPL for our approach, baselines, and oracle methods on test and validation sets of the Room Nav Dataset. Our proposed model (Map Generation + Point Prediction + PointNav + Fine-tune) achieves 0.35 RoomNav-SPL and outperforms all other baselines.

RoomNav model	Validation		Test	
	SPL	Success	SPL	Success
Baseline: Random	0.00	0.00	0.00	0.00
Baseline: Vanilla Room Navigation Policy	0.10	0.15	0.10	0.11
Baseline: Map Generation + Room Navigation Policy	0.16	0.17	–	–
Baseline: Point prediction + PointNav	0.17	0.20	–	–
Baseline: Point prediction + PointNav + Fine-tune	0.21	0.23	–	–
Our: Map Generation + Point Prediction + PointNav	0.31	0.35	–	–
Our: Map Generation + Point Prediction + PointNav + Fine-tune	**0.35**	**0.38**	**0.33**	**0.36**
Oracle: GT Maps + Room Navigation Policy	0.54	0.56	–	–
Oracle: GT Maps + Point Prediction + PointNav	0.61	0.64	–	–
Oracle: GT Maps + Point Prediction + PointNav + Fine-tune	0.67	0.68	–	–
Oracle: GT Point Selection + PointNav	0.83	0.84	0.79	0.82

Vanilla Room Navigation Policy. Here we compare to an approach that does not use semantic maps to model correlations and does not use point navigation. We ablate both the map prediction and point generation components by training room navigation policy from scratch using PPO, similar to the point navigation policy in Sect. 4.3. Instead of a target co-ordinates relative to current state, it takes in the target room ID as input. The agent receives terminal reward $r_T = 2.5$ RoomNav-SPL, and shaped reward $r_t(a_t) = -\Delta_{\text{geo\_dist}} - 0.01$, where $\Delta_{\text{geo\_dist}} = d_t - d_{t-1}$ is the change in geodesic distance to the target room by performing action a_t. The SPL using this baseline is 0.10 on validation and 0.10 on test, significantly worse compared to our approach (SPL 0.35). *This reinforces the effectiveness of our model, specifically the need to generate maps and use point navigation.* Note that this approach mimics an approach that tries to solve room navigation via vanilla ("brute force") reinforcement learning.

Vanilla Room Navigation Policy with Map Generation. We ablate the point prediction model in Sect. 3 and train a room navigation policy to navigate to the target room using RGB images and semantic maps. We use the map generator to predict semantic maps for each room type. We then train a policy to navigate to the target room. The policy is similar to the room navigation policy described above and takes four inputs: the previous action, the target room represented as an ID, the predicted semantic maps for all rooms embedded as in Eq. 6 and the Depth encoding. It is trained the same way as the room navigation policy.

This baseline achieves an SPL of 0.16, which is worse by a large margin of 0.2 when compared to our best performing model (room navigation using Map Generation + Point Prediction). *The improved performance of our*

Table 2. Ablation study of our mapping model and point prediction model.

(a) Map Generation Models Performance. Map Generation is a three-way classification problem. Class-1 consists of regions lying outside the house, Class-2 is regions lying inside the house but not in the room and Class-3 is regions lying in the room.

Mapper	mIoU	Pixel Accuracy %			
		Class-1	Class-2	Class-3	Avg
CNN	25.66	31.84	36.97	5.63	24.81
LSTM(no maps)	32.92	41.44	48.92	10.40	33.59
LSTM(ours)	41.45	56.13	60.92	13.13	43.39

(b) Point Prediction error using Predicted Maps and Ground Truth Maps. With ground truth maps, the prediction error is lower, suggesting scope for improvements.

	Error %
Pred Maps + MLP	39.13
GT Maps + MLP	22.91

best method emphasizes the significance of the point prediction and point navigation modules in our best performing model.

Point Prediction and Point Navigation Policy. We perform room navigation using only the high-level point prediction network and the low-level point navigation controller. We ablate the map generation module and train a modified version of the point prediction network without maps as input. Similar to f_{point} in Eq. 7, it generates $P_t = (x_t, y_t)$ but by using only the image representation $f_i(I_t)$ and target room embedding $f_{emb}(tr)$. The agent then navigates to P_t using the pre-trained point navigation policy as in Sect. 4.3.

This method achieves an SPL of 0.17 when the policy is trained from scratch and an SPL of 0.21 when the policy is fine tuned with points predicted by the point prediction network. Our best model surpasses this by a large margin of ~0.15, which *shows the advantage of using supervision to learn amodal semantic maps that capture correlations.* It also indicates the effectiveness of our map generation network. Since the environments in validation are different from train, we can also conclude that predicting semantic maps allows for better generalization to unseen environments as the RoomNav SPL is a direct indicator of how "quickly" an agent can reach a target room.

Using Ground Truth (GT) Maps. To get a better sense of how well our models can do if we had perfect map generation, we train a few of our baselines with ground truth maps instead of generated maps and report results in Table 1. With GT maps, Vanilla Room Navigation Policy with Map Generation achieves an SPL of 0.54. Adding GT maps to our best model, with and without fine-tuned point navigation policy we achieve SPL of 0.61 and 0.67 respectively. This suggests that there is still a large room for improvements in the Map Prediction module to perform room navigation more effectively. Table 2b reports the prediction error of point prediction model when using generated maps and ground truth maps.

Random. We evaluate using a random agent that takes action randomly among move_forward, turn_left, and turn_right with uniform distribution. It calls stop after 60 steps, which is the average length of an episode in our dataset. This achieves a RoomNav SPL of 0 on both test and validation, which implies that our task is difficult and cannot be solved by random walks.

Using GT Point Selection. We use the pre-trained point navigation policy defined in Sect. 4.3 to navigate to the ground truth target points P^{GT} in the target room defined in Sect. 5. This achieves an SPL of 0.82 and 0.79 on validation and test respectively. It provides an "upper-bound" on the performance that can be achieved by the room navigation policy, as this indicates the maximum RoomNav SPL that can be achieved by the framework in Sect. 4 if the error on point prediction were 0. These numbers are comparable to the SPL values for point navigation on the Matterport-3D dataset in [32], thus indicating our episodes are at least as difficult as the point navigation episodes in [32].

Map Generation Ablations. We also experimented with different semantic map generation models. The results in Table 2a show that the LSTM map generation model described in Sect. 4 performs best with a mean Intersection-over-Union (mIoU) of 41.45 and an Average Pixel Accuracy of 43.39%. The CNN only approach predicts a semantic map from each RGB image without maintaining a memory of the previous maps. This performs poorly and has a mIoU of 25.66 and an Average Pixel Accuracy of 24.81%. We train another LSTM model which doesn't use the semantic maps as input at each time step. This has a mIoU of 32.93 and Average Pixel Accuracy of 33.59.

Trajectory Videos. Qualitative results of our model can be found here. The first image in the first row shows the RGB input. The second and third maps in the first row show the location of the agent in allocentric and egocentric views respectively. The last figure on the first row shows two dots, red indicating the ground truth target point in the target room and green showing the predicted point. When only one dot is visible it indicates the predicted and ground-truth points overlap. There are 5 ground-truth semantic maps for each of the 5 room types we consider. The labels at the bottom indicate the room type being predicted. The second row shows the ground truth semantic maps indicating the location of the rooms in the house. The third row shows the maps predicted by our agent. The target room is mentioned at the very bottom, in this case, "Dining Room". As seen in the video, the model dynamically updates the semantic belief maps and predicts the target point with high precision. The agent is able to detect the room its currently present in and also develop a belief of where other rooms lie. The RoomNav-SPL for this episode is 1.0 as the agent successfully reaches the target room following the shortest path. Additional videos can be found here.

7   Conclusion

In this work, we proposed a novel learning-based approach for Room Navigation which models architectural and stylistic regularities in houses. Our approach consists of predicting the top down belief maps containing room semantics beyond the field of view of the agent, finding a point in the specified target room, and navigating to that point using a point navigation policy. Our model's improved performance (SPL) compared to the baselines confirms that learning to generate amodal semantic belief maps of room layouts improves room navigation performance in unseen environments. Our results using ground truth maps indicate that there is a large scope for improvement in room navigation performance by improving the intermediate map prediction step. We will make our code and dataset publicly available.

Acknowledgements. We thank Abhishek Kadian, Oleksandr Maksymets, and Manolis Savva for their help with Habitat, and Arun Mallya and Alexander Sax for feedback on the manuscript. The Georgia Tech effort was supported in part by NSF, AFRL, DARPA, ONR YIPs, ARO PECASE, Amazon. Prof. Darrell's group was supported in part by DoD, NSF, BAIR, and BDD. The views and conclusions contained herein are those of the authors and should not be interpreted as necessarily representing the official policies or endorsements, either expressed or implied, of the U.S. Government, or any sponsor.

References

1. Anderson, P., et al.: On evaluation of embodied navigation agents. arXiv preprint arXiv:1807.06757 (2018)
2. Anderson, P., et al.: Vision-and-language navigation: interpreting visually-grounded navigation instructions in real environments. In: CVPR (2018)
3. Aydemir, A., Göbelbecker, M., Pronobis, A., Sjöö, K., Jensfelt, P.: Plan-based object search and exploration using semantic spatial knowledge in the real world. In: ECMR (2011)
4. Bailey, T., Durrant-Whyte, H.: Simultaneous localization and mapping (SLAM) Part ii. IEEE Robot. Autom. Mag. **13**, 99–110 (2006)
5. Bengio, S., Vinyals, O., Jaitly, N., Shazeer, N.: Scheduled sampling for sequence prediction with recurrent neural networks. In: Advances in Neural Information Processing Systems (NeurIPS)
6. Bowman, S.L., Atanasov, N., Daniilidis, K., Pappas, G.J.: Probabilistic data association for semantic slam. In: International Conference on Robotics and Automation (ICRA) (2017)
7. Cadena, C., et al.: Past, present, and future of simultaneous localization and mapping: toward the robust-perception age. IEEE Trans. Robot. **32**, 1309–1332 (2016)
8. Carlone, L., Du, J., Kaouk Ng, M., Bona, B., Indri, M.: Active SLAM and exploration with particle filters using Kullback-Leibler divergence. J. Intell. Robot. Syst. **75**(2), 291–311 (2013). https://doi.org/10.1007/s10846-013-9981-9
9. Chang, A., et al.: Matterport3D: Learning from RGB-D data in indoor environments. arXiv preprint arXiv:1709.06158 (2017). matterport3D dataset available at https://niessner.github.io/Matterport/
10. Chen, T., Gupta, S., Gupta, A.: Learning exploration policies for navigation. arXiv preprint arXiv:1903.01959 (2019)
11. Crespo, J., Castillo, J.C., Mozos, O.M., Barber, R.: Semantic information for robot navigation: a survey. Appl. Sci. **10**, 497 (2020)

12. Deng, J., Dong, W., Socher, R., Li, L.J., Li, K., Fei-Fei, L.: ImageNet: a large-scale hierarchical image database. In: IEEE Conference on Computer Vision and Pattern Recognition (CVPR) (2009)
13. Durrant-Whyte, H., Bailey, T.: Simultaneous localization and mapping: part i. IEEE Robot. Autom. Mag. **13**, 99–110 (2006)
14. Fang, K., Toshev, A., Fei-Fei, L., Savarese, S.: Scene memory transformer for embodied agents in long-horizon tasks. In: IEEE Conference on Computer Vision and Pattern Recognition (CVPR) (2019)
15. Fried, D., et al.: Speaker-follower models for vision-and-language navigation. In: Advances in Neural Information Processing Systems (NeurIPS) (2018)
16. Fuentes-Pacheco, J., Ruiz-Ascencio, J., Rendón-Mancha, J.M.: Visual simultaneous localization and mapping: a survey. Artif. Intell. Rev. **43**(1), 55–81 (2012). https://doi.org/10.1007/s10462-012-9365-8
17. Gupta, S., Davidson, J., Levine, S., Sukthankar, R., Malik, J.: Cognitive mapping and planning for visual navigation. In: IEEE Conference on Computer Vision and Pattern Recognition (CVPR) (2017)
18. Hartley, R., Zisserman, A.: Multiple view geometry in computer vision (2003)
19. He, K., Zhang, X., Ren, S., Sun, J.: Deep residual learning for image recognition. In: IEEE Conference on Computer Vision and Pattern Recognition (CVPR) (2016)
20. Hochreiter, S., Schmidhuber, J.: Long short-term memory. Neural Comput. **9**, 1735–1780 (1997)
21. Ioffe, S., Szegedy, C.: Batch normalization: accelerating deep network training by reducing internal covariate shift. arXiv preprint arXiv:1502.03167 (2015)
22. Kingma, D.P., Ba, J.: Adam: a method for stochastic optimization. arXiv preprint arXiv:1412.6980 (2014)
23. Kollar, T., Roy, N.: Trajectory optimization using reinforcement learning for map exploration. Int. J. Robot. Res. **27**, 175–196 (2008)
24. Krishna, R., et al.: Visual genome: connecting language and vision using crowd-sourced dense image annotations. Int. J. Comput. Vis. **123**(1), 32–73 (2017). https://doi.org/10.1007/s11263-016-0981-7
25. LaValle, S.M.: Planning Algorithms. Cambridge University Press, Cambridge (2006)
26. Lillicrap, T.P., et al.: Continuous control with deep reinforcement learning. arXiv preprint arXiv:1509.02971 (2015)
27. Martinez-Cantin, R., de Freitas, N., Brochu, E., Castellanos, J., Doucet, A.: A Bayesian exploration-exploitation approach for optimal online sensing and planning with a visually guided mobile robot. Auton. Robots **27**, 93–103 (2009). https://doi.org/10.1007/s10514-009-9130-2
28. Mirowski, P., et al.: Learning to navigate in complex environments. arXiv preprint arXiv:1611.03673 (2016)
29. Pronobis, A., Jensfelt, P.: Large-scale semantic mapping and reasoning with heterogeneous modalities. In: International Conference on Robotics and Automation (ICRA) (2012)
30. Ronneberger, O., Fischer, P., Brox, T.: U-Net: convolutional networks for biomedical image segmentation. In: Navab, N., Hornegger, J., Wells, W.M., Frangi, A.F. (eds.) MICCAI 2015. LNCS, vol. 9351, pp. 234–241. Springer, Cham (2015). https://doi.org/10.1007/978-3-319-24574-4_28
31. Savinov, N., Dosovitskiy, A., Koltun, V.: Semi-parametric topological memory for navigation. arXiv preprint arXiv:1803.00653 (2018)
32. Savva, M., et al.: Habitat: A platform for embodied AI research. arXiv preprint arXiv:1904.01201 (2019)

33. Schulman, J., Moritz, P., Levine, S., Jordan, M., Abbeel, P.: High-dimensional continuous control using generalized advantage estimation. arXiv preprint arXiv:1506.02438 (2015)
34. Schulman, J., Wolski, F., Dhariwal, P., Radford, A., Klimov, O.: Proximal policy optimization algorithms. arXiv preprint arXiv:1707.06347 (2017)
35. Stachniss, C., Grisetti, G., Burgard, W.: Information gain-based exploration using Rao-Blackwellized particle filters. In: Robotics: Science and Systems (2005)
36. Thrun, S., Burgard, W., Fox, D.: Probabilistic Robotics. MIT Press, Cambridge (2005)
37. Walter, M.R., Hemachandra, S., Homberg, B., Tellex, S., Teller, S.: Learning semantic maps from natural language descriptions. In: Robotics: Science and Systems (2013)
38. Wang, X., et al.: Reinforced cross-modal matching and self-supervised imitation learning for vision-language navigation. In: IEEE Conference on Computer Vision and Pattern Recognition (CVPR) (2019)
39. Wang, X., Xiong, W., Wang, H., Wang, W.Y.: Look before you leap: bridging model-free and model-based reinforcement learning for planned-ahead vision-and-language navigation. In: Ferrari, V., Hebert, M., Sminchisescu, C., Weiss, Y. (eds.) ECCV 2018. LNCS, vol. 11220, pp. 38–55. Springer, Cham (2018). https://doi.org/10.1007/978-3-030-01270-0_3
40. Wang, Z., Zhang, Q., Li, J., Zhang, S., Liu, J.: A computationally efficient semantic SLAM solution for dynamic scenes. Remote Sens. **11**, 1363 (2019)
41. Wijmans, E., et al.: DD-PPO: learning near-perfect pointgoal navigators from 2.5 billion frames. In: International Conference on Learning Representations (ICLR) (2020)
42. Wu, Y., Wu, Y., Gkioxari, G., Tian, Y.: Building generalizable agents with a realistic and rich 3D environment. arXiv preprint arXiv:1801.02209 (2018)
43. Wu, Y., Wu, Y., Tamar, A., Russell, S., Gkioxari, G., Tian, Y.: Bayesian relational memory for semantic visual navigation. arXiv preprint arXiv:1909.04306 (2019)
44. Wu, Y., He, K.: Group normalization. In: Ferrari, V., Hebert, M., Sminchisescu, C., Weiss, Y. (eds.) ECCV 2018. LNCS, vol. 11217, pp. 3–19. Springer, Cham (2018). https://doi.org/10.1007/978-3-030-01261-8_1
45. Xie, S., Girshick, R., Dollár, P., Tu, Z., He, K.: Aggregated residual transformations for deep neural networks. In: IEEE Conference on Computer Vision and Pattern Recognition (CVPR) (2017)
46. Yang, W., Wang, X., Farhadi, A., Gupta, A., Mottaghi, R.: Visual semantic navigation using scene priors. arXiv preprint arXiv:1810.06543 (2018)
47. Zhu, Y., et al.: Target-driven visual navigation in indoor scenes using deep reinforcement learning. In: International Conference on Robotics and Automation (ICRA) (2017)

Learning to Separate: Detecting Heavily-Occluded Objects in Urban Scenes

Chenhongyi Yang[1], Vitaly Ablavsky[2(✉)], Kaihong Wang[1], Qi Feng[1], and Margrit Betke[1]

[1] Boston University, Boston, USA
{hongyi,kaiwkh,fung,betke}@bu.edu
[2] University of Washington, Seattle, USA
vxa@uw.edu

Abstract. While visual object detection with deep learning has received much attention in the past decade, cases when heavy intra-class occlusions occur have not been studied thoroughly. In this work, we propose a Non-Maximum-Suppression (NMS) algorithm that dramatically improves the detection recall while maintaining high precision in scenes with heavy occlusions. Our NMS algorithm is derived from a novel embedding mechanism, in which the semantic and geometric features of the detected boxes are jointly exploited. The embedding makes it possible to determine whether two heavily-overlapping boxes belong to the same object in the physical world. Our approach is particularly useful for car detection and pedestrian detection in urban scenes where occlusions often happen. We show the effectiveness of our approach by creating a model called SG-Det (short for Semantics and Geometry Detection) and testing SG-Det on two widely-adopted datasets, KITTI and CityPersons for which it achieves state-of-the-art performance. Our code is available at https://github.com/ChenhongyiYang/SG-NMS.

1 Introduction

Recent years have witnessed significant progress in object detection using deep convolutional neural networks (CNNs) [1–3]. The approach taken by many state-of-the-art object detection methods [4–8] is to predict multiple bounding boxes for an object and then use a heuristic method such as non-maximum suppression (NMS) to remove superfluous bounding boxes that stem from duplicate detected objects.

The Greedy-NMS algorithm is easy to implement and tends to work well in images where objects of the same class do not significantly occlude each other. However, in urban scenes, where the task is to detect potentially heavily occluded cars or pedestrians, Greedy-NMS does not perform adequately. The decrease in accuracy is due to the fundamental limitation of the NMS algorithm, which uses

---

V. Ablavsky—Work performed at Boston University.

© Springer Nature Switzerland AG 2020
A. Vedaldi et al. (Eds.): ECCV 2020, LNCS 12363, pp. 530–546, 2020.
https://doi.org/10.1007/978-3-030-58523-5_31

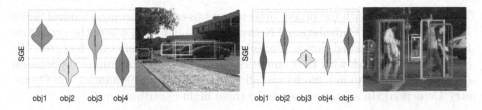

Fig. 1. Learned Semantics-Geometry Embedding (SGE) for bounding boxes predicted by our proposed detector on KITTI and CityPersons images. Heavily overlapped boxes are separated in the SGE space according to the objects they are assigned to. Thus, distance between SGEs can guide Non-Maximum-Suppression to keep correct boxes in heavy intra-class occlusion scenes.

a fixed threshold to determine which bounding boxes to suppress: The algorithm cannot suppress duplicate bounding boxes belonging to the *same* object while preserving boxes belonging to *different* objects, where one object heavily occludes others. Soft-NMS [9] attempts to address this limitation by not removing overlapping boxes but instead lowering their confidence; however, all overlapping boxes are still treated as false positives regardless of how many physical objects are in the image.

The limitation of NMS could be circumvented with an oracle that assigns each bounding box an identifier that corresponds to its physical-world object. Then, a standard NMS algorithm could be applied per set of boxes with the same identifier (but not across identifiers), thus ensuring that false positives from one object do not result in suppression of a true positive from a nearby object.

To approximate such an oracle, we can try to learn a mapping from boxes into a latent space so that the heavily overlapping boxes can be separated in that space. Naively, this mapping can be implemented by learning an embedding for every box based on its region features, e.g., the pooled features after RoIPooling [7]. However, the usefulness of such an embedding would be limited because heavily overlapping boxes tend to yield similar region features, thus would map to nearby points in the embedding space. In this paper, we demonstrate that by considering both the region features and the geometry of each box, we can successfully learn an embedding in a space where heavily overlapping boxes are separated if they belong to different objects. We call the learned embedding Semantics-Geometry Embedding (SGE). We also propose a novel NMS algorithm that takes advantage of the SGE to improve detection recall.

We visualize the concept of a Semantics-Geometry Embedding (SGE) in Fig. 1, where boxes belonging to the same object are mapped to a similar SGE and boxes belonging to different but occluded objects are mapped to SGEs that are far away. Although the embedding algorithm may assign boxes in disparate parts of an image to similar SGEs, this does not negatively impact our SG-NMS algorithm because these boxes can be easily separated based on their intersection-over-union (IoU) score. The SGE is implemented as an associative embedding [10] and learned using two loss functions, *separation* and *group loss*.

To train the SGE with the object detector end-to-end, we propose a novel Serial Region-based Fully Convolutional Network (Serial R-FCN), where the geometric feature of each detected box is precisely aligned with its semantic features. We combine this network with the Semantics-Geometry Non-Maximum-Suppression (SG-NMS) algorithm in a model we call SG-Det (short for Semantics and Geometry Detection). In summary, we make three main contributions:

1. A bounding-box-level *Semantics-Geometry Embedding (SGE)* is proposed, and a Non-Maximum Suppression algorithm, called *Semantics-Geometry Non-Maximum-Suppression (SG-NMS)*, based on this embedding, is derived. The algorithm markedly improves object detection in scenarios with heavy intra-class occlusions.
2. A *serial R-FCN* with self-attention in each head is presented that not only provides the ability to learn the above-mentioned SGE end-to-end, but also improves object detection accuracy.
3. The model *SG-Det* is proposed, which combines the serial R-FCN and the SG-NMS algorithm. *SG-Det* achieves state-of-the-art performance on the tasks of car detection for the KITTI [11] dataset and pedestrian detection for the CityPersons [12] dataset by dramatically improving the detection recall in heavily-occluded scenes.

2   Related Works

Object Detection. CNN-based object detectors can be divided into one-stage and two-stage approaches. One-stage detectors [4,5,13] directly predict the object class and the bounding box regressor by sliding windows on the feature maps. Two-stage object detectors [7,8,14,15], first compute regions of interest (RoIs) [8,16–19] and then estimate the class label and bounding box coordinates for each RoI. Although the two-stage approaches often achieve higher accuracy, they suffer from low computational efficiency. R-FCN [14] addresses this problem by replacing the computation in fully-connected layers with nearly cost-free pooling operations.

Non Maximum Suppression. NMS is widely used in modern object detectors to remove duplicate bounding boxes, but it may mistakenly remove boxes belonging to different objects. Soft-NMS [9] was proposed to address this problem by replacing the fixed NMS threshold with a score-lowering mechanism. However, highly-overlapping boxes are still treated as false positives regardless of the semantic information. In Learning-NMS [20], a neural network is used to perform NMS, but the appearance information is still not considered. The Adaptive-NMS approach [21] learns a threshold with the object detector, but when the threshold is set too high, false positives may be kept. The relation of bounding boxes can also be used to perform NMS by considering their appearance and geometric features [22], but this does not handle intra-class occlusion. The localization quality of each box can be learned to help NMS with keeping accurate boxes [23–26].

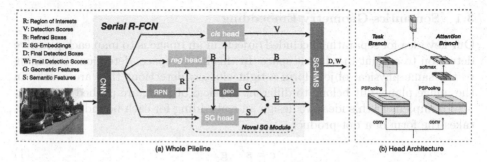

(a) Whole Pileline   (b) Head Architecture

Fig. 2. (a) Overview of our proposed model SG-Det. An input image is first processed by a backbone CNN to yield feature maps. A Region Proposal Network (RPN) [8] is used to extract regions of interests (ROIs). The RoIs will first be refined by the regression head and then fed into the classification head to produce detection scores, making the whole pipeline serial. A novel **Semantics-Geometry Module**, parallel to the classification head, is added to learn the SG embedding for each refined box. Finally, the detected box, detection scores, and SG embeddings are fed into the SG-NMS algorithm to produce final detections. (b) All heads (orange boxes in Fig. 2(a)) share a similar architecture. The feature map computed by the backbone network is processed by two branches to yield two score maps. Then a Position Sensitive RoI-Pooling [14] is applied to produce two grids of k^2 position-sensitive scores – a task-specific score and an attention score. A softmax operation transforms the attention score into a discrete distribution over the k^2 grids. Finally the k^2 task scores are aggregated by the attention distribution to yield the final output scores. (Color figure online)

Other Occlusion Handling Approaches. There are many other methods designed to handle occlusion, including both intra-class or inter-class occlusion. Most of them focus on detecting pedestrians in crowd scenes. Repulsion loss [27] was proposed to prevent boxes from shifting to adjacent objects. The occluded person is detected by considering different body parts separately [17,28–31]. A novel expectation-maximization merging unit was proposed to resolve over-lap ambiguities [32]. Additional annotations such as head position or visible regions have been used [33–35] to create robust person detectors. Although these approaches have been shown to be effective in detecting occluded persons, it is difficult to generalize them to other tasks like car detection.

3 Methodology

In this section, we first introduce the proposed Semantics-Geometry Embedding (Sect. 3.1), then the Semantics-Geometry NMS algorithm (Sect. 3.2), and finally the proposed Serial R-FCN (Sect. 3.3). The overview of the combined proposed model *SG-Det* is shown in Fig. 2.

See also http://www.cs.bu.edu/fac/betke/research/SG-Det.

3.1   Semantics-Geometry Embedding

Our key idea for separating occluded objects in an image is to map each putative detection to a point in a latent space. In this latent space, detections belonging to the same physical object form a tight cluster; detections that are nearby in the image plane, but belong to different physical objects, are pushed far apart.

To implement this idea, we design an embedding for each bounding box that takes the form of a dot-product

$$e = \mathbf{s}^T \cdot \mathbf{g}, \tag{1}$$

where $\mathbf{g}$ is the *geometric feature* and $\mathbf{s}$ is the *semantic feature*. The geometric feature has a fixed form $\mathbf{g} = (x, y, w, h)^T$ with center coordinates (x, y) and width and height (w, h) of the bounding box. We tried different kinds of geometric features (e.g., [22]) and feature vectors with higher dimensions produced by a fully-connected layer, but found that such complexity did not provide further significant improvement.

Unlike the geometric feature, the semantic feature $\mathbf{s}$ is a weight output by a function that yields a vector compatible with $\mathbf{g}$; the function is implemented as a neural network, as shown in the Semantics-Geometry Head in Fig. 2. Note that the SGE is computed by the linear transformation of the geometric feature taking the learned semantic feature as a weight. An interpretation is that the neural network automatically learns how to distinguish the bounding boxes belonging to different objects. Note that a similar idea was proposed that combined geometric and semantic features in a Relation Network [22], but our approach is much simpler and can handle intra-class occlusion effectively.

We train the SGE using the loss function defined in Eq. 1. The training is carried out end-to-end, jointly with the object-detection branch using the loss function defined later (Eq. 6).

The loss function is derived for the SGE by extending the notion of an *associative embedding* [10,36]. Specifically, we use a *group loss* to group the SGEs of boxes belonging to the same object, and use a *separation loss* to distinguish SGEs of boxes belonging to different objects. For one image, the ground-truth boxes are denoted by $B^* = \{b_1^*, b_2^*, ..., b_M^*\}$. For each refined box b_i in the refined box set $B = \{b_i\}$, let b_j^* be the ground truth box with the largest IoU. If $\text{IoU}(b_i, b_j^*) > \theta$, box b_i would be "assigned" to b_j^*. Thus the refined bounding boxes are divided into $M+1$ sets: $B = B_1 \cup B_2 \cup, ..., \cup B_{M+1}$, where B_{M+1} is the set of refined boxes that are not assigned to any ground truth box. Then the *group* and *separation losses* are defined as:

$$L_{\text{group}}(\{e_i\}) = \sum_{j=1}^{M} \sum_{b_i \in B_j} |e_i - e_j^*|, \tag{2}$$

$$L_{\text{sep}}(\{e_i\}) = \sum_i p_i^* \max(0, \sigma - |e_i - \tilde{e}_i|), \tag{3}$$

Algorithm 1. The proposed Semantics-Geometry NMS.

Input: $B = \{b_i\}$: List of detection locations (boxes),
$\quad\quad\quad V = \{v_i\}$: List of detection scores,
$\quad\quad\quad E = \{e_i\}$: List of SGEs,
$\quad\quad\quad N_t$: IoU threshold,
$\quad\quad\quad \Phi(\cdot)$: $\mathbb{R} \to \mathbb{R}$, a monotonically increasing function
begin
$\quad$ D $\leftarrow$ {}; W $\leftarrow$ {} $\quad$ // set D as detected boxes and W as their scores
$\quad$ **while** B $\neq \emptyset$ **do**
$\quad\quad$ $m \leftarrow \arg\max_{i\in\{1...N\}}$ V
$\quad\quad$ D $\leftarrow$ D $\cup \{b_m\}$; W $\leftarrow$ W $\cup \{s_m\}$
$\quad\quad$ B $\leftarrow$ B $\setminus \{b_m\}$; V $\leftarrow$ V $\setminus \{v_m\}$
$\quad\quad$ **for** b_i IN B **do**
$\quad\quad\quad$ $\tau \leftarrow$ IoU(b_m, b_i)
$\quad\quad\quad$ // compare not only IoU but also the embedding distance
$\quad\quad\quad$ **if** $\tau \geq N_t$ AND $d(e_m, e_i) \leq \Phi(\tau)$ **then**
$\quad\quad\quad\quad$ B $\leftarrow$ B $\setminus b_i$; V $\leftarrow$ V $\setminus v_i$
$\quad$ **return** D, W

---

where $e_j{}^*$ is the SGE of ground truth box b_j^*, $\tilde{b}_i^*$ is the ground truth box with the second largest IoU with respect to b_i, and its SGE is $\tilde{e}_i$. We use σ to stabilize the training process by preventing the distances between embeddings to be infinite. We found that the model performance is not sensitive to the actual value of σ. In the definition of the *separation loss*, p_i^* is a indicator variable which is 1 only if $b_i \notin B_{M+1}$ and IoU$(b_i, \tilde{b}_i^*) > \rho$.

Some readers may confuse our loss functions with the Repulsion Loss (RL) [27], which is completely different. The RL was proposed to improve bounding box regression so that the detected bounding boxes better fit ground-truth objects. In contrast, our method does not affect the bounding box regression. The embedding trained through the two loss functions is used to determine if two overlapping boxes belong to the same object. Another difference is that the RL is performed in the box-coordinate space, while our group and separation losses are performed in the latent embedding space.

3.2 Semantics-Geometry Non-Maximum Suppression

We now derive our simple, yet effective NMS algorithm, SG-NMS, which takes advantage of the Semantics-Geometry Embedding. Its pseudo code is given in Algorithm 1.

SG-NMS first selects the box with the highest detection score as the *pivot box*. For each of the remaining boxes, its IoU with the pivot box is denoted by τ, and the box will be kept if the $\tau < N_t$. When $\tau > N_t$, SG-NMS checks the distance between its SGE and the SGEof the pivot box. If the distance is larger than $\Phi(\tau)$, the box will also be kept. Here $\Phi(\cdot)$ is a monotonically increasing function, which means that, as τ increases, a larger distance is required to keep it. In this work, we consider three kinds of SG-NMS algorithms: SG-NMS-Constant, SG-NMS-Linear and SG-NMS-Square, which respectively correspond to:

$$\Phi(\tau) = t_c, \quad \Phi(\tau) = t_l \cdot \tau, \quad \text{and} \quad \Phi(\tau) = t_s \cdot \tau^2, \tag{4}$$

where t_c, t_l, and t_s are hyper-parameters.

3.3 The Proposed Serial R-FCN

In order to compute SGEs that can capture the difference between geometric features of boxes belonging to different objects, we need to align extracted semantic features strictly with the refined boxes after bounding box regression. However, this cannot be achieved by normal two-stage CNN-based object detectors where the pooled feature is aligned with the RoI instead of the refined box because of the bounding-box regression.

To address this problem, we propose *Serial* R-FCN, see Fig. 2(a). In Serial R-FCN, the classification head along with the SG module is placed *after* the class-agnostic bounding box regression head [7]; thus, the whole pipeline becomes a serial structure. The classification head and the SG module use the refined boxes for feature extraction rather than the RoIs. Thus, the pooled features are strictly aligned with the refined boxes.

A light-weight self-attention branch is added into each head, as in Fig. 2(b). The output of the attention head is a discrete distribution over the k^2 position-sensitive grid. The position-sensitive scores are then aggregated through a weighted sum based on that distribution. There are two reasons why we introduced the self-attention in each head: 1). The self-attention helps the network to capture the semantic difference between heavily overlap-ping boxes and hence the SGE can be learned effectively. 2) we suggest that merging the position-sensitive scores by averaging (as done previous work [14]) could be sub-optimal, while adding the self-attention module helps the model to learn how to merge the score better. The idea of our Serial R-FCN is similar to a Cascade R-CNN [15]. However, while Cascade R-CNN stacks multiple classification and regression heads, we here only use one regression head and one classification head, thus do not introduce an extra parameter. Although the serial structure can be used by any two-stage detector, it suits the R-FCN best since no extra operation is added, and so the computation of the refined box is nearly cost free.

Placing the classification head after the regression head can bring us another benefit: It enables us to train the classification head using a higher IoU threshold. This yields more accurate bounding boxes. Without the serial structure, setting the IoU threshold to a very high value would result in the shortage of positive samples. However, in practice, we find that simply adopting the serial structure could easily yield a network that overfits on the training data. The reason is that as training progresses, the regression head becomes more and more powerful so that the classification head cannot receive enough hard negative examples (i.e., boxes whose IoU with the ground truth box is slightly smaller than the training threshold). The result is that the model cannot distinguish these examples and true positives when the model is tested. To alleviate the overfitting

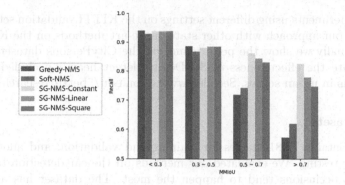

Fig. 3. Detection recall of the proposed SG-NMS and competing NMS algorithms on the KITTI validation set for different levels of occlusion, denoted by the max-mutual-IoU (MMIOU) among ground-truth boxes.

problem, we propose the simple but effective approach to add some noise to the refined bounding box so that the classification head continues to obtain hard false examples. Formally, during training, a box $b = (x, y, w, h)$ is transformed to $b' = (x', y', w', h')$ to train the classification head and the SG module:

$$x' = \sigma_x w + x, \ y' = \sigma_y h + y,$$
$$w' = w \cdot \exp(\sigma_w), \ h' = h \cdot \exp(\sigma_h), \tag{5}$$

where $\sigma_x, \sigma_y, \sigma_w, \sigma_h$ are noise coefficient drawn from a uniform distribution $\prod_{j=1}^{j=k} (-\zeta_k, \zeta_k)$ where the four dimensions correspond to x, y, w, h respectively. In practice we set $\zeta_x = \zeta_y = 0.05$ and $\zeta_w = \zeta_h = 0.2$.

The whole pipeline is trained end-to-end with the loss functions

$$L_{\text{total}} = L_{\text{rpn}} + \alpha L_{\text{det}} + \beta L_{\text{SGE}}, \tag{6}$$

$$L_{\text{rpn}} = L_{\text{cls-anchor}} + L_{\text{reg-anchor}}, \tag{7}$$

$$L_{\text{det}} = L_{\text{cls-rbox}} + L_{\text{reg-roi}}, \tag{8}$$

$$L_{\text{SGE}} = L_{\text{group-rbox}} + L_{\text{sep-rbox}}, \tag{9}$$

where the L_{rpn} is the commonly used loss to train the Region Proposal Network (RPN) [8], L_{det} is object detection loss [7] and L_{SGE} is the loss to train SGE as described in Sect. 3.1. We use two hyper-parameter α and β to balance between losses (Eq. 6). The RPN classification and regression losses are applied to the anchor boxes (Eq. 7), the regression loss to RoIs (Eq. 8), and the classification, group, and separation losses to the refined boxes (Eq. 9).

4 Experiments

We conducted quantitative experiments on two commonly used urban scenes datasets: KITTI [11] and CityPersons [12]. To show the advantage of our *SG-Det model* and also to give deep insights into our approach, we first conducted several experiments using different settings on the KITTI validation set. We then compared our approach with other state-of-the-art methods on the KITTI test set, and finally we show the performance on the CityPersons datasets. Results demonstrate the effectiveness of SG-Det to detect heavily-occluded cars and pedestrians in urban scenes. See also www.cs.bu.edu/~betke/research/SG-Det.

4.1 Datasets

KITTI contains 7,481 images for training and validation, and another 7,518 images for testing. We evaluated our methods on the car detection task where intra-class occlusions tend to happen the most. The dataset has a standard split into three levels of difficulty: Easy, Moderate, and Hard, according to the object scale, occlusion level, and maximum truncation. To further demonstrate how our methods handles intra-class occlusions, we proposed a new difficulty split that divide the dataset into disjoint subsets based on the max-mutual-IoU (MMIoU), denoted by MMIoU, between ground-truth boxes. The max-mutual-IoU of a ground-truth box is defined by its maximum IoU with other ground-truth boxes in the same category. We separate the validation set into three levels: Bare ($0 < $ MMIoU ≤ 0.2), Partial ($0.2 < $ MMIoU ≤ 0.5) and Heavy ($0.5 < $ MMIoU). Average Precision (AP) is used to evaluate performance [11]. Following prior work [37], we randomly held out 3,722 images for validation and use the remaining 3,759 images for training, in which a simple image L_2 similarity metric was adopted to differentiate training and validation images.

CityPersons contains 5,000 images (2,975 for training, 500 for validation, and 1,525 for testing). The log-average Miss Rate (MR) is used to evaluate performance. Following [27], we compare the detection log-average Miss Rate (MR) in different occlusion degrees. Following prior work [27], we separated the data into four subsets according to occlusion degree.

4.2 Implementation Details

We implemented our Serial R-FCN in TensorFlow [38] and trained it on a Nvidia Titan V GPU. For KITTI, we chose a ResNet-101 [2] based on a Feature Pyramid Network (FPN) as the backbone and set the batch size to 4. The model was trained for 100,000 iterations using the Adam [39] optimizer with learning rate of 0.0001. For CityPersons, we chose a ResNet-50 [2] as the backbone network and trained the model for 240,000 iterations with batch size of 4, and the initial learning rate was set to 0.0001 and decreased by a factor of 10 after 120,000 iterations. In all experiments, OHEM is adopted to accelerate convergence [14]. For both datasets, we set θ, σ, ρ to 0.7, 0.3, and 1.0, respectively, and set α and β to 1. Our code is available https://github.com/ChenhongyiYang/SG-NMS.

4.3 Effectiveness of SG-NMS

We report the performance of different NMS algorithms on the KITTI validation set applied to the same initial boxes so that a fair comparison is ensured

Table 1. Average precision (AP) in % of the proposed SG-NMS algorithm and other commonly-used NMS algorithms on the KITTI validation set.

Algorithm	Easy	Moderate	Hard ↑
Greedy-NMS	**97.98**	95.16	90.21
Soft-NMS	97.72	95.13	91.15
SG-Constant	97.56	95.35	92.31
SG-Linear	97.69	**95.41**	**92.54**
SG-Square	97.52	95.14	92.38

Table 2. AP (in %) of NMS algorithms with different thresholds and occlusion levels (highest AP per level in red).

Greedy				SG-Constant				SG-Linear				SG-Square			
N_t	Bare	Partial	Heavy	c	Bare	Partial	Heavy	t_l	Bare	Partial	Heavy	t_s	Bare	Partial	Heavy ↑
Soft	94.63	84.62	54.62	1.2	**93.77**	84.49	57.21	2.0	**93.74**	85.14	58.72	3.0	**93.65**	85.23	60.36
0.3	94.33	76.56	35.10	1.1	93.73	**84.65**	58.74	1.9	93.72	**85.19**	59.17	2.9	93.62	85.16	60.24
0.4	94.03	83.38	40.58	1.0	93.46	84.33	60.02	1.8	93.70	85.11	59.81	2.8	93.61	85.14	60.83
0.5	93.63	**85.15**	50.63	0.9	93.58	84.52	**60.05**	1.7	93.68	85.10	60.19	2.7	93.59	85.17	61.48
0.6	91.56	82.85	55.49	0.8	93.46	84.33	60.02	1.6	93.64	85.06	60.24	2.6	93.58	85.09	62.08
0.7	46.25	27.24	**57.21**	0.7	93.31	83.83	59.05	1.5	93.60	84.97	**61.08**	2.5	93.55	84.98	62.03

(Table 1). For Soft-NMS, we only report the results of the linear version because we find its performance is consistently better than the Gaussian version. All three SG-NMS algorithms outperform the Greedy-NMS and Soft-NMS on the Moderate and Hard levels. In particular, SG-NMS-Linear outperforms Greedy-NMS and Soft-NMS by 2.33 pp and 1.39 pp, respectively, on the Hard level where heavy intra-class occlusions occur. We also explored the efficacy of the Relation Network [22] in occlusion situations, but found that it did not work well due to generating numerous false positive detections in crowded scenes.

We report the detection recall on different MMIoU intervals and show the results in Fig. 3. When MMIoU is less than 0.5, the tested NMS algorithms achieve similar recall scores. When there is severe intra-class occlusion, i.e., MMIoU > 0.5, the recall of Greedy-NMS and Soft-NMS drops significantly. However, all three SG-NMS keep a relatively high recall. When MMIoU > 0.5, the difference in recall among the three SG-NMS algorithms is caused by the different slope of their $\Phi(\cdot)$ function. This result indicates that our SG-NMS improves the detection by promoting detection recall for objects in crowded scenes.

We report how the hyper-parameter t, introduced by our SG-NMS, affects detection performance (Table 2). Overall, the variants of SG-NMS outperform Greedy-NMS and Soft-NMS for the Heavy and Partial occlusion levels, while maintaining high performance for the Bare level. For the Heavy level, the best result, 62.08%, is achieved by SG-NMS-Square, which is 4.87 percent points (pp) higher than the best result of Greedy-NMS and Soft-NMS. Although Greedy-

Table 3. AP for different settings for the proposed SG-Det model and a baseline R-FCN model on car detection on the KITTI validation set. SG stands for SG-NMS; Noise stands for box noise, Attention stands for the self-attention branch used in each head.

Model	SG	Noise	Attention	Easy	Moderate	Hard	Bare	Partial	Heavy ↑
R-FCN (FPN)	–	–	–	95.57	95.08	88.66	92.41	81.83	45.96
Our SG-Det				94.77	94.44	89.30	92.35	80.94	44.03
			✓	95.84	94.55	90.10	93.27	84.87	47.79
		✓		95.04	95.12	90.01	93.11	83.28	43.98
	✓			94.62	94.50	89.54	92.45	81.56	52.62
		✓	✓	**97.98**	95.16	90.21	**93.63**	85.15	50.63
	✓	✓		97.80	**95.24**	91.86	93.15	82.71	51.58
	✓		✓	95.25	94.50	92.30	93.21	84.62	58.43
	✓	✓	✓	97.52	95.14	**92.38**	93.59	**85.17**	**61.48**

NMS can achieve an AP of 55.49% for the Heavy level (when $N_t = 0.6$), the AP in the Bare and Partial levels drops significantly due to the false-positive boxes it generates.

4.4   Ablation Study

We conducted an ablation study that demonstrates how the different model components affect the overall detection performance (Table 3). Our SG-Det model is proposed for detecting occluded objects, thus the analysis is focused on the detection of objects at the Hard difficulty (in the official split) and the Heavy occlusion level.

When the self-attention and bounding box noise are removed from our Serial R-FCN, we obtain a baseline Serial R-FCN that achieves an AP of 89.30% on the Hard and 44.03% on the Heavy occlusion level. When SG-NMS is included, the detection AP on the Heavy level is improved by 8.59 pp. When the self-attention branch is added into each head, the detection AP in the Hard and Heavy levels is lifted by 0.8 pp and 3.76 pp, respectively, compared to the baseline Serial R-FCN. This verifies our assumption that the learnable score aggregation enabled by the self-attention is superior to the naive average aggregation. By adding SG-NMS, the APs are further improved to 92.3% and 58.43%, which indicates that the self-attention head is important in capturing the semantic difference between heavily overlapping boxes. By adding box noise during training, the detection APs for all settings are improved, except for the heavy occlusion level. This means that the box noise can improve the detection precision by alleviating the overfitting problem in the Serial R-FCN, but it cannot help with improving the detection recall for heavily occluded objects. By combining self-attention, box noise, and SG-NMS, the full SG-Det model achieves APs of 92.38% and 61.48% on the Hard difficulty and Heavy occlusion level, respectively.

To conclude, we note that self-attention is useful to capture the semantic difference between heavily overlapping boxes. The box noise can alleviate the

Table 4. Comparison of AP between the proposed Semantics-Geometry Embedding (SGE) , the pure Semantic Embedding (SE) and the pure Geometric Embedding (GE) on the KITTI validation set.

Embedding	Bare	Partial	Heavy ↑
SE	93.31	83.95	55.60
GE	**94.11**	78.53	38.12
SGE	93.59	**85.14**	**61.48**

Table 5. Comparison of AP using different ρ during training.

ρ	0.05	0.10	0.15	0.20	0.25	0.30	0.35	0.40	0.45	0.50
Bare	93.92	94.03	92.50	**94.24**	93.28	93.68	92.98	93.56	93.17	93.07
Partial	81.73	82.89	83.75	84.06	**85.70**	85.10	84.89	82.38	83.24	83.52
Heavy	53.42	54.73	58.37	57.08	60.05	**60.19**	58.93	56.05	55.11	53.25

overfitting problem so that the detection precision is improved and the SG-NMS algorithm can improve the detection performance for heavily occluded objects.

4.5 Discussion

The Importance of Semantics and Geometry. We explored the importance of the semantic and geometric features by removing them from the embedding calculation. We first removed the semantic features by computing a Geometric Embedding (GE) for each box, where the GE is computed using a fixed $\hat{s}$ that is the mean of all the s vectors in the validation set. The performance of GE, shown in Table 4, is inferior than our SGE in occlusion situations, demonstrating the benefit of computing semantic features adaptively. Then we tested the purely-semantic model: for every box, a $1D$ Semantic Embedding (SE) is computed directly from its pooled region feature (Table 4). Our SGE performs better than the SE for all three occlusion levels. In fact, the two loss functions, defined in Sect. 3.1 for the SE, produce very unstable results during training. This means it is difficult for the neural network to learn such an embedding based on semantic features only, and it reveals the benefit of including geometric features.

How to Set ρ When Training SGE? We use a hyper-parameter ρ to determine occlusion during training (Sect. 3.1): For a detected box b, if its second largest IoU with any ground-truth box is larger than ρ, we assert b is occluded or occludes another object. Thus, the value of ρ becomes critical to the performance. In Table 5, we report the AP on different ρ using SG-NMS-Linear with $t_l = 1.7$. The results show that the performance on the bare difficulty level does not depend on ρ, which is reasonable because our SGE and SG-NMS do not affect objects without occlusion. The best ρ for the partial and heavy difficulty levels are 0.25 and 0.3, so we suggest to use a ρ of 0.27. A different value for ρ leads to a decrease in performance. To explain this, we suggest that a low

Table 6. Runtime and AP (%) on the KITTI test set as reported on the KITTI leaderboard. All methods are ranked based on Moderate difficulty.

Model	Runtime ↓	Easy	Moderate	Hard ↑
RRC [37]	3.60 s	95.68	93.40	87.37
SenseKITTI [40]	4.50 s	94.79	93.17	84.38
SDP+RPN [41]	0.40 s	95.16	92.03	79.16
ITVD [42]	0.30 s	95.85	91.73	79.31
SINet+ [43]	0.30 s	94.17	91.67	78.6
Cascade MS-CNN [15]	0.25 s	94.26	91.60	78.84
LTN [44]	0.40 s	94.68	91.18	81.51
Aston-EAS [45]	0.24 s	93.91	91.02	77.93
Deep3DBox [46]	1.50 s	94.71	90.19	76.82
R-FCN (FPN) [14,47]	0.20 s	93.53	89.35	79.35
Ours	0.20 s	95.81	93.03	83.00

value of ρ brings too much noise into the computing of the *group loss*, while a high value of ρ results in the model failing to capture the semantic difference of overlapping boxes that belong to different objects.

4.6   Comparison with Prior Methods

We compared our model with other state-of-the-art models on the KITTI car detection leaderboard (Table 6). Our Serial R-FCN and SG-NMS are ranked at the third place among the existing methods. The respective APs on the Moderate and Hard level are 1.00 pp and 3.84 pp higher than the fourth-place values [41]. Although RRC [37] and sensekitti [40] are ranked higher than ours, the speed of our method is more than ten times faster than theirs. A reason is that our main contribution focuses on the post-processing step rather than the detection pipeline (Fig. 4).

4.7   Experiments on CityPersons

We compare miss rates of NMS algorithms on the CityPersons validation set for different occlusion degrees in Table 7. We also compare our model with existing methods. The NMS hyper-parameters are obtained from a grid search, and we report the best result for each algorithm. With Greedy-NMS, our Serial R-FCN achieves miss rates of 11.7% (reasonable difficulty level) and 52.4% (heavy). Using Soft-NMS yields a slight improvement. SG-NMS-Linear and SG-NMS-Square yield 0.2 and 0.7 pp improvements (reasonable difficulty), but using SG-NMS-Constant harms the performance for this level because a single threshold cannot handle the various complex occlusion situations. All three SG-NMS improve performance on the heavy and partial occlusion levels. Especially, the

(a) Proposed SG-NMS   (b) Greedy-NMS   (c) Soft-NMS

Fig. 4. Visualization of results with true positive (green), false positive (blue) and missed (red) detections. Two failure cases of SG-NMS are shown with a false positive detection ((a) middle), and a missed detection ((a) bottom). (Color figure online)

Table 7. The miss rate (%) on the CityPersons validation set.

Model	Reasonable	Bare	Partial	Heavy ↓
Adapted FasterRCNN [12]	12.8	–	–	–
Repulsion Loss [27]	11.6	7.0	14.8	55.3
OR-CNN [17]	11.0	**5.9**	13.7	51.3
Adaptive-NMS [20]	**10.8**	6.2	11.4	54.0
SR-FCN+Greedy-NMS	11.7	7.5	11.0	52.4
SerialR-FCN+Soft-NMS	11.4	7.1	10.9	51.8
SR-FCN+SG-Constant	11.5	7.4	11.2	52.3
SR-FCN+SG-Linear	11.3	7.3	10.8	51.6
SR-FCN+SG-Square	11.0	7.2	**10.7**	**51.1**

SG-NMS-Square improves the respective miss rate to 10.7% and 51.1% on the partial and heavy occlusion levels, making our methods superior to the state of the art on those two levels. This means our method excels at handling occlusions.

5   Conclusion

In this paper, we presented two contributions, a novel Semantics-Geometry Embedding mechanism that operates on detected bounding boxes and an effective Semantics-Geometry Non-Maximum-Suppression algorithm that improves detection recall for heavily-occluded objects. Our combined model SG-Det achieves state-of-the-art performance on KITTI and CityPersons datasets by dramatically improving the detection recall and excelling in a low run time.

Acknowledgements. We acknowledge partial support of this work by the MURI Program, N00014-19-1-2571 associated with AUSMURIB000001, and the National Science Foundation under Grant No. 1928477.

References

1. Deng, J., Dong, W., Socher, R., Li, L.-J., Li, K., Fei-Fei, L.: ImageNet: a large-scale hierarchical image database. In: 2009 IEEE Conference on Computer Vision and Pattern Recognition, pp. 248–255. IEEE (2009)
2. He, K., Zhang, X., Ren, S., Sun, J.: Deep residual learning for image recognition. In: Proceedings of the IEEE Conference on Computer Vision and Pattern Recognition, pp. 770–778 (2016)
3. Simonyan, K., Zisserman, A.: Very deep convolutional networks for large-scale image recognition (2014)
4. Redmon, J., Divvala, S., Girshick, R., Farhadi, A.: You only look once: unified, real-time object detection. In: Proceedings of the IEEE conference on computer vision and pattern recognition, pp. 779–788 (2016)
5. Liu, W., et al.: SSD: single shot multibox detector. In: Leibe, B., Matas, J., Sebe, N., Welling, M. (eds.) ECCV 2016. LNCS, vol. 9905, pp. 21–37. Springer, Cham (2016). https://doi.org/10.1007/978-3-319-46448-0_2
6. Girshick, R., Donahue, J., Darrell, T., Malik, J.: Rich feature hierarchies for accurate object detection and semantic segmentation. In: Proceedings of the IEEE Conference on Computer Vision and Pattern Recognition, pp. 580–587 (2014)
7. Girshick, R.: Fast R-CNN. In: Proceedings of the IEEE International Conference on Computer Vision, pp. 1440–1448 (2015)
8. Ren, S., He, K., Girshick, R., Sun, J.: Faster R-CNN: towards real-time object detection with region proposal networks. In: Advances in Neural Information Processing Systems, pp. 91–99 (2015)
9. Bodla, N., Singh, B., Chellappa, R., Davis, L. S.: Soft-NMS-improving object detection with one line of code. In: Proceedings of the IEEE International Conference on Computer Vision, pp. 5561–5569 (2017)
10. Newell, A., Huang, Z., Deng, J.: Associative embedding: end-to-end learning for joint detection and grouping. In: Advances in Neural Information Processing Systems, pp. 2277–2287 (2017)
11. Geiger, A., Lenz, P., Urtasun, R.: Are we ready for autonomous driving? The KITTI vision benchmark suite. In: 2012 IEEE Conference on Computer Vision and Pattern Recognition, pp. 3354–3361. IEEE (2012)
12. Zhang, S., Benenson, R., Schiele, B.: CityPersons: a diverse dataset for pedestrian detection. In: Proceedings of the IEEE Conference on Computer Vision and Pattern Recognition, pp. 3213–3221 (2017)
13. Lin, T.-Y., Goyal, P., Girshick, R., He, K., Dollár, P.: Focal loss for dense object detection. In: Proceedings of the IEEE International Conference on Computer Vision, pp. 2980–2988 (2017)
14. Dai, J., Li, Y., He, K., Sun, J.: R-FCN: object detection via region-based fully convolutional networks. In: Advances in Neural Information Processing Systems, pp. 379–387 (2016)
15. Cai, Z., Vasconcelos, N.: Cascade R-CNN: delving into high quality object detection. In: Proceedings of the IEEE Conference on Computer Vision and Pattern Recognition, pp. 6154–6162 (2018)
16. Uijlings, J.R., Van De Sande, K.E., Gevers, T., Smeulders, A.W.: Selective search for object recognition. Int. J. Comput. Vis. **104**(2), 154–171 (2013)
17. Zhang, S., Wen, L., Bian, X., Lei, Z., Li, S.Z.: Occlusion-aware R-CNN: detecting pedestrians in a crowd. In: Proceedings of the European Conference on Computer Vision (ECCV), pp. 637–653 (2018)

18. Zitnick, C.L., Dollár, P.: Edge boxes: locating object proposals from edges. In: Fleet, D., Pajdla, T., Schiele, B., Tuytelaars, T. (eds.) ECCV 2014. LNCS, vol. 8693, pp. 391–405. Springer, Cham (2014). https://doi.org/10.1007/978-3-319-10602-1_26

19. He, K., Zhang, X., Ren, S., Sun, J.: Spatial pyramid pooling in deep convolutional networks for visual recognition. IEEE Trans. Pattern Anal. Mach. Intell. **37**(9), 1904–1916 (2015)

20. Hosang, J., Benenson, R., Schiele, B.: Learning non-maximum suppression. In: Proceedings of the IEEE Conference on Computer Vision and Pattern Recognition, pp. 4507–4515 (2017)

21. Liu, S., Huang, D., Wang, Y.: Adaptive NMS: refining pedestrian detection in a crowd. In: Proceedings of the IEEE Conference on Computer Vision and Pattern Recognition, pp. 6459–6468 (2019)

22. Hu, H., Gu, J., Zhang, Z., Dai, J., Wei, Y.: Relation networks for object detection. In: Proceedings of the IEEE Conference on Computer Vision and Pattern Recognition, pp. 3588–3597 (2018)

23. Tychsen-Smith, L., Petersson, L.: Improving object localization with fitness NMS and bounded IoU loss. In: 2018 IEEE/CVF Conference on Computer Vision and Pattern Recognition, June 2018

24. He, Y. Zhu, C., Wang, J., Savvides, M., Zhang, X.: Bounding box regression with uncertainty for accurate object detection. In: Proceedings of the IEEE Conference on Computer Vision and Pattern Recognition, pp. 2888–2897 (2019)

25. Jiang, B., Luo, R., Mao, J., Xiao, T., Jiang, Y.: Acquisition of localization confidence for accurate object detection. In: Proceedings of the European Conference on Computer Vision (ECCV), pp. 784–799 (2018)

26. Tan, Z., Nie, X., Qian, Q., Li, N., Li, H.: Learning to rank proposals for object detection. In: The IEEE International Conference on Computer Vision (ICCV), October 2019

27. Wang, X., Xiao, T., Jiang, Y., Shao, S., Sun, J., Shen, C.: Repulsion loss: detecting pedestrians in a crowd. In: Proceedings of the IEEE Conference on Computer Vision and Pattern Recognition, pp. 7774–7783 (2018)

28. Zhang, S., Yang, J., Schiele, B.: Occluded pedestrian detection through guided attention in CNNs. In: Proceedings of the IEEE Conference on Computer Vision and Pattern Recognition, pp. 6995–7003 (2018)

29. Noh, J., Lee, S., Kim, B., Kim, G.: Improving occlusion and hard negative handling for single-stage pedestrian detectors. In: Proceedings of the IEEE Conference on Computer Vision and Pattern Recognition, pp. 966–974 (2018)

30. Tian, Y., Luo, P., Wang, X., Tang, X.: Deep learning strong parts for pedestrian detection. In: Proceedings of the IEEE International Conference on Computer Vision, pp. 1904–1912 (2015)

31. Zhou, C., Yuan, J.: Multi-label learning of part detectors for heavily occluded pedestrian detection. In: Proceedings of the IEEE International Conference on Computer Vision, pp. 3486–3495 (2017)

32. Goldman, E., Herzig, R., Eisenschtat, A., Goldberger, J., Hassner, T.: Precise detection in densely packed scenes. In: The IEEE Conference on Computer Vision and Pattern Recognition (CVPR), June 2019

33. Zhou, C., Yuan, J.: Bi-box regression for pedestrian detection and occlusion estimation. In: Proceedings of the European Conference on Computer Vision (ECCV), pp. 135–151 (2018)

34. Pang, Y., Xie, J., Khan, M.H., Anwer, R.M., Khan, F.S., Shao, L.: Mask-guided attention network for occluded pedestrian detection. In: Proceedings of the IEEE International Conference on Computer Vision, pp. 4967–4975 (2019)
35. Zhang, K., Xiong, F., Sun, P., Hu, L., Li, B., Yu, G.: Double anchor R-CNN for human detection in a crowd. arXiv preprint arXiv:1909.09998 (2019)
36. Law, H., Deng, J.: CornerNet: detecting objects as paired keypoints. In: Proceedings of the European Conference on Computer Vision (ECCV), pp. 734–750 (2018)
37. Ren, J., et al.: Accurate single stage detector using recurrent rolling convolution. In: Proceedings of the IEEE Conference on Computer Vision and Pattern Recognition, pp. 5420–5428 (2017)
38. Abadi, M., et al.: TensorFlow: large-scale machine learning on heterogeneous distributed systems. arXiv preprint arXiv:1603.04467 (2016)
39. Kingma, D.P., Ba, J.: Adam: a method for stochastic optimization. arXiv preprint arXiv:1412.6980 (2014)
40. Yang, B., Yan, J., Lei, Z., Li, S.Z.: Craft objects from images. In: Proceedings of the IEEE Conference on Computer Vision and Pattern Recognition, pp. 6043–6051 (2016)
41. Yang, F., Choi, W., Lin, Y.: Exploit all the layers: fast and accurate CNN object detector with scale dependent pooling and cascaded rejection classifiers. In: Proceedings of the IEEE Conference on Computer Vision and Pattern Recognition, pp. 2129–2137 (2016)
42. Liu, W., Liao, S., Hu, W., Liang, X., Zhang, Y.: Improving tiny vehicle detection in complex scenes. In: 2018 IEEE International Conference on Multimedia and Expo (ICME), pp. 1–6. IEEE (2018)
43. Hu, X., et al.: SINet: a scale-insensitive convolutional neural network for fast vehicle detection. IEEE Trans. Intell. Transp. Syst. 20(3), 1010–1019 (2018)
44. Wang, T., He, X., Cai, Y., Xiao, G.: Learning a layout transfer network for context aware object detection. IEEE Trans. Intell. Transp. Syst. (2019)
45. Wei, J., He, J., Zhou, Y., Chen, K., Tang, Z., Xiong, Z.: Enhanced object detection with deep convolutional neural networks for advanced driving assistance. IEEE Trans. Intell. Transp. Syst. (2019)
46. Mousavian, A., Anguelov, D., Flynn, J., Kosecka, J.: 3D bounding box estimation using deep learning and geometry. In: Proceedings of the IEEE Conference on Computer Vision and Pattern Recognition, pp. 7074–7082 (2017)
47. Lin, T.-Y. Dollár, P., Girshick, R., He, K., Hariharan, B., Belongie, S.: Feature pyramid networks for object detection. In: Proceedings of the IEEE Conference on Computer Vision and Pattern Recognition, pp. 2117–2125 (2017)

Towards Causal Benchmarking of Bias in Face Analysis Algorithms

Guha Balakrishnan[1,3]([✉]), Yuanjun Xiong[3], Wei Xia[3], and Pietro Perona[2,3]

[1] Massachusetts Institute of Technology, Cambridge, USA
balakg@mit.edu
[2] California Institute of Technology, Pasadena, USA
[3] Amazon Web Services, Seattle, USA

Abstract. Measuring algorithmic bias is crucial both to assess algorithmic fairness, and to guide the improvement of algorithms. Current bias measurement methods in computer vision are based on *observational* datasets, and so conflate algorithmic bias with dataset bias. To address this problem we develop an *experimental* method for measuring algorithmic bias of face analysis algorithms, which directly manipulates the attributes of interest, e.g., gender and skin tone, in order to reveal causal links between attribute variation and performance change. Our method is based on generating synthetic image grids that differ along specific attributes while leaving other attributes constant. Crucially, we rely on the perception of human observers to control for synthesis inaccuracies when measuring algorithmic bias. We validate our method by comparing it to a traditional observational bias analysis study in gender classification algorithms. The two methods reach different conclusions. While the observational method reports gender and skin color biases, the experimental method reveals biases due to gender, hair length, age, and facial hair. We also show that our synthetic transects allow for more straightforward bias analysis on minority and intersectional groups.

Keywords: Faces · Fairness · Bias · Causality · Counterfactuals · Image synthesis · Generative adversarial networks (GANs)

1 Introduction

Automated machine learning methods are increasingly used to support decisions in industry, medicine and government. While their performance is often excellent, accuracy is not guaranteed, and needs to be assessed through careful measurements. Measuring *biases*, i.e., performance differences, across protected attributes such as age, sex, gender, and ethnicity, is particularly important for decisions that may affect peoples' lives.

Electronic supplementary material The online version of this chapter (https://doi.org/10.1007/978-3-030-58523-5_32) contains supplementary material, which is available to authorized users.

© Springer Nature Switzerland AG 2020
A. Vedaldi et al. (Eds.): ECCV 2020, LNCS 12363, pp. 547–563, 2020.
https://doi.org/10.1007/978-3-030-58523-5_32

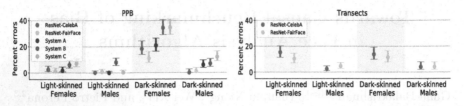

Fig. 1. Algorithmic bias measurements are test set dependent. (Left) Gender classification error rates of three commercial face analysis systems (System A–C) on the Pilot Parliaments Benchmark (PPB) [9] of portrait pictures. Error rates for dark-skinned females were found to be significantly higher than for other groups. We observed the same qualitative behavior when replicating the study with a standard classifier (ResNet-50) on two public face datasets (CelebA, FairFace). (Right) Our experimental investigation using our synthetic Transects dataset, where faces are matched across attributes, reveals a different picture of algorithmic bias (see Fig. 8 for a more complete analysis).

The prevailing technique for measuring the performance of computer vision algorithms is to measure error frequencies on a test set sampled *in the wild* [7,9,26,30] that hopefully mirrors the application domain. Each test image is annotated for attributes of interest, and split into groups that have homogeneous attribute values. Comparing error rates across such groups yields predictions of bias. As an example, Fig. 1-left shows the results of a recent study of algorithmic bias in gender classification of face images from the Pilot Parliaments Benchmark (PPB) dataset. This type of study is called *observational*, because the independent variables (e.g., skin color and gender) are sampled from the environment, rather than controlled by the investigator.

One reason to measure bias is to determine the actions one should take to remove it. For example, based on the results of Fig. 2-left, engineers of systems A-C may decide that incorporating more training examples of dark-skinned women is needed. In this case, measuring bias has one main goal: revealing *causal* connections between attributes of interest and algorithmic performance. Unfortunately, observational studies are ill-suited for drawing such conclusions. When one samples data in the wild, hidden variables that correlate with the variable of interest may have an influence on the performance of the algorithm. As the saying goes, *"correlation does not imply causation"*.

For example, in PPB very few males have long hair and almost no light-skinned females have short hair (Fig. 7, and [37]). The fact that hair length (a variable that may affect gender classification accuracy) is correlated in PPB with skin color (a variable of interest) complicates the analysis. Furthermore, the test dataset used to measure bias is often not representative of the population of interest. E.g., the middle-aged Scandinavians and Africans of PPB are not representative of, say, the broad U.S. Caucasian and African-American population [33]. While observational methods do yield useful information on disparate impact within a given test set population, generalizing observational performance predictions to different target populations is hit-or-miss [51] and can

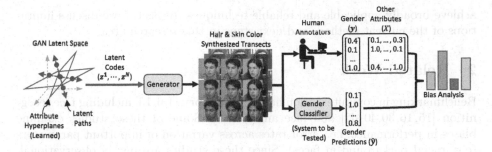

Fig. 2. Method overview. A GAN Generator is used to synthesize "transects," or grids of images along selected attributes by traversing the latent space in specific directions. Human annotations on the transects provide ground truth to be compared with algorithm output to measure errors. Attribute-specific bias measurements are obtained by comparing the algorithm's predictions with human annotations as the attributes are varied. The depicted example may study the question: *Does hair length, skin tone, or any combination of the two have a causal effect on classifier errors?*

negatively impact underrepresented, or minority populations [35,48]. One would want a method that systematically identifies algorithmic bias independent of the peculiarities of specific test sets.

A powerful approach to discovering causal relationships is the *experimental method*, used in other disciplines like medicine and social sciences, which involves manipulating the variable of interest while fixing all the other inputs [5,39]. In this work, we offer a practical way forward to systematically measure bias in computer vision algorithms using the experimental method. Our approach (Fig. 2) generates the test images synthetically, rather than sampling them from the wild, so that they are varied selectively along attributes of interest. This is enabled by recent progress in controlled and realistic image synthesis [22,23], along with methods for collecting large amounts of accurate human evaluations [8] to quantify the perceptual effect of image manipulations. Our synthesis approach can alter multiple attributes at a time to produce grid-like matched samples of images we call *transects*. We quantify the image manipulations with detailed human evaluations which we then compare with algorithm output to estimate algorithmic bias.

We evaluate our methodology with experiments on two gender classification algorithms. We first find that our transect generation strategy creates significantly more balanced data across key attributes compared to "in the wild" datasets. Next, inspired by [9], we use this synthetic data to explore the effects of various attributes on gender classifier errors. Our findings reveal that the experimental method can change the picture of algorithmic bias (Fig. 8), which will affect the strategy of algorithm improvement, particularly concerning groups that are often underrepresented in training and test sets. This work is a first step to developing experimental methods for algorithmic bias testing in computer vision, and so much remains to be done both in design and experimentation to achieve broadly-applicable and reliable techniques. In Sect. 5 we discuss limitations of the current method, and next steps in this research area.

2   Related Work

Benchmarking in computer vision has a long history [4,6,13] including face recognition [15,16,30,40–42] and face analysis [9]. Some of these studies examine biases in performance, i.e., error rates across variation of important parameters (e.g. racial background in faces). Since these studies are purely observational, they raise the question of whether the biases they measure depend on algorithmic bias, or on correlations in the test data. Our work addresses this question.

A dataset is said to be biased when combinations of features are disproportionately represented. Computer vision datasets are often found to be biased [44, 51]. Human face datasets are particularly scrutinized [2,12,26,28,29,36] because methods and models trained on these data can end up being biased along attributes that are protected by the law [27]. Approaches to mitigating dataset bias include collecting more thorough examples [36], using image synthesis to compensate for distribution gaps [29], and example resampling [31].

Studies of face analysis systems [9,26,34] and face recognition systems [17,30] attempt to measure bias across gender and skin-color (or ethnicity). However, the evaluations are based on observational rather than interventional techniques – and therefore any conclusions from these studies should be treated with caution. A notable exception is a recent study [37] using the experimental method to investigate the effect of skin color in gender classification. In that study, skin color is modified artificially in photographs of real faces to measure the effects of differences in skin color, all else being equal. However, the authors observe that generalizing the experimental method to other attributes, such as hair length, is too onerous if one is to modify existing photographs. Our goal is to develop a generally applicable experimental method, where *any* attribute may be studied independently.

Recent work uses generative models to explore face classification system biases. One study explores how variations in pose and lighting affect classifier performance [2,28,29]. A second study uses a generative model to synthesize faces along particular attribute directions [11]. These studies rely on the strong assumption that their generative models can modify one attribute at a time. However, this assumption relies on having unbiased training data, which is almost always not practical. In contrast, our framework uses human annotations to account for residual correlations produced by our generative model.

3   Method

Our framework consists of two components: a technique to synthesize *transects*, or grid-like constructs of synthesized images with control over semantic attributes (Sect. 3.1), and a procedure using these synthesized images, along with human annotators, to analyze biases of a classifier (Sect. 3.2).

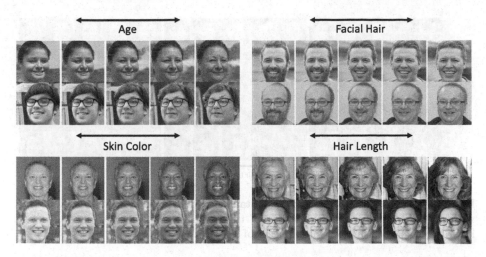

Fig. 3. 1D transects. 1×5 sample transects synthesized by our method for various attributes, with orthogonalization.

3.1 Transects: A Walk in Face Space

We assume a black-box face generator G that can transform a latent vector $\mathbf{z} \in \mathcal{R}^D$ into an image $I = G(\mathbf{z})$, where $p(\mathbf{z})$ is a distribution we can sample from. In our study, G is the generator of a pre-trained, publicly available GAN, "StyleGAN2" [22,23]. We base our approach on a recent study [56] for single attribute traversals in GAN latent spaces. That method trains a linear model to predict a particular image attribute from $\mathbf{z}$, and uses the model to traverse the $\mathbf{z}$-space in a discriminative direction. We generalize this idea to synthesize image grids, i.e., *transects*, spanning arbitrarily many attributes, unlike related work that operate on only one or two attributes at a time [11,47,49,55,56].

Estimating Latents-to-Attributes Linear Models. Let there be a list of N_a image attributes of interest (age, gender, skin color, etc.). We generate an annotated training dataset $\mathcal{D}_\mathbf{z} = \{\mathbf{z}^i, \mathbf{a}^i\}_{i=1}^{N_z}$, where $\mathbf{a}^i$ is a vector of scores, one for each attribute, for generated image $G(\mathbf{z}^i)$. The score for attribute j, $\mathbf{a}^i_j$, may be continuous or binary. We sample a generous number of values of $\mathbf{z}^i$ from $p(\mathbf{z})$ and obtain $\mathbf{a}^i$ from human annotators.

For each attribute j, we use $\mathcal{D}_\mathbf{z}$ to compute a $(D-1)$-dimensional linear hyperplane $h_j = (\mathbf{n}_j, b_j)$, where $\mathbf{n}_j$ is the normal vector and b_j is the offset. For continuous attributes like age or skin color, we train a ridge regression model [20]. For binary attributes we train a support vector machine (SVM) classifier [10].

Multi-attribute Transect Generation. Each hyperplane h_j specifies the subspace of $\mathcal{R}^D$ with boundary values of attribute j, and the normal vector $\mathbf{n}_j$ specifies a direction along which that attribute primarily varies. To construct a one-dimensional, length-L transect for attribute j, we first start with a random

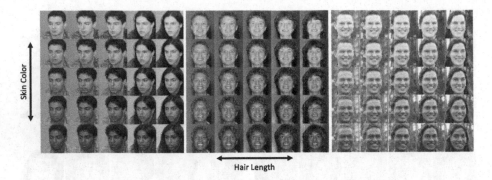

Fig. 4. 2D transects. 5×5 transects simultaneously varying hair length and skin tone, with orthogonalization. Multidimensional transects allow for intersectional analysis, i.e., analysis across the joint distribution of multiple attributes.

point $\mathbf{z}^i$ and project it onto h_j. We then query $L - 1$ evenly-spaced points along $\mathbf{n}_j$, within fixed distance limits on both sides of the h_j. Figure 3 presents some single transect examples (with orthogonalization, a concept introduced in the next section). Further details on querying points are found in Sect. 3.1.

The 1D transect does not allow us to explore the joint space of several attributes, or to fix other attributes in precise ways. We generalize to K-dimensional transects to address this (see supplementary material for algorithm). The main extensions are: (1) we project $\mathbf{z}^i$ onto the intersection of K attribute hyperplanes, and (2) we move in a K-dimensional grid in $\mathbf{z}$-space (see Fig. 4).

Orthogonalization of Traversal Directions. The hyperplane normals $\{\mathbf{n}_j\}_{j=1}^{N_a}$ are not orthogonal to one another. If we set the traversal direction vectors equal to these normal vectors, i.e., $\mathbf{v}_j = \mathbf{n}_j$, we will likely observe unwanted correlations between attributes. We reduce this effect by producing a set of modified direction vectors such that $\mathbf{v}_j \perp \mathbf{n}_k, \forall k \neq j$. See supplementary material for our orthogonalization algorithm and image examples.

Setting Step Sizes and Transect Dimensions. We set L to small values to reduce annotation cost. For example, $L = 5$ for the 1D transects in Fig. 3 and 2D transects in Fig. 4, and $L = 2$ for the 3D transects in Fig. 5. For each attribute j, we manually set min/max signed decision values with respect to h_j, and linearly interpolate L_j points between these extremes. We set per-attribute min/max values so that transects depict a full dynamic range for most random samples.

3.2 Analyses Using Transects

An ideal transect will modify only selected attributes at a time, but in practice, unintended attributes will be accidentally modified. In addition, the degree to which an attribute is altered varies across transects. To measure and control for these factors we annotate each image of each transect, resulting in a second dataset $\mathcal{D}_{transect} = \{I^i, \mathbf{a}^i\}_{i=1}^{N_{images}}$ of images and human annotations.

We assume a target attribute of interest, e.g., gender, and a target attribute classifier C. We denote the ground truth gender of image I^i (as reported by humans) by y^i, and C's prediction by $\hat{y}^i$. For ease of analysis, we discretize the remaining attributes into bins, and assign an independent binary variable to each bin [14]. We denote the vector of concatenated binary covariates for image i by $\mathbf{x}^i$, and the classification error by $e^i = \ell(\hat{y}^i, y^i)$, where $\ell(\cdot, \cdot)$ is an error function.

Our first analysis strategy is to simply compare error rates across different subpopulations [7,9,26,30]. Let E_j^s be the average error of C over test samples for which covariate j is equal to $s \in \{0, 1\}$. If the data is generated from a perfectly randomized or controlled study, the quantity $E_j^1 - E_j^0$ is a good estimate of the "average treatment effect"(ATE) [3,19,38,46] of covariate j on e, a causal measure of the expected change in e when covariate j is flipped from 0 to 1, with other covariates fixed. Note that exactly computing the ATE from an observational dataset is virtually never possible, because we do not observe the counterfactual case(s) for each data point, e.g., the same person with both light and dark skin tones. Though our transects come closer to achieving an ideal study than real datasets do (see Sect. 4.3), there may still be confounding between covariates (see Fig. 10 for an example).

Since any observable confounder may be annotated in $\mathcal{D}_{transect}$, our second strategy is to use a covariate-adjusted ATE estimator [43,45,53]. One simple adjustment approach is to train a linear regression model predicting e^i from $\mathbf{x}^i$: $e^i = \epsilon^i + \beta_0 + \sum_j \beta_j \mathbf{x}_j^i$, where β's are parameters, and ϵ^i is a per-example noise term. β_j captures the ATE, the average change in e given one unit change in $\mathbf{x}_j$ holding all other variables constant, provided: (1) a linear model is a reasonable fit for the relationship between the dependent and independent variables, (2) all relevant attributes are included in the model (i.e., no hidden confounders), and (3) no attributes that are influenced by $\mathbf{x}_j$ are included in the model, otherwise these other factors can "explain away" the impact of $\mathbf{x}_j$. An experimenter can never be completely sure that (s)he has satisfied these conditions but (s)he can strive to do so through careful consideration.

Finally, when the outcome lies in a fixed range, as is the case in our experiments with $e^i \in [0, 1]$, we use logistic instead of linear regression. β_j then represents the expected change in the log odds of e for a unit change in $\mathbf{x}_j$.

3.3 Human Annotation

We collect human annotations on the synthetic faces to construct $\mathcal{D}_\mathbf{z}$ and $\mathcal{D}_{transect}$, using Amazon Mechanical Turk [8] through the AWS SageMaker Ground Truth service [1]. Annotators evaluated each image for seven attributes: gender, facial hair, skin color, age, makeup, smiling, hair length and realism. Each attribute was evaluated on a discrete scale. For complete details about our annotation process, samples of our survey layouts and analysis, please see supplementary material.

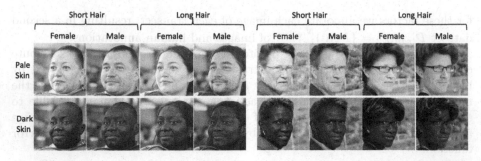

Fig. 5. Examples of transects used in our experiments. We created 1,000 transects spanning pale-to-dark skin tones, short-to-medium hair lengths, and male-to-female genders – two transects are shown here. Other face attributes are approximately held constant. For each image we collected human annotations to measure the perceived attributes. Intended attributes do not always agree with human perception (see Fig. 6, below).

Fig. 6. Human perception of transect attributes. Humans label the first face as a male, though we intended to produce a female. In all our experiments we used human perception, rather than intended attributes, as the ground truth.

4   Experiments

We evaluate our method on benchmarking bias of gender classifiers. The Pilot Parliaments Benchmark (PPB) [9] was the first wild-collected test dataset to balance gender and skin color with the goal of fostering the study of gender classfication bias across both attributes. The authors of that study found a much larger error rate on dark-skinned females, as compared to other groups and conjectured that this is due to bias in the algorithms, i.e., that the performance of the algorithm changes when gender and skin color are changed, all else being equal. Our method allows us to test this hypothesis.

4.1   Gender Classifiers

We trained two research-grade gender classifier models, each using the ResNet-50 architecture [18]. The first was trained on the CelebA dataset [32], and the second on the FairFace dataset [21]. CelebA is the most popular public face dataset due to its size and rich attribute annotations, but is known to have severe imbalances [11]. The FairFace dataset was introduced to mitigate some of these biases. See supplementary material for training details. We decided not

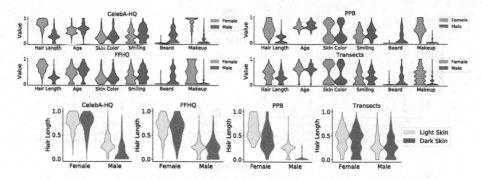

Fig. 7. (Top) Attribute distributions ("violin plots") by gender groups for different datasets. Wild-collected datasets have greater attribute imbalances across gender than synthetic transects, e.g., longer hair and younger ages for women. We designed our transects to mirror PPB's skin color and age distributions, while mitigating hair length imbalance. (Bottom) Hair length distributions by gender and skin color groups. In the wild-collected datasets hair length is correlated with skin color, when gender is held constant. Transects may be designed to minimize this correlation.

to test commercial systems for reproducibility reasons—models we test may be re-implemented and retrained by other researchers, while commercial systems are black boxes which may change unpredictably over time.

4.2   Transect Data

We used the generator from StyleGAN2 trained on Flickr-Faces-HQ (FFHQ) [22, 23]. We use the generator's "style space" as the latent space in our method, because we found it better suited for disentangling semantic attributes than the input noise space. We trained linear regression models to predict age, gender, skin color and hair length attributes from style vectors. For the remaining attributes—facial hair, makeup and smiling—we found that binarizing the ranges and training a linear SVM classifier works best.

We generated 3D transects across subgroups of skin color, hair length, and gender as described in Sect. 3.1. We use a transect size of $2 \times 2 \times 2 \times 1$, with grid decision values (input **c** of transect generation algorithm in supplementary material) spaced to generate pale-to-dark skin colors, short-to-medium hair lengths, male-to-female genders, and adult ages. We set the decision values by trial-and-error, and made them equal for all transects: $(-1.5, 1.7)$ for skin color, $(-0.5, 0)$ for hair length, $(-1.75, 1.75)$ for gender, and 0.5 for age. We generated 1000 such transects, resulting in 8000 total images. Figure 5 presents two example transects.

Not all synthesized images are ideal for our analysis. Some elicit ambiguous human responses or are unrealistic, and others may not belong clearly to the intended category of each attribute. See supplementary material for details on how we prune non-ideal images.

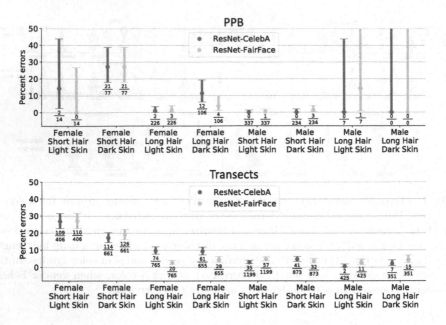

Fig. 8. Algorithmic errors, disaggregated by intersectional groups for wild-collected (PPB, top) and synthetic (transects, bottom). Wilson score 95% confidence intervals [54] are indicated by vertical bars, and the misclassification count and total number of samples are written below each bar. PPB has few samples for several groups, such as short-haired, light-skinned females and long-haired males (see Fig. 7). Synthetic transects provide numerous test samples for all groups. The role of the different attributes in causing the errors is studied in Sect. 4.4 and Fig. 9.

4.3 Comparison of Transects to Real Face Datasets

Figure 7-top analyzes attribute distributions for the CelebA-HQ, FFHQ and PPB datasets, along with our transects, stratified by gender. The wild-collected datasets contain significant imbalances across gender, particularly with hair length and age. In contrast, our transects exhibit more balance across gender. They depict more males with medium-to-long hair, and fewer females with very long hair. Our transects also have a bimodal skin color distribution, and an older population by design, since we are interested in mimicking those population characteristics of PPB. All datasets are imbalanced along the "Beard" and "Makeup" attributes—this is reasonable since we expect these to have strong correlations with gender.

In an ideal matched study, sets of images stratified by a sensitive attribute will exhibit the same distribution over remaining attributes. Figure 7-bottom stratifies by skin color. We see correlations of hair length distributions and skin colors in all the wild-collected data, while the synthetic transects exhibit much better balance.

4.4   Analysis of Bias

We now analyze the performance of the classifiers on PPB and our transects. Our classifiers exhibit similar error patterns to the commercial classifiers already evaluated on PPB [9]. Because PPB only consists of adults, we remove children and teenagers (age <0.4 in the normalized [0, 1] scale) from our transects to make a more direct comparison, leaving us with 5335 total images.

Figure 1 presents classification errors split by gender (M/F) and skin color (L/D). We replicated the reported errors of the commercial classifiers in [9], and report the errors of our classifiers on our in-house version of PPB. All classifiers perform significantly worse on dark-skinned females. Figure 8 presents classification errors, stratified by gender/hair length/skin color combinations.

We can make a number of broad-stroke, qualitative observations. First, the pattern of errors is similar across PPB and transects, with more errors on the left (females) than on the right (males). Second, transect errors are either comparable or higher than in PPB, indicating that synthetic faces can be at least as challenging as real faces. Most significantly, errors are nonzero on males, which allows the study of relative difficulties when attributes are varied. Third, in PPB, there are few males with long hair and few females with short hair and light skin, making measurements unreliable for these categories. This is not a problem with transects. Fourth, transect errors are higher when hair is shorter for women. However, hair length has a negligible effect for males (see Fig. 10 for a possible explanation). Fifth, there is no consistent transect error pattern in skin tone: within homogeneous groups changing skin tone does not seem to affect the performance of either algorithm. Looking at PPB alone, we could not make this observation, since skin tone is so strongly correlated with hair length.

We investigated further by calculating covariate-adjusted causal effects using an $L2$-regularized logistic regression model to predict that classifier's error conditioned on all attributes. See supplementary material for details on how we trained the regression models. Figure 9 presents coefficients for both models, which represent the change in log odds of the classifier's error for a change of one unit of each covariate (see Sect. 3.2). Facial hair, gender, makeup, hair length and age all have significant effects on classification error, while skin color has a negligible effect. We made a simplifying assumption that each covariate has an independent, linear effect on classification error, which we know is not true. Please see supplementary material for further discussion on this topic.

5   Discussion

Our synthesis-based experimental method offers a number of attractive properties over traditional observational methods. First, it generates approximately matched samples along selected attributes, allowing for counterfactual synthesis. Observational image data are almost never matched. Second, image synthesis allows, to a greater degree, uniform sampling of the space of attributes of interest. This is very difficult to do when one relies on images that are sampled from in-the-wild distributions, where some groups are underrepresented. Third, bias

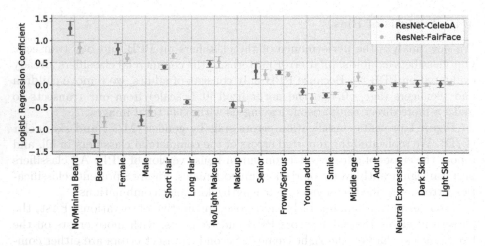

Fig. 9. Logistic regression coefficient values. The regression model is trained to predict absolute errors of the gender classifiers on our transect images given binary attributes as input. Coefficients represent the change in log odds of the error for a change of one unit of each attribute. Larger magnitudes indicate more important attributes, and positive (red)/negative (green) values correspond to attributes that increase/decrease classifier error. We order attributes here by large-to-small coefficient magnitudes. (Color figure online)

may be measured for intersectional groups defined by specific attribute combinations. Single-attribute analysis may conceal biases affecting groups defined by the combination of multiple attributes [25]. Some such combinations are often vastly undersampled in natural data. Fourth, image synthesis is fast and inexpensive, and crowdsourced image annotation is also relatively fast and affordable. By contrast, assembling large datasets of natural images is laborious and expensive. Thus, synthetic data has the potential to democratize testing for bias. And finally, ethical and legal concerns are greatly reduced. Collecting real face images in the wild requires great care to respect the privacy and dignity of individuals, the rights of minors and other vulnerable groups, as well as copyright laws. By contrast, synthetic datasets are freer from such risks because they do not depict real people.

The experimental analysis (transects) and traditional observational analysis (using PPB) diverged most significantly on the effect of skin color, which the observational study flagged as significant and the experimental method found to be not significant in determining algorithmic bias. The experimental method reveals a number of additional sources of bias: age, hair length and facial hair (Fig. 9). The two methods agree on gender.

Bias measurements guide scientists and engineers towards effective corrective measures for improving the performance of their algorithms. It is instructive to view the different predictions of the two methods through this lens. The correlational study based on PPB (Fig. 1) may suggest that, in order to reduce biases

Example 1   Example 2

Fig. 10. Correlated attribute modifications. We found that our method sometimes adds a beard to a male face when attempting to only modify hair length. This is an example of an imprecise intervention which can complicate downstream bias analyses. This bias may be due to the training data itself (men with long hair tend to have facial hair), or injected by the algorithm.

Males   Females

Fig. 11. Hidden confounders. There may always be a hidden confounder lurking in a dataset. As an example, we found that our method tends to add earrings when transitioning from dark-skinned men to dark-skinned women, a cue that a gender classifier might use to perform disproportionately well on the latter group. Interestingly, one male in this image also has an earring; that earring becomes larger for his female counterpart.

in our classifiers, more images of dark-skinned women should be added to their training sets. The experimental method leads engineers in a different direction. First, more training images of long-haired men and short-haired women of all races are needed. Second, correcting age bias requires more training images in the child-teen and, possibly, senior age groups.

Finally, it is important to consider a rich number of attributes and attribute combinations, besides the one(s) of immediate interest. This is for two reasons. First, unobserved confounders can have strong effects and need to be included in the analysis. Second, the combined effect of attributes can be strongly nonlinear (see the interaction of age and gender in supplementary material), and therefore an intersectional analysis [9,24] is necessary. Selecting attributes combinations is as much of an art as a science, and therefore one has to rely on good judgment and on a healthy multidisciplinary debate to progressively reveal missing ones.

6   Limitations and Future Work

Our method can not perfectly eliminate unwanted correlations with annotated variables, nor can it account for hidden confounders [52], and one will need to keep a sharp eye out for both. As an example of the first, we found that our method often adds facial hair to male faces when increasing hair length (see Fig. 10). This is likely a reason for why our classifiers did not have higher error rates for males with longer hair (see Fig. 8). As an example of the second, we found that our method tends to synthesize earrings when modifying a dark-skinned face to look female (see Fig. 11). Depending on culture, earrings may or may not be relevant to the definition of gender. If this is an unwanted correlation, one ought to add earrings to the annotation pipeline so that it may be "orthogonalized away" by the synthesis method. A significant advantage of an approach that is based on synthetic images and human annotation is thus the following: *as soon as one residual correlation is discovered it may be systematically annotated, compensated for in the analysis, and mitigated in the synthesis.*

A number of refinements in face synthesis will make our experimental method more practical and powerful. First, many of the faces we generated contained visible artifacts (see supplementary material), which we eliminated by human annotation – even subtle artifacts can affect classifier outputs, as revealed by the literature on adversarial examples [50]. Second, we do not yet have tools to estimate the sets of physiognomies and attribute combinations that can and cannot be produced by a given generator. Current GANs are known to have difficulties in generating data outside of their training distributions. Third, we observed a bias of StyleGAN2 towards generating Caucasian faces when sampling from its latent distribution. While our method can compensate for biases through carefully oriented traversals calibrated by human annotations, it would be clearly better to start from unbiased synthesis methods.

Our first-order technique for controlling synthesis can also be improved. A better understanding of the geometry of face space will hopefully yield more accurate global coordinate systems. These, in turn, will help reduce residual biases in synthetic transects, which we currently mitigate by having transects annotated by hand. Finally, extending our method beyond gender classification to more complex tasks, such as face recognition, is not straightforward in practice and will require further study.

Acknowledgments. We are grateful to Frederick Eberhardt, Bill Freeman, Lei Jin, Michael Kearns, R. Manmatha, Tristan McKinney, Sendhil Mullainathan, and Chandan Singh for insights and suggestions.

References

1. https://aws.amazon.com/sagemaker/groundtruth/
2. Albiero, V., KS, K., Vangara, K., Zhang, K., King, M.C., Bowyer, K.W.: Analysis of gender inequality in face recognition accuracy. In: Proceedings of the IEEE Winter Conference on Applications of Computer Vision Workshops, pp. 81–89 (2020)
3. Angrist, J.D., Imbens, G.W.: Identification and estimation of local average treatment effects. Technical report, National Bureau of Economic Research (1995)
4. Barron, J.L., Fleet, D.J., Beauchemin, S.S.: Performance of optical flow techniques. Int. J. Comput. Vis. **12**(1), 43–77 (1994)

5. Bertrand, M., Mullainathan, S.: Are Emily and Greg more employable than Lakisha and Jamal? A field experiment on labor market discrimination. Am. Econ. Rev. **94**(4), 991–1013 (2004)
6. Bowyer, K., Phillips, P.J.: Empirical Evaluation Techniques in Computer Vision. IEEE Computer Society Press (1998)
7. Brandao, M.: Age and gender bias in pedestrian detection algorithms. arXiv preprint arXiv:1906.10490 (2019)
8. Buhrmester, M., Kwang, T., Gosling, S.D.: Amazon's mechanical turk: a new source of inexpensive, yet high-quality data? (2016)
9. Buolamwini, J., Gebru, T.: Gender shades: intersectional accuracy disparities in commercial gender classification. In: Conference on Fairness, Accountability and Transparency, pp. 77–91 (2018)
10. Cortes, C., Vapnik, V.: Support-vector networks. Mach. Learn. **20**(3), 273–297 (1995)
11. Denton, E., Hutchinson, B., Mitchell, M., Gebru, T.: Detecting bias with generative counterfactual face attribute augmentation. arXiv preprint arXiv:1906.06439 (2019)
12. Drozdowski, P., Rathgeb, C., Dantcheva, A., Damer, N., Busch, C.: Demographic bias in biometrics: a survey on an emerging challenge. IEEE Trans. Technol. Soc. (2020)
13. Fei-Fei, L., Fergus, R., Perona, P.: Learning generative visual models from few training examples: an incremental Bayesian approach tested on 101 object categories. In: 2004 Conference on Computer Vision and Pattern Recognition Workshop, pp. 178–178. IEEE (2004)
14. Gelman, A., Hill, J.: Data Analysis Using Regression and Multilevel/Hierarchical Models. Cambridge University Press, Cambridge (2006)
15. Grother, P., Ngan, M., Hanaoka, K.: Ongoing face recognition vendor test (FRVT) part 1: verification. Technical report, National Institute of Standards and Technology (2018)
16. Grother, P.J., Ngan, M.L., Hanaoka, K.K.: Ongoing face recognition vendor test (FRVT) part 2: identification. Technical report (2018)
17. Hanaoka, P.G.N.K.: Face recognition vendor test (FRVT) part 3: demographic effects. IR 8280, NIST (2019). https://doi.org/10.6028/NIST.IR.8280
18. He, K., Zhang, X., Ren, S., Sun, J.: Deep residual learning for image recognition. In: Proceedings of the IEEE Conference on Computer Vision and Pattern Recognition, pp. 770–778 (2016)
19. Heckman, J.J., Vytlacil, E.J.: Instrumental variables, selection models, and tight bounds on the average treatment effect. In: Lechner, M., Pfeiffer, F. (eds.) Econometric Evaluation of Labour Market Policies, vol. 13, pp. 1–15. Springer, Heidelberg (2001). https://doi.org/10.1007/978-3-642-57615-7_1
20. Hoerl, A.E., Kennard, R.W.: Ridge regression: biased estimation for nonorthogonal problems. Technometrics **12**(1), 55–67 (1970)
21. Kärkkäinen, K., Joo, J.: FairFace: face attribute dataset for balanced race, gender, and age. arXiv preprint arXiv:1908.04913 (2019)
22. Karras, T., Laine, S., Aila, T.: A style-based generator architecture for generative adversarial networks. In: Proceedings of the IEEE Conference on Computer Vision and Pattern Recognition, pp. 4401–4410 (2019)
23. Karras, T., Laine, S., Aittala, M., Hellsten, J., Lehtinen, J., Aila, T.: Analyzing and improving the image quality of StyleGAN. arXiv preprint arXiv:1912.04958 (2019)

24. Kearns, M., Neel, S., Roth, A., Wu, Z.S.: Preventing fairness gerrymandering: auditing and learning for subgroup fairness. arXiv preprint arXiv:1711.05144 (2017)
25. Kearns, M., Roth, A.: The Ethical Algorithm: The Science of Socially Aware Algorithm Design. Oxford University Press, Oxford (2019)
26. Klare, B.F., Burge, M.J., Klontz, J.C., Bruegge, R.W.V., Jain, A.K.: Face recognition performance: role of demographic information. IEEE Trans. Inf. Forensics Secur. **7**(6), 1789–1801 (2012)
27. Kleinberg, J., Ludwig, J., Mullainathany, S., Sunstein, C.R.: Discrimination in the age of algorithms. Published by Oxford University Press on behalf of The John M. Olin Center for Law, Economics and Business at Harvard Law School (2019). https://academic.oup.com/jla/article-abstract/doi/10.1093/jla/laz001/5476086
28. Kortylewski, A., Egger, B., Schneider, A., Gerig, T., Morel-Forster, A., Vetter, T.: Empirically analyzing the effect of dataset biases on deep face recognition systems. In: Proceedings of the IEEE Conference on Computer Vision and Pattern Recognition Workshops, pp. 2093–2102 (2018)
29. Kortylewski, A., Egger, B., Schneider, A., Gerig, T., Morel-Forster, A., Vetter, T.: Analyzing and reducing the damage of dataset bias to face recognition with synthetic data. In: Proceedings of the IEEE Conference on Computer Vision and Pattern Recognition Workshops (2019)
30. Krishnapriya, K.S., Vangara, K., King, M., Albiero, V., Bowyer, K.: Characterizing the variability in face recognition accuracy relative to race. ArXiv 1904.07325, April 2019
31. Li, Y., Vasconcelos, N.: REPAIR: removing representation bias by dataset resampling. In: Proceedings of the IEEE Conference on Computer Vision and Pattern Recognition, pp. 9572–9581 (2019)
32. Liu, Z., Luo, P., Wang, X., Tang, X.: Deep learning face attributes in the wild. In: Proceedings of International Conference on Computer Vision (ICCV) (2015)
33. Lohr, S.: Facial recognition is accurate, if you're a white guy. New York Times, 9 February 2018. https://nyti.ms/2BNurVq
34. Lu, B., Chen, J.C., Castillo, C.D., Chellappa, R.: An experimental evaluation of covariates effects on unconstrained face verification. IEEE Trans. Biometr. Behav. Identity Sci. **1**(1), 42–55 (2019)
35. Merkatz, R.B., Temple, R., Sobel, S., Feiden, K., Kessler, D.A.: Working group on women in clinical trials: women in clinical trials of new drugs-a change in food and drug administration policy. New Engl. J. Med. **329**(4), 292–296 (1993)
36. Merler, M., Ratha, N., Feris, R.S., Smith, J.R.: Diversity in faces. arXiv preprint arXiv:1901.10436 (2019)
37. Muthukumar, V., et al.: Understanding unequal gender classification accuracy from face images. arXiv preprint arXiv:1812.00099 (2018)
38. Oreopoulos, P.: Estimating average and local average treatment effects of education when compulsory schooling laws really matter. Am. Econ. Rev. **96**(1), 152–175 (2006)
39. Pearl, J.: Causality. Cambridge University Press, Cambridge (2009)
40. Phillips, P.J., Grother, P., Micheals, R., Blackburn, D.M., Tabassi, E., Bone, M.: Face recognition vendor test 2002. In: Proceedings of the 2003 IEEE International SOI Conference (Cat. No. 03CH37443), p. 44. IEEE (2003)
41. Phillips, P.J., Wechsler, H., Huang, J., Rauss, P.J.: The feret database and evaluation procedure for face-recognition algorithms. Image Vis. Comput. **16**(5), 295–306 (1998)

42. Phillips, P.J., et al.: Face recognition accuracy of forensic examiners, superrecognizers, and face recognition algorithms. Proc. Natl. Acad. Sci. **115**(24), 6171–6176 (2018)
43. Pocock, S.J., Assmann, S.E., Enos, L.E., Kasten, L.E.: Subgroup analysis, covariate adjustment and baseline comparisons in clinical trial reporting: current practiceand problems. Stat. Med. **21**(19), 2917–2930 (2002)
44. Ponce, J., et al.: Dataset issues in object recognition. In: Ponce, J., Hebert, M., Schmid, C., Zisserman, A. (eds.) Toward Category-Level Object Recognition. LNCS, vol. 4170, pp. 29–48. Springer, Heidelberg (2006). https://doi.org/10.1007/11957959_2
45. Robinson, L.D., Jewell, N.P.: Some surprising results about covariate adjustment in logistic regression models. Int. Stat. Rev./Revue Inte. Stat. 227–240 (1991)
46. Rubin, D.B.: Matched Sampling for Causal Effects. Cambridge University Press, Cambridge (2006)
47. Shen, Y., Gu, J., Tang, X., Zhou, B.: Interpreting the latent space of GANs for semantic face editing. arXiv preprint arXiv:1907.10786 (2019)
48. Simon, V.: Wanted: women in clinical trials (2005)
49. Singla, S., Pollack, B., Chen, J., Batmanghelich, K.: Explanation by progressive exaggeration. arXiv preprint arXiv:1911.00483 (2019)
50. Szegedy, C., et al.: Intriguing properties of neural networks. arXiv preprint arXiv:1312.6199 (2013)
51. Torralba, A., Efros, A.A., et al.: Unbiased look at dataset bias. In: CVPR, vol. 1, p. 7 (2011)
52. VanderWeele, T.J., Shpitser, I.: On the definition of a confounder. Ann. Stat. **41**(1), 196 (2013)
53. Willan, A.R., Briggs, A.H., Hoch, J.S.: Regression methods for covariate adjustment and subgroup analysis for non-censored cost-effectiveness data. Health Econ. **13**(5), 461–475 (2004)
54. Wilson, E.B.: Probable inference, the law of succession, and statistical inference. J. Am. Stat. Assoc. **22**(158), 209–212 (1927)
55. Xiao, T., Hong, J., Ma, J.: ELEGANT: exchanging latent encodings with GAN for transferring multiple face attributes. In: Proceedings of the European Conference on Computer Vision (ECCV), pp. 168–184 (2018)
56. Zhou, B., Bau, D., Oliva, A., Torralba, A.: Interpreting deep visual representations via network dissection. IEEE Trans. Pattern Anal. Mach. Intell. **41**, 2131–2145 (2018)

Learning and Memorizing Representative Prototypes for 3D Point Cloud Semantic and Instance Segmentation

Tong He, Dong Gong, Zhi Tian, and Chunhua Shen[✉]

The University of Adelaide, Adelaide, Australia
{tong.he,dong.gong,zhi.tian,chunhua.shen}@adelaide.edu.au

Abstract. 3D point cloud semantic and instance segmentation are crucial and fundamental for 3D scene understanding. Due to the complex structure, point sets are distributed off-balance and diversely, appearing as both category and pattern imbalance. It has been proved that deep networks can easily forget the non-dominant cases during training, which influences the model generalization and leads to unsatisfactory performance. Although re-weighting on instances may reduce the influence, it is hard to find a balance between the dominant and the non-dominant cases. To tackle the above issue, we propose a memory-augmented network that learns and memorizes the representative prototypes that encode both geometry and semantic information. The prototypes are shared by diverse 3D points and recorded in a universal memory module. During training, the memory slots are dynamically associated with both dominant and non-dominant cases, alleviating the forgetting issue. In testing, the distorted observations and rare cases can thus be augmented by retrieving the stored prototypes, leading to better generalization. Experiments on the benchmarks, *i.e.*, S3DIS and ScanNetV2, show the superiority of our method on both effectiveness and efficiency, which substantially improves the accuracy not only on the entire dataset but also on non-dominant classes and samples.

Keywords: Point cloud · Instance segmentation · Memory network

1 Introduction

3D scene understanding is important and fundamental for various applications, such as robotics, autonomous driving, and virtual reality. The core tasks include semantic segmentation, and instance segmentation on 3D point clouds, *i.e.*,

---

T. He and D. Gong—Contributed equally.

Electronic supplementary material The online version of this chapter (https://doi.org/10.1007/978-3-030-58523-5_33) contains supplementary material, which is available to authorized users.

A. Vedaldi et al. (Eds.): ECCV 2020, LNCS 12363, pp. 564–580, 2020.
https://doi.org/10.1007/978-3-030-58523-5_33

(a) Input point sets   (b) Instance ground truth   (c) Results wo memory module   (d) Results with memory module

Fig. 1. Comparison of instance segmentation results. The performance of our proposed method shows strong robustness against non-dominant cases.

assigning semantic labels and instance indication labels for each point, respectively. Comparing to the studies on 2D images [2,5,12,13], semantic and instance segmentation on 3D point clouds lag far behind and have just started recently [14,15,33,34,38,39].

Based on the pioneering works of PointNet [24] and PointNet++ [26], directly processing point sets becomes simpler, more memory-efficient and flexible than handling the volumetric grids with 3D convolution [14,21,37]. Some following approaches [33,34,38,39] propose to handle semantic and instance segmentation in an end-to-end network jointly for the fine-grained description of the observation. Specifically, discriminative instance embeddings are learned to measure the instance-level clustering patterns of the points [23,34].

Although existing methods have achieved some impressive results, we still can observe performance bottlenecks on different datasets [1,3], especially on the non-dominant classes with fewer samples (see Fig. 6). It has been proved that deep networks tend to *forget* the rare cases easily while learning on a dataset distributed off balance and diversely [31]. On point cloud data, imbalance issue usually appears as the *category imbalance* and *pattern imbalance*, which is severer than that on 2D images [38]. Defining and measuring the category imbalance is easier, which appears as a significant discrepancy among the proportions of different categories. For example, in an indoor scene (as shown in Fig. 1), the proportions of the points belonging to the background (*e.g.*, wall) are much higher than the objects (*e.g.*, chairs). In S3DIS [1], the total amount of ceiling points is 50 times larger than the chair. The pattern imbalance can be observed on the (non-dominant) rare cases, appearing in both dominant and non-dominant categories, which are usually in the minority of the datasets. It is often caused by complex factors, such as positions, shapes, and relative relationships. For example, chairs are usually placed near a desk, while may occasionally appear with arbitrary positions (e.g., stacking and back-to-back near the cabinet) in an office, as shown in Fig. 1. Conventional methods [38] ignore this issue or simply resort to the focal loss [18], by down weighing the well-learned samples during training. However, it is hard to find a balance between the dominant and non-dominant samples in the dynamic training process.

To address the above issues, we propose to learn and **memorize** the discriminative and representative prototypes covering all the samples, which is implemented as a memory-augmented network, referred to as **MP**Net. The proposed MPNet includes two branches for predicting point-level semantic labels and obtaining per-point embedding for instance grouping, respectively. As shown in Fig. 2, the two branches access the shared memory via two separate memory readers, associating the two tasks via the shared memory. Given an input, MPNet retrieves the most relevant items in the memory for the extracted per-point features and feeds only retrieved embedding to the following segmentation tasks. In MPNet, the memory is maintained as a compact dictionary shared by diverse points. Driven by the task-specific training objectives and the proposed geometry-aware regularization, the compact memory is pushed to record the prototypes encoding the geometric and semantic information that is the most representative for all samples. During training, the memory slots are dynamically associated with both the dominant (common) and non-dominant (rare) categories (and cases) seen in mini-batches, alleviating the example forgetting issue [31]. In testing, the distorted observations and rare cases can thus be augmented by retrieving the stored prototypes, leading to better generalization. Additionally, different from previous methods relying on either pairwise relations computation [33] or KNN based feature aggregation [34], the proposed MPNet is free from complex and time-consuming operations, which is more efficient.

The main contributions are summarized as:

- We propose a memory-augmented network (*i.e.*, MPNet) for point cloud segmentation, by learning and memorizing the discriminative and representative prototypes covering all samples. The memory is dynamically associated with both the dominant (common) and non-dominant (rare) categories and cases seen in mini-batch training, alleviating the forgetting issue of the network and leading to better generalization.
- We propose specific regularizations on the memory to learn meaningful and interpretable prototypes in the memory. The proposed memory module is shared by the semantic and instance segmentation task, which naturally associates the two tasks and facilitates the mutual boost.
- The proposed MPNet achieves state-of-the-art performance on large scale datasets, boosting the performance by a large margin not only on the entire dataset but also on the non-dominant classes and samples, with limited consumption on computation and memory.

2   Related Work

Deep Learning for 3D Point Cloud. Existing methods for extracting features for 3D point cloud can be roughly categorized into three groups, including voxel-based [21,37], multi-view based [4,14,25,29] and point-based [16,24,26,30]. [21,37] are the pioneering works to transfer irregular points to regular volumetric grids, aiming to efficiently extract feature representation with 3D convolution. To reduce irrelevant operation on void places and save runtime memory usage, many works are proposed [10,27]. Multi-view based methods extract features in both 2D and 3D domain. [29] is one of the pioneering multi-view based method, which applies view-pooling over the 2D predictions. 3D-SIS [14], proposed by Hou *et al.*, combines features from 2D and 3D via explicit spatial mapping in an end-to-end trainable network. PointNet [24] is the first deep-learning-based work to operate directly on point sets, which uses shared MLP (multi-layer perceptron) to extract per-point feature. PointNet++ [26] improves the performance by extracting a hierarchical representation. Many following works [16,17,30,32,36] have been proposed to get a better representation of local context. Due to its simplicity, we select PointNet++ as our backbone and leave the choices of other backbones for future work.

Instance Segmentation on Point Cloud. Deep-learning-based instance segmentation for 3D point cloud is rarely studied until huge application potential has been discovered recently. SGPN [33] is the first deep-learning-based method working on this field. It first splits the whole scene into separate blocks. For every single block, per-point grouping candidates are proposed by predicting a similarity matrix that reflects affinity between each pair of points. A block merging algorithm is conducted for post-processing by considering segmentation results of the overlapped area. However, huge memory is needed for storing the pairwise matrix, which makes it memory-consuming for post-processing. In order to solve this, Wang et al. proposed ASIS [34], which utilized a discriminative loss function [2] to encourage points belonging to the same instance are mapped to a metric space with close distances. Moreover, to make the two tasks take advantage of each other, convolution and KNN search are applied for mutual feature aggregation of the two tasks, making it inefficient and time-consuming.

Memory Networks. Memory-based approaches have been discussed for solving various problems. NTM [11] is proposed to improve the generalization ability of the network by introducing an attention-based memory module. Gong et al. [9] proposed a memory augmented auto-encoder for anomaly detection, which is detected by represented the input with prototypical elements of the normal data maintained in a memory module. However, the memory model in [9] only includes a single memory pool in autoencoder for unsupervised representation, which may not work for the other tasks. Prototypical Network [28] maintains a category-wise templates for the problem of few-shot classification. Liu [19] proposed an OLTR algorithm to solve the open-ended and long-tail problem by associating a memory feature that can be transferred to both head and tail classes adaptively. These two methods are designed for the task of classification.

3 The Proposed Method

3.1 Overview of the Proposed MPNet

We propose a memory-augmented network for joint semantic and instance segmentation in the point cloud data, which learns and memorizes the prototypes of the point sets to alleviate the influence of the imbalanced distribution of the data. As shown in Fig. 2, the proposed memory-augmented network (i.e.MPNet) adopts an encoder-decoder architecture, which is free from the specific design of the encoder and decoder. In the proposed MPNet, we use PointNet++ [26] to implement the encoder for per-point feature extraction. Two parallel decoders

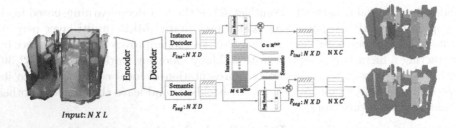

Input: N X L

Fig. 2. The framework of our proposed MPNet, which contains two parallel branches with a shared encoder. A memory module is proposed to memorize representative prototypes that are shared by all samples. Both distorted and rare cases can be augmented by retrieving the stored prototypes.

for instance segmentation and semantic segmentation are built upon the shared encoder. The memory is implemented as a dictionary to record the discriminative and representative prototypes as bases, which are optimized driven by the task-specific objective and the proposed instance-aware geometric regularization.

Given a set of input points $\{\mathbf{p}_i\}_{i=1}^{N}$ with $\mathbf{p}_i \in \mathbb{R}^{L}$, we can formulate the input of the network as a matrix $\mathbf{P} \in \mathbb{R}^{N \times L}$, where L denotes the input feature dimension and N denotes the total number of input points. The input features of each point may consist of both geometry and appearance information, $i.e.$, 3D coordinate (x, y, z) and RGB values. The two decoder branches produce features $\mathbf{F}_{\text{seg}} \in \mathbb{R}^{N \times D}$ and $\mathbf{F}_{\text{ins}} \in \mathbb{R}^{N \times D}$, respectively, where D denotes the dimension of the features. Instead of directly using $\mathbf{F}_{\text{seg}}$ and $\mathbf{F}_{\text{ins}}$ to perform semantic and instance segmentation tasks, respectively, MPNet applies them as queries to retrieve the most relevant prototypes in the memory and then obtains features $\widehat{\mathbf{F}}_{\text{seg}}$ and $\widehat{\mathbf{F}}_{\text{ins}}$, which are delivered to the following semantic classifier and instance embedding module. The memory is randomly initialized and optimized during training. The two branches access the memory with specifically designed reading heads.

3.2 Memory Representation for Prototypes

The *prototype memory* is designed as a matrix $\mathbf{M} \in \mathbb{R}^{M \times D}$, where M is a hyper-parameter that defines the number of memory slots and D is the feature dimension that is identical with the outputs from the two branches. The M memory slots are used to restore the prototypes shared by all instances across all categories. To easily represent the semantic characteristics, we define a *semantic memory* $\mathbf{C} \in \mathbb{R}^{C \times D}$, where C denotes the number of categories in semantic segmentation task and each row of $\mathbf{C}$ represents the summary of a class. $\mathbf{C}$ can be seen as the semantic summary of $\mathbf{M}$ and are generated from $\mathbf{M}$. We equally associate the M memory slots in $\mathbf{M}$ with C categories and thus define $M = M_c \times C$, where M_c is the number of per-category prototypes. As shown in Fig. 2, the i-th row in $\mathbf{C}$ is defined as the average of the i-th subsegment ($i.e.$, rows from $(i - 1) \times M_c + 1$ to $i \times M_c$) in $\mathbf{M}$:

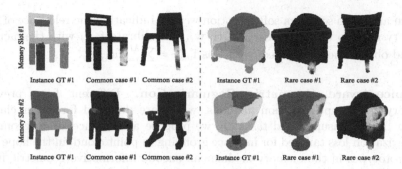

Fig. 3. Visualization of the memory slots in $\mathbf{M}$. We visualize what the memory has learned with the instance segmentation on the PartNet dataset [22], in which the parts (*e.g.*, the chair legs) of the object are treated as instances. For a specific memory slot (*i.e.*, slot #1 and #2 in the figure), we visualize the addressing weights of the points from common and rare cases in pseudo color. The correlation between a specific memory slot and the "visual concepts" (*e.g.*, the components type and relative position) of the most related points are consistent across diverse examples, including common and rare cases, which implies the memory captures meaningful and interpretable semantic and geometric prototypes. Quantitative results are left in the supplementary material.

$$\mathbf{c}_i = \frac{1}{M_c} \sum_{j=(i-1)\times M_c+1}^{i\times M_c} \mathbf{m}_j, \tag{1}$$

where $\mathbf{m}_j$ denotes the j-th row vector of $\mathbf{M}$.

Given the query features $\mathbf{F}_{\text{ins}}$ and $\mathbf{F}_{\text{seg}}$, the instance grouping branch directly accesses the prototypes memory $\mathbf{M}$ and the semantic labeling branch accesses the semantic summary $\mathbf{C}$, with two specifically designed readers. $\mathbf{M}$ can be seen as a dictionary to restore the representative bases shared by all instances, as the instances cross different categories can share some common basic components and characteristics. Because the semantic memory $\mathbf{C}$ is a re-parameterization of $\mathbf{M}$, the two tasks are naturally associated together, without computation-consuming operations as [34].

3.3 Memory-Augmented Instance Embedding

Reading memory for Instance Embedding. Given $\mathbf{F}_{\text{ins}}$, the proposed MPNet reads the most relevant items from $\mathbf{M}$ to obtain instance embedding for instance grouping. For each per-point feature $\mathbf{f}_{\text{ins},i}$ (*i.e.*, the i-th row of $\mathbf{F}_{\text{ins}}$), we calculate the memory addressing weights $\mathbf{w}_i \in \mathbb{R}^M$ according to the similarity between $\mathbf{f}_{\text{ins},i}$ and the prototypes stored in memory $\mathbf{M}$:

$$w_{ij} = \frac{\exp(d(\mathbf{f}_{\text{ins},i}, \mathbf{m}_j))}{\sum_{j=1}^{M} \exp(d(\mathbf{f}_{\text{ins},i}, \mathbf{m}_j))}, \tag{2}$$

where w_{ij} denotes the j-th element of $\mathbf{w}_i$, $\mathbf{m}_j$ is the j-th row in $\mathbf{M}$, and $d(\cdot, \cdot)$ denotes the similarity measurement function. We use cosine similarity as $d(\cdot, \cdot)$.

$\mathbf{w}_i$ can also been seen as a soft-attention weight, indicating the relevance of each memory item to the query $\mathbf{f}_{\text{ins},i}$. With $\mathbf{w}_i$, we calibrate $\mathbf{f}_{\text{ins},i}$ with the memory $\mathbf{M}$ and obtain the augmented feature as $\widehat{\mathbf{f}}_{\text{ins},i} = \sum_{j=1}^{M} w_{ij}\mathbf{m}_j$.

Instance-Aware Geometric Regularization. Different from previous memory-based representation methods, which are designed for either classification [19] or unsupervised tasks [9], we propose an instance-aware geometric regularization loss tailored for instance grouping in point cloud, in the hope that the prototypes in the memory module can encode informative geometric information. To achieve this, we force the memory-augmented features from the same instance to have identical geometric predictions.

We first introduce an instance centroids estimator $G(\cdot)$ that will be trained to predict the instance centroids based on the augmented features as $G(\widehat{\mathbf{f}}_{\text{ins},n})$ and try to enforce the predicted centroids to be grouped around the corresponding geometric centers of the instances. The instance-aware regularization loss R_{ins} is defined as:

$$R_{\text{ins}} = \frac{1}{K} \sum_{k=1}^{K} \frac{1}{N_k} \sum_{n=1}^{N_k} \|G(\widehat{\mathbf{f}}_{\text{ins},n}) - GT_k\|^2, \tag{3}$$

where K is the instance number, N_k is the number of the points of k-th instance, and GT_k denotes the ground truth centroid of the k-th instance. $\widehat{\mathbf{f}}_{\text{ins},n}$ denotes the augmented feature of a point belonging to the k-th instance. $G(\cdot)$ is implemented as an MLP and can be trained in an end-to-end manner.

What are Learnt and Stored in Memory. To have a clear understanding of the learned memory prototypes, we select the category of 'Chair' in PartNet [22] for training and visualization due to its largest number of training samples, as shown in Fig. 3. Quantitative results are presented in the supplementary materials. For each memory item, the points that are addressing it have consistent semantic meaning, implying the capability of the memory module to capture the discriminative and unified representation (for example, position sensitive information) for both common and rare cases.

3.4   Memory-Augmented Semantic Labeling

Reading Memory for Semantic Segmentation. Similar to the instance grouping branch, the semantic branch reads the category prototypes from semantic memory $\mathbf{C}$ for classification. For each $\mathbf{f}_{\text{seg},i}$ from $\mathbf{F}_{\text{seg}}$, we obtain the soft memory addressing weights $\boldsymbol{\gamma}_i \in \mathbb{R}^C$ by calculating the similarity between $\mathbf{f}_{\text{seg},i}$ and each $\mathbf{c}_j$ (*i.e.*, each row of $\mathbf{C}$), similar to Eq. (2). Then we can obtain $\widehat{\mathbf{F}}_{\text{seg}}$ through $\widehat{\mathbf{f}}_{\text{seg},i} = \boldsymbol{\gamma}_i^T \mathbf{C} = \sum_{j=1}^{C} \gamma_{ij}\mathbf{c}_j$, where γ_{ij} denotes the j-th item in $\boldsymbol{\gamma}_i$.

Semantic Memory Regularization. We apply an additional regularization term on the semantic memory to enforce the centroids of different categories (*i.e.*, the semantic summarization c_i's) to be separately distributed. Specifically, R_{seg} is used to encourage the augmented feature close to its corresponding category summary in the memory and far away from others. Given $\hat{f}_{seg,i}$ and its class label y_i, the regularization term R_{seg} is calculated as:

$$R_{seg} = \max(0, \sum_{j=y_i} \|\hat{f}_{seg,i} - c_j\| - \sum_{j \neq y_i} \|\hat{f}_{seg,i} - c_j\| + m), \qquad (4)$$

where m is the margin, which is set as 5 in our implementation. Each c_j performs like an anchor point and pulls the features with identical semantic labels close to it and pushes the features with different semantic labels away from it.

3.5 Loss Functions

Classification Loss. We use the traditional cross-entropy loss L_{ce} for the semantic segmentation task.

Instance Discriminative Loss. Given the per-point memory augmented features $\{\hat{f}_{ins,i}\}_{i=1}^{N}$, point-level embeddings $\{g_{ins,i} \in \mathbb{R}^{c'}\}_{i=1}^{N}$ are generated by a simple MLP layer, where c' is the dimension of the embedding space. Similar to [2,34], we set $c' = 5$ and use the instance discriminative loss for instance grouping. Embeddings from the same instance should be grouped together. A soft margin σ_v is introduced to allow these embeddings distributing on a local manifold rather than having to converge to a single point. Moreover, instance embedding centers are no longer repulsed if their distances are larger than $2\sigma_d$. The instance discriminative loss is formulated as:

$$L_{dis} = \frac{1}{K}\sum_{k=1}^{K}\frac{1}{N_k}\sum_{n=1}^{N_k} \left[\|g_{ins,n} - \mu_k\| - \sigma_v\right]_+^2 + \frac{1}{K(K-1)}\sum_{i=1}^{K}\sum_{\substack{j=1 \\ i \neq j}}^{K} \left[2\sigma_d - \|\mu_i - \mu_j\|\right]_+^2,$$
$$(5)$$

where K is the total instance number, and N_K is the point number of the k-th instance. μ_k is the average embedding of the k-th instance, which is calculated by $\mu_k = \frac{1}{N_k}\sum_{n=1}^{N_k} g_{ins,n}$. σ_v and σ_d in Eq. (5) are the soft margins. During testing, a simple mean shift clustering algorithm is adopted to group the points in the embedding space.

Training Objective. As all operations are differentiable, memory can be updated through back-propagation in an end-to-end manner. By combining the four losses discussed above, the training objective is formulated as:

$$L = L_{ce} + L_{dis} + R_{seg} + \lambda R_{ins}, \qquad (6)$$

where λ is the loss weight for R_{ins}. Moreover, as $\mathbf{C}$ is a re-parameterization of $\mathbf{M}$, the supervisions jointly update $\mathbf{M}$ and then influence the two tasks in turn. The two tasks are thus naturally associated together, free from the complex and time-consuming operation, as introduced in [34].

Table 1. Ablation study on the S3DIS dataset with vanilla Pointnet++ as backbone. **FL** refers to the focal loss. **InsMem** indicates that the memory is updated by the instance information. **SegMem** means the memory is updated by the supervision from semantic segmentation. **Regul** refers to the regularizations used in learning the prototypes memory. Both instance and semantic segmentation results are provided.

Method	FL	InsMem	SegMem	Regul	mPre	mRec
Baseline					52.3	41.4
	✓				55.2	43.0
		✓			58.9	47.0
		✓	✓		60.2	47.2
Ours		✓	✓	✓	62.5	49.0

4   Experiments

To validate the effectiveness of our proposed method, both qualitative and quantitative experiments are conducted on two public datasets: Stanford 3D Indoor Semantic Dataset (S3DIS) [1] and ScanNetV2 [3]. The S3DIS dataset [1] is collected in 6 large-scale indoor areas, including 13 classes. ScanNetV2 consists of 1613 indoor scans from 40 categories. The dataset is split into 1201, 312, and 100 for training, validating, and testing, respectively.

4.1   Evaluation

Following [34] on S3DIS dataset, the results on Area-5 and 6-fold cross-validation are reported in our experiments. For semantic segmentation, we present 1) the overall accuracy (oAcc), which measures the point-level accuracy, 2) the mean class accuracy (mAcc), which calculates the average category-level accuracy, and 3) the instance-wise mean intersection-over-union (mIoU). For instance segmentation, four evaluation metrics are calculated, namely, $mConv$, $mWConv$, $mPrec$, and $mRec$. $mConv$ is defined as the mean instance-wise matching IoU score between the ground truth and the prediction. Instead of treating every instance equally, $mWConv$ is calculated by weighting the size of each instance object. Moreover, traditional $mPrec$ and $mRec$ represent mean precision and recall with the IoU threshold of 0.5, respectively.

4.2   Implementation Details

For the datasets of S3DIS and ScanNetV2, each room is divided into $1m \times 1m$ blocks with a stride of $0.5m$. 4096 points are randomly sampled from each block during the training process. Without special notation, all experiments are conducted using vanilla PointNet++ [26] as backbone (without introducing any multi-scale grouping operation). We utilize the same training setting as ASIS [34]. The whole network is trained in an end-to-end manner for 100 epochs in total. During the inference time, blocks within each room are merged by

Point Cloud   ASIS   Ours   Point Cloud   ASIS   Ours

Fig. 4. Barnes-Hut t-SNE [20] visualization of the instance embedding on S3DIS Area-5 set (Best viewed in color and zoomed in). (Color figure online)

utilizing the semantic and instance results of the overlapped region. Detailed settings of the algorithm are identical with [33].

4.3 Ablation Study

In this section, we study the influence of each integration of the aforementioned components. All the results are tested on S3DIS Area-5 for a fair comparison. We first build a strong baseline which is equivalent to the vanilla ASIS [34]. Building upon the strong baseline, our MPNet surpasses it by a large margin via memorizing representative prototypes. In the following, we provide detailed analyses on different aspects.

Memory M and C. The representative and consistent prototypes are maintained in the prototypes memory **M**, which is universal to represent the shared concepts of all instances. Besides, the semantic memory **C** is served as a semantic summary to represent the category characteristics efficiently. As shown in Table 1, using instance memory **M** alone can boost $mPre$ from 52.3% to 58.9% and $mRec$ from 41.4% to 47.0%. On the other hand, using the semantic memory **C** can bring another 1.3% improvements with the metrics of $mPrec$.

Visualizing the Effects of Memory on Instance Embedding. In Fig. 4, we directly visualize the instance embedding $g_{ins,i}$ to show the positive effects of the memory, which covers both the common and rare scenes, $i.e.$office and lobby. The embeddings are projected to 1-D via Barnes-Hut t-SNE [20] for visualization. In both situations, with the help of the memory module, our MPNet generates more discriminative embedding features than the previous state-of-the-art method ASIS [34], which are critical for separating different instances.

Memory Size. We study the influence of the memory size, $i.e.$, the hyperparameter M or N_c equivalently, to the final performance. The results show that the performance increases as the N_c grows, and becomes stable after when N_c is greater than 200. In all our experiments, N_c is set to 150.

Regularization Loss. To effectively learn representative and discriminative prototypes, regularization losses are proposed in Eq. (3) and Eq. (4), which directly work on the memory-augmented features for instance segmentation and

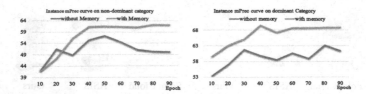

Fig. 5. The training curve on both dominate ("wall") and non-dominant categories ("sofa"). The forgetting issue can be alleviated when associated with our proposed representative memory slots.

Table 2. Instance Segmentation results on S3DIS dataset. Both Area-5 and 6-fold results are reported. All our results are achieved based on a vanilla PointNet++ backbone (without multi-scale grouping) for fair comparison.

Method	Year	mCov	mWCov	mPrec	mRec
Test on area 5					
SGPN [33]	2018	32.7	35.5	36.0	28.7
ASIS [34]	2019	44.6	47.8	55.3	42.4
3D-BoNet [38]	2019	–	–	57.5	40.2
JSNet [40]	2019	48.7	51.5	62.1	46.9
Ours	–	**50.1**	**53.2**	**62.5**	**49.0**
Test on 6-fold					
SGPN [33]	2018	37.9	40.8	31.2	38.2
MT-PNet [23]	2019	–	–	24.9	–
MV-CRF [23]	2019	–	–	36.3	–
ASIS [34]	2019	51.2	55.1	63.6	47.5
3D-BoNet [38]	2019	–	–	65.6	47.6
PartNet [22]	2019	–	–	56.4	43.4
JSNet [40]	2019	54.1	58.0	66.9	**53.9**
Ours	–	**55.8**	**59.7**	**68.4**	53.7

semantic segmentation, respectively. Both of them can be beneficial for both semantic and instance segmentation due to the mutual influence on the memory. As shown in Table 1, the two regularization terms boost the $mPre$ and $mRec$ for about 2.3% and 1.8%, respectively.

Comparing with Focal Loss [18]. The discrepancies among different categories are significant in the 3D point cloud. Focal loss [18] has been widely used in different kinds of vision tasks due to the imbalanced distribution of the training data. It addresses the problem by down-weighting the well-classified samples. However, it only alleviates the category imbalance to some extent and fails to solve the diversely distributed patterns and cases. As shown in Table 1, focal loss can only improve the $mPre$ by 2.9%. Compared with the Focal Loss,

Table 3. Comparison per-class performance of our proposed method with the state-of-the-art methods on the task of semantic segmentation on S3DIS. We use vanilla pointnet++ [26] without multi-scale grouping. Even with a simple backbone, the proposed method surpasses the graph-based method by more than 1% mIOU (reported with 6-fold cross-validation).

	OA	Miou	Ceil	Floor	Wall	Beam	Colu	Wind	Door	Table	Chair	Sofa	Book	Boar	Clut
[24]	78.5	47.6	88.0	88.7	69.3	42.4	23.1	47.5	51.6	54.1	42.0	9.6	38.2	29.4	35.2
[7]	81.1	49.7	90.3	92.1	67.9	44.7	24.2	52.3	51.2	58.1	47.4	6.9	39.0	30.0	41.9
[26]	–	53.2	90.2	91.7	73.1	42.7	21.2	49.7	42.3	62.7	59.0	19.6	45.8	48.2	45.6
[8]	–	58.3	92.1	90.4	**78.5**	37.8	35.7	51.2	65.4	64.0	61.6	25.6	51.6	49.9	53.7
[35]	84.1	56.1	–	–	–	–	–	–	–	–	–	25.6	–	–	–
[16]	85.9	60.0	93.1	95.3	78.2	33.9	**37.4**	56.1	**68.2**	64.9	61.0	34.6	51.5	51.1	**54.4**
Ours	**86.8**	**61.3**	**94.0**	94.1	76.6	**53.4**	33.6	54.2	62.7	**70.2**	60.2	**36.6**	53.4	**54.3**	53.5

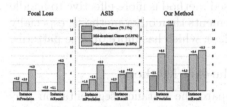

Fig. 6. The comparison of the improvements on both common and uncommon categories. We compare the performance of *mPrec* and *mRec* with Focal Loss [18] and ASIS [34].

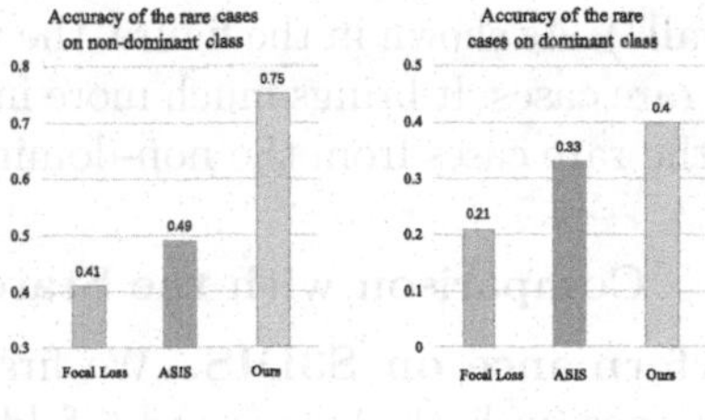

Fig. 7. The instance precision of the rare cases. Both common and uncommon categories are presented. The rare instances are collected as the 20% hardest samples from the baseline model.

our method surpasses the baseline model by a large margin, due to the memorized prototypical patterns and improves *mPre* and *mRec* by 8.6% and 5.9%, respectively.

4.4 Analysis on the Non-dominant Categories and Rare Cases

We study the instance segmentation performance gain brought by the proposed memory network specifically on the non-dominant categories and rare cases.

Analysis on Non-dominant (Rare) Categories. We compare the performance of our proposed MPNet with ASIS [34] on non-dominant classes. We first sort the 13 categories on S3DIS according to their proportions in the training set and split the dataset into three levels: dominant classes (the first 4 classes), mid-dominant classes (the mid 5 classes), and non-dominant classes (the last 4 classes). The amount proportions of the three levels are 79.17%, 16.95%, and 3.88%, respectively. As shown in Fig. 6, we report the improvements with two metrics *mPrec* and *mRec*. Our method not only boosts the overall performances but also brings much more significant improvements to the non-dominant classes than focal loss [18] and ASIS [34].

In Fig. 5, we plot the changes of the instance $mPrec$ scores of the model with or without the memory module during training. The results on both the common category ("wall") and uncommon category ("sofa") from S3DIS are shown. With the proposed memory module, our method can alleviate the forgetting issue on the non-dominant samples.

Analysis on Rare Cases. Analyzing with the rare cases is not as easy as on the rare classes since it is not easy to define. We maintain a set of rare cases from "Area-5" in S3DIS [1] by using the performance of the baseline model as the criterion. Specifically, we evaluate the instance-wise IoU score of vanilla ASIS [34] and collect 20% of the instances with the lowest scores as the rare cases for further studies. in Fig. 7, we show the performance of different methods on the rare cases from both a non-dominant class ("sofa") and a dominant class ("wall"). As shown in the figure, the proposed method is more effective to handle the rare cases. It brings much more improvements than other methods, especially on the rare cases from the non-dominant class, which has more diverse patterns.

4.5   Comparison with the State-of-the-Art Methods

Performance on S3DIS. We first compare the instance segmentation performance on both Area-5 and 6-fold. The results are presented in Table 2. Our proposed MPNet achieves promising results and surpasses the previous state-of-the-art approaches substantially by a large margin. The large improvements are mainly beneficial from the strong ability of the proposed memory module. Qualitative results are shown in Fig. 8. In addition to instance segmentation, we also report the results of semantic segmentation and compare them with other methods. The performance is tested on all areas (6-fold), as shown in Table 3. Although based on a simple PointNet++ backbone, we achieve better results than the other methods which are based on graph neural networks [16,35].

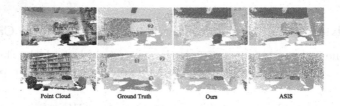

Fig. 8. Qualitative results of our method on S3DIS dataset. From left to right are: input point cloud, instance segmentation ground truth, the results of our method and the results of [34].

Performance on ScanNetV2. In addition to S3DIS, we conduct experiments on ScanNetV2 [3]. The instance segmentation results are reported in Table 4, which is tested on the validation set. To make a fair comparison, we select the methods that are based on PointNet or PointNet++. Our proposed MPNet outperforms previous methods and dominants in many categories.

Table 4. Instance segmentation results on ScannetV2 benchmark (validation set). The results of mAP@0.25 and mAP@0.5 are reported. All methods except [8] are point-based. (Due to the limited space, Table, Toilet, and Sofa are not presented.)

	mAP @0.25	mAP @0.5	bat.	bed	she.	cab.	cha.	cou.	cur.	des	doo	oth.	pic.	ref	sho	sin	sof
[12]	26.1	5.8	33.3	0.2	0.0	5.3	0.2	0.2	2.1	0.0	4.5	2.4	23.8	6.5	0.0	1.4	10.7
[33]	35.1	14.3	20.8	39.0	16.9	6.5	27.5	2.9	6.9	0.0	8.7	4.3	1.4	2.7	0.0	11.2	35.1
[6]	–	24.8	66.7	56.6	7.6	3.5	39.4	2.7	3.5	9.8	9.9	3.0	2.5	9.8	37.5	12.6	60.4
[39]	40.0	23.5	51.3	52.3	12.5	15.2	61.8	0.0	1.5	7.6	29.0	11.7	14.7	25.0	3.7	14.0	34.5
[34]	41.5	24.0	29.9	50.5	0.0	16.7	57.7	0.0	18.4	7.8	14.8	12.9	1.8	12.4	38.0	10.2	36.9
Ours	49.3	31.0	69.4	59.8	2.7	23.7	71.1	4.5	8.4	18.3	11.6	17.3	4.8	21.8	57.0	13.4	27.7

Table 5. Inferencing time comparison on the S3DIS Area-5 set. Forward time is the network running time on GPU, whereas postprocessing time is the BlockMerging algorithm introduced in [33]. ASIS is 45% slower than our method in the forward process due to the usage of KNN, which is extremely time-consuming. The reported time is running on a single 1080ti GPU with 4096 input points.

Method	Backbone	Inference time (ms)			mPre	mRec
		Overall	Forward	Post		
SGPN [33]	PointNet	730	**22**	708	36.0	28.7
ASIS [34]	PointNet2	183	58	**125**	55.3	42.4
Ours	PointNet2	**165**	40	**125**	**62.5**	**49.0**

Speed Analysis. We compare the inference speed with the other two methods: SGPN [33] and ASIS [34]. The whole evaluation process includes two parts: the network forward and instance grouping. The first part is to get per-point semantic labeling and instance embedding. The second part utilizes a grouping algorithm to find out instance groups. SGPN, which is based on PointNet, predicts a pair-wise affinity matrix to group points into instance clusters, requiring a huge memory buffer. Different from SGPN, ASIS utilizes mean-shift for clustering embeddings to instance groups. Meanwhile, ASIS applies KNN for fusing semantic context from a fixed number of neighboring points. This operation is extremely time-consuming. Compared with the above two approaches, our proposed MPNet is free from complex and time-consuming operations, showing superiority in both effectiveness and efficiency (Table 5).

5 Conclusion

In this paper, we propose a memory-augmented network to handle both category and pattern imbalance in the task of point cloud instance and semantic segmentation. Our method shows superiority in both effectiveness and efficiency.

References

1. Armeni, I., et al.: 3D semantic parsing of large-scale indoor spaces. In: Proceedings of the IEEE Conference on Computer Vision and Pattern Recognition (2016)
2. Brabandere, B.D., Neven, D., Gool, L.V.: Semantic instance segmentation with a discriminative loss function. In: Proceedings of the IEEE Conference on Computer Vision and Pattern Recognition (2017)
3. Dai, A., Chang, A.X., Savva, M., Halber, M., Funkhouser, T., Nießner, M.: ScanNet: richly-annotated 3D reconstructions of indoor scenes. In: Proceedings of the IEEE Conference on Computer Vision and Pattern Recognition (2017)
4. Dai, A., Nießner, M.: 3DMV: joint 3D-multi-view prediction for 3D semantic scene segmentation. In: Proceedings of the European Conference on Computer Vision (2018)
5. Dai, J., He, K., Sun, J.: Instance-aware semantic segmentation via multi-task network cascades. In: Proceedings of the European Conference on Computer Vision (2016)
6. Elich, C., Engelmann, F., Kontogianni, T., Leibe, B.: 3D-BEVIS: bird's-eye-view instance segmentation. arXiv preprint arXiv:1904.02199 (2019)
7. Engelmann, F., Kontogianni, T., Hermans, A., Leibe, B.: Exploring Spatial Context for 3D Semantic Segmentation of Point Clouds. In: Proceedings of the IEEE International Conference on Computer Vision Workshops (2017)
8. Engelmann, F., Kontogianni, T., Schult, J., Leibe, B.: Know what your neighbors do: 3D semantic segmentation of point clouds. arXiv:1810.01151 (2018)
9. Gong, D., et al.: Memorizing normality to detect anomaly: memory-augmented deep autoencoder for unsupervised anomaly detection. In: Proceedings of the IEEE International Conference on Computer Vision (2019)
10. Graham, B., Engelcke, M., van der Maaten, L.: 3D semantic segmentation with submanifold sparse convolutional networks. In: Proceedings of the IEEE Conference on Computer Vision and Pattern Recognition (2018)
11. Graves, A., Wayne, G., Danihelk, I.: Neural turing machines. arXiv preprint arXiv:1410.5401 (2014)
12. He, K., Gkioxari, G., Dollár, P., Girshick, R.: Mask R-CNN. In: Proceedings of the IEEE International Conference on Computer Vision (2017)
13. He, T., Shen, C., Tian, Z., Gong, D., Sun, C., Yan, Y.: Knowledge adaptation for efficient semantic segmentation. In: Proceedings of the IEEE Conference on Computer Vision and Pattern Recognition (2019)
14. Hou, J., Dai, A., Nießner, M.: 3D-SIS: 3D semantic instance segmentation of RGB-D scans. In: Proceedings of the IEEE Conference on Computer Vision and Pattern Recognition (2019)
15. Lahoud, J., Ghanem, B., Pollefeys, M., Oswald, M.R.: 3D instance segmentation via multi-task metric learning. arXiv preprint arXiv:1906.08650 (2019)
16. Li, G., Müller, M., Thabet, A., Ghanem, B.: DeepGCNs: can GCNs go as deep as CNNs? In: Proceedings of the IEEE International Conference on Computer Vision (2019)
17. Li, Y., Bu, R., Sun, M., Wu, W., Di, X., Chen, B.: PointCNN: convolution On X-transformed points. In: Proceedings of the Advances in Neural Information Processing Systems (2018)
18. Lin, T.Y., Goyal, P., Girshick, R., He, K., Dollár, P.: Focal loss for dense object detection. In: Proceedings of the IEEE International Conference on Computer Vision (2017)

19. Liu, Z., Miao, Z., Zhan, X., Wang, J., Gong, B., Yu, S.X.: Large-scale long-tailed recognition in an open world. In: Proceedings of the IEEE International Conference on Computer Vision and Pattern Recognition (2019)
20. van der Maaten, L.: Accelerating t-SNE using tree-based algorithms. J. Mach. Learn. Res. **15**, 3221–3245 (2014). http://jmlr.org/papers/v15/vandermaaten14a.html
21. Maturana, D., Scherer, S.: VoxNet: a 3D convolutional neural network for real-time object recognition. In: Proceedings of the IEEE International Conference on Intelligent Robots and Systems (2015)
22. Mo, K., et al.: PartNet: a large-scale benchmark for fine-grained and hierarchical part-level 3D object understanding. In: Proceedings of the IEEE Conference on Computer Vision and Pattern Recognition (2019)
23. Pham, Q.H., Nguyen, D.T., Hua, B.S., Roig, G., Yeung, S.K.: JSIS3D: joint semantic-instance segmentation of 3D point clouds with multi-task pointwise networks and multi-value conditional random fields. In: Proceedings of the IEEE Conference on Computer Vision and Pattern Recognition (2019)
24. Qi, C.R., Su, H., Mo, K., Guibas, L.J.: PointNet: deep learning on point sets for 3D classification and segmentation. In: Proceedings of the IEEE Conference on Computer Vision and Pattern Recognition (2017)
25. Qi, C.R., Su, H., Nießner, M., Dai, A., Yan, M., Guibas, L.J.: Volumetric and multi-view CNNs for object classification on 3D data. In: Proceedings of the IEEE Conference on Computer Vision and Pattern Recognition (2016)
26. Qi, C.R., Yi, L., Su, H., Guibas, L.J.: PointNet++: deep hierarchical feature learning on point sets in a metric space. In: Proceedings of the Advances in Neural Information Processing Systems (2017)
27. Riegler, G., Ulusoy, A.O., Geiger, A.: OctNet: learning deep 3D representations at high resolutions. arXiv preprint arXiv:1611.05009 (2016)
28. Snell, J., Swersky, K., Zemel, R.S.: Prototypical networks for few-shot learning. In: Proceedings of the Advances in Neural Information Processing Systems (2017)
29. Su, H., Maji, S., Kalogerakis, E., Learned-Miller, E.: Multi-view convolutional neural networks for 3D shape recognition. In: Proceedings of the IEEE International Conference on Computer Vision (2015)
30. Thomas, H., Qi, C.R., Deschaud, J.E., Marcotegui, B., Goulette, F., Guibas, L.J.: KPConv: flexible and deformable convolution for point clouds. In: Proceedings of the IEEE International Conference on Computer Vision (2019)
31. Toneva, M., Sordoni, A., Combes, R.T.D., Trischler, A., Bengio, Y., Gordon, G.J.: An empirical study of example forgetting during deep neural network learning. arXiv preprint arXiv:1812.05159 (2018)
32. Wang, L., Huang, Y., Hou, Y., Zhang, S., Shan, J.: Graph attention convolution for point cloud semantic segmentation. In: Proceedings of the IEEE Conference on Computer Vision and Pattern Recognition (2019)
33. Wang, W., Yu, R., Huang, Q., Neumann, U.: SGPN: similarity group proposal network for 3D point cloud instance segmentation. In: Proceedings of the IEEE Conference on Computer Vision and Pattern Recognition (2018)
34. Wang, X., Liu, S., Shen, X., Shen, C., Jia, J.: Associatively segmenting instances and semantics in point clouds. In: Proceedings of the IEEE Conference on Computer Vision and Pattern Recognition (2019)
35. Wang, Y., Sun, Y., Liu, Z., Sarma, S.E., Bronstein, M.M., Solomon, J.M.: Dynamic graph CNN for learning on point clouds. ACM Trans. Graphic **38**, 1–12 (2019)

36. Wu, W., Qi, Z., Fuxin, L.: PointConv: deep convolutional networks on 3D point clouds. In: Proceedings of the IEEE Conference on Computer Vision and Pattern Recognition (2019)
37. Wu, Z., et al.: 3D ShapeNets: a deep representation for volumetric shapes. In: Proceedings of the IEEE Conference on Computer Vision and Pattern Recognition (2015)
38. Yang, B., et al.: Learning object bounding boxes for 3D instance segmentation on point clouds. In: Proceedings of the Advances in Neural Information Processing Systems (2019)
39. Yi, L., Zhao, W., Wang, H., Sung, M., Guibas, L.J.: GSPN: generative shape proposal network for 3D instance segmentation in point cloud. In: Proceedings of the IEEE Conference on Computer Vision and Pattern Recognition (2018)
40. Zhao, L., Tao, W.: JSNet: joint instance and semantic segmentation of 3D point clouds. In: Proceedings of the AAAI Conference on Artificial Intelligence (2020)

Knowledge-Based Video Question Answering with Unsupervised Scene Descriptions

Noa Garcia[(✉)] and Yuta Nakashima

Osaka University, Suita, Japan
{noagarcia,n-yuta}@ids.osaka-u.ac.jp

Abstract. To understand movies, humans constantly reason over the dialogues and actions shown in specific scenes and relate them to the overall storyline already seen. Inspired by this behaviour, we design ROLL, a model for knowledge-based video story question answering that leverages three crucial aspects of movie understanding: dialog comprehension, scene reasoning, and storyline recalling. In ROLL, each of these tasks is in charge of extracting rich and diverse information by 1) processing scene dialogues, 2) generating unsupervised video scene descriptions, and 3) obtaining external knowledge in a weakly supervised fashion. To answer a given question correctly, the information generated by each inspired-cognitive task is encoded via Transformers and fused through a modality weighting mechanism, which balances the information from the different sources. Exhaustive evaluation demonstrates the effectiveness of our approach, which yields a new state-of-the-art on two challenging video question answering datasets: KnowIT VQA and TVQA+.

Keywords: Video question answering · Video description · Knowledge bases

1 Introduction

Robots may not dream of electric sheep yet,[1] but in the last few years, artificial intelligence has shown significant progress towards human-like reasoning. This has been made possible by emulating snippets of human intelligence in constrained tasks [1,11], where machine performance is easily evaluated. Among those tasks, video story question answering [5,17,35] emerged as a testbed to approximate real-world situations, in which not only the spatial relationships between objects are important, but also the temporal coherence between past, present, and future events.

---

[1] 'Do androids dream of electric sheep?' (Philip K. Dick, 1968).

Electronic supplementary material The online version of this chapter (https://doi.org/10.1007/978-3-030-58523-5_34) contains supplementary material, which is available to authorized users.

© Springer Nature Switzerland AG 2020
A. Vedaldi et al. (Eds.): ECCV 2020, LNCS 12363, pp. 581–598, 2020.
https://doi.org/10.1007/978-3-030-58523-5_34

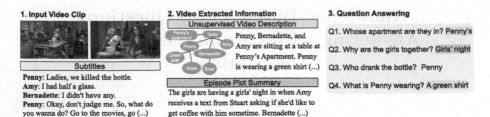

Fig. 1. ROLL performs video story question answering by generating unsupervised descriptions from video scene graphs and obtaining episode summaries.

Video story question answering leverages the structure of video stories, such as movies and TV shows, to formulate questions about specific scenes in a video. Models, then, need to find the correct answer by reasoning over the scene and its underlying plot. However, as the video story unfolds, the details of the plot are often revealed to the spectator over multiple scenes, sometimes far apart from each other. To understand the whole story, humans have the capacity to constantly relate past events with what is currently being shown, acquiring contextual information that forms their story knowledge. We argue that for a full comprehension of video stories, not only what is happening in the current scene has to be considered, but also the knowledge acquired in previous scenes. Some examples are shown in Fig. 1; whereas the answer to Q1, Q3, and Q4 can be guessed from the video scene (and its subtitles), Q2 can only be inferred when the full context is known.

Previous work on video story question answering can be roughly divided into two categories. On one hand, there are models that extract information from the whole video story [14,20,23,35], and use attention mechanisms to find the parts that are relevant to each question. These models obtain contextual representations, which are used to answer general questions about the plot, but barely capture details at the scene level. On the other hand, other models extract detailed information from specific scenes [13,17], without looking at the whole video story. However, relying only on the content of short scenes is insufficient to answer insightful aspects about the story, such as the characters' motivations. To study multiple types of questions about video stories using both contextual and scene-specific information, a knowledge-based video question answering dataset has been recently introduced [5]. The proposed model combines contextual information from external resources with multi-modal representations from specific scenes. However, the contextual data in [5] is obtained from thousands of task-specific human-generated annotations, which are expensive to obtain and difficult to generalise to other domains.

In this paper, we introduce ROLL, <u>R</u>ead, <u>O</u>bserve, and Reca<u>ll</u>, a model that addresses knowledge-based video story question answering with both contextual and scene-specific information using unsupervised scene descriptions and weakly-supervised external knowledge. ROLL consists on a three-branch architecture inspired by three areas of human cognition playing an important role in video understanding: dialog comprehension (*read branch*), scene reasoning (*observe branch*), and storyline recalling (*recall branch*). Whereas the scene-specific details are summarised in the read and observe branches, the recall branch provides a contextual overview about the story using free online resources. To predict the correct answer, the three branches are lately fused through a modality weighting mechanism, which balances the signal from the three different sources.

Contributions: Our contribution is three-fold: 1) we propose a new unsupervised video representation based on video descriptions generated from video scene graphs; 2) we combine specific details from video scenes with weakly supervised external knowledge for a deep understanding of video stories; and 3) we incorporate a modality weighting mechanism to fuse data from different modalities without information loss. Our model is evaluated on two challenging video story question answering datasets: KnowIT VQA and TVQA+, outperforming previous work by more than 6.3% and 1.3%, respectively.

2   Related Work

We develop a model for video story question answering that 1) takes advantage of rich external knowledge sources, and 2) represents video content by generating unsupervised video captions from scene graphs. In the following, we first review work on video story question answering and visual reasoning with external knowledge before discussing scene graphs and methods for video description.

Video Story Question Answering. Video story question answering is a modality in video question answering in which questions are not only related to the visual content of a video, but also to its plot. MovieQA [35] introduced a plot-oriented dataset with questions generated from movie summaries. Most proposed models [14,20,23,35] used frame-level features to represent the entire movie, applying attention mechanisms to find the relevant parts to each question. This provides a high-level overview of the story, but does not consider the details of each scene. Alternatively, PororoQA [15] and TVQA [17] formulated scene-level questions about specific events in the video. Models addressing these datasets described the details of each scene with features [40], captions [15] or visual concepts [13,17,52], but without attending to the ongoing plot in the video story. Recently, KnowIT VQA [5] introduced a combination of detailed questions about scenes and knowledge-based questions about the story. The proposed model relied on human-generated annotations to understand the insights of the plot. On the contrary, our model exploits both specific and general story information without task-specific annotations by using external knowledge bases.

Visual Reasoning with External Knowledge. Using external knowledge in visual reasoning extends the visual question answering task (VQA) to address questions far beyond the visual content of images. Although the acquisition of knowledge depends on the task of interest, structured knowledge bases, such as DBpedia [2] or ConceptNet [34], are commonly used in most methods

[24,25,41,42,45]. However, structured knowledge is usually represented as (subject, predicate, object) triplets, which is a hard constraint on the type of information being processed. Generic solutions [22,31] proposed to exploit unstructured resources in natural language, such as Wikipedia.[2] Following this direction, our model leverages unstructured online data to answer knowledge-based questions about video stories.

Scene Graphs. Scene graphs [12] are structures that represent the objects depicted in an image and their relationships, providing a semantic description of the image. Most scene graph methods consist on an object detector, an attribute classifier and a relationship predictor [19,48,50,54–56]. Scene graphs have been used in multiple vision and language tasks, including image captioning [7,51,54] and VQA [32,36,46]. However, less attention has been paid to generating scene graphs from videos, in which relationships are both spatial and temporal. So far, video scene graphs have been mostly applied to cross-modal retrieval to find video fragments [38,47]. In this work, we rely on video scene graphs to generate unsupervised video scene descriptions.

Video Descriptions. Video captioning aims to describe short video clips using natural language. Most approaches [3,21,26,39,53] use a sequential encoder-decoder framework, in which the input are visual features from multiple frames and the output is the generated sentence. For more detailed descriptions, dense video captioning [16,58] generates multiple sentences describing all the relevant events in the video. However, existing methods require to be trained on large-scale annotated datasets [27,29] with thousands of video-description pairs. We generate rich video scene descriptions in an unsupervised way using the semantic information from video scene graphs.

3   Model Overview

The goal of video story question answering is to understand movies or TV shows in a similar way as we humans do. We argue that there are at least three aspects of human intelligence involved in this task: 1) comprehension of what is being said, 2) comprehension of about what is being watched, and 3) recalling what happened in the story before. Our proposed model, ROLL, emulates each of those aspects in a three branch architecture, as shown in Fig. 2. Each branch in ROLL (read, observe, and recall) represents as text data the information from a different cognitive task, and encodes it through a Transformer with several self-attention layers. Then, the outputs from each Transformer are fused through a modality weighting mechanism to predict the correct answer.

Task Definition. We address video story question answering as a knowledge-based multiple-choice task. For each sample, the input is: 1) a question, 2) N_{ca} candidate answers, 3) a video scene, and 4) the subtitles associated with the

---

[2] https://www.wikipedia.org/.

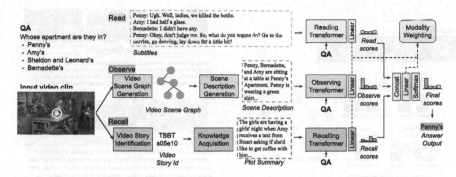

Fig. 2. ROLL overview. Each branch estimates a relevance score for each of the candidate answers based on different information. The read branch relies on subtitles, the observe branch generates unsupervised video descriptions, and the recall branch obtains external knowledge as plot summaries. To predict the correct answer, the three outputs are fused through a modality weighting mechanism.

scene. The output is the index of the predicted answer. As a knowledge-based task, models can access external resources to retrieve contextual information.

Introduction to Transformers. Transformers [37] are sequence-to-sequence modelling architectures that entirely rely on self-attention mechanisms. They have rapidly become the state-of-the-art in many natural language processing tasks. ROLL incorporates three independent Transformers to model the language data extracted from each branch, which is represented by the input string:

$$s_m{}^c = [\text{CLS}] + \text{context}_m + [\text{SEP}] + \text{choice}_m{}^c + [\text{SEP}], \qquad (1)$$

where m indicates the branch, context_m is an input sentence defined for each branch, $\text{choice}_m{}^c$ is a sentence for the c-th candidate answer with $c = 1, \cdots, N_{ca}$, [CLS] is the classification token used to obtain the output representation, [SEP] is the separator token for differentiating sentences, and $+$ is string concatenation. For each sample, N_{ca} input strings are generated, one per candidate answer.

The input string $s_m{}^c$ is tokenised into a sequence of n tokens $\mathbf{x}^c = [x_1, \cdots, x_n]$, and fed into a Transformer network. For each token x_i in $\mathbf{x}^c$, the Transformer creates an input embedding, $\mathbf{h}_i^0 \in \mathbb{R}^{D_h}$ with D_h hidden size, by adding the word, segment, and position embeddings. For each self-attention layer $l = 1, \cdots, N_L$ in the Transformer, denoted by $\text{TBlock}^l(\cdot)$, the contextualised word representation for position i in the sequence is computed as:

$$\mathbf{h}_i^l = \text{TBlock}^l(\mathbf{h}_i^{l-1}) \qquad (2)$$

The encoded representation of the input string $s_m{}^c$ is then obtained as the output of the position of the [CLS] token in the last layer:

$$\mathbf{y}_m{}^c = \mathbf{h}_0^L \in \mathbb{R}^{D_h} \qquad (3)$$

Our Transformers are the BERT$_\text{BASE}$ model [4] with $N_L = 12$ and $D_h = 768$.

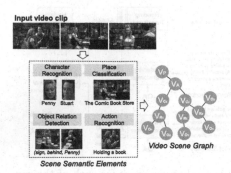

Fig. 3. Video scene graphs are generated from recognising the semantic elements of the scene (characters V_C, places V_P, objects V_O, relations V_R, and actions V_A) and connecting them.

Fig. 4. Examples of generated scene description. Although not natural, they accurately represents the semantics in the video scene.

4   Read Branch

In the read branch, ROLL extracts information from the dialogues of the video scene, which are obtained from the subtitles. The input string for this branch is:

$$s_r{}^c = [\text{CLS}] + subs + q + [\text{SEP}] + a^c + [\text{SEP}], \tag{4}$$

where $subs$ are the subtitles, q the question, and a^c with $c = 1, \cdots, N_{ca}$ each of the candidate answers. Each input string $s_r{}^c$ is fed into the Reading Transformer to obtain $\mathbf{y}_r{}^c$, which is forwarded into a single output linear layer with $\mathbf{w}_r$ weights and b_r bias, to compute a *read score* per candidate answer:

$$\alpha_r{}^c = \mathbf{w}_r{}^\top \cdot \mathbf{y}_r{}^c + b_r \tag{5}$$

5   Observe Branch

In the observe branch, ROLL summarises the semantics of the video scene into a video description. Generating descriptions from video is a challenging problem [49]. Standard video captioning models [3,21,39] require to be trained on large-scale datasets with annotated video and description pairs. As video story question answering datasets commonly do not provide such annotations, training a model for our task is impractical. Similarly, relying on pre-trained models may lead to poor results, as the generated descriptions will miss important information about the story (e.g., character names or frequent locations). Alternatively, we propose to generate unsupervised video descriptions by first creating a video scene graph. The descriptions are then fed into the Observing Transformer to predict a *observe score* for each candidate answer. Below, we first describe the video scene graph generation process, then we provide the details for the unsupervised video description, and finally we summarise the observing Transformer.

5.1   Video Scene Graph Generation

Figure 3 shows the video scene graph generation process, which is built on top of state-of-the-art image and video recognition techniques. We use four modules to detect the most relevant details in the scene: character recognition, place classification, object relation detection, and action recognition. The video scene graph is then generated by building connections between the detected elements.

Character Recognition. This module identifies the characters that appear in the scene using a face recognition classifier trained with images from the cast. We download about 10 images for each of the most common N_{C_T} characters based on IMDb.[3] We extract $f \in \mathbb{R}^{128}$ face representations with FaceNet [30] and train a k-nearest neighbour (kNN) classifier, where $k = N_{C_T}$. At test time, the trained kNN classifier returns a score for the predicted character. If the score is below a threshold, we assigned it to the *unknown* class. Finally, we apply a spatio-temporal filter to remove mispredictions and duplicate characters. Details are provided in the suppl. material. As output, we obtain a set of N_C characters appearing in the scene $C = \{C_i | i = 1, \cdots, N_C\}$, and their bounding boxes.

Place Classification. The place classification module detects where the scene is located. To learn the frequent locations in the video story, we fine-tune the pre-trained Places365 [57] network with ResNet50 [8] backbone in a weakly supervised way. To obtain place annotations, we use video transcripts from specialised websites.[4] We extract the locations that appear at least 10 times in the training set scripts, and include an *unknown* category for the rest. Training is performed at the frame level, i.e., each frame is considered as an independent image. For prediction, we accumulate the scores of the top 5 predicted classes for each frame in a video scene and output the most scored place, P.

Object Relation Detection. This module detects the objects in the scene and their relations. We use the large-scale visual relationship understanding (VRU) [55] pre-trained on the VG200 dataset [48], with 150 object and 50 relation categories. For each frame, VRU returns a list of subject-relation-object triplets, their bounding boxes, and a prediction score for each triplet. We replace the objects and subjects assigned to a person class[5] with its corresponding character name by finding the overlap between the bounding boxes. We only keep triplets assigned to known characters and we filter out duplicates. After discarding the bounding boxes and scores, we obtain a list of N_T triplets, $T = \{T_i | i = 1, \cdots, N_T\}$ with $T_i = (S_i, R_i, O_i)$ and S_i, R_i, and O_i as subject, relation, and object.

Action Recognition. The action recognition module detects the main action in the video scene. We use Long-Term Feature Banks (LFB) [44] pre-trained on the Charades dataset [33] with 157 action categories. LFB extracts information over the entire span of the video scene, improving performance with respect to

---

[3] https://www.imdb.com/.

[4] For example, https://bigbangtrans.wordpress.com/.

[5] Boy, girl, guy, lady, man, person, player, woman.

Table 1. Sentence generation from the video scene graph, $G = (V, E)$ with $V = \{V_C, V_P, V_O, V_R, V_A\}$. We define $e_{R_k, O} = \{e_{R_k, O_j}\}$ with $j \in [1, ..., |V_O|]$. $e_{R_k, C}$, e_{O, R_k}, and e_{C, R_k} are defined likewise.

Graph Condition	Generated Sentence	Example
$|V_C| = 0$ & $|V_P| = 0$	Someone is V_A.	Someone is lying on the floor.
$|V_C| = 1$ & $|V_P| = 0$	V_C is V_A.	Leonard is smiling.
$|V_C| > 1$ & $|V_P| = 0$	$V_{C_1}, ..., V_{C_{|V_C|-1}}$ and $V_{C_{|V_C|}}$ are V_A.	Penny and Amy are holding a bag.
$|V_C| = 0$ & $|V_P| = 1$	Someone is V_A at V_P.	Someone is walking at the street.
$|V_C| = 1$ & $|V_P| = 1$	V_C is V_A at V_P.	Sheldon is smiling at the bedroom.
$|V_C| > 1$ & $|V_P| = 1$	$V_{C_1}, ..., V_{C_{|V_C|-1}}$ and $V_{C_{|V_C|}}$ are V_A at V_P.	Amy and Raj are talking at the room.
$e_{C_i, R_k} \in E$ & $e_{R_k, O_j} \in E$ & $|e_{R_k, O}| = 1$	$V_{C_i} \ V_{R_k} \ V_{O_j}$.	Penny wearing shorts.
$e_{C_i, R_k} \in E$ & $e_{R_k, O_j} \in E$ & $|e_{R_k, O}| > 1$	$V_{C_i} \ V_{R_k} \ V_{O_1}, ..., V_{O_{|V_O|-1}}$ and $V_{O_{|V_O|}}$.	Raj holding bottle and book.
$e_{R_k, C_i} \in E$ & $e_{O_j, R_k} \in E$ & $|e_{O, R_k}| = 1$	$V_{O_j} \ V_{R_k} \ V_{C_i}$.	Board behind Sheldon.
$e_{R_k, C_i} \in E$ & $e_{O_j, R_k} \in E$ & $|e_{O, R_k}| > 1$	$V_{O_1}, ..., V_{O_{|V_O|-1}}$ and $V_{O_{|V_O|}} \ V_{R_k} \ V_{C_i}$.	Chair, table and door behind Penny.

using short 2–3 s clips. We input the entire scene into the network, and we obtain a predicted action as a result, A.

Graph Generation. The video scene graph, $G = (V, E)$, semantically describes the visual contents of the scene by using a collection of nodes V, and edges E. We consider the following types of nodes:

- *Character nodes*, $V_C \subseteq V$, representing the characters in the scene. If C do not contain any *unknown* character, $V_C = C$. Otherwise, we remove the *unknown* characters $\{\text{UNK}_C\}$, as $V_c = C - \{\text{UNK}_C\}$.
- *Place nodes*, $V_P \subseteq V$, representing the location where the video scene occurs. $V_P = \{P\}$ if $P \neq unknown$, otherwise $V_P = \emptyset$.
- *Object nodes*, $V_O \subseteq V$, representing the objects in the scene, which are obtained from the subjects and objects in the triplets that are not a character, as $V_O = Z - (Z \cap C)$ with $Z = S \cup O$.
- *Relation nodes*, $V_R \subseteq V$, representing the relation between subjects and objects in the triplets, $V_R = R$.
- *Action nodes*, $V_A \subseteq V$, representing the action in the scene as $V_A = \{A\}$, with $|V_A| = 1$.

We use 6 types of directed edges:

- $e_{P,A} = (V_P, V_A) \in E$ between the place node V_P and the action node V_A.
- $e_{A,C_j} = (V_A, V_{C_j}) \in E$ between the action node V_A and each character V_{C_j}.
- $e_{C_i, R_j} = (V_{C_i}, V_{R_j}) \in E$ between a character node V_{C_i} and a relation node V_{R_j} when $V_{C_i} = S_k$ and $V_{R_j} = R_k$ in the triplet $T_k = (S_k, R_k, O_k)$.
- $e_{R_i, C_j} = (V_{R_i}, V_{C_j}) \in E$ between a relation node V_{R_i} and a character node V_{C_j} when $V_{R_i} = R_k$ and $V_{C_j} = O_k$ in the triplet $T_k = (S_k, R_k, O_k)$.
- $e_{O_i, R_j} = (V_{O_i}, V_{R_j}) \in E$ between an object node V_{O_i} and a relation node V_{R_j} when $V_{O_i} = S_k$ and $V_{R_j} = R_k$ in the triplet $T_k = (S_k, R_k, O_k)$.
- $e_{R_i, O_j} = (V_{R_i}, V_{O_j}) \in E$ between a relation node V_{R_i} and an object node V_{C_j} when $V_{R_i} = R_k$ and $V_{C_j} = O_k$ in the triplet $T_k = (S_k, R_k, O_k)$.

with i, j, and k being the index for a certain object in a set.

5.2   Scene Description Generation

Scene descriptions are generated from the video scene graph according to the set of rules in Table 1 in an unsupervised manner. For each true condition in Table 1, a single sentence is generated. The final scene description is the concatenation of all the generated sentences, which serves as a representation of the semantic content in the video scene. Examples are shown in Fig. 4.

5.3   Observing Transformer

The generated description, d, is used in the input string for the observe branch:

$$s_o{}^c = [\text{CLS}] + d + q + [\text{SEP}] + a^c + [\text{SEP}], \tag{6}$$

Each $s_o{}^c$ is fed into the Observing Transformer to obtain $\mathbf{y}_o{}^c$, which is forwarded into a single output linear layer to compute the *observe score*:

$$\alpha_o{}^c = \mathbf{w}_o{}^\top \cdot \mathbf{y}_o{}^c + b_o \tag{7}$$

6   Recall Branch

In the recall branch, ROLL emulates the human experience of watching a TV show by recalling the events that occurred previously. This is inspired by the human evaluation on [5], which provides some insights on human behaviour. In [5], evaluators were asked to answer questions about a popular sitcom under different conditions. Interestingly, the reported performance dropped dramatically when humans were not exposed to the videos. We speculate that this is because humans indirectly used the scene to remember the whole episode and answer questions about the plot. The recall branch imitates this behaviour by first identifying the video and then acquiring knowledge about the story plot.

6.1   Knowledge Acquisition

Differently from previous work [5], in which the external knowledge to answer each question is specifically annotated by humans, we rely on publicly available resources[6] and build a knowledge base (KB) using plot summaries from the Internet.[7] Given a video scene, we first identify the video story it belongs to as in video retrieval [6]. Frames are represented by the output of the second-to-last layer of a pre-trained ResNet50 [8]. We compute the cosine similarity between each frame representation in the scene and all frames in the dataset, keeping the video of the most similar frame. As a result, we obtain an identifier of the most voted video, which is used to query the KB and we obtain a document p with

---

[6] For example, https://the-big-bang-theory.com/.

[7] Generating video plot summaries automatically from the whole video story is a challenging task by itself and out of the scope of this work. However, it is an interesting problem that we aim to study as a our future work.

the plot. In this way, ROLL acquires external knowledge about the video story in an weakly supervised way as 1) the questions and the external knowledge base have not been paired in any way during their generation, 2) the model does not know if there is corresponding text in the external knowledge base that can be useful for a given question, 3) the model is not directly trained with ground-truth episode labels, and 4) the model is not trained with ground-truth text location.

6.2   Recalling Transformer

The document p is fed into the Recalling Transformer to predict a *recall score* for each candidate answer. As many documents exceed the maximum number of words the Transformer can take as input,[8] we adopt a sliding window approach [9,10] to slice p into multiple overlapping segments. To produce the segments k_j with $j = 1, \cdots, N_{s\text{MAX}}$, we slide a window of length W_l with a stride r over the document p, obtaining $N_s = \lceil \frac{L_d - W_l}{r} \rceil + 1$ segments, where L_d is the number of words in the document. For training multiple samples in a minibatch, we set all the documents to have the same number of segments $N_{s\text{MAX}}$, discarding segments if $N_s > N_{s\text{MAX}}$, and zero-padding if $N_s < N_{s\text{MAX}}$. We encode the plot segments along with the question and candidate answers into multiple input strings:

$$s_{ll_j}^c = [\text{CLS}] + q + [\text{SEP}] + a^c + k_j + [\text{SEP}] \tag{8}$$

Each $s_{ll_j}^c$ is fed into the Recalling Transformer to obtain $\mathbf{y}_{ll_j}^c$, which is forwarded into a single output linear layer to compute a score for an answer-segment pair:

$$\alpha_{ll_j}^c = \mathbf{w}_{ll}^\top \cdot \mathbf{y}_{ll_j}^c + b_{ll} \tag{9}$$

Then, the final recall score for each of the candidate answers α_{ll}^c is:

$$\alpha_{ll}^c = \max(\alpha_{ll_j}^c) \quad \text{with } j = 1, \cdots, N_{s\text{MAX}} \tag{10}$$

7   Final Prediction

To output the final prediction score, the model concatenates the output of the three branches into a score vector $\boldsymbol{\alpha}^c = [\alpha_r^c, \alpha_o^c, \alpha_{ll}^c]$, which is input into a single layer classifier. The predicted answer $\hat{a}$ is then:

$$\omega^c = \mathbf{w}_c^\top \cdot \boldsymbol{\alpha}^c + b_c \tag{11}$$

$$\hat{a} = a^{\arg\max_c \omega} \quad \text{with } \boldsymbol{\omega} = [\omega^1, \cdots, \omega^{N_{ca}}]^\top \tag{12}$$

---

[8] In The Big Bang Theory, the longest summary contains 1,605 words.

Table 2. Evaluation on KnowIT VQA test set.

Method	Encoder	Data			Accuracy				
		Dialog	Vision	Know.	Vis.	Text.	Temp.	Know.	All
Rookies [5]	–	–	–	No	0.936	0.932	0.624	0.655	0.748
Masters [5]	–	–	–	Yes	0.961	0.936	0.857	0.867	0.896
TVQA [17]	LSTM	Subs	Concepts	–	0.612	0.645	0.547	0.466	0.522
ROCK Img [5]	BERT	Subs	ResNet	Human	0.654	0.681	0.628	0.647	0.652
ROCK Cpts [5]	BERT	Subs	Concepts	Human	0.654	0.685	0.628	0.646	0.652
ROCK Faces [5]	BERT	Subs	Characters	Human	0.654	0.688	0.628	0.646	0.652
ROCK Caps [5]	BERT	Subs	Captions	Human	0.647	0.678	0.593	0.643	0.646
ROLL-human	BERT	Subs	Descriptions	Human	0.708	**0.754**	0.570	0.567	0.620
ROLL	BERT	Subs	Descriptions	Summaries	**0.718**	0.739	**0.640**	**0.713**	**0.715**

Modality Weighting Mechanism. Wang et al. [43] have shown that multi-modality training often suffers from information loss, degradating performance with respect to single modality models. To avoid losing information when merging the three branches in ROLL, we use a modality weighting (MW) mechanism. First, we ensure that each Transformer learns independent representations by training them independently. The multi-class cross-entropy loss is computed as:

$$\mathcal{L}(\boldsymbol{\delta}, c^*) = -\log \frac{\exp(\delta^{c^*})}{\sum_c \exp(\delta^c)} \tag{13}$$

where c^* is the correct answer, and $\boldsymbol{\delta} = [\delta^1, \cdots, \delta^{N_{ca}}]$ the vector with the scores of the candidate answers. Next, the Transformers are frozen and the three branches are fine-tuned together. To ensure the multi-modal information is not lost, the model is trained as a multi-task problem with $\beta_r + \beta_o + \beta_{ll} + \beta_\omega = 1$:

$$\mathcal{L}_{\mathrm{MW}} = \beta_r \mathcal{L}(\boldsymbol{\alpha}_r, c^*) + \beta_o \mathcal{L}(\boldsymbol{\alpha}_o, c^*) + \beta_{ll} \mathcal{L}(\boldsymbol{\alpha}_{ll}, c^*) + \beta_\omega \mathcal{L}(\boldsymbol{\omega}, c^*)$$

8 Evaluation

Datasets. We evaluate ROLL on the KnowIT VQA [5] and the TVQA+ [18] datasets. KnowIT VQA is the only dataset for knowledge-based video story question answering, containing 24,282 questions about 207 episodes of The Big Bang Theory TV show. Questions in the test set are divided into four categories: visual-based, textual-based, temporal-based, and knowledge-based, and each question is provided with $N_{ca} = 4$ candidate answers. Accuracy is computed as the number of correct predicted answers over the total number of questions. Even though our model is specifically designed for leveraging external knowledge, we also evaluate its generalisation performance on non knowledge-based video story question answering. For this purpose, we use the TVQA+ dataset, in which questions are compositional and none of them requires external knowledge. TVQA+ contains 29,383 questions, each with $N_{ca} = 5$ candidate answers.

Table 3. Evaluation on the TVQA+ val set. No external knowledge is used.

Method	Vision	Lang.	Acc
TVQA [17]	Concepts	LSTM	62.28
TVQA [17]	Regional	LSTM	62.25
STAGE [18]	Regional	GloVe	67.29
STAGE [18]	Regional	BERT	68.31
ROLL	Description	BERT	**69.61**

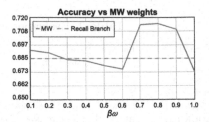

Fig. 5. ROLL accuracy according to β_ω on KnowIT VQA test set.

Implementation Details. We use the BERT uncased base model with pre-trained initialisation for our three Transformers. The maximum number of tokens is set to 512. For the single branch training, transformers are fine-tuned following the details in [4]. For the joint model training, we use stochastic gradient descent with momentum 0.9 and learning rate 0.001. In the observe branch, we extract the frames for the Character Recognition, Place Classification and Object Relation Detection modules at 1 fps, and for the Action Recognition module at 24 fps. In total, we use 17 characters, 32 places, 150 objects, 50 relations, and 157 action categories. In the recall branch, we use a window length $W_l = 200$, stride $r = 100$, and maximum number of segments $N_{s_{MAX}} = 5$. In the modality weighting mechanism, we set $\beta_r = 0.06$, $\beta_o = 0.06$, $\beta_{ll} = 0.08$, and $\beta_\omega = 0.80$ unless otherwise stated.

Evaluation on KnowIT VQA. We compare ROLL against the latest reported results on the KnowIT VQA dataset: TVQA and four different models in ROCK. TVQA [17] is based on a two-stream LSTM encoder for subtitles and visual concepts, whereas ROCK [5] uses task-specific human annotations to inform a BERT based model with external knowledge. ROCK reports results using four different visual representations: ResNet features, visual concepts, list of characters, and generated captions. For a more complete comparison, we also report results of ROLL using the human annotations from [5] as external knowledge (ROLL-human). Results are found in Table 2. Main findings are summarised as:

1. Overall, ROLL outperforms previous methods in all the question categories by a large margin, with 6.3% improvement on the overall accuracy with respect to the best performing ROCK.
2. When comparing the visual representations, our proposed video descriptions contain more semantic information than previous methods, improving visual-based questions by at least 6.4%. Specially, the boost in performance in visual-based questions with respect to standard captioning (ROCK Captions) or visual concepts (ROCK Concepts, TVQA) validates our unsupervised video descriptions as the best representation for this task.
3. Additional evidence of the superior performance of our proposed unsupervised descriptions is shown when ROLL uses human annotations as external knowledge. Although the overall performance is lower because ROLL-human

Table 4. ROLL ablation study.

Branch	Vis.	Text.	Temp.	Know.	All
Read	0.656	**0.772**	0.570	0.525	0.584
Observe	0.629	0.424	0.558	0.514	0.530
Recall	0.624	0.620	0.570	**0.725**	0.685
Read-Observe	0.695	0.732	0.570	0.527	0.590
Observe-Recall	0.712	0.601	0.628	0.704	0.691
Read-Recall	**0.722**	0.732	0.628	0.708	0.711
Full Model	0.718	0.739	**0.640**	0.713	**0.715**

Table 5. Fusion methods comparison.

Method	Vis.	Text.	Temp.	Know.	All
Average	0.726	0.710	0.628	0.648	0.672
Maximum	0.685	0.757	0.593	0.678	0.686
Self-att	**0.737**	**0.761**	**0.651**	0.641	0.677
QA-att	0.736	0.743	0.605	0.637	0.670
FC w/o MW	0.728	0.743	0.616	0.637	0.669
FC w/ MW	0.718	0.739	0.640	**0.713**	**0.715**

is not optimised to exploit this kind of information, on the visual-based questions our method improves best previous work by 5.4%. As the same source of knowledge is used, the superior performance can only be due to the contribution of our proposed visual representations.

4. On the knowledge-based samples, our method based on plot summaries outperforms task-specific human annotations by 6.7%, even when less annotations are required. This implies that the proposed slicing mechanism in the recall branch successfully extracts the relevant information from the long documents provided as external knowledge.

5. When compared against human performance, ROLL is 18% behind masters accuracy (humans that have watched the show) and it is closer to rookies non-knowledge accuracy (humans that have never watched the show). This shows how challenging this task is, with still plenty room for improvement.

Evaluation on TVQA+. To show ROLL generalisation performance even when external knowledge is not necessary, we additionally evaluate it on the TVQA+ dataset. For a fair comparison against previous work, 1) we remove the recall branch in ROLL and only use the read and observe branches, i.e., no external knowledge is used, and 2) we compare ROLL against models that use the answer labels as the only supervision for training, i.e., no extra annotations such as timestamps or spatial bounding boxes are used. Results are found in Table 3. Consistent with the results on the KnowIT VQA dataset, ROLL based on scene descriptions outperforms models based on other visual representations, such as visual concepts or Faster R-CNN [28] regional features, by at least 1.3%.

Ablation Study. We perform an ablation study to measure the contribution of each branch. Results when using one, two, or the three branches on the KnowIT VQA dataset are reported in Table 4. When a single branch is used, the observe branch gets the worst overall accuracy and the recall branch performs the best. This is consistent with the types of questions in the dataset, with 22% being visual-based and 63% being knowledge-based. The read branch gets the best performance in the text-based questions (i.e., about the subtitles), and the recall branch gets the best accuracy in the knowledge-based questions (i.e., about the storyline). When the observe branch is combined with other branches it consistently contributes to improve the results. Again, this result strongly suggests

Generated Description: *Sheldon, Leonard, Howard and Raj are sitting at a table at the Caltech cafeteria.* Leonard wearing shirt, jacket and glass. Leonard holding bottle. Hand, mouth, face and head of Leonard. Leonard has hair. Howard has hair and head. Door behind Howard. Howard at table. Shirt on Howard. Door, window and tree behind Sheldon. Pant, shirt, glass and short on Sheldon. Sheldon at table. Sheldon holding bottle. Sheldon has hand, hair, head and ear. Face of Sheldon. Chair behind Raj. Shirt and jacket on Raj. Raj wearing sock and glass. Raj has mouth, head, ear, face and hair.

Q: Where are Sheldon, Raj, Howard, and Leonard having a meal?	
A1: In their car.	A2: At their house.
A3: In their lab.	**A4: <u>The Caltech lunch room.</u>**

Generated Description: Penny and Sheldon are holding a cup/glass/ bottle of something at the main building. Door, sign, building and light behind Sheldon. Pant, shirt, glove and short on Sheldon. Sheldon wearing sock. Sheldon in room. Sheldon has hand, finger, hair and arm. Sheldon holding bottle, book, cup and handle. Head of Sheldon. Girl near Sheldon. Door behind Penny. Shirt on Penny. Penny wearing tie and short. Penny has eye, hair and head. Face of Penny. Penny holding flag. Penny near man.

Q: What is on Sheldon's sheets?	
A1: The periodic table.	A2: Batman.
A3: <u>Star Wars.</u>	A4: Superman.

Fig. 6. ROLL visual results. Underline/colour for correct/predicted answers. The relevant part for the question in the generated description is highlighted in blue. (Color figure online)

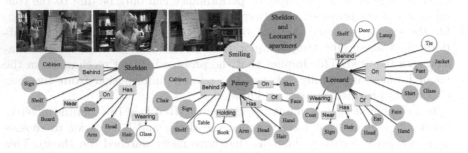

Fig. 7. Generated scene graph. Solid for correct and bordered for incorrect nodes.

that the generated scene descriptions do contain meaningful information for the task. The full model combining the three branches performs the best.

Fusion Methods Comparison. We also study the performance of our proposed MW mechanism and compare it against several fusion methods. Results are reported in Table 5. Given the three prediction scores from each of the branches, Average and Maximum compute the average and maximum score, respectively. The Self-att method implements a self-attention mechanism based on the Transformer outputs, and the QA-att mechanism attends each of the modality predictions based on the BERT representation of the question and candidate answers. The FC w/o MW predicts the answer scores by concatenating the scores of the three branches and feeding them into a linear layer, and FC w/ MW builds our proposed MW mechanism on top. The results show that most of the methods fail at properly fusing the information from the three branches, i.e., the overall performance is lower than the best single branch (recall, as reported in Table 4). This is probably because the fusion of the different modalities incurs in information loss. Our MW mechanism, in contrast, successfully balances the contribution from the three branches. Figure 5 compares different values of β_ω in the MW against the best performing single modality, with β_r, β_o, and β_{ll}

uniformly distributed. When the MW is not used ($\beta_\omega = 1$) the model obtains the worst performance. Likewise, when the loss contribution from the final prediction is too weak ($\beta_\omega < 0.6$), the model is not able to fuse the information correctly.

Qualitative Results. We visually inspect ROLL results to understand the strengths and weaknesses of our model. An example of scene graph can be seen in Fig. 7, whereas results on visual-based questions are provided in Fig. 6. ROLL performs well on questions related to the general content of the scene, such as places, people, or objects, but fails in detecting fine-grained details. Performance of the individual video modules is reported in the supplementary material.

9   Conclusion

We introduced ROLL, a model for knowledge-based video story question answering. To extract the visual information from videos, ROLL generates video descriptions in an unsupervised way by relying on video scene graphs. This new video representation led the model to an important increase of accuracy on visual-based questions on two datasets. Moreover, unlike previous work, ROLL leverages information from external knowledge without specific annotations on the task, easing the requirements of human labelling. This came without a drop in performance. On the contrary, as ROLL successfully fuses specific details from the scene with general information about the plot, the accuracy in KnowIT VQA and TVQA+ datasets was improved by more than 6.3% and 1.3%, respectively. Finally, by incorporating a modality weighting mechanism, ROLL avoided the information loss that comes from fusing different sources.

Acknowledgement. This work was supported by a project commissioned by the New Energy and Industrial Technology Development Organization (NEDO), and JSPS KAKENHI Nos. 18H03264 and 20K19822. We also would like to thank the anonymous reviewers for they insightful comments to improve the paper.

References

1. Antol, S., et al.: VQA: visual question answering. In: Proceedings of the ICCV, pp. 2425–2433 (2015)
2. Auer, S., Bizer, C., Kobilarov, G., Lehmann, J., Cyganiak, R., Ives, Z.: DBpedia: a nucleus for a web of open data. In: Aberer, K., et al. (eds.) ASWC/ISWC -2007. LNCS, vol. 4825, pp. 722–735. Springer, Heidelberg (2007). https://doi.org/10.1007/978-3-540-76298-0_52
3. Chen, Y., Wang, S., Zhang, W., Huang, Q.: Less is more: picking informative frames for video captioning. In: Proceedings of the ECCV, pp. 358–373 (2018)
4. Devlin, J., Chang, M.W., Lee, K., Toutanova, K.: BERT: pre-training of deep bidirectional transformers for language understanding. In: Proceedings of the NAACL, pp. 4171–4186 (2019)
5. Garcia, N., Otani, M., Chu, C., Nakashima, Y.: KnowIT VQA: answering knowledge-based questions about videos. In: Proceedings of the AAAI (2020)

6. Garcia, N., Vogiatzis, G.: Asymmetric spatio-temporal embeddings for large-scale image-to-video retrieval. In: BMVC (2018)
7. Gu, J., Joty, S., Cai, J., Zhao, H., Yang, X., Wang, G.: Unpaired image captioning via scene graph alignments. In: Proceedings of the ICCV (2019)
8. He, K., Zhang, X., Ren, S., Sun, J.: Deep residual learning for image recognition. In: Proceedings of the CVPR, pp. 770–778 (2016)
9. Hewlett, D., Jones, L., Lacoste, A., Gur, I.: Accurate supervised and semi-supervised machine reading for long documents. In: Proceedings of the EMNLP, pp. 2011–2020 (2017)
10. Hu, M., Peng, Y., Huang, Z., Li, D.: Retrieve, read, rerank: towards end-to-end multi-document reading comprehension. In: Proceedings of the ACL, pp. 2285–2295 (2019)
11. Johnson, J., Hariharan, B., van der Maaten, L., Fei-Fei, L., Zitnick, L.C., Girshick, R.: CLEVR: a diagnostic dataset for compositional language and elementary visual reasoning. In: Proceedings of the CVPR, pp. 2901–2910 (2017)
12. Johnson, J., et al.: Image retrieval using scene graphs. In: Proceedings of the CVPR, pp. 3668–3678 (2015)
13. Kim, J., Ma, M., Kim, K., Kim, S., Yoo, C.D.: Progressive attention memory network for movie story question answering. In: Proceedings of the CVPR, pp. 8337–8346 (2019)
14. Kim, K.M., Choi, S.H., Kim, J.H., Zhang, B.T.: Multimodal dual attention memory for video story question answering. In: Proceedings of the ECCV, pp. 673–688 (2018)
15. Kim, K.M., Heo, M.O., Choi, S.H., Zhang, B.T.: DeepStory: video story QA by deep embedded memory networks. In: Proceedings of the IJCAI, pp. 2016–2022 (2017)
16. Krishna, R., Hata, K., Ren, F., Fei-Fei, L., Carlos Niebles, J.: Dense-captioning events in videos. In: Proceedings of the ICCV, pp. 706–715 (2017)
17. Lei, J., Yu, L., Bansal, M., Berg, T.L.: TVQA: localized, compositional video question answering. In: Proceedings of the EMNLP, pp. 1369–1379 (2018)
18. Lei, J., Yu, L., Berg, T.L., Bansal, M.: TVQA+: spatio-temporal grounding for video question answering. arXiv preprint arXiv:1904.11574 (2019)
19. Li, Y., Ouyang, W., Zhou, B., Wang, K., Wang, X.: Scene graph generation from objects, phrases and region captions. In: Proceedings of the ICCV, pp. 1261–1270 (2017)
20. Liang, J., Jiang, L., Cao, L., Li, L.J., Hauptmann, A.G.: Focal visual-text attention for visual question answering. In: Proceedings of the CVPR, pp. 6135–6143 (2018)
21. Liu, S., Ren, Z., Yuan, J.: SibNet: sibling convolutional encoder for video captioning. In: Proceedings of the ACM Multimedia, pp. 1425–1434 (2018)
22. Marino, K., Rastegari, M., Farhadi, A., Mottaghi, R.: OK-VQA: a visual question answering benchmark requiring external knowledge. In: Proceedings of the CVPR, pp. 3195–3204 (2019)
23. Na, S., Lee, S., Kim, J., Kim, G.: A read-write memory network for movie story understanding. In: Proceedings of the ICCV, pp. 677–685 (2017)
24. Narasimhan, M., Lazebnik, S., Schwing, A.: Out of the box: reasoning with graph convolution nets for factual visual question answering. In: Proceedings of the NIPS, pp. 2659–2670 (2018)
25. Narasimhan, M., Schwing, A.G.: Straight to the facts: learning knowledge base retrieval for factual visual question answering. In: Proceedings of the ECCV, pp. 451–468 (2018)

26. Pan, Y., Mei, T., Yao, T., Li, H., Rui, Y.: Jointly modeling embedding and translation to bridge video and language. In: Proceedings of the CVPR, pp. 4594–4602 (2016)
27. Pini, S., Cornia, M., Bolelli, F., Baraldi, L., Cucchiara, R.: M-VAD names: a dataset for video captioning with naming. Multimed. Tools Appl. **78**(10), 14007–14027 (2019)
28. Ren, S., He, K., Girshick, R., Sun, J.: Faster R-CNN: towards real-time object detection with region proposal networks. In: Proceedings of the NIPS, pp. 91–99 (2015)
29. Rohrbach, A., Rohrbach, M., Tandon, N., Schiele, B.: A dataset for movie description. In: Proceedings of the CVPR, pp. 3202–3212 (2015)
30. Schroff, F., Kalenichenko, D., Philbin, J.: FaceNet: a unified embedding for face recognition and clustering. In: Proceedings of the CVPR, pp. 815–823 (2015)
31. Shah, S., Mishra, A., Yadati, N., Talukdar, P.P.: KVQA: knowledge-aware visual question answering. In: Proceedings of the AAAI (2019)
32. Shi, J., Zhang, H., Li, J.: Explainable and explicit visual reasoning over scene graphs. In: Proceedings of the CVPR, pp. 8376–8384 (2019)
33. Sigurdsson, G.A., Varol, G., Wang, X., Farhadi, A., Laptev, I., Gupta, A.: Hollywood in homes: crowdsourcing data collection for activity understanding. In: Proceedings of the ECCV, pp. 510–526 (2016)
34. Speer, R., Chin, J., Havasi, C.: ConceptnNet 5.5: an open multilingual graph of general knowledge. In: Proceedings of the AAAI (2017)
35. Tapaswi, M., Zhu, Y., Stiefelhagen, R., Torralba, A., Urtasun, R., Fidler, S.: MovieQA: understanding stories in movies through question-answering. In: Proceedings of the CVPR, pp. 4631–4640 (2016)
36. Teney, D., Liu, L., van den Hengel, A.: Graph-structured representations for visual question answering. In: Proceedings of the CVPR, pp. 1–9 (2017)
37. Vaswani, A., et al.: Attention is all you need. In: Proceedings of the NIPS, pp. 5998–6008 (2017)
38. Vicol, P., Tapaswi, M., Castrejon, L., Fidler, S.: MovieGraphs: towards understanding human-centric situations from videos. In: Proceedings of the CVPR, pp. 8581–8590 (2018)
39. Wang, B., Ma, L., Zhang, W., Jiang, W., Wang, J., Liu, W.: Controllable video captioning with POS sequence guidance based on gated fusion network. In: Proceedings of the ICCV, pp. 2641–2650 (2019)
40. Wang, B., Xu, Y., Han, Y., Hong, R.: Movie question answering: remembering the textual cues for layered visual contents. In: Proceedings of the AAAI (2018)
41. Wang, P., Wu, Q., Shen, C., Dick, A., van den Hengel, A.: FVQA: fact-based visual question answering. IEEE Trans. PAMI **40**(10), 2413–2427 (2018)
42. Wang, P., Wu, Q., Shen, C., Dick, A., Van Den Henge, A.: Explicit knowledge-based reasoning for visual question answering. In: Proceedings of the IJCAI, pp. 1290–1296 (2017)
43. Wang, W., Tran, D., Feiszli, M.: What makes training multi-modal networks hard? In: Proceedings of the CVPR, pp. 12695–12705 (2020)
44. Wu, C.Y., Feichtenhofer, C., Fan, H., He, K., Krahenbuhl, P., Girshick, R.: Long-term feature banks for detailed video understanding. In: Proceedings of the CVPR, pp. 284–293 (2019)
45. Wu, Q., Wang, P., Shen, C., Dick, A., van den Hengel, A.: Ask me anything: free-form visual question answering based on knowledge from external sources. In: Proceedings of the CVPR, pp. 4622–4630 (2016)

46. Xiong, P., Zhan, H., Wang, X., Sinha, B., Wu, Y.: Visual query answering by entity-attribute graph matching and reasoning. In: Proceedings of the CVPR, pp. 8357–8366 (2019)
47. Xiong, Y., Huang, Q., Guo, L., Zhou, H., Zhou, B., Lin, D.: A graph-based framework to bridge movies and synopses. In: Proceedings of the ICCV, pp. 4592–4601 (2019)
48. Xu, D., Zhu, Y., Choy, C.B., Fei-Fei, L.: Scene graph generation by iterative message passing. In: Proceedings of the CVPR, pp. 5410–5419 (2017)
49. Xu, J., Mei, T., Yao, T., Rui, Y.: MSR-VTT: a large video description dataset for bridging video and language. In: Proceedings of the CVPR, pp. 5288–5296 (2016)
50. Yang, J., Lu, J., Lee, S., Batra, D., Parikh, D.: Graph R-CNN for scene graph generation. In: Proceedings of the ECCV, pp. 670–685 (2018)
51. Yang, X., Tang, K., Zhang, H., Cai, J.: Auto-encoding scene graphs for image captioning. In: Proceedings of the CVPR, pp. 10685–10694 (2019)
52. Yang, Z., Garcia, N., Chu, C., Otani, M., Nakashima, Y., Takemura, H.: BERT representations for video question answering. In: Proceedings of the WACV (2020)
53. Yao, L., et al.: Describing videos by exploiting temporal structure. In: Proceedings of the ICCV, pp. 4507–4515 (2015)
54. Zellers, R., Yatskar, M., Thomson, S., Choi, Y.: Neural motifs: scene graph parsing with global context. In: Proceedings of the CVPR, pp. 5831–5840 (2018)
55. Zhang, J., Kalantidis, Y., Rohrbach, M., Paluri, M., Elgammal, A., Elhoseiny, M.: Large-scale visual relationship understanding. In: Proceedings of the AAAI, vol. 33, pp. 9185–9194 (2019)
56. Zhang, J., Shih, K.J., Elgammal, A., Tao, A., Catanzaro, B.: Graphical contrastive losses for scene graph parsing. In: Proceedings of the CVPR, pp. 11535–11543 (2019)
57. Zhou, B., Lapedriza, A., Khosla, A., Oliva, A., Torralba, A.: Places: A 10 million image database for scene recognition. IEEE Trans. PAMI **40**, 1452–1464 (2017)
58. Zhou, L., Zhou, Y., Corso, J.J., Socher, R., Xiong, C.: End-to-end dense video captioning with masked transformer. In: Proceedings of the CVPR, pp. 8739–8748 (2018)

Transformation Consistency Regularization – A Semi-supervised Paradigm for Image-to-Image Translation

Aamir Mustafa[✉] and Rafał K. Mantiuk

Department of Computer Science and Technology, University of Cambridge,
Cambridge, UK
{am2806,rafal.mantiuk}@cl.cam.ac.uk

Abstract. Scarcity of labeled data has motivated the development of semi-supervised learning methods, which learn from large portions of unlabeled data alongside a few labeled samples. Consistency Regularization between model's predictions under different input perturbations, particularly has shown to provide state-of-the art results in a semi-supervised framework. However, most of these method have been limited to classification and segmentation applications. We propose *Transformation Consistency Regularization*, which delves into a more challenging setting of image-to-image translation, which remains unexplored by semi-supervised algorithms. The method introduces a diverse set of geometric transformations and enforces the model's predictions for unlabeled data to be invariant to those transformations. We evaluate the efficacy of our algorithm on three different applications: image colorization, denoising and super-resolution. Our method is significantly data efficient, requiring only around 10–20% of labeled samples to achieve similar image reconstructions to its fully-supervised counterpart. Furthermore, we show the effectiveness of our method in video processing applications, where knowledge from a few frames can be leveraged to enhance the quality of the rest of the movie.

1 Introduction

In recent past, deep neural networks have achieved immense success in a wide range of computer vision applications, including image and video recognition [19,24,25], object detection [17,37], semantic segmentation [8,29] and image-to-image (I2I) translation [12,22,48]. However, a fundamental weakness of the existing networks is that they owe much of this success to large collections of labeled datasets. In real-world scenarios creating these extensive datasets is

Codes are made public at https://github.com/aamir-mustafa/Transformation-CR.

Electronic supplementary material The online version of this chapter (https://doi.org/10.1007/978-3-030-58523-5_35) contains supplementary material, which is available to authorized users.

© Springer Nature Switzerland AG 2020
A. Vedaldi et al. (Eds.): ECCV 2020, LNCS 12363, pp. 599–615, 2020.
https://doi.org/10.1007/978-3-030-58523-5_35

expensive requiring time-consuming human labeling, e.g. expert annotators, as in case of medical predictions and artistic reconstructions. As we enter the age of deep learning, wide-spread deployment of such models is still constrained for many practical applications due to lack of time, expertise and financial resources required to create voluminous labeled datasets.

Conceptually situated between supervised and unsupervised learning, Semi-Supervised Learning (SSL) [7] aims at addressing this weakness by leveraging large amounts of unlabeled data available alongside smaller sets of labeled data to provide improved predictive performance. Lately extensive research has been done in SSL and has shown to work well in the domain of image [4,5,32,39,40] and text classification [44]. However, it would be highly desirable to create I2I translation networks that can take advantage of the abundance of unlabeled data while requiring only a very small portion of the data to be labeled. For example, to colorize a black and white movie, we may want an artist to colorize only 1–5% of the frames and rest is done by the network. For capturing video in low light, we may want to capture a few reference frames on a tripod with long exposure times (therefore low noise) and use those to remove noise from the rest of the video. We may also want to design a camera, which captures only every n-th frame at a higher resolution (as the sensor bandwidth is constrained) and use those frames to enhance the resolution of the rest of the video. Unsupervised I2I translation methods have shown to generate compelling results, however, although unpaired, they still require large datasets from both the input and output domains to train the network [13,20,28,31,38,45,51]. For example to train an unsupervised super-resolution model we still require huge amounts of high resolution images as in [6,46]. On the contrary an SSL method would require only low-resolution images and a few low-high resolution image pairs for training.

In this work, we draw insights from Consistency Regularization (CR) – that has shown state-of-the art performance in classification tasks – to leverage unlabeled data in a semi-supervised fashion in a more challenging setting i.e. image-to-image translation. CR enforces a model's prediction to remain unchanged for an unsupervised sample when the input sample is perturbed [3,26,39]. However, applying CR in I2I domain is not straightforward, because images with varied transformations should have different predictions, unlike in the case of image classification. We derive our motivation for our approach from *a) smoothness assumption*, which in the case of image classification states that if two sample points are close enough to each other in the input space, then their predicted labels must be same and *b) manifold assumption*, which states that natural images lie on a low-dimensional manifold [52].

This paper introduces a regularization term over the unsupervised data called *Transformation Consistency Regularization* (TCR), which makes sure the model's prediction for a geometric transform of an image sample is consistent with the geometric transform of the model's reconstruction of the said image. In other words, we propose a modification to the *smoothness assumption* [52] for image reconstruction models postulating that if two input images $x, \hat{x} \in \mathcal{X}$ that are geometric transformations of each other and close by in the input space, then

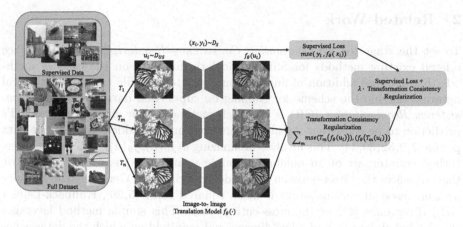

Fig. 1. An illustration of our training scheme using the Transformation Consistency Regularization ($\mathcal{L}_{us}$) over the unlabeled data for image colorization. The same method is used for Image Denoising and Single Image Super Resolution.

their corresponding predictions $y, \acute{y}$ must be equally spaced in the output manifold. Our training objective is inspired by the *transformation invariance loss* [14] introduced to design stable CNNs aiming at removing temporal inconsistencies and artifacts in video sequences. Our method is, however, fundamentally different, since the proposed loss formulation and the notion of using unlabeled data in designing near perfect mapping functions (CNNs) have not been previously explored for image-to-image translation settings. The proposed *TCR* over unlabeled data works jointly with the model's supervised training to effectively reconstruct better and visually appealing images than its supervised counterpart. To the best of our knowledge this is the first work to study the applicability of semi-supervised learning in an image-to-image setting.

The main contributions of this paper are as follows:

1. Through extensive empirical evaluations, we show the efficacy of our semi-supervised training scheme in regularizing model's predictions to be invariant to input transformations for three image-to-image translation problems, viz. image colorization, denoising and single image super-resolution.
2. We hypothesize that addition of unsupervised data during training makes a model generic enough to better remap images from one image manifold to other. We provide validation for our manifold assumption (see Sect. 5).
3. We provide analysis of how much unsupervised data is ideal for training a semi-supervised model in contrast to its supervised counterpart per batch.
4. Using less than 1% of labeled data, we perform colorization, denoising and super-resolution of movie clips. Our semi-supervised scheme achieves an absolute average gain of 6 dB in PSNR than its supervised counterpart, which uses the same percentage of labeled data.

2   Related Work

To set the stage for *Transformation Consistency Regularization* we introduce related existing methods for SSL, particularly focusing on the class of methods that propose addition of an additional loss term while training the neural network, leaving the scheme for training on supervised data unchanged. *Consistency Regularization* (CR) is one such method which enforces the model's prediction to be consistent if realistic perturbations are added to the input data points [2,9,32,39,44]. This involves minimizing $d(f_\theta(u), f_\theta(\acute{u}))$, where $\acute{u}$ is a perturbed counterpart of an unlabeled sample u and $d(\cdot, \cdot)$ is a distance metric that measures the offset between the model's predictions. Generally the distance measure used in various works is mean squared error [5,26], Kullback-Leibler (KL) divergence [44] or the cross-entropy loss. This simple method leverages the unlabeled data to find a low-dimensional manifold on which the dataset lies and has shown to provide state-of-the art performance in image classification tasks. Different CR techniques choose different forms of perturbations added to unlabeled data, most common form include domain-specific data augmentation [5,26,39,44]. More recently, Unsupervised Data Augmentation (UDA) [44], ReMixMatch [4] and FixMatch [40] enforce consistency among strongly augmented image samples making use of artificial labels generated for weakly augmented samples. Virtual Adversarial Training [32] adds small amounts of input noise a.k.a adversarial perturbations, which are carefully crafted to significantly alter the models predictions. [26,35,41] used dropout noise as a noise injection module to enforce consistency between the model's output predictions. Grandvalet *et al.* introduced *Entropy Minimization* [18] making use of unlabeled data to ensure that classes are well separated by adding an additional loss term making sure the model outputs confident (low-entropy) predictions for unsupervised samples. Interpolation Consistency Training [42] builds on this idea and enforces the predictions at an interpolation of unlabeled images to be consistent with the interpolation of the model's predictions for those images. This achieves the decision boundary to lie on low density regions of the class distribution. MixMatch [5] combined the ideas of entropy minimization and consistency regularization for better generalization. *Self training* is another such technique that uses the labeled data to initially train a model, which is then used to generate pseudo-labels for unlabeled data. If the model's prediction for an unlabeled sample (u) is above a certain threshold, then the sample and its pseudo-label are added to the training set as supervised samples [7,33,43,47]. However, the method relies heavily on the threshold value; a larger value results in a small set of pseudo-labeled samples, preventing model's performance to reach its full potential, whereas a smaller threshold may harm the performance with significant amount of erroneous labels. Most of these noise injection methods, however, are designed for image classification problems, with very little work being done in more challenging settings like image-to-image translation.

We in our work broaden the scope of CR to image-to-image translation, which remains untouched by current SSL algorithms. Our aforementioned regularization term and geometric transformations bear closest resemblance to

Eilertsen *et al.'s* [14] work on colorization of video sequences in designing temporally stable CNNs. They enforce temporal stability by regularizing the various types of motion artifacts that occur between frames in a fully supervised fashion. Our method is, however, fundamentally different, since the proposed loss formulation and the notion of using unlabeled data in designing near perfect mapping functions (CNNs) have not been previously explored for I2I translation settings.

3 Our Approach

3.1 Fully Supervised Perspective

In a traditional supervised image-to-image learning protocol, we are provided with a finite collection of B image pairs $D_s = \{(x_i, y_i) : i \in (1, \ldots, B)\}$ per batch, where each data point $x_i \in \mathcal{X}$ is sampled from an input distribution $\mathcal{X}$ and $y_i \in \mathcal{Y}$ belongs to a separate target space $\mathcal{Y}$. The goal is to train a regression model, say a Convolutional Neural Network (CNN) $f_\theta(\cdot)$ parameterized by θ, which promotes an accurate mapping between the input images x and the ground truth images y by minimizing the loss:

$$\sum_i \mathcal{L}\Big(f_\theta(x_i), \, y_i \Big) \tag{1}$$

The loss used to train a CNN could be mean squared error (L_2), L_1 loss or perceptual loss based on the main objective of the network.

3.2 Transformation Consistency Regularization

In semi-supervised learning, we are provided with an additional set of data points, sampled from the same input distribution $\mathcal{X}$. Let $D_{us} = \{(u_i) : i \in (1, \ldots, rB)\}$ be a batch of rB unlabeled data, where r is the ratio of unlabeled to labeled data per batch used in training and is a hyper-parameter.

Our goal is to leverage the unsupervised data to learn more about the inherent structure of this distribution and thereby regularize the image-mapping network $f_\theta(\cdot)$. We propose a modification to the *smoothness assumption*, postulating that if two input images $x, \acute{x} \in \mathcal{X}$ are close enough in the input space, then their corresponding predictions $y, \acute{y}$ must be equally spaced. Our approach is also motivated by *manifold assumption* [52], which states that natural images lie on a low-dimensional manifold. These assumptions form the basis of semi-supervised learning algorithms in image classification domain [15]. If natural images lie on a low-dimensional manifold, this explains why low-dimensional intermediate feature representations of deep networks accurately capture the structure of real image datasets. Image-to-image translation models, such as an image denoising network, approximate the input image manifold and help in remapping them to the output natural image manifold. Figure 2 shows a low dimensional manifold of noisy and recovered images. Noisy image samples of real world datasets sampled from a distribution $\mathcal{X}$, can be considered to lie on

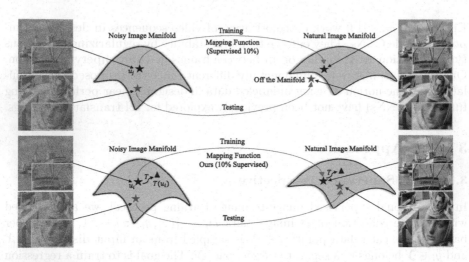

Fig. 2. The figure illustrates the mapping of image samples from noisy to original image manifold while training and testing. Model trained using only labeled data provides imperfect mapping resulting in reconstructed images lying off-the natural image manifold. Making use of large chunks of unlabeled data for training the same underlying model can provide a better mapping function.

a separate manifold. A trained image denoising model learns to map these data samples to an output natural image manifold. However, insufficient amount of data results in some output images that lie off the manifold. In this work we propose that using additional unlabeled images can go a long way in learning a near perfect mapping function resulting in remapping *off-the-manifold* output images. We provide a detailed validation for our proposition in Sect. 5.

The foundation of our semi-supervised learning algorithm are the following propositions:

Proposition 1: *Since both labeled and unlabeled data are sampled from the same underlying data distribution $p(x)$ over the input space, unlabeled data can be used to extract information about $p(x)$ and thereby about the posterior distribution $p(y|x)$.*

Proposition 2: *The unsupervised data D_{us} provides additional insights about the shape of the data manifold, which results in better reconstruction of images in an image-to-image setting.*

Proposition 3: *In an image-to-image translation setting, for a diverse set of geometric transformations $T(\cdot)$, if a model predicts a reconstructed image $\hat{u}_i$ for an ith unlabeled input sample $u_i \in D_{us}$, then the model's prediction for its transformation $T(u_i)$ must be $T(\hat{u}_i)$.*

In other words, we propose that a model's output predictions in an image-to-image setting should be transformed by the same amount as the input image is transformed before being fed to the CNN. The loss function for our semi-supervised learning protocol includes two terms *a)* a *supervised loss* term $\mathcal{L}_s$ applied on the labeled data D_s and *b)* a *transformation consistency regularization* (TCR) term $\mathcal{L}_{us}$ applied on the combination of D_s and D_{us}.[1] For illustration see Fig. 1. Specifically, $\mathcal{L}_s$ is the mean-squared loss on the labeled examples:

$$\mathcal{L}_s(x,y) = \frac{1}{B} \sum_{i=1}^{B} \| f_\theta(x_i) - y_i \|_2^2 \tag{2}$$

During the training process, each mini-batch includes both labeled and unlabeled images. The labeled data is trained using a typical I2I translation approach. Additionally, we introduce TCR on both labeled as well as the unlabeled data. We make use of a series of geometric transformations $T(\cdot)$ to enforce consistency between the reconstructed images for a given unlabeled data sample u_i. Specifically our loss function is given by:

$$\mathcal{L}_{us}(u) = \frac{1}{rB} \sum_{i=1}^{rB} \left(\frac{1}{M} \sum_{m=1}^{M} \| T_m(f_\theta(u_i)) - f_\theta(T_m(u_i)) \|_2^2 \right) \tag{3}$$

Here $f_\theta(\cdot)$ is a parametric family of I2I mappings and is a non-linear function, therefore $T_m(f_\theta(u_i))$ and $f_\theta(T_m(u_i))$ will indeed have different values. Here r is the ratio of amount of unlabeled data per mini-batch to that of labeled data. The loss function in Eq. 3 leverages unsupervised data and thereby helps regularize the model's predictions over varied forms of geometric transformations. The loss in Eq. 3 can appear similar to data augmentation with a random transformation but it should be noted that it is fundamentally different. The loss in data augmentation is measured between the outputs and their corresponding ground truths and is typically expressed as $\| T_m(f_\theta(x_i)) - (T_m(y_i)) \|$. TCR loss on the other hand enforces consistency between the predictions of transformed inputs without any knowledge of the ground truth.

The overall loss is composed of the original objective of the CNN and our transformation consistency regularization term as:

$$\mathcal{L} = \mathcal{L}_s(x,y) + \lambda \left(\mathcal{L}_{us}(u) + \mathcal{L}_{us}(x) \right) \tag{4}$$

The scalar λ is to control the weight given to the regularization term. Our proposed semi-supervised approach is, however, not restricted to L_2 loss and works equally well for L_1 and perceptual loss [16,23] (see supplementary material).

3.3   Transformations

In a traditional consistency regularization (CR) framework for image classification, a noise operation is employed that adds perturbations to the input image

---

[1] We include all labeled data, without using their labels, alongside the unlabeled data in our transformation consistency regularization term.

u to generate its counterpart $\acute{u}$. An additional regularization term then aims at minimizing the L_2 loss, cross entropy loss or KL divergence between the output predictions $f_\theta(u)$ and $f_\theta(\acute{u})$ [26,32,41,44]. These noise injection methods include random augmentations, adversarial perturbations, Gaussian noise, dropout noise etc [10,11]. CR has shown to provide state-of-the art performance in classification domain by enforcing that u and $\acute{u}$ lie on the same image manifold [34].

However, choosing the noise operations in an image-to-image translation model is a challenging task as the output predictions for an unsupervised image sample and, say, its augmented version are no longer same. In this paper we generate unsupervised image pairs using a series of geometric transformations $T(\cdot)$ that capture the occurrences in real world scenarios. These include rotation, translation, scaling and horizontal flipping. We choose a specified range for the degree of transformations so as to prevent images going off their natural manifold. The degree of rotation is uniformly drawn from a range of values between $-45°$ and $45°$. For translation and zooming the range is set between -30 px and 30 px and between $0.9\times$ and $1.1\times$ respectively.

For a particular geometric transformation T_m, unlike image classification, the model's predictions for an input image pair u_i and $T_m(u_i)$ require modifications to be equivalent. The additional loss term for TCR is computed between the model's prediction for $T_m(u_i)$, denoted by $f_\theta(T_m(u_i))$ and the transformation of model's output for u_i, i.e. $T_m(f_\theta(u_i))$. These transformations over unsupervised data regularize the model and force the reconstructed images to lie on the output image manifold (see Fig. 2).

4   Experiments

In this section, we evaluate the efficacy of TCR on a variety of image-to-image translation tasks where CNNs are employed. Specifically, we perform experiments with varying amounts of labeled data for image colorization, image denoising and single image super-resolution. For all the cases, we provide results for baseline models (using only supervised data), models trained with addition of our TCR over labeled data and finally the models trained using our semi-supervised paradigm. The codes used for our experiments are based on PyTorch. In our experiments, both the supervised and unsupervised data are supplied together in each mini-batch. To maintain their ratio r (see Eq. 2 and 3), we use separate data loaders for labeled and unlabeled data.

In the following sections we provide details about the dataset, model and training schemes employed in different image-to-image translation applications. Finally in Sect. 5, we provide validation for our manifold assumption that is built on the hypothesis that unsupervised data, sampled from the same data distribution as supervised data, can be leveraged to learn a generic image-to-image mapping function.

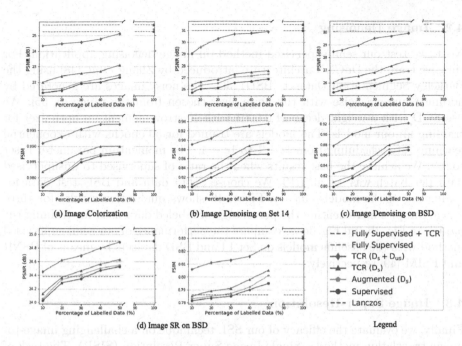

(a) Image Colorization   (b) Image Denoising on Set 14   (c) Image Denoising on BSD

(d) Image SR on BSD

Legend
- - - Fully Supervised + TCR
- - - Fully Supervised
—◆— TCR ($D_s + D_{us}$)
—▲— TCR (D_s)
—●— Augmented (D_s)
—●— Supervised
- - - Lanczos

Fig. 3. The plots provide results for baseline models (using only supervised data), models trained with addition of our TCR but using only labeled data i.e. TCR (D_s), models trained with image augmentation over the supervised data i.e. Augmented (D_s) and finally the models trained using our semi-supervised paradigm i.e. TCR ($D_s + D_{us}$). The addition of unsupervised data while training provides substantial improvement in image reconstruction. (Color figure online)

4.1   Image Colorization

To begin, we compare our semi-supervised method on image colorization. For this, we train the architecture introduced by Iizuka et al. [21] with varying amounts of labeled data (image pairs) from the Places dataset [50]. We use the SGD with Nestrov momentum optimizer for all our experiments. The models are trained for 45 epochs with the initial learning rate set to 0.1 with scheduling by a factor of 10 after epoch 30. The momentum parameter was set to 0.9. We achieved the best results when the ratio of supervised to unsupervised data per batch was set at 1:10. Figure 4a shows qualitative results for three image samples incrementing the percentage of labeled data used for training the model. In Fig. 3a we compare the performance of our semi-supervised method with the baseline models by means of PSNR and Feature Similarity Index (FSIMc) [49]. We use FSIMc, rather than SSIM, as it is a more modern metric that was shown to correlate much better with subjective image quality assessment data [36] and can operate on both gray-scale and color images.

4.2   Image Denoising

Next, we test our semi-supervised method on image denoising application. For this, we train the DnCNN architecture introduced by Zhang *et al.* [48] using the Berkeley Segmentation Dataset (BSD) [30]. The noisy images are generated by adding Gaussian noise with the noise level chosen from the range [0, 55]. We use SGD with a weight decay of 0.0001 with Nestrov momentum optimizer for training all our models. The models are trained for 50 epochs with exponential learning rate scheduling from $1e-1$ to $1e-4$. The momentum parameter is set to 0.9. We achieve the best results when the ratio of supervised to unsupervised data per batch was set at 1:10. We choose Set14 data and BSD test data for evaluating of the models efficacy. Figure 4b shows qualitative results for three image samples incrementing the percentage of labeled data used for training the model. In Fig. 3b and Fig. 3c we compare the performance of our semi-supervised method with the baseline models on Set 14 and BSD datasets by means of PSNR and FSIM [49] respectively.

4.3   Image Super-Resolution

Finally, we evaluate the efficacy of our SSL technique on a challenging image-to-image translation method – Single Image Super-Resolution (SISR). The task of SISR is to estimate a High Resolution (HR) image given a Low Resolution (LR) image. The LR image is first generated by downsampling the original HR image by a factor of τ. For this application we use state-of-the art SISR architecture – Enhanced Deep Super-Resolution (EDSR) proposed by [27], which uses a hierarchy of such residual blocks. We train the models on BSD 500 dataset, choosing 400 images for training and 100 for testing. The upscaling factor τ is chosen to be 3. Each model is trained for 500 epochs with an initial learning rate of 0.01 with gradual scheduling. We use the PSNR and FSIM [49] as performance metrics to compare the performance of our SSL scheme with its supervised counterpart (see Fig. 3d). Additional results are included in the supplementary material.

4.4   Results and Discussions

Comparisons: The goal of our TCR is to reconstruct better images using the combination of D_s and D_{us}, than what have been obtained using only D_s. This is a very common scenario in I2I translation as we typically have abundance of input images, but much fewer corresponding (labeled) output images. As illustrated in Fig. 3, we show a substantial performance boost across all applications. We report an absolute PSNR gain of 3.1 dB and 3.0 dB for image colorization and denoising using our semi-supervised scheme with only 10% of labeled data. A similar trend can be observed for FSIM results, suggesting a substantial gain in the perceived quality of translation. We compare the performance of our SSL method for three applications with the supervised baseline, which uses only the labeled data and the same underlying models for training. We further evaluate the performance of our TCR loss term when applied on the labeled data only,

(a) Image Colorization for three image samples from Places dataset

(b) Image Denoising for three image samples from Set 14 dataset

Fig. 4. Results showing comparison between reconstructed images using our model and supervised baseline models. The column title indicates the percentage of data used for training. The last column shows our results where we use only 10% of the entire dataset as labeled and rest in an unsupervised fashion.

as done in [14]. This is shown as green lines with triangle markers in Fig. 3. We observe that TCR improves the reconstruction quality even when trained only on the labeled data. Finally, we compare all tested approaches to data augmentation, shown as blue lines with pentagon markers in Fig. 3. We used the same set of transformations for data augmentation as for TCR. The results show that augmentation improves the performance as compared to the baseline, however, its still inferior to performance achieved using TCR under supervised settings.

Movie Applications: In this section, we show the potential of our SSL method in applications like movie colorization and movie super-resolution. In movie colorization, we use Blender Foundation's open source short film 'Big Buck Bunny' [1]. We divide the movie into train and test set with each comprising of 510 and 43 seconds respectively. In our SSL setting, we make use of only 1% of the total training frames in supervised fashion, while rest were fed into the TCR term. We compare our method with its supervised counterpart (using 1% of total training frames), and achieve an absolute gain of 6.1 dB in PSNR. Following a similar setting, we further evaluate the efficacy of TCR for denoising and enhancing the resolution of video clips in our experiments.

We believe our semi-supervised learning technique can go a long way in colorization of old movie clips requiring an artist to colorize only a few frames. In a similar fashion, we can take advantage of the abundance of unlabeled data in removing noise and enhancing the quality and resolution of movie clips.

5   Manifold Assumption Validation

In this paper we propose that input image samples (e.g. a noisy image) and their reconstructed counterparts (e.g. denoised image) lie on different manifolds and large portions of unsupervised data can better help in regularizing the remapping function $f_\theta(\cdot)$. Addition of this data while training promotes the reconstructed image to be mapped to the natural image manifold as illustrated in Fig. 2.

To validate this assumption for image denoising, we fine-tune an ImageNet pre-trained Inception-v3 model as a binary classifier using 10,000 pairs of clean (class 0) and noisy (class 1) images, 299×299 pixels each, drawn from ILSVRC validation set. The noisy samples are generated using two kinds of distributions namely, Gaussian and Poisson. For Gaussian noise we randomize the standard deviation $\sigma \in [0, 50]$ for each image. For Poisson noise, which is a dominant source of noise in photographs, we randomize the noise magnitude $\lambda \in [0, 50]$. With the learning rate reduced by a factor of 10, we retrain the top-2 blocks while freezing the rest. The global average pooling layer of the model is followed by a batch normalization, drop-out and two dense layers (1024 and 1 nodes, respectively). This model now efficiently leverages the subtle difference between clean images and their noisy counterparts and separates the two with a very high accuracy (99.0%). To validate our manifold assumption we test the denoised images generated using *a)* supervised model using 10% labeled data and *b)* model trained using TCR with same percentage of labeled data in semi-supervised fashion. The classifier labels only 44.3% of the denoised images generated using the supervised model as clean, whereas 97.9% of denoised images generated using our method are labeled as clean. This confirms that vast majority of the images restored using our method are remapped to the clean image manifold.

Figure 5a, shows a plot of the features extracted from the penultimate layer of the binary classifier to visualize our manifold assumption validation. We employ t-SNE to reduce the dimensionality of features to 3 for visualization.

To perform a similar validation for image colorization, we use an ResNet-50 model[2] pre-trained on the Places 365 dataset [50] to test the efficacy of our colorization model compared to the supervised counterpart. We convert 100 image samples into gray-scale from a particular class of the validation set (houses

---

[2] https://github.com/CSAILVision/places365.

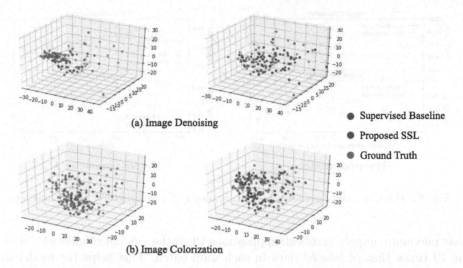

(a) Image Denoising

● Supervised Baseline

● Proposed SSL

● Ground Truth

(b) Image Colorization

Fig. 5. Manifold Assumption Validation for Denoising and Colorization. On the top we show 3D t-SNE plots of the intermediate features extracted from denoised images generated using our semi-supervised model (blue) and the baseline supervised model (red) compared with the ground truth noise-free images (green). The bottom plots show the features extracted from colored images. The plots clearly show that our method provides an accurate mapping of images back onto the the natural i.e. the ground truth image manifold than the baseline model. (Color figure online)

in our case) and generate colored images using both the aforementioned methods. In Fig. 5b, we plot the features extracted after the Global Pooling Layer of the classifier to visualize our manifold assumption validation. We employ t-SNE to reduce the dimensionality of features from 1024 to 3. The plots show that TCR helps to project images in each class to the ground truth manifold.

6 Ablation Studies

To achieve the best performance, we need to select the values for the hyper-parameters: ratio of unsupervised to supervised data per batch (r), and the weight of the regularization term (λ). We show that the optimal values of both hyper-parameters are consistent across the problems and datasets.

6.1 The Amount of Unsupervised Data We Need

First, we study the effect of the ratio of unsupervised to supervised data per batch (r) of training in a semi-supervised learning settings. In Fig. 6a, we plot the PSNR values of the reconstructed images versus different ratios for image colorization and denoising. We perform an ablation study using 10% and 20% of the total data as labeled. We observe a significant increase in the quality of images by using large amounts of unlabeled data. Our proposed SSL technique

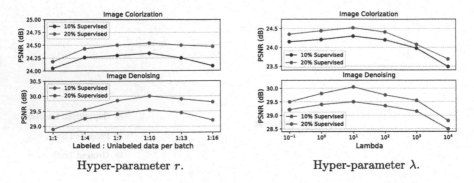

Hyper-parameter r.   Hyper-parameter λ.

Fig. 6. Ablation Study over two hyper-parameters for colorization and denoising.

has maximum impact across all applications when the amount of unlabeled data is 10 times that of labeled data in each mini-batch. This helps the model to learn over all possible distributions thereby resulting in a near perfect mapping function using only a small fraction of labeled data.

6.2   Lambda

We conduct a hyper-parameter search over the scalar λ used to control the weight given to our unsupervised loss term (see Eq. 4). We searched over values in $\{10^{-1}, 10^0, 10^1, 10^2, 10^3, 10^4\}$. Figure 6b plots the PSNR value versus λ for image colorization and denoising. Again in the above settings only 10% and 20% of the total data is labeled. In our experiments we found the maximum performance boost of our technique for $\lambda = 10$.

7   Conclusions

There has been an immense surge in semi-supervised learning techniques, however, none of the methods have addressed image-to-image translation. We introduce Transformation Consistency Regularization, a simple yet effective method that fills in this gap and provides great boosts in performance using only a small percentage of supervised data. The strategy proposes an additional regularization term over unlabeled data and does not require any architectural modifications to the network. We also show the efficacy of our semi-supervised method in colorization, denoising and super-resolution of movie clips by using only a few frames in supervised fashion, while rest been fed to our unsupervised regularization term while training. On the whole, we believe that our method can be used in diverse video applications where knowledge from a few frames can be effectively leveraged to enhance the quality of the rest of the movie.

Acknowledgements. This project has received funding from the European Research Council (ERC) under the European Union's Horizon 2020 research and innovation programme (grant agreement N° 725253–EyeCode).

References

1. Big buck bunny (code-named project peach), open source movie by blender foundation. https://peach.blender.org/
2. Athiwaratkun, B., Finzi, M., Izmailov, P., Wilson, A.G.: There are many consistent explanations of unlabeled data: why you should average. arXiv preprint arXiv:1806.05594 (2018)
3. Bachman, P., Alsharif, O., Precup, D.: Learning with pseudo-ensembles. In: Advances in Neural Information Processing Systems, pp. 3365–3373 (2014)
4. Berthelot, D., et al.: ReMixMatch: semi-supervised learning with distribution alignment and augmentation anchoring. arXiv preprint arXiv:1911.09785 (2019)
5. Berthelot, D., Carlini, N., Goodfellow, I., Papernot, N., Oliver, A., Raffel, C.A.: MixMatch: a holistic approach to semi-supervised learning. In: Advances in Neural Information Processing Systems, pp. 5050–5060 (2019)
6. Bulat, A., Yang, J., Tzimiropoulos, G.: To learn image super-resolution, use a GAN to learn how to do image degradation first. In: Proceedings of the European Conference on Computer Vision (ECCV), pp. 185–200 (2018)
7. Chapelle, O., Scholkopf, B., Zien, A.: Semi-supervised learning. IEEE Trans. Neural Netw. 20(3), 542 (2009). (Chapelle O., et al. (eds.) 2006)[book reviews]
8. Chen, L.C., Papandreou, G., Kokkinos, I., Murphy, K., Yuille, A.L.: DeepLab: semantic image segmentation with deep convolutional nets, atrous convolution, and fully connected CRFs. IEEE Trans. Pattern Anal. Mach. Intell. 40(4), 834–848 (2018)
9. Clark, K., Luong, M.T., Manning, C.D., Le, Q.V.: Semi-supervised sequence modeling with cross-view training. arXiv preprint arXiv:1809.08370 (2018)
10. Cubuk, E.D., Zoph, B., Mane, D., Vasudevan, V., Le, Q.V.: AutoAugment: learning augmentation strategies from data. In: Proceedings of the IEEE Conference on Computer Vision and Pattern Recognition, pp. 113–123 (2019)
11. Cubuk, E.D., Zoph, B., Shlens, J., Le, Q.V.: RandAugment: practical data augmentation with no separate search. arXiv preprint arXiv:1909.13719 (2019)
12. Dong, C., Loy, C.C., He, K., Tang, X.: Image super-resolution using deep convolutional networks. IEEE Trans. Pattern Anal. Mach. Intell. 38(2), 295–307 (2015)
13. Dong, H., Neekhara, P., Wu, C., Guo, Y.: Unsupervised image-to-image translation with generative adversarial networks. arXiv preprint arXiv:1701.02676 (2017)
14. Eilertsen, G., Mantiuk, R.K., Unger, J.: Single-frame regularization for temporally stable CNNs. In: Proceedings of the IEEE Conference on Computer Vision and Pattern Recognition, pp. 11176–11185 (2019)
15. van Engelen, J., Hoos, H.: A survey on semi supervised learning. Mach. Learn. 109, 373–440 (2020)
16. Gatys, L., Ecker, A.S., Bethge, M.: Texture synthesis using convolutional neural networks. In: Advances in Neural Information Processing Systems, pp. 262–270 (2015)
17. Girshick, R.: Fast R-CNN. In: Proceedings of the IEEE International Conference on Computer Vision, pp. 1440–1448 (2015)
18. Grandvalet, Y., Bengio, Y.: Semi-supervised learning by entropy minimization. In: Advances in Neural Information Processing Systems, pp. 529–536 (2005)
19. He, K., Zhang, X., Ren, S., Sun, J.: Deep residual learning for image recognition. In: Computer Vision and Pattern Recognition (2016)
20. Huang, X., Liu, M.Y., Belongie, S., Kautz, J.: Multimodal unsupervised image-to-image translation. In: Proceedings of the European Conference on Computer Vision (ECCV), pp. 172–189 (2018)

21. Iizuka, S., Simo-Serra, E., Ishikawa, H.: Let there be color!: joint end-to-end learning of global and local image priors for automatic image colorization with simultaneous classification. ACM Trans. Graph. (Proc. SIGGRAPH 2016) **35**(4), 110:1–110:11 (2016)
22. Isola, P., Zhu, J.Y., Zhou, T., Efros, A.A.: Image-to-image translation with conditional adversarial networks. In: Proceedings of the IEEE Conference on Computer Vision and Pattern Recognition, pp. 1125–1134 (2017)
23. Johnson, J., Alahi, A., Fei-Fei, L.: Perceptual losses for real-time style transfer and super-resolution. In: Leibe, B., Matas, J., Sebe, N., Welling, M. (eds.) ECCV 2016. LNCS, vol. 9906, pp. 694–711. Springer, Cham (2016). https://doi.org/10.1007/978-3-319-46475-6_43
24. Karpathy, A., Toderici, G., Shetty, S., Leung, T., Sukthankar, R., Fei-Fei, L.: Large-scale video classification with convolutional neural networks. In: Proceedings of the IEEE Conference on Computer Vision and Pattern Recognition, pp. 1725–1732 (2014)
25. Krizhevsky, A., Sutskever, I., Hinton, G.E.: ImageNet classification with deep convolutional neural networks. In: Advances in Neural Information Processing Systems, pp. 1097–1105 (2012)
26. Laine, S., Aila, T.: Temporal ensembling for semi-supervised learning. arXiv preprint arXiv:1610.02242 (2016)
27. Lim, B., Son, S., Kim, H., Nah, S., Mu Lee, K.: Enhanced deep residual networks for single image super-resolution. In: Proceedings of the IEEE Conference on Computer Vision and Pattern Recognition Workshops, pp. 136–144 (2017)
28. Liu, M.Y., Breuel, T., Kautz, J.: Unsupervised image-to-image translation networks. In: Advances in Neural Information Processing Systems, pp. 700–708 (2017)
29. Long, J., Shelhamer, E., Darrell, T.: Fully convolutional networks for semantic segmentation. In: Proceedings of the IEEE Conference on Computer Vision and Pattern Recognition, pp. 3431–3440 (2015)
30. Martin, D., Fowlkes, C., Tal, D., Malik, J.: A database of human segmented natural images and its application to evaluating segmentation algorithms and measuring ecological statistics. In: Proceedings of the 8th International Conference on Computer Vision, vol. 2, pp. 416–423, July 2001
31. Mejjati, Y.A., Richardt, C., Tompkin, J., Cosker, D., Kim, K.I.: Unsupervised attention-guided image-to-image translation. In: Advances in Neural Information Processing Systems, pp. 3693–3703 (2018)
32. Miyato, T., Maeda, S.I., Koyama, M., Ishii, S.: Virtual adversarial training: a regularization method for supervised and semi-supervised learning. IEEE Trans. Pattern Anal. Mach. Intell. **41**(8), 1979–1993 (2018)
33. Nguyen, N.V., Rigaud, C., Burie, J.C.: Semi-supervised object detection with unlabeled data. In: International Conference on Computer Vision Theory and Applications (2019)
34. Oliver, A., Odena, A., Raffel, C.A., Cubuk, E.D., Goodfellow, I.: Realistic evaluation of deep semi-supervised learning algorithms. In: Advances in Neural Information Processing Systems, pp. 3235–3246 (2018)
35. Park, S., Park, J., Shin, S.J., Moon, I.C.: Adversarial dropout for supervised and semi-supervised learning. In: Thirty-Second AAAI Conference on Artificial Intelligence (2018)
36. Ponomarenko, N., et al.: Image database TID2013: peculiarities, results and perspectives. Signal Process.: Image Commun. **30**, 57–77 (2015). https://doi.org/10.1016/j.image.2014.10.009. http://linkinghub.elsevier.com/retrieve/pii/S0923596514001490

37. Ren, S., He, K., Girshick, R., Sun, J.: Faster R-CNN: towards real-time object detection with region proposal networks. In: Advances in Neural Information Processing Systems, pp. 91–99 (2015)

38. Royer, A., et al.: XGAN: unsupervised image-to-image translation for many-to-many mappings. In: Singh, R., Vatsa, M., Patel, V.M., Ratha, N. (eds.) Domain Adaptation for Visual Understanding, pp. 33–49. Springer, Cham (2020). https://doi.org/10.1007/978-3-030-30671-7_3

39. Sajjadi, M., Javanmardi, M., Tasdizen, T.: Regularization with stochastic transformations and perturbations for deep semi-supervised learning. In: Advances in Neural Information Processing Systems, pp. 1163–1171 (2016)

40. Sohn, K., et al.: FixMatch: simplifying semi-supervised learning with consistency and confidence. arXiv preprint arXiv:2001.07685 (2020)

41. Tarvainen, A., Valpola, H.: Mean teachers are better role models: weight-averaged consistency targets improve semi-supervised deep learning results. In: Advances in Neural Information Processing Systems, pp. 1195–1204 (2017)

42. Verma, V., Lamb, A., Kannala, J., Bengio, Y., Lopez-Paz, D.: Interpolation consistency training for semi-supervised learning. arXiv preprint arXiv:1903.03825 (2019)

43. Wang, K., Yan, X., Zhang, D., Zhang, L., Lin, L.: Towards human-machine cooperation: self-supervised sample mining for object detection. In: Proceedings of the IEEE Conference on Computer Vision and Pattern Recognition, pp. 1605–1613 (2018)

44. Xie, Q., Dai, Z., Hovy, E., Luong, M.T., Le, Q.V.: Unsupervised data augmentation. arXiv preprint arXiv:1904.12848 (2019)

45. Yi, Z., Zhang, H., Tan, P., Gong, M.: DualGAN: unsupervised dual learning for image-to-image translation. In: Proceedings of the IEEE International Conference on Computer Vision, pp. 2849–2857 (2017)

46. Yuan, Y., Liu, S., Zhang, J., Zhang, Y., Dong, C., Lin, L.: Unsupervised image super-resolution using cycle-in-cycle generative adversarial networks. In: Proceedings of the IEEE Conference on Computer Vision and Pattern Recognition Workshops, pp. 701–710 (2018)

47. Zhai, X., Oliver, A., Kolesnikov, A., Beyer, L.: S4L: self-supervised semi-supervised learning. In: Proceedings of the IEEE International Conference on Computer Vision, pp. 1476–1485 (2019)

48. Zhang, K., Zuo, W., Chen, Y., Meng, D., Zhang, L.: Beyond a Gaussian denoiser: residual learning of deep CNN for image denoising. IEEE Trans. Image Process. **26**(7), 3142–3155 (2017)

49. Zhang, L., Zhang, L., Mou, X., Zhang, D.: FSIM: a feature similarity index for image quality assessment. IEEE Trans. Image Process. **20**(8), 2378–2386 (2011)

50. Zhou, B., Lapedriza, A., Khosla, A., Oliva, A., Torralba, A.: Places: a 10 million image database for scene recognition. IEEE Trans. Pattern Anal. Mach. Intell. **40**, 1452–1464 (2017)

51. Zhu, J.Y., Park, T., Isola, P., Efros, A.A.: Unpaired image-to-image translation using cycle-consistent adversarial networks. In: Proceedings of the IEEE International Conference on Computer Vision, pp. 2223–2232 (2017)

52. Zhu, X., Goldberg, A.B.: Introduction to semi-supervised learning. Synth. Lect. Artif. Intell. Mach. Learn. **3**(1), 1–130 (2009)

LIRA: Lifelong Image Restoration from Unknown Blended Distortions

Jianzhao Liu[iD], Jianxin Lin, Xin Li[iD], Wei Zhou[iD], Sen Liu[iD],
and Zhibo Chen[✉][iD]

CAS Key Laboratory of Technology in Geo-spatial Information Processing
and Application System, University of Science and Technology of China,
Hefei 230027, China
{jianzhao,linjx,lixin666,weichou}@mail.ustc.edu.cn,
elsen@iat.ustc.edu.cn, chenzhibo@ustc.edu.cn

Abstract. Most existing image restoration networks are designed in a disposable way and catastrophically forget previously learned distortions when trained on a new distortion removal task. To alleviate this problem, we raise the novel lifelong image restoration problem for blended distortions. We first design a base fork-join model in which multiple pre-trained expert models specializing in individual distortion removal task work cooperatively and adaptively to handle blended distortions. When the input is degraded by a new distortion, inspired by adult neurogenesis in human memory system, we develop a neural growing strategy where the previously trained model can incorporate a new expert branch and continually accumulate new knowledge without interfering with learned knowledge. Experimental results show that the proposed approach can not only achieve state-of-the-art performance on blended distortions removal tasks in both PSNR/SSIM metrics, but also maintain old expertise while learning new restoration tasks.

Keywords: Image restoration · Blended distortions · Lifelong learning

1 Introduction

Image restoration, which is a highly ill-posed problem, has been studied for a long time due to its high demand in different application scenarios such as surveillance imaging [42,51] and medical imaging [8,10]. Most existing image restoration methods tend to treat different degradation factors individually and design a dedicated model for each task, which is inefficient and impractical in real world since images are often degraded by various factors with unknown mixture

---

J. Liu and J. Lin—Authors contributed equally to this work.

Electronic supplementary material The online version of this chapter (https://doi.org/10.1007/978-3-030-58523-5_36) contains supplementary material, which is available to authorized users.

© Springer Nature Switzerland AG 2020
A. Vedaldi et al. (Eds.): ECCV 2020, LNCS 12363, pp. 616–632, 2020.
https://doi.org/10.1007/978-3-030-58523-5_36

ratios and strengths. Recent studies [41,47] have been raised for the blended distortion removal task. However, these approaches are designed in a disposable way and lack the ability to maintain previously learned knowledge when dealing with new distortions.

Unlike deep neural networks that are prone to catastrophic forgetting [28,29], human can gradually accumulate knowledge without seriously perturbing past memories [1] due to special neurophysiological and biological mechanism of learning and memorizing. One explanation about human memory system is stability-plasticity dilemma [30], stating that plasticity is required for the integration of new knowledge and stability is required for avoiding perturbation of old memory.

To imitate human memory system and alleviate catastrophic forgetting in image restoration networks, we design an expansible network for lifelong image restoration, in order to accommodate to new blended distortions without forgetting previous restoration tasks. Meanwhile, assuming that we cannot access the old training data while training a new task, we leverage a Generative Adversarial Network (GAN) [11] to replay learned experience to ensure model stability.

Starting from three typical degradation factors (Gaussian blur, Gaussian noise and JPEG compression), we train a base network in a fork-join manner. In the fork stage, three expert models specializing in specific distortion removal are trained. Then in the join stage, we build the base network by aggregating features from different experts, which helps the three experts work collaboratively to complete a blended distortions removal task.

When the input is degraded by another new distortion (e.g. haze or darkness), we develop a neural growing strategy where our pre-trained base network can incorporate a new expert branch to form an expanded network and continually accumulate new knowledge. The neural growing strategy is inspired by adult neurogenesis studies [3,16,37], which reveals that new neurons are preferentially recruited in response to behavioral novelty and integrated into existing neuronal circuits in selective regions.

The base network and the expanded network have shared parameters and constitute an incremental network. The following question is how to adapt the shared parameters to both the old task (dealing with blended distortions formed by the superposition of three degradation factors) and the new task (dealing with blended distortions formed by the superposition of the three degradation factors and another new degradation factor) simultaneously, avoiding the learned knowledge from being overwritten by the new learning knowledge. Studies about Complementary Learning Systems (CLS) [27] illustrated that recent experience is encoded in the hippocampus and is consolidated in the neocortex through replays of the encoded experience. Considering that old training data may not be allowed to stored for a long time or shared with others due to storage or privacy issues in real scenarios, we train a GAN to generate pseudo old samples to imitate the memory replay process. The pseudo samples are paired with the corresponding responses of the pre-trained base network, serving as supervision of the old task. More interestingly, we find that the memory replay strategy can even boost the performance of the old task rather than simply maintain the performance, which means the old task can benefit from incremental learning of new task through our method.

The contributions of this work can be summarized as follows:

- We handle the blended distortions in a fork-join manner. Complex tasks are assigned to different expert models. The experimental results show that the expert models dealing with heterogeneous problems can work adaptively in the case of blended distortions.
- We consider lifelong learning in image restoration. We incorporate a new expert branch into the pre-trained base network and leverage a generative adversarial network to replay the learned experience to achieve a trade-off between plasticity and stability.

2   Related Work

2.1   Image Restoration

Image restoration tasks, which aim to reconstruct uncorrupted images from corrupted low-quality images, have been widely studied with the rapidly developed CNN-based methods, such as image denoising [19,50], image dehazing [33,49] and image deblurring [31,43]. Liu et al. [24] proposed a "dual residual connection" which exploits the potential of paired operations. Work [23] proposed DuRB-M which can handle different degradation factors and demonstrated that multi-task learning of diverse restoration tasks can bring about synergetic performance improvement. However, images in the real world are usually degraded by complicated blended distortions with unknown mixture ratios. Unlike previous image restoration methods which treat different degradation factors individually and design a dedicated model for each task, some works [41,47] have been proposed to restore images with blended distortions. Yu et al. [47] first proposed a reinforcement learning based approach that can progressively restore the corrupted images. Then Suganuma et al. [41] presented an operation-wise attention layer, which performs various operations in parallel and weights these operations through an attention mechanism. Although these methods have shown preliminary successes in image restoration from blended distortions, they neglect the problem of how to tackle a novel distortion. It would be important to have an intelligent restorer that could continuously learn to address new distortions, without forgetting how to deal with existing ones. In this paper, we take the first step towards lifelong image restoration from blended distortions, in order to mitigate catastrophic forgetting in image restoration networks.

2.2   Lifelong Learning

There have been many works proposed to address the catastrophic forgetting in high-level computer vision tasks such as classification. Regularization-based approaches, such as EWC [18], HAT [38] and MAS [4], add a regularization term that discourages the alteration to weights important to previous tasks, which effectively prevents old knowledge from being erased or overwritten. Li et al. [21], Castro et al. [7] and Dhar et al. [9] also proposed to employ a distillation loss to encourage the responses of the original and the new network to be similar. The rehearsal based methods [25,32,34,45] refer to relearning representative samples selected from the previous task while learning a new task, which can protect the learned knowledge from disruption by new knowledge. However, due to complexity, privacy and legal issues, previously learned data may not be allowed to be stored for a long period in the real world [45], which limits the application scenarios of rehearsal based methods. The pseudo-rehearsal based methods [15,40,44,45] utilize generative models to generate pseudo-samples for modeling the data distribution of previous tasks and replay learned experiences to consolidate the old task knowledge. Dynamic architecture methods [35,46] dynamically accommodate new branches or increase the number of trainable parameters to adapt to new tasks. In this paper, we introduce a fork-join model for lifelong image restoration, which has not been explored before. Unlike lifelong learning for classification aiming at increasing the number of recognizable classes in the network output, our work aims at handling more degradation factors in the input and produce the same clean output. For new distortion coming in, our model only needs to incorporate a new expert branch and continually accumulates new learning knowledge without losing previously learned memory.

3   Our Approach

In the following subsections, we will first introduce the base fork-join network dealing with blended distortions formed by the superposition of N degradation factors and then demonstrate how to extend the existing base network in the presence of a new degradation factor. Finally, we will describe the training strategy for lifelong image restoration.

3.1   Base Network Design

The overall architecture of the proposed network is illustrated in Fig. 1. The entire base network consists of three parts: a feature extractor, a feature amalgamator and a decoder. The feature extractor consists of convolutional layers and produces two-scale feature maps, while the decoder has two upsampling layers that enlarge the spatial size of feature maps. The feature amalgamator consists of N experts, a set of gate layers, an attention layer, an aggregation layer and two Bidirectional Sequential Gateing Units (Bi-SGUs). Each expert is composed of M Selective Kernel Dual Residual Blocks (SK-DuRBs). The SK-DuRB is modified from DuRB-M [23] by replacing the improved SE-ResNet module with SK-Block [20] because the DuRB-M has been proved to be robust with different degradation factors and the SK-Block has the capability of adaptively adjusting the receptive fields according to the input. We first train N experts each of which specializes in individual distortion removal. Specially, these experts are trained sharing a feature extractor and a decoder. Next, with weight parameters

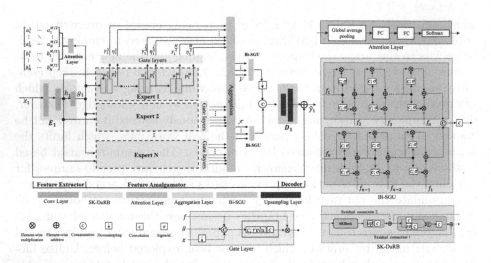

Fig. 1. Framework of the base network. It consists of three parts: a feature extractor, a feature amalgamator and a decoder. Expertises of N experts specializing in individual distortion removal are dynamically aggregated for blended distortions removal. FC means Fully Connected layer; se means squeeze-and-excitation block [12]; up means PixelShuffle [39] modules with convolutional operations. In the Gate layer, $f \in \{\{s_i^j\}_{j=1}^{M/2}, \{q_i^j\}_{j=1}^{M/2}\}_{i=1}^N$, $g \in \{h_1, g_1\}$ and x denotes input image. Zooming in for better viewing.

of the N experts loaded, amalgamated features are generated to the decoder for eliminating blended distortions.

Let's denote the outputs of the j-th SK-DuRB for the i-th expert as u_i^j and p_i^j and take the outputs of two strided convolutional layers in the feature extractor as h_1 and g_1. h_1 and g_1 can be seen as the global descriptor for the input images thus we refer to them as global features. Each expert can be depicted as follows:

$$u_i^j, p_i^j = B_i^j(u_i^{j-1}, p_i^{j-1}), 1 \le j \le M; 1 \le i \le N \tag{1}$$

where $u_i^0 = h_1$, $p_i^0 = g_1$ and $B_i^j(\cdot)$ refers to the jth SK-DuRB of the ith expert model. Let $s_i = \{u_i^{2j}\}_{j=1}^{M/2}$ and $q_i = \{p_i^{2j}\}_{j=1}^{M/2}$ denote features chosen from multiple SK-DuRBs of each expert. As illustrated in [14,36], models trained with gates applied to multiple layers implicitly learn to suppress irrelevant regions in an input image while highlighting salient features for a specific task. Therefore, in order to make more use of image information, we employ gate layers (as shown in Fig. 1) on the intermediate features s_i and q_i for each expert. Meanwhile, gate layers among different experts are expected to learn distortion-specific features. The gate layers produce gated features γ_i^j and η_i^j:

$$\gamma_i^j = G(h_1, x_1, s_i^j) \odot s_i^j, \tag{2}$$
$$\eta_i^j = G(g_1, x_1, q_i^j) \odot q_i^j, \tag{3}$$

where G takes as input the intermediate feature s_i^j or q_i^j, the global feature h_1 or g_1 and the input image x_1, and generate a pixel-level feature map for s_i^j or q_i^j. We then utilize attention layers to dynamically adjust the mixture ratios of distortion-specific features extracted from different experts thus obtaining two sets of aggregated features $\mathcal{V}$ and $\mathcal{F}$:

$$\mathcal{V} = \{v_j\}_{j=1}^{M/2} = \{\sum_{i=1}^{N} a_i^j \gamma_i^j\}_{j=1}^{M/2}, \tag{4}$$

$$\mathcal{F} = \{f_j\}_{j=1}^{M/2} = \{\sum_{i=1}^{N} b_i^j \eta_i^j\}_{j=1}^{M/2}, \tag{5}$$

where a_i^j and b_i^j are the (i,j) element of $(a_i^j)_{N \times M/2}$ and $(b_i^j)_{N \times M/2}$ generated by the attention layers fitted on the global feature g_1. $\sum_{i=1}^{N} a_i^j = 1$ and $\sum_{i=1}^{N} b_i^j = 1$.

$\mathcal{V}$ and $\mathcal{F}$ are composed of features from multiple layers, which contains multi-level information. In order to more effectively leverage multi-level information, we develop a Bidirectional Sequential Gating Unit (Bi-SGU) inspired by the Sequential Gating Unit [22] and BLSTM [52]. Given an activate input f_a and a passive input f_b, equation depicting the SGU unit is given as below:

$$SGU(\mathbf{f}_a, \mathbf{f}_b) = \sigma(conv(\mathbf{f}_a)) * \mathbf{f}_a + \sigma(conv(\mathbf{f}_a)) * \mathbf{f}_b, \tag{6}$$

where $\sigma(\cdot)$ is a sigmoid activation function and $conv$ denotes convolutional operation. Then the Bi-SGU can be defined as:

$$\begin{aligned} Bi\text{-}SGU(\mathcal{V}) = conv([SGU(v_{M/2}, SGU(v_{M/2-1}, (\cdots))), \\ SGU(v_1, SGU(v_2, (\cdots))))]), \end{aligned} \tag{7}$$

where $[\cdot]$ refers to concatenation operation. The final reconstructed image is:

$$\hat{y}_1 = D_1(conv([Bi\text{-}SGU(\mathcal{V}), Bi\text{-}SGU(\mathcal{F})])) + x_1, \tag{8}$$

where $D_1(\cdot)$ corresponds to the decoder part in Fig. 1.

In this way, we obtain a base network that can handle N degradation factors. Expertises of different experts can be dynamically aggregated for the blended distortions removal task.

3.2   Incremental Network Design

For a new degradation factor, we add a new expert branch and first train it together with the original N experts whose parameters are frozen, sharing a new feature extractor E_2 and a new decoder D_2. As Fig. 2 shows, we assign E_2 and D_2 to the new task. All parts except Bi-SGUs in the amalgamator of the base network are treated as a whole and named as an old expert afterwards. Now we describe the expanded network in detail below.

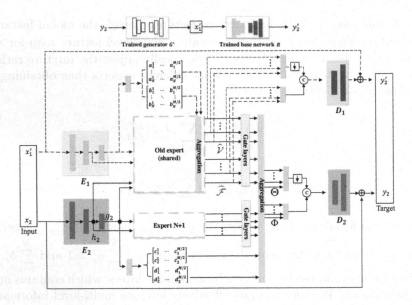

Fig. 2. Framework of the incremental network. The base network and the expanded network share an old expert and constitute the incremental network. Pseudo-samples x_1' paired with the corresponding responses of the trained base network y_2' are used to optimize the parameters of the base network. The new training samples x_2 and y_2 are used to train the expanded network.

For new input image x_2, the global features extracted from the new feature extractor E_2 are denoted as h_2 and g_2 and the new expert branch can be modeled as:

$$u_{N+1}^j, p_{N+1}^j = B_{N+1}^j(u_{N+1}^{j-1}, p_{N+1}^{j-1}), 1 \leq j \leq M \qquad (9)$$

where $u_{N+1}^0 = h_2$ and $p_{N+1}^0 = g_2$. Similarly, let $\hat{s}_1 = \widehat{\mathcal{V}} = \{\hat{v}_j\}_{j=1}^{M/2}$ and $\hat{q}_1 = \widehat{\mathcal{F}} = \{\hat{f}_j\}_{j=1}^{M/2}$ denote features chosen from the old expert where $\widehat{\mathcal{V}}$ and $\widehat{\mathcal{F}}$ represent aggregated features of the old expert for x_2. Let $\hat{s}_2 = \{u_{N+1}^{2j}\}_{j=1}^{M/2}$ and $\hat{q}_2 = \{p_{N+1}^{2j}\}_{j=1}^{M/2}$ denote features chosen from the new expert branch. The aggregated features Θ and Φ for the expanded network are:

$$\Theta = \{\theta_j\}_{j=1}^{M/2} = \{\sum_{i=1}^{2} c_i^j \xi_i^j\}_{j=1}^{M/2}, \qquad (10)$$

$$\Phi = \{\varphi_j\}_{j=1}^{M/2} = \{\sum_{i=1}^{2} d_i^j \psi_i^j\}_{j=1}^{M/2}, \qquad (11)$$

where c_i^j and d_i^j are the (i, j) element of $(c_i^j)_{2 \times M/2}$ and $(d_i^j)_{2 \times M/2}$ generated by the attention layers fitted on the global feature g_2. $\sum_{i=1}^{2} c_i^j = 1$ and $\sum_{i=1}^{2} d_i^j = 1$. ξ_i^j and ψ_i^j are given by the formulas below:

$$\xi_i^j = G(h_2, x_2, \widehat{s}_i^j) \odot \widehat{s}_i^j, \tag{12}$$

$$\varphi_i^j = G(g_2, x_2, \widehat{q}_i^j) \odot \widehat{q}_i^j. \tag{13}$$

The final reconstructed image for the new task is given as follow:

$$\widehat{y}_2 = D_2(conv([Bi\text{-}SGU(\Theta), Bi\text{-}SGU(\Phi)])) + x_2. \tag{14}$$

In this way, the new expert branch can be assimilated into the base network, forming an expanded network to handle new blended distortions. We finally get an incremental network consisting of the base network and the expanded network. It should be noticed that old expert's parameters are shared by the base network and the expanded network and are dynamically optimized for both old and new tasks.

3.3   Training Strategy for Lifelong Image Restoration

Our training strategy is inspired by the Complementary Learning System(CLS) theory [27], which illustrates that experience encoded in the hippocampus is consolidated in the neocortex through memory replay. We leverage a generative adversarial network to model the data distribution of old training samples, playing the role of memory replay. As shown at the top of Fig. 2, we first learn a mapping function G [53] aiming to convert clean images y_1 to blended distortion images x_1 of the old task. We denote the data distribution as $p_{data}(x_1)$ and $p_{data}(y_1)$. The adversarial loss and the cycle consistency loss can be depicted as follows:

$$\mathcal{L}_{GAN} = \mathbb{E}_{x_1 \sim p_{data}(x_1)}[\log D_{X_1}(x_1)]$$
$$+ \mathbb{E}_{y_1 \sim p_{data}(y_1)}[\log(1 - D_{X_1}(G(y_1)))], \tag{15}$$

$$\mathcal{L}_{cyc} = \mathbb{E}_{y_1 \sim p_{data}(y_1)}[\|R(G(y_1)) - y_1\|_1], \tag{16}$$

where G learns to generate blended distorted images, D_{X_1} is the discriminator network leaning to distinguish real distorted images and generated ones. R is the previously trained base network on the old task. The full objective is:

$$\mathcal{L} = \mathcal{L}_{GAN} + \lambda_1 \mathcal{L}_{cyc}, \tag{17}$$

where λ_1 controls the relative importance of the two objectives.

After obtaining the trained generator G^*, we can generate pseudo old training samples $x_1' = G^*(y_2)$ and $y_2' = R(G^*(y_2))$ using training samples $\{x_{2_i}, y_{2_i}\}_{i=1}^T$ of the new task where T is the number of the new task training samples. In the following step, we train the incremental network with new training samples and pseudo old training samples. Let's denote the base network as $\widetilde{R}$ with parameters $\theta_{Base} = \{\theta_S, \theta_{P_1}\}$ and denote the expanded network as H with parameters $\theta_{Expanded} = \{\theta_S, \theta_{P_2}\}$. θ_S refers to the shared parameters (the parameters of the old expert) between the base network and the expanded network, while θ_{P_1} and

θ_{P_2} denote specific parameters of the base network and the expanded network respectively.

First, the base network $\widetilde{R}$ is initialized with the previously trained base network R and the expanded network is initialized with the trained $(N + 1)$th expert. Then we jointly optimize θ_{Base} and $\theta_{Expanded}$ to minimize loss for both two tasks:

$$\mathcal{L}_{incremental} = \mathcal{L}_{new}(y_2, \widehat{y}_2) + \lambda_2 \mathcal{L}_{old}(y_2', \widehat{y}_2'), \tag{18}$$

where $\widehat{y}_2 = H(x_2; \theta_{Expanded})$, $\widehat{y}_2' = \widetilde{R}(x_1'; \theta_{Base})$ and λ_2 is a loss balance weight to balance the old task performance and the new task performance. $\mathcal{L}_{new}$ and $\mathcal{L}_{old}$ are both l_1 loss functions.

Through this training strategy, the incremental network can continually accumulate new knowledge while consolidating previously learned knowledge through memory replay, without accessing the old training samples.

4 Experiments

4.1 Datasets

Following the experimental settings of works [41,47], we use the DIV2K dataset [2] which contains 800 images (750 images for training and 50 images for testing). The images are corrupted by a sequence of Gaussian blur, Gaussian noise and JPEG compression with random levels and are then cropped into 64×64 sub-images, resulting in total 230,139 training images and 3,584 testing images. The images for training experts that specialize in individual distortion removal, are generated in the same way and individual degradation factor is randomly added. In order to verify the effectiveness of our incremental network, we add another two degradation factors (haze and darkness) respectively to the above three degradation factors with random levels and form another two training sets with 230,139 images and two testing sets with 1,827 images. Specially, to further verify the generalization ability of our base network on images from different datasets and in different resolutions, we randomly add blended distortions on five public benchmark datasets: BSDS100 [5], MANGA109 [26], SET5 [6], SET14 [48] and URBAN100 [13] and randomly crop the images into 64×64, 128×128 and 256×256 respectively, generating the additional testing sets.

The standard deviations of Gaussian blur and Gaussian noise are uniformly distributed in [0, 5] and [0, 50] respectively. We use the "imwrite" function in MATLAB to apply JPEG compression, where the parameter "quality" within the range of [0,100] controls the compression quality. We randomly choose the "quality" from the range of [10]. For haze degradation, we randomly choose the scattering coefficient within [0.05, 0.25]. For darkness degradation, we use the "imadjust" function in MATLAB to randomly apply a gamma adjustment within the range of [1.5, 2.5]. The images can be classified into three groups based on the applied degradation levels: mild, moderate and severe. To test the generalization ability of the network, the training is performed on the moderate group and testing is performed on all three groups (DIV2K dataset). BSDS100, MANGA109, SET5, SET14 and URBAN100 are added with moderate-level distortions.

4.2   Implementation Details

In our basic network, we set the number of experts N to 3 and the number of SK-DuRBs M to 6. For model training, we use Adam solver [17] with parameters $\beta_1 = 0.9$ and $\beta_2 = 0.999$. We adopt cosine annealing strategy to adjust the initial learning rate $1e^{-4}$. For the adversarial generator used to generate pseudo-samples, we adopt the same architecture as the adversarial generator G with 9 residual blocks in [53] and replace the adversarial generator F in [53] with our trained base network. We set λ_1 in Eq. 17 to 10.0 and λ_2 in Eq. 18 to 0.2. Specially, in order to discourage the alteration to the parameters closely related to the old task, we lower down the learning rate of θ_{Base} and set $lr(\theta_{P_2}) = \mu$ and $lr(\theta_S, \theta_{P_1}) = \rho\mu$, where $lr(\cdot)$ represents the learning rate and ρ is a small number within the range of $(0, 1)$. We set μ, and ρ to $1e^{-4}$ and $1/1000$ respectively. We take 5 epochs to warm up the network (freeze the parameters of the old expert and only train the expanded network) and then jointly optimize all parameters.

4.3   Performance of Base Network

We compare our base network with RL-Restore [47] and Operation-wise Attention Network (OWAN) [41] under PSNR and SSIM metrics. RL-Restore and OWAN are state-of-the-art models that are specially designed for blended distortions. Table 1 shows the PSNR and SSIM values of the three methods at different degradation levels on DIV2K test sets. It can be seen that our method outperforms the compared methods at all degradation levels in both the PSNR and SSIM metrics.

Table 1. Quantitative results of RL-Restore, OWAN and our base network on DIV2K test sets. The best results are highlighted in bold.

Method	Mild(unseen)		Moderate		Severe(unseen)	
	PSNR	SSIM	PSNR	SSIM	PSNR	SSIM
RL-Restore	28.04	0.6498	26.45	0.5587	25.20	0.4777
OWAN	28.33	0.7455	27.07	0.6787	25.88	0.6167
Ours	**28.61**	**0.7496**	**27.24**	**0.6832**	**25.93**	**0.6219**

We also show restored images obtained by different methods along with their input images and ground truths in Fig. 3. It can be seen that the visual results are consistent with quantitative results, where our base network can achieve the best visual quality and tends to retain more image detail information. Furthermore, we carry out extensive experiments on five public benchmark datasets: BSDS100, MANGA109, SET5, SET14 and URBAN10 and the quantitative results are shown in Table 2. We can see that our network has the best generalization ability on different datasets and has better robustness in terms of resolution variations.

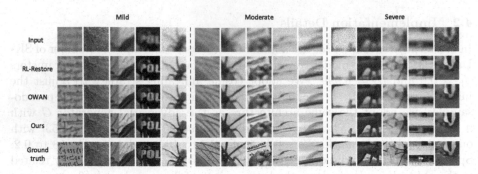

Fig. 3. Qualitative comparisons with RL-Restore and OWAN on DIV2K test sets.

Table 2. Quantitative results of RL-Restore, OWAN and our base network on BSDS100, MANGA109, SET5, SET14 and URBAN100. The best results are highlighted in bold.

Method	Image size	BSDS100		MANGA109		SET5		SET14		URBAN100	
		PSNR	SSIM	PSNR	SSIM	PSNR	SSIM	PSNR	SSIM	PSNR	SSIM
RL-Restore	64 × 64	25.57	0.6074	25.70	0.7292	25.46	0.6549	24.68	0.6198	25.01	0.6354
OWAN		25.99	0.6261	26.68	0.7648	26.14	0.6874	25.39	0.6567	25.64	0.6721
Ours		**26.21**	**0.6298**	**27.69**	**0.7751**	**26.49**	**0.6955**	**25.78**	**0.6652**	**25.97**	**0.6783**
RL-Restore	128 × 128	24.60	0.6108	24.58	0.7263	25.98	0.7005	24.30	0.6725	23.87	0.6227
OWAN		23.70	0.6127	21.66	0.7143	23.52	0.7005	22.53	0.6650	22.65	0.6341
Ours		**25.24**	**0.6377**	**26.13**	**0.7741**	**27.58**	**0.7637**	**25.10**	**0.7079**	**24.72**	**0.6641**
RL-Restore	256 × 256	23.56	0.5724	23.15	0.6908	28.86	0.7881	24.58	0.6974	22.54	0.6077
OWAN		22.50	0.5745	19.44	0.6549	20.66	0.6726	21.73	0.6686	21.37	0.6136
Ours		**24.10**	**0.6038**	**24.45**	**0.7428**	**28.86**	**0.8098**	**25.33**	**0.7328**	**23.39**	**0.6564**

4.4 Ablation Studies

In this section, we conduct several ablation studies to investigate the effectiveness of pre-trained experts, the SK-DuRB, the gate layer, the attention layer and the Bi-SGU, which are all the basic components in our base network. So total six variant networks are designed for comparison: (a) To validate the effectiveness of the pre-trained experts, we train the base network without the initialization of experts. (b) To validate the effectiveness of the SK-DuRB, we replace the SK-DuRB with the DuRB-M [23]. (c) To verify the effectiveness of the gate layer, we remove the gate layers applied to s_i^j in Eq. 2. Then we further remove all gate layers applied to q_i^j in Eq. 3. (d) To verify the effect of the attention layer, we set all attention weights equal to 1. (e) To verify the effect of the Bi-SGU, we replace the Bi-SGU with the concatenation operation followed by a convolutional layer. We train these networks using the same strategy as aforementioned in Sect. 4.2. Table 3 shows that the original base network achieves the best performance among these variants at all degradation levels in terms of PSNR and SSIM (on DIV2K testing set).

Table 3. Ablation study of our base network.

Model	Mild(unseen)		Moderate		Severe(unseen)	
	PSNR	SSIM	PSNR	SSIM	PSNR	SSIM
w/o Experts	27.28	0.7009	25.95	0.6370	24.72	0.5832
w/o SK-Blocks	28.14	0.7232	26.76	0.6588	25.41	0.6010
w/o Half of Gate layers	28.51	0.7437	27.14	0.6783	25.85	0.6191
w/o All Gate layers	27.96	0.7186	26.64	0.6560	25.38	0.6018
w/o Attention layer	28.10	0.7236	26.74	0.6591	25.39	0.6027
w/o Bi-SGUs	28.52	0.7435	27.18	0.6789	25.91	0.6209
Ours	**28.61**	**0.7496**	**27.24**	**0.6832**	**25.93**	**0.6219**

4.5 Performance of Incremental Network

We have three settings for training the incremental network:

Setting A. Only real new training samples $\{x_{2_i}, y_{2_i}\}_{i=1}^{T}$ are used to train the expanded network H. We discourage alteration to the old expert's parameters by reducing the learning rate.

Setting B (Joint Learning). Real new training samples $\{x_{2_i}, y_{2_i}\}_{i=1}^{T}$ and real old training samples $\{x_{1_i}, y_{1_i}\}_{i=1}^{K}$ are used to optimize the parameters of the expanded network H and the base network $\widetilde{R}$ respectively, with the learning rate of the base network lowered. K is the number of the old task training samples.

Setting C (LIRA). As illustrated in Sect. 3.3, real new training samples $\{x_{2_i}, y_{2_i}\}_{i=1}^{T}$ and pseudo old training samples $\{x'_{1_i}, y'_{2_i}\}_{i=1}^{T}$ are used to optimize the parameters of the expanded network H and the base network $\widetilde{R}$ respectively, with the learning rate of the base network lowered.

We present the training curves of the three settings on mild DIV2K test group in Fig. 4. It should be mentioned that we initialize the expanded network H with trained parameters obtained in the warm-up step under the three settings, thereby eliminating random initialization interference under the three settings. We can see that the performance on the old task continually drops under setting A. In contrast, Setting B and LIRA can even boost the performance of the old task rather than simply maintain the performance, since the new task can provide complementary knowledge for the old task by updating the shared parameters. It should be noted that setting B can be seen as an upper bound on the performance of LIRA since an adversarial generator cannot fully resemble the data distribution of the real old samples.

In addition to the above three settings, we also compare LIRA with the following baselines: **(a) Elastic Weight Consolidation (EWC)** [18]: Add a regularization term to discourage the alteration to weights important to the old task. We implement EWC on the old expert's parameters to find which weights

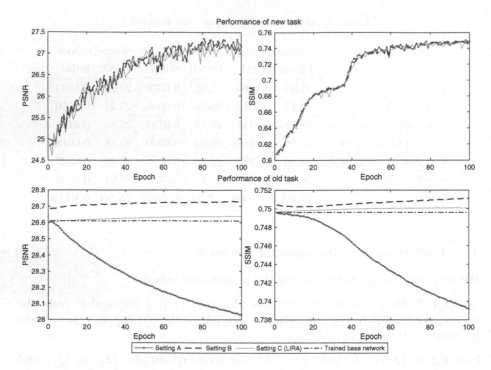

Fig. 4. PSNR and SSIM curves with respect to training epochs of 3 different settings.

are important for the old task when training the new task. **(b) Progressive Neural Networks (PNN)** [35]: Allocate a new column whose weights are randomly initialized for each new task while freezing the weights of previous columns. Transfer of old knowledge is enabled via lateral connections to features of previously learned columns. We implement PNN by freezing the old expert's weights and randomly initializing the weights of the new expert branch. **(c) Learning without Forgetting (LWF)** [21]: Responses to the new task from the previously trained network are used to optimize old task's performance while new training data is used to optimize new task's performance. We implement LWF by using new training samples $\{x_{2_i}, y_{2_i}\}_{i=1}^{T}$ and responses to the new task from trained base network $\{x_{2_i}, R(x_{2_i})\}_{i=1}^{T}$ to train the expanded network H and the base network $\widetilde{R}$ respectively, where R denotes previously trained base network. **(d) Fine-tuning**: Modify the parameters of an existing network to train a new task. To implement this method, we train the expanded network H using only real new training samples $\{x_{2_i}, y_{2_i}\}_{i=1}^{T}$. The difference between Fine-tuning and Setting A is that we don't specially lower down the learning rate of the old expert and all parameters are optimized with the same learning rate. **(e) OWAN** [41]: We merge old and new training samples together to train OWAN, whose parameters are initialized with trained weights for old task.

The comparison results of old and task performance regarding haze degradation factor are shown in Table 4. We can see that EWC cannot effectively prevent performance degradation of old task. Although PNN can guarantee the old task performance, it limits the new task performance because the old expert's parameters are frozen. LWF can achieve relatively good performance on old task but it still causes the old task performance to slightly drop. For Fine-tuning, the descending rate of the old task performance is faster compared with Setting A, because there is no constraints to the old expert's learning rate and the shared parameters will be more adaptive to the new task. Moreover, since the data distribution of the two training sets is different, it is difficult to find a local minimum that is most suitable for both old and new tasks for OWAN. Therefore, both the old and the new task performances of OWAN are worse than our model. Compared with the above methods, LIRA can maintain old expertise while accumulating new knowledge without accessing old training samples.

Table 4. Quantitative results of old and new tasks regarding haze degradation factor. The top three results are highlighted in red, blue and cyan colors respectively in each column.

Method	Mild(unseen)		Moderate		Severe(unseen)	
	old PSNR/SSIM	new PSNR/SSIM	old PSNR/SSIM	new PSNR/SSIM	old PSNR/SSIM	new PSNR/SSIM
EWC	22.05/0.4327	27.16/0.7485	20.86/0.3449	25.79/0.6750	19.61/0.2644	23.92/0.5795
PNN	28.61/0.7496	24.78/0.6119	27.24/0.6832	24.21/0.6037	25.93/0.6219	22.68/0.4706
LWF	28.56/0.7496	27.23/0.7493	27.13/0.6804	25.97/0.6767	25.81/0.6191	24.30/0.5887
Setting A	28.03/0.7394	27.33/0.7502	26.50/0.6633	26.06/0.6781	25.13/0.5950	24.31/0.5899
Fine-tuning	20.25/0.3959	27.51/0.7566	19.13/0.3127	25.90/0.6783	17.88/0.2311	24.26/0.5907
OWAN	27.57/0.7459	27.18/0.7539	26.30/0.6785	25.71/0.6731	25.02/0.6097	24.28/0.5891
Joint learning	28.73/0.7511	27.39/0.7527	27.25/0.6837	26.11/0.6793	25.90/0.6228	24.36/0.5917
LIRA	28.63/0.7501	27.37/0.7509	27.24/0.6830	26.05/0.6772	25.93/0.6223	24.36/0.5901

5 Conclusions

In this paper, we take the first step toward lifelong image restoration in the presence of blended distortions. To alleviate the catastrophic forgetting in image restoration networks, our model LIRA is designed by imitating human memory system. Meanwhile, we leverage a GAN to replay learned experience to ensure model stability. As a result, LIRA can continually accumulate new knowledge and retain previous capabilities without accessing old training samples. Extensive experiment results show that LIRA achieves state-of-the-art performance on both ordinary image restoration tasks and lifelong image restoration tasks.

Acknowledgements. This work was supported in part by NSFC under Grant U1908209, 61632001 and the National Key Research and Development Program of China 2018AAA0101400.

References

1. Abraham, W.C., Robins, A.: Memory retention-the synaptic stability versus plasticity dilemma. Trends Neurosci. **28**(2), 73–78 (2005)
2. Agustsson, E., Timofte, R.: Ntire 2017 challenge on single image super-resolution: dataset and study. In: Proceedings of the IEEE Conference on Computer Vision and Pattern Recognition Workshops, pp. 126–135 (2017)
3. Aimone, J.B., Deng, W., Gage, F.H.: Resolving new memories: a critical look at the dentate gyrus, adult neurogenesis, and pattern separation. Neuron **70**(4), 589–596 (2011)
4. Aljundi, R., Babiloni, F., Elhoseiny, M., Rohrbach, M., Tuytelaars, T.: Memory aware synapses: learning what (not) to forget. In: Proceedings of the European Conference on Computer Vision (ECCV), pp. 139–154 (2018)
5. Arbelaez, P., Maire, M., Fowlkes, C., Malik, J.: Contour detection and hierarchical image segmentation. IEEE Trans. Pattern Anal. Mach. Intell. **33**(5), 898–916 (2010)
6. Bevilacqua, M., Roumy, A., Guillemot, C., Alberi-Morel, M.L.: Low-complexity single-image super-resolution based on nonnegative neighbor embedding (2012)
7. Castro, F.M., Marín-Jiménez, M.J., Guil, N., Schmid, C., Alahari, K.: End-to-end incremental learning. In: Proceedings of the European Conference on Computer Vision (ECCV), pp. 233–248 (2018)
8. Chen, Y., Shi, F., Christodoulou, A.G., Xie, Y., Zhou, Z., Li, D.: Efficient and accurate MRI super-resolution using a generative adversarial network and 3D multi-level densely connected network. In: Frangi, A.F., Schnabel, J.A., Davatzikos, C., Alberola-López, C., Fichtinger, G. (eds.) MICCAI 2018. LNCS, vol. 11070, pp. 91–99. Springer, Cham (2018). https://doi.org/10.1007/978-3-030-00928-1_11
9. Dhar, P., Singh, R.V., Peng, K.C., Wu, Z., Chellappa, R.: Learning without memorizing. In: Proceedings of the IEEE Conference on Computer Vision and Pattern Recognition, pp. 5138–5146 (2019)
10. Gondara, L.: Medical image denoising using convolutional denoising autoencoders. In: 2016 IEEE 16th International Conference on Data Mining Workshops (ICDMW), pp. 241–246. IEEE (2016)
11. Goodfellow, I., et al.: Generative adversarial nets. In: Advances in Neural Information Processing Systems, pp. 2672–2680 (2014)
12. Hu, J., Shen, L., Sun, G.: Squeeze-and-excitation networks. In: Proceedings of the IEEE Conference on Computer Vision and Pattern Recognition, pp. 7132–7141 (2018)
13. Huang, J.B., Singh, A., Ahuja, N.: Single image super-resolution from transformed self-exemplars. In: Proceedings of the IEEE Conference on Computer Vision and Pattern Recognition, pp. 5197–5206 (2015)
14. Jetley, S., Lord, N.A., Lee, N., Torr, P.H.: Learn to pay attention. arXiv preprint arXiv:1804.02391 (2018)
15. Kamra, N., Gupta, U., Liu, Y.: Deep generative dual memory network for continual learning. arXiv preprint arXiv:1710.10368 (2017)
16. Kee, N., Teixeira, C.M., Wang, A.H., Frankland, P.W.: Preferential incorporation of adult-generated granule cells into spatial memory networks in the dentate gyrus. Nat. Neurosci. **10**(3), 355 (2007)
17. Kingma, D.P., Ba, J.: Adam: a method for stochastic optimization. arXiv preprint arXiv:1412.6980 (2014)

18. Kirkpatrick, J., et al.: Overcoming catastrophic forgetting in neural networks. Proc. Nat. Acad. Sci. **114**(13), 3521–3526 (2017)
19. Lefkimmiatis, S.: Universal denoising networks: a novel CNN architecture for image denoising. In: Proceedings of the IEEE Conference on Computer Vision and Pattern Recognition, pp. 3204–3213 (2018)
20. Li, X., Wang, W., Hu, X., Yang, J.: Selective kernel networks (2019)
21. Li, Z., Hoiem, D.: Learning without forgetting. IEEE Trans. Pattern Anal. Mach. Intell. **40**(12), 2935–2947 (2017)
22. Lin, J., Zhou, T., Chen, Z.: Multi-scale face restoration with sequential gating ensemble network. In: Thirty-Second AAAI Conference on Artificial Intelligence (2018)
23. Liu, X., Suganuma, M., Okatani, T.: Joint learning of multiple image restoration tasks. arXiv preprint arXiv:1907.04508v1 (2019)
24. Liu, X., Suganuma, M., Sun, Z., Okatani, T.: Dual residual networks leveraging the potential of paired operations for image restoration. In: Proceedings of the IEEE Conference on Computer Vision and Pattern Recognition, pp. 7007–7016 (2019)
25. Lopez-Paz, D., Ranzato, M.: Gradient episodic memory for continual learning. In: Advances in Neural Information Processing Systems, pp. 6467–6476 (2017)
26. Matsui, Y., et al.: Sketch-based manga retrieval using manga109 dataset. Multimed. Tools Appl. **76**(20), 21811–21838 (2017)
27. McClelland, J.L., McNaughton, B.L., O'Reilly, R.C.: Why there are complementary learning systems in the hippocampus and neocortex: insights from the successes and failures of connectionist models of learning and memory. Psychol. Rev. **102**(3), 419 (1995)
28. McCloskey, M., Cohen, N.: Catastrophic interference in connectionist networks: the sequential learning problem. Psychol. Learn. Motiv. Adv. Res. Theory **24**(C), 109–165 (1989). https://doi.org/10.1016/S0079-7421(08)60536-8
29. McCloskey, M., Cohen, N.J.: Catastrophic interference in connectionist networks: the sequential learning problem. In: Psychology of Learning and Motivation, vol. 24, pp. 109–165. Elsevier (1989)
30. Mermillod, M., Bugaiska, A., Bonin, P.: The stability-plasticity dilemma: investigating the continuum from catastrophic forgetting to age-limited learning effects. Front. Psychol. **4**, 504 (2013)
31. Nah, S., Hyun Kim, T., Mu Lee, K.: Deep multi-scale convolutional neural network for dynamic scene deblurring. In: Proceedings of the IEEE Conference on Computer Vision and Pattern Recognition, pp. 3883–3891 (2017)
32. Rebuffi, S.A., Kolesnikov, A., Sperl, G., Lampert, C.H.: iCaRL: incremental classifier and representation learning. In: Proceedings of the IEEE Conference on Computer Vision and Pattern Recognition, pp. 2001–2010 (2017)
33. Ren, W., et al.: Gated fusion network for single image dehazing. In: Proceedings of the IEEE Conference on Computer Vision and Pattern Recognition, pp. 3253–3261 (2018)
34. Robins, A.: Catastrophic forgetting, rehearsal and pseudorehearsal. Connect. Sci. **7**(2), 123–146 (1995)
35. Rusu, A.A., et al.: Progressive neural networks. arXiv preprint arXiv:1606.04671 (2016)
36. Schlemper, J., et al.: Attention gated networks: learning to leverage salient regions in medical images. Med. Image Anal. **53**, 197–207 (2019)
37. Schmidt-Hieber, C., Jonas, P., Bischofberger, J.: Enhanced synaptic plasticity in newly generated granule cells of the adult hippocampus. Nature **429**(6988), 184 (2004)

38. Serrà, J., Surís, D., Miron, M., Karatzoglou, A.: Overcoming catastrophic forgetting with hard attention to the task. arXiv preprint arXiv:1801.01423 (2018)
39. Shi, W., et al.: Real-time single image and video super-resolution using an efficient sub-pixel convolutional neural network. In: The IEEE Conference on Computer Vision and Pattern Recognition (CVPR), June 2016
40. Shin, H., Lee, J.K., Kim, J., Kim, J.: Continual learning with deep generative replay. In: Advances in Neural Information Processing Systems, pp. 2990–2999 (2017)
41. Suganuma, M., Liu, X., Okatani, T.: Attention-based adaptive selection of operations for image restoration in the presence of unknown combined distortions. In: Proceedings of the IEEE Conference on Computer Vision and Pattern Recognition, pp. 9039–9048 (2019)
42. Svoboda, P., Hradiš, M., Maršík, L., Zemcík, P.: CNN for license plate motion deblurring. In: 2016 IEEE International Conference on Image Processing (ICIP), pp. 3832–3836. IEEE (2016)
43. Tao, X., Gao, H., Shen, X., Wang, J., Jia, J.: Scale-recurrent network for deep image deblurring. In: Proceedings of the IEEE Conference on Computer Vision and Pattern Recognition, pp. 8174–8182 (2018)
44. Wu, C., Herranz, L., Liu, X., van de Weijer, J., Raducanu, B., et al.: Memory replay GANs: learning to generate new categories without forgetting. In: Advances In Neural Information Processing Systems, pp. 5962–5972 (2018)
45. Wu, Y., et al.: Incremental classifier learning with generative adversarial networks. arXiv preprint arXiv:1802.00853 (2018)
46. Yoon, J., Yang, E., Lee, J., Hwang, S.J.: Lifelong learning with dynamically expandable networks. arXiv preprint arXiv:1708.01547 (2017)
47. Yu, K., Dong, C., Lin, L., Change Loy, C.: Crafting a toolchain for image restoration by deep reinforcement learning. In: Proceedings of the IEEE Conference on Computer Vision and Pattern Recognition, pp. 2443–2452 (2018)
48. Zeyde, R., Elad, M., Protter, M.: On single image scale-up using sparse-representations. In: Boissonnat, J.-D., et al. (eds.) Curves and Surfaces 2010. LNCS, vol. 6920, pp. 711–730. Springer, Heidelberg (2012). https://doi.org/10.1007/978-3-642-27413-8_47
49. Zhang, H., Patel, V.M.: Densely connected pyramid dehazing network. In: Proceedings of the IEEE Conference on Computer Vision and Pattern Recognition, pp. 3194–3203 (2018)
50. Zhang, K., Zuo, W., Zhang, L.: FFDNet: toward a fast and flexible solution for CNN-based image denoising. IEEE Trans. Image Process. **27**(9), 4608–4622 (2018)
51. Zhang, L., Zhang, H., Shen, H., Li, P.: A super-resolution reconstruction algorithm for surveillance images. Sig. Process. **90**(3), 848–859 (2010)
52. Zhang, S., Zheng, D., Hu, X., Yang, M.: Bidirectional long short-term memory networks for relation classification. In: Proceedings of the 29th Pacific Asia Conference on Language, Information and Computation, pp. 73–78 (2015)
53. Zhu, J.Y., Park, T., Isola, P., Efros, A.A.: Unpaired image-to-image translation using cycle-consistent adversarial networks. In: Proceedings of the IEEE International Conference on Computer Vision, pp. 2223–2232 (2017)

HDNet: Human Depth Estimation for Multi-person Camera-Space Localization

Jiahao Lin[✉][iD] and Gim Hee Lee[iD]

Department of Computer Science, National University of Singapore,
Singapore, Singapore
{jiahao,gimhee.lee}@comp.nus.edu.sg

Abstract. Current works on multi-person 3D pose estimation mainly focus on the estimation of the 3D joint locations relative to the root joint and ignore the absolute locations of each pose. In this paper, we propose the Human Depth Estimation Network (HDNet), an end-to-end framework for absolute root joint localization in the camera coordinate space. Our HDNet first estimates the 2D human pose with heatmaps of the joints. These estimated heatmaps serve as attention masks for pooling features from image regions corresponding to the target person. A skeleton-based Graph Neural Network (GNN) is utilized to propagate features among joints. We formulate the target depth regression as a bin index estimation problem, which can be transformed with a soft-argmax operation from the classification output of our HDNet. We evaluate our HDNet on the root joint localization and root-relative 3D pose estimation tasks with two benchmark datasets, *i.e.*, Human3.6M and MuPoTS-3D. The experimental results show that we outperform the previous state-of-the-art consistently under multiple evaluation metrics. Our source code is available at: https://github.com/jiahaoLjh/HumanDepth.

Keywords: Human Depth Estimation · Multi-person pose estimation · Camera coordinate space

1 Introduction

Human pose estimation is one of the active research topics in the community of computer vision and artificial intelligence due to its importance in many applications such as camera surveillance, virtual/augmented reality, and human-computer interaction, *etc.* Extensive research has been done for human pose estimation in both 2D image space and 3D Cartesian space, respectively. Great successes have been achieved in the single-person 2D/3D pose estimation tasks thanks to the rapid development of deep learning techniques and the emergence of large-scale human pose datasets [1,12,16]. On the other hand, multi-person 2D/3D pose estimation tasks are more challenging due to the unknown number of persons in the scene. To mitigate this problem, the multi-person pose estimation task is typically tackled in a two-stage scheme that decouples the estimation of the number of persons and the pose of each person, *i.e.*, the top-down [3,5,8,10,28] or bottom-up [2,22,24] scheme. In recent years, large-scale

© Springer Nature Switzerland AG 2020
A. Vedaldi et al. (Eds.): ECCV 2020, LNCS 12363, pp. 633–648, 2020.
https://doi.org/10.1007/978-3-030-58523-5_37

Fig. 1. Top-down multi-person 3D pose estimation pipeline. Camera-space root joint coordinate is estimated for each detected person bounding box, followed by a root-relative 3D pose estimation, to obtain the absolute 3D poses and locations.

multi-person 3D pose datasets such as MuPoTS-3D [20] are created to facilitate the research of multi-person 3D pose estimation. However, most of the existing works [4,20,26,27] focus on the estimation of 3D pose relative to the root joint of each person in the scene. Global absolute locations of the respective 3D poses with respect to the camera coordinate space are ignored.

Estimating the absolute 3D location of each pose in an image is essential for understanding human-to-human interactions. Recently, Moon *et al.* [21] propose a multi-stage pipeline for the task of multi-person 3D pose estimation in the camera coordinate space as shown in Fig. 1. The pipeline adopts the top-down scheme which predicts a bounding box for each person in the first stage. This is followed by estimation of the absolute root joint location for the person in each bounding box. Finally, the global pose of each person is recovered by applying single-person 3D pose estimation to get the relative location of other joints with respect to the root joint. The root joint localization framework proposed in [21] estimates the depth of root joint for each person based on the size of the bounding box. Despite showing promising results, the approach relies on the size of bounding box for root joint localization, and hence is not sufficiently effective due to two reasons: (1) The compactness of bounding boxes varies from person to person and also between different object detectors. (2) Sizes of bounding boxes carries no direct information about the size of the particular person due to the variation of poses.

In this paper, we propose an end-to-end Human Depth Estimation Network (HDNet) to address the problems of root joint depth estimation and localization. We adopt the same top-down pipeline for the task of multi-person absolute 3D pose estimation. Our key observation is that we can estimate the depth of a person in a monocular image with considerably high accuracy by leveraging on the prior knowledge of the typical size of the human pose and body joints. Inspired by this observation, we propose to jointly learn the 2D human pose and the depth estimation tasks in our HDNet. More specifically, we utilize the heatmaps of the joints from the human pose estimation task as attention masks to achieve pose-aware feature pooling in each joint type. Subsequently, we put the pose-aware features of the joints into a skeleton-based Graph Neural Network (GNN), where information are effectively propagated among body joints to enhance depth estimation. Following a recent work on scene depth estimation [7], we formulate our depth estimation of the root joint as a classification task, where the target depths are discretized into a preset number of bins. We also adopt a soft-argmax operation on the bins predicted by our HDNet for faster convergence during training and better performance without losing precision compared to direct numerical depth regression. Our approach outperforms previous state-of-the-art [21] on the task of root joint localization on two benchmark datasets, i.e., Human3.6M [12] and MuPoTS-3D [20] under multiple evaluation metrics. Experimental results also show that accurate root localization benefits the task of root-aligned 3D human pose estimation.

Our contributions in this work are:

- An end-to-end Human Depth Estimation Network (HDNet) is proposed to address the problem of root joint localization for multi-person 3D pose estimation in the camera-space.
- Several key components are introduced in our framework: (1) pose heatmaps are used as attention masks for pose-aware feature pooling; (2) a skeleton-based GNN is designed for effective information propagation among the body joints; and (3) depth regression of the root joint is formulated as a classification task, where the classification output is transformed to the estimated depth with a soft-argmax operation to facilitate accurate depth estimation.
- Quantitative and qualitative results show that our approach consistently outperforms the state-of-the-art on multiple benchmark datasets under various evaluation metrics.

2   Related Works

Human pose estimation has been an interesting yet challenging problem in computer vision. Early methods use a variety of hand-crafted features such as silhouette, shape, SIFT features, HOG for the task. Recently, with the power of deep neural networks and well-annotated large-scale human pose datasets, increasing learning-based approaches are proposed to tackle this challenging problem.

Single-Person 2D Pose Estimation. Early works, such as Stacked Hourglass [23], Convolutional Pose Machines [30], etc., have been proposed to use deep convolutional neural networks as feature extractors for 2D pose estimation. Heatmaps of joints are the commonly used representation to indicate the presence of joints at spatial locations with Gaussian peaks. More recent works including RMPE [5], CFN [10], CPN [3], HRNet [28], etc., introduce various framework designs to improve the joint localization precision.

Single-Person 3D Pose Estimation. Approaches for 3D pose estimation can be generally categorized into two groups. Direct end-to-end estimation of 3D pose from RGB images regresses both 2D joint locations and the z-axis root-relative depth for each joint. [25,29] extend the notion of heatmap to the 3D space, where estimation is performed in a volumetric space. Another group of approaches decouples the task into a two-stage pipeline. The 2D joint locations are first estimated, followed by a 2D-to-3D lifting. [6,18] utilize Multi-Layer Perceptron (MLP) to learn the mapping.

Multi-person 2D Pose Estimation. Top-down [3,5,8,10,28] and bottom-up [2,22,24] approaches have been proposed to estimate poses for multiple persons. Top-down approaches utilize a human object detector to localize the bounding box, followed by a single-person pose estimation pipeline with image patch cropped from the bounding box. Bottom-up approaches detect human joints in a person-agnostic way, followed by a grouping process to identify joints belonging to the same person. Top-down approaches usually estimate joint locations more precisely because bounding boxes of different sizes are scaled to the same size in the single-person estimation stage. However, top-down approaches tend to be more computationally expensive due to the redundancy in bounding box detections.

Multi-person 3D Pose Estimation. Several works [4,20,26,27,31] have been conducted on multi-person 3D pose estimation. Rogez *et al.* [26] propose a LCR-Net which consists of localization, classification, and regression parts and estimates each detected human with a classified and refined anchor pose. Mehta *et al.* [20] propose a bottom-up approach which estimates a specially designed occlusion-robust pose map and readout the 3D poses given 2D poses obtained with Part Affinity Fields [2]. Dabral *et al.* [4] propose to incorporate hourglass network into Mask R-CNN detection heads for better 2D pose localization, followed by a standard residual network to lift 2D poses to 3D. Zanfir *et al.* [31] design a holistic multi-person sensing pipeline, *i.e.* MubyNet, to jointly address the problems of multi-person 2D/3D skeleton/shape-based pose estimation. However, these works only estimate and evaluate the 3D pose after root joint alignment and ignore the global location of each pose. Recently, Moon *et al.* [21] propose a multi-stage pipeline for multi-person camera-space 3D pose estimation. The pipeline follows the top-down scheme and consists of a RootNet which localizes the root joint for each detected bounding box. We also adopt the top-down scheme pipeline and estimate the camera-space root joint location and 3D pose for each detected bounding box. To our best knowledge, [21] and our work are the only two works that focus on the estimation and evaluation of multi-person root joint locations. Compared to [21] which relies on the size of detected bounding box, we utilize the underlying features and design a human-specific pose-based root joint depth estimation framework to significantly boost the root localization performance.

3   Our Approach

3.1   Overview

Given a 2D image with an unknown number of persons, the task of camera-space multi-person 3D pose estimation is to: (1) identify all person instances,

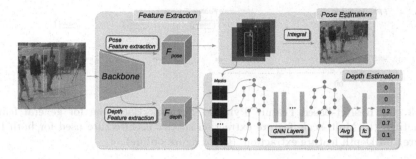

Fig. 2. Our HDNet architecture. The framework takes an image together with the bounding box of a target person as input. A Feature Pyramid Network backbone is used for general feature extraction followed by separated multi-scale feature extraction for the tasks of pose and depth estimation. Estimated heatmaps are used as attention masks to pool depth features. A Graph Neural Network is utilized to propagate and aggregate features for the target person depth estimation.

(2) estimate the 3D pose with respect to the root joint, *i.e.*, pelvis, for each person, and (3) localize each person by estimating the 3D coordinate of root joint in the camera coordinate space.

Following the top-down approaches in the literature of multi-person pose estimation, we assume that the 2D human bounding boxes for each person in the input image are available from a generic object detector. Given the person instances and detected bounding boxes, we propose an end-to-end depth estimation framework to localize the root joint of each person in the camera coordinate space as illustrated in Fig. 2. The root joint localization is decoupled into two sub-tasks: (1) localization of the root joint image coordinate (u, v), and (2) estimation of the root joint depth Z in the camera frame, which is then used to back-project (u, v) to 3D space. We use an off-the-shelf single-person 3D pose estimator to estimate the 3D joint locations of each person with respect to the root joint. The final absolute 3D pose of each person in the camera coordinate system is obtained by the transformation of each joint location with the absolute location of the root joint.

The details of our proposed root joint localization framework are introduced in Sect. 3.2. The choices of specific object detector and single-person 3D pose estimator used in our experiments are given in the implementation details in Sect. 4.2.

3.2 Root Localization Framework

Our framework for monocular image single/multi-person depth estimation is shown in Fig. 2. The framework consists of a Feature Pyramid Network (FPN)-based backbone, a heatmap-based human pose estimation branch, and a Graph Neural Network (GNN)-based depth estimation branch.

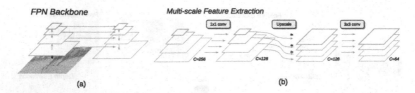

Fig. 3. (a) ResNet-based Feature Pyramid Network Backbone for general feature extraction. (b) Multi-scale feature extraction subnet architecture used for both Pose feature and Depth feature extraction. (Color figure online)

Backbone Network. We choose FPN [15] as our backbone network due to its capability of explicitly handling features of multiple scales in the form of feature pyramids. Hence, it is suitable for perceiving the scale of human body parts and consequently enhances depth estimation of the human pose in an image. The FPN network consists of a ResNet-50 [9] with feature blocks of four different scales C_2, C_3, C_4, C_5 (cyan layers in Fig. 3(a)), where a reversed hierarchy of feature pyramid P_5, P_4, P_3, P_2 is built upon (orange layers in Fig. 3(a)). Each of the four scales encodes hierarchical levels of feature representations, which are then passed through two consecutive convolutional layers as shown in Fig. 3(b). An upsampling operation with corresponding upsample scale factor is applied between the two convolutional layers to ensure matching spatial resolution from the output of the four scales. Batch Normalization [11] and ReLU operations are used after each convolution layer. Weights are not shared across scales. Blocks of all scales are then concatenated to form the final feature block $\mathbf{F}$. Since we find that the downstream tasks of pose estimation and depth estimation are not collaboratively correlated, we split the multi-scale feature processing from the output feature pyramid P_5, P_4, P_3, P_2 of the backbone into two parallel branches without shared weights as shown in Fig. 2. We denote the features as $\mathbf{F}_{\text{pose}}$ and $\mathbf{F}_{\text{depth}}$, respectively.

2D Pose Estimation Branch. We propose to use estimated 2D pose as a guide to aggregate information from useful feature regions to effectively distil information from the image and discard irrelevant areas such as the background. We first regress N_J heatmaps $\hat{\mathbf{H}}$ that correspond to the N_J joints with a 1×1 convolution from feature block $\mathbf{F}_{\text{pose}}$. Each of the N_J heatmaps are normalized across all spatial locations with a softmax operation. A direct read out of the coordinate from the local maximum limits the precision of the joint location estimation due to the low resolution of the output heatmap (4x downsample from input image in ResNet backbone). To circumvent this problem, we follow the idea of "soft-argmax" in [29] and compute the "integral" version of estimated coordinate $(\hat{u}, \hat{v})$ for each joint j using the weighted sum of coordinates:

$$(\hat{u}_j, \hat{v}_j) = \sum_{(u,v)=(0,0)}^{(W-1, H-1)} \hat{\mathbf{H}}_{u,v}^{(j)} \cdot (u, v), \tag{1}$$

where W and H are the width and height of output heatmap. The softmax operation guarantees that the weights $\hat{\mathbf{H}}_{u,v}$ form a valid distribution which sum up to 1 over all spatial locations. To supervise the heatmap regression, we generate a ground truth heatmap $\mathbf{H}^{(j)\mathrm{GT}}$ for each joint j. A Gaussian peak is created around the ground truth joint location (u_j, v_j) with a preset standard deviation that controls the compactness of the Gaussian peak. We use standard Mean Squared Error (MSE) as the heatmap regression loss and $L1$ loss for the pose after soft-argmax as follows:

$$\mathcal{L}_{\mathrm{hm}} = \frac{1}{N_J HW} \sum_{j}^{N_J} \sum_{(u,v)=(0,0)}^{(W-1,H-1)} \left\| \mathbf{H}_{u,v}^{(j)\mathrm{GT}} - \hat{\mathbf{H}}_{u,v}^{(j)} \right\|^2, \tag{2}$$

$$\mathcal{L}_{\mathrm{pose}} = \frac{1}{N_J} \sum_{j}^{N_J} \left(\left| u_j^{\mathrm{GT}} - \hat{u}_j \right| + \left| v_j^{\mathrm{GT}} - \hat{v}_j \right| \right). \tag{3}$$

To deal with multiple persons in the image, we focus on a target person by zeroing out the regions of the heatmap outside the bounding box of that person from the object detector.

Depth Estimation Branch. After we obtain the heatmaps, we use them as attention masks to guide the network into focusing on specific regions of the image related to the target person. More specifically, we only care about features from pixel locations that are close to the joints of the target person. The intuition behind our design choice is that joint locations contain more scale-related information than the larger yet less discriminative areas such as the whole upper body trunk. Attention-guided feature pooling is also adopted in other tasks such as action recognition [17] and hand pose estimation [13]. We compute the weighted sum feature vector $\mathbf{d}$ for each joint j from the feature block $\mathbf{F}_{\mathrm{depth}}$ as:

$$\mathbf{d}^{(j)} = \sum_{(u,v)=(0,0)}^{(W-1,H-1)} \hat{\mathbf{H}}_{u,v}^{(j)} \cdot \mathbf{F}_{\mathrm{depth}_{u,v}}. \tag{4}$$

To effectively aggregate features corresponding to different joint types, we formulate a standard Graph Neural Network (GNN) where each node represents one joint type, e.g., elbow, knee, etc. The aggregated features $\mathbf{d}^{(j)}$ for each joint type j is fed into the corresponding node $X_{in}^{(j)}$ in the graph as input. Each layer of the GNN is defined as:

$$X_{out}^{(i)} = \sigma\left(\tilde{a}_{ii} f_{self}(X_{in}^{(i)}; \Theta_{self}) + \sum_{j \neq i} \tilde{a}_{ij} f_{inter}(X_{in}^{(j)}; \Theta_{inter}) \right). \tag{5}$$

The feature of each input node X_{in} undergoes the linear mappings $f_{self}(.)$ and $f_{inter}(.)$ that are parametrized by Θ_{self} and Θ_{inter}, respectively. The output of the node, i.e. $X_{out}^{(i)}$ is computed from a weighted aggregation of $f_{self}(.)$ and $f_{inter}(.)$ of all other nodes. The weighting factor $\tilde{a}_{ij}$ is an element of the normalized adjacency matrix $\tilde{A} \in \mathbb{R}^{N_j \times N_j}$ that controls the extent of influence of the nodes on each other. The original adjacency matrix is $A \in \{0, 1\}^{N_j \times N_j}$; an element a_{ij} equals 1 if there is a skeletal link between joint i and j, e.g. left knee to left ankle, or otherwise 0. $\tilde{A} \in \mathbb{R}^{N_j \times N_j}$ is obtained by applying $L1$-normalization on each row of A. The non-linearity function $\sigma(.)$ is implemented with a Batch Normalization followed by a ReLU. We stack L GNN layers in total. After the last GNN layer, we merge the feature output from each node with an average pooling operation.

Target Output Formulation. Inspired by the work [7] for scene depth estimation, we formulate the depth estimation as a classification problem instead of directly regressing the numerical value of depth. We follow the practice in [7] to discretize the log-depth space into a preset number of bins, $N_\mathbf{B}$. We compute:

$$b(d) = \frac{\log d - \log \alpha}{\log \beta - \log \alpha} \cdot (N_\mathbf{B} - 1), \tag{6}$$

where $\lfloor b \rfloor$ gives the bin index of the depth, and the depth d of a pose is assumed to be within the range $[\alpha, \beta]$. Here $\lfloor . \rfloor$ is the round-off to the nearest integer operator. To eliminate quantization errors, we assign non-zero values to two consecutive bins i and $i + 1$, where $i \leq b < i + 1$. This operation is similar to the weights in bi-linear interpolation. For example, the ground truth values of the bins are given by $\mathbf{B} = [0, 0, 0.6, 0.4, 0]$ for $N_\mathbf{B} = 5$ and $b = 2.4$. Consequently, $\mathbf{B}$ is a 1D heatmap that can achieve any level of precision with a sufficiently accurate categorical estimation on the bins.

Since a different focal length of the camera affects the scale of a target person in the image, it is unrealistic to estimate the absolute depth from images taken by any arbitrary camera. To alleviate this problem, we normalize out the camera intrinsic parameters by replacing the target d with $\hat{d} = d/f$, where f is the focal length of camera. We approximate with $\hat{d} = d/\sqrt{f_x \cdot f_y}$ in our experiments since the focal lengths in x and y directions are usually very close. Finally, we add a fully connected layer after the pooled feature from the last GNN layer to regress the $N_\mathbf{B}$ values of the bins $\hat{\mathbf{B}}$. Softmax operation is used to normalize the output into a valid distribution. We transform $\hat{\mathbf{B}}$ back to the estimated depth d of the root joint by:

$$d = \exp\left[\frac{\hat{b}}{N_\mathbf{B} - 1} \cdot (\log \beta - \log \alpha) + \log \alpha\right] \cdot \sqrt{f_x \cdot f_y}, \text{ where } \hat{b} = \sum_{i=0}^{N_\mathbf{B}-1} \hat{\mathbf{B}}_i \cdot i. \tag{7}$$

Similar to the soft-argmax operation used to transform heatmaps to joint locations, $\hat{b}$ is the weighted sum of the bin indices with the estimated heatmap $\hat{\mathbf{B}}$. To supervise the learning of the depth estimation branch, we adopt cross-entropy loss on the estimated bins $\hat{\mathbf{B}}$ and $L1$ loss on $\hat{b}$ as follows:

$$\mathcal{L}_{\text{bins}} = - \sum_{i=0}^{N_B-1} \mathbf{B}_i^{\text{GT}} \cdot \log \hat{\mathbf{B}}_i, \text{ and } \mathcal{L}_{\text{idx}} = \left| b^{\text{GT}} - \hat{b} \right|. \tag{8}$$

We train the whole framework with losses from the pose estimation and depth estimation branches:

$$\mathcal{L} = \lambda_{\text{hm}}\mathcal{L}_{\text{hm}} + \lambda_{\text{pose}}\mathcal{L}_{\text{pose}} + \lambda_{\text{bins}}\mathcal{L}_{\text{bins}} + \lambda_{\text{idx}}\mathcal{L}_{\text{idx}}. \tag{9}$$

4 Experiments

4.1 Datasets and Evaluation Metrics

Human3.6M Dataset. Human3.6M dataset [12] is currently the largest publicly available dataset for human 3D pose estimation. The dataset consists of 3.6 million video frames captured by MoCap system in a constrained indoor studio environment. 11 actors performing 15 activities are captured from 4 camera viewpoints. 3D ground truth poses in world coordinate system and camera extrinsic (rotation and translation with respect to world coordinate) and intrinsic parameters (focal length and principal point) are available. We follow previous works that five subjects (S1, S5, S6, S7, S8) are used in training and two subjects (S9 and S11) are used for evaluation. We use every 5th and 64th frames in each video for training and evaluation respectively. No extra 2D pose dataset is used to augment the training. We follow the metric Mean Root Position Error (MRPE) proposed in [21] to evaluate the root localization accuracy. Specifically, we consider the Euclidean distance between the estimated and the ground truth 3D coordinate of the root joint.

MuCo-3DHP and MuPoTS-3D Datasets. MuCo-3DHP and MuPoTS-3D are two datasets proposed by Mehta *et al.* [20] to evaluate multi-person 3D pose estimation performance. The training set MuCo-3DHP is a composite dataset which merges randomly sampled 3D poses from single-person 3D human pose dataset MPI-INF-3DHP [19] to form realistic multi-person scenes. The test set MuPoTS-3D is a markerless motion captured multi-person dataset including both indoor and outdoor scenes. We use the same set of MuCo-3DHP synthesized images from [21] for a fair comparison. No extra 2D pose dataset is used to augment the training. For evaluation of multi-person root joint localization, we follow [21] to report the average precision and recall of 3D root joint location under different thresholds. A root joint with a smaller distance to the matched ground truth root joint location than a threshold is considered a true positive estimation. We follow [21] to report 3DPCK$_{abs}$ for evaluation of the root-aware 3D pose estimation, where 3DPCK (3D percentage of correct keypoints) for the estimated poses is evaluated without root alignment. 3DPCK treats an estimated joint as correct if it is 15 cm distance from the matched ground truth joint. Although our framework does not focus on root-relative 3D pose estimation, we also report the root-aligned 3DPCK$_{rel}$ to show that accurate root localization also benefits the precision of 3D pose estimation.

Table 1. MRPE results comparison with state-of-the-arts on the Human3.6M dataset. $MRPE_x$, $MRPE_y$, and $MRPE_z$ are the average errors in x, y, and z axes, respectively.

Method	MRPE	$MRPE_x$	$MRPE_y$	$MRPE_z$
Baseline	267.8	27.5	28.3	261.9
Baseline w/o limb joints	226.2	24.5	24.9	220.2
Baseline with RANSAC	213.1	24.3	24.3	207.1
RootNet [21]	120.0	23.3	23.0	108.1
Ours	**77.6**	**15.6**	**13.6**	**69.9**

4.2   Implementation Details

Following previous work [21], we use Mask R-CNN [8] as our person detector due to its high performance and generalizability to in-the-wild images. For single-person 3D pose estimation, we use the volumetric-based 3D pose estimator by [29]. Instead of cropping out areas of interest using bounding boxes, we keep the original scale of image and crop out a fixed size patch centered around the bounding box, or the principal point if no bounding box is provided in the single-person scenario. The cropped out image is then rescaled to 256×256 and used as input to our network. The output resolution of the heatmap is 64×64. We use 2 layers of GNN operations in the depth estimation branch. We set the standard deviation of the Gaussian peak in the ground truth heatmap to be 0.75, and the bin range of d/f to $[\alpha = 1.0, \beta = 8.0]$ for a reasonably sufficient range of the depth. We do not see much performance change when different number of bins N_B are used. All results of the experiments shown in the paper are obtained with $N_B = 71$. We set λ in Eq. 9 to balance the four loss terms to same order of magnitudes. For training, we use Adam optimizer [14] with learning rate $1e-4$ and batch size 16. We train the model for 200k steps and decay the learning rate with a factor of 0.8 at every 20k steps. The evaluation of each image takes around 7ms with our root joint localization HDNet.

4.3   Results on Human3.6M

The root joint localization results on Human3.6M dataset are shown in Table 1. The baselines reported in the top 3 rows follow a two-stage approach, where 2D pose [29] and 3D pose [18] are estimated separately, and an optimization process is adopted to obtain the global root joint location that minimizes the reprojection error. "w/o limb joints" refers to optimization using only head and body trunk joints. "with RANSAC" refers to randomly sampling the set of joints used for optimization with RANSAC. The baseline results are taken from the figures reported in [21]. We also compare with the state-of-the-art approach [21]. It can be seen from Table 1 that optimization-based methods can achieve reasonable results, but with limited accuracy due to the errors from both the 2D and 3D estimation stages. Our root joint localization framework 69.9 mm

Table 2. Root joint localization accuracy comparison in average precision and recall with state-of-the-arts on MuPoTS-3D dataset.

Method	AP_{25}^{root}	AP_{20}^{root}	AP_{15}^{root}	AP_{10}^{root}	AR_{25}^{root}	AR_{20}^{root}	AR_{15}^{root}	AR_{10}^{root}
RootNet [21]	31.0	21.5	10.2	2.3	55.2	45.3	31.4	15.2
Ours	**39.4**	**28.0**	**14.6**	**4.1**	**59.8**	**50.0**	**35.9**	**19.1**

Table 3. Sequence-wise $3DPCK_{abs}$ comparison with state-of-the-arts on MuPoTS-3D dataset. Accuracy is measured on matched ground-truths.

Method	S1	S2	S3	S4	S5	S6	S7	S8	S9	S10	−
RootNet [21]	**59.5**	**45.3**	51.4	**46.2**	**53.0**	**27.4**	23.7	26.4	39.1	23.6	−
Ours	21.4	22.7	**58.3**	27.5	37.3	12.2	**49.2**	**40.8**	**53.1**	**43.9**	−
Method	S11	S12	S13	S14	S15	S16	S17	S18	S19	S20	Avg
RootNet [21]	18.3	14.9	38.2	**29.5**	36.8	23.6	14.4	20.0	18.8	25.4	31.8
Ours	**43.2**	**43.6**	**39.7**	28.3	**49.5**	**23.8**	**18.0**	**26.9**	25.0	**38.8**	**35.2**

depth estimation error in $MRPE_z$ with a 35% improvement over [21]. Since our approach uses the original scale image without scaling to person bounding box size which limits the 2D (u, v) localization precision, we also adopt a state-of-the-art 2D pose estimator CPN [3] within the person bounding box area to further refine the (u, v) localization. Our MRPE for root joint achieves an overall performance 77.6 mm which significantly outperforms the state-of-the-art.

4.4 Results on MuPoTS-3D

Root Joint Localization. To evaluate our root joint localization performance on the multi-person MuPoTS-3D dataset, we estimate the root joint 3D coordinate for each bounding box detected from the object detector. All root joint candidates are matched with the ground truth root joints, and only candidates with distance to the matched ground truth lesser than a threshold are considered as an accurate estimate. We then analyze the average precision and recall over the whole dataset under various settings of thresholds ranging 25 cm to 10 cm. The results are shown in Table 2. Our method achieves much higher AP and AR consistently across all levels of thresholds compared to the state-of-the-art approach [21].

Camera-Space Absolute 3D Pose Estimation. We also evaluate the camera-space absolute 3D pose estimation performance with $3DPCK_{abs}$. $3DPCK_{abs}$ compares the estimated 3D pose with the matched ground truth pose in the camera coordinate space without root alignment, thus requires highly accurate root joint localization. We use the same 3D pose estimator [29] as the state-of-the-art root joint localization method [21] for a fair comparison. Results

Table 4. Joint-wise 3DPCK$_{abs}$ comparison with state-of-the-arts on MuPoTS-3D dataset. Accuracy is measured on matched ground-truths.

Method	Head	Neck	Shoulder	Elbow	Wrist	Hip	Knee	Ankle	Avg
RootNet [21]	37.6	35.6	34.0	34.1	30.7	30.6	31.3	25.3	31.8
Ours	**38.3**	**37.8**	**36.2**	**37.4**	**34.0**	**34.9**	**36.4**	**29.2**	**35.2**

Table 5. Sequence-wise 3DPCK$_{rel}$ comparison with state-of-the-arts on MuPoTS-3D dataset. Accuracy is measured on matched ground-truths.

Method	S1	S2	S3	S4	S5	S6	S7	S8	S9	S10	–
Rogez et al. [26]	69.1	67.3	54.6	61.7	74.5	25.2	48.4	63.3	69.0	78.1	–
Mehta et al. [20]	81.0	65.3	64.6	63.9	75.0	30.3	65.1	61.1	64.1	83.9	–
Rogez et al. [27]	88.0	73.3	67.9	74.6	81.8	50.1	60.6	60.8	78.2	89.5	–
RootNet [21]	**94.4**	78.6	79.0	82.1	86.6	72.8	**81.9**	75.8	**90.2**	90.4	–
Ours	**94.4**	**79.6**	**79.2**	**82.4**	**86.7**	**73.0**	81.6	**76.3**	90.1	**90.5**	–

Method	S11	S12	S13	S14	S15	S16	S17	S18	S19	S20	Avg
Rogez et al. [26]	53.8	52.2	60.5	60.9	59.1	70.5	76.0	70.0	77.1	81.4	62.4
Mehta et al. [20]	72.4	69.9	71.0	72.9	71.3	83.6	79.6	73.5	78.9	**90.9**	70.8
Rogez et al. [27]	70.8	74.4	72.8	64.5	74.2	84.9	85.2	78.4	75.8	74.4	74.0
RootNet [21]	**79.4**	**79.9**	75.3	81.0	**81.1**	90.7	**89.6**	83.1	81.7	77.3	82.5
Ours	77.9	79.2	**78.3**	**85.5**	**81.1**	**91.0**	88.5	**85.1**	**83.4**	90.5	**83.7**

in Table 3 show that our method consistently outperforms the state-of-the-art in most of the test sequences and achieves a 35.2% average 3DPCK (3.4% improvement). The performance breakdown of all joint types is shown in Table 4.

Root-Relative 3D Pose Estimation. The state-of-the-art root-relative 3D pose estimator [29] adopts a volumetric output representation and estimates the root-relative depth for each joint. Absolute root joint depth has to be available to recover the 3D pose through back-projection. We follow [21] and use our estimated root depth to back-project the 3D pose and evaluate the root-relative 3D pose estimation accuracy with 3DPCK$_{rel}$ after root joint alignment. Results are shown in Table 5, where our method outperforms the previous best performance by 1.2% average 3DPCK. This demonstrates that more accurate root localization also benefits the precise 3D pose estimation in volumetric-based approaches [21,25,29].

4.5   Ablation Studies

We conduct ablation studies to show how each component in our framework affects the root joint localization accuracy. We evaluate the depth estimation

Table 6. Ablation studies on components of the framework. Depth error MRPE_z (mm) on Human3.6M dataset and $\text{AP}_{25}^{\text{root}}$ (%) on MuPoTS-3D dataset are measured.

Method	$\text{MRPE}_z(\downarrow)$	$\text{AP}_{25}^{\text{root}}(\uparrow)$
RootNet [21]	108.1	31.0
Ours direct regression	94.5	27.3
Ours shared feature branch	72.0	31.9
Ours w/o GNN	72.9	32.7
Ours w/o HM pooling	71.8	26.0
Ours (full)	**69.9**	**39.4**

accuracy MRPE_z on Human3.6M dataset and the root joint localization $\text{AP}_{25}^{\text{root}}$ on MuPoTS-3D dataset with different variants of our framework in Table 6. The state-of-the-art approach [21] is also included for comparison.

- "Ours direct regression": Performance drop (24.6 mm and 12.1%) with directly regressing target depth instead of performing classification over binning shows the effectiveness of formulating the depth estimation as a classification task.
- "Ours shared feature branch": One single multi-scale feature branch is kept after FPN, which means $\mathbf{F}_{\text{pose}}$ and $\mathbf{F}_{\text{depth}}$ use the same feature representation. This setting causes performance to drop (2.1 mm and 7.5%), and thus demonstrates that the features used for pose estimation and depth estimation are not highly correlated.
- "Ours w/o GNN": We replace the GNN layers in our depth estimation branch with same number of fully-connected layers and observe a performance drop (3 mm and 6.7%), showing the effectiveness of the graph neural network in propagating and refining the features extracted for different types of joints.
- "Ours w/o HM pooling": We remove feature pooling with estimated heatmaps as mask in the depth estimation branch and instead apply a global average pooling to obtain a single feature vector. The GNN layers are replaced with fully-connected layers since we do not explicitly differentiate between different joint types. We observe a performance drop (1.9 mm and 13.4%), which demonstrates the effectiveness of utilizing estimated pose as attention mask for useful feature aggregation.

4.6 Discussions

We analyze the root joint localization results on the challenging multi-person dataset MuPoTS-3D and observe several sources of large errors as shown in Fig. 4: (1) Bounding boxes for two persons tend to have overlapping areas when the person closer to the camera partially occludes the other person farther away (Fig. 4(a)). Masking the heatmaps with bounding box cannot effectively remove undesired regions of information and consequently the depth estimation for both

Fig. 4. Typical errors in multi-person root localization. (a) Close and overlapping bounding box regions. (b) Different sizes of target persons.

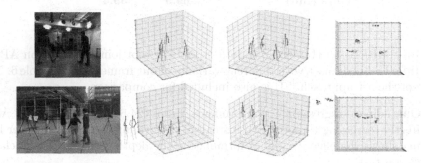

Fig. 5. Qualitative results on MuPoTS-3D dataset. Columns are: (1) image with bounding boxes (2) left-front view (3) right-front view (4) top-down view

persons are affected. The problem of fine-grained target person segmentation will be of interest for future research. (2) Since monocular depth estimation relies on prior knowledge such as typical scale of human bodies, estimation tends to be erroneous when the size of target person is far away from the "average" size, *e.g.*, the target is a child or a relatively short person (Fig. 4(b)). Research on person 3D size estimation may complement our depth estimation task and improve the generalizability to persons of different sizes.

5   Conclusions

In this work, we proposed the Human Depth Estimation Network (HDNet), an end-to-end framework to address the problem of accurate root joint localization for multi-person 3D absolute pose estimation. Our HDNet utilizes deep features and demonstrates the capability to precisely estimate depth of root joints. We designed a human-specific pose-based feature aggregation process in the HDNet to effectively pool features from regions of human body joints. Experimental results on multiple datasets showed that our framework significantly outperforms the state-of-the-art in both root joint localization and 3D pose estimation.

References

1. Andriluka, M., Pishchulin, L., Gehler, P., Schiele, B.: 2D human pose estimation: new benchmark and state of the art analysis. In: CVPR (2014)
2. Cao, Z., Simon, T., Wei, S.E., Sheikh, Y.: Realtime multi-person 2D pose estimation using part affinity fields. In: CVPR (2017)
3. Chen, Y., Wang, Z., Peng, Y., Zhang, Z., Yu, G., Sun, J.: Cascaded pyramid network for multi-person pose estimation. In: CVPR (2018)
4. Dabral, R., Gundavarapu, N.B., Mitra, R., Sharma, A., Ramakrishnan, G., Jain, A.: Multi-person 3D human pose estimation from monocular images. In: 3DV, pp. 405–414. IEEE (2019)
5. Fang, H.S., Xie, S., Tai, Y.W., Lu, C.: RMPE: regional multi-person pose estimation. In: ICCV (2017)
6. Fang, H., Xu, Y., Wang, W., Liu, X., Zhu, S.C.: Learning pose grammar to encode human body configuration for 3D pose estimation. In: AAAI (2018)
7. Fu, H., Gong, M., Wang, C., Batmanghelich, K., Tao, D.: Deep ordinal regression network for monocular depth estimation. In: CVPR, pp. 2002–2011 (2018)
8. He, K., Gkioxari, G., Dollár, P., Girshick, R.: Mask R-CNN. In: ICCV (2017)
9. He, K., Zhang, X., Ren, S., Sun, J.: Deep residual learning for image recognition. In: CVPR, pp. 770–778 (2016)
10. Huang, S., Gong, M., Tao, D.: A coarse-fine network for keypoint localization. In: ICCV (2017)
11. Ioffe, S., Szegedy, C.: Batch normalization: accelerating deep network training by reducing internal covariate shift. arXiv preprint arXiv:1502.03167 (2015)
12. Ionescu, C., Papava, D., Olaru, V., Sminchisescu, C.: Human3.6M: large scale datasets and predictive methods for 3D human sensing in natural environments. TMAPI **36**(7), 1325–1339 (2014)
13. Iqbal, U., Molchanov, P., Breuel, T., Gall, J., Kautz, J.: Hand pose estimation via latent 2.5D heatmap regression. In: Ferrari, V., Hebert, M., Sminchisescu, C., Weiss, Y. (eds.) ECCV 2018. LNCS, vol. 11215, pp. 125–143. Springer, Cham (2018). https://doi.org/10.1007/978-3-030-01252-6_8
14. Kingma, D.P., Ba, J.: Adam: a method for stochastic optimization. arXiv preprint arXiv:1412.6980 (2014)
15. Lin, T.Y., Dollár, P., Girshick, R., He, K., Hariharan, B., Belongie, S.: Feature pyramid networks for object detection. In: CVPR, pp. 2117–2125 (2017)
16. Lin, T.-Y., et al.: Microsoft COCO: common objects in context. In: Fleet, D., Pajdla, T., Schiele, B., Tuytelaars, T. (eds.) ECCV 2014. LNCS, vol. 8693, pp. 740–755. Springer, Cham (2014). https://doi.org/10.1007/978-3-319-10602-1_48
17. Luvizon, D.C., Picard, D., Tabia, H.: 2D/3D pose estimation and action recognition using multitask deep learning. In: CVPR, pp. 5137–5146 (2018)
18. Martinez, J., Hossain, R., Romero, J., Little, J.J.: A simple yet effective baseline for 3D human pose estimation. In: ICCV, vol. 1, p. 5. IEEE (2017)
19. Mehta, D., et al.: Monocular 3D human pose estimation in the wild using improved CNN supervision. In: 3DV. IEEE (2017)
20. Mehta, D., et al.: Single-shot multi-person 3D pose estimation from monocular RGB. In: 3DV. IEEE (2018)
21. Moon, G., Chang, J., Lee, K.M.: Camera distance-aware top-down approach for 3D multi-person pose estimation from a single RGB image. In: ICCV (2019)
22. Newell, A., Huang, Z., Deng, J.: Associative embedding: end-to-end learning for joint detection and grouping. In: NeurIPS (2017)

23. Newell, A., Yang, K., Deng, J.: Stacked hourglass networks for human pose estimation. In: Leibe, B., Matas, J., Sebe, N., Welling, M. (eds.) ECCV 2016. LNCS, vol. 9912, pp. 483–499. Springer, Cham (2016). https://doi.org/10.1007/978-3-319-46484-8_29

24. Papandreou, G., Zhu, T., Chen, L.-C., Gidaris, S., Tompson, J., Murphy, K.: PersonLab: person pose estimation and instance segmentation with a bottom-up, part-based, geometric embedding model. In: Ferrari, V., Hebert, M., Sminchisescu, C., Weiss, Y. (eds.) Computer Vision – ECCV 2018. LNCS, vol. 11218, pp. 282–299. Springer, Cham (2018). https://doi.org/10.1007/978-3-030-01264-9_17

25. Pavlakos, G., Zhou, X., Derpanis, K.G., Daniilidis, K.: Coarse-to-fine volumetric prediction for single-image 3D human pose. In: CVPR, pp. 1263–1272. IEEE (2017)

26. Rogez, G., Weinzaepfel, P., Schmid, C.: LCR-Net: localization-classification-regression for human pose. In: CVPR, pp. 3433–3441 (2017)

27. Rogez, G., Weinzaepfel, P., Schmid, C.: LCR-Net++: multi-person 2D and 3D pose detection in natural images. TPAMI **42**, 1146–1161 (2019)

28. Sun, K., Xiao, B., Liu, D., Wang, J.: Deep high-resolution representation learning for human pose estimation. In: CVPR (2019)

29. Sun, X., Xiao, B., Wei, F., Liang, S., Wei, Y.: Integral human pose regression. In: Ferrari, V., Hebert, M., Sminchisescu, C., Weiss, Y. (eds.) ECCV 2018. LNCS, vol. 11210, pp. 536–553. Springer, Cham (2018). https://doi.org/10.1007/978-3-030-01231-1_33

30. Wei, S.E., Ramakrishna, V., Kanade, T., Sheikh, Y.: Convolutional pose machines. In: CVPR (2016)

31. Zanfir, A., Marinoiu, E., Zanfir, M., Popa, A.I., Sminchisescu, C.: Deep network for the integrated 3D sensing of multiple people in natural images. In: NeurIPS, pp. 8410–8419 (2018)

SOLO: Segmenting Objects by Locations

Xinlong Wang[1], Tao Kong[2], Chunhua Shen[1(✉)], Yuning Jiang[2], and Lei Li[2]

[1] The University of Adelaide, Adelaide, Australia
chunhua.shen@adelaide.edu.au
[2] ByteDance AI Lab, Beijing, China

Abstract. We present a new, embarrassingly simple approach to instance segmentation. Compared to many other dense prediction tasks, *e.g.*, semantic segmentation, it is the arbitrary number of instances that have made instance segmentation much more challenging. In order to predict a mask for each instance, mainstream approaches either follow the "detect-then-segment" strategy (*e.g.*, Mask R-CNN), or predict embedding vectors first then use clustering techniques to group pixels into individual instances. We view the task of instance segmentation from a completely new perspective by introducing the notion of "instance categories", which assigns categories to each pixel within an instance according to the instance's location and size, thus nicely converting instance segmentation into a single-shot classification-solvable problem. We demonstrate a much simpler and flexible instance segmentation framework with strong performance, achieving *on par* accuracy with Mask R-CNN and outperforming recent single-shot instance segmenters in accuracy. We hope that this simple and strong framework can serve as a baseline for many instance-level recognition tasks besides instance segmentation. Code is available at https://git.io/AdelaiDet.

Keywords: Instance segmentation · Location category

1 Introduction

Instance segmentation is challenging because it requires the correct separation of all objects in an image while also semantically segmenting each instance at the pixel level. Objects in an image belong to a fixed set of semantic categories, but the number of instances varies. As a result, semantic segmentation can be easily formulated as a dense per-pixel classification problem, while it is challenging to predict instance labels directly following the same paradigm.

To overcome this obstacle, recent instance segmentation methods can be categorized into two groups, *i.e.*, top-down and bottom-up paradigms. The former approach, namely 'detect-then-segment', first detects bounding boxes and then

Electronic supplementary material The online version of this chapter (https://doi.org/10.1007/978-3-030-58523-5_38) contains supplementary material, which is available to authorized users.

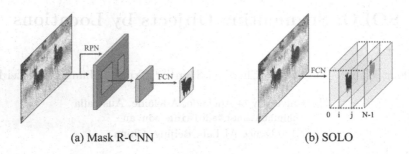

(a) Mask R-CNN   (b) SOLO

Fig. 1. Comparison of the pipelines of Mask R-CNN and the proposed SOLO.

segments the instance mask in each bounding box. The latter approach learns an affinity relation, assigning an embedding vector to each pixel, by pushing away pixels belonging to different instances and pulling close pixels in the same instance. A grouping post-processing is then needed to separate instances. Both these two paradigms are step-wise and *indirect*, which either heavily rely on accurate bounding box detection or depend on per-pixel embedding learning and the grouping processing (Fig. 1).

In contrast, we aim to directly segment instance masks, under the supervision of full instance mask annotations instead of masks in boxes or additional pixel pairwise relations. We start by rethinking a question: *What are the fundamental differences between object instances in an image?* Take the challenging MS COCO dataset [16] for example. There are in total 36,780 objects in the validation subset, 98.3% of object pairs have center distance greater than 30 pixels. As for the rest 1.7% of object pairs, 40.5% of them have size ratio greater than 1.5×. To conclude, in most cases two instances in an image either have different center locations or have different object sizes. This observation makes one wonder whether we could directly distinguish instances by the center locations and object sizes?

In the closely related field, semantic segmentation, now the dominate paradigm leverages a fully convolutional network (FCN) to output dense predictions with N channels. Each output channel is responsible for one of the semantic categories (including background). Semantic segmentation aims to distinguish different semantic categories. Analogously, in this work, we propose to distinguish object instances in the image by introducing the notion of *"instance categories"*, *i.e.*, the quantized center locations and object sizes, which enables to segment objects by locations, thus the name of our method, **SOLO.**

Locations. An image can be divided into a grid of $S \times S$ cells, thus leading to S^2 center location classes. According to the coordinates of the object center, an object instance is assigned to one of the grid cells, as its center location category. Note that grids are used conceptually to assign location category for each pixel. Each output channel is responsible for one of the center location categories, and the corresponding channel map should predict the instance mask of the object belonging to that location. Thus, structural geometric information is naturally preserved in the spatial matrix with dimensions of height by width. Unlike Deep-Mask [24] and TensorMask [4], which run in a dense sliding-window manner and segment an object in a fixed local patch, our method naturally outputs accurate masks for all scales of instances without the limitation of (anchor) box locations and scales.

In essence, an instance location category approximates the location of the object center of an instance. Thus, by classification of each pixel into its instance location category, it is equivalent to predict the object center of each pixel in the latent space. The importance here of converting the location prediction task into classification is that, with classification it is much more straightforward and easier to model varying number of instances using a fixed number of channels, at the same time not relying on post-processing like grouping or learning embeddings.

Sizes. To distinguish instances with different object sizes, we employ the feature pyramid network (FPN) [14], so as to assign objects of different sizes to different levels of feature maps. Thus, all the object instances are separated regularly, enabling to classify objects by "instance categories". Note that FPN was designed for the purposes of detecting objects of different sizes in an image.

In the sequel, we empirically show that FPN is one of the core components for our method and has a profound impact on the segmentation performance, especially objects of varying sizes being presented.

With the proposed SOLO framework, we are able to optimize the network in an end-to-end fashion for the instance segmentation task using mask annotations solely, and perform pixel-level instance segmentation out of the restrictions of local box detection and pixel grouping. For the first time, we demonstrate a very simple instance segmentation approach achieving *on par* results to the dominant "detect-then-segment" method on the challenging COCO dataset [16] with diverse scenes and semantic classes. Additionally, we showcase the generality of our framework via the task of instance contour detection, by viewing the instance edge contours as a one-hot binary mask, with almost no modification SOLO can generate reasonable instance contours. The proposed SOLO only needs to solve two pixel-level classification tasks, thus it may be possible to borrow some of the recent advances in semantic segmentation for improving SOLO. *The embarrassing simplicity and strong performance of the proposed SOLO method may predict its application to a wide range of instance-level recognition tasks.*

1.1 Related Work

We review some instance segmentation works that are closest to ours.

Top-down Instance Segmentation. The methods that segment object instance in a priori bounding box fall into the typical top-down paradigm. FCIS [13] assembles the position-sensitive score maps within the region-of-interests (ROIs) generated by a region proposal network (RPN) to predict instance masks. Mask R-CNN [9] extends the Faster R-CNN detector [25] by adding a branch for segmenting the object instances within the detected bounding boxes. Based on Mask R-CNN, PANet [19] further enhances the feature representation to improve the accuracy, Mask Scoring R-CNN [10] adds a mask-IoU branch to predict the quality of the predicted mask and scoring the masks to improve the performance. HTC [2] interweaves box and mask branches for a joint multi-stage processing. TensorMask [4] adopts the dense sliding window paradigm to segment the instance in the local window for each pixel with a predefined number of windows and scales. In contrast to the top-down methods above, our SOLO is totally box-free thus not being restricted by (anchor) box locations and scales, and naturally benefits from the inherent advantages of FCNs.

Bottom-up Instance Segmentation. This category of the approaches generate instance masks by grouping the pixels into an arbitrary number of object instances presented in an image. In [22], pixels are grouped into instances using the learned associative embedding. A discriminative loss function [7] learns pixel-level instance embedding efficiently, by pushing away pixels belonging to different instances and pulling close pixels in the same instance. SGN [18] decomposes the instance segmentation problem into a sequence of sub-grouping problems. SSAP [8] learns a pixel-pair affinity pyramid, the probability that two pixels belong to the same instance, and sequentially generates instances by a cascaded graph partition. Typically bottom-up methods lag behind in accuracy compared to top-down methods, especially on the dataset with diverse scenes. Instead of exploiting pixel pairwise relations SOLO directly learns with the instance mask annotations solely during training, and predicts instance masks end-to-end without grouping post-processing.

Direct Instance Segmentation. To our knowledge, no prior methods directly train with mask annotations solely, and predict instance masks and semantic categories in one shot without the need of grouping post-processing. Several recently proposed methods may be viewed as the 'semi-direct' paradigm. AdaptIS [26] first predicts point proposals, and then sequentially generates the mask for the object located at the detected point proposal. PolarMask [28] proposes to use the polar representation to encode masks and transforms per-pixel mask prediction to distance regression. They both do not need bounding boxes for training but are either being step-wise or founded on compromise, *e.g.*, coarse parametric representation of masks. Our SOLO takes an image as input, directly outputs instance masks and corresponding class probabilities, in a fully convolutional, box-free and grouping-free paradigm.

2   Our Method: SOLO

2.1   Problem Formulation

The central idea of SOLO framework is to reformulate the instance segmentation as two simultaneous category-aware prediction problems. Concretely, our system divides the input image into a uniform grids, *i.e.*, $S \times S$. If the center of an object falls into a grid cell, that grid cell is responsible for 1) predicting the semantic category as well as 2) segmenting that object instance.

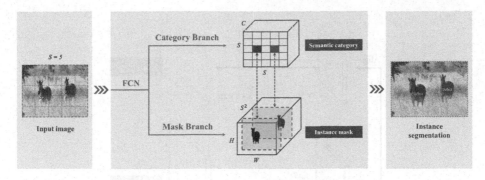

Fig. 2. SOLO framework. We reformulate the instance segmentation as two sub-tasks: category prediction and instance mask generation problems. An input image is divided into a uniform grids, *i.e.*, $S \times S$. Here we illustrate the grid with $S = 5$. If the center of an object falls into a grid cell, that grid cell is responsible for predicting the semantic category (top) and masks of instances (bottom). We do not show the feature pyramid network (FPN) here for simpler illustration.

Semantic Category. For each grid, our SOLO predicts the C-dimensional output to indicate the semantic class probabilities, where C is the number of classes. These probabilities are conditioned on the grid cell. If we divide the input image into $S \times S$ grids, the output space will be $S \times S \times C$, as shown in Fig. 2 (top). This design is based on the assumption that each cell of the $S \times S$ grid must belong to one individual instance, thus only belonging to one semantic category. During inference, the C-dimensional output indicates the class probability for each object instance.

Instance Mask. In parallel with the semantic category prediction, each positive grid cell will also generate the corresponding instance mask. For an input image I, if we divide it into $S \times S$ grids, there will be at most S^2 predicted masks in total. We explicitly encode these masks at the third dimension (channel) of a 3D output tensor. Specifically, the instance mask output will have $H_I \times W_I \times S^2$ dimension. The kth channel will be responsible to segment instance at grid (i, j), where $k = i \cdot S + j$ (with i and j zero-based)[1]. To this end, a one-to-one correspondence is established between the semantic category and class-agnostic mask (Fig. 2).

A direct approach to predict the instance mask is to adopt the fully convolutional networks, like FCNs in semantic segmentation [20]. However the conventional convolutional operations are *spatially invariant* to some degree. Spatial invariance is desirable for some tasks such as image classification as it introduces robustness. However, here we need a model that is *spatially variant*, or in more precise words, position sensitive, since our segmentation masks are conditioned on the grid cells and must be separated by different feature channels.

---

[1] We also show an equivalent and more efficient implementation in Sect. 4.

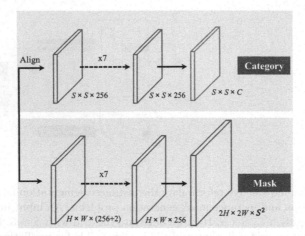

Fig. 3. SOLO Head architecture. At each FPN feature level, we attach two sibling sub-networks, one for instance category prediction (top) and one for instance mask segmentation (bottom). In the mask branch, we concatenate the x, y coordinates and the original features to encode spatial information. Here numbers denote spatial resolution and channels. In the figure, we assume 256 channels as an example. Arrows denote either convolution or interpolation. 'Align' means bilinear interpolation. During inference, the mask branch outputs are further upsampled to the original image size.

Our solution is very simple: at the beginning of the network, we directly feed normalized pixel coordinates to the networks, inspired by 'CoordConv' operator [17]. Specifically, we create a tensor of same spatial size as input that contains pixel coordinates, which are normalized to $[-1, 1]$. This tensor is then concatenated to the input features and passed to the following layers. By simply giving the convolution access to its own input coordinates, we add the spatial functionality to the conventional FCN model. It should be noted that CoordConv is not the only choice. For example the semi-convolutional operators [23] may be competent, but we employ CoordConv for its simplicity and being easy to implement. If the original feature tensor is of size $H \times W \times D$, the size of new tensor becomes $H \times W \times (D + 2)$, in which the last two channels are x-y pixel coordinates. For more information on CoordConv, we refer readers to [17].

Forming Instance Segmentation. In SOLO, the category prediction and the corresponding mask are naturally associated by their reference grid cell, *i.e.*, $k = i \cdot S + j$. Based on this, we can directly form the final instance segmentation result for each grid. The raw instance segmentation results are generated by gathering all grid results. Finally, non-maximum-suppression (NMS) is used to obtain the final instance segmentation results. No other post processing operations are needed.

2.2 Network Architecture

SOLO attaches to a convolutional backbone. We use FPN [14], which generates a pyramid of feature maps with different sizes with a fixed number of channels (usually 256-d) for each level. These maps are used as input for each prediction head: semantic category and instance mask. Weights for the head are shared across different levels. Grid number may varies at different pyramids. Only the last conv is not shared in this scenario.

To demonstrate the generality and effectiveness of our approach, we instantiate SOLO with multiple architectures. The differences include: (a) the *backbone* architecture used for feature extraction, (b) the network *head* for computing the instance segmentation results, and (c) training *loss function* used to optimize the model. Most of the experiments are based on the *head* architecture as shown in Figure 3. We also utilize different variants to further study the generality. We note that our instance segmentation heads have a straightforward structure. More complex designs have the potential to improve performance but are not the focus of this work.

2.3 SOLO Learning

Label Assignment. For category prediction branch, the network needs to give the object category probability for each of $S \times S$ grid. Specifically, grid (i, j) is considered as a positive sample if it falls into the *center region* of any ground truth mask, Otherwise it is a negative sample. Center sampling is effective in recent works of object detection [12,27], and here we also utilize a similar technique for mask category classification. Given the mass center (c_x, c_y), width w, and height h of the ground truth mask, the center region is controlled by constant scale factors ϵ: $(c_x, c_y, \epsilon w, \epsilon h)$. We set $\epsilon = 0.2$ and there are on average 3 positive samples for each ground truth mask.

Besides the label for instance category, we also have a binary segmentation mask for each positive sample. Since there are S^2 grids, we also have S^2 output masks for each image. For each positive samples, the corresponding target binary mask will be annotated. One may be concerned that the order of masks will impact the mask prediction branch, however, we show that the most simple row-major order works well for our method.

Loss Function. We define our training loss function as follows:

$$L = L_{cate} + \lambda L_{mask}, \tag{1}$$

where L_{cate} is the conventional Focal Loss [15] for semantic category classification. L_{mask} is the loss for mask prediction:

$$L_{mask} = \frac{1}{N_{pos}} \sum_k \mathbb{1}_{\{\mathbf{p}^*_{i,j} > 0\}} d_{mask}(\mathbf{m}_k, \mathbf{m}^*_k), \tag{2}$$

Here indices $i = \lfloor k/S \rfloor, j = k \bmod S$, if we index the grid cells (instance category labels) from left to right and top to down. N_{pos} denotes the number of positive samples, $\mathbf{p}^*$ and $\mathbf{m}^*$ represent category and mask target respectively. $\mathbb{1}$ is the indicator function, being 1 if $\mathbf{p}^*_{i,j} > 0$ and 0 otherwise.

We have compared different implementations of $d_{mask}(\cdot, \cdot)$: Binary Cross Entropy (BCE), Focal Loss [15] and Dice Loss [21]. Finally, we employ Dice Loss for its effectiveness and stability in training. λ in Eq. (1) is set to 3. The Dice Loss is defined as

$$L_{Dice} = 1 - D(\mathbf{p}, \mathbf{q}), \tag{3}$$

where D is the dice coefficient which is defined as

$$D(\mathbf{p}, \mathbf{q}) = \frac{2 \sum_{x,y} (\mathbf{p}_{x,y} \cdot \mathbf{q}_{x,y})}{\sum_{x,y} \mathbf{p}^2_{x,y} + \sum_{x,y} \mathbf{q}^2_{x,y}}. \tag{4}$$

Here $\mathbf{p}_{x,y}$ and $\mathbf{q}_{x,y}$ refer to the value of pixel located at (x, y) in predicted soft mask $\mathbf{p}$ and ground truth mask $\mathbf{q}$.

2.4   Inference

The inference of SOLO is very straightforward. Given an input image, we forward it through the backbone network and FPN, and obtain the category score $\mathbf{p}_{i,j}$ at grid (i, j) and the corresponding masks $\mathbf{m}_k$, where $k = i \cdot S + j$. We first use a confidence threshold of 0.1 to filter out predictions with low confidence. Then we select the top 500 scoring masks and feed them into the NMS operation. We use a threshold of 0.5 to convert predicted soft masks to binary masks.

Maskness. We calculate maskness for each predicted mask, which represents the quality and confidence of mask prediction maskness $= \frac{1}{N_f} \sum_i^{N_f} \mathbf{p}_i$. Here N_f the number of foreground pixels of the predicted soft mask $\mathbf{p}$, *i.e.*, the pixels that have values greater than threshold 0.5. The classification score for each prediction is multiplied by the maskness as the final confidence score.

3   Experiments

We present experimental results on the MS COCO instance segmentation benchmark [16], and report ablation studies by evaluating on the 5k `val2017` split. For our main results, we report COCO mask AP on the `test-dev` split, which has no public labels and is evaluated on the evaluation server.

Training Details. SOLO is trained with stochastic gradient descent (SGD). We use synchronized SGD over 8 GPUs with a total of 16 images per minibatch. Unless otherwise specified, all models are trained for 36 epochs with an initial learning rate of 0.01, which is then divided by 10 at 27th and again at 33th epoch. Weight decay of 0.0001 and momentum of 0.9 are used. All models are initialized from ImageNet pre-trained weights. We use scale jitter where the shorter image side is randomly sampled from 640 to 800 pixels, following [4].

Table 1. Instance segmentation mask AP (%) on the COCO `test-dev`. All entries are *single-model* results. Here we adopt the "6×" schedule (72 epochs), following [4]. Mask R-CNN* is our improved version with scale augmentation and longer training time. D-SOLO means Decoupled SOLO as introduced in Sect. 4.

	Backbone	AP	AP_{50}	AP_{75}	AP_S	AP_M	AP_L
Two-stage							
MNC [5]	Res-101-C4	24.6	44.3	24.8	4.7	25.9	43.6
FCIS [13]	Res-101-C5	29.2	49.5	–	7.1	31.3	50.0
Mask R-CNN [9]	Res-101-FPN	35.7	58.0	37.8	15.5	38.1	52.4
MaskLab+ [3]	Res-101-C4	37.3	59.8	39.6	16.9	39.9	53.5
Mask R-CNN*	Res-101-FPN	37.8	59.8	40.7	20.5	40.4	49.3
One-stage							
TensorMask [4]	Res-50-FPN	35.4	57.2	37.3	16.3	36.8	49.3
TensorMask [4]	Res-101-FPN	37.1	59.3	39.4	17.4	39.1	51.6
YOLACT [1]	Res-101-FPN	31.2	50.6	32.8	12.1	33.3	47.1
PolarMask [28]	Res-101-FPN	30.4	51.9	31.0	13.4	32.4	42.8
Ours							
SOLO	Res-50-FPN	36.8	58.6	39.0	15.9	39.5	52.1
SOLO	Res-101-FPN	37.8	59.5	40.4	16.4	40.6	54.2
D-SOLO	Res-101-FPN	38.4	59.6	41.1	16.8	41.5	54.6
D-SOLO	Res-DCN-101-FPN	40.5	62.4	43.7	17.7	43.6	59.3

Table 2. The impact of **grid number and FPN**. FPN significantly improves the performance thanks to its ability to deal with varying sizes of objects.

Grid number	AP	AP_{50}	AP_{75}	AP_S	AP_M	AP_L
12	27.2	44.9	27.6	8.7	27.6	44.5
24	29.0	47.3	29.9	10.0	30.1	45.8
36	28.6	46.3	29.7	9.5	29.5	45.2
Pyramid	35.8	57.1	37.8	15.0	38.7	53.6

3.1 Main Results

We compare SOLO to the state-of-the-art methods in instance segmentation on MS COCO `test-dev` in Table 1. SOLO with ResNet-101 achieves a mask AP of 37.8%, the state of the art among existing *two-stage* instance segmentation methods such as Mask R-CNN. SOLO outperforms all previous *one-stage* methods, including TensorMask [4]. Some SOLO outputs are visualized in Fig. 6, and more examples are in the supplementary.

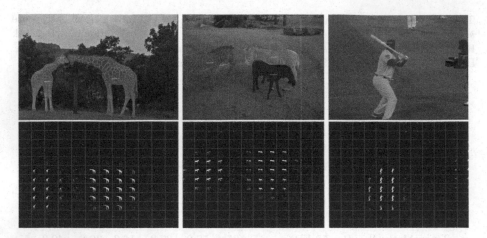

Fig. 4. SOLO behavior. We show the visualization of soft mask prediction. Here $S = 12$. For each column, the top one is the instance segmentation result, and the bottom one shows the mask activation maps. The sub-figure (i, j) in an activation map indicates the mask prediction results (after zooming out) generated by the corresponding mask channel.

3.2   How SOLO Works?

We show the network outputs generated by $S = 12$ grids (Fig. 4). The sub-figure (i, j) indicates the soft mask prediction results generated by the corresponding mask channel. Here we can see that different instances activates at different mask prediction channels. By explicitly segmenting instances at different positions, SOLO converts the instance segmentation problem into a position-aware classification task. Only one instance will be activated at each grid, and one instance may be predicted by multiple adjacent mask channels. During inference, we use NMS to suppress these redundant masks.

3.3   Ablation Experiments

Grid Number. We compare the impacts of grid number on the performance with single output feature map as shown in Table 2. The feature is generated by merging C3, C4, and C5 outputs in ResNet (stride: 8). To our surprise, $S = 12$ can already achieve 27.2% AP on the challenging MS COCO dataset. SOLO achieves 29% AP when improving the grid number to 24. This results indicate that our single-scale SOLO can be applicable to some scenarios where object scales do not vary much.

Multi-level Prediction. From Table 2 we can see that our single-scale SOLO could already get 29.0 AP on MS COCO dataset. In this ablation, we show that the performance could be further improved via multi-level prediction using FPN [14]. We use five pyramids to segment objects of different scales (details in supplementary). Scales of ground-truth masks are explicitly used to assign them

Table 3. Conv vs. CoordConv. CoordConv can considerably improve AP upon standard convolution. Two or more layers of CoordConv are not necessary.

#CoordConv	AP	AP_{50}	AP_{75}	AP_S	AP_M	AP_L
0	32.2	52.6	33.7	11.5	34.3	51.6
1	35.8	57.1	37.8	15.0	38.7	53.6
2	35.7	57.0	37.7	14.9	38.7	53.3
3	35.8	57.4	37.7	15.7	39.0	53.0

Table 4. Different loss functions may be employed in the mask branch. The Dice loss (DL) leads to best AP and is more stable to train.

Mask loss	AP	AP_{50}	AP_{75}	AP_S	AP_M	AP_L
BCE	30.0	50.4	31.0	10.1	32.5	47.7
FL	31.6	51.1	33.3	9.9	34.9	49.8
DL	35.8	57.1	37.8	15.0	38.7	53.6

to the levels of the pyramid. From P2 to P6, the corresponding grid numbers are [40, 36, 24, 16, 12] respectively. Based on our multi-level prediction, we further achieve 35.8 AP. As expected, the performance over all the metrics has been largely improved.

CoordConv. Another important component that facilitates our SOLO paradigm is the *spatially variant* convolution (CoordConv [17]). As shown in Table 3, the standard convolution can already have spatial variant property to some extent, which is in accordance with the observation in [17]. As also revealed in [11], CNNs can implicitly learn the absolute position information from the commonly used zero-padding operation. However, the implicitly learned position information is coarse and inaccurate. When making the convolution access to its own input coordinates through concatenating extra coordinate channels, our method enjoys 3.6 absolute AP gains. Two or more CoordConvs do not bring noticeable improvement. It suggests that a single CoordConv already enables the predictions to be well spatially variant/position sensitive.

Loss Function. Table 4 compares different loss functions for our mask optimization branch. The methods include conventional Binary Cross Entropy (BCE), Focal Loss (FL), and Dice Loss (DL). To obtain improved performance, for Binary Cross Entropy we set a mask loss weight of 10 and a pixel weight of 2 for positive samples. The mask loss weight of Focal Loss is set to 20. As shown, the Focal Loss works much better than ordinary Binary Cross Entropy loss. It is because that the majority of pixels of an instance mask are in background, and the Focal Loss is designed to mitigate the sample imbalance problem by decreasing the loss of well-classified samples. However, the Dice Loss achieves the best results without the need of manually adjusting the loss hyper-parameters. Dice Loss views the pixels as a whole object and could establish the right bal-

Table 5. Different head depth. We use depth being 7 in other experiments, as the performance becomes stable when the depth grows beyond 7.

Head depth	4	5	6	7	8
AP		34.6	35.2	35.5	35.8	35.8

ance between foreground and background pixels automatically. Note that with carefully tuning the balance hyper-parameters and introducing other training tricks, the results of Binary Cross Entropy and Focal Loss may be considerably improved. However the point here is that with the Dice Loss, training typically becomes much more stable and more likely to attain good results without using much heuristics. To make a fair comparison, we also show the results of Mask R-CNN with Dice loss in the supplementary, which performs worse (-0.9AP) than original BCE loss.

Alignment in the Category Branch. In the category prediction branch, we must match the convolutional features with spatial size $H \times W$ to $S \times S$. Here, we compare three common implementations: interpolation, adaptive-pool, and region-grid-interpolation. (a) Interpolation: directly bilinear interpolating to the target grid size; (b) Adaptive-pool: applying a 2D adaptive max-pool over $H \times W$ to $S \times S$; (c) Region-grid-interpolation: for each grid cell, we use bilinear interpolation conditioned on dense sample points, and aggregate the results with average. From our observation, there is no noticeable performance gap between these variants (± 0.1AP), indicating that the alignment process does not have a significant impact on the final accuracy.

Different Head Depth. In SOLO, instance segmentation is formulated as a pixel-to-pixel task and we exploit the spatial layout of masks by using an FCN. In Table 5, we compare different head depth used in our work. Changing the head depth from 4 to 7 gives 1.2 AP gains. The results show that when the depth grows beyond 7, the performance becomes stable. In this paper, we use depth being 7 in other experiments.

Previous works (*e.g.*, Mask R-CNN) usually adopt four conv layers for mask prediction. In SOLO, the mask is conditioned on the spatial position and we simply attach the coordinate to the beginning of the head. The mask head must have enough representation power to learn such transformation. For the semantic category branch, the computational overhead is negligible since $S^2 \ll H \times W$.

3.4   SOLO-512

Speed-wise, the Res-101-FPN SOLO runs at 10.4 FPS on a V100 GPU (all post-processing included), vs. TensorMask's 2.6 FPS and Mask R-CNN's 11.1 FPS. We also train a smaller version of SOLO designed to speed up the inference. We use a model with smaller input resolution (shorter image size of 512 instead of 800). Other training and testing parameters are the same between SOLO-512 and SOLO (Table 6).

Table 6. SOLO-512. SOLO-512 uses a model with smaller input size. All models are evaluated on val2017. Here the models are trained with "6×" schedule.

	Backbone	AP	AP$_{50}$	AP$_{75}$	fps
SOLO	ResNet-50-FPN	36.0	57.5	38.0	12.1
SOLO	ResNet-101-FPN	37.1	58.7	39.4	10.4
SOLO-512	ResNet-50-FPN	34.2	55.9	36.0	22.5
SOLO-512	ResNet-101-FPN	35.0	57.1	37.0	19.2

With 34.2 mask AP, SOLO-512 achieves a model inference speed of 22.5 FPS, showing that SOLO has potentiality for real-time instance segmentation applications. The speed is reported on a single V100 GPU by averaging 5 runs.

3.5 Error Analysis

To quantitatively understand SOLO for mask prediction, we perform an error analysis by replacing the predicted masks with ground-truth values. For each predicted binary mask, we compute IoUs with ground-truth masks, and replace it with the most overlapping ground-truth mask. As reported in Table 7, if we replace the predicted masks with ground-truth masks, the AP increases to 68.1%. This experiment suggests that there are still ample room for improving the mask branch. We expect techniques developed (a) in semantic segmentation, and (b) for dealing occluded/tiny objects could be applied to boost the performance.

Table 7. Error analysis. Replacing the predicted instance mask with the ground-truth ones improves the mask AP from 37.1 to 68.1, suggesting that the mask branch still has ample room to be improved. The models are based on ResNet-101-FPN.

	AP	AP$_{50}$	AP$_{75}$	AP$_S$	AP$_M$	AP$_L$
Baseline	37.1	58.7	39.4	16.0	41.1	54.2
w/gt mask	68.1	68.3	68.2	46.1	75.0	78.5

4   Decoupled SOLO

Given a predefined grid number, e.g., $S = 20$, our SOLO head outputs $S^2 = 400$ channel maps. However, the prediction is somewhat redundant as in most cases the objects are located sparsely in the image. In this section, we further introduce an equivalent and significantly more efficient variant of the vanilla SOLO, termed **Decoupled SOLO**, shown in Fig. 5.

In Decoupled SOLO, the original output tensor $M \in \mathbb{R}^{H \times W \times S^2}$ is replaced with two output tensors $X \in \mathbb{R}^{H \times W \times S}$ and $Y \in \mathbb{R}^{H \times W \times S}$, corresponding two

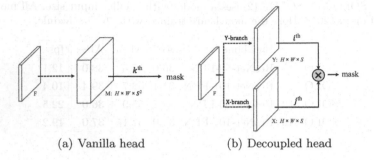

(a) Vanilla head   (b) Decoupled head

Fig. 5. Decoupled SOLO head. F is input feature. Dashed arrows denote convolutions. $k = i \cdot S + j$. '$\otimes$' denotes element-wise multiplication.

axes respectively. Thus, the output space is decreased from $H \times W \times S^2$ to $H \times W \times 2S$. For an object located at grid location (i, j), the mask prediction of that object is defined as the element-wise multiplication of two channel maps:

$$\mathbf{m}_k = \mathbf{x}_j \otimes \mathbf{y}_i, \tag{5}$$

where $\mathbf{x}_j$ and $\mathbf{y}_i$ are the jth and i^{th} channel map of X and Y after sigmoid operation. The motivation behind this is that the probability of a pixel belonging to location category (i, j) is the joint probability of belonging to i^{th} row and j^{th} column, as the horizontal and vertical location categories are independent.

We conduct experiments using the same hyper-parameters as vanilla SOLO. As shown in Table 1, Decoupled SOLO even achieves slightly better performance (0.6 AP gains) than vanilla SOLO. With DCN-101 [6] backbone, we further

Fig. 6. Visualization of instance segmentation results using the Res-101-FPN backbone. The model is trained on the COCO train2017 dataset, achieving a mask AP of 37.8 on the COCO test-dev.

achieve 40.5 AP, which is considerably better than current dominant approaches. It indicates that the Decoupled SOLO serves as an efficient and equivalent variant in accuracy of SOLO. Note that, as the output space is largely reduced, the Decoupled SOLO needs considerably less GPU memory during training and testing.

5 Conclusion

In this work we have developed a direct instance segmentation framework, termed SOLO. Our SOLO is end-to-end trainable and can directly map a raw input image to the desired instance masks with constant inference time, eliminating the need for the grouping post-processing as in bottom-up methods or the bounding-box detection and RoI operations in top-down approaches. Given the simplicity, flexibility, and strong performance of SOLO, we hope that our SOLO can serve as a cornerstone for many instance-level recognition tasks.

Acknowledgement. We would like to thank Dongdong Yu and Enze Xie for the discussion about maskness and dice loss. We also thank Chong Xu and the ByteDance AI Lab team for technical support. Correspondence should be addressed to CS. This work was in part supported by ARC DP 'Deep learning that scales'.

References

1. Bolya, D., Zhou, C., Xiao, F., Lee, Y.J.: YOLACT: real-time instance segmentation. In: Proceedings of the IEEE International Conference on Computer Vision (2019)
2. Chen, K., et al.: Hybrid task cascade for instance segmentation. In: Proceedings of the IEEE Conference on Computer Vision and Pattern Recognition (2019)
3. Chen, L.C., Hermans, A., Papandreou, G., Schroff, F., Wang, P., Adam, H.: MaskLab: instance segmentation by refining object detection with semantic and direction features. In: Proceedings of the IEEE Conference on Computer Vision and Pattern Recognition (2018)
4. Chen, X., Girshick, R., He, K., Dollar, P.: TensorMask: a foundation for dense object segmentation. In: Proceedings of the IEEE International Conference on Computer Vision (2019)
5. Dai, J., He, K., Sun, J.: Instance-aware semantic segmentation via multi-task network cascades. In: Proceedings of the IEEE Conference on Computer Vision and Pattern Recognition (2016)
6. Dai, J., et al.: Deformable convolutional networks. In: Proceedings of the IEEE International Conference on Computer Vision (2017)
7. De Brabandere, B., Neven, D., Van Gool, L.: Semantic instance segmentation with a discriminative loss function. arXiv:1708.02551 (2017)
8. Gao, N., et al.: SSAP: single-shot instance segmentation with affinity pyramid. In: Proceedings of the IEEE International Conference on Computer Vision (2019)
9. He, K., Gkioxari, G., Dollár, P., Girshick, R.B.: Mask R-CNN. In: Proceedings of the IEEE International Conference on Computer Vision (2017)

10. Huang, Z., Huang, L., Gong, Y., Huang, C., Wang, X.: Mask scoring R-CNN. In: Proceedings of the IEEE Conference on Computer Vision and Pattern Recognition (2019)
11. Islam, M.A., Jia, S., Bruce, N.D.B.: How much position information do convolutional neural networks encode? In: Proceedings of the International Conference on Learning Representations (2020)
12. Kong, T., Sun, F., Liu, H., Jiang, Y., Li, L., Shi, J.: FoveaBox: beyond anchor-based object detector. IEEE Trans. Image Process. **29**, 7389–7398 (2020)
13. Li, Y., Qi, H., Dai, J., Ji, X., Wei, Y.: Fully convolutional instance-aware semantic segmentation. In: Proceedings of the IEEE Conference on Computer Vision and Pattern Recognition (2017)
14. Lin, T., Dollár, P., Girshick, R.B., He, K., Hariharan, B., Belongie, S.J.: Feature pyramid networks for object detection. In: Proceedings of the IEEE Conference on Computer Vision and Pattern Recognition (2017)
15. Lin, T.Y., Goyal, P., Girshick, R., He, K., Dollár, P.: Focal loss for dense object detection. In: Proceedings of the IEEE International Conference on Computer Vision (2017)
16. Lin, T.-Y., et al.: Microsoft COCO: common objects in context. In: Fleet, D., Pajdla, T., Schiele, B., Tuytelaars, T. (eds.) ECCV 2014. LNCS, vol. 8693, pp. 740–755. Springer, Cham (2014). https://doi.org/10.1007/978-3-319-10602-1_48
17. Liu, R., et al.: An intriguing failing of convolutional neural networks and the CoordConv solution. In: Proceedings of the Advances in Neural Information Processing Systems (2018)
18. Liu, S., Jia, J., Fidler, S., Urtasun, R.: Sequential grouping networks for instance segmentation. In: Proceedings of the IEEE International Conference on Computer Vision (2017)
19. Liu, S., Qi, L., Qin, H., Shi, J., Jia, J.: Path aggregation network for instance segmentation. In: Proceedings of the IEEE Conference on Computer Vision and Pattern Recognition (2018)
20. Long, J., Shelhamer, E., Darrell, T.: Fully convolutional networks for semantic segmentation. In: Proceedings of the IEEE Conference on Computer Vision and Pattern Recognition (2015)
21. Milletari, F., Navab, N., Ahmadi, S.A.: V-Net: fully convolutional neural networks for volumetric medical image segmentation. In: Proceedings of the International Conference on 3D Vision (2016)
22. Newell, A., Huang, Z., Deng, J.: Associative embedding: end-to-end learning for joint detection and grouping. In: Proceedings of the Advances in Neural Information Processing Systems (2017)
23. Novotny, D., Albanie, S., Larlus, D., Vedaldi, A.: Semi-convolutional operators for instance segmentation. In: Ferrari, V., Hebert, M., Sminchisescu, C., Weiss, Y. (eds.) ECCV 2018. LNCS, vol. 11205, pp. 89–105. Springer, Cham (2018). https://doi.org/10.1007/978-3-030-01246-5_6
24. Pinheiro, P.H.O., Collobert, R., Dollár, P.: Learning to segment object candidates. In: Proceedings of the Advances in Neural Information Processing Systems (2015)
25. Ren, S., He, K., Girshick, R.B., Sun, J.: Faster R-CNN: towards real-time object detection with region proposal networks. In: Proceedings of the Advances in Neural Information Processing Systems (2015)
26. Sofiiuk, K., Barinova, O., Konushin, A.: AdaptIS: adaptive instance selection network. In: Proceedings of the IEEE International Conference on Computer Vision (2019)

27. Tian, Z., Shen, C., Chen, H., He, T.: FCOS: fully convolutional one-stage object detection. In: Proceedings of the IEEE International Conference on Computer Vision (2019)
28. Xie, E., et al.: PolarMask: single shot instance segmentation with polar representation. In: Proceedings of the IEEE Conference on Computer Vision and Pattern Recognition (2020)

Learning to See in the Dark with Events

Song Zhang[1,3], Yu Zhang[3,4(✉)], Zhe Jiang[3], Dongqing Zou[3], Jimmy Ren[3], and Bin Zhou[1,2(✉)]

[1] State Key Laboratory of Virtual Reality Technology and Systems, School of Computer Science and Engineering, Beihang University, Beijing, China
{zhangs,zhoubin}@buaa.edu.cn
[2] Peng Cheng Laboratory, Shenzhen, China
zhangyulb@gmail.com, {zhangyu1,jiangzhe,zoudongqing,
rensijie}@sensetime.com
[3] SenseTime Research, Beijing, China
[4] Tsinghua University, Beijing, China

Abstract. Imaging in the dark environment is important for many real-world applications like video surveillance. Recently, the development of *Event Cameras* raises promising directions in solving this task thanks to its High Dynamic Range (HDR) and low requirement of computational sources. However, such cameras record sparse, asynchronous intensity changes of the scene (called *events*), instead of canonical images. In this paper, we propose learning to see in the dark by translating HDR events in low light to canonical sharp images as if captured in day light. Since it is extremely challenging to collect paired event-image training data, a novel unsupervised domain adaptation network is proposed that explicitly separates domain-invariant features (*e.g.* scene structures) from the domain-specific ones (*e.g.* detailed textures) to ease representation learning. A detail enhancing branch is proposed to reconstruct day light-specific features from the domain-invariant representations in a residual manner, regularized by a ranking loss. To evaluate the proposed approach, a novel large-scale dataset is captured with a DAVIS240C camera with both day/low light events and intensity images. Experiments on this dataset show that the proposed domain adaptation approach achieves superior performance than various state-of-the-art architectures.

Keywords: Domain adaptation · Event camera · Image reconstruction · Low light imaging

1 Introduction

Event cameras [5,23], a kind of bio-inspired vision sensors that mimic the human eye in receiving the visual information, have gained more and more attention in

---

S. Zhang, Y. Zhang, and Z. Jiang—Equal contribution.

Electronic supplementary material The online version of this chapter (https://doi.org/10.1007/978-3-030-58523-5_39) contains supplementary material, which is available to authorized users.

© Springer Nature Switzerland AG 2020
A. Vedaldi et al. (Eds.): ECCV 2020, LNCS 12363, pp. 666–682, 2020.
https://doi.org/10.1007/978-3-030-58523-5_39

Fig. 1. Motivation of the proposed approach. (a) In low light the conventional camera fails to capture the scene clearly. (b) Event cameras, on the other side, can perceive the scene due to its high dynamic range, though with noisy measurements. (c) In low light, even the state-of-the-art event-to-image translation approach [30] fails reconstructing a clean image. (d) Despite the large domain gap, the proposed domain adaptation approach learns to reconstruct high-quality day-light image by observing low-light events. (e)(f) Domain translation results of strong baselines [28,38], respectively.

computer vision community. Different from traditional cameras capturing intensity frames at a fixed rate, event cameras transmit the changes of scene intensities asynchronously, which are called events. Endowed with microsecond temporal resolution, low power assumption and a high dynamic range (e.g., 140 dB compared to 60 dB of most consumer-level standard cameras), event cameras have promising applications in various scenes with broad illumination range.

Reconstructing canonical images from asynchronous events has been explored in various research [21,27,30,32]. Early works, which were primarily inspired by the physical formation model of events, are prone to real-world imperfectness and often generate noisy reconstructions. Recent deep reconstruction model [30] has demonstrated impressive performance through being trained on large amounts of simulated image/event pairs. Despite their successes, few works pay sufficient attention to event-based imaging in low-light environment. Like conventional cameras, events in low light have their own distributions. The sparity and noisiness of low-light events render their distribution rather different with that in normal light. As a result, even the state-of-the-art event-to-image translation approach [30] fails generating clean reconstruction due to the domain gap, as shown in Fig. 1(c). Besides, collecting large datasets with events and reference sharp images in low light is hardly practical.

Motivated by recent advances of deep domain adaptation methods [11,16, 17,24,28] and the HDR property of event cameras, a novel domain adaptation method is proposed to tackle the problem of generating clear intensity images from events in the dark. Specifically, features are extracted from the low-light event domain and transferred to the day-light one, in which the reference sharp images are much easier to collect. To this end, previous domain adaptation methods usually project features from different domains into a common feature space, which can be supervised with available rich labels [9,16,28,34]. In this manner,

discriminative cross-domain attributes are preserved while the distractive ones
are neglected. While this is fine for high-level downstream tasks such as classification and segmentation, it may not be the best choice for the proposed task. In
fact, the domain-specific "distractors", such as the high frequency scene details
and textures, largely affects the reconstruction quality and should be complementary to the domain-shared attributes (e.g. scene structures). Therefore, we
propose distangled representation learning: we decompose the day-light domain
features into a domain-shared part and a domain-specific part. The domain-
shared part is expected to encode scene-level information such as structures,
intensity distributions, etc., which could also be perceived from the low-light
events. The domain-specific part contains day-light exclusive patterns, e.g. high-
frequency details, which may not be easily recovered from low-light events. A
dedicated residual detail enhancing network is incorporated to recover plausible
day-light details from the common representations in a generative manner. We
show that such decomposition can be regularized with a simple ranking loss. To
evaluate the proposed approach, we capture a large dataset with real-world low-
light/day-light events and images, with a DAVIS240C camera [5]. Experiments
on this dataset show that the proposed architecture outperforms various state-
of-the-art domain adaptation and event-based image reconstruction approaches.

Our contributions are: 1) we propose a novel domain-adaptation approach for
event-based intensity image reconstruction in low light. Our approach achieves
the best reconstruction quality among various state-of-the-art architectures; 2)
we propose to learn decomposed representations to smooth the adaptation process, which can be regularized by a simple ranking loss; 3) a large dataset with
low-light/day-light events and images of various real-world scenes is compiled,
and will be made publicly available to facilitate future research.

2   Related Work

Intensity Image Reconstruction from Events. Dynamic Vision Sensors
(DVS) [5,23] are a kind of biology inspired sensors which sends signals when the
scene exhibits illumination changes. An "on" signal is generated when the pixel
brightness goes up to a certain degree, or an "off' signal otherwise. The signals,
often referred as *events*, can be denoted with a tuple (u, v, t, p), where u, v are
the pixel coordinates of the event, t is the timestamp, and $p \in \{-1, +1\}$ is the
polarity, i.e, representing "on" and "off" states.

Early attempts on reconstructing the intensity image from events are inspired
by their physical imaging principles. Generally, the spatiotemporal formation of
events are formulated with various forms of pixel tracking models. For example, Kim et al. [20] estimate scene movements and gradients, and integrate
them in a particle filter to generate final intensity reconstructions. Barua et al.
[2] employs patch-based sparse dictionary learning on events to address image
reconstruction. Bardow et al. [1] integrates optical flow estimation with intensity
reconstruction into a cost function with various spatiotemporal regularizations.
Scheerlinck et al. [32] updates each pixel location with incoming events via a
complementary ordinary differential equation filter.

Recent research directs to learning deep event-to-image translation models from data [21,30]. Generally, these approaches achieve much improved reconstructions due to deep networks. However, a challenge is how to collect clean training pairs of events and reference images. Wang *et al.* [21] captures such data with off-the-shelf hybrid event camera simultaneously recording events and images with time calibration. Rebecq *et al.* applies a carefully designed event simulator [29] to synthesize event data from existing large-scale image datasets. However, in low-light scenario, while the former fails due to the limited dynamic range of image sensor, the latter may not well model real noise distributions.

Unsupervised Domain Adaptation. To overcome the above difficulty, we treat low-light image reconstruction from events as an unsupervised domain adaptation task. Unsupervised domain adaptation has been widely adopted for image classification [4,11,13,24,33,35], segmentation [9,16] and synthesis [3,15,28]. In the line of pixel-level adaptation networks, Hoffman *et al.* [15] proposes an early attempt that obtains domain-agnostic features. Ghifary *et al.* [13] and Murez *et al.* [28] develop further constraints (*e.g.* decodable shared features, cycle-consistent mapping originating from [38]), to further improve the robustness of adaptation. Based on these constraints, further extensions are proposed [9,34] by exploring structural cues, tailored for segmentation. However, restricting the inference on domain-shared features may neglect domain-specific high-frequency information, which is not problematic or even desired for high-level tasks such as classification, but may sacrifice the sharpness and realistics for image generation. Bousmalis *et al.* [4] propose to explicitly decompose the features into domain-shared and domain-specific ones. However, their purpose is to obtain more discriminative and clean domain-shared features for classification, instead of leveraging domain-specific high-frequency details to improve reconstruction.

Deep Image Enhancement. Recently, impressive results were shown for training deep image enhancing models on enhancing bayer raw [7,8] and HDR [6,12,19,25,37] images. However, collecting the reference aligned images for low-light events are extremely difficult. A widely adopted alternative is unpaired training, via modality translation paradigm [10,17,18,38]. Nevertheless, the proposed task involves modality translation as a task (*i.e.* event-to-image translation), in two separate domains (*i.e.* low light and day light). In addition, the domain difference in the proposed task involves various factors like color, noise, scene textures and details. This introduces further challenges as opposed with existing success, for which domain difference often lies in color distribution [10,18]. In fact, we find that existing domain translation approaches cannot well address the proposed task, as illustrated in Fig. 1 and our experimental section.

3   The Proposed Approach

Our approach assumes that a dataset with events and their reference intensity images are given in day light environment, while events and images are

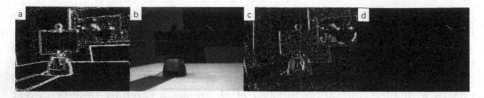

Fig. 2. Events and intensity images of the same scene in day-light and low-light environment, captured with the DAVIS240C camera in a lab room with controlled lighting. The first two columns show events (a) and corresponding intensity image (b) in day-light, respectively. The last two columns (c) (d) illustrate them in the dark.

time-calibrated. Such data could be easily collected with modern off-the-shelf hybrid cameras with calibrated sensors, *e.g.* DAVIS240C [5]. In addition, a set of captures of events in the target low-light environment are also required, but this time the reference images are allowed absent. The scene coverage in day light and low light do not need to be aligned. On this dataset, our objective is to train an event-to-image translation model that generalizes to the target domain. We start with introducing the high-level design of the proposed architecture, then goes into the technical details of each component.

3.1   Learning Architecture

As described previously, event cameras capture changes of scene intensities on each pixel location. Therefore, the distributions of events are closely related to lighting conditions, like standard frame-based cameras. In Fig. 2 we shown sample events and the corresponding image captures of the same scene. We can observe that 1) the low-light and day-light events distributions differ in a number of aspects including noise, event density and the sharpness of details. The events captured in low light are obviously much noisier and less sharper; 2) however, due to the high dynamic range of event camera, the coarse scene structures are still preserved in different lightings. Therefore, an ideal domain adaptation approach should be capable of 1) extracting the shared scene-level features in two domains; 2) "hallucinating" plausible day light-specific details that are missing in the low-light domain. While the former has been verified as a common rule for domain adaptation methods [28], the latter and how it can be integrated into the whole pipeline for image reconstruction, is still less explored.

Keeping this in mind, we propose a novel training architecture for domain-adapted low-light event-to-image translation, as summarized in Fig. 3. It consists of a day light-specific private encoder $\mathbf{E}_p$, a domain-shared encoder $\mathbf{E}_c$, a shared decoder $\mathbf{R}$ for image reconstruction, a domain discriminator $\mathbf{D}$, and a detail enhancing branch $\mathbf{T}_e$ which will be explained shortly after.

Given a source (day-light) domain event representation $\mathbf{x}^s$, we concatenate it with a spatial noise channel $\mathbf{z}$, and feed them into the source-private encoder $\mathbf{E}_p$, obtaining domain-private features $\mathbf{x}_p^f = \mathbf{E}_p\left(\mathbf{x}^s, \mathbf{z}; \boldsymbol{\theta}_p\right)$. Conditioning on both the real samples and a noise channel helps improve the generalization of the model,

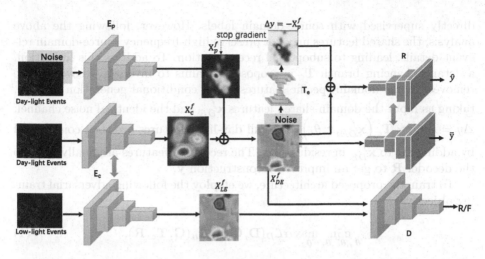

Fig. 3. The proposed framework. The day-light events and low-light events are fed into a shared encoder $\mathbf{E}_c$ to extract their representations $\mathbf{x}_c^f$ and $\mathbf{x}_{LE}^f$. Meanwhile, the day-light events are also fed into a private encoder $\mathbf{E}_p$ along with a noise channel, yielding source domain-specific residuals $\mathbf{x}_p^f$. By adding operation, the modulated day-light features $\mathbf{x}_{DE}^f = \mathbf{x}_c^f + \mathbf{x}_p^f$ and the low-light features $\mathbf{x}_{LE}^f$ lie in a domain-shared feature space, guaranteed by adversarial training with a discriminator $\mathbf{D}$. The detail enhancement branch $\mathbf{T}_e$ reconstructs day-light domain-specific residuals from the shared features. Finally, a shared decoder $\mathbf{R}$ reconstructs the intensity images using both the domain-specific and shared representations. "R/F" represents Real or Fake logits.

which is verified in previous works [16] and our experiments. Meanwhile, the source events $\mathbf{x}^s$ itself also go through the domain-shared encoder $\mathbf{E}_c$, yielding domain-public features $\mathbf{x}_c^f = \mathbf{E}_c(\mathbf{x}^s; \boldsymbol{\theta}_c)$. On the other side, the target domain (low-light) event representation $\mathbf{x}_{LE}^f$ is sent to the same shared encoder to get its encoding $\mathbf{x}_{LE}^f = \mathbf{E}_c(\mathbf{x}_{LE}^t, \boldsymbol{\theta}_c)$. Here, $\boldsymbol{\theta}_p$ and $\boldsymbol{\theta}_c$ are trainable parameters.

We expect that the domain-private features $\mathbf{x}_p^f$ encode day-light exclusive patterns such as high-frequency details, which may not be easily captured by the shared features $\mathbf{x}_{LE}^t$ extracted from the low-light events. On the other hand, the domain-public features $\mathbf{x}_c^f$ from day-light events may still contain source-domain relevant attributes. We perform an "add" operation between $\mathbf{x}_c^f$ and $\mathbf{x}_p^f$, obtaining a modulated feature $\mathbf{x}_{DE}^f = \mathbf{x}_c^f + \mathbf{x}_p^f$, which lies in a shared feature space with $\mathbf{x}_{LE}^f$ as if extracted from the low-light domain. In this manner, $\mathbf{x}_p^f$ can be deemed as the "negative" complementary features of source domain. To guarantee the desired behavior, we feed $\mathbf{x}_{LE}^f$ and $\mathbf{x}_{DE}^f$ to a domain discriminator $\mathbf{D}(\cdot; \boldsymbol{\theta}_D)$, which serves to distinguishing between the transformed feature map $\mathbf{x}_{DE}^f$ from the day-light domain, and the features $\mathbf{x}_{LE}^f$ extracted from the real low-light domain.

The features $\mathbf{x}_{DE}^f$ projected into the domain-shared space can be then fed into the decoder $\mathbf{R}(\cdot; \boldsymbol{\theta}_R)$ to get the reconstruction result $\bar{\mathbf{y}}$, which can be directly supervised with source-domain labels. However, following the above analysis, the shared features may not preserve high-frequency source-domain relevant details, leading to suboptimal reconstruction. To address this limitation, a detail enhancing branch $\mathbf{T}_e$ is proposed. It aims to inversely recovering the removed source domain-specific features $\mathbf{x}_p^f$ in conditional generation manner, taking as input the domain-shared features $\mathbf{x}_{DE}^f$ and the identical noise channel: $\Delta y = -\mathbf{x}_p^f \approx \mathbf{T}_e\left(\mathbf{x}_{DE}^f, \mathbf{z}; \boldsymbol{\theta}_t\right)$. The full day-light features are then constructed by adding Δy to $\mathbf{x}_{DE}^f$ in residual way. The recovered features are finally fed into the decoder $\mathbf{R}$ to get an improved reconstruction $\hat{\mathbf{y}}$.

To train the proposed architecture, we employ the following adversarial training objective:

$$\min_{\boldsymbol{\theta}_g, \boldsymbol{\theta}_r, \boldsymbol{\theta}_t} \max_{\boldsymbol{\theta}_d} \alpha \mathcal{L}_D(\mathbf{D}, \mathbf{G}) + \mathcal{L}_R(\mathbf{G}, \mathbf{T}_e, \mathbf{R}), \tag{1}$$

where α balances the interaction of different terms, and is set to 0.1. For the domain loss $\mathcal{L}_D(\mathbf{D}, \mathbf{G})$, we define it as a common logistic loss in training GANs:

$$\begin{aligned}
\mathcal{L}_D(\mathbf{D}, \mathbf{G}) = &\mathcal{E}_{\mathbf{x}^t}[\log \mathbf{D}(\mathbf{x}^t; \boldsymbol{\theta}_d)] + \\
&\mathcal{E}_{\mathbf{x}^s, \mathbf{z}}[\log(1 - \mathbf{D}(\mathbf{G}(\mathbf{x}^s, \mathbf{z}, \boldsymbol{\theta}_g); \boldsymbol{\theta}_d))],
\end{aligned} \tag{2}$$

where $\mathcal{E}$ denotes expectation over distribution.

Our reconstruction related loss $\mathcal{L}_R$ can be decomposed into three terms:

$$\mathcal{L}_R(\mathbf{G}, \mathbf{T}_e, \mathbf{R}) = L_p(\hat{\mathbf{y}}, \bar{\mathbf{y}}, \mathbf{y}^g) + \beta L_t(\Delta \mathbf{y}, \mathbf{x}_p^f) + \gamma L_r(\hat{\mathbf{y}}, \bar{\mathbf{y}}, \mathbf{y}^g), \tag{3}$$

where $\bar{\mathbf{y}}$ and $\hat{\mathbf{y}}$ are the reconstructions from shared and inversely recovered features (see Fig. 3), $\mathbf{y}^g$ is the groundtruth reconstruction available on the source day-light domain. The balancing weights β and γ are empirically set to 0.1.

The first term is the ℓ_1 photometric reconstruction loss:

$$L_p(\hat{\mathbf{y}}, \bar{\mathbf{y}}, \mathbf{y}^g) = \frac{1}{|\mathbb{I}^s|} \sum_{i=1}^{\mathbb{I}^s} \left(\|\hat{y}_i - y_i^g\|_1 + \|\bar{y}_i - y_i^g\|_1\right), \tag{4}$$

in which $\mathbb{I}^s$ denotes the set of all the image pixels in source domain.

With the second term, we train the detail enhancing branch to regress the source domain-specific residuals:

$$L_t(\Delta \mathbf{y}, \mathbf{x}_p^f) = \frac{1}{|\mathbb{F}^s|} \sum_{i=1}^{\mathbb{F}^s} \|\Delta y_i + (\mathbf{x}_p^f)_i\|_1. \tag{5}$$

The set $\mathbb{F}_s$ denotes all the pixel locations on the immediate feature maps $\mathbf{x}_p^f$. Note that the target labels in (5) are not fixed but dynamically changed with network training. To avoid contaminating representation learning in early training stage, gradients from L_t to the private encoder $\mathbf{E}_p$ is blocked, as illustrated in Fig. 3. It effectively cancels the loops in information flows and gets rid of mode collapse.

Finally, we add further regularizations between $\bar{\mathbf{y}}$, the reconstructions from the shared-space features $\mathbf{x}_{DE}^f$, and $\hat{\mathbf{y}}$, reconstructions from the recovered domain-specific features $\mathbf{x}_{DE}^f + \Delta\mathbf{y}$. We employ a ranking loss, preferring stronger performance for the latter:

$$L_r(\hat{\mathbf{y}}, \bar{\mathbf{y}}, \mathbf{y}^g) = \max\left(\frac{1}{|\mathbb{I}^s|}\sum_{i=1}^{\mathbb{I}^s}\|\hat{y}_i - y_i^g\|_1 - \frac{1}{|\mathbb{I}^s|}\sum_{i=1}^{\mathbb{I}^s}\|\bar{y}_i - y_i^g\|_1 + \epsilon, 0\right), \quad (6)$$

where $\epsilon > 0$ is a predefined margin. Eq. (6) encourages the reconstruction $\hat{\mathbf{y}}$ to have a lower loss compared with that of $\bar{\mathbf{y}}$ so as to probe a better reconstruction. This is somewhat counter-intuitive, since that (6) can achieve the same minimizer by degrading $\bar{\mathbf{y}}$ instead of improving $\hat{\mathbf{y}}$. The rationale behind this regularization is that the domain loss L_D and the reconstruction loss L_R in (1) do not always have consistent objectives, especially in perceptual image synthesis tasks. Actually, as advocated by recent studies [26,34], adversarial generation and domain adaptation without regularization tends to align the most discriminative (instead of many) modes of two distributions. To this end, it may underplay important distortion-critic features, which are often not the most discriminative ones between domains but crucial for reconstruction quality. It shares with a similar finding from GAN-based image superresolution [22,36]. Including GAN objective may damages distortion-based metrics (though perceptual quality can be improved by training on carefully created, high-quality clean datasets). In our setting, the large distribution gap between day-light and low-light domains can readily guide the discriminator to focus on several main differences, $e.g.$ noise and texture distributions, deviating from reconstructing other details and damages reconstruction quality. Through the regularization (6), it would be not a great issue if this happens, as the detail enhancing branch can encode the lost less discriminative but important information to promote a better reconstruction. In the experimental section, we show that such regularization indeed improves the results in both quantitative and subjective comparisons.

In Fig. 4 we visualize the reconstructed images with and without the ranking-based regularization. Without the ranking loss, the reconstruction in Fig. 4(a) comes from the shared feature representations. In this case, the inter-domain adversarial loss dominates the training process, which focuses on discriminative distributional differences but neglects tiny scene details. Adding the ranking loss guides the detail enhancing network to recover such details while leaving the discriminative adaptation unaffected. As the loss itself does not increases network capacity, it serves as $self\text{-}regularization$ that smooths the training process by distangling representation learning for different network components.

3.2   Implementation Details

Event Representation. Events are asynchronous signals that cannot be trivially processed by image-specialized networks like CNN. To overcome this issue, we employ the stacked event frame representations [21]. Specifically, we divide

Fig. 4. Effectiveness of rank regularization. Reconstruction without and with rank regularization are visualized in (a) and (b), respectively. (c) The reference image. Reconstruction error maps of (a) and (b) are shown in (d) and (e).

the events captured from a scene into equal-size temporal chunks. In each chunk, we sum up event polarities triggered in the time interval of this chunk at per pixel level. For each pixel $\mathbf{p}$, this writes to

$$\Phi(\mathbf{p}) = \sum_{\mathbf{e}=(e_{\mathbf{p}},e_t,e_l)\in\mathbb{E}} \mathbb{1}(e_{\mathbf{p}} = \mathbf{p} \wedge e_t \in [t, t+\tau]) \cdot e_l, \qquad (7)$$

where $e_{\mathbf{p}}$ and e_t represent the pixel location and time stamp when the event is triggered, and $e_l \in \{-1, 1\}$ denotes the polarity of event. The indicator function $\mathbb{1}(\cdot)$ takes 1 if the input condition holds, or 0 otherwise. In our implementary, we employ 4 chunks, where the time length τ of each chunk spans across 1.25 ms. In addition, we concatenate the time stamps of the events (normalized into $[0, 1]$) with the event frames, resulting into a 8-channel representation.

Noise Channel. The noise channel is a 1-dimensional map with the same spatial resolution with that of event frames. Values of this noise map are randomly sampled from a spatial Gaussian distribution whose mean is zero and standard deviation is 1.

Layer Configurations. The encoder and decoder forms the generator that aims to synthesizing plausible real-like images. Thus, they are implemented with residual networks. For encoder $\mathbf{E}_p$ and $\mathbf{E}_c$ in Fig. 3, the concatenated event frames and noise map go through a 7×7 convolution without striding, and two 3×3 convolutions with stride $= 2$. The feature dimensions are scaled up to 32, 64 and 128, sequentially. These features are then fed into 9 identical residual blocks, which has two 3×3 convolutions. For the decoder $\mathbf{R}$, it has 9 the same residual blocks, then upsample the features with two 3×3 deconvolution layers. The feature dimension reduces from 128 to 64 and 32 during upsampling. Finally a 7×7 convolution layer fuses the features to a single output map. The discriminator $\mathbf{D}$ consists of three 5×5 convolutions with stride $= 2$ and two 5×5 convolutions with stride $= 1$. The feature dimensions go up to 32, 64, 128 and 256, finally fused to a single output logit channel. Average pooling is taken to compute final score. All the convolution and deconvolution layers are interleaved with Instance Normalization and Leaky ReLU activation.

The detail enhancing branch $\mathbf{T}_e$ consumes encoded feature maps and the same noise channel as input of the private encoder $\mathbf{E}_p$. To make spatial dimensions consistent, the noise map individually goes through three 3×3 convolution layers with 32 output dimensions. The output are then concatenated with the input features, followed by a 3×3 processing convolution layer and 9 residual identity blocks. Again, instance normalization and ReLU (or leaky ReLU) activation are used for interleaving the convolution layers. Please refer to our supplementary material for more detailed layer/parameter configurations.

Testing on Low-Light Events. In testing phase, low-light event representations are passed through the shared encoder $\mathbf{E}_c$ to get $\mathbf{x}_{LE}^f$. A noise channel is sampled then combined with $\mathbf{x}_{LE}^f$, which are fed into the detail enhancing branch $\mathbf{T}_e$ to get the residual features $\Delta \mathbf{y}$. Finally, the shared decoder $\mathbf{R}$ consumes $\mathbf{x}_{LE}^f + \Delta \mathbf{y}$ to obtain the final reconstruction. Note that the private encoder $\mathbf{E}_p$ and the discriminator $\mathbf{D}$ are not involved in testing phase.

4   Experiments

4.1   Experimental Settings

Data Collection. To the best of our knowledge, there does not exist a mature dataset for evaluating event-based low-light imaging. Thus, we introduce a large novel dataset to evaluate the proposed approach. The dataset is captured by a DAVIS240C camera [5], which has an Active Pixel Sensor (APS) to record intensity image sequences as well as an event sensor recording event streams. These two modalities are calibrated in spatial and temporal axis. The dataset consists of image scenes of urban environment at resolution 180×240, captured in both day-light and low-light conditions with various kinds of camera/scene movement. In summary, there are 8820 scenes captured in day time, and 8945 ones in the night. For each scene, the event stream spans across roughly 100 ms. Note that in day light, the reference scenes and lightings are carefully chosen to avoid saturated pixels (though some still remains in high-light area), so that the images captured by APS sensors can serve as groundtruth. In the night, however, we do not control lighting and there is no high-quality reference images available. Across the experiments, this dataset is referred as DVS-Dark.

Baseline Settings. To compare the proposed domain adaptation framework with state-of-the-art frameworks, we carefully choose and implement 4 representative baselines, as described as follows:

1) Pix2Pix [17,21]. Pix2Pix is a successful conditional generative framework for image-to-image translation, while Wang et al. [21] extend it to event-to-image translation. We carefully follow Wang et al. [21] to set this baseline, and train it on the day-light event/image pairs. The trained network can be applied to low-light test. Thus, no adaptation is performed in this baseline.
2) CycleGAN [38]. The milestone unpaired domain translation framework is proposed by Zhu et al. [38]. We adapt this framework to translate between low-light events and day-light images. However we find the naive implementation hardly working in practice, potentially due to the large domain gap between low-light events and day-light images. Thus, we instead adopt a modification

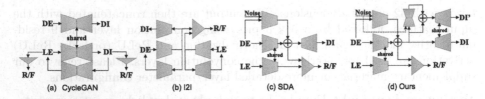

Fig. 5. Illustration of baseline architectures. The red, blue, and green graphics denote the encoder, decoder, and discriminator, respectively. Symbols "D" and "L" represents day-light and low-light, while "E" and "I" refer to events and intensity images. (Color figure online)

in semi-supervised setting: the paired events and images are used in supervised manner to train the forward/backward mapping networks, together with unpaired domain translation.

3) I2I [28]. In the image-to-image translation architecture proposed by Murez *et al.* [28], the day-light and low-light events are projected into a shared feature space. A decoder then learns to reconstruct day-light images from the shared features, which are directly supervised by paired day-light events and images. Besides, two domain-specific decoders (weights are not shared) are involved, responsible for reconstructing the shared features to the original event representations.

4) SDA [16]. The structured domain adaptation framework proposed by Hong *et al.* [16] is originally for semantic segmentation, which we tailor to address image reconstruction. In this framework, the source samples are passed through a source-private encoder to learn source-specific residuals, and also through a shared encoder to get immediate shared features. The immediate shared features then cancels the residuals, yielding domain-shared features that lie in the same space with the features extracted from the target domain. Note that compared with our architecture, SDA does not address detail recovery, and the final reconstruction still comes from the shared feature space.

An illustration of baseline architectures except Pix2Pix is referred to Fig. 5. In addition, we also compare with state-of-the-art event-based image reconstruction approaches CIE [32] and RIRM [27] and E2V [30]. They directly reconstruct intensity images from events via physical or learning-based rules.

Training Details. We train the network for 200 epochs, using a mini-batch of 4 images and 10^{-4} as initial learning rate. The learning rate linearly decays to zero from the 100th to the 200th epoch. As for CIE, RIRM and E2V, we directly adopts the author's open source code.

4.2   Comparisons with State-of-the-Art Models

Quantitative Evaluation Through Simulated Experiments. Since groundtruth is only available for day-light data, we perform simulated experiments by artificially translating a random half of day-light subset to low light

Table 1. PSNR and SSIM numbers on simulated data.

	RIRM	CIE	E2V	E2V$^+$	CGAN	Pix2Pix	I2I	SDA	Ours
PSNR	11.28	13.30	12.16	17.15	14.20	22.65	23.36	24.93	**26.03**
SSIM	0.29	0.45	0.33	0.63	0.27	0.68	0.69	0.75	**0.77**

Table 2. Inception scores(higher is better) and Frechet Inception Distance(lower is better) on real low-light data. Numbers in the parentheses denote standard deviation.

	CIE	RIRM	E2V	CGAN	Pix2Pix	I2I	SDA	Ours
IS	3.36	2.28	3.21	2.89	3.86	2.44	3.75	**3.87**
	($\pm$0.47)	($\pm$0.37)	($\pm$0.61)	($\pm$0.15)	($\pm$0.24)	($\pm$0.15)	($\pm$0.24)	($\pm$0.21)
FID	267.8	208.5	210.42	163.28	109.08	177.59	110.18	**104.21**

to conduct quantitative analysis. In details, we apply a pixel-valued S-shape tone curve to adjust the day-light image brightness. This curve globally shrinks the pixel brightness by roughly 50%, and further shrinks high-light pixels by 60%. To account for event distribution gap between lightings, we discard the captured day-light events and regenerate them with the ESIM simulator [29] for day-light and artificially created low-light images. We use 20% data for testing, and the left for training.

In Table 1 we summarize PSNR and SSIM numbers of different approaches. We note that since RIRM, CIE and E2V all reconstruct intensity frames purely from events, they are not aware of dataset-specific color distributions thus their PSNRs and SSIMs are not meaningful. For fair comparison we propose the variant E2V$^+$, which is obtained by finetuning E2V on the day-light images of DVS-Dark, using a similar process as described in [30] to simulate events. However, it still falls behind our approach in training and testing phase. Among domain translation/adaptation approaches, the proposed approach achieves much improved performance by additionally considering detail recovery from the shared feature space. In contrast, conventional reconstruction-from-shared-space paradigm, as adopted by I2I and SDA, gets worse results.

Comparisons on Real Low-Light Data. Due to the lack of groundtruth reference images in low light, we measure the performance via two widely adopted perceptual measures, the Inception Scores (IS) [31] and the Frechet Inception Distance (FID) [14], as summarized in Table 2. The event-based reconstruction approaches CIE, RIRM and E2V do not get satisfactory perceptual score. Among domain adaptation approaches, the proposed approach still achieves the best performance in both metrics, with the best perceptual quality. In Fig. 6, we provide representative results generated by different approaches. The event-based reconstruction methods CIE, RIRM and E2V recover plenty of scene details, but

Fig. 6. Representative reconstruction results generated by different approaches from real low-light event data on the DVS-Dark dataset. More results are referred to our supplementary material. Best viewed in color with zoom.

Table 3. Ablation studies of different network components.

Noise	Detail	Rank	PSNR	SSIM
✗	✗	✗	20.69	0.63
✓	✗	✗	21.24	0.66
✓	✓	✗	24.91	0.75
✓	✓	✓	26.03	0.77

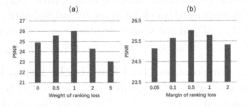

Fig. 7. PSNR as function of the weight (a) and margin (b) of ranking loss.

tend to be noisy. The domain adaptation approaches effectively addresses the noise but may miss tiny details/structures of the scene. The proposed approach preserves scene details best while successfully suppressing the noise.

Fig. 8. Effectiveness of rank regularization. (a) w/o regularization; (b) w/ regularization, $\gamma = 5.0$; (c) w/ regularization, $\gamma = 1.0$; (d) reference image.

4.3 Performance Analysis

Analysing Different Components. We analyse the contributions of different network components on the simulated datasets, as summarized in Table 3. Here, "noise.", "detail." and "rank." denote the noise map augmentation, detail enhancing branch, and ranking mechanism, respectively. From the results, we observe that incorporating noise channel improve the results significantly as it implicitly augments the data by introducing randomness into the inputs. The detail enhancing branch restores domain-specific details that are not well-modelled in shared feature learning, and leads to a substantial improvement. By further combining the rank regularization, the network is guided to learn both domain-discriminative and domain-exclusive features to promote better reconstruction.

Effectiveness of the Rank Regularization. The ranking loss (6) can effectively regularize the domain adaptation process of the proposed network. However, improper parameters may overwhelm the contribution of other loss terms, deteriorating the reconstruction quality in order to satisfy the ranking constraints. Figure 7 shows that the optimal choices of the loss weight and margin on the DVS-Dark dataset are 1.0 and 0.5 respectively.

Besides recovering small details as analysed in Sect. 3.1, the rank regularization also leads to more smooth reconstruction. As shown in Fig. 8, without regularization, the discriminative adaptation would dominate the training process. The results are sharp, but with false textures. With over-regularization ($\gamma = 5.0$), false textures vanish but the results tend to be blurry. Proper strength of regularization ($\gamma = 1.0$) leads to sharp and clean results.

5 Conclusion

We present in this work a deep domain adaptation method for intensity image reconstruction from events captured in low light. The model explicitly learns the shared representation inter domains and domain-specific features via novel detail enhancing mechanism regularized by relative ranking. Our method outperforms related existing methods of image generation from events and unsupervised domain adaptation methods, in both quantitative and qualitative comparisons.

Acknowledgement. We thank the anonymous reviewers for their valuable comments. This work was supported in part by National Natural Science Foundation of China (U1736217 and 61932003), National Key R&D Program of China (2019YFF0302902), Pre-research Project of the Manned Space Flight (060601), and Beijing Postdoctoral Research Foundation (ZZ-2019-89).

References

1. Bardow, P., Davison, A., Leutenegger, S.: Simultaneous optical flow and intensity estimation from an event camera. In: IEEE Conference on Computer Vision and Pattern Recognition (CVPR), pp. 884–892 (2016)
2. Barua, S., Miyatani, Y., Veeraraghavan, A.: Direct face detection and video reconstruction from event cameras. In: IEEE Winter Conference on Applications of Computer Vision (WACV), pp. 1–9 (2016)
3. Bousmalis, K., Silberman, N., Dohan, D., Erhan, D., Krishnan, D.: Unsupervised pixel-level domain adaptation with generative adversarial networks. In: IEEE Conference on Computer Vision and Pattern Recognition (CVPR), pp. 3722–3731 (2017)
4. Bousmalis, K., Trigeorgis, G., Silberman, N., Krishnan, D., Erhans, D.: Domain separation networks. In: Advances in Neural Information Processing Systems (NIPS), pp. 343–351 (2016)
5. Brandli, C., Berner, R., Yang, M., Liu, S., Delbruck, T.: A 240 × 180 130 db 3 μs latency global shutter spatiotemporal vision sensor. IEEE J. Solid State Circ. **49**(10), 2333–2341 (2014)
6. Cai, J., Gu, S., Zhang, L.: Learning a deep single image contrast enhancer from multi-exposure images. IEEE Trans. Image Process. (TIP) **27**(4), 2049–2062 (2018)
7. Chen, C., Chen, Q., Do, M.N., Koltun, V.: Seeing motion in the dark. In: IEEE International Conference on Computer Vision (ICCV), pp. 3185–3194 (2019)
8. Chen, C., Chen, Q., Xu, J., Koltun, V.: Learning to see in the dark. In: IEEE Conference on Computer Vision and Pattern Recognition (CVPR), pp. 3291–3300 (2018)
9. Chen, C., Dou, Q., Chen, H., Qin, J., Heng, P.A.: Synergistic image and feature adaptation: towards cross-modality domain adaptation for medical image segmentation. AAAI Conf. Artif. Intell. (AAAI) **33**, 865–872 (2019)
10. Chen, Y.S., Wang, Y.C., Kao, M.H., Chuang, Y.Y.: Deep photo enhancer: unpaired learning for image enhancement from photographs with GANs. In: IEEE Conference on Computer Vision and Pattern Recognition (CVPR) (June 2018)
11. Ganin, Y., et al.: Domain-adversarial training of neural networks. J. Mach. Learn. Res. (JMLR) **17**(1), 2096–2030 (2016)
12. Gharbi, M., Chen, J., Barron, J., Hasinoff, S., Durand, F.: Deep bilateral learning for real-time image enhancement. ACM Trans. Graph. (TOG) **36**(4), 1–12 (2017)
13. Ghifary, M., Kleijn, W.B., Zhang, M., Balduzzi, D., Li, W.: Deep reconstruction-classification networks for unsupervised domain adaptation. In: Leibe, B., Matas, J., Sebe, N., Welling, M. (eds.) ECCV 2016. LNCS, vol. 9908, pp. 597–613. Springer, Cham (2016). https://doi.org/10.1007/978-3-319-46493-0_36
14. Heusel, M., Ramsauer, H., Unterthiner, T., Nessler, B., Hochreiter, S.: GANs trained by a two time-scale update rule converge to a local Nash equilibrium. In: Advances in Neural Information Processing Systems (NIPS), pp. 6626–6637 (2017)

15. Hoffman, J., Wang, D., Yu, F., Darrell, T.: FCNs in the wild: Pixel-level adversarial and constraint-based adaptation. arXiv preprint arXiv:1612.02649 (2016)

16. Hong, W., Wang, Z., Yang, M., Yuan, J.: Conditional generative adversarial network for structured domain adaptation. In: IEEE Conference on Computer Vision and Pattern Recognition (CVPR), pp. 1335–1344 (2018)

17. Isola, P., Zhu, J., Zhou, T., Efros, A.A.: Image-to-image translation with conditional adversarial networks. In: IEEE Conference on Computer Vision and Pattern Recognition (CVPR), pp. 5967–5976 (2017)

18. Jiang, Y., et al.: EnlightenGAN: Deep light enhancement without paired supervision. arXiv preprint arXiv:1906.06972 (2019)

19. Kalantari, N.K., Ramamoorthi, R.: Deep high dynamic range imaging of dynamic scenes. ACM Trans. Graph. (TOG) 36(4), 1–12 (2017)

20. Kim, H., Handa, A., Benosman, R., Ieng, S.H., Davison, A.: Simultaneous mosaicing and tracking with an event camera. Br. Mach. Vis. Conf. (BMVC) 43, 566–576 (2008)

21. Wang, L., Mostafavi, S.M., Ho, Y.S., Yoon, K.J.: Event-based high dynamic range image and very high frame rate video generation using conditional generative adversarial networks (2019)

22. Ledig, C., et al.: Photo-realistic single image super-resolution using a generative adversarial network. In: IEEE Conference on Computer Vision and Pattern Recognition (CVPR), pp. 4681–4690 (2017)

23. Lichtsteiner, P., Posch, C., Delbruck, T.: A 128×128 120 db 15µs latency asynchronous temporal contrast vision sensor. IEEE J. Solid State Circ. 43(2), 566–576 (2008)

24. Long, M., Cao, Y., Wang, J., Jordan, M.: Learning transferable features with deep adaptation networks. In: International Conference on Machine Learning (ICML) (2015)

25. Lore, K.G., Akintayo, A., Sarkar, S.: LLNet: a deep autoencoder approach to natural low-light image enhancement. Pattern Recogn. 61, 650–662 (2017)

26. Mao, Q., Lee, H.Y., Tseng, H.Y., Ma, S., Yang, M.H.: Mode seeking generative adversarial networks for diverse image synthesis. In: IEEE Conference on Computer Vision and Pattern Recognition (CVPR), pp. 1429–1437 (2019)

27. Munda, G., Reinbacher, C., Pock, T.: Real-time intensity-image reconstruction for event cameras using manifold regularisation. Int. J. Comput. Vis. (IJCV) 126(12), 1381–1393 (2018)

28. Murez, Z., Kolouri, S., Kriegman, D., Ramamoorthi, R., Kim, K.: Image to image translation for domain adaptation. In: IEEE Conference on Computer Vision and Pattern Recognition (CVPR), pp. 4500–4509 (2018)

29. Rebecq, H., Gehrig, D., Scaramuzza, D.: ESIM: an open event camera simulator. In: Conference on Robotics Learning (CoRL) (2018)

30. Rebecq, H., Ranftl, R., Koltun, V., Scaramuzza, D.: Events-to-video: bringing modern computer vision to event cameras. In: IEEE Conference on Computer Vision and Pattern Recognition (CVPR) (2019)

31. Salimans, T., Goodfellow, I., Zaremba, W., Cheung, V., Radford, A., Chen, X.: Improved techniques for training GANs. In: Advances in Neural Information Processing Systems (NIPS), pp. 2234–2242 (2016)

32. Scheerlinck, C., Barnes, N., Mahony, R.: Continuous-time intensity estimation using event cameras. In: Jawahar, C.V., Li, H., Mori, G., Schindler, K. (eds.) ACCV 2018. LNCS, vol. 11365, pp. 308–324. Springer, Cham (2019). https://doi.org/10.1007/978-3-030-20873-8_20

33. Sun, B., Feng, J., Saenko, K.: Return of frustratingly easy domain adaptation. In: AAAI Conference on Artificial Intelligence (AAAI) (2016)
34. Tsai, Y.H., Sohn, K., Schulter, S., Chandraker, M.: Domain adaptation for structured output via discriminative patch representations. In: IEEE International Conference on Computer Vision (ICCV), pp. 1456–1465 (2019)
35. Tzeng, E., Hoffman, J., Darrell, T., Saenko, K.: Simultaneous deep transfer across domains and tasks. In: IEEE International Conference on Computer Vision (ICCV), pp. 4068–4076 (2015)
36. Wang, X., et al.: ESRGAN: enhanced super-resolution generative adversarial networks. In: Leal-Taixé, L., Roth, S. (eds.) ECCV 2018. LNCS, vol. 11133, pp. 63–79. Springer, Cham (2019). https://doi.org/10.1007/978-3-030-11021-5_5
37. Wu, S., Xu, J., Tai, Y.-W., Tang, C.-K.: Deep high dynamic range imaging with large foreground motions. In: Ferrari, V., Hebert, M., Sminchisescu, C., Weiss, Y. (eds.) ECCV 2018. LNCS, vol. 11206, pp. 120–135. Springer, Cham (2018). https://doi.org/10.1007/978-3-030-01216-8_8
38. Zhu, J.Y., Park, T., Isola, P., Efros, A.A.: Unpaired image-to-image translation using cycle-consistent adversarial networks. In: International Conference on Computer Vision (ICCV) (2017)

Trajectron++: Dynamically-Feasible Trajectory Forecasting with Heterogeneous Data

Tim Salzmann[1], Boris Ivanovic[1(✉)], Punarjay Chakravarty[2], and Marco Pavone[1]

[1] Autonomous Systems Lab, Stanford University, Stanford, USA
{timsal,borisi,pavone}@stanford.edu
[2] Ford Greenfield Labs, Palo Alto, USA
pchakra5@ford.com

Abstract. Reasoning about human motion is an important prerequisite to safe and socially-aware robotic navigation. As a result, multi-agent behavior prediction has become a core component of modern human-robot interactive systems, such as self-driving cars. While there exist many methods for trajectory forecasting, most do not enforce dynamic constraints and do not account for environmental information (e.g., maps). Towards this end, we present *Trajectron++*, a modular, graph-structured recurrent model that forecasts the trajectories of a general number of diverse agents while incorporating agent dynamics and heterogeneous data (e.g., semantic maps). *Trajectron++* is designed to be tightly integrated with robotic planning and control frameworks; for example, it can produce predictions that are optionally conditioned on ego-agent motion plans. We demonstrate its performance on several challenging real-world trajectory forecasting datasets, outperforming a wide array of state-of-the-art deterministic and generative methods.

Keywords: Trajectory forecasting · Spatiotemporal graph modeling · Human-robot interaction · Autonomous driving

1 Introduction

Predicting the future behavior of humans is a necessary part of developing safe human-interactive autonomous systems. Humans can naturally navigate through many social interaction scenarios because they have an intrinsic "theory of

---

T. Salzmann and B. Ivanovic—Equal contribution.

T. Salzmann—Work done as a visiting student in the Autonomous Systems Lab.

Electronic supplementary material The online version of this chapter (https://doi.org/10.1007/978-3-030-58523-5_40) contains supplementary material, which is available to authorized users.

© Springer Nature Switzerland AG 2020
A. Vedaldi et al. (Eds.): ECCV 2020, LNCS 12363, pp. 683–700, 2020.
https://doi.org/10.1007/978-3-030-58523-5_40

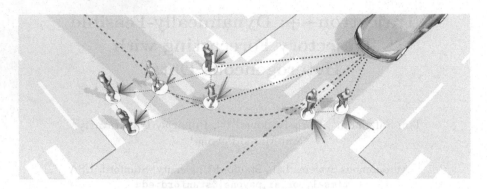

Fig. 1. Exemplary road scene depicting pedestrians crossing a road in front of a vehicle which may continue straight or turn right. The graph representation of the scene is shown on the ground, where each agent and their interactions are represented as nodes and edges, visualized as white circles and dashed black lines, respectively. Arrows depict potential future agent velocities, with colors representing different high-level future behavior modes.

mind," which is the capacity to reason about other people's actions in terms of their mental states [14]. As a result, imbuing autonomous systems with this capability could enable more informed decision making and proactive actions to be taken in the presence of other intelligent agents, e.g., in human-robot interaction scenarios. Figure 1 illustrates a scenario where predicting the intent of other agents may inform an autonomous vehicle's path planning and decision making. Indeed, multi-agent behavior prediction has already become a core component of modern robotic systems, especially in safety-critical applications like self-driving vehicles which are currently being tested in the real world and targeting widespread deployment in the near future [49].

There are many existing methods for multi-agent behavior prediction, ranging from deterministic regressors to generative, probabilistic models. However, many of them were developed without directly accounting for real-world robotic use cases; in particular, they ignore agents' dynamics constraints, the ego-agent's own motion (important to capture the interactive aspect in human-robot interaction), and a plethora of environmental information (e.g., camera images, lidar, maps) to which modern robotic systems have access. Table 1 provides a summary of recent state-of-the-art approaches and their consideration of such desiderata.

Accordingly, in this work we are interested in developing a multi-agent behavior prediction model that (1) accounts for the dynamics of the agents, and in particular of ground vehicles [27,35]; (2) produces predictions possibly conditioned on potential future robot trajectories, useful for intelligent planning taking into account human responses; and (3) provides a generally-applicable, open, and extensible approach which can effectively use heterogeneous data about the surrounding environment. Importantly, making use of such data would allow for the incorporation of environmental information, e.g., maps, which would enable producing predictions that differ depending on the structure of the scene (e.g.,

Table 1. A summary of recent state-of-the-art pedestrian (left) and vehicle (right) trajectory forecasting methods, indicating the desiderata addressed by each approach.

Method	GNA	CD	HD	FCP	OS	Method	GNA	CD	HD	FCP	OS
DESIRE [31]	✓		✓			IntentNet [8]	✓		✓		
Trajectron [20]	✓				✓	PRECOG [39]			✓	✓	✓
S-BiGAT [28]	✓		✓			MFP [18]	✓		✓		✓
DRF-Net [23]	✓		✓			NMP [51]	✓		✓		
MATF [53]	✓		✓		✓	SpAGNN [7]	✓		✓		
Our work	✓	✓	✓	✓	✓	Our work	✓	✓	✓	✓	✓

Legend: GNA = General Number of Agents, CD = Considers Dynamics, HD = Heterogeneous Data, FCP = Future-Conditional Predictions, OS = Open Source.

interactions at an urban intersection are very different from those in an open sports field!). One method that comes close is the Trajectron [20], a multi-agent behavior model which can handle a time-varying number of agents, accounts for multimodality in human behavior (i.e., the potential for many high-level futures), and maintains a sense of interpretability in its outputs. However, the Trajectron only reasons about relatively simple vehicle models (i.e., cascaded integrators) and past trajectory data (i.e., no considerations are made for added environmental information, if available).

In this work we present *Trajectron++*, an open and extensible approach built upon the Trajectron [20] framework which produces dynamically-feasible trajectory forecasts from heterogeneous input data for multiple interacting agents of distinct semantic types. Our key contributions are twofold: First, we show how to effectively incorporate high-dimensional data through the lens of encoding semantic maps. Second, we propose a general method of incorporating dynamics constraints into learning-based methods for multi-agent trajectory forecasting. *Trajectron++* is designed to be tightly integrated with downstream robotic modules, with the ability to produce trajectories that are optionally conditioned on future ego-agent motion plans. We present experimental results on a variety of datasets, which collectively demonstrate that *Trajectron++* outperforms an extensive selection of state-of-the-art deterministic and generative trajectory prediction methods, in some cases achieving 60% lower average prediction error.

2   Related Work

Deterministic Regressors. Many earlier works in human trajectory forecasting were deterministic regression models. One of the earliest, the Social Forces model [16], models humans as physical objects affected by Newtonian forces (e.g., with attractors at goals and repulsors at other agents). Since then, many approaches have been applied to the problem of trajectory forecasting, formulating it as a time-series regression problem and applying methods like Gaussian Process Regression (GPR) [38,48], Inverse Reinforcement Learning (IRL) [32], and Recurrent Neural Networks (RNNs) [1,34,47] to good effect. An excellent review of such methods can be found in [40].

Generative, Probabilistic Approaches. Recently, generative approaches have emerged as state-of-the-art trajectory forecasting methods due to recent advancements in deep generative models [12,44]. Notably, they have caused a shift from focusing on predicting the single best trajectory to producing a *distribution* of potential future trajectories. This is advantageous in autonomous systems as full distribution information is more useful for downstream tasks, e.g., motion planning and decision making where information such as variance can be used to make safer decisions. Most works in this category use a deep recurrent backbone architecture with a latent variable model, such as a Conditional Variational Autoencoder (CVAE) [44], to explicitly encode multimodality [11,20,21,31,39,42], or a Generative Adversarial Network (GAN) [12] to implicitly do so [13,28,41,53]. Common to both approach styles is the need to produce position distributions. GAN-based models can directly produce these and CVAE-based recurrent models usually rely on a bivariate Gaussian Mixture Model (GMM) to output position distributions. However, both of these output structures make it difficult to enforce dynamics constraints, e.g., non-holonomic constraints such as those arising from no side-slip conditions. Of these, the Trajectron [20] and MATF [53] are the best-performing CVAE-based and GAN-based models, respectively, on standard pedestrian trajectory forecasting benchmarks [33,37].

Accounting for Dynamics and Heterogeneous Data. There are few works that account for dynamics or make use of data modalities outside of prior trajectory information. This is mainly because standard trajectory forecasting benchmarks seldom include any other information, a fact that will surely change following the recent release of autonomous vehicle-based datasets with rich multisensor data [6,9,26,50]. As for dynamics, current methods almost exclusively reason about positional information. This does not capture dynamical constraints, however, which might lead to predictions in position space that are unrealizable by the underlying control variables (e.g., a car moving sideways). Table 1 provides a detailed breakdown of recent state-of-the-art approaches and their consideration of these desiderata.

3   Problem Formulation

We aim to generate plausible trajectory distributions for a time-varying number $N(t)$ of interacting agents $A_1, \ldots, A_{N(t)}$. Each agent A_i has a semantic class S_i, e.g., Car, Bus, or Pedestrian. At time t, given the state $\mathbf{s} \in \mathbb{R}^D$ of each agent and all of their histories for the previous H timesteps, which we denote as $\mathbf{x}$, $\mathbf{x} = \mathbf{s}_{1,\ldots,N(t)}^{(t-H:t)} \in \mathbb{R}^{(H+1) \times N(t) \times D}$, as well as additional information available to each agent $I_{1,\ldots,N(t)}^{(t)}$, we seek a distribution over all agents' future states for the next T timesteps $\mathbf{y} = \mathbf{s}_{1,\ldots,N(t)}^{(t+1:t+T)} \in \mathbb{R}^{T \times N(t) \times D}$, which we denote as $p(\mathbf{y} \mid \mathbf{x}, I)$.

We also assume that geometric semantic maps are available around A_i's position, $M_i^{(t)} \in \mathbb{R}^{\lceil C/r \rceil \times \lceil C/r \rceil \times L}$, with context size $C \times C$, spatial resolution r,

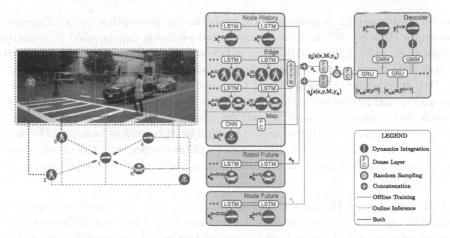

Fig. 2. Left: Our approach represents a scene as a directed spatiotemporal graph. Nodes and edges represent agents and their interactions, respectively. **Right:** The corresponding network architecture for Node 1.

and L semantic channels. Depending on the dataset, these maps can range in sophistication from simple obstacle occupancy grids to multiple layers of human-annotated semantic information (e.g., marking out sidewalks, road boundaries, and crosswalks).

We also consider the setting where we condition on an ego-agent's future motion plan, for example when evaluating responses to a set of motion primitives. In this setting, we additionally assume that we know the ego-agent's future motion plan for the next T timesteps, $\mathbf{y}_R = \mathbf{s}_R^{(t+1:t+T)}$.

4 Trajectron++

Our approach[1] is visualized in Fig. 2. At a high level, a spatiotemporal graph representation of the scene in question is created from its topology. Then, a similarly-structured deep learning architecture is generated that forecasts the evolution of node attributes, producing agent trajectories.

Scene Representation. The current scene is abstracted as a spatiotemporal graph $G = (V, E)$. Nodes represent agents and edges represent their interactions. As a result, in the rest of the paper we will use the terms "node" and "agent" interchangeably. Each node also has a semantic class matching the class of its agent (e.g., Car, Bus, Pedestrian). An edge (A_i, A_j) is present in E if A_i influences A_j. In this work, the ℓ_2 distance is used as a proxy for whether agents are influencing each other or not. Formally, an edge is directed from A_i to A_j if $\|\mathbf{p}_i - \mathbf{p}_j\|_2 \leq d_{S_j}$ where $\mathbf{p}_i, \mathbf{p}_j \in \mathbb{R}^2$ are the 2D world positions of agents A_i, A_j,

---

[1] All of our source code, trained models, and data can be found online at https://github.com/StanfordASL/Trajectron-plus-plus.

respectively, and d_{S_j} is a distance that encodes the perception range of agents of semantic class S_j. While more sophisticated methods can be used to construct edges (e.g., [47]), they usually incur extra computational overhead by requiring a complete scene graph. Figure 2 shows an example of this scene abstraction.

We specifically choose to model the scene as a directed graph, in contrast to an undirected one as in previous approaches [1,13,20–22,47], because a directed graph can represent a more general set of scenes and interaction types, e.g., asymmetric influence. This provides the additional benefit of being able to simultaneously model agents with different perception ranges, e.g., the driver of a car looks much farther ahead on the road than a pedestrian does while walking on the sidewalk.

Modeling Agent History. Once a graph of the scene is constructed, the model needs to encode a node's current state, its history, and how it is influenced by its neighboring nodes. To encode the observed history of the modeled agent, their current and previous states are fed into a Long Short-Term Memory (LSTM) network [19] with 32 hidden dimensions. Since we are interested in modeling trajectories, the inputs $\mathbf{x} = \mathbf{s}_{1,...,N(t)}^{(t-H:t)} \in \mathbb{R}^{(H+1)\times N(t)\times D}$ are the current and previous D-dimensional states of the modeled agents. These are typically positions and velocities, which can be easily estimated online.

Ideally, agent models should be chosen to best match their semantic class S_i. For example, one would usually model vehicles on the road using a bicycle model [27,35]. However, estimating the bicycle model parameters of another vehicle from online observations is very difficult as it requires estimation of the vehicle's center of mass, wheelbase, and front wheel steer angle. As a result, in this work pedestrians are modeled as single integrators and wheeled vehicles are modeled as dynamically-extended unicycles [29], enabling us to account for key non-holonomic constraints (e.g., no side-slip constraints) [35] without requiring complex online parameter estimation procedures – we will show through experiments that such a simplified model is already quite impactful on improving prediction accuracy. While the dynamically-extended unicycle model serves as an important representative example, we note that our approach can also be generalized to other dynamics models, provided its parameters can either be assumed or quickly estimated online.

Encoding Agent Interactions. To model neighboring agents' influence on the modeled agent, *Trajectron++* encodes graph edges in two steps. First, edge information is aggregated from neighboring agents of the same semantic class. In this work, an element-wise sum is used as the aggregation operation. We choose to combine features in this way rather than with concatenation or an average to handle a variable number of neighboring nodes with a fixed architecture while preserving count information [3,21,22]. These aggregated states are then fed into an LSTM with 8 hidden dimensions whose weights are shared across all edge instances of the same type, e.g., all Pedestrian-Bus edge LSTMs share the same weights. Then, the encodings from all edge types that connect to the modeled node are aggregated to obtain one "influence" representation vector, representing the effect that all neighboring nodes have. For this, an additive attention module is used [2]. Finally, the node history and edge influence encodings are concatenated to produce a single node representation vector, e_x.

Incorporating Heterogeneous Data. Modern sensor suites are able to produce much more information than just tracked trajectories of other agents. Notably, HD maps are used by many real-world systems to aid localization as well as inform navigation. Depending on sensor availability and sophistication, maps can range in fidelity from simple binary obstacle maps, i.e., $M \in \{0,1\}^{H \times W \times 1}$, to HD semantic maps, e.g., $M \in \{0,1\}^{H \times W \times L}$ where each layer $1 \leq \ell \leq L$ corresponds to an area with semantic type (e.g., "driveable area," "road block," "walkway," "pedestrian crossing"). To make use of this information, for each modeled agent, *Trajectron++* encodes a local map, rotated to match the agent's heading, with a Convolutional Neural Network (CNN). The CNN has 4 layers, with filters $\{5,5,5,3\}$ and respective strides of $\{2,2,1,1\}$. These are followed by a dense layer with 32 hidden dimensions, the output of which is concatenated with the node history and edge influence representation vectors.

More generally, one can include further additional information (e.g., raw LIDAR data, camera images, pedestrian skeleton or gaze direction estimates) in this framework by encoding it as a vector and adding it to this backbone of representation vectors, e_x.

Encoding Future Ego-Agent Motion Plans. Producing predictions which take into account future ego-agent motion is an important capability for robotic decision making and control. Specifically, it allows for the evaluation of a set of motion primitives with respect to possible responses from other agents. *Trajectron++* can encode the future T timesteps of the ego-agent's motion plan y_R using a bi-directional LSTM with 32 hidden dimensions. A bi-directional LSTM is used due to its strong performance on other sequence summarization tasks [5]. The final hidden states are then concatenated into the backbone of representation vectors, e_x.

Explicitly Accounting for Multimodality. *Trajectron++* explicitly handles multimodality by leveraging the CVAE latent variable framework [44]. It produces the target $p(y \mid x)$ distribution by introducing a discrete Categorical latent variable $z \in Z$ which encodes high-level latent behavior and allows for $p(y \mid x)$ to be expressed as $p(y \mid x) = \sum_{z \in Z} p_\psi(y \mid x, z) p_\theta(z \mid x)$, where $|Z| = 25$ and ψ, θ are deep neural network weights that parameterize their respective distributions. z being discrete also aids in interpretability, as one can visualize which high-level behaviors belong to each z by sampling trajectories.

During training, a bi-directional LSTM with 32 hidden dimensions is used to encode a node's ground truth future trajectory, producing $q_\phi(z \mid x, y)$ [44].

Producing Dynamically-Feasible Trajectories. After obtaining a latent variable z, it and the backbone representation vector e_x are fed into the decoder, a 128-dimensional Gated Recurrent Unit (GRU) [10]. Each GRU cell outputs the parameters of a bivariate Gaussian distribution over control actions $u^{(t)}$ (e.g., acceleration and steering rate). The agent's system dynamics are then integrated with the produced control actions $\mathbf{u}^{(t)}$ to obtain trajectories in position space [25,46]. The only uncertainty at prediction time stems from *Trajectron++*'s output. Thus, in the case of linear dynamics (e.g., single integrators, used in this work to model pedestrians), the system dynamics are linear Gaussian. Explicitly, for a single integrator with control actions $\mathbf{u}^{(t)} = \dot{\mathbf{p}}^{(t)}$, the position mean at $t + 1$ is $\mu_{\mathbf{p}}^{(t+1)} = \mu_{\mathbf{p}}^{(t)} + \mu_{\mathbf{u}}^{(t)} \Delta t$, where $\mu_{\mathbf{u}}^{(t)}$ is produced by *Trajectron++*. In the case of nonlinear dynamics (e.g., unicycle models, used in this work to model vehicles), one can still (approximately) use this uncertainty propagation scheme by linearizing the dynamics about the agent's current state and control. Full mean and covariance equations for the single integrator and dynamically-extended unicycle models are in the appendix. In contrast to existing methods which directly output positions, our approach is uniquely able to guarantee that its trajectory samples are dynamically feasible by integrating an agent's dynamics with the predicted controls.

Output Configurations. Based on the desired use case, *Trajectron++* can produce many different outputs. The main four are outlined below.

1. *Most Likely (ML)*: The model's deterministic and most-likely single output. The high-level latent behavior mode and output trajectory are the modes of their respective distributions, where

$$z_{\text{mode}} = \arg\max_z p_\theta(z \mid \mathbf{x}), \qquad \mathbf{y} = \arg\max_{\mathbf{y}} p_\psi(\mathbf{y} \mid \mathbf{x}, z_{\text{mode}}). \qquad (1)$$

2. z_{mode}: Predictions from the model's most-likely high-level latent behavior mode, where

$$z_{\text{mode}} = \arg\max_z p_\theta(z \mid \mathbf{x}), \qquad \mathbf{y} \sim p_\psi(\mathbf{y} \mid \mathbf{x}, z_{\text{mode}}). \qquad (2)$$

3. *Full*: The model's full sampled output, where z and y are sampled sequentially according to

$$z \sim p_\theta(z \mid \mathbf{x}), \qquad \mathbf{y} \sim p_\psi(\mathbf{y} \mid \mathbf{x}, z). \qquad (3)$$

4. *Distribution*: Due to the use of a discrete latent variable and Gaussian output structure, the model can provide an analytic output distribution by directly computing $p(\mathbf{y} \mid \mathbf{x}) = \sum_{z \in Z} p_\psi(\mathbf{y} \mid \mathbf{x}, z) p_\theta(z \mid \mathbf{x})$.

Training the Model. We adopt the InfoVAE [52] objective function, and modify it to use discrete latent states in a conditional formulation (since the model uses a CVAE). Formally, we aim to solve

$$\max_{\phi,\theta,\psi} \sum_{i=1}^N \mathbb{E}_{z \sim q_\phi(\cdot \mid \mathbf{x}_i, \mathbf{y}_i)} \big[\log p_\psi(\mathbf{y}_i \mid \mathbf{x}_i, z) \big]$$
$$- \beta D_{KL}\big(q_\phi(z \mid \mathbf{x}_i, \mathbf{y}_i) \parallel p_\theta(z \mid \mathbf{x}_i)\big) + \alpha I_q(\mathbf{x}; z), \qquad (4)$$

where I_q is the mutual information between $\mathbf{x}$ and z under the distribution $q_\phi(\mathbf{x}, z)$. To compute I_q, we follow [52] and approximate $q_\phi(z \mid \mathbf{x}_i, \mathbf{y}_i)$ with

$p_\theta(z \mid \mathbf{x}_i)$, obtaining the unconditioned latent distribution by summing out $\mathbf{x}_i$ over the batch. Notably, the Gumbel-Softmax reparameterization [24] is not used to backpropagate through the Categorical latent variable z because it is not sampled during training time. Instead, the first term of Eq. (4) is directly computed since the latent space has only $|Z| = 25$ discrete elements. Additional training details can be found in the appendix.

5   Experiments

Our method is evaluated on three publicly-available datasets: The ETH [37], UCY [33], and nuScenes [6] datasets. The ETH and UCY datasets consist of real pedestrian trajectories with rich multi-human interaction scenarios captured at 2.5 Hz ($\Delta t = 0.4$ s). In total, there are 5 sets of data, 4 unique scenes, and 1536 unique pedestrians. They are a standard benchmark in the field, containing challenging behaviors such as couples walking together, groups crossing each other, and groups forming and dispersing. However, they only contain pedestrians, so we also evaluate on the recently-released nuScenes dataset. It is a large-scale dataset for autonomous driving with 1000 scenes in Boston and Singapore. Each scene is annotated 2 Hz ($\Delta t = 0.5$ s) and is 20 s long, containing up to 23 semantic object classes as well as HD semantic maps with 11 annotated layers.

Trajectron++ was implemented in PyTorch [36] on a desktop computer running Ubuntu 18.04 containing an AMD Ryzen 1800X CPU and two NVIDIA GTX 1080 Ti GPUs. We trained the model for 100 epochs ($\sim$3 h) on the pedestrian datasets and 12 epochs ($\sim$8 h) on the nuScenes dataset.

Evaluation Metrics. As in prior work [1,13,20,28,41,53], our method for trajectory forecasting is evaluated with the following four error metrics:

1. *Average Displacement Error (ADE):* Mean ℓ_2 distance between the ground truth and predicted trajectories.
2. *Final Displacement Error (FDE):* ℓ_2 distance between the predicted final position and the ground truth final position at the prediction horizon T.
3. *Kernel Density Estimate-based Negative Log Likelihood (KDE NLL):* Mean NLL of the ground truth trajectory under a distribution created by fitting a kernel density estimate on trajectory samples [20,45].
4. *Best-of-N (BoN):* The minimum ADE and FDE from N randomly-sampled trajectories. We compare our method to an exhaustive set of state-of-the art deterministic and generative approaches.

Deterministic Baselines. Our method is compared against the following deterministic baselines: (1) *Linear*: A linear regressor with parameters estimated by minimizing least square error. (2) *LSTM*: An LSTM network with only agent history information. (3) *Social LSTM* [1]: Each agent is modeled with an LSTM and nearby agents' hidden states are pooled at each timestep using a proposed social pooling operation. (4) *Social Attention* [47]: Same as [1], but all other agents' hidden states are incorporated via a proposed social attention operation.

Generative Baselines. On the ETH and UCY datasets, our method is compared against the following generative baselines: (1) *S-GAN* [13]: Each agent is modeled with an LSTM-GAN, which is an LSTM encoder-decoder whose outputs are the generator of a GAN. The generated trajectories are then evaluated against the ground truth trajectories with a discriminator. (2) *SoPhie* [41]: An LSTM-GAN with the addition of a proposed physical and social attention module. (3) *MATF* [53]: An LSTM-GAN model that leverages CNNs to fuse agent relationships and encode environmental information. (4) *Trajectron* [20]: An LSTM-CVAE encoder-decoder which is explicitly constructed to match the spatiotemporal structure of the scene. Its scene abstraction is similar to ours, but uses undirected edges.

On the nuScenes dataset, the following methods are also compared against: (5) *Convolutional Social Pooling (CSP)* [11]: An LSTM-based approach which explicitly considers a fixed number of movement classes and predicts which of those the modeled agent is likely to take. (6) *CAR-Net* [42]: An LSTM-based approach which encodes scene context with visual attention. (7) *SpAGNN* [7]: A CNN encodes raw LIDAR and semantic map data to produce object detections, from which a Graph Neural Network (GNN) produces probabilistic, interaction-aware trajectories.

Evaluation Methodology. For the ETH and UCY datasets, a leave-one-out strategy is used for evaluation, similar to previous works [1,13,20,28,41,53], where the model is trained on four datasets and evaluated on the held-out fifth. An observation length of 8 timesteps (3.2 s) and a prediction horizon of 12 timesteps (4.8 s) is used for evaluation. For the nuScenes dataset, we split off 15% of the train set for hyperparameter tuning and test on the provided validation set.

Throughout the following, we report the performance of *Trajectron++* in multiple configurations. Specifically, *Ours* refers to the base model using only node and edge encoding, trained to predict agent velocities and Euler integrating velocity to produce positions; *Ours+∫* is the base model with dynamics integration, trained to predict control actions and integrating the agent's dynamics with the control actions to produce positions; *Ours+∫, M* additionally includes the map encoding CNN; and *Ours+∫, M*, $\mathbf{y}_R$ adds the robot future encoder.

5.1   ETH and UCY Datasets

Our approach is first evaluated on the ETH [37] and UCY [33] Pedestrian Datasets, against deterministic methods on standard trajectory forecasting metrics. It is difficult to determine the current state-of-the-art in deterministic methods as there are contradictions between the results reported by the same authors in [13] and [1]. In Table 1 of [1], Social LSTM *convincingly* outperforms a baseline LSTM without pooling. However, in Table 1 of [13], Social LSTM is actually *worse* than the same baseline on average. Thus, when comparing against Social LSTM we report the results summarized in Table 1 of [13] as it is the most recent work by the same authors. Further, the values reported by Social Attention in [47] seem to have unusually high ratios of FDE to ADE. Nearly every

Table 2. (a) Our model's deterministic Most Likely output outperforms other deterministic methods on displacement error metrics, even if it was not originally trained to do so. (b) Our model's probabilistic Full output significantly outperforms other methods, yielding accurate predictions even in a small number of samples. Lower is better. Bold indicates best.

Dataset	(a) ADE/FDE (m)					
	Linear	LSTM	S-LSTM [13]	S-ATTN [47]	Ours (ML)	Ours+$\int$ (ML)
ETH	1.33/2.94	1.09/2.41	1.09/2.35	**0.39**/3.74	0.71/**1.66**	0.71/1.68
Hotel	0.39/0.72	0.86/1.91	0.79/1.76	0.29/2.64	**0.22/0.46**	**0.22/0.46**
Univ	0.82/1.59	0.61/1.31	0.67/1.40	**0.33**/3.92	0.44/1.17	0.41/**1.07**
Zara 1	0.62/1.21	0.41/0.88	0.47/1.00	**0.20/0.52**	0.30/0.79	0.30/0.77
Zara 2	0.77/1.48	0.52/1.11	0.56/1.17	0.30/2.13	**0.23/0.59**	**0.23/0.59**
Average	0.79/1.59	0.70/1.52	0.72/1.54	**0.30**/2.59	0.38/0.93	0.37/**0.91**

Dataset	(b) ADE/FDE, Best of 20 Samples (m)					
	S-GAN [13]	SoPhie [41]	Trajectron [20]	MATF [53]	Ours (Full)	Ours+$\int$ (Full)
ETH	0.81/1.52	0.70/1.43	0.59/1.14	1.01/1.75	**0.39/0.83**	0.43/0.86
Hotel	0.72/1.61	0.76/1.67	0.35/0.66	0.43/0.80	**0.12/0.21**	**0.12**/0.19
Univ	0.60/1.26	0.54/1.24	0.54/1.13	0.44/0.91	**0.20/0.44**	0.22/**0.43**
Zara 1	0.34/0.69	0.30/0.63	0.43/0.83	0.26/0.45	**0.15/0.33**	0.17/**0.32**
Zara 2	0.42/0.84	0.38/0.78	0.43/0.85	0.26/0.57	**0.11/0.25**	0.12/**0.25**
Average	0.58/1.18	0.54/1.15	0.47/0.92	0.48/0.90	**0.19/0.41**	0.21/**0.41**

Legend: $\int$ = Integration via Dynamics, M = Map Encoding, $\mathbf{y}_R$ = Robot Future Encoding.

other method (including ours) has FDE/ADE ratios around 2–3× whereas Social Attention's are around 3–12×. Social Attention's errors on the Univ dataset are especially striking, as its FDE of 3.92 is 12× its ADE of 0.33, meaning its prediction error on the other 11 timesteps is essentially zero. We still compare against the values reported in [47] as there is no publicly-released code, but this raises doubts of their validity. To fairly compare against prior work, neither map encoding nor future motion plan encoding is used. Only the node history and edge encoders are used in the model's encoder. Additionally, the model's deterministic ML output scheme is employed, which produces the model's most likely single trajectory. Table 2(a) summarizes these results and shows that our approach is competitive with state-of-the-art deterministic regressors on displacement error metrics (outperforming existing approaches by 33% on mean FDE), even though our method was not originally trained to minimize this. It makes sense that the model performs similarly with and without dynamics integration for pedestrians, since they are modeled as single integrators. Thus, their control actions are velocities which matches the base model's output structure.

To more concretely compare generative methods, we use the KDE-based NLL metric proposed in [20,45], an approach that maintains full output distributions and compares the log-likelihood of the ground truth under different methods' outputs. Table 3 summarizes these results and shows that our method significantly outperforms others. This is also where the performance improve-

Table 3. Mean KDE-based NLL for each dataset. Lower is better. 2000 trajectories were sampled per model at each prediction timestep. Bold indicates the best values.

Dataset	KDE NLL			
	S-GAN [13]	Trajectron [20]	Ours (Full)	Ours+$\int$ (Full)
ETH	15.70	2.99	1.80	**1.31**
Hotel	8.10	2.26	−1.29	**−1.94**
Univ	2.88	1.05	−0.89	**−1.13**
Zara 1	1.36	1.86	−1.13	**−1.41**
Zara 2	0.96	0.81	−2.19	**−2.53**
Average	5.80	1.79	−0.74	**−1.14**

Legend: $\int$ = Integration via Dynamics, M = Map Encoding, $\mathbf{y}_R$ = Robot Future Encoding.

ments brought by the dynamics integration scheme are clear. It yields the best performance because the model is now explicitly trained on the distribution it is seeking to output (the loss function term $p_\psi(\mathbf{y}|\mathbf{x}, z)$ is now directly over positions), whereas the base model is trained on velocity distributions, the integration of which (with no accounting for system dynamics) introduces errors. Unfortunately, at this time there are no publicly-released models for SoPhie [41] or MATF [53], so they cannot be evaluated with the KDE-based NLL metric. Instead, we evaluate *Trajectron++* with the Best-of-N metric used in their works. Table 2(b) summarizes these results, and shows that our method *significantly* ourperforms the state-of-the-art [53], achieving 55–60% lower average errors.

Map Encoding. To evaluate the effect of incorporating heterogeneous data, we compare the performance of *Trajectron++* with and without the map encoder. Specifically, we compare the frequency of obstacle violations in 2000 trajectory samples from the Full model output on the ETH - University scene, which provides a simple binary obstacle map. Overall, our approach generates colliding predictions 1.0% of the time with map encoding, compared to 4.6% without map encoding. We also study how much of a reduction there is for pedestrians that are especially close to an obstacle (i.e. they have at least one obstacle-violating trajectory in their Full output), an example of which is shown in the appendix. In this regime, our approach generates colliding predictions 4.9% of the time with map encoding, compared to 21.5% without map encoding.

5.2   nuScenes Dataset

To further evaluate the model's ability to use heterogeneous data and simultaneously model multiple semantic classes of agents, we evaluate it on the nuScenes dataset [6]. Again, the deterministic ML output scheme is used to fairly compare with other single-trajectory predictors. The trajectories of both Pedestrians and Cars are forecasted, two semantic object classes which account for most of the 23

Table 4. [nuScenes] (a): Vehicle-only FDE across time for *Trajectron++* compared to that of other single-trajectory and probabilistic approaches. Bold indicates best. (b): Pedestrian-only FDE and KDE NLL across time for *Trajectron++*.

(a) Vehicle-only

Method	FDE (m)			
	@1s	@2s	@3s	@4s
Const. Velocity	0.32	0.89	1.70	2.73
S-LSTM* [1,7]	0.47	–	1.61	–
CSP* [7,11]	0.46	–	1.50	–
CAR-Net* [7,42]	0.38	–	1.35	–
SpAGNN* [7]	0.36	–	1.23	–
Ours (ML)	0.18	0.57	1.25	2.24
Ours+$\int$,M (ML)	**0.07**	**0.45**	**1.14**	**2.20**

(b) Pedestrian-only

Method	KDE NLL				FDE (m)			
	@1s	@2s	@3s	@4s	@1s	@2s	@3s	@4s
Ours (ML)	−2.69	−2.46	−1.76	−1.09	0.03	0.17	0.37	0.60
Ours+$\int$,M (ML)	−5.58	−3.96	−2.77	−1.89	0.01	0.17	0.37	0.62

*We subtracted 22–24 cm from these reported values (their detection/tracking error [7]), as we do not use a detector/tracker. This is done to establish a fair comparison. Legend: $\int$ = Integration via Dynamics, M = Map Encoding, y_R = Robot Future Encoding.

possible object classes present in the dataset. To obtain an estimate of prediction quality degradation over time, we compute the model's FDE at $t = \{1, 2, 3, 4\}$ s for all tracked objects with at least 4 s of available future data. We also implement a constant velocity baseline, which simply maintains the agent's heading and speed for the prediction horizon. Table 4(a) summarizes the model's performance in comparison with state-of-the-art vehicle trajectory prediction models. Since other methods use a detection/tracking module (whereas ours does not), to establish a fair comparison we subtracted other methods' detection and tracking error from their reported values. The dynamics integration scheme and map encoding yield a noticeable improvement with vehicles, as their dynamically-extended unicycle dynamics now differ from the single integrator assumption made by the base model. Note that our method was only trained to predict 3 s into the future, thus its performance at 4 s also provides a measure of its capability to generalize beyond its training configuration. Other methods do not report values at 2 s and 4 s. As can be seen, *Trajectron++* outperforms existing approaches without facing a sharp degradation in performance after 3 s. Our approach's performance on pedestrians is reported in Table 4(b), where the inclusion of HD maps and dynamics integration similarly improve performance as in the pedestrian datasets.

Table 5. [nuScenes] (a): Vehicle-only prediction performance for ablated versions of our model. (b): The same, but excluding the ego-robot from consideration (as it is being conditioned on). This shows that our model's robot future conditional performance does not arise from merely removing the ego-vehicle.

(a) Including the Ego-vehicle												
Ablation	KDE NLL				FDE ML (m)				B. Viol. (%)			
$\int$ M $\mathbf{y_R}$	@1s	@2s	@3s	@4s	@1s	@2s	@3s	@4s	@1s	@2s	@3s	@4s
– – –	0.81	0.05	0.37	0.87	0.18	0.57	1.25	2.24	0.2	0.6	2.8	6.9
✓ – –	−4.28	−2.82	−1.67	−0.76	0.07	0.45	1.13	2.17	0.2	0.7	3.2	8.1
✓ ✓ –	−4.17	−2.74	−1.62	−0.70	0.07	0.45	1.14	2.20	0.3	0.6	2.8	7.6
(b) Excluding the Ego-vehicle												
Ablation	KDE NLL				FDE ML (m)				B. Viol. (%)			
$\int$ M $\mathbf{y_R}$	@1s	@2s	@3s	@4s	@1s	@2s	@3s	@4s	@1s	@2s	@3s	@4s
✓ ✓ –	−4.26	−2.86	−1.76	−0.87	0.07	0.44	1.09	2.09	0.3	0.6	2.8	7.6
✓ ✓ ✓	−3.90	−2.76	−1.75	−0.93	0.08	0.34	0.81	1.50	0.3	0.5	1.6	4.2

Legend: $\int$ = Integration via Dynamics, M = Map Encoding, $\mathbf{y_R}$ = Robot Future Encoding.

Ablation Study. To develop an understanding of which model components influence performance, a comprehensive ablation study is performed in Table 5. As can be seen in the first row, even the base model's deterministic ML output performs strongly relative to current state-of-the-art approaches for vehicle trajectory forecasting [7]. Adding the dynamics integration scheme yields a drastic reduction in NLL as well as FDE at all prediction horizons. There is also an associated slight increase in the frequency of road boundary-violating predictions. This is a consequence of training in position (as opposed to velocity) space, which yields more variability in the corresponding predictions. Additionally including map encoding maintains prediction accuracy while reducing the frequency of boundary-violating predictions.

The effect of conditioning on the ego-vehicle's future motion plan is also studied, with results summarized in Table 5(b). As one would expect, providing the model with future motion plans of the ego-vehicle yields significant reductions in error and road boundary violations. This use-case is common throughout autonomous driving as the ego-vehicle repeatedly produces future motion plans at every timestep by evaluating motion primitives. Overall, dynamics integration is the dominant performance-improving module.

Qualitative Comparison. Figure 3 shows trajectory predictions from the base model, with dynamics integration, and with dynamics integration + map encoding. In it, one can see that the base model (predicting in velocity space) undershoots the turn for the red car, predicting that it will end up in oncoming traffic. With the integration of dynamics, the model captures multimodality in the agent's action, predicting both the possibility of a right turn and continuing straight. With the addition of map encoding, the predictions are not only more accurate, but nearly all probability mass now lies within the correct side of the road. This is in contrast to versions of the model without map encoding which predict that the red car might move into oncoming traffic.

Online Runtime. A key consideration in robotics is runtime complexity. As a result, we evaluate the time it takes *Trajectron++* to perform forward inference on commodity hardware. The results are summarized in the appendix, and confirm that our model scales well to scenes with many agents and interactions.

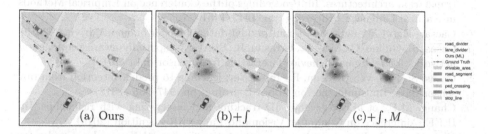

Fig. 3. [nuScenes] The same scene as forecast by three versions of *Trajectron++*. **(a)** The base model tends to under-shoot turns, and makes overly-confident predictions. **(b)** Our approach better captures position uncertainty with dynamics integration, producing well-calibrated probabilities. **(c)** The model is able to leverage the additional information that a map provides, yielding accurate predictions.

6 Conclusion

In this work, we present *Trajectron++*, a generative multi-agent trajectory forecasting approach which uniquely addresses our desiderata for an open, generally-applicable, and extensible framework. It can incorporate heterogeneous data beyond prior trajectory information and is able to produce future-conditional predictions that respect dynamics constraints, all while producing full probability distributions, which are especially useful in downstream robotic tasks such as motion planning, decision making, and control. It achieves state-of-the-art prediction performance in a variety of metrics on standard and new real-world multi-agent human behavior datasets.

Acknowledgment. This work was supported in part by the Ford-Stanford Alliance. This article solely reflects the opinions and conclusions of its authors.

References

1. Alahi, A., Goel, K., Ramanathan, V., Robicquet, A., Fei-Fei, L., Savarese, S.: Social LSTM: human trajectory prediction in crowded spaces. In: IEEE Conference on Computer Vision and Pattern Recognition (2016)

2. Bahdanau, D., Cho, K., Bengio, Y.: Neural machine translation by jointly learning to align and translate. In: International Conference on Learning Representations (2015)
3. Battaglia, P.W., Pascanu, R., Lai, M., Rezende, D., Kavukcuoglu, K.: Interaction networks for learning about objects, relations and physics. In: Conference on Neural Information Processing Systems (2016)
4. Bowman, S.R., Vilnis, L., Vinyals, O., Dai, A.M., Jozefowicz, R., Bengio, S.: Generating sentences from a continuous space. In: Proceedings of the Annual Meeting of the Association for Computational Linguistics (2015)
5. Britz, D., Goldie, A., Luong, M.T., Le, Q.V.: Massive exploration of neural machine translation architectures. In: Proceedings of the Conference on Empirical Methods in Natural Language Processing, pp. 1442–1451 (2017)
6. Caesar, H., et al.: nuScenes: a multimodal dataset for autonomous driving (2019)
7. Casas, S., Gulino, C., Liao, R., Urtasun, R.: SpAGNN: spatially-aware graph neural networks for relational behavior forecasting from sensor data (2019)
8. Casas, S., Luo, W., Urtasun, R.: IntentNet: learning to predict intention from raw sensor data. In: Conference on Robot Learning, pp. 947–956 (2018)
9. Chang, M.F., et al.: Argoverse: 3D tracking and forecasting with rich maps. In: IEEE Conference on Computer Vision and Pattern Recognition (2019)
10. Cho, K., et al.: Learning phrase representations using RNN encoder-decoder for statistical machine translation. In: Proceedings of the Conference on Empirical Methods in Natural Language Processing, pp. 1724–1734 (2014)
11. Deo, M.F., Trivedi, J.: Multi-modal trajectory prediction of surrounding vehicles with maneuver based LSTMs. In: IEEE Intelligent Vehicles Symposium (2018)
12. Goodfellow, I., et al.: Generative adversarial nets. In: Conference on Neural Information Processing Systems (2014)
13. Gupta, A., Johnson, J., Li, F., Savarese, S., Alahi, A.: Social GAN: socially acceptable trajectories with generative adversarial networks. In: IEEE Conference on Computer Vision and Pattern Recognition (2018)
14. Gweon, H., Saxe, R.: Developmental cognitive neuroscience of theory of mind, chap. 20. In: Neural Circuit Development and Function in the Brain, pp. 367–377. Academic Press (2013). https://doi.org/10.1016/B978-0-12-397267-5.00057-1. http://www.sciencedirect.com/science/article/pii/B9780123972675000571
15. Hallac, D., Leskovec, J., Boyd, S.: Network lasso: clustering and optimization in large graphs. In: ACM International Conference on Knowledge Discovery and Data Mining (2015)
16. Helbing, D., Molnár, P.: Social force model for pedestrian dynamics. Phys. Rev. E **51**(5), 4282–4286 (1995)
17. Higgins, I., et al.: β-VAE: learning basic visual concepts with a constrained variational framework. In: International Conference on Learning Representations (2017)
18. Ho, J., Ermon, S.: Multiple futures prediction. In: Conference on Neural Information Processing Systems (2019)
19. Hochreiter, S., Schmidhuber, J.: Long short-term memory. Neural Comput. **9**, 1735–1780 (1997)
20. Ivanovic, B., Pavone, M.: The trajectron: probabilistic multi-agent trajectory modeling with dynamic spatiotemporal graphs. In: IEEE International Conference on Computer Vision (2019)
21. Ivanovic, B., Schmerling, E., Leung, K., Pavone, M.: Generative modeling of multimodal multi-human behavior. In: IEEE/RSJ International Conference on Intelligent Robots & Systems (2018)

22. Jain, A., Zamir, A.R., Savarese, S., Saxena, A.: Structural-RNN: deep learning on spatio-temporal graphs. In: IEEE Conference on Computer Vision and Pattern Recognition (2016)
23. Jain, A., et al.: Discrete residual flow for probabilistic pedestrian behavior prediction. In: Conference on Robot Learning (2019)
24. Jang, E., Gu, S., Poole, B.: Categorial reparameterization with Gumbel-Softmax. In: International Conference on Learning Representations (2017)
25. Kalman, R.E.: A new approach to linear filtering and prediction problems. ASME J. Basic Eng. **82**, 35–45 (1960)
26. Kesten, R., et al.: Lyft Level 5 AV Dataset 2019 (2019). https://level5.lyft.com/dataset/
27. Kong, J., Pfeifer, M., Schildbach, G., Borrelli, F.: Kinematic and dynamic vehicle models for autonomous driving control design. In: IEEE Intelligent Vehicles Symposium (2015)
28. Kosaraju, V., et al.: Social-BiGAT: multimodal trajectory forecasting using bicycle-GAN and graph attention networks. In: Conference on Neural Information Processing Systems (2019)
29. LaValle, S.M.: Better unicycle models. In: Planning Algorithms, p. 743. Cambridge University Press (2006)
30. LaValle, S.M.: A simple unicycle. In: Planning Algorithms, pp. 729–730. Cambridge University Press (2006)
31. Lee, N., et al.: DESIRE: distant future prediction in dynamic scenes with interacting agents. In: IEEE Conference on Computer Vision and Pattern Recognition (2017)
32. Lee, N., Kitani, K.M.: Predicting wide receiver trajectories in American football. In: IEEE Winter Conference on Applications of Computer Vision (2016)
33. Lerner, A., Chrysanthou, Y., Lischinski, D.: Crowds by example. Comput. Graph. Forum **26**(3), 655–664 (2007)
34. Morton, J., Wheeler, T.A., Kochenderfer, M.J.: Analysis of recurrent neural networks for probabilistic modeling of driver behavior. IEEE Trans. Pattern Anal. Mach. Intell. **18**(5), 1289–1298 (2017)
35. Paden, B., Čáp, M., Yong, S.Z., Yershov, D., Frazzoli, E.: A survey of motion planning and control techniques for self-driving urban vehicles. IEEE Trans. Intell. Veh. **1**(1), 33–55 (2016)
36. Paszke, A., et al.: Automatic differentiation in PyTorch. In: Conference on Neural Information Processing Systems - Autodiff Workshop (2017)
37. Pellegrini, S., Ess, A., Schindler, K., Gool, L.: You'll never walk alone: modeling social behavior for multi-target tracking. In: IEEE International Conference on Computer Vision (2009)
38. Rasmussen, C.E., Williams, C.K.I.: Gaussian Processes for Machine Learning (Adaptive Computation and Machine Learning), 1st edn. MIT Press, Cambridge (2006)
39. Rhinehart, N., McAllister, R., Kitani, K., Levine, S.: PRECOG: prediction conditioned on goals in visual multi-agent settings. In: IEEE International Conference on Computer Vision (2019)
40. Rudenko, A., Palmieri, L., Herman, M., Kitani, K.M., Gavrila, D.M., Arras, K.O.: Human motion trajectory prediction: a survey (2019). https://arxiv.org/abs/1905.06113

41. Sadeghian, A., Kosaraju, V., Sadeghian, A., Hirose, N., Rezatofighi, S.H., Savarese, S.: SoPhie: an attentive GAN for predicting paths compliant to social and physical constraints. In: IEEE Conference on Computer Vision and Pattern Recognition (2019)
42. Sadeghian, A., Legros, F., Voisin, M., Vesel, R., Alahi, A., Savarese, S.: CAR-Net: Clairvoyant attentive recurrent network. In: Ferrari, V., Hebert, M., Sminchisescu, C., Weiss, Y. (eds.) ECCV 2018. LNCS, vol. 11215. Springer, Cham (2018). https://doi.org/10.1007/978-3-030-01252-6_10
43. Schöller, C., Aravantinos, V., Lay, F., Knoll, A.: What the constant velocity model can teach us about pedestrian motion prediction. IEEE Robot. Autom. Lett. **5**, 1696–1703 (2020)
44. Sohn, K., Lee, H., Yan, X.: Learning structured output representation using deep conditional generative models. In: Conference on Neural Information Processing Systems (2015)
45. Thiede, L.A., Brahma, P.P.: Analyzing the variety loss in the context of probabilistic trajectory prediction. In: IEEE International Conference on Computer Vision (2019)
46. Thrun, S., Burgard, W., Fox, D.: The extended Kalman filter. In: Probabilistic Robotics, pp. 54–64. MIT Press (2005)
47. Vemula, A., Muelling, K., Oh, J.: Social attention: modeling attention in human crowds. In: Proceedings of the IEEE Conference on Robotics and Automation (2018)
48. Wang, J.M., Fleet, D.J., Hertzmann, A.: Gaussian process dynamical models for human motion. IEEE Trans. Pattern Anal. Mach. Intell. **30**(2), 283–298 (2008)
49. Waymo: Safety report (2018). https://waymo.com/safety/. Accessed 9 Nov 2019
50. Waymo: Waymo Open Dataset: An autonomous driving dataset (2019). https://waymo.com/open/
51. Zeng, W., et al.: End-to-end interpretable neural motion planner. In: IEEE Conference on Computer Vision and Pattern Recognition (2019)
52. Zhao, S., Song, J., Ermon, S.: InfoVAE: balancing learning and inference in variational autoencoders. In: Proceedings of the AAAI Conference on Artificial Intelligence (2019)
53. Zhao, T., et al.: Multi-agent tensor fusion for contextual trajectory prediction. In: IEEE Conference on Computer Vision and Pattern Recognition (2019)

Context-Gated Convolution

Xudong Lin[1]([✉]), Lin Ma[2], Wei Liu[2], and Shih-Fu Chang[1]

[1] Columbia University, New York, USA
{xudong.lin,shih.fu.chang}@columbia.edu
[2] Tencent AI Lab, Bellevue, USA
forest.linma@gmail.com, wl2223@columbia.edu

Abstract. As the basic building block of Convolutional Neural Networks (CNNs), the convolutional layer is designed to extract local patterns and lacks the ability to model global context in its nature. Many efforts have been recently devoted to complementing CNNs with the global modeling ability, especially by a family of works on global feature interaction. In these works, the global context information is incorporated into local features before they are fed into convolutional layers. However, research on neuroscience reveals that the neurons' ability of modifying their functions dynamically according to context is essential for the perceptual tasks, which has been overlooked in most of CNNs. Motivated by this, we propose one novel Context-Gated Convolution (CGC) to explicitly modify the weights of convolutional layers adaptively under the guidance of global context. As such, being aware of the global context, the modulated convolution kernel of our proposed CGC can better extract representative local patterns and compose discriminative features. Moreover, our proposed CGC is lightweight and applicable with modern CNN architectures, and consistently improves the performance of CNNs according to extensive experiments on image classification, action recognition, and machine translation. Our code of this paper is available at https://github.com/XudongLinthu/context-gated-convolution.

Keywords: Convolutional neural network · Context-gated convolution · Global context information

1 Introduction

Convolutional Neural Networks (CNNs) have achieved remarkable successes on various tasks, e.g., image classification [21,26], object detection [16,44], image translation [58], action recognition [5], sentence/text classification [30,56],

X. Lin—This work was done when Xudong Lin interned at Tencent AI Lab.

Electronic supplementary material The online version of this chapter (https://doi.org/10.1007/978-3-030-58523-5_41) contains supplementary material, which is available to authorized users.

© Springer Nature Switzerland AG 2020
A. Vedaldi et al. (Eds.): ECCV 2020, LNCS 12363, pp. 701–718, 2020.
https://doi.org/10.1007/978-3-030-58523-5_41

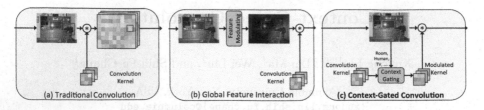

Fig. 1. (a) Traditional convolution only composes local information. (b) Global feature interaction methods modify input feature maps by incorporating global information. (c) Our proposed CGC, in a fundamentally different manner, modulates convolution kernels under the guidance of global context. ⊛ denotes convolution.

machine translation [14], etc. However, the sliding window mechanism of convolution makes it only capable of capturing local patterns, limiting its ability of utilizing global context. Taking the 2D convolution on the image as one example, as Fig. 1(a) shows, the traditional convolution only operates on the local image patch and thereby composes local features.

According to the recent research on neuroscience [15,32], neurons' awareness of global context is important for us to better interpret visual scenes, stably perceive objects, and effectively process complex perceptual tasks. Many methods [3,4,8,25,33,39,48,49,51] have been recently proposed to introduce global context modeling modules into CNN architectures. As Fig. 1(b) shows, these methods, which are named as global feature interaction methods in this paper, modulate intermediate feature maps by incorporating the global context into the local feature representation.

However, as stated in [15], "*rather than having a fixed functional role, neurons should be thought of as adaptive processors, changing their function according to the behavioural context*". Therefore, the context information should be utilized to explicitly modulate the convolution kernels for "*changing the structure of correlations over neuronal ensembles*" [15]. However, to the best of our knowledge, such a modulating mechanism has not been exploited in CNNs yet, even though it is one efficient and intuitive way. Motivated by this, we will model convolutional layers as "*adaptive processors*" and explore how to leverage global context to guide the composition of local features in convolution operations.

In this paper, we propose Context-Gated Convolution (CGC), as shown in Fig. 1(c), a new perspective of complementing CNNs with the awareness of the global context. Specifically, our proposed CGC learns a series of mappings to generate gates from the global context feature representations to modulate convolution kernels accordingly. With the modulated kernels, the traditional convolution is performed on input feature maps, which enables convolutional layers to dynamically capture representative local patterns and compose local features of interest under the guidance of global context. Our contributions lie in three-fold.

- To the best of our knowledge, we make the first attempt of introducing the context-awareness to convolutional layers by modulating their weights according to the global context.

- We propose a novel lightweight Context-Gated Convolution (CGC) to effectively generate gates for convolution kernels to modify the weights with the guidance of global context. Our CGC consists of a Context Encoding Module that encodes context information into latent representations, a Channel Interacting Module that projects them into the space of output dimension, and a Gate Decoding Module that decodes the latent representations to produce the gate.
- Our proposed CGC can better capture local patterns and compose discriminative features, and consistently improve the generalization of traditional convolution with a negligible complexity increment on various tasks including image classification, action recognition, and machine translation.

2   Related Works

There have been many efforts in augmenting CNNs with context information. They can be roughly categorized into three types: first, adding backward connections in CNNs [47,54,55] to model the top-down influence [15] like humans' visual processing system; second, modifying intermediate feature representations in CNNs according to the attention mechanism [4,8,48,51,52]; third, dynamically generating the parameters of convolutional layers according to local or global information [9,12,28,29,35,53,59].

For the first category of works, it is still unclear how the feedback mechanism can be effectively and efficiently modeled in CNNs. For example, Yang *et al.* [54] proposed an Alternately Updated Clique to introduce feedback mechanisms into CNNs. However, compared to traditional CNNs, the complex updating strategy increases the difficulty for training them as well as the latency at the inference time. The second category of works is the global feature interaction methods. They [3,4,8,25,39,48,49,51,52] were proposed recently to modify local features according to global context information, usually by a global correspondence, i.e., the self-attention mechanism. There are also works on reducing the complexity of the self-attention mechanism [10,40]. However, this family of works only considers changing the input feature maps.

The third type of works is more related to our work. Zhu *et al.* [59] proposed to adaptively set the offset of each element in a convolution kernel and the gate value for each element in the input local feature patch. However, the mechanism only changes the input to the convolutional layer. The weight tensor of the convolutional layer is not considered. Wu *et al.* [53] proposed to dynamically generate the weights of convolution kernels. However, it is specialized for Lightweight Convolution [53] and only takes local segments as inputs. Another family of works on dynamic filters [28,29,35] also belongs to this type. They generate weights of convolution kernels using features extracted from input images by another CNN feature extractor. The expensive feature extraction process makes it more suitable for generating a few filters, e.g., in the case of low-level image processing. It is impractical to generate weights for all the layers in a deep CNN model in this manner.

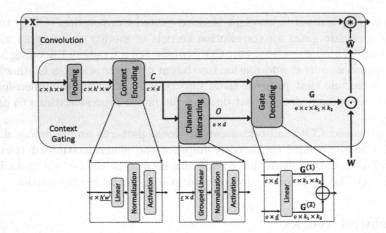

Fig. 2. Our proposed CGC consists of three components, namely the Context Encoding Module, the Channel Interacting Module, and the Gate Decoding Module. The Context Encoding Module encodes global context information into a latent representation C; the Channel Interacting Module transforms C to O with output dimension o; the Gate Decoding Module produces $\mathbf{G}^{(1)}$ and $\mathbf{G}^{(2)}$ from C and O to construct the gate $\mathbf{G}$. ⊛ and ⊙ denote convolution and element-wise multiplication operations, respectively. ⊕ is shown in Eq. (1). The dimension transformed in each linear layer is underlined.

3 Context-Gated Convolution

3.1 Preliminaries

Without loss of generality, we consider one sample of 2D case. The input to a convolutional layer is a feature map $\mathbf{X} \in \mathbb{R}^{c \times h \times w}$, where c is the number of channels, and h, w are respectively the height and width of the feature map. In each convolution operation, a local patch of size $c \times k_1 \times k_2$ is collected by the sliding window to multiply with the kernel $\mathbf{W} \in \mathbb{R}^{o \times c \times k_1 \times k_2}$ of this convolutional layer, where o is the number of output channels, and k_1, k_2 are respectively the height and width of the kernel. Therefore, only local information within each patch is extracted in one convolution operation. Although in the training process, the convolution kernels are learned from all the patches of all the images in the training set, the kernels are not adaptive to the current context during the inference time.

3.2 Module Design

In order to handle the aforementioned drawback of traditional convolution, we propose to incorporate the global context information during the convolution process. Different from the existing approaches that modify the input features according to the context, e.g., a global correspondence of feature representations, we attempt to directly modulate the convolution kernel under the guidance of the global context information.

One simple and straightforward way of modulating the convolution kernel $\mathbf{W}$ with global context information is to directly generate a gate $\mathbf{G} \in \mathbb{R}^{o \times c \times k_1 \times k_2}$ of the same size as $\mathbf{W}$ according to the global context. Assuming that we generate the gate from a context vector $v \in \mathbb{R}^l$ using a linear layer without the bias term, the number of parameters is $l \times o \times c \times k_1 \times k_2$, which is extremely catastrophic when we modulate the convolution kernel of every convolutional layer. For modern CNNs, o and c can be easily greater than 100 or even 1,000, which makes $o \times c$ the dominant term in the complexity. Inspired by previous works on convolution kernel decomposition [11,24], we propose to decompose the gate $\mathbf{G}$ into two tensors $\mathbf{G}^{(1)} \in \mathbb{R}^{c \times k_1 \times k_2}$ and $\mathbf{G}^{(2)} \in \mathbb{R}^{o \times k_1 \times k_2}$, so that the complexity of $o \times c$ can thereby significantly be broken down.

However, directly generating these two tensors is still impractical. Supposing that we generate them with two linear layers, the number of parameters is $l \times (o + c) \times k_1 \times k_2$, which is of the same scale as the number of parameters of the convolution kernel itself. The bottleneck now is jointly modeling channel-wise and spatial interactions, namely l and $(o + c) \times k_1 \times k_2$, considering that $v \in \mathbb{R}^l$ is encoded from the input feature map $\mathbf{X} \in \mathbb{R}^{c \times h \times w}$. Inspired by depth-wise separable convolutions [11,24], we propose to model the spatial interaction and the channel-wise interaction separately to further reduce the complexity.

In this paper, we propose one novel Context-Gated Convolution (CGC) to incorporate the global context information during the convolution process. Specifically, our proposed CGC consists of three modules: the Context Encoding Module, the Channel Interacting Module, and the Gate Decoding Module. As shown in Fig. 2, the Context Encoding Module encodes global context information in each channel into a latent representation C via spatial interaction; the Channel Interacting Module projects the latent representation to the space of output dimension o via channel-wise interaction; the Gate Decoding Module produces $\mathbf{G}^{(1)}$ and $\mathbf{G}^{(2)}$ from the latent representation C and the projected representation O to construct the gate $\mathbf{G}$ via spatial interaction. The detailed information is described in the following.

Context Encoding Module. To extract contextual information, we first use a pooling layer to reduce the spatial resolution to $h' \times w'$ and then feed the resized feature map to the Context Encoding Module. It encodes information from all the spatial positions for each channel, and extracts a latent representation of the global context. We use a linear layer with weight $\mathbf{E} \in \mathbb{R}^{h' \times w' \times d}$ to project the resized feature map in each channel to a latent vector of size d. Inspired by the bottleneck structure from [21,25,48,51], we set $d = \frac{k_1 \times k_2}{2}$ to extract informative context, when not specified. The weight $\mathbf{E}$ is shared across different channels. A normalization layer and an activation function come after the linear layer. There are c channels, so the output of the Context Encoding Module is $C \in \mathbb{R}^{c \times d}$. Since the output is fed into two different modules, we accordingly apply two individual Normalization layers so that different information can be conveyed if needed.

Channel Interacting Module. It projects the feature representations $C \in \mathbb{R}^{c \times d}$ to the space of output dimension o. Inspired by [19], we use a grouped linear layer $\mathbf{l} \in \mathbb{R}^{\frac{c}{g} \times \frac{o}{g}}$, where g is the number of groups. The weight $\mathbf{l}$ is shared among different dimensions of d and different groups. A normalization layer and an activation function come after the linear layer. The final output of the Channel Interacting Module is $O \in \mathbb{R}^{o \times d}$.

Gate Decoding Module. It takes both C and O as inputs, and decodes the latent representations to the spatial size of convolution kernels. We use two linear layers whose weights $\mathbf{D}_c \in \mathbb{R}^{d \times k_1 \times k_2}$ and $\mathbf{D}_o \in \mathbb{R}^{d \times k_1 \times k_2}$ are respectively shared across different channels in C and O. Then each element in the gate $\mathbf{G}$ is produced by:

$$\mathbf{G}_{h,i,j,k} = \sigma(\mathbf{G}^{(1)}_{i,j,k} + \mathbf{G}^{(2)}_{h,j,k}) = \sigma((C\mathbf{D}_c)_{i,j,k} + (O\mathbf{D}_o)_{h,j,k}), \tag{1}$$

where $\sigma(\cdot)$ denotes the sigmoid function. Now we have $\mathbf{G}$ with the same size of the convolution kernel $\mathbf{W}$, which is generated from the global context by our lightweight modules. Then we can modulate the weight of a convolutional layer by element-wise multiplication to incorporate rich context information:

$$\hat{\mathbf{W}} = \mathbf{W} \odot \mathbf{G}. \tag{2}$$

With the modulated kernel, a traditional convolution process is performed on the input feature maps, where the context information can help the kernel capture more representative patterns and also compose features of interest.

Complexity. The computational complexity of our three modules is $O(c \times d \times h' \times w' + c \times o/g + c \times d \times k_1 \times k_2 + o \times d \times k_1 \times k_2 + o \times c \times k_1 \times k_2)$, where h', w' can be set independent of h, w. It is negligible compared to convolution's $O(o \times c \times k_1 \times k_2 \times h \times w)$. Except the linear time of pooling, the complexity of these three modules is independent of the input's spatial size. The total number of parameters is $O(d \times h' \times w' + c \times o/g^2 + d \times k_1 \times k_2)$, which is negligible compared to traditional convolution's $O(o \times c \times k_1 \times k_2)$. Therefore, we can easily replace the traditional convolution with our proposed CGC with a very limited computation and parameter increment, and enable convolutional layers to be adaptive to global context.

3.3 Discussions

We are aware of the previous works on dynamically modifying the convolution operation [12,28,29,35,53,59]. As discussed before, [59] essentially changes the input to the convolutional layer but not the weight tensor of the convolutional layer. Dynamic Convolution [53] is specialized for the Lightweight convolution [53] and only adaptive to local inputs. The family of work on dynamic filters [28,29,35] generates weights of convolution kernels using features extracted from input images by another CNN feature extractor. It is too expensive to generate weights for all the layers in a deep CNN model in this manner. In contrast, our CGC takes feature maps of a convolutional layer as input and makes it possible to dynamically modulate the weight of each convolutional layer, which systematically improves CNNs' global context modeling ability.

Both global feature interaction methods[3,4,8,25,39,48,49,51] modifying feature maps and our proposed CGC modulating kernels can incorporate the global context information into CNN architectures, which can boost the performance of CNNs. However, 1) with our CGC, the complexity of modulating kernels does not depend on input size, but global feature interaction methods, e.g., Non-local, may suffer from a quadratic computational complexity w.r.t. the input size; 2) our CGC can be easily trained from scratch and improve the training stability of CNNs according to our experiments (Sects. 4.2 and 4.3); 3) by modulating kernels, our CGC can dynamically create kernels with specialized functions according to context (Sect. 4.2) and thus enable CNNs to accordingly capture discriminative information as adaptive processors, which cannot be realized by modifying feature maps. Moreover, our CGC is also somewhat complementary to global feature interaction methods (Sects. 4.2 and 4.3) and we can further improve CNN's performance by applying both CGC and global feature interaction methods.

4 Experiments

In this section, we demonstrate the effectiveness of our proposed CGC in incorporating 1D, 2D, and 3D context information in 1D, 2D, and (2+1)D convolutions. We conduct extensive experiments on image classification, action recognition, and machine translation, and observe that our CGC consistently improves the performance of modern CNNs with a negligible parameter increment on six benchmark datasets: ImageNet [45], CIFAR-10 [31], ObjectNet [2], Something-Something (v1) [18], Kinetics [5], and IWSLT'14 De-En [6].

4.1 Implementation Details

All of the experiments are based on PyTorch [41]. All the linear layers are without bias terms. We follow common practice to use Batch Normalization [27] for computer vision tasks, and Layer Normalization [1] for natural language processing tasks, respectively. We use ReLU [36] as the activation function for all the experiments in this paper. We use average pooling with $h' = k_1$ and $w' = k_2$, when not specified. Note that we only replace the convolution kernels with a spatial size larger than 1. For those point-wise convolutions, we take them as linear layers and do not modulate them. To reduce the size of I, we fix $c/g = 16$ when not specified. We initialize all these layers as what [20] did for computer vision tasks and as what [17] did for natural language processing tasks, when not specified.

Table 1. Image classification results on ImageNet and CIFAR-10. Param indicates the number of parameters in the model. ΔMFLOPs is the increment of the number of multiplication-addition operations compared to ResNet-50 (R50, 4 GFLOPs) for ImageNet models and ResNet-110 (R110, 256 MFLOPs) for CIFAR-10 models. Bold indicates the best result.

Dataset	Training setting	Model	Param	ΔMFLOPs	Top-1 (%)	Top-5 (%)
ImageNet	-	R50 + GloRe [8]	30.5M	1200	78.4	-
		DCNv2-R50 [59]	27.4M	200	78.2	94.0
		GC-R50 [4]	28.08M	100	77.70	93.66
	Default	SE-R50 [25]	28.09M	8	77.18	93.67
		BAM-R50 [39]	25.92M	83	76.90	93.40
		GC-R50 [4]	28.11M	8	73.90	91.70
		DCNv2-R50 [59]	27.4M	200	77.21	93.69
		SK-R50 [33]	37.25M	1837	77.15	93.54
		R50 [21]	25.56M	-	76.16	92.91
		R50 + CGC (Ours)	**25.59M**	6	**77.48**	**93.81**
		CBAM-R50 [52]	28.09M	15	77.34	**93.69**
		CBAM-R50 + CGC (Ours)	28.12M	21	**77.68**	93.68
	Advanced	DCNv2-R50 [59]	27.4M	200	78.89	94.60
		SE-R50 [25]	28.09M	8	78.79	94.52
		R0 [21]	25.56M	-	78.13	94.06
		R50 + CGC (Ours)	**25.59M**	6	**79.54**	**94.78**
		CBAM-R50 [52]	28.09M	15	78.86	94.58
		CBAM-R50 + CGC (Ours)	28.12M	21	**79.74**	**94.83**
CIFAR-10		R110 [22]	1.73M	-	93.96	99.73
		R110 + CGC (Ours)	**1.80M**	2	**94.86**	**99.82**

4.2   Image Classification

Experimental Setting. Following previous works [21] on ImageNet [45], we train models on the ImageNet 2012 training set, which contains about 1.28 million images from 1,000 categories, and report the results on its validation set, which contains 50,000 images. We replace all the convolutions that are not 1×1 in ResNet-50 [21] with our CGC and train the network from scratch. Note that for the first convolutional layer, we use $I \in \mathbb{R}^{3 \times 64}$ for the Channel Interacting Module. We conduct experiments in two settings: Default and Advanced. For the Default setting, we follow common practice [21] and apply minimum training tricks. For the Advanced setting, we borrow training tricks from [23] to validate that our CGC can still improve the performance, even under a strong baseline. CIFAR-10 contains 50K training images and 10K testing images in 10 classes. We follow common practice [22] to train and evaluate the models. We take ResNet-110 [22] (with plain blocks) as the baseline model. All the compared methods are trained based on the same training protocol. The details are provided in the supplementary material. For evaluation, we report Top-1 and Top-5 accuracies of a single crop with the size 224×224 for ImageNet and 32×32 for CIFAR-10, respectively.

ObjectNet [2] is a new challenging evaluation dataset for image classification. There are 113 classes out of 313 ObjectNet classes, which overlap with ImageNet classes. We follow [2] to evaluate models trained on ImageNet on the overlapped classes.

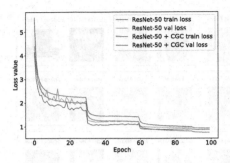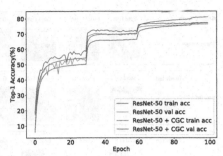

Fig. 3. The training curves of ResNet-50 and ResNet-50 + CGC (ours) on ImageNet under the default training setting.

Table 2. Image classification results on ObjectNet. Bold indicates the best result.

Model	Top-1 (%)	Top-5 (%)
R50 [21]	29.35	48.42
SE-R50 [25]	29.48	45.55
DCNv2-R50 [59]	29.74	48.83
CBAM-R50 [52]	29.56	48.68
R50 + CGC (Ours)	**31.53**	**50.16**

Performance Results. As Table 1 shows, our CGC significantly improves the performances of baseline models on both ImageNet and CIFAR-10. On ImageNet, our CGC improves the Top-1 accuracy of ResNet-50 under the Advanced setting by 1.41% with only 0.03M more parameters and 6M more FLOPs, which verifies our CGC's effectiveness of incorporating global context and its efficiency. We observe that our CGC outperforms DCN-v2 [59], SK-Net [33] and CBAM [52], which indicates the superiority of modulating kernels. We also observe that our CGC can also improve the performance of CBAM-ResNet-50 [52] consistently under both settings, which indicates that our proposed CGC is applicable with state-of-the-art global feature interaction methods. CBAM-ResNet-50 + CGC even reaches 79.74% Top-1 accuracy, which outperforms the other compared methods by a large margin.

We also find that GC-ResNet-50 is hard to train from scratch unless using the fine-tuning protocol reported by [4], which indicates that modifying features may be misleading in the early training process. Although our CGC introduces a few new parameters, our model converges faster and more stably compared to vanilla ResNet-50, as shown in Fig. 3. We conjecture that this is because the adaptiveness to global context improves the model's generalization ability and the gating mechanism reduces the norm of gradients back-propagated to the convolution kernels, which leads to a smaller Lipschitz constant and thus better training stability [42, 46].

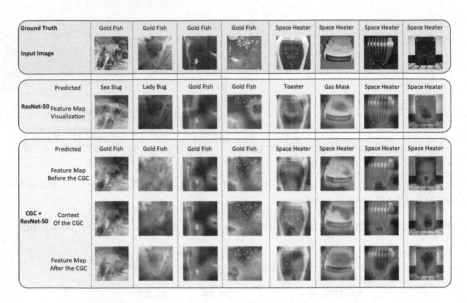

Fig. 4. Visualization of the feature maps produced by ResNet-50 and CGC-ResNet-50 from the ImageNet validation set images. (Best viewed on a monitor when zoomed in.)

To further validate the generalization ability of our CGC, we use Object-Net to evaluate models with good performances on ImageNet. ObjectNet [2] is recently proposed to push image recognition models beyond their current limit of generalization. The dataset contains images "in many rotations, on different backgrounds, from multiple viewpoints", which makes it hard for models trained on ImageNet to correctly classify these images. As Table 2 shows, our CGC significantly improves the generalization ability of the ResNet-50 baseline. The improvement (2.18%) is even larger than that on ImageNet validation set.

Visualization. To understand how CGC helps the model capture more informative features under the guidance of context information, we visualize the feature maps of ResNet-50 and our CGC-ResNet-50 by Grad-CAM++ [7]. As Fig. 4 shows, overall, the feature maps (After the CGC) produced by our CGC-ResNet-50 cover more informative regions, e.g., more instances or more parts of the ground-truth object, than vanilla ResNet-50.

Specifically, we visualize the feature maps before the last CGC in the model, the context information used by CGC, and the resulting feature maps after the CGC. As is clearly shown in Fig. 4, the proposed CGC extracts the context information from representative regions of the target object and successfully refines the feature maps with comprehensive understanding of the whole image and the target object. For example, in Gold Fish 1, the heads of the fishes are partially visible. Vanilla ResNet-50 mistakes this image as Sea Slug, because it only pays attention to the tails of the fishes, which are similar to sea slugs. However, our CGC utilizes the context of the whole image and guides the convolution with information from the entire fishes, which helps the model classify this image correctly.

Analysis of the Gate. To further validate that our CGC uses context information of the target objects to guide the convolution process, we calculate the average modulated kernel (in the last CGC of the model) for images of each class in the validation set. Then we calculate inter-class $L - 2$ distance between every two average modulated kernels, i.e., class centers, and the intra-class $L - 2$ distance (mean distance to the class center) for each class. As is shown in the supplementary material, we visualize the difference matrix between inter-class distances and intra-class distances. In more than 93.99% of the cases, the inter-class distance is larger than the corresponding intra-class distance, which indicates that there are clear clusters of these modulated kernels and the clusters are aligned very well with the classes.

This observation strongly supports that our CGC successfully extracts class-specific context information and effectively modulates the convolution kernel to extract representative features. Meanwhile, the intra-class variance of the modulated kernels indicates that our CGC dynamically modulates convolution kernels according to different input contexts.

Ablation Study. In order to demonstrate the effectiveness of our module design, ablation studies are conducted on CIFAR-10, as illustrated in Table 3a. Specifically, we ablate many variants of our CGC and find that our default setting is a good trade-off between parameter increment and performance gain. The experiments on the combination of $\mathbf{G}^{(1)}$ and $\mathbf{G}^{(2)}$ show that our decomposition approach in Eq. (1) is a better way to construct the gate. For channel interacting, we find that using a full linear model with $g = 1$ achieves better performance with more parameters, as is expected. We try removing the bottleneck structure and set $d = k_1 \times k_2$, and the performance drops, which validates the necessity of the bottleneck structure.

Shared Norm indicates using the same Normalization layer for the following two branches. For Two **E**s, we learn another **E** to encode C only for the Channel Interacting Module. We also try sharing **D** for generating $\mathbf{G}^{(1)}$ and $\mathbf{G}^{(2)}$, using larger resized feature maps and using max pooling instead of average pooling. All the results support our default setting. We also test different numbers of layers to replace traditional convolutions with our CGC. The result indicates that the more, the better. We select 3 variants with a similar number of parameters and performance on CIFAR-10 and further perform ablation studies for them on ImageNet. As Table 3b shows, we observe the same performance rankings of these variants on ImageNet as those on CIFAR-10.

4.3   Action Recognition

Baseline Methods. For the action recognition task, we adopt three baselines to evaluate the effectiveness of our CGC: TSN [50], P3D-A [43] (details are in the supplementary material), and TSM [34]. Because our CGC's effectiveness of introducing 2D spatial context to CNNs has been verified on image classification, in this part, we focus on its ability of incorporating 1D temporal context and 3D spatiotemporal context. For the 1D case, we apply our CGC to temporal convolutions in every P3D-A block. For the 3D case, we apply our CGC to

Table 3. Ablation studies on CIFAR-10 and ImageNet. Param denotes the number of parameters in the model. ΔMFLOPs is the increment of the number of multiplication-addition operations compared to ResNet-110 (256 MFLOPs). Bold indicates our default setting. Top-1 Accuracy (%) (average of 3 runs) is reported.

(a) CIFAR-10

Model	Param	ΔMFLOPs	Top-1 (%)
ResNet-110 [22]	1.73M	-	93.96
ResNet-110 + CGC	1.80M	1.681	94.86
only $G^{(1)}$	1.75M	1.447	94.53
only $G^{(2)}$	1.78M	1.472	94.41
$G^{(1)} * G^{(2)}$	1.80M	1.681	94.59
$g = 1$	1.96M	1.681	94.97
$d = k_1 \times k_2$	1.81M	1.741	94.61
Shared Norm	1.79M	1.681	94.72
Two Es	1.80M	1.871	94.53
Shared D	1.79M	1.681	94.78
$h' = 2k_1, w' = 2k_2$	1.81M	1.681	94.80
MaxPool	1.80M	1.681	94.44
(res1,2,3)	1.80M	1.678	94.55
(res2,3)	1.78M	1.052	94.43
(res3)	1.76M	0.622	94.26

(b) ImageNet

Model	Top-1 (%)
ResNet-50 [21]	76.16
ResNet-50 + CGC	77.48
Shared Norm	77.21
Shared D	77.28
$h' = 2k_1, w' = 2k_2$	77.34

spatial convolutions in P3D-A or 2D convolutions in TSN or TSM; the pooling layer produces $c \times k \times k \times k$ cubes, the Context Encoding Module encodes $k \times k \times k$ feature maps into a vector of length $k^3/2$, and the Gate Decoding Module generates $o \times c \times t \times k \times k$ gates. Note that for the first convolutional layer, we use $I \in \mathbb{R}^{3 \times 64}$ for the Channel Interacting Module.

Experimental Setting. The Something-Something (v1) dataset has a training split of 86,017 videos and a validation split of 11,522 videos, with 174 categories. We follow [42] to train on the training set and report evaluation results on the validation set. We follow [34] to process videos and augment data. Since we only use ImageNet for pretraining, we adapt the code base of TSM but the training setting from [42]. We train TSN- and TSM-based models for 45 epochs (50 for P3D-A), starting from a learning rate of 0.025 (0.01 for P3D-A) and decreasing it by 0.1 at 26 and 36 epochs (30, 40, 45 for P3D-A). The Kinetics [5] dataset has 400 action classes and 240K training samples. We follow [34] to train and evaluate all the compared models.

For TSN- and TSM-based models, the batch size is 64 for 8-frame models and 32 for 16-frame models, and the dropout rate is set to 0.5. P3D-A takes 32 continuously sampled frames as input and the batch size is 64, and the dropout ratio is 0.8. We use the evaluation setting of [34] for TSN- and TSM-based models and the evaluation settings of [51] for P3D-A. All the models are trained with 8-GPU machines.

Performance Comparisons. As Table 4 and Table 5 show, our CGC significantly improves the performance of baseline CNN models, compared to Non-local [51]. As aforementioned, Non-local modules modify the input feature maps of convolutional layers by reassembling local features according to the global correspondence. We apply Non-local blocks in the most effective way as is reported by [51]. However, we observe that its performance gain is not consistent when training the model from scratch. When applied to TSM on the

Table 4. Action recognition results on Something-Something (v1). Backbone indicates the backbone network architecture. Param indicates the number of parameters in the model. Frame indicates the number of frames used for evaluation. Bold indicates the best result.

Model	Backbone	Param	Frame	Top-1 (%)	Top-5 (%)
TRN [57]	BNInception	18.3M	8	34.4	-
TRN [34]	ResNet-50	31.8M	8	38.9	68.1
ECO [60]	BNInc+Res18	47.5M	8	39.6	-
ECO [60]	BNInc+Res18	47.5M	16	41.4	-
ECO$_{En}$Lite [60]	BNInc+Res18	150M	92	46.4	-
TSN [50]	ResNet-50	23.86M	8	19.00	44.98
TSN + Non-local [51]	ResNet-50	31.22M	8	25.73	55.17
TSN + CGC (Ours)	ResNet-50	24.07M	8	**32.58**	**60.06**
P3D [43]	ResNet-50	25.38M	32 × 30	45.17	74.61
lslsP3D + Non-local [51]	ResNet-50	32.73M	32 × 30	45.88	74.94
P3D + CGC 1D (Ours)	ResNet-50	25.39M	32 × 30	46.14	75.92
P3D + CGC 3D (Ours)	ResNet-50	25.61M	32 × 30	46.35	75.97
P3D + CGC 1D & 3D (Ours)	ResNet-50	25.62M	32 × 30	**46.73**	**76.04**
TSM [34]	ResNet-50	23.86M	8	44.65	73.94
TSM + Non-local [51]	ResNet-50	31.22M	8	43.91	72.18
TSM + CGC (Ours)	ResNet-50	24.07M	8	**46.00**	**75.11**
TSM [34]	ResNet-50	23.86M	16	46.61	76.18
TSM + CGC (Ours)	ResNet-50	24.09M	16	**47.87**	**77.22**

Table 5. Action recognition results on Kinetics. Backbone indicates the backbone network architecture. Param indicates the number of parameters in the model. Bold indicates the best result.

Model	Backbone	Param	Top-1 (%)	Top-5 (%)
TSM [34]	ResNet-50	23.86M	74.12	91.21
TSM + Non-local [51]	ResNet-50	31.22M	75.60	92.15
TSM + CGC (Ours)	ResNet-50	24.07M	76.06	**92.50**
TSM + Non-local + CGC (Ours)	ResNet-50	31.43M	**76.40**	**92.50**

Something-Something dataset, it even degrades the performance. Our proposed CGC consistently improves the performances of all the baseline models. We also observe that on Kinetics, our CGC and Non-local are somewhat complementary to each other since applying both of them to the baseline achieves the highest performance. This is consistent with the observation of the combination of CBAM and our CGC in Sect. 4.2.

4.4 Machine Translation

Baseline Methods. The LightConv proposed by [53] achieves better performance with a lightweight convolutional model, compared to Transformer [48]. We

Table 6. Machine translation results on IWSLT'14 De-En. Param indicates the number of parameters in the model. Bold indicates the best result.

Model	Param	BLEU-4
Deng *et al.* [13]	-	33.08
Transformer [48]	39.47M	34.41
LightConv [53]	38.14M	34.84
LightConv + Dynamic Encoder [53]	38.44M	35.03
LightConv + CGC Encoder (Ours)	**38.15M**	**35.21**

take it as the baseline model and augment its Lightweight Convolution with our CGC. Note that the Lightweight Convolution is a grouped convolution $\mathbf{L} \in \mathbb{R}^{H \times k}$ with weight sharing, so we remove the Channel Interacting Module since we do not need it to project latent representations. We resize the input sequence $S \in \mathbb{R}^{c \times L}$ to $\mathbb{R}^{H \times 3k}$ with average pooling. For those sequences shorter than $3k$, we pad them with zeros. Since the decoder decodes translated words one by one at the inference time, it is unclear how to define global context for it. Therefore, we only replace the convolutions in the encoder.

Experimental Setting. We follow [53] to train all the compared models with 160K sentence pairs and 10K joint BPE vocabulary. We use the training protocol of DynamicConv [53] provided in [37]. The widely-used BLEU-4 [38] is reported for evaluation of all the models. We find that it is necessary to set beam width to 6 to reproduce the results of DynamicConv reported in [53], and we fix it to be 6 for all the models.

Performance Comparisons. As Table 6 shows, replacing Lightweight Convolutions in the encoder of LightConv with our CGC significantly outperforms LightConv and LightConv + Dynamic Encoder by 0.37 and 0.18 BLEU, respectively, yielding the state-of-the-art performance. As was discussed previously, Dynamic Convolution leverages a linear layer to generate the convolution kernel according to the input segment, which lacks the awareness of global context. This flaw may lead to sub-optimal encoding of the source sentence and thus the unsatisfying decoded sentence. However, our CGC incorporates global context of the source sentence and helps significantly improve the quality of the translated sentence. Moreover, our CGC is much more efficient than Dynamic Convolution because of our module design. Our CGC only needs 0.01M extra parameters, but Dynamic Convolution needs 30× more.

5   Conclusions

In this paper, motivated by the neuroscience research on neurons as *"adaptive processors"*, we proposed a lightweight Context-Gated Convolution (CGC) to incorporate global context information into CNNs. Different from previous works which usually modify input feature maps, our proposed CGC directly modulates convolution kernels under the guidance of global context information. In specific, we proposed three modules to efficiently generate a gate to modify the kernel. As such, our CGC is able to extract representative local patterns according to global context. The extensive experimental results show consistent performance improvements on various tasks with a negligible computational complexity and parameter increment. In the future, our proposed CGC can be incorporated into the searching space of Neural Architecture Search (NAS) to further improve the performance of NAS models.

References

1. Ba, J.L., Kiros, J.R., Hinton, G.E.: Layer normalization. arXiv preprint arXiv:1607.06450 (2016)
2. Barbu, A., et al.: ObjectNet: a large-scale bias-controlled dataset for pushing the limits of object recognition models. In: Advances in Neural Information Processing Systems, pp. 9448–9458 (2019)
3. Bello, I., Zoph, B., Vaswani, A., Shlens, J., Le, Q.V.: Attention augmented convolutional networks. arXiv preprint arXiv:1904.09925 (2019)
4. Cao, Y., Xu, J., Lin, S., Wei, F., Hu, H.: GCNet: Non-local networks meet squeeze-excitation networks and beyond. arXiv preprint arXiv:1904.11492 (2019)
5. Carreira, J., Zisserman, A.: Quo Vadis, action recognition? A new model and the kinetics dataset. In: CVPR (2017)
6. Cettolo, M., Niehues, J., Stüker, S., Bentivogli, L., Federico, M.: Report on the 11th IWSLT evaluation campaign, IWSLT 2014 (2015)
7. Chattopadhay, A., Sarkar, A., Howlader, P., Balasubramanian, V.N.: Grad-CAM++: generalized gradient-based visual explanations for deep convolutional networks. In: 2018 IEEE Winter Conference on Applications of Computer Vision (WACV), pp. 839–847. IEEE (2018)
8. Chen, Y., Rohrbach, M., Yan, Z., Shuicheng, Y., Feng, J., Kalantidis, Y.: Graph-based global reasoning networks. In: Proceedings of the IEEE Conference on Computer Vision and Pattern Recognition, pp. 433–442 (2019)
9. Cheng, C., et al.: Dual skipping networks. In: Proceedings of the IEEE Conference on Computer Vision and Pattern Recognition, pp. 4071–4079 (2018)
10. Child, R., Gray, S., Radford, A., Sutskever, I.: Generating long sequences with sparse transformers. arXiv preprint arXiv:1904.10509 (2019)
11. Chollet, F.: Xception: deep learning with depthwise separable convolutions. In: Proceedings of the IEEE Conference on Computer Vision and Pattern Recognition, pp. 1251–1258 (2017)
12. Dai, J., et al.: Deformable convolutional networks. In: Proceedings of the IEEE International Conference on Computer Vision, pp. 764–773 (2017)
13. Deng, Y., Kim, Y., Chiu, J., Guo, D., Rush, A.: Latent alignment and variational attention. In: Advances in Neural Information Processing Systems, pp. 9712–9724 (2018)
14. Gehring, J., Auli, M., Grangier, D., Yarats, D., Dauphin, Y.N.: Convolutional sequence to sequence learning. In: Proceedings of the 34th International Conference on Machine Learning, vol. 70, pp. 1243–1252. JMLR.org (2017)
15. Gilbert, C.D., Li, W.: Top-down influences on visual processing. Nat. Rev. Neurosci. **14**(5), 350 (2013)

16. Girshick, R., Donahue, J., Darrell, T., Malik, J.: Rich feature hierarchies for accurate object detection and semantic segmentation. In: CVPR (2014)
17. Glorot, X., Bengio, Y.: Understanding the difficulty of training deep feedforward neural networks. In: Teh, Y.W., Titterington, M. (eds.) Proceedings of the 13th International Conference on Artificial Intelligence and Statistics. Proceedings of Machine Learning Research, Chia Laguna Resort, Sardinia, Italy, 13–15 May 2010, vol. 9, pp. 249–256. PMLR. http://proceedings.mlr.press/v9/glorot10a.html
18. Goyal, R., et al.: The "something something" video database for learning and evaluating visual common sense. In: The IEEE International Conference on Computer Vision (ICCV) (October 2017)
19. Ha, D., Dai, A., Le, Q.V.: Hypernetworks. arXiv preprint arXiv:1609.09106 (2016)
20. He, K., Zhang, X., Ren, S., Sun, J.: Delving deep into rectifiers: surpassing human-level performance on ImageNet classification. In: 2015 IEEE International Conference on Computer Vision (ICCV) (December 2015). https://doi.org/10.1109/iccv.2015.123
21. He, K., Zhang, X., Ren, S., Sun, J.: Deep residual learning for image recognition. In: CVPR (2016)
22. He, K., Zhang, X., Ren, S., Sun, J.: Identity mappings in deep residual networks. In: Leibe, B., Matas, J., Sebe, N., Welling, M. (eds.) ECCV 2016. LNCS, vol. 9908, pp. 630–645. Springer, Cham (2016). https://doi.org/10.1007/978-3-319-46493-0_38
23. He, T., Zhang, Z., Zhang, H., Zhang, Z., Xie, J., Li, M.: Bag of tricks for image classification with convolutional neural networks. In: Proceedings of the IEEE Conference on Computer Vision and Pattern Recognition, pp. 558–567 (2019)
24. Howard, A.G., et al.: MobileNets: Efficient convolutional neural networks for mobile vision applications. arXiv preprint arXiv:1704.04861 (2017)
25. Hu, J., Shen, L., Sun, G.: Squeeze-and-excitation networks. In: Proceedings of the IEEE Conference on Computer Vision and Pattern Recognition, pp. 7132–7141 (2018)
26. Huang, G., Liu, Z., Van Der Maaten, L., Weinberger, K.Q.: Densely connected convolutional networks. In: CVPR (2017)
27. Ioffe, S., Szegedy, C.: Batch normalization: accelerating deep network training by reducing internal covariate shift. In: ICML (2015)
28. Jia, X., De Brabandere, B., Tuytelaars, T., Gool, L.V.: Dynamic filter networks. In: Advances in Neural Information Processing Systems, pp. 667–675 (2016)
29. Jo, Y., Wug Oh, S., Kang, J., Joo Kim, S.: Deep video super-resolution network using dynamic upsampling filters without explicit motion compensation. In: Proceedings of the IEEE Conference on Computer Vision and Pattern Recognition, pp. 3224–3232 (2018)
30. Kim, Y.: Convolutional neural networks for sentence classification. arXiv preprint arXiv:1408.5882 (2014)
31. Krizhevsky, A., et al.: Learning multiple layers of features from tiny images. Technical report. Citeseer (2009)
32. Li, W., Piëch, V., Gilbert, C.D.: Perceptual learning and top-down influences in primary visual cortex. Nat. Neurosci. **7**(6), 651–657 (2004)
33. Li, X., Wang, W., Hu, X., Yang, J.: Selective kernel networks. In: Proceedings of the IEEE Conference on Computer Vision and Pattern Recognition, pp. 510–519 (2019)
34. Lin, J., Gan, C., Han, S.: Temporal shift module for efficient video understanding. arXiv preprint arXiv:1811.08383 (2018)

35. Mildenhall, B., Barron, J.T., Chen, J., Sharlet, D., Ng, R., Carroll, R.: Burst denoising with kernel prediction networks. In: Proceedings of the IEEE Conference on Computer Vision and Pattern Recognition, pp. 2502–2510 (2018)
36. Nair, V., Hinton, G.E.: Rectified linear units improve restricted Boltzmann machines. In: Proceedings of the 27th International Conference on Machine Learning, ICML 2010, pp. 807–814 (2010)
37. Ott, M., et al.: fairseq: a fast, extensible toolkit for sequence modeling. In: Proceedings of NAACL-HLT 2019: Demonstrations (2019)
38. Papineni, K., Roukos, S., Ward, T., Zhu, W.J.: BLEU: a method for automatic evaluation of machine translation. In: Proceedings of the 40th Annual Meeting on Association for Computational Linguistics, pp. 311–318. Association for Computational Linguistics (2002)
39. Park, J., Woo, S., Lee, J.Y., Kweon, I.S.: BAM: Bottleneck attention module. arXiv preprint arXiv:1807.06514 (2018)
40. Parmar, N., et al.: Image transformer. arXiv preprint arXiv:1802.05751 (2018)
41. Paszke, A., Gross, S., Chintala, S., et al.: Automatic differentiation in PyTorch (2017)
42. Qiao, S., Wang, H., Liu, C., Shen, W., Yuille, A.: Weight standardization. arXiv preprint arXiv:1903.10520 (2019)
43. Qiu, Z., Yao, T., Mei, T.: Learning spatio-temporal representation with pseudo-3D residual networks. In: 2017 IEEE International Conference on Computer Vision (ICCV), pp. 5534–5542. IEEE (2017)
44. Ren, S., He, K., Girshick, R., Sun, J.: Faster R-CNN: towards real-time object detection with region proposal networks. In: NIPS (2015)
45. Russakovsky, O., et al.: ImageNet large scale visual recognition challenge. Int. J. Comput. Vis. 115(3), 211–252 (2015)
46. Santurkar, S., Tsipras, D., Ilyas, A., Madry, A.: How does batch normalization help optimization? In: Advances in Neural Information Processing Systems, pp. 2483–2493 (2018)
47. Stollenga, M.F., Masci, J., Gomez, F., Schmidhuber, J.: Deep networks with internal selective attention through feedback connections. In: Advances in Neural Information Processing Systems, pp. 3545–3553 (2014)
48. Vaswani, A., et al.: Attention is all you need. In: Advances in Neural Information Processing Systems, pp. 5998–6008 (2017)
49. Wang, F., et al.: Residual attention network for image classification. In: Proceedings of the IEEE Conference on Computer Vision and Pattern Recognition, pp. 3156–3164 (2017)
50. Wang, L., et al.: Temporal segment networks: towards good practices for deep action recognition. In: Leibe, B., Matas, J., Sebe, N., Welling, M. (eds.) ECCV 2016. LNCS, vol. 9912, pp. 20–36. Springer, Cham (2016). https://doi.org/10.1007/978-3-319-46484-8_2
51. Wang, X., Girshick, R., Gupta, A., He, K.: Non-local neural networks. arXiv preprint arXiv:1711.07971 10 (2017)
52. Woo, S., Park, J., Lee, J.-Y., Kweon, I.S.: CBAM: convolutional block attention module. In: Ferrari, V., Hebert, M., Sminchisescu, C., Weiss, Y. (eds.) ECCV 2018. LNCS, vol. 11211, pp. 3–19. Springer, Cham (2018). https://doi.org/10.1007/978-3-030-01234-2_1
53. Wu, F., Fan, A., Baevski, A., Dauphin, Y.N., Auli, M.: Pay less attention with lightweight and dynamic convolutions. arXiv preprint arXiv:1901.10430 (2019)

54. Yang, Y., Zhong, Z., Shen, T., Lin, Z.: Convolutional neural networks with alternately updated clique. In: Proceedings of the IEEE Conference on Computer Vision and Pattern Recognition, pp. 2413–2422 (2018)
55. Zamir, A.R., et al.: Feedback networks. In: Proceedings of the IEEE Conference on Computer Vision and Pattern Recognition, pp. 1308–1317 (2017)
56. Zhang, X., Zhao, J., LeCun, Y.: Character-level convolutional networks for text classification. In: Advances in Neural Information Processing Systems, pp. 649–657 (2015)
57. Zhou, B., Andonian, A., Oliva, A., Torralba, A.: Temporal relational reasoning in videos. In: Ferrari, V., Hebert, M., Sminchisescu, C., Weiss, Y. (eds.) ECCV 2018. LNCS, vol. 11205, pp. 831–846. Springer, Cham (2018). https://doi.org/10.1007/978-3-030-01246-5_49
58. Zhu, J.Y., Park, T., Isola, P., Efros, A.A.: Unpaired image-to-image translation using cycle-consistent adversarial networks. In: 2017 IEEE International Conference on Computer Vision (ICCV) (2017)
59. Zhu, X., Hu, H., Lin, S., Dai, J.: Deformable ConvNets v2: more deformable, better results. In: Proceedings of the IEEE Conference on Computer Vision and Pattern Recognition, pp. 9308–9316 (2019)
60. Zolfaghari, M., Singh, K., Brox, T.: ECO: efficient convolutional network for online video understanding. In: Ferrari, V., Hebert, M., Sminchisescu, C., Weiss, Y. (eds.) ECCV 2018. LNCS, vol. 11206, pp. 713–730. Springer, Cham (2018). https://doi.org/10.1007/978-3-030-01216-8_43

Polynomial Regression Network
for Variable-Number Lane Detection

Bingke Wang, Zilei Wang$^{(\boxtimes)}$, and Yixin Zhang

University of Science and Technology of China, Hefei, China
{wbkup,zhyx12}@mail.ustc.edu.cn, zlwang@ustc.edu.cn

Abstract. Lane detection is a fundamental yet challenging task in autonomous driving and intelligent traffic systems due to perspective projection and occlusion. Most of previous methods utilize semantic segmentation to identify the regions of traffic lanes in an image, and then adopt some curve-fitting method to reconstruct the lanes. In this work, we propose to use polynomial curves to represent traffic lanes and then propose a novel polynomial regression network (PRNet) to directly predict them, where semantic segmentation is not involved. Specifically, PRNet consists of one major branch and two auxiliary branches: (1) polynomial regression to estimate the polynomial coefficients of lanes, (2) initialization classification to detect the initial retrieval point of each lane, and (3) height regression to determine the ending point of each lane. Through the cooperation of three branches, PRNet can detect variable-number of lanes and is highly effective and efficient. We experimentally evaluate the proposed PRNet on two popular benchmark datasets: TuSimple and CULane. The results show that our method significantly outperforms the previous state-of-the-art methods in terms of both accuracy and speed.

Keywords: Lane detection · Polynomial curve · Deep neural network · Polynomial regression

1 Introduction

The past decade has witnessed the great progress of autonomous driving and intelligent transport systems in academia and industry. In these systems, lane detection is one of the fundamental tasks to fully understand the traffic environment, in which the road lanes represent some kind of traffic rules made by human being. Currently, lane detection is still challenging due to the diversity of lane appearance (*e.g.*, colors, line types) and complexity of traffic environmental conditions (*e.g.*, various weathers, lights, and shadows). For example, it is quite difficult to detect lanes in crowded traffic conditions even for human being due to heavy occlusion by vehicles.

Electronic supplementary material The online version of this chapter (https://doi.org/10.1007/978-3-030-58523-5_42) contains supplementary material, which is available to authorized users.

© Springer Nature Switzerland AG 2020
A. Vedaldi et al. (Eds.): ECCV 2020, LNCS 12363, pp. 719–734, 2020.
https://doi.org/10.1007/978-3-030-58523-5_42

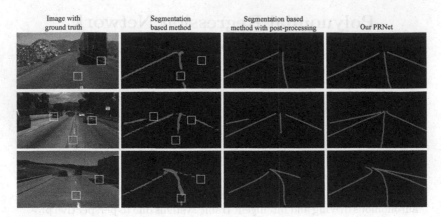

Fig. 1. Comparison of lane detection results between the segmentation based method and our PRNet. Typically, the segmentation based method would suffer the noisy points and intermittent lane segments which need post-processing methods to handle, while PRNet can avoid them due to polynomial representation of traffic lanes.

Many lane detection methods have been proposed to tackle these challenges. Traditional methods [1,2,9,14] usually utilize hand-crafted low-level features to detect the edges or colors, which cannot handle complex conditions. In recent years, some works try to employ the popular deep neural networks to solve this problem [6,10,12,16–18]. Typically, most of these methods treat lane detection as a semantic segmentation task, where each image pixel is classified if it belongs to one of lanes. However, the segmentation based methods often suffer discontinuous and noisy detection results due to thinness of traffic lanes, as shown in Fig. 1. To alleviate this issue, these methods usually use some curve-fitting strategy to filter the noise points [12,16] or cluster the intermittent lane segments [12]. Here we argue that it is unnecessary to explicitly produce semantic segmentation maps for lane detection because such a task essentially targets to get the curves of traffic lanes in an image.

In this paper, we propose to use polynomial curves to represent the traffic lanes and a novel polynomial regression network (PRNet) to directly predict them, in which no semantic segmentation is performed. The key idea of PRNet is to use a piecewise curve to represent a traffic lane rather than a set of image pixels in the previous works. Following this idea, we decompose lane detection into one major subtask and two auxiliary subtasks, *i.e.*, polynomial regression, initialization classification, and height regression, as shown in Fig. 2. Here polynomial regression is used to estimate the polynomial coefficients of lane segments in an image. Initialization classification is used to detect the point to retrieve the initial polynomial coefficients of each lane. Height regression is used to predict the height of ending point for each lane, which together with the estimated polynomial curves determines the ending point of a traffic lane. In this work, we particularly define the initial retrieval point of one lane as the lane point closest to the bottom boundary of image. Evidently, the initial retrieval points

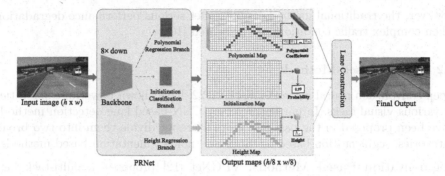

Fig. 2. Illustration of our proposed PRNet. The input image is first transformed into low-resolution feature maps by a backbone network. Then three branches, *i.e.*, polynomial regression, initialization classification, and height regression, take the feature maps as input to predict the polynomial curves of traffic lanes. Finally, the lanes are constructed by fusing the information from three branches. Best viewed in color. (Color figure online)

of different lanes in an image are usually far apart from each other according to the traffic rules.

Different from the segmentation based methods that assign the pixels of different lanes different semantic labels, PRNet identifies a lane by detecting its initial retrieval point. Thus PRNet can detect variable-number lanes, like object detection. Moreover, the curve representations of traffic lanes are inherently smooth, and thus no extra post-processing is needed in constructing lane curves.

The contributions of this work are summarized as:

- We propose to use polynomial curves to represent a traffic lane in images, and then formulate lane detection into three subtasks, *i.e.*, polynomial regression, initialization classification, and height regression.
- We propose a novel polynomial regression network (PRNet) to efficiently perform the three subtasks by three branches, in which low-resolution feature maps having global receptive field at the input images are shared.
- We experimentally verify the effectiveness of our proposed PRNet, and the results on both TuSimple and CULane well demonstrate the superiority of our method to other state-of-the-art methods.

2 Related Work

2.1 Traditional Methods

Traditional methods generally use hand-crafted features to detect traffic lanes. For example, the Gaussian filter [1], Steerable filter [14,15], and Gabor filter [26] are adopted to extract the edge features for lane detection. The color features [9] and histogram based features [7] are also exploited to achieve more accurate lane detection results. For these methods, Hough Transformation (HT) [2] is often employed to perform the lane fitting as a post-processing technique. In practice, however, the traditional methods would suffer serious performance degradation when complex traffic conditions are presented [16].

2.2   CNN-Based Methods

Deep convolution neural networks [8,11,21,22] have shown powerful capabilities in various visual tasks. In particular, many CNN-based lane detection methods have been proposed in the past few years. Here we divide them into two broad categories: segmentation based methods and non-segmentation based methods.

Segmentation Based Methods. VPGNet [12] proposes a multi-task network to jointly handle lane and road marking detection under the guidance of vanishing point. Spatial CNN (SCNN) [17] generalizes the traditional deep layer-by-layer convolutions to slice-by-slice convolutions within feature maps, which contribute to detecting long continuous slender structure or large objects. LaneNet [16] proposes to formulate lane detection into an instance segmentation problem and then predict a perspective transformation matrix for better fitting lanes. Embedding-loss GAN (EL-GAN) [6] introduces a GAN framework to make the produced semantic segmentation maps more realistic or better structure-preserving. Self Attention Distillation (SAD) [10] allows a model to learn from itself and gains substantial improvement without any additional supervision or labels. Different from these methods, our proposed method in this work does not involve semantic segmentation.

Non-segmentation Based Methods. Inspired by Faster RCNN [19], Li *et al.*proposed Line-CNN that utilizes line proposals as references to locate traffic curves [13]. Line-CNN need generate a large number of line proposals to achieve good performance. FastDraw [18] proposes to estimate the joint distribution of neighboring pixels belonging to the same lane and draw the lane in an iterative way, in which a binary segmentation map is needed as guidance. 3D-LaneNet [5] directly predicts the 3D layout of lanes in a road scene from a single image through an end-to-end network, which uses the anchor-based lane representation similar to Line-CNN. [24] proposes to estimate lane curvature parameters by solving a weighted least-squares problem in-network, whose weights are generated by a deep network conditioned on the input image. However, the method needs to generate the segmentation-like weight map for each lane separately, and thus can only detect a fixed number of lanes. In addition, the involved huge matrix operation for solving the weighted least-squares problem is time-consuming.

Most of previous methods need to perform some post-processing method to obtain the final traffic curves in practice. For the segmentation based methods, the clustering method (*e.g.*, DBSCAN [3]) or line fitting method (*e.g.*, RANSAC [4]) is often required. In addition, Line-CNN needs to employ NMS [19] to eliminate the redundant line proposals. Evidently, the post-processing in these methods would involve extra computational cost. On the contrary, our proposed network can directly produce the traffic curves and the number of lanes in an image is not required to be fixed.

3   Our Approach

Traffic lanes belong to the man-made objects that are used to specify the traffic rules. In general, lanes are drawn on the roads with a shape of line or curve. So we propose to use the intrinsic curves to represent the traffic lanes in images, and it is expected that such curves can be directly predicted by some network. Following this idea, we particularly propose polynomial curves to represent traffic lanes.

Due to the perspective projection, a lane in images may present a complicated shape that is hard to be accurately represented by one single polynomial curve. To tackle the issue, we propose to use the piecewise polynomials with different coefficients to represent one lane curve. As a result, each lane in an image can be represented by

$$
\begin{cases}
[a_0^1, a_1^1, \cdots, a_{n-1}^1, a_n^1] & \text{if } h_r - H \leqslant x \leqslant h_r \\
[a_0^2, a_1^2, \cdots, a_{n-1}^2, a_n^2] & \text{if } h_r - 2H \leqslant x < h_r - H \\
\cdots \\
[a_0^k, a_1^k, \cdots, a_{n-1}^k, a_n^k] & \text{if } h_e \leqslant x < h_r - (k-1)H
\end{cases}
\tag{1}
$$

where n is the polynomial order, k is the number of polynomials, and $\{a_i^j\}_{i=0}^n$ are the polynomial coefficients of the jth polynomial piece. In addition, h_r is the height of the initial retrieval point, H is a hyper-parameter that denotes the height of each polynomial piece, and h_e is the height of the ending point. Obviously, we have $k = \lceil \frac{h_r - h_e}{H} \rceil$. Different from the splines to represent lanes by identifying the control points, our proposed piecewise polynomials target to get the polynomial coefficients directly.

According to the above formulation, our task turns to predict the polynomial coefficients $\{a_i^j\}_{i=0}^n$ for each lane segment. Here our main challenges lie in how to model all polynomial pieces in an image so that each lane curve can be effectively constructed, and how to design efficient implementation. To this end, we propose a novel Polynomial Regression Network (PRNet) in this paper, as shown in Fig. 2. Specifically, we formulate lane detection into three subtasks, *i.e.*, polynomial regression, initialization classification, and height regression, and complete them by three branches with sharing the input features. Here polynomial regression is the major task that is used to estimate the polynomial coefficients of lane segments. Initialization classification is used to detect the initial point of each lane for retrieving the coefficients of the first segment from polynomial map. Height regression is to estimate the height of the ending point for each lane, which determines the ending point together with the estimated polynomial curve. Once the results of three branches are obtained, we can directly construct the curve representation of lanes, where each lane consists of k polynomials.

More specifically, a backbone network is employed to extract the shared features of three branches of PRNet. Here the down-sampled features with global receptive field at the input images are used, *i.e.*, the decoder in the segmentation-based methods is eliminated, since the information of encoded features is enough

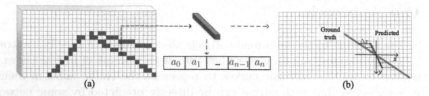

(a)   (b)

Fig. 3. Illustration of Polynomial Regression. (a) Polynomial map, where the red points denote the used points during training, namely, *polynomial points*. (b) One polynomial piece, where the red line denotes the predicted one from a polynomial point, and the green one corresponds to the ground truth. The differences of sampled points on two lines are used to calculate the loss. Best viewed in color. (Color figure online)

for PRNet. Such a design makes PRNet very efficient. In our implementation, 8× down-sampling is particularly adopted that can achieve a good trade-off between efficiency and effectiveness. Note that we design the output maps of three branches to have the same spatial size, in which the points of three maps at the same position together represent the polynomial curve of a lane segment. In the following, we elaborate on the important components of PRNet.

3.1 Polynomial Regression

The polynomial regression branch is used to predict the polynomial coefficients of all lane segments in an image. For such a task, we design the output to be a $(n + 1)$-channel map with the same size as the input features, which is called *polynomial map*. One point in the polynomial map denotes a n-order polynomial. In our implementation, only a part of points are chosen to represent the lane segments, which are called *polynomial points* in this paper. Particularly, the points lying on the lanes are used to calculate the loss during training, *e.g.*, the red points in Fig. 2. Each polynomial point is to perform regression of the closest lane segment. More specifically, we segment traffic lanes in images along the vertical orientation, *i.e.*, the height H is used to denote the length of polynomial pieces. In our implementation, this branch only contains one convolutional layer and thus is highly efficient.

Formally, let $[a_0, a_1, \cdots, a_n]$ and $[\bar{a}_0, \bar{a}_1, \cdots, \bar{a}_n]$ denote the predicted polynomial coefficients and corresponding ground truth for one lane segment. To supervise training of the network, we propose to transform each polynomial segment into some sampling points in images. Particularly, we first sample m points uniformly along the vertical orientation for each lane segment, and then compute the corresponding horizontal coordinates by applying them to the involved polynomial. As a result, we can get $\{(x_p^1, y_p^1), (x_p^2, y_p^2), \cdots, (x_p^m, y_p^m)\}$ and $\{(x_{gt}^1, y_{gt}^1), (x_{gt}^2, y_{gt}^2), \cdots, (x_{gt}^m, y_{gt}^m)\}$ corresponding to the predicted polynomial piece and ground truth. Obviously, $y_p^i = y_{gt}^i$. In this work, the polynomials are enforced to fit the ground truth that are inherently continuous and we use the differences of sampled points on the x-coordinate to define the loss, *i.e.*,

$$L_{poly}(x_p, x_{gt}) = \frac{1}{m} \sum_{i=1}^{m} smooth_{L_1}(x_p^i - x_{gt}^i), \qquad (2)$$

where

$$smooth_{L_1}(x) = \begin{cases} \dfrac{0.5(x)^2}{\beta} & if \ |x| < \beta \\ |x| - 0.5\beta & otherwise \end{cases} \qquad (3)$$

It can be seen that when $|x_p^i - x_{gt}^i| < \beta$, the predicted point is considered near the traffic lane and the L_2 loss is adopted. The computation of polynomial regression loss is illustrated in Fig. 3.

3.2 Initialization Classification

The initialization classification branch is used to detect the initial retrieval points of all lanes in an image. Through this subtask, we can identify arbitrary number of lanes in principle as each point represents one traffic lane. Here we particularly define the initial retrieval point of a lane by its closest point to the bottom boundary of image. Considering the perspective projection of car cameras, such points are usually far apart from each other, which makes accurate detection easier than the dense points. Note that the initial retrieval points are mainly used to retrieve the polynomial coefficients from the polynomial map rather than to determine their starting points in image. Here the standard cross entropy loss is adopted, and a probability map with the same size as input features would be produced, which is called *initialization map*. Similar to polynomial regression, this branch only contains one convolutional layer. During inference, we get the initial retrieval points by scanning the initialization map. The points whose probability is local maximum and greater than the threshold are considered as the initial retrieval points. Here no post-processing technique are applied in our implementation.

3.3 Height Regression

An intuitive approach to get the ending point of each lane is to directly detect them, as in initialization classification. However, the ending points of traffic lanes in an image are often close to each other due to perspective projection. Consequently, it is difficult to accurately localize them and match them with traffic lanes. Instead, we propose to estimate the height of ending point for each traffic lane, as in [24], which together with the estimated polynomial curve can exactly produce the ending point.

Similar to the polynomial regression branch, this branch regresses the heights of ending points of all traffic lanes, and produces an one-channel *height map* with the same size as the input features. One point in the height map gives the estimated height of ending point of the traffic lane it belongs to. Specifically, only the points lying on traffic lanes are used in training the network, *e.g.*, the yellow points in Fig. 2. Here the smooth $L1$ loss [19] in Eq. (3) is adopted. Similarly, the branch only contains one convolutional layer.

Algorithm 1: Lane Construction

Input: L_1: Initial retrieval point whose height is h_r; P_{map}: Polynomial map;
　　　　H_{map}: Height map; H: Height of lane segments.

Output: Polynomial coefficients and height of the lane.

1 Retrieve the polynomial coefficients of first polynomial piece
$A_1 = [a_0^1, a_1^1, \cdots, a_n^1]$ from P_{map} at L_1;

2 Retrieve the initial ending height h_{end}^1 from the height map H_{map} at L_1 ;

3 $i = 1$; $h_e = h_{end}^1$;

4 **while** $h_e < h_r - i * H$ **do**

5 　　i=i+1;

6 　　Get L_i by applying the y-coordinate $(h_r - (i-1)H)$ to A_{i-1};

7 　　Retrieve the polynomial coefficients of ith polynomial piece
　　　　$A_i = [a_0^i, a_1^i, \cdots, a_n^i]$ from P_{map} at L_i;

8 　　Retrieve the ith height h_{end}^i from H_{map} at L_i;

9 　　Update h_e by voting over $\{h_{end}^1, \cdots, h_{end}^i\}$;

10 **Return** Polynomial coefficients $\{A_i\}_{i=1}^k$ and the height h_e.

3.4 Lane Construction

The three branches of PRNet produce the polynomial coefficients, initial retrieval points, and heights of ending points. Here we explain how to construct each traffic lane in an image using the produced information. Algorithm 1 gives the procedure to construct one of traffic lanes, and Fig. 4 illustrates it. Note that the maps produced by three branches have the same size, implying that they can naturally match with each other.

Specifically, we first get all initial retrieval points by scanning the initialization map, each of which represents one traffic lane. Then we construct the traffic lanes one-by-one by connecting multiple lane pieces belonging to the same lane and at the same time calculating the height of the ending point. For a single traffic lane, the initial retrieval point is used to retrieve the polynomial coefficients of first polynomial piece and initial height of ending point, and additionally its height is considered as the height of the starting point. For next lane segment, we first use the vertical interval H to get the y-coordinate of retrieval point and then get the x-coordinate by applying it to the current polynomial piece. That is, the ending point of current polynomial piece is regarded as the retrieval point of next polynomial piece. For each iteration, we would update the estimated height of ending point. Here a voting strategy is particularly adopted over the currently obtained height values, *i.e.*, the most often value is selected as the estimated height. Note that the height values are discretized with an interval of ten pixels in our implementation. In our experimental evaluation, the lanes are represented by the sampled points from polynomials which inherently form the continuous lane curves.

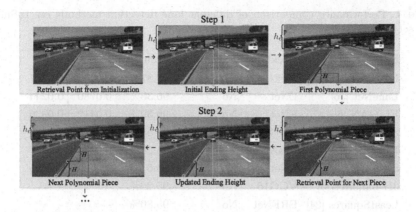

Fig. 4. Illustration of lane construction. We first get the initial retrieval point by scanning initialization map, and use it to retrieve the initial height from height map and polynomial coefficients of first polynomial piece from polynomial map. Then we can get the ending point of current polynomial piece, which is used as the retrieval point of next polynomial piece. The procedure is repeated until the ending point of the traffic lane is reached. Here the polynomial pieces are connected to form a traffic lane and the height is updated iteratively. Best viewed in color. (Color figure online)

4 Experiment

In this section, we experimentally evaluate our proposed PRNet on two popular benchmark datasets: TuSimple [23] and CULane [17]. The representative lane detection methods are used for comparison, including Line-CNN [13], LaneNet [16], EL-GAN [6], SCNN [17], FastDraw [18], 3D-LaneNet [5], SAD [10], and LeastSquares [24]. For each dataset, the reported results of methods in the original literatures are adopted for performance comparison, and one method would not be involved if it does not offer the corresponding results.

4.1 Experimental Setup

To show the generalization of our PRNet, we choose the BiSeNet [25] with ResNet18 [8] and ERFNet [20] as the backbone. Both of them are efficient and their features have a global receptive field at the input image. Specifically, we replace the FFM module of BiSeNet and the decoder module of ERFNet with one convolutional layer followed by a SCNN_D block [17], which can effectively extract discriminative features for lane detection. All the networks are implemented in PyTorch, and we run experiments on NVIDIA GTX1080Ti GPUs. The model pretrained on the ImageNet is used for initialization. For PRNet, we train the three branches jointly and the loss weights of the three branches are set to 1, 1 and 0.1 respectively. Adam optimizer is adopted for optimization. The learning rate is set to 0.0001. The hyper-parameters m, β mentioned in Sect. 3.1 are set to 20 and 0.005, which are determined empirically by cross validation.

Table 1. Performance comparison of different lane detection methods on TuSimple (test set).

Method	Backbone	Extra data	Accuracy	FP	FN
Line-CNN [13]	ResNet50	-	96.87%	0.0442	0.0197
LaneNet [16]	ENet	No	96.38%	0.0780	0.0244
EL-GAN [6]	ENet	No	96.39%	0.0412	0.0336
SCNN [17]	VGG16	Yes	96.53%	0.0617	0.0180
FastDraw [18]	ResNet50	No	95.2%	0.076	0.045
SAD [10]	ENet	No	96.64%	0.0602	0.0205
3D-LaneNet [5]	VGG16	No	95.20%	-	-
LeastSquares [24]	ERFNet	No	95.80%	-	-
PRNet	BiSeNet	No	**97.18%**	**0.0397**	**0.0172**
PRNet	ERFNet	No	97.00%	0.0491	0.0209

Throughout the experiments, images in TuSimple and CULane datasets are first resized to 256×512 and 256×768 respectively. Three types of data augmentation strategies are adopted, including randomly flipped, randomly rotated, and randomly varying brightness.

4.2   Results on TuSimple

Dataset. TuSimple [23] is a popular dataset for lane detection in recent years. It includes $3,268$ images for training, 358 images for validation, and $2,782$ images for test. The sizes of these images are all 720×1280. The annotations of traffic lanes are given in the form of polylines of lane markings, which have a fixed height-interval of 10 pixels. For each image, only the current (ego) lanes and left/right lanes are annotated in both the training and test set. When a lane is crossed, a 5th lane would be added to avoid confusion, which means that each image contains at most 5 lanes.

Evaluation Metrics. We follow the official evaluation metrics ($Acc/FP/FN$). The accuracy is defined as $Acc = \frac{C_{pred}}{T_{gt}}$, where C_{pred} is the number of lane points correctly predicted by the network and T_{gt} is the total number of lane points in ground truth. FP and FN are defined as $FP = \frac{F_{pred}}{N_{pred}}$ and $FN = \frac{M_{pred}}{N_{gt}}$, where F_{pred} is the number of wrongly predicted lanes, N_{pred} is the total number of predicted lanes, and M_{pred} is the number of missed lanes, and N_{gt} is the number of all groundtruth lanes.

Performance Comparison. Table 1 reports the performance comparison of our PRNet against the previous representative methods, where the test set of TuSimple is adopted for evaluation. It can be seen that our method outperforms the previous state-of-the-art methods on all three metrics, which implies that PRNet can detect the lanes more accurately with less wrong prediction and lane missing. Note that no extra data are used for training our PRNet.

Table 2. Performance (F_1-measure) of different lane detection methods on CULane (test set). Here $*$ denotes that the backbone is BiSeNet with ResNet18 and † denotes that the backbone is ERFNet. For crossroad, only FP is reported for fair comparison. The second column denotes the proportion of each scenario in the test set.

Category	Proportion	SCNN [17]	FastDraw [18]	SAD [10]	**PRNet***	**PRNet†**
Normal	27.7%	90.6	85.9	90.7	90.8	**92.0**
Crowded	23.4%	69.7	63.6	70.0	72.3	**74.7**
Night	20.3%	66.1	57.8	66.3	69.2	**70.5**
No line	11.7%	43.4	40.6	43.5	47.6	**51.7**
Shadow	2.7%	66.9	59.9	67.0	70.6	**76.0**
Arrow	2.6%	84.1	79.4	84.4	85.2	**87.8**
Dazzle light	1.4%	58.5	57.0	59.9	64.2	**68.4**
Curve	1.2%	65.7	65.2	65.7	67.2	**70.0**
Crossroad	9.0%	1990	7013	2052	**1113**	2114
Total	–	71.6	–	71.8	74.8	**76.4**

4.3 Results on CULane

Dataset. CULane [17] is a large lane detection dataset which contains about $130k$ images. The dataset is divided into the training set with $88,880$ images, validation set with $9,675$ images, and test set with $34,680$ images. The images are collected at the urban, rural, and highways in Beijing. All images in CULane datest have the same resolution of 590×1640. For each image, only at most 4 lanes are annotated: the current (ego) lanes and left/right lanes. The format of annotations are same with the TuSimple dataset. In general, the CULane dataset is considered more challenging than the TuSimple dataset.

Evaluation Metrics. Following SCNN [17], we extend the predicted lanes to a width of 30 pixels and then calculate the intersection-over-union (IoU) between the ground truth and prediction. True positives (TP) are the number of predicted lanes whose IoUs are greater than a certain threshold, and false positives (FP) are opposite. Here we choose 0.5 as the preset threshold by following [17]. False negatives (FN) are the number of missed lanes. Then we adopt the F_1-measure to evaluate the methods, which is defined as $F_1 = \frac{2 \times Precision \times Recall}{Precision + Recall}$, where $Precision = \frac{TP}{TP+FP}$ and $Recall = \frac{TP}{TP+FN}$.

Performance Comparison. As the CULane dataset is more challenging than the TuSimple dataset, the results on CULane can better demonstrate the capacity of different methods. Table 2 gives the detection performance of different methods, and we have the following observations. First, our method can always get better results than the previous state-of-the-art methods for each category. Second, the performance improvement of our method is more significant for complex scenarios, *e.g.*, *crowded*, *dazzle light*, and *no line*, which well demonstrates the robustness of PRNet to the traffic conditions. We also provide the visualization results of some examples in Fig. 5, which intuitively show the performance of different lane detection methods.

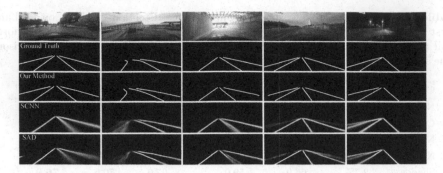

Fig. 5. Visualization of different methods. The segmentation based methods often fail to predict some lanes when the scenarios are complex. Our PRNet could handle the complex scenarios well.

Table 3. Performance comparison between segmentation and regression. Here $*$ denotes that the backbone is BiSeNet with ResNet18 and $\dagger$ denotes that the backbone is ERFNet. Here the accuracy and F_1-measure are reported for TuSimple and CULane respectively, and the test set is used.

Method	TuSimple	CULane
Segmentation*	96.11%	70.7
Regression (ours)*	**97.18%**	**74.8**
Segmentation†	95.92%	73.6
Regression (ours)†	**97.00%**	**76.4**

4.4 Ablation Study

Regression _vs_ Segmentation. The key idea of our PRNet is to use polynomial regression to complete lane detection rather than semantic segmentation in previous works. Here we particularly explore the advantages of regression by fairly comparing them with the same backbone and settings. Specifically, we construct a semantic segmentation head (producing lane markings) and a lane classification head (judging existence of lane markings) by following SCNN [17], and then append them to the backbone of PRNet, like the polynomial regression head. Moreover, the segmentation results will be fitted as splines for evaluation. We conduct the experiments on both TuSimple and CULane datasets, and Table 3 provides the results. Evidently, the experimental results well demonstrate the superiority of our proposed regression to semantic segmentation, especially for more challenging CULane.

Polynomial Order and Piece Height of Polynomials. In PRNet, the polynomial order n and piece height of polynomials H are two main hyperparameters, which represent the ability and complexity to describe lane curves. In principle, a smaller piece height requires a lower polynomial order since shorter lane segments are easier to be fitted by curves. Here we study the effects of dif-

Table 4. Detection performance of different polynomial orders and piece heights. Here the F_1-measure on CULane (validation set) is particularly reported.

Order (n) /Height (H)	8	16	32	64
1	77.31	76.85	76.82	76.66
2	77.48	**77.72**	77.18	76.80
3	77.66	77.70	77.59	77.23

Table 5. Run-time performance of different methods. Here * denotes that we run the model provided by authors on the used platform, and otherwise the result reported in the original paper is directly adopted.

Method	Platform	Input size	FPS
Line-CNN (Res50) [13]	Titan X	288 × 512	30
SCNN (VGG16)* [17]	GTX1080Ti	208 × 976	20
LaneNet (ENet) [16]	GTX1080Ti	256 × 512	52
FastDraw (Res50) [18]	GTX1080	128 × 256	90
SAD (ENet)* [10]	GTX1080Ti	208 × 976	79
LeastSquares (ERFNet)* [24]	GTX1080Ti	256 × 512	50
Ours (BiSeNet)*	GTX1080Ti	256 × 512	**110**
Ours (ERFNet)*	GTX1080Ti	256 × 512	81

Table 6. The statistics of failure cases on two datasets. Here the four failure categories are adopted.

Category	IRPM	IRPW	PRI	HRI
TuSimple	134	502	9	26
CULane	28457	16104	669	334

ferent combinations of polynomial order and piece height. Particularly, CULane is adopted due to its challenging and the BiSeNet with ResNet18 is chosen as the backbone of PRNet. Table 4 shows the results, and we have the following observations. First, for a large piece height (*e.g.*, 64), a higher polynomial order is better since more powerful ability of fitting is required. Second, a low order is enough to achieve good detection performance for some reasonable piece height (*e.g.*, 16). Considering the complexity, we finally set the polynomial order $n = 2$ and piece height $H = 16$ throughout the experiments.

Run-Time Performance. Here we evaluate the run-time performance of different lane detection methods. Particularly, one GTX1080Ti GPU is used for fair comparison. Table 5 gives the run-time performance. It can be seen that our proposed PRNet achieves a speed of 110 FPS for the backbone of BiSeNet with ResNet18, which is very competitive to other state-of-the-art methods.

Fig. 6. Visualization of Failure cases. Top Row: images with ground truth. Bottom Row: the results produced by our PRNet. Here the four categories of failures are shown from left to right, including initial retrieval point missing (IRPM), wrong prediction of initial retrieval points (IRPW), inaccuracy of polynomial regression (PRI), and inaccuracy of height regression (HRI).

Failure Cases Analysis. Here we analyse the failure cases of our PRNet on both TuSimple and CULane datasets. We classify failure cases into four categories: initial retrieval point missing (IRPM), wrong prediction of initial retrieval points (IRPW), inaccuracy of polynomial regression (PRI), and inaccuracy of height regression (HRI). Table 6 shows the statistics over the four categories of failure cases, and Fig. 6 gives the visualization of typical failure cases. From the results, it can be seen that most of failures are about initial retrieval points, including wrong prediction and detection missing. Furthermore, we visualize many failure cases to deeply analyse the failure causes. We find that detection missing is mainly due to irregular scenes, *e.g.*, dark light, crowded vehicles, and no lane markings, and wrong prediction is mainly due to deceptive or confused scenes, *e.g.*, lane-like lines and unlabeled lanes (both datasets limit the number of lanes to annotate according to their protocols). For the complex scenes, however, our PRNet performs much better than other methods, as shown in Table 2. To further address these issues, we plan to introduce the structure information in the future works, *e.g.*, embedding the layout of lanes into network.

5   Conclusion

In this paper, we propose to use the in-network polynomial curves to represent the traffic lanes in images, and then propose a novel polynomial regression network (PRNet) for variable-number lane detection. Specifically, PRNet consists of three cooperative branches: polynomial regression, initialization classification, and height regression. The experimental results on two benchmark datasets show our proposed method significantly outperforms the previous state-of-the-art methods, and achieves competitive run-time performance. In particular, our PRNet presents better robustness to complex traffic conditions than other methods.

Acknowledgment. This work is supported by the National Natural Science Foundation of China under Grant 61673362 and 61836008, Youth Innovation Promotion Association CAS (2017496), and the Fundamental Research Funds for the Central Universities.

References

1. Aly, M.: Real time detection of lane markers in urban streets. In: IV (2008)
2. Borkar, A., Hayes, M., Smith, M.T.: Polar randomized hough transform for lane detection using loose constraints of parallel lines. In: ICASSP (2011)
3. Ester, M., Kriegel, H.P., Sander, J., Xu, X., et al.: A density-based algorithm for discovering clusters in large spatial databases with noise. In: KDD (1996)
4. Fischler, M.A., Bolles, R.C.: Random sample consensus: a paradigm for model fitting with applications to image analysis and automated cartography. Commun. ACM **24**, 381–395 (1981)
5. Garnett, N., Cohen, R., Pe'er, T., Lahav, R., Levi, D.: 3D-lanenet: end-to-end 3D multiple lane detection. In: CVPR (2019)
6. Ghafoorian, M., Nugteren, C., Baka, N., Booij, O., Hofmann, M.: EL-GAN: embedding loss driven generative adversarial networks for lane detection. In: Leal-Taixé, L., Roth, S. (eds.) ECCV 2018. LNCS, vol. 11129, pp. 256–272. Springer, Cham (2019). https://doi.org/10.1007/978-3-030-11009-3_15
7. Gonzalez, J.P., Ozguner, U.: Lane detection using histogram-based segmentation and decision trees. In: IEEE Intelligent Transportation Systems (2000)
8. He, K., Zhang, X., Ren, S., Sun, J.: Deep residual learning for image recognition. In: CVPR (2016)
9. He, Y., Wang, H., Zhang, B.: Color-based road detection in urban traffic scenes. In: IEEE Transactions on Intelligent Transportation Systems, pp. 309–318 (2004)
10. Hou, Y., Ma, Z., Liu, C., Loy, C.C.: Learning lightweight lane detection CNNs by self attention distillation. In: ICCV (2019)
11. Krizhevsky, A., Sutskever, I., Hinton, G.E.: Imagenet classification with deep convolutional neural networks. In: Advances in Neural Information Processing Systems, pp. 1097–1105 (2012)
12. Lee, S., et al.: Vpgnet: Vanishing point guided network for lane and road marking detection and recognition. In: ICCV (2017)
13. Li, X., Li, J., Hu, X., Yang, J.: Line-CNN: end-to-end traffic line detection with line proposal unit. IEEE Trans. Intell. Transp. Syst. **21**, 248–258 (2019)
14. McCall, J.C., Trivedi, M.M.: An integrated, robust approach to lane marking detection and lane tracking. In: IV (2004)
15. McCall, J.C., Trivedi, M.M.: Video-based lane estimation and tracking for driver assistance: survey, system, and evaluation (2006)
16. Neven, D., De Brabandere, B., Georgoulis, S., Proesmans, M., Van Gool, L.: Towards end-to-end lane detection: an instance segmentation approach. In: IV (2018)
17. Pan, X., Shi, J., Luo, P., Wang, X., Tang, X.: Spatial as deep: Spatial cnn for traffic scene understanding. In: AAAI (2018)
18. Philion, J.: FastDraw: addressing the long tail of lane detection by adapting a sequential prediction network. In: CVPR (2019)
19. Ren, S., He, K., Girshick, R., Sun, J.: Faster R-CNN: towards real-time object detection with region proposal networks. In: Advances in Neural Information Processing Systems (2015)

20. Romera, E., Alvarez, J.M., Bergasa, L.M., Arroyo, R.: ERFNet: efficient residual factorized convNet for real-time semantic segmentation. IEEE Trans. Intell. Transp. Syst. **19**, 263–272 (2017)
21. Simonyan, K., Zisserman, A.: Very deep convolutional networks for large-scale image recognition. arXiv preprint arXiv:1409.1556 (2014)
22. Szegedy, C., et al.: Going deeper with convolutions. In: CVPR (2015)
23. TuSimple: http://benchmark.tusimple.ai/#/t/1. Accessed 08 Sept 2018
24. Van Gansbeke, W., De Brabandere, B., Neven, D., Proesmans, M., Van Gool, L.: End-to-end lane detection through differentiable least-squares fitting. In: ICCV Workshop (2019)
25. Yu, C., Wang, J., Peng, C., Gao, C., Yu, G., Sang, N.: BiSeNet: bilateral segmentation network for real-time semantic segmentation. In: Ferrari, V., Hebert, M., Sminchisescu, C., Weiss, Y. (eds.) ECCV 2018. LNCS, vol. 11217, pp. 334–349. Springer, Cham (2018). https://doi.org/10.1007/978-3-030-01261-8_20
26. Zhou, S., Jiang, Y., Xi, J., Gong, J., Xiong, G., Chen, H.: A novel lane detection based on geometrical model and Gabor filter. In: IV (2010)

Structural Deep Metric Learning for Room Layout Estimation

Wenzhao Zheng[1,2,3], Jiwen Lu[1,2,3(✉)], and Jie Zhou[1,2,3,4]

[1] Department of Automation, Tsinghua University, Beijing, China
`zhengwz18@mails.tsinghua.edu.cn`, {`lujiwen,jzhou`}`@tsinghua.edu.cn`
[2] State Key Lab of Intelligent Technologies and Systems, Beijing, China
[3] Beijing National Research Center for Information Science and Technology,
Beijing, China
[4] Tsinghua Shenzhen International Graduate School, Tsinghua University,
Beijing, China

Abstract. In this paper, we propose a structural deep metric learning (SDML) method for room layout estimation, which aims to recover the 3D spatial layout of a cluttered indoor scene from a monocular RGB image. Different from existing room layout estimation methods that solve a regression or per-pixel classification problem, we formulate the room layout estimation problem from a metric learning perspective where we explicitly model the structural relations across different images. We propose to learn a latent embedding space where the Euclidean distance can characterize the actual structural difference between the layouts of two rooms. We then minimize the discrepancy between an image and its ground-truth layout in the learned embedding space. We employ a metric model and a layout encoder to map the RGB images and the ground-truth layouts to the embedding space, respectively, and a layout decoder to map the embeddings to the corresponding layouts, where the whole framework is trained in an end-to-end manner. We perform experiments on the widely used Hedau and LSUN datasets and achieve state-of-the-art performance.

Keywords: Deep metric learning · Room layout estimation · Structured prediction

1 Introduction

Room layout estimation has attracted great attention in recent years, since it serves as a basic step to provide strong priors for a variety of applications such as indoor navigation [1,30,55], augmented reality [26,27,46], and scene understanding [8,9,13]. The goal of room layout estimation is to find a projection of a 3D box onto the image which best fits the actual layout of the scene, as described in Fig. 1. A major challenge of this task is the presence of clutter, where different kinds of furniture like beds, sofas, and tables may totally or partially occlude

© Springer Nature Switzerland AG 2020
A. Vedaldi et al. (Eds.): ECCV 2020, LNCS 12363, pp. 735–751, 2020.
https://doi.org/10.1007/978-3-030-58523-5_43

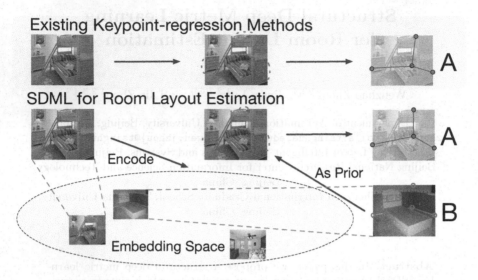

Fig. 1. Comparison of the proposed SDML method with existing methods for room layout estimation which simply regress the locations of keypoints (e.g. point of intersection among the left wall, the front wall, and the ceiling). The keypoints are often occluded by clutter like beds or sofas, making it hard to directly determine the exact locations of keypoints. Instead, we propose to learn a latent embedding space to explicitly model the structural relations across different images and use the globally similar images as prior to better infer the precise keypoints of the query image.

the boundary between two surfaces of the box, making it difficult to recover the underlying layout of the room.

A main characteristic of room layout estimation is that the output is of a structured form, as indoor scenes typically satisfy the "Manhattan world assumption" [2]. Conventional methods on room layout estimation represent the room layout by a set of parameters and solve a structured prediction problem [16,24,36]. With the success of deep learning, recent works begin to employ deep convolutional neural networks (CNNs) [15,20,37,41] to extract features from images. The first category of works obtain a segmentation map for each image and then perform a search process for the best legitimate layout [3,29,49,51]. The second category of works define a set of keypoints to describe the room layout and regress the locations of those keypoints [17,23]. However, all these previous works fail to consider the structural correlations among different images and cannot well capture the global features of the layouts. On the other hand, being able to globally consider the whole indoor scene is important for room layout estimation due to the occlusion problem caused by clutter. For instance, as demonstrated by scene A in Fig. 1, it is difficult to directly find the point of intersection of the floor, the left wall, and the right wall, yet it is simpler to say it is similar to scene B in terms of their layouts. Using the predicted layout of scene B as prior, it is easier to determine the keypoints and layout of scene A.

Motivated by the above example, in this work, we formulate the room layout estimation problem from a metric learning perspective and propose a structural deep metric learning (SDML) method which can be trained in an end-to-end manner. We explicitly model the structural relations across images by learning a mapping from the image space to a latent embedding space where the Euclidean distance can reflect the similarity of the underlying layouts of two images, i.e., a smaller distance between two images with more similar layouts and vice versa. We further propose a dense structural loss to enforce such a continuous constraint, which generalizes beyond previous triplet-based loss and enjoys the advantage of efficient sampling and faster convergence. We propose a layout antoencoder with a layout encoder to obtain a representation for each layout in the embedding space and a layout decoder to generate layouts from the embedding space. We then minimize the difference of the embeddings of an image and its corresponding ground-truth layout as well as the reconstruction cost between the original layout and the reconstructed layout. In the test phase, we simply connect the metric model and the layout decoder to obtain an estimated layout for an image. Extensive experiments on the widely used Hedau [16] and LSUN [50] datasets demonstrate the effectiveness of the proposed approach.

2 Related Work

Room Layout Estimation: The problem of room layout estimation was first formally introduced by Hedau *et al.* [16] and has attracted constant attention since then. Most conventional methods used structured prediction learning algorithms (e.g., structured SVMs [42]) for room layout estimation [16,24,35,36,43]. For example, Hedau *et al.* [16] proposed to iteratively localize visible objects and refit the box with structured SVMs in order to be more robust to clutter. Lee *et al.* [24] generated room hypotheses from line segments [25] and eliminated invalid hypotheses by volumetric reasoning before ranking them. Wang *et al.* [43] introduced latent variables to implicitly describe clutters and proposed to parameterize a layout by four factors inferred using structured SVMs.

Recent methods took advantage of deep networks and employed fully convolutional networks (FCNs) [28] to extract features, improving the performance of conventional methods dramatically. Mallya *et al.* [29] and Ren *et al.* [32] extracted per-pixel feature maps using FCNs and then ranked layout proposals based on them. Dasgupta *et al.* [3] and Zhao *et al.* [51] proposed to perform inference through optimization instead of proposal ranking, but they still require a two-step procedure to obtain the layout. Lee *et al.* [23] proposed an end-to-end framework for room layout estimation by simultaneously predicting the room type of a scene and regressing a set of pre-defined keypoints. However, existing methods lack a global understanding of the scene which is important to infer the underlying layout, especially for a cluttered room. Differently, the proposed SDML method explicitly models the structural relations among images which provide strong priors for more exact and robust room layout inference.

Metric Learning: Metric learning aims to learn a good distance function to measure the similarities between images with a common objective of minimizing intra-class distances and maximizing inter-class distances. Conventional metric learning methods [4,11,44] usually learn a Mahalanobis distance as the distance metric. Recently a variety of deep metric learning methods [7,34,38,39,54] have been proposed and demonstrated promising results. They usually employ deep CNNs to compute an embedding for each image and then adopt the Euclidean distance between embeddings to model their similarities, where the loss function is carefully designed to enforce a certain relational constraint within a structured tuple. For example, Schroff et $al.$ [34] proposed to separate the distances between the positive pair and the negative pair in a triplet by a fixed margin. Sohn [38] considered an $(N + 1)$-tuple each time and required the metric to recognize one positive from $N - 1$ negatives. Kim et $al.$ [19] extended the triplet loss and required the distance ratio in the continuous label space to be preserved in the learned embedding space.

A straightforward application of metric learning is image retrieval as we only need to find the nearest neighbors of the query image under the learned metric. We can easily extend it to the classification problem by utilizing a K-nearest neighbor classifier [10]. However, it is still unclear how to use deep metric learning for classifying with structured output labels. We move a step forward and propose a SDML method for end-to-end structured room layout estimation.

3 Proposed Approach

In this section, we first present our method for structural layout distance learning and then describe the design of the layout autoencoder. Lastly, we detail the proposed SDML approach for relational room layout inference.

3.1 Structural Layout Distance Learning

Room layout estimation aims to obtain a boxy representation of the underlying layout of an indoor scene. Formally, let $\mathbf{X} = [\mathbf{x}_1, \mathbf{x}_2, \cdots, \mathbf{x}_N]$ be a set of indoor scene images and $\mathbf{Z} = [\mathbf{z}_1, \mathbf{z}_2, \cdots, \mathbf{z}_N]$ be their corresponding ground-truth underlying layouts. Our objective is to assign a label (left wall, front wall, right wall, ceiling, and floor) to each pixel of the image indicating the surface it belongs to. For an image $\mathbf{x}$ of size $[h, w, 3]$, we estimate its layout on the resolution of $[\frac{h}{8}, \frac{w}{8}]$ following the protocol in previous work [23].

Existing room layout estimation methods fail to consider the structural relations across different images, which are actually of great value to obtain the estimated room layout, especially for a cluttered scene. Motivated by this, we propose to explore the correlations among different scenes by learning an embedding space where the Euclidean distance can reflect the actual structural distance between layouts.

To achieve this, we employ a CNN network to construct the mapping $\mathbf{f}(\mathbf{x}; \theta_f) = \mathbf{y}$ from the image space to the latent embedding space, where θ_f is the parameters of the metric network. We first extract a tensor $\mathbf{M} \in \mathbb{R}^{\frac{h}{8} \times \frac{w}{8} \times 2048}$

from the last convolutional layer of the CNN. We then use a 1×1 convolution to obtain a feature map and flatten it to an n-dimension embedding that globally represents the layout. The learned distance is then defined as the Euclidean distance between the corresponding embeddings of two images:

$$D(\mathbf{x}_i, \mathbf{x}_j; \theta_f) = ||\mathbf{f}(\mathbf{x}_i; \theta_f) - \mathbf{f}(\mathbf{x}_j; \theta_f)||_2, \tag{1}$$

where $|| \cdot ||_2$ denotes the L2-norm.

We use the pixelwise surface label difference between two layouts as the layout distance [16], where we first employ the Hungarian algorithm [21] to find the matching surfaces. We then learn the distance metric to approximate the layout distance $d(\mathbf{z}_i, \mathbf{z}_j)$ with the Euclidean distance $D(\mathbf{x}_i, \mathbf{x}_j)$ so that images with similar room layouts are clustered together. The objective function of the metric learning problem can be formulated as:

$$L(\mathbf{x}_i, \mathbf{x}_j) = (D(\mathbf{x}_i, \mathbf{x}_j) - d(\mathbf{z}_i, \mathbf{z}_j))^2, \tag{2}$$

where $\mathbf{z}_i$ and $\mathbf{z}_j$ are the underlying room layouts of $\mathbf{x}_i$ and $\mathbf{x}_j$, respectively.

Directly imposing such a constraint lacks flexibility and leads to inferior performance, and Kim $et\ al.$ [19] instead minimizes the difference of log distance ratios in the two spaces based on triplets:

$$L(\mathbf{x}_a, \mathbf{x}_i, \mathbf{x}_j) = (\log \frac{D(\mathbf{x}_a, \mathbf{x}_i)}{d(\mathbf{z}_a, \mathbf{z}_i)} - \log \frac{D(\mathbf{x}_a, \mathbf{x}_j)}{d(\mathbf{z}_a, \mathbf{z}_j)})^2. \tag{3}$$

The goal of the structural layout distance learning is to obtain a metric to accurately represent the structural distance between layouts. Exploiting more relations imposes tighter constraints on the metric and is expected to perform better. Kim $et\ al.$ [19] shows that densely sampling more triplets in one minibatch improves the performance. However, there exist $O(b^3)$ triplets that can be sampled from one minibatch where b is the batch size. Naively sampling all triplets and directly applying (3) will greatly increase the time complexity. To move a step forward, we propose a dense structural loss which includes not only all the triplets but also all the quadruplets in the minibatch:

$$L_{dense} = \frac{1}{2} \sum_{\{i,j\} \neq \{k,l\}} (\log \frac{D(\mathbf{x}_i, \mathbf{x}_j)}{d(\mathbf{z}_i, \mathbf{z}_j)} - \log \frac{D(\mathbf{x}_k, \mathbf{x}_l)}{d(\mathbf{z}_k, \mathbf{z}_l)})^2, \tag{4}$$

where $i, j, k, l \in \mathbf{B}$. Note that the summands in (4) contain all the triplets when $i = k$. The proposed loss actually exploits all the triplet-wise relations and can be seen as a generalization of (3).

Still, directly computing (4) is computationally infeasible. Using Lagrange's identity [45], we can rewrite (4) as:

$$L_{dense} = \frac{b(b-1)}{2} \sum_{i<j \in \mathbf{B}} (\log \frac{D(\mathbf{x}_i, \mathbf{x}_j)}{d(\mathbf{z}_i, \mathbf{z}_j)})^2 - (\sum_{i<j \in \mathbf{B}} \log \frac{D(\mathbf{x}_i, \mathbf{x}_j)}{d(\mathbf{z}_i, \mathbf{z}_j)})^2, \tag{5}$$

where $\mathbf{B}$ denotes a minibatch of indices of training images and b is the batch size. We can efficiently compute the pairwise squared distance matrix $\mathbf{D}^2$ by

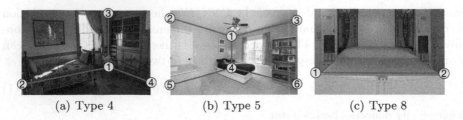

(a) Type 4   (b) Type 5   (c) Type 8

Fig. 2. Examples of the keypoint-based parameterization of room layouts. The LSUN dataset [50] defines 11 room types with a total of 48 keypoints, where each room type may contain different numbers of keypoints. We can generate a boxy layout by linking the keypoints with straight lines based on a predefined rule specific to each room type.

matrix operations $\mathbf{D}^2 = \tilde{\mathbf{y}}\mathbf{1}^{\mathrm{T}} + \mathbf{1}^{\mathrm{T}}\tilde{\mathbf{y}} - 2\mathbf{Y}\mathbf{Y}^{\mathrm{T}}$, where $D_{ij} = D(\mathbf{x}_i, \mathbf{x}_j)$, $\tilde{\mathbf{y}} = [\{\|\mathbf{y}_i\|_2^2\}_{i\in\mathbf{B}}]^{\mathrm{T}} \in \mathbb{R}^{b\times 1}$, $\mathbf{Y} = [\{\mathbf{y}_i\}_{i\in\mathbf{B}}] \in \mathbb{R}^{b\times m}$, and m is the embedding size.

We see that the proposed dense structural loss (5) takes $O(b^2)$ time and can take full advantage of the minibatch. It explicitly constrains each pair to have the same ratio of distances with every other pair in one minibatch, which generalizes (3) and can exploit more information without substantially increasing the computing overhead.

3.2   Layout Autoencoder

Having obtained the latent embedding space, we can effectively measure the structural distance between two scenes. We minimize the discrepancies between the images and ground-truth labels by calculating L2 loss in this space. To achieve this, we propose to learn a layout autoencoder composed of a layout encoder $\mathbf{g}(\mathbf{z}; \theta_g)$ to map a layout $\mathbf{z}$ to the embedding space and a decoder $\mathbf{h}(\mathbf{y}; \theta_h)$ to map an embedding $\mathbf{y}$ back to the layout space.

We use the keypoint-based parameterization to describe a layout as specified in the LSUN dataset [50]. See Fig. 2 for a demonstration. They define 11 types of layouts with a total of 48 keypoints which include most pictures taken with standard cameras in a cuboid room. Each room type contains a sequence of different keypoints defined in a specific order. Connecting these keypoints following predefined rules will generate the final layout. The location of each keypoint can be expressed by a 2D Gaussian distribution heatmap centered at the keypoint, which is a more effective form as the input and output of CNNs [23]. Therefore, we can represent all of the 48 keypoints by a 3D tensor $\mathbf{K} \in \mathbb{R}^{\frac{h}{8} \times \frac{w}{8} \times 48}$ called keypoint tensor with each channel as the corresponding keypoint heatmap. A room layout is then a combination of room type and the keypoint tensor $\mathbf{z} = \{l, \mathbf{K}\}$, where $l \in \{0, 1, \cdots, 10\}$.

The layout encoder $\mathbf{g}$ is a two-layer convolutional network followed by a 1×1 convolutional layer, which shares the parameters with that in the metric model. The encoder then flattens it to obtain the layout embedding. The input of the encoder is a ground-truth keypoint tensor $\mathbf{K}^g$, where channels corresponding to keypoints that do not appear in the ground-truth room type are set to zeros.

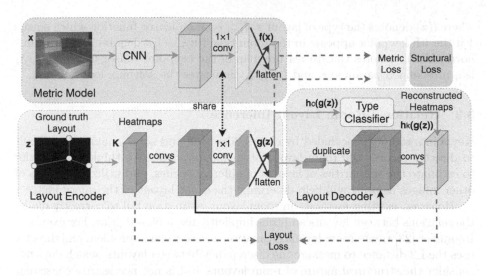

Fig. 3. An illustration of the proposed SDML method for room layout estimation. The metric model employs a CNN backbone network and a 1×1 convolutional layer to obtain a feature map of the query image and flattens it to obtain the embedding. The layout encoder first represents a ground-truth layout by the keypoint tensor and then uses two convolutional layers and one 1×1 convolutional layer to obtain the feature map. The 1×1 convolutional layer shares parameters with that in the metric model and we similarly flatten it to obtain the layout embedding. The layout decoder duplicates the layout embedding and concatenates them with the tensor from the encoder to aggregate global and local information. We then employ two convolutional layers to reconstruct the keypoint tensor and use a softmax classifier to select correct heatmaps.

The layout decoder $\mathbf{h}$ is composed of two parts: the type classifier $\mathbf{h}_c$ and the keypoint estimator $\mathbf{h}_k$. The type classifier $\mathbf{h}_c$ is a fully connected layer which takes as input an embedding $\mathbf{y}$ and outputs an 11-dimension vector $\mathbf{h}_c(\mathbf{y})$ representing the predicted room type possibilities. The keypoint estimator $\mathbf{h}_k$ first duplicates the embedding to obtain a tensor of size $\mathbb{R}^{\frac{h}{8} \times \frac{w}{8} \times d}$ and concatenates them to the feature maps outputted by the CNN in the layout encoder to aggregate both global and local information. We then use a two-layer convolutional network to obtain the estimated keypoint tensor $\mathbf{h}_k(\mathbf{y})$.

We use the reconstruction cost between the original and reconstructed ground-truth label as the objective to train the layout autoencoder. Similar to RoomNet [23], we measure the discrepancy of two keypoint tensors by the L2 loss of only the channels corresponding to keypoints that appear in the *original* room type $l(\mathbf{z})$. We use the softmax loss to train the type classifier to reconstruct the original room type. The layout reconstruction loss is formulated as:

$$L_{layout} = \sum_{i=0}^{47} I_{l(\mathbf{z})}(i) \|\mathbf{h}_k(\mathbf{g}(\mathbf{z}))_i - \mathbf{K}(\mathbf{z})_i\|_F + \lambda_s L_{softmax}(\mathbf{h}_c(\mathbf{g}(\mathbf{z})), l(\mathbf{z})), \quad (6)$$

where $l(\mathbf{z})$ denotes the type of layout $\mathbf{z}$, $I_{l(\mathbf{z})}$ is an indicator function which equals 1 if the ith keypoint appears in the room type $l(\mathbf{z})$, $||\cdot||_F$ denotes the Frobenius norm, $\mathbf{K}(\mathbf{z})_i$ denotes the ith feature map of the keypoint tensor of layout $\mathbf{z}$, λ_s is a pre-defined parameter, and $L_{softmax}$ indicates the softmax loss function.

3.3   Relational Room Layout Inference

Keypoints are often occluded by clutter like beds and sofas, making it difficult to directly recover the room layout of an indoor scene. We think it is simpler to estimate the similarities of layouts of different scenes, and utilizing layouts of other images as prior will help determine the exact layout of the query image.

However, existing methods on room layout estimation fail to correctly model the relations between layouts and thus implicitly use a biased prior. For example, RoomNet [23] treats room layout estimation as a regression problem and directly uses the L2 distance to measure the discrepancy between layouts, which does not consider the structural nature of room layouts and is not necessarily consistent with the actually proper metric like the pixel error [16]. In this case, nearest neighbors may not have the most similar underlying layouts and the correlations cannot be used as a good prior. The core problem is that such a structural metric is usually non-differential. We address this by learning an embedding space where the L2 distance reflects the structural distance. The correlation between images now is an unbiased prior and the nearest samples in the embedding space can be correctly decoded to similar layouts.

To achieve this, we first review an image $\mathbf{x}$ from a global perspective and consider its structural relation with other layouts to obtain the corresponding embedding $\mathbf{f}(\mathbf{x})$ in the latent embedding space. Then we employ the layout decoder $\mathbf{h}$ to map the embedding to the final estimated layout $\mathbf{h}(\mathbf{f}(\mathbf{x}))$. The overall framework of the proposed SDML is illustrated in Fig. 3.

To constrain the metric model $\mathbf{f}$ to map an image to the embedding corresponding to its ground-truth layout, we minimize the discrepancy between the predicted layout and the ground-truth layout using a simple L2 loss in the embedding space. The structural prediction loss is formulated as:

$$L_{str} = ||\mathbf{f}(\mathbf{x}) - \mathbf{g}(\mathbf{z})||^2. \tag{7}$$

We can also regard (7) as a distribution matching loss [18] to minimize the shift between the distributions of the image embeddings and layout embeddings.

Since we obtain an estimated layout by reconstruction from the embedding space, the embedding of an image should be as close to the corresponding layout embedding as possible. A shift from the correct layout embedding will cause an error to the estimated layout and thus harms the accuracy of layout estimation. Therefore, we tailor the proposed dense structural loss to further strengthen the connections between image embeddings and layout embeddings. We propose to constrain the log distance ratio between images and labels to be equal in the embedding space and the layout space:

$$L_{metric} = b(b-1) \sum_{i \neq j \in \mathbf{B}} (\log \frac{D(\mathbf{x}_i, \mathbf{z}_j)}{d(\mathbf{z}_i, \mathbf{z}_j)})^2 - (\sum_{i \neq j \in \mathbf{B}} \log \frac{D(\mathbf{x}_i, \mathbf{z}_j)}{d(\mathbf{z}_i, \mathbf{z}_j)})^2, \qquad (8)$$

where $D(\mathbf{x}_i, \mathbf{z}_j) = ||\mathbf{f}(\mathbf{x}_i) - \mathbf{g}(\mathbf{z}_j)||_2$ is the Euclidean distance between the mapped image $\mathbf{f}(\mathbf{x}_i)$ and the encoded layout $\mathbf{g}(\mathbf{z}_j)$ in the embedding space. Note that (8) has the same computation complexity as the original dense structural loss since we can similarly compute an asymmetric distance matrix. Moreover, the proposed metric loss integrates the constraints on the relations of image embeddings and layout embeddings and thus is more adaptive to the proposed SDML method. We only use (8) to update the metric function $\mathbf{f}$ in order to make the training more stable.

The entire framework of the proposed SDML approach is composed of three parts, a metric model $\mathbf{f}$ to map a scene to its embedding, a layout encoder $\mathbf{g}$ to map a layout to its embedding and a layout decoder $\mathbf{h}$ to map an embedding to its corresponding layout. Our SDML method can be trained end-to-end where we simultaneously learn the three parts $\mathbf{f}$, $\mathbf{g}$, and $\mathbf{h}$. The overall objective of the proposed approach can be formulated as:

$$\min_{\theta_f, \theta_g, \theta_h} L = \min_{\theta_f} L_{metric} + \lambda_1 \min_{\theta_f, \theta_g} L_{str} + \lambda_2 \min_{\theta_g, \theta_h} L_{layout}, \qquad (9)$$

where λ_1 and λ_2 are parameters to balance the contributions of different losses.

L_{metric} constructs an embedding space where the Euclidean distance can approximate the actual layout distance. L_{str} minimizes the difference between the predicted layout and the ground-truth layout by the L2 loss in the embedding space. L_{layout} learns a layout autoencoder to connect corresponding points in the embedding space and structured layout space.

In the test phase, we can directly estimate the room layout of an image $\mathbf{x}$ by $\mathbf{h}(\mathbf{f}(\mathbf{x}))$. Having obtained the keypoint tensor, we can find the corresponding keypoint heatmaps based on the estimated room type and extract the locations of each keypoint by an argmax operation on the heatmap. We then link the keypoints following the pre-defined protocol to obtain the final boxy representation of the estimated room layout.

4   Experiments

In this section, we conducted a variety of experiments to evaluate the performance of the proposed SDML for room layout estimation. Our experiments demonstrate the superiority of the proposed dense structural loss on the room layout retrieval task and analyze the origins of performance improvement through the ablation study. For room layout estimation, we employ the pixel error (PE) and the keypoint error (KPE) as performance metrics. The pixel error measures the average classification error of the predicted surface label of each pixel. The keypoint error computes the average Euclidean distance between the estimated keypoint locations and the ground-truth keypoint locations normalized by the image diagonal length.

4.1 Datasets

We followed existing methods [3,23,29] and evaluated our method on the widely used Hedau [16] and Large-scale Scene Understanding Challenge (LSUN) [50] room layout benchmark datasets. The Hedau dataset [16] consists of 367 images collected from the Internet using LabelMe [33], including 209 training images, 53 validation images, and 105 test images. The LSUN dataset [50] consists of 5,394 images sampled from the SUN database [47], including 4000 training images, 394 validation images, and 1000 test images.

For all the experiments, we trained our model from scratch on the training split of the LSUN dataset. In the training phase, we resized each image to a scale of 320×320 and estimated its room layout at a scale of 40×40. In the test phase, we obtained the estimated layouts at the scale of 40×40, but rescale it to the original image scale.

4.2 Implementation Details

We conducted all the experiments using the PyTorch package. We instantiated the metric function f by a ResNet-50 [15] backbone model with the dilated network strategy [48] similar to that used by PSPNet [52], which takes as input an 320×320 image and outputs a $40 \times 40 \times 2048$ tensor. The following 1×1 convolutional layer further maps it to a feature map of size $40 \times 40 \times 1$, rendering 1600 as the dimension of the embedding space. The layout encoder takes as input a keypoint tensor of size $40 \times 40 \times 48$ and employs two convolutional layers with 512 1×1 kernels and 2048 3×3 kernels to obtain a $40 \times 40 \times 2048$ tensor. The following 1×1 convolutional layer shares parameters with that in the metric model. The layout decoder is composed of two convolutional layers with 512 3×3 kernels and 48 1×1 kernels. The classifier is a softmax layer with the input of a 1600-dimension embedding and the output of an 11-dimension vector indicating the estimated probabilities of the room type. We performed a random horizontal mirror of images during training for data augmentation. We fixed the batch size to 15 due to the limited physical memory of the GPU card. We used the Adam optimizer and set the learning rate to 10^{-4}. We set λ_1, λ_2, and λ_s to 5, 1, and 0.3, respectively, to balance the effect of different losses.

4.3 Results and Analysis

Evaluation of the Dense Structural Loss: We first conducted an experiment on the task of room layout retrieval to verify the effectiveness of the proposed dense structural loss. The goal is to retrieve a set of images with similar underlying layouts given a query image. We measure the distance between layouts by the pixelwise surface label difference. We obtain the retrieved images by searching for the nearest neighbors in the learned latent embedding space, which requires the learned metric to accurately reflect the actual layout similarity. We adopted the mean label distance (smaller is better) and the modified version of the normalized discounted cumulative gain (nDCG) [19,22] (larger is better) as

Table 1. Room layout retrieval results (%) on the LSUN dataset compared with other methods. Red and bold numbers denote the best and second-best results, respectively.

Method	Mean label distance					nDCG				
Number of retrievals	1	2	4	8	16	1	2	4	8	16
Contrastive [14]	31.2	31.6	32.7	33.0	33.4	92.0	93.2	94.7	95.2	96.9
Triplet [34]	27.2	27.4	27.7	28.3	29.5	94.1	94.7	95.6	96.3	97.1
N-pair [38]	29.5	29.9	30.8	31.1	32.1	93.1	93.9	95.7	98.5	97.3
RoomNet [23]	27.9	28.2	29.7	30.9	32.5	93.3	93.8	94.9	96.0	96.5
Log-ratio (Random) [19]	25.2	26.4	27.1	28.7	30.5	95.3	96.0	96.9	97.4	97.8
Log-ratio (Dense) [19]	23.3	24.1	25.3	26.5	27.9	96.1	96.6	97.6	98.3	98.9
Dense structural	20.7	21.5	22.9	**24.1**	**25.3**	**97.2**	97.5	**98.6**	98.8	**99.1**
Dense structural (Layout)	**20.9**	**21.6**	**23.0**	24.0	25.1	97.4	97.5	**98.5**	**98.7**	99.2

Table 2. Performance of the proposed SDML method using different losses on the LSUN dataset.

Setting	PE (%)	KPE (%)
SDML without L_{metric}	9.93	6.42
SDML without L_{str}	15.36	10.23
SDML with L_{dense}	8.56	5.86
SDML	**6.95**	**5.29**

Table 3. Performance of the proposed SDML method under different model settings on the LSUN dataset.

Setting	PE (%)	KPE (%)
RoomNet [23]	10.46	6.95
RoomNet + L_{dense}	9.23	6.27
Keypoints regression	9.97	6.60
SDML	**6.95**	**5.29**

the evaluation metrics. We employed 60 images as queries in the validation split of the LSUN dataset. We refer readers to previous work [19] for more details.

We compared our dense structural loss with several baseline methods including the contrastive loss [14], the triplet loss [34], the N-pair loss [38] and the state-of-the-art method log-ratio loss with dense sampling [19]. The first three methods aim to pull closer samples from the same class and push away samples from different classes. For each sample, we chose its 30 nearest neighbors as positive samples, and others as negative samples to perform training. We evaluated all the losses using the same metric model as described in Sect. 3.1. We also tested the framework of RoomNet [23] and used the 512-dimension vector in the room type prediction module as the embedding for room layout retrieval. For our method, Dense structural denotes using the proposed dense structural loss (5) and Dense structural (Layout) denotes using the proposed metric loss (8) with a layout encoder.

Table 1 shows the results of room layout retrieval on the LSUN dataset. We see that the proposed loss outperforms the other baseline methods by a large margin. In particular, our method performs better than the state-of-the-art log-ratio loss with dense sampling. This is because our loss exploits full information from the batch and imposes more structural constraints on the metric.

Table 4. Experimental results of the proposed SDML method compared with existing methods. Red and bold numbers denote the best and second-best results, respectively.

Method	Hedau dataset	LSUN dataset		Time (s/image)
	PE (%)	PE (%)	KPE (%)	
Hedau et al. [16]	21.20	24.23	15.48	-
Del Pero et al. [5]	16.30	-	-	720
Gupta et al. [12]	16.20	-	-	-
Zhao et al. [53]	14.50	-	-	-
Ramalingam et al. [31]	13.34	-	-	6
Schwing et al. [35]	12.8	-	-	0.15
Del Pero et al. [6]	12.7	-	-	900
Mallya et al. [29]	12.83	16.71	11.02	-
DeLay [3]	9.73	10.63	8.20	30
CFILE [32]	8.67	9.31	7.95	-
RoomNet [23]	12.19	10.46	6.95	0.052
RoomNet recurrent 3-iter [23]	8.36	9.86	6.30	0.17
LayoutNet [56]	9.69	-	-	**0.039**
Hirzer et al. [17]	7.44	7.79	5.84	0.086
Zhang et al. [49]	**7.36**	6.58	5.17	150.18
SDML	7.21	**6.95**	**5.29**	0.017

Ablation Study: We first evaluated the effect of the three losses in (9). Table 2 shows the performance of the proposed SDML method using different losses. SDML without L_{metric} indicates training the SDML framework without explicitly modeling the structural relations across different scenes. The degraded performance verifies the advantage of considering the correlations among room layouts. SDML without L_{str} denotes training our framework without the structural loss, which constrains the metric model and the label encoder to map an image to the same embedding corresponding to its ground-truth label. It achieves inferior results since the label decoder might decode an inconsistent layout embedding leading to mistaken estimation. However, the reduction of the performance is not too large because L_{metric} still has an effect of decreasing the distribution shift. In addition, replacing the metric loss L_{metric} with the dense structural loss L_{dense} also decreases the performance of the proposed SDML method. This demonstrates L_{metric} is more adaptive with our method since it reinforces the connections between image embeddings and layout embeddings.

Moreover, we evaluated the proposed SDML method under different model settings, as shown in Table 3. We modified RoomNet [23] by further applying the dense structural loss L_{dense} to the 512-dimension vector in the room type classifier in RoomNet. We see that RoomNet + L_{dense} outperforms the original RoomNet. This is because the dense structural loss encourages the decoder to

Fig. 4. Qualitative results of our SDML in comparison with RoomNet [23].

consider the structural relations between images which further constrain it to encode more global information relative to the underlying layout. We also tested the performance of our model when removing the layout decoder and simply regress the positions of keypoints similar to RoomNet. We observe that the performance is only slightly better than RoomNet. This shows the performance improvement of our method mainly results from the relation modeling.

Quantitative Results: We compared the proposed SDML framework with existing methods on room layout estimation, where we summarize the results on the Hedau dataset and LSUN dataset in Table 4. We see that our method is the fastest with comparable performance with Zhang *et al.* [49]. In particular, while Zhang *et al.* [49] achieves slightly better results on the LSUN dataset than the proposed framework, we want to emphasize that it is a two-step method which is orders of magnitude slower than our method. Note that our method was only trained on the training split of the LSUN dataset without using external data. Zhao *et al.* [51] achieved better results (5.29% pixel error and 3.84% keypoint error on the LSUN dataset), yet they exploited more information by additionally training a model on the SUBRGBD dataset [40] on a 37-class semantic segmentation task to better describe clutter.

The proposed SDML framework exploits the learned structural relations as global prior to infer the underlying room layout of a scene, leading to superior performance. Our method has the advantage of balanced performance and cost and has the potential to be further applied to other structural prediction tasks.

Qualitative Results: Figure 4 shows the visualization of the room layout estimation results of our method. We provide a comparison with the current state-of-the-art method RoomNet [23], which directly regresses the locations of key-

Fig. 5. Some ambiguous cases where SDML and RoomNet [23] both predict incorrectly.

points. We observe that our method can more robustly estimate the locations of the keypoints even when they are occluded by clutter like beds or tables, which intuitively demonstrates the effectiveness of the proposed approach.

Figure 5 demonstrates some ambiguous scenes where the predictions of the proposed SDML method and RoomNet both fail to match the ground-truth annotations. Still, we see that our method produces better estimations than RoomNet since we explicitly model the structural relations among different scenes and use them as prior to assist the room layout estimation process.

5 Conclusion

In this paper, we have presented a structural deep metric learning framework (SDML) for room layout estimation, which formulates the problem from a metric learning perspective. We learn a latent embedding space to explicitly model the relations across different indoor scenes and utilize a layout autoencoder to connects the embedding space and the underlying layout in order to perform relational room layout inference. We have performed experimental evaluations on two widely used datasets which have verified the effectiveness of our method. In the future, it is interesting to further extend our method to the general structural prediction problem and apply SDML to more tasks such as human pose estimation and hand pose estimation.

Acknowledgements. The authors would like to thank Yangyang Song for his kind support and helpful discussions. This work was supported in part by the National Key Research and Development Program of China under Grant 2017YFA0700802, in part by the National Natural Science Foundation of China under Grant 61822603, Grant U1813218, Grant U1713214, and Grant 61672306, in part by Beijing Natural Science

Foundation under Grant No. L172051, in part by Beijing Academy of Artificial Intelligence (BAAI), in part by a grant from the Institute for Guo Qiang, Tsinghua University, in part by the Shenzhen Fundamental Research Fund (Subject Arrangement) under Grant JCYJ20170412170602564, and in part by Tsinghua University Initiative Scientific Research Program.

References

1. Boniardi, F., Valada, A., Mohan, R., Caselitz, T., Burgard, W.: Robot localization in floor plans using a room layout edge extraction network. In: Proceedings of the IROS, pp. 5291–5297 (2019)
2. Coughlan, J.M., Yuille, A.L.: The manhattan world assumption: regularities in scene statistics which enable Bayesian inference. In: Proceedings of the NIPS, pp. 845–851 (2001)
3. Dasgupta, S., Fang, K., Chen, K., Savarese, S.: DeLay: robust spatial layout estimation for cluttered indoor scenes. In: Proceedings of the CVPR, pp. 616–624 (2016)
4. Davis, J.V., Kulis, B., Jain, P., Sra, S., Dhillon, I.S.: Information-theoretic metric learning. In: Proceedings of the ICML, pp. 209–216 (2007)
5. Del Pero, L., Bowdish, J., Fried, D., Kermgard, B., Hartley, E., Barnard, K.: Bayesian geometric modeling of indoor scenes. In: Proceedings of the CVPR, pp. 2719–2726 (2012)
6. Del Pero, L., Bowdish, J., Kermgard, B., Hartley, E., Barnard, K.: Understanding Bayesian rooms using composite 3D object models. In: Proceedings of the CVPR, pp. 153–160 (2013)
7. Duan, Y., Zheng, W., Lin, X., Lu, J., Zhou, J.: Deep adversarial metric learning. In: Proceedings of the CVPR, pp. 2780–2789 (2018)
8. Eigen, D., Fergus, R.: Predicting depth, surface normals and semantic labels with a common multi-scale convolutional architecture. In: Proceedings of the ICML, pp. 2650–2658 (2015)
9. Eigen, D., Puhrsch, C., Fergus, R.: Depth map prediction from a single image using a multi-scale deep network. In: Proceedings of the NIPS, pp. 2366–2374 (2014)
10. Fix, E., Hodges Jr., J.L.: Discriminatory analysis-nonparametric discrimination: consistency properties. Technical report, California Univ Berkeley (1951)
11. Globerson, A., Roweis, S.T.: Metric learning by collapsing classes. In: Proceedings of the NIPS, pp. 451–458 (2006)
12. Gupta, A., Hebert, M., Kanade, T., Blei, D.M.: Estimating spatial layout of rooms using volumetric reasoning about objects and surfaces. In: Proceedings of the NIPS, pp. 1288–1296 (2010)
13. Gupta, S., Arbeláez, P., Girshick, R., Malik, J.: Aligning 3D models to RGB-D images of cluttered scenes. In: Proceedings of the CVPR, pp. 4731–4740 (2015)
14. Hadsell, R., Chopra, S., LeCun, Y.: Dimensionality reduction by learning an invariant mapping. In: Proceedings of the CVPR, pp. 1735–1742 (2006)
15. He, K., Zhang, X., Ren, S., Sun, J.: Deep residual learning for image recognition. In: Proceedings of the CVPR, pp. 770–778 (2016)
16. Hedau, V., Hoiem, D., Forsyth, D.: Recovering the spatial layout of cluttered rooms. In: Proceedings of the ICCV, pp. 1849–1856 (2009)
17. Hirzer, M., Roth, P.M., Lepetit, V.: Smart hypothesis generation for efficient and robust room layout estimation. In: Proceedings of the WACV, pp. 2912–2920 (2020)

18. Huang, J., Gretton, A., Borgwardt, K., Schölkopf, B., Smola, A.J.: Correcting sample selection bias by unlabeled data. In: Proceedings of the NIPS, pp. 601–608 (2007)
19. Kim, S., Seo, M., Laptev, I., Cho, M., Kwak, S.: Deep metric learning beyond binary supervision. In: Proceedings of the CVPR., pp. 2288–2297 (2019)
20. Krizhevsky, A., Sutskever, I., Hinton, G.E.: ImageNet classification with deep convolutional neural networks. In: Proceedings of the NIPS, pp. 1097–1105 (2012)
21. Kuhn, H.W.: The Hungarian method for the assignment problem. Naval Res. Logistics Q. **2**(1–2), 83–97 (1955)
22. Kwak, S., Cho, M., Laptev, I.: Thin-slicing for pose: learning to understand pose without explicit pose estimation. In: Proceedings of the CVPR, pp. 4938–4947 (2016)
23. Lee, C.Y., Badrinarayanan, V., Malisiewicz, T., Rabinovich, A.: RoomNet: end-to-end room layout estimation. In: Proceedings of the ICCV, pp. 4865–4874 (2017)
24. Lee, D.C., Gupta, A., Hebert, M., Kanade, T., Blei, D.M.: Estimating spatial layout of rooms using volumetric reasoning about objects and surfaces. In: Proceedings of the NIPS, pp. 1288–1296 (2010)
25. Lee, D.C., Hebert, M., Kanade, T.: Geometric reasoning for single image structure recovery. In: Proceedings of the CVPR, pp. 2136–2143 (2009)
26. Lin, C., Li, C., Furukawa, Y., Wang, W.: Floorplan priors for joint camera pose and room layout estimation. arXiv abs/1812.06677 (2018)
27. Liu, C., Schwing, A.G., Kundu, K., Urtasun, R., Fidler, S.: Rent3D: floor-plan priors for monocular layout estimation. In: Proceedings of the CVPR, pp. 3413–3421 (2015)
28. Long, J., Shelhamer, E., Darrell, T.: Fully convolutional networks for semantic segmentation. In: Proceedings of the CVPR, pp. 3431–3440 (2015)
29. Mallya, A., Lazebnik, S.: Learning informative edge maps for indoor scene layout prediction. In: Proceedings of the ICCV, pp. 936–944 (2015)
30. Mirowski, P., et al.: Learning to navigate in complex environments. In: Proceedings of the ICLR (2017)
31. Ramalingam, S., Pillai, J.K., Jain, A., Taguchi, Y.: Manhattan junction catalogue for spatial reasoning of indoor scenes. In: Proceedings of the CVPR, pp. 3065–3072 (2013)
32. Ren, Y., Li, S., Chen, C., Kuo, C.-C.J.: A coarse-to-fine indoor layout estimation (CFILE) method. In: Lai, S.-H., Lepetit, V., Nishino, K., Sato, Y. (eds.) ACCV 2016. LNCS, vol. 10115, pp. 36–51. Springer, Cham (2017). https://doi.org/10.1007/978-3-319-54193-8_3
33. Russell, B.C., Torralba, A., Murphy, K.P., Freeman, W.T.: LabelMe: a database and web-based tool for image annotation. Int. J. Comput. Vision **77**(1–3), 157–173 (2008)
34. Schroff, F., Kalenichenko, D., Philbin, J.: FaceNet: a unified embedding for face recognition and clustering. In: Proceedings of the CVPR, pp. 815–823 (2015)
35. Schwing, A.G., Hazan, T., Pollefeys, M., Urtasun, R.: Efficient structured prediction for 3D indoor scene understanding. In: Proceedings of the CVPR, pp. 2815–2822 (2012)
36. Schwing, A.G., Urtasun, R.: Efficient exact inference for 3D indoor scene understanding. In: Fitzgibbon, A., Lazebnik, S., Perona, P., Sato, Y., Schmid, C. (eds.) ECCV 2012. LNCS, vol. 7577, pp. 299–313. Springer, Heidelberg (2012). https://doi.org/10.1007/978-3-642-33783-3_22
37. Simonyan, K., Zisserman, A.: Very deep convolutional networks for large-scale image recognition. In: Proceedings of the ICLR (2015)

38. Sohn, K.: Improved deep metric learning with multi-class N-pair loss objective. In: Proceedings of the NIPS, pp. 1857–1865 (2016)
39. Song, H.O., Xiang, Y., Jegelka, S., Savarese, S.: Deep metric learning via lifted structured feature embedding. In: Proceedings of the CVPR, pp. 4004–4012 (2016)
40. Song, S., Lichtenberg, S.P., Xiao, J.: Sun RGB-D: a RGB-D scene understanding benchmark suite. In: Proceedings of the CVPR, pp. 567–576 (2015)
41. Szegedy, C., et al.: Going deeper with convolutions. In: Proceedings of the CVPR, pp. 1–9 (2015)
42. Tsochantaridis, I., Joachims, T., Hofmann, T., Altun, Y.: Large margin methods for structured and interdependent output variables. J. Mach. Learn. Res. 6(Sep), 1453–1484 (2005)
43. Wang, H., Gould, S., Koller, D.: Discriminative learning with latent variables for cluttered indoor scene understanding. In: Daniilidis, K., Maragos, P., Paragios, N. (eds.) ECCV 2010. LNCS, vol. 6314, pp. 497–510. Springer, Heidelberg (2010). https://doi.org/10.1007/978-3-642-15561-1_36
44. Weinberger, K.Q., Saul, L.K.: Distance metric learning for large margin nearest neighbor classification. J. Mach. Learn. Res. 10(2), 207–244 (2009)
45. Weisstein, E.W.: CRC Concise Encyclopedia of Mathematics. Chapman and Hall/CRC, New York (2002)
46. Xiao, J., Furukawa, Y.: Reconstructing the world's museums. Int. J. Comput. Vision 110(3), 243–258 (2014)
47. Xiao, J., Hays, J., Ehinger, K.A., Oliva, A., Torralba, A.: Sun database: large-scale scene recognition from Abbey to Zoo. In: Proceedings of the CVPR, pp. 3485–3492 (2010)
48. Yu, F., Koltun, V.: Multi-scale context aggregation by dilated convolutions. In: Proceedings of the ICLR (2016)
49. Zhang, W., Zhang, W., Gu, J.: Edge-semantic learning strategy for layout estimation in indoor environment. TCYB (2019)
50. Zhang, Y., Yu, F., Song, S., Xu, P., Seff, A., Xiao, J.: Large-scale scene understanding challenge: room layout estimation. In: CVPR Workshop (2015)
51. Zhao, H., Lu, M., Yao, A., Guo, Y., Chen, Y., Zhang, L.: Physics inspired optimization on semantic transfer features: an alternative method for room layout estimation. In: Proceedings of the CVPR, pp. 10–18 (2017)
52. Zhao, H., Shi, J., Qi, X., Wang, X., Jia, J.: Pyramid scene parsing network. In: Proceedings of the CVPR, pp. 2881–2890 (2017)
53. Zhao, Y., Zhu, S.C.: Scene parsing by integrating function, geometry and appearance models. In: Proceedings of the CVPR, pp. 3119–3126 (2013)
54. Zheng, W., Chen, Z., Lu, J., Zhou, J.: Hardness-aware deep metric learning. In: Proceedings of the CVPR, pp. 72–81 (2019)
55. Zhu, F., Zhu, L., Yang, Y.: Sim-real joint reinforcement transfer for 3D indoor navigation. In: Proceedings of the CVPR, pp. 11388–11397 (2019)
56. Zou, C., Colburn, A., Shan, Q., Hoiem, D.: LayoutNet: reconstructing the 3D room layout from a single RGB image. In: Proceedings of the CVPR, pp. 2051–2059 (2018)

Adaptive Task Sampling
for Meta-learning

Chenghao Liu[1](✉), Zhihao Wang[2], Doyen Sahoo[3], Yuan Fang[1], Kun Zhang[4],
and Steven C. H. Hoi[1,3]

[1] Singapore Management University, Singapore, Singapore
{chliu,yfang}@smu.edu.sg
[2] South China University of Technology, Guangzhou, China
ptkin@outlook.com
[3] Salesforce Research Asia, San Francisco, USA
{dsahoo,shoi}@salesforce.com
[4] Carnegie Mellon University, Pittsburgh, USA
kunz1@cmu.edu

Abstract. Meta-learning methods have been extensively studied and
applied in computer vision, especially for few-shot classification tasks.
The key idea of meta-learning for few-shot classification is to mimic
the few-shot situations faced at test time by randomly sampling classes
in meta-training data to construct few-shot tasks for episodic training.
While a rich line of work focuses solely on how to extract meta-knowledge
across tasks, we exploit the complementary problem on how to generate
informative tasks. We argue that the randomly sampled tasks could be
sub-optimal and uninformative (e.g., the task of classifying "dog" from
"laptop" is often trivial) to the meta-learner. In this paper, we propose
an adaptive task sampling method to improve the generalization performance. Unlike instance based sampling, task based sampling is much
more challenging due to the implicit definition of the task in each episode.
Therefore, we accordingly propose a greedy class-pair based sampling
method, which selects difficult tasks according to class-pair potentials.
We evaluate our adaptive task sampling method on two few-shot classification benchmarks, and it achieves consistent improvements across
different feature backbones, meta-learning algorithms and datasets.

1 Introduction

Deep neural networks have achieved great performance in areas such as image
recognition [17], machine translation [9] and speech synthesis [51] when large

C. Liu and Z. Wang—contributed equally, and completed most of this work when working at the School of Information Systems, Singapore Management University (SMU).
Steven C.H. Hoi is currently with Salesforce Research Asia and on leave from SMU.

Electronic supplementary material The online version of this chapter (https://
doi.org/10.1007/978-3-030-58523-5_44) contains supplementary material, which is
available to authorized users.

© Springer Nature Switzerland AG 2020
A. Vedaldi et al. (Eds.): ECCV 2020, LNCS 12363, pp. 752–769, 2020.
https://doi.org/10.1007/978-3-030-58523-5_44

amounts of labelled data are available. In stark contrast, human intelligence naturally possesses the ability to leverage prior knowledge and quickly learn new concepts from only a handful of samples. Such fast adaptation is made possible by some fundamental structures in human brains such as the "shape bias" to learn the learning procedure [25], which is also known as *meta-learning*. The fact that deep neural networks fail in the small data regime formulates a desirable problem for understanding intelligence. In particular, leveraging meta-learning algorithms to solve few-shot learning problems [24, 38] has recently gained much attention, which aims to close the gap between human and machine intelligence by training deep neural networks that can generalize well from very few labelled samples. In this setup, meta-learning is formulated as the extraction of cross-task knowledge that can facilitate the quick acquisition of task-specific knowledge from new tasks.

In order to compensate for the scarcity of training data in few-shot classification tasks, meta-learning approaches rely on an *episodic training* paradigm. A series of few-shot tasks are sampled from meta-training data for the extraction of transferable knowledge across tasks, which is then applied to new few-shot classification tasks consisting of unseen classes during the meta-testing phase. Specifically, optimization-based meta-learning approaches [12, 46] aim to find a global set of model parameters that can be quickly and effectively fine-tuned for each individual task with just a few gradient descent update steps. Meanwhile, metric-based meta-learning approaches [37, 47] learn a shared distance metric across tasks.

Despite their noticeable improvements, these meta-learning approaches leverage uniform sampling over classes to generate few-shot tasks, which ignores the intrinsic relationships between classes when forming episodes. We argue that exploiting class structures to construct more informative tasks is critical in meta-learning, which improves its ability to adapt to novel classes. For example, in the midst of the training procedure, a randomly sampled task of classifying dogs from laptops may have little effect on the model update due to its simpleness. Furthermore, in the conventional classification problem, prioritizing challenging training examples [42, 43] to improve the generalization performance has been widely used in various fields, ranging from AdaBoost [14] that selects harder examples to train subsequent classifiers, to Focal Loss [28] that adds a soft weighting scheme to emphasize harder examples.

A natural question thus arises: Can we perform adaptive task sampling and create more difficult tasks for meta-learning? Compared to the traditional instance-based adaptive sampling scheme, one key challenge in task sampling is to define the difficulty of a task. A naïve solution is to choose the difficult classes since each task is constructed by multiple classes. However, the difficulty of a class, and even the semantics of a class, is dependent on each other. For instance, the characteristics to discriminate "dog" from "laptop" or "car" are relatively easier to uncover than those for discriminating "dog" from "cat" or "tiger". In other words, the difficulty of a task goes beyond the difficulty of individual classes, and adaptive task sampling should consider the intricate relationships between different classes.

In this work, we propose a **class-pair based** adaptive task sampling method for meta-learning with several appealing qualities. First, it determines the task selection distribution by computing the difficulty of all class-pairs in it. As a result, it could capture the complex-structured relationships between classes in a multi-class few-shot classification problem. Second, since the cost of computing the task selection distribution for K-way classification problem is ($|\mathbb{C}_{tr}|$ choose K) or $O(|\mathbb{C}_{tr}|^K)$, where $|\mathbb{C}_{tr}|$ is the number of classes in the meta-training data, we further propose a **greedy class-pair based** adaptive task sampling method which only requires $O(K)$ time. Meanwhile, it can be formally established that the proposed greedy approach in fact samples from a distribution that is identical to that in the non-greedy version. Lastly, our method could be applied to any meta-learning algorithms that follow episodic training and works well with different feature backbones. In summary, our work makes the following contributions. **(1)** We propose a class-pair based adaptive task sampling approach for meta-learning methods, to improve the generalization performance on unseen tasks. **(2)** We further develop a greedy class-pair based approach that not only significantly reduces the complexity of task distribution computation, but also guarantees the generation of an identical distribution as that in the non-greedy approach. **(3)** We study the impact of the adaptive task sampling method by integrating it with various meta-learning approaches and performing comprehensive experiments on the miniImageNet and CIFAR-FS few-shot datasets, which quantitatively demonstrates the superior performance of our method. **(4)** We also conduct an extensive investigation of different sampling strategies, including class-based method, easy class-pair based method and uncertain class-pair based method. The results show that hard class-pair based sampling consistently leads to more accurate results.

2   Related Work

Meta-learning. The original idea of meta-learning, training a meta-model to learn a base model, has existed for at least 20 years [35,48]. Recently, the meta-learning framework has been used to solve few-shot classification problems. One typical work is the optimization based method. [38] uses the LSTM-based meta-learner to replace the SGD optimizer in the base model. MAML [12] and its variants [4,27] aim to learn a good model initialization so that the model for new tasks can be learned with a small number of samples and gradient update steps. Another category of work is the metric based method. It learns a set of embedding functions such that when represented in this space, images are easy to be recognized using a non-parametric model like nearest neighbor [37,44,50]. All of these methods follow the uniform sampling scheme to generate tasks at each episode. Besides, [46] considers a heuristic sampling method, which uses memory to store all the failure classes from k continuous tasks, and then constructs a hard task from them. [29,49] utilize pre-defined class structure information to

Fig. 1. The episodic training paradigm for meta-learning few-shot classification.

construct tasks in both meta-training and meta-testing phases. In this way, the experiment setting could more closely resemble realistic scenarios. In contrast, our work, inspired by importance sampling in stochastic optimization, aims to adaptively update task generating distribution in the meta-training phase, and this, in turn, improves its ability to adapt to novel classes with few training data in the meta-testing phase. We also present a theoretical analysis of the generalization bound to justify our approach.

Adaptive Sampling. Instance-based sampling is ubiquitous in stochastic optimization. Generally, it constantly reevaluates the relative importance of each instance during training. The most common paradigm is to calculate the importance of each instance based on the gradient norm [1], bound on the gradient norm [20], loss [31], approximate loss [21] or prediction probability [7]. One typical line of research work is to leverage adaptive sampling for fast convergence [2,55]. Researchers also consider improving the generalization performance rather than speeding up training [30]. Specifically, [5] considers instances that increase difficulty. Hard example mining methods also prioritize challenging training examples [28,43]. Some other researchers prioritize uncertain examples that are close to the model's decision boundary [7,45]. In this work, we also evaluate easy sampling and uncertain sampling at the task level, but experimental results show that hard sampling performs better. There also exists work for sampling mini-batches instead of a single instance [11,18]. [52,53] consider sampling diverse mini-batches via the repulsive point process. Nonetheless, these methods are not designed for meta-learning and few-shot learning.

3 Preliminaries

In this section, we review the *episodic training* paradigm in meta-learning and the vanilla instance-based adaptive sampling method for SGD.

3.1 Episodic Training

In the meta-learning problem setting, the goal is to learn models that can learn new tasks from small amounts of data. Formally, we have a large meta-training dataset $\mathbb{D}_{tr}$ (typically containing a large number of classes) and a meta-test dataset $\mathbb{D}_{test}$, in which their respective category sets $\mathbb{C}_{tr} = \{1, \ldots, |\mathbb{C}_{tr}|\}$ and $\mathbb{C}_{test} = \{|\mathbb{C}_{tr}| + 1, \ldots, |\mathbb{C}_{tr}| + \mathbb{C}_{test}\}$ are disjoint. We aim to learn a classification model on $\mathbb{D}_{tr}$ that can generalize to unseen categories $\mathbb{C}_{test}$ with one or few training examples per category.

The success of existing meta-learning approaches relies on the *episodic train-ing* paradigm [50], which mimics the few-shot regime faced at test time during training on $\mathbb{D}_{tr}$. Particularly, meta-learning algorithms learn from a collection of K-way-M-shot classification tasks sampled from the amply labelled set $\mathbb{D}_{tr}$ and are evaluated in a similar way on $\mathbb{D}_{test}$. In each episode of meta-training, we first sample K classes $\mathbb{L}^K \sim \mathbb{C}_{tr}$. Then, we sample M and N labelled images per class in $\mathbb{L}^K$ to construct the support set $\mathbb{S} = \{(s_m, y_m)_m\}$ and query set $\mathbb{Q} = \{(q_n, y_n)_n\}$, respectively. The episodic training for few-shot learning is achieved by minimizing, for each episode, the loss of the prediction for each sample in the query set, given the support set. The model is parameterized by θ and the loss is the negative loglikelihood of the true class of each query sample: $\ell(\theta) = \mathop{\mathbb{E}}\limits_{(S,Q)} [-\sum_{(q_n, y_n) \in Q} \log p_\theta(y_n | q_n, S)]$, where $p_\theta(y_n | q_n, S)$ is the classifica-tion probability based on the support set. The model then back-propagates the gradient of the total loss $\nabla \ell(\theta)$. Different meta-learning approaches differ in the manner in which this conditioning on the support set is realized. To better explain how it works, we show its framework in Fig. 1.

3.2 Instance-Based Adaptive Sampling for SGD

Let $\mathbb{D} = \{(x_i, y_i)_i\}$ indicate the training dataset. The probability of selecting each sample is equal at the initial stage (i.e., $p_0(i|\mathbb{D}) = \frac{1}{|\mathbb{D}|}$). To emphasize difficult examples while applying SGD, we adaptively update the selection prob-ability $p^{t+1}(i)$ for instance i at iteration $t + 1$ according to the current predic-tion probability $p(y_i | x_i)$ and the selection probability at previous iteration $p^t(i)$, $p^{t+1}(i) \propto (p^t(i))^\tau e^{\alpha(1 - p(y_i|x_i))}$, where the hyperparameters τ is a discounting parameter and α scales the influence of current prediction. This multiplicative update method has a close relation to maximum loss minimization [42] and AdaBoost [15], which can result in improved generalization performance, espe-cially when only a few "rare" samples exist. Moreover, when the gradient update is weighted by the inverse sampling probability, we obtain an unbiased gradient estimation that improves the convergence by reducing its variance [16,55].

4   Adaptive Task Sampling for Meta-Learning

In this section, we first propose the class-based adaptive task sampling method which is a straightforward extension of the instance-based sampling. Then, we discuss its defect and present the class-pair based sampling method. Finally, we propose the greedy class-pair based sampling method, which significantly reduces the computation cost while still generating the identical task distribution as that in the non-greedy approach.

Class-Based Sampling. A major challenge of adaptive task sampling for meta-learning is the implicit definition of the task, which is randomly generated by sampling K classes in each episode. Although direct task based sampling is infeasible, we can adaptively sample classes for each K-way classification task. With this goal in mind, we propose a **class-based sampling (c-sampling)** approach that updates the class selection probability $p_C^{t+1}(c)$ in each episode. Given $\mathbb{S}^t$ and $\mathbb{Q}^t$ at episode t, we could update the class selection probability for each class in current episode $c \in \mathbb{L}_K^t$ in the following way,

$$p_C^{t+1}(c) \propto (p^t(c))^\tau e^{\alpha \frac{\sum_{(q_n, y_n) \in \mathbb{Q}^t} \mathbb{I}[c \neq y_n] p(c|q_n, \mathbb{S}^t) + \mathbb{I}[c = y_n](1 - p(c|q_n, \mathbb{S}^t))}{NK}}. \tag{1}$$

Note that we average the prediction probability of classifying each query sample n into incorrect classes in $\mathbb{L}_K^t$. Then we can sample K classes without replacement to construct the category set $\mathbb{L}_K^{t+1}$ for the next episode.

Despite its simplicity, such a sampling approach does suffer from an important limitation. It implicitly assumes that the difficulty of each class is independent. Therefore, it updates the class selection probability in a decoupled way. In concrete words, suppose we have two different tasks: discerning "corgi", "Akita" and "poodle" and discerning "corgi", "car" and "people". Obviously, it is quite hard to tell "corgi" in the first task while it could be easy in the second one. This would be a challenging aspect for updating the class selection probability as the class-based sampling is agnostic to the context of the task and could accidentally assign contradictory scores to the same class. Secondly, even if the class selection probability is updated correctly, it cannot ensure that difficult tasks are generated properly. That is, assembling the most difficult classes do not necessarily lead to a difficult task.

Class-Pair Based Sampling. To address the above issue, we further propose a **class-pair based sampling (cp-sampling)** approach that exploits the pairwise relationships between classes. This idea is commonly used in the multi-class classification that constructs binary classifiers to discriminate between each pair of classes [3], as two-class problems are much easier to solve. Recently, it has also been considered to extract the pairwise relationships between classes for task-dependent fast adaptation in few-shot learning [40]. In this work, we formulate the task selection probability by leveraging the Markov random field [10] over class pairs. Formally, the probability of choosing a category set $\mathbb{L}_K^{t+1}$ at episode $t+1$ is defined as:

$$p_{CP}^{t+1}(\mathbb{L}_K^{t+1}) \propto \prod_{(i,j) \in \mathbb{L}_K^{t+1}} C^t(i,j), \qquad \text{s.t. } i, j \in \mathbb{C}_{tr} \tag{2}$$

where $C^t(i,j)$ is a potential function over class pair (i,j) at episode t. Notice that the classes in $\mathbb{C}_{tr}$ form a complete and undirected graph. The category set $\mathbb{L}_K^{t+1}$ that have a relatively high probability to be selected are those K-cliques with large potentials. Similarly, we adaptively update the potential function $C^{t+1}(i,j)$ according to

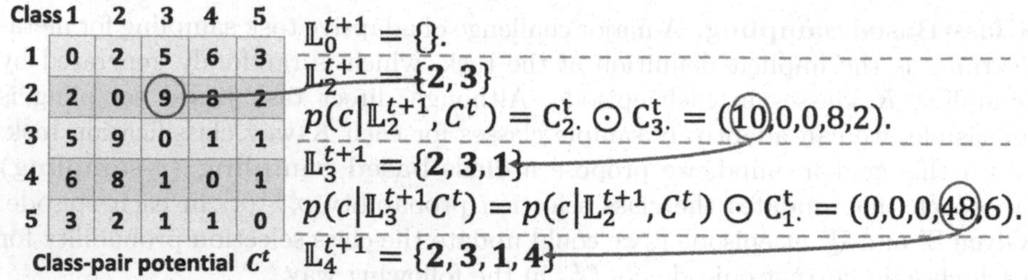

Class 1	2	3	4	5
1	0	2	5	6	3
2	2	0	9	8	2
3	5	9	0	1	1
4	6	8	1	0	1
5	3	2	1	1	0

Class-pair potential C^t

$\mathbb{L}_0^{t+1} = \{\}.$

$\mathbb{L}_2^{t+1} = \{2, 3\}$

$p(c|\mathbb{L}_2^{t+1}, C^t) = C_2^t \odot C_3^t. = (10, 0, 0, 8, 2).$

$\mathbb{L}_3^{t+1} = \{2, 3, 1\}$

$p(c|\mathbb{L}_3^{t+1} C^t) = p(c|\mathbb{L}_2^{t+1}, C^t) \odot C_1^t. = (0, 0, 0, 48, 6).$

$\mathbb{L}_4^{t+1} = \{2, 3, 1, 4\}$

Fig. 2. A toy example to illustrate how greedy class-pair based sampling chooses 4-class category set $\mathbb{L}_4^{t+1}$ from 5 classes. The left correlation matrix indicates the class-pair potentials C^t and the right part denotes the state of each step in sequential sampling. The blue number on the right denotes the chosen class and the red circle highlights the highest unnormalized class selection probability. $\odot$ denotes the element-wise multiplication. (Color figure online)

$$C^{t+1}(i,j) \leftarrow (C^t(i,j))^\tau e^{\alpha \bar{p}((i,j)|\mathbb{S}^t, \mathbb{Q}^t)}, \quad i \neq j \qquad (3)$$

where $\bar{p}((i,j)|\mathbb{S}^t, \mathbb{Q}^t)$ denotes the average prediction probability that classifies query samples in class j into its incorrect class i or vice versa. Specifically, we define it as

$$\bar{p}((i,j)|\mathbb{S}^t, \mathbb{Q}^t) = \frac{\sum_{(q_n, y_n = j) \in \mathbb{Q}^t} p(c = i|q_n, \mathbb{S}^t)}{N} + \frac{\sum_{(q_n, y_n = i) \in \mathbb{Q}^t} p(c = j|q_n, \mathbb{S}^t)}{N}.$$
$$(4)$$

Greedy Class-Pair Based Sampling. It is important to note that class-pair based sampling has the disadvantage that $\binom{K}{2} \cdot \binom{|\mathbb{C}_{tr}|}{K}$ multiplication operations need to be performed for calculating $p_{CP}^{t+1}(\mathbb{L}_K^{t+1})$ for different combinations of K-class in the category set. To significantly reduce the complexity, we now design a **greedy class-pair based sampling (gcp-sampling)** method, which samples not only at the cost $O(K)$ but also from a distribution identical to that in Eq. (2), due to the independence of the potential function $C^t(i,j)$ over class pairs. In particular, we sequentially sample classes in $K - 1$ steps based on the previous results. At episode t, we first sample two classes based on class-pair potential function $C^t(i,j)$. Then we iteratively sample a new class based on the already sampled classes. Figure 2 gives an example to illustrate the process. Formally, the task selection probability is defined as

$$p_{GCP}^{t+1}(\mathbb{L}_{k+1}^{t+1}) \propto \begin{cases} C^t(i,j), & k = 1 \\ p(c|\mathbb{L}_k^{t+1}, C^t), & k > 1 \end{cases} \qquad (5)$$

where $p(c = i|\mathbb{L}_k^{t+1}, C^t) \propto \prod_{j \in \mathbb{L}_k^{t+1}} C^t(i,j)$. It considers the joint probability over class pairs between the chosen class i and every sampled class j in the category set $\mathbb{L}_k^{t+1}$. Compared to the distribution in Eq. (2), the greedy sampling approach in Eq. (5) has a different normalization constant in each step k. However, for the

Algorithm 1. gcp-sampling: Greedy Class-Pair based Sampling in K-Way-M-Shot

Require: meta-training data $\mathbb{D}_{tr}$, hyperparameters α, τ, T

1: Randomly initialize meta model parameter θ. Initialize class-pair potentials C by ones
2: **for** $t = 1, \ldots, T$ **do**
3:   Initialize $\mathbb{L}_0^t$ by an empty set. Initialize $p(c|\mathbb{L}_0^t, C^{t-1})$ by $\frac{1}{|C_{tr}|}$
4:   Sample class pair $(i, j) \propto C(i, j)$, add class i and j into $\mathbb{L}_0^t$
5:   **for** $k = 2, \ldots, K - 1$ **do**
6:     Update $p(c = i|\mathbb{L}_k^t, C^{t-1}) \propto \prod_{j \in \mathbb{L}_k^t} C^{t-1}(i, j)$
7:     Sample class c based on $p(c|\mathbb{L}_k^t, C^{t-1})$, add class c into $\mathbb{L}_{k+1}^t$
8:   **end for**
9:   Construct support set $\mathbb{S}^t$ and query set $\mathbb{Q}^t$ by sampling M and N image per class in category set $\mathbb{L}_K^t$, respectively
10:   Update meta model θ based on support set and query set
11:   Update class-pair potentials C according to Eq. (3)
12: **end for**
13: **return** θ_T

---

evaluation of task selection distribution, the unnormalized joint probability over the class pairs of a specific category set is identical which makes the distribution in Eq. (5) exactly the same as that in Eq. (2), which we prove in Proposition 1.

Proposition 1. *The greedy class-pair based sampling strategy in Eq. (5) is identical to the class-pair based sampling in Eq. (2).*

Proof. We present a proof by induction. It is obvious that $p_{GCP}^{t+1}(\mathbb{L}_2^{t+1}) = p_{CP}^{t+1}(\mathbb{L}_2^{t+1})$ since $p_{GCP}^{t+1}(\mathbb{L}_2^{t+1}) \propto C^t(i, j)$. Now let us consider a general case where we have previously sampled k classes with $\mathbb{L}_k^{t+1}$ and are about to sample the $(k+1)$-th class. Suppose we sample a new class l to generate $\mathbb{L}_{k+1}^{t+1}$, according to Eq. (5), we have

$$p_{GCP}^{t+1}(\mathbb{L}_{k+1}^{t+1}) = p_{GCP}^{t+1}(\mathbb{L}_k^{t+1})p(c = l|\mathbb{L}_k^{t+1}, C^t) \propto \prod_{(i,j) \subset \mathbb{L}_k^{t+1}} C^t(i, j) \prod_{j \in \mathbb{L}_k^{t+1}} C^t(l, j)$$

$$= \prod_{(i,j) \subset \mathbb{L}_{k+1}^{t+1}} C^t(i, j) = p_{CP}^{t+1}(\mathbb{L}_{k+1}^{t+1}). \tag{6}$$

The pseudocode of the proposed gcp-sampling algorithm is given in Algorithm 1. Due to the space limitation, we leave the theoretical analysis of the proposed gcp-sampling method in terms of its generalization ability to the supplementary material.

5   Experiments

In this section, we evaluate the proposed adaptive task sampling method on two few-shot classification benchmarks: miniImageNet [50] and CIFAR-FS [6]. We first introduce the datasets and settings, and then present a comparison to state-of-the-art methods, followed by a detailed evaluation of the compatibility when integrating with different meta-learning algorithms and the efficacy of different sampling strategies. Finally, we demonstrate qualitative results to characterize the gcp-sampling.

5.1   Datasets and Implementation Details

Datasets. We conduct experiments to evaluate our method on two few-shot classification benchmarks. Firstly, **miniImageNet** [50] is widely used for few-shot learning, which is constructed based on the ImageNet dataset [39] and thus has high diversity and complexity. This dataset has 100 classes with 600 84×84 images per class. These classes are divided into 64, 16 and 20 classes for meta-training, meta-validation and meta-test, respectively, as suggested earlier [12,38,46]. Secondly, **CIFAR-FS** is another recent few-shot image classification benchmark [6] constructed by randomly sampling from the CIFAR-100 dataset [23] using the same criteria as the miniImageNet, and has the same number of classes and samples. The limited resolution of 32×32 makes the task still difficult. We also use the 64/16/20 divisions for consistency with previous studies [6,26].

Evaluation Metric. We report the mean accuracy (%) of 1000 randomly generated episodes as well as the 95% confidence intervals on the meta-test set. In every episode during meta-test, each class has 15 queries.

Network Architectures. We conduct experiments with 2 different feature extractor architectures, Conv-4 and ResNet-12. **Conv-4** is a shallow embedding function proposed by [50] and widely used [4,12,36,44]. It is composed of 4 convolutional blocks, each of which comprises a 64-filter 3×3 convolution, batch normalization (BN) [19], a ReLU nonlinearity and a 2×2 max-pooling layer. We also adopt a deep backbone **ResNet-12** [17], which achieves significant improvement in recent works [33,34,37]. It consists of 4 residual blocks, each of which has three 3×3 convolutional layers and a 2×2 max-pooling layer. The number of filters starts from 64 and is doubled every next block. There is also a mean-pooling layer compressing the feature maps to a feature embedding in the end. In our experiments, we integrate gcp-sampling with PN, MetaOptNet-RR and MetaOptNet-SVM with ResNet-12 to compare with state of the arts. We follow the settings of [26] and use SGD with Nesterov momentum of 0.9 and weight decay of 0.0005. Besides, we use Conv-4 to evaluate the compatibility when integrating with different meta-learning algorithms and the efficacy of different sampling strategies. We follow the settings of [8] and use Adam [22] optimizer with an initial learning rate of 0.001.

5.2   Results and Analysis

Comparison with State-of-the-Art. Tables 1 and 2 present the 5-way 1-shot and 5-way 5-shot results on miniImageNet and CIFAR-FS datasets, respectively. Note that it shows the highest accuracies for which the iterations are chosen by

Table 1. Average 5-way, 1-shot and 5-shot classification accuracies (%) on the mini-ImageNet dataset. * denotes the results from [26].

Methods	Backbone	5-way-1-shot	5-way-5-shot
Matching network [50]	CONV-4	43.44 ± 0.77	55.31 ± 0.73
Relation network [47]	CONV-4	50.44 ± 0.82	65.32 ± 0.70
PN [44]	CONV-4	49.42 ± 0.78	68.20 ± 0.66
MAML [12]	CONV-4	48.70 ± 1.84	63.11 ± 0.92
MAML++ [4]	CONV-4	52.15 ± 0.26	68.32 ± 0.44
MAML++, AS (ours)	CONV-4	52.34 ± 0.81	69.21 ± 0.68
Bilevel Programming [13]	ResNet-12	50.54 ± 0.85	64.53 ± 0.68
MetaGAN [54]	ResNet-12	52.71 ± 0.64	68.63 ± 0.67
SNAIL [33]	ResNet-12	55.71 ± 0.99	68.88 ± 0.92
adaResNet [34]	ResNet-12	56.88 ± 0.62	71.94 ± 0.57
TADAM [37]	ResNet-12	58.50 ± 0.30	76.70 ± 0.30
MTL [46]	ResNet-12	61.2 ± 1.8	75.5 ± 0.8
PN* [26]	ResNet-12	59.25 ± 0.64	75.60 ± 0.48
PN with gcp-sampling	ResNet-12	**61.09 ± 0.66**	**76.80 ± 0.49**
MetaOptNet-RR [26]	ResNet-12	61.41 ± 0.61	77.88 ± 0.46
MetaOptNet-RR with gcp-sampling	ResNet-12	**63.02 ± 0.63**	**78.91 ± 0.46**
MetaOptNet-SVM [26]	ResNet-12	62.64 ± 0.61	78.63 ± 0.46
MetaOptNet-SVM with gcp-sampling	ResNet-12	**64.01 ± 0.61**	**79.78 ± 0.47**

Table 2. Average 5-way, 1-shot and 5-shot classification accuracies (%) on the CIFAR-FS dataset. * denotes the results from [26].

Methods	Backbone	5-way-1-shot	5-way-5-shot
Relation network [47]	CONV-4	55.0 ± 1.0	69.3 ± 0.8
PN* [44]	CONV-4	55.5 ± 0.7	72.0 ± 0.6
MAML* [12]	CONV-4	58.9 ± 1.9	71.5 ± 1.0
GNN [41]	CONV-4	61.9	75.3
R2D2 [26]	CONV-4	65.3 ± 0.2	79.4 ± 0.1
PN* [26]	ResNet-12	72.2 ± 0.7	84.2 ± 0.5
PN with gcp-sampling	ResNet-12	**74.1 ± 0.7**	**84.5 ± 0.5**
MetaOptNet-RR [26]	ResNet-12	72.6 ± 0.7	84.3 ± 0.5
MetaOptNet-RR with gcp-sampling	ResNet-12	**74.2 ± 0.7**	**85.1 ± 0.4**
MetaOptNet-SVM [26]	ResNet-12	72.0 ± 0.7	84.2 ± 0.5
MetaOptNet-SVM with gcp-sampling	ResNet-12	**73.9 ± 0.7**	**85.3 ± 0.5**

The results in Table 2 demonstrate that using gcp-sampling for meta-learning methods consistently improves the 1-shot classification performance. More over, the performance improvement is more significant for 1-shot classification than 5-shot classification.

Table 3. Average 5-way classification accuracies (%) on miniImageNet and CIFAR-FS. All methods use shallow feature backbone (Conv-4). [†] denotes the local replication results. We run PN without oversampling the number of ways.

Model	miniImageNet		CIFAR-FS	
	1-shot	5-shot	1-shot	5-shot
Matching network [†]	48.26 ± 0.76	62.27 ± 0.71	53.14 ± 0.85	68.16 ± 0.76
Matching network with gcp-sampling	$\mathbf{49.61 \pm 0.77}$	$\mathbf{63.23 \pm 0.75}$	$\mathbf{54.72 \pm 0.87}$	$\mathbf{69.28 \pm 0.74}$
PN [†]	44.15 ± 0.76	63.89 ± 0.71	54.87 ± 0.72	71.64 ± 0.58
PN with gcp-sampling	$\mathbf{47.13 \pm 0.81}$	$\mathbf{64.75 \pm 0.72}$	$\mathbf{56.12 \pm 0.81}$	$\mathbf{72.77 \pm 0.64}$
Reptile [†]	46.12 ± 0.80	63.56 ± 0.70	55.86 ± 1.00	71.08 ± 0.74
Reptile with gcp-sampling	$\mathbf{47.60 \pm 0.80}$	$\mathbf{64.56 \pm 0.69}$	$\mathbf{57.25 \pm 0.99}$	$\mathbf{71.69 \pm 0.71}$
MAML [†]	48.25 ± 0.62	64.09 ± 0.70	56.93 ± 0.99	72.10 ± 0.74
MAML with gcp-sampling	$\mathbf{49.65 \pm 0.85}$	$\mathbf{65.37 \pm 0.70}$	$\mathbf{57.62 \pm 0.97}$	$\mathbf{72.51 \pm 0.72}$
MAML++ [†]	50.60 ± 0.82	68.24 ± 0.68	58.87 ± 0.97	73.86 ± 0.76
MAML++ with gcp-sampling	$\mathbf{52.34 \pm 0.81}$	$\mathbf{69.21 \pm 0.68}$	$\mathbf{60.14 \pm 0.97}$	$\mathbf{73.98 \pm 0.74}$

Table 4. Average 5-way classification accuracies (%) on miniImageNet and CIFAR-FS. Using MAML++ on a Conv-4 backbone, we compare different sampling methods: random, c-sampling with hard class, gcp-sampling with hard/uncertain/easy class.

Sampling strategy	miniImageNet		CIFAR-FS	
	5-way-1-shot	5-way-5-shot	5-way-1-shot	5-way-5-shot
Random sampling	50.60 ± 0.82	68.24 ± 0.68	58.87 ± 0.97	73.36 ± 0.76
c-sampling with hard class	51.43 ± 0.75	68.74 ± 0.67	58.61 ± 0.92	73.98 ± 0.72
gcp-sampling with easy class	50.88 ± 0.88	68.22 ± 0.72	58.73 ± 1.14	73.41 ± 0.76
gcp-sampling with uncertain class	51.73 ± 0.87	69.01 ± 0.72	59.43 ± 1.02	73.84 ± 0.82
gcp-sampling with hard class	$\mathbf{52.34 \pm 0.81}$	$\mathbf{69.21 \pm 0.68}$	$\mathbf{60.14 \pm 0.97}$	$\mathbf{74.58 \pm 0.74}$

validation. For our approach, we integrate gcp-sampling with PN, MON-RR and MON-SVM, which are strong baselines. For all cases, we achieve comparable performance surpassing prior methods by a meaningful margin. For example, PN with gcp-sampling outperforms the PN with ResNet-12 by around 1.84 and 1.2% points in miniImageNet and 1.89 and 1.0% points in CIFAR-FS. It is worth noting that the adaptive task sampling method is orthogonal to the meta-learning algorithm. Moreover, even for a deep feature backbone, our approach is still able to preserve the performance gain.

Compatibility with Different Meta-learning Algorithms. Next, we study the impact of gcp-sampling when integrating with different types of meta-learning algorithm. We consider gradient-based meta-learning methods: MAML, Reptile and MAML++, and metric-based meta-learning methods: PN and MN. The results in Table 3 demonstrate that using gcp-sampling for meta-learning methods consistently improves the few-shot classification performance. Moreover, the performance improvement is more significant for 1-shot classification than 5-shot classification.

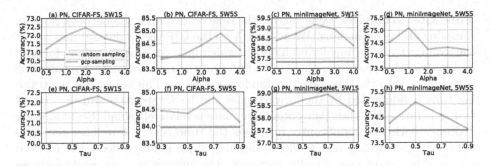

Fig. 3. Impact of hyperparameters α and τ. First row (a–d): we fix the discounting factor $\tau = 0.5$ and tune the updating factor α; Second row (e–g): we fix $\alpha = 1$ and tune τ.

Efficacy of Different Adaptive Task Sampling Strategies. In literature, there exist contradicting ideas in adaptive sampling strategies which work well in different scenarios [7]. Preferring easier samples may be effective when solving challenging problems containing noise or outliers. The opposite hard sample mining strategy may improve the performance since it is more likely to be minority classes. Therefore, we explore different sampling strategies for meta-learning for few-shot classification. As defined in Eq. (4) for hard class, the probability of easy class is $1 - \bar{p}(i,j)$ and uncertain class is $(1 - \bar{p}(i,j))(\bar{p}(i,j))$, respectively. We report the results in Table 4. We observe that gcp-sampling with hard or uncertain class outperforms that with random sampling, but uncertain sampling offers a smaller improvement. We also compare gcp-sampling with c-sampling, in which c-sampling achieves similar performance as random sampling, verifying the efficacy of using class pairs to represent task difficulty.

Impact of Hyperparameters α and τ. In the proposed gcp-sampling, the hyperparameter α controls the aggressiveness of the update while the hyperparameter τ controls the degree of forgetting past updates. Here we adopt PN with **ResNet-12** backbone and report the effect of α and τ on the testing performance in Fig. 3.

Table 5. Time cost comparison between random sampling and gcp-sampling. All the experiments are conducted with PN on the CIFAR-FS dataset.

	Random sampling	gcp-sampling	Factor
5-way-1-shot, Conv-4	235.4	251.8	1.070
5-way-1-shot, ResNet-12	531.2	554.6	1.044
5-way-5-shot, Conv-4	342.2	367.3	1.073
5-way-10-shot, Conv-4	471.4	491.0	1.042
5-way-15-shot, Conv-4	617.2	634.6	1.028
10-way-1-shot, Conv-4	411.3	451.7	1.098
15-way-1-shot, Conv-4	624.9	723.5	1.158
20-way-1-shot, Conv-4	816.8	992.5	1.215

Time Cost Analysis. Table 5 shows the time cost comparison between random sampling and gcp-sampling. We adopt PN on the CIFAR-FS dataset and report the average training time for each epoch, which includes task sampling, forward and backward propagation phases. We find that the time taken by gcp-sampling is comparable to the time taken by random-sampling. This is because the training time is dominated by the forward pass and backward pass and the cost of task generation and class-pair potential update is relatively small. Besides, using a deeper backbone significantly increases the time cost but reduces the ratio between gcp-sampling and random-sampling, since it only affects the forward pass and backward pass. Finally, increasing the number of ways would increase the time cost while increasing the number of shots will not. This is because the complexity of gcp-sampling scales linearly to the number of ways.

Visual Analysis of Adaptive Task Sampling. To qualitatively characterize adaptive task sampling, we visualize the prototype of each class generated by the training procedure of PN with gcp-sampling and random sampling. We use the t-SNE [32] method to convert the prototypes into two-dimensional vectors by preserving the cosine similarity between them. As shown in Fig. 4, the classes sampled by random sampling achieve better clustering results than gcp-sampling. This is because gcp-sampling tends to sample classes with highly overlapping embeddings, which is much more difficult to learn for meta-learner.

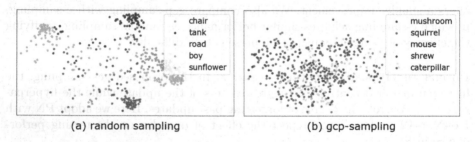

(a) random sampling   (b) gcp-sampling

Fig. 4. Feature embedding of the classes sampled by (a) random sampling and (b) task adaptive sampling. The dimension reduction is performed based on all 64 training classes of CIFAR-FS, while we show only the 5 selected classes in each sub-figure for better visualization.

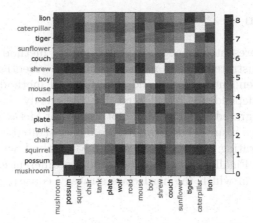

Fig. 5. Correlation matrix w.r.t. class-pair potentials. Each element indicates the class-pair potential. The higher the correlation weight (i.e., the darker the color), the higher the probability of this two-class combination being sampled. The green and red colors denote the classes sampled by random sampling and adaptive sampling, respectively. (Color figure online)

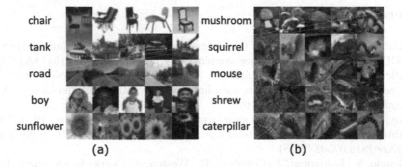

Fig. 6. Sample images from classes by (a) random sampling and (b) gcp-sampling.

We also visualize the class-pair potentials constructed by gcp-sampling in Fig. 5. We show 16 classes of CIFAR-FS, where the green and red colors denote the classes sampled by random sampling and gcp-sampling, respectively. We can see that the classes sampled by random sampling are often easier to distinguish, which leads to inefficient training, while the gcp-sampling tends to sample the classes that, when combined with other classes, display greater difficulty. We also randomly select some sampled images from each class for observation. As shown in Fig. 6, the classes sampled by random sampling do vary greatly (e.g., with unique shapes or colors) and are easier to recognize, while the classes sampled by gcp-sampling are visually more confusing (e.g., small animals or insects in the wild) and much more difficult to distinguish.

6   Conclusion

In this paper, we presented an adaptive task sampling method for meta-learning. Our results demonstrated that in meta-learning it is essential for the sampling process to be dependent on tasks, and the proposed method naturally models and exploits this dependence. We showed that the greedy class-pair based sampling method, integrated with PN, MetaOptNet-RR or MetaOptNet-SVM, could achieve competitive results. Furthermore, we demonstrated consistent improvement when integrating the proposed sampling method with different meta-learning methods. Finally, we explore and evaluate different sampling strategies for gcp-sampling, in which the hard class strategy consistently leads to more accurate results.

Acknowledgment. This research is supported by the National Research Foundation, Singapore under its AI Singapore Programme (AISG Award No: AISG-RP-2018-001). Any opinions, findings and conclusions or recommendations expressed in this material are those of the author(s) and do not reflect the views of National Research Foundation, Singapore.

References

1. Alain, G., Lamb, A., Sankar, C., Courville, A., Bengio, Y.: Variance reduction in SGD by distributed importance sampling. arXiv preprint arXiv:1511.06481 (2015)
2. Allen-Zhu, Z., Qu, Z., Richtárik, P., Yuan, Y.: Even faster accelerated coordinate descent using non-uniform sampling. In: International Conference on Machine Learning, pp. 1110–1119 (2016)
3. Aly, M.: Survey on multiclass classification methods. Neural Netw. **19**, 1–9 (2005)
4. Antoniou, A., Edwards, H., Storkey, A.: How to train your MAML. arXiv preprint arXiv:1810.09502 (2018)
5. Bengio, Y., Louradour, J., Collobert, R., Weston, J.: Curriculum learning. In: Proceedings of the 26th Annual International Conference on Machine Learning, pp. 41–48. ACM (2009)
6. Bertinetto, L., Henriques, J.F., Torr, P.H., Vedaldi, A.: Meta-learning with differentiable closed-form solvers. arXiv preprint arXiv:1805.08136 (2018)
7. Chang, H.S., Learned-Miller, E., McCallum, A.: Active bias: training more accurate neural networks by emphasizing high variance samples. In: Advances in Neural Information Processing Systems, pp. 1002–1012 (2017)
8. Chen, W., Liu, Y., Kira, Z., Wang, Y.F., Huang, J.: A closer look at few-shot classification. In: Proceedings of the 7th International Conference on Learning Representations, ICLR 2019, New Orleans, LA, USA, 6–9 May (2019). https://openreview.net/forum?id=HkxLXnAcFQ
9. Cho, K., et al.: Learning phrase representations using RNN encoder-decoder for statistical machine translation. arXiv preprint arXiv:1406.1078 (2014)
10. Cross, G.R., Jain, A.K.: Markov random field texture models. IEEE Trans. Pattern Anal. Mach. Intell. **5**(1), 25–39 (1983)
11. Csiba, D., Richtárik, P.: Importance sampling for minibatches. J. Mach. Learn. Res. **19**(1), 962–982 (2018)

12. Finn, C., Abbeel, P., Levine, S.: Model-agnostic meta-learning for fast adaptation of deep networks. In: Proceedings of the 34th International Conference on Machine Learning, vol. 70, pp. 1126–1135 (2017). JMLR.org
13. Franceschi, L., Frasconi, P., Salzo, S., Grazzi, R., Pontil, M.: Bilevel programming for hyperparameter optimization and meta-learning. In: International Conference on Machine Learning, pp. 1563–1572 (2018)
14. Freund, Y., Schapire, R.: A short introduction to boosting. J. Jpn. Soc. Artif. Intell. **14**(771–780), 1612 (1999)
15. Freund, Y., Schapire, R.E.: A decision-theoretic generalization of on-line learning and an application to boosting. J. Comput. Syst. Sci. **55**(1), 119–139 (1997)
16. Gopal, S.: Adaptive sampling for SGD by exploiting side information. In: International Conference on Machine Learning, pp. 364–372 (2016)
17. He, K., Zhang, X., Ren, S., Sun, J.: Deep residual learning for image recognition. In: Proceedings of the IEEE Conference on Computer Vision and Pattern Recognition, pp. 770–778 (2016)
18. Horváth, S., Richtárik, P.: Nonconvex variance reduced optimization with arbitrary sampling. arXiv preprint arXiv:1809.04146 (2018)
19. Ioffe, S., Szegedy, C.: Batch normalization: accelerating deep network training by reducing internal covariate shift. In: International Conference on Machine Learning, pp. 448–456 (2015)
20. Katharopoulos, A., Fleuret, F.: Biased importance sampling for deep neural network training. arXiv preprint arXiv:1706.00043 (2017)
21. Katharopoulos, A., Fleuret, F.: Not all samples are created equal: deep learning with importance sampling. arXiv preprint arXiv:1803.00942 (2018)
22. Kingma, D.P., Ba, J.: Adam: a method for stochastic optimization. arXiv preprint arXiv:1412.6980 (2014)
23. Krizhevsky, A., Hinton, G., et al.: Learning multiple layers of features from tiny images. Technical report, Citeseer (2009)
24. Lake, B.M., Salakhutdinov, R., Tenenbaum, J.B.: Human-level concept learning through probabilistic program induction. Science **350**(6266), 1332–1338 (2015)
25. Landau, B., Smith, L.B., Jones, S.S.: The importance of shape in early lexical learning. Cogn. Dev. **3**(3), 299–321 (1988)
26. Lee, K., Maji, S., Ravichandran, A., Soatto, S.: Meta-learning with differentiable convex optimization. In: Proceedings of the IEEE Conference on Computer Vision and Pattern Recognition, pp. 10657–10665 (2019)
27. Li, Z., Zhou, F., Chen, F., Li, H.: Meta-SGD: Learning to learn quickly for few-shot learning. arXiv preprint arXiv:1707.09835 (2017)
28. Lin, T.Y., Goyal, P., Girshick, R., He, K., Dollár, P.: Focal loss for dense object detection. In: Proceedings of the IEEE International Conference on Computer Vision, pp. 2980–2988 (2017)
29. Liu, L., Zhou, T., Long, G., Jiang, J., Zhang, C.: Learning to propagate for graph meta-learning. arXiv preprint arXiv:1909.05024 (2019)
30. London, B.: A PAC-Bayesian analysis of randomized learning with application to stochastic gradient descent. In: Advances in Neural Information Processing Systems, pp. 2931–2940 (2017)
31. Loshchilov, I., Hutter, F.: Online batch selection for faster training of neural networks. arXiv preprint arXiv:1511.06343 (2015)
32. van der Maaten, L., Hinton, G.: Visualizing data using t-SNE. J. Mach. Learn. Res. **9**(86), 2579–2605 (2008)
33. Mishra, N., Rohaninejad, M., Chen, X., Abbeel, P.: A simple neural attentive meta-learner. In: Proceedings of the ICLR (2017)

34. Munkhdalai, T., Yuan, X., Mehri, S., Trischler, A.: Rapid adaptation with conditionally shifted neurons. In: International Conference on Machine Learning, pp. 3661–3670 (2018)

35. Naik, D.K., Mammone, R.J.: Meta-neural networks that learn by learning. In: Proceedings of the International Joint Conference on Neural Networks, IJCNN 1992, vol. 1, pp. 437–442. IEEE (1992)

36. Nichol, A., Achiam, J., Schulman, J.: On first-order meta-learning algorithms. arXiv preprint arXiv:1803.02999 (2018)

37. Oreshkin, B., López, P.R., Lacoste, A.: TADAM: task dependent adaptive metric for improved few-shot learning. In: Advances in Neural Information Processing Systems, pp. 721–731 (2018)

38. Ravi, S., Larochelle, H.: Optimization as a model for few-shot learning. In: Proceedings of the ICLR (2016)

39. Russakovsky, O., et al.: ImageNet large scale visual recognition challenge. Int. J. Comput. Vision 115(3), 211–252 (2015)

40. Rusu, A.A., et al.: Meta-learning with latent embedding optimization. In: Proceedings of the 7th International Conference on Learning Representations, ICLR 2019, New Orleans, LA, USA, 6–9 May (2019). https://openreview.net/forum?id=BJgklhAcK7

41. Satorras, V.G., Bruna, J.: Few-shot learning with graph neural networks. In: Proceedings of the ICLR (2018)

42. Shalev-Shwartz, S., Wexler, Y.: Minimizing the maximal loss: how and why. In: Proceedings of the ICML, pp. 793–801 (2016)

43. Shrivastava, A., Gupta, A., Girshick, R.: Training region-based object detectors with online hard example mining. In: Proceedings of the IEEE Conference on Computer Vision and Pattern Recognition, pp. 761–769 (2016)

44. Snell, J., Swersky, K., Zemel, R.: Prototypical networks for few-shot learning. In: Advances in Neural Information Processing Systems, pp. 4077–4087 (2017)

45. Song, H., Kim, S., Kim, M., Lee, J.G.: Ada-boundary: accelerating the DNN training via adaptive boundary batch selection (2018)

46. Sun, Q., Liu, Y., Chua, T.S., Schiele, B.: Meta-transfer learning for few-shot learning. In: Proceedings of the IEEE Conference on Computer Vision and Pattern Recognition, pp. 403–412 (2019)

47. Sung, F., Yang, Y., Zhang, L., Xiang, T., Torr, P.H., Hospedales, T.M.: Learning to compare: relation network for few-shot learning. In: Proceedings of the IEEE Conference on Computer Vision and Pattern Recognition, pp. 1199–1208 (2018)

48. Thrun, S., Pratt, L.: Learning to learn: introduction and overview. In: Thrun, S., Pratt, L. (eds.) Learning to Learn, pp. 3–17. Springer, Boston (1998). https://doi.org/10.1007/978-1-4615-5529-2_1

49. Triantafillou, E., et al.: Meta-dataset: A dataset of datasets for learning to learn from few examples. arXiv preprint arXiv:1903.03096 (2019)

50. Vinyals, O., Blundell, C., Lillicrap, T., Kavukcuoglu, K., Wierstra, D.: Matching networks for one shot learning. In: Advances in Neural Information Processing Systems, pp. 3630–3638 (2016)

51. Ze, H., Senior, A., Schuster, M.: Statistical parametric speech synthesis using deep neural networks. In: Proceedings of the 2013 IEEE International Conference on Acoustics, Speech and Signal Processing, pp. 7962–7966. IEEE (2013)

52. Zhang, C., Kjellstrom, H., Mandt, S.: Determinantal point processes for mini-batch diversification. arXiv preprint arXiv:1705.00607 (2017)

53. Zhang, C., Öztireli, C., Mandt, S., Salvi, G.: Active mini-batch sampling using repulsive point processes. In: Proceedings of the AAAI Conference on Artificial Intelligence, vol. 33, 5741–5748 (2019)
54. Zhang, R., Che, T., Ghahramani, Z., Bengio, Y., Song, Y.: MetaGAN: an adversarial approach to few-shot learning. In: Advances in Neural Information Processing Systems, pp. 2365–2374 (2018)
55. Zhao, P., Zhang, T.: Stochastic optimization with importance sampling for regularized loss minimization. In: International Conference on Machine Learning, pp. 1–9 (2015)

Deep Complementary Joint Model for Complex Scene Registration and Few-Shot Segmentation on Medical Images

Yuting He[1], Tiantian Li[1], Guanyu Yang[1,2](✉), Youyong Kong[1,2], Yang Chen[1,2], Huazhong Shu[1,2], Jean-Louis Coatrieux[2,3], Jean-Louis Dillenseger[2,3], and Shuo Li[4]

[1] Laboratory of Image Science and Technology, Southeast University, Nanjing 210096, China
yang.list@seu.edu.cn

[2] Centre de Recherche en Information Biomédicale Sino-Français (CRIBs), Rennes, France

[3] Univ Rennes, Inserm, LTSI - UMR1099, 35000 Rennes, France

[4] Department of Medical Biophysics, University of Western Ontario, London, ON, Canada
slishuo@gmail.com

Abstract. Deep learning-based medical image registration and segmentation joint models utilize the complementarity (augmentation data or weakly supervised data from registration, region constraints from segmentation) to bring mutual improvement in complex scene and few-shot situation. However, further adoption of the joint models are hindered: 1) the diversity of augmentation data is reduced limiting the further enhancement of segmentation, 2) misaligned regions in weakly supervised data disturb the training process, 3) lack of label-based region constraints in few-shot situation limits the registration performance. We propose a novel Deep Complementary Joint Model (DeepRS) for complex scene registration and few-shot segmentation. We embed a perturbation factor in the registration to increase the activity of deformation thus maintaining the augmentation data diversity. We take a pixel-wise discriminator to extract alignment confidence maps which highlight aligned regions in weakly supervised data so the misaligned regions' disturbance will be suppressed via weighting. The outputs from segmentation model are utilized to implement deep-based region constraints thus relieving the label requirements and bringing fine registration. Extensive experiments on the CT dataset of MM-WHS 2017 Challenge [42] show great advantages of our DeepRS that outperforms the existing state-of-the-art models.

1 Introduction

Deep learning-based medical image segmentation models and registration models [11,19,23] are limited in complex scene and few-shot situation. In complex

© Springer Nature Switzerland AG 2020
A. Vedaldi et al. (Eds.): ECCV 2020, LNCS 12363, pp. 770–786, 2020.
https://doi.org/10.1007/978-3-030-58523-5_45

scene which has complex but task-unconcerned backgrounds, the unsupervised registration models [2,7] pay equal attention to all regions for overall alignment so that the performance on regions of interest (ROIs) will be limited by background. In few-shot situation which lacks labels, the segmentation models [24,29] will over-fit [30,40] due to the lack of supervision information.

The registration and segmentation tasks has great complementarity which will bring mutual improvement in complex scene and few-shot situation. As shown in Fig. 1, the registration model provides diverse augmentation data (warped images and labels) or weakly supervised data (fixed images and warped labels) for segmentation model [37,40] during the training process, thus reducing the requirement of labels and enhancing the segmentation generalization in few-shot situation. The segmentation model feeds back region constraints [6,22,37] so that additional attention on ROIs is paid for finer registration in complex scene.

Unfortunately, further exploiting of this complementary topology are hindered [6,22,37,40] due to: **Limitation 1:** *Degradation of data augmentation capability* (Fig. 1(a)). During the training of registration model, it learns the deformation rule that matches real situation and generates diverse warped images as augmentation data to improve the segmentation generalization ability [36,37]. However, the similarity between warped and fixed images increa-

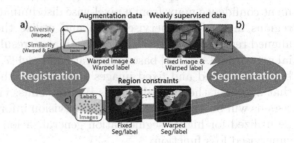

Fig. 1. The complementary topology and limitations of registration and segmentation tasks. Registration provides augmentation data and weakly supervised data for segmentation for higher generalization in few-shot situation, the segmentation feeds back region constraints for finer registration on ROIs in complex scene. a), b), c) illustrate the limitations in the utilization of this complementary topology.

ses and tends to become stable, and the diversity of warped images is gradually reduced as the similarity stabilizes. Therefore, in the later training stage of registration network, the identical warped images are generated in different epochs, resulting in the reduction of augmentation data diversity. Thus, the data augmentation ability of registration model is degraded and the further enhancement of segmentation will be limited. **Limitation 2:** *Misaligned regions in weakly supervised data* (Fig. 1(b)). The weakly supervised data enlarges the labeled dataset and provide additional supervision information for the segmentation model. However, large misaligned regions in these data will produce incorrect optimization targets and it will disturb the training process leading to serious mis-segmentation if used directly [37]. **Limitation 3:** *Lack of label-based region constraints* (Fig. 1(c)). Region constraints provide specific alignment information for regions bringing finer registration optimization. However, in few-shot situation, the label-based region constraints [6,14,22,37] are lacked with few labels. Thus if in complex scene, the registration model [2,7,36] will take rough optimization and the complex backgrounds will limit the registration performance on ROIs.

Solution 1 for the degradation of data augmentation capability: we embed a random perturbation factor in the registration to increase the activity of deformation for sustainable data augmentation capability. The registration process is a displacement of structure information, and the adjustment of deformation degree is the sampling of the structure information on this displacement path [12,20]. Therefore, our perturbation factor adjusts the deformation degree randomly to sample the structure information which is consistent with the real distribution to produce diverse and real augmentation data for the segmentation model.

Solution 2 to suppress the misaligned regions' disturbance: we extract alignment confidence maps from a pixel-wise discriminator to suppress the misaligned regions in weakly supervised data and utilize the supervision information in aligned regions. The pixel-wise discriminator, resulting in a generative adversarial network (GAN) [10] based registration model [7,8,13,38], learns the similarity between warped and fixed images and outputs the alignment confidence maps that highlight the aligned regions [15,26]. Thus, via these maps ,the misaligned regions will be suppressed and the supervision information in aligned regions will be utilized for higher segmentation generalization when calculating the weakly supervised loss function.

Solution 3 to cope with the lack of label-based region constraints: we build deep-based region constraints that calculate the loss value via the warped and fixed segmentations from the segmentation model so that fine registration optimization targets are available. Therefore, 1) label requirements of label-based region constraints are freed in few-shot situation, 2) different regions are independently optimized to avoid the misalignment of each region and 3) region attention on the ROIs is paid for finer registration.

In this paper, we propose a *Deep Complementary Joint Model (DeepRS)* that minimizes background interference in complex scene for finer registration on ROIs, and greatly reduces the label requirements of segmentation in few-shot situation for higher generalization ability. In short, the contributions of our work are summarized as follows:

- To the best of our knowledge, we build a novel complementary topology of registration and segmentation for the first time, and propose the DeepRS model utilizing the data generation ability of registration for few-shot segmentation, and the label-free region constraint ability of segmentation for complex scene registration.
- We propose a deep structure sampling (DSS) block adding a random perturbation factor to the registration for sustainable data augmentation ability.
- We propose an alignment confidence map (ACM) method which efficiently utilizes the supervision information in weakly supervised data thus bringing powerful segmentation generalization.

- We propose a deep-based region constraint (DRC) strategy which frees up the label requirements of label-based methods achieving finer registration on ROIs.

2   Related Works

2.1   Registration and Segmentation Joint Models

Registration and segmentation tasks have great complementarity, thus building a registration and segmentation joint model has the potential of mutual improvement. The registration provides augmentation data and weakly supervised data for the segmentation [35,37,40], and the segmentation feeds back additional region constraints [6,14,22]. Zhao et al. [40] took a pre-trained registration model to generate augmentation data for more powerful segmentation ability. Li et al. [22] made a hybrid framework that took the label-based region constraints from labels and segmentations for finer registration. Similarity, Xu et al. [37] designed a semi-supervised method that combined registration and segmentation models bringing the mutual improvement in knee and brain images.

However, these existing methods only took the advantage of partial complementarity which hardly gives full play to their potential. The convergence of the registration model limits the diversity of the augmentation data and prevents further enhancement of the segmentation model [40]. Misaligned regions in weakly supervised data disturb the training of segmentation models, and if used directly, it will lead to serious mis-segmentation [37]. In few-shot situation, label-based region constraints are lacked due to the small labeled dataset [22], thus with inaccurate optimization targets, complex backgrounds will limit registration performance on ROIs in complex scene.

2.2   Data Augmentation

Data augmentation [30], generating bigger dataset, has the ability to improve learning models [31], especially in few-shot situation. Some data augmentation strategies (random cropping, mirroring, rotation, flipping, etc.) are often used for higher generation ability, while inappropriate strategy combinations will generate unreasonable data which will weaken the model performance [30]. Learning-based data augmentation strategies [4,16,21,27] learn the augmentation methods from dataset for real augmentation data. Registration learns transformation rules of structure information from the images [12,20,37,40] so that the augmentation images with real structure information are obtained.

Disappointingly, the registration-based augmentation ability will degrade due to the reduction of deformation diversity. As the registration model converges, the moving image is stably aligned onto the fixed image and the identical warped images in different epochs are generated, resulting in the reduction of augmentation data diversity and limiting the further improvement of segmentation.

2.3 Weakly-Supervised Learning

Weakly-supervised learning [14,15,18,26,28,33] utilizing non-precisely labeled data is a strategy for labeled data limitation. It has three typical types according to the weakly supervised data types [41]: 1) incomplete supervision where part of the dataset without labels [15,26], 2) inexact supervision where data with coarse-grained labels [14,18] and 3) inaccurate supervision where data with inaccurate labels [32,33]. In registration and segmentation tasks, the warped labels and fixed images from registration model make up weakly supervised data leading to inaccurate supervision which will improve the segmentation performance with appropriate strategy. Unfortunately, if the weakly supervised data is used directly, the misaligned regions will brings inaccurate optimization target, thus disturbing the training process and lead to mis-segmentation.

2.4 Generative Adversarial Networks

Generative adversarial networks (GANs) [9,10,39], consisting of a generator G and a discriminator D, learns a similarity metric of the generated and real images. The discriminator learns to distinguish the real or generated images and the generator takes the adversarial loss from the discriminator to improve the authenticity of the generated image to deceive the discriminator. GAN-based registration models [7,8,13,38] take global discriminator to learn image-wise similarity metric

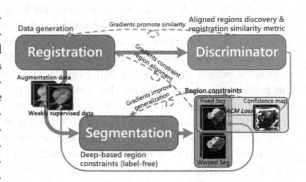

Fig. 2. The overview of our DeepRS. The data generation ability of the registration, the deep-based region constraint of the segmentation, the aligned regions discovery ability and the learned similarity metric of discriminator interact in the alternating training process.

of warped and fixed images which can be used to evaluate the weakly supervised data in our task.

However, the image-wise similarity has no ability to evaluate the regional similarity and in our segmentation task (pixel-wise), it will still introduce the error information in the weakly supervised data. Patch-GANs utilize pixel-wise discriminator [15,26] consisting of a full convolution network to learn the pixel-wise similarity and output confidence maps which highlight task-beneficial regions. Thus, a patch-GAN is used in our model for alignment confidence maps to suppress the misaligned regions and utilize the supervision information in weakly supervised data.

3   Methodology

Our DeepRS model (Figs. 3 and 2), which consists of registration, pixel-wise discriminator and segmentation models, leverages their complementarity for complex scene registration and few-shot segmentation (Sect. 3.1) bringing mutual improvement. The registration generates diverse augmentation data via randomly adjusting the deformation field in a DSS block (Sect. 3.1) and provides weakly supervised data for the segmentation network to reduce the labeled data requirements in few-shot situation. The pixel-wise discriminator provides ACMs (Sect. 3.1) for the segmentation network for supervision information utilization in weakly supervised data. The segmentation network provides DRC (Sect. 3.1) for the registration network for finer registration on ROIs in complex scene. The joint strategy (Sect. 3.2) maximizes the complementarity via alternating training.

3.1   DeepRS for Stronger Registration and Segmentation

The proposed DeepRS model leverages the complementarity of registration and segmentation tasks via the DSS block, ACM method and DRC strategy.

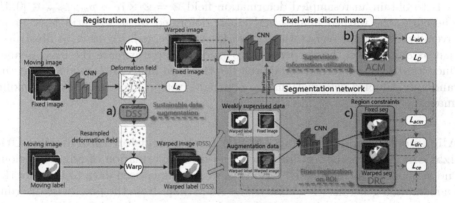

Fig. 3. In detail of our DeepRS model, we design a DSS block, a ACM method and a DRC strategy cleverly dealing with the limitations. a) The DSS block maintains the diversity of warped images bringing sustainable data augmentation ability. b) The ACM method utilizes the supervision information in weakly supervised data. c) The DRC strategy provides region attention on ROIs for finer registration.

Deep Structure Sampling (DSS) for Sustainable Data Augmentation.
DSS block generates diverse augmentation data sustainedly via embedding a random perturbation factor in the deformation field to increase the uncertainty of the warped images and labels. The registration process is the displacement of image structure information, and the perturbation of deformation degree realizes the sampling of information on this displacement path [12,20]. Therefore,

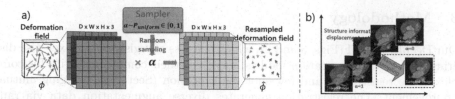

Fig. 4. The DSS block for sustainable data augmentation. a) A perturbation factor $\alpha \in [0, 1]$ from uniform distribution adjusts the deformation field making the sampling process. b) Illustration of the sampling that registration makes the structure information displacement and our DSS samples the information on its displacement path.

the DSS block brings two advantages: 1) Sustainable data augmentation. The perturbation factor controls the deformation degree so that the registration network is guaranteed to generate diverse augmentation data sustainedly. 2) Real distribution. Sampling structure information from its displacement path generates the augmentation data more matching real distribution than other manual augmentation methods.

As shown in Fig. 4(a), a deformation field $\varnothing$ from registration network is multiplied by a random perturbation factor α from uniform distribution $p_{uniform} \in [0, 1]$ to obtain an resampled deformation field $\hat{\varnothing} = \varnothing \times \alpha \backsim p_{uniform} \in [0, 1]$. Therefore, the warped images and labels deformed by it will still have great diversity, even if the registration network has converged. Figure 4(b) illustrates that as α increases, the warped images gradually approximate the fixed images since its structure information approaches the fixed image. It is evident that the randomly sampled deformations are non-rigid, yet produce realistically-looking images.

Alignment Confidence Map (ACM) for Supervision Information Utilization. ACM method utilizes the supervision information of aligned regions and suppresses the misaligned regions in weakly supervised data to improve the segmentation generalization ability. The ACM maps from the pixel-wise discriminator evaluate the pixel-wise similarity between warped and fixed images and will highlight the aligned regions. Thus, these maps will be taken to weight the loss of weakly supervised data to utilize its supervision information in aligned regions, as illustrated in Eq. 1:

$$\mathcal{L}_{acm} = -D(W(x_m, \hat{\varnothing}), x_f)W(y_m, \hat{\varnothing}) \log S(x_f) \tag{1}$$

where x_m, y_m, x_f and $\hat{\varnothing}$ are the moving image, moving label, fixed image and resampled deformation field from DSS block. As shown in Fig. 3, $W(\cdot, \cdot)$ is the 'warp' block which deforms the moving images and labels to the fixed images for warped images and labels following the spatial transformation layer in [2]. The pixel-wise discriminator $D(\cdot, \cdot)$ measures the similarity between warped and fixed images for the ACMs to weight the cross-entropy loss between warped labels and fixed segmentation (seg-) masks $S(x_f)$. Therefore, the loss value in misaligned region will get low weight and the disturbance will be suppressed.

The contribution of the weakly supervised data is increasing during the training. In early training stage, the powerful discriminator outputs weak maps, so that the loss from weakly supervised data is suppressed greatly and the optimization target of the segmentation network is dominated by the loss $\mathcal{L}_{ce}$ from augmentation data. As the training progresses, the registration network defeats the discriminator and obtains high responsive maps, thus increasing the contribution of ACM loss $\mathcal{L}_{acm}$, so that the segmentation generalization ability will be further enhanced.

Deep-Based Region Constraint (DRC) for Finer Registration on ROIs. DRC strategy guides the attention on the ROIs for finer registration via constraints between the fixed and warped seg-masks from the segmentation network. This deep-based region constraint takes the alignment of the corresponding regions in warped and fixed images as the optimization target, so that 1) label requirements of label-based region constraints is freed in few-shot situation, 2) different regions are independently optimized to avoid the misalignment between each other and 3) additional region attention on the ROIs is paid for finer registration.

As shown in Fig. 3(c), the warped image and the fixed image are input into the segmentation network respectively for the warped and the fixed seg-masks firstly. Then a mean square error loss between these two seg-masks is calculated as is illustrated in Eq. 2:

$$\mathcal{L}_{drc} = -(S(W(x_m, \hat{\varnothing})) - S(x_f))^2 \tag{2}$$

where x_m, x_f and $\hat{\varnothing}_n$ are the moving image, fixed image and deformation field from the DSS block. $W(\cdot, \cdot)$ is the deformation process in registration network and $S(\cdot)$ is the segmentation network. Each ROI is calculated in different channels obtaining independent fine optimization, while the task unconcerned regions are calculated in a background channel together. Thus, fine registration on ROIs is available and inter-regional error registration is avoided.

3.2 Joint Learning Strategy Exerts Complementarity

The registration network, segmentation network and pixel-wise discriminator in our DeepRS model (Fig. 3) are trained by different loss function combinations to coordinate the training process and achieve mutual improvement.

Registration Network. The registration network is optimized by four different targets. An adversarial loss $\mathcal{L}_{adv}$ [7] from the pixel-wise discriminator provides the similarity metric between warped and fixed images. The DRC loss $\mathcal{L}_{drc}$ from the segmentation network brings registration attention on ROIs. A local cross-correlation (CC) [2] $\mathcal{L}_{cc}$ maintains the stability of the training process, and a smooth loss [2] $\mathcal{L}_R$ penalizes local spatial variations in deformation field. Therefore, the total loss function $\mathcal{L}_{reg}$ is:

$$\mathcal{L}_{reg} = \lambda_{adv}\mathcal{L}_{adv} + \lambda_{drc}\mathcal{L}_{drc} + \lambda_{cc}\mathcal{L}_{cc} + \lambda_R\mathcal{L}_R \qquad (3)$$

Segmentation Network. The loss function of the segmentation network $\mathcal{L}_{seg}$ consists of two components. One is the ACM loss $\mathcal{L}_{acm}$ that adds the weakly supervised data to the training for higher segmentation generalization ability. The other is cross-entropy loss $\mathcal{L}_{ce}$ between the warped images and labels that maintains the right optimization target:

$$\mathcal{L}_{seg} = \lambda_{acm}\mathcal{L}_{acm} + \lambda_{ce}\mathcal{L}_{ce} \qquad (4)$$

Pixel-wise Discriminator. The training strategy of pixel-wise discriminator follows [7]: well-registered image pairs consisting of reference images x_r and fixed images x_f as positive cases and misaligned images consisting of warped images x_w and fixed images x_f as negative cases. The reference image x_r is a fusion of a moving image x_m and a fixed image x_f according to the formula $x_r = \beta * x_m + (1 - \beta) * x_f$. Thus, the loss for the discriminator $\mathcal{L}_D$ is:

$$\mathcal{L}_D = -\log(D(x_r, x_f)) - \log(1 - D(x_w, x_f)) \qquad (5)$$

4   Experiments

Extensive experimental results show that our DeepRS model enhances the performance of complex scene registration and few-shot segmentation tasks on cardiac CT data which has complex task-unconcerned backgrounds.

4.1   Evaluation Settings

Dataset. We validated the superiority of our DeepRS model on the whole heart registration and segmentation tasks on the CT dataset of *MM-WHS 2017 Challenge* [42] which has complex backgrounds (lung, rib cage, etc.). This dataset consists of 20 labeled and 40 unlabeled CT images. Our experiments aim to register and segment seven cardiac structures including the ascending aorta, left atrial cavity (LA), left ventricular cavity(LV), myocardium of the left ventricle (Myo), pulmonary artery (PA), right atrial cavity (RA) and right ventricular cavity (RV). We first crop the rectangular regions containing the hearts for affine transformation and resample them to $128 \times 128 \times 96$. Then the labeled images are randomly split into 5 parts, 1 part (4 images) is used in training set as the moving images for few-shot situation and the remaining 4 parts (16 images) in testing set resulting in 5-folds evaluation. We put 40 unlabeled images as the fixed images in the training set leading to 160 data pairs together with the moving images, and the 16 images in the testing set are paired separately leading to 240 data pairs. Following the [2], we use Elastix[1] to perform affine transformation so that our model only needs to pay attention to the deformation registration process.

---

[1] https://www.elastix.org/.

Implementation. The segmentation network and pixel-wise discriminator follow the same 3D U-Net [3] structure. The registration network follows the VoxelMorph-2 [2] structure. We use RMSprop [34] to train the registration network and the discriminator for stable process [1], and Adam [17] to train the segmentation network for fast convergence. These models share the same learning rate of $2e^{-4}$ and training batch size of 1 due to the limitation of memory. According to extensive experiments, we finally set $\lambda_{adv} = 1$, $\lambda_{drc} = 10$, $\lambda_{cc} = 1$, $\lambda_R = 1$, $\lambda_{acm} = 1$ and $\lambda_{ce} = 1$. The models were implemented via Keras[2] with a Tensorlow[3] backend and were trained on a single NVIDIA TitanX GPU with 12 GB memory.

Comparison Settings. The comparison demonstrates the advancement on segmentation and registration of our DeepRS model. We compare our model's segmentation performance with three general segmentation networks (3D U-Net [3], V-Net [25], 3D FCN [24]) to illustrate the enhancement brought by registration. The 3D U-Net augmented by manual strategies (random rotate in $[-10°, 10°]$, random mirroring and random flipping) is compared with to show the advantages of registration-based data augmentation. We also compare two unsupervised registration models (VoxelMorph-2 [2], Adv-Reg [7]) to illustrate the superiority of our deep-based region constraints in complex scene. In addition, two registration and segmentation joint models (DeepAtlas [37], HybridCNN [22]) are compared with to demonstrate the superiority of the DeepRS brought by our DSS block, ACM method and DRC strategy. What's more, our proposed DeepRS is also evaluated on different data amount to illustrate its excellent generalization ability in few-shot segmentation. Finally, an ablation study is used to analyse the contributions of each our innovation.

Evaluation Metric. We evaluate the registration and segmentation methods with dice coefficient [5]. The dice coefficient ($[\%]$) is a metric that measures the coincidence degree between two sets according to $Dice(G, P) = \frac{2|G \cap P|}{|G| + |P|}$ where the G is the ground truth and the P is the predicted mask. It is suitable to evaluate the agreement between the predicted segmentation/registration and the ground truth. The Dice coefficients of the corresponding seven cardiac structures are calculated, and presented as *mean ± std*.

4.2   Results

Extensive experimental results on cardiac CT dataset show that with merely 4 training labels, our proposed DeepRS appears to be a strong superiority both in the quantitative comparison and in visual. The experiments on different label amounts illustrate that our DeepRS greatly reduces the label dependence of the segmentation model.

---

[2] https://github.com/keras-team/keras.
[3] https://github.com/tensorflow/tensorflow.

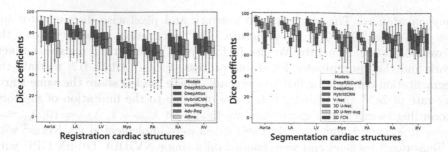

Fig. 5. Our DeepRS achieves excellent dice coefficients on each structure. The box plots shows the proposed DeepRS (red box) model achieves the state-of-the-art performance in complex scene registration (*Left*) and few-shot segmentation (*Right*). (Color figure online)

Quantitative Comparison. As shown in Table 1, our DeepRS model achieves the state-of-the-art performance in both registration and segmentation tasks whose mean dice coefficients of all cardiac structures are 77.6% and 85.7%. Figure 5 illustrates that the proposed DeepRS achieves excellent dice coefficients on each structure in complex scene registration and few-shot segmentation.

On the registration task, the registration network gets the deep-based region constraints from the segmentation network, bringing finer registration on ROIs in complex scene than other registration models. VoxelMorph-2 lacks region constraints, thus the dice is 5.9% lower than ours. The Adv-Reg takes a GAN whose training process is unstable to learn a similarity metric and gets worse results than VoxelMorph-2. DeepAtlas utilizes weakly supervised data directly, thus the misaligned regions disturbs the training process of the segmentation model finally in turn disturbing the registration performance (71.3%). HybridCNN lacks label-based region constraints in our few-shot situation thus the influence of the misaligned regions are more pronounced (69.2%).

Table 1. The proposed DeepRS model achieves the state-of-the-art performance both in registration (R) and segmentation (S) tasks on cardiac CT data.

Method	R-Dice	S-Dice
Affine only	64.6 ± 10.7	-
VoxelMorph-2 [2]	71.7 ± 10.6	-
Adv-Reg [7]	68.8 ± 10.7	-
3D U-Net [3]	-	78.8 ± 9.2
3D U-Net-aug [3]	-	80.0 ± 12.0
3D FCN [24]	-	71.4 ± 11.3
V-Net [25]	-	69.8 ± 10.9
DeepAtlas [37]	71.3 ± 10.5	81.8 ± 7.5
HybridCNN [22]	69.2 ± 10.3	78.8 ± 7.9
DeepRS (Ours)	$\mathbf{77.6 \pm 7.9}$	$\mathbf{85.7 \pm 7.7}$

On the segmentation task, the segmentation network in our DeepRS model effectively utilizes the augmentation data and weakly supervised data from the registration network, thus achieving much higher dice coefficient than 3D U-Net, 3D FCN and V-Net. Although the 3D U-Net augmented by manual strategies has get 1.2% dice improvement compared with non-augmentation, our DeepRS model has even greater advantage by 5.7%. Due to the influence of the misaligned regions in weakly supervised data, the HybridCNN only gets 78.8% dice. Similarly, the DeepAtlas takes the augmentation data from registration, but the misaligned regions still limits the enhancement which make it get only 81.8% dice.

Visual Superiority. Visually, our DeepRS model brings higher segmentation generalization ability with few labels, and achieves finer registration performance on ROIs in complex scene.

As illustrated in Fig. 6, our DeepRS brings finer registration on ROIs making 5 structures in moving image look more similar to these structures in fixed image. The HybridCNN uses weakly supervised data directly and lacks label-based region constraints. Therefore the misaligned regions interrupt the segmentation training process and in turn weaken the registration performance bringing serious region correspondence errors. The Adv-Reg is optimized by the unstable GAN making the warped image messy and rough in detail.

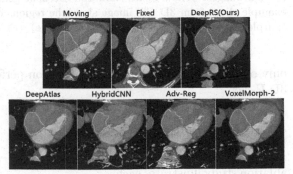

Fig. 6. Our DeepRS gets finer registration on ROIs. The example slices from 3D CT image show the overlaid boundaries of the LV (green), RA (yellow), RV (purple), LV (blue) and Myo (pink). Our model makes these structures in moving image alike structures in fixed image. (Color figure online)

As shown in Fig. 7, our DeepRS model brings much higher segmentation generalization ability trained on merely 4 labeled images. Case 1 shows the excellent generalization ability and the yellow boxes show the performance in detail. Our DeepRS has achieved fine segmentation, while the 3D U-Net, 3D FCN, 3D U-Net-aug and V-Net have many mis-segmentation regions. The HybridCNN and DeepAtlas has more mis-segmentation regions than others due to the misaligned regions in weakly supervised data. Case 2 shows the fine segmentation capability in another perspective and sample. The 3D U-Net ,3D FCN and V-Net are limited by small dataset leading to various serious mis-segmentation.

DeepRS for Few-Shot Segmentation. In few-shot situation, the segmentation (S) network in our DeepRS model achieves higher mean dice coefficients of all structures than 3D U-Net as illustrated in Fig. 8. The effectiveness of our DeepRS is evaluated on randomly-sampled labeled data whose amount is 1, 4, 7 and 10 respectively. 3D U-Net is used for comparison and the mean dice coefficients of all structures are calculated. As the labeled data decreases, the superiority of our DeepRS on segmentation task becomes more prominent. When

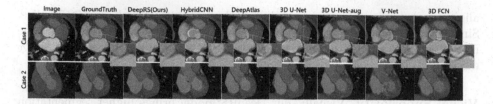

Fig. 7. Our DeepRS brings higher segmentation generalization ability trained on 4 labeled images. Yellow boxes show the excellent generalization ability in detail. The example slices from 3D CT image show the regions of Aorta (red), RA (yellow), RV (purple), Myo (pink), LV (green) and LV (blue). (Color figure online)

only one label is available, our segmentation performance is 18.1% higher than 3D U-Net.

4.3 Ablation Study

As shown in Table 2, an ablation study illustrates each great advantage brought by our innovations. The directly joint model only utilizes the registration's data augmentation ability thus the segmentation gets 80.5% dice and the registration gets 72.9% dice. Our DSS block embeds a random perturbation factor in the registration to maintain the diversity of augmentation data (**Solution 1**), thus bringing 3.4% segmentation dice growth. The ACM method adds the supervision

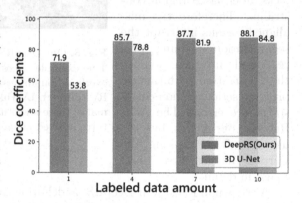

Fig. 8. Especially in few-shot situation, the segmentation network in our DeepRS model achieves much higher mean dice coefficients of all structures than 3D U-Net [3].

information in weakly supervised data (**Solution 2**) to segmentation network so that it gets 3.6% segmentation dice improvement. The DRC strategy builds deep-based region constraints instead of label-based methods (**Solution 3**) via the warped and fixed segmentations increasing the direct joint model by 3% registration dice. We find that the segmentation and registration models achieve further promotion in our final DeepRS model owing to their complementarity, thus finally achieving 77.6% registration dice and 85.7% segmentation dice which are increased by 4.7% and 5.2% respectively.

5   Conclusion

This paper presents a *Deep Complementary Joint Model (DeepRS)* for complex scene registration and few-shot segmentation. Our proposed *DSS block* adjusts deformation fields randomly via a perturbation factor, thus increasing the activity of the warped images and labels and achieving sustainable data augmentation capability. Our pro-

Table 2. The ablation study analyses the contributions of our innovations.

R	S	DSS	ACM	DRC	R-Dice	S-Dice
✓					72.2 ± 10.3	-
	✓				-	78.8 ± 9.2
✓	✓				72.9 ± 10.4	80.5 ± 10.2
✓	✓	✓			72.9 ± 9.6	83.9 ± 8.3
✓	✓		✓		72.5 ± 10.1	84.1 ± 8.3
✓	✓			✓	75.9 ± 9.1	82.5 ± 9.2
✓	✓	✓	✓	✓	**77.6 ± 7.9**	**85.7 ± 7.7**

posed *ACM method* efficiently utilizes the supervision information in weakly supervised data via alignment confidence maps from a pixel-wise discriminator bringing higher segmentation generalization. Our proposed *DRC strategy* constructs label-free loss between the warp and fixed images from the segmentation model resulting in finer registration on ROIs. We train our proposed DeepRS model on the cardiac CT dataset which has complex background with few labels with merely 4 labels and shows great advantages in registration and segmentation tasks compared to existing methods.

Our work greatly reduces the requirement of a large labeled dataset and provides the fine optimization targets, thus the registration and segmentation accuracy are improved and the cost is greatly saved. Especially, our DeepRS model has great potential in some situations where the labeling is difficult, the scene is complex or the dataset is small.

Acknowledgments. This research was supported by the National Natural Science Foundation under grants (61828101,31571001,31800825), the Short-Term Recruitment Program of Foreign Experts (WQ20163200398), and Southeast University-Nanjing Medical University Cooperative Research Project (2242019K3DN08). We thank the Big Data Computing Center of Southeast University for providing the facility support on the numerical calculations in this paper.

References

1. Arjovsky, M., Chintala, S., Bottou, L.: Wasserstein gan. arXiv preprint arXiv:1701.07875 (2017)
2. Balakrishnan, G., Zhao, A., Sabuncu, M.R., Guttag, J., Dalca, A.V.: An unsupervised learning model for deformable medical image registration. In: Proceedings of the IEEE Conference on Computer Vision and Pattern Recognition, pp. 9252–9260 (2018)

3. Çiçek, Ö., Abdulkadir, A., Lienkamp, S.S., Brox, T.: 3D U-Net: learning dense volumetric segmentation from sparse annotation. In: Ourselin, S., Joskowicz, L., Sabuncu, M., Unal, G., Wells, W. (eds.) Medical Image Computing and Computer-Assisted Intervention – MICCAI 2016. Lecture Notes in Computer Science, vol. 9901, pp. 424–432. Springer, Cham (2016). https://doi.org/10.1007/978-3-319-46723-8_49

4. Cubuk, E.D., Zoph, B., Mane, D., Vasudevan, V., Le, Q.V.: Autoaugment: learning augmentation strategies from data. In: Proceedings of the IEEE Conference on Computer Vision and Pattern Recognition, pp. 113–123 (2019)

5. Dice, L.R.: Measures of the amount of ecologic association between species. Ecology 26(3), 297–302 (1945)

6. Estienne, T., et al.: U-ReSNet: ultimate coupling of registration and segmentation with deep nets. In: Shen, D., et al. (eds.) Medical Image Computing and Computer Assisted Intervention – MICCAI 2019, vol. 11766, pp. 310–319. Springer, Cham (2019). https://doi.org/10.1007/978-3-030-32248-9_35

7. Fan, J., Cao, X., Wang, Q., Yap, P.T., Shen, D.: Adversarial learning for mono-or multi-modal registration. Med. Image Anal. 48, 101545 (2019)

8. Fan, J., Cao, X., Xue, Z., Yap, P.T.: Adversarial similarity network for evaluating image alignment in deep learning based registration. In: Frangi, A., Schnabel, J., Davatzikos, C., Alberola-López, C., Fichtinger, G. (eds.) Medical Image Computing and Computer Assisted Intervention – MICCAI 2018. Lecture Notes in Computer Science, vol. 11070, pp. 739–746. Springer, Cham (2018). https://doi.org/10.1007/978-3-030-00928-1_83

9. Ge, R., Yang, G., Xu, C., Chen, Y., Luo, L., Li, S.: Stereo-correlation and noise-distribution aware ResVoxGAN for dense slices reconstruction and noise reduction in thick low-dose CT. In: Shen, D., et al. (eds.) Medical Image Computing and Computer Assisted Intervention – MICCAI 2019. Lecture Notes in Computer Science, vol. 11769, pp. 328–338. Springer, Cham (2019). https://doi.org/10.1007/978-3-030-32226-7_37

10. Goodfellow, I., Pouget-Abadie, J., Mirza, M., Xu, B., Warde-Farley, D., Ozair, S., Courville, A., Bengio, Y.: Generative adversarial nets. In: Advances in Neural Information Processing Systems, pp. 2672–2680 (2014)

11. Haskins, G., Kruger, U., Yan, P.: Deep learning in medical image registration: a survey. arXiv preprint arXiv:1903.02026 (2019)

12. Hauberg, S., Freifeld, O., Larsen, A.B.L., Fisher, J., Hansen, L.: Dreaming more data: class-dependent distributions over diffeomorphisms for learned data augmentation. Artif. Intell. Stat. 342–350 (2016)

13. Hu, Y., et al.: Adversarial deformation regularization for training image registration neural networks. In: Frangi, A., Schnabel, J., Davatzikos, C., Alberola-López, C., Fichtinger, G. (eds.) Medical Image Computing and Computer Assisted Intervention – MICCAI 2018. Lecture Notes in Computer Science, vol. 11070, pp. 774–782. Springer, Cham (2018). https://doi.org/10.1007/978-3-030-00928-1_87

14. Hu, Y., et al.: Weakly-supervised convolutional neural networks for multimodal image registration. Med. Image Anal. 49, 1–13 (2018)

15. Hung, W.C., Tsai, Y.H., Liou, Y.T., Lin, Y.Y., Yang, M.H.: Adversarial learning for semi-supervised semantic segmentation. arXiv preprint arXiv:1802.07934 (2018)

16. Jackson, P.T., Atapour-Abarghouei, A., Bonner, S., Breckon, T., Obara, B.: Style augmentation: data augmentation via style randomization. arXiv preprint arXiv:1809.05375 (2018)

17. Kingma, D.P., Ba, J.: Adam: a method for stochastic optimization. arXiv preprint arXiv:1412.6980 (2014)

18. Kolesnikov, A., Lampert, C.H.: Seed, expand and constrain: three principles for weakly-supervised image segmentation. In: Leibe, B., Matas, J., Sebe, N., Welling, M. (eds.) Computer Vision – ECCV 2016. Lecture Notes in Computer Science, vol. 9908, pp. 695–711. Springer, Cham (2016). https://doi.org/10.1007/978-3-319-46493-0_42

19. Lateef, F., Ruichek, Y.: Survey on semantic segmentation using deep learning techniques. Neurocomputing **338**, 321–348 (2019)

20. Learned-Miller, E.G.: Data driven image models through continuous joint alignment. IEEE Trans. Pattern Anal. Mach. Intell. **28**(2), 236–250 (2005)

21. Lemley, J., Bazrafkan, S., Corcoran, P.: Smart augmentation learning an optimal data augmentation strategy. IEEE Access **5**, 5858–5869 (2017)

22. Li, B., et al.: A hybrid deep learning framework for integrated segmentation and registration: evaluation on longitudinal white matter tract changes. In: Shen, D., et al. (eds.) Medical Image Computing and Computer Assisted Intervention – MICCAI 2019. Lecture Notes in Computer Science, pp. 645–653. Springer, Cham (2019). https://doi.org/10.1007/978-3-030-32248-9_72

23. Litjens, G., Kooi, T., Bejnordi, B.E., Setio, A.A.A., Ciompi, F., Ghafoorian, M., Van Der Laak, J.A., Van Ginneken, B., Sánchez, C.I.: A survey on deep learning in medical image analysis. Med. Image Anal. **42**, 60–88 (2017)

24. Long, J., Shelhamer, E., Darrell, T.: Fully convolutional networks for semantic segmentation. In: Proceedings of the IEEE Conference on Computer Vision and Pattern Recognition, pp. 3431–3440 (2015)

25. Milletari, F., Navab, N., Ahmadi, S.A.: V-Net: fully convolutional neural networks for volumetric medical image segmentation. In: 2016 Fourth International Conference on 3D Vision (3DV), pp. 565–571. IEEE (2016)

26. Nie, D., Gao, Y., Wang, L., Shen, D.: ASDNet: attention based semi-supervised deep networks for medical image segmentation. In: Frangi, A., Schnabel, J., Davatzikos, C., Alberola-López, C., Fichtinger, G. (eds.) Medical Image Computing and Computer Assisted Intervention – MICCAI 2018. Lecture Notes in Computer Science, vol. 11073, pp. 370–378. Springer, Cham (2018). https://doi.org/10.1007/978-3-030-00937-3_43

27. Nielsen, C., Okoniewski, M.: GAN data augmentation through active learning inspired sample acquisition. In: Proceedings of the IEEE Conference on Computer Vision and Pattern Recognition Workshops, pp. 109–112 (2019)

28. Papandreou, G., Chen, L.C., Murphy, K.P., Yuille, A.L.: Weakly-and semi-supervised learning of a deep convolutional network for semantic image segmentation. In: Proceedings of the IEEE International Conference on Computer Vision, pp. 1742–1750 (2015)

29. Ronneberger, O., Fischer, P., Brox, T.: U-Net: convolutional networks for biomedical image segmentation. In: Navab, N., Hornegger, J., Wells, W., Frangi, A. (eds.) Medical Image Computing and Computer-Assisted Intervention – MICCAI 2015. Lecture Notes in Computer Science, vol. 9351, pp. 234–241. Springer, Cham (2015). https://doi.org/10.1007/978-3-319-24574-4_28

30. Shorten, C., Khoshgoftaar, T.M.: A survey on image data augmentation for deep learning. J. Big Data **6**(1), 60 (2019)

31. Sun, C., Shrivastava, A., Singh, S., Gupta, A.: Revisiting unreasonable effectiveness of data in deep learning era. In: The IEEE International Conference on Computer Vision (ICCV) (2017)

32. Tang, M., Djelouah, A., Perazzi, F., Boykov, Y., Schroers, C.: Normalized cut loss for weakly-supervised CNN segmentation. In: Proceedings of the IEEE Conference on Computer Vision and Pattern Recognition, pp. 1818–1827 (2018)

33. Tang, M., Perazzi, F., Djelouah, A., Ayed, I.B., Schroers, C., Boykov, Y.: On regularized losses for weakly-supervised CNN segmentation. In: Ferrari, V., Hebert, M., Sminchisescu, C., Weiss, Y. (eds.) Computer Vision – ECCV 2018. Lecture Notes in Computer Science, vol. 11220, pp. 524–540. Springer, Cham (2018). https://doi.org/10.1007/978-3-030-01270-0_31

34. Tieleman, T., Hinton, G.: Lecture 6.5-rmsprop, coursera: Neural networks for machine learning. University of Toronto, Technical report (2012)

35. Vakalopoulou, M., et al.: ATLASNet: multi-atlas non-linear deep networks for medical image segmentation. In: Frangi, A., Schnabel, J., Davatzikos, C., Alberola-López, C., Fichtinger, G. (eds.) Medical Image Computing and Computer Assisted Intervention – MICCAI 2018. Lecture Notes in Computer Science, vol. 11073, pp. 658–666. Springer, Cham (2018). https://doi.org/10.1007/978-3-030-00937-3_75

36. de Vos, B.D., Berendsen, F.F., Viergever, M.A., Sokooti, H., Staring, M., Išgum, I.: A deep learning framework for unsupervised affine and deformable image registration. Med. Image Anal. **52**, 128–143 (2019)

37. Xu, Z., Niethammer, M.: DeepatLas: joint semi-supervised learning of image registration and segmentation. arXiv preprint arXiv:1904.08465 (2019)

38. Yan, P., Xu, S., Rastinehad, A.R., Wood, B.J.: Adversarial image registration with application for MR and TRUS image fusion. In: Shi, Y., Suk, H.I., Liu, M. (eds.) Machine Learning in Medical Imaging. MLMI 2018. Lecture Notes in Computer Science, vol. 11046, pp. 197–204. Springer, Cham (2018). https://doi.org/10.1007/978-3-030-00919-9_23

39. Yi, X., Walia, E., Babyn, P.: Generative adversarial network in medical imaging: a review. Med. Image Anal. **58**, 101552 (2019)

40. Zhao, A., Balakrishnan, G., Durand, F., Guttag, J.V., Dalca, A.V.: Data augmentation using learned transformations for one-shot medical image segmentation. In: Proceedings of the IEEE Conference on Computer Vision and Pattern Recognition, pp. 8543–8553 (2019)

41. Zhou, Z.H.: A brief introduction to weakly supervised learning. Nat. Sci. Rev. **5**(1), 44–53 (2017)

42. Zhuang, X., Shen, J.: Multi-scale patch and multi-modality atlases for whole heart segmentation of mri. Med. Image Anal. **31**, 77–87 (2016)

Improving Multispectral Pedestrian Detection by Addressing Modality Imbalance Problems

Kailai Zhou, Linsen Chen, and Xun Cao$^{(\boxtimes)}$

Nanjing University, Nanjing, China
{calayzhou,linsen}@smail.nju.edu.cn, caoxun@nju.edu.cn

Abstract. Multispectral pedestrian detection is capable of adapting to insufficient illumination conditions by leveraging color-thermal modalities. On the other hand, it is still lacking of in-depth insights on how to fuse the two modalities effectively. Compared with traditional pedestrian detection, we find multispectral pedestrian detection suffers from modality imbalance problems which will hinder the optimization process of dual-modality network and depress the performance of detector. Inspired by this observation, we propose Modality Balance Network (MBNet) which facilitates the optimization process in a much more flexible and balanced manner. Firstly, we design a novel Differential Modality Aware Fusion (DMAF) module to make the two modalities complement each other. Secondly, an illumination aware feature alignment module selects complementary features according to the illumination conditions and aligns the two modality features adaptively. Extensive experimental results demonstrate MBNet outperforms the state-of-the-arts on both the challenging KAIST and CVC-14 multispectral pedestrian datasets in terms of the accuracy and the computational efficiency. Code is available at https://github.com/CalayZhou/MBNet.

Keywords: Multispectral pedestrian detection · Modality imbalance problems · Multimodal feature fusion

1 Introduction

Recent years have witnessed increasing researches towards object detection among vision community by taking the advantages of multi-modal inputs, such as RGB + thermal, RGB + depth, RGB + LiDAR and so on [1,15,17,19]. Compared than traditional single-modal RGB images, which present great challenges at complex scenarios (e.g.. dim environment, face spoofing detection [47], autonomous driving [24,39], etc.), the introducing of another modality dramatically benefits the tasks of object detection. For instances, spectral images are

Electronic supplementary material The online version of this chapter (https://doi.org/10.1007/978-3-030-58523-5_46) contains supplementary material, which is available to authorized users.

© Springer Nature Switzerland AG 2020
A. Vedaldi et al. (Eds.): ECCV 2020, LNCS 12363, pp. 787–803, 2020.
https://doi.org/10.1007/978-3-030-58523-5_46

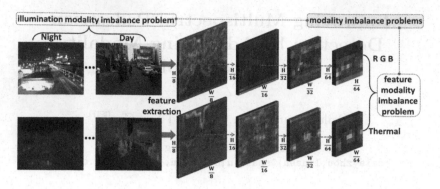

Fig. 1. The modality imbalance problems which consist of two parts: the illumination modality imbalance problem and the feature modality imbalance problem.

able to detect the optical radiation of matter and reveal the essential color properties of target object, to avoid the metamerism ambiguity. Thermal images can be captured based on the heat radiation difference of the object, which does not rely on external light sources. Time-of-flight (TOF) or LiDAR sensors provide additional depth information of the target scene, which has been widely used as data representation for many vision applications. Even with these remarkable benefits, however, how to effectively fuse multi-modal information in the context of advanced algorithms, like convolutional neural network, still remains much to be studied.

As for ordinary optimization process of object detection from multi-modality inputs, the imbalance problems [33] are crucial. The most known imbalance problem is the foreground-to-background imbalance [26]. This drawback is caused by an extremely inequality between the number of positive examples and negative ones. Nevertheless, the imbalance problems are not limited to the class imbalance. For instance, in multi-task losses minimization, the imbalance problems exist since the norms of gradients are different and the ranges of loss functions vary [14]. The common solution is to add coefficients upon each loss function to guide a balanced optimization process. Similarly, the modality imbalance issue in multispectral detection has a substantial influence on the algorithm performance.

The traditional Caltech [10] and CityPersons [45] pedestrian detection datasets only have RGB modality images captured during the day, so as shown in Fig. 1, modality imbalance problems existing in multispectral pedestrian detection datasets can be divided into two categories: the illumination modality imbalance and the feature modality imbalance. Illumination modality imbalance means the difference of illumination conditions between the daytime and the night images. Intuitively, pedestrians in RGB images have clearer texture features than thermal images in daytime. Comparatively, thermal images can provide more distinct pedestrian shapes than RGB images during night time. The RGB modality branch and the thermal modality branch tend to obtain different confidence scores and have uneven contributions to the object losses under diverse illumination conditions. It is expected that the RGB modality branch and the thermal modality branch should be optimized adaptively according to illumination conditions [4,23].

Feature modality imbalance problem signifies that the misalignment and inadequate integration of different modalities can lead to an uneven contribution and representation of the features. On the one hand, as the visualization results shown in Fig. 1, it is obvious that the RGB and thermal modality features are diverse in terms of pedestrian morphology, texture and properties in the two independent backbone networks. In RGB modality, the complexion and hair of the pedestrian can be some important hints of the pedestrian characteristics [6], but none of the thermal images has such cues. It is necessary to sufficiently incorporate the cross-modality complementarity to generate robust features. On the other hand, the misalignment between the RGB and thermal modalities will cause unbalanced modality feature representation in the fixed receptive fields of a convolution kernel. Both the balance and the integration of different modalities are the cornerstone we should consider in multispectral pedestrian detection. Unfortunately, existing RGB-Thermal based detection methods simply fuse the RGB and the thermal input/features by concatenation [22,23,37,42]. The inherent complementary is not fully exploited yet between different modalities.

To address the modality imbalance problems above, we investigate the impact and explore solutions in this paper. First, we construct the Modality Balance Network (MBNet) based on SSD [29] to extract the characteristics of two modalities separately. Then for the purpose of fully fusing features at different scales in the network, Differential Modality Aware Fusion (DMAF) module is proposed to tap the difference between RGB and thermal feature maps which brings more complementary information at each channel. Finally we design an illumination aware feature alignment module to align two modality features and induce the network to be optimized adaptively according to illumination conditions.

The main contributions of this paper are as follows: (1) We present modality imbalance problems specific to multispectral pedestrian detection, and analyse that modality imbalance problems will affect the performance of the detector due to the modality inconsistency in the optimization of the network; (2) We propose a one-stage detector named Modality Balance Network (MBNet) which consists of Differential Modality Aware Fusion (DMAF) module and illumination aware feature alignment (IAFA) module to address the modality imbalance problems. With DMAF module and IAFA module, the contribution of each feature map from two modalities will be explicitly integrated and balanced. In addition, The MBNet backbone (ResNet embedded with DMAF) may also do a favor to other computer vision communities; (3) MBNet achieves state-of-the-art results on both the challenging KAIST and CVC-14 multispectral pedestrian datasets in terms of the accuracy while maintaining the fastest speed.

2   Relate Work

2.1   Multispectral Pedestrian Detection

CNN-based pedestrian detection has achieved notable progress in recent years with methods of occlusion handling [32,38,46], cascaded detection systems

[2,30], semantic attention [3,49], anchor-free approach [31], etc. Nevertheless, current pedestrian detectors using single RGB modality may fail under the insufficient illumination condition.The KAIST multispectral pedestrian detection dataset [17] provides a new way to solve this problem by combining RGB modality and thermal modality. The initial baseline F + T + THOG is extended from Aggregated Channel Features (ACF) [9] with the thermal channel added. As the popularization of deep learning, the CNN-based methods [7,34,37,40] greatly reduce the miss rate of multispectral pedestrian detection. Inspired by [41], Boosted Decision Trees classifier [21] is built on high-resolution RPN feature maps to reduce potential false positive detections. MSDS RCNN [22] is learned by jointly optimizing pedestrian detection and semantic segmentation tasks.

How to fuse the information of two modalities is the common concerned problem in multispectral pedestrian detection. Liu et al. [27] design four distinct fusion architectures that integrate two modality branches on different DNNs stages and reveal the Halfway Fusion model provides the best performance. GFD-SSD [48] proposes two variations of novel Gated Fusion Units (GFU) that learn the combination of feature maps generated by the two SSD middle layers. Zhang et al. [42] explore the cross-modality disparity problem in multispectral pedestrian detection and propose a novel region feature alignment module to solve this problem. CIAN [43] makes the middle-level feature maps of two streams converge to a unified one under the guidance of cross-modality interactive attention and adopts the context enhancement blocks (CEBs) to further augment contextual information. Illumination-aware Faster R-CNN [23] adaptively merges color and thermal sub-networks to obtain the final confidence scores via a gate function defined over the illumination value. As the most popular solution, the two-stream architecture with concatenating RGB-Thermal feature maps has achieved significant improvements. Nevertheless, direct concatenation will inevitably introduce redundant features and a selection module is required to unveil the relation of modality complementary features.

2.2   Imbalance Problems in Object Detection

Oksuz et al. [33] present a comprehensive review of the imbalance problems in object detection and group these problems in a taxonomic tree with four main types: spatial imbalance, objective imbalance, class imbalance and scale imbalance. Spatial imbalance and objective imbalance focus on spatial properties of the bounding boxes and multiple loss functions respectively. Class imbalance occurs due to the significant inequality among different classes of training data. RetinaNet [26] addresses class imbalance by means of reshaping the standard cross entropy loss to prevent the vast number of easy negatives from overwhelming the detector. AP-Loss [5] and DR Loss [35] also provide ideas of designing loss function to solve the class imbalance problem. Scale imbalance occurs when certain sizes of the object bounding boxes are over-represented in the network. For instances, SSD [29] makes independent predictions from features at different layers. Since abstractness of information varies among different layers, it is unreliable to make predictions directly from different layers of the backbone network.

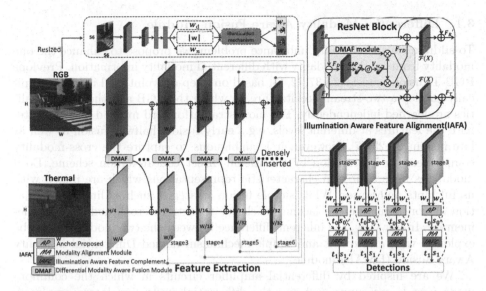

Fig. 2. Overview framework of the Modality Balance Network (MBNet). The MBNet consists of three parts: feature extraction module, illumination aware feature alignment module and illumination mechanism. The feature extraction module adopts ResNet-50 [16] as the backbone network and embeds DMAF module to supplement modality information. Illumination mechanism is designed to acquire illumination values which will assign weights to two modality streams. Illumination aware feature alignment module plays the role of adapting the model to different illumination conditions and aligning the two modality features in the region proposal stage.

Feature Pyramid Network [25] exploits an additional top-down pathway in order to have a balanced mixed of features from different scales. FPN can be further enhanced [28] by integrating and refining pyramidal feature maps.

In addition to the integration balance of different level, we argue that the integration of different modality features should also be balanced in the two-stream network. In other words, different modality features should be fully integrated and represented in order to have a balanced modality optimization in the training.

3 Approach

The overall architecture of the proposed method is shown in Fig. 2. The MBNet extends the framework of SSD [29] and it consists of three parts: feature extraction module, illumination aware feature alignment module and illumination mechanism. Details of DMAF module are introduced in Sect. 3.1. The design of illumination aware feature alignment module is introduced in Sect. 3.2.

3.1 Differential Modality Aware Fusion Module

To address feature modality imbalance problem, we propose to enhance the one modality from another modality with differential modality information. Previous RGB-T fusion models [22,37,42,43] based on deep convolutional networks typically employ a two-stream architecture, in which the RGB and thermal modalities are learned independently. The most straightforward method is to concatenate the features at different levels, e.g.., early fusion, halfway fusion as well as late fusion [21,23,27]. However, it is ambiguous to capture the cross-modality complementary information by traditional direct concatenation scheme. Both modalities have their own characteristic representations which are mixed with useful hints and noises. While simple fusion strategies such as linear combination or concatenation are lacking in clarity to extract cross-modality complementary. In our view, the inherent difference between the two modalities can be exploited with an explicit and simple mechanism named Differential Modality Aware Fusion (DMAF) module.

We are inspired by differential amplifier circuits in which the common-mode signals are suppressed and the differential-mode signals are amplified. Our DMAF module retains the original features and compensates according to differential features. The RGB convolution feature map F_R and the thermal convolution feature map F_T can be represented with common modality part and differential modality part at each channel as follows:

$$
\begin{aligned}
F_T &= \frac{F_T + F_T}{2} + \frac{F_R - F_R}{2} = \frac{F_R + F_T}{2} + \frac{F_T - F_R}{2} \\
F_R &= \frac{F_R + F_R}{2} + \frac{F_T - F_T}{2} = \frac{F_R + F_T}{2} + \frac{F_R - F_T}{2}
\end{aligned}
\tag{1}
$$

where the common modality part reflects the common features and the differential modality part reflects the unique features captured by two modalities. Equation 1 illustrates the principle of splitting which is same behind differential amplifier circuits and DMAF module. The key idea of our DMAF module is acquiring complementary features from another modality with channel-wise differential weighting. We expect the learning of complementary features to be enhanced by explicitly modeling modality interdependencies, so that the network sensitivity to informative features from another modality can be increased.

In order to make sufficient use of cross-modality complements, the DMAF module is densely inserted in each ResNet block. As the top right corner of Fig. 2 shows, we obtain the differential feature F_D by direct subtraction of two modalities first. Then we squeeze global spatial information F_D into a global differential vector which contains channel-wise differential statistics with global average pooling. The global differential vector can be interpreted as a channel descriptor whose statistics are expressive for the discrepancy between RGB and thermal modality. The tanh activation function ranging from -1 to 1 is applied for the global differential vector to obtain the fusion weight vector V_w. The two modality features F_T and F_R are recalibrated by the fusion weight vector V_w with channel-wise multiplication. The recalibration results F_{RD}, F_{TD} will be

Fig. 3. Feature map visualization of one channel in stage3 (shown in Fig. 2) before and after DMAF module. The two modality feature maps are remedied with the differential information from each other.

added to the original modality path as complementary information. After the enhancement from another modality with DMAF module, the more informative and robust features are generated and sent to the next ResNet block in the following step. The whole procedure of DMAF module can be formulated as:

$$
\begin{aligned}
F'_T = F_T + \mathcal{F}\left(F_T \oplus F_{RD}\right) = F_T + \mathcal{F}\left(F_T \oplus \left(\sigma\left(GAP\left(F_D\right)\right) \odot F_R\right)\right) \\
F'_R = F_R + \mathcal{F}\left(F_R \oplus F_{TD}\right) = F_R + \mathcal{F}\left(F_R \oplus \left(\sigma\left(GAP\left(F_D\right)\right) \odot F_T\right)\right)
\end{aligned}
\tag{2}
$$

where $\mathcal{F}(\mathcal{X})$ is considered as the residual function. σ refers to the tanh function, GAP refers to Global Average Pooling, and $\oplus$, $\odot$ represent element-wise sum and element-wise multiplication respectively. It is noteworthy that the F_{RD}, F_{TD} are added to the residual branch which formulates the complementary feature learning as residual learning inspired by RFBNet [8]. With residual mapping, the complementary feature would not directly impact the modality-specific stream. The DMAF module acts as a part of residual function in the ResNet block.

The visualization result of DMAF module is illustrated in Fig. 3. Due to the differences in the characteristics of two modalities, thermal and RGB modalities have certain limitations respectively in capturing pedestrian and background features. As the CNN goes deeper, pedestrian features gradually become salient and background features are re-integrated. The integration of background features means useful background information is refined and noisy background information is eliminated as much as possible. The DMAF module which effectively combine modality features can contribute to the integration of background information and make pedestrian features prominent from low level to high level. In our opinion, the DMAF module facilitates modality interaction in the network which reduces the learning of redundancy and conveys more information (refer the detailed analysis in appendix). In terms of no extra parameters and low computational complexity, the MBNet backbone (ResNet embedded with DMAF) may do a favor to other computer vision communities such as RGB-Depth tasks, stereo image SR, RGB-LiDAR tasks, etc.

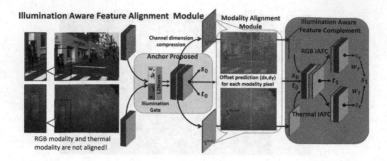

Fig. 4. The structure of illumination aware feature alignment module. Anchor Proposed (AP) stage generates an approximate location and Illumination Aware Feature Complement (IAFC) stage predicts based on the results of AP stage with the illumination aware balance of the two modality features. Modality Alignment (MA) module fixes the misalignments between the RGB modality and the thermal modality.

3.2   Illumination Aware Feature Alignment Module

Illumination Aware Feature Alignment module plays the role of adapting the model to different illumination conditions and aligning the two modality features in the region proposal stage. As the top of Fig. 2 shows, we design a tiny neural network to capture the illumination values in which only the RGB images are used because the thermal images are difficult to reflect the environment illumination condition. In order to reduce computational complexity, the RGB images are resized to 56×56 and sent into the illumination aware module which consists of two convolutional layers and three fully-connected layers. The ReLU activation function and a 2×2 maxpooling layer are followed after the convolutional layer to compress and extract features. The network is optimized by minimizing cross entropy loss function between the predicted illumination values and the true labels. The illumination loss L_I is formulated as:

$$L_I = -\widehat{w}_d \cdot \log(w_d) - \widehat{w}_n \cdot \log(w_n)$$
$$w_r = (\frac{w_d - w_n}{2}) \cdot (\alpha_w \cdot |w| + \gamma_w) + \frac{1}{2} \quad w_t = 1 - w_r \tag{3}$$

where w_d and w_n are the softmax output of full connection layers. $\widehat{w}_d$ and $\widehat{w}_n$ represent the true labels of the day and night. To be self-adaptable in the network, w_d, w_n are readjusted in the illumination mechanism in which $|w| \in [0,1]$ is the independent prediction of the bias from 0.5 and α_w, γ_w are two learnable parameters initialized with 1, 0. Then the re-scaled results w_r, w_t are embedded into the network to have a balanced optimization during different illumination conditions. We tailor an illumination gate to control the weight of thermal modality stream and RGB modality stream before the Anchor Proposal (AP) stage. By element-wise multiplying with illumination value, the feature maps from two modalities have different scales after reweighting, and we use L2-normalization to rescale their norms to 10.

Considering that RGB and thermal cameras are not always captured at the same time, there are slight misalignments between RGB and thermal modality as shown in Fig. 4. In the fixed receptive field of a convolution kernel, the modality misalignments will cause the unbalanced feature representations and contributions of the two modalities. We contrive a Modality Alignment (MA) module which predicts offsets (dx, dy) for every pixel (x, y) of each modality. Channel dimensions are compressed because the rearrangement of two modality features according to the learned offsets is time-consuming. Since (dx, dy) is the float type, we adopt bilinear interpolation to obtain the final pixel value (x + dx, y + dy) from the four neighborhood pixels.

Due to the vague and sparse pedestrian distribution, we employ a cascade architecture inspired by [18,20,30]. Figure 4 shows the cascade region proposal module which consists of two stages, Anchor Propose (AP) stage and Illumination Aware Feature Complement (IAFC) stage. First, the reweighted RGB and thermal feature maps are combined to generate an approximate location estimation by AP stage. The predicted regression offsets t_0 are used to propose the deformable anchors as the basic reference for position prediction in the next IAFC stage. Then the deformable anchors and confidence scores are further fine-tuned through the IAFC stage. The confidence scores s_r, s_t predicted by RGB and thermal feature maps are reweighted according to illumination values. The final confidence scores s_{final} and regression offsets t_{final} are computed as follows:

$$s_{final} = s_0 \times s_1 = s_0 \times (w_r \cdot s_r + w_t \cdot s_t) \qquad t_{final} = t_0 + t_1 \qquad (4)$$

The multiplication in confidence scores is to encourage the final score only if two-stage scores s_0, s_1 are both high. While for regression offsets, summation is adopted to progressively approach the pedestrian bounding boxes. Inspired by [26], we append the focal weight in classification loss L_{cls} to address the positive-negative imbalance. The L_{cls} is formulated as:

$$L_{cls} = -\alpha \sum_{i \in S_+} (1 - s_i)^\gamma \log (s_i) - (1 - \alpha) \sum_{i \in S_-} s_i^\gamma \log (1 - s_i) \qquad (5)$$

S_+, S_- are the positive and negative anchor boxes. As suggested in [26], we experimentally set $\alpha = 0.25$ and $\gamma = 2$. s_i is the positive probability of samples i. The total loss is the sum of illumination loss L_I, classification loss L_{cls} and regression loss L_{reg}, where the regression loss L_{reg} is the smooth L1 loss raised by Faster-RCNN [36]. The total loss function L is as follows:

$$L = L_I + L_{cls0} + L_{cls1} + [y = 1]L_{reg0} + [y = 1]L_{reg1} \qquad (6)$$

With the progressive detection of AP stage and IAFC stage, more positive cases are generated to benefit bounding box regression in the second IAFC stage. The adaptive illumination aware feature alignment of RGB modality and thermal modality provides a solution to feature modality imbalance problems by aligning two modality features, meanwhile, it also makes the detector more robust to the illumination variation.

Table 1. Comparisons with the state-of-the-art methods on the KAIST reasonable subset in terms of MR^{-2} [17] with different thresholds of IoU. In addition, Comparisons of running time are also provided.

Methods	MR^{-2} (IoU = 0.5)			MR^{-2} (IoU = 0.75)			Plateform	Speed (s)
	All	Day	Night	All	Day	Night		
ACF [17]	47.32	42.57	56.17	88.79	87.70	91.22	MATLAB	2.73
Halfway Fusion [27]	25.75	24.88	26.59	81.29	78.43	86.80	TITAN X	0.43
Fusion RPN + BF [21]	18.29	19.57	16.27	72.97	68.14	81.35	MATLAB	0.80
IAF R-CNN [23]	15.73	14.55	18.26	75.50	72.34	81.12	TITAN X	0.21
IATDNN + IASS [13]	14.95	14.67	15.72	76.69	76.46	77.05	TITAN X	0.25
RFA [42]	14.61	16.78	10.21	-	-	-	TITAN X	0.08
CIAN [43]	14.12	14.77	11.13	74.45	71.42	80.16	1080 Ti	**0.07**
MSDS-RCNN [22]	11.34	10.53	12.94	70.57	67.36	79.25	TITAN X	0.22
AR-CNN [44]	9.34	9.94	8.38	64.22	57.87	76.82	1080 Ti	0.12
MBNet(ours)	**8.13**	**8.28**	**7.86**	**60.12**	**54.90**	**68.34**	1080 Ti	**0.07**

4 Experiments

In this section, we first introduce the KAIST dataset [17] and CVC-14 dataset [12]. Then we show implementation details and experiment results to compare logMR and runtime of the proposed MBNet with the state-of-the-art methods. The evaluation is based on the reasonable setup (55 pixel or taller under partial or no occlusion) unless otherwise mentioned. Finally, we will carry out ablation studies for the proposed method on the KAIST dataset.

4.1 Datasets

Our approach is evaluated on the KAIST dataset [17] and CVC-14 dataset [12].

KAIST. The KAIST dataset [17] contains 95,328 aligned color-thermal image pairs, with a total of 103,128 bounding boxes covering 1,182 unique pedestrians. Due to the problematic annotations in original training data, we adopt the annotations improved by Zhang et al. [44] for training. The test set consists of 2,252 frames sampled every 20th frame from video, among which 1,455 images are captured during daytime and the rest 797 images are during nighttime. The evaluation metric follows the standard KAIST evaluation [17]: log-average Miss Rate over False Positive Per Image (FPPI) range of $[10^{-2}, 10^0]$ (denoted as MR^{-2}). We evaluate the detection performance on the KAIST test set with annotations improved by Liu et al. [27] and report the runtime of the proposed MBNet using a single NVIDIA GTX 1080Ti GPU for fair comparison with the state-of-the-art methods before.

CVC-14. The CVC-14 dataset [12] contains visible (grayscale) and thermal paired images. It was recorded in various scenes at day and night by on-board

Table 2. Evaluations on the KAIST dataset under all nine test subscts and ablation experiments are also provided. The lowest MR are highlighted with bold font.

Methods	Rea.	Day	Night	Near	Medium	Far	None	Partial	Heavy
ACF [17]	47.32	42.57	56.17	28.74	53.67	88.20	62.94	81.40	88.08
Halfway Fusion [27]	25.75	24.88	26.59	8.13	30.34	75.70	43.13	65.21	74.36
FusionRPN + BF [21]	18.29	19.57	16.27	0.04	30.87	88.86	47.45	56.10	72.20
IAF R-CNN [23]	15.73	14.55	18.26	0.96	25.54	77.84	40.17	48.40	69.76
IATDNN + IASS [13]	14.95	14.67	15.72	0.04	28.55	83.42	45.43	46.25	64.57
CIAN [43]	14.12	14.77	11.13	3.71	19.04	55.82	30.31	41.57	62.48
MSDS-RCNN [22]	11.63	10.60	13.73	1.29	16.19	63.73	29.86	38.71	63.37
AR-CNN [44]	9.34	9.94	8.38	0.00	16.08	69.00	31.40	38.63	**55.73**

MBNet (ours)

IAFC	DMAF	Aligned								
			11.93	12.51	10.86	0.00	20.08	63.70	34.16	44.10	65.11
	✓		10.96	11.13	10.48	0.00	20.33	61.25	33.50	39.80	62.68
✓			10.53	11.00	9.75	0.00	16.50	58.47	29.39	40.25	59.13
✓	✓		9.36	9.72	8.63	0.00	16.18	**54.66**	28.02	38.19	60.70
✓	✓	✓	**8.13**	**8.28**	**7.86**	**0.00**	**16.07**	55.99	**27.74**	**35.43**	59.14

color and thermal cameras 10 Hz. The training and testing set contains 7, 085 and 1, 433 frames, respectively. Annotations are individually provided in each modality since the cameras are not well calibrated.

4.2 Implementation Details

Our MBNet detector uses ResNet-50 [16] as the backbone network, which is pretrained on ImageNet unless otherwise stated. The training IoU of the AP stage is set to {0.3, 0.5} and the IAFC stage is set to {0.5, 0.7}. The Xavier method [11] is used to randomly initialize other convolutional layers. In the training, we crop a patch with the size of [0.3, 1] of the input image and resize it to 640 × 512, then each image is randomly color distorted and horizontal flipped with a probability of 0.5 to increase the diversity. The whole network is trained by adam optimizer for 7 epoches with a learning rate of 0.0001 and a batch size of 10. Followed by [22], the width of initial anchors are set to [25.84, 29.39], [33.81, 38.99], [44.47, 52.54], [65.80, 131.40] for stage3 to stage6 (shown in Fig. 2) with a single anchor ratio of 0.41.

4.3 Evaluation on the KAIST Dataset

We show the superiority of our method from both aspects of miss rate (MR) and speed.

Miss Rate. Our proposed approach achieves 8.28 MR, 7.86 MR, and 8.13 MR on the reasonable day, night and all-day subset respectively under the IoU threshold of 0.5, all of them are lower than the previous best competitor AR-CNN [44]. In the case of a stricter IoU threshold of 0.75, Table 1 shows our proposed method achieves about 4.10% lower on MR^{-2} which implies that the MBNet has a substantially better localization accuracy compared with AR-CNN. The larger the IoU threshold is set, the harder the predicted bounding boxes are considered to be True Positives (TP) in the evaluation. In order to have a comprehensive understanding of detector performance, we also make an evaluation under all nine subsets including the pedestrian distances and the occlusion levels. Table 2 shows MBNet outperforms other methods under most subsets with no extra treatment to handle the small and occlusion pedestrians, especially on the none subset (27.74 vs. 29.86 MR) and partial subset (35.43 vs. 38.63 MR).

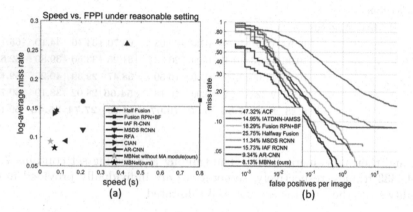

Fig. 5. (a) Log-average miss rate versus the running time of each detector. (b) Performance comparisons with the state-of-the-art methods on the KAIST dataset under reasonable subset.

Speed. We also compare the running time of MBNet with state-of-the-art methods. MBNet directly takes 640 × 512 multispectral images as input without image up-scaling. Since modality alignment module is time-consuming, we draw the Speed vs. FPPI results of MBNet and MBNet without MA module respectively in Fig. 5(a). The MBNet without MA module reaches the fastest speed of 20 fps and has a comparable performance with AR-CNN [44]. MBNet achieves the state-of-art performance on the test annotations improved by Liu et al. [27] while maintaining high computational efficiency.

Overall, the Speed vs. FPPI and performance comparisons under reasonable setting are shown in Fig. 5(a) and (b). The result indicates that MBNet is an attractive multispectral pedestrian detector in both accuracy and speed.

4.4   Evaluation on the CVC-14 Dataset

We fine-tune from the KAIST pretrained model in the training of CVC-14 dataset. We follow the protocol in [34] to conduct the evaluation experiments

Table 3. Evaluation results on the CVC-14 dataset. The first column refers to input modalities of the approach.

Visible	Methods	MR^{-2}			Visible + Thermal	Methods	MR^{-2}		
		Day	Night	All			Day	Night	All
	SVM [12]	37.6	76.9	-		MACF [34]	61.3	48.2	60.1
	DPM [12]	25.2	76.4	-		Choi et al. [7]	49.3	43.8	47.3
	Random Forest [12]	26.6	81.2	-		Halfway Fusion [34]	38.1	34.4	37.0
	ACF [34]	65.0	83.2	71.3		Park et al. [34]	31.8	30.8	31.4
	Faster R-CNN [34]	43.2	71.4	51.9		AR-CNN [44]	24.7	18.1	22.1
						MBNet (ours)	**24.7**	**13.5**	**21.1**

and adopt the strategy in [44]. Specifically, there exsiting serious misalignments between thermal and RGB modalities, so we consider pedestrians in the RGB modality as the training target, and pedestrians in the thermal modality act as a reference. It can be observed from Table 3 that MBNet can still achieve good results even in the case of serious modality misalignments, which demonstrates that modality balance strategy improves the robustness to the position shift problem.

4.5   Ablation Study

Ablation experiments are performed on the KAIST dataset for a detailed analys in this section. The baseline is initialized from two-stream SSD [29] and adopts simple concatenation to fuse two modalities as most previous methods do. We show how to construct the MBNet with the principle of modality balance.

Differential Modality Aware Fusion. Simple concatenation modulates the complementary part implicitly and has a large computational cost. It encourages us to seek a better integration of modalities. DMAF modules are inserted in each ResNet block and act as a part of residual function to be learned which will impel a deep integration of RGB and thermal modalities. From the ablation experiment results in Table 2, we could see that the MBNet generates more accurate detection results with DMAF module added. The DMAF module capture the complementary modality information in a more explicit way with no extra parameters and very little computation, so that the MBNet can maintain a fast speed. The feature representation can be more balanced after the two modalities are fully integrated.

Modality Alignment. It could be seen from the Table 4 that the MBNet with modality alignment module has a much lower miss rate on the reasonable subset. It indicates that the MA module with bilinear interpolation used can locate pedestrians more precisely especially in the deeper network which has a smaller resolution of the feature maps. After the MA module, the RGB and thermal modality features are rearranged and the aligned features can make a balanced contribution at one position. In addition, it can be observed that MA module narrows the gap of miss rates between day and night subsets. The experiment results demonstrate that the illumination modality imbalance problem will be

Table 4. Ablation experiments of illumination aware feature alignment module evaluated on KAIST reasonable set.

Component	Choice				
RGB IAFC	✓		✓	✓	✓
Thermal IAFC		✓	✓	✓	✓
Illumination gate	✓	✓		✓	✓
Modality alignment					✓
MR (%)	Day	9.80	10.05	10.74	9.72	8.28
	Night	10.17	9.76	9.54	8.63	7.86
	All	9.86	9.89	10.27	9.36	8.13

mitigated with the alignment of feature modalities. The illumination and feature modality imbalance problems exist side by side and play a part together.

Illumination Aware Feature Alignment. To have a deep insight into the effectiveness of illumination aware feature alignment module, we investigate the performance of different design choices in Table 4. RGB IAFC and thermal IAFC represent complementing the prediction results of AP stage with RGB and thermal modality features according to the illumination conditions. It is observed that performance gains can generally be achieved by illumination aware feature complement. RGB modality is beneficial to the pedestrian detection during the day while thermal modality is beneficial to the night. By introducing the illumination gate which applies weights to the RGB and thermal stream, MBNet has a more balanced performance under different illumination conditions. This demonstrates that the detection performance can be further improved by illumination aware mechanism, since it helps the network mitigate illumination modality imbalance problem.

5   Conclusion

In this work, we explore a one-stage detector named MBNet to alleviate the modality imbalance problems in multispectral pedestrian detection. Specifically, the DMAF module is densely inserted in the ResNet block to fully integrate features and the MA module aligns two modalities so that the RGB and thermal features can have an equal contribution and representation. Meanwhile, the illumination gate embedded in the backbone network and the adaptive illumination aware feature complement in the region proposal stage make the detector robust to the variant illumination. We argue the modality imbalance problems are not limited to multispectral pedestrian detection. They are widespread in multimodal computer vision task to which the balance and integration of different modality features should be paid attention. We will further study how to reconcile the balance and reduce the learning of redundancy between different modalities in other computer vision task in the future.

Acknowledgment. This work was supported in part by the National Natural Science Foundation of China (Nos. 61627804).

References

1. Behley, J., et al.: A dataset for semantic scene understanding of lidar sequences. In: Proceedings of the IEEE/CVF International Conf. on Computer Vision (ICCV), vol. 3 (2019)
2. Brazil, G., Liu, X.: Pedestrian detection with autoregressive network phases. In: Proceedings of the IEEE Conference on Computer Vision and Pattern Recognition, pp. 7231–7240 (2019)
3. Brazil, G., Yin, X., Liu, X.: Illuminating pedestrians via simultaneous detection & segmentation. In: Proceedings of the IEEE International Conference on Computer Vision, pp. 4950–4959 (2017)
4. Cao, Y., Guan, D., Wu, Y., Yang, J., Cao, Y., Yang, M.Y.: Box-level segmentation supervised deep neural networks for accurate and real-time multispectral pedestrian detection. ISPRS J. Photogram. Remote Sens. **150**, 70–79 (2019)
5. Chen, K., et al.: Towards accurate one-stage object detection with AP-loss. In: Proceedings of the IEEE Conference on Computer Vision and Pattern Recognition, pp. 5119–5127 (2019)
6. Chi, C., Zhang, S., Xing, J., Lei, Z., Li, S.Z., Zou, X.: Relational learning for joint head and human detection. arXiv preprint arXiv:1909.10674 (2019)
7. Choi, H., Kim, S., Park, K., Sohn, K.: Multi-spectral pedestrian detection based on accumulated object proposal with fully convolutional networks. In: 2016 23rd International Conference on Pattern Recognition (ICPR), pp. 621–626. IEEE (2016)
8. Deng, L., Yang, M., Li, T., He, Y., Wang, C.: RFBNet: deep multimodal networks with residual fusion blocks for RGB-D semantic segmentation. arXiv preprint arXiv:1907.00135 (2019)
9. Dollár, P., Appel, R., Belongie, S., Perona, P.: Fast feature pyramids for object detection. IEEE Trans. Pattern Anal. Mach. Intell. **36**(8), 1532–1545 (2014)
10. Dollar, P., Wojek, C., Schiele, B., Perona, P.: Pedestrian detection: an evaluation of the state of the art. IEEE Trans. Pattern Anal. Mach. Intell. **34**(4), 743–761 (2011)
11. Glorot, X., Bengio, Y.: Understanding the difficulty of training deep feedforward neural networks. In: Proceedings of the Thirteenth International Conference on Artificial Intelligence and Statistics, pp. 249–256 (2010)
12. González, A., et al.: Pedestrian detection at day/night time with visible and fir cameras: a comparison. Sensors **16**(6), 820 (2016)
13. Guan, D., Cao, Y., Yang, J., Cao, Y., Yang, M.Y.: Fusion of multispectral data through illumination-aware deep neural networks for pedestrian detection. Inf. Fusion **50**, 148–157 (2019)
14. Guo, M., Haque, A., Huang, D.A., Yeung, S., Fei-Fei, L.: Dynamic task prioritization for multitask learning. In: Ferrari, V., Hebert, M., Sminchisescu, C., Weiss, Y. (eds.) Computer Vision – ECCV 2018. Lecture Notes in Computer Science, vol. 11220, pp. 282–299. Springer, Cham (2018). https://doi.org/10.1007/978-3-030-01270-0_17
15. Ha, Q., Watanabe, K., Karasawa, T., Ushiku, Y., Harada, T.: MFNet: towards real-time semantic segmentation for autonomous vehicles with multi-spectral scenes. In: 2017 IEEE/RSJ International Conference on Intelligent Robots and Systems (IROS), pp. 5108–5115. IEEE (2017)

16. He, K., Zhang, X., Ren, S., Sun, J.: Deep residual learning for image recognition. In: The IEEE Conference on Computer Vision and Pattern Recognition (CVPR) (2016)
17. Hwang, S., Park, J., Kim, N., Choi, Y., So Kweon, I.: Multispectral pedestrian detection: benchmark dataset and baseline. In: Proceedings of the IEEE Conference on Computer Vision and Pattern Recognition, pp. 1037–1045 (2015)
18. Jang, H.D., Woo, S., Benz, P., Park, J., Kweon, I.S.: Propose-and-attend single shot detector. arXiv preprint arXiv:1907.12736 (2019)
19. Ju, R., Ge, L., Geng, W., Ren, T., Wu, G.: Depth saliency based on anisotropic center-surround difference. In: 2014 IEEE International Conference on Image Processing (ICIP), pp. 1115–1119. IEEE (2014)
20. Kong, T., Sun, F., Liu, H., Jiang, Y., Shi, J.: Consistent optimization for single-shot object detection. arXiv preprint arXiv:1901.06563 (2019)
21. Konig, D., Adam, M., Jarvers, C., Layher, G., Neumann, H., Teutsch, M.: Fully convolutional region proposal networks for multispectral person detection. In: Proceedings of the IEEE Conference on Computer Vision and Pattern Recognition Workshops, pp. 49–56 (2017)
22. Li, C., Song, D., Tong, R., Tang, M.: Multispectral pedestrian detection via simultaneous detection and segmentation. arXiv preprint arXiv:1808.04818 (2018)
23. Li, C., Song, D., Tong, R., Tang, M.: Illumination-aware faster R-CNN for robust multispectral pedestrian detection. Pattern Recogn. **85**, 161–171 (2019)
24. Li, X., et al.: A unified framework for concurrent pedestrian and cyclist detection. IEEE Trans. Intell. Transp. Syst. **18**(2), 269–281 (2016)
25. Lin, T.Y., Dollár, P., Girshick, R., He, K., Hariharan, B., Belongie, S.: Feature pyramid networks for object detection. In: Proceedings of the IEEE Conference on Computer Vision and Pattern Recognition, pp. 2117–2125 (2017)
26. Lin, T.Y., Goyal, P., Girshick, R., He, K., Dollár, P.: Focal loss for dense object detection. In: Proceedings of the IEEE International Conference on Computer Vision, pp. 2980–2988 (2017)
27. Liu, J., Zhang, S., Wang, S., Metaxas, D.N.: Multispectral deep neural networks for pedestrian detection. arXiv preprint arXiv:1611.02644 (2016)
28. Liu, S., Qi, L., Qin, H., Shi, J., Jia, J.: Path aggregation network for instance segmentation. In: Proceedings of the IEEE Conference on Computer Vision and Pattern Recognition, pp. 8759–8768 (2018)
29. Liu, W., et al.: SSD: single shot multibox detector. In: Leibe, B., Matas, J., Sebe, N., Welling, M. (eds.) Computer Vision – ECCV 2016. Lecture Notes in Computer Science, vol. 9905, pp. 21–37. Springer, Cham (2016). https://doi.org/10.1007/978-3-319-46448-0_2
30. Liu, W., Liao, S., Hu, W., Liang, X., Chen, X.: Learning efficient single-stage pedestrian detectors by asymptotic localization fitting. In: Ferrari, V., Hebert, M., Sminchisescu, C., Weiss, Y. (eds.) Computer Vision – ECCV 2018. Lecture Notes in Computer Science, vol. 11218, pp. 643–659. Springer, Cham (2018). https://doi.org/10.1007/978-3-030-01264-9_38
31. Liu, W., Liao, S., Ren, W., Hu, W., Yu, Y.: High-level semantic feature detection: a new perspective for pedestrian detection. In: Proceedings of the IEEE Conference on Computer Vision and Pattern Recognition, pp. 5187–5196 (2019)
32. Noh, J., Lee, S., Kim, B., Kim, G.: Improving occlusion and hard negative handling for single-stage pedestrian detectors. In: Proceedings of the IEEE Conference on Computer Vision and Pattern Recognition, pp. 966–974 (2018)
33. Oksuz, K., Cam, B.C., Kalkan, S., Akbas, E.: Imbalance problems in object detection: a review. arXiv preprint arXiv:1909.00169 (2019)

34. Park, K., Kim, S., Sohn, K.: Unified multi-spectral pedestrian detection based on probabilistic fusion networks. Pattern Recogn. **80**, 143–155 (2018)
35. Qian, Q., Chen, L., Li, H., Jin, R.: DR loss: improving object detection by distributional ranking. arXiv preprint arXiv:1907.10156 (2019)
36. Ren, S., He, K., Girshick, R., Sun, J.: Faster R-CNN: towards real-time object detection with region proposal networks. In: Advances in Neural Information Processing Systems, pp. 91–99 (2015)
37. Wagner, J., Fischer, V., Herman, M., Behnke, S.: Multispectral pedestrian detection using deep fusion convolutional neural networks. In: ESANN (2016)
38. Wang, X., Xiao, T., Jiang, Y., Shao, S., Sun, J., Shen, C.: Repulsion loss: detecting pedestrians in a crowd. In: Proceedings of the IEEE Conference on Computer Vision and Pattern Recognition, pp. 7774–7783 (2018)
39. Wu, B., Iandola, F., Jin, P.H., Keutzer, K.: Squeezedet: unified, small, low power fully convolutional neural networks for real-time object detection for autonomous driving. In: Proceedings of the IEEE Conference on Computer Vision and Pattern Recognition Workshops, pp. 129–137 (2017)
40. Xu, D., Ouyang, W., Ricci, E., Wang, X., Sebe, N.: Learning cross-modal deep representations for robust pedestrian detection. In: Proceedings of the IEEE Conference on Computer Vision and Pattern Recognition, pp. 5363–5371 (2017)
41. Zhang, L., Lin, L., Liang, X., He, K.: Is faster R-CNN doing well for pedestrian detection? In: Leibe, B., Matas, J., Sebe, N., Welling, M. (eds.) Computer Vision – ECCV 2016. Lecture Notes in Computer Science, vol. 9906, pp. 443–457. Springer, Cham (2016). https://doi.org/10.1007/978-3-319-46475-6_28
42. Zhang, L., Liu, Z., Chen, X., Yang, X.: The cross-modality disparity problem in multispectral pedestrian detection. arXiv preprint arXiv:1901.02645 (2019)
43. Zhang, L., et al.: Cross-modality interactive attention network for multispectral pedestrian detection. Inf. Fusion **50**, 20–29 (2019)
44. Zhang, L., Zhu, X., Chen, X., Yang, X., Lei, Z., Liu, Z.: Weakly aligned cross-modal learning for multispectral pedestrian detection. In: Proceedings of the IEEE International Conference on Computer Vision, p. 5127–5137 (2019)
45. Zhang, S., Benenson, R., Schiele, B.: Citypersons: a diverse dataset for pedestrian detection. In: Proceedings of the IEEE Conference on Computer Vision and Pattern Recognition, pp. 3213–3221 (2017)
46. Zhang, S., Yang, J., Schiele, B.: Occluded pedestrian detection through guided attention in CNNs. In: Proceedings of the IEEE Conference on Computer Vision and Pattern Recognition, pp. 6995–7003 (2018)
47. Zhang, S., et al.: CASIA-SURF: a large-scale multi-modal benchmark for face anti-spoofing. arXiv preprint arXiv:1908.10654 (2019)
48. Zheng, Y., Izzat, I.H., Ziaee, S.: GFD-SSD: gated fusion double ssd for multispectral pedestrian detection. arXiv preprint arXiv:1903.06999 (2019)
49. Zhou, C., Wu, M., Lam, S.K.: SSA-CNN: semantic self-attention CNN for pedestrian detection. arXiv preprint arXiv:1902.09080 (2019)

Author Index

Printed in the United States
By Bookmasters